1970s

1970	Clean Air Act	
	First Earth Day	
1972	DDT banned	USA
	Clean Water Act	
	Marine Mammal Protection Act	
	U.N. Environment Program created	
1973	Endangered Species Act (ESA)	
	OPEC oil embargo	
1975	Convention on the International Trade in Endangered Species (CITES)	
1976	National Forest Management Act	
	Resource Conservation and Recovery Act	
	Toxic Substances Control Act	
1978	Endangered Species Act Amendment (requiring designation of critical habitat)	

1980s

1980	Superfund Act	
1982	Moratorium on whaling	
1984	Union Carbide plant explosion	Bhopal, India
1986	Chernobyl nuclear reactor explosion	Chernobyl, Russia
1987	American alligator downlisted from “endangered” to “threatened”	
	Montreal Protocol Agreement	
1988	Elephant Conservation Act	
1989	Moratorium on ivory	
	Exxon Valdez oil spill	Alaska

1990s

1990	Clean Air Act reauthorized	
1992	U.N. Earth Summit	Rio de Janeiro
	Agenda 21 adopted by U.N.	
1994	Gray wolf reintroduced into Yellowstone National Park	Wyoming, Montana, and Idaho
	Death Valley National Park	California and Nevada
	International Conference on Population and Development	Cairo
1995	Bald Eagle downlisted from “endangered” to “threatened” (ESA)	
1996	Red-legged frog listed as “threatened” (first new listing since ESA moratorium ended)	
	California condor experimental release in the Grand Canyon	Arizona
1996	World Food Summit	Rome

Modified from The American Museum of Natural History's exhibit, *Endangered! Exploring a World at Risk* (1997)

Environmental Science

Environmental Science

The Way the World Works

Sixth Edition

Bernard J. Nebel

Catonsville Community College

Richard T. Wright

Gordon College

PRENTICE HALL
Upper Saddle River, New Jersey 07458

Library of Congress Cataloging-in-Publication Data

Nebel, Bernard J.
Environmental science : the way the world works / Bernard J.
Nebel, Richard T. Wright. — 6th ed.
p. cm.
Includes bibliographical references and index.
ISBN 0–13–835331–X
1. Environmental sciences. I. Wright, Richard T. II. Title.
GE105.N42 1998 97–17388
363.7—dc21 CIP

Executive Editor: Sheri L. Snavely
Acquisitions Editor: Teresa K. Ryu
Editor in Chief: Paul F. Corey
Editorial Director: Tim Bozik
Development Editor: Carol Stone
Assistant Vice President of Production and Manufacturing: David W. Riccardi
Executive Managing Editor: Kathleen Schiaparelli
Production Editor: Rose Kernan
Marketing Manager: Jennifer Welchans
Creative Director: Paula Maylahn
Art Director: Heather Scott
Interior Design: Lorraine Mullaney
Cover Photo: David Muench, David Muench Photography, Inc.
Art Manager: Gus Vibal
Manufacturing Manager: Trudy Pisciotti
Associate Editor: Mary Hornby
Photo Editor: Lori Morris-Nantz/Melinda Reo
Photo Research: Tobi Zausner/Beaura Ringrose
Illustrations: Academy ArtWorks
Insert Illustrations: Renaissance Studios, Inc.
Copy Editors: Brian Baker/Margo Quinto/Rose Kernan
Editorial Assistants: Lisa Tarabokjia/Nancy Gross/David Stack
Proofreader: Carolyn M. Gauntt
Text Composition/Prepress: KR Publishing Services

Simon & Schuster/A Viacom Company
Upper Saddle River, New Jersey 07458

Printed in the United States of America
10 9 8 7 6 5 4 3 2 1

ISBN 0-13-835331-X

Prentice-Hall International (UK) Limited, *London*
Prentice-Hall of Australia Pty. Limited, *Sydney*
Prentice-Hall Canada Inc., *Toronto*
Prentice-Hall Hispanoamericana, S. A., *Mexico*
Prentice-Hall of India Private Limited, *New Delhi*
Prentice-Hall of Japan, Inc., *Tokyo*
Simon & Schuster Asia Pte. Ltd., *Singapore*
Editora Prentice-Hall do Brasil, Ltda., *Rio de Janeiro*

Total 10% Recycled Paper

All Post-Consumer Waste

ABOUT THE AUTHORS

Bernard J. Nebel is professor emeritus of Biology at Catonsville Community College in Maryland. He earned his Bachelor of Arts from Earlham College and his Ph.D. from Duke University. Nebel taught environmental science for 24 continuous years and is a member of the American Association for the Advancement of Science, the Institute of Biological Sciences, and the National Association of Science Teachers. He served as the 1990 Earth Day Coordinator for Maryland Colleges, and he strives to make a difference on the environment in his personal life as well. He walks or bikes to activities, he is experimenting with the organic production of fruits, vegetables, and fish in his backyard, and he volunteers his efforts for Habitat for Humanity, and supports the Natural Resources Defense Fund and the World Wildlife Fund, and other environmental organizations.

Richard T. Wright is the chairman of the Division of Natural Sciences, Mathematics, and Computer Science at Gordon College in Massachusetts, where he has taught environmental science for the past 25 years. He earned a Bachelor of Arts from Rutgers and a Master of Arts and Ph.D. from Harvard. Wright has received research grants from the National Science Foundation (NFS) for his work in aquatic microbiology and, in 1981, was a founding faculty member of Au Sable Institute of Environmental Studies in Michigan, of which he is now Academic Chairman. He is a member and Fellow of the American Association for the Advancement of Science, the American Society for Microbiology, and the American Society for Limnology and Oceanography. In his personal life, Wright strives to have as light an impact on our environment as possible by biking to work, by planting trees and a vegetable garden in his yard, and by working with such groups as Bread for the World, Essex County Greenbelt, and the Union of Concerned Scientists.

BRIEF CONTENTS

CONTENTS

7 Addressing the Population Problem 165

8 The Production and Distribution of Food 187

P R E F A C E

Consider the following:

- World population reaches 6.0 billion people during 1999—4.8 billion living in less developed countries and 1.2 billion in more developed countries. Thirty-eight percent of the world's population live in only two countries, China and India.
- The bulk of Earth's inhabitants live in rural settings and 34% (2 billion) of all people are under the age of 15.
- The globalization of popular culture and lifestyles is diminishing folk ways and native livelihoods and damaging Earth's life-sustaining environment with increasing severity.
- Some 15% of Earth's living species could be extinct by the year 2000 if trends of habitat destruction, soil losses, and pollution continue. More than 25% of new medicines are derived from plants, yet deforestation and loss of plant species continues at a record pace.
- The Intergovernmental Panel on Climate Change (IPCC) reached a scientific consensus in 1992 and 1996 that global changes in climates, air temperatures, ocean temperatures, plant and animal habitats, disintegration of Antarctic ice shelves, and a rising sea level are the result of human activities.
- Reserves of petroleum and natural gas continue to decline, forcing the consideration of renewable alternatives—solar, thermal, and electric, wind, energy efficiency and conservation, and small-scale hydroelectric technologies.
- A recent analysis of the Clean Air Act (1970, 1977, 1990) found that total implementation costs were $436 billion whereas the mean estimate of direct benefits exceeded $6.8 trillion—a net benefit of $6.4 trillion in improved human health, reduced environmental damage, and other avoided costs!

In *Environmental Science: The Way the World Works* issues are approached by starting at a basic level and then building in a logical, systematic way toward a comprehensive understanding. Thus, without assuming a science background and without deluging the student in complexity, *Environmental Science* brings the student to an understanding of each issue in a reader-friendly way. The same understanding enables the student to evaluate different courses of action in terms of sustainability as they are discussed. This approach is appropriate for both majors and nonmajors.

By its attention to developing an understanding of the factors underlying environmental issues, *Environmental Science* provides students with a framework of knowledge into which they can readily integrate additional information for a lifetime of continued learning.

Civilization is at a crossroads and in a predicament: We cannot simply halt centuries of technological progress because economic chaos would result, nor can we continue to expend Earth's resources in a wasteful manner because ecological ruin would result. We cannot stop, yet we cannot continue in our present nonsustainable ways.

Environmental science stands at the interface between humans and Earth and explains the interactions and relations that must be considered in virually all future decision-making. Some past civilizations adapted to crisis, whereas other failed. The new millennium lies ahead as a stage on which this one-time, real time human-environment experiment is being played. This text considers the full spectrum of views in an effort to establish a middle ground of understanding and a sustainable formula for the future.

A Guide to the Sixth Edition of Environmental Science

The sixth edition of *Environmental Science* includes a thorough updating of text, tables, and maps to reflect the latest information. We live on a planet where almost all places are accessible on the Internet and its World Wide Web. Students can easily find the threads of environmental information that can be woven into a vast planetary fabric of understanding. Occasionally the text will refer the student to these resources. The Internet icons have been provided throughout the book to refer students to these resources.

Chapter 1 explores the development of the environmental movement, its successes and failures, and the reasons for the current environmental backlash. The concept of sustainability is developed by contrasting global trends that are *not* sustainable with directions for advancing civilization that *are* sustainable. The objective of creating a sustainable base on which human civilization can continue to develop technologically and socially is a principal theme of the text.

Part One (Chapters 2–5) explores natural ecosystems, what they are, how they function, how balances are maintained, and how they evolve and change. This examination, in addition to providing an appreciation of how the natural world functions, brings out four basic principles that underlie sustainability (see Table 4-1, page 104). Sections at the end of Chapters 2–4, *Implications for Humans,* help us realize that our environmental problems basically stem from the failure of humans to abide by these basic principles. Conversely, the principles serve as bench marks to evaluate the sustainability of various courses of action presented in following chapters.

Part Two (Chapter 6–11) confronts issues concerning a growing population and the pressures on soil and water resources that are fundamental to supporting agriculture. Presented more fully is the growing recognition that population, poverty, and environmental degradation are irrevocably interrelated issues; resolving any one of these issues demands simultaneous action on the other two. The focus of the 1994 Cairo Population Conference and other programs aimed at developing this required multifaceted approach are described. All of these chapters are enhanced with revised content to include the latest available statistics.

Part Three (Chapters 12–17) investigates the many forms of pollution, ranging from eutrophication and sewage, through contamination of ground water and hazardous wastes, to air pollution and major atmo-spheric changes, acid rain, depletion of ozone, and the greenhouse effect. By devoting a full chapter to each of five major categories of pollution, the student is given an understanding of sources, consequences, legislation, and control strategies related to different pollution problems. In this section, each chapter has been revised to reflect new information, legislation, and advances in control technologies. Pollution avoidance, as opposed to pollution control, is discussed. Chapter 17, now titled *Pollution and Public Policy,* gives a treatment of how scientific investigation, public perception, and economics all come into play in risk assessment and setting the course of government action toward mitigation. New art enhances the presentation.

Part Four (Chapters 18–24) addresses resources, and the issues inherent in finding a suitable balance between obtaining economic values from natural systems, and their value and need for conservation as intact systems in Chapter 18 and 19. The most recent and remarkable advances in recycling are addressed in Chapter 20. The outlook for the sustainability of U.S. reliance on crude oil (diminishing supplies) and the obvious prospects for solar energy are presented in view of the most recent statistics and developments in Chapters 21–23. The option of nuclear power is presented although problems of fuel availability, cost, nuclear waste storage and disposal, lack of fuel reprocessing, and inherent danger remain potent obstacles. Chapter 24, *Lifestyle and Sustainability*, continues to focus on how environmental problems are made worse and how inner cities have deteriorated as a result of migration to the suburbs and urban sprawl. Discussion emphasizes that the plight of inner cities is a problem of economic exclusion. Making cities livable again, a prerequisite for sustainability, will depend on bolstering the economy of the inner city. Chattanooga, Tennessee, which has become a national model in this respect, is presented.

Individual Text Elements

Chapter Opening: Each chapter begins with a set of *Key Issues and Questions.* Each item consists of a succinct statement regarding a key aspect of the issue being covered and a question inviting the student to explore that statement.

Chapter Outline: Chapter outlines seen in previous editions may still be found in the Table of Contents. Importantly, the text of each chapter is organized according to a logical outline of first, second, and third order headings to assist student outlining, note taking, and learning.

Review Questions: Each chapter concludes with a set of *Review Questions* addressing each aspect of the topic covered. Of course, these questions may serve as learning objectives, test items, or for review.

Thinking Environmentally: A set of questions, *Thinking Environmentally*, is included at the end of each chapter. These questions invite the student to make connections between knowledge gleaned from the chapter and other areas of the environmental arena, and to apply knowledge gained to specific environmental problems. These items may be used also for testing or to focus class discussion.

Vocabulary: Each new term will be found in **bold face** type where it is first introduced and defined. All such items are found in the Glossary at the end of the book.

Boxes: *Environmental Science* features three kinds of boxes: Earth Watch, Ethics, and Global Perspective. Lists of all boxes are found inside the front cover of the book.

Earth Watch boxes, similar to the *FYI* boxes in the fifth edition, provide further information that enhances the student's understanding of particular aspects of the topic being covered.

Ethics boxes focus on the fact that many environmental issues do not involve clear-cut rights or wrongs, but present ethical dilemmas.

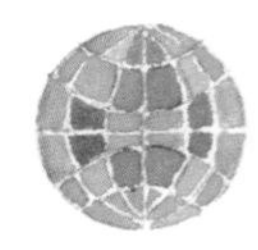

Global Perspective boxes, help the student appreciate the global nature and extent of the topic in question.

Making a Difference: We believe that no amount of text-based learning about the environment truly becomes useful until students challenge themselves and those around them to begin *making a difference*. With this in mind, each of the four parts of the text concludes with a section that suggests courses of action that each student can take to bring about the needed changes to foster sustainability.

Video Case Studies: Selected from the archives of ABC news, these timely and relevant video segments offer students an overview of a particular environmental issue or controversy. Case study material is found directly after the end of the text, but has direct application to particular chapters in the text. We have provided a brief synopsis of each video and have included a list of discussion questions in the hopes of stimulating healthy classroom debate and discussion of these topics. A list of the video segments is found in the Video Case Study section, as well as on the inside front cover of the book.

Internet icons allow students to further explore a topic on the Web. They can be found throughout the text and in boxes.

Appendices

At many points in the text, reference is made to the work being done by various environmental organizations. A listing of major national environmental organizations, as well as many Web addresses, is given in Appendix A. Remember that most of these organizations and agencies have a home page on the Internet.

A conversion chart for various English and metric units is found in Appendix B.

A discussion of atoms, molecules, atomic bonding, and other basic elements of chemistry is provided in Appendix C.

A discussion of science and the scientific method, is in Appendix D.

Acknowledgments

Completion of this text represents the fruition of devoted labors and contributions from countless people. We wish to gratefully acknowledge and express our sincere thanks and appreciation to all those who contributed to this volume in so many ways.

In particular, our many thanks go to those who reviewed all or various chapters of the text and offered comments, suggestions, and constructive criticisms.

Narayanaswamy Bharathan
Northern State University–Aberdeen

Roger G. Bland
Central Michigan University

Jack L. Butler
University of South Dakota

Ann S. Causey
Auburn University

Robert W. Christopherson
American River College

Lynnette Danzl-Tauer
Rock Valley College

Phil Evans
East Carolina University, Pitt Community College

Gian Gupta
University of Maryland Eastern Shore

John P. Harley
Eastern Kentucky University

Vern Harnapp
University of Akron

Stanley Hedeen
Xavier University

Clyde W. Hibbs
Ball State University

John C. Jahoda
Bridgewater State College

Karolyn Johnston
California State University–Chico

Guy R. Lanza
East Tennessee State University

John Mathwig
College of Lake County

Richard J. McCloskey
Boise State University

SuEarl McReynolds
San Jacinto College–Central

Eric Pallant
Allegheny College

David J. Parrish
Virginia Polytechnic Institute

Carol Skinner
Edinboro University of Pennsylvania

And many thanks to those colleagues and associates who provided special assistance with particular chapters or sections. In particular Bernard Nebel gives thanks to: Walter Boynton for information regarding Chesapeake Bay; Ree Brennin for information regarding marine food chains; George Farrant for providing information for and reviewing Chapter 14; Werner Fornos, President Population Institute for help with and review of Chapter 7; Ron Klauda for information regarding Chesapeake Bay; Tom Matey for help with energy information; Bernie Noeller for help with geological aspects; Don Sprinkle and Dan Ward for information regarding sewage treatment; Peter Vorster for help regarding Mono Lake; and Monica Yost for supplying information regarding human evolution.

We also thank all the dedicated people at Prentice Hall who had a hand in the production of this volume. In particular, we thank our Editors, Teresa Ryu and Sheri Snavely, for constant enthusiasm and support; our Production Editor, Rose Kernan, always good natured, who managed untold details of putting the text together; our Developmental Editor, Carol Stone, who edited the entire manuscript and has made it an exceptionally reader-friendly text; our photo researchers, Tobi Zausner and Beaura Ringrose for tracking down appropriate photos; Clark Adams, Texas A&M University for writing the Study Guide; John Peck, St. Cloud State University for a fine job with the Instructors Guide; and Michele Hluchy and Gordon Godshalk, Alfred University for revising the Test Bank.

Finally, Bernard Nebel gives great thanks again to Richard Wright for continuing to share in the work of creating this volume, help and support which it would be impossible to do without; Ava Chitwood for typing up the bibliography; and my son, Christopher, who has helped with computer problems. Most of all I wish to thank my late wife Janet, who passed away April 26, 1994 after a long ordeal with cancer, but whose memory and love continue to sustain me. Many other friends and colleagues who, while not being mentioned individually, contributed in untold ways.

Richard Wright again wishes to express his appreciation for the opportunity of working with Bernard Nebel. This has been a fruitful collaboration, and the mutual respect and help that has been built over the years have been essential to the success of this new edition. I also thank my colleagues at Gordon College—Jane Andrus, Russ Camp, and Grace Ju—who have tolerated my frequent preoccupation with book work, and who have continually encouraged me in spite of inconveniences to them. Finally, I would like to thank my wife Ann, who has been with me since the beginning of my career in biology and has provided the emotional base and companionship without which I would be far less of a person and a biologist. Her love and patience have sustained me in immeasurable ways.

For the Instructor

Instructor's Manual (0-13-769423-7)
By John Peck, St. Cloud State University
This manual features a unique annotated lecture outline with presentation suggestions and topics for classroom discussion and debate.

Test Item File (0-13-769449-0)
By Michele Hluchy and Gordon Godshalk, Alfred University. Contains over 1800 test questions, including multiple choice, short answer, and essay questions all with text page references and coded by level of difficulty.

Prentice Hall Custom Test IBM (013-769456-3)
Prentice Hall Custom Test Macintosh (0-13-769464-4)
Prentice Hall Custom Test is based on the powerful testing technology developed by Engineering Software Associates, Inc. (ESA). Available for Windows, Macintosh, and DOS, Prentice Hall Custom Text allows educators to create exams specifically for their own needs. With the Online Testing option, exams can also be administered online and data can then be automatically transferred for evaluation. A comprehensive desk reference guide is included, along with on-line assistance.

Transparency Pack (0-13-769472-5)
A selection of 125 four-color transparencies of images from the text as well as 50 blackline transparency masters.

Slides (0-13-769480-6)
A selection of 125 slides. Same images as transparency acetates available in slide format.

The ABC News/Prentice Hall Video Library
Volume III (0-13-769498-9)
This unique video series contains nine new (3–10 minute) segments from award-winning shows such as 20/20, World News Tonight, and the American Agenda. Selected from the archives of ABC news, each video includes a written summary that ties the video segment to particular sections of the text, making it easier to enhance your classroom presentation with timely and relevant video segments.
Volume 1 (0-13-285578-7) and **Volume II** (0-13-381450-5)
Twenty-eight video segments from the previous edition are also available.

Presentation Manager CD-ROM Image Bank for Environmental Science
This unique image bank contains all illustrations from the Sixth Edition of Environmental Science as well as animations and videos in a digitized format for use in the classroom. The CD-ROM includes a navigational tool to allow instructors to customize lecture presentations. Additional features include keyword searches and the ability to incorporate lecture notes based on custom presentations. The Microsoft Internet Explorer is included on the Presentation Manager CD-ROM for insturctors to browse the Web.

College Newslink
This subscription service delivers discipline-specific news from more than 40 national and international news sources daily to your e-mail in box. Visit http://www.ssnewslink.com for more information or to register for a free trial subscription.

For the Student

The New York Times *Themes of the Times* Supplement
Coordinated by Linda Butler, University of Texas at Austin.
This unique newspaper-format supplement brings together a collection of recent environmental articles from the pages of *The New York Times*. This free supplement, available in quantity through your local representative, encourages students to make connections between the classroom and the world around them.

Study Guide (0-13-761362-8)
By Clark Adams, Texas A&M University
An excellent review tool offering both concept and content review exercises.

Environmental Science World Wide Web Home Page
(http://www/prenhall.com/nebel)
This unique tool is designed to launch student exploration of environmental science resources on the Web. This page is regularly updated and linked specifically to text chapters.

Biology on the Internet: A Student's Guide (0-13-890120-1)
Andrew Stull
The perfect guide to help your students take advantage of our *Environmental Science* home page on the World Wide Web. This unique resource gives clear steps to access our regularly updated *Environmental Science* resource area as well as an overview of general navigation and research strategies.

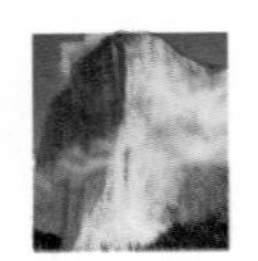

YOU CAN MAKE A DIFFERENCE

Simply by purchasing this text you can have a positive impact on the environment. The authors of this text are pleased to contribute a portion of their royalties to one of the environmental organizations listed below. We are proud that together we can make a difference.

Directions: Please complete the information below and then check the box next to the organization to which you would like us to make a contribution. Carefully cut this page from your text and mail it to: Biology Editor, Prentice Hall, One Lake Street, Upper Saddle River, NJ 07458.

Name: ______________________________

School: ______________________________

Year in School: FR SO JR SR Graduate Student Other

Your Instructor's Name: ______________________________

Please send my contribution to the organization I have inidcated:

☐ **Bread for the World**

☐ **Freedom from Hunger**

☐ **Habitat for Humanity International**

☐ **Natural Resource Defense Council**

☐ **Nature Conservancy**

☐ **World Wildlife Fund**

Environmental Science

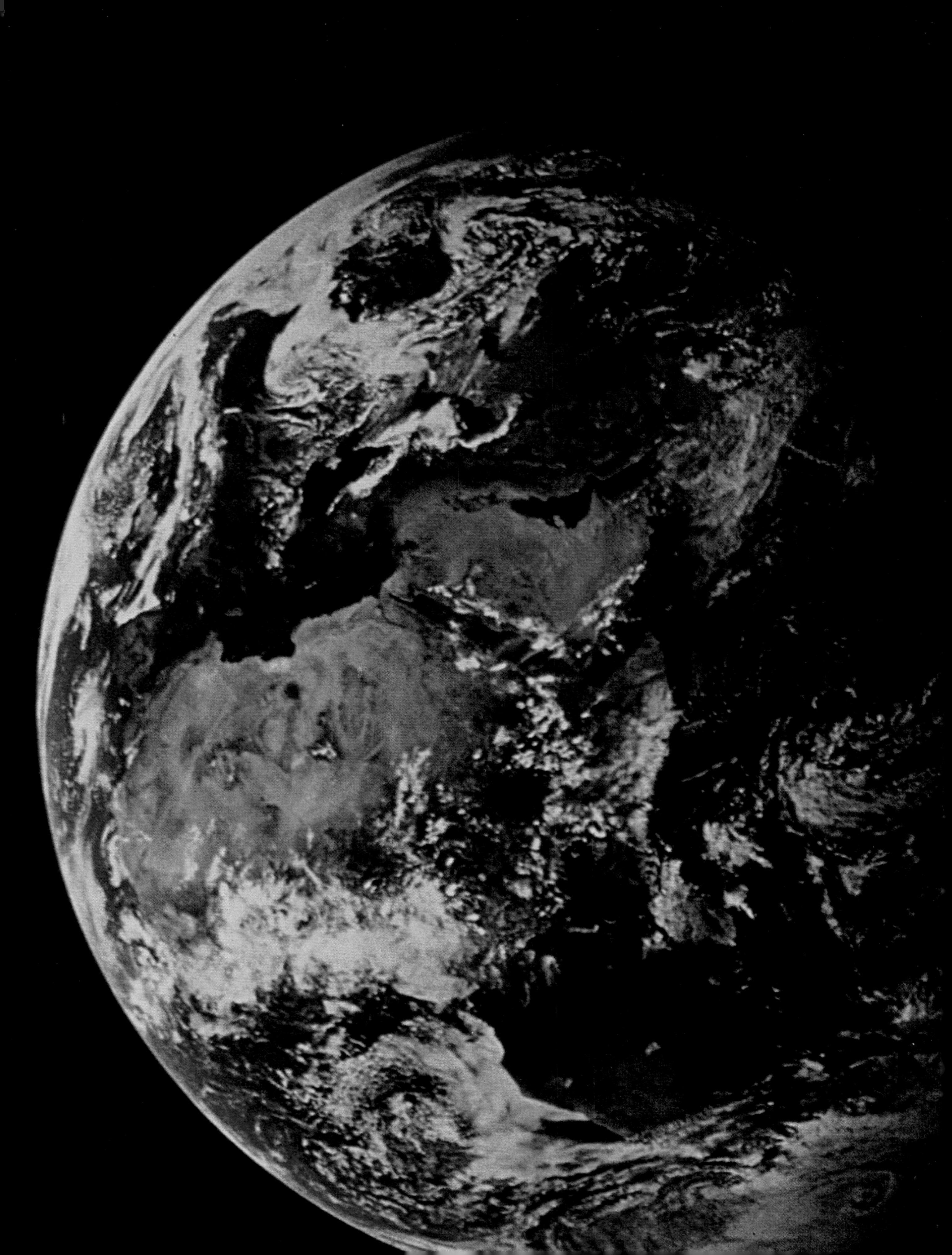

CHAPTER 1

Introduction: Environmental Science and Sustainability

Key Issues and Questions

1. Sharp contrasts exist in the way in which different peoples of the world interact with their environments. Give a range of contrasting "pictures" supporting this statement.
2. History is a saga of rises and falls of civilizations. What are the factors that have brought about the collapse of past civilizations? Are there any parallels in the present?
3. Human concern for the environment has evolved and grown over the years. How have these concerns, the number of people involved, and their responses changed over the years? What are the grounds upon which environmental critics rest their arguments?
4. Environmentalism has won many local battles, but it is still losing the war on a worldwide scale. Describe some of the local battles environmentalism has won? Which global trends indicate that we still might be losing the war?
5. The key concept for the future is *sustainability*. What is meant by sustainability? For a sustainable society, what are the principal prerequisites?
6. A new concept of *sustainable development* is in sharp contrast with the traditional concept of development. Contrast and compare sustainable and traditional development concepts.
7. Present-day environmentalism includes people from the business, economics, and religious communities, as well as world leaders. Discuss the common interests that are bringing all these people together? What is the prospect for the future? Assess the future of the Earth–human relation.

Imagine that you are assigned the task of traveling throughout the world to document human interactions with the environment. Armed with a camera, you start your trip in South America. Boating down the Amazon River through Peru's rain forest, you might photograph small clearings along the shore where there are a few small houses. Constructed of poles cut from the forest, lashed together with vines, and covered with a thatch of braided palm leaves, these primitive dwellings provide simple but adequate shelter in a climate where temperatures range from 75° to 85° F (24° to 29° C) year round (Fig. 1-1).

Earth photo taken by astronauts on the way to the Moon.
[NASA photo.]

People living here have no running water or sewers, no electricity or telephones, and there are no shops or markets of any kind. The rain forest and the river provide all needs: fish, game, fruits, and even medicines for those who learn which plants to use. People here have lived this way for centuries and might continue living in this manner for untold centuries more—except that their way of life is being upset by the impact of modern civilization, population expansion, forests destroyed for pasture, and mining and oil exploration.

In Tanzania, in contrast, you would observe a much harsher life: Women and girls must trudge miles across a denuded, eroded landscape each day to collect all the water they will use for drinking, cooking, and washing. But even these supplies of water may be polluted, and amounts are dwindling. Similar treks of increasing length must be taken to collect the firewood that is the only source of fuel for cooking (Fig. 1-2). Food is mostly coarse grains such as sorghum, and the amounts people get of it are minimal.

Back in the United States and other more developed countries (MDCs), life for many appears much easier. People adjust the indoor temperature to their liking, and any amount of safe hot or cold water is available with the turning of a faucet handle. Travelers go nearly anywhere they wish in the air-conditioned comfort of private cars. Their greatest concern about food is that they not gain weight. News and entertainment from around the world are displayed on several television sets in each home, and many families own computers with rapidly growing interactive capabilities. There is an appearance of a civilization that is detached from its environment unlike those living in less developed countries (LDCs) such as Brazil and Tanzania. Yet in MDCs, we see more of the natural environment giving way to human developments of one kind or another (Fig. 1-3).

Human societies interact with nature with varying intensity in the ways they exploit the natural environment. We can look to past civilizations for better understanding. On Easter Island, a remote spot of land in the South Pacific, we find giant stone heads standing as sentinels with their backs toward the sea (Fig. 1-4). These statues are evidence of a once sophisticated civilization. Yet, Easter Island natives encountered by 19th-century explorers were living at a primitive level, scratching out a meager existence on a desolate island. When asked about the great stone heads, the natives could say only, "Our legends say that our ancestors made them, but we don't know how or why." The past culture and civilization of the island was not sustained.

Working from the legends that natives told, and conducting excavations for evidence, archeologists have pieced together the following chronology of events. Where the original inhabitants of Easter Island came from remains a mystery. However, they probably arrived sometime in the 13th century. The evidence from pollen grains found in the soil and in artifacts

FIGURE 1–1
People living in the rain forest along the banks of the Amazon River in Peru depend on the forest and river for all their needs. Canoes are made from hollowed out larger logs. Food is the daily catch. The river serves as both water supply and sewer. The native people are living in greater harmony with the environment than we see in a modern city. (Photographs by BJN.)

FIGURE 1–2
For one-third of the world population the only source of fuel for cooking is firewood. Many women in less developed countries must spend several hours each day gathering wood. Treks become longer and more difficult as the landscape is increasingly denuded. See Figure 1–12 for a solution to this problem. (The Stock Market.)

show that these early arrivals found an island abundantly forested with a wide variety of trees, including palms, conifers, and mahogany. As their population grew and flourished, they cut trees for agriculture and structural materials, even clearing entire forests. Without plant roots, the cleared land failed to hold water, and the soil washed into the sea, killing the fish and shellfish near the shore. The eroded soil baked hard and dry after rains, offering little support for agriculture. More forestland was cleared, and the population continued to grow. Also, scientists are discovering vast linkages in the global climate system. Changes in expected precipitation and temperature might have hastened the failure of this society.

As the forest was depleted and soil and water resources were degraded, the work necessary for existence became harder and the rewards fewer. The gap between the ruling and worker classes widened, apparently becoming intolerable. In 1678, there was a sudden revolt of the workers. In the great war that ensued, virtually the entire ruling class was killed. Still, the situation worsened. Anarchy broke out among the workers, who splintered into groups and continued to fight among themselves. In the struggle, there was no thought of preserving forests or maintaining agriculture; the remaining trees were cut, and more soil washed into the sea. Starvation and disease became epidemic. A population that had numbered about 8000 at the time of the revolt was down to a few hundred persons by the mid-1800s.

The lesson of Easter Island is all too clear: When a society fails to care for the environment that sustains it, when its population increases beyond the capacity of the land and water to provide adequate food for all, and when the disparity between haves and have-nots widens into a gulf of social injustice, the result is disaster. The civilization collapses.

Are there some parallels between Easter Island and the late 20th century? Is there evidence that we are making some of the same mistakes made by the Easter Islanders? To answer these questions, let us examine the basic thinking of some individuals and groups known as environmentalists. Everyone alive participates with and depends on environmental systems for energy, food, oxygen, water, and waste processing, but many do not recognize the implications. **Environmentalists** are people who pay close attention to the connections between people and their environment.

FIGURE 1–3
Natural ecosystems giving way to development. Increasing production requires a massive reorganization and exploitation of natural resources. Here we see that stripping away of a mature forest in the northeastern United States. (M. Kirk/Peter Arnold Inc.)

FIGURE 1–4
Easter Island. The great stone heads and other artifacts found on Easter Island provide evidence of a once-prosperous culture. The present barren, eroded landscape indicates that the civilization collapsed as a result of overexploitation of forest and soil resources. Is the story of Easter Island a parable for modern civilization? (Jan Halaska/Photo Researchers, Inc.)

Environmentalism

Concern about the degradation of the natural world has led many to become environmentalists. Although what we think of now as the environmental movement began less than 50 years ago, it had its roots in the late 19th century, when some people realized that the unique, wild aspects of the United States were disappearing. The 1890 U.S. census demonstrated the "closing of the frontier," an event that was noted with some sadness. No longer could any area of the country be classified as totally uninhabited. Around that time, several groups devoted to conservation formed, among them the National Audubon Society and the National Wildlife Federation. The Sierra Club was founded in California by naturalist John Muir, who helped popularize the idea of wilderness. The national parks system was created. The national environmental consciousness was stirring.

In the scientific era that followed World War I, ideas of wilderness and conservation began to seem quaint and old fashioned. Chemical fertilizers and pesticides were developed, and deserts were irrigated, changing agricultural productivity. Western lands were converted from unfenced grazing areas to neat, fenced wheat farms. New medical knowledge increased human longevity by controlling outbreaks of infectious diseases, and plastics were developed, replacing natural materials for many purposes. Huge machines replaced farm animals and simple plows. An earlier conservation ethic gave way to daring technology and the engineering disciplines.

Technological achievements, however, eventually helped create an environmental crisis, the Dust Bowl of the 1930s, when wheat failed to hold soil in place and the topsoil eroded (Fig. 1-5). During the Great Depression (1930–1936), conservation suddenly became fashionable again as a means both of restoring the land and providing work for the unemployed. Many trails, erosion control projects, and other improvements in national parks and forests were originally installed by the Civilian Conservation Corps (CCC). This program played a major role in pulling the country out of the Depression.

Before the budding conservation ethic of the 1930s could become firmly established, World War II broke out. As a nation at war, the United States was forced to utilize its resources, not save them for future generations; fuel, metal, and other natural resources were thus directed toward the war effort.

Silent Spring, Loud Reaction

The two decades following World War II (1945–65) were full of optimism. We had won the "greatest" of all wars, in part because of technology. The tremendous productive capacity built up during the war, and new developments ranging from rocket science to computers and from pesticides to antibiotics, could be redirected to peacetime applications. Except for the tensions of the Cold War during those years, and some apprehension about atomic energy, it seemed as if nothing but opportunity and prosperity lay ahead.

FIGURE 1–5
The Dust Bowl and an abandoned farm near Guyman, Oklahoma, in 1937. In the 1930s, what were then modern farming practices combined with drought conditions and a century of overgrazing to create an environmental disaster during which farmland throughout the central U.S. was largely destroyed by wind erosion. For a time this disaster spurred a rising environmental awareness. Has the lesson been forgotten? (The Bettmann Archives.)

Although economic expansion enabled most families to have a home, a car, and other possessions, certain problems became obvious. The air in and around cities was becoming murky and irritating to people's eyes and respiratory systems. Street lights from St. Louis to Pittsburgh were left on during the day because of the smoke from coal-burning industries. In winter, even freshly fallen snow soon turned gray as the soot fell from city chimneys. Rivers and beaches were increasingly fouled with raw sewage, garbage, and chemical wastes. The effects of air, land, and water pollution (contamination) affected all living systems: Conspicuous declines occurred in many bird populations—including our national symbol, the bald eagle, aquatic species, and other animals.

It was easy to identify some culprits—belching industrial smokestacks, open burning dumps, and municipal and industrial sewers discharging raw sewage and chemical wastes into waterways. The decline of the bald eagle and other bird populations was found to be caused by the accumulation in their bodies of DDT, the long-lasting pesticide that had been used in large amounts since the 1940s. In short, it was clear that we were seriously contaminating our environment.

In 1962, biologist Rachel Carson wrote *Silent Spring*, presenting her scenario of a future with no songbirds and with other dire consequences if pollution of the environment with DDT and other pesticides continued. Carson's voice was soon joined by others, many of whom formed organizations to focus and amplify the voices of thousands more in demanding a cleaner environment. This was the beginning of the modern **environmental movement**, in which a newly militant citizenry demanded the curtailment of pollution, the cleanup of polluted environments, and the protection of still-pristine areas. It is significant that the environmental movement began as a grassroots initiative, and it maintains its momentum and force today only by continuing to command public interest and support.

Members of the environmental movement of the 1960s joined forces with older organizations, such as the National Audubon Society and the National Parks Conservation Association, that had a considerable history of dedication to preserving wildlife and natural habitats. There is an obvious overlap between the goal of reducing pollution and that of protecting wildlife. Wildlife cannot survive in polluted surroundings, and preserving an uncontaminated environment presupposes that the plant and animal species within that environment will be protected. Therefore, established wildlife-preservation organizations and their members became significant players in the blossoming environmental movement, along with newly formed organizations such as the Environmental Defense Fund, the Natural Resources Defense Council, Greenpeace, and Zero Population Growth. This dual focus continues today. As the broad category of "environmentalists" includes everyone concerned with reducing pollution and/or protecting wildlife, it is no wonder that at least 70% of the population identify themselves as environmentalists.

Early Results Were Encouraging

By almost any measure, the environmental movement has been successful. Pressured by a concerned citizenry, Congress created the Environmental Protection Agency in 1970 and passed numerous laws for pollution control and wildlife protection—among them the Endangered Species Act of 1973, the Marine Mammals Protection Act of 1972, the Clean Air Act of 1970, the Clean Water Act of 1972, and the Safe Drinking Water Act of 1974. You will read more about these laws and their effects in subsequent chapters. For now, let us simply say that a number of species probably have been saved from extinction—at least for the present. In the area of pollution abatement alone, industry has spent hundreds of billions of dollars, both on installing pollution control devices and on redesigning procedures and products so that less pollution is created. The pollution control devices on our cars are an example. Governments have spent additional billions on upgrading sewage treatment, on refuse disposal, and on other measures to reduce pollution. As a result of these expenditures, the air in most cities and the water in numerous lakes and rivers are cleaner now than they were in the late 1960s. Without a doubt, they are immensely cleaner than they would be if the environmental movement had not come into existence.

Expenditures for pollution control are continuing. In its various aspects ranging from law enforcement to the design, manufacture, and operation of equipment, pollution abatement has grown to be a major sector of our economy, providing jobs that did not even exist in 1960. Anyone contemplating becoming a professional environmentalist may do well to consider careers in areas such as environmental law, environmental economics, or environmental engineering, and geographic information systems (GIS) for example.

Environmentalism Acquires Its Critics

In the early stages of the environmental movement, the sources of the problems were specific and visible. That made it easy to identify certain polluters as the "bad guys" and take the side of the environmental movement. Likewise, the solutions seemed relatively straightforward—install waste treatment and pollution control equipment and ban the use of DDT, substituting safer pesticides. Despite the obvious pollution and needed solutions, researchers like Rachael Carson were severely attacked by industry in the media.

Importantly, it was possible in many cases to achieve significant improvements in air and water quality with only modest expenditures by addressing these **point sources** (specific producers of pollution).

Given this beginning at cleaning up major point sources of pollution, further improvements in air quality demand addressing **diffuse sources**—the innumerable small sources such as automobiles and transportaion, home fireplaces, barbecue grills, gas lawn mowers, and the runoff from agricultural fields and every homeowner's lawn and garden. Although no one individual is responsible for much of this pollution, the sum total from everyone is significant. In other words the "bad guys" are no longer "them," but "us," and many of us do not take well to this notion: We perceive restrictions on such things as lawn mowers as unacceptable infringements on our personal liberty. Yet, the improper disposal of auto crankcase oil annually exceeds that from tanker spills in the ocean!

Another early effort—the banning of DDT to save the bald eagle—caused little disturbance because alternative pesticides were available, and after all, our national symbol was at stake. Indeed, populations of

FIGURE 1–6
Spotted owl and old growth forests of the Northwest. The survival of the spotted owl is about the survival of the forests. The spotted owl requires undisturbed old growth forests to survive. Thus, its listing as an endangered species has been used as a means to curtail logging and to preserve the last-remaining stands of old-growth forests. (background photo G. C. Kelly/Photo Researchers, Inc.; owl Tom and Pat Leeson/Photo Researchers, Inc.)

FIGURE 1–7
Antagonism toward environmentalists and environmental regulations is becoming more common as people feel that their livelihoods are threatened. This scene is part of a crowd of 15 to 20 thousand demonstrators at a pro-logging rally in Victoria, B.C., Canada, in March 1994. (Chip Vinai/Gamma Liaison.)

bald eagles and certain other fish-eating birds have recovered markedly since DDT was banned in the early 1970s. However, populations of bald eagles in the Northwest are now declining again for as yet undetermined causes. In the future, continuing to save the bald eagle may lead to more difficult choices. If the cause of the eagle's decline is found to be something having a positive value for the economy, banning it may prove controversial.

The bitterness of the conflict between environmentalists and those with special business interests is illustrated by the case of the spotted owl, which inhabits old-growth forests of the Northwest (Fig. 1-6). Saving the spotted owl from extinction requires putting large tracts of old-growth forests off limits to logging. Indeed, the central interest of environmentalists is to save some of these old-growth forests in their natural state. However, curtailing logging of the last-remainingin old-growth forests is portrayed as threatening the jobs of many loggers, some logging companies, and even whole communities whose economies are based solely on logging. Many loggers are hostile toward environmentalists who would effectively trade their jobs for "some bird no one ever heard of before the controversy started" (Fig. 1-7). Similarly, many landowners become hostile toward environmentalists as they find themselves prevented from developing their property because it is home to an endangered species or is subject to some other environmental restriction. They may also become hostile toward the government in general, since the government is responsible for enforcing these regulations. Although we might ask if the forestry industry had ever heard of Easter Island.

Many people are concerned about regulations that may threaten their well-being and even their economic survival. This makes them easily swayed by critics of the environmental movement who want to be free of regulations protecting the environment. A significant contingent of such critics is made up of the *Wise-Use* movement. Leaders of this movement are the oil corporations, polluting industries, ranchers, executives of logging companies, officials of extractive mining corporations, and so on. Their common interest is that they wish to continue to exploit the Earth's resources, with no restrictions or controls on their activities. Today, we see that even loggers recognize that policies and practices of the logging companies cost them more jobs than environmental regulations. New alliances of loggers and environmentalists are forming to promote sustainable forestry.

Has the environmental movement accomplished everything it can? Is environmentalism merely a passing fad? Has it served its purpose? Along with most other professionals in the environmental arena, we feel that our future will depend on expanding and deepening the role of environmentalism, rather than weakening it.

The Global Environmental Picture

As we have seen in the preceding, environmentalism has won many local battles, but is still losing the war on a worldwide scale. Four global trends are of particular concern: (a) population growth and increasing consumption per person, (b) degradation of soils, (c) global atmospheric changes, and (d) loss of biodiversity. Each of these issues will be explored in greater depth in later chapters.

Population Growth

The world's human population, 6.0 billion persons in 1998, has grown by 2 billion in just the last 25 years. It is continuing to grow more rapidly than at any other time in history, adding nearly 88 million persons per year. Even though the growth rate is gradually slowing, world population is projected to reach about 10 billion by the year 2050 (Fig. 1-8). Each person creates a certain demand on Earth's resources, and the demand tends to increase with greater affluence. Consider, for example, the resources required to support a typical

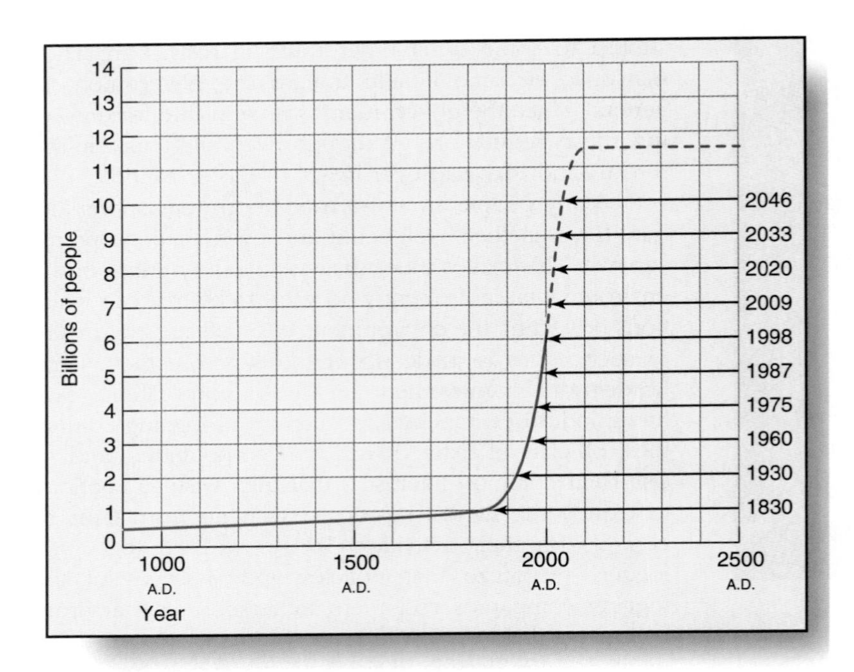

FIGURE 1–8
World population started a rapid growth phase in the early 1800s and has grown sixfold in the last 200 years. It continues to grow by nearly 88 million persons per year. (See Chapter 6.)

American or Canadian lifestyle, compared with those required to support Native Americans living along the banks of the Peruvian Amazon.

Vital resources are stressed by the dual demands of increasing population and increasing consumption per person. Around the world we see groundwater supplies being depleted, agricultural soils being degraded, oceans being overfished, oil reserves being drawn down, and forests being cut faster than they can regrow. Despite efforts to improve this situation, growing numbers of people are suffering—as is amply demonstrated by the African women and girls who must trek long distances across barren land to fetch water and firewood (Fig. 1-2).

How can Earth support a near doubling of its human population over the next 50 years, as is projected, and still increase standards of living at the same time?

Degradation of Soils

Fertile soil is the foundation for plant growth and food production. Yet, around the world, soils are being degraded by erosion (Fig. 1-9a), grazing lands are turning into deserts, irrigated lands are becoming too salty to support crops, water supplies for irrigation are being depleted, and additional millions of acres of agricultural land are being sacrificed for development (Fig. 1-9b).

Global Atmospheric Changes

Historically, pollution has been a relatively local problem, affecting a given river, lake, bay, or the air in a city. Today, scientists are analyzing pollution on a global scale. A case in point is the danger of global warming. An unavoidable byproduct of burning fossil fuels—gasoline and other liquid fuels from crude oil, coal, and natural gas—is carbon dioxide (CO_2).

Carbon dioxide is a natural component of the lower atmosphere, along with nitrogen and oxygen. It is required by plants for photosynthesis and is important to the Earth-atmosphere energy budget. Carbon dioxide is transparent to incoming light from the Sun but absorbs infrared (heat) energy radiated from Earth's surface, thus delaying its loss to space. This process warms the lower atmosphere in a process known as the *greenhouse effect*. Although the concentration of carbon dioxide is a small percentage of the atmosphere, even slight increases in its volume affects temperatures.

Because of the large amount of fossil fuels being burned today, carbon dioxide levels in the atmosphere have grown from about 280 parts per million (ppm), or 0.028 percent, in 1900 to over 370 ppm as we approach the end of the century. It is increasing at .4% per year and is expected to double during the next century. The latest conclusion of the Intergovernmental Panel on Climate Change (IPCC), published in 1995, stated that

> Human activities, including the burning of fossil fuels...are increasing the atmospheric concentrations of greenhouse gases. These changes...are projected to change regional and global climate and climate related parameters such as temperature, precipitation, soil moisture, and sea level.

Figure 1-10 graphs air temperatures from 1880 to the present and illustrates an overall warming trend—more on this in a later chapter. Carbon dioxide is thought to be responsible for almost 60% of the global warming trend. Even considering a margin of error, a

(a)

(B)

FIGURE 1–9
(a) Throughout much of the world, agricultural soils are degraded by erosion. (Jeff Lepore/Photo Researchers, Inc.) (b) Around every metropolitan area in North America, as well as in other parts of the world, agricultural land is consumed by development causing serious loss of millions of acres of soil from

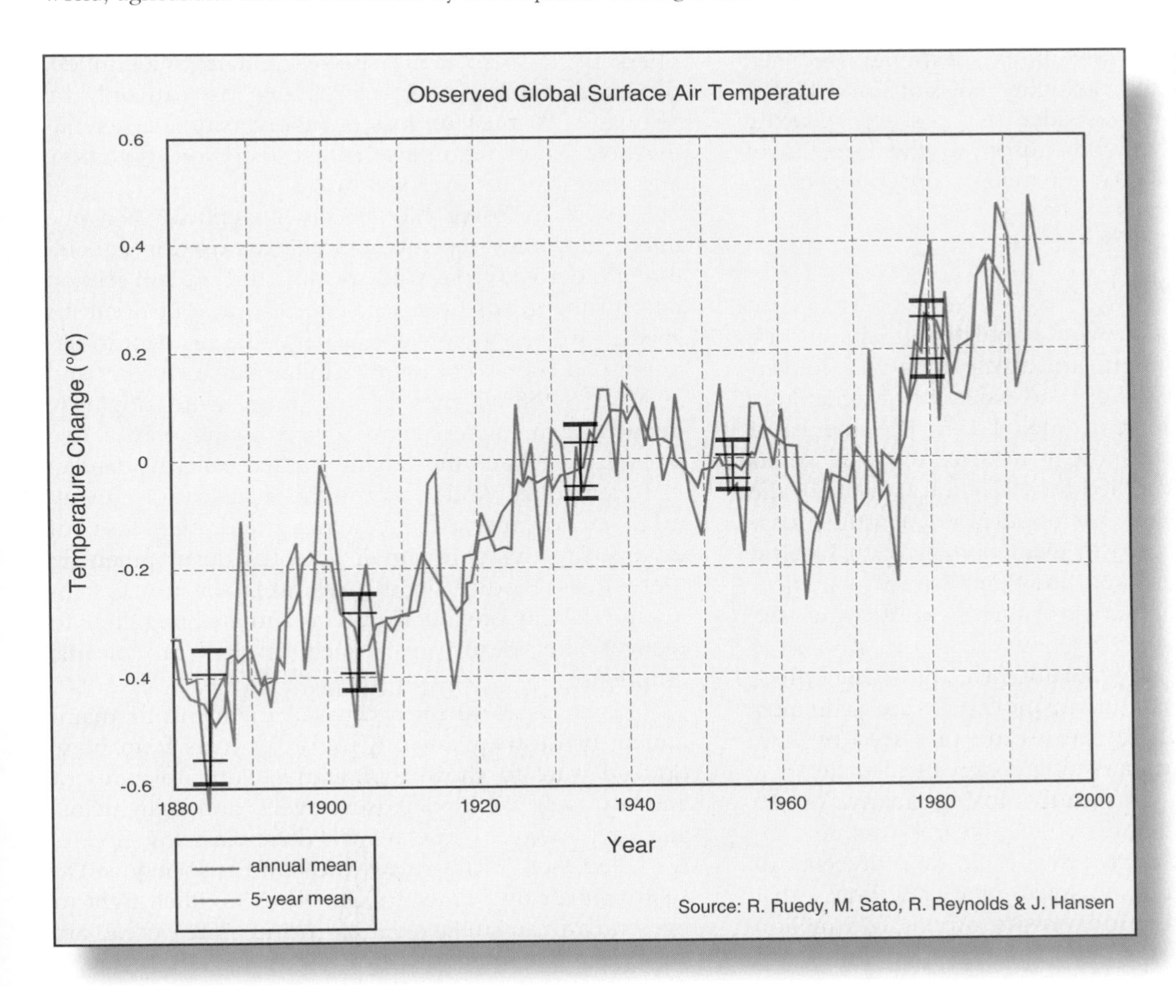

FIGURE 1–10
Global temperature trends from 1880 to 1995. The 0 baseline represents the 1950–1980 global average. Note the cooling effect of the Mount Pinatubo volcanic eruption in 1991. Global temperatures quickly recovered setting a new record in 1995.

FIGURE 1–11
The loss of biodiversity is a serious problem. Record numbers of plant and animal species face extinctions. This photo shows massive clearing of tropical rainforests in Malaysia for sugar plantations. (Fletcher and Baylis/Photo Researchers, Inc.)

reduction in fossil fuel use will produce benefits beyond slowing the greenhouse warming (reduced energy costs, reduced pollution emissions and tanker spills). Society must consider the cost of reducing greenhouse gases in comparison to the benefits of avoiding some of the financial consequences of increased temperatures.

Loss of Biodiversity

Rapidly increasing human populations, along with increasing consumption, are accelerating the conversion of forests, grasslands, and wetlands to agriculture and urban development (Fig. 1-11). The inevitable result is the extermination of most of the wild plants and animals that occupied those natural habitats. If the species involved have no populations at other locations, they are doomed to extinction by such **habitat alteration**. Pollution alters additional habitats—particularly aquatic and marine habitats—destroying the species they house. Also, hundreds of species of mammals, reptiles, amphibians, fish, birds, butterflies, and innumerable plants are exploited for their commercial value. Even where species are protected by law, hunting, killing, and marketing continues illegally.

Thus, Earth is rapidly losing many of its species—as many as 17,500 per year by some estimates. The term used to refer to the total diversity of living things—plants, animals, and microbes—that inhabit the planet is **biodiversity**. About 1.75 million species have been described and classified, but scientists estimate that up to 100 million still remain unidentified. Because so many species remain unidentified, the exact number of species being lost can only be estimated. At present, loss of biodiversity is accelerating because of mounting habitat alteration, pollution, and pressures for exploitation.

Why is losing biodiversity so critical? For one thing, all domestic plants and animals used in agriculture are derived from wild species, and we still rely on introducing genes from wild species into our domestic species to keep them vigorous and able to adapt to different conditions. For another thing, between 1959 and 1980, 25% of all prescription drugs were originally derived from higher plants, even though only a few percent have been thoroughly studied from this medicinal viewpoint. Biodiversity is the mainstay of agricultural crops and of medicines, and the loss of biodiversity can only curtail potential development in these areas. Biodiversity is a critical factor in maintaining the stability of natural systems and enabling them to recover after disturbances such as fires or volcanic eruptions.

There are aesthetic and moral arguments for maintaining biodiversity, also. Forty or 50 years from now, do you want to show your grandchildren pictures of animals such as rhinoceroses, tigers, and orangutans, and have to say, "These animals don't exist any more—we killed them"? The moral question for society is, Do such animals and other species have as much right to live on Earth as humans do? More and more people are answering this moral question with a yes.

Ethics

Are We in the Process of a Major Paradigm Shift?

Very rare but significant events in human history are what scientists call *paradigm shifts*. They are major changes in the way humans view the world and their place and role in it. They can also be called major shifts in world view. They tend to be fraught with controversy and conflict at the time they are first presented, but then they usher in a whole new era in the advancement of knowledge and understanding. An example will help illustrate the concept.

Prior to the time of Nicolaus Copernicus (1473–1543), most people believed that Earth was the center of the Universe. The Sun, Moon, planets, and stars were thought to all revolved about Earth. In 1512, Copernicus presented his theory that the Sun was the center of a system, and Earth and the other planets revolved around the Sun. Steeped in the old world view, not only did people ignore the new theory, but anyone who suggested that it had merits was vigorously attacked by the existing power structure, dominated by the Catholic Church that had a vested interest in maintaining the old beliefs. Indeed, about 100 years later, Galileo faced an inquisition and was prohibited by the church from doing further scientific work for daring to present hard evidence supporting the Copernican theory. However, other scientists extended Galileo's observations, thus opening up a whole new era of advancing understanding about the universe, an era that continues, of course.

The question we now pose to you is, Are we in the midst of another, even more significant, paradigm shift regarding an old world view and that of environmentalism? The old world view, dating from early Judeo-Christian times, at least in Western culture, is that the world's plants, animals, minerals, etc., exist for the express purpose of benefiting humans. Therefore, treating these things simply as resources to be exploited not only is acceptable, but is the right and proper course of action. It is further implied and assumed in this view that the "bounty" awaiting exploitation is infinite, making conservation or preservation needless. Given this world view, it is perhaps understandable that the entire economy and lifestyle of Western civilization came to be what it is.

The world view presented and promoted by environmentalists is the antithesis of this old world view in virtually all respects. According to environmentalists, the world is not infinite. Continued exploitation is not sustainable. The continued well-being of humans will depend on the conservation of wild plants and animals and the protection of air and water. In short, the new view amounts to a paradigm shift from seeing humans as the center of things, free to ride over nature in any manner possible, to seeing nature and humans intricately linked in life processes and global systems. Perhaps the reluctance of many to make this shift and the consequent vehemence they express toward environmentalists is similar in ways to what Copernicus and Galileo faced.

Of course, we want to consider the evidence presented by environmentalists in support of their new world view. But if it stands up to scrutiny, can we afford to ignore it? There is more riding on this environmental paradigm shift than on the shift in the old view of the solar system. With environmentalism, we are saying the future of humanity in the years immediately ahead is at risk. Moreover, environmentalism will demand more than just an academic awareness. To have any meaning, it must come down to our personal lives, our lifestyles, and how we personally affect the environment. Thus, beyond an academic belief, environmentalism will require an ethical and moral commitment to the stewardship of Earth that will engender concrete actions.

Toward the Future

Increasing numbers of people in all walks of life—scientists, economists, business people, and world leaders, as well as professional environmentalists—are recognizing that the global trends just outlined are *not sustainable*. Common sense dictates that these trends are all on a collision course not only with basic human needs, but also with fundamental systems that maintain our planet as a tolerable place to live. A finite planet cannot go on adding nearly 90 million additional persons annually, nor can we tolerate current losses of soil, atmospheric changes, losses of species, and depletion of water resources without arriving at a point where resources are no longer adequate to support the human population and civil order breaks down.

Few, if any, disagree with the idea that extrapolating current trends far enough into the future leads to a dismal scenario. The debate concerns what kinds of actions we need to take to avoid that scenario. Population versus food production serves to illustrate two distinct schools of thought. In 1798, the English economist Thomas Malthus realized that the human population, then still fewer than one billion persons, was embarked on exponential growth whereas agricultural production would be limited. He forecast that massive famines would occur in the early 1800s.

Contrary to his dire predictions, technological innovations have enabled agricultural production to grow even more rapidly than population. Today, as the human population reaches 6 billion, agricultural surpluses still exist but their margin has narrowed. (Famines, of course, do occur, but they result from the difficulty of distributing food to needy areas, not from the total amount of food available.) Malthus's prediction was based on the faulty assumption that food production methods would remain the same; he failed to realize that technology would change the picture markedly.

Is the situation any different today? Environmentalists believe that there are two differences today: (a) Stresses on the environment are growing faster than at any time in history, and (b) we are reaching certain limits on technological solutions.

As for technology, since Malthus's time, humans have mechanized agriculture, irrigated dry land, added chemical fertilizers to soil, developed new and more productive varieties of crops, and increased catches from the sea by using new fishing methods. Environmentalists see evidence that all of these techniques have nearly reached their full potential. For example, because of limited water supplies, irrigation can no longer expand in some areas; there are no new varieties of plants that may double crop yields again waiting to be introduced; most major fishing areas have already been overfished; and so on. In fact, world food production per person seems to be leveling off and even declining.

On the other side, critics of environmentalism (sometimes called **cornucopians**) have an unbounded faith in the human capacity to devise a technological solution to any problem that arises, environmental or otherwise. Indeed, some critics insist on casting environmentalists as pessimists and doomsayers whose predictions should be ignored, or even in the role of "the enemy" that is standing in the way of human progress.

Most environmentalists counter that it is not their intent to diminish or underestimate the technological capacity of humans in any way. Rather, they only point out that in the past, unbridled optimism and faith in technology have led to disaster as well as success. Consider, for example, the "last words" of the captain of the *Titanic* shortly before it sank: "Damn the icebergs. Full steam ahead!" In short, environmentalists see the issues we have just outlined, and many others as well, as figurative icebergs in our path. Environmentalists are firmly convinced that ignoring these dangers and going full speed ahead is not in our best interests. They believe that the way to a prosperous future is through studying, understanding, and evaluating the environmental issues for exactly the problems they are and using our technological capability to navigate accordingly.

Sustainable Development

To say that a system or process is **sustainable** is to say that it can be continued indefinitely without depleting any of the material or energy resources required to keep it running. The term was first applied in relation to the idea of **sustainable yields** in human endeavors such as forestry and fisheries. Trees, fish, and other biological species are able to grow and reproduce at rates faster than that required just to keep their populations stable. This built-in capacity allows every species to increase or replace a population following some natural disaster.

Thus, it is possible to harvest a certain percentage of trees or fish every year without depleting the forest or reducing the fish population below a certain base number. As long as the number harvested stays within the capacity of the population to grow and replace itself, the practice can be continued indefinitely. The harvest then represents a *sustainable yield*. It becomes nonsustainable only when cutting or catches exceed the capacity for reproduction and growth. The concept of sustainable yield also is applied to freshwater supplies, soils, and the ability of natural systems such as the atmosphere or a river to absorb pollutants without being damaged. In contrast, the global trends we pointed out (population growth, loss of biodiversity, etc.) can all be seen as examples of going beyond "sustainable yields;" they are *not* sustainable.

Extending the concept of sustainability further, we can speak of a **sustainable society** as a society that continues, generation after generation, neither depleting its resource base by exceeding sustainable yields nor producing pollutants in excess of nature's capacity to absorb them.

Therefore, when the concept of sustainability is applied to modern society, it takes on added dimensions. Beyond just having our species survive, the sustainability of society implies preserving the capacity to explore, reflect on, and understand new things. Exploring frontiers—whether they be the far reaches of outer space, the depths of subatomic space, or the workings of the human mind—requires increasingly sophisticated technology and instruments—the Hubble Space Telescope, modern computers, and genetic research as examples. To build, launch, and maintain such technology, we need a phenomenal base of science education, technological skills, and manufacturing capability, as well as an economy that is willing and able to pay for it. Sometimes the technology required is

Earth Watch

Agenda 21

Agenda 21 is the official document signed by world leaders representing 98 percent of the global population at the United Nations Earth Summit in Rio de Janeiro, Brazil, in June of 1992. The following is an excerpt from an abridged version edited by Daniel Sitarz:[1]

> *Agenda 21* is, first and foremost, a document of hope. . . . [I]t is the principal global plan to confront and overcome the economic and ecological problems of the late 20th century. It provides a comprehensive blueprint for humanity to use to forge its way into the next century by proceeding more gently upon the Earth. . . .
>
> Humanity is at a crossroads of enormous consequence. Never before has civilization faced an array of problems as critical as the ones now faced. As forbidding and portentous as it may sound, what is at stake is nothing less than the global survival of humankind.
>
> . . . Where once nature seemed forever the dominant force on Earth, evidence is rapidly accumulating that human influence over nature has reached a point where natural forces may soon be overwhelmed. Only very recently have the citizens of Earth begun to appreciate the depth of the potential danger of the human impact on our planet. . . . Scientists around the world, in every country on Earth, are documenting the hazard of ignoring our dependence upon the natural world. . . . For the first time in history, humanity must face the risk of unintentionally destroying the foundation of life on Earth. . . . To prevent such a collapse is an awesome challenge for the global community. . . .
>
> *Agenda 21* is not a static document. It is a plan of action. It is meant to be a hands-on instrument to guide the development of the Earth in a sustainable manner. . . . It is based on the premise that sustainable development of the Earth is not simply an option: it is a requirement—a requirement increasingly imposed by the limits of nature to absorb the punishment which humanity has inflicted upon it. *Agenda 21* is also based on the premise that sustainable development of the Earth is entirely feasible.
>
> The bold goal of *Agenda 21* is to halt and reverse the environmental damage to our planet and to promote environmentally sound and sustainable development in all countries on Earth. It is a blueprint for action in all areas relating to the sustainable development of our planet into the 21st century. . . . It includes concrete measures and incentives to reduce the environmental impact of the industrialized nations, revitalize development in developing nations, eliminate poverty world-wide and stabilize the level of human population.
>
> *Agenda 21* provides a myriad of opportunities. Suggestions are furnished to develop new industries, pioneer innovative technologies, evolve fresh techniques and institute novel trade arrangements.

Various meetings involving leaders of both governments and non-governmental organizations are continuing to take place around the world to develop, refine, and implement strategies to confront the world's environmental problems. "Gradually it is being understood that the issues of poverty, population growth, industrial development, depletion of natural resources and destruction of the environment are all very closely interrelated." The global society will have to solve all of them—or it will wind up solving none of them.

[1]Daniel Sitartz, *Agenda 21* (Boulder, CO: Earth Press, 1993), pp. 1–5.

simple and inexpensive. For example, contrast the women walking miles to gather wood for cooking fires (see Figure 1-2) with women in Kenya using their solar-panel cookers, which do not require wood scavenging (Figure 1-12).

Finally, a prerequisite underlying all of these ideas is an environment that has clean air, clean water, and healthy and functioning natural systems. An impoverished environment only supports impoverished people. Philip Handler, past president of the U.S. National Academy of Sciences, once said,

> I cannot believe that the principal objective of humanity is to establish experimentally how many human beings the planet can just barely sustain. But I can imagine a remarkable world in which a limited population can live in abundance, free to explore the full extent of man's imagination and spirit.

It is easy to see why there has been a growing recognition in recent years that traditional environmental issues are irrevocably intertwined with social and economic issues.

A New Meaning of Development

Sustainable development is a term that was first brought into common use by the World Commission on Environment and Development, a group appointed by

FIGURE 1–12
Kenyan women in training to use their solar-panel cookers. Using simple solar cookers made of reflective cardboard and black cook pots, villagers are able to cook meals and sanitize drinking water with only the sun for fuel energy. In less-developed countries the pressing need is for decentralized energy sources appropriate in scale to everday needs. (Photo by Solar Cookers International, Sacramento.)

the United Nations. The commission made sustainable development the theme of its final report, *Our Common Future*, published in 1987. They defined the term as a form of development or progress that "meets the needs of the present without compromising the ability of future generations to meet their own needs."

Of course, in the traditional sense, we still use the word *development* to refer to the clearing of natural areas to make room for more shopping malls, housing tracts, or agricultural land—a process that is conspicuously nonsustainable in the long term. Therefore, the idea of sustainable development strikes many people as an oxymoron—a self-contradictory concept.

We need to think of development in a broader sense, of moving toward protecting and enhancing those aspects of the environment and social justice which are necessary conditions for the sustainability of continuing gains in knowledge and understanding. For example, can we think of development in terms of protecting farm soil from erosion? In terms of saving natural areas and the wildlife they support, for their aesthetic, recreational, and scientific value? In terms of stabilizing world population? Can we define development in terms of bettering our physical, mental, and emotional health? In terms of reducing the underlying factors that lead to crime and corruption? In terms of improving relations among different peoples and nations? In terms of increasing people's access to educational and career opportunities? In terms of discovering and implementing new technologies for recycling, reducing pollution, and harnessing solar energy?

In short, can we think of development as learning to be stewards of Earth, not just to protect wild species, but to enhance the general well-being and security of human life for generations to come?

Importantly, this concept of sustainable development should not be confused with the idea of returning to the status of a primitive culture "living in harmony with nature." First, the idea of primitive cultures living happily and peacefully in harmony with nature is largely imaginary. Their lives actually included much suffering, discomfort, and pain, high infant mortality, and early death. They also had wars with neighboring tribes. In reality, humans have always been territorial and covetous of their neighbor's resources. Living in balance with nature was just a result of lacking the understanding and technology to manipulate nature more to their advantage and a result of a belief system that tended to prevent innovation. Also, we must point out again that the world is replete with the ruins of civilizations that did not sustain themselves and became extinct.

Thus, we are not talking about moving back to a state that used to be. We are talking about moving forward to a different human–Earth relation (see "Ethics" box, page 13). It will require a special level of

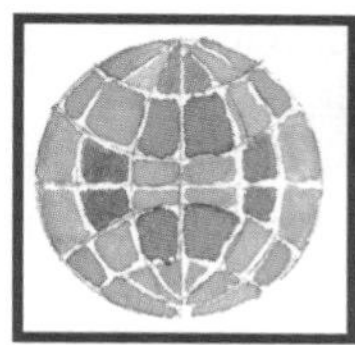

Global Perspective

Mapping Human Impacts on the Biosphere

As the Earth's human population swells, there is an obvious demand to use more and more land for agriculture. This map, created in 1983 by Dr. Elaine Matthews from satellite data and published reports, gives an overview of the extent to which the Earth's surface had been converted to cultivating crops. For the purposes of mapping, a grid cell of approximately 60 by 60 miles was used, and the color indicates the percentage of each area under cultivation.

The extent of human impacts on Earth is considerably greater than the map indicates for three reasons: (1) The world population has grown by another billion people or 20% since the data was compiled which has forced agriculture into new areas and increased the percentage of cultivated land in existing agricultural areas; (2) Areas cleared for or otherwise degraded by grazing, specifically large portions of eastern South America, Africa, and Australia are not shown; (3) Areas deforested for wood products, comprising large portions of Latin America, Asia, and Africa, are not shown.

The kinds of natural habitats (ecosystems) that are being sacrificed can be determined by comparing this map with the map showing the distribution of natural biomes, Fig. 2-4 on page 27. The loss of biodiversity occurring is a function of the biological richness of the ecosystem and its degree of conversion. Importantly, major land areas that appear unaffected according to this map are actually deserts or tundra, which house relatively little biodiversity. Thus, biologists predict massive extinctions as human pressures continue to mount.

Map and information: Courtesy of Dr. Elaine Matthews, NASA Goddard Institute for Space Studies, New York.

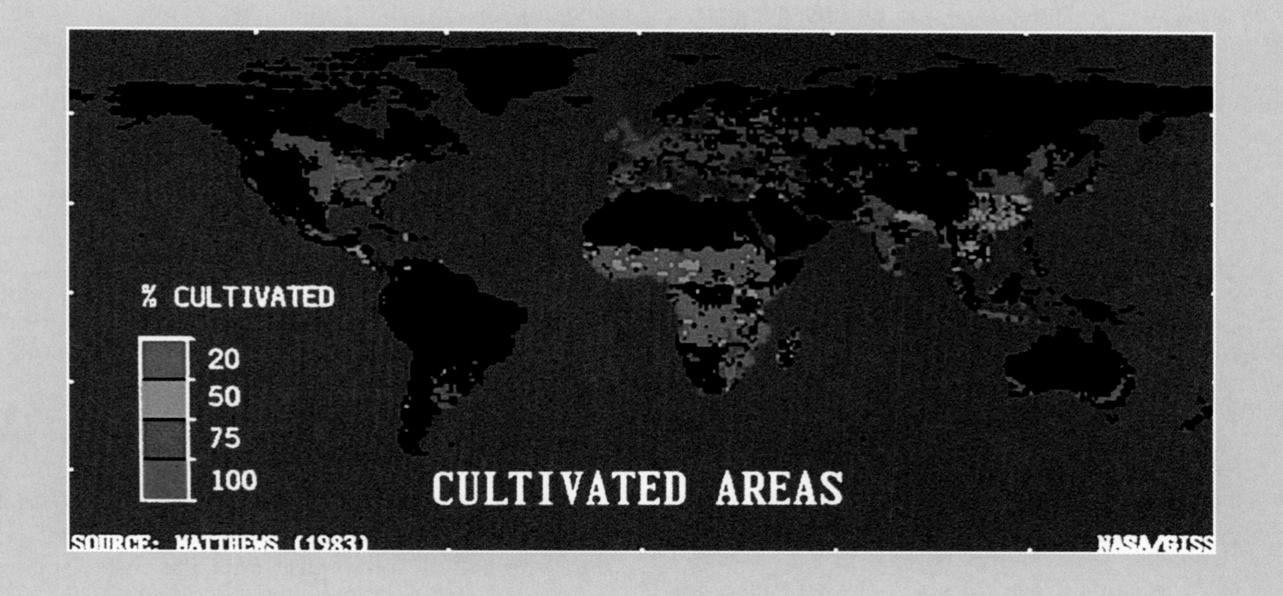

dedication, commitment, and caring for one another as citizens of a global community, as well as being stewards of Earth and other species we share it with. If we actually achieve the transition to a sustainable society, it will be a unique event in human history; no civilization on Earth has ever done it. Achieving a sustainable society will make possible, the continuing growth of a scientific, cultural, and spiritual understanding of ourselves and our place in the universe.

The New Environmentalists

The good news is that, critics notwithstanding, growing numbers of people in all walks of life—scientists, sociologists, workers and executives, economists, government leaders, and clergy, as well as traditional environmentalists—are recognizing that "business as usual" is not sustainable. These caring people are beginning to play an important role in changing society's treatment of Earth. For example, adding their voices to the literally hundreds of traditional environmental and professional organizations devoted to controlling pollution and protecting wildlife, people in business have formed the Business Council for Sustainable Development, economists have formed the International Society for Ecological Economics, religious leaders have formed the National Religious Partnership for the Environment, and philosophers are speaking out for a new ethic of "caring for creation." And for the first time in 1994, the U.S. Department of Commerce began to consider the consumption of environmental assets in their calculation of the gross domestic product (GDP).

Sustainable development was the primary focus of a 1992 world summit meeting of leaders and representatives from 180 nations—the United Nations Conference on Environ-ment and Development, held in Rio de Janeiro, Brazil. The outcome of this conference, a "blueprint" intended to guide development in sustainable directions into and through the next century, is now published in book form as *Agenda 21* (See "Earth Watch" box, page 15).

Thus, we are seeing a melding of environmentalists with many other individuals and groups that appreciate the problems jeopardizing sustainability. Together, they are working to bring about corrective measures. With this breadth of participation, it is difficult today to define an environmentalist, except perhaps as "anyone who cares for creation, is interested in human sustainability, and is making efforts to achieve it."

In this light, the textbook you are holding is our own "best effort" to contribute to the cause, from our perspective as scientists and educators in biology and ecology. Given the vastness of what is now the environmental arena, we are the first to say that this text is far from exhaustive, nor should it be taken as the last word. Of course, viewpoints are subject to change as human experience and understanding increase. Our basic premise, however, is that sound policy in the long run must be based on sound science.

In the first part of the text (Chapters 2 through 5) we describe the basics of how natural systems function and perpetuate themselves, and their limits in terms of adaptability to changing conditions. In addition to giving us some appreciation for natural systems, this study will reveal certain basic principles underlying sustainability. We contend that, to achieve sustainability, we must incorporate these principles into the functioning of our own society. Subsequent chapters will address issues of population and development, pollution in its various forms, and issues relating to resources, energy and land. In each case, we will attempt to give you a deeper understanding of just what the problems are and how far we have come, and a view of the path ahead toward sustainability.

Review Questions

1. Describe the range of ways people in different parts of the world live and interact with their environments.
2. What factors brought about the collapse of the Easter Island civilization? Draw parallels and counterparallels between the current global situation and the prelude to the collapse of Easter Island civilization.
3. How and when did environmentalism arise?
4. What changes in issues, concerns, and responses occurred in the post-World War II period?
5. What are the successes of the environmental movement?
6. What are the grounds and concerns of present critics of environmentalism?
7. Are all the environmental problems under control? Cite four global trends that bespeak of "still losing the environmental war."
8. Define *sustainability* and a *sustainable society*.
9. Distinguish between the survival and the sustainability of a modern civilization.
10. Beyond survival, what features of society do we wish to preserve?
11. What are the prerequisites for the sustainability of modern society?
12. Define and give features of *sustainable development*.
13. Contrast our traditional concept of development with the concept of sustainable development.
14. What aspects might or should sustainable development focus on?
15. What are the new directions, new focus, and new adherents of modern environmentalism?

Thinking Environmentally

1. Have a class debate between people representing environmentalists and environmental critics. Characterize the two sides in terms of long-term vs. short-term viewpoint, personal interests vs. interests of society in general, and local vs. global perspective.
2. Some people say that the concept of sustainable development either is an oxymoron or represents going back to some kind of primitive living. Present an argument demonstrating that neither is the case, but that sustainable development is the only course that will allow the continued advancement of civilization.
3. List all the prerequisites for a sustainable society that can continue to advance in gaining further understanding of the universe and our place in it. Why is it necessary that people from all walks of life be involved? Give the roles (in general) that each needs to play in achieving a sustainable society.
4. How can you contribute to a sustainable society?

PART ONE

Ecosystems and How They Work

Tropical rain forests—humid, warm, dense, full of unusual plants and animals. These are the impressions that our senses bring to us when we take the first few steps into a rain forest. However, our senses will not tell us that the tropical rain forests are home to a greater diversity of living things than anywhere else on Earth, that they are the result of millions of years of adaptive evolution, or that they are the site of storage of more carbon than the entire atmosphere. This information comes from the work of many scientists who have been asking the basic questions of how such natural systems work, how they came to be, and how they relate to the rest of the natural world.

How important is the information that comes from the work of such scientists? We hope in this first Part to convince you that information about how the natural world works is absolutely crucial to the human enterprise of living on Planet Earth. Ecosystems—which is the name we give to natural units like the tropical rain forest—are not just the backdrop for human activities, they are the basic context of life on Earth, including human life. These are systems that are self-sustaining, and if we accept the notion that the human enterprise should also be self-sustaining (and currently is not), then perhaps we can learn how to construct a sustainable society by studying ecosystems. Environmental science begins by understanding how the natural world works.

Rain forest. Photo by Michael Melford/The Image Bank.

CHAPTER 2

Ecosystems: Units of Sustainability

Key Issues and Questions

1. Life can exist only in the context of an ecosystem. What is an ecosystem?
2. The biosphere is a matrix of interconnected ecosystems. What are the major biomes? How are they linked together?
3. The organisms in every ecosystem can be assigned to key categories. What are the categories? How do they function together to make a sustainable system? What is the general movement of chemicals and energy throughout the system?
4. Ecosystems are differentiated largely by environmental factors. How do precipitation and temperature patterns influence the variations in ecosystems?
5. The human perception of being independent from nature is erroneous. How are humans still dependent on nature? When and how did this erroneous perception arise?
6. Balances exist between natural ecosystems. What factors hold natural ecosystems in balance? What is the failure in relationships between humans and nature?
7. To be sustainable the human system will need ecological improvement in several basic respects. What are the deficiencies? What is the potential for improvement?

Can we learn to interact with the natural world and conduct our own endeavors in ways that are sustainable—that is, "meet the needs of the present without compromising the ability of future generations to meet their own needs?"

Study of natural ecosystems—those groupings of plants, animals, and other organisms that we think of as unspoiled forests, grasslands, coral reefs, ponds or other such entities—can serve several objectives.

First, our agriculture and technology notwithstanding, we remain dependent on the natural world and its biodiversity for such basic needs as clean air, clean water, a suitable climate for growing food, as well as countless less obvious ways. Study of natural ecosystems will help us understand these intricate relationships between the environment and living things and between the natural world and ourselves. From this we can more clearly understand how the impact of humans is influencing the natural world and the consequences this may have. By

Ocean floor. Photo by Jim Zipp/Photo Researchers, Inc.

the same token, gaining understanding of natural ecosystems will help us understand how to make our human–Earth relation more sustainable.

Second, natural ecosystems are models of sustainability. The individual kinds of trees and other plants, animals, and microbes making up a forest, for example, are known to have propagated themselves over many tens of thousands, or even millions, of years, with little change. Our premise is that bringing out the basic principles that enable the perpetuation or sustainability of natural ecosystems will provide insights into the ways in which we need to direct human development in order to achieve a sustainable future.

Finally, it is our hope that the study of natural ecosytems will lead you to a greater appreciation of the amazing diversity and beauty of our home planet and a heightened desire to care for it. Thus, our study begins with an examination of natural ecosystems.

What are Ecosystems?

The grouping or assemblage of plants, animals, and microbes we observe when we study a natural forest, grassland, pond, coral reef, or other undisturbed area is referred to as the area's **biota** (*bio*, living) or **biotic community**. Importantly, the plant portion of the biotic community includes all vegetation, from large trees down through microscopic algae. Likewise, the animal portion includes everything from large mammals, birds, reptiles, and amphibians through earthworms, tiny insects, and mites. Microbes encompass a large array of microscopic bacteria, fungi, and protozoans. Thus, one may speak of the biotic community as comprising a plant community, an animal community, and a microbial community.

The particular kind of biotic community that we witness in a given area is, in large part, determined by **abiotic** (meaning nonliving chemical and physical) factors, such as the amount of water or moisture present, temperature, salinity, and soil type. These abiotic conditions both support and limit the particular community. For example, a relative lack of available moisture prevents the growth of most species of plants, but supports certain species, such as cacti; we recognize such areas as deserts. Land with plenty of available moisture and suitable temperature supports forests. Obviously, the presence of water is the major factor that sustains aquatic communities.

The first step in investigating a biotic community may be to simply catalogue all the *species* present. **Species** are the different kinds of plants, animals, and microbes. Each species includes all those individuals that have a very strong similarity in appearance to one another, and which are distinct in appearance from other such groups (Robins and Mallard ducks for example.) The similarity in appearance implies a close genetic relationship. Indeed, the biological definition of a species is the entirety of a population that can interbreed and produce fertile offspring, whereas members of different species generally do not interbred or, if they do, fertile offspring are not produced. Breeding is often impractical or impossible to observe, however, so for purposes of identification the aspect of appearance suffices.

In cataloguing the species of a community, one will observe that each species is represented by a certain **population**—that is, by a certain number of individuals that make up the interbreeding, reproducing group. The distinction between *population* and *species* is that the term *population* is used to refer only to those individuals of a certain species that live within a given area, whereas the term *species* is all inclusive—it refers to all the individuals of a certain kind, even though they may exist in different populations in widely separated areas.

Continuing our study, along with the incredible variety of species and communities it is impressive that the species within a community depend on and support one another in many ways. Most evident, certain animals will not be present unless particular plants that provide their necessary food and shelter are also present. Thus, the plant community supports (or limits by its absence) the animal community. Additionally, every plant and animal species is adapted to cope with the abiotic factors of the region. For example, every species that lives in temperate regions is adapted in one way or another to survive the winter season, which includes a period of freezing temperatures (Fig. 2-1). We shall explore these interactions among organisms and their environments later. For now, the point is that the populations of different species within a biotic community are constantly interacting with each other and the abiotic environment.

This brings us to the concept of an *ecosystem*. An **ecosystem** is both the biotic community *and* the abiotic conditions in which the biotic community members live. Additionally, it includes considerations of the ways populations interact with each other and the abiotic environment to reproduce and perpetuate the entire grouping. In one sentence, an ecosystem is a grouping of plants, animals, and microbes interacting with each other and with their environment in such a way as to perpetuate the grouping. For study purposes, an ecosystem may be taken to be any more or less distinctive biotic community living in a certain environment. Thus, a forest, a grassland, a wetland, a marsh, a pond, a beach, and a coral reef, each with its respective species in a particular environment, can be studied as distinct ecosystems.

Since no organism can live apart from its environment or from interactions with other species, ecosystems are the functional units of sustainable life on Earth. The study of ecosystems and the interactions that occur among organisms and between organisms and their environment is the science of **ecology**, and the investigators who conduct such studies are called **ecologists**.

While it is convenient to divide the living world into different ecosystems, any investigation soon reveals that there are seldom distinct boundaries between ecosystems, and they are never totally isolated from one another. Many species will occupy (and thus be a part of) two or more ecosystems at the same time. Or, they may move from one ecosystem to another at different times, as in the case of migrating birds. In passing from one ecosystem to another, one may observe only a gradual decrease in the populations of one biotic community and an increase in the populations representing another. Thus, one ecosystem may gradate into the next through a transitional region known as an **ecotone**, which shares many of the species and characteristics of the two adjacent ecosystems (Fig. 2-2).

The ecotone between adjacent systems may also include unique conditions that support distinctive plant and animal species. Consider, for example, the marshy

FIGURE 2–1
Many trees and other plants of temperate forests are so adapted to the winter season that they actually require a period of freezing temperature in order to recommence growth in the spring. (Photograph by BJN.)

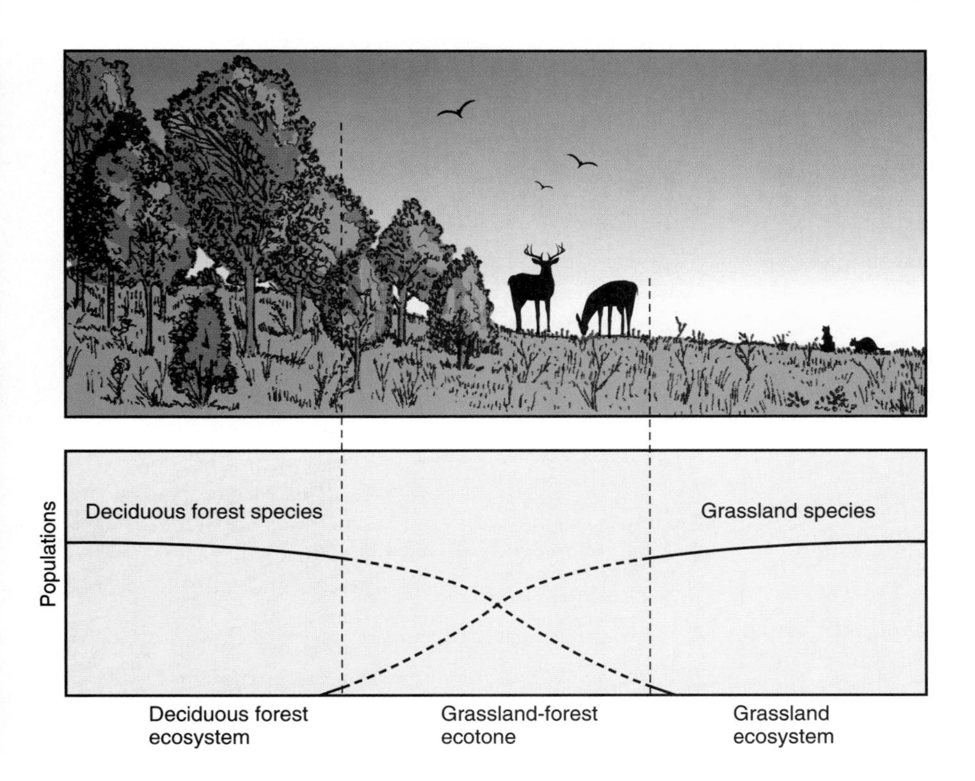

FIGURE 2–2
Ecosystems are not isolated from one another. One ecosystem blends into the next through a transitional region, an ecotone, that contains many species common to the two adjacent systems.

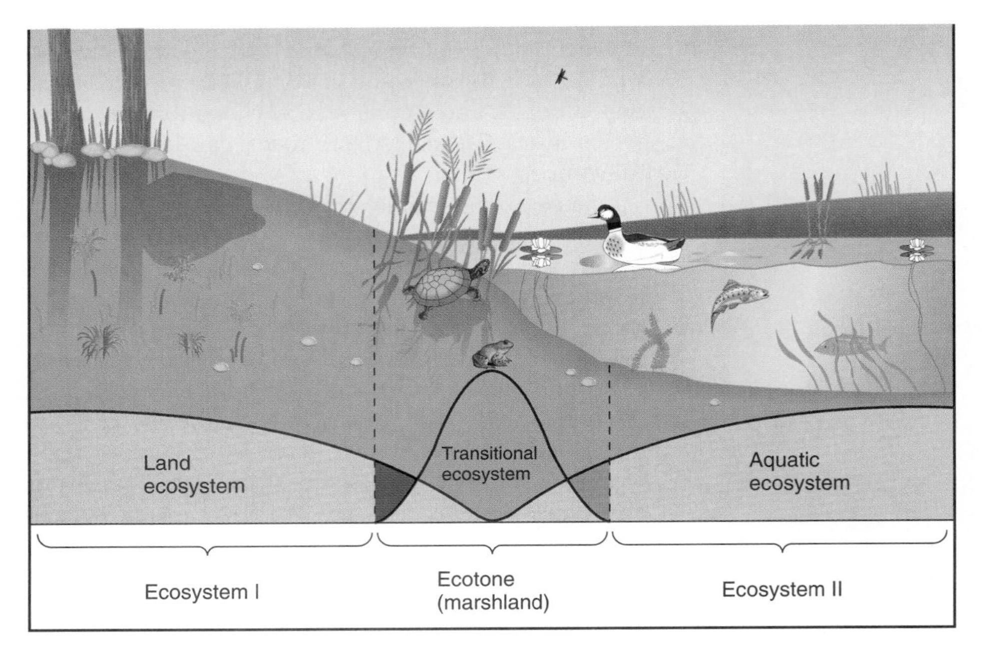

FIGURE 2–3
An ecotone may create a unique habitat that harbors specialized species not found in either of the ecosystems bordering it.

area that often occurs between the open water of a lake and dry land (Fig. 2-3). Ecotones may be studied as distinct ecosystems in their own right.

Furthermore, what happens in one ecosystem will definitely affect other ecosystems. For example, losses and fragmentation of forests have disrupted migration lanes and resulted in drastic declines in the populations of certain North American songbirds. How the loss of all these birds will affect various ecosystems is a question we cannot answer at this time.

Similar or related ecosystems are often grouped together to form major kinds of ecosystems called **biomes**. Tropical rain forests, grasslands, and deserts are examples. While more extensive than an ecosystem in its breadth and complexity, a biome is still basically a certain biotic community supported and limited by certain abiotic environmental factors. Names and brief descriptions of some of the major terrestrial biomes are given on the front of the foldout following page 42. Their distribution around the globe as a consequence of abiotic (namely, climatic) factors is shown in Fig. 2-4. Again, there are generally no distinct boundaries between biomes, but one grades into the next through transitional regions. Indeed, there is no general agreement among ecologists as to whether certain kinds of ecosystems should be lumped within one major biome or considered as separate biomes. Therefore, you may well see maps similar to Fig. 2-4 that depict more or fewer biomes.

Likewise, there are a large variety of aquatic and wetland ecosystems that are determined primarily by the depth, salinity, and permanence of water. A few are shown on the back of the foldout following page 42. Then there are various marine (ocean) ecosystems

FIGURE 2–4
World distribution of the major terrestrial biomes. (After Robert Christopherson, *Geosystems,* 2/e. ©1994, p.624–25. Adopted by permission of Prentice Hall, Upper Saddle River, New Jersey.)

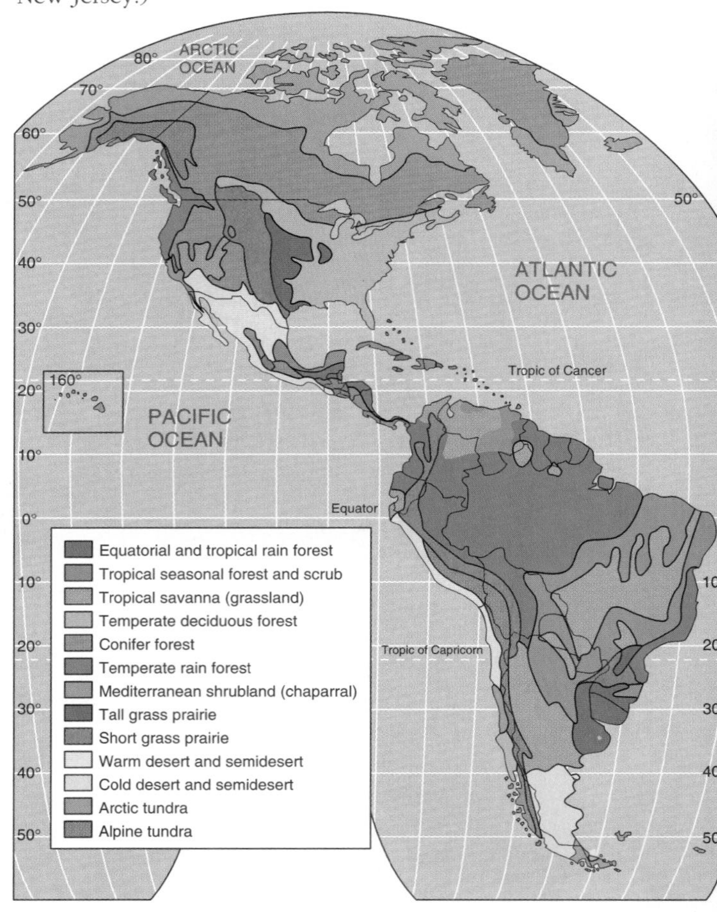

that are determined by depth, texture of the bottom (mud vs. rock ledges), and nutrient levels, as well as by water temperature. Thus, marine ecosystems are determined more by specific environmental factors at the location than by general climatic factors, as is the case for terrestrial biomes. Therefore, we generally speak of marine environments and not marine biomes.

Regardless of how we choose to divide (or group) and name different ecosystems, it is important to recognize that they all remain interconnected and interdependent. Terrestrial biomes are connected by the flow of rivers between them and by migrating animals. Sediments and nutrients washing from the land may nourish or pollute the ocean. Seabirds and mammals connect the oceans with the land, and all biomes share a common atmosphere and water cycle.

Therefore, all the species on Earth, along with all their environments, can be seen as one vast ecosystem, which is called the **biosphere** (see Chapter 1, opening photo). Although the separate local ecosystems are the individual units of sustainability, they are all interconnected to form the biosphere. The concept is analogous to the idea that the cells of our bodies are the units of living systems, but are all interconnected to form the whole body. Carrying the analogy further, to what degree may individual ecosystems be upset or destroyed before the entire biosphere is affected? And to what degree can basic global parameters such as atmosphere and temperature be altered before all ecosystems on Earth are affected? Let us now begin a more intensive study of ecosystems to discover the principles of their sustainability.

The key terms introduced in this section are summarized in Table 2–1.

The Structure of Ecosystems

Let us examine the *structure* of ecosystems. Structure refers to parts and the way they fit together to make the whole. There are two key aspects to every ecosystem: the biota or biotic community and the abiotic environmental factors. The way different categories of organisms fit together is referred to as the **biotic structure**.

Biotic Structure

Despite the diversity of ecosystems, all have a similar biotic structure, based on feeding relationships. That is,

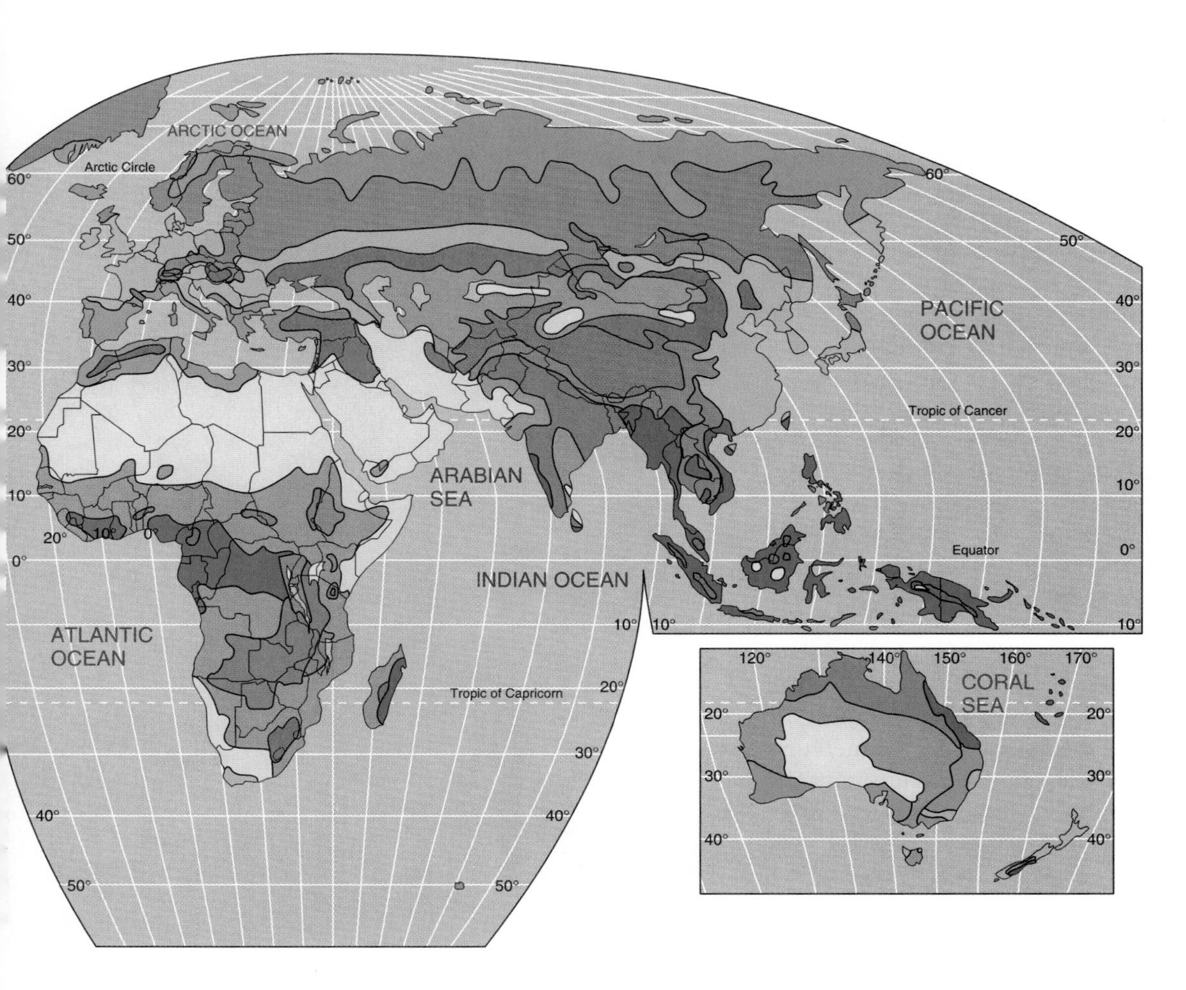

TABLE 2–1
Important Terms

Species	All the members of a specific kind of plant, animal, or microbe; a kind given by similarity of appearance and/or capacity for interbreeding, and producing fertile offspring.
Population	All the members of a particular species occupying a given area.
Biotic community	All the populations of different plants, animals, and microbes occupying a given area.
Abiotic factors	All the factors of the physical environment: moisture, temperature, light, wind, pH, soil type, salinity, etc.
Ecosystem	The biotic community together with the abiotic factors; all the interactions among the members of the biotic community and between the biotic community and the abiotic factors.
Biome	A grouping of all the ecosystems of a similar type, e.g., tropical forests, grasslands, etc.
Biosphere	All species and physical factors on Earth functioning as one mammoth ecosystem.

all ecosystems have the same three basic categories of organisms that interact in the same ways.

Categories of Organisms. The major categories of organisms are (1) *producers*, (2) *consumers,* and (3) *detritus feeders and decomposers*. Together, these groups produce food, pass it along food chains, and return the starting materials to the abiotic parts of the environment.

Producers. **Producers** are mainly green plants, which use light energy from the Sun to convert carbon dioxide (absorbed from air or water) and water to a sugar called glucose and release oxygen as a by-product. This chemical conversion, which is driven by light energy, is called **photosynthesis**. Plants are able to manufacture all the complex molecules that make up their bodies from the glucose produced in photosynthesis, plus a few additional *mineral nutrients* such as nitrogen, phosphorus, potassium, and sulfur, which they absorb from the soil or from water (Fig. 2-5).

The molecule that plants use to capture light energy for photosynthesis is **chlorophyll**, a green pigment. Hence, plants that carry on photosynthesis are easily identified by their green color. (In some cases, the green may be overshadowed by additional red or brown pigments. Thus, red algae and brown algae also carry on photosynthesis.) Producers range in diversity from microscopic, single-celled algae through medium-sized plants such as grass, daisies, and cacti to gigantic trees. Every major ecosystem, both aquatic and terrestrial, has its particular producers carrying on photosynthesis.

The term **organic** is used to refer to all those materials that make up the bodies of living *organ*isms—

FIGURE 2–5
The producers in all major ecosystems are green plants, because they contain the green pigment, chlorophyll. Chlorophyll absorbs light energy, which is then used to produce the sugar glucose from carbon dioxide and water, releasing oxygen as a by-product. The glucose, along with a few additional mineral nutrients from the soil, is used in the production of all plant tissues, leading to growth.

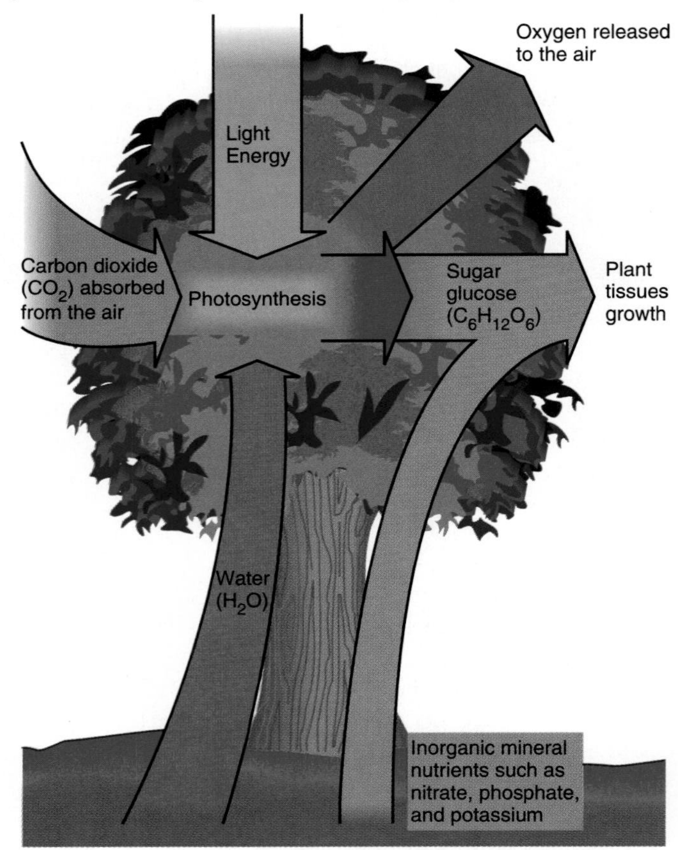

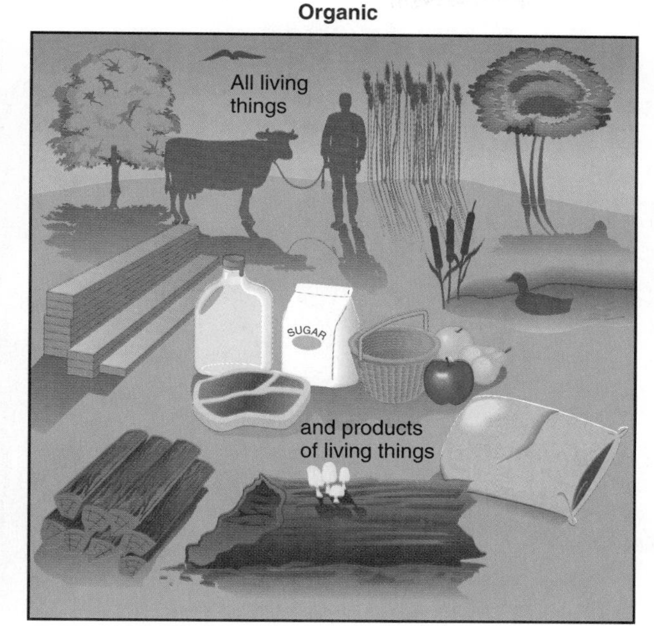

FIGURE 2–6
Organic and inorganic. Water and the simple molecules found in air and in rocks and soils are *inorganic*. The complex molecules that make up plant and animal tissues are *organic*. Producers, using energy derived from light, convert inorganic substances to organic substances. Organic materials then are broken down to inorganic materials again by burning or digestion, which also releases energy. Chemically, organic compounds involve carbon-carbon and carbon-hydrogen bonds not found in inorganic materials.

molecules such as proteins, fats or lipids, and carbohydrates. Likewise, materials that are specific products of living organisms, such as dead leaves, leather, sugar, and wood, are considered *organic*. On the other hand, materials and chemicals of air, water, rocks and minerals, which exist apart from the activity of living organisms, are considered **inorganic** (Fig. 2-6). The key feature of *organic* materials and molecules is that they are in large part constructed from bonded carbon and hydrogen atoms, a structure that is not found among *inorganic* materials. This carbon-hydrogen structure has its origins in the process of photosynthesis. Hydrogen atoms taken from water molecules and carbon atoms from carbon dioxide are joined together to form organic compounds in the process of photosynthesis. Green plants use light as the energy source to produce all the complex organic molecules their bodies need from the simple inorganic chemicals (carbon dioxide, water, mineral nutrients) present in the environment. As this conversion from inorganic to organic occurs, some of the energy from light is stored in the organic compounds.

Now, all organisms in the ecosystem *other than green plants* feed on organic matter as their source of both energy and nutrients. These include not only all animals, but also **fungi** (mushrooms, molds, and similar organisms), most bacteria, and even a few higher plants, such as Indian pipe (Fig. 2-7), that do not have

FIGURE 2–7
Indian pipe, a flowering plant that is not a producer. It does not carry on photosynthesis, but derives its energy from other organic matter, as do animals and other consumers (heterotrophs). (Photograph by BJN.)

FIGURE 2–8
Gray wolves have brought down an elk. (Tom McHugh/ Photo Researchers, Inc.)

chlorophyll and that therefore cannot carry on photosynthesis.

Thus, green plants, which carry on photosynthesis, are absolutely essential to every ecosystem. Their photosynthesis and growth constitute the production of organic matter, which sustains all other organisms in the ecosystem.

Indeed, all organisms in the biosphere can be divided into two categories, *autotrophs* and *heterotrophs*, on the basis of whether they do or do not produce the organic compounds they need to survive and grow. Those organisms such as green plants, which produce their own organic material from inorganic constituents in the environment using an external energy source, are **autotrophs** (*auto*, self; *troph*, feeding). As previously mentioned, the most important and common autotrophs by far are green plants, which use chlorophyll to capture light energy for photosynthesis. However, a few bacteria use a purple pigment for photosynthesis, and some other bacteria acquire their energy from certain high-energy inorganic chemicals. All other organisms, however, which must *consume* organic material to obtain energy and nutrients, are **heterotrophs** (*hetero*, other). Heterotrophs may be divided into numerous subcategories, the two major categories being **consumers** (which eat living prey), and **detritus feeders** and **decomposers** both of which feed on dead organisms or their products.

Consumers. Consumers encompass a wide variety of organisms ranging in size from microscopic bacteria to blue whales and include such diverse groups as protozoans, worms, fish and shellfish, insects, reptiles, amphibians, birds, and mammals (including humans).

For the purpose of understanding ecosystem structure, consumers are divided into various subgroups according to their food source. Animals, be they as large as elephants or as small as mites, that feed directly on producers are called **primary consumers**. They are also called **herbivores** (*herb*, grass).

Animals that feed on primary consumers are called **secondary consumers**. Thus, elk, which feed on vegetation, are primary consumers, whereas wolves, because they feed on elk, are secondary consumers (Fig. 2-8). There may also be third, fourth, or even higher levels of consumers, and certain animals may occupy more than one position on the consumer scale. For instance, humans are primary consumers when they eat vegetables, secondary consumers when they eat beef, and third-level consumers when they eat fish that feed on smaller fish that feed on algae. Secondary and higher order consumers are also called **carnivores** (*carni*, meat). Consumers that feed on both plants and animals are called **omnivores** (*omni*, all).

In a relationship in which one animal attacks, kills, and feeds on another, the animal that attacks and kills is called the **predator**; the animal that is killed is called the **prey**. Together, the two animals are said to have a **predator-prey** relationship.

Parasites are another important category of consumers. Parasites are organisms—either plants or animals—that become intimately associated with their "prey" and feed on it over an extended period of time, typically without killing it (at least not immediately), but often weakening it so that it becomes more prone to being killed by other predators or by adverse conditions. The plant or animal that is fed upon is called the **host**; thus, we speak of a **host-parasite** relationship.

A tremendous variety of organisms may be parasitic. Various worms are well-known examples, but certain protozoans, insects, and even certain mammals (vampire bats) and plants (dodder) (Fig. 2-9a) are also parasites. Many serious plant diseases and some animal diseases (such as athlete's foot) are caused by parasitic fungi. Indeed, virtually every major group of organisms has at least some members that are parasitic. Parasites may live inside or outside their host, as the examples shown in Fig. 2-9 illustrate.

In medical circles, a distinction is generally made between bacteria and viruses, which cause disease, and parasites, which are usually larger organisms. Ecologically, however, there is no real distinction. Bacteria are foreign organisms and viruses are organism-like entities feeding on and multiplying in their hosts over a period of time and doing the same damage as do other parasites. Therefore, disease-causing bacteria and viruses can be considered highly specialized parasites. Representative examples of producers and consumers, and feeding relationships among them, are shown in Fig. 2-10.

Detritus Feeders and Decomposers. Dead plant material such as fallen leaves, branches and trunks of dead trees, dead grass, the fecal wastes of animals, and occasional dead animal bodies is called **detritus** (pronounced di-TRI-tus). Many organisms are specialized to feed on detritus, and we refer to such consumers as **detritus feeders** or *detritivores*. Examples include earthworms, millipedes, crayfish, termites, ants, and wood beetles. As with regular consumers, one can identify ***primary* detritus feeders** (those that feed directly on detritus), ***secondary* detritus feeders** (those that feed on primary detritus feeders), and so on.

An extremely important group of primary detritus feeders is the **decomposers**, namely, fungi and bacteria. Much of the detritus in an ecosystem—particularly dead leaves and the wood of dead trees or branches—does not appear to be eaten as such, but rots away. Rotting is the result of the metabolic activity of fungi and bacteria. These organisms secrete digestive enzymes that cause the breakdown of wood, for example, into simple sugars that the fungi or bacteria then absorb for their nourishment. Thus, the rotting we observe is really the result of material being consumed by fungi and bacteria. Even though fungi and bacteria are called decomposers because of their unique behavior, we group them with detritus feeders because their function in the ecosystem is the same. In turn, decomposers are fed upon by such secondary detritus feeders as protozoans, mites, insects, and worms (Fig. 2-11). When a fungus or other decomposer dies, its body becomes part of the detritus and the source of energy and nutrients for still more detritus feeders and decomposers.

In summary, despite the apparent diversity of ecosystems, they all have a similar *biotic structure*. They can all be described in terms of autotrophs, or producers, which produce organic matter that becomes the source of energy and nutrients for heterotrophs, which are various categories of consumers and detritus feeders and decomposers (Fig. 2-12).

Feeding Relationships: Food Chains, Food Webs, and Trophic Levels. In describing the biotic structure of ecosystems, it is evident that major interactions

FIGURE 2–9
Diversity of parasites. Nearly every major biological group of organisms has at least some members that are parasitic on others. Shown here are (a) dodder, a plant parasite. The orange "strings" are the dodder stems, which suck sap from the host plant; dodder has no leaves and no chlorophyll. (b) Nematode worms *(Ascaris lumbricoides)*, the largest of the human parasites, reach a length of 14 inches (35 cm). (c) Lamprey attached to a salmon. Lampreys are parasitic on fish. (a) Jeff Lepore/Photo Researchers, Inc.; b) Sinclair Stammers/Science Photo Library/Photo Researchers, Inc.; c) Runk/Schoenberger/Grant Heilman Photography.)

(a)

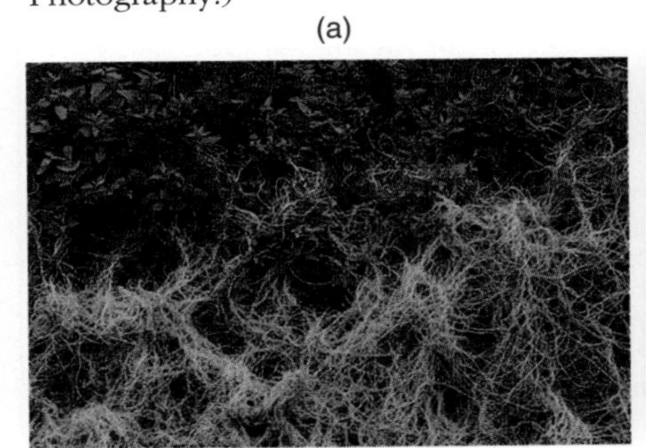

(b)

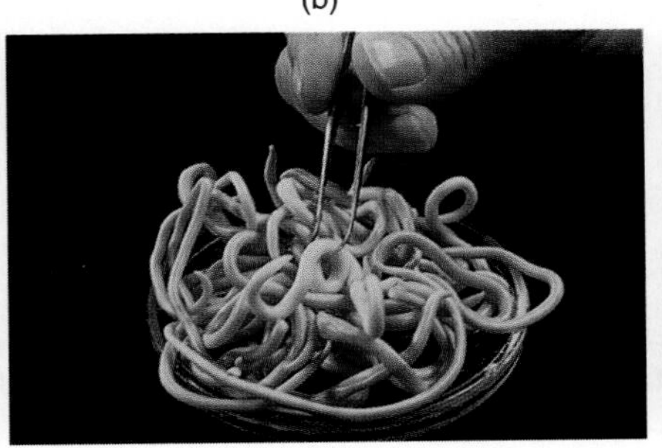

(c)

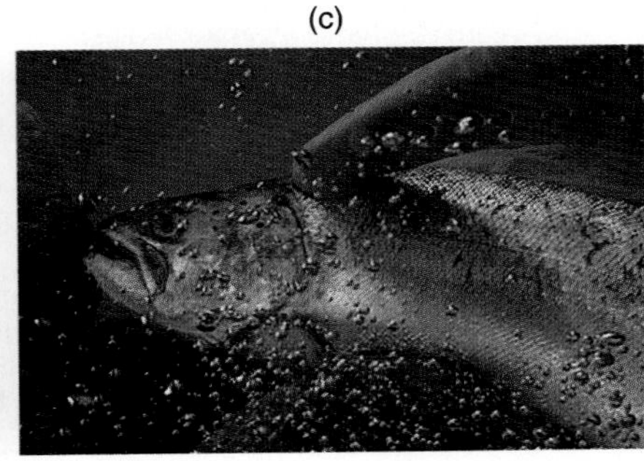

FIGURE 2–10
Common feeding (trophic) relationships among producers and consumers.

FIGURE 2–11
The feeding (trophic) relationships among primary detritus feeders, secondary detritus feeders, and consumers. Organisms that feed on detritus support many other organisms living in the soil, and these, in turn, may be eaten by larger consumers.

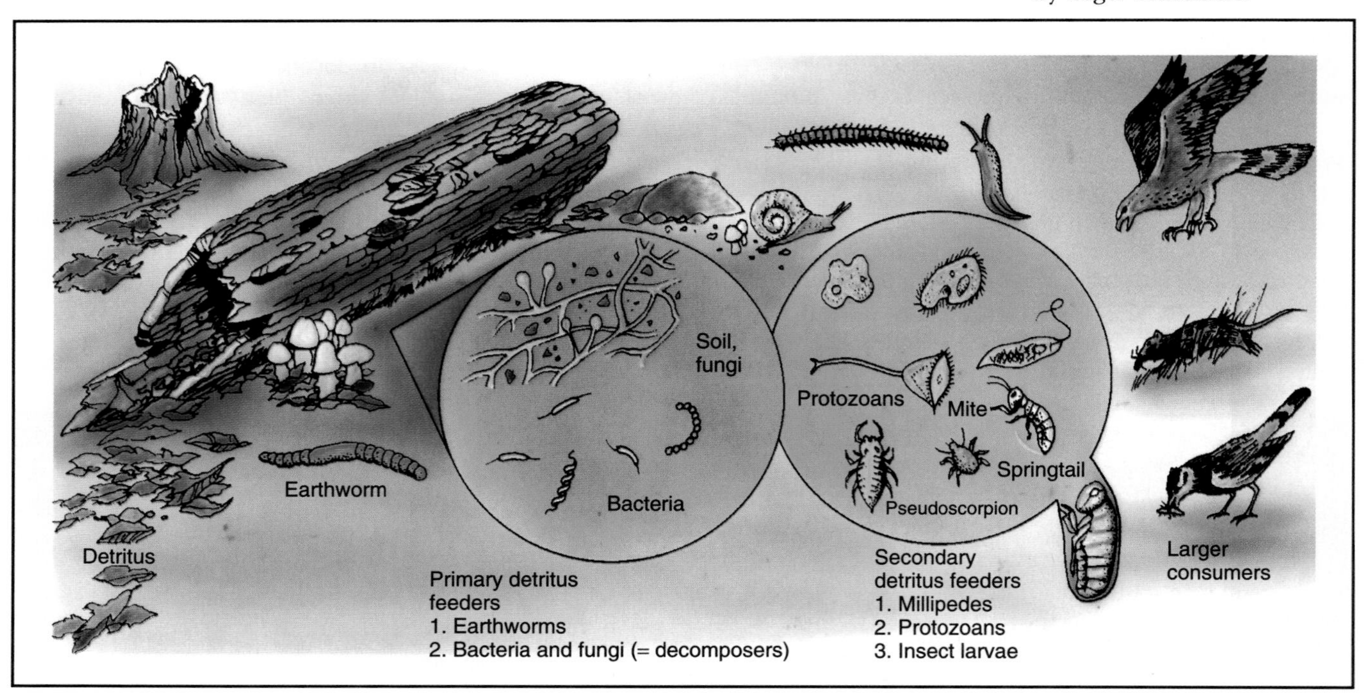

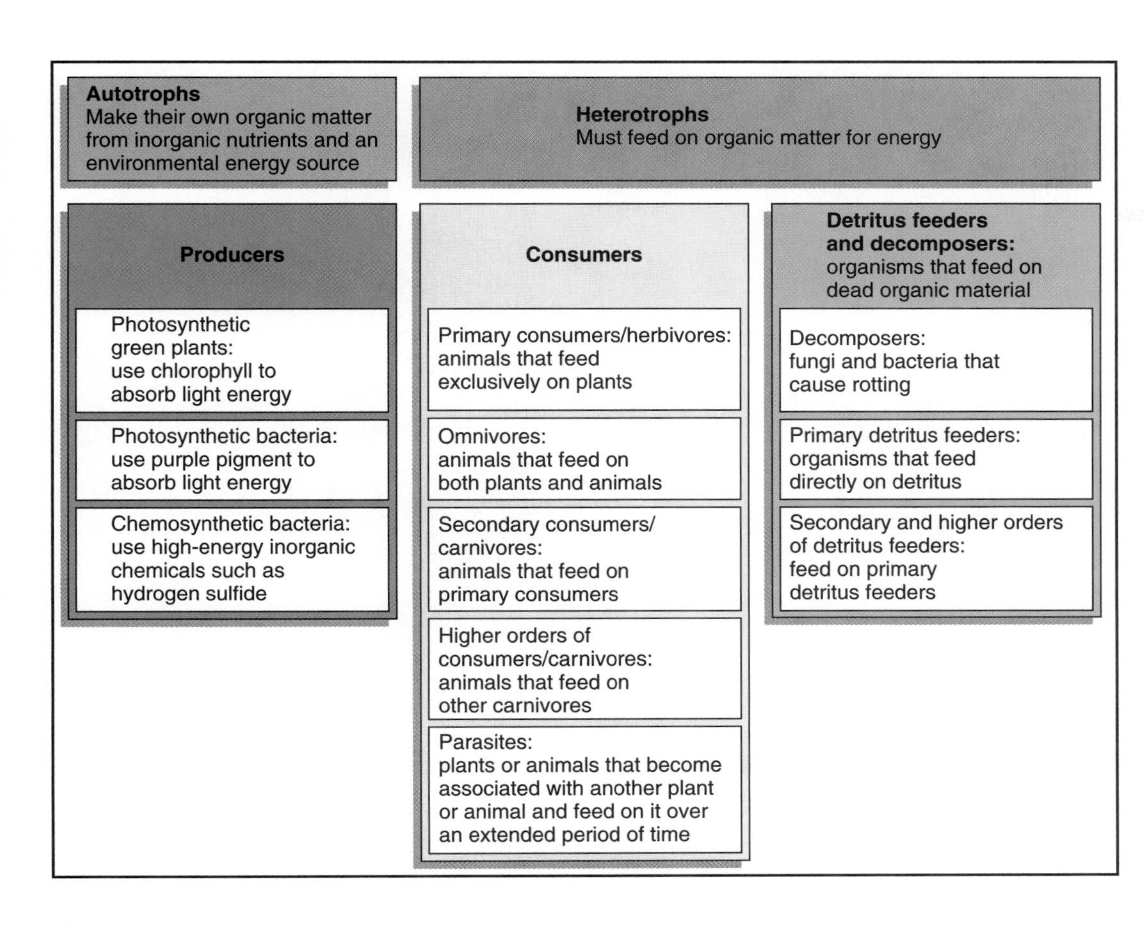

FIGURE 2–12
A summary of how living organisms are ecologically categorized according to feeding attributes.

among organisms involve feeding relationships. We can identify innumerable pathways where one organism is eaten by a second, which is eaten by a third, and so on. Each such pathway is called a **food chain**.

While it is interesting to trace such pathways, it is important to recognize that food chains seldom exist as isolated entities. An herbivore population feeds on several kinds of plants and is preyed upon by several secondary consumers or omnivores. Consequently, virtually all food chains are interconnected and form a complex *web* of feeding relationships. Indeed, the term **food web** is used to denote the complex network of interconnected food chains.

Despite the number of theoretical food chains and the complexity of food webs, there is a simple overall pattern: All food chains basically lead through a series of steps or levels—namely, from producers to primary consumers (or primary detritus feeders) to secondary consumers (or secondary detritus feeders), and so on. These *feeding levels* are called **trophic levels** (*trophic*, feeding). All producers belong to the first trophic level; all primary consumers (in other words, all herbivores), whether feeding on living or dead producers, belong to the second trophic level; organisms feeding on these herbivores belong to the third level, and so on.

Whether we visualize the biotic structure of an ecosystem in terms of food chains, food webs, or trophic levels, we should see, through each feeding step, that there is a fundamental movement of the chemical nutrients and stored energy they contain from one organism or level to the next. These movements of energy and nutrients will be described in more detail later. A diagrammatic comparison of a food chain, a food web, and trophic levels is shown in Fig. 2-13a. A marine food web is shown in Fig. 2-13b.

How many trophic levels are there? Usually, no more than three or four in any ecosystem. This answer comes from straightforward observations. The **biomass**, or total combined (net dry) weight, of all the organisms at each trophic level can be estimated by collecting (or trapping) and weighing suitable samples. In terrestrial ecosystems, the biomass is about 90–99 percent less at each higher trophic level. If the biomass of producers in a grassland is 10 tons (20,000 lb) per acre, the biomass of herbivores will be no more than 2000 pounds and that of carnivores no more than 200 pounds. Clearly, you can't go through very many trophic levels before the biomass approaches zero. Depicting this graphically gives rise to what is commonly called a **biomass pyramid** (Fig. 2-14).

The biomass decreases so much at each trophic level largely because much of the food that is consumed by a heterotroph is not converted to the body tissues of the heterotroph; rather, it is broken down so

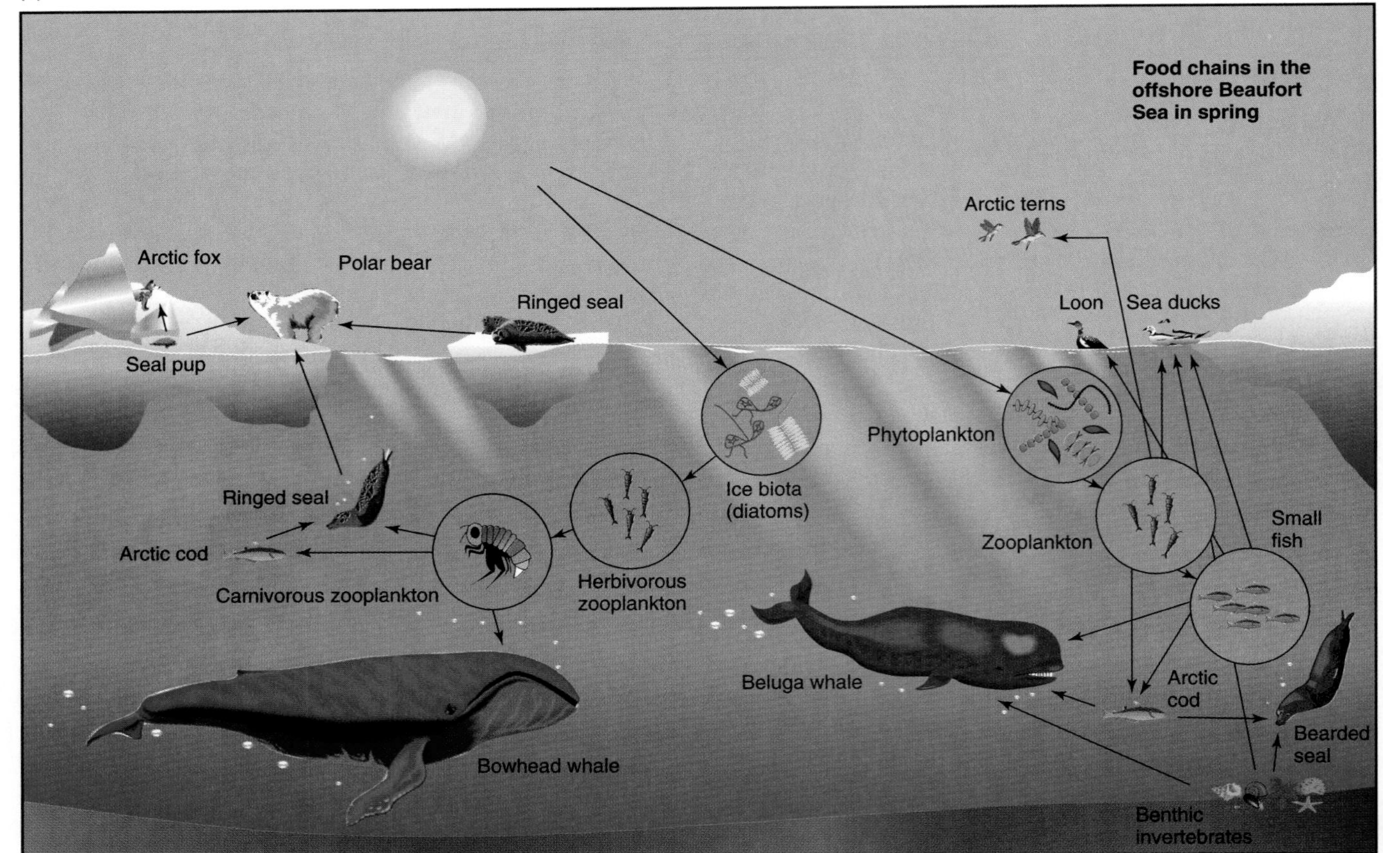

FIGURE 2–13

(a) Three ways of representing the movement of food through an ecosystem. Specific pathways such as from nuts to squirrels to foxes (shown by green arrows) are referred to as *food chains*. A *food web* refers to the collective consideration of all food chains, which are invariably interconnected (all arrows). Trophic levels, indicated by shading at the left, stresses the general pattern that food always flows from producers to herbivores to carnivores. (b) A marine food web.

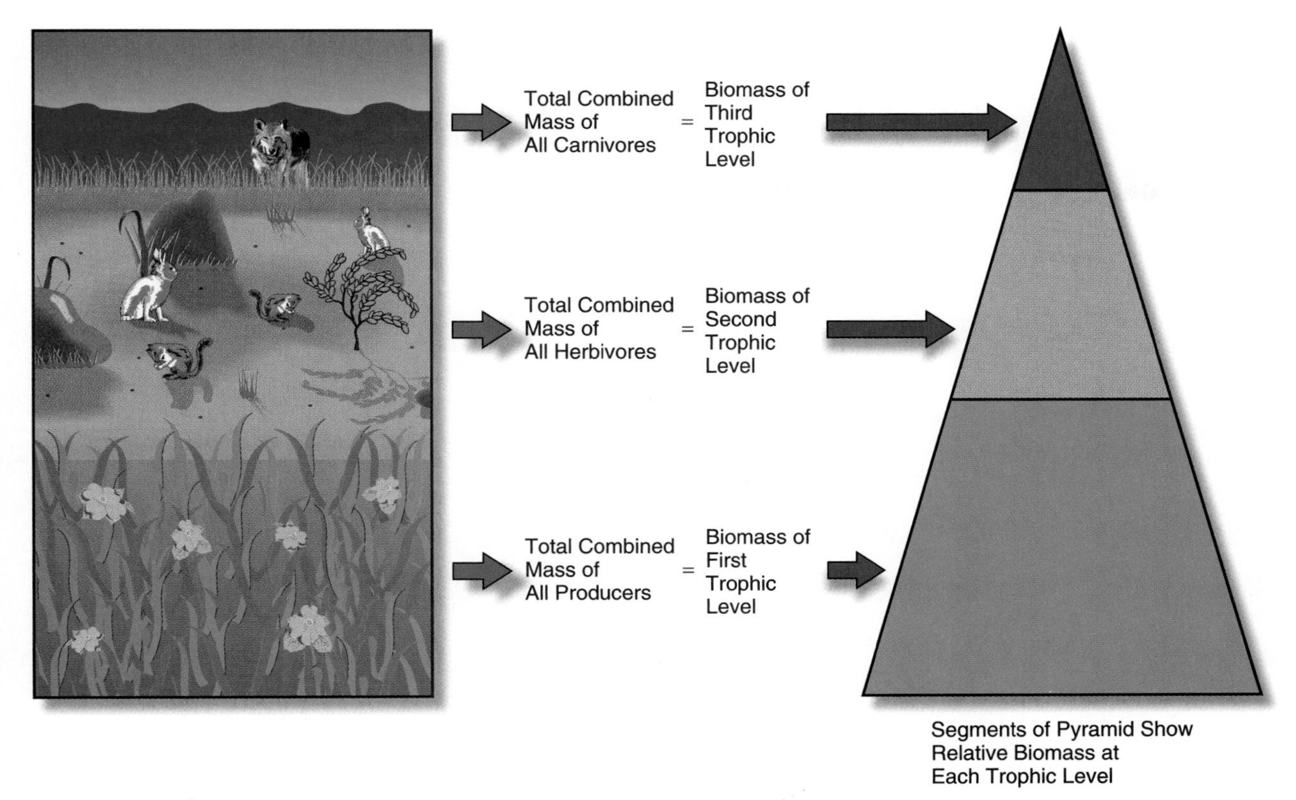

FIGURE 2–14
Biomass pyramid. A graphic representation of the biomass (the total combined mass of organisms) at successive trophic levels has the form of a pyramid.

that the stored energy it contains can be released and used by the heterotroph. Hence, there is an inevitable loss of biomass with the movement to higher trophic levels. It is very significant to observe that all heterotrophs depend on the continual input of fresh organic matter produced by the autotrophs (green plants). Without such input, the heterotrophs would all run out of food and starve as the organic matter was broken down to release its stored energy.

As this breakdown of organic matter occurs, the chemical elements are released back to the environment in the inorganic state, where they may be reabsorbed by autotrophs (producers). Thus, there is a continuous cycle of nutrients from the environment through organisms and back to the environment. The spent energy, on the other hand, is lost as heat is given off from bodies (Fig. 2-15). These concepts of recycling of chemical nutrients and the flow of energy will be discussed in greater detail in Chapter 3.

In summary, all food chains, food webs, and trophic levels *must start with producers*, and producers must have suitable environmental conditions to support their growth. Populations of all heterotrophs, including humans, are ultimately limited by what plants produce, in accordance with the concept of the biomass pyramid. Should any factor cause the productive capacity of green plants to be diminished, all other organisms at higher trophic levels will be diminished accordingly.

Nonfeeding Relationships. ***Mutually Supportive Relationships.*** The overall structure of ecosystems is dominated by feeding relationships, as we have just seen. In any feeding relationship, we generally think of one species benefiting and the other being harmed to a greater or lesser extent. However, there are many relationships that provide a mutual benefit to both species. This phenomenon is called **mutualism**. A common example is the relationship between flowers and insects: The insects benefit by obtaining nectar from the flowers, and the plants benefit by being pollinated in the process. Another example is observed in tropical seas: Clownfish are immune to the toxin in the tentacles of sea anemones, which the anemones use to immobilize their prey. Thus, these fish are able to feed on detritus around the anemones, at the same time receiving protection from would-be predators that are *not* immune. The anemones benefit by being cleaned (Fig. 2-16).

In some cases, the mutualistic relationship has become so close that the species involved are no longer capable of living alone. A classic example is the group

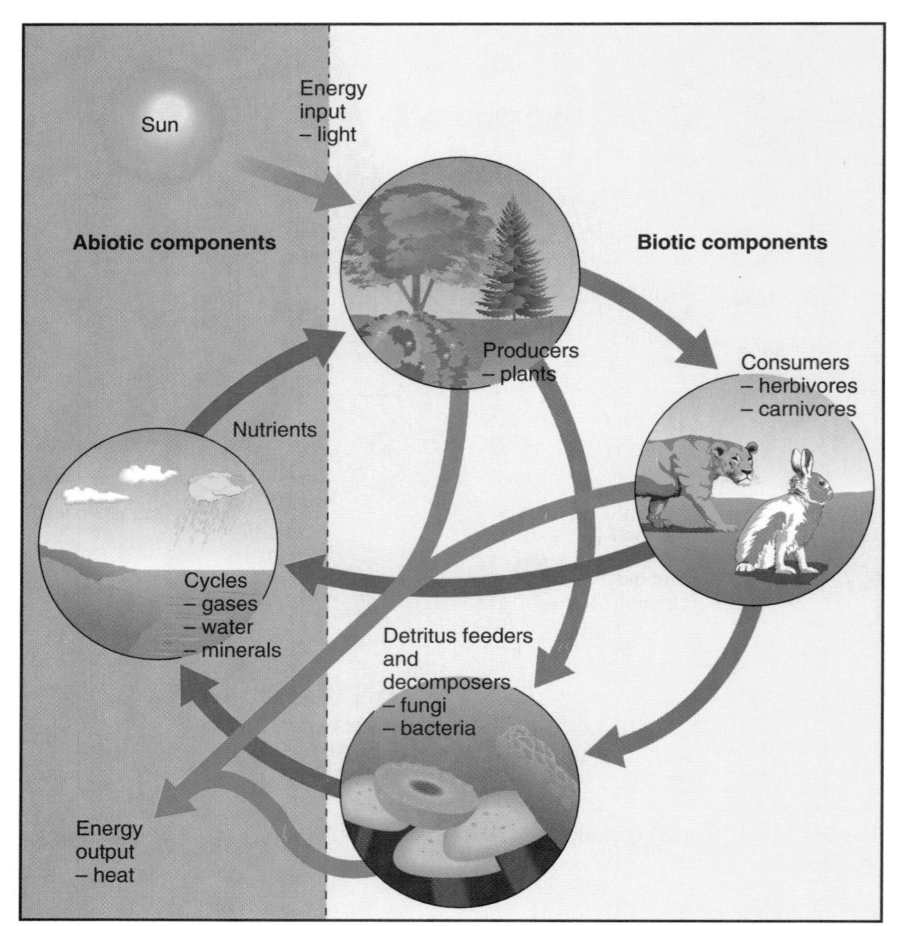

FIGURE 2–15
The movement of nutrients (blue arrows) and energy (red arrows) and both (brown arrows) through the ecosystem. Nutrients follow a cycle, being used over and over. Light energy absorbed by producers is released and lost as heat energy as it is "spent." (From Robert Christopherson, *Geosystems,* 2/e. ©1994, p. 587. Reprinted by permission of Prentice Hall, Upper Saddle River, New Jersey. After B.J. Nebel, *Environmental Science,* 1/e 1980, Prentice Hall, Upper Saddle River, New Jersey.)

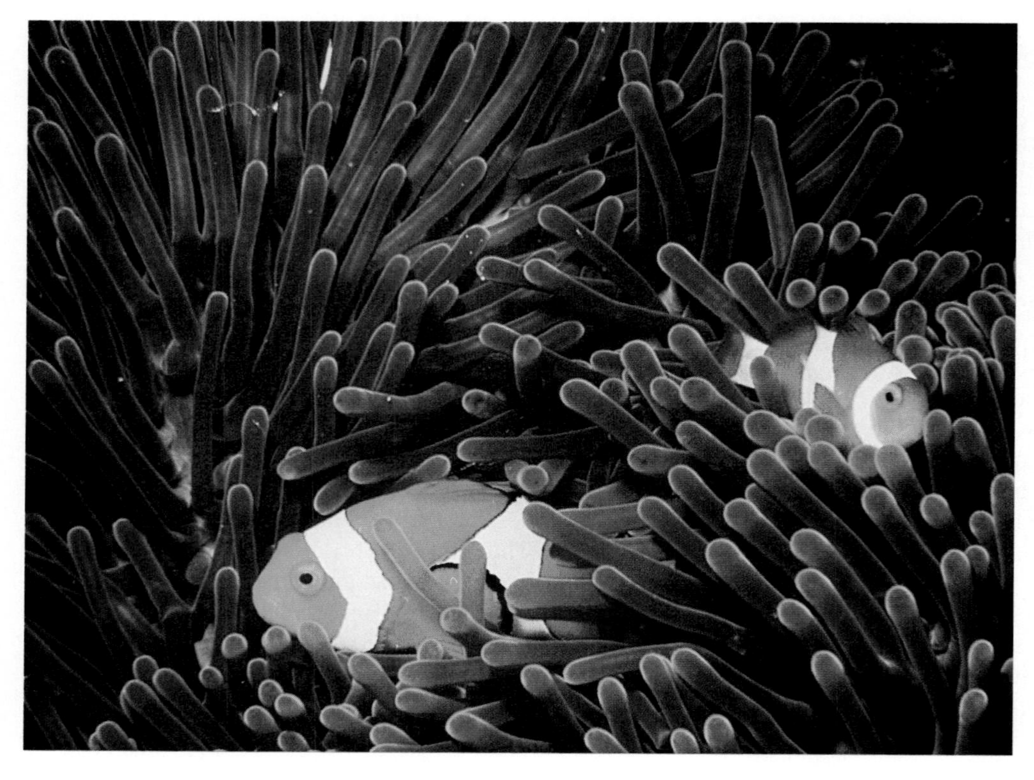

FIGURE 2–16
A mutualistic relationship—one in which both species benefit—is seen between clownfish and sea anemones. The fish, protected by the anemones, can forage without risk of predation; the anemones are cleaned of detritus. (Greg Dimijan/Photo Researchers, Inc.)

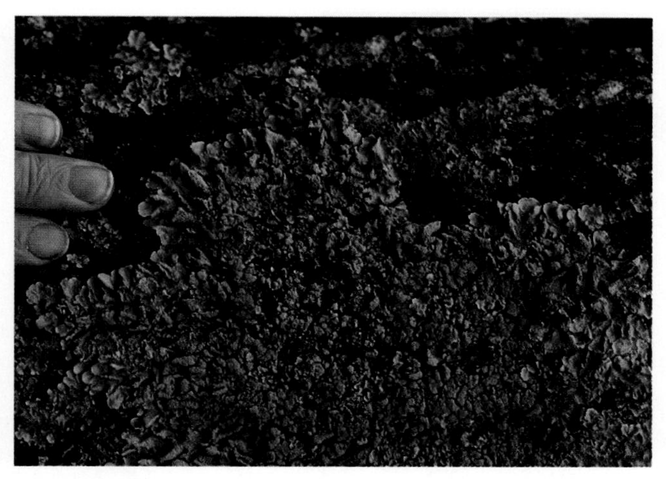

FIGURE 2–17
Lichens. The crusty-appearing "plants" commonly seen growing on rocks or the bark of trees are actually composed of a fungus and an alga growing in a symbiotic relationship. (Photograph by BJN.)

of plants known as *lichens* (Fig. 2-17). A lichen actually comprises two organisms: a fungus and an alga. The fungus provides protection for the alga, enabling it to survive in dry habitats where it could not live by itself, and the alga, which is a producer, provides food for the fungus, which is a heterotroph. Two species living together in close union are said to have a *symbiotic* relationship. However, **symbiosis** by itself simply refers to the fact of "living together" in close union (*sym*, together; *bio*, living); it does not specify a mutual benefit or harm. Therefore, symbiotic relationships may include parasitic relationships as well as mutualistic relationships.

While not categorized as mutualistic, countless relationships in an ecosystem may be seen as aiding its overall sustainability. For example, plant detritus provides most of the food for decomposers and soil-dwelling detritus feeders such as earthworms. Thus, these organisms benefit from plants, but the plants also benefit because the activity of the organisms is instrumental in releasing nutrients from the detritus and in returning them to the soil where they can be reused by the plants. In another example, insect-eating birds benefit from vegetation by finding nesting materials and places among trees, while the plant community benefits because the birds feed on and reduce the populations of many herbivorous insects. Even in predator-prey relationships, some mutual advantage may exist. The killing of individual prey that are weak or diseased may benefit the population at large by keeping it healthy. Predators and parasites may also prevent herbivore populations from becoming so abundant that they overgraze their environment.

Competitive Relationships. Given the concept of food webs, it might seem that species of animals would be in a great "free-for-all" competition with each other. In fact, fierce competition rarely occurs, because each species tends to be specialized and adapted to its own *habitat* and/or *niche*.

Habitat refers to the kind of place—defined by the plant community and the physical environment—where a species is biologically adapted to live. For example, a deciduous forest, a swamp, and an open grassy field denote types of habitats. Types of forests (e.g., conifer vs. deciduous) provide markedly different habitats and support a variety of wildlife.

Even when different species occupy the same habitat, competition may be slight or nonexistent, for the most part, because each species has its own *niche*. An animal's **niche** refers to what it feeds on, where it feeds, when it feeds, where it finds shelter, and where it nests. Seeming competitors can coexist in the same habitat but have separate niches. For example, woodpeckers, which feed on insects in dead wood, are not in competition with birds that feed on seeds. Many species of songbirds coexist in forests because they feed on insects from distinct levels in the trees (Fig. 2-18). Bats and swallows both feed on flying insects, but they are not in competition, because bats feed at night and swallows feed during the day.

There is often interspecies competition where different habitats or niches overlap. If two species do compete directly in every respect, as sometimes occurs when a species is introduced from another continent, one of the two generally perishes in the competition—this is the *competitive exclusion principle*.

All green plants require water, nutrients, and light, and where they are growing in the same location, one species may eliminate others through competition. (Hence, maintaining flowers and vegetables against the advance of weeds is a constant struggle.) However, different plant species are also adapted and specialized to particular conditions. Thus, each species is able to hold its own against competition where conditions are well suited to it. The same concepts hold true for species in aquatic and marine ecosystems.

Abiotic Factors

We now turn to the *abiotic* side of the ecosystem. As noted before, the environment involves the interplay of many physical and chemical factors, or **abiotic factors**, the major ones being rainfall (amount and distribution over the year and/or available moisture in the soil), temperature (extremes of heat and cold, as well as average,) light, wind, chemical nutrients, pH (acidity), salinity (saltiness), and fire. Within aquatic systems, the key abiotic factors are salinity (fresh vs. salt water), temperature, chemical nutrients, texture of the bottom (rocky vs. silty), depth and turbidity (cloudiness) of water (determining how much, if any, light reaches the

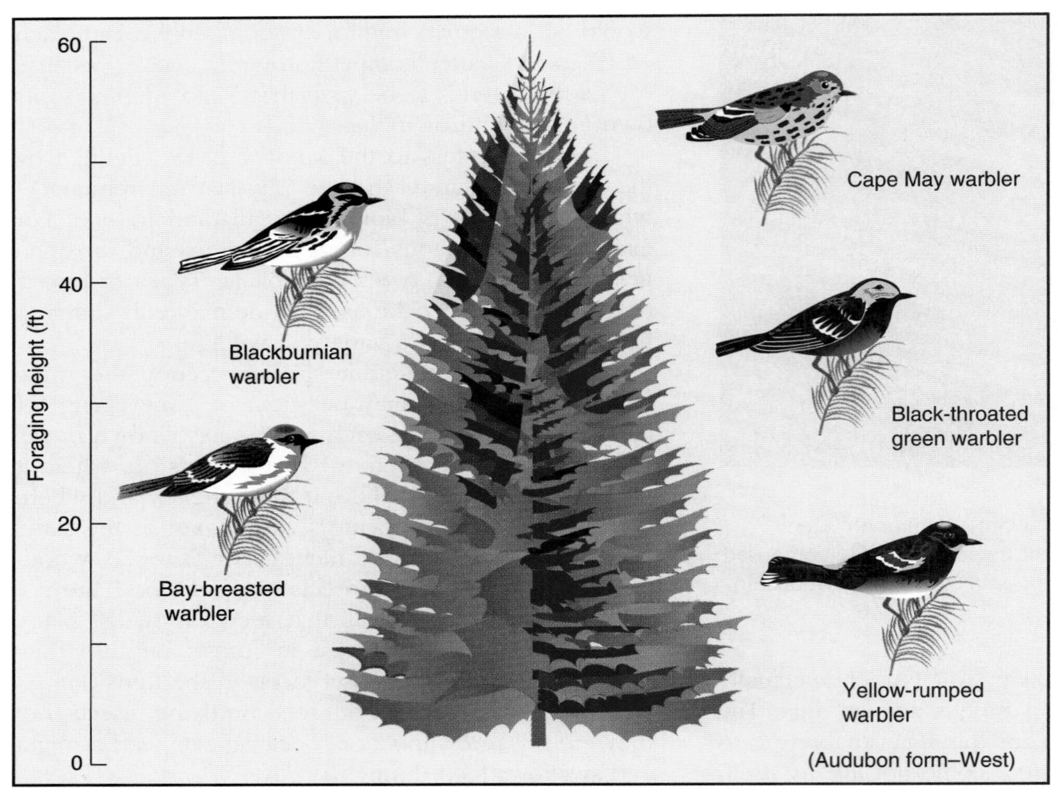

FIGURE 2–18
Five species of North American warblers reduce the competition among themselves by feeding at different levels and on different parts of the trees.

bottom), and currents. The degree to which each factor is present or absent and high or low profoundly affects the ability of organisms to survive. However, each species may be affected differently by each factor. We shall find that this difference in response to environmental factors determines which species may or may not occupy a given region or particular area within a region. In turn, which organisms do or do not survive determines the nature of a given ecosystem.

Optimum, Zones of Stress, and Limits of Tolerance. In any study of ecology, a primary observation is that *different species thrive under different conditions.* This principle applies to all living things, both plants and animals. Some survive where it is very wet; others relatively dry. Some thrive in warmth; others do best in cooler situations. Some tolerate freezing; others do not. Some require bright sun; others do best in shade. Aquatic systems are divided into fresh and salt water, each with its respective fish and other organisms.

Laboratory experiments clearly bear out the fact that different species are best adapted to different conditions. Organisms can be grown under controlled conditions where one factor is varied while other factors are held constant. Such experiments demonstrate that for every factor there is an **optimum**, a certain level at which the organisms do best. At higher or lower levels the organisms do less well, and at further extremes they may not be able to survive at all. This concept is shown graphically in Fig. 2-19. Temperature is shown as the variable in the figure, but the idea pertains to any abiotic factor that might be tested.

The point at which the best response occurs is called the optimum, but since this often occurs over a range of several degrees, it is common to speak of an *optimal range*. The entire span that allows any growth at all is called the **range of tolerance**. The points at the high and low ends of the range of tolerance are called the **limits of tolerance**. Between the optimal range and the high or low limit of tolerance, there are **zones of stress**. That is, as the factor is raised or lowered from the optimal range, the organisms experience increasing stress, until, at either limit of tolerance, they cannot survive.

Of course, not every species has been tested for every factor; however, the consistency of such observations leads us to conclude that the following is a fundamental biological principle: *Every species (both plant and animal) has an optimum range, zones of stress, and limits of tolerance with respect to every abiotic factor.*

This line of experimentation also demonstrates that different species vary in characteristics with respect to the values at which the optimum and limits of tolerance occur. For instance, what may be an optimal amount of water for one species may stress a second and result in the death of a third. Some plants cannot tolerate any freezing temperatures, others can tolerate slight but not intense freezing, and some actually require several weeks of freezing temperatures in order to complete their life cycles. Also, some species have a very broad range of tolerance, whereas others have a much narrower range. While optimums and limits of tolerance may differ from one species to another, there may be great overlap in their ranges of tolerance.

The concept of a range of tolerance does not just affect the growth of individuals; insofar as the health and vigor of individuals affect reproduction and survival of the next generation, the population is also influenced. That is, the population density (individuals per unit area) of a species will be greatest where all conditions are optimal, and population density will decrease as any one or more conditions depart from the optimum. Can you begin to relate this to the existence of ecotones, the grading of one ecosystem or biome into another, as described earlier in the chapter?

Law of Limiting Factors. There is an optimum and limits of tolerance for every single abiotic factor. Therefore, it follows that any one factor being outside the optimal range will cause stress and limit the growth, reproduction, or even the survival of the population. A factor that limits growth is called the **limiting factor**. The preceding observation is referred to as the **law of limiting factors**.

Also, keep in mind that the limiting factor may be a problem of "too much," as well as a problem of "too little." For example, plants may be stressed or killed not only by underwatering or underfertilizing, but also by overwatering or overfertilizing, which are common pitfalls for beginning gardeners. Note also that the limiting factor may change from one time to another. For instance, in a single growing season, temperature may

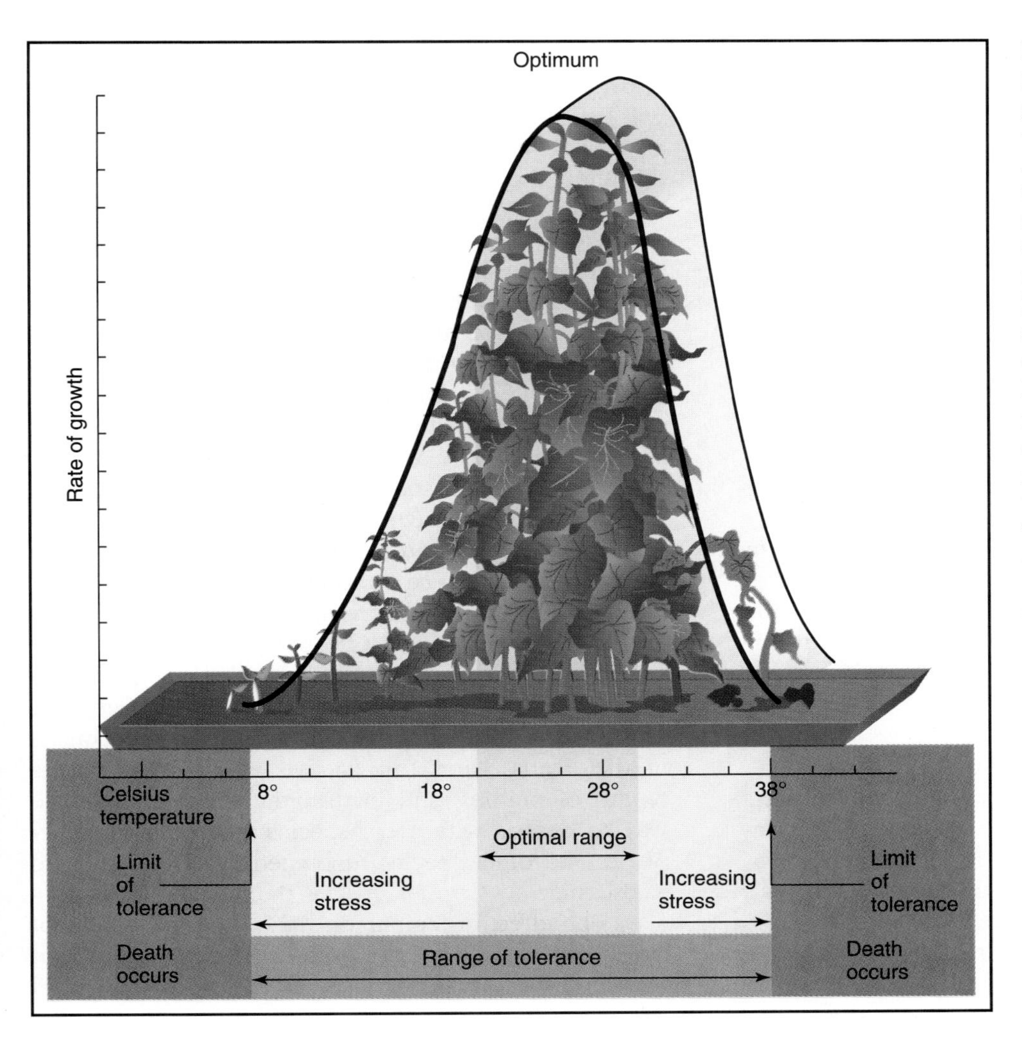

FIGURE 2–19
For every factor influencing growth, reproduction, and survival, there is an optimum level. Above and below the optimum, there is increasing stress, until survival becomes impossible at the limits of tolerance. The total range between the high and low limits is the range of tolerance. Levels at which the optimum, zones of stress, and limits of tolerance occur are different for each species and are a function of the genetic makeup and variability within the species population.
The genetic makeup is the basis of a species' adaptation to its environment. Not only are individuals more robust at the optimums, but they are also more numerous.

be limiting in the early spring, nutrients may be limiting later, and then water may be limiting if a drought occurs. Also, if one limiting factor is corrected, growth will increase only until another factor comes into play. Of course, the organism's genetic potential is an ultimate limiting factor: A daisy will never grow to be the height of a tree, nor a mouse to the bulk of an elephant, regardless of optimal environmental factors.

The law of limiting factors was first presented by Justus von Liebig in 1840 in connection with his observations regarding the effects of chemical nutrients on plant growth. He observed that restricting any one of the many different nutrients at any given time had the same effect: It limited growth. Thus, this law is also called Liebig's law of minimums.

Observations made since Liebig's time show that his law has a much broader application: Growth may be limited not only by abiotic factors, but also by biotic factors. Thus, the limiting factor for a population may be competition or predation from another species. This is certainly the case with our agricultural crops, where it is a constant struggle to keep them from being limited or even eliminated by weeds and "pests."

Finally, while one factor may be determined to be limiting at a given time, several factors outside the optimum may combine to cause additional stress or even death. Particularly, pollutants may act in a way that causes organisms to become more vulnerable to disease or drought. Such cases are examples of **synergistic effects**, or **synergisms**, which are defined as two or more factors interacting in a way that causes an effect much greater than one would anticipate from the effects of each of the two acting separately.

Why Do Different Regions Support Distinct Ecosystems?

We can now use the concepts of optimums and limiting factors to gain a better understanding of why different regions or even localized areas may have distinct biotic communities, creating a variety of ecosystems and biomes (Figure 2-3).

Climate

Weather, of course, is defined mainly by the temperatures and precipitation that occur on any particular day. Climate is derived from weather in the following way: The weather that has occurred on January 1 of each year for the past 30 years or more is averaged to give a "normal" expectation of weather for January 1. The same is done for each day of the year. Thus, the **climate** of a given region is a description of the average temperature and precipitation that may be expected on each day throughout the entire year.

Climates in different parts of the world vary widely. In general, the temperature of equatorial regions is continuously warm, with high rainfall and no discernible seasons. Above and below the equator, the temperatures become increasingly seasonal (warm or hot summers and cool or cold winters); the farther we go toward the poles, the longer and colder the winters become, until at the poles it is perpetually winter. Likewise, colder temperatures are found at higher elevations, so that there are a number of snowcapped mountains on or near the equator.

Annual precipitation in any area also may vary greatly, from virtually zero to well over 100 inches (250 cm) per year. It may be evenly distributed throughout the year or may all occur in certain months, dividing the year into wet and dry seasons.

Different temperature and rainfall conditions may occur in almost any combination to give a wide variety of climates. In turn, any climate will support only those species that find the temperature and precipitation levels optimal or at least within the range of tolerance. As indicated in Fig. 2-19, population densities will be greatest where conditions are optimal and will decrease as any condition departs from the optimum. A species will be excluded from a region (or local areas) where any condition is beyond its limit of tolerance. How will this affect the biotic community?

To illustrate, let us consider three major types of ecosystems: forests, grasslands, and deserts. These three types of ecosystems are determined mainly by a single limiting factor: amount of rainfall. For example, annual precipitation over most of the eastern United States is 40–60 inches (100–150 cm). It tapers off to 20–30 inches (50–75 cm) per year over the Plains States and declines still more in the Southwest, with large regions receiving an average of less than 10 inches (25 cm) per year. For most temperate tree species, 46–56 inches (115–140 cm) per year is optimal. Below about 40 inches (100 cm) per year, many trees begin to be stressed, and most reach their limit of tolerance at about 30 inches (75 cm) per year. Grasses, however, have a much lower limit of tolerance, about 10 inches (25 cm) per year, and many species of cacti and other specialized desert plants do well with as little as 2–4 inches (5–10 cm) per year. Thus, the eastern United States supports forests, while in the Plains States forests give way to the grasses typical of prairies, and in the Southwest grasses give way to deserts with a sparse coverage of cacti, sagebrush, and other species renowned for their tolerance to drought.

The effect of temperature, the other parameter of climate, is largely superimposed on that of rainfall. That is, 36 inches (90 cm) or more of rainfall per year will support a forest, but temperature will determine the

Earth Watch

A Dose of Limiting Factor

Environmental factors—particularly temperature and rainfall—are always changing. What does it mean to say that rainfall, for example, is a limiting factor when weather is almost always changing?

Consideration of a limiting factor should always include the concept of *dose*. **Dose** is defined as the level of exposure multiplied by the length of time over which the exposure occurs. You have probably experienced this kind of thing for yourself. For example, you can enter a walk-in freezer or expose yourself to the intense heat from an oven for a few seconds without particular discomfort, whereas prolonged exposure to such heat or cold would be extremely damaging, if not fatal. You might not be hurt by breathing noxious fumes for a short time, but longer exposures might well kill you.

Similarly, a limiting factor that causes the dieoff of a species involves both the intensity of the factor and the duration of exposure to it. For example, most plants can tolerate drought to some degree. It is more a matter of how long they can tolerate it. Cacti and other desert plants can tolerate much longer periods of drought than nondesert species. Likewise, it is not a question of whether plants can tolerate flooding. Rather, it is a question of how long they can tolerate being flooded. Marsh plants, of course, thrive on perpetually flooded land; in contrast, more than a day or so of flooding will kill many other terrestrial species.

Thus, the balance between ecosystems is not a perfect steady-state balance with a difference in average rainfall as a sharp dividing line. What actually occurs, where forest meets grassland, for example, is that trees encroach into grasslands during years of normal rainfall, but then a severe drought occurs, and trees in the grassland are killed back, and then the cycle starts again. Hence, there is a seesaw effect between adjoining ecosystems that depends on the occurrence of limiting doses of one or another factor. This seesaw effect is a particularly important consideration in the face of global warming.

There is no question that most species could probably tolerate the average temperature being a few degrees higher if this were the only factor involved. The problem is, how will the slightly higher average temperature affect the occurrence of heat waves and the duration of droughts and floods? Such warming could create limiting doses that have far-reaching and unpredictable effects. Scientists are already tracking changing habitats attributable to global warming.

kind of forest. For example, broadleaf evergreen species, which are extremely vigorous and fast growing but cannot tolerate freezing temperatures, predominate in the tropics. By dropping their leaves and going into dormancy each autumn, deciduous trees are well adapted to freezing temperatures. Therefore, wherever rainfall is sufficient, deciduous forests predominate in temperate latitudes. Most deciduous trees, however, cannot tolerate the extremely harsh winters and short summers that occur at higher latitudes and higher elevations. Therefore, northern regions and high elevations are dominated by spruce-fir coniferous forests, which are better adapted to these conditions.

Temperature by itself limits forests only when it becomes low enough to cause **permafrost** (permanently frozen subsoil). Permafrost prevents the growth of trees because roots cannot penetrate deeply enough to provide adequate support. However, a number of grasses, clovers, and other small flowering plants can grow in the topsoil above permafrost. Consequently, where permafrost sets in, coniferous forests give way to tundra. Of course, at still colder temperatures, the tundra gives way to permanent snow and ice cover.

The same relationship of rainfall effects being primary and temperature effects secondary applies in deserts. Any region receiving less than about 10 inches (25 cm) of rain per year will be a desert, but the plant and animal species found in hot deserts are different from those found in cold deserts.

Temperature also exerts considerable influence on an ecosystem by its effect on the rate of evaporation of water. Higher temperatures effectively reduce the amount of available water because more is lost through evaporation. Consequently, the transitions from deserts to grasslands and from grasslands to forests are found at higher precipitation levels in hot regions than in cold regions.

A summary of these temperature and rainfall conditions and the biomes they support and preclude is given in Fig. 2-20. The average temperature for a region varies with both latitude and altitude, as shown in Fig. 2-21.

Microclimate and Other Abiotic Factors

A specific site may have temperature and moisture conditions that are significantly different from the overall climate of the region in which it is located, which is necessarily an average. For example, a south-facing slope, which receives more direct sunlight, will be rel-

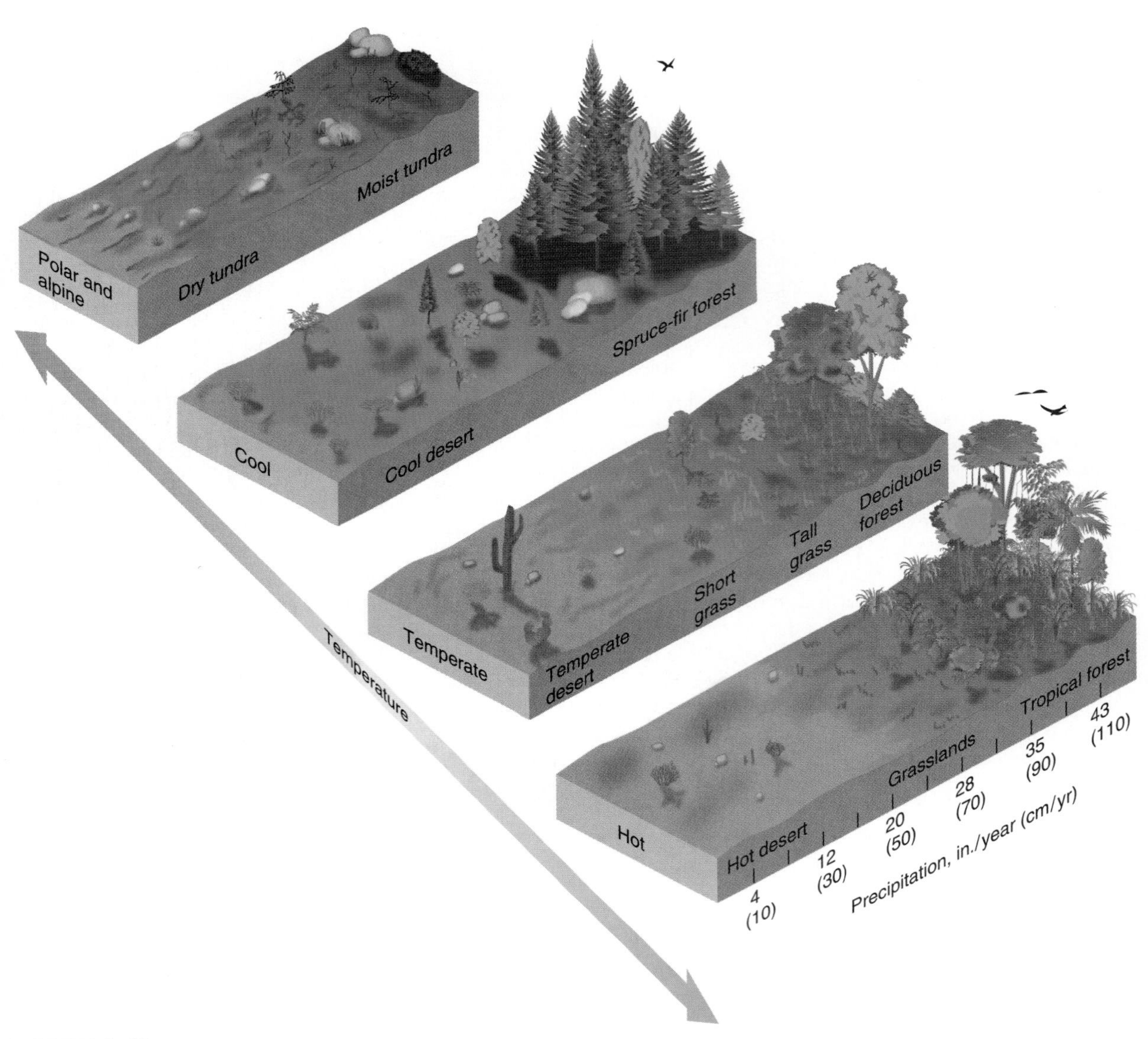

FIGURE 2–20
Climate and major biomes. Moisture is generally the overriding factor determining the type of biome that may be supported in a region. Given adequate moisture, an area will generally support a forest. Temperature, however, determines the *kind* of forest. The situation is similar for grasslands and deserts. At cooler temperatures, there is a shift toward less precipitation because lower temperatures reduce evaporative water loss. Temperature becomes the overriding factor only when it is low enough to sustain permafrost. (From Robert Christopherson, *Geosystems*, 2/e. ©1994, p.594. Reprinted by permission of Prentice Hall, Upper Saddle River, New Jersey. After B.J. Nebel, *Environmental Science*, 1/e. ©1980, Prentice Hall, Upper Saddle River, New Jersey.)

atively warmer and hence also drier than a north-facing slope. Similarly, temperature range in a sheltered ravine will be narrower than that in a more exposed location, and so on. The conditions found in a specific localized area are referred to as the **microclimate** of that location. In the same way that different climates determine the major biome of the region, different microclimates result in variations of the biotic community within the biome.

Soil type and topography may also contribute to the diversity found in a biome because these two factors affect the availability of moisture. For example, in the eastern United States, oaks and hickories generally predominate on rocky, sandy soils and on hilltops, which retain little moisture, whereas beeches and maples are found on richer soils, which hold more moisture, and red maples and cedars inhabit low, swampy areas. In the transitional region between desert

Coniferous Forests

Primary Regions: Northern portions of North America, Europe and Asia, extending southward at high elevations (glaciated terrain).
Climate and Soils: Seasonal, winters usually long and cold. Precipitation often light in winter, heavier in summer. Soils acidic and humus-rich, much litter.
Major Vegetation: Coniferous trees (spruce, fir, pine, hemlock), smaller amounts of deciduous trees (birch, maple). Poor understory.
Animals: Large herbivores such as mule deer, moose, elk, caribou; smaller herbivores such as mice, hares, red squirrels; predators like lynx, foxes, bears, wolverines, fisher, marten; important nesting area for many migratory warblers, thrushes and others.

Environmental concerns: *Pesticide spraying to control forest insect damage can lead to poisoning of the food chain and loss of predatory hawks, owls and eagles (Ch. 10). Damming for hydroelectric and water supply drowns forestlands in the north (Ch. 11, 23). Coniferous forests downwind of industrial complexes are heavily damaged by ozone and acid deposition (Ch. 15, 16). Harvesting of old growth coniferous forests destroys crucial habitat for endangered species (Ch. 19).*

Tundra

Primary Regions: North of the coniferous forest in the Northern Hemisphere and extending southward at elevations above the coniferous forest.
Climate and Soils: Bitter cold except for an 8-to-10 week growing season having long days and moderate temperatures. Precipitation low, less than 10 inches annually. Soils thin and underlain by permanent frost.
Major Vegetation: Low-growing lichens, mosses, grasses, sedges, and dwarf shrubs.
Animals: Year-round: lemmings, arctic hares, ptarmigan, arctic fox, lynx, grizzly bears, snowy owls, gyrfalcons; large herbivores such as caribou, reindeer, musk ox and mountain sheep migrate in and out of tundra. Summers: many geese, ducks, sandpipers, and other waterfowl migrate in to raise their young. Insects and other invertebrates dense during the short summers.

Environmental concerns: *Harsh conditions and low productivity prevent most human exploitation of this biome. Oil exploration and development disrupts wilderness and can lead to long-term contamination of affected areas and decline of large animals (Ch. 21).*

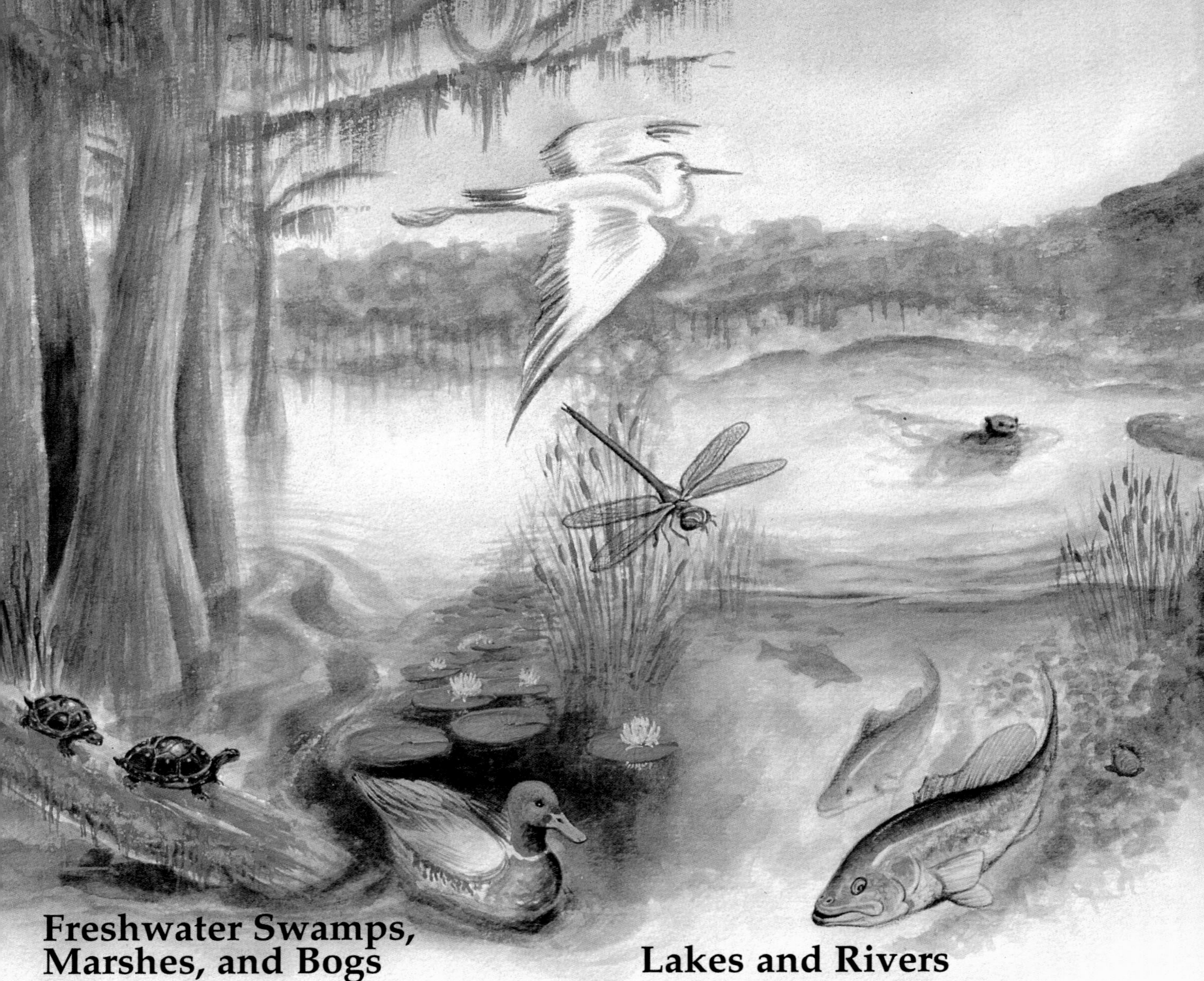

Freshwater Swamps, Marshes, and Bogs

Locations: Wetlands in poorly drained regions receiving moderate to heavy precipitation; often occupy sites of lakes and ponds that have filled in.
Environmental parameters: Shallow systems, sometimes only seasonally wet. Sediment is black and rich, often anaerobic below the surface. Nutrients usually abundant, except in acidic bogs.
Vegetation: Marshes usually heavily vegetated with cattails, sedges and reeds. Swamps vegetated by water tolerant trees such as red maple and cedars. Bogs occupied by sphagnum moss and low shrubs like leatherleaf.
Animals: Amphibians and reptiles, small fish, numerous invertebrates; wading birds, ducks and geese; raccoons. Alligators in warmer regions.

Environmental concerns: Drainage from irrigated lands into wetlands can lead to accumulation of toxic chemicals (Ch. 9). Wetlands are often drained and converted into home sites or agricultural lands (Ch. 12, 19). Wetlands used as toxic waste dumps and landfills can cause human health problems (Ch. 14, 20).

Lakes and Rivers

Locations: Lakes and ponds: physical depressions that allow precipitation and groundwater to accumulate; rivers and streams: water flows by gravity toward oceans or large lakes.
Environmental parameters: Lower concentration of dissolved solids than in ocean, determined primarily by soils around water body. Seasonal vertical stratification in lakes separates water masses; one-way currents in rivers and streams.
Vegetation: Microscopic algae suspended in water (phytoplankton) or on rocks and sediment (periphyton); higher plants rooted on bottom and submerged or emergent (macrophytes).
Animals: Microscopic crustaceans and rotifers suspended in water (zooplankton); many invertebrates, esp. insect larvae; reptiles and amphibians common; many kinds of fish feeding on other animal and plant life; otters, raccoons, wading birds, ducks, geese and swans.

Environmental concerns: Toxic chemicals and other pollutants affect water quality, kill wildlife and create human health problems (Ch. 3, 11, 14, 21). Eutrophication from excessive nutrients creates unwanted growth of vegetation (Ch. 3, 12, 13). Introduced species like water hyacinth (Ch. 4) and zebra mussels (Ch. 18) kill native species and choke waterways. Acid deposition leads to acidification of bodies of water, killing fish life (Ch. 16). Erosion changes riverbeds and causes flooding and loss of aquatic habitat (Ch. 11, 12).

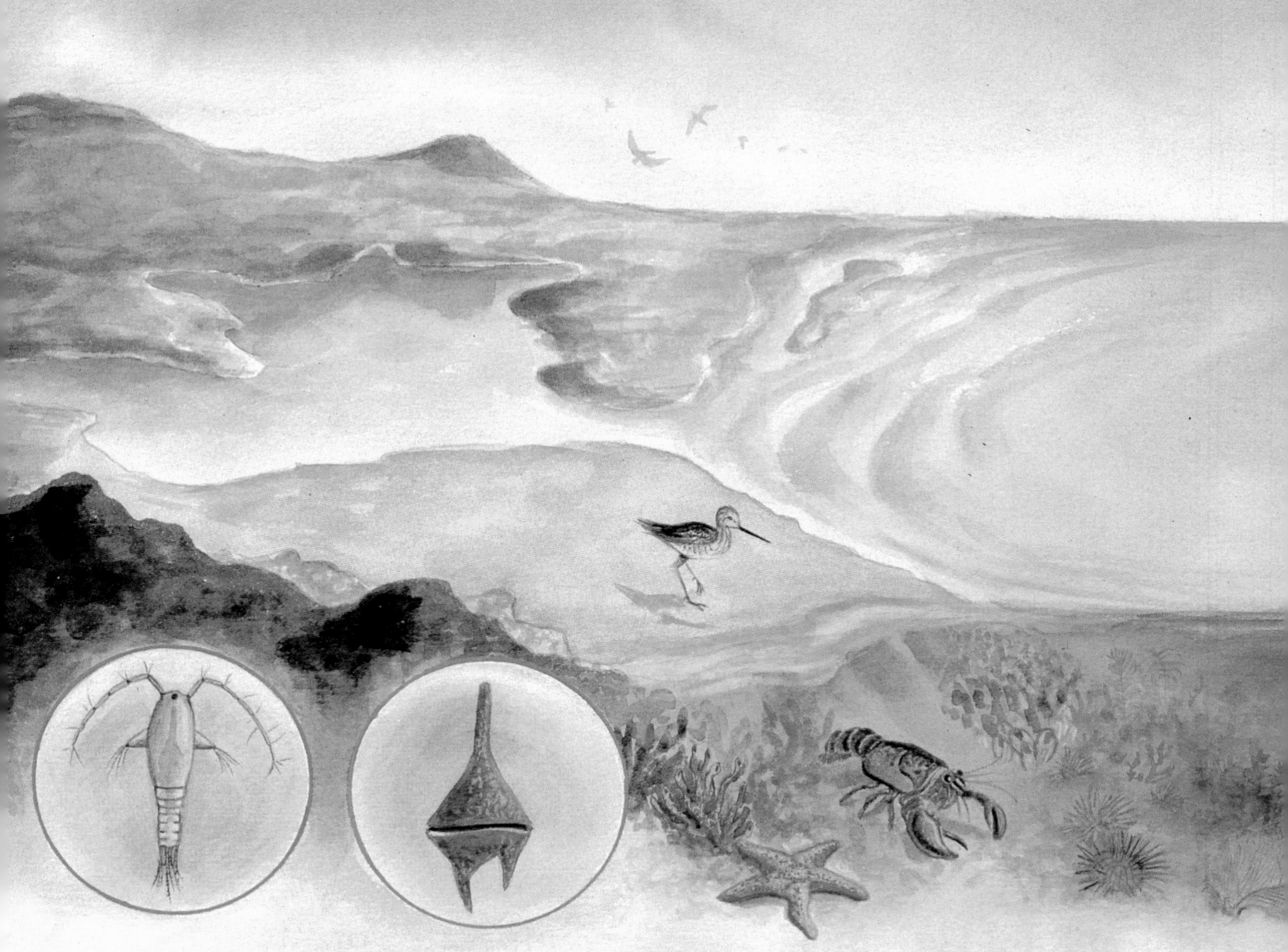

Estuaries

Locations: Coastal regions where rivers meet the ocean; may form bays behind outer sandy barrier island.
Environmental parameters: Variable salinity due to mixing of fresh and salt water on a gradient from freshwater to the ocean; tides create two-way currents, promote mixing. Often rich in nutrients and suspended sediments. Bottom sediments often anaerobic just below the surface.
Vegetation: Phytoplankton in water column, rooted aquatics such as eelgrass and kelps; saltmarsh grasses growing intertidally and forming unique saltmarsh environments; in tropics, mangrove swamps form, with many salt-tolerant species of trees and shrubs.
Animals: Zooplankton in water column; rich shellfish, crustacean and worm fauna on and in bottom sediments; abundant fish, some larvae of oceanic species; wading birds, ducks and geese abundant.

Environmental concerns: *Most estuaries are affected by eutrophication, leading to unwelcome changes in vegetation and animal life (Ch. 12, 13). Estuarine saltmarshes are filled, dredged and bulkheaded to provide residences and commercial buildings (Ch. 12, 19). Estuaries receive poorly treated sewage, rendering the shellfish unfit for consumption (Ch. 13). Shellfish and fish are readily overharvested, leading to the tragedy of the commons (Ch. 19).*

Inter-tidal Zone

Locations: Land margins along the ocean wherever tides occur.
Environmental parameters: Cyclical exposure to air and inundation with seawater due to tides; strong wave action on outward facing land. If substratum is sandy, beaches will occur; otherwise, rocky surfaces will occur, where competition for space between attached algae and shellfish is intense.
Vegetation: Microscopic algae on wet sand beaches; large red and brown algae such as kelp, irish moss and rockweed attached to rocky surfaces where wave action is strong.
Animals: Small crustaceans and molluscs in sand beaches; snails, bivalve molluscs, barnacles, anemones, sea urchins and starfish abundant on rocky intertidal, especially in tide pools; wading birds, gulls, terns, loons, ducks, grebes.

Environmental concerns: *Human use of beaches prevents some endangered bird species from breeding (Ch. 18). Building jetties and bulkheads to stem storm and wave damage leads to further coastal damage (Ch. 18). Oil spills create unsightly and toxic conditions on coastlines (Ch. 21).*

Coastal Ocean

Locations: From coastline outward, often over a continental shelf, to a depth of 200 meters. In tropics, coral reefs are major shallow coastal forms.
Environmental parameters: High productivity due to coastal upwelling and transport of nutrients from estuaries; water column mixes to bottom except where seasonal vertical stratification develops. Tidal currents promote mixing.
Vegetation: Microscopic phytoplankton algae dominate productivity; some large benthic plants where water clarity permits. In coral reefs, symbiotic algae living in coral animals, other large algae and turtle grass predominate.
Animals: Abundant microscopic zooplankton in water column; rich bottom fauna of worms, shellfish and crustaceans; great variety of fauna on coral reefs; diverse and abundant fish fauna; jellyfish, turtles, gulls, terns; diving ducks, gannets, puffins, cormorants and other fish-feeding birds abundant; seals, sea lions, penguins, dolphins and whales regionally abundant.

Environmental concerns: *Pollutants from estuaries and coastal cities contaminate shellfish and fish (Ch. 13, 14). Rising sea levels will inundate many low-lying coastal areas (Ch. 16). Overfishing of coastal fisheries causes loss of breeding stock and changes in ecology (Ch. 19). Past whaling in coastal areas has depleted the stocks of most species (Ch. 19).*

Open Ocean

Locations: Covering 70% of the earth's surface, from the edge of the continental shelf outwards.
Environmental parameters: Great depths (to 11,000 meters), all but upper 200 meters without light and cold. Nutrient-poor, except where vertical currents bring deep water to surface (upwelling).
Vegetation: Exclusively phytoplankton (coccolithophorids, diatoms, dinoflagellates predominate). Varies according to nutrient availability.
Animals: Diverse zooplankton fauna together with fish fauna adapted to different depths. Bottom fauna sparse except in regions of deep hydrothermal vents. Seabirds like petrel, shearwater, albatross. Whales, dolphins, tuna, sharks, flying fish, squid. Unique deep-sea fish with bioluminescence.

Environmental concerns: *Ozone shield depletion will kill phytoplankton in the Antarctic, affecting the entire food chain (Ch. 16). Drift-netting on the high seas depletes fisheries and kills ocean birds, turtles and mammals (Ch. 19). Whaling has led to steep declines in most whale species, which are still not out of danger in spite of a moratorium (Ch. 19).*

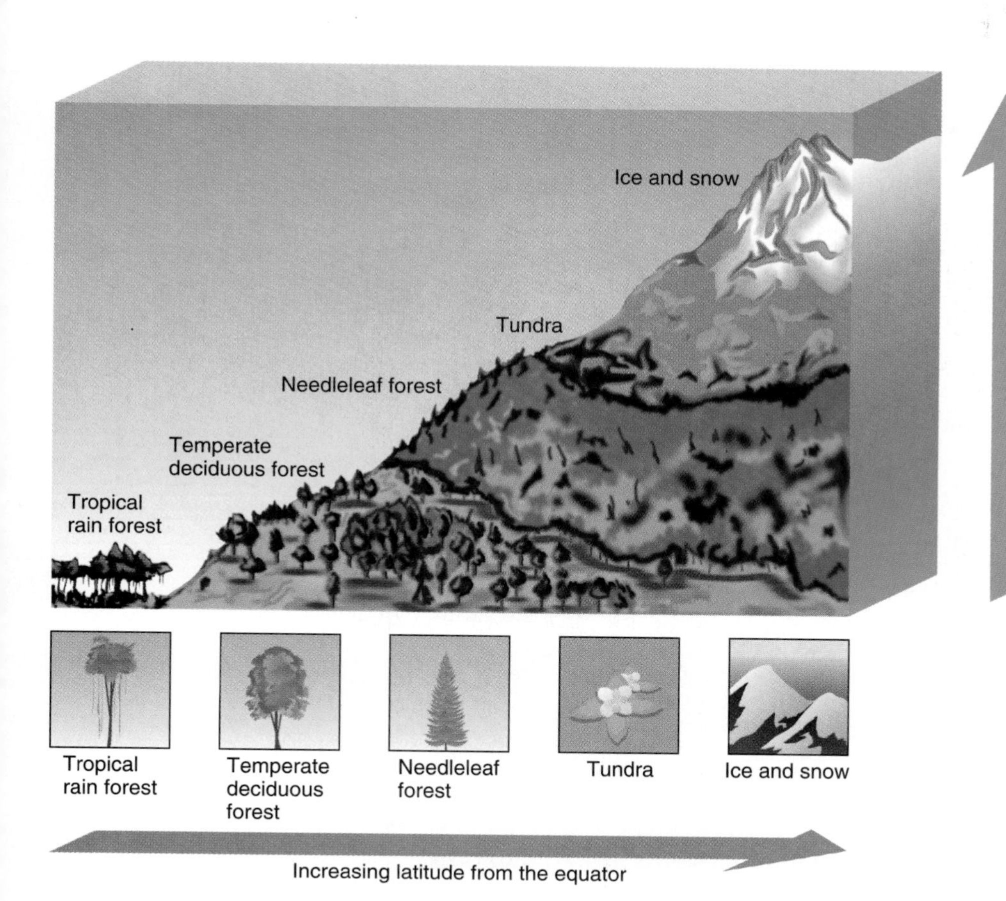

FIGURE 2–21
Decreasing temperatures that result in the biome shifts noted in Fig. 2-20 occur both with increasing latitude (distance from the equator) and increasing altitude. (From Robert Christopherson, *Geosystems,* 2/e. ©1994, p. 595. Adapted by permission of Prentice Hall, Upper Saddle River, New Jersey.)

and grassland [10–20 inches (25–50 cm) of rainfall per year], a soil with good water-holding capacity will support grass, but a sandy soil with little ability to hold water will support only desert species (Fig. 2-22).

In certain cases, an abiotic factor other than rainfall or temperature may be the primary limiting factor. For example, the strip of land adjacent to a coast frequently receives a salty spray from the ocean, a factor that relatively few plants can tolerate. Consequently, this strip is frequently occupied by a community of salt-tolerant plants (Fig. 2-23). Relative acidity or alkalinity (pH) may also have an overriding effect on a plant or animal community. This fact is particularly significant in view of acid rain.

Biotic Factors

Limiting factors may also be biotic, that is, caused by other species. Grasses thrive when rainfall is more than 30 inches (75 cm). However, when the rainfall is great enough to support trees, increased shade may limit grasses. Thus, the factor that limits grasses from taking over high-rainfall regions is biotic: overwhelming competition from taller species. The distribution of plants may also be limited by the presence of certain herbivores, particularly insect species and parasitic fungi.

The concept of limiting factors also applies to animals. As with plants, the limiting factor may be abiotic—cold temperatures or lack of open water, for example—but it is more frequently biotic, in the form of a lack of a plant community that provides suitable food or habitat or both.

Physical Barriers

A final factor that may limit species to a particular region is the existence of a physical barrier, such as an ocean, desert, or mountain range, that the species are unable to cross. Thus, species making up the communities on separate continents or remote islands are usually quite different despite having similar climates.

When such barriers are overcome—for example, by humans transporting a species from one continent to another—the introduced species may make a successful "invasion." However, a successful invasion by a foreign species may cause an ecological disaster, because the invader often displaces existing species through competition. Examples of such problems with imported species will be explored further in Chapter 4. Note also that humans erect barriers—dams, roadways, cities, and

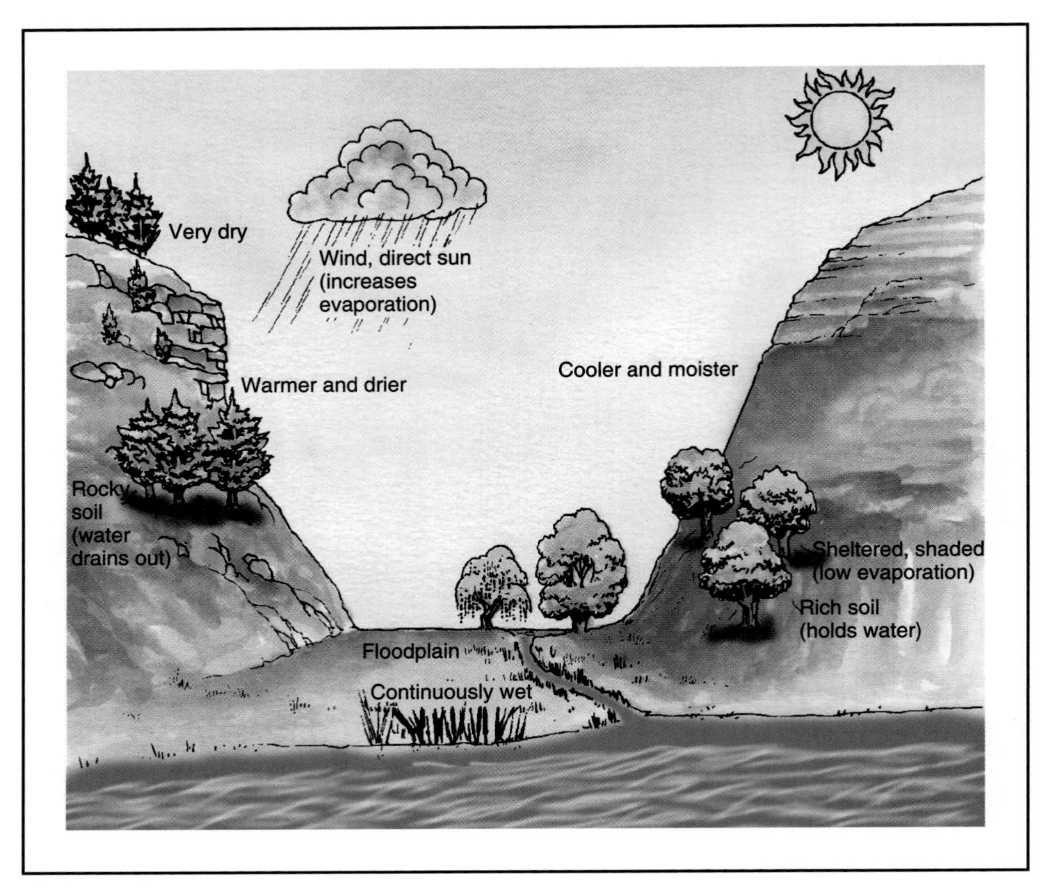

FIGURE 2–22
Abiotic factors such as terrain, wind, and type of soil create different microclimates by influencing temperature and moisture in localized areas.

farms—that may block the normal movement of populations and cause their demise.

In summary, the biosphere consists of a great variety of environments, both aquatic and terrestrial. In each environment we find plant, animal, and microbial species that are adapted to all the abiotic factors. In addition, they are adapted to each other, in various feeding and nonfeeding relationships. Each environment supports a more or less unique grouping of organisms interacting with each other and with the environment in a way that perpetuates or sustains the entire group. That is, each environment, together with the species it supports, is an ecosystem. Every ecosystem is tied with others through species that migrate from one system to another, and through exchanges of air, water, and minerals common to the whole planet. At the same time, each species, and as a result, each ecosystem, is kept within certain bounds by limiting factors. The spread of each species is at some point limited by its not being able to tolerate particular conditions, compete with other species, or cross some physical barrier. Species distribution is always due to one or more limiting factors.

Implications for Humans

In the preceding discussion, we have given you an understanding of what ideal ecosystems are like apart from human impacts. Again, we emphasize that natural ecosystems have existed and perpetuated themselves on Earth for hundreds of millions of years, while humans are relative newcomers on this scene.

Throughout this discussion, you have probably been wondering about how we humans—with all our associated agriculture, industry, and transportation—interrelate with natural ecosystems. Is our human system an ecosystem? Does it function like one? What does the study of natural systems really have to do with the sustainability of our human system? We shall attempt to give some insight regarding these questions.

FIGURE 2–23
Specific abiotic factors determine the type of vegetation found in a given location. Along the Pacific Coast in central California, the salt spray limits the growth of most vegetation; but certain succulent, salt-tolerant species thrive there. (Photographs by BJN.)

Ethics

Can Ecosystems Be Restored?

The human capacity for destroying ecosystems is well established. To some degree, however, we also have the capacity to restore them. In many cases, restoration involves simply stopping the abuse. For example, it has been found that water quality improves, and fish and shellfish gradually return to previously polluted lakes, rivers, and bays, after the input of pollutants is curtailed. Similarly, forests may gradually return to areas that have been cleared. Humans can speed up this process by seeding, planting seedling trees, and reintroducing populations of fish and animals that have been eliminated.

Under mandate of the Endangered Species Act the U.S. Fish and Wildlife Service has a number of ongoing programs aimed at building up populations of endangered species—captive breeding is commonly used—and then reintroducing populations into their original native habitat. The reintroduction of wolves to Yellowstone National Park in the fall of 1994 is an example. But this reintroduction is not without critics. Ranchers, who were largely responsible for extermination of the wolves in the first place, are particularly concerned that the wolves will leave the park and attack their livestock. Others simply don't see such efforts as a worthwhile expenditure of taxpayer's money. What do you think?

An additional dimension of this problem should be the recognition that the potential for restoration of any ecosystem rests on three assumptions: (a) abiotic factors must have remained unaltered, or can be returned to their original state; (b) viable populations of the species involved remain in existence, and (c) the ecosystem has not been upset by introduction of one or more foreign species that cannot be eliminated, and that may preclude the survival of reintroduced native species. As you study further chapters you will find that there are countless cases where one or more of these assumptions has been violated, cutting off any chance of complete restoration of the original ecosystem. Furthermore, human impacts are increasingly taking this direction.

What *are* our values, considering that we seem content to let the forces of extinction and irreversible ecosystem alteration go forward? Can such values be analyzed in terms of allowing the self-interests of a few to supersede the interests of society as a whole? As members of society, do we agree that this is right, or do we have a moral right, even an obligation, to stand up for other values?

In light of these questions it appears curious that the 104th Congress terminated the National Biological Survey. This agency was preparing an inventory of biodiversity in the United States.

Evidence gained through archaeology and anthropology shows that hominid ancestry goes back at least 4.4 million years. Hominids include all extinct human-like creatures as well as present-day humans The evidence indicates that several different hominid species were involved on the evolutionary pathway from our primate ancestors to our present-day human species, which emerged about 100,000 years ago (Fig. 2-24).

Early hominids survived in small tribes as hunter-gatherers, which means that they "lived off the land," catching wildlife and gathering seeds, nuts, roots, berries, and other plant foods. (Fig. 2-25). Settlements were never large and were of relatively short duration because, as one area was "picked over," the tribe was forced to move on. As hunter-gatherers, hominids were much like other omnivorous consumers in natural ecosystems. Populations could not expand beyond the sizes that natural food sources supported, and deaths from predators, disease, and famine were no doubt common.

About 10,000 years ago, however, a highly significant change occurred: Humans began to develop agriculture—the domestication of wild species. Animal husbandry and agriculture are processes of taking particular animal and plant species out of the wild, clearing space, and providing other conditions to grow them preferentially. Plants are protected from competition (weeds) and other would-be consumers, and additional nutrients (fertilizer) and water may be provided. Animals are protected from predators and given food for optimal growth. Over the years, nearly all agricultural plants and animals have been modified greatly through selective breeding, so that now they are quite different from their wild ancestors, but the basic processes of providing conditions to grow certain plants and animals preferentially remains unchanged.

The development of agriculture provided a more abundant and reliable food supply, but it created a turning point in human history for other reasons as well. Conducting agriculture does not just allow, but *requires*, permanent (or at least long-term) settlements and the specialization of labor. Some members of the settlement specialize in tending crops and producing food, freeing others to specialize in other endeavors. With this specialization of labor in permanent settlements, there is more incentive and potential for technological development: better tools, better dwellings,

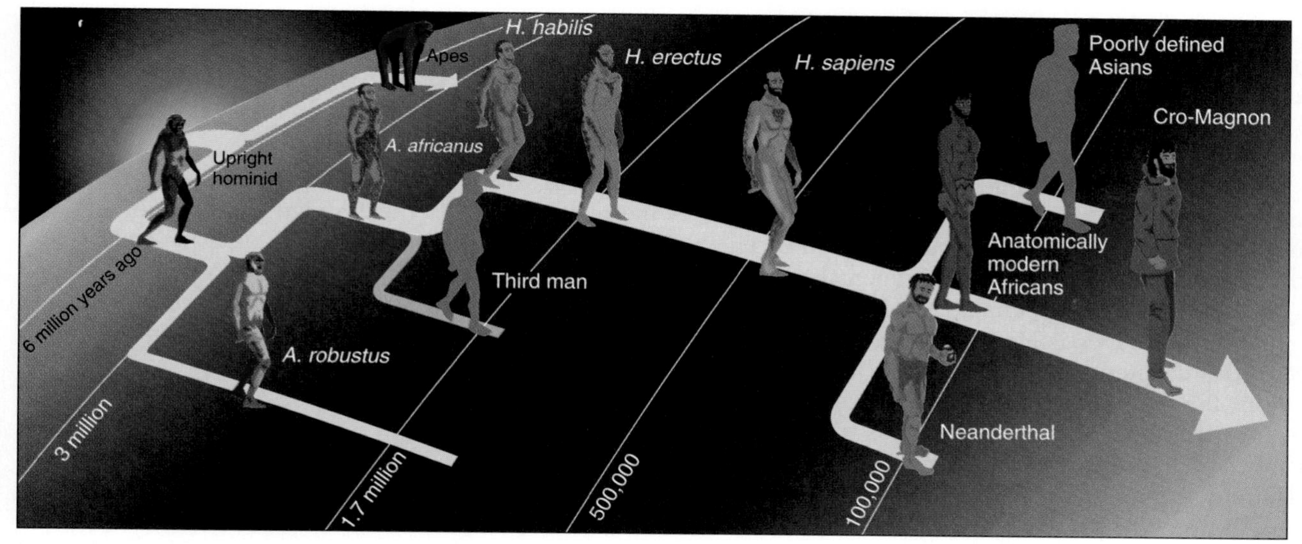

FIGURE 2–24
Human evolution. Fossil evidence shows that several hominid (upright human-like) species occurred between the origin of the oldest known hominid 4.4 million years ago and anatomically modern humans, which first appeared about 100,000 years ago. The diagram shows the time periods of their existence and the assumed relationship between them. Causes for extinction of other hominid species (where bars terminate) are unknown.

and better means of transporting water and other materials. Trade with other settlements begins, and thus commerce is born. Also, living in settlements enables better care and protection for everyone, and thus the number of deaths is reduced. This reduced mortality rate, coupled with more reliable food production, supports population growth, which in turn is supported by expanding agriculture. In short, civilization had its origins in the advent of agriculture about 10,000 years ago. We can see a continuing historical trend of a growing population creating bigger and bigger settlements (cities) supported by ever-expanding agricultural and industrial revolution.

One can easily envision how attitudes toward nature may have shifted with the development of agriculture and civilization. As long as humans lived as hunter-gatherers nature was necessarily in the position of being the provider of all materials and sustenance. This being the case, it is not surprising that various aspects of nature were frequently deified by primitive cultures. With the advent of agriculture (including animal husbandry), however, humans gained what seemed like an independence from nature. It became right and proper to convert natural ecosystems into agriculture and other human developments to support growing human populations. Furthermore, it was right and proper to attempt to exterminate "natural enemies" (weeds, insects, and predators) that interfered with agriculture, using any means available. Finally, supported by agriculture, it became feasible to exploit other species, even into extinction, purely for profit, without paying real immediate consequences.

In short, we can see the attitude toward nature as something to conquer, exploit, or push aside having its origin with the development of agriculture. Human technologcal "progress" has, to a large extent, involved learning to carry on these attacks against nature with ever greater efficiency.

FIGURE 2–25
Before the advent of agriculture about 10,000 years ago, humans subsisted by gathering seeds, nuts, berries, and edible roots from wild plants and by snaring or hunting whatever wildlife they could. (The Bettmann Archive.)

Many will point out that if humans had not exploited and overcome nature in these ways, we would still be living in caves and chasing wild animals with stone-tipped spears. This is undoubtedly so. However, does the fact that overcoming nature has been a necessary phase of human advancement in the past mean that the same trends can be continued in the future without consequence? This question is the center of the debate between environmentalists and Cornucopians discussed in Chapter 1. But now, let's look at it from the ecological perspective we have gained in this chapter.

From the ecological perspective, it can be seen that every species propagates and spreads to the limits of its capacity. It is contained only by physical barriers and limiting factors beyond its range of tolerance. Humans, however, have been able to overcome the usual barriers and limiting factors. Thus, humans have been able to exploit every biome and marine environment on Earth, a process that is continuing and even accelerating. Additional ecosystems are being damaged by pollution. Note that from an ecological point of view, humans are not behaving "unnaturally." It is simply that we have gained a capacity to overcome barriers and limiting factors that restrict other species.

In other words, humans, with their agriculture and technology, have developed into super-consumers fully capable of usurping every natural ecosystem on Earth. What are the problems in continuing with this trend? They are aesthetic, ethical, and ecological.

First, is there any real advantage in pursuing a trend toward more and more people and human developments and less and less nature? Is a purely artificial world, assuming one is possible, what we want to create for ourselves and future generations? Would such a world offer greater amounts of the things we really value? Increasingly, people are answering these questions in the negative. The aesthetic value of unspoiled nature is gaining increased recognition. Even more, for some it is becoming a moral imperative: "Other species have as much right to live on this planet as we do."

But, if such aesthetic or ethical arguments are not persuasive, perhaps the ecological argument will be. As noted, we see natural systems being displaced by the **human system**, a term we will use to refer to our total system including animal husbandry, agriculture, and all human developments. If the human system functioned as a true ecosystem there might be an argument for its sustainability. We could see the process of human domination of the environment as one ecosystem displacing another, a phenomenon that does occur occasionally in nature.

Indeed, the human system does have some features in common with natural ecosystems, the series of trophic levels from crop producers to human consumers, is an example. But, in other respects it is far off the mark—failing to break down and recycle its "detritus" (trash, chemical wastes), and other byproducts and suffering the consequences of pollution as one example. Remaining highly dependent on fossil fuels, and thereby suffering the buildup of carbon dioxide in the atmosphere is another.

However, the attitude that we are independent of natural ecosystems and the biosphere is even more erroneous. Even as we get increasing proportions of our food and materials from agriculture, agriculture remains dependent on the periodic infusion of genes from wild species to maintain its vigor. Many scientists and agricultural experts doubt that we can maintain a viable agricultural system without the backup from natural biodiversity. (We will explore the reasons for this more fully in Chapter 5.)

We have seen how all ecosystems are interconnected to make the biosphere and that the biosphere, by maintaining predictable climates, atmospheric composition, and so on, supports all ecosystems. The human system cannot divorce itself from these interactions. As we begin to affect the natural world to the extent that basic parameters of the biosphere are altered—global warming and depletion of the stratospheric ozone shield to name just two examples—we will affect not just one ecosystem in a limited area; we will inevitably alter the balances between all species and ecosystems on Earth. We cannot predict what these changes may comprise, much less the final outcome as they ramify through the entire biosphere.

Importantly, we do have the knowledge and technological expertise to create an ecologically sustainable human system, one that can flourish in harmony with natural ecosystems. The question is this: Can we rely on our basic biological instincts, and on economic forces, both geared toward maximizing short-term consumption and discounting long-term consequences, to achieve sustainability? Or, at this stage in human development, do we need to muster a higher level of consciousness and caring in order to achieve sustainable development? With further study of how ecosystems function, we will proceed to unveil the principles of sustainability in Chapter 3.

Review Questions

1. Define and compare the terms: species, population, biotic community, abiotic environmental factors, ecosystem.
2. In what ways are the biosphere, biomes, ecosystems, and ecotones the same, and how are they different?
3. Identify the biotic components of the biome of your region.
4. Identify and describe the abiotic components of the biome of your region.
5. Name the three main categories of organisms that make up the biotic structure of every ecosystem, and describe the role that each plays.
6. Give four categories of consumers present in ecosystems and the role that each plays.
7. Give similarities and differences between detritus feeders and decomposers in terms of what they do, how they do it, and the kinds of organisms involved in each category.
8. All organisms can be divided into two major categories. Name and describe the attributes of these two categories.
9. What is the distinction between food chains, food webs, and trophic levels?
10. Describe three nonfeeding relationships that exist among organisms.
11. How is competition among different species of consumers in an ecosystem reduced?
12. List five abiotic factors affecting ogranisms. What is the effect on a population when any abiotic factor shifts from the optimum to the limit of tolerance and beyond?
13. How are the distribution of species and the formation of ecotones determined by limiting factors?
14. What is climate, and how does it affect the distribution of biomes?
15. What is a microclimate, and how does it affect the diversity of species within an ecosystem?
16. What things in addition to abiotic factors may act as limiting factors?
17. How was the development of agriculture a major turning point in human history?
18. How have humans overcome the limits that tend to keep natural ecosystems within certain bounds?
19. How is the human system similar to and different from natural ecosystems?

Thinking Environmentally

1. Select a woods, field, pond, or other natural area near where you live, and discuss how it functions as an ecosystem. Organize your discussion by assigning each species that is present to a biotic category and describing the role the species plays in the system.
2. From local, national, and international news, compile a list of the many ways humans are altering abiotic and biotic factors on local, regional, and global scales. Analyze the ways in which local changes may affect ecosystems on larger levels and the ways in which global changes may affect local levels.
3. Write a scenario of what would happen to an ecosystem or to the human system in the event of one of the following: (a) All producers were killed through losses of soil fertility or toxic contamination (b) all parasites were eliminated; (c) decomposers/detritus feeders were eliminated. Support all of your statements with reasons drawn from your understanding of the way ecosystems function.
4. Consider the various kinds of relationships humans have with other species, both natural and domestic. Give examples of relationships that benefit humans but harm the other species; that benefit both humans and the other species; that benefit the other species but harm humans. Give examples in which the relationship may be changing—for instance, from exploitation to protection. Discuss the ethical issues involved in changing relationships.
5. Can the human system be modified into a sustainable ecosystem in balance with (i.e., preserving) other natural ecosystems without losing the benefits of modern civilization?

Ecosystems: How They Work

Key Issues and Questions

1. All the elements that comprise living things come from the environment. What are these key elements? Where is each found?
2. Life processes can be seen in terms of the assembly and disassembly of complex molecules. What molecules are formed in growth? In the breakdown of organic compounds?
3. Changes in matter cannot be separated from changes in energy. Relate inputs and outputs of energy to various changes in matter.
4. Photosynthesis and cell respiration are the two fundamental biological processes. What matter and energy changes occur in these two processes? Relate them to the dynamics of ecosystems.
5. Recycling of elements and flow of energy are fundamental aspects of all ecosystems. Describe specifically how various atoms are recycled in, and how energy flows through, ecosystems.
6. A third fundamental feature of all ecosystems is the biomass pyramid. What three factors make such a pyramid inevitable? What occurs if higher trophic levels are increased?
7. Sustainability of natural ecosystems is supported by three principles. Name these principles. Relate problems regarding sustainability of the human system to these principles.

In this chapter, we explore how ecosystems work at the fundamental level of chemicals and energy. Our look at this basic level will reveal underlying principles that enable natural ecosystems to be sustainable, and it will provide insight into the pathways we must take to make our human system sustainable. Also, understanding at this level will provide a background for understanding agricultural problems, pollution, global warming, and other issues covered in the text.

Masai Mara National Park, Kenya. [Photo by M.P. Kahl/Photo Researchers, Inc.]

Elements, Life, and Energy

The basic building blocks of all **matter** (all gases, liquids, and solids in both living and nonliving systems) are **atoms**. Only 92 different kinds of atoms occur in nature, and these are known as the 92 naturally occurring **elements**. In addition, physicists have created 14 more in the laboratory, but all of these break down again into the naturally occurring elements such as carbon, hydrogen, oxygen, and iron (see Table C-1, p. 639).

How can the innumerable materials that make up our world, including the tissues of living things, be made of just 92 elements? More specifically, 99% of Earth's crust is composed of only eight of these natural elements.

Elements are analogous to Lego® blocks: From a small number of basic kinds of blocks, we can build innumerable different things. Also, like blocks, nature's materials can be taken apart into their separate constituent atoms, and the atoms can then be reassembled into different materials. All chemical reactions, whether they occur in a test tube, in the environment, or inside living things, and whether they occur very slowly or very fast, involve rearrangements of atoms to form different kinds of matter.

Atoms do not change during the disassembly and reassembly of different materials. A carbon atom, for instance, will always remain a carbon atom. Furthermore, atoms are not created or destroyed during any chemical reactions. This constancy of atoms is regarded as a fundamental natural law, the *law of conservation of matter.*

On the chemical level, then, the cycle of growth, reproduction, death, and decay of organisms can be seen as a continuous process of taking various atoms from the environment, assembling them into living organisms (growth) and then disassembling them (decay) and repeating the process. Of course, in nature, there is no one visible doing the assembling and disassembling; it occurs according to the atoms' chemical nature and to flows of energy. Nonetheless, the simplicity of the concept does not diminish the wonder of it.

Which atoms make up living organisms? Where are they found in the environment? How do they become part of living organisms? We answer these questions next.

Organization of Elements in Living and Nonliving Systems

A more detailed discussion of atoms—how they differ from one another, how they bond to form various gases, liquids, and solids, and how we use chemical formulas to describe different chemicals—is given in Appendix C (page 639). Studying that appendix first may give you a better comprehension of the material we are about to cover. At the very least, the definitions of two terms are essential: *molecule* and *compound.*

A **molecule** refers to *any* two or more atoms bonded together in a specific way. The properties of a material are dependent on the specific way in which atoms are bonded to form molecules as well as on the atoms themselves. Similarly, a **compound** refers to any *two or more different kinds* of atoms bonded together. Note the distinction that a molecule may consist of two or more of the *same kind*, as well as different kinds, of atoms bonded together. A compound always implies that at least two different kinds of atoms are involved. For example, the fundamental units of oxygen gas, which consist of two oxygen atoms bonded together, are molecules but not a compound. Water, on the other hand, can be referred to as either molecules or a compound, since the fundamental units are two hydrogen atoms bonded to an oxygen atom. Some further distinctions are given in Appendix C.

The key elements in living systems (and their chemical symbols) are carbon (C), hydrogen (H), oxygen (O), nitrogen (N), phosphorus (P), and sulfur (S). You can remember them by the acronym N. CHOPS. These six elements are the building blocks of all the organic molecules that make up the tissues of plants, animals, and microbes. We have said that growth and decay can be seen as a process of atoms moving from the environment into living things and returning to the environment. By looking at the chemical nature of air, water, and minerals, we shall see where our six key elements and others occur in the environment (Table 3-1).

The lower atmosphere is a mixture of molecules of three important gases—oxygen (O_2), nitrogen (N_2), and carbon dioxide (CO_2)—along with trace amounts of several other gases that have no immediate biological importance. Also generally present in air are variable amounts of polluting materials and water vapor. The three main gases found in air are shown in Figure 3-1. Note three of the key elements among these molecules. Thus, air is a source of carbon, oxygen, and nitrogen for all organisms.

TABLE 3-1
Elements Found in Living Organisms and Locations in the Environment

		Biologically Important Molecule or Ion in Which the Element Occurs[a]		*Location in the Environment*		
Element (Kind of Atom)	***Symbol***	***Name***	***Formula***	***Air***	***Dissolved in Water***	***Some Rock and Soil Minerals***
Carbon	C	Carbon dioxide	CO_2	X	X	X(CO_3^-)
Hydrogen	H	Water	H_2O		(Water itself)	
Atomic oxygen (required in respiration)	O	Oxygen gas	O_2	X	X	
Molecular oxygen (released in photosynthesis)	O_2	Water	H_2O		(Water itself)	
Nitrogen	N	Nitrogen gas	N_2	X	X	Via fixation
		Ammonium ion	NH_4^+		X	X
		Nitrate ion	NO_3^-		X	X
Sulfur	S	Sulfate ion	SO_4^{2-}		X	X
Phosphorus	P	Phosphate ion	PO_4^{3-}		X	X
Potassium	K	Potassium ion	K^+		X	X
Calcium	Ca	Calcium ion	Ca^{2+}		X	X
Magnesium	Mg	Magnesium ion	Mg^{2+}		X	X
Trace Elements[b]						
Iron	Fe	Iron ion	Fe^{2+}, Fe^{3+}		X Fe^{2+} only	X
Manganese	Mn	Manganese ion	Mn^{2+}		X	X
Boron	B	Boron ion	B^{3+}		X	X
Zinc	Zn	Zinc ion	Zn^{2+}		X	X
Copper	Cu	Copper ion	Cu^{2+}		X	X
Molybdenum	Mo	Molybdenum ion	Mo^{2+}		X	X
Chlorine	Cl	Chloride ion	Cl^-		X	X

NOTE: These elements are found in *all* living organisms—plants, animals, and microbes. Some organisms require certain elements in addition to the ones given. For example, humans require sodium and iodine.
[a] A molecule is a chemical unit of two or more atoms bonded together. An ion is a single atom or group of bonded atoms that has acquired a positive or negative charge as indicated.
[b] Only small or trace amounts of these elements are required.

Saying that air is a **mixture** means that there is no chemical bonding between the molecules involved. Indeed, it is this lack of connection between molecules that results in air being gaseous. Attraction, or bonding, between molecules results in liquid or solid states.

The source of the key element hydrogen is water. Each molecule of water consists of two hydrogen atoms bonded to an oxygen atom, as indicated by the formula for water: H_2O. A weak attraction between water molecules is known as *hydrogen bonding*. At temperatures below freezing, hydrogen bonding holds the molecules in position with respect to one another, and the result is a solid (ice or snow). At temperatures above freezing, but below vaporization (evaporation), hydrogen bonding still holds the molecules close, but allows them to move around one another, producing the liquid state. Vaporization occurs as hydrogen bonds break and

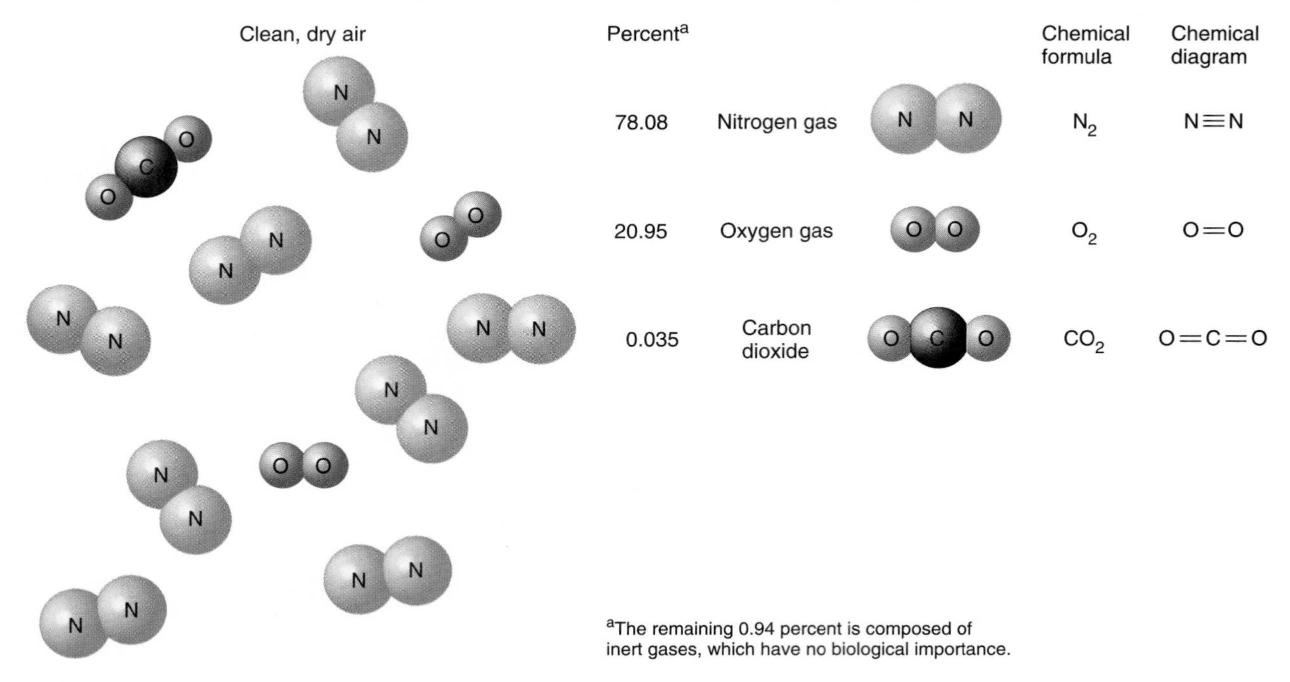

FIGURE 3-1
From a biological point of view, the three most important gases of the lower atmosphere are nitrogen, oxygen, and carbon dioxide.

water molecules move into the air independently. With a lowering of temperature, all these changes in state go in the reverse direction (Fig. 3-2). We reemphasize that, regardless of the changes in state, the water molecules themselves retain their basic structure of two hydrogen atoms bonded to an oxygen atom. It is only the relationship between the molecules that changes.

All the other elements required by living organisms, as well as the 72 or so elements that are not required, are found in various rock and soil minerals. A **mineral** refers to any hard, crystalline, inorganic material of a given chemical composition. Most rocks are made up of relatively small crystals of two or more minerals, and soil generally consists of particles of many different minerals. Each mineral is made up of dense clusters of two or more kinds of atoms bonded together by an attraction between positive and negative charges on the atoms as explained in Appendix C and Fig. 3-3.

There are simple but significant interactions between air, water, and minerals. Gases from the air and ions (charged atoms) from minerals may dissolve in water. Therefore, natural water is inevitably a *solution* containing variable amounts of dissolved gases and minerals. This solution is constantly subject to change, as any dissolved substances may be removed from it by various processes, or additional materials may dissolve in it. Molecules of water enter the air by evaporation and leave it by means of condensation and precipitation. (See the water cycle, p. 264). Thus, the amount of moisture in air is constantly fluctuating. Wind may carry a certain amount of dust or mineral particles, and this amount is also changing constantly, since the particles gradually settle out from the air. The various interactions are summarized in Fig. 3-4.

By contrast to the relatively simple molecules that occur in the environment (for example, CO_2, H_2O, N_2), in living organisms we find the key atoms (C, H, O, N, P, S) bonded into very large, complex molecules known as proteins, carbohydrates (sugars and starches), lipids (fatty substances), and nucleic acids. Some of these molecules may contain millions of atoms, and their potential diversity is infinite. Indeed, the diversity of living things is a reflection of the diversity of such molecules.

The molecules that make up the tissues of living things are constructed mainly from carbon atoms bonded together into chains with hydrogen atoms attached. Oxygen, nitrogen, phosphorus, and sulfur may be present also, but the key common denominator is carbon-carbon and/or carbon-hydrogen bonds (Fig. 3-5). Recall (page 28) that material making up the tissues of living organisms is referred to as *organic*. Hence, these *carbon-based molecules, which make up the tissues of living organisms*, are called **organic mol-**

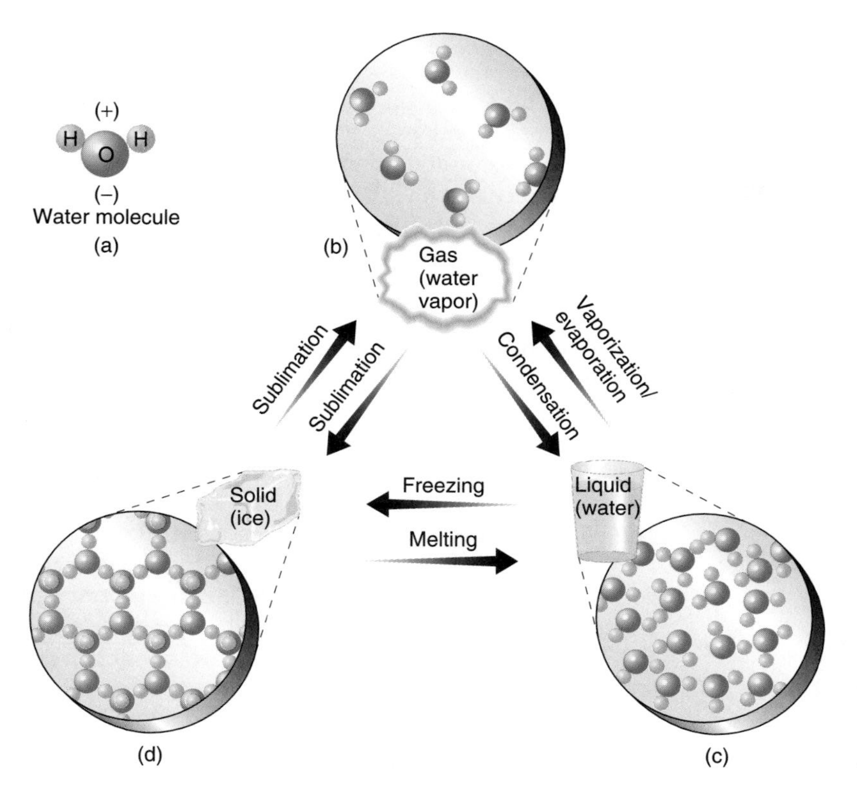

FIGURE 3-2
(a) Water consists of molecules, each of which is formed by two hydrogen atoms bonded to an oxygen atom (H_2O). (b) In water vapor, the molecules are separate and independent. (c) In liquid water, the weak attraction between water molecules known as hydrogen bonding gives the water its liquid property. (d) At freezing temperatures, hydrogen bonding holds the molecules firmly, giving the solid state—ice. (After Robert Christopherson, *Geosystems,* 2/e. ©1994, p. 186. Adapted by permission of Prentice Hall, Upper Saddle River, New Jersey. After B.J. Nebel, *Environmental Science,* 2/e. ©1987, p. 46. Prentice Hall, Upper Saddle River, New Jersey.)

ecules. (Don't miss the similarity between the words *organic* and *organism*.) **Inorganic**, then, refers to molecules or compounds with neither carbon-carbon nor carbon-hydrogen bonds.

Causing some confusion is the fact that all plastics and countless other human-made compounds are based on carbon-carbon bonding and are, chemically speaking, organic compounds—although they have nothing to do with living systems. Where there is doubt, we resolve this confusion by referring to the compounds of living organisms as **natural organic compounds** and the human-made ones as **synthetic organic compounds**.

FIGURE 3-3
Minerals (hard crystalline compounds) are composed of dense clusters of atoms of two or more elements. The atoms of most elements gain or lose one or more electrons, becoming negative (−) or positive (+) ions. The ions are held together by an attraction between positive and negative charges.

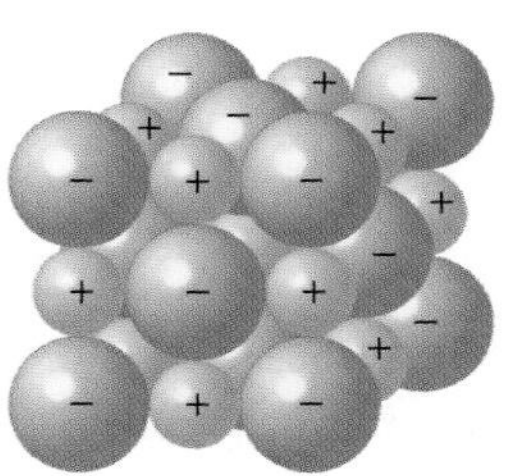

In conclusion, we can see that the elements essential to life (C, H, O, and so on) are present in air, water, or minerals in relatively simple molecules. In living *organ*isms, on the other hand, they are *organ*ized into very complex *organ*ic molecules. These organic compounds in turn make up the various parts of cells, which make up the tissues and organs of the body (Fig. 3-6). Growth, then, may be seen as using the atoms from simple molecules in the environment to construct the complex organic molecules of an organism. Decomposition and decay may be seen as the reverse process. We shall look at each of these processes in more detail later in the chapter; first, however, we must consider another factor: *energy*.

Energy Considerations

In addition to the rearrangement of atoms, chemical reactions also involve the absorption or release of energy. To grasp this concept, let us examine the distinction between matter and energy.

Matter and Energy. The universe is made up of *matter* and *energy*. A more technical definition of **matter** than the one given earlier in this chapter is, *anything that occupies space and has mass*—that is, can

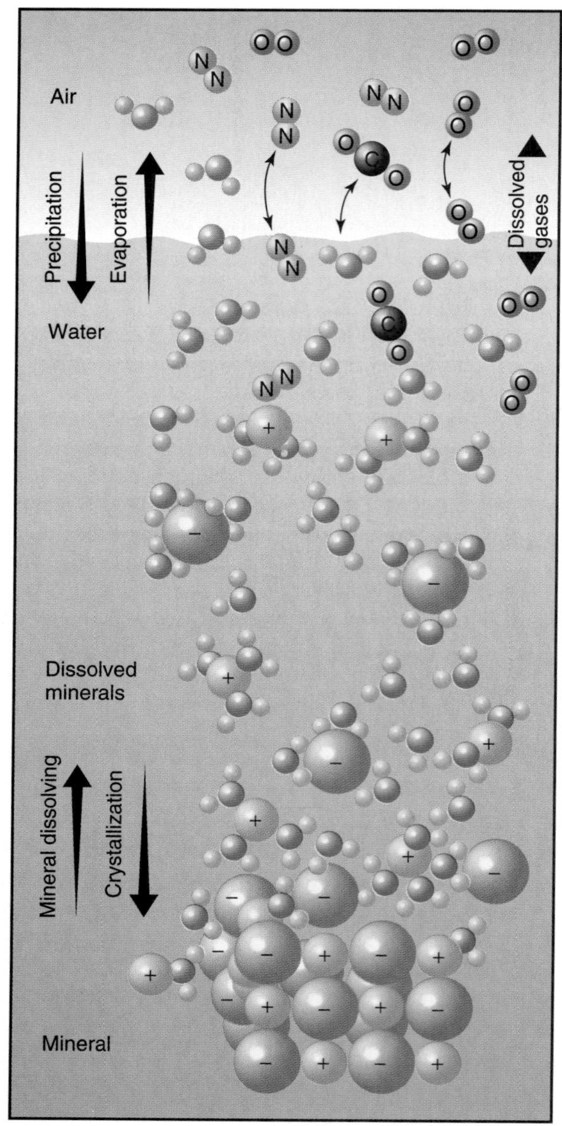

FIGURE 3-4
Interrelationship among air, water, and minerals. Minerals and gases dissolve in water, forming solutions. Water evaporates into air, causing humidity. These processes are all reversible: Minerals in solution recrystallize, and water vapor in the air condenses to form liquid water.

be weighed when gravity is present. This definition obviously covers all solids, liquids, and gases, and living as well as nonliving things.

Atoms are made up of protons, neutrons, and electrons, which in turn are made of still smaller particles. Thus, physicists debate what the most basic unit of matter is. However, since atoms are the basic units of all elements and remain unchanged during chemical reactions, it is practical to consider them as the basic units of matter.

Light, heat, movement, and *electricity*, on the other hand, do not have mass, nor do they occupy space. (Note that heat, as used here, refers not to a hot object but to the heat energy we can feel radiating from the hot object.) These are the common forms of *energy* which we experience continually—or perhaps their lack is a more significant experience. What do forms of energy have in common? They *affect* matter, causing changes in its *position* or its *state*. For example, the release of energy in an explosion causes things to go flying, a change in position. Heating water causes it to boil and change to steam, a change in state. On a molecular level, changes in state may be seen as movements of atoms or molecules. For example, the degree of heat energy is actually a measure of the relative vibrational motion of the atoms and molecules of the substance. Therefore, **energy** is the ability to move matter.

Energy is commonly divided into two major categories: *kinetic* and *potential* (Fig. 3-7). **Kinetic energy** is *energy in action or motion*. Light, heat energy, physical motion, and electrical current are all forms of kinetic energy. **Potential energy** is energy in storage. A substance or system with potential energy has the capacity, or *potential*, to release one or more forms of kinetic energy. A stretched rubber band, for example, has potential energy; it can send a paper clip flying. Numerous chemicals, such as gasoline and other fuels, release kinetic energy—heat energy, light, and movement—when ignited. The potential energy contained in such chemicals and fuels is called **chemical energy**.

Energy may be changed from one form to another in innumerable ways. How many examples can you think of in addition to those shown in Fig. 3-8? Besides seeing that potential energy may be converted to kinetic

FIGURE 3-5
The organic molecules making up living organisms are larger and more complex than the inorganic molecules found in the environment. Glucose and cystine show this relative complexity. (Do *not* memorize these formulas; they are here just to give you a sense of the complexity we are describing.)

```
  O  OH  H  OH OH  H
  ||  |   |   |   |   |
H—C—C—C—C—C—C—H
      |   |   |   |   |
      H  OH  H   H  OH
```
Glucose, a sugar

```
      H   H   O
      |   |   ||
HS—C—C—C—OH
      |   |
      H  NH2
```
Cystine, an amino acid occurring in proteins

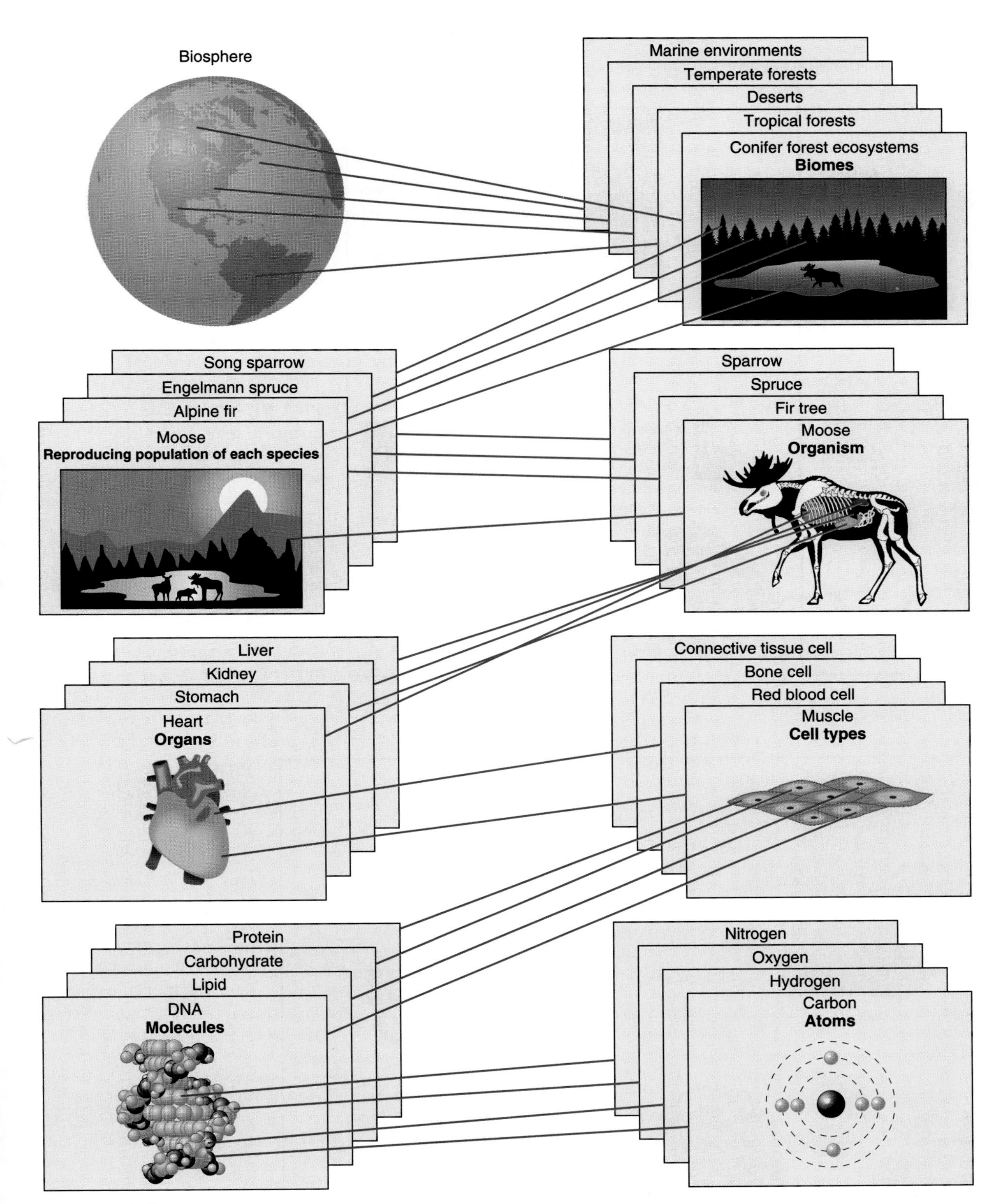

FIGURE 3-6
Life can be seen as a hierarchy of organization of matter. In the inorganic sphere, elements are arranged simply in molecules of the air, water, and minerals. In living organisms, they are arranged in very complex organic molecules, which, in turn, make up cells that constitute tissues, organs, and, thus, the whole organism. Levels of organization continue up through populations, species, ecosystems, and, finally, the whole biosphere.

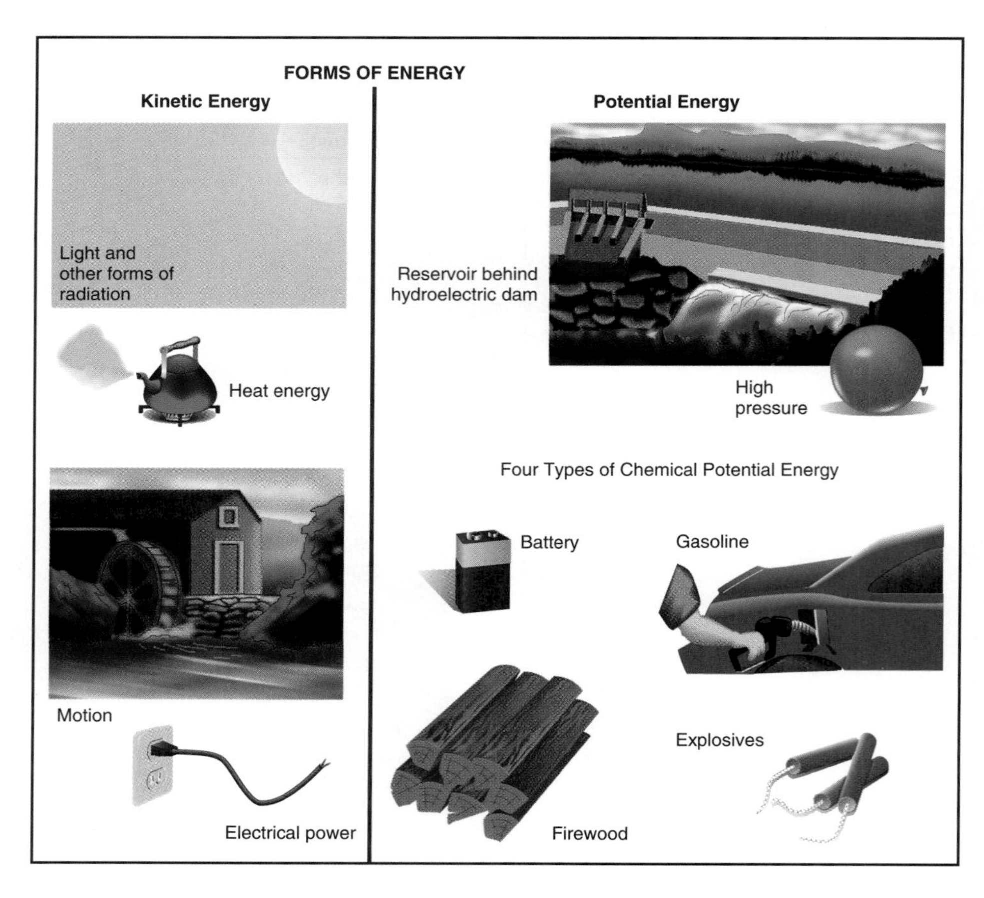

FIGURE 3-7
Energy is distinct from matter in that energy neither has mass nor occupies space. It has the ability to act on matter, changing the position of the matter and/or its state. Kinetic energy is energy in one of its active forms. Potential energy refers to systems or materials that have the potential to release kinetic energy. In this text, we use the term *heat energy* to refer to thermal infrared energy.

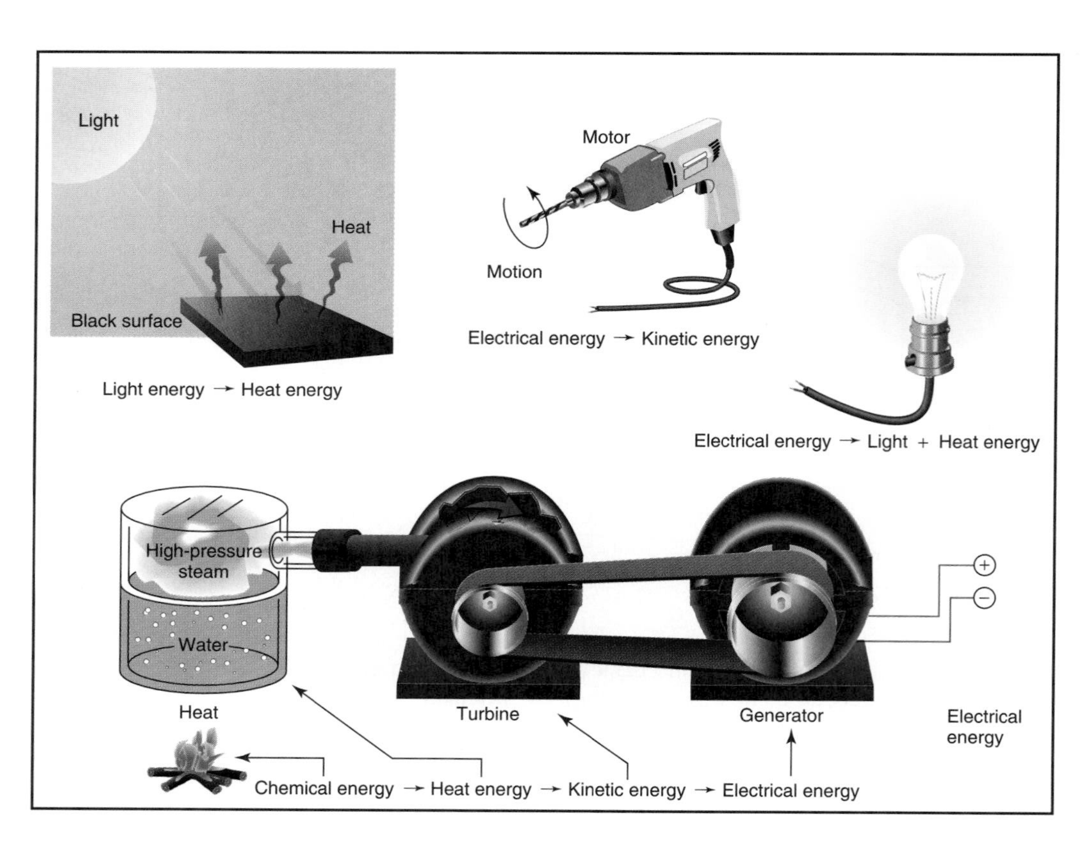

FIGURE 3-8
Any form of energy can be converted to any other form, except that heat energy can be transferred only to something cooler. Heat is a form of energy that flows from one system or object to another because the two are at different temperatures.

energy, it is especially important to recognize that kinetic energy may be converted to potential energy, as in charging a battery or pumping water into a high-elevation reservoir. We shall see shortly that photosynthesis is another such process.

Because energy does not have mass or occupy space, it cannot be measured in units of weight or volume, but it can be measured in other kinds of units. One of the most common units is the **calorie**, which is defined as the *amount of heat required to raise the temperature of 1 gram (1 milliliter) of water 1 degree Celsius.* Since this is a very small unit, it is frequently more convenient to speak in terms of kilocalories (1 kilocalorie = 1,000 calories), the amount of heat required to raise 1 liter (1000 milliliters) of water 1 degree Celsius. Kilocalories are sometimes denoted as "Calories" with a capital "C." Food Calories, which are a measure of how much energy our bodies can derive from given foods, are actually kilocalories. Any form of energy can be measured in calories by converting it to heat energy and measuring that heat in terms of a rise in the temperature of water. Temperature is a measurement of the molecular motion in a substance caused by the kinetic energy present.

We define energy as the ability to move matter. Conversely, no change in the movement of matter can occur *without* the absorption or release of energy. Indeed, no change in matter—from a few atoms joining together or coming apart in a chemical reaction to a major volcanic eruption—can be separated from respective changes in energy.

Energy Laws: Laws of Thermodynamics. Knowing that energy can be converted from one form to another has led numerous would-be inventors over the years to try to build machines or devices that would produce more energy than they consumed. A common idea that occurs to many students is to use the output from a generator to drive a motor that, in turn, drives the generator to keep the cycle going and yields additional power in the bargain. Unfortunately, all such devices have one feature in common: They don't work. When all the inputs and outputs of energy are carefully measured, they are found to be *equal.* There is no net gain or loss in total energy. This observation is now accepted as a fundamental natural law, **the law of conservation of energy**, also called the **first law of thermodynamics:** *Energy is neither created nor destroyed, but may be converted from one form to another.* The law is also commonly stated as "You can't get something for nothing."

Fanciful "energy generators" fail for two reasons: First, in every energy conversion, a portion of the energy is converted to heat energy (thermal infrared). Second, heat always flows toward cooler surroundings. There is no way of trapping and recycling heat energy, since it can flow only "downhill" toward a cooler place. Consequently, in the absence of energy inputs, any and every system will sooner or later come to a stop as its energy is converted to heat and lost. This is now accepted as another natural law, the **second law of thermodynamics**. Basically, the second law says that, *in any energy conversion, you will end up with* less *usable energy than you started with.* So, not only can you not get something for nothing (the first law), you can't even break even.

A principle that underlies the loss of heat is the principle of increasing *entropy.* **Entropy** refers to the degree of *disorder: Increasing entropy means increasing disorder.* The principle is that, without energy inputs, everything goes in one direction only: toward increasing entropy. This principle of ever-increasing entropy is most readily apparent in the fact that all human-made things tend to deteriorate. We never observe the reverse—a run-down building renovating itself, for example. Students often like to speak of the increasing disorder of their dormitory rooms as the semester wears on as an example of entropy.

The conversion of energy and the loss of heat are both aspects of increasing entropy. Heat energy is the result of the random vibrational motion of atoms and molecules. Thus, it is the lowest (most disordered) form of energy, and its flow to cooler surroundings is a way for that disorder to spread. Therefore, the second law of thermodynamics is nowadays more generally stated as: *Systems will go spontaneously in one direction only; toward increasing entropy.* The second law also says that systems will go spontaneously only toward *lower* potential energy, a direction that releases heat from the systems. (Fig. 3-9).

Very important in the statement of the second law is the word *spontaneously.* It is possible to pump water uphill, charge a battery, stretch a rubber band, compress air, or otherwise increase the potential energy of some system. However, inherent in such words as *pump, charge, stretch*, and *compress* is the fact that energy is being put into the system; in contrast, the opposite direction, which releases energy, occurs spontaneously.

The conclusion is that whenever you see something gaining potential energy, you should realize that that energy is being obtained from somewhere else (the first law). Moreover, the amount of energy lost from that somewhere else is greater than the amount gained (the second law). Let us now relate these concepts of matter and energy to organic molecules, organisms, ecosystems, and the biosphere.

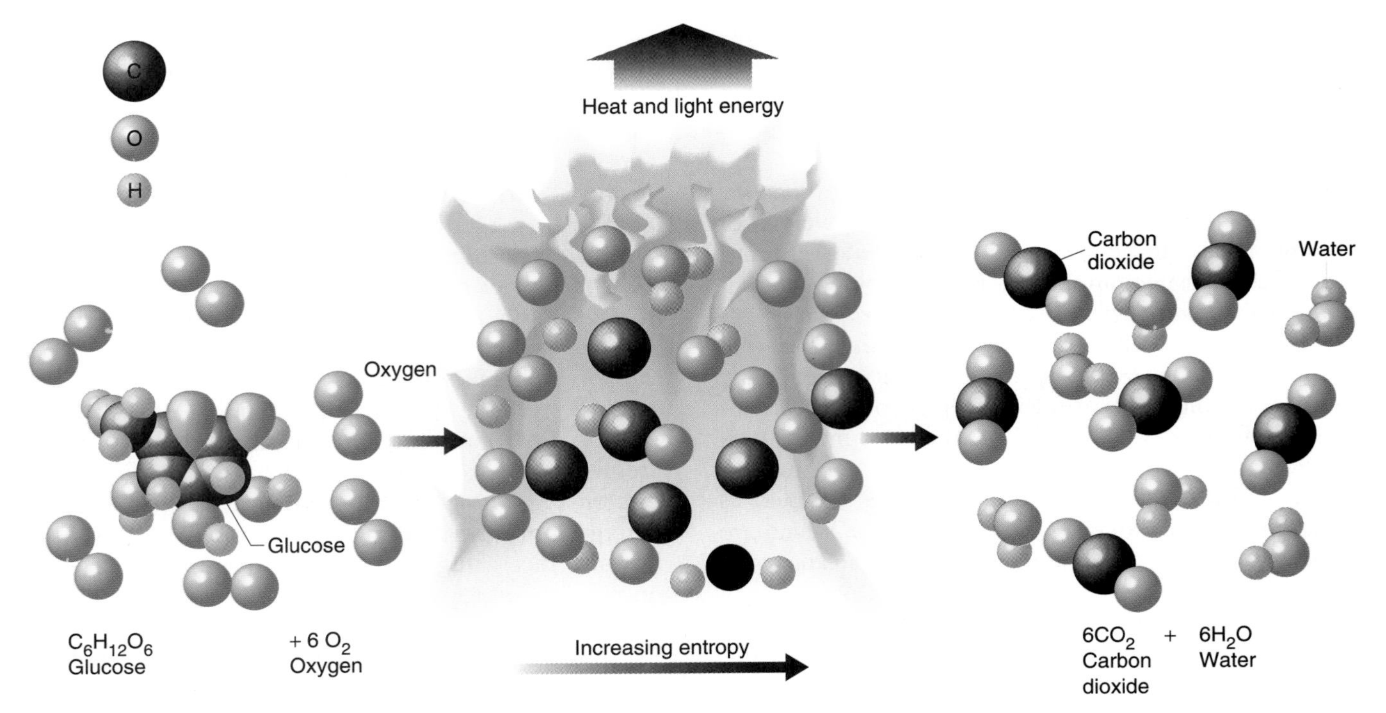

FIGURE 3-9
Systems go spontaneously only in the direction of increasing entropy. When glucose, the building-block molecule of wood, is burned, heat is released, and the atoms become more and more disordered, both aspects of increasing entropy. The fact that wood will burn but not form spontaneously is an example of the second law of thermodynamics.

Matter and Energy Changes in Organisms and Ecosystems

All organic molecules, which make up the tissues of living organisms, contain *high potential energy*. This is evident from the simple fact that they burn: The heat and light of the flame are their potential energy being released as kinetic energy. On the other hand, try as you might, you will not be able to get energy by burning inorganic molecules, such as carbon dioxide, water, or mineral compounds that occur in nature. Indeed, many of these materials are used as fire extinguishers. This extreme nonflammability is evidence that such materials have very *low potential energy*. Thus, the production of organic material from inorganic material involves a *gain* in potential energy. Conversely, the breakdown of organic matter involves a *release* of energy.

In this relationship between the formation and breakdown of organic matter and the gain and release of energy, we can see the energy dynamics of ecosystems. Producers (green plants) play the role of making high-potential-energy organic molecules for their bodies from low-potential-energy raw materials in the environment—namely, carbon dioxide, water, and a few dissolved compounds of nitrogen, phosphorus, and other elements. Such "uphill" conversion is made possible by the light energy absorbed by chlorophyll. On the other hand, all consumers, detritus feeders, and decomposers obtain their energy requirement for movement and other body functions from feeding on and breaking down organic matter (Fig. 3-10). Let us now look at this energy flow in somewhat more detail for each category of organisms.

Producers. Recall from Chapter 2 that producers are green plants, which use light energy in the process of *photosynthesis* to make sugar (glucose, stored chemical energy) from carbon dioxide and water and release oxygen gas as a by-product. The process is expressed by the following formula:

PHOTOSYNTHESIS

$$6\,CO_2 + 6\,H_2O \longrightarrow C_6H_{12}O_6 + 6\,O_2$$

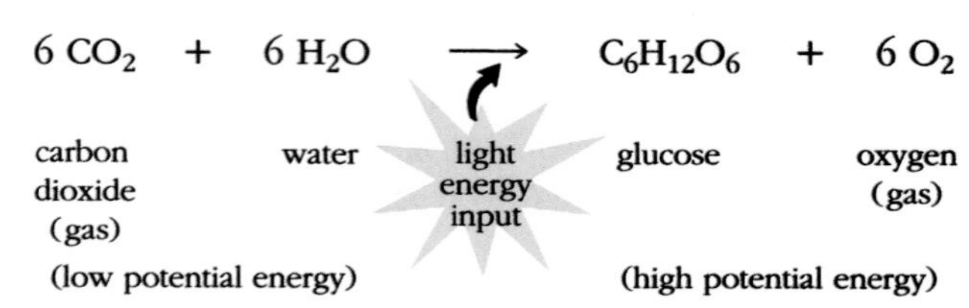

The kinetic energy of light is absorbed by chlorophyll in the cells of the plant and used to remove the hydrogen atoms from water (H_2O) molecules. The hydrogen atoms are transferred to carbon atoms coming from carbon dioxide as the carbons are joined in a chain to begin forming a glucose molecule. After the removal of hydrogen from water, the oxygen atoms that remain combine with each other to form oxygen gas, which is released into the air.

Each molecule of glucose is constructed from 6 carbon atoms, 12 hydrogen atoms, and 6 oxygen atoms—hence its formula, $C_6H_{12}O_6$. Thus, the construction of one molecule of glucose requires 6 molecules of carbon dioxide to provide the 6 carbon atoms and 6 molecules of water to provide the 12 hydrogen atoms. Among these molecules of carbon dioxide and water are 18 oxygen atoms, but only 6 are needed. The extra oxygen atoms are given off as molecules of oxygen gas (O_2), 6 molecules for every molecule of glucose formed. This accounting, based on careful quantitative measurements, supports the law of conservation of matter. Note that oxygen gas, which is essential for the respiration of animals, is a *waste product* of photosynthesis.

The key energy steps in photosynthesis are removing the hydrogen from water molecules and joining carbon atoms together to form the high-potential-energy carbon-carbon and carbon-hydrogen bonds of glucose in place of the low-potential-energy bonds in water and carbon dioxide molecules. But the laws of thermodynamics are not violated or even strained in this process. Careful measurements show that the rate of photosynthesis (which determines the amount of glucose formed) is proportional to the intensity of light, and only 2–5 calories' worth of sugar is formed for each 100 calories' worth of light energy falling on the plant. Thus, plants are not particularly efficient "machines" in performing this conversion of light energy to chemical energy.

The glucose produced in photosynthesis plays three roles in the plant. First, either by itself or along

FIGURE 3-10

Storage and release of potential energy. (a) A simple physical example of the storage and release of potential energy. With suitable energy input, water can be pumped to a higher elevation, thus capturing a portion of the energy input. A portion of the potential energy can then be harnessed to do useful work by letting the water flow back to low potential energy over a turbine. (b) The same principle applies to ecosystems. Through photosynthesis, light energy builds up elements from a low-potential-energy state in inorganic materials to a high-potential-energy state in the form of molecules in organic materials. The breakdown of these molecules releases the energy, which then drives all the active functions of organisms.

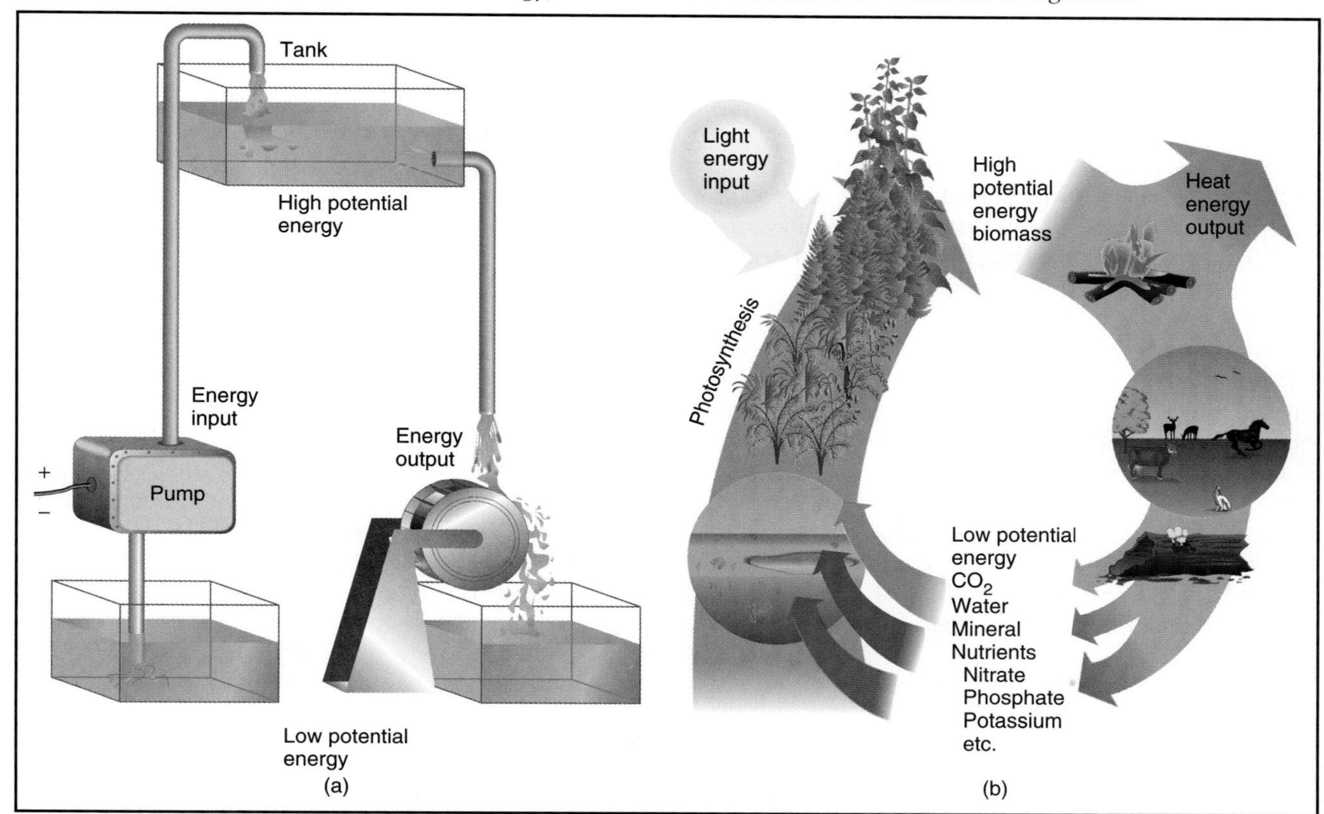

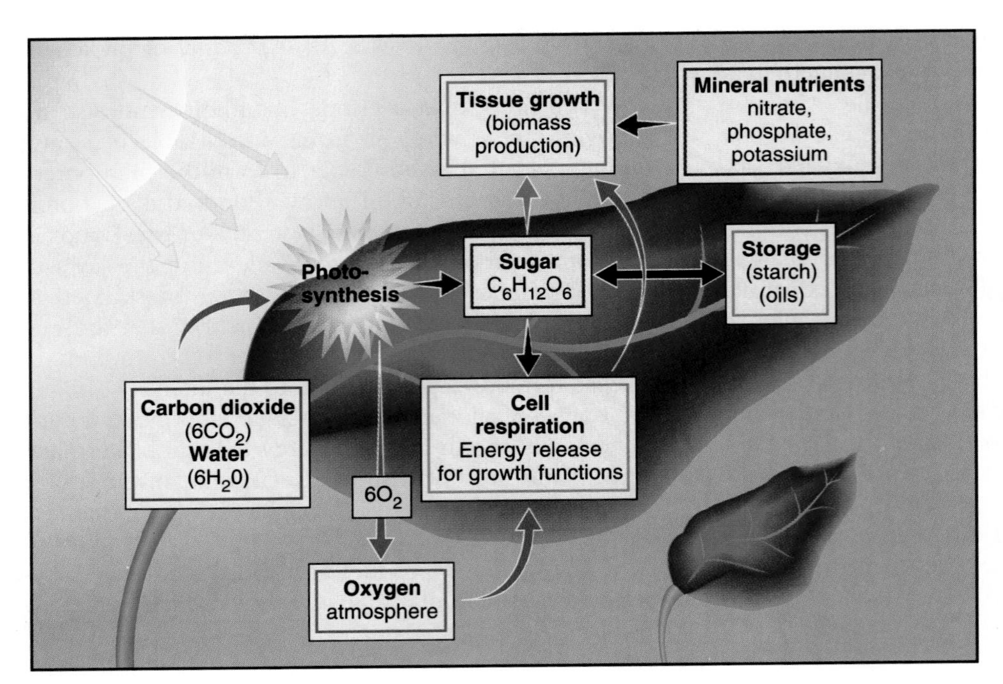

FIGURE 3-11
Producers are remarkable chemical factories. Using light energy from the Sun, they make glucose from carbon dioxide and water, releasing oxygen as a byproduct. Breaking down some of the glucose to provide additional chemical energy, they combine the remaining glucose with certain nutrients from the soil to form other complex organic molecules that the plant then uses for growth.

with nitrogen, phosphorus, sulfur, and other mineral nutrients absorbed from the soil or water surrounding the plant's roots, glucose is the raw material used for making all the other organic molecules (proteins, carbohydrates, and so on) that make up the stems, roots, leaves, flowers, and fruits of the plant. Second, the synthesis of all these organic molecules requires additional energy, as does the plant's absorption of nutrients from the soil and certain other functions. This energy is provided when the plant breaks down a portion of the glucose to release its stored energy in a process called *cell respiration,* which will be discussed shortly. Third, a portion of the glucose produced may be stored for future use. For storage, the glucose is generally converted to starch, as in potatoes, or to oils, as in seeds. These conversions are summarized in Fig. 3-11.

Consumers and Other Heterotrophs—*Energy of Food.* Obviously, consumers need energy to move about and to perform such bodily functions as pumping blood. In addition, consumers need energy to synthesize all the molecules required for growth, maintenance, and repair of the body. Where does this energy come from? It comes from the breakdown of organic molecules of food (or of the body's own tissues if food is not available). About 60–90 percent of the food that we or other consumers eat and digest acts as "fuel" to provide energy.

First, the starches, fats, and proteins that you eat are digested in the stomach and/or intestine, which means that they are broken into simpler molecules—starches into sugar (glucose), for example. These simpler molecules are then absorbed from the intestine into the bloodstream and transported to individual cells of the body.

Inside each cell, organic molecules may be broken down through a process called **cell respiration** to release the energy required for the work done by that cell. Most commonly, cell respiration involves the breakdown of glucose, and the overall chemistry is the reverse of that for photosynthesis:

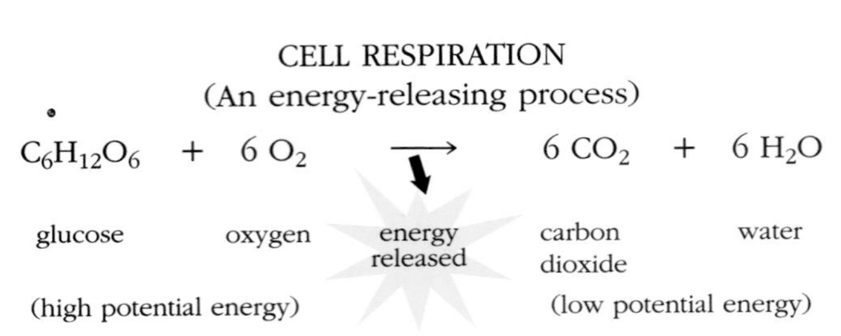

CELL RESPIRATION
(An energy-releasing process)

$$C_6H_{12}O_6 + 6\,O_2 \longrightarrow 6\,CO_2 + 6\,H_2O$$

glucose (high potential energy) + oxygen → energy released; carbon dioxide + water (low potential energy)

Again, the key point of cell respiration is to release the potential energy contained in organic molecules to perform the activities of the organism. However, other aspects of the chemistry are also significant. Note that oxygen is *released* in photosynthesis, but in cell respiration it is *used* to complete the breakdown of glucose to carbon dioxide and water. Oxygen is absorbed through the lungs with every inhalation (or through the gills, in the case of fish) and is transported to all cells via the circulatory system. Carbon dioxide, which is formed as a waste product, moves from the cells into the circulatory system and is eliminated through the lungs (or gills) with every exhalation. The other byproduct, water, serves any of the body's needs for water, which reduces the need to drink water. A number of desert animals, which are adapted to conserve water, do not need to drink any water because that produced by cell respiration is sufficient. However, the bodies of most animals, including ourselves, are less conserving of water; therefore, drinking additional water is necessary. Such water loss from a plant is called *transpiration*.

Again in keeping with the laws of thermodynamics, the energy conversions involved in the body's using the potential energy from glucose to do work are not 100 percent efficient. Considerable waste heat is produced, and this is the source of body heat. This heat output can be measured in cold-blooded animals and in plants, as well as in warm-blooded animals. It is more noticeable in warm-blooded animals only because they produce extra heat, via cell respiration, to maintain their warm-body temperature.

The basis of weight gain or loss should become evident here also. Organic matter is broken down in cell respiration only as it is needed to meet the energy demands of the body; this is why your breathing rate, the outer reflection of cell respiration, varies with changes in your level of exercise and activity. If you consume more calories from food than your body needs, the excess is converted to fat and stored, and the result is a gain in weight. Conversely, the principle of dieting is to eat less and exercise more, to create an energy demand that exceeds the amount of energy contained in the food being consumed. This imbalance forces the body to break down its own tissues to make up the difference, and the result is a weight loss. Of course, carried to an extreme, this imbalance leads to *starvation* and death when the body runs out of anything expendable to break down for its energy needs.

The overall reaction for cell respiration is the same as that for simply burning glucose. Thus, it is not uncommon to speak of "burning" our food for energy. Such a breakdown of molecules is also called **oxidation**. The distinction between burning and cell respiration is that in cell respiration the oxidation takes place in about 20 small steps, so that the energy is released in small "packets" suitable for driving the functions of each cell. If all the energy from glucose molecules were released in a single "bang," as occurs in burning, it would be like heating and lighting a room with large firecrackers—energy, yes, but hardly useful energy.

We have learned that in addition to containing carbon and hydrogen, many organic molecules contain nitrogen, phosphorus, sulfur, and other elements. When such molecules are broken down in cell respiration, the waste byproducts include compounds of nitrogen, phosphorus, and any other elements present, in addition to the usual carbon dioxide and water. These byproducts are excreted in the urine (or as similar waste in other kinds of animals) and returned to the environment, where they may be reabsorbed by plants (Fig. 3-12). Here you can see the movement of elements

FIGURE 3-12
Animal wastes are plant fertilizer. When consumers burn food to obtain energy, the waste products are the inorganic nutrients needed by plants. Here, dog urine has been deposited on a lawn. The ring of dark green grass is where the urine has been diluted to optimal concentration; the grass in the center has been killed by overfertilization with concentrated urine. (Photograph by BJN.)

in a cycle between the environment and living organisms. We expand on these cycles shortly.

Also, you can visualize a flow of energy that enters as light and exits as heat. Finally, recall the *biomass pyramid* (Fig. 2-14). At each trophic level, the amount of biomass inevitably decreases by the amount that is oxidized to provide energy for the consuming organisms.

Nutritive Role of Food. Whereas 60–90 percent of the food that consumers eat, digest, and absorb is oxidized for energy, the remaining 10–40 percent, which is converted to the body tissues of the consumer, is no less important. This is the fraction that enables the body to grow, as well as to maintain and repair itself.

Carbohydrates (sugars and starches), and fats can be oxidized easily by the body to provide energy. Body growth, maintenance, and repair, however, require particular nutrients—namely, the various vitamins, minerals, and proteins—that are not present in carbohydrates or fats. If any one or more of the specific nutrients is absent from the diet, various diseases associated with **malnutrition** will develop. Thus arises the problem of overconsumption of highly processed "junk foods" such as potato chips, sodas, candies, various baked goods, and alcohol. Rich in fat or sugar or both, these items are very high in calories, but they contain little or none of the necessary nutrients. Consequently, a diet high in

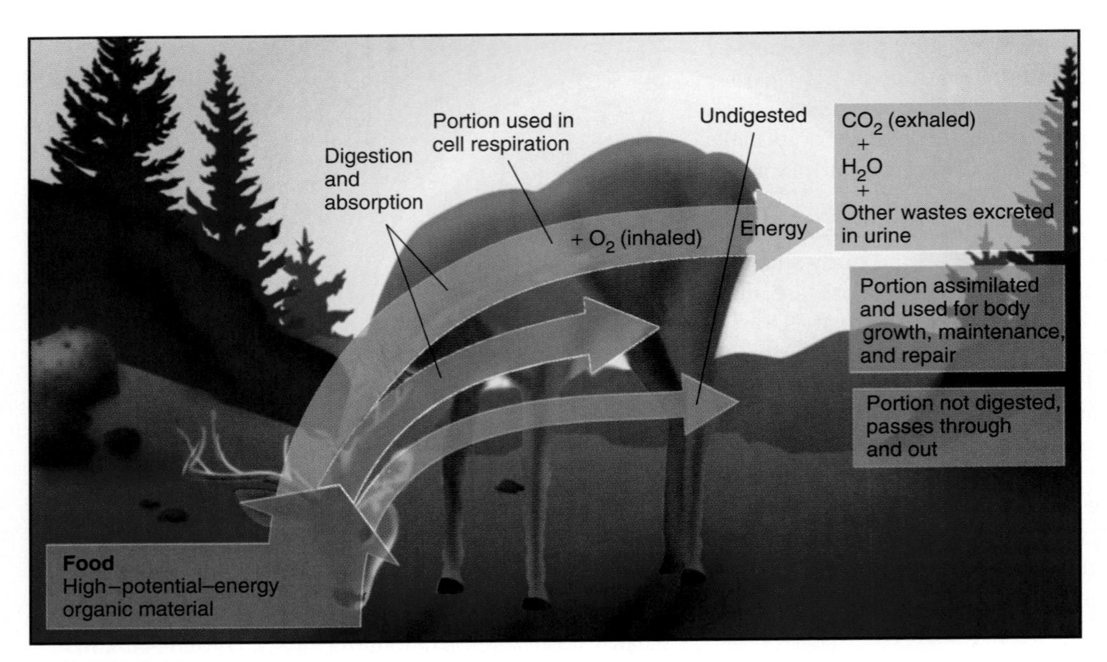

FIGURE 3-13
Consumers. Only a small portion of the food ingested by a consumer is assimilated into body growth, maintenance, and repair. A larger amount is used in cell respiration to provide energy for assimilation, movements, and other functions; waste products are carbon dioxide, water, and various mineral nutrients. A third portion is not digested, but instead passes through, becoming fecal waste.

such items may easily oversupply calories and under supply essential nutrients, causing a weight gain and serious disorders of malnutrition at the same time. Yes, many fat people are also malnourished.

Material Consumed, but Not Digested. A portion of what is ingested by consumers is not digested (broken down so that it can be absorbed), but simply passes through the digestive system and out as fecal wastes. For consumers that eat plants, such waste is largely the material of plant cell walls, **cellulose**. We often refer to it as *fiber, bulk*, or *roughage*. Some fiber is a necessary part of the diet in order for the intestines to have something to push through to keep clean and open.

In summary, organic material (food) eaten by any consumer follows one of three pathways: (1) More than 60 percent of what is digested and absorbed is oxidized to provide energy, and waste byproducts are released back to the environment; (2) the remainder of what is digested and absorbed goes into body growth, maintenance and repair, or storage (fat); and (3) the portion that is not digested or absorbed passes out as fecal waste (Fig. 3-13). Recognize that in an ecosystem, it is only that portion of the food that becomes body tissue of the consumer that becomes food for the next organism in the food chain. Of course, the still-organic fecal waste becomes food for detritus feeders and/or decomposers.

Detritus Feeders and Decomposers—The Detritivores. Recall that detritus is mostly dead leaves, the woody parts of plants, and animal fecal wastes. As such, it is largely cellulose, which is unusable by most consumers because they are unable to digest it. Nevertheless, it is still organic and high in potential energy for those organisms that can digest it—namely, the decomposers we learned about in Chapter 2, various species of fungi and bacteria, and a few other microbes. Beyond having this ability to digest cellulose, decomposers act as any other consumer, using the cellulose as a source of both energy and nutrients. Termites and some other detritus feeders can digest woody material by virtue of maintaining decomposer microorganisms in their guts in a mutualistic symbiotic relationship. The termite (a detritus feeder) provides a cozy home for the microbes (decomposers) and takes in the cellulose, which the microbes digest for both their own and the termites' benefit.

Most decomposers make use of cell respiration. Thus, the detritus is broken down to carbon dioxide, water, and mineral nutrients. Likewise, there is a release of waste heat, which you may observe as the "steaming" of a manure or compost pile on a cold day.

Some decomposers (certain bacteria and yeasts) can meet their energy needs through the partial oxidation of glucose that can occur in the absence of oxygen. This modified form of cell respiration is called **fermentation**. It results in such end products as ethyl alcohol (C_2H_6O), methane gas (CH_4), and vinegar (acetic acid, $C_2H_4O_2$). The commercial production of

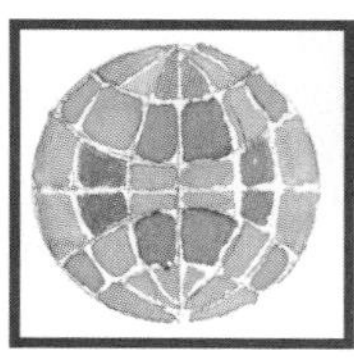

Global Perspective

Light and Nutrients, the Controlling Factors in Marine Ecosystems

Even though running on solar energy and recycling nutrients are basic principles of sustainability, they are limiting factors in some ecosystems. Indeed, although the availability of moisture is a primary determining factor in terrestrial ecosystems, the availability of light and/or nutrients is a primary determining factor in marine ecosystems.

First, light gets dimmer and dimmer as water depth increases because even clear water absorbs light to some extent. The layer of water from the surface down to the greatest depth at which there is adequate light for photosynthesis is known as the **euphotic zone**. Below the euphotic zone, by definition, photosynthesis does not occur. In clear water, the euphotic zone may be as deep as 600 feet (200 meters), but in very turbid (cloudy) water, it may be a matter of only a few centimeters. If the euphotic zone extends to the bottom, the bottom may support abundant plant life—that is, aquatic vegetation attached to or rooted in the bottom sediments. If the euphotic zone does not extend to the bottom, however, the bottom will be barren of plant life.

That the euphotic zone does not extend to the bottom does not preclude an ecosystem from existing in it. Many species of phytoplankton—algae and photosynthetic bacteria that grow as single cells or in small groups of cells—can maintain themselves close to the surface in the euphotic zone. Phytoplankton may support a diverse food web, including many species of fish and sea mammals (such as whales).

Also, an entire ecosystem operates in the cold, dark depths below the euphotic layer nourished by detritus precipitation from above, and closer to the ocean floor, by vents and fissures that produce mineral-rich water and warmth.

In a phytoplankton-based system, nutrients dissolved in the water become critically important. If the water contains too few dissolved nutrients such as phosphorus or nitrogen compounds, the growth of phytoplankton and, hence, the rest of the ecosystem, will be limited. If the bottom receives light, it may support vegetation despite nutrient-poor water because such vegetation draws nutrients from the bottom materials. Indeed, nutrient-rich water may be counterproductive to bottom vegetation, because the dissolved nutrients support the growth of phytoplankton, which makes the water turbid and shades out the bottom vegetation.

Let us put these concepts together to understand particular marine environments. The most productive areas of the ocean—the areas supporting the most abundant marine life of all sorts—are mostly found within 200 miles (300 km) of shorelines. This is true because either the bottom is within the euphotic zone and thus supports abundant bottom vegetation, or nutrients washing in from the land support an abundant primary production of phytoplankton.

In the open ocean, there is less and less marine life as one moves farther from shore. Indeed, marine biologists speak of most of the open ocean as being a "biological desert." The lack of life occurs both because the bottom is well below the euphotic zone and because the water is nutrient poor.

The nutrients carried to the bottom with the settling detritus are released into solution by decomposers, thus making the bottom water nutrient rich. This nutrient-rich bottom water may be carried along by ocean currents. Where the currents hit underwater mountains or continental rims, the nutrient-rich water is brought to the surface. Phytoplankton flourish in these areas of **upwelling** (rising) nutrient-rich water and support a rich diversity of fish and marine mammals.

In sum, the world's oceans are far from uniformly stocked with fish. By far the richest marine fishing areas are continental shelves and regions of upwelling as shown on the accompanying map. Unfortunately, however, many of these areas are now being depleted by overfishing.

(Image is courtesy of Jane A. Elrod and Gene Feldman. NASA/Goddard Space Flight Center.)

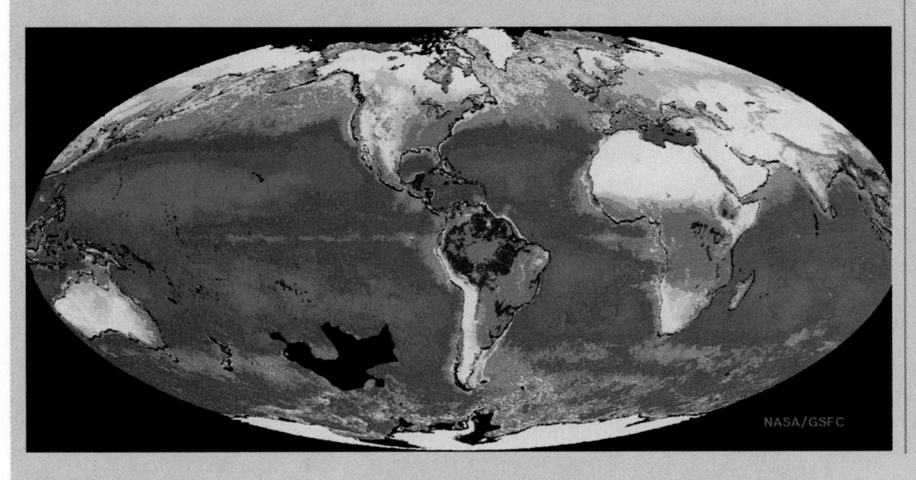

Magenta—mid oceans: lowest productivity (0.1 mg chlorophyll/m^3 or less).
Red/orange—along coasts: highest productivity (10 mg chlorophyll/m^3 or more).

these compounds is achieved by growing the particular organism on suitable organic matter in a vessel without oxygen. In nature, **anaerobic**, or *oxygen-free*, environments commonly exist in the sediment at the bottom of marshes or swamps, buried deep in the earth, and in the guts of animals where oxygen does not penetrate readily. Methane gas is commonly produced in such locations. A number of large grazing animals, including cattle, maintain fermenting bacteria in their digestive systems in a mutualistic, symbiotic relationship similar to that just described for termites. Both cattle and termites produce methane as a result.

For simplicity, our orientation in this chapter is directed toward terrestrial ecosystems. It is important to realize that exactly the same processes occur in aquatic ecosystems. As aquatic plants and algae absorb dissolved carbon dioxide and mineral nutrients from the water, their photosynthetic production becomes the food and dissolved oxygen that sustain consumers and other heterotrophs. Likewise, aquatic heterotrophs return carbon dioxide and mineral nutrients to the aquatic environment. Of course, aquatic and terrestrial systems are never entirely isolated from one another, and exchanges between them go on all the time.

Principles of Ecosystem Function

The preceding examination of how ecosystems function reveals that three common denominators underlie them all: (a) recycling of nutrients, (b) using sunlight as a basic energy source, and (c) populations are such that overgrazing does not occur. In turn, these common features reveal basic principles underlying the sustainability of ecosystems. Let us examine this further.

Nutrient Cycling

Looking at the various inputs and outputs of producers, consumers, detritus feeders, and decomposers, you should be impressed by how they fit together. The products and byproducts of each group are the food and/or essential nutrients for the other. Specifically, the organic material and oxygen produced by green plants are the food and oxygen required by consumers and other heterotrophs. In turn, the carbon dioxide and other wastes generated when heterotrophs break down their food are exactly the nutrients needed by green plants. Such recycling is fundamental, for two reasons: (a) It prevents wastes, which would cause problems, from accumulating. (b) It assures that the ecosystem will not run out of essential elements. Thus, we uncover the **first basic principle of ecosystem sustainability:**

For sustainability, ecosystems dispose of wastes and replenish nutrients by recycling all elements.

If we reconsider the natural law of conservation of matter which says that atoms cannot be created, destroyed, nor changed, we can see that recycling is the only possible way to maintain a dynamic system, and the biosphere has mastered this to a profound degree. We can see this even more clearly by focusing on the pathways of three key elements: carbon, phosphorus, and nitrogen. Because these pathways all lead in circles, they are known as the *carbon cycle*, the *phosphorus cycle*, and the *nitrogen cycle*. (Note that energy is not recycled; it must be renewably supplied by sunlight.)

The Carbon Cycle. For descriptive purposes, it is convenient to start the carbon cycle (Fig. 3-14) with the "reservoir" of carbon dioxide molecules present in the air and dissolved in water. Through photosynthesis and further metabolism, carbon atoms from carbon dioxide become the carbon atoms of all the organic molecules making up the plant's body. Through food chains, the carbon atoms then move into and become part of the tissues of all the other organisms in the ecosystem. However, it is unlikely that a particular carbon atom will be passed through many organisms in any one cycle, because at each step there is a considerable chance that the consumer will break down the organic molecule in cell respiration. As this occurs, the carbon atoms are released back to the environment in molecules of carbon dioxide, which completes one cycle, but, of course, starts another. Likewise, burning organic material returns the carbon atoms locked up in the material to the air in carbon dioxide molecules.

No two successive cycles of a particular carbon atom are likely to be the same. Nor are the two cycles likely to be within the same ecosystem, because carbon in the atmosphere will be carried around the globe by wind. By calculating the total amount of carbon dioxide in the atmosphere and the amount of primary production (photosynthesis) occurring in the biosphere, scientists have concluded that about a third of the total atmospheric carbon dioxide is taken up in photosynthesis in a year, but an equal amount is returned to the atmosphere through cell respiration. This means that, on the average, a carbon atom makes a cycle from the atmosphere through one or more living things and back to the atmosphere every three years. What does this mean in terms of sharing earth's supply of carbon atoms with every other living thing that exists (or has existed) on Planet Earth? This fact is not without ethical implications. (See Ethics box, "Who Are You?" on page 73.)

The Phosphorus Cycle. The phosphorus cycle is representative of the cycles for all the mineral nutrients—

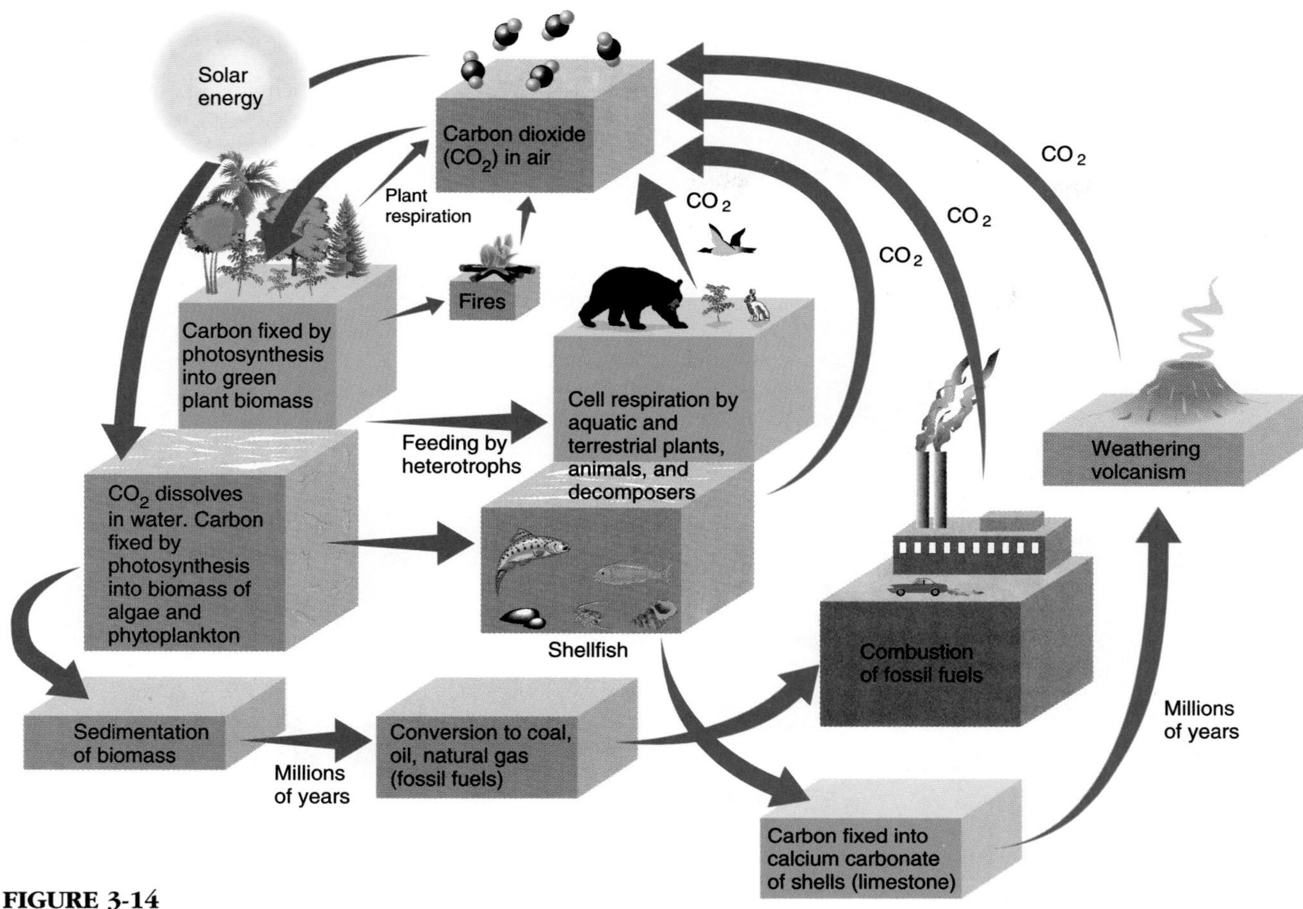

FIGURE 3-14
The carbon cycle.

those required elements that have their origin in rock and soil minerals (see Table 3-1). We focus on phosphorus for both simplicity and because its shortage tends to be a limiting factor in a number of ecosytems.

The phosphorus cycle is illustrated in Fig. 3-15. Phosphorus exists in various rock and soil minerals as the inorganic ion *phosphate* (PO_4^{3-}). As rock gradually breaks down, phosphate and other ions are released. Phosphate dissolves in water, but does not enter the air. Plants absorb phosphate from the soil or from a water solution, and when it is bonded into organic compounds by the plant, it is referred to as **organic phosphate**. Moving through food chains, organic phosphate is transferred from producers to the rest of the ecosystem. As with carbon, at each step there is a high likelihood that the organic compounds containing phosphate will be broken down in cell respiration, releasing inorganic phosphate in urine or other waste. The phosphate may then be reabsorbed by plants to start another cycle.

There is an important difference between the carbon cycle and the phosphorus cycle. No matter where carbon dioxide is released, it will mix into and maintain the concentration of carbon dioxide in the air. Phosphate, however, which does not have a gas phase, is recycled only if the wastes containing it are deposited on the soil *from which it came*. The same holds true for other mineral nutrients. Of course, in natural ecosystems wastes (urine, detritus) are deposited in the same area so that recycling occurs efficiently. Humans have been extremely prone to interrupt this cycle, however.

A very serious case of humans disrupting the phosphorus cycle is the cutting of tropical rain forests. This type of ecosystem is supported by a virtually 100-percent-efficient recycling of nutrients. There are little or no reserves of nutrients in the soil. When the forest is cut and burned, the nutrients that were stored in the organisms and detritus are readily washed away by the heavy rains, and the land is thus rendered unproductive. Another human effect on the cycle is that much phosphate from agricultural crop lands makes its way into waterways—either directly, in runoff from the crop lands, or indirectly, in sewage effluents. Because there is essentially no return of phosphate from water to soil, this addition results in overfertilization of bodies of water, which in turn leads to a severe pollution problem

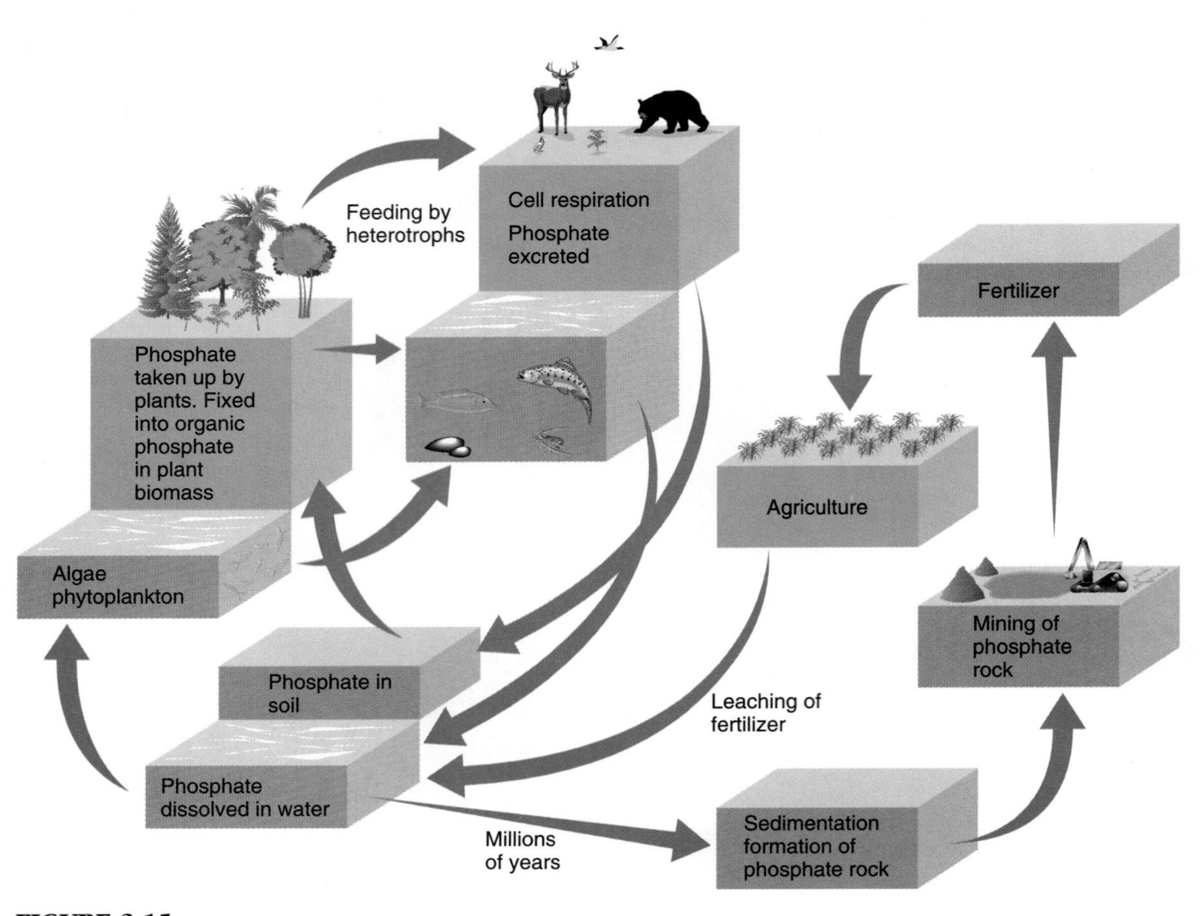

FIGURE 3-15
The phosphorus cycle.

known as eutrophication. (See Chapter 12.) Meanwhile, the lost phosphorus must be replaced on the crop lands by mining phosphate rock—a process that will ultimately result in depletion of the phosphate.

When humans use manure, compost (rotted plant wastes), or sewage sludge (see Chapter 13) on crops, lawns, or gardens, the foregoing natural cycle is duplicated. But in too many cases it is not, and the applied chemical fertilizers end up being leached (carried by water seepage) into waterways, resulting in eutrophication.

The Nitrogen Cycle. The nitrogen cycle (Fig. 3-16) is unique; it has aspects of both the carbon cycle and the phosphorus cycle. The main reservoir of nitrogen is the air, which is about 78% nitrogen gas (N_2). Plants and animals cannot utilize nitrogen gas directly from the air. Instead, the nitrogen must be in mineral form, such as ammonium ions (NH_4^+) or nitrate ions (NO_3^-). A number of bacteria and cyanobacteria (bacteria that contain chlorophyll; formerly referred to as blue-green algae) can convert nitrogen gas to the ammonium form, a process called biological **nitrogen fixation**. For terrestrial ecosystems, the most important among these nitrogen-fixing organisms is a bacterium called *Rhizobium*, which lives in nodules on the roots of legumes, the plant family that includes peas and beans (Fig. 3-17). This is another example of mutualistic symbiosis: The legume provides the bacteria with a place to live and with food (sugar), and gains a source of nitrogen in return.

From the legumes, nitrogen is passed down whatever food chains exist. At each step, as we have observed before, the nitrogen-containing compounds may be broken down in cell respiration. As this occurs, nitrogen-compounds are returned to the soil with excrements and may be absorbed by other plants. Thus, after it is fixed, nitrogen may be recycled in a manner similar to phosphorus and other mineral nutrients. However, nitrogen does not remain in this "mineral phase" of the cycle indefinitely. Other kinds of bacteria in the soil gradually change the nitrogen compounds back to nitrogen gas (See Fig. 3-16). Consequently, nitrogen will not accumulate in the soil. Additionally, some nitrogen gas may enter the cycle by being converted to the ammonium form by discharges of light-

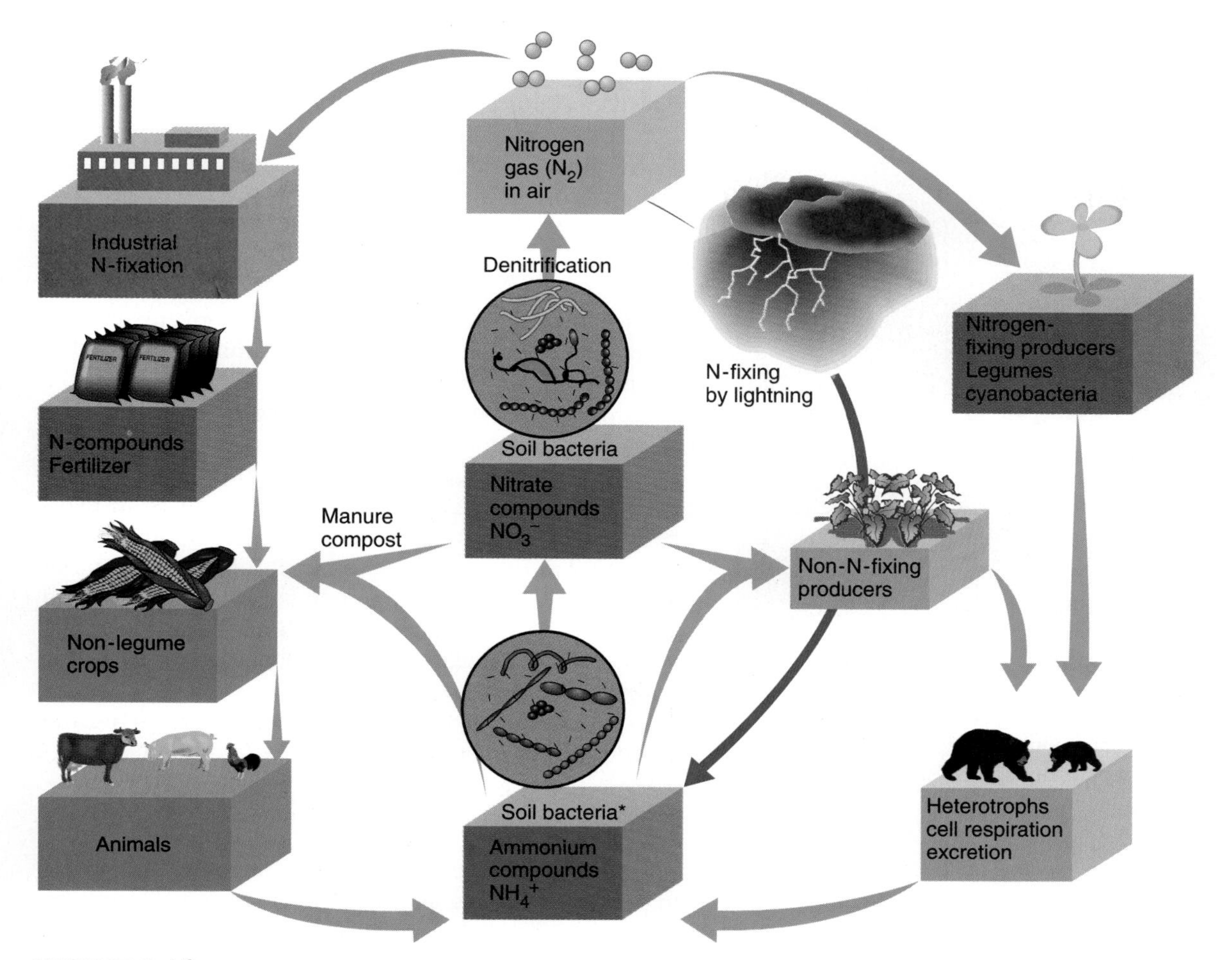

FIGURE 3-16
The nitrogen cycle.

ning in a process known as *atmospheric nitrogen fixation* and by coming down with rainfall. This pathway is estimated to be only about 10 percent of the biological pathway.

All natural ecosystems, then, depend on nitrogen-fixing organisms; legumes, with their symbiotic bacteria are, by far, the most important. The legume family includes a huge diversity of plants, ranging from clovers (common in grasslands) through desert shrubs to many trees. Every major terrestrial ecosystem, from tropical rain forest to desert and tundra, has its representative legume species, and legumes are generally the first plants to recolonize a burned-over area. Without them, all production would be sharply impaired because of lack of available nitrogen—precluding the formation of proteins, nucleic acids, and other building blocks of life.

The nitrogen cycle in aquatic ecosystems is similar. There, cyanobacteria are the most significant nitrogen fixers.

Only humans have been able to bypass the necessity for legumes when growing nonlegume crops such as corn, wheat, and other grains. We do this by fixing nitrogen in chemical factories (industrial nitrogen fixing). Synthetically produced ammonium and nitrate compounds are major constituents of fertilizer. However, the intensive use of such chemical fertilizer is not without disadvantages as regards maintaining a productive soil structure. Therefore, some farmers are readopting the natural process of enriching the soil by alternating legumes with nonlegume crops—that is, by **crop rotation**. This topic will be discussed further in Chapter 9.

While we have focused on the cycles of carbon, phosphorus, and nitrogen, it should be evident that cycles exist for oxygen, hydrogen, and all the other elements that play a role in living things. Also, while the routes taken by distinct elements may differ, it should be evident that all are going on simultaneously and that all come together in the tissues of living things.

From studying these cycles it might appear that every element is always recycled 100 percent. In fact there are some significant losses, or sinks, in the cycles. For example, carbon dioxide is removed from the cycle we have discussed by clams, oysters, and other marine organisms that use it in making their shells, which are

FIGURE 3-17
Nitrogen fixation. Conversion of nitrogen gas in the atmosphere to forms that can be used by plants, is carried out by bacteria that live in the root nodules of legumes. This process allows nitrogen gas to be converted into compounds needed for life. (USDA.)

calcium carbonate (Ca_2CO_3). Deposits of this calcium carbonate later convert to limestone, and thus the carbon atoms may remain trapped for hundreds of millions of years. Additionally, huge amounts of carbon were sidetracked millions of years ago in the formation of what are now our fossil fuels (see Chapter 21).

In the meantime, however, there is venting of carbon dioxide, as well as other gases, from volcanoes. On the millions-of-years time scale sea beds may be uplifted and carbon dioxide released from limestone by acidic leaching. Therefore, the cycles we have discussed are tied to much longer-term geological cycles; but over the millennia equilibrium is maintained. Will humans upset this equilibrium by burning the fossil fuels releasing the carbon dioxide over a relatively short period of time?

Running on Solar Energy

We have seen that no system can run without an input of energy, and living systems are no exception. For all major ecosystems, both terrestrial and aquatic, the initial source of energy is *sunlight* absorbed by green plants through the process of photosynthesis. (The only exceptions are ecosystems near the ocean floor or in dark caves, where the producers are bacteria that derive energy from the oxidation of hydrogen sulfide in those locations. These bacteria use that energy to make organic compounds, in a manner similar to that of higher plants. The process is called *chemosynthesis,* because it runs on chemical energy rather than light.)

Using sunlight as the basic energy source is fundamental to sustainability for two reasons: it is both *nonpolluting* and *nondepletable.*

Nonpolluting. Light from the Sun is a form of pure energy; it contains no substance that can pollute the environment. All the matter and pollution involved in the production of light energy are conveniently left behind on the Sun some 93 million miles (150 million kilometers) away in space.

Nondepletable. The Sun's energy output is constant. How much or how little of this energy is used on Earth will not influence, much less deplete, the Sun's output. For all practical purposes, the sun is an everlasting source of energy. True, astronomers tell us that the Sun will burn out in another 3–5 billion years, but we need to put this figure in perspective. One thousand is only 0.0001 percent of a billion. Thus, even the passing of millennia is hardly noticeable on this time scale.

Hence, we uncover the **second basic principle of ecosystem sustainability**:

> For sustainability, ecosystems use sunlight as their source of energy.

From the preceding discussion, you should be impressed with the importance of chemical nutrients and light as major prerequisites for the functioning of every ecosystem. Yet, we observed in Chapter 2 that rainfall and temperature (climate) are the primary limiting factors determining different terrestrial biomes. Basically, this is because every region that is above water receives abundant light, and most soils contain a modicum of nutrients or retain their nutrients through recycling. Therefore, light and nutrients are generally not limiting factors on land. But the situation is different in aquatic and marine environments, where light and nutrients dominate as the determining factors. (See the Global Perspective box, page 65.)

Prevention of Overgrazing

We now return to the concept of the food or biomass pyramid presented in Chapter 2 (Fig. 2-14). We have seen that a consumer cannot gain an amount of weight equal to what it eats because first, 60–90 percent of what is consumed is broken down for energy and second, another portion passes through without being digested. These two facts by themselves would result in the declining biomass at each higher trophic level—the biomass pyramid observed. However, there is a third reason for the observed decline. It is the following.

In a grazing situation it is readily apparent that if

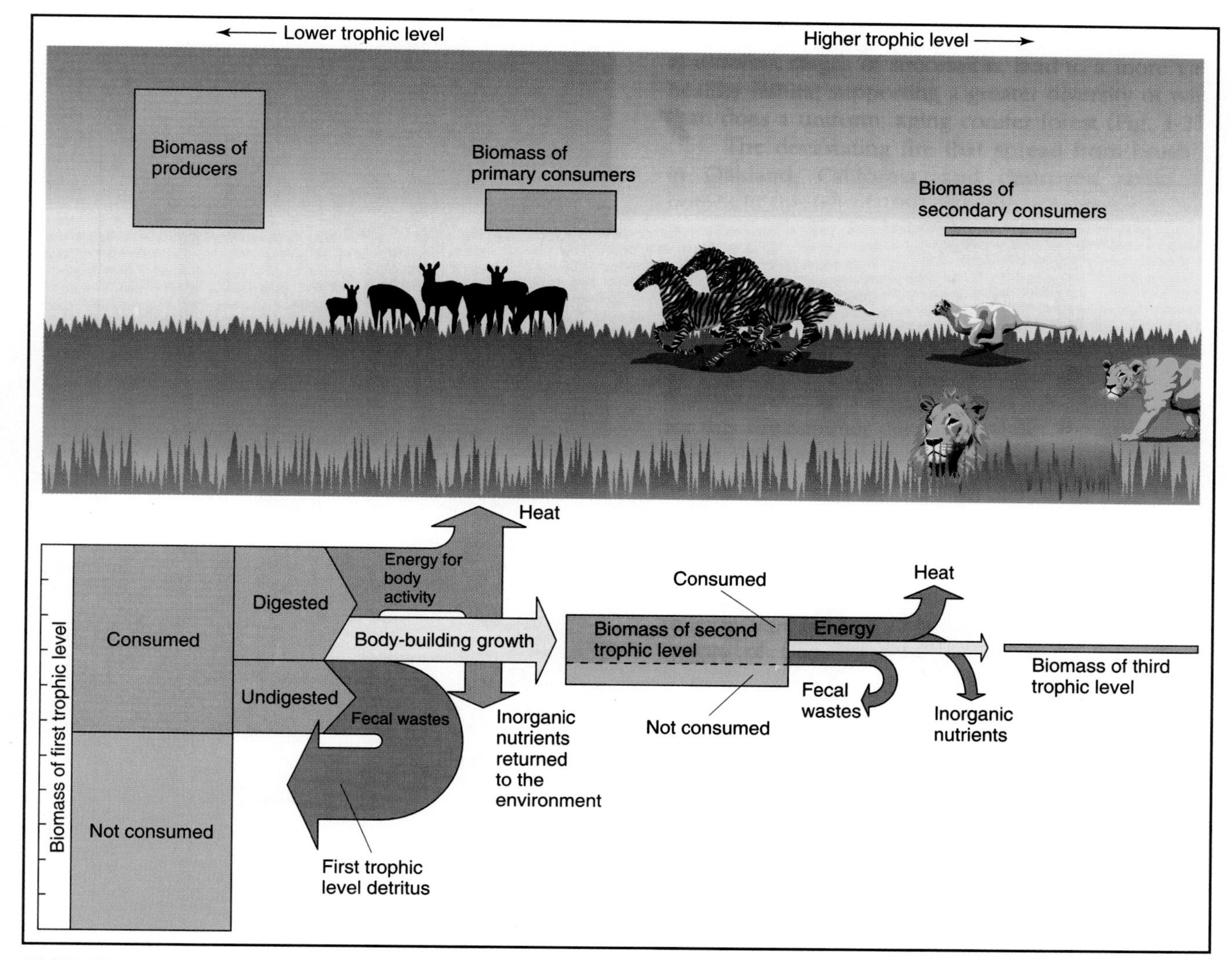

FIGURE 3-18
Decreasing biomass at higher trophic levels. The decrease results from three facts: (1) Much of the preceding trophic level is standing biomass and so is not available for consumption; (2) much of what is consumed is broken down in order to release energy; and (3) some of what is consumed passes through the organism.

the animals eat the grass faster than the grass can regrow, sooner or later the grass will be destroyed, and all the animals will starve. This situation is known as **overgrazing**. The same holds true in the case of carnivores and their prey. Basically, a sustainable situation demands that, on average, consumption cannot exceed production. It follows, then, that a large portion of the producer must remain intact to maintain that production. This portion, or population, that is not consumed, and which must remain intact to assure continued production, is called the **standing biomass**. In natural ecosystems, which are sustainable, we observe that consumers eat no more than a small proportion of the total biomass available; most is left as standing biomass (Fig. 3-18). We can readily see that this is another feature that is fundamental to sustainability. Hence, here is the **third basic principle of ecosystem sustainability**:

> For sustainability, the size of consumer populations is maintained so that overgrazing or other overuse does not occur.

How populations are regulated in nature to prevent such overuse will be the subject of Chapter 4. For now,

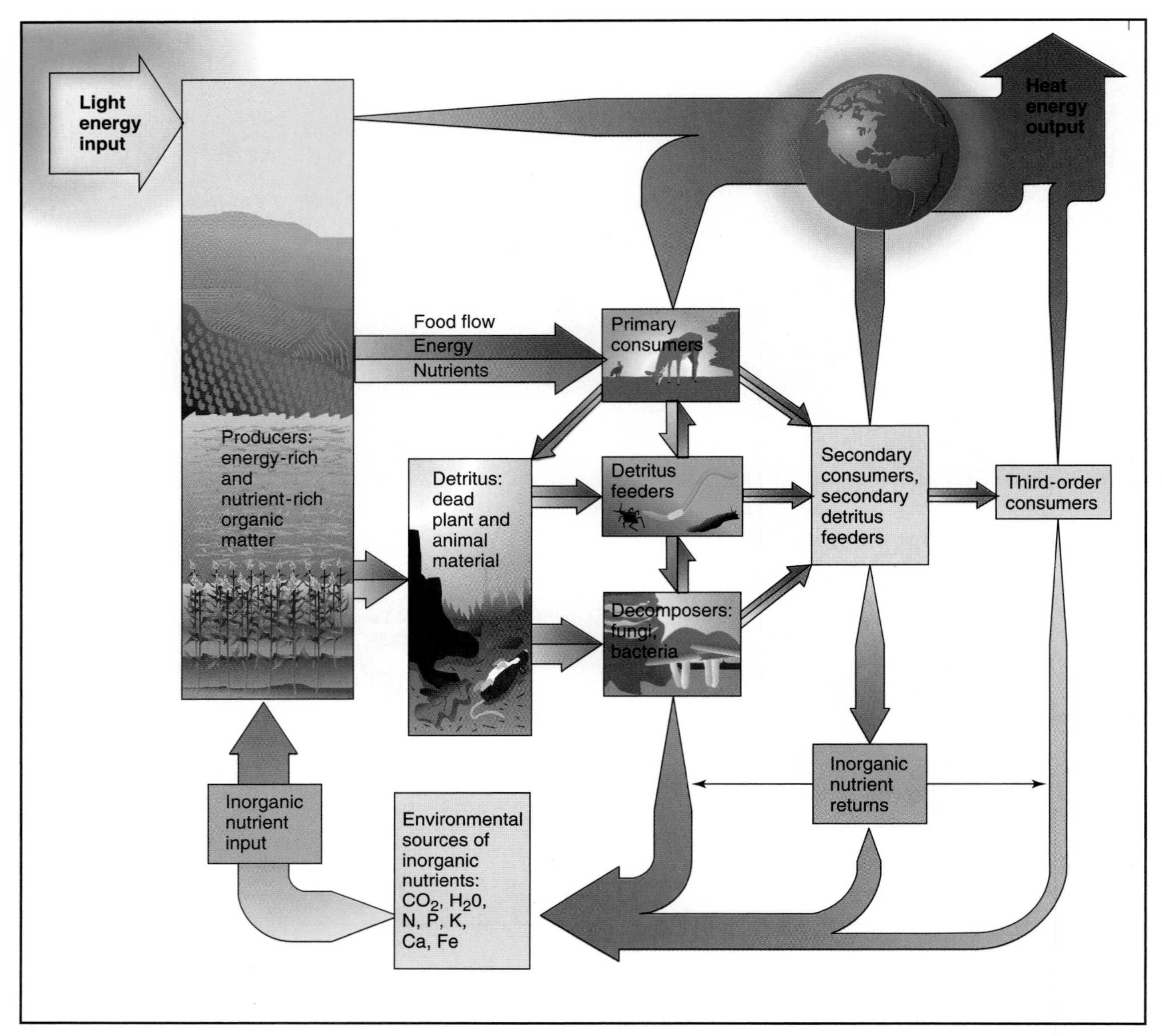

FIGURE 3-19
Nutrient cycling in and energy flowing through an ecosystem. Arranging organisms by feeding relationships and depicting the energy and nutrient inputs and outputs of each relationship show a continuous recycling of nutrients (blue) in the ecosystem, a continuous flow of energy through it (red), and a decrease in biomass.

it is sufficient to appreciate that such regulation is mandatory for sustainability. The three basic principles of ecosystem sustainability are illustrated in Figure 3-19.

Recognizing these principles of ecosystem sustainability, many students attempt to make their own artificial ecosystems in sealed bottles, with varying degrees of success. One artificial ecosystem big enough to maintain people, Biosphere 2, has been built in Arizona. ("Earth Watch" box, Biosphere 2, page 75).

Implications for Humans

We have said that a major part of our purpose in studying natural ecosystems lies in the fact that they are models of sustainability. We have said that if we can elucidate the principles that underlie their sustainability, we may be able to apply those principles toward our own efforts to achieve a sustainable society. Impor-

Ethics

Who Are You?

The cycling of carbon and other elements from the environment through organisms and back to the environment is more than just theory. Atoms can be made radioactive (see Appendix C) so that they behave chemically exactly like normal atoms, except that they give off radiation, which can be detected with suitable monitors. Introducing radioactive compounds into a cycle is analogous to planting a small radio transmitter on a vehicle to monitor where the vehicle goes. By this technique, scientists have verified that carbon atoms move from carbon dioxide in the atmosphere into glucose and then into various macromolecules making up plant tissues. Similarly, scientists can observe the progress of the radioactive atoms down food chains and their return to the atmosphere via cell respiration.

Through these studies, a striking fact becomes evident: All tissues of animal bodies, including our own, "turn over" quite rapidly. That is, although we maintain our general appearance from year to year, every tissue in our body is constantly being broken down, oxidized, and replaced with newly made molecules derived from the food we eat. Atom for atom, molecule for molecule, our bodies are entirely replaced about every four years. Thus, all forms of life, including human, are constantly participating in the cycles of all the elements.

Imagine having a microscope powerful enough to see the individual carbon atoms in the protein of the skin on your hand. Focus on a single carbon atom. Where did it come from? From the food you ate a few weeks ago? Before that? From carbon dioxide in the atmosphere that was incorporated into a plant by photosynthesis? Where will it go? In a few weeks, it will in all likelihood be back in the atmosphere as carbon dioxide after the top layer of your skin is sloughed off and oxidized by microorganisms. Will this be the end of the carbon atom's travels? No, it will no doubt go on to additional cycles.

If this or any other atom in your body could tell you its "life history," the story might go something like this:

> I have existed since the formation of Earth. In my countless cycles from the air through living things and back, I have participated in the bodies of virtually every species that has ever existed anywhere on Earth, including trees and animals of the forests; seaweeds, fishes, and other creatures of the oceans; and the dinosaurs that roamed the land 100 million years ago. In more recent times, my travels along food chains have led through humans of all races, as well as other plants and animals that share your environment. This is my fate till the end of time.

In a very real and verifiable way, all life is interconnected through sharing and recycling a common pool of atoms. Generations pass, but atoms remain the same. Does the scientific fact that we are all continually sharing the same atoms with every human and other living creature on Earth have ethical and moral implications?

---

tantly, discovering and applying principles is different from simple copying. For example, birds were models for the ability to fly. However, human flight was not achieved by copying birds; indeed, such efforts failed. Flight was attained by studying birds and discovering principles of aerodynamic lift and guidance. By applying these principles to "human-built machines," we not only achieved flight, but surpassed the capability of birds manyfold.

The reverse holds true also. Forging ahead in a way that is out of keeping with basic principles, whether from ignorance, arrogance, or stupidity, assuredly leads to problems and perhaps much worse. Any number of disasters have resulted from people's failure to give adequate attention to principles of engineering, physics, chemistry, and so on.

We have discussed three of four specific principles regarding sustainability of living systems. (The fourth will be covered in Chapter 4.) Can we see our environmental problems in terms of failing to abide by these principles? Even more, can we see these principles as providing guideposts for achieving sustainable development? The following is a synopsis contrasting our human system with these first three principles of sustainability.

First Principle of Sustainability. *For sustainability, ecosystems dispose of wastes and replenish nutrients by recycling all elements.*

In contrast to the remarkable recycling seen in natural ecosystems, we have constructed our human system, in large part, on the basis of a *one-directional flow* of elements. We have already noted that the fertilizer-nutrient phosphate, which is mined from deposits, ends up going into waterways with effluents from sewage treatment. The same one-way-flow can be seen in such metals as aluminum, mercury, lead, and

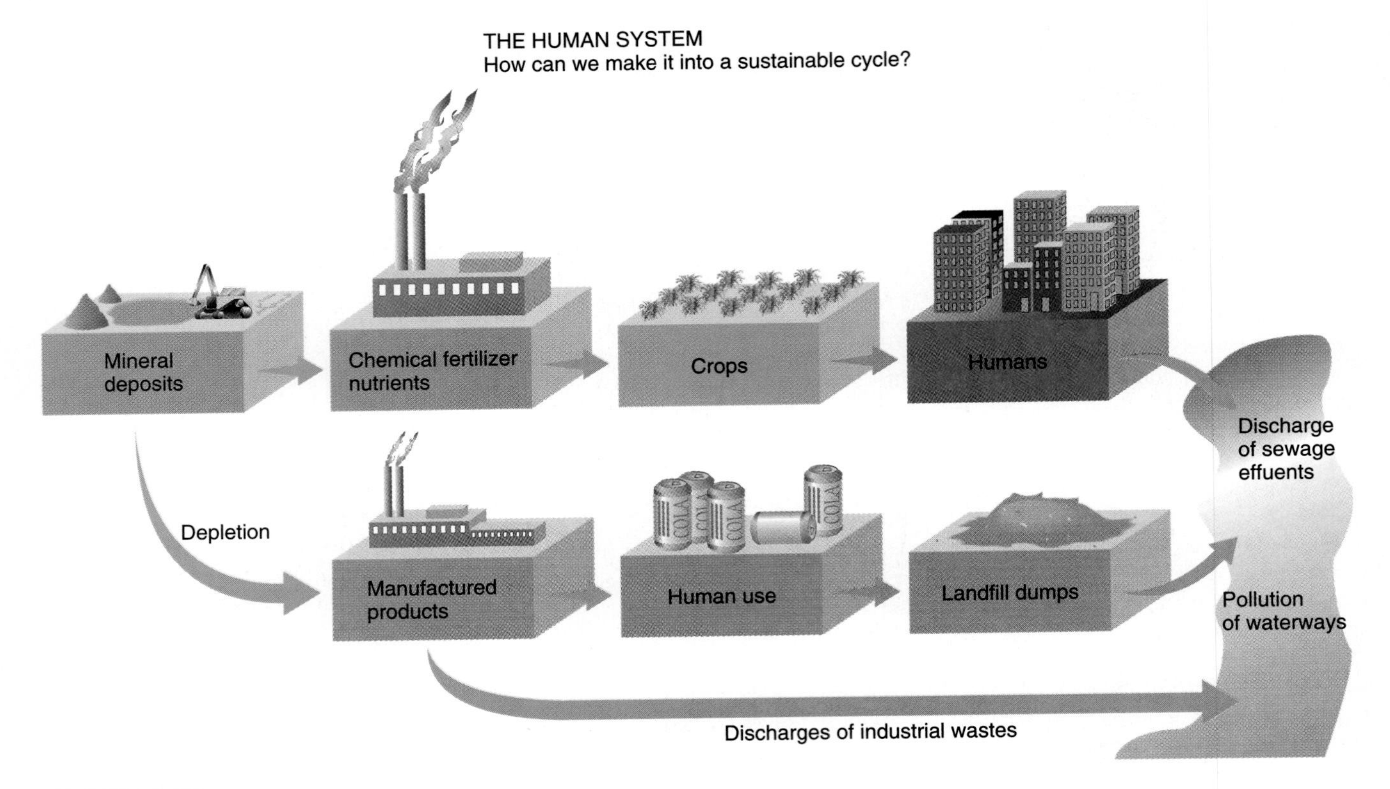

FIGURE 3-20
In contrast to applying the ecological principle of nutrient recycling, human society has developed a pattern of one-directional nutrient flow. There are increasing problems at both ends. This drawing illustrates one-way flow for phosphorus, but the scheme also applies to all other elements we use in our daily lives.

cadmium, which are the "nutrients" of our industry. At one end, they are mined from the earth; at the other, they end up in dumps and landfills, as items containing them are discarded. Is it any wonder that there are depletion problems at one end and pollution problems at the other (Fig. 3-20)? Actually, pollution problems are most significant at the present time. The earth has vast deposits of most minerals, but the capacity of ecosystems (even the whole biosphere) to absorb wastes without being disturbed is comparatively limited. This limitation is aggravated even more by the fact that many of the products we use are nonbiodegradable. Conversely, can you see the rationale for expanding the concept of recycling to include, not just paper, bottles, and cans, but everything from sewage to industrial wastes as well?

Second Principle of Sustainability. *For sustainability, ecosystems use sunlight as their source of energy.*

In contrast to running on solar energy, which is nonpolluting and nondepletable, we have constructed a human system that is heavily dependent on fossil fuels—coal, natural gas, and crude oil. Crude oil is the base for refinement of all liquid fuels: gasoline, diesel fuel, fuel oil, and so on. Even in the production of food, which is fundamentally supported by sunlight and photosynthesis, it is estimated that we use about 10 calories of fossil fuel for every calorie of food consumed. This additional energy is used in the course of field preparation, fertilizing, controlling pests, harvesting, processing, preserving, transportation, and finally cooking.

Again, the most pressing problem in connection with consuming these fuels is the limited capacity of the biosphere to absorb the waste byproducts produced from burning them. Air pollution problems, including urban smog, acid rain, and the potential for global warming, are the result of these byproducts. Also problems stemming from depletion, particularly that of crude oil, are on the horizon. You see why most people concerned about sustainability are also solar-energy advocates. Solar energy is extremely abundant (Fig. 3-21). Just as important, we do have the technology to obtain most, if not all, of our energy needs from sunlight and the forces it causes such as wind (Fig. 3-22; see also Chapter 23).

Third Principle of Sustainability. *For sustainability, the sizes of consumer populations are maintained so that overgrazing or other overuse does not occur.*

We have seen that, in natural ecosystems, a suit-

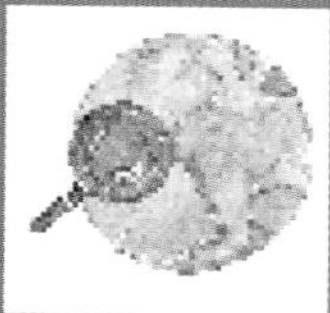

Earth Watch

Biosphere 2

The proof of a theory lies in testing it. If the biosphere functions as we have described—running on solar energy and recycling all the elements from the environment through living organisms and back to the environment—then it should be possible to create an artificial biosphere that functions similarly. Indeed, students commonly conduct an exercise of creating a "biosphere in a bottle": Some photosynthetic and compatible consumer organisms are sealed in a bottle and kept in the light. Varying degrees of success are achieved, however; such systems usually do not sustain themselves beyond a few weeks for various reasons.

The largest such experiment to date is Biosphere 2, constructed in Arizona 30 miles north of Tucson. Biosphere 2 was developed entirely with private venture capital, with a view toward gaining information and experience that might be used in creating permanent space stations on the Moon or other planets or for long-distance space travel. Additionally, Biosphere 2 is expected to yield information that will further our understanding of our own biosphere—Biosphere 1.

Biosphere 2 is a supersealed "greenhouse," including seals underneath, enclosing an area of 2.5 acres (1 ha). Entry and exit is through a double air lock. Different environmental conditions within the containment support several ecosystems. Accordingly, there is an area of tropical rain forest, savannah, scrub forest, desert, fresh- and saltwater marshes, and a miniocean complete with a coral reef, each stocked with respective species—over 4000 in all. There is an agricultural area and living quarters for a crew of up to ten "Biospherians."

Water vapor from evaporation and transpiration of plants is condensed to create high rainfall over the tropical rain forest. The water trickles back toward the marshes and the ocean through soil filters, providing a continuous supply of fresh water for both humans and ecosystems. Carbon

(Gonzalo Arcila/Decisions Investments Corp.)

dioxide from respiration is reabsorbed and oxygen is replenished through photosynthesis. Thus, these "natural ecosystems" provide the basic ecological stability for the atmosphere and hydrosphere of Biosphere 2. All wastes, including human and animal excrements, are treated, decomposed, and recycled to support the growth of plants.

Biosphere 2 is not self-sufficient as regards solar energy falling on the structure. Biosphere 2's energy demands for machinery are such that an additional (30) acres (12 ha) of solar collectors would be required. (Actually, external natural gas–driven generators are used.)

The first crew of four men and four women, a crew with a variety of skills and academic backgrounds, completed the first two-year mission in September 1993. In addition to monitoring and collecting data on the natural systems, their main occupation was intensive organic agriculture (no chemical pesticides) to produce plant foods both for themselves and for feeding a few goats and chickens, which produced eggs, milk, and a little meat. Some fish farming supplied additional protein. The living quarters include the comforts and conveniences of modern living, but all communication with the outside world was via electronics. Their environment with other plants and animals is totally sealed from that of the outside world. In such a closed system, the water soil, and nutrients they had when they began were the same when they finished, having cycled inumerable times within the system.

At the end of their two-year experiment sojourn, the "Biospherians" emerged somewhat thinner, but all in good health. The overall conclusion is that it worked: It is possible to build a biosphere, that includes humans, and have it function within tolerable limits.

Not everything went perfectly. At one point additional oxygen had to be introduced, because the amount of oxygen that was absorbed by decomposers in the rich organic soil was underestimated. Excessive carbon dioxide was absorbed through chemical reactions with exposed concrete surfaces. A considerable number of species that were introduced, especially insect pollinators, became extinct, necessitating pollinating many plants by hand.

Some writers have suggested that if we trash our own planet, we may end up living in Biosphere 2-like enclosures, although the costs for such escape from a polluted environment would be prohibitive. An important lesson from Biosphere 2 is an appreciation of the operational complexity of Biosphere 1. If we fail to maintain our natural biophere, there will be no alternative for survival.

Many scientific experiments continue within the sealed structure under the supervision of Columbia University's Lamont–Doherty Earth Observatory.

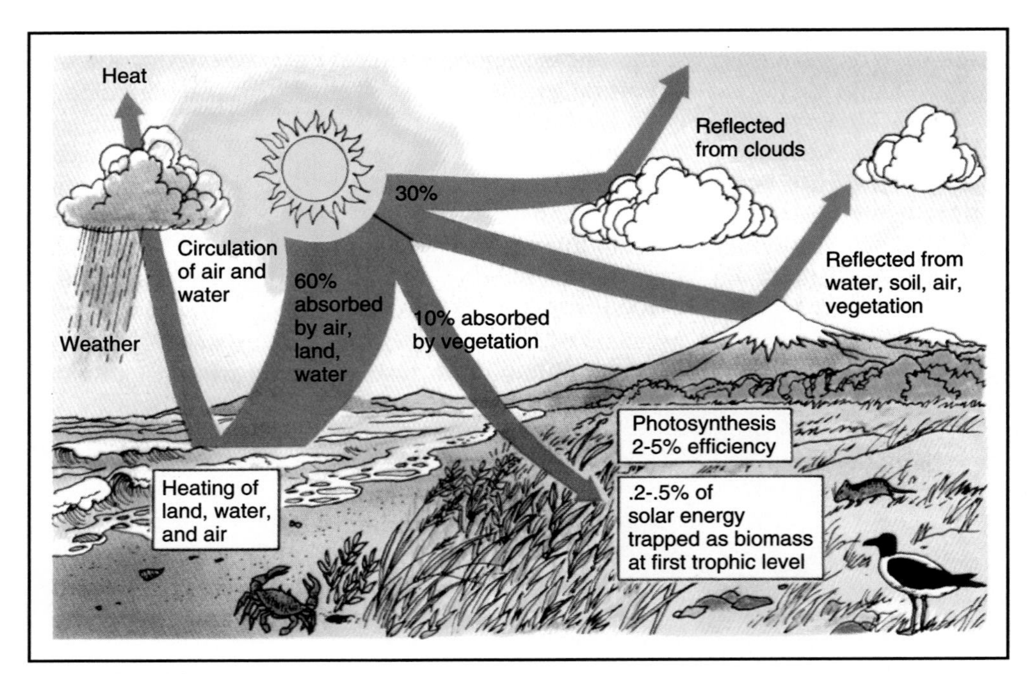

FIGURE 3-21
Of the total solar energy reaching Earth, 30% is reflected from clouds, water, and land surfaces. Sixty percent is absorbed by, and goes into heating land, water, and air. This heating causes evaporation of water and the circulation of air and water, resulting in weather. Only 10% of the solar energy is absorbed by vegetation. (Of course these percentages vary greatly among each ecosystem, the season of the year, and location on Earth.) Given the low efficiency of photosynthesis, only 2–5% of the 10% absorbed by vegetation (0.2–0.5% of the total solar energy) is trapped as biomass at the first trophic level, but this amount of energy supports all the rest of the ecosystem. It is estimated that just 0.1% of the solar energy arriving at Earth's surface would supply all human needs and not affect the dynamics of the biosphere.

able standing biomass is maintained so that future production is assured. Even a slight familiarity with such problems as the world's loss of biodiversity, loss of tropical rainforests, overfishing of the oceans, overgrazing of range lands, or other examples of overuse is enough to show that we are wanting with respect to this principle.

Nor are the causes of this overuse hard to identify—the demands of a rapidly growing human population and increasing per capita consumption. Human population has increased more than sixfold in the past 200 years, and it is continuing to increase at a rate of nearly 88 million people per year—10 times faster than in 1800 (Chapter 6). Can you see why the success or failure of efforts toward stabilizing population will have profound implications regarding sustainability?

The necessity of stabilizing population takes on even more importance when we consider another trend—increasing per capita consumption. Given the same basic human system, better lives translate into more consumption of virtually everything. A particular case in point is the human fondness for meat. A trend that parallels increasing affluence in every country observed is increasing meat consumption. Because of the principles involved in the biomass pyramid, it takes about 10 pounds of grain to grow a pound of meat [more for beef, less for chicken (Fig. 3-23)]. Therefore, for every increment of increase in meat consumption, there is a tenfold increase in the demand on plant production and on all the land use, fertilizer, pesticides, energy, and pollution that that increase entails. Of course, the reverse is also true. Dropping down a trophic level alleviates the demand proportionally. The implication of this is profound when you consider that half the cultivated acreage in the U.S. produces animal feed!

In this brief assessment of our human system in relation to the principles of sustainability, we see that we are missing the mark by a wide margin, and the environmental troubles we face can be seen as a mounting consequence. We can use an awareness of our present circumstance as a guide to show the directions to be taken to attain a sustainable future. Specific things we can do will become obvious as we discuss the issues in later chapters.

FIGURE 3-22
Wind turbines generating electrical power in Palm Springs, CA. More than 50,000 wind turbines have been installed worldwide since 1974—15,000 in California alone. By the middle of the next century, wind-generated electricity will be a necessity, along with other renewable energy sources, conservation, and energy efficiency. (Matt Meadows/Peter Arnold, Inc.)

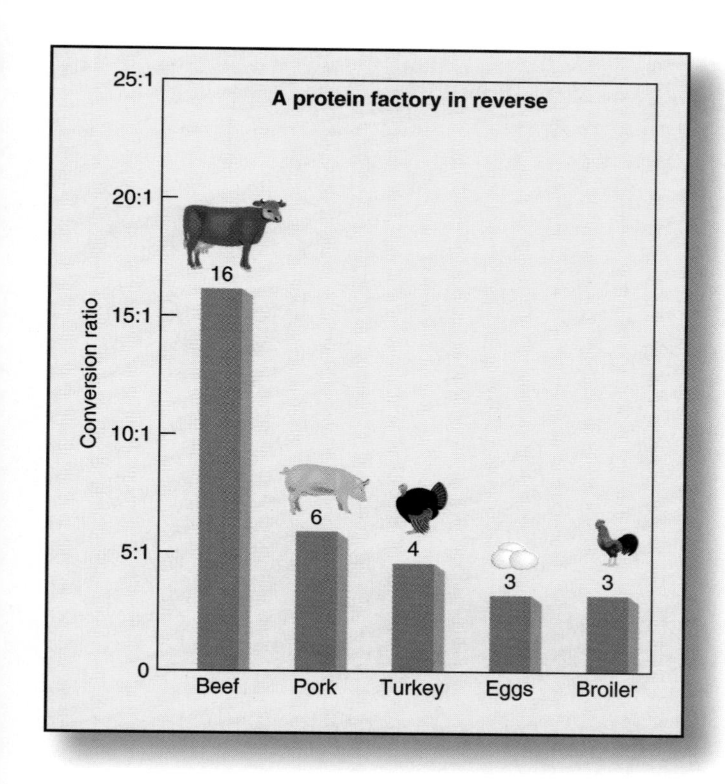

FIGURE 3-23
To obtain one pound of meat, poultry, or eggs, farmers must invest these many pounds of grains and soy feeds. To get one pound of beef requires an expenditure of 16 pounds of feed. Said another way, the grain consumed to support one person eating meat could support 16 persons eating the grains directly.

Review Questions

1. What are the six key elements of living organisms, and where does each occur in the environment? In four cases, identify the specific molecule that the element comes from.
2. Give a simple description of what is happening to the six key elements in the course of growth and decay. What is the "common denominator" that distinguishes organic and inorganic molecules?
3. State the definitions of matter and energy. Name three main categories of matter, and list four forms of kinetic energy.
4. Give five examples that demonstrate different conversions among forms of kinetic energy. Give another five examples that demonstrate conversions between kinetic and potential energy.
5. Name the two energy laws and apply each, describing the relative amounts and forms of energy going into and coming out of the conversions listed in Question 4.
6. What is chemical energy? What energy changes are involved in the formation and breakdown of organic molecules?
7. Describe the process of photosynthesis in terms of what is happening to specific atoms (matter) and where energy is coming from and going to. Do the same for cell respiration.
8. Food ingested by a consumer follows three different pathways. Describe each pathway in terms of what happens to the food involved and what products and byproducts are produced in each case.
9. Define and contrast starvation and malnutrition, giving the causes and results of each.
10. Compare and contrast the decomposers with other consumers in terms of matter and energy changes that they perform.
11. Where do carbon, phosphorus, and nitrogen exist in the environment, and how do they move into and through organisms and back to the environment?
12. In what forms does energy enter, move through, and leave ecosystems?
13. What three factors account for decreasing biomass at higher trophic levels—that is, the food pyramid?
14. What is meant by standing biomass? What does it imply regarding consumer populations?
15. What are the three principles of ecosystem sustainability?
16. Contrast our human system with each of the principles of sustainability. What environmental problems result from not abiding by the principles? What would you recommend regarding directions to achieve sustainability?

Thinking Environmentally

1. Describe the consumption of fuel by a car in terms of the laws of conservation of matter and energy. That is, what are the inputs and outputs of matter and energy? (*Note*: Gasoline is an organic carbon-hydrogen compound.)
2. Relate your level of exercise, breathing hard, and "working up an appetite" to cell respiration in your body. What materials are consumed, and what products and byproducts are produced?
3. Using your knowledge of photosynthesis and cell respiration, create an illustration showing the hydrogen cycle and the oxygen cycle.
4. Write a short essay supporting or contradicting the statement, "Waste is a human concept and invention; it does not exist in natural ecosystems!"
5. Tundra and desert ecosystems support a much smaller biomass of animals than do tropical rainforests. Give two reasons for this fact.
6. Evaluate the sustainability of parts of the human system, such as transportation, manufacturing, agriculture, and waste disposal, by relating them to specific principles of sustainability. How can such things be modified to make them more sustainable?
7. In Chapter 2, we observed that humans have overcome the *usual* limits and barriers that restrict other species. Does this mean that we can disregard basic ecological principles? What is the difference between "usual limits" that restrict other species and the "basic principles" revealed here?

Ecosystems In and Out of Balance

Key Issues and Questions

1. Ecosystem sustainability demands maintaining a balance between all the populations in the biotic community. Explain what is meant by this statement.
2. Stable natural populations involve balances between biotic potential and environmental resistance. Define these terms and explain how a balance occurs.
3. Various biotic and abiotic factors play a role in maintaining balances between herbivores and their natural enemies, herbivores and vegetation, and between competing plant species. What are the main factors involved in maintaining balances among the groups noted?
4. The introduction of a foreign species frequently has disruptive ecological results. Explain why this is the case. Give examples.
5. Natural ecosystems may undergo gradual succession until they reach a climax, or more stable state of ongoing adaptation. What is meant by succession, and what factors are responsible?
6. Fire is a significant natural abiotic factor. How does fire play a role in the balance of some ecosystems? Can fire be used in ecosystem management?
7. The major ecological imbalance on Earth is between the human species and the rest of the Earth's biota. What will be the probable consequences of not achieving a balance? What directions can be taken toward achieving a balance?

In protected areas of Africa, Antarctica, deep tropical forests, tundra, or other remote areas that have escaped human impacts, we find nature very much as it was described by the earliest explorers. The same species of plants, fungi, arthropods, mollusks, reptiles, mammals, and birds are still there. On the other hand, every individual plant or animal, because of its limited life span, is many generations removed from those first seen and described. We have pointed out that this is the essence of sustainability, and we have shown that such sustainability is supported by the continuous recycling of all nutrients and the use of sunlight as the primary energy source.

Yellow Stone National Park. [Photo by Stan Osolinski/The Stock Market.]

Still, the most difficult question of sustainability is left unanswered. Given the host of different species that exist in a natural ecosystem and all the predatory and competitive relationships among them in their food web, how is it that one species does not eliminate the others? How do they all manage to coexist over many generations?

In Chapter 3, we noted that a standing biomass of lower trophic (feeding) levels must be maintained to assure continued production. Thereby, we derived our third basic principle of sustainability—for sustainability, populations of consumers are maintained such that overgrazing or other forms of overuse do not occur. It is evident that if this were not the case, the lower member of the food chain would be entirely devoured, and the consumer would perish as a result. However, this does not answer the question of how such overuse is prevented in natural systems.

Some people like to think that a natural law or force (such as gravity) acting from the outside causes all ecosystems to keep in balance. Casual observation, however, shows the falsity of this notion. The fossil record is filled with examples of the rise, fall, and extinction of past species. Now, with humans on the scene, the rate of extinctions is increasing. We are observing the decline and demise of species because of many factors: pollution, the introduction of new species that displace native species, the elimination of predators, alterations in water levels, and changes in nutrient levels in lakes, to say nothing of the direct pressures of overhunting and the outright destruction of ecosystems for human developments.

The fact that we must go to a very remote or protected area to find things as they were centuries ago implies that essentially all ecosystems have been altered by human influence. Indeed, it is doubtful whether *any* ecosystem on Earth remains totally free of human impact, and numerous ecosystems have been essentially obliterated, losing many species. Further, such impacts are increasing, and the loss of species is accelerating; a recent conservative estimate places extinction rates at one every 30 minutes; other estimates range to one every 16 minutes, or 30,000 per year! The actual extinction rate could be much higher.

Therefore, far from retaining the notion that nature is stable and will take care of itself, we must realize that an ecosystem is a system of dynamic balances—balances among the populations of different species and between each species and the abiotic environment. Changing any factor, biotic or abiotic, may shift these balances and cause sometimes unpredictable consequences.

Two interrelated issues are important here: First, we cannot hope to sustain natural ecosystems and many of the species in them unless we understand and preserve these balances; second, we cannot hope to sustain our own civilization unless we establish suitable balances between our human system and the natural world. Our objective in this chapter is to examine some of the mechanisms of balance that underlie the sustainability of all ecosystems including human systems and the rest of the biosphere.

Ecosystem Balance Is Population Balance

Each species in an ecosystem exists as a population—that is to say, a reproducing group. For an ecosystem to remain stable (to retain the same mix of populations of different species) over a long period of time, the population of each species in the ecosystem must remain more or less constant in size and geographic distribution. In short, on average deaths must equal births; otherwise the population would shrink or grow accordingly. We speak of an equilibrium between births and deaths as **population balance**. In turn, ecosystems sustainability or balance boils down to a problem of how population balance is maintained for all the species that comprise the ecosystem.

Biotic Potential versus Environmental Resistance

First among the factors for increasing population size is **biotic potential**, the number of offspring (live births, eggs laid, or seeds or spores set in plants) that a species may produce under ideal conditions. Looking at different species you can readily see that biotic potential varies tremendously, averaging from less than one birth per year in certain mammals and birds to many millions per year in the case of many plants, invertebrates, and fish. However, to have any effect on the size of subsequent generations, the young must survive and reproduce in turn. Survival through the early growth stages to become part of the breeding population is called **recruitment**. Recruitment is the second factor in population growth.

Considering differences in biotic potential and recruitment, you can see two different **reproductive**

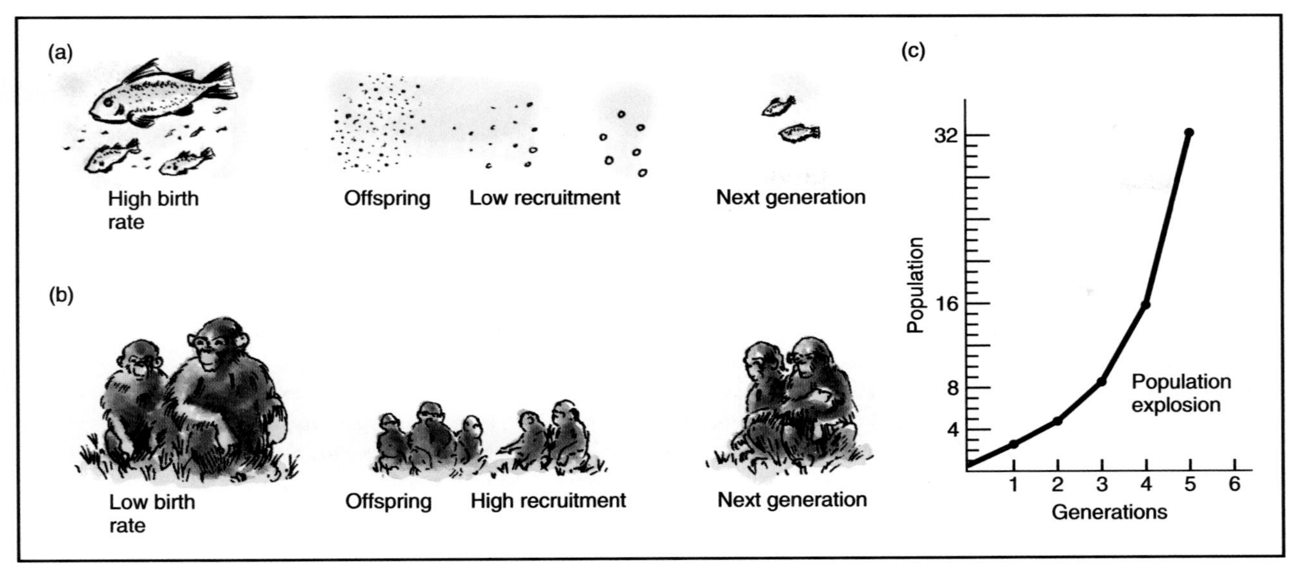

FIGURE 4-1
Two reproductive strategies. (a) High birth rate–low recruitment and (b) low birth rate–high recruitment. Both accomplish survival. (c) If recruitment is high, even a relatively low birth rate will lead to rapid population growth. Shown here, a birth rate of just four offspring per couple will double the population each generation if all offspring survive and reproduce. Such exponential growth is called a *population explosion.*

strategies (Fig. 4-1). The first strategy is to produce massive numbers of young, but then leave survival to the whims of nature. This strategy results in very low recruitment. Thus, despite high biotic potential, population increase may be nil because recruitment is so low. (Note that "low recruitment" is a euphemism for high mortality of the young.) The second strategy is to have a much lower reproductive rate, but then provide care and protection to the young so as to enhance recruitment.

Additional factors that influence population growth and geographic distribution are the ability of animals to migrate, or of seeds to disperse, to similar habitats in other regions; the ability to adapt to and invade new habitats; defense mechanisms; and resistance to adverse conditions and to disease.

Taking all these factors for population growth together, we find that every species has the capacity to increase its population when conditions are ideal. Furthermore, growth of a population under ideal conditions will be *exponential.* For example, a pair of rabbits producing 20 offspring, 10 of which are female, may grow by a factor of 10 each generation: 10, 100, 1000, 10,000, (10^1, 10^2, 10^3) and so on. Such a series is called an **exponential increase**. When this occurs in a population, it is commonly called a **population explosion** (Fig. 4-1c). A basic feature of such an exponential increase is that the numbers increase faster and faster as the population doubles and redoubles, with each doubling occurring in the same amount of time.

Population explosions are seldom seen in natural ecosystems, however, because a large number of both biotic and abiotic factors tends to decrease population. Among the biotic factors are predators, parasites, competitors, and lack of food. Among abiotic factors are unsuitable temperature, moisture, light, salinity, pH, and lack of nutrients. The combination of all the abiotic and biotic factors that may limit population increase is referred to as **environmental resistance**.

You may already foresee the result of the interplay between the factors promoting population growth and those leading to population decline. Conditions are always changing. When they are favorable, populations may increase. When they are unfavorable, populations decrease.

In general, the reproductive rate of a species remains fairly constant, because that rate is part of the genetic endowment of the species. What varies substantially is recruitment. It is in the early stages of growth that individuals (plants or animals) are most vulnerable to predation, disease, lack of food (or nutrients) or water, and other adverse conditions. Consequently, environmental resistance effectively reduces recruitment. Of course, some adults also perish, particularly the old or weak. If recruitment is at the **replacement level**, that is, just enough to replace these adults, then the size of the population will remain constant. If recruitment is not sufficient to replace losses in the breeding population, at that point, the size of the population will decline.

Earth Watch

Maximum versus Optimum Population

We define later in the chapter that *carrying capacity* is the *maximum* population a habitat will support without being degraded over the long term. This population level, however, is not a simple fixed number that can be arrived at easily, and it will vary from year to year in any given ecosystem. Carrying capacity may be considerably higher in wet years than in dry years, which inhibit the growth of vegetation. In addition, the line between degrading and not degrading the habitat over the long term is anything but clear. Degradation may take place slowly and be hardly perceptible from year to year; yet it does occur.

Therefore, managers nowadays focus more on what is the *optimal* rather than the maximum population. The optimal population is large enough to ensure a healthy breeding stock, yet considerably less than the theoretical maximum. It allows flexibility between good years and bad years without upsetting the ecosystem.

The optimum population is not simple or fixed either. It too can be influenced by habitat management. For example, planting certain species, thinning a forest, and starting new trees will increase the amount of browse available and enhance the carrying capacity for deer. Creation of ponds and marshes will attract and support waterfowl.

What is the minimum size of habitat that is required to support a breeding population? Natural areas isolated from surrounding similar areas proceed to lose species. The smaller the isolated area, the greater the number of species lost. Thus fragmentation of forests by intervening development is leading to many extinctions because the fragments are insufficient to support critical numbers even though the ecosystems are otherwise preserved. It is being found that the loss of species may be reduced to some extent by keeping the remaining fragments connected by "green corridors."

Some 300 formal *biosphere reserves* are recognized worldwide. These areas contain a core in which genetic material is protected from outside disturbances, surrounded by a buffer zone, that is, in turn surrounded by a transition area. In an ideal sense, these reserves are meant to slow extinction rates and the fragmenting of natural sites.

In certain situations, environmental resistance may affect reproduction as well as causing mortality directly. For example, the loss of suitable habitat often prevents animals from breeding. Also, certain pollutants adversely affect reproduction. However, we can still view these situations as environmental resistance that either blocks a population's growth or causes its decline.

In sum, whether a population grows, remains stable, or decreases is the result of an *interplay between its biotic potential and environmental resistance* (Fig. 4-2). In general, biotic potential remains constant; it is shifts in environmental resistance that allow populations to increase or cause them to decrease. For example, a number of favorable years (low environmental resistance) will allow a population to increase; then a drought or other unfavorable conditions may cause it to die back, and the cycle may be repeated.

We emphasize that population balance is a **dynamic balance**, which implies that additions (births) and subtractions (deaths) are occurring continually and the population may fluctuate around a median. Some fluctuate very little; others fluctuate widely, but as long as decreased populations restore their numbers, the system may be said to be balanced. Still, the questions remain: What maintains the balance within a certain range? What prevents a population from "exploding" or, conversely, becoming extinct? Indeed, as balances are upset, these events can and do occur.

Density Dependence and Critical Numbers

In general, the size of a population remains within a certain range because most factors of environmental resistance are *density dependent*. That is, as **population density** (the number of individuals per unit area) increases, environmental resistance becomes more intense and causes such an increase in mortality, that population growth ceases or declines. Conversely, as population density decreases, environmental resistance is generally mitigated, allowing the population to recover. This balancing act will become clearer as we discuss specific mechanisms of population balance in the next section.

But again, there are no guarantees that a population *will* recover from low numbers. Extinctions can and do occur in nature. For example, where are the dinosaurs? The survival and recovery of a population depends on a certain minimum population base, which is referred to as the **critical number**. You can see the idea of critical number at work in terms of a herd of deer, a pack of wolves, a flock of birds, or a school of fish. Often, the group is necessary to provide protection and support for its members. In some cases the critical

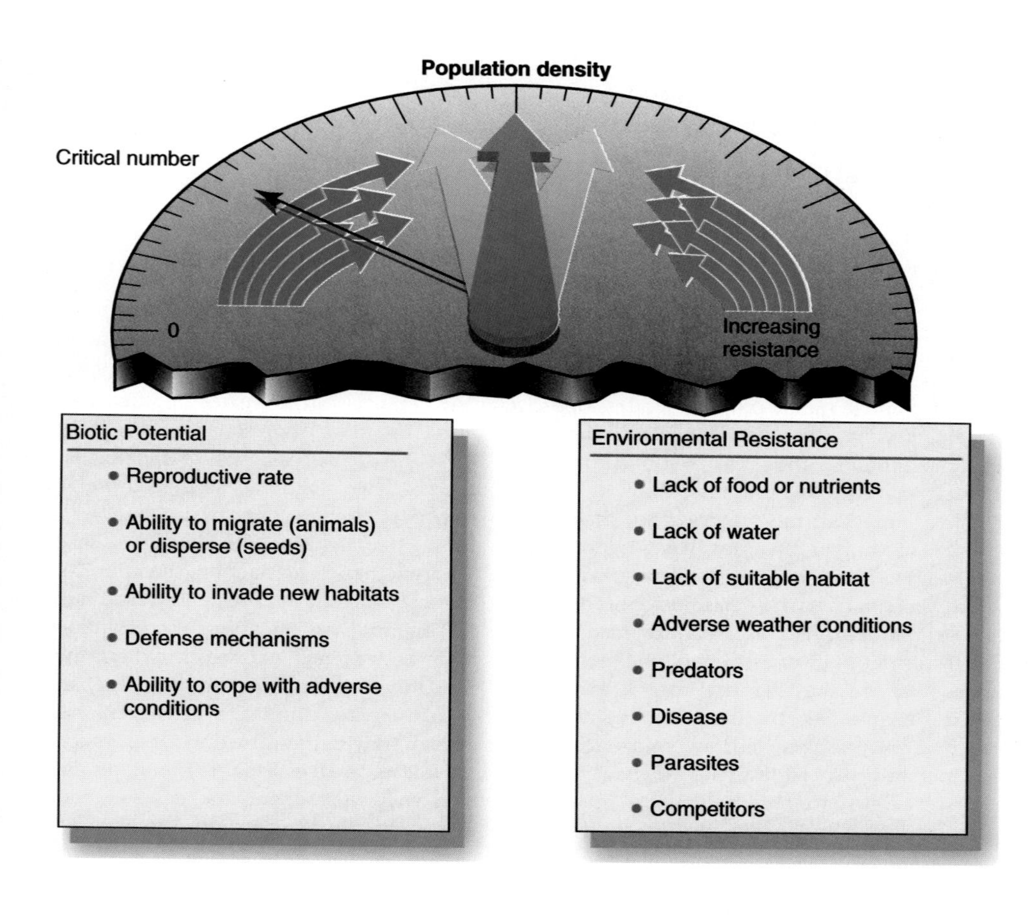

FIGURE 4-2
A stable population in nature is the result of a balance between factors tending to increase population (biotic potential) and factors tending to decrease population (environmental resistance). A balance results because many factors of environmental resistance become more intense with increasing population.

number is larger than a single pack or flock, because interactions between groups may be necessary as well. In any case, if a population is depleted to below the critical number needed to provide such supporting interactions, the surviving members actually become more vulnerable, breeding fails, and extinction is virtually inevitable. Thus, we must recognize that the density-dependent decline and recovery of a population occurs well above the critical number for the population.

Why are human activities causing so many extinctions? It is simply that human impacts, such as alterating habitats, pollution, hunting, and other forms of exploitation are not density dependent; they can even intensify as populations decline toward extinction. (See "Tragedy of the Commons," page 496.)

Environmentalism represents an attempt to create a new kind of balance through using acquired wisdom and intelligence. Scientists are monitoring many populations. Species whose populations are declining rapidly because of human impacts are classified as **threatened**. If the population is near what scientists believe to be its critical number, the species may be classified as **endangered**. These definitions, when officially assigned by the U.S. Fish and Wildlife Service, set into motion a number of actions aimed at stemming the negative impacts on, and providing protection for and even artificial breeding of, the species in question. Nongovernmental organizations, such as the World Wildlife Fund, are also playing a great role in protecting threatened and endangered species. The actions on behalf of such species are covered in Chapter 18. (See the "Earth Watch" box, "A Beach for Birds," on page 462.)

Mechanisms of Population Balance

With our general understanding of population balance as a dynamic interplay between biotic potential and environmental resistance, we now turn our attention to some specific kinds of population-balancing interactions. (Keep in mind, however, that in the natural world a population is subjected to the total array of all the biotic and abiotic environmental factors around it. Single factors are seldom totally responsible for the regulation of a given population; rather, regulation results from many factors acting together.)

Ethics

Hunting versus Animal Rights

Hunting and animal rights are increasingly the topic of moral and ethical debates, with hunting enthusiasts and animal rights activists at the extremes of the issue. Certainly each individual has a right to choose for himself or herself whether or not to engage in legal hunting, eating meat, wearing leather, or otherwise being a party to killing animals. But which point of view will best serve the long-term interests of society and the biosphere?

Proponents on both sides of the question could profit from a greater ecological understanding and awareness. If maintaining the biodiversity and ecological balances in the biosphere is taken as the primary aim, then there is a rationale for both sides. If we are looking at an endangered species, condoning its further hunting for any reason seems unconscionable because such hunting can only hasten its extinction, causing a permanent loss for both society and the biosphere. This reasoning can be extended beyond hunting to include habitat destruction for development or other purposes, a process that is also causing the extinction of species. Where animal rights projects lend support to saving endangered species and their habitats, the efforts of the people involved are commendable.

On the other hand, if we are looking at a species that is overpopulating and overgrazing because its natural enemies have been removed (e.g., deer) or because it is an introduced species without natural enemies, then some would argue for human intervention. Continuing to allow such animals to overpopulate and overgraze can lead only to widespread damage to the ecosystem, damage that will include the death and even extinction of other animals dependent on the same vegetation for food and habitat. It will lead also to the death of the animals that are overpopulating because they will eventually deplete their food supply and die of starvation or disease.

Is keeping their population in check by hunting less cruel? Does allowing animals to overgraze in any way serve the value of preserving the integrity of the biosphere? Does one particular animal have rights that exceed those of others or of the ecosystem as a whole?

In conclusion, it makes little sense to argue, much less act on, the ethics of animal rights or hunting as issues in and of themselves. Meaningful resolution can be reached, however, when the debate or action is put in terms of particular species and the ecosystems in which they exist and when preservation of that ecosystem is made the determining value. Preserving the ecosystem is the only way to preserve biodiversity. As for the actual act of killing an animal, this remains a personal decision based on individual values.

Predator-Prey and Host-Parasite Balances

The best known mechanism of population balance is regulation of a population by a predator, that is, a *predator-prey balance*. An example well documented in nature is the interaction between wolves and moose on Isle Royale, a 45-mile-long island in Lake Superior not far from the Canadian shore.

During a hard winter early in this century, a small group of moose crossed the ice to the island and stayed. Their population grew considerably in the absence of predators. Then, in 1948, a pair of wolves also managed to reach the island. The populations of the two species have been carefully tracked by wildlife biologists since 1958 (Fig. 4-3). As seen in the figure, the rise in the moose population is followed by a rise in the wolf population, and the decline in the moose population is followed by a decline in the wolf population. The data can be interpreted as follows: A paucity of wolves represents low environmental resistance for the moose, so the moose population increases. Then, the abundance of moose representing optimal conditions (low environmental resistance) for the wolves, the wolf population increases. The growing wolf population means higher predation on the moose, so the moose population falls. The decline in the moose population is followed by a decline in the wolf population, because now there are fewer prey. We can see how this cycle can be repeated indefinitely, providing a dynamic balance between the populations of moose and wolves.

The most recent data from Isle Royale show another cycle in progress, but the wolf population has not increased as much as should be expected. Apparently, inbreeding and disease are additional factors affecting the viability of the wolf population. Also the most recent dramatic fall in the moose population cannot be attributed entirely to predation by the small number of wolves. Therefore, the situation may be less straighforward than implied above. Ecologists generally agree that the rise and fall of the predator population is in response to the availability of prey. But there is debate as to whether the predator is the *primary* or *only* cause of the leveling and decline of the prey population. Other factors are generally involved as well. For example, the shortage of vegetation that occurs as the herbivore population grows may stress the animals, especially the old, sick, and young, and make them more vulnerable to predation. It has been

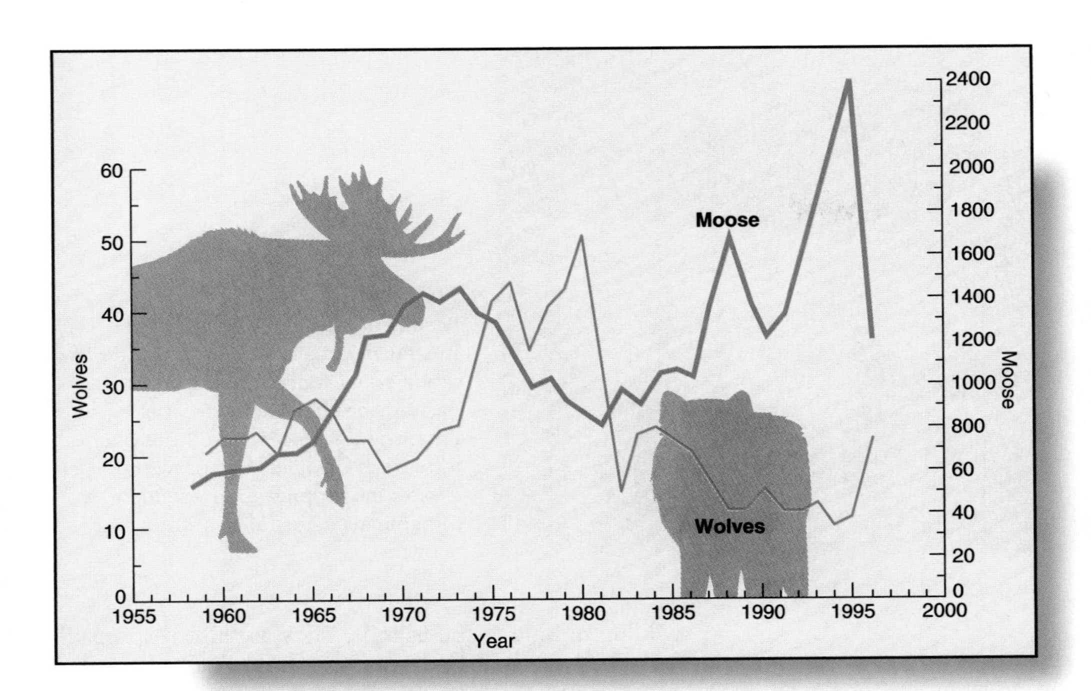

FIGURE 4-3
The predator-prey relationship creates a balance between predator and prey populations. The data from direct observations of the moose and wolves on Isle Royale shows that the moose population declined as the wolf population increased and increased as the wolf population decreased. During the winter of 1996, moose were dying at 2 to 3 times normal rates. Reasons for this population decline are deep snow, ticks, and increasing wolf population. (Rolf O. Peterson, Professor of Forestry, Michigan Technological University.)

observed, for instance, that wolves are incapable of bringing down an adult moose in good physical condition. The animals they kill are the young and those weakened by another factor—which is not the case when the predators are human hunters.

The observation that predators are incapable of killing individuals of their prey that are mature and in good physical condition is extremely significant. This is what puts the brakes on a predator eliminating its prey. As the prey population is culled down to those healthy individuals that can escape attack, the predator population has no choice but to starve back to a lower level. Meanwhile, the survivors of the prey population are the most healthy of the stock and can readily procreate the next generation.

Much more abundant and ecologically important than predators in population control are a huge diversity of *parasitic organisms.* Recall from Chapter 2 that these organisms range from tapeworms, which may be a foot or more in length (Fig. 2-9), to microscopic disease-causing protozoans, fungi, bacteria, and viruses. All species of plants and animals, and even microbes themselves, may be infected with parasites.

In terms of population balance, parasitic organisms act in the same way as large predators do. As the population density of the host organism increases, parasites and their *vectors* (agents that carry the parasites from one host to another), such as disease-carrying insects, have little trouble finding new hosts, and infection rates increase, causing dieoff. Conversely, when the population density of the host is low, transfer of infection is impeded, and there is a great reduction in levels of infection, a condition that allows the population to recover.

You can readily see how a parasite can work in conjunction with a large predator in the control of a given herbivore population. Parasitic infection breaks out in a dense population. Individuals weakened by infection are readily removed by predators, leaving a smaller but healthier population.

In a food web, a population of any given organism is affected by a number of predators and parasites simultaneously. Consequently, the balance among them can be thought of more broadly as a balance between the population of an organism and its **natural enemies**.

Balances between an organism and several natural enemies are generally much more stable and less prone to wide fluctuations than when only a single natural enemy is involved, because different natural enemies come into play at different population densities. Also, when the preferred prey is at a low density, the population of the natural enemy may be supported by its feeding on something else. Thus, the lag time between an increase in the prey population and that of the natural enemy is diminished. These factors have a great damping effect on the rise and fall of the prey population. The wide swings observed in the populations of moose and wolves on Isle Royale are seen to occur in very simple ecosystems involving relatively few species. (This relationship between simplicity and greater instability has a bearing on sustainability that will be discussed later.)

FIGURE 4-4
Results of rabbits overgrazing the Australian ecosystem. On one side of a rabbit-proof fence, there is lush pasture; on the other side, it is barren. Rabbit-proof fences were built over thousands of miles, but proved unsuccessful in stopping the movement of the animals. (Australian Information Service.)

When Balances Are Absent

Another way of seeing that relationships among species are indeed delicate balances is to observe how vulnerable they are to the introduction of species from a foreign ecosystem. Consider the following examples.

In 1859, rabbits were introduced to Australia from England, to be used for sport shooting. The Australian environment proved favorable to the rabbits, and contained no carnivore or other natural enemies capable of controlling them. The result was that the rabbit population exploded and devastated vast areas of rangeland by overgrazing (Fig. 4-4). The devastation was extremely damaging to both native wildlife and sheep ranching. It was finally brought under control, at least to some extent, by the introduction of a rabbit disease virus that provided a host-parasite balance.

Prior to 1900, the dominant tree in the Eastern deciduous forests of the United States was the American chestnut, which was highly valued for both its high-quality wood and its prolific production of chestnuts, which were eaten by wildlife as well as people. In 1904, however, a fungal disease called the chestnut blight was accidentally introduced when some Chinese chestnut trees that were carriers were planted in New York. The fungus spread through the forests, killing nearly every American chestnut tree by 1950. Although oaks filled in where the chestnuts died, the ecological and commercial loss was incalculable. Much research has been done in an attempt to breed a resistant tree or otherwise control the blight but success has been elusive.

Many of the most severe insect pests in crop lands and forests—Japanese beetles, fire ants, and gypsy moths, for example—are species introduced from other ecosystems. Domestic cats introduced into island ecosystems have often proved to be overly effective predators and have exterminated many species of wildlife unique to the islands. They are also responsible for greatly diminished songbird populations in urban and suburban areas, including parks. Goats introduced onto islands, because of their voracious appetites, have been devastating to both native plants and animals dependent on that vegetation.

Such unfortunate introductions are not all in the past. With expanding world trade and travel, the problem is increasing. In 1986, the zebra mussel was introduced to the Great Lakes with the discharge of ballast water from European ships (Fig. 4-5a). It is now spreading through the Mississippi River basin and may go much further, causing untold ecological and commercial damage as it displaces other species and clogs water intake pipes. (See "Earth Watch" box, p. 476.) Of course, such problems may be exported as well as imported. In 1982, a jellyfishlike animal known as the ctenophore was similarly transported from the east coast of the United States to the Black and Azov Seas in eastern Europe. The ctenophore has cost Black Sea fisheries an estimated $250 million and has totally shut down those in the Azov, for it both kills fish directly and deprives them of food (Fig. 4-5b).

The list of introductions goes on and on, totaling in the tens of thousands. Still, not all of them have turned out badly. The ring-necked pheasant and the day lily, for instance, have found places in new ecosystems, adding apparent charm. However, we may be unaware of native species displaced in the process.

The ecological lesson to be learned from our experience of unfortunate introductions is twofold. First, it should emphasize again that regulation of populations is a matter of delicately balanced interactions among the members of biotic community. Second, and just as important, the balanced relationships are specific to the organisms in each particular ecosystem. Therefore, when we transport a species over a physical

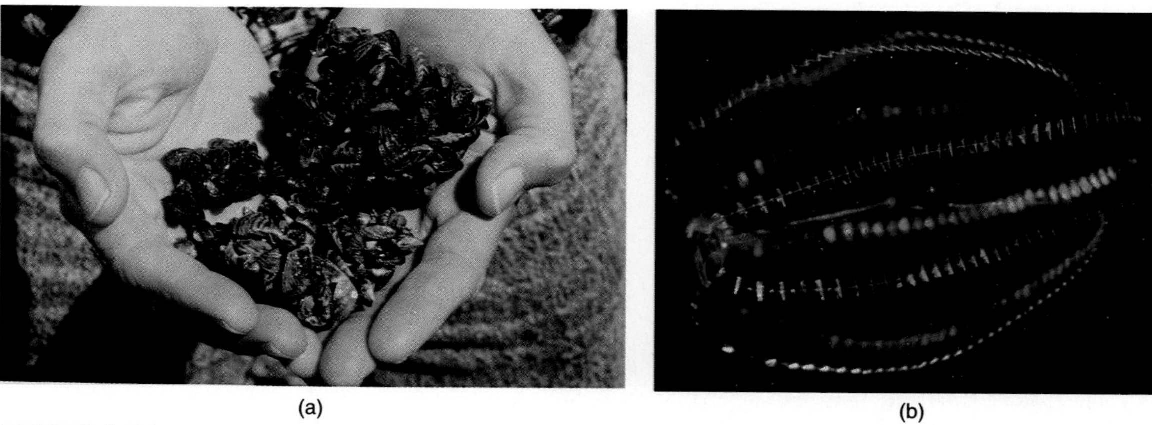

FIGURE 4-5
Species introduced from foreign ecosystems may find the new physical environment favorable and natural enemies lacking. Such introduced species cause great damage to the receiving ecosystem as their populations explode and they outcompete native species for food and living space. (a) Zebra mussels introduced from Europe are now proliferating throughout the Great Lakes and the Mississippi valley. (b) Ctenophores, originating on the South Atlantic Coast of the United States and introduced in Europe, have destroyed fishing in the Black and Azov Seas (Oxford Scientific Films/Peter Parks. Animals Animals/Earth Science.)

barrier from one ecosystem to another, it may find the environment favorable, but it is unlikely to fit into the framework of balances in the new biotic community. Actually, in the best case, it finds the environmental resistance of the new system too severe and dies out. No harm is then done: The receiving system continues unaltered. In the worst case, the transported species finds physical conditions and a food supply that are hospitable, together with an insufficient number of natural enemies to stop its population growth. The damage is done as its population explodes, and it drives out native species by out-competing them for space, food, or other resources, if not by predation. Such upsets may be caused by any category of organism—plant, herbivore, carnivore, or parasite, large or small. (We will give examples of upsets caused by introductions of plants in a later section.)

You may be wondering why the balances in different ecosystems should differ. It is because ecosystems on different continents or remote islands have been isolated by physical barriers for millions of years. Consequently, the species within each system have developed adaptations that provide balance with other species *within their own ecosystem, and these are independent of those that have developed in other ecosystems*. We shall address the question of how species adapt to and develop such balances with each other in Chapter 5.

The seemingly obvious solution to the takeover by a "foreign invader" is to introduce a natural enemy. Indeed, this approach has been used in a number of cases; rabbit control in Australia is one. Others are discussed in connection with the biological control of pests in Chapter 10. However, such control is more easily said than done. Recall that balance is achieved as a result of the interplay among *all* the factors of environmental resistance, which often include several natural enemies, as well as *all* the abiotic factors. Thus, a single natural enemy that will control a pest simply may not exist. There is no guarantee that the natural enemy, when introduced into the new ecosystem, will focus its attention on the target pest. Control of the rabbit population in Australia was initially attempted by introducing foxes. However, the foxes soon learned that they could catch other Australian wildlife more easily than rabbits and thus went their own way. In short, a great amount of research needs to be done before introducing a natural enemy, to prevent doing more harm than good.

Territoriality

In discussing predator-prey balances, we said that in lean times the excess carnivore population—the wolf, for instance—simply starved. Actually, another factor is often involved in the control of carnivore and some herbivore populations of a great number of species ranging from fish to birds and mammals: *territoriality*. **Territoriality** refers to individuals or groups such as a pack of wolves defending a territory against the encroachment of others of the *same species*. For example, the males of many species of songbirds claim a territory at the time of nesting. Their song has the function of warning other males to keep away (Fig. 4-6). The males of many carnivorous mammals, including dogs, "stake out" a territory by spotting it with urine, the smell of which warns others to stay away. If there is encroachment, there may be a fight, but in natural species a large part of the battle is intimidation—an actual fight rarely results in death.

FIGURE 4-6
Territoriality. Breeding members of many species "stake out" a territory so that they will have sufficient food resources to successfully raise a brood. Since members unable to hold a territory do not breed, this trait prevents the population of the species from exceeding a number that the ecosystem can support. For songbirds, such as the Sedge Wren shown here, the song is actually a way of telling competing males, "This is my territory, stay away." (Maslowski Photo/Photo Researchers, Inc.)

In territoriality what is really being defended, or sought after by the "invader," is claim to an area from which adequate food resources can be obtained in order to rear a brood successfully. Hence the territory is only defended against others that would cause a direct competition for those resources. As a consequence of territoriality some members of the population are able to gain access to sufficient food resources to rear a well-fed next generation. Thus, a healthy population of the species survives. If, instead, there were an even rationing of inadequate resources to all the members, all of them trying to raise broods, the entire population would become malnourished and might perish. By territoriality, breeding is restricted to only those individuals that are capable of claiming and defending territory, and thus population growth is curtailed.

Individuals unable to claim a territory are in large part the young of the previous generation(s). Some may hang out on the fringes and seize their opportunity as they become more mature, and older members with territories weaken. Some fall prey to one or another factors of environmental resistance as they are continually chased out of one territory after another. Finally, some may be driven to migrate. Of course, it is an open question whether such migration will lead them to another region where they can successfully breed, or whether it will lead them to perish in conditions beyond their limit of tolerance along the way. In any case, territoriality may be seen as a powerful force behind migration as well as population stabilization.

Plant-Herbivore Balance

In previous discussion we observed that herbivore populations are commonly held in check by various natural enemies. In turn, keeping the herbivore population in check prevents it from growing to the extent that overgrazing occurs. Let us examine this relationship further.

The best way to appreciate the role of natural enemies in preventing a herbivore population from overgrazing is to observe what occurs when natural enemies are *not* present. A classic example with good documentation is the case of reindeer on St. Matthew Island, a 128-square-mile island in the Bering Sea midway between Alaska and Russia. In 1944, a herd of 29 reindeer (5 males and 24 females) was introduced onto the island, where they had no predators. From these 29 animals, the herd multiplied some two-hundred-fold over the next 19 years. Early in the cycle, the animals were observed to be healthy and well nourished, as supporting vegetation was abundant. However, by 1963, when the size of the herd had reached an estimated 6,000, the animals were conspicuously malnourished. Lichens, an important winter food source, had been virtually eliminated and replaced by unpalatable sedges and grasses. During the winter of 1963–64, this factor combined with harsh weather and resulted in death by starvation of nearly the entire herd—there were only 42 surviving animals in 1966 (Fig. 4-7).

The lesson is that no population can escape ultimate limitation by environmental resistance. But the form of environmental resistance and the consequences may differ. If a population is not held in a reasonable balance, it may explode, overgraze, and then crash as a result of starvation. It is also crucial to note that the consequence is not just to the herbivore in question. One or more types of vegetation may be eliminated and replaced by other forms or not replaced at all, leaving a "desert." Other herbivores that were dependent on the original vegetation, and secondary and higher levels of consumers dependent on them, also are eliminated as food chains are severed. Innumerable extinctions have occurred among the unique flora and fauna of islands because goats were introduced by sailors to create a convenient food supply for return trips.

Eliminating predators or other natural enemies upsets basic plant-herbivore balances in the same way as introducing an animal without natural enemies does. Examples of this type of folly abound as well. For example, in much of the United States, deer populations were originally controlled by wolves, mountain lions, and black bear, most of which have been killed because they were felt to be a threat to livestock and even humans. Now, were it not for human hunting in place of these natural predators, deer populations in most areas would increase to the point of overgrazing. Indeed, drastic population increases do occur where hunting is prevented. Similarly, prairie dogs and other small rodents are becoming an increasing problem in the western United States as a result of a reduction in the number of their predators, such as coyotes. In this case, hunting is obviously not practical—who wants to go hunting for mice?

A second factor influencing plant-herbivore balance is that large herbivores—bison in the American West or elephants in Africa, for example—were originally able to roam vast regions. As forage was reduced in one area, a herd would simply move on before overgrazing could occur. Because humans have fenced such regions for agriculture and cattle ranching, however, wild herbivores are increasingly confined to areas such as parks and reserves. Overgrazing in such presumably "protected" areas is become an increasing danger.

Carrying Capacity

The preceding examples lay open another basic concept. There is a definite upper limit to the population of any particular animal that an ecosystem can support without being upset and degraded. This limit is known as the *carrying capacity*. More precisely, **carrying capacity** is defined as the maximum population of an animal that a given habitat will support without the habitat being degraded over the long term. The concept of carrying capacity is being given more and more consideration by wildlife managers. (See "Earth Watch" box, page 84.) Also, many are asking, What is the carrying capacity of the biosphere for humans? Unfortunately, carrying capacity is subject to many variables—especially when the subject is humans—and is not easy to determine. Keep this in mind when we get to Chapters 6 and 7.

To this point, our discussion has focused on large grazing animals, but such animals are really a small fraction of the total biota. It is significant to observe that plants are attacked by an incredible variety of insects and other arthropods; these, in turn, both support and are controlled by a vast variety of arthropod-eating organisms, including carnivorous insects, spiders, birds, amphibians, reptiles, and some mammals, such as bats and tree mice.

Balances among Competing Plants

A natural ecosystem may contain hundreds or even thousands of species of green plants, all competing for nutrients, water, and light. How is the balance among competing plants maintained? What prevents one plant

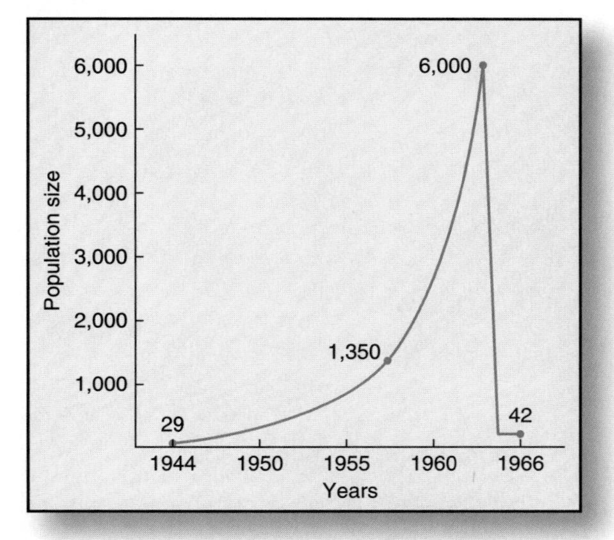

FIGURE 4-7
In a favorable habitat and without natural predators, a population of 29 reindeer introduced onto St. Matthew Island was observed to increase exponentially to about 6,000 and then plummet as the habitat was overgrazed and the animals starved. (Photograph by Fran Allan/Animals Animals/Earth Science.)

FIGURE 4-8
The balance between competing plant communities is often maintained by differing amounts of available moisture. Many regions receive rainfall sufficient to support grasslands, but not forests. However, in such regions trees do grow along streams and rivers where more moisture is available, giving rise to what are called riparian woodlands as shown in this photograph taken in the Tallgrass Prairie Preserve, Oklahoma. (Grant Heilman/Grant Heilman Photography.)

species from driving out others in competition—again, an event that may occur in the case of introduced species?

First, we observed in Chapter 2 that, because of differences in topography, soil type, and so on, the environment is far from uniform. It is actually comprised of numerous microclimates or microhabitats. That is to say, the specific conditions of moisture, temperature, light, and so on differ from location to location. Thus, the adaptation of a species to specific conditions enables it to thrive and overcome its competitors in one location but not in another. An example is the distribution of trees along streams and rivers. In the Great Plains states, trees only grow along waterways, because elsewhere the environment is too dry. This creates what are called **riparian** woodlands (Fig. 4-8). In the eastern United States sycamore and/or cedar, which can thrive in water-saturated soil, grow along river banks. Oaks, which require well-drained soil, occupy higher elevations. In the West, white alder, willow, and cottonwoods can survive in water saturated soils.

A second factor affecting the balance among competing plant species is the fact that a single species generally cannot utilize all of the resources in a given area. Therefore, any resources that remain may be gathered by other species having different adaptations. For example, grasslands contain both grasses, which have a fibrous root system, and plants that have tap roots (Fig. 4-9). These different root systems enable the plants to coexist because they get their water and nutrients from different layers of the soil. Also, trees in a forest, while competing with each other for light in the canopy (the "layer" of treetops), leave lots of space near the ground, and this space may be occupied by plants (ferns and mosses, for example) that can tolerate the reduced light intensity. Another adaptation is the plethora of spring wild flowers that inhabit temperate deciduous forests. Sprouting from perennial roots or bulbs in the early part of that season, these plants take advantage of the light that can reach the forest floor before the trees grow leaves. In warm, humid climates, the branches of trees are often covered with **epiphytes**, or air plants (Fig. 4-10). Such plants are not parasitic; indeed there may be mutualistic symbiosis involved. There is some evidence that the epiphytes help to gather the minute amounts of nutrients that come with rainfall (e.g., nitrogen compounds fixed by lightning, page 69) and make them accessible to the tree on which the epiphytes are located.

A third and very important factor in multiple-plant balance is called *balanced herbivory*. It is easiest to understand if we start from the point of view of a

FIGURE 4-9
Plants having fibrous roots may coexist with plants having tap roots because each is drawing water and nutrients from a different part of the soil. (Photograph by BJN.)

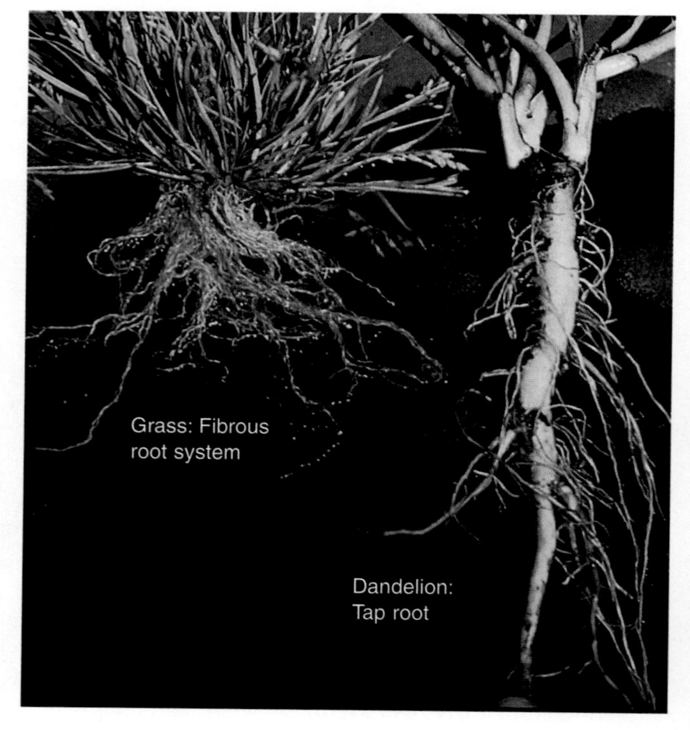

FIGURE 4-10
Epiphytes in the Amazon region. The plants (bromeliads) growing on the tree branches are epiphytes. They are not parasitic but perch on the branches of trees to gain access to light. (Photograph by BJN.)

monoculture—the growth of a single species over a wide area, a practice commonly followed in agriculture and forestry for economic efficiency.

Experience shows that monocultures are exceedingly vulnerable to insects, fungal diseases, or other pests, while diverse ecosystems are much more resistant to them. To understand why this should be, consider the following. First, insects, fungal diseases, and other parasites are, for the most part, **host specific**. That is, they will attack only one species and, possibly, its close relatives. They are unable to attack species unrelated to their specific host. Second, such organisms have an enormous biotic potential. An individual often produces thousands of offspring—even millions in the case of fungal spores—and they have a generation time of only a few days or weeks.

Now, a monoculture may be seen as a continuous, lush food supply for its particular host-specific attacker, a situation highly conducive to supporting a population explosion. Indeed, the pest population may explode so fast that its natural enemies cannot keep up with it even if they are present. Only virtual elimination of the monoculture halts the multiplication of the pest—a scenario not unlike that of the reindeer described earlier. (It is for this reason that many farmers and forest managers feel obliged to use chemical pest controls despite recognizing that they may present certain environmental hazards. See Chapter 10.)

On the other hand, in a diverse ecosystem—one consisting of a mixture of many different species of plants growing together—the host-specific attacker has trouble reaching its next host. With this limitation, most of the pest's offspring may perish, and the surviving population may be held in check by its natural enemies.

To conclude, visualize a monoculture developing in a natural situation. Its being largely wiped out by an outbreak of its host-specific pest would leave space that might be invaded by another plant species, which in turn might be largely wiped out by an outbreak of *its* pest, leaving space that might be occupied by a third plant species, and so on. The end result of this process would be a diversified plant community, with each species held to a low density by its specific herbivore(s) and the herbivores held in check by their natural enemies (Fig. 4-11a). Thus, a **balanced herbivory** may be defined as a balance among competing plant populations being maintained by herbivores feeding on the respective populations.

A prime example of a balanced herbivory is seen in the tropical rain forests of the Amazon River basin in Brazil and Peru (Fig. 4-11b). A single acre may contain a hundred or more species of trees, but often no more than a single individual of each. The next individual of the same species may be as much as a half mile away. Evidence that this diversity is maintained by a balanced herbivory is seen in that attempts to create plantations of single species—rubber plantations, for example—met with failure because outbreaks of various pests proved uncontrollable in the monoculture situation. (However, rubber plantations proved successful in Malaysia, where climatic and soil conditions are different enough to limit pests while still supporting the rubber trees.)

Again, the balances among competing plant populations, and the ecosystem's vulnerability to upset, are demonstrated by the havoc raised when plant species are introduced from foreign ecosystems. Examples abound.

In 1884, the water hyacinth, a plant originally from South and Central America, was introduced into Florida as an ornamental flower. It soon escaped into waterways, where it had little competition and few natural enemies. By now, it has proliferated to the extent of making navigation difficult or impossible on the waters in which it thrives (Fig. 4-12a). Millions of dollars have been spent attempting to get rid of this weed, but with little success.

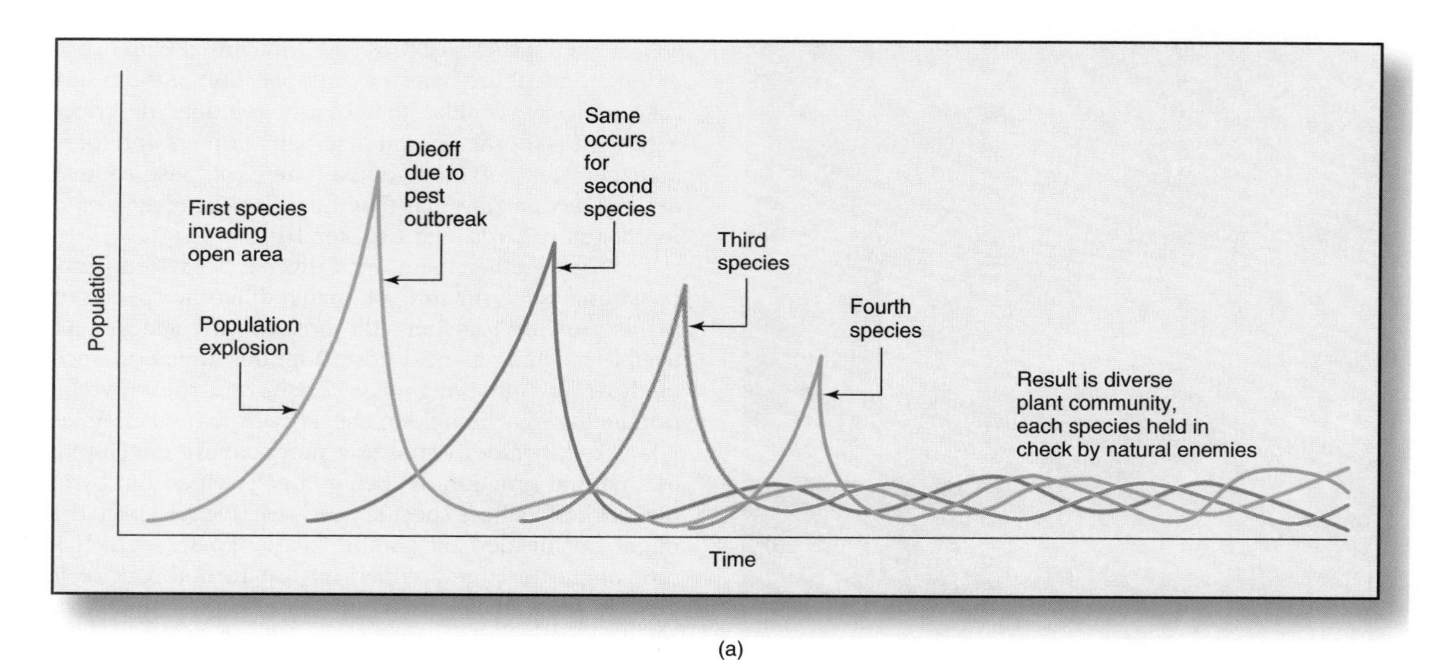

(a)

(b)

FIGURE 4-11
(a) Balanced herbivory. A plant species may experience a population explosion as it invades an open area. If it dies back and is held down by an herbivore, space is opened up for a second invader, which may experience the same fate. Repeating this process for additional species results in a plant community of many species, each held in check by its specific herbivores a patchwork mosaic operating in constant adaptation. (b) View from canopy walkway overlooking the Amazonian rain forest near Iquitos, Peru. Nearly every tree in view is a different species. (Photograph by BJN.)

Kudzu, a vigorous vine introduced from Japan in 1876, was widely planted on farms throughout the southeastern United States, with the idea of using it for cattle fodder and also for erosion control. From wherever it is planted, however, kudzu invades and climbs over adjacent forests, smothering everything (Fig. 4-12b). Considerable efforts are now being exerted by the Forest Service to eradicate it. People have been heard to say "kudzu aint't good for nuthin!"

Spotted knapweed (Fig. 4-12c), which was probably unwittingly introduced into this country with alfalfa seed imported from Europe in the 1920s, is taking over vast areas of range lands in the northwestern United States and southwestern Canada. Totally inedible, knapweed is endangering wildlife such as elk and rendering the lands worthless for grazing by domestic cattle as it displaces native plants and grasses.

Wetlands throughout temperate regions of the United States and Canada, already reduced by over 50 percent by development pressures, are now being further degraded by the invasion of purple loosestrife, which was introduced from Europe in the 1800s as an ornamental and medicinal plant (Fig. 4-12d). By outcompeting and eliminating native wetland vegetation and being inedible itself, purple loosestrife is threatening many wildlife species, including waterfowl. Control of such weeds by the introduction of plant-eating insects is being investigated. However, it is hard to find an insect (or other organism) that will control the target weed, but not attack desired species. (See Chapter 10.)

The balance among competing plant species may also be upset by the introduction of an herbivore,

FIGURE 4-12
Introduced species that find a favorable environment without natural enemies may overgrow and crowd out natural species. (a) Water hyacinth overgrowing waterways. (John Pontier/Animals Animals/Earth Science) (b) Kudzu overgrowing forests. (Leonard Lee Rue III/Photo Researchers, Inc.) (c) Spotted knapweed, unwittingly introduced from Europe, is taking over range land in the northwestern United States and southwestern Canada, displacing native plants and diminishing the grazing of both cattle and wildlife. (Robert Bornemann/Photo Researchers, Inc.) (d) Purple loosestrife, a native of Eurasia, introduced into North America 200 years ago. This is a specific problem in the northeastern United States and Canada. (John Mitchell/Photo Researchers, Inc.)

which invariably attacks some plant species but not others. For example, millions of acres of pasture and range lands are now dominated by inedible weeds and shrubs as cattle have eaten back the grass and allowed the competing weeds to flourish (Fig. 4-13).

Population Growth Curves

When the size of a population is plotted over time, two basic kinds of curve can be seen: *S*-curves and *J*-curves (Fig. 4-14). For example, suppose some abnormally severe years have reduced a population to a low level, but then conditions return to normal. Once normal conditions return, the population may increase exponentially for a time, but then either of two things may occur. One is that natural enemies come into play and cause the population to level off and continue in a dynamic balance with those enemies. This pattern is known as the *S*-curve (the green-and-yellow curve in Fig. 4-14). In the absence of natural enemies, however, the population keeps growing exponentially until it exhausts essential resources—usually food—and then there is a precipitous dying off caused by famine and, perhaps, diseases related to malnutrition. This pattern is known as a *J*-curve (the blue-and-purple curve in Fig. 4-14). The case of the reindeer on St. Matthew Island is an example of the latter pattern.

An important characteristic of *J*-curve growth is the rapidity with which a population can go from modest levels to the peak and then crash. Consider, for example, that an insect population is doubling (hence doubling the amount it eats) *each week*. Suppose it has taken a given insect population eight weeks to devour one-half of a crop. How long will it take to devour the second half? The answer is *1 week*! In any doubling sequence, the last doubling necessarily includes one-half of the total. You may test this for yourself by taking a sheet of

FIGURE 4-13
The range on the right side of the fence has been lightly grazed; the range on the left side has been heavily grazed. Grazing tips the balance among grasses and other plant species such that plants resistant to grazing gain dominance. (United States Department of Agriculture.)

graph paper and blackening first one square, then two, then four, then eight, and so on, doubling the number each time. You will find that, regardless of the size of the paper or the size of the squares, the last turn will involve blackening half the paper. If you double the size of the paper, how many more turns will that provide? Imagine in human systems, if a specific carrying capacity were exceeded, what such a crash would be like.

What follows the *J*-curve crash? Any one of three scenarios may unfold. First, if the ecosystem has not been too seriously damaged, the producers may recover, allowing a recovery of the herbivore population, and the *J*-curve may be repeated. This scenario is seen in periodic outbreaks of certain pest insects, even in natural ecosystems. Second, after the initial *J*, natural enemies may come into the picture as the ecosystem recovers and bring the population into an *S*-balance. There is evidence that such a balance is being established in the eastern United States as regards the introduced gypsy moth. Stands of oak trees that were devastated by the initial invasion of gypsy moths a few years ago are now recovering, and the gypsy moth is remaining at low levels. In the third scenario, damage to the ecosystem may be so severe that recovery does not occur, but small surviving populations eke out an existence in a badly degraded environment.

The outstanding feature of natural ecosystems—ecosystems that are more or less undisturbed by human activities—is that they are made up of populations that are in the dynamic balances represented by *S*-curves. *J*-curves come about when there are unusual disturbances, such as the introduction of a foreign species, the elimination of a predator, or the sudden alteration of a habitat. Nowadays such disturbances are increasingly caused by humans. In short, humans are progres-

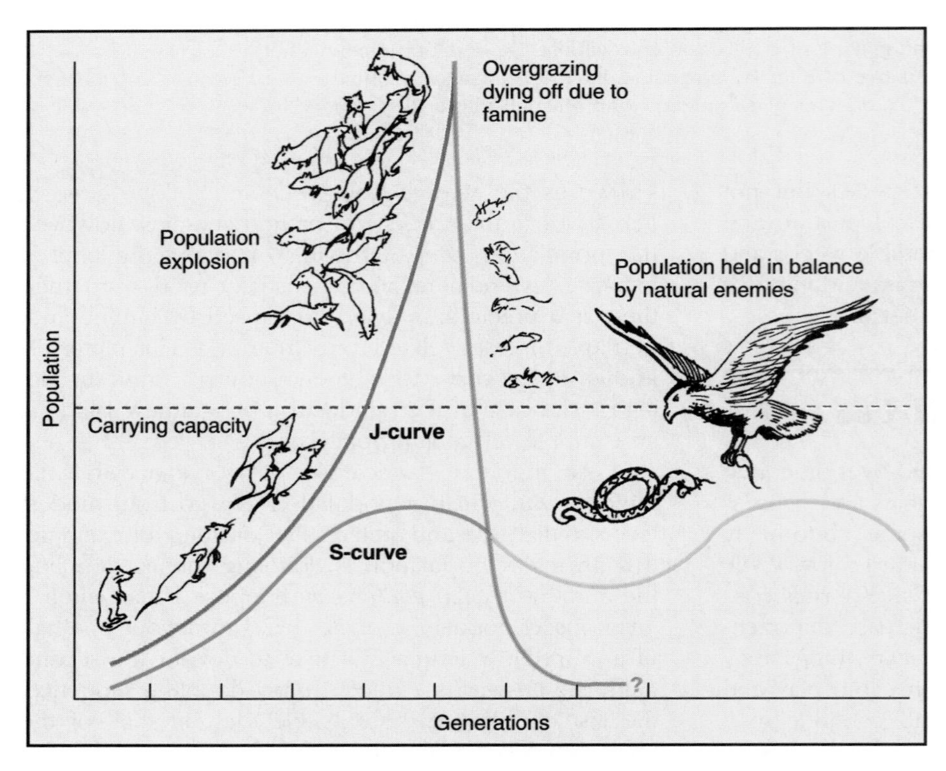

FIGURE 4-14
Two types of growth curves, the *S*-curve (green and yellow) and the *J*-curve (blue and purple). Given optimal conditions, the biotic potential of any species will result in a rapid exponential increase. This initial increase is typical of both curves, but then the curves diverge. In *S*-curve growth, with effective natural enemies, the population is brought into balance within the carrying capacity of the system. The dynamic population balance may continue indefinitely (yellow). In *J*-curve growth, in the absence of natural enemies, the exponential growth continues until overgrazing results in a precipitous dying off of the population due to famine and malnutrition (purple).

sively upsetting the natural population balances that have existed for millennia and are causing *J*-curves. The consequences of these changes are still unfolding.

Again, we should visualize the dynamics behind the third principle of ecosystem sustainability: Populations are maintained such that overgrazing and other forms of overuse do not occur. The balances inherent in *S*-curves are generally provided by herbivore populations being controlled by natural enemies. On the other hand, a *J*-curve is a picture of the non-sustainability of a population growing without control. The continuing exponential increase of one species implies the displacement of other species in the ecosystem. The exponential increase cannot be sustained in animal populations, there is an inevitable dying off as resources are exhausted.

Ecological Succession

Given the prevalence of human impacts on the environment, you may be led to think that all ecological changes are caused by humans. However, there are situations in nature where, over the course of years, we observe one biotic community gradually giving way to a second, the second perhaps to a third, and even the third to a fourth. This phenomenon of *orderly transition* from one biotic community to another is called **ecological**, or **natural**, **succession**. Natural succession occurs because the physical environment may be gradually modified by the growth of the biotic community itself, such that the area becomes more favorable to another group of species and less favorable to the present occupants.

The succession of species does not go on indefinitely, however. A stage of equilibrium is reached during which there is a dynamic balance between all species and the physical environment. This final state is referred to as a **climax ecosystem**, or assemblage of species in constant adaptation, each striving to achieve an optimal range and low environmental stress. The major biomes and natural ecosystems discussed up to this point in the text are climax ecosystems. Of course, it remains important to keep in mind that all balances are relative to the *current* biotic community and the *existing* climatic conditions. Therefore, even climax systems are subject to change if climatic conditions change, if new species are introduced, or old ones are removed. Nevertheless, natural succession may be seen as a progression toward a relatively more stable climax situation. Note, there may be several "final" stages, or a polyclimax condition, with adjoining ecosystems in the same environment at different stages. The following are three classic examples.

Primary Succession

If the area has not been occupied previously, the process of initial invasion and then progression from one biotic community to the next is termed **primary succession**. An example is the gradual invasion of a bare rock surface by what eventually becomes a climax forest ecosystem. Bare rock is an inhospitable environment. There are few places for seeds to lodge and germinate, and if they do, the seedlings are killed by lack of water, or by exposure to wind and sun on the rock surface. However, certain species of moss are uniquely adapted to this environment. Their tiny spores, specialized cells that function reproductively, can lodge and germinate in minute cracks, and moss can withstand severe drying simply by becoming dormant. With each bit of moisture, it grows and gradually forms a mat that acts as a sieve, catching and holding soil particles as they are broken from the rock or as they blow or wash by. Thus, a layer of soil, held in place by the moss, gradually accumulates (Fig. 4-15). The mat of moss and

FIGURE 4-15
Primary succession on bare rock. Moss invades bare rock and acts as a collector, accumulating a layer of soil sufficient for additional plants to become established. (Photograph by BJN.)

Global Perspective

Protecting the World's Biodiversity

Safeguarding the world's diversity of species is integral to a sustainable future. Yet, as humans dominate and alter increasing portions of the planet, increasing numbers of species are threatened with extinction for various reasons, one of which is simple short-term monetary gain. Illegal trade in wildlife is an estimated $2–3 billion a year business, with profit margins comparable to the drug trade. Highlighted on the map are 10 representative animal and plant species particularly threatened by illegal trade, and in some cases legal international trade that is excessive and in great need of further protective regulation.

The World Wildlife Fund and Conservation International are two nongovernmental organizations (NGO's) dedicated to conservation on a worldwide basis (see Appendix A). Their efforts involve both field and policy work. They work with governments and local NGOs to help with programs aimed at creating conservation areas and species protection. At the same time, they address the economic needs of people living in and around protected areas, by stepping up anti-poaching activities, initiating educational programs to engender greater appreciation on the part of local people for their wildlife, and providing training for local field staff, park guards, and villagers. The protection of endangered species and ecosystems is explored further in Chapter 18.

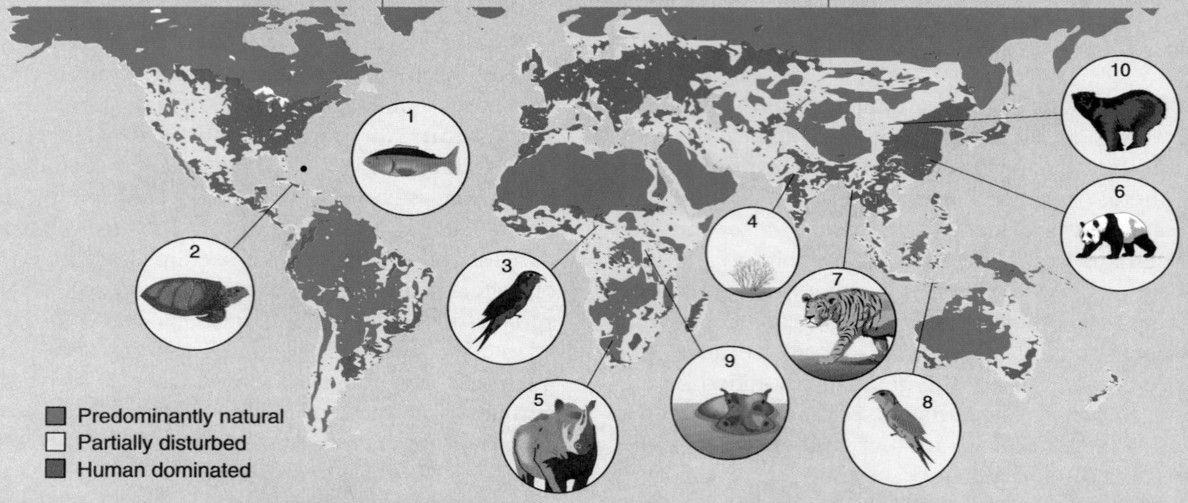

1. Atlantic Bluefin Tuna. Largest fish in the Atlantic, migrates thousands of miles each year, reaching speeds up to 55 miles per hour. Its population has declined 80% over the last 20 years because of overfishing to supply the international luxury-food market.
2. Hawksbill Sea Turtle. Found around tropical reefs, these three-foot long turtles are the principal source of what is called "tortoise shell." While officially protected, poaching and illegal trade threatens them with extinction.
3. African Gray Parrot. Capture for the pet trade has reduced the status of this bird, prized for its ability to mimic human speech, from common to threatened over much of its range in central and west Africa.
4. Himalayan Yew Tree. The species is threatened by general deforestation and collection of its leaves and bark for the anti-cancer drug, Taxol.
5. Black Rhino. Populations have dropped to fewer than 8000—a decline of more than 80%—since 1979. Now protected, they are still illegally slaughtered for the value of their horns; which are prized material for dagger handles in Yemen and believed to have medicinal value by some Asians although no such value has been documented. White Rhinos in their subSaharan, northern range, number only two dozen from an original population of thousands.
6. Giant Panda. Now reduced to fewer than 1000 animals, and not successfully bred in captivity, the giant panda is still hunted illegally.
7. Tiger. With a range once extending from India and southeast Asia to Siberia, the total population of this great cat is now estimated to be less than 6000. Its greatest threat is illegal hunting for the worldwide black-market trade. Recently, poaching for their bones and other body parts, used in Oriental medicines, has accelerated. Again, no medicinal value has been documented.
8. Red and Blue Lory. Existing only in a small range covering a few Indonesian islands, collectors have wiped out an estimated one-third of the population original to supply a sudden fad in the pet trade. Fewer than 2000 remain.
9. Hippopotamus. While not yet considered endangered, growing use of hippo ivory (in the form of its teeth) as a substitute for elephant ivory is decimating populations throughout its range in sub-Saharan Africa.
10. Asiatic Black Bear. Illegal killing to obtain their gall bladders for use in Oriental medicines imperils this and a number of other bear species throughout the world. Again, no medicinal value has been documented.

Sources: Human distribution map. Conservation International, *Annual Report*, 1992.Threatened Species. World Wildlife Fund, *Focus*, (Nov/Dec, 1994).

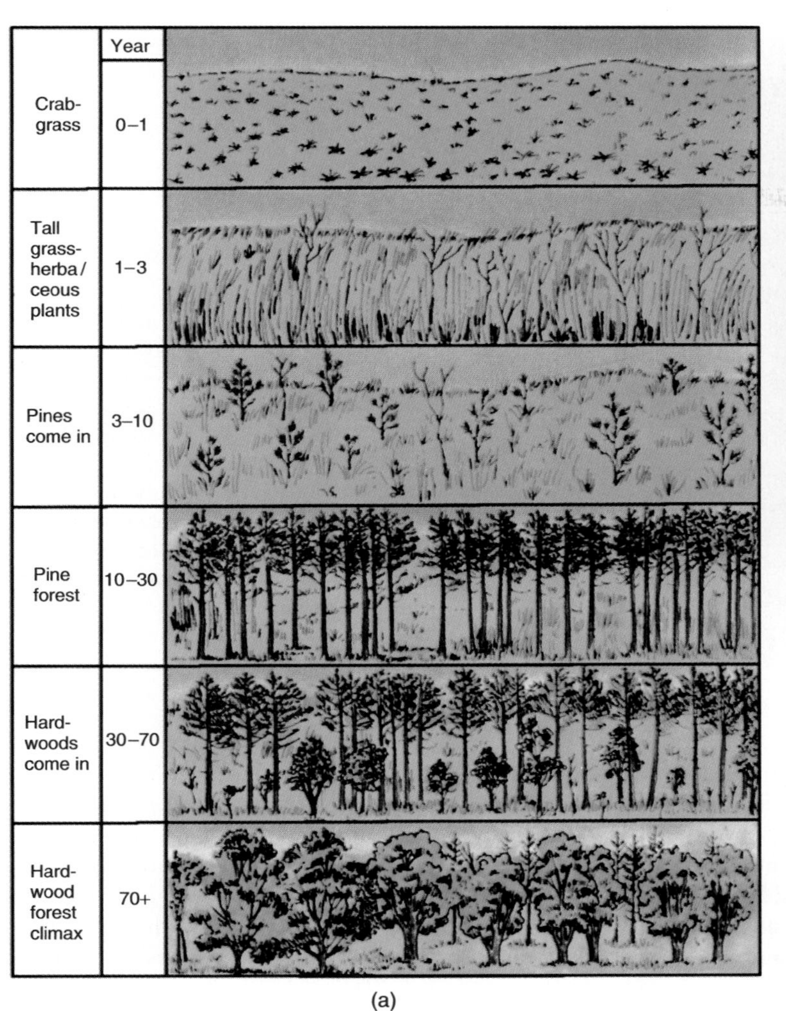

(a)

(b)

FIGURE 4-16
Secondary succession. (a) Reinvasion of an agricultural field by a forest ecosystem occurs in the stages shown. (b) Hardwoods (species of oak) growing up underneath and displacing pines in eastern Maryland. (Photograph by BJN.)

soil provides a suitable place for seeds of larger plants to lodge, and the greater amount of water held by the mat supports their germination and growth. The larger plants in turn collect and build additional soil, and eventually there is enough soil to support shrubs and trees. In the process, the fallen leaves and other litter from the larger plants smother and eliminate the moss and most of the smaller plants that initiated the process. Thus, there is a gradual succession from moss through small plants and finally to trees that form a climax forest ecosystem. The nature of the climax ecosystem, of course, differs according to the prevailing abiotic factors of the region, thus giving us the biomes typical of different climatic regions, as described in Chapter 2.

Because bare substrate can be reexposed by erosion, earthquakes, landslides, and volcanic eruptions, there are always places for primary succession to start anew.

Secondary Succession

When an area has been cleared by fire or artificial means and then left alone, the surrounding ecosystem may gradually reinvade the area, not at once, but through a series of distinct stages termed **secondary succession**. The major difference between primary and secondary succession is that secondary succession starts with the preexisting soil substrate. Therefore, the early, prolonged stages of soil building are bypassed. Still, as you can readily experience, a clear area has a microclimate quite the opposite from the cool, moist, shaded conditions beneath a forest canopy. Those plant species that propagate themselves in the microclimate of the forest floor cannot tolerate the harsh conditions of the clearing. Hence, the process of reinvasion necessarily begins with different species and proceeds accordingly. The steps leading from abandoned agricultural fields in the eastern United States back to deciduous forests provide a classic example of secondary succession (Fig. 4-16a).

On an abandoned agricultural field, crabgrass is predominant among the initial invaders. Crabgrass is particularly well adapted to invading bare soil. Its seeds germinate in the spring, and it grows and spreads rapidly by means of runners; moreover, it is exceptionally resistant to drought. In spite of its

FIGURE 4-17
Ravages of erosion in Nepal. Erosion washes away topsoil with seeds and seedlings before they can gain a foothold and leaves an inhospitable subsoil surface that fails to support any new growth. Therefore, severe erosion such as that seen here prevents secondary succession from getting started and may continue indefinitely. (George Turner/Photo Researchers.)

vigor on bare soil, crabgrass is easily shaded out by taller plants. Consequently, taller weeds and grasses, which take a year or more to develop, eventually take over from the crabgrass. Next, young pine trees, which are well adapted to thrive in the direct sunlight and heat of open fields, gradually develop and shade out the smaller, sun-loving weeds and grasses, eventually forming a pine forest. But pine trees also shade out their own seedlings, which need bright sun to grow. Thus, the seedlings of deciduous trees, not pines, develop in the cool shade beneath the pine trees (Fig. 4-16b). Consequently, as the pines die off (their life span is 40 to 100 years), they are replaced by oaks, hickories, beeches, maples, and other species of hardwoods that characterize Eastern deciduous forests. The seedlings of the latter continue to flourish beneath the cover of their parents, providing a stable balance—the climax deciduous forest ecosystem.

We emphasize again that secondary succession requires a suitable soil base to start. If this base is lost by erosion, even the initial plants cannot get a start. Consequently, the bare subsoil may continue to erode and degrade indefinitely, preventing even the possibility of reforestation (Fig. 4-17).

Aquatic Succession

Another example of natural succession is seen as lakes or ponds are gradually filled and taken over by the surrounding terrestrial ecosystem. This process occurs because a certain quantity of soil particles is inevitably eroded from the land and settles out in ponds or lakes, gradually filling them. Aquatic vegetation produces detritus that also contributes to the filling process. As the buildup occurs, terrestrial species can advance, and aquatic species move farther out into the lake. In short, the shoreline gradually advances toward the center of the lake until, finally, the lake disappears altogether (Fig. 4-18).

Succession and Biodiversity

In order for natural succession to occur, the spores and seeds of the various invading plants and the breeding populations of the various invading animals must already be present in the vicinity. Ecological succession is not a matter of new species developing, or even old species adapting, to new conditions. It is a matter of populations of existing species taking advantage of a new area as conditions become favorable. If certain species have been eliminated, natural succession will be blocked or modified. For example, beginning with the early colonization by Norsemen in the 11th and 12th centuries, the forests of Iceland were cut for fuel, a process that accelerated with further European colonization in the 18th and 19th centuries. By 1850, not a tree was left standing, and Iceland remained a barren, tundralike habitat because there was no remaining source of seeds to foster natural regeneration (Fig. 4-19). Tree seedlings are now being imported and planted in Iceland.

In short, natural succession depends on maintaining the biodiversity of the surrounding area.

(a)

FIGURE 4-18
Aquatic succession. (a) Ponds and lakes are gradually filled and invaded by the surrounding land ecosystem. (b) In this photograph, taken in Banff National Park in the Canadian Rockies, you can visualize the lake that used to exist in the low-level area. It is now filled with sediment and covered by scrub willow. Spruce and fir forest is gradually encroaching. (Photograph by BJN.)

(b)

FIGURE 4-19
Iceland. Forests that originally covered much of this island nation were totally stripped for fuel in the 18th and 19th centuries. Forests could not regenerate through succession because no source of seeds remained. Therefore, Iceland has remained barren and tundralike, as seen here. Efforts toward reforestation are underway. (George Holton/Photo Researchers, Inc.)

Fire and Succession

Fire is an abiotic factor that has particular relevance to succession. About 75 years ago, forest managers interpreted the potential destructiveness of fire to mean that all fire is bad and embarked on fire prevention programs that eliminated fires from many areas. Unexpectedly, fire prevention did not preserve all ecosystems in their existing state. In pine forests of the southeastern United States, for instance, economically worthless scrub oaks and other broadleaf species began to displace the more valuable pines. Grasslands were gradually taken over by scrubby, woody species that hindered grazing. Pine forests of the western United States that were once clear and open became cluttered with the trunks and branches of trees that had died in the normal aging process. This deadwood became the breeding ground for wood-boring insects that proceeded to attack live trees. In California, the regeneration of redwood seedlings began to be blocked by the proliferation of broadleaf species.

Scientists now recognize that fire, which is often started by lightning, is a natural abiotic factor. As with all abiotic factors, different species have different degrees of tolerance to fire. In particular, grasses and pines have their growing buds located deep among the leaves or needles, where they are protected from fire, whereas broadleaf species such as oaks have their buds exposed, where they are sensitive to damage from fire. Consequently, where these species exist in competition, periodic fires are instrumental in maintaining a balance in favor of pines, grasses, or redwood trees. In relatively dry ecosystems, where natural decomposition is slow, fire may also play a role in releasing nutrients from dead organic matter. Some plant species actually depend on fire. The cones of lodgepole pine, for example, will not release their seeds until they have been scorched by fire.

Ecosystems that depend on the recurrence of fire to maintain the existing balance are now referred to as **fire climax ecosystems**. This category includes various grasslands and pine forests. Fire is now increasingly used as a tool in the management of such ecosystems. In pine forests, if ground fires occur every few years, there is relatively little accumulation of deadwood. With only small amounts of fuel, fires usually just burn along the ground harming neither pines nor wildlife significantly (Fig. 4-20). In forests where fire has

FIGURE 4-20
Ground fire. Far from being harmful, periodic ground fires are necessary to preserve the balance of pine forests. Such fires remove excessive fuel and kill competing species. (USDA–Soil Conservation Service.)

FIGURE 4-21
Fire is a natural abiotic factor, and many species are well adapted to it. This photograph was taken one year after the 1988 fire in Yellowstone National Park. Note the prolific regeneration of healthy vegetation. Other species will come back in the process of ecological succession. Periodic burning of sections of pine forests actually results in a more diverse and healthy ecosystem. (Renee Lynn/Photo Researchers.)

not occurred for more than 60 years, however, so much deadwood has accumulated that, if a fire does break out, it will almost certainly become a crown fire. That is, so much heat is generated that entire living trees are ignited and destroyed. This long-term lack of fire was a major factor in the fires in Yellowstone National Park in the summer of 1988.

Crown fires do occur in nature since not every area is burned on a regular basis, and exceedingly dry conditions can make a forest vulnerable to crown fires even when large amounts of deadwood are not present. Thus, humans have only promoted the potential for crown fires by fire prevention programs. Even crown fires, however, serve to clear the deadwood and sickly trees that provide a breeding ground for pests, to release nutrients, and to provide for a fresh ecological start. Burned areas soon become productive meadows as secondary succession starts anew. We have seen this in Yellowstone National Park. Thus, periodic crown fires, by creating a patchwork of meadows and forests at different stages of succession, lead to a more varied, healthy habitat supporting a greater diversity of wildlife than does a uniform, aging conifer forest (Fig. 4-21).

The devastating fire that spread from brush land in Oakland, California, and destroyed some 1,000 homes in the fall of 1991 is another example of lack of fire management. The grasslands on the hills east of San Francisco are a fire climax ecosystem. Preventing fire in this area for many years allowed a superabundance of woody shrubs to grow, not the least of which was eucalyptus that had been introduced, which has a particularly flammable wood. Homeowners created additional fuel by planting a lush, forested landscape, unnatural for this dry-summer climatic regime. The result, finally, was a fire that could not be controlled (Fig. 4-22).

In summary, the concept that is most important to recognize is that the sustainability of ecosystems is not only dependent upon maintaining existing balances among populations of all the species in the biotic community; it is also dependent upon maintaining existing relationships between the biotic community and abiotic factors of the environment. We have seen that the natural biotic community itself may induce changes in abiotic factors that, in turn, result in changes in the biotic

FIGURE 4-22
The Oakland, California, fire in 1991 destroyed about 1,000 homes. This tragedy could have been prevented with proper fire management of surrounding natural areas. (Michael A. Jones/Sygma.)

FIGURE 4-23
Increasing levels of carbon dioxide in the atmosphere will shift balances between competing plant species, because some are stimulated by increased levels of CO_2 and others are not. Through experiments shown here that artificially alter carbon dioxide levels, scientists are attempting to determine the kinds of effects that may be anticipated. (Keith Lewin/Brookhaven Lab.)

community (succession). Given this dynamic succession it should be clear that if any one or more physical factors of the environment are shifted, the biotic community may again be pushed into a state of flux in which certain species that are stressed by the new conditions die out and other species that are better suited to the new conditions thrive and become more abundant.

The sobering reality is that humans are unwittingly embarked on changing basic abiotic parameters on a global scale with no knowledge as to what the effects may be. The foremost example of this is our elevating the atmospheric content of carbon dioxide. Most of the discussion pertaining to elevated levels of carbon dioxide centers on the fact that the additional carbon dioxide is likely to lead to global warming, which may upset all ecosystems on Earth, including our human system, for agriculture is affected also. (See Chapter 16.) However, plant scientists point out that, since carbon dioxide is a basic nutrient for photosynthesis, elevated carbon dioxide levels will have an impact quite apart from any change in temperature, because some species of plants are stimulated by higher levels and other species are not. Scientists at Duke University in North Carolina and elsewhere are currently conducting experiments to gain insight into how the higher levels expected will affect an entire ecosystem (Fig. 4-23). For other species, the increasing temperatures associated with global warming are causing a *forced succession* as habitats change.

The Fourth Principle of Ecosystem Sustainability

Throughout this chapter you may have noted a great importance on maintaining a diversity of species; that is, biodiversity. To reiterate some of the major points: The most stable population balances are achieved by a diversity of natural enemies. The most stable ecosystems are those with a high degree of biodiversity. Simple systems, especially monocultures, are inherently unstable. Most or all succession depends on a preservation of biodiversity, and succession underlies the ability of an ecosystem to recover from damage (e.g., fire, the loss of the American chestnut due to blight) and its ability to accommodate to changing conditions. Thus we begin to see another basic principle of sustainability, the fourth. Very simply, the **fourth principle of ecosystem sustainability** may be stated:

For sustainability, biodiversity is maintained.

In Chapter 5 we will find that biodiversity is also required in order to sustain a much longer-term, evolutionary adaptation of species to changing conditions that are inevitable, even without human interference.

Our four principles of ecosystem sustainability are stated in Table 4-1 for review purposes.

TABLE 4-1
Principles of Ecosystem Sustainability

For sustainability
• **Ecosystems dispose of wastes and replenish nutrients by recycling all elements.**
• **Ecosystems use sunlight as their source of energy.**
• **The size of consumer populations is maintained such that overgrazing and other forms of overuse do not occur.**
• **Biodiversity is maintained.**

Implications for Humans

The central lesson of this chapter is the understanding that all species on Earth are sustainable only insofar as balances are maintained among members of the biotic community and between the biotic community and the abiotic factors of the environment. From the examples discussed, you can create a list for yourself summarizing how ecosystems have been upset through altering a biotic factor—that is, introducing or eliminating one or another species—or altering an abiotic factor.

How can we use the information presented to create a sustainable future? The answer to this question has two aspects. One is protecting or managing the natural environment to maintain the beauty, interest, biodiversity, and other intrinsic values of the natural world. The second aspect is establishing a balance between our own species and the rest of the biosphere. The two aspects are interrelated.

Focusing on the first aspect, the basic rule for anyone interested in wildlife management and ecosystem protection is to think in terms of balances that must be maintained, or reestablished in cases where they have been upset. In ecosystems that have not been upset, simple protection from adverse human impacts and avoidance of the introduction of foreign species may be adequate for protection. However, if an ecosystem has been upset, restorative measures may involve a number of options, depending on what the upset is. Observe that restoring balances may involve anything from the painstaking reintroduction of a predator or parasite to the hunting of a species that is multiplying beyond the carrying capacity of the system. Observe that simplistic values such as "hunting is right" or "hunting is wrong" will not suffice, because what is necessary to reestablish a balance will differ in different situations.

Likewise, preventing alteration of the environment generally is necessary for the protection of ecosystems, but in some cases alterations may be done to encourage certain kinds of wildlife. This is called **ecological restoration**: How do you restore some semblance of a natural ecosystem after an area has been close to or totally altered by such activities as development, agriculture, or mining? Restoration begins with the creation of the desired physical environment and introduction of appropriate plants to support the desired animals. An example is the creation of ponds and wetlands to encourage waterfowl.

In the long run, any efforts to protect natural ecosystems will be overwhelmed and thwarted by growing pressures from the human system if current trends of population growth and exploitation of natural ecosystems continue.

You may observe that a graph of the human population over the last 200 years (see page 10) has a distinct similarity to the upsweeping portion of the *J*-curve for the reindeer population on St. Matthew Island. We can also draw parallels in that humans have overcome, for the most part, those factors that generally keep natural populations in balance—disease, parasites, and predators. Therefore, we may observe that the human population is behaving much like any population in the absence of natural enemies.

But what lies ahead? If we were being viewed by an alien from afar who made the assumption that we were just "dumb" animals like the reindeer, the alien could predict the peak and crash portions of the *J*-curve with total confidence. After all, the alien would see that we are "overgrazing" our environment as we cut forests, overfish the seas, deplete water resources, and cause the extinction of ever more species even as our population continues to grow. It is only a matter of time, the alien would reason, before essential resources are exhausted and the human population crashes.

The glaring fallacy in the alien's scenario, of course, is that we are not dumb animals. We have options not available to the reindeer. But this, by itself, does not rule out the scenario: Smart people are still capable of making dumb decisions and doing dumb things, particularly en masse. We need to be not only smart enough to *recognize* the critical need to establish a balance between our human system and the rest of the biosphere, but also wise enough to *do so*.

What is the urgency? In discussing the reindeer population on St. Matthew Island, we noted the rapidity with which an exponentially growing population can deplete its resource base. Half of the original resource base is consumed in the last doubling. This is to say that half of the original resource base is needed just to accomplish the last doubling, and then the crash will occur anyhow because no resources remain to sustain the high population, much less to allow it to grow further (Fig. 4-24).

Now, the human population doubled from 2.5 billion in 1950 to 5 billion in 1987 (37 years). Reaching 6.0 billion in 1998 and still growing at the rate of 1.52%, nearly 88 million persons per year, it is expected to reach 10 billion in the 2040s. But in many populous regions of the world, forests, water, fish, fertile soil, and other resources have already been exploited well beyond the 50-percent level needed for the final doubling. Are there enough resources to support another doubling? Are there enough to sustain a population of 10 billion, should that level be reached? What level of resource consumption per person will see the human carrying capacity exceeded? If the answer to either of the first two questions is no, and the exploitation or

FIGURE 4-24
The ruins of Chichen-Itza on the Yukatán Peninsula of Mexico are but one of innumerable examples of civilizations that arose, prospered, and then passed into oblivion. Much evidence indicates that the reason for collapse of ancient civilizations was a failure to maintain a suitable balance between their human population and environmental resources; that is, the J curve. Will we do better? (Ulrike Welsch/Photo Researchers, Inc.)

consumption of resources continues as in the past, we will find ourselves facing an impoverished future.

As natural biota and ecosystems are increasingly displaced by the human system, we are effectively replacing very complex systems containing great biodiversity with a relatively simple system. Over 50% of our planetary food demand is supported by four basic grains—wheat, corn, rice, and soybeans—compared with about 7000 plant species that have been gathered for food throughout human history, and some 30,000 plants that have edible parts. We have noted that simple systems are very vulnerable to upset. Many agriculturists are quite concerned that an outbreak of a pest could destroy a critical portion of the world's food supply at any time. What does this bode for the future?

The scenario of a major population crash would not just involve "those people over there." In all likelihood, all countries and people would be drawn into the turmoil, and it could end civilization as we know it. Nor would the result be benevolent for wildlife. Every natural resource even remotely usable would be consumed to exhaustion or extinction in the desperate struggle for survival at the peak and during the crash of the population. (People who are starving cannot be expected to be concerned about saving endangered species.) Again, this drastic scenario is painted only to illustrate a path we wish to avoid.

We need to return again to the question: What is the carrying capacity of our planet for human beings? Many experts have considered and debated this question. Unfortunately, there is no uniform conclusion because humans are capable of such diverse behavior, but let us consider the major points. Most significantly, we observe that major environmental parameters such as forest, fisheries, soils, water resources, and atmosphere are in a state of decline under the pressures of the present population of 6.0 billion. Therefore, by definition we are already over the Earth's carrying capacity. How much over?

Human pressures on the envrionment vary with affluence as higher incomes both enable higher levels of consumption and produce higher levels of pollution as waste or byproducts from consumption are inevitably disposed. Only 20 percent of the world's population or 1.2 billion people live at a level of affluence typical of the United States. The rest live at considerably lower levels. It is calculated that the Earth's carrying capacity (sustainability) for people living a typical American lifestyle would be about 2.0 billion. On the other hand, the Earth's carrying capacity for people living at minimal levels of resource consumption, as most do in China for example, is estimated to be in order of 12 billion.

Neither of these end figures, however, consider the potential of implementing new technologies, most particularly in the areas of solar energy, total product and waste recycling, and improvements in crop production and soil conservation. If ideal states could be reached in all these areas, it is conceivable that the Earth might sustain 8–10 billion people at a reasonably comfortable lifestyle. Thus, we do not have to imagine either reducing the world's population by 70 percent or all having to live in abject poverty. There is realistic hope that the overall condition of humanity and the environment can be improved through a combination of stabilizing population growth and implementaion of technologies noted. But this cannot be expected to happen automatically through "business as usual." It will require using our intelligence and wisdom to understand and appreciate ecological principles and in using our technological skills to bring our human system into compliance with those principles.

In relation to the concepts presented in this chapter, anything that we are doing or that we may do toward preserving or reestablishing natural balances among biota and protecting natural biota from destructive human impacts may be seen as progress toward sustainability. Likewise, protecting ecosystems, to say nothing of the total biosphere, from unnatural changes in abiotic factors such as pollution or atmospheric alterations is equally important. We must work toward stabilizing the human population in morally acceptable ways and husbanding resources such that they remain available to future generations. A concern for posterity is a valid motivating factor.

Chapters 6–8 will focus on the problems of stabilizing the human population, and later chapters will focus on husbanding resources. Before proceeding with those issues, however, we wish to examine the question of how species become adapted to one another and their abiotic environment in such a way as to form balanced, sustainable ecosystems.

Review Questions

1. What factors are involved in biotic potential and environmental resistance?

2. What is the distinction between reproduction and recruitment, and what occurs in all populations if conditions are ideal?

3. How does population density relate to a balance between biotic potential and environmental resistance?

4. How are herbivore populations in nature generally controlled? Carnivore populations?

5. What is meant by territoriality, and how does it control certain populations in nature?

6. What events occur if herbivore populations are not controlled by natural enemies?

7. What are the three main ways that enable different plant species to coexist in the same region?

8. What may occur when plants, herbivores, carnivores, or parasites from foreign regions are introduced into an ecosystem?

9. What is ecological succession? a climax ecosystem?

10. What role may fire play in ecological succession, and how may fire be used in the management of certain ecosystems?

11. What are the two fundamental kinds of population growth curves? What are the causes and consequences of each?

12. What is the fourth principle of ecosystem sustainability?

13. What are some potential consequences of not maintaining biodiversity?

14. Do humans need to control their population? If so, what methods are available that are not available to all other animals?

Thinking Environmentally

1. Describe, in terms of biotic potential and environmental resistance, how the human population is affecting natural ecosystems.

2. Analyze the population-balancing mechanisms that are operating among various plants, animals, and other organisms present in a natural area near you.

3. Choose one species (plant or animal) in a natural area near you, and predict what will happen to it if two or three other species native to the area are removed. Then predict what will happen to it if two or three foreign species are introduced into the area.

4. Evaluate such practices as legal hunting, controlling pests with chemical sprays, using or preventing fires, and poaching of endangered species in terms of supporting sustainable balances.

5. Make an argument, pro or con, regarding sustainability of the human system in terms of the concepts you have learned in this chapter. What new directions, if any, do humans need to take to achieve sustainability?

CHAPTER 5

Ecosystems: Adapting to Change

Key Issues and Questions

1. In nature, we find that each species is remarkably adapted to the factors of the ecosystem in which it exists. What does this imply about the "moldability" of species?
2. The characteristics of a population can be modified by differential reproduction. Give an example of this in terms of selective breeding; in terms of natural selection.
3. Mutations and differential reproduction lead to inevitable changes in the gene pool of a species. Explain how this occurs.
4. Natural selection can lead to the development of new species. How does this occur? What are the prerequisites and limitations?
5. All ecosystems have the same basic groups of organisms, and they function in the same way. Explain this result in terms of selection at the ecosystem level.
6. In the face of environmental changes, insects and some plants are most likely to survive, whereas large mammals and birds are more likely to become extinct. What attributes influence the survival of a species?
7. Loss of biodiversity undercuts the ability of species, ecosystems, and agriculture to adapt to changing conditions. Why is this the case?

In Chapter 4, we showed how the sustainability of ecosystems depends on maintaining balances among the populations of species that make up the biotic community. Further, we observed that such balances are frequently upset by the introduction of a species from a foreign ecosystem. These upsets led us to conclude that balances among populations are not "automatic," but apparently evolve as species live together over many millennia. The evolved strengths and weaknesses of a species that provide for population balances within its native ecosystem, then, differ from the strengths and weaknesses of a similar species in a different ecosystem, and ecological upsets may occur when a species moves from one ecosystem to another.

Snow geese. Snow geese show two distinct color patterns, the white phase and the blue phase as seen in this group. Birds of the two colors interbreed freely but the colorations remain distinct. The genes for the two colorations are apparently held in equilibrium in the population by the fact that the white phase has a survival advantage when the ground is covered by snow while the blue phase is better camouflaged when the gound is snow free. [Photo by Gilbert Grant/Photo Researchers, Inc.]

This simple explanation implies an important concept—that species are gradually "molded" to form balanced relationships with other members of the biotic community and with the abiotic environment (Fig. 5-1). How does such "molding," or **adaptation**, of species occur? Further, how do new species arise, an event that is evident in the fossil record, as are extinctions? Then, if the generation of new species and the development of balances are natural processes, can we assume that nature will take care of the upsets caused by humans and nullify the consequences? Or can humans understand these processes sufficiently to manipulate them toward a desired end?

Our objective in this chapter is to examine the process by which species may change and thereby adapt to their abiotic environments and other members of the biotic community with which they coexist. By understanding this process, you may also understand its prerequisites and limitations. Finally, the importance of maintaining biodiversity—our fourth principle of sustainability—will become abundantly clear.

Selection by the Environment

First, consider our modern understanding that all the inborn characteristics of an individual organism (plant, animal, or microbe) are expressions of its genetic (DNA) makeup. That is, each inborn characteristic is, in one way or another, the manifestation of certain genes or DNA the individual is born with. Second, we are familiar with the fact that any two individuals (save identical twins, which will be considered later) can be identified or distinguished on the basis of their DNA, as well as on the basis of their physical appearance. In other words, the differences between individuals that we see on the physical level correspond to measurable differences on the genetic level. We speak of these genetic differences that exist among individuals as **genetic variation** in the population. We find genetic variation in the populations of *all* species.

Third, imagine that we take a DNA sample from every individual of the species. This would give us what is referred to as the **gene pool** of the species. Whereas, we have not sampled every individual, we have sampled enough to find that the DNA gene pool of one species differs from that of another species, reflecting the degree of physical differences between the species.

As the members of a species reproduce, the genetic makeup of the species is passed from one generation to the next. Can the gene pool of the species change in the process? It can be mathematically demonstrated that if breeding among all the members of a species were totally random, and if each member produced the same number of offspring, each having the same chance of survival, the gene pool would remain constant in its genetic makeup. However, observation shows us that such constancy is not the case. Invariably, we observe some individuals reproducing more than others, a phenomenon called **differential reproduction**, which, itself, is a product of individual adaptation.

As a result of differential reproduction, the particular genes carried by the prolific individuals are reproduced and passed on and genes uniquely carried by individuals that reproduce little or not at all will become increasingly rare or nonexistent in the population. In this way a gene pool can change over generations. Differential reproduction will lead to a gradual modification in the gene pool as some genes become increasingly common in the population and other genes diminish in frequency of occurrence. A change in the gene pools of species over the course of generations is the essence of **biological evolution**.

The changes in the genetic characteristics of populations that can occur through differential reproduction is most dramatically illustrated through *selective breeding* of domestic plants and animals.

Change through Selective Breeding

Virtually all agricultural and domestic breeds of both plants and animals have been derived from wild populations through the process of **selective breeding**. The primary technique behind all such breeding is the following. Breeders first envision the traits they would like to achieve in a given species: a dog with short, squat legs that is able to wiggle into animal burrows to aid hunters, for example. The breeders then examine the existing population of dogs and *select* those individuals that show the sought-after trait a little more than other members of the population. The selected individuals are then bred. The offspring tend to be like the parents, but some offspring express the particular trait *more* than the parents do, and others express it *less*. Those offspring that show the trait most are *selected* to be the parents for the next generation, whereas the other offspring are prevented from breeding. Repeating this process of selection and breeding over many generations gradually yields the traits desired by the breeder, as illustrated in Fig. 5-2. In this example the desired outcome is seen in the dachshund.

The fact that all the different breeds of dogs, from

FIGURE 5-1
Every species is adapted to cope with the environment and with other species in the biotic community in which it exists. Modifications of body shape and color to allow the species to blend into the background and thus provide protection from predation are among the most amazing. Shown here: (a) Thomisus spider, (b) Span worm, (c) Leaf katydid, (d) Giant Australian stick. (a) Nuridsany et Perennou/Photo Researchers, Inc. b) J. H. Robinson/Photo Researchers, Inc. c) Michael Fogden/Animals Animals/Earth Science. d) R. H. Armstrong/Animals Animals/Earth Science.)

Great Danes to Chihuahuas, are derived from the same original species of wild dog demonstrates that a gene pool of a species may be molded in any number of different ways through selective breeding. It should be noted, however, that all breeds of dogs are still considered the same species, because they can interbreed and produce fertile offspring.

The development or enhancement of desired traits through selective breeding has been practiced since antiquity, many centuries before there was any knowledge about genes or genetics. Breeders made their selections of which animals or plants to breed and

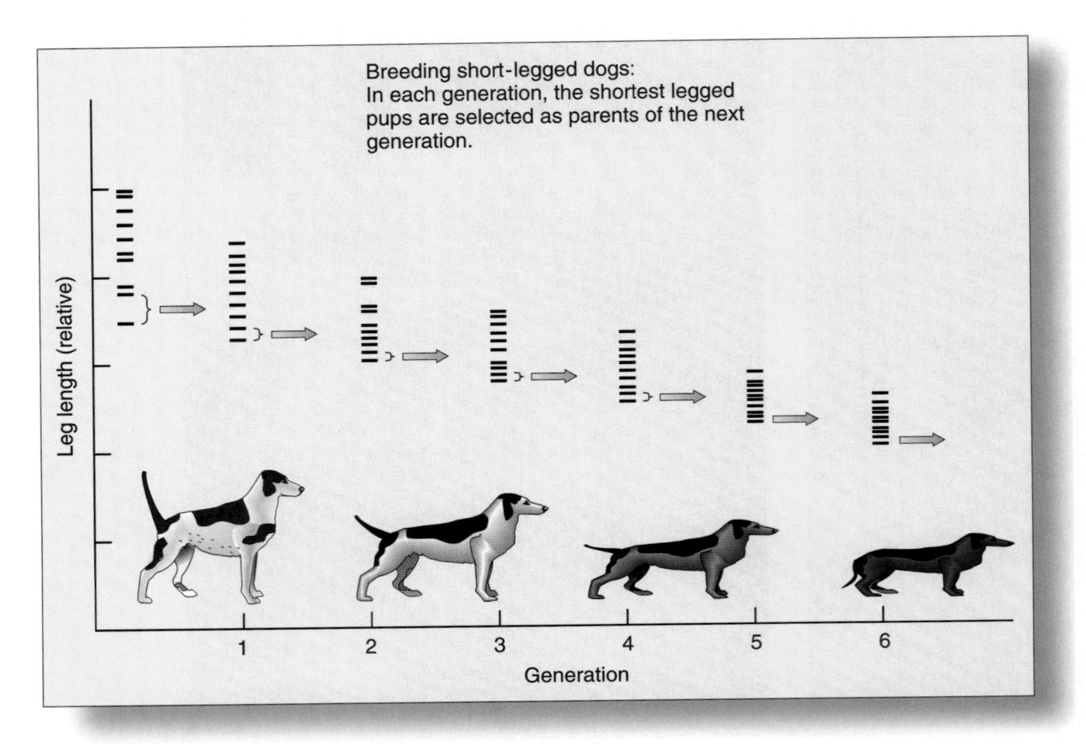

FIGURE 5-2
Selective breeding is a crucial technique in the development of all agricultural and domestic plants and animals. It entails selecting from the existing population those individuals showing the desired features to the highest degree, breeding them, and preventing others from breeding. As this process is repeated over many generations, the desired features are gradually developed, as shown here for the development of the dachshund.

which to exclude purely on the basis of visible characteristics, and it worked. Only now do we recognize that the visible characteristics are actually the manifestation of certain genes. In selecting particular individuals that express a certain trait more than others, the breeder is in fact selecting to increase the proportion of certain genes in the gene pool and to decrease others. As we can see, the results are profound (Fig. 5-3).

Change through Natural Selection

Does selective breeding or differential reproduction occur in nature? We note that in nature most offspring do not survive; rather, they fall victim to various factors of environmental resistance, as described in Chapter 4. These various factors of environmental resistance—predators, parasites, drought—can now be seen as

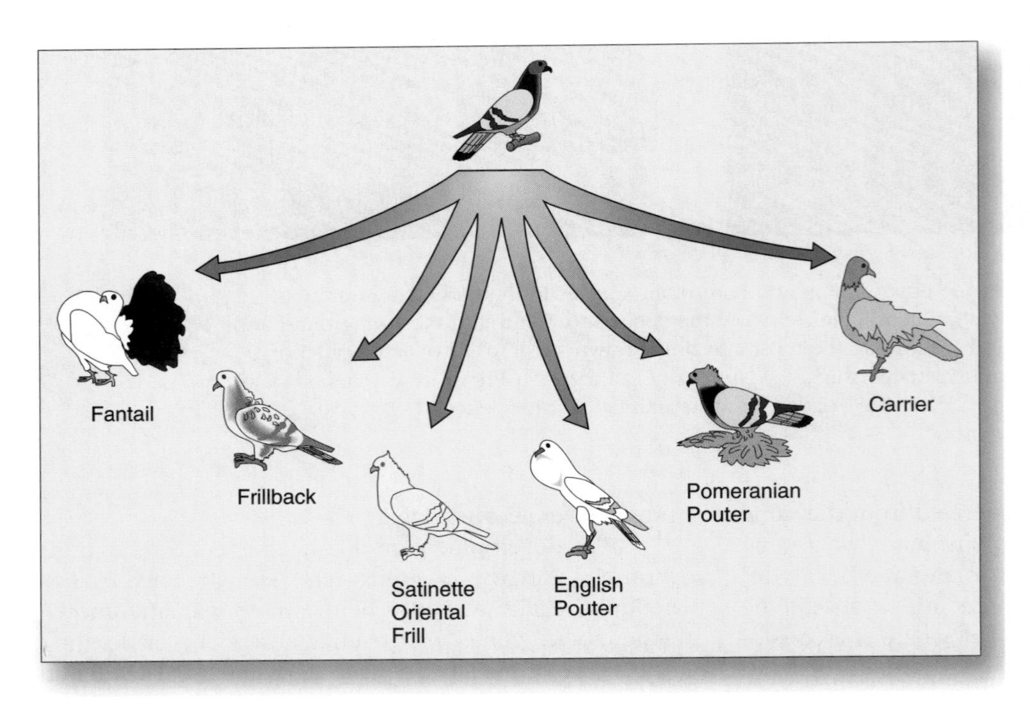

FIGURE 5-3
Breeds of pigeons developed by selective breeding. The wild Rock Pigeon of Europe, top, is thought to be the ancestor of all the domesticated breeds shown here. (Redrawn from W. W. Levi, *The Pigeon*, Sumter, SC.: Levi Publishing Co., 1957.)

Ethics

Selection—Natural and Unnatural

The fact that the gene pool of a species changes only through differential reproduction has tremendous practical, ethical, and moral implications for humans. First, we must restate a thoroughly discredited but still commonly held notion called the *inheritance of acquired characteristics*. This is the *false* notion that characteristics acquired by parents through training, learning, or accident will somehow be inherited by the offspring. Numerous experiments and common experience, show that such acquired characteristics are not inherited, because they absolutely do *not* affect the DNA of the parent. The offspring will inherit the same genes no matter what kind of skills the parents have developed.

A special aspect of this false notion concerns genetic disorders. Many congenital or genetic diseases, defined as conditions that are caused by certain mutant genes, can be held in check with medicines or treatments. Certain forms of diabetes, sickle cell anemia, and hemophilia are examples. Medicines do not correct the defective gene, however. Thus, any offspring of those affected may inherit the disorder and require the same treatment.

You may have heard a statement, "The human race is evolving toward bigger brains because we are required to use them," or "The human race is evolving toward smaller legs because we don't use them." These statements are basically expressions of the discredited notion of the inheritance of acquired characteristics. The only way such events would come to pass is if short-legged people consistently had more children than people with normal-sized legs. The way to arrive at an answer to such questions regarding the future evolution of the human race is to reflect upon the following question: Are people with this characteristic consistently reproducing more than average? If the answer is no, then there is no basis for projecting the change.

In short, the only way the genetics of a population will change from one generation to the next is through differential reproduction. The population will become increasingly dominated by the characteristics of those who reproduce the most, and a smaller and smaller proportion will bear the genetic traits of those who reproduce the least.

Let us speculate about what might be done with the human species if we could selectively breed humans. The study of improving the genetic future of humans is known as *eugenics* and has a cadre of serious advocates, but implementing any eugenics program collides with moral and ethical issues. Consider, for example, the ethics of one group dictating how many children another group should or should not produce. Who is to make the decisions, and how are they to be implemented? In China, families are allowed only one child—a direct effort to slow population growth. Traditionally, a male child is valued more than a female. If more males reach adulthood, will there be enough females for marriage? Take a moment and assess the outcome of this situation.

Yet, through suitable education, information, and understanding, some people may choose to practice a form of eugenics on a personal level. Prospective parents may obtain counseling regarding the genetic basis of any defect they may have and the probability of its being passed to an offspring. They can then choose whether or not to bear their own children. Also, it is now possible to perform DNA analyses on a fetus and determine the presence or absence of a number of genetic defects. Parents can then make the decision whether or not to have the fetus aborted. Of course, this assumes that having an abortion is a legal and acceptable option, a moral dilemma in itself.

A practice already widely used in animal breeding is to collect sperm samples from certain males—a prize racehorse or bull, for example. The sperm is kept frozen and then used to artificially inseminate females whenever and wherever desired. Thus, the genes of the prize animal are passed on to a much larger portion of the next generation than would be possible through natural breeding. How do you feel about applying this practice to human breeding? Artificially inseminating women with sperm from an unknown man is already commonly practiced.

Through advances in genetic engineering, it may be possible in the relatively near future to manipulate and "correct" genes in the developing embryo. To what degree should such manipulation be conducted? Who should make the decision, and who will pay the costs?

In short, technologies for identifying, manipulating, and propagating particular genes is becoming increasingly available. More and more, we will have to face the ethical dilemmas involved in choosing to use or not to use these technologies.

selective pressures. That is, each factor exerts a certain force in determining which individuals will survive and reproduce, and which will be eliminated. For example, if a predator is present, prey animals having traits that protect them or allow them to escape from their enemies (such as coloration that blends in with the background) tend to survive and reproduce, and those without such traits tend to become the predator's dinner. Any individual with a gene that manifests itself as a handicap will be eaten. Thus, predators may be seen as a selective pressure favoring the reproduction of genes that enhance the prey's ability to escape or protect itself and causing the elimination of any genes handicapping those functions. The need for food is a selective pressure acting on the predator, enhancing those characteristics benefiting predation—such as keen eyesight and sharp claws.

Every factor of environmental resistance is a selective pressure resulting in the survival and reproduction of only those individuals with a genetic endowment enabling them to cope with their surroundings. In nature there is undeniably a constant selection and, consequently, a modification of the species' gene pool, toward features that enhance survival and reproduction *within the existing biotic community and environment.* Since the process occurs naturally, it is known as **natural selection**.

These are the concepts presented by Charles Darwin in his famous book, *The Origin of Species by Natural Selection* (1859). The phrase "survival of the fittest" means the survival of those individuals having traits that best enable them to cope with the biotic and abiotic factors of their environment. The modification of the gene pool that occurs through natural selection over the course of many generations is the sum and substance of *biological evolution.* Darwin deserves great credit for constructing his theory purely from empirical evidence, without any knowledge of genes or genetics, information that wasn't discovered until several decades later. Today, our modern understanding of DNA, mutations, and genetics fully supports Darwin's theory of evolution by natural selection. Fully understanding this theory has many moral and ethical implications. (See "Ethics" box, p 113.)

Adaptations to the Environment

The key distinction between the genetic changes that can be brought about through selective breeding by humans and those brought about through natural selection lies in the end point. For human breeders, the end point toward which selection is carried on is some image in their minds. In natural selection, the criteria for selection in every generation is simply survival and reproduction.

Under the selective pressures exerted by the factors of environmental resistance, the gene pool of each population is modified such that the population becomes increasingly well adapted for survival and reproduction in the particular biotic community and environment in which it exists. Indeed, virtually all traits of any organism can be looked at in terms of features that adapt the organism for survival and reproduction. Essentially all characteristics of organisms can be grouped as follows:

- Adaptations for coping with climatic and other abiotic factors.
- Adaptations for obtaining food and water in the case of animals, and nutrients, energy, and water in the case of plants.
- Adaptations for escaping from or protecting against predation and for resistance to disease-causing or parasitic organisms.
- Adaptations for reproduction—for finding or attracting mates in animal populations and for pollination and setting seed in plant populations.
- Adaptations for migration in the case of animals and dispersal of seeds in the case of plants.

The fundamental question about any trait is, Does it support survival and reproduction of the organism? If the answer is yes, the trait will be maintained through natural selection. Consequently, various organisms have evolved different traits to accomplish the same function. For example, the ability to run fast, to fly, or to burrow, and protective features such as quills, thorns, and an obnoxious smell or taste, all support the function of reducing predation and are seen in various organisms (Fig. 5-4).

You may have observed that in our discussion we speak of survival and reproduction together. This is because in the process of breeding the two cannot be separated. It is obvious that an organism will not reproduce if it fails to survive. However, it should be equally obvious that survival by itself is not enough. In terms of passing its genes on to future generations, an individual that lives to a "ripe old age" without ever reproducing is no different from one that died in infancy. Any effect on the genetics of future generations is accomplished only through reproduction. Furthermore, the genetic effect of an individual on the next generation is directly proportional to the number of offspring that it produces that survive and reproduce in turn.

Selection of Traits and Genes

Selection invariably takes place at the level of the *individual organism*—that is, it is the individual organism that survives and reproduces, or ends up as somebody's dinner. Yet, by differential reproduction—weeding out certain individuals while others reproduce—it is the *population* that gradually becomes adapted to its particular physical environment and the species with which it interacts. Because the genetic makeup of the individual often is the determining factor in its success or failure to survive and reproduce, differential reproduction leads to a change in the gene pool of the population.

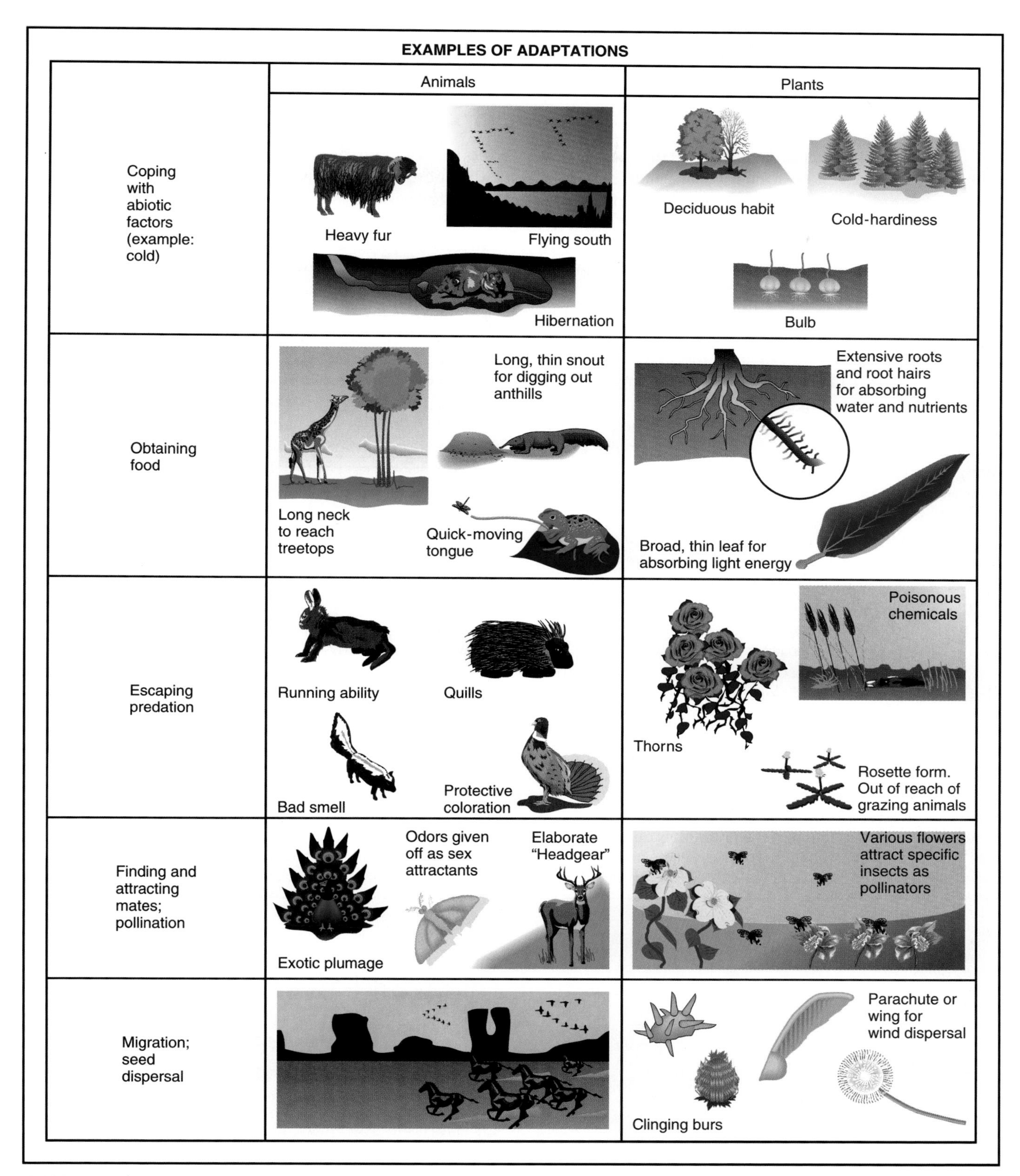

FIGURE 5-4
Adaptation for survival and reproduction. The five categories listed at the left are the basic essentials for the continuation of every species. Each feature of every species can be seen in terms of an adaptation that enables the species to meet its need in one or more of these categories. In each case, a multitude of features will accomplish the same function. Thus, a tremendous diversity of species exists, each adapted in its own special way.

In this section, we explore more fully how characteristics are determined by genes, how genes are passed from one generation to the next, and how they may be altered. We begin with a brief consideration of the first issue.

Describe an organism in every detail, and you are describing its *traits*. A **trait** is any particular characteristic of the individual. Such characteristics include all features of physical appearance, metabolism, aptitude, and behavior. Body height, shape of the nose, eye color, body build, lung capacity, and size of the appendix are examples of physical traits. Metabolic traits include such things as sensitivity to allergens, digestive capacity, tolerance to heat or cold, and resistance to disease. Traits pertaining to aptitude refer to natural abilities such as running, swimming, jumping, and, in humans, additional things such as talent in mathematics or music. Behavioral traits are the instinctive ways organisms act, such as a spider spinning a web, a bird flying south for the winter, and gentle versus aggressive behavior.

Many traits may be modified or developed more fully by experience, learning, or training. However, the underlying basis of traits is *hereditary*, or *genetic*. That is to say, each individual (plant, animal, or microbe) is born with a certain potential for each trait as a result of the transfer of genetic material (genes) from the parents to the offspring in the process of reproduction.

Since the 1940s, research has made it abundantly clear that, regardless of species, the genetic material responsible for all traits is the chemical known as deoxyribonucleic acid (DNA), which resides in the cells of every organism (save some viruses that have a related chemical, RNA). DNA is a long, chainlike molecule, and the sequence of four subunits (nucleotides) along its length provides a "code" analogously to the way the dots and dashes of the Morse code provide a code for various letters. The genetic code on the DNA is translated through the cell's metabolism into the production of proteins.

Some of the proteins specified by DNA, called *structural proteins*, determine the physical structure of the body—not just the overall structure determining whether the body is that of a mouse, an elephant, a human, or a turnip, but down to such details as the length of whiskers, shape of the nose, and whether hair is straight or curly. Another set of proteins, called *enzymes*, catalyzes or speeds up all the chemical reactions that go on in the body. Other proteins may serve as hormones or control the production of hormones that in turn regulate overall body functions such as growth, development, metabolism, and sex. In short, DNA specifies proteins, and the proteins determine everything else (Fig. 5-5). One **gene** can be thought of as a segment of DNA that codes for one particular protein. The physical structure and functioning of every individual organism are the result of the coordinated interaction of many thousands of genes. The number of genes for humans is estimated to be between 50,000 and 100,000.

FIGURE 5-5
DNA, genes, and traits. The hereditary information for every organism is chemically encoded on the chain-like molecule, DNA. Segments along its length code for synthesis of specific proteins. Specific proteins, in turn, determine and control physical structure, metabolism and other hereditary attributes of the organism, as indicated. In addition, the DNA molecule is constructed so that it can be chemically replicated, and the coded information may be passed from one generation to the next.

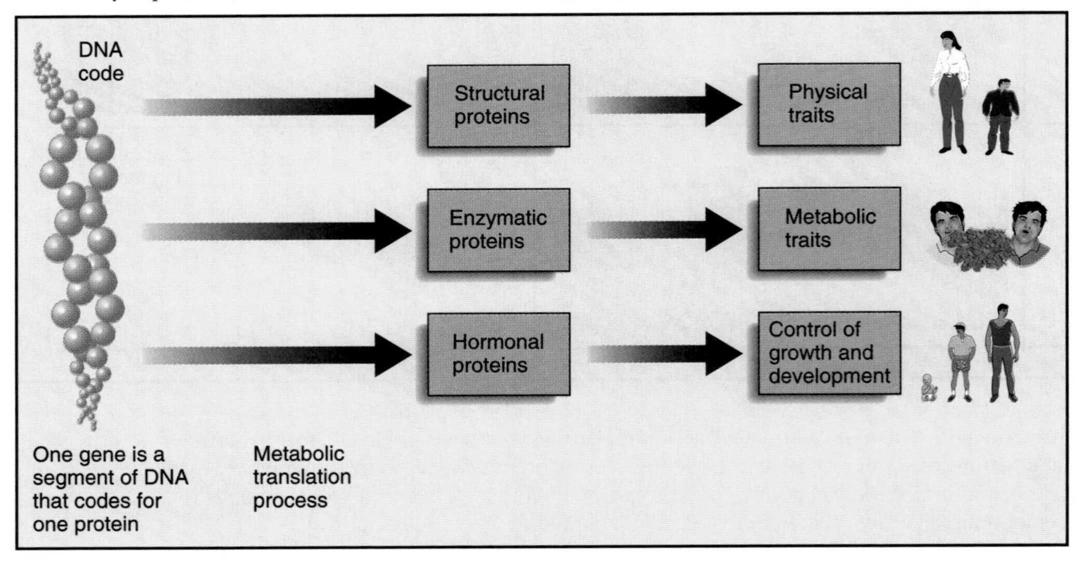

For every individual, this interaction actually involves interactions among *two sets of genes*, one set received from the male parent by way of a sperm cell and the other received from the female parent by way of an egg cell. In this way, each individual manifests characteristics of both parents and also has certain unique traits stemming from the new combinations of genes. The genetic makeup of the new individual is determined when the sperm enters the egg and the two sets of genes combine in the process of fertilization. The fertilized egg divides and redivides. As the cells continue to divide and grow, they gradually develop through various stages into the embryo and so on. In each cell division leading to growth of the body, all the genes are replicated (copied) exactly, and each cell resulting from division receives a complete copy of both sets of genes.

Genetic Variation and Gene Pools

It is easy to see that there is *variation* among individuals of the same species. The human species is a good example. We come in a wide range of sizes, shapes, and colors; our bodies have different chemical abilities, such as the ability to digest certain foods; we have different tolerances to various conditions; and we have widely different abilities quite apart from any training or teaching we may receive. Similar variation exists among the individuals of all other species: dogs, elephants, oak trees, mushrooms, flies, and on and on. The only reason we tend to think of all houseflies, for example, as identical is that we do not examine them closely. When we do, it is not hard to find differences in size, color, or any other trait we choose to focus on.

These variations imply that there are variations in the genes controlling each particular aspect of the individual, and indeed, there are. Variations of a single trait—blue and brown eye color, for example—result from two different forms of a particular gene for eye color. Different forms of a gene are called **alleles**. Thus, we say that there are two major alleles for eye color: blue and brown. Other, less common alleles for eye color exist as well, giving rise to other colors sometimes seen. Similarly, differences in natural hair color, height, blood type, and so on all fundamentally result from different versions or alleles of the particular genes in question. Likewise hereditary diseases such as sickle-cell anemia, hemophilia, cystic fibrosis, and Huntington's disease are the result of alleles of particular genes that fail to function as the normal gene should.

Since each individual has two sets of genes—one from the father and one from the mother—each individual carries two alleles for each gene. The two alleles for any one gene may be different from each other, or they may be identical. How they interact determines the trait that the individual shows. You are probably familiar with the ideas of dominance and recessiveness as they apply to eye color. The allele for brown eye color dominates over that for blue. Hence, brown eyes result from either the presence of two alleles for brown eyes or the presence of just one allele for brown and one for blue. Blue eyes, the recessive condition, result only when the allele for brown is absent, that is, when both alleles are for blue. By the same token, can you see how a person may carry an allele for a genetic disease without showing any symptoms of that disease?

Understanding that each individual has two alleles (which may be the same or different) for each characteristic is important because, in the formation of an egg or sperm cell, there is a segregation process such that the egg or sperm receives *one and only one* allele for each characteristic. Furthermore, if the alleles are different, there is a 50-50 chance as to which of the two will end up in any given egg or sperm. Said another way, half the sex cells will carry one of the alleles, and half will carry the other. Then, as egg and sperm unite in fertilization, there is a recombination of the alleles such that, again, there will be two for each gene, but one will have come from each parent. Can you see from this how two brown-eyed parents may have a blue-eyed child? (See Fig. 5-6.)

Such examples show how a particular characteristic may be determined by the respective alleles of just one gene. These cases provide a basic understanding of the process involved. However, many characteristics, such as overall body stature, may be influenced by numerous genes, each with different alleles playing a greater or lesser role in the final manifestation. Thus, one may have any of the gradations between the extremes of the particular characteristic, as well as the extremes themselves. Other exceptions with which you may be familiar are the so-called sex-linked characteristics such as color blindness. The alleles for such characteristics are located on the X chromosome and do not have a counterpart in the male, which has only one X chromosome. Also, keep in mind that an individual is always the sum total of all its many thousands of traits, each being determined by respective genes.

In summary, it is the different combination of alleles in each individual that is responsible for the genetic variation among the individuals.

Let us now relate this idea of different alleles to the concepts of DNA and the gene pool of a species discussed earlier. Each different allele is a slightly different piece of DNA coding for a slightly different protein, which causes some variation of the trait in question. The total of all the alleles in the population represents the gene pool of the population. Note again that some alleles will be very widespread in the population, while other alleles may be rare, meaning that only a few individuals carry or have the particular traits associated with them.

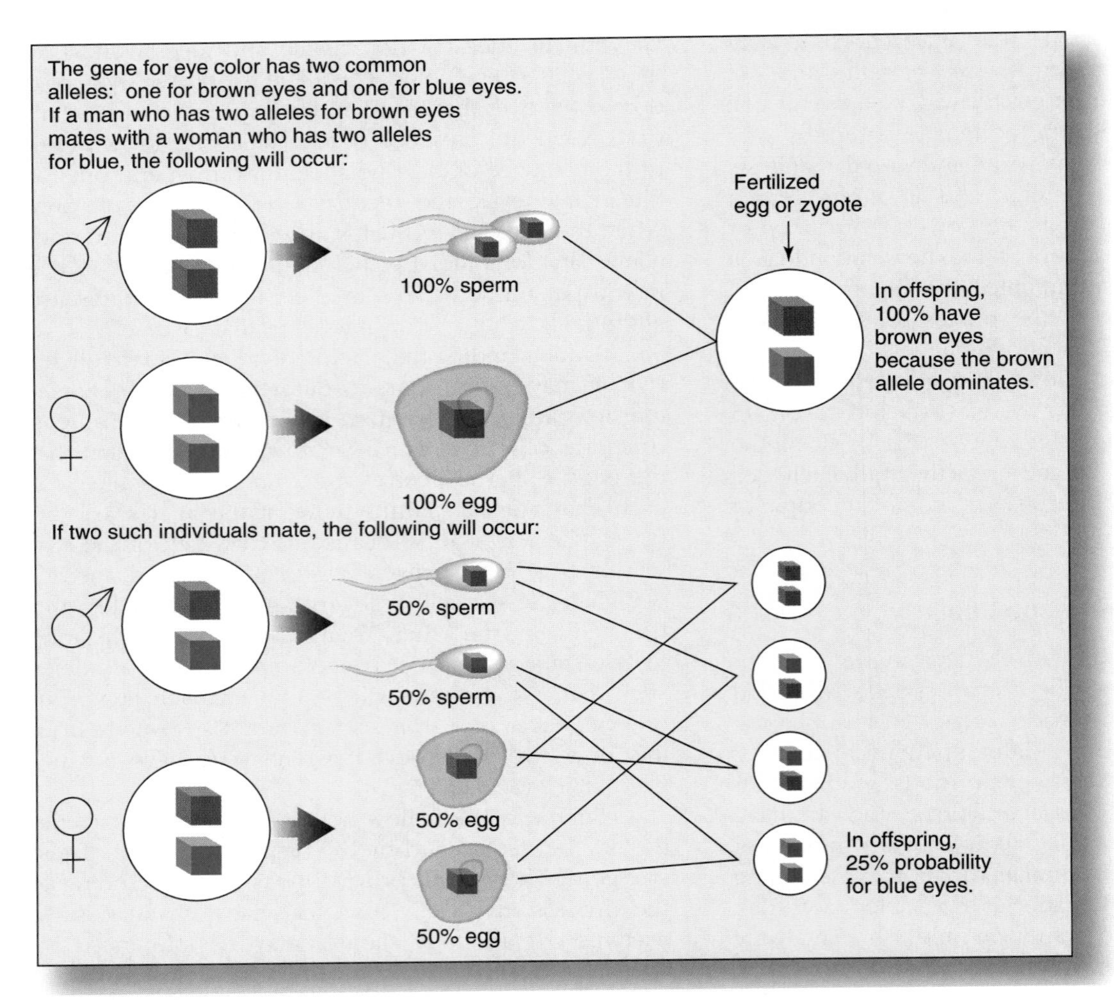

FIGURE 5-6
Inheritance of eye color.

Thus, we should try to visualize the reproduction of a species as far more than popping out additional copies in a cookie cutter fashion. We should visualize the reproduction of a population over generations as a constant mixing of all the alleles within its gene pool. With so many genes and alleles involved, it is more than likely that each individual will be genetically unique (that is, have a unique combination of alleles). But obvious exceptions exist, namely, identical twins and *clones*.

Identical twins arise in the following way: Instead of the fertilized egg developing into a single embryo, the two cells formed by the first division separate, and each goes on to form an embryo and, hence, two individuals. We have noted that in regular cell division, all the DNA (the set of genes) is replicated exactly. Therefore, the two cells derived from the same fertilized egg and subsequently the two individuals who develop from that egg, are genetically identical.

Clones, which are populations of genetically identical individuals, are propagated by asexual reproduction. The most common example is the propagation of plants by means of cuttings or graftings. Such propagation involves only the regular cell divisions in which all genes are replicated exactly. Therefore each "offspring" is an exact genetic copy of the "parent." For example, all Macintosh apples are the same, because all Macintosh apple trees are a clone propagated from cuttings from the original tree. But how did that original Macintosh variety occur? We address this question in the next section.

Mutations—the Source of New Alleles

We have noted that the forces of environmental resistance acting on a population serve to "weed out" individuals with any traits that handicap any aspect of survival or reproduction. Thus, those alleles and com-

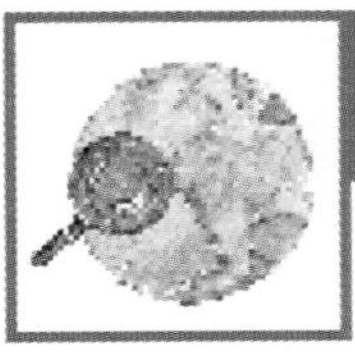

Earth Watch

What Is a Species?

A species was originally, and to a considerable extent still is, defined simply as a distinct kind of organism. The problem is, What constitutes a "distinct kind"? For example, two populations of tree frogs, one in New England and one in Georgia, appear to be distinct enough to be classified as different species. In the intervening states, however, we find tree frog populations that, when all put together, create a smooth transition of variations from the New England to the Georgia populations. Should all these frogs be classified as one large species with much variation, as two species, or as many very similar species? This problem exists with numerous kinds of plants and animals, and there is no clear-cut answer.

Taxonomists, the term for those who make a profession of classifying organisms, fall into two philosophical camps: splitters and lumpers. Splitters recommend dividing groups, such as the frogs just mentioned, into multiple species; lumpers recommend putting them all into one species.

In an attempt to remedy the situation, the definition of *species* was amended to include the aspect of interbreeding. If interbreeding occurs among individuals, and the offspring are fertile, then they are of one species, regardless of how different they may appear. Conversely, if interbreeding does not occur between two groups, they should be considered separate species. Or, if two different populations interbreed but produce sterile offspring, as in the case of horses and donkeys producing mules, which are sterile, the two populations are likewise different species.

This definition may help in an intellectual sense, but often does not help in a practical sense, mainly because interbreeding is often impractical or impossible to test or observe. Also, natural interbreeding may not occur between two populations of similar animals, which would indicate that they were separate species. When members of the two populations are placed together *in captivity*, however, they may interbreed readily. This situation is even more problematic in plants that generally do not cross-pollinate in nature, but can be readily crossed by artificial pollination. The offspring from populations that generally do not interbreed in nature, but that can be induced to do so, are termed **hybrids**.

So what is a species, then? We can conclude only that the concept of a species as a particular kind of organism distinct from all others, is a hypothetical construct of the human mind. (We like to categorize things and put them in discrete pigeonholes.) However, if species are undergoing a process of separation and change as we have described, then examples such as those described are what we should expect to observe. For the sake of discussion, we still need to use the term *species* to refer to kinds of organisms. However, the problem inherent in defining precisely what a species is should be taken as reinforcement of the concept that species are always in the process of changing and creating new species.

binations of alleles responsible for such traits are removed from the population, while those that enhance survival and reproduction become more abundant. If this weeding out of less desirable and reproduction of more desirable traits were the only process at work, however, we should expect a limit to be reached as the most advantageous combination of alleles is attained.

But experience gained through selective breeding shows that such limits do not occur. Breeders find that regardless of how far the development of a given trait is pursued, they still observe variation that provides the basis for further enhancement of the trait in question. Thus, chickens having tail feathers 15 feet (5 meters) long have been bred, and still the breeding for longer feathers continues.

The potential for genetic change seems to have no boundaries, except insofar as it may lead to a nonfunctional organism. For example, breeders have found that they can produce a dachshund with legs so short that it is unable to walk. That breeders still easily skid down that slippery slope, is evidenced by the many overbred dogs with hip dysplasias and other anatomical problems.

Modern knowledge of DNA allows us to understand this potential for unlimited genetic change. In the course of cell division and reproduction, the DNA molecules are generally copied exactly, but occasionally mutations occur. A **mutation** is any change in the DNA molecule. A small change causes the protein coded by the changed segment to be altered. In turn, the trait determined by that protein will be modified in one way or another. In other words, mutations introduce into the population new alleles and, hence, new modifications of traits. If the mutation involves a large change in the DNA, one or more proteins are so altered that the organism fails to function and dies as a result. Mutations that result in death are termed **lethal mutations**.

Very important is the fact that *mutations are random events*. There is no way, so far as is known, either in nature or through human technology, of causing a specific mutation to cope with a particular selective pressure. Indeed, like randomly turning screws in an engine, most mutations result in modifications that are harmful; only rarely is one beneficial.

How does nature cope with the problem of sorting out the good mutations from the bad? Again, it acts through the process of natural selection. Any individual having a mutation resulting in a handicap will usually perish before it can reproduce. Thus, the harmful allele is eliminated from the gene pool. The individual with the rare mutation (allele) that enhances survival, however, is likely to produce offspring. As these offspring, which inherit the new trait, reproduce in their turn, the population will comprise more and more individuals that carry the new allele. Mutations that result in neither benefit nor harm are called neutral. Such neutral alleles continue to be reproduced at the same proportion in the population and manifest themselves as the variations we see among the individuals of the population.

Changes in Species and Ecosystems

Understanding the process of how the gene pool of a population may be changed through mutations, breeding, and natural selection enables us to comprehend how new species evolve. We need only recognize, as did Darwin, that members of the same population are invariably in competition with one another for survival and that more offspring are produced than can survive. Individuals with any characteristic that provides an advantage over their cohorts stand a better chance of survival and reproduction. Thus, alleles for the characteristic are differentially reproduced, while contrary alleles are eliminated from the population. For example, if a population takes to browsing on leaves overhead as its main source of food, it will be the individuals with the longest necks that get the most food and leave the most offspring; shorter individuals perish. As such natural selection occurs over many generations, the end result, as you may guess in this case, will be the giraffe (Fig. 5-7). Similarly, a population that takes to feeding on ants and termites will, over generations, be modified to be more and more proficient in this regard, and we will have the giant anteater of Brazil (Fig.5-8).

Speciation

The infusion of new variations from mutations and the pressures of natural selection serve to adapt a species to the biotic community and the environment *in which it exists.* In this process of adaptation, the final "product"—giraffe, anteater, redwood tree—may be so different from the population that started the process, that it is considered a different species. This is one aspect of the process of **speciation**.

The same process also may result in two or more species developing from one. There are only two prerequisites. The first is that the original population separate into smaller populations that do not interbreed with one another. This reproductive isolation is significant, because if the subpopulations continue to interbreed all the alleles will continue to mix through the entire population, keeping it as one species. The second prerequisite is that separated subpopulations be exposed to different selective pressures. As the separated populations adapt to these different selective pressures they may gradually become so different as to be considered different species and be unable to interbreed with one another, even if they later come together again.

For example, consider our present Arctic and gray foxes. It is assumed that some thousands of years ago a subpopulation, perhaps under pressure of territoriality as discussed in Chapter 4, broke away from the main population in central North America and migrated northward. In the Arctic, selective pressures favor individuals that have heavier fur, a shorter tail, legs, ears, and nose (all of which help conserve body heat), and a white color (which helps animals hide in snow). In the southern regions, selective pressures regarding adaptation to temperature and background color are the reverse. Mutations creating alleles adaptive for the Arctic would actually be harmful in southern animals, because in warmer climates animals need a thinner coat to dissipate excessive body heat, and a white coat would make the animals more conspicuous. The result of this selection over many generations is the two species we see today (Fig. 5-9).

Tundra plants provide striking examples of speciation within the plant kingdom. They have a short, dense, mosslike vegetative structure, yet, as the similarity of their flowers reveals, are closely related to taller plants at lower elevations (Fig. 5-10). It is not hard to visualize ice crystals driven by strong winds—common at high elevations and in the arctic—being the selective pressure toward the compact form of tundra plants, as such crystals destroy any plant taller and less dense.

A renowned example of speciation, is one first observed by Darwin himself, which had a considerable influence in his formulation of the concept of natural selection. These are the fourteen species of finches living on the Galápagos Islands, located in the Pacific Ocean about 650 miles (1000 kilometers) west of Ecuador (South America). By similarities in their overall body structure, Darwin could see that all were clearly members of the finch family. Yet, each having a different beak structure adapting it to a specialized feeding niche demarked it as a separate species (Fig. 5-11).

Darwin's reasoning, which still holds, is that at some time in the past a population of migrating finches from the South American mainland was blown west-

FIGURE 5-7
(a) Selective pressure. A selective pressure is any factor that enables individuals with a particular variation to survive and reproduce more than individuals without that variation. In the evolution of the giraffe, we can readily visualize that once the behavior of browsing on trees started, there was a strong selective pressure for longer necks in that longer necked individuals got to eat and shorter necked individuals did not. When a longer neck is of no additional benefit, selective pressure and further modification cease. (b) Masai giraffe in its native habitat. (Stephen J. Krasemann/Photo Researchers, Inc.)

ward by a freak storm and became the first terrestrial birds to inhabit the relatively new (10,000-year-old) volcanic islands. As the initial population grew and members faced increased competition from each other, groups dispersed to nearby islands, where they were separated from the main population. These subpopulations encountered different selective pressures and became specialized for feeding on different things. In time, when these changed populations dispersed back to their original islands, they were different enough from the parent species to be distinguishable as new species and different enough that interbreeding among them did not occur. Since the process of speciation is a

FIGURE 5-8
Giant anteater. The major herbivores (in terms of biomass consumption) on the savannas of Brazil are termites, which build large earthen mounds as nests (see Fig. 5-14). The giant anteater, actually a termite eater, has evolved huge forelimbs and claws that it uses to break open the mounds and a head and mouth structure adapted to feeding on the termites. (Photograph by BJN.)

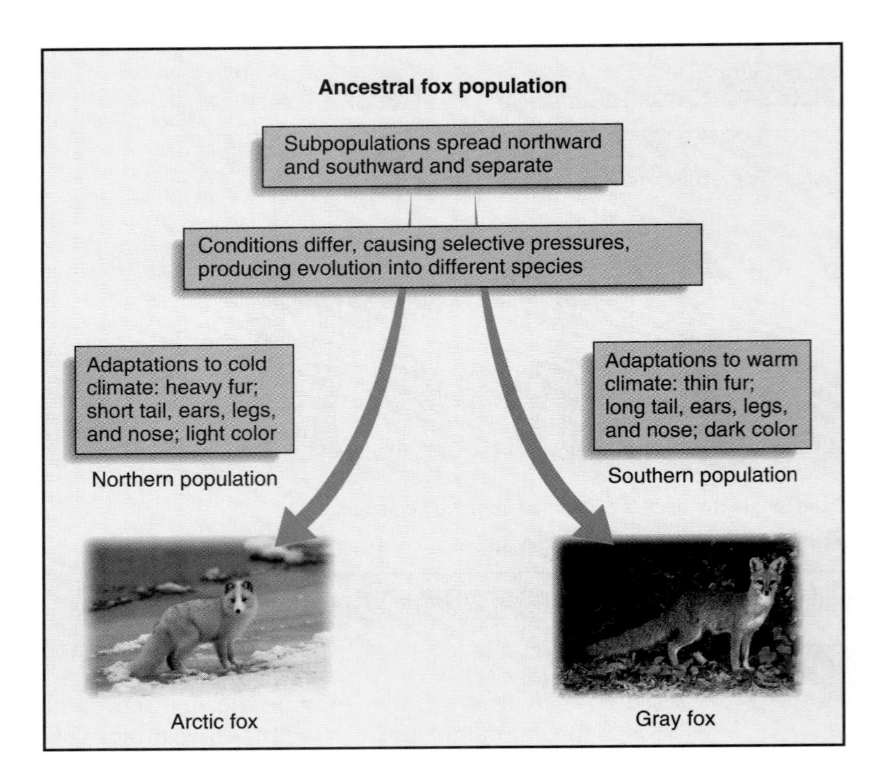

FIGURE 5-9
A population spread over a broad area may face various selective pressures. If the population splits so that interbreeding among the subpopulations does not occur, the different selective pressures may result in the subpopulations evolving into new species, as shown here for the Arctic fox and the gray fox. (a) Jeff Lapore, (b) C. K. Lorenz/Photo Researchers, Inc.)

gradual change occurring over many generations, you might ask, At what stage in the process do the separated populations become different enough to be considered different species? Indeed this presents a problem for those who name species. (See the "Earth Watch" Box, page 119.)

FIGURE 5-10
Frequently subjected to the harsh abrasive force of ice and sand particles driven by high winds, taller vegetation is destroyed giving way to tundra plants that have low, dense, moss-like habits like the mountain avens shown here. (Photograph by BJN.)

In sum, the key feature of our understanding of speciation is that new species are not formed from scratch; they are formed only by the gradual modification of existing species. Additionally, the gene pool of a species may be molded in many different directions by taking different populations within the species and subjecting them to different selective pressures. This concept is entirely consistent with the fact that in nature we generally find groups of closely related species, rather than a distinct species with no close relatives. It is also consistent with and explains the fact that all living things can be classified into a relatively few major groups—mammals and insects, for example—all the members of which can be seen as variations on the same theme and, hence, derived from a common ancestral population.

Some object to this concept and see nature as having more profound modifications than anything achieved by breeders. This objection is overcome by the fact that nature has been working at the process for many *hundreds of millions of years*. Another important point is that we observe that natural selection works to adapt species only to the biotic community and the environment in which they currently exist. There is no way by which species can foresee and preadapt to different conditions that may occur in the future.

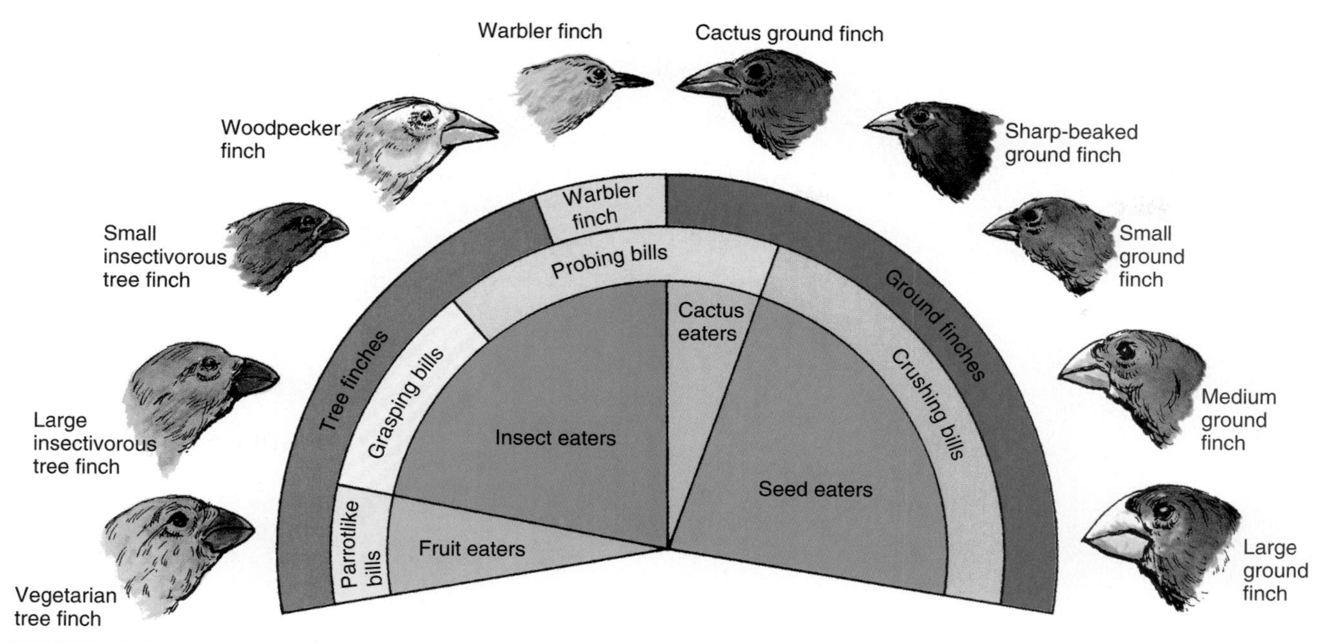

FIGURE 5-11
Some of Darwin's finches. The similarities among these birds attest to their common ancestor. Selective pressures to feed on different foods have caused modification and speciation in adapting subpopulations. (Adapted from Raven, P. H. and G. Johnson. 1988. *Biology*, 2d ed. St. Louis, Mosby-Year Book, Inc. Original illustration by George Venable.)

Developing Ecosystems

To understand the concept of change through natural selection, it is necessary to focus on how one species may be modified in a particular way by a particular selective pressure; we have employed such a focus in the preceding section. However, each organism is always living and reproducing within the context of the entire ecosystem, and therefore, it is simultaneously adapting to all the biotic and abiotic factors present in the ecosystem. By the same token, all the species in the ecosystem are *simultaneously adapting to each other*! Does this mean, then, that there is a process of natural selection acting at the ecosystem level as well as at the species level?

Only balanced relationships are sustainable. Any unbalanced feeding relationship will lead to extinction of the eaten, and also of the eater, as it runs out of food. Any unbalanced competitive relationship between species will lead to the extinction of both the loser and the other organisms dependent on it. Finally, any ecosystem that does not maintain suitable movements of nutrients and energy will be eliminated altogether. Therefore, at the same time that populations are being molded by the specific selective pressures of individual biotic and abiotic factors, there is also a selective pressure at an all-inclusive level for an overall functional system—any dysfunction such as an unbalanced relationship is eliminated.

This recognition gives us another insight regarding why different ecosystems or biomes (forests, grassland, tundra) have such dramatically different biotic communities, yet all function in basically the same way, with producers, consumers, detritus feeders, decomposers, and so on. Species have necessarily evolved in ways that adapt them to particular abiotic factors (such as temperature, moisture, and light). But, at the same time, they have evolved in ways that adapt them to particular roles in the entire system as a result of selective pressures at the ecosystem level.

But why have species developed so differently on different continents and remote islands, even though the climates and other environmental factors in these regions may be very similar? This is explained by the fact that nature can modify only preexisting species. Thus, a land mass will bear species that are modifications of what was there or what arrived as the land mass was isolated from other land masses. For example, when Australia broke away from a central land mass some 65 million years ago (shifting plates of Earth's crust, see Figure 5-16), marsupials (pouched animals) were present, but there were no higher mammals that existed elsewhere. Therefore, the animal life on Australia today is marked by modifications (many species) of marsupials—in particular kangaroos (Fig. 5-12)—while other continents have modifications of the higher mammals.

FIGURE 5-12
In Australia, the marsupials present 65 million years ago evolved into the prime grazing animals of today. Nowadays, kangaroos and other marsupials, dominate the grasslands. (Australian Information Center.)

Likewise, giant tortoises occupying the grazing niche on the Galápagos Islands, are a reflection of the fact that turtles were present when the islands were first formed and other animals were not (Fig. 5-13). Similarly, the grasslands of southern Brazil have essentially no large herbivores at all; the primary consumer niche is occupied largely by termites (Fig. 5-14), and giant anteaters (which actually eat termites) are the major "carnivore."

Recognizing that each ecosystem, irrespective of the species that are present in it, evolves in such a way as to establish its own operational integrity should make it abundantly clear why species introduced from one ecosystem into another may cause gross upsets.

How Rapid is Evolution?

Since natural selection necessarily acts on each generation, Darwin pictured evolution as slow, but continuous, modifications of species to become ever better adapted to their surroundings. However, a more detailed picture of the fossil record, developed since Darwin's time, shows that evolution plotted against time has been a stepwise process. There have been events that caused massive extinctions of many species, followed by the development of many new species derived from the survivors and then long periods of relative stability, ending with another event causing mass extinction. This is the theory of *punctuated equilibrium*.

The most recent mass extinction occurred about 65 million years ago, when dinosaurs were at the height of their dominance. Evidence indicates that this extinction was caused by a giant asteroid crashing into Earth. The collision and its aftermath resulted in the extinction of not only the dinosaurs, but many other species as well. Among the survivors, however, were some of the earliest mammals. The record shows that, following the demise of the dinosaurs, there was a relatively rapid (rapid, geologically speaking—several *million* years in reality) proliferation and speciation of these mammals, which came to dominate the niches occupied by large animals; relative stability has prevailed since then. This finding does not change the basic concept of the evolution of species through natural selection.

We have seen that ecosystems tend to develop toward sustainable dynamic balances among species. As this occurs, selective pressures on species within the system reach an equilibrium. After a species becomes well adapted for survival and reproduction in its particular ecosystem, there is no great selective pressure for further adaptation, especially since specialized adaptations have their disadvantages as well as advantages. For example, a giraffe with a neck longer than what is necessary to reach the treetops has no advantage over other giraffes, and may in fact suffer a disadvantage: It is harder for that giraffe to bend over for a drink of water than it is for its shorter necked companions.

In summary, selective pressures acting on the gene pools of all species gradually lead to the development of an ecosystem that is in dynamic balance. As a dynamically balanced status is reached, the selective

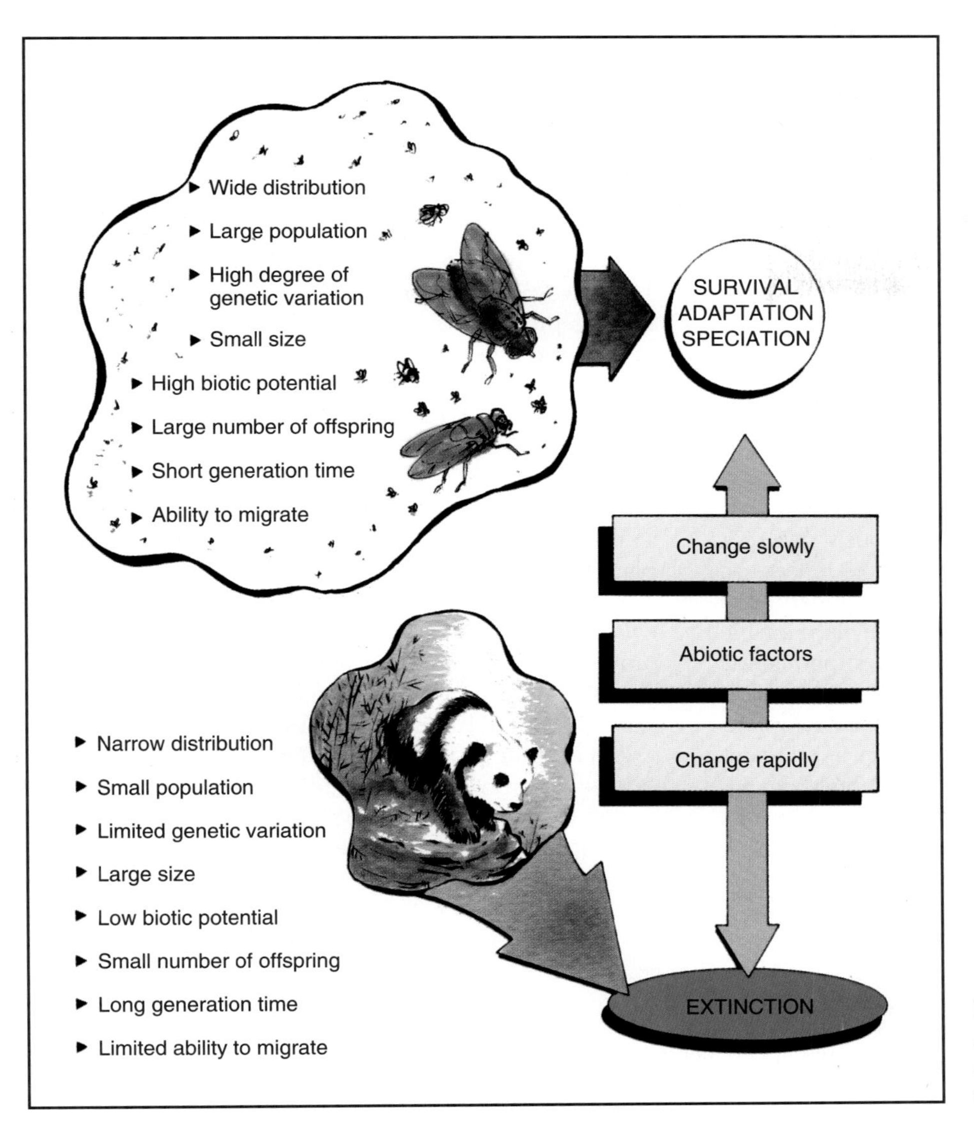

FIGURE 5-15
A summary of factors supporting the survival and adaptation of species, as opposed to their extinction.

message. We have a slow generation time, which speaks to slow genetic adaptation. But our limitation in this regard is more than outweighed by our wide distribution and our capacity to adapt to almost any environment through our technology. Therefore, in all likelihood, the human species will survive almost any calamity. The greater questions are, What level of civilization will we sustain and what kind of world will the survivors be living in?

Global Changes

Many scientists are working on understanding the causes behind shifting weather patterns and long-term changes in climate so that we can predict what the future climate of any given region will be as a result of either human or natural causes. To date, the ability to make exact predictions is still elusive. However, everyone agrees that over the long term change is inevitable. One slow but spectacular global change is that continents are continually on the move toward different relative positions on Earth. Let us look further at this movement of continental land masses.

The interior of Earth is molten rock kept hot by the radioactive decay of unstable isotopes remaining from the time when the Solar System was formed about 5 billion years ago. Earth's crust, which includes the bottom of oceans as well as the continents, is a relatively thin (no more than 65 miles [100 kilometers] thick) layer that can be visualized as huge slabs of rock floating on a plastic (rock that flows under heat and pressure) layer beneath, much like crackers float on a bowl of soup. These slabs of rock are called **tectonic plates**; about half a dozen major plates and two dozen minor ones make up the Earth's crust.

Tectonic Plates

Tectonic plates are not stationary. Within Earth's semi-molten interior, convection currents exist because hotter material rises toward the surface and spreads out at some locations, while cooler material sinks toward the interior at other locations. Riding atop these currents, the plates move slowly but inexorably with respect to one another, much like crackers might move if the soup below were gently stirred.

The average rate of movement can be up to 3 inches (6 centimeters) per year, but over 100 million years, this adds up to almost 5000 miles (8000 kilometers) in the fastest-moving segments. Movement of the crust is not gradual, for the plate boundaries are locked by friction. You can understand this process by pressing your thumb firmly on a tabletop and simultaneously sliding it across the surface. You will find that it moves in jerks as it sticks, jumps, sticks, and jumps. A plate boundary may not move for decades and then snap suddenly in dramatic events—namely, earthquakes and volcanic eruptions—as pressures from the movement gradually build until a break occurs. The earthquake or volcanic eruption releases the pressures and is thus followed by a quiescent period during which pressures gradually build again, eventually to cause another break or eruption. You can see how misleading is an average rate of motion.

Adjacent tectonic plates may move with respect to each other in four basic ways. First, where rising convection currents of molten material reach the surface, plates are forced apart, and the gap between them is filled with molten material that solidifies. Such regions are presently seen as mid-ocean ridges, and there is considerable volcanic activity along these ridges. Various lines of evidence indicate that about 225 million years ago all the continents were positioned as one major continent, which we now call *Pangaea* (Fig. 5-16). The spreading process has brought the continents to their present positions and is the basic driving force behind the other interactions between tectonic plates.

In the second kind of interaction, two plates may gradually slide past each other. An example of this sliding can be seen in the United States in the San Andreas fault system in California, that marks the boundary between the Pacific plate, which is moving in a northwesterly direction, and the North American plate, which forms the bulk of the United States (Fig. 5-17). Along this fault line, there is a major earthquake every 50 to 100 years as the pressures built up by the movement of 1–2 centimeters per year is suddenly released in a jump of 1–2 meters. The last two major slippages on the central California stretch of the San Andreas fault were the San Francisco earthquake of 1906 and the Loma Prieta earthquakes of 1989. Therefore, you can understand the dread of the "big one" happening at any time.

In the third kind of interaction, one plate may slide under another. This is currently taking place in the

FIGURE 5-16
Pangaea. (a) Similarities of rock types, the distribution of fossil species, and other lines of evidence indicate that 225 million years ago all the present continents were formed into one huge land mass, which we now call Pangaea. (b) Slow but steady movement of the tectonic plates over the intervening time caused the breakup of Pangaea and has brought the continents to their present positions. (Robert S. Dietz and John C. Holden, *Journal of Geophysical Research* 75, no. 26 September 10, 1970:4939–56, ©The American Geophysical Union.)

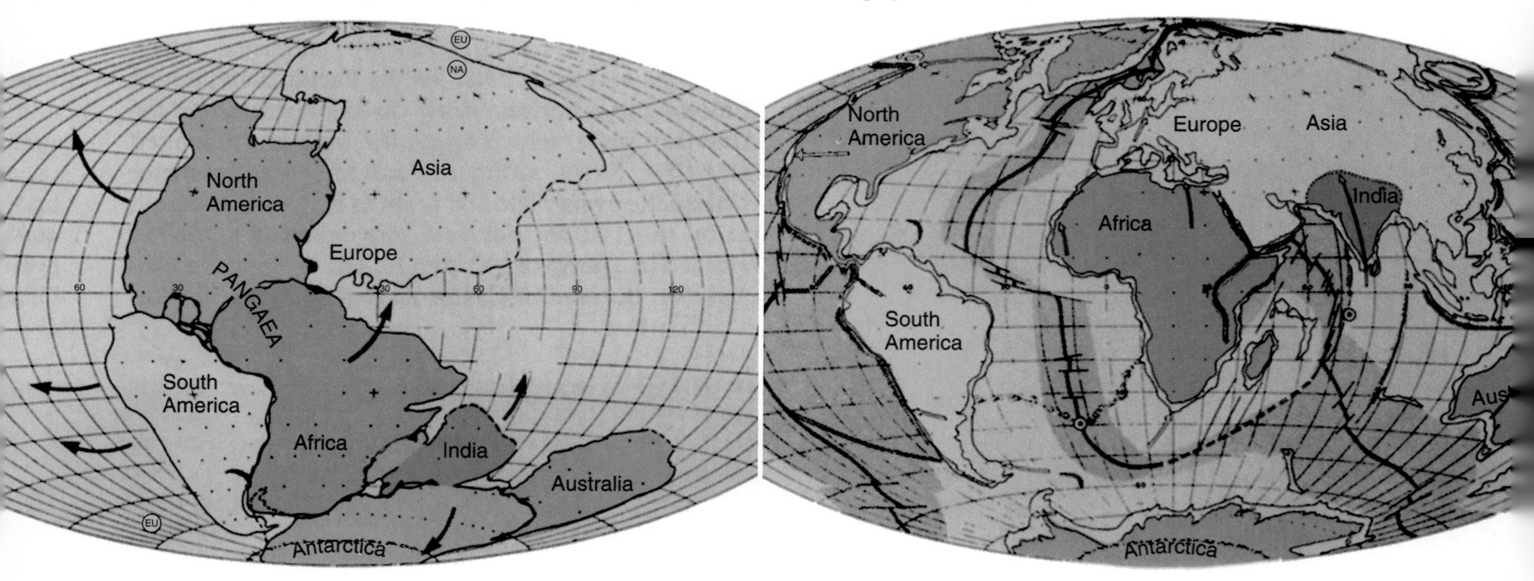

FIGURE 5-17
The narrow valleylike scar running from the top to the bottom of the picture is the San Andreas fault as it crosses the Carrizo Plain about 300 miles south of San Francisco. Every 50 to 100 years, mounting pressures of plate movement cause the fault to rupture in a major earthquake and the left side moves away from the viewer a distance of 3 to 6 feet relative to the right side. Hills on either side of the fault are pressure ridges formed as a result of many movements. (David Parker/Science Photo Library/Photo Researchers, Inc.)

northwestern United States, as part of the Pacific plate slides under Oregon and Washington. There are periodic earthquakes because the sliding is not smooth, but in this case another dramatic event also occurs. As the edge of the plate that is sliding under is forced down beneath the continental crust, it melts and periodically erupts to the surface. Thus, there is a volcanic mountain chain from northern California through Oregon and Washington. The most recent major eruption in the Cascade Mountains was Mount Saint Helens in 1980 (Fig. 5-18).

The fourth kind of interaction between tectonic plates is a continental-plate collision. Again, this produces periodic earthquakes, but the gradual result is the crumpling and uplifting of the plates into mountain ranges similar to the way in which the hoods of colliding cars crumple. An example of this is the Himalayas, which resulted from the plate that bears India moving northward and crashing into the Eurasian plate.

In addition to the periodic catastrophic destruction that may be caused in localized regions by earthquakes and volcanic eruptions, tectonic movements may gradually lead to major shifts in climate in three ways. First, as continents gradually move to different positions on the globe, their climates change accordingly. Second, the movement of continents alters the direction and flow of ocean currents, which in turn have an effect on climate. Third, the uplifting of mountains alters the movement of air currents, which also affect climate. (For example, see the rain shadow effect in Fig. 11-6). Additionally, there appear to be other factors involved in climatic change that are not well understood.

The fact that volcanic eruptions and earthquakes continue to occur is evidence that the tectonic plates are continuing to move today much as they have over billions of years and other conditions are changing accordingly. The fact that we find every region on Earth occupied by well-adapted living organisms makes it self-evident that they have been able to evolve, speciate, and adapt to these geological changes as they have occurred. Let us briefly examine the evolution of species as it is revealed in the fossil record.

An Overview of Evolution

An overview of the major events in the evolution of species, as it is revealed in the fossil record, is shown in Fig. 5-19. One way to appreciate the relative time periods involved in the different episodes of evolution is to make the entire geological history of Earth, 4.6 billion (4,600 million) years, equivalent to one calendar

FIGURE 5-18
Volcanic eruptions, such as Mount Saint Helens in southern Washington state, on May 18, 1980, are a result of tectonic movement in which one plate off the coast is sliding under the continental plate. (John H. Meehan/Science Source/Photo Researchers, Inc.)

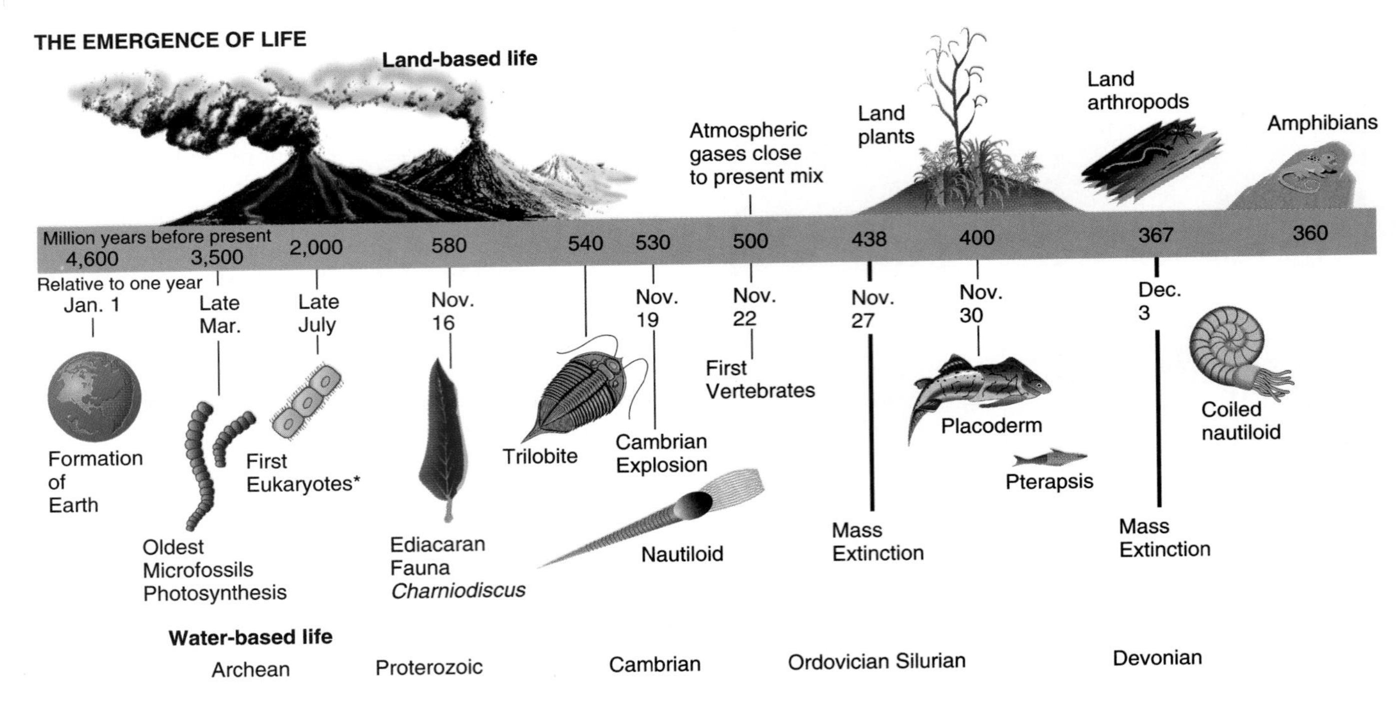

FIGURE 5-19
Contrasting the geological time scale with a single year gives an appreciation for the relative amount of time taken for various evolutionary stages. Note that two-thirds of the time is taken in the development of cells; then the pace quickens. Humans developed in just the last 8 hours, civilization since the advent of agriculture occurred in the last 2 minutes, and progress since the Industrial Revolution occupies only the last 2 seconds of the year. (*Eukaryotes are cells that possess a membrane-enclosed nucleus that contains chromosomes of DNA and proteins.)

year. Dividing by 365, each day on our calendar represents about 12.6 million years. At first, the planet was entirely molten, and it took several hundred million years—till the end of February on our calendar—for the surface to cool sufficiently to allow life, but by the end of March (3.5 billion years before the present [BP]) bacterialike organisms are in evidence. The first photosynthesis began in cyanobacteria at about this time, beginning the accumulation of atmospheric oxygen gas. Nothing other than bacteria existed for another four months (1.5 billion years). Then, in late July (2 billion years BP), the first more complex cells typical of today's higher life forms are seen. Still, these cells seem to have done little more than reproduce as single cells or filaments of cells for nearly another four months (1.5 billion years).

During the third week of November (600 to 500 million years BP), an episode known as the Cambrian explosion occurs in which there appear the representatives of all the major groups of organisms, including the first primitive forerunners of vertebrates. The last few days of November and the first few of December (436 to 367 million years BP) are marked by the development of the primitive vertebrates into many species of fishes and the evolution of land plants and land arthropods (organisms with an external skeleton). The development of amphibians, reptiles (leading to dinosaurs), and insects follows in close order over the second week of December (350–250 million years BP). During the third week of December (250–150 million years BP), reptiles become the dominant animals of Earth; also, the first primitive mammals and birds are seen. The dinosaurs dominated until December 26 (65 million years BP), when the Earth was apparently struck by an asteroid causing their rapid demise, but mammals survived and proliferated.

It is not until about 3 P.M. of December 31 (4.4 million years BP) that hominid creatures with an upright body posture appear. Even then, hominids and then early humans existed as hunter-gatherers, much like other animals in the ecosystem, for most of the 4.4 million years till the present. It was only a little over a minute ago on our calendar (10,000 years BP) that humans developed agriculture and civilization started. Finally, the explosion of technology and knowledge of the last 200 years is represented by the last 2 *seconds* of our geological year. Thus, relatively speaking, humans have barely opened their eyes on the evolutionary scene.

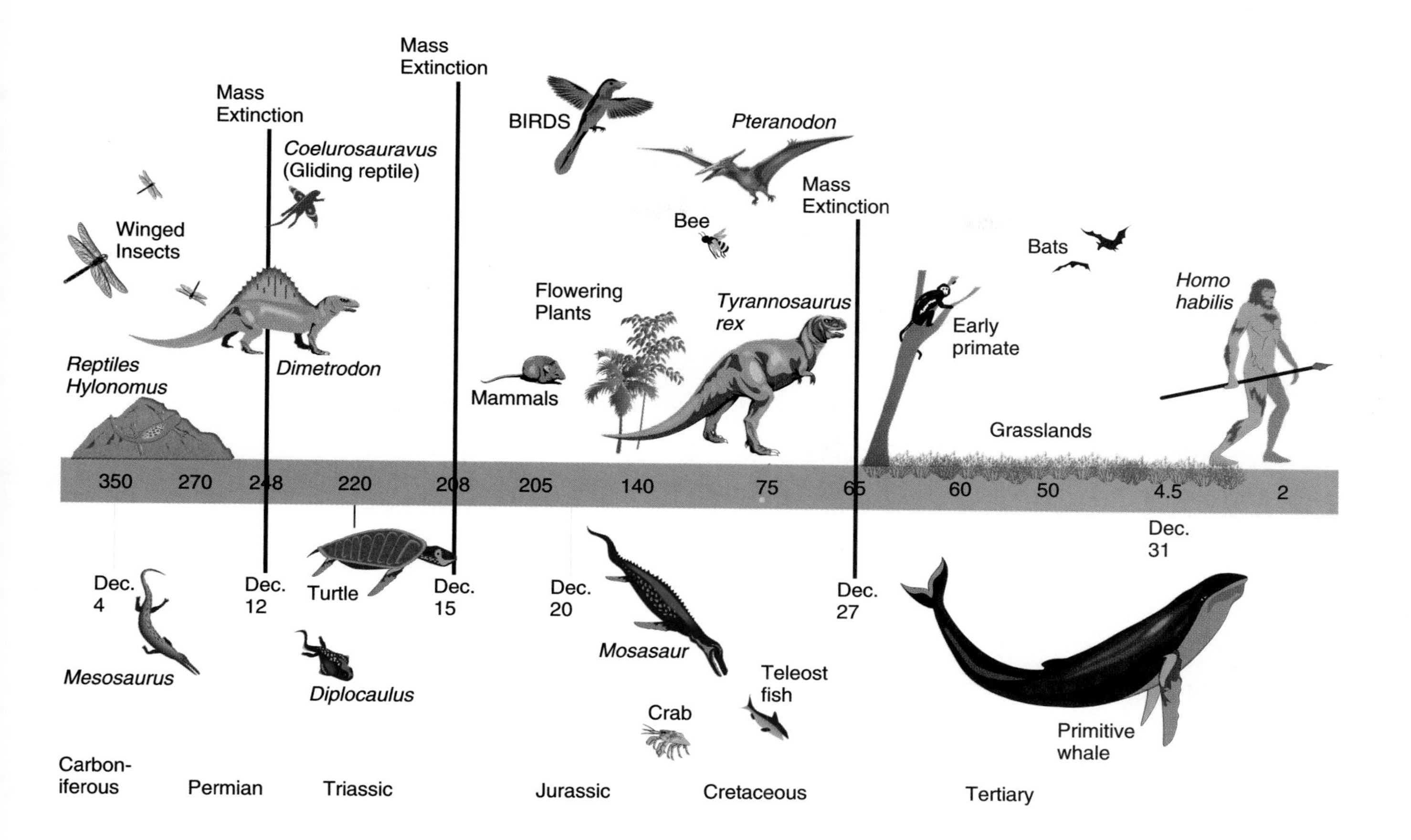

The Human Animal

There is no question among scientists that our ancestors of some 5 million years ago were a population of primates that speciated into both hominids and present-day members of the great ape family. Evidence for our common ancestry is the fact that over 98 percent of human DNA is identical to chimpanzee DNA, which is to say that differences between us and chimpanzees entail a divergence of only 2 percent of the gene pool—a most significant 2 percent to be sure, but still, only 2 percent. Even more clinching is the discovery reported recently (1994) of a 4.4 million-year-old fossil with distinct characteristics of both hominids and apes.

The selective pressures that caused the divergence of one part of the ancestral primate population toward hominids while the other part remained apelike are a matter of continuing speculation. One hypothesis is that the initial change was a change in behavior, which can spread rapidly through a population by learning. The hypothetical change in behavior was from the entire troop continually moving and feeding, as present chimpanzees do, to maintaining a home base from which some went out to forage, hunt, and gather and bring food back to share with others who stayed at the base caring for the young and protecting possessions. There is a selective advantage in this behavior, as it affords better protection and survival of the young. There are other traits that greatly enhance the ability to carry food back to share—in particular, moving on two legs to free the forelimbs for carrying; developing a greater intellect and manipulative skills to make containers for use in carrying; and developing greater vocalization skills for communicating ideas and feelings about hunting places and dangers. The early control and use of fire appears important in the maintenance of landscapes in an intermediate stage of succession, providing abundant food for game. The significant point is that, starting with an apelike population, we can visualize selective pressures that would logically lead to the modifications that distinguish humans from other primates.

In most cases, natural selection leads a species to become increasingly well adapted to a particular niche within the ecosystem, and in becoming adapted, the species also becomes specialized to that particular niche. For example, the giraffe is beautifully adapted to browsing on the tops of trees, but it is also specialized for that niche: It can no longer compete with animals grazing on the ground. In the case of humans, however, the naturally selected features—intellect and moving on

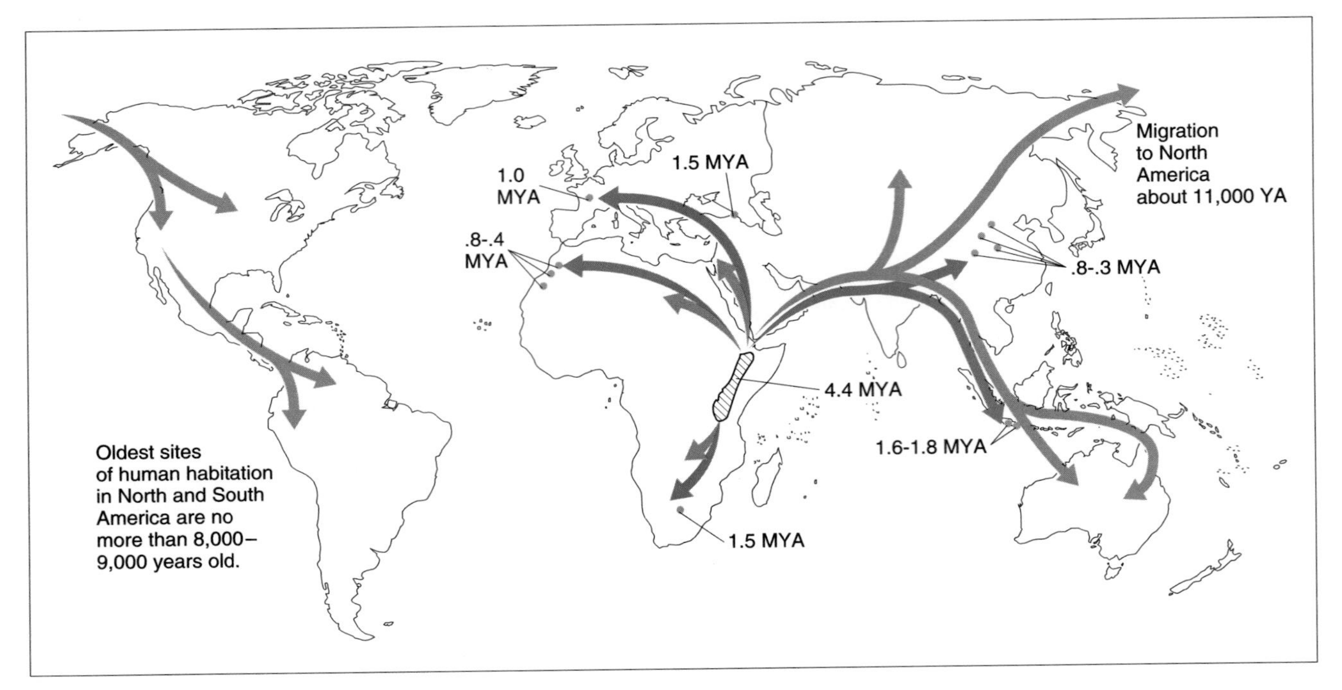

FIGURE 5-20
Fossil evidence indicates that the first hominids developed in East Africa (hatched area) 5 to 4 million years ago (MYA). Early hominid fossils found in other parts of Africa, Europe, Asia, and Indonesia (dates of such fossils noted on map) indicate that there were extensive migrations from the African center with some speciation occurring in the process (blue arrows). (See also Fig. 2-24). However, recent studies comparing the DNA of all races of present-day humans indicate that we all come from the same relatively small ancestral population, existing about 100 thousand years ago. The interpretation is that speciation to anatomically modern humans also occurred in Africa about 100 thousand years ago. In a second great wave of migration (red arrows), modern humans spread around the world, probably causing the extinction of earlier hominid species such as the Neanderthals in the process. The new world was inhabited only as a relatively recent part of the second wave, as humans were able to walk across the Bering land bridge that existed due to the lower sea level during the last ice age.

two feet to free the arms for carrying and manipulation—lead to a very generalized capability. Far from evolving into a specialized niche in the ecosystem, humans have became capable of out-competing virtually everything and subsequently migrating across the globe (Fig. 5-20).

Humans are going to determine the future course of evolution of life on Earth. This statement is not meant to be prophetic or say anything profound about humans; it is simply a statement of fact. Indeed, humans have already altered the surface of the planet, caused the extinction of numerous species, induced the migration of species from one place to another, and so on, to such an extent that the course of evolution unquestionably has been altered already from what it would have been without the presence of humans.

To be sure, every species, by its very existence, presents an array of both positive and negative selective pressures on other species in its biotic community. Therefore, every species may be seen as having a certain effect on the evolution of other species. But with humans, the impacts are more massive and sudden, and, most striking, for the first time in evolutionary history they are *conscious*. Never before in evolution has a species been conscious of its impact on the environment, much less been able to exercise a conscious choice as to what that impact will be.

One might spend a long time dwelling on the philosophical or spiritual implications of an evolutionary process that has progressed by means of random mutations and unconscious selective pressures to finally give rise to an organism that can recognize the process of its own evolution and can hold the future of that process in its own hands. Nevertheless, this is the station at which we humans have arrived whether we like it or not.

The choice before us is not *whether* we will alter the future course of evolution; we are already doing so. The choice is only in regard to the degree and the direction of our intervention. The overall global trend we see is a continuing sacrifice of natural ecosystems in favor of agricultural systems to support our growing

Earth Watch

Preserving Genes for Agriculture

There are two basic approaches to agriculture: You can select and grow plants suited to the environmental conditions, or you can change the conditions to suit the plants. Having limited ability to control the conditions, premodern cultures selected and cultivated plants that had the ability and tenacity to produce even in unfavorable conditions, without the benefits of irrigation, pesticides, or fertilizer.

Along the Missouri River in North Dakota, Native Americans grew numerous species of corn and squash. On the dry, wind-shifting sands of the Colorado Plateau, Hopi farmers tended sunflowers. In the dry lands of the Southwest, Indians grew hundreds of species of beans. By sending roots as much as six feet into the soil to find moisture, these beans had a high yield despite scorching 115°F (46°C) heat and almost no rain. In Peru, each valley grew its particular variety of potato or other tuber. Thus, ancient people survived for millennia by growing remarkably adapted crops. These crops are often referred to as heirloom vegetables by horticulturists, because they were grown only in one particular locality and their seeds were passed down through the generations.

Modern agriculture, however, has taken the approach of modifying environmental conditions to fit the plant. It has focused on using relatively few high-yielding varieties and obtaining maximum yields through the intensive use of pesticides, fertilizer, and irrigation. Agriculturalists are well aware of the vulnerability of this monocultural system. In 1980, for example, 20 percent of the U.S. corn crop was lost to a fungus disease because the disease was able to spread so rapidly through the monoculture. Resistance to disease and pests and vigor of modern varieties can be improved through the infusion of genes from the old heirloom varieties, but only so long as those varieties continue to survive—recall that new genes cannot be "invented." Scientists have mastered an ability to do new DNA makeups. Such is the emerging field of biotechnology in the development of new plant varieties.

Navajo cornfield in Canyon de Chelly in northern Arizona shows how native seeds thrive without irrigation. (Bill Steen, Elgin, Az.)

In the process of shifting to modern agriculture, many of the old varieties and species have been abandoned, and many have been lost forever. Of all the food plants that were grown by Native Americans at the time of Columbus, roughly 75 percent no longer exist, and many more are at risk. The loss of these species and the alleles they contain puts modern agriculture increasingly at risk. Aware of this problem, a worldwide network of "seed banks"—institutions devoted to the collection and preservation of seeds—has been established under the auspices of the World Bank's Consultative Group on International Agricultural Research. Also, a number of nongovernmental organizations, such as Seed Savers Exchange of Decorah, Iowa, Native Seed/SEARCH of Phoenix, Arizona, and the National Gardening Association, operate seed exchanges whereby members can obtain, grow, and keep alive heirloom varieties. The viability of future agriculture may depend on the efforts of those who are creating and maintaining these species repositories.

From Gary Paul Nabhan, Asst. Director of Desert Botanical Gardens, Phoenix, and cofounder of Native Seed/SEARCH.

---

population. Thus, the real alternative is whether we will continue this trend and use our consciousness simply to chronicle the loss of more and more species, or whether we will use our consciousness and our will to preserve a biologically diverse planet capable of meeting the needs of the future.

In Chapter 4 we brought out the fourth principle of ecosystem sustainability—*For sustainability, biodiversity is maintained.* The information presented in this chapter enables us to comprehend the significance of this principle. First, notice that biodiversity is really an expression of the gene pools and the genetic variation within those gene pools for all the species that inhabit the planet. We have seen that the ability of a species to adapt to new conditions depends, to a large extent, upon the genetic variation within its gene pool. Therefore, as we reduce the size of surviving populations—which we are doing with countless large mammals, birds, and other species—we inevitably reduce the genetic variation within their gene pools and thereby undercut their potential for future adaptation. Small surviving populations are increasingly vulnerable to any number of factors. Even if a population is brought back from the brink of extinction and restored in numbers, it will still carry the legacy of the genetic uniformity of that small surviving population for many generations. Therefore it will remain more vulnerable. For example, an outbreak of a disease may eliminate an entire population because the variations that would have conferred disease resistance to some members may have been lost. Of course, when a species becomes extinct its gene pool and everything it might have given rise to is lost forever.

This connection between lack of genetic variation and inability to adapt to new conditions has particular relevance in agriculture. Around the world, a major trend of modern agriculture, particularly in the last 40 years, has been to replace a tremendous diversity of local varieties of crop plants and animal breeds with a single high-yielding variety or breed. This practice has increased production and has been a major factor in the world's being able to support a growing human population. (See the discussion of the Green Revolution, Chapter 8). However, as discussed in Chapter 4, such genetically uniform monocultures are extremely vulnerable to outbreaks of pests and disease. Therefore, not only are we creating a monocultural system with dubious stability and sustainability, but with the loss of biodiversity represented by all the local varieties and breeds, future options for adapting agricultural species to changing conditions will be sharply curtailed. (See the "Earth Watch" box, page 133.)

Some argue that biotechnology or genetic engineering, the transferring of genes from one species to another, may be our saving grace. But the potential of biotechnology depends on a source of those genes to transfer. We are still a long way from being able to synthesize a gene for a specific characteristic, much less a new species. Naturally occurring species are the source of the genes on which biotechnology depends. In short, the loss of biodiversity will handicap the potential of biotechnology as well as the traditional methods of breeding.

Since it undercuts the potential of future adaption to new conditions, reducing biodiversity seems dubious enough in itself, because we have seen that climatic changes are inevitable. But humans also appear to be embarked on causing global climatic changes at an unprecedented rate. For example, if projections of global warming come to pass (Chapter 16), the climatic changes we cause in the next 100 years will be greater than all the climatic changes that have occurred over the past 10,000 years. That is, they will be occurring at a rate 100 times faster! We have seen that most species are capable of adapting to slow changes, but only those rapidly reproducing species with short generation times may be able to adapt to rapid changes.

Another comparison is seen in the Hawaiian ecosystem. Before human transport, this ecosystem was invaded by a new species from the mainland perhaps once every 10,000 years. By contrast, in the past 200 years several hundred species of various plants and animals have been introduced to the islands. Is it any wonder that the natural Hawaiian ecosystem is being overwhelmed and exterminated by the new invaders (Fig. 5-21)?

FIGURE 5-21
Because of isolation from other ecosystems, the ecosystem on the Hawaiian Islands developed thousands of species that are found nowhere else. More than 500 of these unique species are in danger of extinction as a result of the introduction of more vigorous competing species. Shown here is the silversword. (Stan Goldblatt/Photo Researchers, Inc.)

An additional argument for preserving biodiversity that may have even more immediate importance is that about a quarter of all prescription drugs marketed in the United States have active ingredients derived from plants found in the wild. Thus, it is speculated that an additional bonanza of wonder drugs awaits discovery since 98 percent of tropical flora is still to be screened for medicinal properties. If this flora is destroyed before such screening occurs, we may lose medical properties of incalculable value.

Still, nonenvironmental business interests fail to be concerned about losing future values that have not been concretely proven when there is money to be made by short term exploitation. An argument used to support continued exploitation is that early in the environmental movement it was frequently stated that, since all living things were intricately connected in food webs, extinction of a single species might cause the collapse of the entire web—that is the entire biosphere. But many species have become extinct, and we are obviously still here, rendering this notion false.

However, does the fact that the biosphere has tolerated some loss of species without undue upset say that it can tolerate any amount of loss? How many examples can you think of where a little damage or injury is tolerable, but too much is disastrous? Perhaps it is more appropriate to see the loss of species, even if they seem insignificant, like the canaries used by miners in the early days of coal mining. Extra sensitive to poisonous gases, the death of the canaries served as an early warning of "bad air" in the mine. Unlike the miners, however, we cannot stay out of the "mine."

But all these arguments miss the most fundamental point. That is, what kind of world do we ultimately wish to create and leave to our progeny? As we have said, The reins of future evolution are in our hands. Doubts regarding its sustainability aside, do we wish to lead the process of evolution to a climax that consists almost entirely of humans and the agricultural systems devoted to supporting them? Or, would we rather see natural ecosystems with a richness of biodiversity flourishing and evolving naturally beside developing human culture? Probably few, if any, of us really desire the former outcome over the latter. The problem is that most of us have simply assumed that nature would always be there and would take care of itself, despite human activities. In the last few decades the choice has become clear; we need to make a conscious, concerted, unified effort to avoid exterminating a major portion of the biodiversity on the planet and perhaps ourselves in the process. Otherwise, we will be taking the former pathway by default, as the present trend of mounting loss of biodiversity testifies.

On the basis of this foundation, we now turn to the actual "mechanics" of how we may achieve the goal of sustainable development. We begin with a more in-depth study of human population.

Review Questions

1. What is meant by the gene pool of a species? By differential reproduction?
2. How is it possible to change the gene pool and the characteristics of a species?
3. What is selective pressure, and how does it relate to differential reproduction in nature?
4. What are the similarities and differences between selective breeding and natural selection in terms of the process and the outcome?
5. What is the chemical structure of genes, and what is the relationship of genes to alleles and to observable traits or characteristics?
6. How can natural selection lead to the development of new species? What is the process called? What is the role of mutations? What are the prerequisites and limitations?
7. Does natural selection act on a higher level than the species?
8. How may natural selection lead to the development of a balanced ecosystem?
9. What factors are responsible for the similarities and differences among ecosystems?
10. What factors determine whether a species can adapt to, or will be forced into extinction, by a change?
11. Which species are most likely, and which are least likely, to survive changes and be able to adapt to new conditions? Can species adapt to changes before they actually occur?
12. What factors are making future changes in the biosphere inevitable?
13. What are the implications of diminishing biodiversity for the future of agriculture? For the future survival of wildlife?
14. Contrast the "Ages" of fish, amphibians, and reptiles with the "Age of Humans."
15. How long is the "Age of Humans" likely to last? What factors will determine its length?

Thinking Environmentally

1. Select a particular species, and describe how its various traits support its survival and reproduction and how these traits may have developed by, and are maintained through, natural selection.
2. How do strains of disease-causing organisms become resistant to medicines used against them?
3. Why is it biologically impossible for a "Ninja turtle" to arise from a single mutation?
4. To what extent is differential reproduction occurring in the human population, and what may be the long-term results?
5. Speculate as to the future course of evolution, including the evolution of humans, if substantial biodiversity is lost.
6. Discuss how we humans will determine the length of the "Age of Humans," in terms of our ability to abide by each of the four basic principles of ecosystem sustainability.

Making a Difference—PART ONE: Chapters 1, 2, 3, 4, 5

1. Begin to find out where and how your school, college, community or city gets its water and power, how it disposes of sewage and refuse, and what is the trend of land development and preservation around the area.
2. Find out if there is an environmental organization or club on your campus or in your community. Join if there is, or create one if there isn't.
3. Learn the common names of trees and other plants, birds and mammals found in your area. This knowledge is a prerequisite to understanding and describing an actual ecosystem. Then, pass your knowledge on to children, who are often curious about the natural world.
4. Take courses offered by colleges or local environmental organizations that give you a greater appreciation for the natural world; amaze and influence your friends by becoming a local expert in some taxonomic group, such as birds, wildflowers, butterflies.
5. Consider small animals, birds, butterflies and other wildlife that might be present in your yard, school or campus grounds were it not for human-created limiting factors. Investigate and begin a project to create natural habitats that will attract and support additional wildlife.
6. Create a list of the things you do and the things that you use and throw away. Write a brief evaluation of each in terms of the principles of sustainability. Consider how you might begin to change any of these habits to move toward a more sustainable option. Begin by making one such change.
7. Rather than using chemical pesticides and fertilizers to maintain your lawn, which is an "unnatural" monoculture, introduce clover and other low-growing flowering plants that will create a sustainable balance with only mowing.
8. Read local papers, make contact with environmental organizations and become aware of efforts to protect natural areas locally. Support efforts to protect such areas locally, regionally and globally.
9. Select a current environmental issue related to biodiversity or endangered species, and write your congresspersons to express your concern and ask for their support for legislation that effectively addresses the issue.
10. Support and join efforts and organizations, such as the World Wildlife Fund and Conservation International, that are devoted to protecting endangered and threatened species.
11. Continue your education toward a profession that is important for environmental concerns, and make your life work one that helps rather than hinders the environmental revolution.
12. Use your citizenship to support environmentally sound practices and policies by voting in your local, state and national elections for candidates who are clearly environmentally aware.

PART TWO

Finding Balance Among Population, Soil, Water, and Agriculture

A mountainside on the island of Bali is terraced for rice cultivation; tall-grass prairies of Illinois and Iowa are plowed under to raise corn; tropical rain forests in Brazil are converted to cattle pasture; dry woodlands in Kenya are grazed by goats. Wherever there are people, the environment is brought into service to produce food. By March 1999, six billion people will depend on the environment to bring them their daily bread. Soil, water, nutrients, sunlight—the basis for productivity in natural ecosystems—are diverted toward meeting human needs through agricultural enterprises or through direct harvesting from natural systems. Many of these same systems also produce fuel, building materials, fabrics, and much else for human use. The global economy is largely based on extracting food and products from the natural world.

The population continues to increase by nearly 88 million per year and, thankfully, soil, water, trees, crops, and animals are all "renewable" resources—if this were not so, we would long since have run out of them. Yet their capacity to renew is limited, and in our efforts to put nature to use, we can easily take faster than these systems can give. A moment's thought is enough to conclude that at some point, the human population must come into a basic balance with resources for food and other needs. In Part Two, we make serious application of the principles of sustainability, as we consider how many of us there are and how we have used and misused the natural world, and what we must do to move to a sustainable future.

Mountainside on the Island of Bali. [Photo by Michele Burgess/the Stock Market.

FILM·CAMERAS
Sale
EXCLUSIVE
to 20th APRIL
CHHAYA
BALUJA
SHOES
saree emporium
CHHAYA
SAPNA
Superlon
SUITINGS
SHIRTINGS
KARAM CHAND & CO.

The Global Human Population Explosion: Causes and Consequences

Key Issues and Questions

1. The human population is undergoing an explosion. When did it start? What are its causes? What is the current growth rate?
2. The world comprises high, middle, and low-income nations. Identify the nations or regions, and describe representative lifestyles in these three groups.
3. The most rapid population growth is occurring in developing countries, or less developed countries as they are sometimes called. What are the social and environmental consequences of such growth for developing countries? For developed countries?
4. Population profiles may be used to predict future populations. What is a population profile? How are future populations predicted?
5. Population profiles for developed and developing nations are fundamentally different. What are the differences?
6. Populations of developing countries will continue growing for some time, even after fertility is reduced to replacement rates. Why? What is the phenomenon called?
7. Birth rates and death rates have been observed to undergo a change as a country develops. What is the change in each? How is it related to the population explosion? To future stability?
8. There are serious fallacies in the notion that the world can reach a sustainable state through development alone. What are these fallacies?

The third basic principle of ecosystem sustainability, first set forth in Chapter 4, is that the size of consumer populations is maintained so that overgrazing or other forms of overuse do not occur. Yet the global human population is undergoing what many term an explosion. It has grown some sixfold in the last 200 years, and while the rate of growth is slowing somewhat, the increase in absolute numbers continues to be greater than at any time in history—nearly 88 million people per year (births minus deaths). Also, there has been and continues to be an average trend toward higher per capita consumption of both energy and materials, not the least of which is land. In the United States alone, about 2 million

India. [Photo by Dillip Mehta/The Stock Market.]

acres (0.8 million hectares) per year of natural habitat are being sacrificed to satisfy human need or greed. Thus, the demise of natural ecosystems and the additional stresses caused by pollution around the world may be seen as a direct consequence of the global human population explosion together with increasing consumption.

What do these trends portend for sustainability? Many environmentalists see the impacts of global human population on the planet as overgrazing and overuse. They see human population as on a *J*-curve (Chapter 4, p. 96), much like the reindeer on St. Matthew Island and perhaps not too far from the peak at which critical resources are exhausted, causing a breakdown of the system. The recent episodes of anarchy, social collapse, and killing in Somalia and Rwanda might be taken as an indication of what is in store on a larger scale if population is not stabilized.

Yet, specific efforts to reduce population growth collide with various ethical and moral values, creating an impasse. Moreover, not everyone agrees that the population issue is serious. The major counterarguments are as follows:

- Some claim that population growth is beneficial in that more people provide more ideas, creativity, and work. This belief is supported by the fact that the greatest technological advances and improvements in living standards have occurred in parallel with the population explosion of the past 200 years.
- Some take the position that any artificial interference in the reproductive process (including the use of sex education, contraceptives, and especially abortions) is immoral. Therefore, for them, any thought of altering the course of population growth, except by abstinence, is not debatable.
- Some argue that population growth is not the issue as much as consumption is, at least for the present. What needs to be achieved more than reducing growth in numbers, they say, is adopting conservation measures that will reduce consumption.
- Others take the position that population growth will level off by itself well within the capacity to support the population. This view is supported by the facts that we have been able to expand agricultural production even faster than population growth and that the average number of births per woman is coming down.
- Still others are disturbed by the fact that population programs often seem to have the trappings of social engineering—the rich trying to get rid of the poor or minorities by preventing them from having children.

Our objective in this chapter is to gain an understanding of the dynamics of population growth and its social and environmental consequences. Ultimately, all individuals will have to reach their own conclusions on the population issue and act accordingly.

The Population Explosion and Its Cause

Considering all the thousands of years of human history, the explosion of the global human population is a recent and unique event—a phenomenon of just the past 200 years (Fig. 6-1). Let us look at this event and why it occurred more closely.

The Explosion

From the dawn of human history until the beginning of the 1800s, population increased slowly and variably with periodic setbacks. It was roughly 1830 before world population reached the 1 billion mark. But by 1930, just 100 years later, the population had doubled to 2 billion. Barely 30 years later, in 1960, it reached 3 billion, and in only 15 more years, by 1975, it had climbed to 4 billion. Thus, population doubled in just 45 years, from 1930 to 1975. Then, 12 years later, in 1987, it crossed the 5 billion mark! In early 1999, world population is projected to pass 6.0 billion, as it continues to grow at the rate of nearly 88 million people per year. This rate is equivalent to fitting into the world every year the combined populations of New York, Los Angeles, Chicago, Philadelphia, Detroit, Dallas, Boston, and ten other U.S. metropolitan areas.

Based on current trends, the Population Reference Bureau (a private educational organization) projects that world population will pass the 7 billion mark in

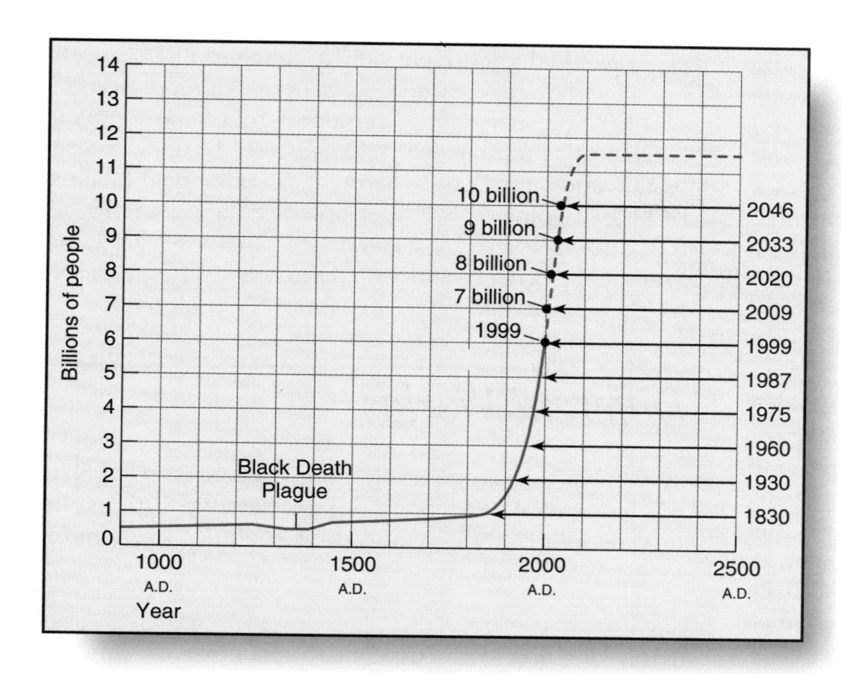

FIGURE 6-1
The world population explosion. For most of human history, population grew slowly, but in modern times it has suddenly "exploded." Further, the gradually declining but still high birth rates will carry the population to about 12 billion by the end of the next century. This is simply a mathematical projection based on current population statistics. It does not address the question of whether the biosphere will be able to support 12 billion people and at what lifestyle. (From Joseph A. McFalls, Jr., "Population: A Lively Introduction," *Population Bulletin*, 46, no. 2 [Washington, DC: Population Reference Bureau, Inc., Oct. 1991], p. 4.)

2009, the 8 billion mark in 2020, the 9 billion mark in 2033, and the 10 billion mark in 2046, before population finally levels off between 11 and 12 billion people by the end of the next century (Fig. 6-1).

Reasons for the Explosion

The main reason for slow and fluctuating population growth prior to the early 1800s was the prevalence of diseases, such as smallpox, diphtheria, measles, and scarlet fever, that were often fatal. These diseases hit infants and children particularly hard. It was not uncommon for a woman who had seven or eight live births to have only one or two children reach adulthood. In addition, epidemics of diseases such as typhoid fever and cholera, and the black plague of the 14th century, would eliminate large numbers of adults. Famines also were not unusual.

Biologically speaking, prior to the 1800s, the population was essentially in a dynamic balance with natural enemies—mainly diseases—and other aspects of environmental resistance. High reproductive rates were largely balanced by high mortality, especially among infants and children. With high birth and death rates, the population growth rate was low in these preindustrial societies.

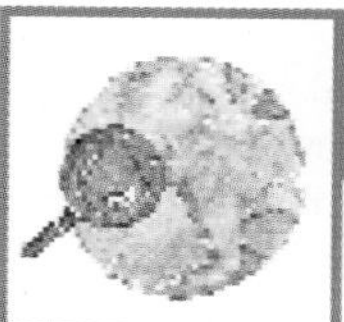

Earth Watch

We Are Living Longer

It is commonly said that, with the introduction of modern medical technology and disease control, average longevity increased from about 40 to 65 years of age. While mathematically correct, this statement may be misleading. Many people take it to mean that nearly everyone used to die around age 40 and now everyone lives to around 65. However, the years between 35 and 45 are generally the healthiest period of human life. With or without modern medicine, a relatively small portion of the population dies in this age range. The "average longevity of 40" before modern techniques of disease control was a function of a large fraction of the population dying in childhood, thus counterbalancing another fraction that lived into their sixties and beyond.

Through disease control, most of those who once would have died in childhood now live past 50. Extending the life span of this group raises the average age of death of the population. But the basic life span of the human species has changed little, if at all, as a result of modern medicine. All modern medicine has done is to increase the proportion of people who get to or near the maximum age.

In the 1800s, Louis Pasteur and others made the major discovery that diseases were caused by infectious agents (now identified as various bacteria, viruses, and parasites) and that these were transmitted via water and food, insects, and other vermin. With these discoveries came major improvements in sanitation and personal hygiene. Then, techniques of providing protection by means of vaccinations came into play. Later, in the 1930s, the discovery of penicillin, the first in a long line of antibiotics, resulted in cures for otherwise often-fatal diseases such as pneumonia. Improvements in nutrition began to be significant as well. In short, better sanitation, medicine, and nutrition brought about spectacular reductions in mortality, especially among infants and children, while birth rates remained high.

Again from the biological point of view, the human population entered into exponential growth, as does any natural population on being freed from natural enemies and other environmental restraints. Note the rapidly declining death rates must be compared to birth rates that remained high.

In the last few decades, the average fertility rate—that is, the average number of babies born to a woman over her lifetime—has declined, resulting in a decreasing rate of growth of population. Still, as the number of people of reproductive age has expanded with the increasing numbers of children growing up, even the lower fertility rates continue to add absolute numbers faster than at any other time in history (Fig. 6-2). Extrapolating the trend of lower fertility rates leads to the projection that the global human population will level off at around 12 billion toward the end of the next century.

The projected leveling off at around 12 billion is what leads some to believe that the population situation will take care of itself. However, such complacency fails to recognize that the currently declining fertility rates are a reflection of already significant efforts at reducing births, including what many consider an extremely coercive one-child-per-couple policy exercised in China.

Even more, the future global population projection of 12 billion in no way considers any of the looming ecological questions of whether the biosphere can sustain such numbers. Where are the additional billions of people going to live, and how are they going to be fed, clothed, housed, educated, and otherwise cared for? Will enough energy and material resources be available for them to fulfill a poor, medium, or rich lifestyle?

Different Worlds

To answer the foregoing questions, we must recognize the tremendous economic disparity among nations. In fact, people in wealthy and poor countries live almost in separate worlds.

Rich Nations and Poor Nations

The World Bank, an arm of the United Nations, divides the countries of the world into three main economic categories according to average per capita gross national product (Figure 6–3):

1. *High-income, highly developed, industrialized countries*. This group includes the United States, Canada, Japan, Australia, New Zealand, and the countries of western Europe and Scandinavia. (1994 gross national product per capita = $22,745 average; Japan first at $34,630, United States second at $25,860 and Germany third at $25,580—the three highest per capita income countries as well.
2. *Middle-income, moderately developed countries:* These are mainly the countries of Latin America (Mexico, Central America, and South America), northern and western Africa, eastern Asia, eastern

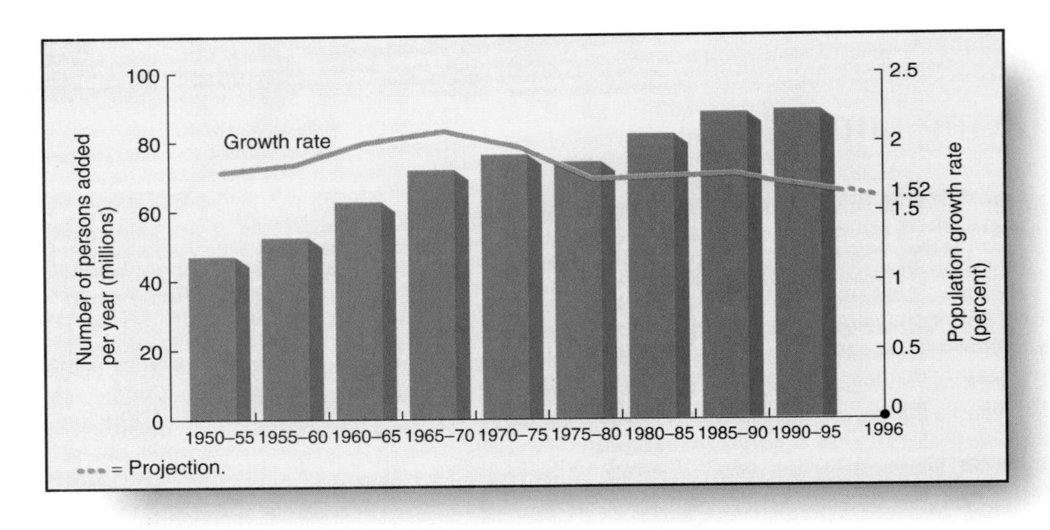

FIGURE 6-2
World population growth rate and absolute growth. Declining fertility rates in the last three decades have resulted in a decreasing rate of growth. However, with a larger population base, absolute numbers are growing faster than ever. (Shiro Horiuchi, "World Population Growth Rate," *Population Today*, June 1993. Washington, DC: Population Reference Bureau, Inc., June 1993, p. 7 and June/July 1996, p.1.)

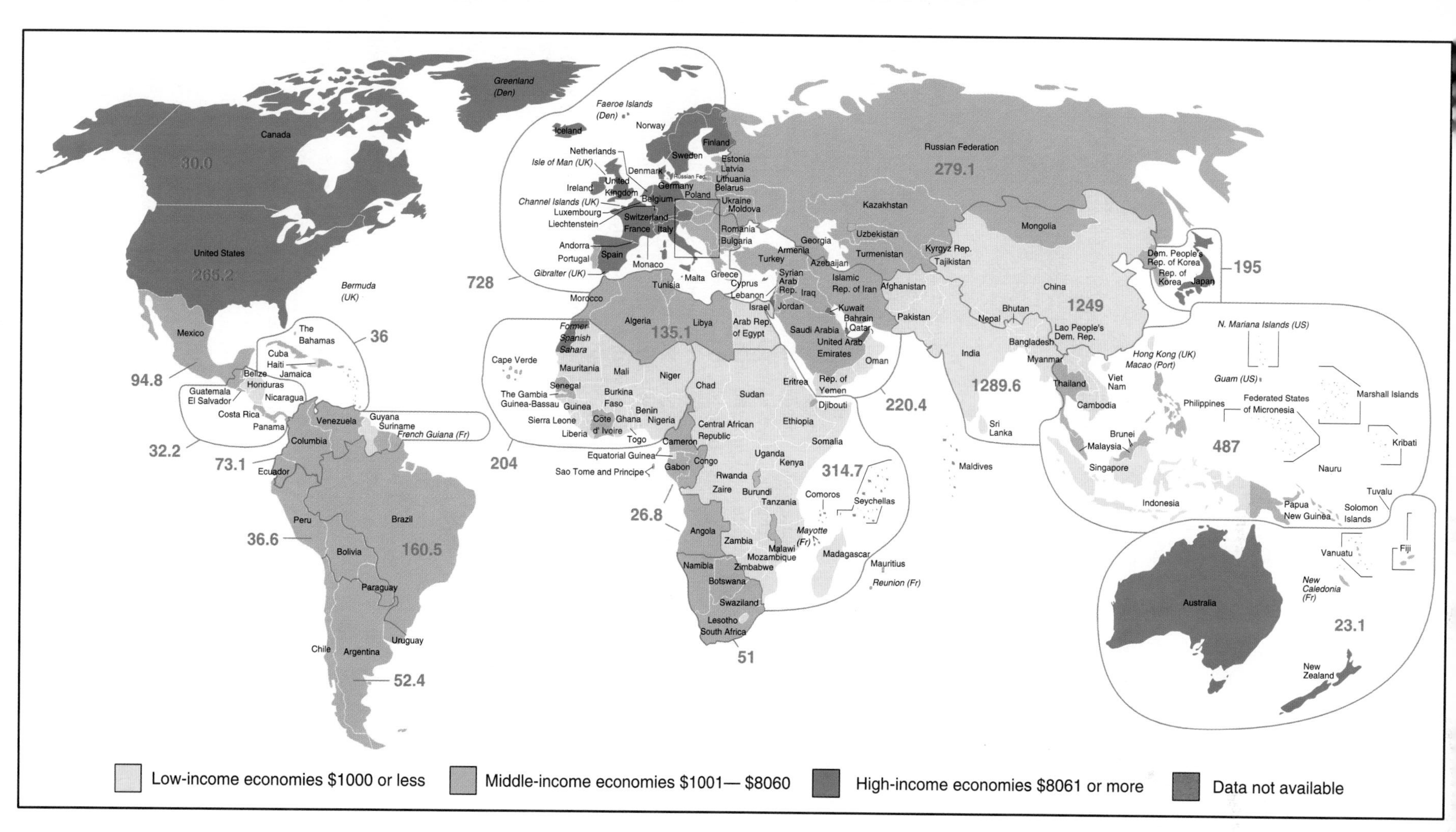

FIGURE 6-3
Nations of the world grouped according to gross national product per capita, a general indicator of standard of living. The population in millions of various regions is also shown by magenta lines and numbers. (From the *World Development Report*, 1994. Copyright © 1994 by the International Bank for Reconstruction and Development/The World Bank. Reprinted by permission of Oxford University Press, Inc., New York. Population and per capita gross national product data from the Population Reference Bureau, *World Population Data Sheet*, 1996.)

TABLE 6-1
World Consumption Classes, 1992

Category of consumption	*Consumers (1.1 billion)*	*Middle (3.4 billion)*	*Poor (1.3 billion)*
Diet	Meat, packaged food, soft drinks	Grain, clean water	Insufficient grain, unsafe water
Transport	Private cars	Bicycles Buses	Walking
Materials	Throwaways	Durables	Local biomass

Source: Alan Durning, *How Much Is Enough*, Washington, DC: Worldwatch Inst., 1992, p. 27; updated United Nations, 1996.

Europe, and countries of the former USSR. (1994 gross national product per capita range from $1001–8060)

3. *Low-income countries.* This group comprises the countries of eastern and central Africa, India, and other countries of central Asia. (1994 gross national product per capita = less than $1000.) Note that the People's Republic of China is still placed in this category with a per capita GNP of $530 (1994), but it may soon move into category 2.

The high-income nations are commonly referred to as **developed countries**, whereas the middle- and low-income countries are often grouped together and referred to as **developing countries**. The terms *highly developed countries* (HDCs) and *Third World countries* are being phased out, although you will still hear them used. (The Second World was the former Communist bloc, which no longer exists. Therefore, referring to the developing countries as Third World countries is obsolete.)

The disparity in distribution of wealth among the countries of the world is mind boggling. More developed countries hold just 21% of the world's population, yet they control about 80% of the world's wealth. Thus, developing countries, which have 79% of the world's population, have only about 20% of the world's wealth. Of course, the distribution of wealth *within* each country is also disproportionate. Between 10% and 15% of the people in more developed countries are recognized as poor (unable to afford adequate food, shelter, and/or clothing), while about 90% of those in developing countries are.

The disparity of wealth is difficult to understand just by looking at monetary figures. Therefore, Allen Durning of the WorldWatch Institute describes the relative wealth for different peoples of the world in terms of their access to different kinds of food, drinks, and transportation, as is shown in Table 6-1. It is sobering that, although some people in developed nations such as the United States may feel poor, they are still well within the rich category in terms of representative living standards for the world.

The people in the lower-middle and poor categories, predominantly in low- and middle-income countries, live along a bare "subsistence margin," with little beyond the minimum requirements for survival. Roughly half of these—somewhat over a billion people—live in a condition of "absolute poverty," defined by Robert McNamara, president of the World Bank in 1978, as, "A condition of life so limited by malnutrition, illiteracy, disease, squalid surroundings, high infant mortality, and low-life expectancy as to be beneath any reasonable definition of human decency."

The present disparity between high- and low-income countries is in part a legacy of the colonialism that existed in the 18th and 19th centuries. France, England, Portugal, Spain, and other European countries effectively took over Asia, Africa, and Latin America and made them into colonies. Land ownership was taken away from the native people. Without other means of sustenance, they worked for minimal wages growing cash crops such as cotton and tea for the colonial parent. These countries were prevented from industrializing. Of course, the United States started out in this way, too, but Native Americans refused to be "enslaved" and were largely killed off or segregated onto reservations instead. Then, the early colonizers rebelled, became independent, and joined the Industrial Revolution. Most of the countries of the now developing world gained independence only shortly before or after World War II. Starting with no industrial base and with little experience or skill in self-governing, they have, with few exceptions, remained poor. Nonetheless, the major portion of population growth is occurring in these developing countries.

Population Growth in Rich and Poor Nations

The population growth shown in Figs. 6-1 and 6-2 is for the world as a whole. If we look at population growth in more developed versus developing countries, we find that about 90% of the population growth is occurring in the developing countries alone. Why is this so?

In the absence of high mortality, the major determining factor for population growth is **total fertility rate**—the average number of children each woman has over her lifetime. It follows that a total fertility rate of 2.0 will give a stable population in that two children per woman will just replace the parents when they eventually die. Fertility rates greater than 2.0 will give a growing population in that each generation is replaced by a larger one. Conversely, barring immigration, a total fertility rate less than 2.0 will lead to a declining population because each generation will eventually be replaced by a smaller one. Because infant and childhood mortality are not in fact zero, **replacement fertility**—that fertility rate which will just replace the population of parents—is slightly higher: 2.03 for developed countries and 2.16 for developing countries, which have higher infant and childhood mortality.

TABLE 6-2
Population Data for Selected Countries

Country	*Total Fertility*	*Doubling Time (Years)*
World	3.0	46
Developing Countries		
Average (excluding China)	4.0	31
Egypt	3.6	31
Kenya	5.4	25
Madagascar	6.1	22
India	3.4	37
Iraq	6.7	19
Viet Nam	3.7	30
Haiti	4.8	30
Brazil	2.8	41
Mexico	3.1	32
Developed Countries		
Average	1.6	501
United States	2.0	114
Canada	1.6	116
Japan	1.5	315
Denmark	1.8	462
Germany	1.3	—*
Italy	1.2	—*
Spain	1.2	1155

Data from 1996 World Population Data Sheet. Population Reference Bureau, Inc.
* Population is declining.
Internet: http://www.prb.org/prb
http://sunsite.unc.edu/lunarbin/worldpop

To argue that a higher fertility rate in developing nations is *necessary* because more children die in such countries makes little sense and is not supported by facts. In a humane society, our focus must be on mitigating the conditions that cause infant and childhood mortality. As this mitigation occurs, replacement fertility rates will come closer to 2.0.

For reasons we shall discuss later, total fertility rates in developed countries have declined over the past several decades to the point where they are now almost all below 2.0. The one major exception is the United States, which has a total fertility rate of 2.0 (1996). In developing countries, on the other hand, although fertility rates have come down considerably in recent years, they are still mostly in the range of 3.0 to 6.0, rates that will cause the populations of those countries to double in just 20 to 40 years (Table 6-2). At the very least, this means that the populations of poor countries will continue growing, while the populations of more developed countries will stabilize or even decline. Therefore, the percentage of the world's population living in developing countries—already 79%—will climb steadily to over 90% by 2075 (Fig. 6-4). However, this is not to say that only the developing countries have a population problem.

Different Populations Present Different Problems

Today increasing numbers of people put increasing demands on the environment, both through demands for resources including food and water, and through the production of wastes. However, it should also be clear that the demand each individual makes on the environment depends on how much and what that individual consumes. Each additional thing purchased represents a certain additional demand on resources for its production, as well as additional wastes produced in the course of its production, use, and, finally, disposal. Therefore, negative effects on the environment also increase dramatically as consumption increases.

For example, it is estimated that because of differences in consumption, the average American causes at least 20 times the demand on Earth's resources, including its ability to absorb pollutants, as does the average person in Bangladesh, a poor Asian country. Major world pollution problems, including depletion of the ozone shield, the implications of global warming, and the accumulation of toxic wastes in the environment, are almost exclusively the consequence of the high consumption associated with affluent lifestyles. For instance, it is people who drive cars and heat and cook with fossil fuels that contribute most significantly to rising levels of carbon dioxide in the atmosphere. Likewise, much of the global deforestation and loss of biodiversity is a consequence of consumer demands in developed countries.

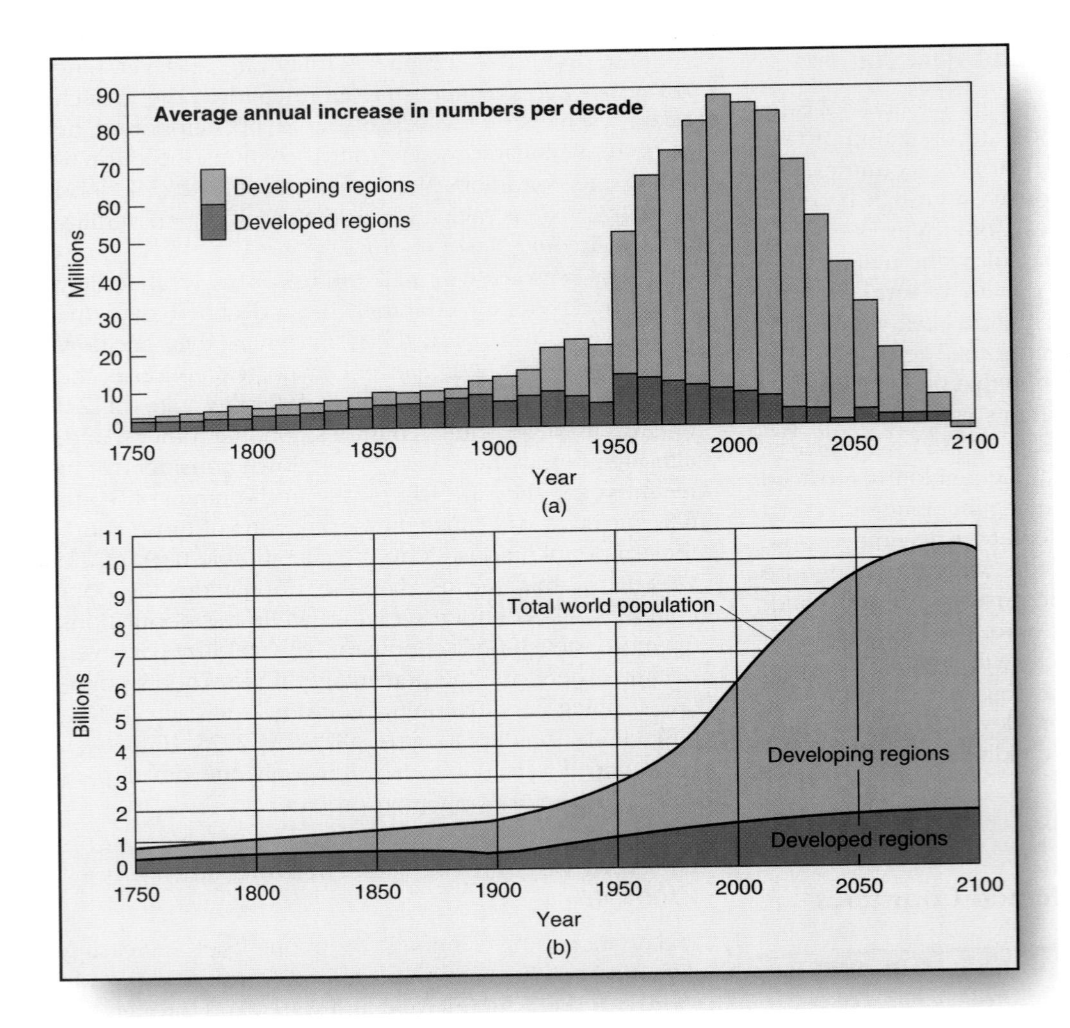

FIGURE 6-4
Population increase in developed and developing countries. (a) Because of higher populations and higher birth rates, most of the population growth is occurring and will continue to occur in developing countries. (b) Because of higher growth rates, developing countries represent a larger and larger share of the world's population. If current trends continue, developing nations will have 90% of the world population by 2075. (From Thomas W. Merrick, "World Population in Transition," *Population Bulletin*, 41, no. 2 [Washington, DC: Population Reference Bureau, Inc., Jan. 1988, reprint], p. 4.)

Environmental impacts of affluent lifestyles may be moderated to a large extent by a factor we call **environmental regard**. For example, suitable attention to wildlife conservation, pollution control, energy conservation and efficiency, and recycling may offset, to some extent, the negative impact of a consumer lifestyle. One might make a good argument that a life devoted to conservation or other aspects of environmental regard might entirely offset the negatives and have a very positive effect overall. The relationship between these factors may be expressed as

$$\text{Negative environmental impact} \propto \frac{\text{Population size} \times \text{Consumptiveness of lifestyle}}{\text{Environmental regard}}$$

This proportionality should be read as "Negative environmental impact is proportional to the population multiplied by the consumptiveness of the population's lifestyle, moderated by the environmental regard of the population." Note that it is not a strictly mathematical proportionality, because meaningful numbers cannot be assigned to any of the factors except population.

You may hear debates among people, one side arguing that population growth is the main problem, the other arguing that our highly consumption-oriented lifestyle is the main problem. Our contention in this text is that in order to reach sustainability, all *three* areas must be addressed: Stabilize population, decrease consumption, and increase environmental regard. For the present, we wish to maintain our focus on the factors and consequences of population growth, mainly in developing countries.

Environmental and Social Impacts of Growing Populations and Affluence

Both growing populations and increasing affluence have numerous environmental and social implications for the whole world.

The Growing Populations of Developing Countries

Prior to the Industrial Revolution, most of the human population survived through subsistence agriculture. That is, families lived on the land and produced enough food for their own consumption and perhaps enough extra to barter for other essentials. Natural forests provided firewood, structural materials for housing, and wild game for meat. With a small, stable population, this system was basically sustainable. As the older generation passed away, the land and natural systems could still support the next generation. Indeed, many cultures sustained themselves in this way over thousands of years, and considerable areas of Latin America, Africa, and Asia maintain this tradition today.

After World War II, modern medicines—chiefly vaccines and antibiotics—were introduced to developing nations, whereupon death rates plummeted, and population growth increased. What are the impacts of rapid growth on a population that is largely engaged in subsistence agriculture? Five basic alternatives are possible, all of which are being played out to various degrees by people in these societies.

1. Subdivide farms among the children of the next generation and/or intensify cultivation of existing land to increase production per unit area.
2. Open up new land to farm.
3. Move to cities and seek employment.
4. Engage in illicit activities for income.
5. Emigrate to other countries legally or illegally.

In addition, rapid population growth especially affects women and children.

Let us look at each of the preceding alternatives and their consequences in a little more detail.

Subdividing Farms and Intensifying Cultivation. Over wide areas of Asia and Africa, plots of land have been divided and redivided to the point that the United Nations Food and Agriculture Organization (FAO) estimates that over a billion rural people live in households that have too little land to meet even their own meager needs for food and fuel, much less producing extra for income from barter.

Adding to the problem is the quest for firewood. Some 3 billion people, or 60% of the world's population, do not have gas, electric, or even kerosene stoves; hence, they depend on firewood in the preparation of their daily food. Forests are being cut faster than they can regenerate. Much of East Africa, Nepal, and Tibet, as well as many slopes of the Andes in South America, have been deforested as a result. Already, women in many developing countries spend a large portion of each day on increasingly long treks to gather firewood (Fig. 6-5). As many as 3 billion poor people face acute shortages of firewood. Worse, in the long term, as the forests are removed, massive soil erosion occurs. In Chapter 1 we discussed a simple and inexpensive solution to this problem—solar panel cookers (see Figure 1-12).

FIGURE 6-5
Like the Nepalese woman seen here, some 50% of the world's population still depends on gathering firewood for cooking and other fuel needs. The resulting deforestation is causing both ecological and human tragedy. (Zviki-Eshet/ The Stock Market.)

The consequences of soil erosion are manifold. The loss of topsoil diminishes the future productivity of the land. It also results in more water running off rather than soaking into the ground, a situation that aggravates flooding in the lowlands and diminishes groundwater. Soil washing into streams and rivers destroys fisheries, clogs channels, and also aggravates flooding. (These effects are explained further in Chapter 11.)

With regard to the intensification of cultivation, the introduction of more highly productive varieties of basic food grains has had a dramatic positive effect in supporting the growing population, but is not without some

FIGURE 6-6
Millions of acres of rain forest in Central and South America are being cut down each year to make room for agriculture, as shown in this photograph from Peru. Unfortunately, the agricultural benefits are meager, because a very thin topsoil is soon washed away, leaving only a hard, nutrient-poor subsoil that is nearly impossible to till and yields little produce. (Asa C. Thresen/Photo Researchers, Inc.)

concerns, as will be described in Chapter 8. However, intensification of cultivation also means simply working the land harder. For example, traditional subsistence farming in Africa involved rotating cultivation among three plots. In that way, after being cultivated for one year, the soil in each plot had two years to regenerate. With pressures for increasing productivity, plots have been put into continuous production, with no time off. The results have been deterioration of soil, decreased productivity (ironically), and erosion.

In addition, the increasing intensity of grazing is damaging the land. (Desertification, the extremely serious consequence of overgrazing, will be considered further in Chapter 9.) Given the countertrends of rapidly increasing population and deterioration of land from overcultivation, food production per capita in Africa, for example, is currently on a downward course.

Opening Up New Lands for Agriculture. With respect to opening up new lands for agriculture, first consider that there really is no such thing as "new land." Instead, it is always a matter of converting the land of natural ecosystems to agricultural production, which means increasing pressure on wildlife and an unavoidable loss of biodiversity. Even then, converted land may not be well suited for agriculture. (Most good agricultural land is already in production.) For example, it is estimated that two thirds of the tropical deforestation that is occurring in Brazil and Central America is for the purpose of increasing agricultural production (Fig. 6-6). Much of this deforestation is done by poor, young people who are seeking an opportunity to get ahead, but are unskilled and untrained in the unique requirements of maintaining tropical soils. Consequently, beyond the loss of biodiversity, we have the additional problem that between a third and a half of the cleared land becomes unproductive within three to five years, again leaving the people in absolute poverty.

Additional "new lands" include steep slopes, which suffer horrendous erosion when plowed, and areas with minimal rainfall that turn to deserts under cultivation.

Migration to Cities. Faced with the poverty and hardship of the countryside, many hundreds of millions of people in developing nations continue to migrate to cities in search of employment and a better life. The result is that a number of cities of the developing world are now among the world's largest and are still growing rapidly (Fig. 6-7a). Opportunities in many cities have not expanded fast enough to handle the influx. People are forced to live in sprawling, wretched squatter set-

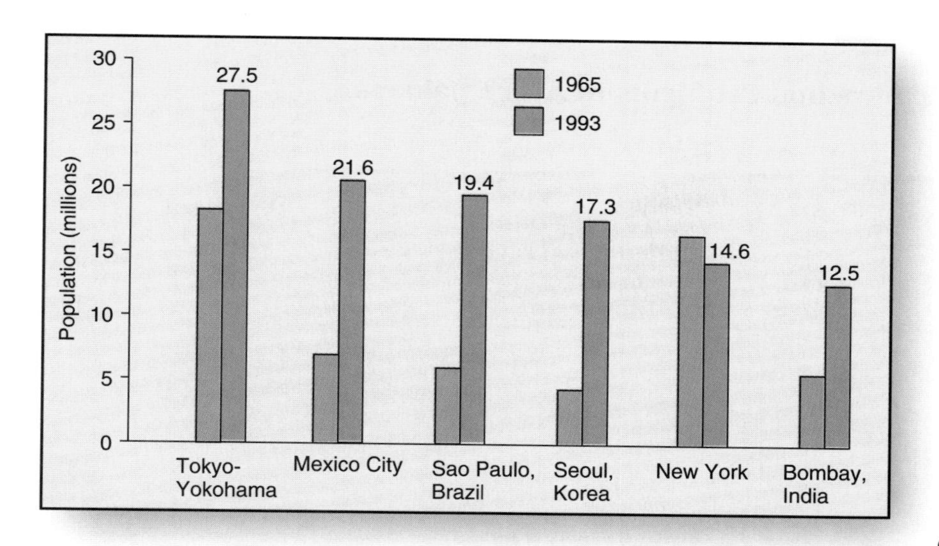

(a)

(b)

FIGURE 6-7
(a) Growth of some major world metropolitan areas. Since 1965, cities in the developing world have grown phenomenally, and a number of them are now among the world's largest. (b) Unfortunately, in many cases, the growth rate has far surpassed the city's ability to provide adequate housing, sewer and water systems, and schools, leaving many people to live in despicable squatter settlements and slums as seen here on the outskirts of São Paulo, Brazil where 32% of the city's population live. By contrast, the growth of cities in the developed world has been more modest. ((a) From the *Christian Science Monitor,* Dec. 11, 1990, p. 12 updated from U.S. Bureau of the Census, *International Data Base,* 1993. (b) Nickelsberg/Gamma-Liaison, Inc.)

tlements and slums that do not even provide adequate water and sewers, much less other services (Fig. 6-7b, Table 6–3). Consider the following description of São Paulo, the industrial center of Brazil:

> Too many people in the wrong places. The shacks and shanty towns surround São Paulo in concentric circles—called the rings of misery—of millions of people living below the poverty level, trying to earn, beg or steal a living with virtually no hope of aid from a government that feels it has a long way to go before it can consider social welfare programs. And most of those people are young; each year three million Brazilians enter the job market. They come to the cities because, bad as life is there, it is better than in the desolate rural area where many were born. It is expected that the population of São Paulo—as well as of Rio de Janeiro, with its five million people—will double by the end of the century. Brazil's economy must grow rapidly if the country is to keep from eating itself alive.[1]

Consider also this, said of Bombay, India:

> More than a million people crouch cheek by jowl along a maze of suffocating narrow footpaths. Naked children play in black puddles of stagnant sewage beside tents of rags and corrugated tin huts. A choking stench, unending noise and festering disease are everywhere.[2]

[1]Brian Kelly & Mark Landon, *Amazon.* (New York: Harcourt Brace Jovanovich, 1983, p. 19.)

[2]Robert Benjamin, *Baltimore Sun.* June 16, 1991.

TABLE 6-3
Incidence of Household Environmental Problems in Accra, Jakarta, and São Paulo

Environmental Indicator	*Accra, Ghana*	*Incidence of Problem (percentage of all households surveyed)* Jakarta, Indonesia	*São Paulo, Brazil*
Water			
No water source at residence	46	13	5
No drinking water source at residence	46	33	5
Sanitation			
Toilets shared with more than 10 households	48	14–20	<3
Solid waste			
No home garbage collection	89	37	5
Waste stored indoors in open container	40	27	14
Indoor air			
Wood or charcoal is main cooking fuel	76	2	0
Mosquito coils used	45	28	8
Pests			
Flies observed in kitchen	82	38	17
Rats/mice often seen in home	61	82	25

Note: Sample sizes were as follows: Accra, $N = 1{,}000$; Jakarta, $N = 1{,}055$; and São Paulo, $N = 1{,}000$. Missing values, never more than 3% of the sample, are omitted. *Missing values* refers to cases where the question under consideration was not answered. It includes both "no response" and "not applicable." In some cases, questions were slightly different for the three cities, which accounts for the use of intervals.
From: McGranahan & Songsore, "Health, Wealth, and the Urban Household," *Environment*, July–Aug 1994, p. 9.

It is not coincidental that three epidemics of cholera have occurred in the developing world in recent years. Cholera is caused by a bacterium spread via sewage that gets into drinking water. The disease causes extreme vomiting and diarrhea, which result in great loss of body fluids; it is frequently fatal if not properly treated.

Worse, these cities do not provide the jobs people seek. Indeed, the high numbers of rural immigrants in the cities dilute the value of the one thing they have to sell—their labor. A common wage for a day's unskilled work is often equivalent to no more than a dollar or two, not enough for food, much less housing, clothing, and other amenities. Thousands, including many children, make their living by scavenging in dumps to find items they can salvage, repair, and sell. Many survive by begging—or worse.

Illicit Activities. Anyone who doesn't have a way to grow sufficient food must gain enough income to buy it—and sometimes, desperate people break the law to do this. Of course, it is difficult to draw the line between need and the greed that also draws people into illicit activities. However, it is undeniable that the shortage of adequately paying employment exacerbates the problem. Besides the rampant petty thievery and corruption that pervade many developing countries, income is also obtained from the following illegal activities.

Illicit drugs. A small peasant farmer with too little land to make a living growing food can make a decent income growing the various crops from which illicit drugs are made. Of course, those who synthesize and sell the drugs make even greater profits.

Poaching of wildlife. A person in a tropical developing country can make a considerable income hunting or trapping various fish, birds, reptiles, and other animals and selling them into the "pet" trade. Illegal trade involving endangered species has grown steadily over the years; in black market activity, it is second only to drugs. Since few of the animals collected survive, and fewer yet are put into situations in which they will breed, poaching is a major factor in pushing many species toward extinction.

Given the poverty in developing countries, can you see why these activities persist despite efforts to

Ethics

The Dilemma of Immigration

For people trapped by poverty or lack of opportunity in their homeland, emigration to another country has always seemed a way to achieve a better life. As the New World opened up, many millions of people recognized this dream by emigrating to the United States and other countries. The United States and Canada are largely countries composed of immigrants and their descendants. Until 1875, all immigration into the United States was legal; all who could manage to arrive could stay and become citizens. This openness was inscribed on the Statue of Liberty, which reads, in part, "Give me your tired, your poor, your huddled masses yearning to breathe free, the wretched refuse of your teeming shore. . . . Send these, the homeless, tempest-tost to me."

Emigration from the Old World created a flood of migrants immigrating to the New World. This relieved population pressures in European countries and aided in the development of the New World. It is apparent, however, that a totally open policy toward immigration would be untenable today. The United States, with its current population of 265 million, is no longer a vast open land awaiting development; yet hundreds of millions of people would immigrate if they could. How much immigration should be permitted, and should some groups be favored over others? For example, in 1882, the U.S. Congress passed the Chinese Exclusion Act, which barred the immigration of Chinese laborers, but not Chinese teachers, diplomats, students, merchants, or tourists. This act remained in effect until 1943, when China and the United States became allies in World War II. However, the current immigration policy still makes it relatively easy for trained people to gain citizenship and relatively difficult for untrained people to do so. This policy created what is commonly referred to as a "brain drain." Brain power, many pointed out, is the "export" that developing nations can least afford.

Under the Immigration Reform Act of 1995, the United States can now accept up to 675,000 new immigrants per year, a number larger than we have received at any time since the 1920s and larger than is accepted by all other countries combined. Further, a lottery system was adopted for selection. Under the act, immigration accounts for about 30% of U.S. population growth, which is presently 2.7 million per year. The remainder of the population growth is called natural increase, births exceeding deaths. If the fertility rate remains low, immigration will account for a growing proportion of population growth.

The preceding deals, of course, with legal immigration; illegal immigration is another matter. Hundreds of thousands, unable to gain access through legal channels, seek ways to illegally enter the country. The United States maintains an active border patrol, especially along the border with Mexico, which several thousand people try to cross each night. Most are caught and returned, but an undetermined number slip through.

The Haitian boat people who made the headlines in 1991 and 1992 presented a different issue—that of granting asylum. Under terms of a 1951 United Nations convention, all European Community countries and the United States pledged to grant asylum to persons who can show a "well-founded fear of being persecuted for reasons of race, religion, nationality, membership in a particular social group or political opinion." People qualifying for asylum may stay in their host countries regardless of other restrictions on immigration. It was anticipated that the number seeking asylum would be trivial compared with those seeking a better life; however, often the distinction is not clear.

Haiti, an island republic in the Caribbean, is the poorest country in the Western Hemisphere. Environmental stresses there are severe. Less than 2% of the original forest cover remains. Soil erosion has already severely undercut productivity and remains grave. The crystal blue of the Caribbean near the Haitian shore is often brown with eroding soil, and mud settling on the coral has destroyed fishing in many areas. Yet Haiti's population of 7.3 million has a fertility rate of 4.8. The country's environmental and population disaster stems in no small part from a long line of repressive dictatorial regimes. Hopes brightened with the election of a democratic government in 1991, but the forceful takeover by yet another military dictatorship drove thousands of Haitians to despair. They crowded into anything that would float and headed for the United States seeking asylum. They argued that their plight was political and that they would face further political repression if they were returned to Haiti.

Back in February 1992, the U.S. Supreme Court ruled that the Haitian boat people were *economic* refugees—that is, they were only seeking a better life—and they were deported back to Haiti to face a dubious future. While some people in the United States lauded this ruling, others decried it as utterly lacking in compassion and making a mockery of what we say we stand for.

In 1994, the Clinton administration launched Project Hope, a diplomatic military intervention which removed the dictatorial regime and restored the exiled democratically elected leader, Aristide. New, free elections were held in 1996. The broader objective and subsequent economic aid, is to enable Haitians to earn a decent living at home, thus relieving the pressure to emigrate. Humanitarian and economic aid to other foreign countries might be seen in the same light. However, there is mounting pressure currently to cut such aid in view of our budget deficit.

As population pressures in developing countries continue to mount, the question of how many immigrants to accept, from what countries, and where to draw the line regarding asylum seems certain to become more and more pressing, as will needs for aid. In addition to compassion, the social, economic, and environmental consequences, both national and global, of the alternatives must be weighed in making the final decision. Where do you stand?

stop them? Of course, it shouldn't escape your notice that the affluent consumer at the end of the chain provides the basic incentive for such activities.

Emigration and Immigration. Facing the stresses and limited opportunities in their own countries, many people of developing countries see emigration to developed countries as their best hope for a brighter future. Historically, the New World was colonized by the overflow population from Europe, which experienced its population explosion early in the Industrial Revolution, and certainly, immigrants have contributed immensely to the development and economic growth of the United States, Canada, and other countries. However, is it feasible for the United States or any other developed nation simply to open its doors to all who would flee from poverty and lack of opportunity in the developing world?

The Worldwatch Institute estimates that there are already over 23 million environmental refugees—people living outside their homeland because they can't make a living there—and the number is growing rapidly. The current (1996) population of the United States—265 million—is growing at about 2.6 million per year. About one third of this growth is new immigrants.

Can the U.S. or any other country simply open its doors to all who would like to enter? You don't need this book to tell you that the problem of immigration, both legal and illegal, is becoming an increasingly contentious issue.

Impoverished Women and Children. The hardships and deprivation of poverty fall most heavily on women and children. Men are more free to roam and pick up whatever work is available, and they may keep their wages for themselves. Some men take no responsibility at all for the women they make pregnant, much less the children that are produced. Even many married men, under the stress of poverty, abandon wives and children. Numerous developing countries have no welfare system that will provide care in such situations. Too often, the women cannot cope; children are abandoned and women turn to begging, stealing, or prostitution.

And what happens to their children? If they survive at all, it is by begging, scrounging through garbage, stealing, and finding shelter in any hole or crevice they can find (Fig. 6-8). The problem is great: Nearly every sizable developing world city has thousands of these "stray" children, on the order of 20 million in all by some estimates, and their numbers are growing. Forced child labor, child prostitution, and selling children for adoption are additional problems that exist in no small measure. (See "Global Perspective" box, p. 154.) One can speculate as to the kinds of adults these children become as they grow up. At the very least, all of the factors tend to lock the poor into the vicious cycle of illiteracy and squalid conditions that defines absolute poverty.

A summary of these factors is given in Fig. 6-9. It is most notable that population growth, poverty, and environmental degradation are not separate issues. They are very much interrelated.

Effects of Increasing Affluence

Increasing the average wealth of a population affects the environment both positively and negatively. An affluent country certainly can and does provide such things as safe drinking water, sanitary sewage systems and sewage treatment, and collection and disposal of refuse. Thus, in terms of the most immediate forms of human wastes, pollution decreases, and the environment we live in improves with increasing affluence. In addition, if we can afford gas and electricity, we are not destroying our parks and woodlands for firewood. In short, we are able to afford conservation and manage-

FIGURE 6-8
In cities of the developing world, many poor people, including mothers and children, subsist only by scrounging through refuse for bits of food and items they can resell. (Jerry Cooke/Photo Researchers, Inc.)

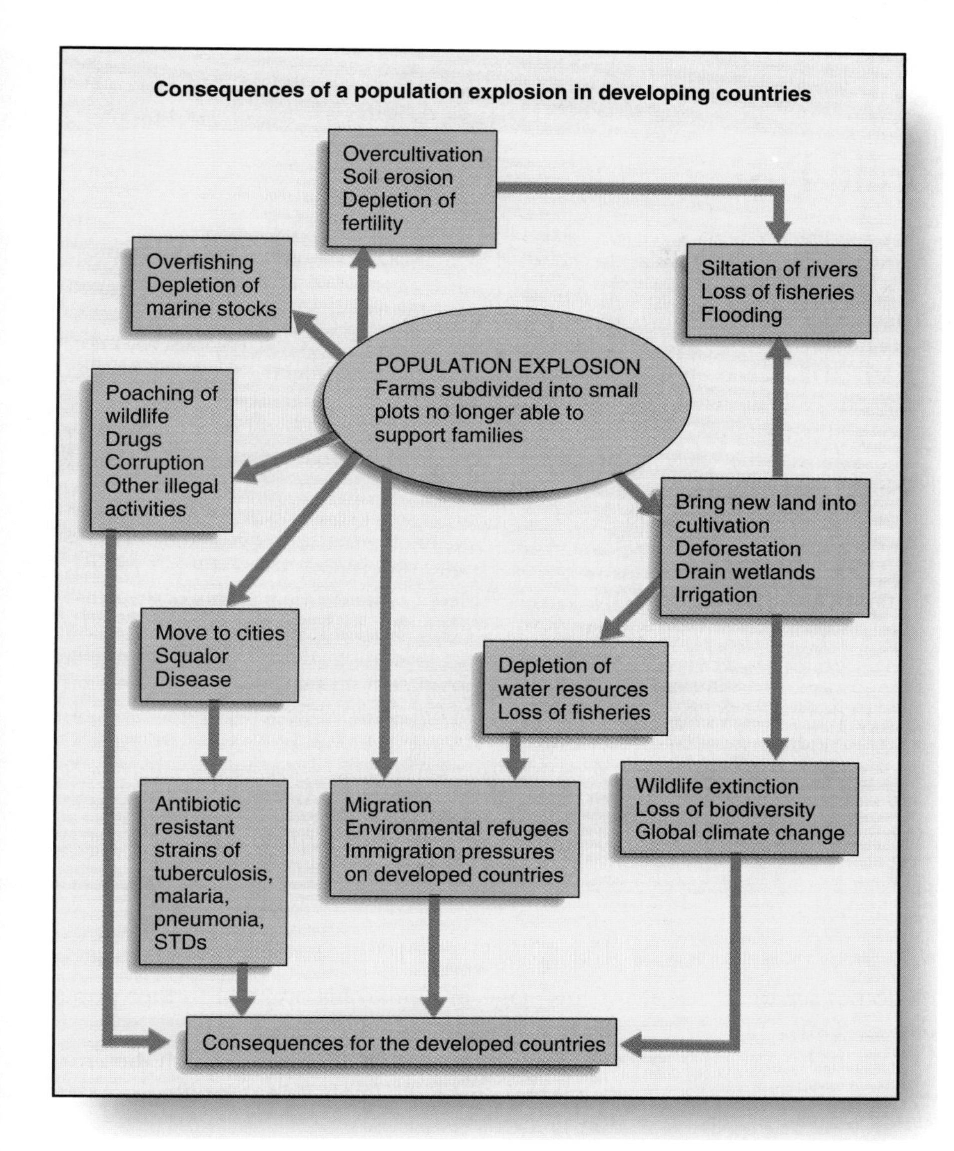

FIGURE 6-9
The diagram shows the numerous connections between unchecked population growth and social and environmental problems.

ment, better agricultural practices, and so on, that in many respects improve our environment.

But affluence does have its negative aspects as well. For example, by using such large quantities of fossil fuel (coal, oil, and natural gas) to drive our cars, heat and cool our homes, generate electricity, and so on, the United States is responsible for a large share of the production of carbon dioxide, which is the inevitable byproduct of burning these fuels. Specifically, with about 4.5% of the world's population, the United States is responsible for about 25% of the emissions of carbon dioxide that may be changing global climate. Similarly, emissions of chlorofluorocarbons (CFCs), which degrade the ozone shield, emissions of chemicals that cause acid rain, and emissions of hazardous chemicals and nuclear wastes are all primarily byproducts of affluent societies.

Economic factors place further demands on the environment. For example, many rain forests in Brazil have been cut in order to convert the land into cattle pasture, serving the appetites of the affluent for meat. A large portion of the oceans have been overfished for the same reason. Currently, old-growth forests in southern South America are being clearcut and turned into chips to make fax paper. Oil spills are a "byproduct" of our appetite for energy. We have already mentioned the pressure—perhaps to the point of extinction—being put on endangered species from people willing to pay exorbitant prices for exotic "pets" or trinkets made from animal parts. As increasing numbers of people strive for and achieve greater affluence, it seems more than likely that these and similar pressures will mount.

One way of generalizing the effect of affluence is to say that it enables humans to clean up their immediate environment by disposing of their wastes to more distant locations. It also allows them to obtain resources

Global Perspective

Repeating a Sordid Past

In the 19th century, Charles Dickens wrote novels and magazine stories describing the horrors and inhumanity of the child labor sweatshops that developed in the early part of the Industrial Revolution. Young children were forced to work at machines or perform simple repetitive tasks for 12–14 hours a day, seven days a week. Often, they ate and slept under their worktables. When they grew old enough to become rebellious, they were simply turned out onto the streets—with no education and no skills—to face a world they had never seen.

In the early years of the 20th century, the public's outrage at learning of such atrocities led to the child labor laws and the generally improved working conditions we in the developed world have come to expect. Thus, we would like to think that the conditions Dickens wrote about have passed into history. Yet, in 1989, the United Nations held a "Convention on the Rights of the Child" that unanimously adopted resolutions outlawing child labor and trafficking in children.

Why did the United Nations find such a convention necessary in 1989? Because conditions every bit as horrendous as what Dickens described were (and still are) occurring in many places in the developing and also the developed world. According to the Anti-Slavery Society, 200 million "child slaves" are working in sweatshops around the world. In Tamil Nadu, India, 45,000 children—some of them only 4 or 5 years old—work in the fireworks and match industry. Many of the beautiful, intricate, handmade rugs from Asia, which one can buy at a "very reasonable price" are made by children forced to work. In Bangkok, Thailand, according to the Thai Center for the Protection of Children's Rights, there are some 800,000 girl prostitutes between the ages of 12 and 15. As recently as 1990, children were being sold near Bangkok's main railway station.[1]

Even in developed countries there are incidences of sweat-shop conditions. The United States still has not ratified an International Rights of the Child treaty, still pending since 1991.

Consider the relationships among population growth, poverty, and social injustice, particularly the impact on children.

[1] Jonathan Power, "Where Death Squads Hunt Stray Children," *Baltimore Sun*, Oct. 25, 1991, p. 15A.

from more distant locations such that they do not see or feel the impacts of obtaining those resources. Thus, in many respects, the affluent isolate themselves and may become totally unaware of the environmental stresses they cause with their consumption-oriented lifestyles. Since the world is not flat, however, the "more distant locations" come back around, and we discover that we are having a global impact.

The most important aspect of affluence, however, is that it provides us with choices and opportunities. It may be true, for example, that an affluent society puts more pressure on ecosystems to graze more beef. However, as individuals, we can decide whether to eat beef or not, and we can join groups dedicated to saving rainforests. In conclusion, the point is not to decry or feel guilty about our affluence, but to recognize that we can use it for constructive purposes.

Summing Up

In summing up this section, we revisit the different points of view outlined in the introduction of this chapter (page 140). We simply pose questions and let you consider and debate the answers for yourself.

1. At this stage in human history, will further population growth be beneficial?
2. Can we count on technology to solve all the environmental problems implicit in unbridled population growth?
3. Are all moral values on an equal level? How do values that lead to the denial or obstruction of family planning compare with those necessary for sustainability?
4. Will population taper off by itself at a level that will enable all to enjoy an affluent lifestyle?
5. Is population growth only a social problem, only an environmental problem, or both? Is it a problem that will affect only developing countries, or will developed countries be affected as well?
6. Is a *J*-curve for the human population out of the question, possible, or inevitable? Will how we direct our efforts make a difference?

In Chapter 7, we shall discuss avenues and issues regarding how population might be stabilized. Before that, however, it is necessary to present some concepts and terminology that are common in discussing population issues.

Dynamics of Population Growth

In considering population growth, we consider more than just the increase in numbers, which is simply births minus deaths; we also consider how the numbers of births ultimately affect the entire population over the **longevity**, or lifetimes, of the individuals.

Population Profiles

A **population profile** is a bar graph showing the number of people (males and females separately) at each age for a given population. The population profile of the United States in 1991 is shown in Fig. 6-10. The data are collected through a census of the entire population, a process in which each household is asked to fill out a questionnaire concerning the status of each of its members. Various estimates are made for those who do not maintain regular households. For the United States and most other countries a detailed census is taken every 10 years. Between censuses, the population profile may be adjusted by using data regarding births, deaths, immigration, and, of course, the aging of the population. The field of collecting, compiling, and presenting information about populations is called **demography**; the people engaged in this work are **demographers**.

A population profile shows the **age structure** of the population, that is, the proportion of people in each age group at a given date. However, leaving out the complication of emigration and immigration for the moment, recognize that each bar in the profile started out as a cohort of babies at a given point in the past, and that cohort has only been diminished by deaths as it has aged. In developed countries such as the United States, the proportion of people who die before age 60 is relatively small. Therefore, the population profile below age 60 is an "echo" of past events insofar as those events affected birth rates. For example, in Figure 6-10 you can observe that smaller numbers of people were born between 1931 and 1936. This is a reflection of lower birth rates during the Great Depression. The

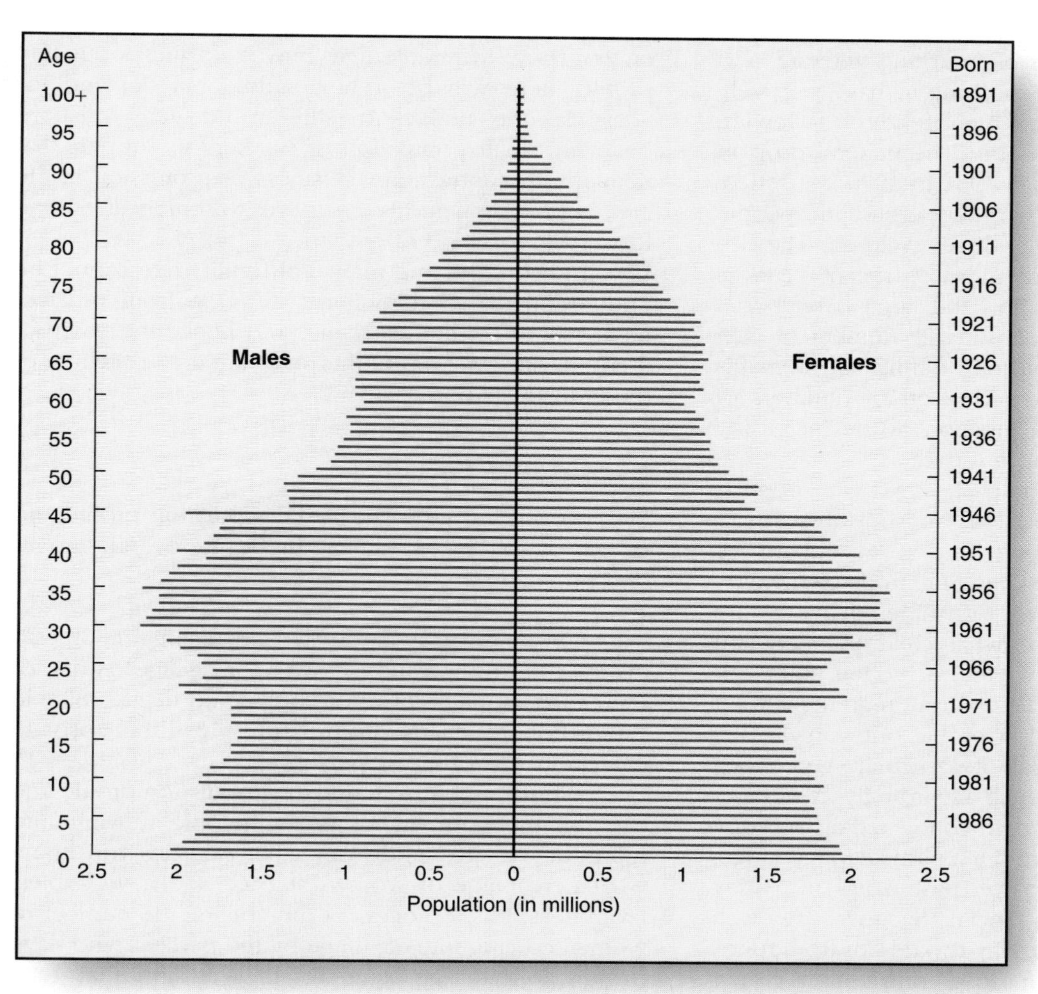

FIGURE 6-10
Population profile of the United States, 1991. (*Population Today*, Sept. 1993, p. 10. Washington, DC: Population Reference Bureau, Inc.)

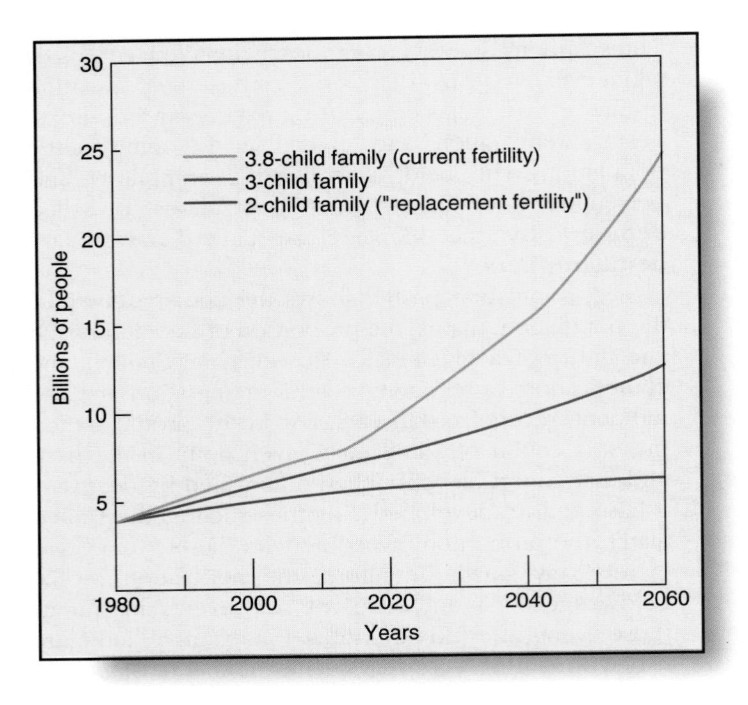

FIGURE 6-11
Projected world population, using different total fertility rates. As can be seen, the different rates will lead to vastly different population figures over the course of 80 years. (From Elaine Murphy, *World Population: Toward the Next Century* [Washington, DC: Population Reference Bureau, Inc., Oct. 1985], p. 7.)

dramatic increase in people born in 1946 and in following years is a reflection of returning veterans and others starting families and choosing to have relatively large numbers of children (high fertility) following World War II—the "baby boom." The general drop in numbers of people born from 1961 to 1976 is a reflection of sharply declining fertility rates, with people choosing to have significantly fewer children. The rise in numbers of people born in more recent years is termed the "baby boom echo," the large baby boom generation producing a similarly large number of children even though the actual total fertility rate remains near two. Of course, the tapering off of numbers of people beyond age 65 is a function of the increasing number of deaths that occur in old age.

More than a view of the past, however, a population profile provides governments and businesses with a means of realistic planning for future demand for various goods and services, ranging from elementary schools to retirement homes. Consumer demands are largely age specific; that is, what children need and want differs from what teenagers, or young adults, or older adults, or finally people entering retirement want and need. Using a population profile one can literally see the projected populations of particular age cohorts and plan to expand or retrench accordingly. A number of industries expanded and then contracted as the baby boom generation moved through a particular age range, and this phenomenon is not yet past. You can readily see that concerns about Social Security are well founded, because Social Security is paid not from funds contributed by retirees, but from contributions being made by current workers. As the large baby boom generation, now in middle age, moves up the population profile there will be a huge additional demand for Social Security outlays to retirees, and there will be a relatively small population of workers to support that demand. In contrast, any business or profession in the area of providing goods or services to seniors is looking forward to a period of growth.

In short, the importance of fertility rate cannot be underestimated. Fertility rates differing from replacement will affect the economy, and hence the environment, in one way or another over the entire lifetime of the individuals born.

Population Projections

Most significantly we can use a population profile and certain other data to estimate the future overall growth of a population.

The Forecasting Technique. The technique of estimating future population growth (or decline) is one of estimating numbers of future births and deaths. Simple subtraction of deaths from births gives the absolute change in size of the population.

Estimating births: Using the population profile you can see the numbers of young people moving into reproductive ages, and from other statistics you know what percentage of women at each age have babies. From here, it is a process of multiplying the number of women coming into each age by the percent who have babies at that age, totaling, and adding that total number of babies as a new bar at the bottom of the profile.

Estimating deaths: From statistics regarding deaths, you know the percentage of people who die at each age. These are the actuarial tables used by insurance companies to determine life insurance rates. Using the population profile you can see the number of people moving into each age. Particularly significant are the numbers of people moving into older ages, where death rates are highest. Again, multiplying the number of people moving into each age by the percentage of people who die at that age, and totaling, gives the overall deaths for the year, and all the bars on the profile are adjusted accordingly.

Repeating this process over and over allows demographers to project births and deaths and, by subtraction, change in population as far into the future as desired. You can see the need for many calculations, but computers make the process relatively straightforward. Of course, such projections become increasingly subject to error as they are extended farther into the future, because one has only current statistics and trends to work from, and many things may occur to either increase or decrease longevity. Just as troublesome in making long-term projections are changes in fertility rates, which are basically a function of how many babies women choose to have at any given age. As we noted at the beginning of this chapter, the projection that the world population will level off at around 12 billion is based on the assumption that fertility rates will continue a gradually declining trend. However, fertility rates tend to rise and fall, for reasons that are not always well understood or predictable. We have already seen how fertility rates rose after World War II, producing the baby boom generation, and then declined resulting in fewer numbers of people following.

Therefore, different projections are made on the basis of different assumptions regarding future fertility rates. Thus, the Population Reference Bureau gives three different projections of future world population (Fig. 6-11). Still imponderable, of course, are events that may drastically increase death rates. Some look toward such an event as the way to stabilize human population; it is, after all, nature's way. This may be true, but it is also the *J*-curve. The social upheaval and human suffering associated with such an event are exactly what we wish to make every effort to avoid.

Simplified Population Projections. Demographers use statistics and computers to make projections, because they need those projections to be as accurate as possible. However, we may gain a reasonable picture of future population by a more simplified procedure in which population profiles are presented in a "condensed" form lumping people into five-year age groups. (See Fig. 6-12.) In this version, each one of the bars moves up one space every five years. Since in modern human societies relatively few deaths occur before old age, the population loss from deaths can be approximated simply by removing the uppermost bar. To determine the number of births, observe that the average child-bearing age range is 20–24. Therefore, we simply multiply the number of *women* (only the left side of the bar) by the total fertility rate and obtain an approximation of the number of babies born in the relevant five-year period. Again, the process can be repeated over and over to project the population into the future as far as desired, and we can make different projections based on different total fertility rates.

Population Projections for a More Developed Country. The 1990 population profile of Denmark, a developed country in northern Europe, is shown in Fig. 6-12a. It is more or less columnar in shape because Danish women have had a low fertility rate for some time. Assuming that the 1991 total fertility rate of 1.6 remained constant for the next 20 years, then, following the technique just described, we obtain the profiles presented in Figs. 6-12b through e. The profiles show an increase in the numbers of older people and, with one exception, a marked decline in the number of children and young people.

The profiles also show that, for the next 20 years, Denmark's total population will remain relatively constant because the number of deaths (removal of top bars) will be nearly the same as the number of births (addition of bottom bars). But the population will be **graying**, a term used to indicate that the proportion of elderly people is increasing. We can also predict a considerable population decline following 2010, as the large elderly population suffers mortality while fewer and fewer children are born.

What opportunities and risks does the changing population profile imply for Denmark? If you were an advisor to the Danish government, what would your advice be for the short term? For the longer term? Unless a smaller population is the goal, our advice here might well be to encourage, and provide incentives for, Danish couples to bear more children. We might also advocate allowing more immigration. But allowing the declining numbers of Danish people to be replaced by an immigrant population requires one to reflect on the effect that would have on the Danish culture, religion, etc. (In fact, Denmark's total fertility has increased some since 1991—1.8 in 1996.)

The very low fertility rate and prospective declining population seen in Denmark are typical of nearly all highly developed nations. Therefore, the preceding analysis and questions can be applied to any of them. The one exception is the United States: In contrast to other developed countries, the U.S. fertility rate reversed directions in the late 1980s and started back

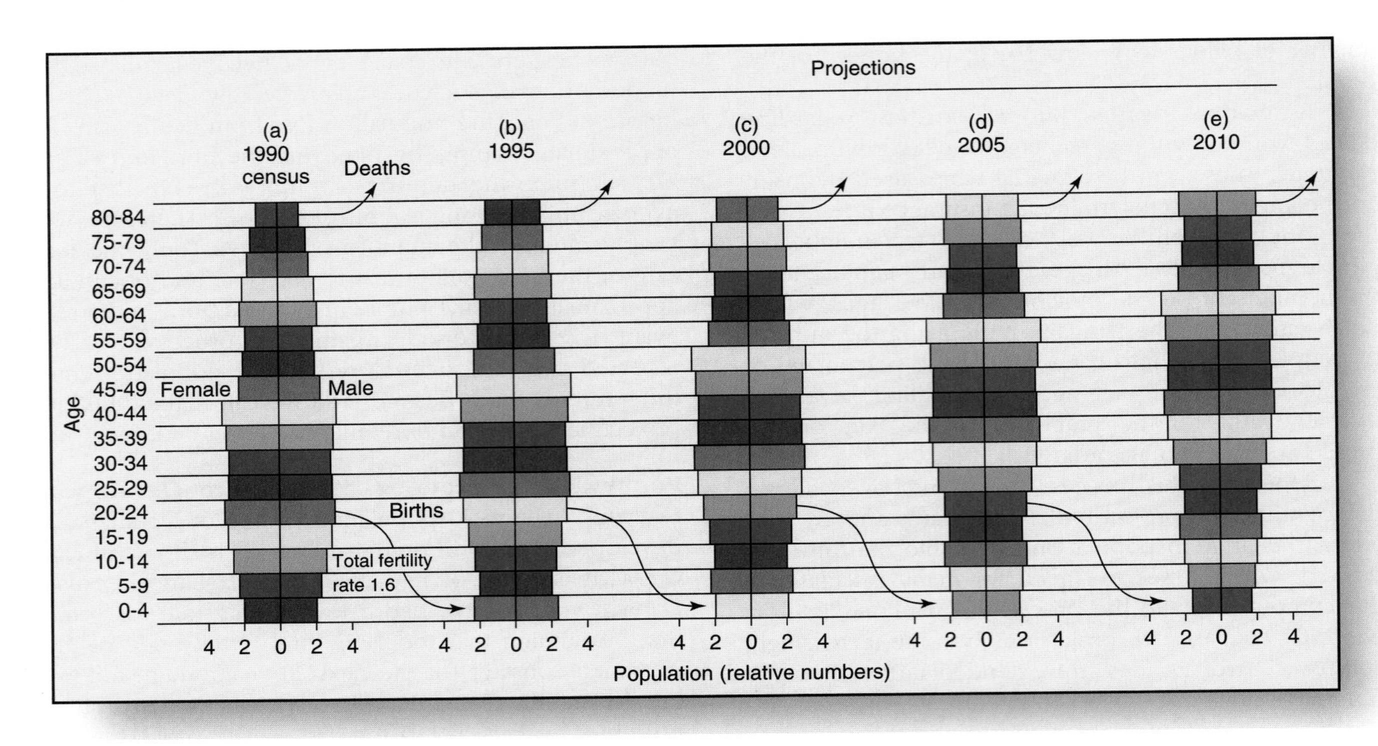

FIGURE 6-12
(a) A population profile representative of a highly developed country, Denmark, in 1990. (b–e) Projections of Denmark's population made as described in the text, assuming that the total fertility rate remained at its 1991 value of 1.6. Note how larger numbers of persons are moving into older age groups and the number of children is diminishing. (Profile shown in [a] from Joseph A. McFalls, Jr., "Population: A Lively Introduction," *Population Bulletin*, 46, no. 2 [Washington, DC: Population Reference Bureau, Inc., Oct. 1991], p. 4.)

up. Based on the lower fertility rate, the U.S. population had been projected to stabilize at between 290 and 300 million toward the middle of the next century. With a higher fertility rate of 2.0, the U.S. population is projected to be 392 million by 2050 and to continue growing indefinitely (Fig. 6-13). Immigration will add even more to the figures, especially in that many newly arrived immigrants tend to have fertility rates well above the average of the rest of the population.

These projections for the United States show what a profound effect on population slight differences in the total fertility rate make when they are extrapolated 50 or more years forward. What do you think the population policy of the United States should be?

Population Projections for a Less Developed Country. Developing countries are in a vastly different situation from that of developed countries. In developing countries, while fertility rates are generally declining, they are still well above 2.1. (The average is currently 4.0, excluding China.) Because of even higher past fertility rates, the population profiles of developing countries have a pyramidal shape. For example, the 1990 population profile for the African country of Kenya, which had a fertility rate of 6.7, is shown in Fig. 6-14a. Assuming

FIGURE 6-13
Population projections for the United States. Projections shift drastically with changes in fertility. Contrast the 1988 projection, based on a fertility rate of 1.8, with the 1993 projection, based on the increased fertility rate of 2.13. (Data from Kelvin M. Pollard, *Population Stabilization No Longer in Sight for U.S. Population Today* [Washington, DC: Population Reference Bureau, Inc., May 1994], p. 1.)

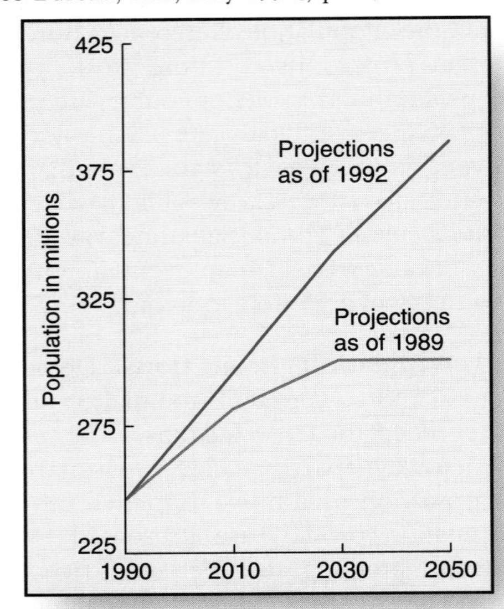

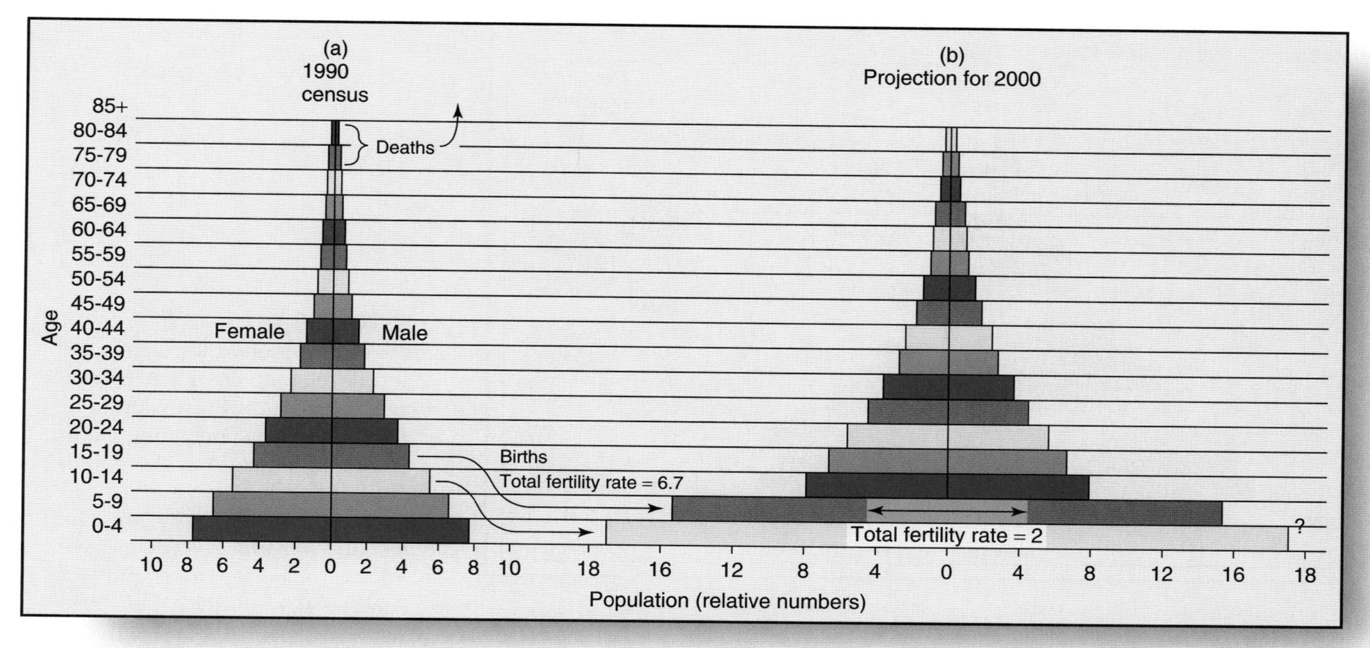

FIGURE 6-14
(a) The 1990 population profile of Kenya, a developing country. (b) Projection for the year 2000, based on the assumption that the total fertility rate will continue at 6.7. Even if the total fertility rate immediately dropped to 2, the number of births (hatched inner portion of the bottom two bars) would still greatly exceed the number of deaths because so few persons are in the upper age groups. (Profile shown in [a] from Joseph A. McFalls, Jr., "Population: A Lively Introduction," *Population Bulletin*, 46, no. 2 [Washington, DC: Population Reference Bureau, Inc., Oct. 1991], p. 4.)

that this fertility rate remains constant, projecting Kenya's population ahead just 10 years gives us the profile shown in Fig. 6-14b. Note that even the relatively high child mortality rate, which is unacceptable in itself, comes nowhere close to offsetting the high fertility. Thus, the pyramidal form of the age structure remains the same because each rising generation of young adults produces an even larger generation of children. The pyramid just gets wider and wider, leading to a doubling and redoubling of the population. Kenya's total fertility has dropped slightly—5.4 in 1996—but at the lower rate the population will still double in just 25 years.

Another factor exacerbating population growth in Kenya and other developing countries is a shorter generation time. Many women in developing countries start bearing children in their early teens, whereas women in developed countries generally wait until their late twenties or even longer.

While highly developed countries are facing the problems of a graying population, the high fertility rates in developing countries maintain an exceedingly young population. An "ideal" population structure with equal numbers of persons in each age group and a longevity of 75 years would have one fifth, or 20%, of the population in each 15-year age group. By comparison, in many developing countries, 40%–50% of the population is below 15 years of age. In contrast, in most developed countries, less than 20% of the population is below the age of 15 (Table 6-4).

Consider what all this means in terms of the percentage of the population that is available to build new

TABLE 6-4
Populations by Age Group

Region/ country	*Percent of Population in Specific Age Groups*		
	<15	*15 to 65*	*>65*
Africa	44	53	3
Kenya	48	49	3
Latin America	35	60	5
Asia	32	63	5
Europe	19	67	14
Denmark	17	68	15
Japan	16	69	15
United States	22	65	13

Source: Population Reference Bureau, Population Data Sheet 1996.

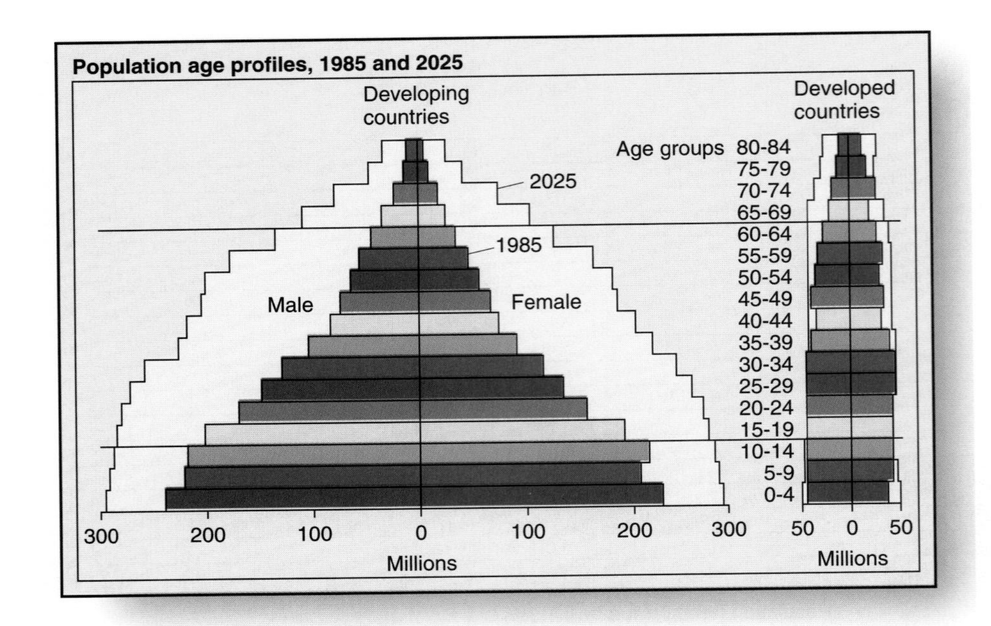

FIGURE 6-15
Population profiles for developed and developing countries, projected to the year 2025. Even with the assumption of a continuing gradual decline in fertility rate, the populations of developing countries will grow enormously. (*Population and the Environment: The Challenges Ahead*, United Nations Population Fund, 1991.)

schools, housing units, hospitals, roads, sewage collection and treatment facilities, telephones, etc. A young population is bound to be largely untrained and unskilled. Yet, if a country such as Kenya is simply to maintain its current standard of living, the amount of housing and all other facilities (to say nothing of food production) must be doubled in as little as 20 years. Accordingly, it is not difficult for a developing country's efforts to get ahead to be more than nullified by its own population growth.

A comparison of present and projected population profiles for developed and developing countries is shown in Fig. 6-15. The figure shows that, while little growth will occur in developed countries over the next 35 years, enormous growth is in store for the developing world. What is worse, this is so assuming that fertility rates in the developing world continue their current downward trend.

Again, these or any other population projections should not be confused with predicting the future. They are intended only to show where we will end up if we continue the present course. According to an old saying, "If you don't like where you are going, change direction." In other words, if we feel that the projected population growth is not desirable, we can try to bring fertility rates down faster. However, even bringing the fertility rates of developing countries down to 2.0 will not stop their growth immediately. This is because of a phenomenon known as *population momentum*, which we examine next.

Population Momentum

Countries with a pyramid-shaped population profile, such as Kenya, will continue to grow for 50–60 years, even after the total fertility rate is reduced to the replacement level. This phenomenon, called **population momentum**, occurs because such a small portion of the population is in the upper age groups, where most deaths occur, and many children are entering their reproductive years. Even if these rising generations have only two children per woman, the number of births will far exceed the number of deaths. The imbalance will continue until the current children reach the Kenyan limits of longevity—50 to 60 years.

To acknowledge population momentum is not, of course, to say that efforts to stabilize population are fruitless. It is, however, to say that fertility rates *must* be reduced before a crisis point is reached.

Changing Fertility Rates and the Demographic Transition

The concept of a stable, nongrowing global human population based on people freely choosing to exercise a lower fertility rate is possible because it is already the fact in developed countries. If we can understand the factors that have caused the decline in fertility rates in developed countries, then perhaps we can make those factors operative in developing countries.

Early demographers observed that the economic development of a nation brings about more than just a lower death rate resulting from better health care; a decline in fertility rate also occurs as people choose to limit the size of their families (Fig. 6-16). Thus, as development occurs, human societies move from a primitive population stability, in which high birth rates are offset by high infant and childhood mortality, to a modern population stability, in which low infant and childhood mortality are balanced by low birth rates. This gradual

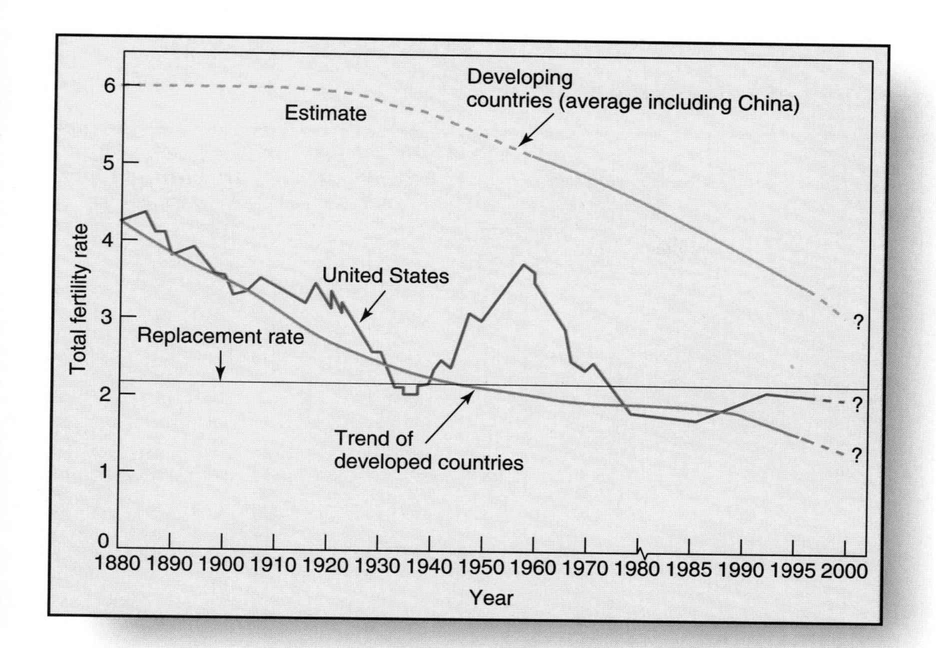

FIGURE 6-16
Change in fertility rates. Historically, high fertility rates were the norm for the human population. In more modern times, the trend has been toward lower fertility rates. In developed countries, the trend started earlier and, except for the anomalous post-World War II baby boom, has progressed to near replacement level. In developing countries, the trend toward lower fertility started later and has not progressed as far. (Data from the Population Reference Bureau, Inc., Washington, DC.)

shift from the primitive to the modern condition that is correlated with development is called **demographic transition**.

To picture the demographic transition, we need to introduce two new terms: the **crude birth rate (CBR)** and **crude death rate (CDR)**. These terms are defined as the number of births or deaths per 1000 of population per year. By giving the data in terms of "per 1000," populations of different countries can be compared regardless of their total size. The term *crude* is used because no consideration is given to what proportion of the population is old or young, male or female. Subtracting the CDR from the CBR gives the increase (or decrease) per 1000 per year. Dividing this result by 10 then puts it in terms of "per 100," or percent.

CBR		CDR		Natural increase
Number of		Number of		(or decrease)
births per	−	*deaths* per	=	in population
1000		1000		per 1000
per year		per year		per year
				÷ 10 = %

Of course, a stable population is achieved if, and only if, the CBR and CDR are equal.

The *doubling time*, or the number of years it will take a population growing at a constant percent per year to double, is calculated by dividing the percentage rate of growth into 70. The answer is the number of years it will take the population to double. (The 70 has nothing to do with population; it is simply the number that works to give the result.) CBR, CDR, and doubling times of various countries are shown in Table 6–5.

The demographic transition has four phases, as shown in Fig. 6–17. **Phase I** is the primitive stability resulting from a high CBR being offset by an equally high CDR. **Phase II** is marked by a declining CDR brought about by the reduction of infant and childhood mortality. Because fertility and, hence, the CBR remain high during Phase II, this is a phase of accelerating population growth. **Phase III** is a phase of declining CBR resulting from a declining fertility rate. Finally, **Phase IV** is reached, where modern stability is achieved by a continuing low CDR, but an equally low CBR.

Basically, developed countries have completed the demographic transition. The population momentum from higher birth rates of the past is still causing some growth, but stability will soon be reached as this bulge in the population profile passes. Developing countries, on the other hand, are still in Phase III. Death rates (infant and childhood mortality) have declined markedly, and fertility and birth rates are declining, but are still considerably above replacement levels. Therefore, populations in developing countries are still growing rapidly.

It is on the basis of the demographic transition that some argue that we not worry about population; it will stabilize by itself as developing countries reach Phase IV. Therefore, the argument goes, we need only focus on free world trade and other factors that will speed economic growth in developing countries. But major flaws in this argument loom. First, population growth itself is so rapid in many developing nations, that it is undercutting economic development in that any advances are continually being divided among more people. Second, we emphasize again that present stresses on the biosphere and loss of biodiversity are largely a consequence of the consumption-oriented lifestyles of the current 1.2 billion people in developed

TABLE 6-5
Crude Birth and Death Rates for Selected Countries

Country	*Crude Birth Rate*	*Crude Death Rate*	*Annual Rate of Increase (%)*	*Doubling Time (Years)*
World	24	9	1.5	46
Developing Nations				
Average (excluding China)	31	10	2.2	32
Egypt	30	7	2.2	31
Kenya	40	13	2.7	25
Madagascar	44	12	3.2	22
India	29	10	1.9	37
Iraq	44	7	3.7	19
Viet Nam	30	7	2.3	30
Haiti	35	12	2.3	30
Brazil	25	8	1.7	41
Mexico	27	5	2.2	32
Developed Nations				
Average	12	10	0.1	501
United States	15	9	0.6	114
Canada	13	7	0.6	116
Japan	10	7	0.2	315
Denmark	13	12	0.2	462
Germany	9	11	−0.1	—*
Italy	9	10	0.0	—*
Spain	9	9	0.1	1155

Data from 1996 World Population Data Sheet. Population Reference Bureau, Inc.
*Population is declining.

nations. And recognizing that severe stresses are being caused by the lifestyles of 1.2 billion people makes any notion of a world with 12 billion people living with the same lifestyle utterly absurd.

Finally, and most importantly, the demographic transition really only shows a *correlation* between development and changing birth and death rates; it does not prove that development is the *cause* of the changing demographic picture. Indeed, much evidence has accumulated in recent years that the gross national product (GNP) of a country (or gross domestic product, GDP, as it is called in the United States), the usual indicator of economic development, has little bearing on fertility rates. Other factors are much more significant.

In Chapter 7, we investigate the factors that have a more direct bearing on how many babies people have and what we can do to provide incentives for reducing that number.

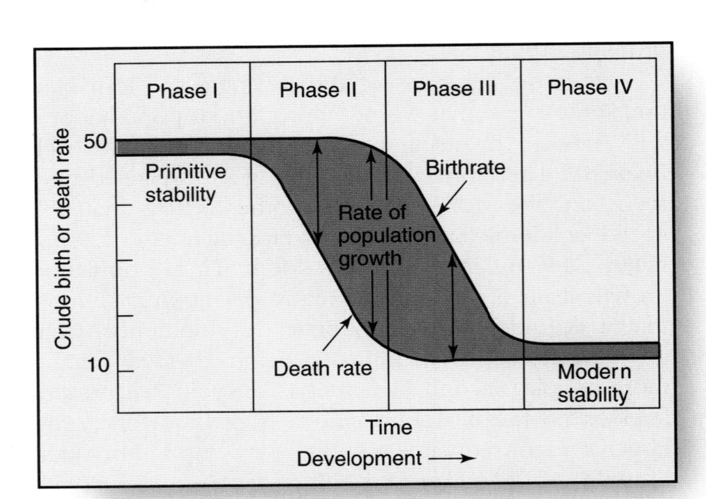

FIGURE 6-17
The demographic transition. Phase I: Before any modernization, a primitive stability is based on high birth rates being balanced by high infant and childhood mortality. Phase II: As modernization occurs, first the death rates drop while birth rates remain high, causing accelerating population growth. Phase III: Later, as development progresses, birth rates also come down, slowing population growth. Phase IV: Finally, a modern stability is reached in which a low death rate is balanced by a similarly low birth rate. Modern developed countries have already reached Phase IV, while developing countries are still in various stages of Phase III. (From Joseph A. McFalls, Jr., "Population: A Lively Introduction," *Population Bulletin*, 46, no. 2 [Washington, DC: Population Reference Bureau, Inc., Oct. 1991], p. 33.)

Review Questions

1. How has the global human population changed from early times to 1800? from 1800 until the present? What is projected over the next 50 years?
2. How is the world divided in terms of relative per capita incomes? Fertility rates? Population growth rates?
3. What two factors are multiplied to give total environmental impact? Are developed nations exempt from environmental impact? Why not?
4. What third factor affects environmental impact, and how does it do so?
5. What are the social and environmental consequences of rapid population growth in developing countries? How may developed countries be affected?
6. What information is given by a population profile? How is the information presented?
7. How are numbers of future deaths and future births predicted from a population profile? What simplifying assumptions can be made to enable you to approximate a projection of the future size and structure of a population?
8. How do the population profiles, fertility rates, and population projections of developed countries differ from those of developing countries?
9. How might future population goals of developed and developing countries contrast?
10. What is meant by population momentum, and what is its cause?
11. Define crude birth rate (CBR) and crude death rate (CDR).
12. Describe how the CBR and CDR are used to calculate the percent rate of growth and the doubling time of a population.
13. What is meant by the demographic transition? What are its four phases?
14. How do developed and developing nations differ regarding their current positions in the demographic transition?
15. Can future sustainability be attained through development alone? Why or why not?

Thinking Environmentally

1. Make a cause-and-effect "map" showing the many social and environmental consequences linked to unabated population growth. Include crossovers between the developed and developing worlds.
2. It has been proposed that excess human populations be accommodated by building orbiting space stations that would house about 10,000 persons each. Each station would be able to produce its own food. How many space stations would be required to accommodate the world's projected population growth over the next 10 years? What kind of population policy would have to be enforced on the space stations? Are space stations a logical solution to the population problem?
3. Starting with a hypothetical population of 14,000 people and an even age distribution (1000 in each five-year age group from 0–4 to 65–70), assume that this population initially has a total fertility rate of 2 and an average longevity of 70 years. Project how the population will change over the next 60 years under each of the following conditions:
   a. Total fertility rate and longevity remain constant.
   b. Total fertility rate changes to 4; longevity remains constant.
   c. Total fertility rate changes to 1; longevity remains constant.
   d. Total fertility rate remains at 2; longevity increases to 100.
   e. Total fertility rate remains at 2; age at which people die decreases to 50.
4. Contrast the long-term effect on population growth caused by changes in the total fertility rate with that caused by changes in longevity.
5. A country such as Denmark, which faces a decreasing population, has the following options:
   a. Accept the decreasing population.
   b. Encourage women to have more children.
   c. Accept more immigration.
   Discuss the long-term (50 years) economic and social implications of each option.
6. From the 1996 crude birth and crude death rates, calculate the rate of population growth and the population doubling time for each of the following countries (p. 161).

	CBR	CDR
Algeria	30	6
Ethiopia	46	16
Argentina	20	8
Iran	36	7
Russia	9	15
France	12	9
Australia	14	7

डाकटर की सलाह

Addressing the Population Problem

Key Issues and Questions

1. The factors that influence fertility rates are more specific than development in general. What are the factors that actually influence the number of children desired?
2. The vicious cycle of poverty, high fertility, and environmental degradation continues over much of the globe despite past efforts at development. Why have such efforts missed the mark?
3. To break the vicious cycle, development efforts must focus on five areas simultaneously. What are these areas? How and by what means may they be addressed?
4. Family planning services are considered essential to enable people to achieve a desired family size. What are the key aspects of family planning?
5. In 1994, world leaders met in Cairo, Egypt, at the United Nations Conference on Population and Development. What was the significance of this meeting? What are the agreed-upon strategies for addressing the problems of poverty, excessive population, and environmental degradation?

In September 1994, some 15,000 leaders and representatives from 179 nations and nearly 1000 nongovernmental organizations (NGOs) met in Cairo, Egypt, at the United Nations Conference on Population and Development (Fig. 7-1). Before the delegates was a draft document of a "Program of Action" to address the world's persistent problems of poverty and population. Norway's Prime Minister Gro Harlem Brundtland, in her keynote address to the Convention (Fig. 7-2) stated:

> Population growth is one of the most serious obstacles to world prosperity and sustainable development. . . .
>
> We may soon be facing new famine, mass migration, destabilization, and even armed struggle as people compete for ever more scarce land and water resources. . . .
>
> Today's newborns will be facing the ultimate collapse of vital resource bases. . . . Only when people have the right to take part in the shaping of society by participating in democratic political processes will changes be politically sustainable. Only then can we fulfill the hopes and aspirations of generations yet unborn.

Realizing that increasing population is causing a large segment of the population to become locked in poverty, India is actively promoting family planning. The poster is extolling the virtues of a two-child family. [Photo by Paolo Koch/Photo Researchers, Inc.]

Some objections were raised to the draft document by the Vatican and by Moslem countries regarding the wording of certain sections alluding to birth control methods and abortions. However, differences were resolved, and in the end all 179 nations, large and small, rich and poor, signed the final document committing themselves to achieve basic goals set forth in the document by the year 2015.

The historic significance of this event is that for the first time in history, the political, religious, and scientific communities of the world reached a consensus on the population issue. Some outspoken individuals are still voicing arguments to the effect that population is not a critical issue. However, such voices are now overwhelmed by worldwide agreement that the intertwined issues of excessive population, poverty, and environmental degradation are jeopardizing the future of the planet. If they are not mitigated over the next 20 years, the world is more than likely to see a future of incalculable biologic and human impoverishment.

This is not to say that the world is just awakening to the problems of population, poverty, and environmental degradation. Indeed, these problems have been long recognized, and many billions of dollars have been directed toward their solution through governmental, charitable, and United Nations organizations, and marked accomplishments have occurred. However, the persistence of the problems speaks to the requirement for improvement. So what is different now? Can we really expect policy changes in the future, or will those who do not believe there to be a population problem prevent change and a conscious control of birth rates.

Our first objective in this chapter is to examine the relationship between fertility rates and development (the demographic transition) in more detail. Second, we examine various policies and programs that were used in the past to improve development and lower fertility, and determine whether they have succeeded or failed. Finally, we examine the goals of the Cairo Conference in terms of this experience.

Population and Development

In 1798, a British economist, Thomas Malthus, pointed out that populations tend to grow exponentially, but there are ultimate limits to the expansion of agriculture. Thus, at the very beginning of the population explosion, Malthus foresaw a world headed toward calamity if something was not done to control population. But Malthus could foresee neither the tremendous expansion of agriculture that would come with the Industrial Revolution nor the demographic transition—the fact that fertility rates would decline with the industrial revolution (Fig. 6-17, page 162).

Therefore, from early on there were two basic schools of thought regarding the growth of population: (a) we need to focus on controlling population and (b) a focus on development will take care of the situation

FIGURE 7-1
World leaders at the U.N. Conference on Population and Development congratulate each other as they reach a concensus regarding a "Program of Action."

FIGURE 7-2
Prime Minister Gro Harlem Brundtland of Norway, the keynote speaker at the Cairo Conference. (Chiaki Tsukumo/AP/Wide World Photos.)

and bring about a balance "automatically." These two schools of thought were reflected at two previous United Nations Population Conferences, the first in Bucharest, Romania, in 1974 and the second in Mexico City in 1984. (The 1994 conference in Cairo was the third.) At the Bucharest conference, the United States was a strong advocate of population control, while the developing nations argued that "development was the best contraceptive." Their resistance to family planning was also bolstered by feelings that the developed world's promotion of population control was another form of economic imperialism or even genocide.

At the second conference in Mexico City, the sides were reversed. Developing nations facing real problems of excessive population growth were asking for more assistance with family planning, whereas the United States, under pressure from "right-to-life" advocates, took the position that development was the answer and terminated all contributions to international family planning efforts, a policy that remained in effect until 1993. This difference in objectives between developed and developing countries largely stymied any coordinated focus on population. But the development of poor nations has been consistently promoted over the past 50 years. Let us look at it more closely.

Promoting the Development of Low-Income Countries

As described in Chapter 6, the absence of development and the impoverished societies in what we now call developing countries is largely a legacy of 18th- and 19th-century colonialism, which persisted well into this century. Thus, reversing colonial policy and fostering the development of these countries can be amply justified on all grounds: humanitarian, political, and economic. By providing better jobs and incomes, development fosters improved standards of living, which, in turn, create

Ethics
Forgiving Debts

In the text, we pointed out that there is currently a net flow of $50 billion a year going from the developing to the developed world. Despite these payments, the debt of the developing world is growing because these countries have not been meeting their interest payments. Some policymakers argue that the best thing we can do for development, humanity, and the forests and wildlife that are being exploited to pay the interest is to forgive these debts. Such a policy makes sense, they argue, because the original loan amounts have been more than repaid in full.

Since no one seriously believes that most developing countries ever will be able to pay the amount they owe, the losses have already been absorbed and written off in a practical sense. IOUs for debt are bought and sold much like stocks on the stock market. The value of the IOU depends on the buyer's evaluation of the debtor's ability or intention to pay. Thus, much of the debt of developing countries is currently traded on the "debt market" for 10–20 cents on the dollar. That is, a speculator can buy (or sell) a $1000 IOU from a developing country for $100–$200. This effectively means that 80–90% of the face value of the debt has already been written off and absorbed by those who traded the IOU on the way down. What do you think? Should debts of developing nations be forgiven?

expanded markets for the developed world. Finally, development fosters trade, cooperation, and peace among nations. If fertility rates decline to replacement levels in the process, then, according to the theory, we can look forward to a future in which a stable world population lives in a state of relative affluence.

In 1944, during World War II, delegates from around the world met in Bretton Woods, New Hampshire, and conceived a vision of development for poor countries. They established the International Bank for Reconstruction and Development (or, as it is more commonly known now, the **World Bank**). The World Bank now functions as a special agency within the United Nations. With deposits from governments and commercial banks in the developed world, the World Bank lends money to developing nations for a variety of projects at interest rates somewhat below the going market rates. Effectively, the World Bank helps governments of developing countries (the Bank loans only to governments) borrow large sums of money for projects they otherwise could not afford.

Loans from the World Bank have climbed steadily, from $1.3 billion in 1949 to $23.7 billion in 1993. With the power to approve or disapprove loans, and through the amount of money it lends, the World Bank has been the major instrument in determining the course of development in poor countries for the past 45 years.

Successes and Failures of the World Bank

The World Bank has been successful in many areas. The gross national products of some countries have increased as much as fivefold, bringing them from the low to the medium-income countries; and some medium-income nations are now close to high-income status. In conjunction with efforts from other branches of the United Nations, such as the World Health Organization (WHO), Food and Agricultural Organization (FAO), U.N. Educational, Scientific, and Cultural Organization (UNESCO) and United Nations Children's Fund (UNICEF), from various government programs, and from innumerable charitable organizations, great strides have been made in social progress. Literacy rates, the percentage of the population with access to clean drinking water and sanitary sewers, and other social indicators of development, generally speaking, have improved (Table 7-1). Further, in keeping with the concept of the demographic transition, the fertility rates of most developing countries have declined, although not to the replacement level (Table 7-2).

However, these successes are dulled by the facts described in Chapter 6: A fifth of the world population, 1.3 billion people, remains in absolute poverty, illiterate, and without access to clean water or adequate nutrition (an increase of 300 million over 1990); envi-

TABLE 7-1
Improvement in Development Indicators 1970–1990

	Low-Income Countries (excluding India and China)[1]		*Upper/Middle-Income Countries*	
	1970	*1990*	*1970*	*1990*
Population with Access to:				
Safe drinking water (percent)	22	51	59	87
Sanitary sewers (percent)	29	38	61	86
Age Group Enrolled In Education:				
Primary (percent of total)	55	79	94	105[2]
females per 100 males	61	77	95	(No data)
Secondary (percent of total)	13	28	32	54
females per 100 males	44	66	95	(No data)
Infant Mortality: (Per 1000 live births)	114	73	70	40
Life Expectancy at Birth:				
Female	47	57	64	72
Male	46	55	59	66

[1]See map on page 143
[2]Adults in addition to children enrolled.
Source: Data from World Bank, *World Development Report*, 1994.

TABLE 7-2
Decline in Total Fertility Rate

	1981	*1985*	*1990*	*1995*
Africa	6.4	6.3	6.2	5.8
Latin America and Caribbean	4.4	4.2	3.5	3.1
Asia (excluding China)	5.5	4.6	4.1	3.5
China	2.3	2.1	2.3	1.9
Developed Countries	2.0	2.0	2.0	1.6

ronmental degradation is rampant; and fertility rates remain unacceptably high. In short, the vicious cycle of high fertility, poverty, and environmental degradation seem about to overwhelm development efforts and commence a cycle of decline. Indeed, there is mounting evidence that this turning point has been reached in some countries. For example, per capita grain production in Africa has been declining since 1980 (see Fig. 8-7, p. 195).

Not only have World Bank efforts fallen short of overcoming poverty, but critics point to many examples where the Bank's projects have actually exacerbated the cycle of poverty and environmental decline. For example, the World Bank lent India nearly a billion dollars to create a huge electric-generating facility at Singraali consisting of five coal-burning power plants and to develop open-pit coal mines to support the plants. However, the increased power does little for the poor, who cannot afford electrical hookups. The project displaced over 200,000 rural poor people, who had farmed the fertile soil of the region for generations, and moved them to a much less fertile area without allowing them any say in the matter and giving them little, if any, compensation. In addition, the project has caused extensive air and water pollution. Hydroelectric dams in a number of countries have similarly displaced people and worsened their poverty.

Nowhere are the failures and environmental destructiveness of large-scale projects more evident than in agriculture. The World Bank funneled $1.5 billion into Latin America from 1963 to 1985 for clearing millions of acres of tropical forests. Most of the cleared land was given over to large cattle operations for producing beef for export. However, no type of agriculture requires less labor per acre than ranching. Spreads of more than a million acres are run by millionaire cattle barons checking their herds by aircraft (Fig. 7-3). Meanwhile, the poor are pushed into more marginal lands or into cities, as described in Chapter 6. Because the soil of cleared tropical forests is so poor, some of the ranches have already been abandoned, and

FIGURE 7-3
Cattle ranch in eastern Brazil. Large World Bank loans went to clear rain forest and convert the land into rangeland, as seen here. The ranches provide almost no jobs and, because of poor soil, are only marginally profitable. (Photograph by BJN.)

Ethics

Additional Incentives for Reducing Fertility

What are our options as we foresee the limits of Earth's carrying capacity for the human species? What are the ethical and moral implications of each option?

1. We can argue, as cornucopians do, that the problems do not really exist—that technology will always have a solution to make life better and better for more and more people. Does global environmental evidence support this position, or is it a means of blinding oneself to reality? Is it simply a means of delaying making tough choices? Is it creating an even greater crisis for the future, as more people will have to support themselves with ever more depleted or degraded resources?

   An extreme of the cornucopian argument is the future export of people to space stations or other planets. We would have to be exporting 88 million persons per year just to stabilize population. The astronimcal costs of orbiting small numbers of people who can afford tickets makes this a silly alternative.
2. Perhaps a disease, famine, or a natural disaster—or even war—will come along and take care of the situation for us in "nature's way." Again, this argument seems to lack an appreciation for the magnitude of the numbers involved. For example, the current world death toll from AIDS, tragic as it is, is about 1.5 million—less than 2% of what would actually stabilize population. Therefore, looking for a "death" solution is looking for untold human suffering. It is doubtful that anyone would escape the ramifications of that suffering. The entire human endeavor has been to try to avoid such calamities. Are we going to change that?
3. Some consider forced sterilizations or abortions as a solution to the population problem. This option immediately implies that some person or group is deciding who is going to reproduce and who is not. The backlash from this kind of endeavor in the few cases where it has been tried has been so severe, that the agencies or governments which have tried it have been summarily thrown out of power.
4. The most acceptable option—the one that has been the theme of this chapter—is to create an economic and social climate in which people of their own volition will desire to have fewer children and then provide the means (family planning) to enable them to meet that choice. This, of course, has happened "unconsciously" in developed countries, although we are confronted by dilemmas. How far can one go in manipulating the social or economic environment before choice becomes coercion?

With its current population of 1.2 billion (a fifth of the world's people), China provides the most comprehensive example of a country that offers extensive economic incentives and disincentives for reducing population growth. Some years ago, China's leaders recognized that, unless population growth was stemmed, the country would be unable to live within the limits of its resources. Because of inevitable population momentum, the leaders felt that the country could not even afford a total fertility rate of 2.0; they set a goal of a one-child family, and to achieve that goal, they instituted an elaborate array of incentives and deterrents. The prime incentives were as follows:

- Paid leave to women who have fertility-related operations—namely, sterilization or abortion procedures.
- A monthly subsidy to one-child families.
- Job priority for only children.
- Additional food rations for only children.
- Housing preferences for single-child families.
- Preferential medical care to parents whose only child is a girl. (There is a strong preference for sons in China, and parents generally wish to have children until at least one son is born.)

Penalties for an excessive number of children in China included the following:

- Repayment to the government of bonuses received for the first child if a second is born.
- Payment of a tax for having a second child.
- Payment of higher prices for food for a second child.
- Maternity leave and paid medical expenses only for the first child.

Along with improving economic opportunities, these incentives and deterrents have helped China achieve a precipitous drop in its fertility rate, from about 4.5 in the mid-1970s to 1.8 in 1996. (The one-child policy has not been consistently promoted; a higher fertility rate in rural areas offsets a rate of less than 2 in cities.) However, the population of China is still growing because of momentum (a large percentage of the population is still at, or below, reproduction age).

Is China's population policy morally just or unjust? If we do not succeed in stabilizing population with less stringent measures, can we all look forward to more coercive measures being implemented? Or should we just ignore the whole situation, which by default puts us back at the first, second, or third option?

FIGURE 7-4
Reynosa, Mexico, two miles from the United States border. U.S. firms have built factories and employ Mexicans for 60–70 cents per hour to make consumer items, mostly for the U.S. market. On this wage, the workers can afford to live only in the shacks and slum conditions seen here. (Nubar Alexandrian/ Woodfin Camp & Associates.)

most of the rest are only marginally profitable. Innumerable projects in other countries have emphasized growing cash crops for export, fostering huge mechanized plantations while leaving the poor likewise marginalized.

Development or Exploitation of the Poor?

In many instances of industrial plants built in developing nations, there is much debate as to whether the projects should be classified as development or as exploitation of the poor. Developing nations lack strong labor unions. Consequently, they do not have the wages, fringe benefits, or other protections that unions in developed nations have won for workers over the years. Also, developing countries lack or fail to enforce pollution regulations.

An example of the result of establishing such plants may be seen in Mexico, just across the border with the United States. Here, more than 1000 U.S. firms have set up plants (with the government of Mexico holding a 20% share) to assemble items such as television sets, which are mostly returned to the U.S. market for sale. An example, General Motors has 25 such plants employing over 25,000. Collectively, all these U.S. industrial facilities are know as *Maquiladoras*.

Around the plants people, paid less than $10.00 per day, live several per room in cinder block huts with no indoor plumbing or electricity. Water is obtained from a single community pipe. Outdoor privies commonly overflow. Air and water are often fouled by pollution from the plant (Fig. 7-4).

Why do these people put up with such deplorable conditions? The answer is the poverty-population problem. The workers are mostly young people desperate for work of any kind. Many are children in their early teens who lie about their age (working is technically illegal in Mexico till age 16), quit school, and go to work to help their struggling and often fatherless families. According to the Mexican Center for Children's Rights, there are some 12 million such working children in Mexico.

The Debt Crisis

Another consequence of promoting development through World Bank loans is similar to enticing people to buy on credit. Borrowers become overwhelmed by interest payments. Theoretically, development projects were intended to generate additional revenues that would be sufficient for the recipient to pay back the loan with interest. However, a number of things have gone wrong with this theory, not the least of which are corruption, mismanagement, and, perhaps, honest miscalculations. In their eagerness to obtain a loan for a billion-dollar project, government officials often overestimate the virtues of and expected revenues from the project.

In any case, far from paying off loans, developing countries as a group have become increasingly indebted. Their total debt reached $1.8 trillion in 1994 and is still climbing. Of course, interest obligations climb accordingly, and any failure to pay interest gets added to debt, increasing the interest owed—the typical credit–debt trap. Many developing countries are now paying a substantial amount of their export earnings in interest, and the situation in Africa is worsening. (Fig. 7-5).

Because of the rising interest being paid by developing countries, there is now a net flow of capital from developing to developed countries in the amount of some $50 billion per year, and there is no question that

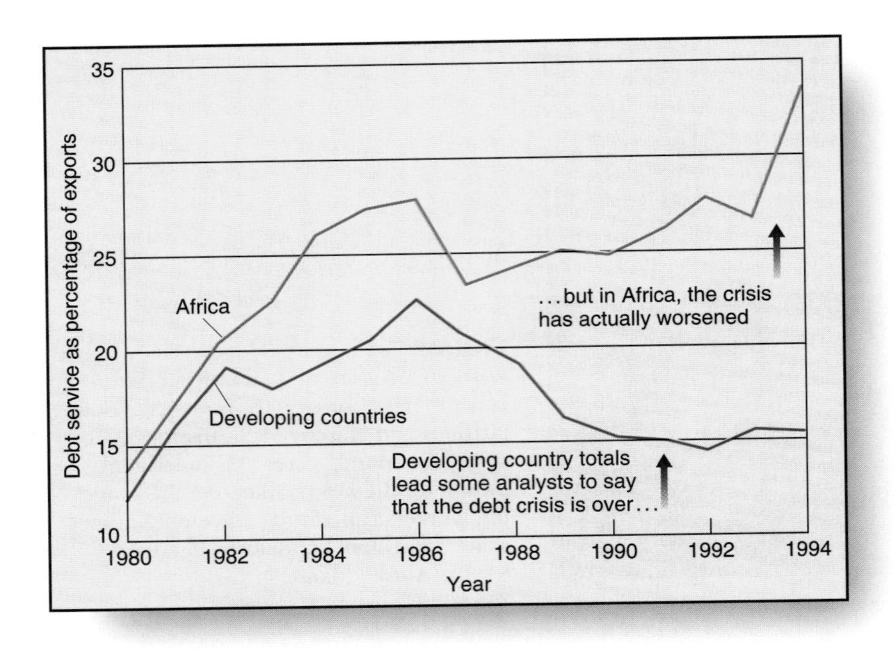

FIGURE 7-5
Loans aimed at promoting development have caught many countries in a debt trap. (Gary Gardener, "Third World Debt Is Still Growing," *WorldWatch*, Jan./Feb. 1995, p. 38.)

income disparities between rich and poor countries are widening. Ironically, then, who is aiding the development of whom?

This debt situation continues to be an economic, social, and ecological disaster for many developing countries. In order to keep up even partial interest payments, poor countries have done one or more of the following:

1. Focused agriculture on large-scale growing of cash crops for export. This has occurred at the expense of peasant farmers growing food. Thus, hunger and malnutrition have increased, and so has poverty, as peasants have been pushed from the land.
2. Adopted austerity measures. Government expenditures have been drastically reduced so that income can go to pay interest. But what is cut? Usually, it is schools, health clinics, police protection in poor areas, building and maintenance of roads in rural areas, and other goods and services that benefit not only the poor, but the country as a whole.
3. Invited the rapid exploitation of natural resources (e.g., logging of forests and extraction of minerals) for quick cash. With the emphasis on quick cash, few, if any, environmental restrictions are imposed. Thus, the debt crisis has caused disaster for the environment. Ironically, this forced exploitation of material resources has placed such oversupplies of commodities on the market, that prices have been severely depressed, so that actual earnings are low.

In essence, these are examples of liquidating capital assets to raise cash for short term needs. Clearly they do not represent sustainability. Also, it is clear that the brunt of these measures falls on the poor. Thus, many observers point out that it is the poor, who gained nothing from the development, who are now being required to pay for it. As is typical with the credit trap, many countries have paid back in interest many times what they originally borrowed; yet the debt remains. Is there any point—humanitarian, ecological, or economic—to keeping such debts in place? (See "Ethics" box, page 167.)

In sum, the concept of fostering the development of poor countries through massive loans for large-scale projects—whatever the advantages in enhancing gross national products—have not broken the cycle of excess population, poverty, and environmental degradation. The reason for this failure is that policymakers at the World Bank saw development in terms of the hallmarks of the industrialized world: energy—in particular, electrification through centralized plants; transportation—in particular, building roads and cars; and mechanization of agriculture. Thus, about three fourths of the money loaned by the World Bank has been for projects in these three areas. Moreover, the policymakers at the Bank were mainly economists and engineers, and the traditional training in these two professions has a strong cornucopian bias, viewing Earth as yielding endless resources and seeing humans as able to manipulate anything to their benefit. Thus, such professionals, lacking any ecological training, have been notoriously slow to appreciate the environmental ramifications of the projects they propose. We now see, however, that superimposing these large-scale projects on a culture that is basically engaged in subsistence agriculture bypasses and perhaps worsen the lives of the poor. The people who benefited most from the projects were those who were already relatively well off and in power. In terms of electrification, development efforts were building large, capital-intensive, centralized generation plants to serve a labor-intensive, decentralized energy need—a mismatch in scale and concept.

These past failures, however, should not be taken as a reason to give up and cut off all efforts to help. The combined problems of poverty, population, and environmental degradation do threaten our future. The lesson should be that breaking the cycle demands focusing on the problems of the poor specifically. Let us begin with a reassessment of the demographic transition.

Reassessing the Demographic Transition

In retrospect, it should be self-evident that no one makes a decision whether to have a child based on the average GNP of the home country. Indeed, plotting fertility rate against GNP per capita shows a weak correlation (Fig. 7-6). Yet there is no question that fertility rates in industrialized countries have declined with development. Therefore, we need to determine the specific aspects of development that influence childbearing.

Factors Influencing Family Size

Many students in developed countries find it difficult to understand why poor women in developing countries have large numbers of children. It is obvious from our perspective that more children spread a family's income more thinly and handicap efforts to get ahead economically. "Why," many ask, "do poor people behave so irrationally?"

What we fail to recognize is that the poor in developing countries live in a very different sociocultural situation. When we understand that situation, we find that their choices for larger families are quite

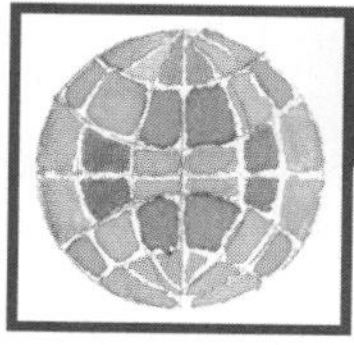

Global Perspective
Fertility and Literacy

Illiteracy, particularly among women, is one of the prime indicators of poverty and high fertility. The following map depicts countries according to female illiteracy rate and fertility rate. The strong correlation is evident. A conclusion which may be drawn is that improving women's literacy will have an impact on lowering fertility.

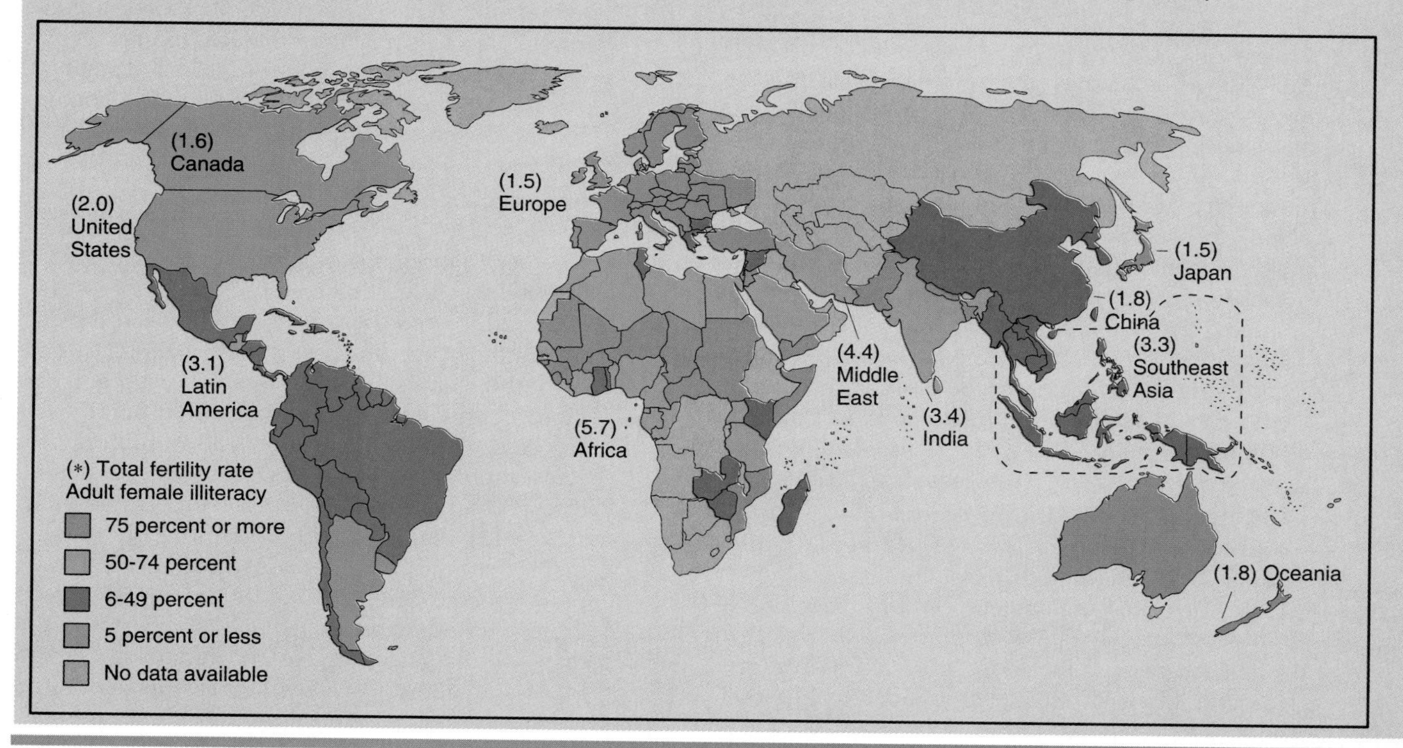

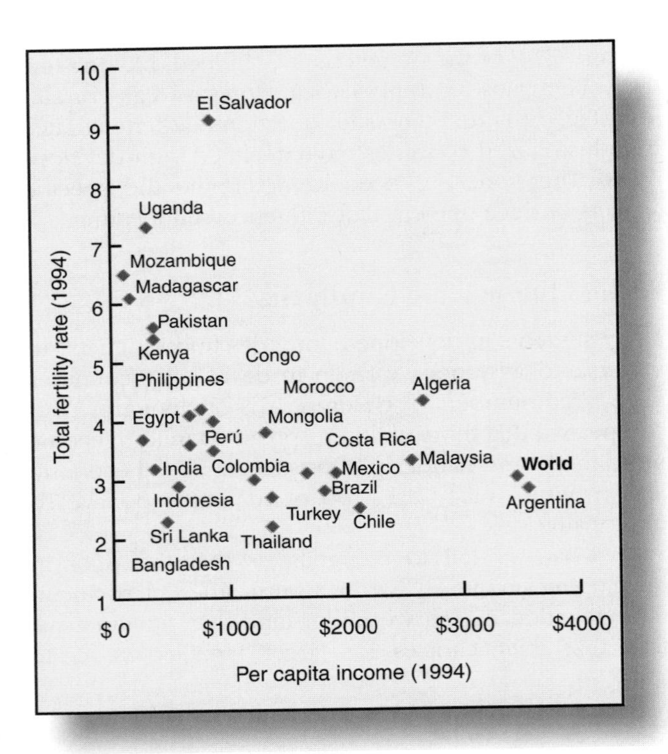

FIGURE 7-6
Fertility rate in relation to income for selected developing countries. There is a weak correlation between income and lower total fertility. Factors affecting fertility more directly are health care, education for women, and the availability of contraceptive information and services. (From the *World Tables 1994*. Copyright 1994 by the World Bank and *World Population Data Sheet 1996,* Population Reference Bureau.)

logical. Numerous studies and surveys reveal the following as primary reasons that the poor in developing countries desire large numbers of children:

1. **Security in one's old age.** The traditional custom and expectation in developing countries is that old people will be cared for by their children. Social security, welfare, medicare, retirement, and nursing homes are all relatively new developments, found in high-income nations alone. Such things are not available to the poor of developing countries. Therefore, a primary reason given by poor women in developing nations for desiring many children is "to assure my care in old age."
2. **Infant and childhood mortality.** Closely coupled with security in one's old age is a high infant and childhood mortality. According to UNICEF, 42,000 children below one year of age die every day in the developing world. This high infant mortality is the most profound indicator of the conditions of squalor and poverty in which people are living; it is unacceptable on any moral, ethical, or religious grounds. Nor does it serve to stabilize population, for the following reason: The common and often personal experience of children dying leads people to desire additional children as an "insurance policy" for security in their old age. Therefore, high infant mortality actually leads to higher fertility rates. It is only when there is a very high likelihood that children will survive (low childhood mortality) that people feel secure in having just two children or even one child.
3. **Children: an economic asset or a liability?** A third reason given by women of developing nations for desiring many children is "to help me with my work." In the subsistence-agriculture societies of the developing world, it has been and remains traditional for women to do most of the work relating to the direct care and support of the family. Clearing fields and turning the soil in preparation for planting is done largely by men, but all of the rest of the work, from planting, weeding, and harvesting to going to the market and gathering firewood and water, falls to the women (Fig. 7-7). A child as young as 5 can begin to help with many of these chores, and 12-year-olds can do an adult's work (Fig. 7-8). In short, children are seen as an economic asset. As environmental degradation occurs, and more time is required to fetch water, collect firewood, and tend crops, the asset of many little hands becomes even greater (Fig. 7-9). It is only in an urban setting in a developed nation that opportunities for children to contribute to the economic welfare of the family become extremely limited, that the costs of feeding, clothing, and educating children are prolonged, and that the economic burden of children is acutely felt.
4. **Importance of education.** The importance given by a society to education is closely allied to whether children are seen as an economic asset or liability. If it is felt that children do not need to be educated, then it is easy to cast them in the role of simply being the "many little hands" to help with the chores of everyday survival, and the more the better. If, on the other hand, it is required that they go to school, they not only are removed from the labor force, but also need additional economic support for suitable clothing, school supplies, and so on. In short, the requirement that children be educated changes their position from being seen as an economic asset to being viewed as an economic liability and influences fertility rates. Again, in traditional, subsistence-agriculture societies, education has been

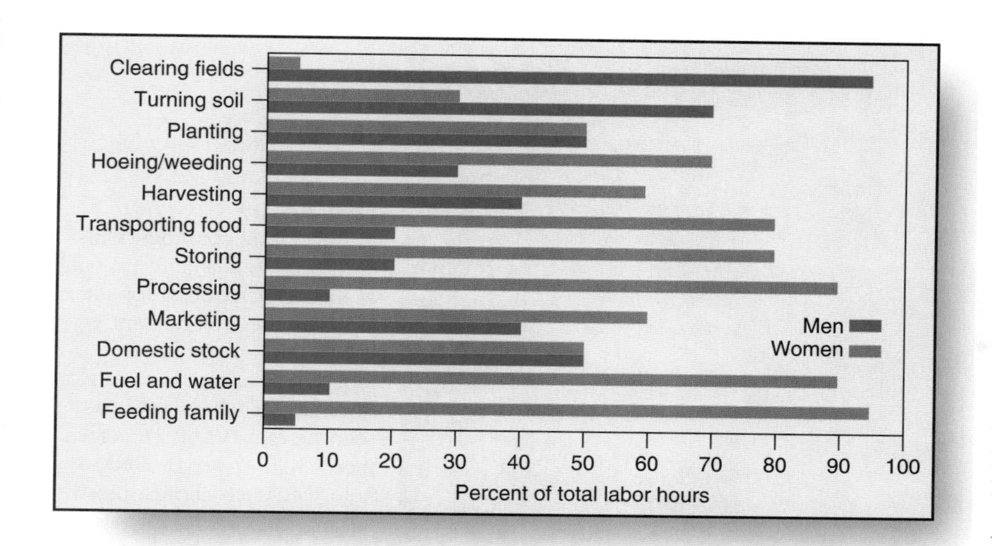

FIGURE 7-7
In developing countries—especially Africa—women do most of the work relating to care and maintenance of the family, including heavy farming tasks. (Jodi Jacobson, "Gender Bias: Roadblock to Sustainable Development," *Worldwatch Paper 110*, 1992. Washington, DC: Worldwatch Institute.)

deemed unnecessary, and this remains the case for many children in the developing world, especially girls. Thus, the continued high fertility of parents is supported by a belief that educating girls is unnecessary. Of course, education does much more than keep children from working; it opens up any number of additional opportunities that also affect fertility rates. (See "Global Perspective" box, p. 173.)

5. **Status of women: opportunities for women's education and careers.** The traditional social structure in many developing countries still discourages and in many cases bars women from obtaining higher education, owning businesses, owning land, and pursuing many careers. Such discrimination against women forces them into doing what only they can do: bear children. Worse, in many of these countries, the male's respect for a woman is only proportional to the number of children she bears. Breaking down such barriers of discrimination so that girls are educated and gain status outside the context of raising children has probably contributed more than anything toward the very low fertility rates seen in developed countries today. Indeed, studies show that women with just an eighth-

FIGURE 7-8
Children working with adults in the fields in Bali, Indonesia. In most developing countries, children perform adult work and thus contribute significantly to the income of the family. Far from being an expense, in such situations children are seen as an economic asset. (Bruno J. Zehnder/Peter Arnold, Inc.)

FIGURE 7-9
In developng countries, children provide a significant part of the labor required for survival. As the environment is degraded more labor is required to secure basic necessities such as water, as seen here in Brazil. Therefore, more children are desired to help with such chores, promoting the cycle of population, poverty and environmental degradation. (J. L. Bulcao/Gamma Liaison, Inc.)

grade education have, on average, only half the number of children as their uneducated counterparts.

6. **Availability of contraceptives.** There can be no doubt regarding the importance of contraceptives in achieving a lower fertility rate. Studies show a strong correlation between lower fertility rates and the percentage of couples using contraception (Fig. 7-10). In the developed world, we take the availability of contraceptives almost for granted. Perhaps the most profound finding in surveys of women in the developing world is that large numbers state that they want to delay having their next child or that they do not want any more children. Yet many of these women are not using contraceptives. Women in rural areas report that contraceptives are frequently not available. Women in cities also have trouble getting them in spite of free clinics; the clinics may be too far away or crowded, or they may run out of contraceptives.

Students frequently raise the point that religious beliefs play a role in determining family size. To some extent this may be true, but it becomes less of a factor as educational and other barriers to women are broken down. For example, Italy, a strongly Catholic country, has the lowest fertility rate of any country. Obviously, most Italian Catholics are not taking the Pope's admonitions seriously. The same is true for Mexico, host country for the 1984 World U.N. Population Conference.

Conclusions

Reflecting on the foregoing items, we can see that the factors supporting large families are common to preindustrialized, agrarian societies, while those conducive to raising small families (or no children) generally appear with industrialization and development. Fertility rates in developing countries remain high not because people in those countries are behaving irrationally, but because the sociocultural climate in which they live favors high fertility and, often, contraceptives are not available. Furthermore, we should be able to understand how poverty, environmental degradation, and high fertility drive one another in a vicious cycle (Fig. 7-11).

One may ask how the now-industrialized nations came through the demographic transition without getting caught in the poverty-population trap. Two points are significant in this regard. First, the improvements in disease control that lowered death rates occurred gradually, through the 1800s and early 1900s. Industrialization, which introduced factors that lowered fertility, occurred over the same period. Therefore, there was never a huge discrepancy between birth and death rates (Fig. 7-12a). Second, and perhaps even more significant, surplus population from European nations could and did readily emigrate to the United States, Canada, Latin America, New Zealand, and Australia.

In contrast, modern medicine was introduced to the developing world relatively suddenly, bringing

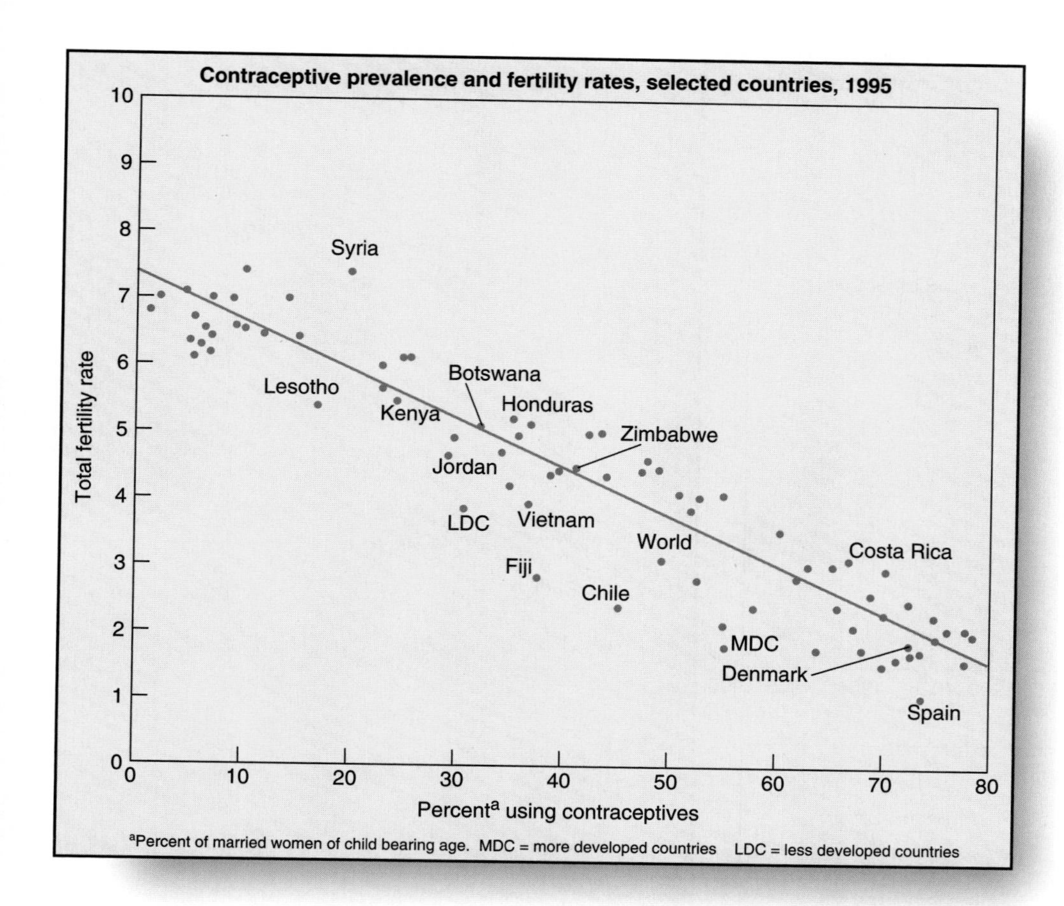

FIGURE 7-10
More than any other single factor, lower fertility rates are correlated with the percent of the population using contraceptives. (Population Data Sheet, 1996, Population Reference Bureau, Inc.)

about a precipitous decline in death rates, while the fertility-lowering effects of development were delayed (Fig. 7-12b). Even more, development actually served to *increase* birth rates in some ways. (Observe the rise in the birth rate after World War II in the figure.) For example, breast-feeding babies tends to inhibit ovulation. Therefore, nursing a baby may prevent another pregnancy for up to three years. Despite this knowledge, companies of the industrialized world have introduced and vigorously promoted infant formulas and bottle-feeding ever since World War II. Worse yet, mixed with nonsterile water, the formula adds to disease and mortality among infants.

All these observations point to the fact that it is not economic development by itself that leads to declining fertility rates. Rather, fertility rates decline insofar as development provides (1) security in one's old age apart from ministrations of children; (2) lower infant and childhood mortality; (3) mandatory education for children; (4) opportunities for higher education and careers for women; and (5) unrestricted access to contraceptives. It should be clear why traditional development efforts are failing to bring about the projected

FIGURE 7-11
Poverty, environmental degradation, and high fertility rates become locked in a self-perpetuating vicious cycle.

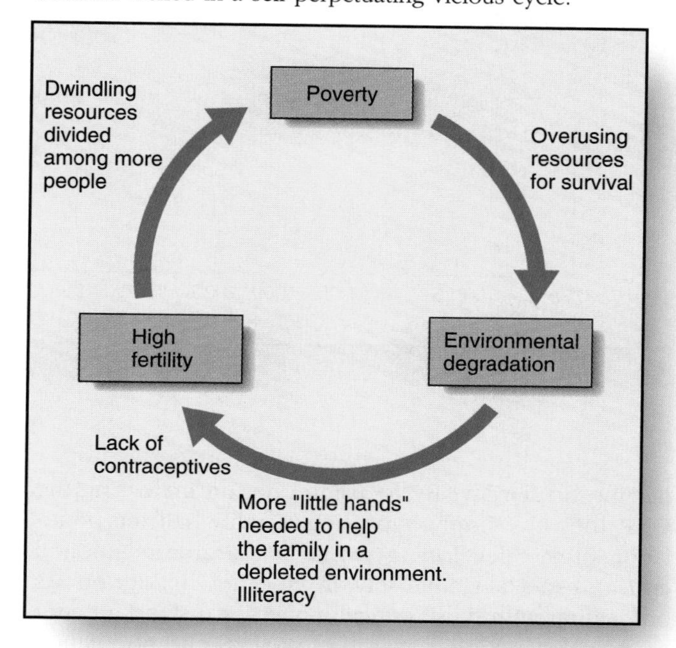

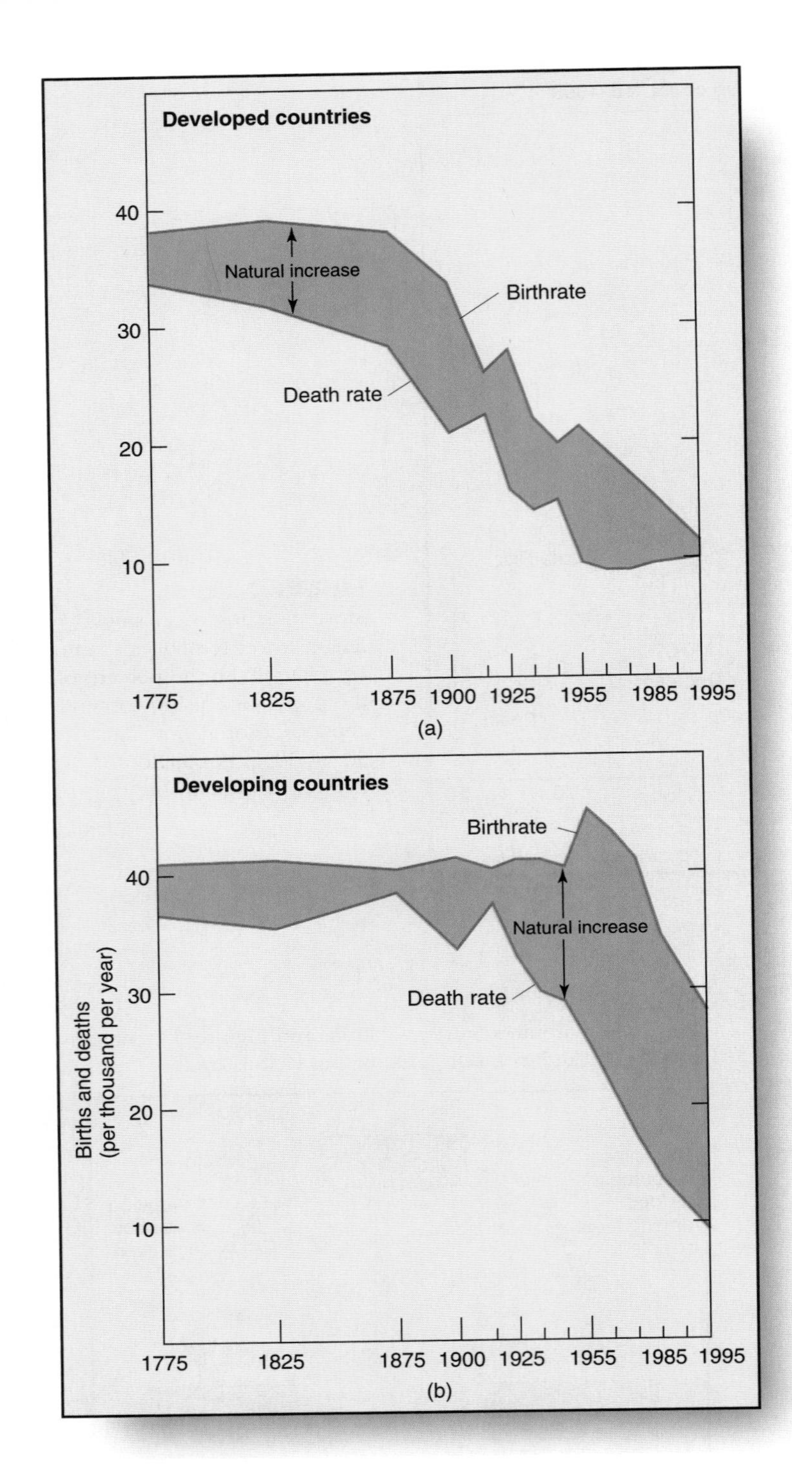

FIGURE 7-12
The increase in world population has resulted from a drop in the death rate, not from an increase in the birth rate. The increasing rate of population growth is seen as the gap between the birth rate and the death rate. (a) In developed countries, the decrease in birth rates proceeded soon after and along with the decrease in death rates, so very rapid population growth never occurred. (b) In developing countries, both birth and death rates remained high until the early 1900s. Then a precipitous decline in death rates was caused by the rapid introduction of modern medicine, whereas birth rates remained high, causing very rapid population growth. (Redrawn with permission of Population References Bureau, Inc., Washington, DC.)

declines in fertility: By focusing only on increasing the GNP, they are simply bypassing roughly half the population of the developing world in the aforementioned critical respects. Conversely, if we can focus our efforts on dealing with these critical matters, we stand a better chance of breaking the cycle of poverty and population with minimal expenditure of resources.

A New Direction for Development

To break the cycle of poverty, high fertility, and environmental degradation, efforts need to be made on

Earth Watch

An Integrated Approach to Alleviating the Conditions of Poverty

Freedom from Hunger is a nongovernmental organization that is pioneering an integrated approach toward improving the lives of the poor described in the text. Freedom from Hunger's Credit with Education program provides opportunities for women in extreme poverty to invest in their own small businesses and to save for emergency needs. These financial services are linked to education for better health, nutrition, and family planning—always respecting local beliefs and culture.

Women interested in receiving loans come together to form credit associations, composed of about 20–30 members, the great majority of whom are very poor. The members guarantee repayment of each other's loans, so they must agree that each woman is capable of making a sufficient profit from her proposed income-earning activity. After training the new members to manage their own association within specified rules, Freedom from Hunger makes a four- to six-month loan to the credit association. The members then break the large loan into small loans averaging $64 per individual. Women invest in activities in which they are already skilled and need no technical assistance, such as baking and selling food, raising chickens, operating a small shop, and making or buying and selling clothing.

Initially attracted by the offer of credit, the women of the credit association become engaged in weekly "learning sessions" to discuss and plan how to provide more and better food for their children. In addition, learning about family planning—with natural or artificial methods that prevent or postpone conception (rather than with postconception methods)—is critically important, because multiple, closely spaced births often lead to maternal and child malnutrition, poor health, and even death.

Freedom from Hunger has found that poor women who enter the Credit with Education program are good credit risks; their repayment currently stands at 99% over three or more years. The impact of Credit with Education goes beyond augmenting income and gaining some knowledge. The women increase their confidence and self-esteem, and they go on to provide more active leadership in their communities.

Freedom from Hunger's desire is to have its program adopted by others. The organization is actively enlisting the direct support of banks and other financial and nonfinancial institutions in each country and training them to deliver Credit with Education. Thus, the program is designed to become self-financing and therefore can be expanded to reach very large numbers of women and families in each country. Started in 1989, Credit with Education programs now exist in six countries: Bolivia, Burkina Faso, Ghana, Honduras, Mali, and Thailand.

Freedom from Hunger, Davis, CA 95617

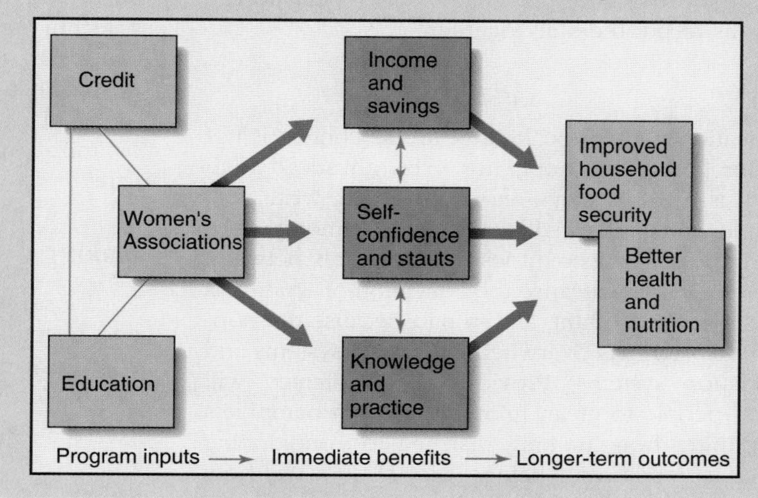

behalf of the poor, with particular emphasis on the following:

1. Education—especially improving literacy and educating girls and women equally with boys and men
2. Improving health—especially lowering infant mortality
3. Making contraceptives available
4. Enhancing income
5. Improving resource management (reversing environmental degradation)

In all of these areas, the focus should be on women because they not only bear the children, but are also the primary providers of nutrition, child care, hygiene, and early education. In short, it is women who are most relevant in determining the number and welfare of subsequent generations.

Fortunately, the world is by no means starting

FIGURE 7-13
Providing education in developing countries can be very cost-effective, since any open space can suffice as a classroom, and few materials are required. This is a class in Bombay, India. (Jagdish Agarwal/Uniphoto Picture Agency.)

from scratch in any of the preceding areas. Numerous programs, both private and government funded, have been going on for many years, and a wealth of experience and knowledge has been gained. We shall look at each area in somewhat more detail, and then we shall return to the program set forth at the Cairo Conference, which aims to put them all together.

Education

The education we are speaking of in this context is not college or graduate school, nor even advanced high school. It is basic literacy—learning to read, write, and do simple calculations. Illiteracy rates among poor women in developing countries are commonly between 50 and 70%, in part because the education of women is not considered important, and in part because the population explosion has overwhelmed school systems and transportation systems. Providing basic literacy will empower people to glean information from pamphlets on everything from treating diarrhea to conditioning soils with compost and baking bread. Consider how much more efficiently information can be distributed with written materials as opposed to one-on-one oral instruction, which is required if people cannot read.

Any number of organizations, both governmental and private, serve in the area of education by sponsoring teachers and contributing materials. Cost-effectiveness is high because often the "classroom" may consist of a group of eager children and adults sitting under a tree with the teacher (Fig. 7-13).

Improving Health

Like education, the health care required most by poor developing-world communities is not high-tech bypass surgery or chemotherapy; rather, it is the basics of good nutrition and hygiene—steps such as boiling water to avoid the spread of disease, and proper treatment of infections and common ailments such as diarrhea. (In developing countries, diarrhea is a major killer of young children, but is easily treated by giving suitable liquids, a technique called oral rehydration therapy.) Health care in the developing world must emphasize pre- and postnatal care of the mother, as well as that of the children. Many governmental, charitable, and religious organizations are involved in providing basic health care.

Making Contraceptives Available

For those who can pay, information on contraceptives and related materials and treatments are readily available from private doctors and health-care institutions. The poor, however, must depend on family planning agencies, which are supported by a combination of private donations, government funding, and small amounts the clients may be able to afford.

The stated policy of family planning agencies (or private services) is, as the name implies, to enable people to plan their own family size—that is, to have children only if and when they want them. In addition to helping people avoid unwanted pregnancies, planning often involves determining and overcoming fertility problems for those couples who are having reproductive difficulties. More specifically, family planning services include the following:

- Counseling and education for singles, couples, and groups regarding the reproductive process, the hazards of sexually transmitted diseases (AIDS in particular), and the benefits and risks of various contraceptive techniques.

- Counseling and education on achieving the best possible pre- and postnatal health for mother and child. The emphasis is on good nutrition, sanitation, and hygiene.
- Counseling and education to avoid high-risk pregnancies. Pregnancies that occur when a woman is too young or too old, and pregnancies that follow too closely on a previous pregnancy, are considered high risk; they seriously jeopardize the health and even the life of the mother. Any existing children are also at risk if the mother suffers injury or death.
- Provision of contraceptive materials and/or treatments after people have been properly instructed about all alternatives.

The vigorous promotion and provision of contraceptives has proven all by itself to have a decided effect in lowering fertility rates, as seen in Fig. 7-10.

Abortions, by definition, are terminations of unwanted pregnancies. Nearly everyone agrees that an abortion is the least desirable way to avoid having an unwanted child—especially in view of the other alternatives available. Indeed, the document resulting from the Cairo Conference explicitly states that abortions should *never* be used as a means of family planning. Therefore, it is particularly important to observe that the primary functions of family planning are education and services directed at avoiding unwanted or high-risk pregnancies. If family planning services were universally available and people availed themselves of them, there would be virtually no unwanted pregnancies, and hence few abortions. The recourse to abortions should be seen as a consequence of the lack of family planning education and services. All studies show that cutbacks in family planning services result in more unwanted pregnancies and more demand for abortions, not less. This occurred following the cut off of family planning monies by the Reagan administration in 1984—abortions increased an estimated 70,000 a year. An ironic twist because of the vocal antiabortion rhetoric at the time.

Furthermore, if abortions are not legal in a particular country, they tend to be provided illegally, or women will attempt to induce them on their own. Both of these practices have tremendous adverse health consequences. It is estimated that 500,000 women die annually from pregnancy-related problems. Most of these deaths result from illegal and self-induced abortions.

Planned Parenthood, which operates clinics throughout the world, is probably the best known family planning agency. Another significant player is the United Nations Population Fund (UNFPA), which provides financial and technical assistance to developing countries at their request. The emphasis of UNFPA is on combining family planning services with maternal and child health care and expanding the delivery of such services in rural and marginal urban areas. Yet U.S. support of UNFPA was cut off in response to right-to-life demands. Funding was in part restored in 1997.

Enhancing Income

The bottom line of any economic system is the exchange of goods and services. At its simplest level, this entails a barter economy in which people agree on direct exchanges of certain things and/or services. Barter economies are still widespread in the developing world.

The introduction of a cash economy facilitates the exchange of a wider variety of goods and services, and everyone may prosper, as they have a wider market for what they can provide and a wider choice of what they can get in return. In a poor community, the ironic twist is that everyone may have the potential to provide certain goods or services and may want other things in return, but there is no money to get the system off and running. In a going economy, people who wish to start a new business venture generally begin by obtaining a bank loan to "set up shop." However, the poor are considered very poor credit risks, they may want a smaller loan than what a commercial bank wants to deal with, and many are women who are denied credit because of gender discrimination alone. For these three reasons, poor communities are handicapped in getting start-up capital.

In 1976, Muhammad Yunus, an economics professor in Bangladesh, conceived and created a new kind of bank (now known as the Grameen Bank) that would engage in **microlending** to the poor. As the name implies, microloans are small—they average just 67 dollars—and they are short term, usually just four to six months. Nevertheless, they provide such basic things as seed and fertilizer for a peasant farmer to start growing tomatoes, some pans for a baker to start baking bread, a supply of yarn for a weaver, some tools for an auto mechanic, and so on.

Yunus secured his loans by having the recipients form **credit associations**, groups of several people who agreed to be responsible for each other's loans. With this arrangement, the Grameen Bank experienced an exceptional rate of payback: Less than 3% of the people defaulted on their loans. Loans from the Grameen Bank have had outstanding results when applied to small-scale agriculture. In a rural area of Bangladesh, small loans, along with horticultural advice, are now enabling peasant farmers to raise tomatoes and other vegetables for sale to the cities. These people have doubled their incomes in three years.

Microlending has been found to have the greatest

social benefits when focused on women because, as Yunus observed, "When women borrow, the beneficiaries are the children and the household. In the case of a man, too often the beneficiaries are himself and his friends." Note that the credit associations also create another level of cooperation and mutual support within the community, particularly when the loans are directed toward women.

The unqualified success of microlending in stimulating the economic activity and enhancing the incomes of people within poor communities has been so remarkable, that the concept has been adopted with various modifications by a considerable number of private organizations dedicated to alleviating hunger and poverty—Oxfam and Freedom from Hunger among them. (See Appendix A.) Freedom from Hunger, which has projects in six countries around the world, combines its lending with "problem-solving education" in a program called Credit with Education. (See "Earth Watch" box, page 179.)

Recently, the U.S. Agency for International Development (AID) has entered the picture by providing grants to other organizations that wish to do microlending. Even the World Bank is expressing the need to be more sensitive to local communities in its lending practices and is making 200 million dollars available for microlending.

FIGURE 7-14
A major step in enhancing incomes and protecting the environment is to encourage better resource management. Here local people are learning the skills to raise tree seedlings that will later be transplanted in a reforestation project. This is part of the Greenbelt Movement in Kenya. (William Campbell/Peter Arnold, Inc.)

Improving Resource Management

The world's poor, almost by definition, are dependent on local resources, particularly water, soil for growing food, and forests for firewood. We have amply described how the pressures of excessive population are degrading these resources. However, a considerable part of the problem lies in poor utilization of resources—failing to replant trees and prevent erosion of the soil, for example. Conversely, a major factor in enhancing income is providing the technical skills and information necessary to manage those basic resources more effectively (Fig. 7-14). This in itself can greatly slow or even reverse the tide of environmental degradation. Allowing impoverishment of the environment can only lead to further impoverishment of the people.

Putting It All Together

Each of the five components just described both depends on and supports the other components. For example, better health and nutrition support better economic productivity, better economic productivity supports obtaining a better education, and a better education leads to a delay in marriage and the desire to have fewer children. The availability of family planning services is essential to realizing the desire of parents to have fewer children. In short, all the components work together in harmony to alleviate the conditions of poverty, reverse environmental degradation, and stabilize population (Fig. 7-15). Conversely, the lack of any component—especially family planning services—will undercut the ability to achieve all other components.

It is not too hard to imagine putting these components together without undue expense. Most importantly, doing so involves the local people uplifting themselves. For example, Zimbabwe is training 5000 women to be preventive health care workers in their own communities. Already a number of developing countries have itinerant "nurses" who make the rounds of rural villages, treating illnesses and injuries, giving advice regarding better health maintenance, and dispensing contraceptive supplies. Another one or two persons traveling with the nurses might conduct classes to improve literacy, provide technical skills, and dispense microloans.

The importance of putting the five components together into a single "package" may seem self-evident. However, probably the biggest reason that United Nations, governmental, and charitable programs have not achieved greater success is because they have been largely focused on only a single component in their approach. This brings us back to the significance of the Cairo Conference.

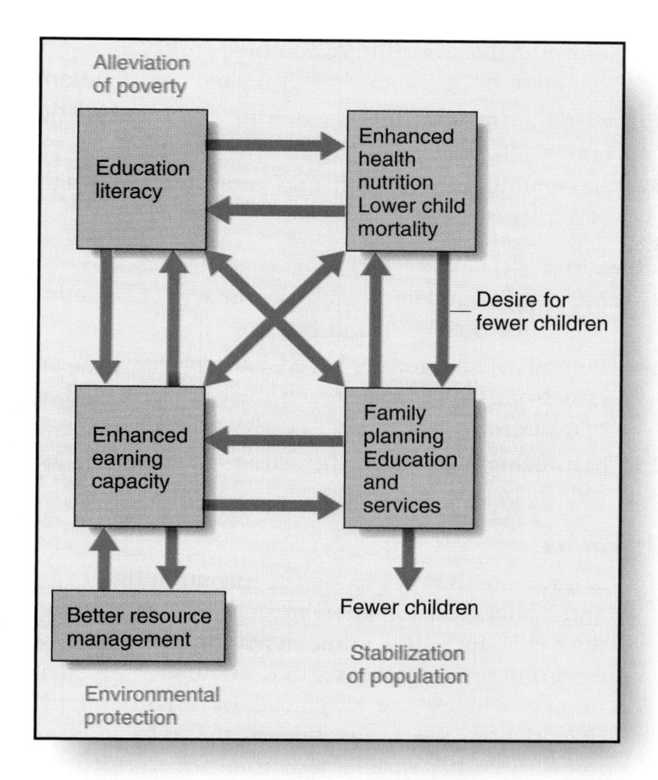

FIGURE 7-15
Five main aspects of enhancing the well-being of the poor are mutually supporting and dependent on one another as illustrated. If any one component is missing, the success of the others will be undercut. Together, they create an integrated "package" not only enhancing people's life potential, but also leading to lower fertility and protecting the environment. The entire system contributes to sustainability.

The Cairo Conference

The decades since World War II must be recognized as a time in which a substantial portion of world leaders held views to the effect that technological progress could always support more people and prevent environmental degradation, or that with more general development, the population problem would take care of itself. They were also decades in which we were gaining knowledge and experience regarding the effectiveness of different kinds of programs. Therefore, it is perhaps understandable why the world has not reached a consensus before now.

Recognizing this diversity of opinions makes the consensus reached at the Cairo Conference in 1994 all the more significant. Effectively, all nations agreed that population is an issue of crisis proportions that must be confronted forthrightly. This sentiment was summed up in the words of Lewis Preston, president of the World Bank: "Putting it bluntly, if we do not deal with rapid population growth, we will not reduce poverty—and development will not be sustainable." Not just the words themselves but hearing them emanate from the World Bank, which traditionally has been a proponent of the "development will take care of it" view, is a reflection of an attitudinal shift of major proportions on the part of numerous leaders and organizations.

The document that nations have signed, and thus, given their commitment to achieve, is technically known as the *1994 International Conference on Population and Development Program of Action* (1994 ICPD Program of Action). It sets forth various principles and goals to be achieved by 2015.

Importantly, the goals of the 20-year Program of Action are not cast simply in terms of reducing fertility or population control per se, policies that have proven unacceptable. For example, one government of India was summarily voted out of power because of resentment growing out of its policy of promoting male sterilization. Instead, the 1994 ICPD document asks that "interrelationships between population, resources, the environment and development should be fully recognized, properly managed and brought into a harmonious, dynamic balance." Thus, the goals are set in terms of creating an economic, social, and cultural environment in which *all people,* regardless of race, gender, or age, can equitably share a state of well-being. In this context of providing opportunities such that people can improve their quality of life, it is assumed that they will choose to have fewer children, as a result, and population growth will level off to the medium projection shown in Figure 6-11. This means that no more than 7.5 billion persons will have to share the world by 2015, and 9.8 billion by the year 2050.

The basic premises that underlie the Program of Action include the following:

Every child should be a wanted child.
Every person should have access to family planning services.
Every person should have access to basic education.
Every person should have access to good nutrition.
Every person should have access to basic health care.
Every person should have rights to own and manage property.
Every person should have access to employment.
No person should fear for his or her care and support in old age.

We have observed that the burdens of poverty fall unduly on women and children because women are often denied access to education and opportunities for business and professional careers. Therefore, the phrase "every person" in the foregoing premises has special significance to women—so much so, that the Program has been widely reported as a document for the empowerment of women.

By signing the Program of Action, the governments of 179 nations have indicated their commitment to achieving, over the next 20 years, a host of objectives, including the following, in seven different categories:

General

1. Implementing strategies for income generation and employment, especially for the rural poor and those living within or on the edge of fragile ecosystems. Strategies should include maintaining and enhancing the productivity of natural resources.

Empowerment of women

2. Eliminating gender discrimination in hiring, wages, benefits, training, and job security, with a view toward eradicating gender-biased disparities in income.
3. Changing customs and laws, where necessary, such that women can buy, hold, and sell property and land, obtain credit, and negotiate contracts equally with men.
4. Promoting the full involvement of women in community life, with speaking and voting rights equal to those of men.

Family

5. Promoting the full involvement of men in family life and creating policies to ensure men's responsibility to and financial support for their children and families.
6. Assuring that programs are created and administered in ways that encourage keeping family members together.

Reproductive and Basic Health

7. Making access to basic health care and health maintenance central strategies for reducing mortality and morbidity, especially among infants, children, and childbearing women.
8. Ensuring community participation in planning health policy.
9. Strengthening education and communication regarding health and nutrition.
10. Making reproductive health accessible to all. Included are counseling on family planning; education dealing with the consequences of unsafe abortions, sexually transmitted diseases, and other reproductive health conditions; and the dissemination of other health-related information.
11. Placing special emphasis on controlling the spread of AIDS.
12. Integrating basic health and reproductive health services.

Education

13. Ensuring complete access to primary school education for both girls and boys.
14. Removing any gender-based barriers that prevent girls from going on to reach their full potential in education.
15. Sensitizing parents to the value of education for girls.

Migration

16. Addressing the root causes of migration into cities and emigration to other areas. This includes the development and implementation of effective environmental management strategies so that people will not be displaced by environmental degradation; the redistribution and relocation of industries and businesses from urban to rural areas; and conflict resolution and other strategies that may help people find opportunities where they are, rather than feeling forced to move.

International Cooperation

17. The developed countries are called upon to cooperate in transferring technology such that the need for contraceptives and other basic items can be met by local production.
18. Further, each member of the developed world is called upon to set aside 0.7% of its GNP for the achievement of the Program of Action's objectives throughout the world.

What is to prevent the 1994 ICPD Program of Action document from simply being forgotten? It is significant that the program does not spell out exactly how the objectives should be met; much less does it have any power of enforcement. It is left to each nation to decide how it will meet the objectives. The need for a forum to facilitate the exchange of information and garner support for the program is conspicuous. This is where nongovernmental organizations come into play. Since the Cairo meeting, the Population Institute (see Appendix A) has been organizing and holding high-level consultations around the world to discuss the implementation of the Program for Action and engender further commitment. The gatherings are drawing high-level representatives from nearly all countries and many nongovernmental organizations.

Still, the bottom line of commitment and support will be funding the implementation of the various programs as they are designed. With the 0.7% GNP requirement, the United States, with by far the world's largest GNP ($6,738.4 billion in 1994), will be a major player (or a major holdout). Which path the U.S. takes in this matter will be largely dependent on public sentiment, pro or con.

The amount of money is significant, but it should be put in the context of other issues and priorities. Rightly and properly, we place a high priority on our national security. This is evidenced by the fact that 4.4% of our GNP currently goes for national defense, and there is considerable pressure to raise the percentage. Our point is not to question whether national security is worth this percentage of our GNP. However, we should ask, What are the threats to our national security, and are our expenditures in line with the relative threats? If we do not diffuse the cycle of population, poverty, and environmental decline, what are the future threats and costs likely to be? Also, while our focus has been on the developing world in this chapter, we should not lose sight of the fact that major environmental impacts, including issues of global pollution, are mainly the result of the consumption-oriented lifestyles in the industrialized world.

Review Questions

1. What has been the major agency and mechanism for promoting development in poor nations over the past 45 years?
2. Why have fertility rates not come down with development, as was expected in the demographic transition?
3. What is meant by the developing-world debt crisis?
4. What are the specific factors that influence the number of children desired?
5. How is the number of children desired influenced by infant mortality? the importance given to education? the status of women? environmental degradation?
6. How does the availability of family planning influence the number of children?
7. Describe how poverty, environmental degradation, and high fertility rates drive one another in a vicious cycle?
8. What are the five components that must be addressed simultaneously to break the vicious cycle mentioned in Question 7?
9. How can each of the following be addressed in a cost-effective way: income enhancement? reduction of infant mortality? education? environmental degradation?
10. What is meant by microlending?
11. What is meant by improving resource management? Give examples.
12. What are the key aspects of family planning, and why is family planning of critical importance to all other aspects of development?
13. What was the significance of the 1994 Cairo conference? What are the agreed-upon strategies for addressing the problems of poverty, excessive population, and environmental degradation?

Thinking Environmentally

1. Is the world population below, at, or above the optimum? Defend your answer by pointing out things that may get better and things that may get worse by increasing population.
2. Suppose you are the head of an island nation with a poor, growing population, and the natural resources of the island are being degraded. What kinds of programs would you initiate, and what help would you ask for to try to provide a better, sustainable future for your nation's people?
3. Suppose you are the head of an agency responsible for the development of a poor nation. Describe major aspects of the program you would initiate to get the best overall improvement in the lives of people with the least cost to your agency.
4. List and discuss the benefits and harms of writing off debts owed by developing nations.
5. Describe what you think will be long-term results (for people, the society, and the environment) of limiting family planning services in developing nations.
6. List what you think are the threats to our national security, and give them a priority ranking according to how large, likely, and close at hand the threat is.
7. What priority do you give to the cycle of population, poverty, and environmental decline? Give reasons pro and con as to your ranking. What proportion of national security expenditures would you allocate toward addressing the population-poverty issue?

CHAPTER 8

The Production and Distribution of Food

Key Issues and Questions

1. In industrialized societies, an agricultural revolution has taken place that has radically affected the practice of farming and its environmental impact. What is industrialized agriculture, how did it develop, and what are its environmental costs?
2. The agricultural revolution has been transferred to the developing world in a process called the Green Revolution. What are the origins and impacts of the Green Revolution?
3. Agriculture in most of the developing world is still practiced in traditional ways—called subsistence agriculture. How important is subsistence agriculture?
4. Continued population growth puts pressure on agricultural practice to keep producing more food. What are the prospects for increasing food production?
5. There is a lively and important world trade in foodstuffs. What are the global patterns of food trade, and what are its consequences?
6. Hunger and malnutrition still plague human societies. What is the extent of hunger, malnutrition, and undernutrition in the world?
7. Famines continue to occur. What are the causes of famine, and which geographical areas are affected?
8. Food aid is distributed to countries all over the world. Is food aid necessary? How is aid distributed?
9. Agricultural sustainability is a desirable goal. How can the four principles of ecosystem sustainability be applied to agriculture—both in developed and developing countries?

There is a land where hills and valleys were covered with green forests and fields. Now the hillsides are cut by erosion gullies that have long since lost their soil. Trees are almost totally gone. The remaining vegetation on the arid landscape is continually grazed by more than a million goats and sheep. Over half of the original cropland has been lost to erosion. To produce charcoal for fuel, people cut the mangroves along the coast and dig up the tree stumps left on the deforested land.

Marketplace in South America. [Photo by David Ball/The Stock Market.]

Even with substantial food aid, the average person here receives only 80% of daily food needs. Severe malnutrition and even starvation are becoming common. The per capita gross national product is $220, and it is declining every year. Unemployment and inflation are in the double digits. Until recently, a corrupt military regime maintained control over a human population of 7.3 million having an annual growth rate of 2.3%. The people fled their country by the thousands in makeshift boats, refugees from political and economic oppression. Behind them, they left a land that can only be described as an environmental wasteland. That land is Haiti (Figure 8-1).

The sad plight of the Haitians is reflected in countries throughout the world, where two decades of rapid population growth and declining food harvests have left hundreds of millions dependent on food aid. As the world population continues its relentless rise (see Figure 6-1), no resource is more directly affected than food. Can the world's farmers and herders produce food fast enough to keep up with population growth? After all, our planet holds only a finite amount of all the resources needed for food production: suitable land, water, energy, and fertilizer.

By many measures, human societies have done very well at putting food on the table. Optimistic observers note that more people are being fed than ever before, with more nutritional food. In the last 25 years, world food production has more than doubled. A lively world trade in foodstuffs forms the bulk of economic production for many nations. Although pockets of chronic hunger and malnutrition exist—and even occasional famines—these are the exceptions to an otherwise remarkable accomplishment. Some experts are convinced that, when the world population levels off (at 12 billion?), enough food will still be produced to sustain it. Others question whether that is at all possible, considering that we are not doing well at feeding half that number.

This chapter takes a look at food production systems, problems surrounding the lack of food, and questions of how to build sustainability, as well as justice, into the agricultural enterprise.

Crops and Animals: Major Patterns of Food Production

Whereas the growth of the human population is part of the food-supply story, lack of food cannot be blamed simply on overpopulation. It can also be traced to social and economic conditions.

The Development of Modern Industrialized Agriculture

Until 150 years ago, the majority of people in the United States lived and worked on small farms. Human and animal labor turned former forests and grasslands into systems that produced enough food to supply a robust and growing nation (Fig. 8-2a). Farmers used traditional approaches to combat pests and soil erosion: Crops were rotated regularly, many different crops were grown, and animal wastes were returned to the soil. The land was good, and farming was efficient enough to allow a substantial segment of the population to leave the farm and join the growing ranks of merchants and workers living in cities and towns. This was important, because in the mid-1800s the Industrial Revolution came to the United States, and it had a major impact on farming.

The Transformation of Traditional Agriculture. The Industrial Revolution contributed to a revolution in agriculture so profound, that today less than 3% of the

FIGURE 8-1
Decades of rapid population growth and intense poverty have left the Haitian landscape in a state of ruin, as forests have been cut down, grasslands overgrazed, and croplands eroded. The land is no longer capable of growing the crops needed to feed all of the Haitian people. (J.C. Francolon/Gamma-Liaison, Inc.)

FIGURE 8-2
(a) Traditional farming practiced in Arkansas. For hundreds of years, traditional practices on American farms supported the growing U.S. population. (b) Modern agricultural practice illustrated by a combine harvesting wheat in America's midwest. The use of machinery and other technologies has promoted an agricultural revolution in the developed world that allows a small percentage of the people to feed the entire population and also produce a surplus. (a) Garry D. McMichael/Photo Researchers, Inc. b) E. R. Degginger/Earth Science)

United States work force produces enough food for all the nation's needs plus a substantial amount for trade on world markets (Fig. 8-2b). Indeed, this revolution has achieved such gains in production, that the United States has had to formulate policies to cope with surpluses of many crops.

The agricultural revolution involved the following developments:

- shifting from animal labor to machinery powered by fossil fuels
- bringing additional land into cultivation
- increasing the use of chemical fertilizers and pesticides
- increasing the use of irrigation
- substituting new varieties of crops.

Virtually every industrialized nation has experienced an agricultural revolution. The pattern of developments in United States agriculture could just as well describe that of France, Australia, or Japan. The combination of these developments in agricultural practice has raised crop production to new heights, doubling or tripling yields per acre (Fig. 8-3). However, each development carries an environmental cost. According to many agricultural experts, expanding production by these methods has reached, or even exceeded, sustainable limits. Let us examine the costs.

Machinery. The shift from animal labor to machinery has created a dependency on fossil fuel energy that adds significantly to the energy demands of the industrial societies (Chapter 21). For example, calculations indicate that 4% of total energy use in the United States is farm related. Further, continued use of farm machines for plowing, planting and harvesting causes soil compaction (Chapter 9).

Land Under Cultivation. Before 1960, much of the increased production in the United States came from bringing new land into production. Since then,

FIGURE 8-3
U.S. corn yield in kilograms per hectare (8000 kg/hectare = 3.5 tons/acre). This graph demonstrates two phenomena: the long-term rise in yields and the effects of droughts in 1970, 1973, 1980, 1983, and 1988. (Worldwatch Institute, Washington, DC.)

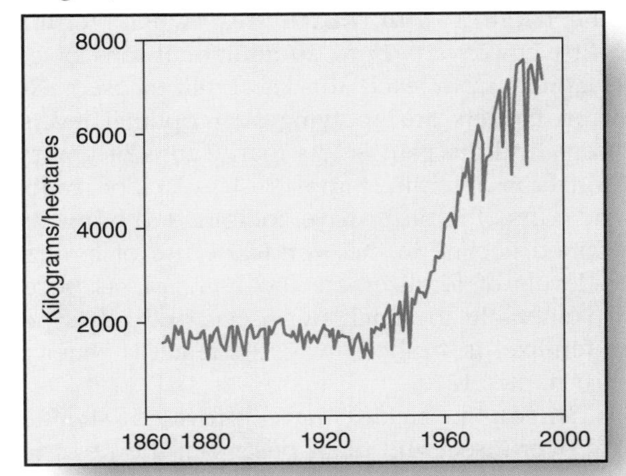

Earth Watch

The Promise of Biotechnology

Biotechnology has raised the prospect of making some remarkable advances in food production. The application of genetic engineering to living plants and animals has already created the following products: (1) the Flavr Savr, a tomato that can be vine ripened and subsequently brought to market and kept fresh much longer than the locally produced products; (2) cotton plants with built-in resistance to insects that comes from genes taken from a bacterium; (3) bacteria that produce bovine somatotropin (BST), a hormone that induces greater milk production in cows. However, these are hardly earth-shattering advances. This technology to produce more food can help the developing world. Biotech crop research is proceeding at a rapid pace in developing countries. Products under development include the incorporation of insect resistance into corn and potatoes, virus resistance into sweet potatoes, melons, and papaya, vitamin A production into rice, protein enhancement into corn and soybeans, and drought tolerance into sorghum and corn. The potential for *transgenic* crops and animals—that is, organisms with genes from another species—seems almost unlimited.

There is a down side to biotechnology, however. Because the main purpose of agricultural biotechnology in the developed countries is to reduce costs, some of the biotech products could hurt the economies of developing countries as substitutes for their cash crops are engineered. For example, common oil-producing plants could be engineered to produce the more exotic coconut and palm kernel oils, important exports of many tropical countries.

In spite of these drawbacks, a look to the future almost certainly will include major advances in food production from biotechnology. If food production is to keep pace with population growth, such advances will be essential, in the view of most observers.

attempts have been made to increase the land used to raise grain, but these new land areas were not well suited for agriculture and are now being abandoned because erosion or depletion of water resources has rendered them no longer productive. Current farm policy (the Conservation Reserve Program, CRP) now reimburses farmers for "retiring" erosion-prone land and planting it to produce trees or grasses. CRP acres may be later returned to cultivation. Essentially all of the good cropland in the United States is now under cultivation or held in short-term reserve. Worldwide, cropland on a per capita basis is on the decline as population continues to rise. Any significant expansion in cropland will come at the expense of forests and wetlands, which are both economically important and ecologically fragile.

Fertilizers and Pesticides. When fertilizers were first employed, 15 to 20 additional tons of grain were gained from each ton of fertilizer used. Now, however, farmers are applying near optimal levels of fertilizers, and the gain is less than 2 tons of grain per additional ton of fertilizer applied. This lack of response and the costs of fertilizer have combined to bring about a sustained decline in the worldwide use of fertilizers. When levels of fertilizer are too high, plants become more vulnerable to attack by insects and other pests; when fertilizer is washed away, the result is water pollution (Chapter 12).

Chemical pesticides have provided significant control over insect and plant pests, but the pests have become resistant to most of the pesticides as a result of natural selection. Also, there are efforts to reduce the use of pesticides because of side effects to human and environmental health. Many of the chemicals in use have not been adequately tested in terms of human concerns and their potential to cause genetic defects. As we shall see in Chapter 10, progress is being made toward developing natural means of control that will be environmentally safe, but these new methods will be unlikely to increase yields.

Irrigation. Worldwide, irrigated acreage increased about 2.6 times from 1950 to 1980 and by 1993 represented 17% of all cropland—some 248 million hectares. It is still expanding, but at a much slower pace because of limits on water resources. More ominous, much present irriga-tion is not sustainable because groundwater resources are being depleted. In addition, production is being adversely affected on as much as one third of the world's irrigated land because of waterlogging and accumulation of salts in the soil, consequences of irrigating where there is poor drainage (Chapter 9).

High-yielding Plant Varieties. Several decades ago, plant geneticists developed new varieties of wheat, corn, and rice that gave yields of double to triple those of traditional varieties. As these new varieties were introduced throughout the world, production soared. However, most of their potential has now been realized, and plant geneticists have no more "super-high-yielding" varieties waiting in the wings to repeat the performance. The widespread use of genetically identical crops has given rise to major pest damage, as pests have become adapted to the new varieties and resistant

FIGURE 8-4
Comparison of (a) an old variety of wheat, shown growing in Rwanda, with (b) a new, high-yielding variety of dwarf wheat growing in Mexico. The new varieties have short stalks and large heads and are much more responsive to fertilizers. (Dr. Nigel Smith/Animals Animals/Earth Science.)

to pesticides. Even breakthroughs in genetic engineering, such as introducing pest-resistance genes into crops, may enable farmers to reduce their costs for pesticides, but they will not dramatically increase yields.

The Green Revolution. The same technologies that gave rise to the agricultural revolution in the industrialized countries were eventually introduced into the developing world. There they gave birth to the remarkable increase in rice and wheat production called the **Green Revolution**.

In 1944, The Rockefeller Foundation sent agricultural expert Norman Borlaug and three other United States agricultural scientists to Mexico, with the objective of exporting U.S. agricultural technology to a less developed nation that had serious food problems. Their aim was to improve the traditional crops grown in Mexico, especially wheat. Mexican wheat was well adapted to the subtropical climate, but it gave low yields and responded to fertilization by growing very tall stalks that were easily blown over. Using wheat from other areas of the world, Borlaug and his coworkers bred a dwarf hybrid with a large head and a thick stalk; the hybrid did well in warm weather when provided with fertilizer and sufficient water (Figure 8-4). The program was highly successful. By the 1960s, Mexico had closed the gap between food production and food needs, wheat production had tripled, and Mexican wheat appeared on the export market.

Research workers with the Consultative Group on International Agricultural Research (CGIAR) extended the work done in Mexico, introducing both high-yielding wheat and high-yielding rice to other developing countries. To cite just one success, India imported hybrid Mexican wheat seed in the mid-1960s, and in six years, India's wheat production tripled. Within a few years, many of the world's most populous countries turned the corner from being grain importers to achieving stability, and, in some cases, even becoming grain exporters. Thus, while the world population was increasing at its highest rate (2% per year), rice and wheat production underwent increases in yields of 4% or more per year. This Green Revolution has probably done more than any other single scientific or other achievement to prevent hunger and malnutrition. Norman Borlaug was awarded the Nobel Peace Prize in 1970 in recognition of his contribution.

The high-yielding grains are now cultivated throughout the world and have become the basis of food production in China, Latin America, the Middle East, southern Asia, and, of course, the industrialized nations. But as remarkable as it was, the Green Revolution is not a panacea for all of the world's food-population difficulties, for the following reasons:

1. Most of its potential has been realized; many of the most populous countries are reaching a plateau in their grain production and in acreage planted to high-yielding varieties, while their populations continue to increase.
2. Without irrigation, it does not work in drought-prone lands. It also requires constant inputs of fer-

tilizers, pesticides, and energy-using mechanized labor, all of which are often in short supply in developing countries.

3. Because it is patterned after agriculture in the developed world, Green Revolution agriculture tends to benefit larger land-holders. More food is raised by a smaller farm work force, causing many farm laborers and small land-holders to become displaced and migrate to the cities, joining the ranks of the unemployed.
4. The most important African food crops (sorghum, millet, and yams) are not commonly used in the developed world, so they have not benefited from the Green Revolution technology. The fact is, the Green Revolution has had little impact on the large part of the developing world where another kind of agriculture—subsistence agriculture—is practiced.

Subsistence Agriculture in the Developing World

In most of the developing world, plants and animals continue to be raised for food by *subsistence farmers*, using traditional agricultural methods. These farmers represent the great majority of the rural populations. **Subsistence farmers** live on small parcels of land that provide them with the food for their households and, it is hoped, a small cash crop. From the point of view of the modern world, such farmers are very poor, although some do not consider themselves to be so. Like past agricultural practice in the United States, subsistence farming is labor intensive and lacks practically all of the inputs of industrialized agriculture. Also, it is often practiced on marginally productive land (Fig. 8-5).

Typically, a family owns a small parcel of land for growing food and maintains a few goats, chickens, or cattle. Such a family is making the best use of very limited resources, and very often the people are adapted well enough to the prevailing social and environmental conditions to provide a livelihood for a household. An important fact to remember, however, is that subsistence agriculture is practiced in regions experiencing the most rapid population growth, even though this kind of agriculture is best suited for low population densities. An estimated 1.4 billion people in Latin America, Asia, and Africa—over one third of the people there—are sustained by subsistence agriculture.

The pressures of population and the diversion of better land to industrialized agriculture often lead to practices that are at best nonsustainable and at worst ecologically suicidal. In many regions in developing countries, woodlands and forests are cleared for agriculture or removed for firewood and animal fodder, leaving the soil susceptible to erosion and forcing the gatherers to travel farther and farther from their homes. The scarcity of firewood leads the residents to burn animal dung for cooking and heat, thus diverting nutrients from the land. Erosion-prone land suited only to growing grass or trees is planted to annual crops. Good land is forced to produce multiple crops instead of being left fallow to recover nutrients. Growing populations force the continued subdivision of existing land, which diminishes the land's ability to support each

FIGURE 8-5
Mexican subsistence farmer plowing marginally productive land with a team of oxen. Subsistence farming feeds more than 1.4 billion people in the developing world. (Dick Davis/Photo Researchers, Inc.)

FIGURE 8-6
Countryside in Mindanao, Philippine Islands, showing the results of traditional slash-and-burn agriculture. Cultivated fields are interspersed with trees and natural areas in a diverse ecosystem. (Roland Seitre/Peter Arnold, Inc.)

household. All these factors tend to increase the poverty characteristic of populations supported by subsistence agriculture, and, in a relentless circle, the added poverty in turn puts increased pressures on the land to produce food and income.

Because subsistence agricultural practice varies with the local climate, it is difficult to draw sweeping generalities. In some areas, subsistence agriculture involves shifting cultivation within tropical forests—often called slash-and-burn agriculture. Research has shown that such a practice can be sustainable (Fig. 8-6). The cultivators create highly diverse ecosystems, where the cleared land supports a few years of crops and gradually shifts into agroforestry—a system of tree plantations with different ground crops arising as the trees grow.

In other areas, subsistence farmers are showing remarkable success in adapting to the changing needs of local societies as they are forced to support expanding populations on the same land. For example, in Kenya, land in the Machakos region was seriously degraded during the 1930s. Now, the land has recovered and supports a population six times larger, as the 1.5 million people there have diversified their agriculture and practiced soil and water conservation.

Animal Farming and its Consequences

Raising livestock—sheep, goats, cattle, buffalo, and poultry—has many parallels to raising crops (Table 8-1), and there are also direct connections between the two. Fully one fourth of the world's croplands are used to feed nonhuman animals; in the United States alone, 70% of the grain crop goes to animals—half the cultivated acreage. Indeed, in many developing countries farmers are switching from food crops to feed crops, to satisfy the growing export-market demand for meat. In essence, this practice removes much of the land's potential to alleviate hunger and malnutrition.

An estimated 15 billion domestic animals on Earth are tended by a human population one third that number. The care, feeding, and "harvesting" of all those animals constitute one of the most important economic activities on the planet. The primary force driving this livestock economy is the growing number of the world's people who enjoy eating meat and dairy products—primarily, most of the developed world and the affluent people in less developed nations.

A thriving livestock economy does not mean that all is well, however. Animal farming affects the environment in a host of nonsustainable ways. Since so much of the plant crop is fed to animals, all of the problems of industrialized agriculture apply to animal farming. In addition, rangelands are susceptible to overgrazing, either because of mismanagement of prime grazing land or because the land used for grazing is marginal dry grasslands, used in that manner because the better lands have been converted to producing crops. For example, overstocking on the rangelands of the western United States has reduced the carrying capacity by an estimated 50%. Much of the western rangeland is public land leased at subsidized fees that

TABLE 8-1
Parallels between Plant and Animal Farming

	Plant	***Animal***
Major products	Grains, fruits, and vegetables for food	Meat, dairy products, and eggs for food
Other important products	Oils, fabrics, rubber, specialty crops (spices, nuts, etc.)	Labor, leather, wool, manure, lanolin
Modern practices	Industrialized agriculture on former grasslands and forests	Ranching, dairy farming, and stall feeding
Traditional practices	Subsistence agriculture on marginally productive lands	Pastoral herding on nonagricultural land
Current global land use	3.7 billion acres (1.5 billion ha) (11% of land surface)	7.6 billion acres (3.1 billion ha) (25% of land surface)

easily lead to overgrazing. (Grazing fees on public land are 80% lower than fees on private land.)

In Latin America, more than 49 million acres (20 million ha) of tropical rainforests have been converted to cattle pasture. Even though most of this land is best suited for growing rainforest trees, some of it could support a rural population of subsistence farmers producing a diversity of crops. Instead, it is held by relatively few ranchers who own huge spreads.

The burning of rainforests throughout the world has released an estimated 1.4 billion tons of carbon to the atmosphere, contributing a significant amount of carbon dioxide to the greenhouse effect. Because their digestive process is anaerobic, cows annually belch and eliminate some 80 million tons of methane, another greenhouse gas. Anaerobic decomposition of manure leads to an additional 35 million tons of methane per year. All this methane released by livestock makes up about 3% of the gases causing global warming (discussed in Chapter 16).

Such significant environmental costs make it clear that animal farming needs to be brought into an ecologically sustainable balance. Essentially, humans are violating the third principle of sustainability: We are maintaining populations of herbivores that are overgrazing the land.

Prospects for Increasing Food Production

Figure 8-7 gives some idea of the race between food and population. The data show that per capita food production is rising in some regions, but not in others (Fig. 8-7a). The one region that stands out as being in serious trouble is Africa. On balance, food production per capita in Africa has been on a downward course since 1970, a consequence of rapid population growth and poor harvests. On a world basis, per capita grain production has been declining since 1984 (Fig.8-7b); 1993 recorded a 7% decline. Nonetheless, there is no shortage of food in the world as a whole: Food production appears to be keeping up with demand. The greatest concern continues to be about the future: How will we manage to produce enough food for twice as many people in a few decades?

Because the end of the increases in yield brought about by the Green Revolution is in sight, and essentially all arable land in the world either is now or has recently been in cultivation, we have only two prospects for increasing food production: (1) Continue to increase crop yields; (2) begin growing food crops on land that is now used for feedstock crops or cash crops.

Grain yields in the most important producing areas are showing signs of reaching a plateau; the slowdown is affecting all of the major grains and has been attributed to limitations of climate and a decline in the response of grain plants to fertilizer. Increasing yields further will be difficult for two reasons. First, a significant percentage of cropland, especially in the less developed countries, has been seriously degraded by erosion. Second, weather conditions in years to come may be unfavorable for farming. The climate between 1940 and 1980 was exceptionally stable and ideal for agriculture in most parts of the world. Then three severe droughts occurred, drastically affecting harvests in North America in 1980, 1983, and 1988. Just the opposite—cool, rainy weather—in the United States and northeast Asia brought a decline in grain harvest in 1993. Fortunately, the world, and particularly the United States, had ample stocks from previous surpluses. A year of ideal weather in North America in 1994 produced bumper crops, driving prices down and prompt-

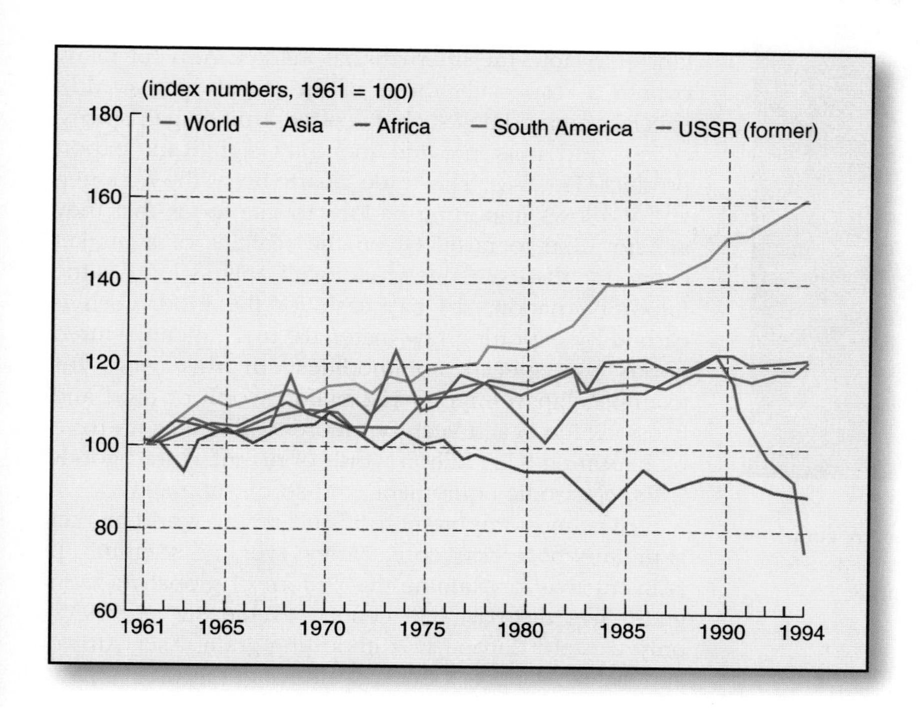

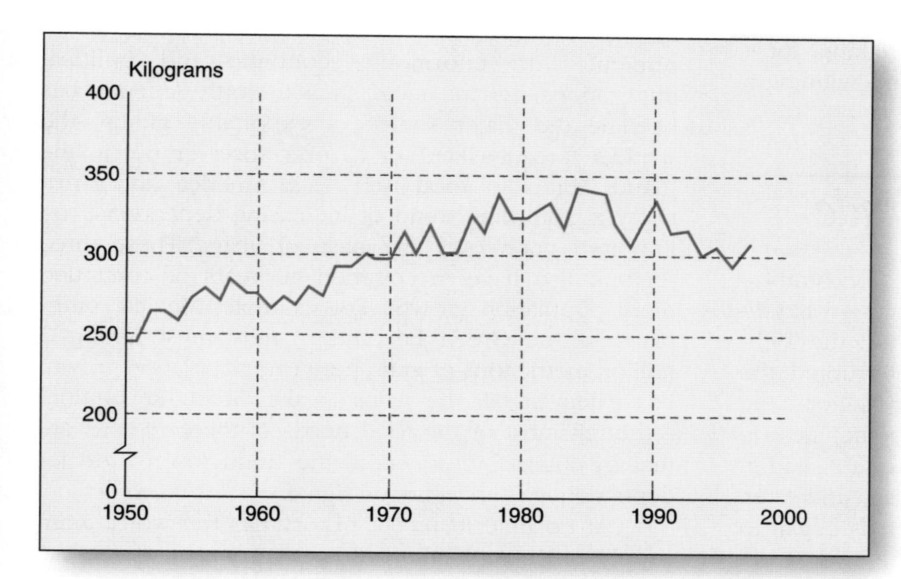

FIGURE 8-7
(a) Changes in per capita food production by region (1961 = 100). (World Resources, 1996–1997; copyright Oxford University Press 1996.) (b) World per capita grain production, 1950–1996. (WorldWatch Institute.)

ing farmers to put their grains in storage. This seesaw harvest situation points to the instability in food production that makes future planning practically impossible.

On the basis of computer models, some respected climatologists are predicting that droughts will become increasingly commonplace as the climate warms due to the greenhouse effect (Chapter 16). At best, climate cannot do more than return to the "ideal" of the 1940–1980 period. Thus, expecting climatic change to increase yields is unrealistic, and the crucial question is, will there be more droughts in major areas of the world in the years to come?

Many observers have pointed out the inefficiency of using grain to feed livestock and then eating the livestock or their products (milk, eggs, butter). Fully 70% of domestic grain in the United States is used to feed livestock; the percentages drop over other regions of the world, in proportion to the economic level of the region. Sub-Saharan Africa and India, for example, use only 2% of their grain for livestock feed. Feed grain can be considered a buffer against world hunger; if the food supply becomes critical, it might force more of the world's people to eat lower on the food chain (less meat, more grain). As we have seen, however, the trend is in exactly the opposite direction.

FIGURE 8-8
A coffee harvest in East Java. Coffee is one of many commodity crops that produce important income for nations of the developing world. (Sam Abell/Woodfin Camp & Associates.)

Converting land use from cash crops to food crops is possible, but is a complex undertaking, for it involves such issues as land reform and maintaining a balance of trade.

Food Distribution and Trade

For centuries, the general rule for basic foodstuffs—grains, vegetables, meat, and dairy products—was *self-sufficiency.* Whenever climate, blight (as in the 19th-century Irish potato famine), or war interrupted the agricultural production of a nation or region, the inevitable result was famine and death, sometimes on the scale of millions. Once colonies were established in the New World, timber, furs, tobacco, fish, and later, sugar, coffee, cotton, and other raw materials began to flow back to the Old World. In turn, the Old World exported manufactured goods, which helped to transform the colonies into societies much like the European ones that had given birth to them. With the Industrial Revolution, trade between nations intensified, and soon it became economically feasible to ship basic foodstuffs from one part of the world to another. In time, a lively and important world trade in foodstuffs arose; as it did, the need for self-sufficiency in food diminished.

Patterns in Food Trade

Today, agricultural production systems do much more than supply a country's internal food needs. For some nations (such as the United States and Canada), the capacity to produce more basic foodstuffs than the home population needs represents an extremely important entry into the international market. And for many countries (especially those of the developing world), special commodities such as coffee, fruit, sugar, spices, cocoa, and nuts provide the only significant export product (Fig. 8-8). This trade clearly helps the exporter, and it allows importing nations to use foods that they are not able to raise. Given the realities of a market economy, the exchange works well only as long as the importing nation can pay cash for the food. Cash is earned by exporting raw materials, fuels, manufactured goods, or special commodities. In this way, for example, Japan imports $22 billion worth of food and livestock feed each year, but more than makes up for it by exporting $223 billion worth of manufactured goods (cars, electronic equipment, and so on) annually.

The most important foodstuff on the world market is grain: wheat, rice, corn, barley, rye, and sorghum. It is instructive to examine the pattern of global trade in grain over the past half century (Table 8-2). In 1935, only western Europe was importing grain; Asia, Africa, and Latin America were self-sufficient. By 1950, new patterns were emerging, and today the trade in grains—as well as other basic foodstuffs—represents a development with enormous economic and political implications. As the table shows, North America has become the major source of exportable grains—the world's "breadbasket," or as one observer put it, the "Saudi Arabia" of food. Asia, Latin America, and Africa show a disturbing trend of increasing dependence on imported grain over the past 40 years. These three regions also have in common 40 years of continued rapid population growth. For example, Mexico, birthplace of the Green Revolution, now must import 7 million metric tons of grain per year; population growth has eaten up all the gains of the Green Revolution. Although most of the food needs of these regions are met by internal production, the trend toward greater dependency is an ominous signal.

At no time in recent history has the world grain supply run out; on the average, carryover stocks (the amount in storage as a new harvest begins) are enough to supply two years' worth of international trade and aid. In other words, enough food is produced to satisfy the world market and keep a healthy surplus in storage. Why, then, are there people in every nation who are hungry and malnourished? Shouldn't every nation make an all-out effort to provide food for its people? And if it can't, shouldn't the rest of the world assume some of the responsibility for providing food to a hungry nation? Where does the responsibility lie for meeting the need for this most basic resource?

Levels of Responsibility in Supplying Food

To begin to answer the question of who is responsible for meeting food needs, it is helpful to examine Fig.8-9.

TABLE 8-2
World Grain Trade since 1935

Region	*Amount Exported or Imported (million metric tons)*[1]					
	1935	*1950*	*1960*	*1970*	*1980*	*1990*
North America	5	23	39	56	131	123
Latin America	9	1	0	4	−10	−12
Western Europe	−24	−22	−25	−30	−16	28
Eastern Europe and former USSR	5	0	0	0	−46	−38
Africa	1	0	−2	−5	−15	−31
Asia	2	−6	−17	−37	−63	−82
Australia and New Zealand	3	3	6	12	19	15

[1]No sign in front of a figure indicates net export; a minus sign in front of a figure indicates net import.
Sources: 1935 through 1980: U.N. Food and Agriculture Organization, *FAO Production Yearbook* (Rome: various years); U.S. Department of Agriculture, Foreign Agricultural Service, *World Rice Reference Tables* (unpublished printout) (Washington, D.C.: June 1988); USDA, FAS, *World Wheat and Coarse Grains Reference Tables* (unpublished printout) (Washington, D.C.: June 1988). 1990: *World Resources, 1994–95* (New York, 1994).

The figure displays three major levels of responsibility for meeting food needs: the family, the nation and the globe. At each level, the players are part of a cash economy as well as a sociopolitical system. In the cash economy, food flows in the direction of economic demand. Need is not taken into consideration. In the event that there are hungry cats and hungry children, the food will go to the cats if the owners of the cats have money and the children's parents don't. In other words, the cash economy, following the rules of the market, provides the *opportunity* to purchase food, but not the food itself.

Where the economic status of the player (a destitute breadwinner, a poor country) is very low, the sociopolitical system may be able to provide the needed purchasing power or the food. In the United

Ethics

National Food Security

Is there a national mandate for food security in most societies? Such a mandate, if it exists, will be seen in a variety of policies:

1. There will be safety net programs, such as those most commonly seen in the wealthier countries and those with centrally planned (socialistic) economies. Farm policies in these nations will promote soil conservation and other sustainable practices.
2. In countries with limited resources (developing countries, in particular), policy needs will be directed toward encouraging agricultural production, especially in rural areas populated by poor people.
3. Additional incentives that encourage food security are land reform programs (seeing that the land is distributed broadly and not concentrated in the hands of a wealthy few), effective family planning and health programs, the encouragement of a lively market economy (instead of a controlled, centrally planned economy), and old age security programs.

Unfortunately, few countries come close to achieving these policies, for a variety of reasons. One of the most important is the direction of massive national resources to military purposes. Another is the tendency for power in the society to be in the hands of a wealthy elite, whose main interest is maintaining their power and wealth. For most of the countries in which hunger is chronic, natural resources and wealth are simply scarce, and consequently, these countries look to the international level for assistance in meeting their food and development needs.

Family	Country	Globe

Goal: Personal and family food security
Policies:
—Employment security
—Adequate land or livestock
—Good health and nutrition
—Adequate housing
—Effective family planning

Goal: Self-sufficiency in food and nutrition
Policies:
—Just land distribution
—Support of sustainable agriculture
—Effective family planning
—Promotion of market economy
—Avoidance of militarization
—Effective safety net

Goal: Sustainable food and nutrition for all countries
Policies:
—Food aid for famine relief
—Appropriate technology in development aid
—Aid for sustainable agricultural development
—Debt refief
—Fair trade
—Disarmament

FIGURE 8-9
Goals and strategies for meeting food needs at three levels of responsibility: the family, the country and the globe. (a) Susan Kulkin, b) John Spragens, Jr./Photo Researchers, Inc. c) European Space Agency/Science Photo Library/Photo Researchers, Inc.)

States, this help is described as the "safety net," and it is represented by a variety of welfare measures such as the Food Stamp Program, Aid to Families with Dependent Children, and the Supplemental Security Income program. (http://www.bread.org/)

The most important level of responsibility is the microlevel—the family. The *goal* at this level is **food security**: the ability to meet the food needs of everyone in the family at a nutritional level that grants freedom from hunger and malnutrition.

For an individual, there are three legitimate options for attaining food security: Purchase the food; raise the food or gather it from natural ecosystems; or have it provided by someone (dependency), usually in the context of the family.

In the event of economic or agricultural failure at the family level, the third option implies that there is an effective safety net—that at the national (or local) level there exist policies with the objectives of meeting the food security needs of all individuals in the society. Thus, an appropriate goal at the national level would be *self-sufficiency in food*—enough food to satisfy the nutritional needs of all of a country's people. The nation can either produce all the food its people need or buy it on the world market. This goal implies policies to eliminate chronic hunger and malnutrition in the society.

Many nations are not self-sufficient in food, however, and must turn to the global community for unmet food needs. According to the Food and Agriculture Organization (FAO) of the United Nations (UN), some 14.0 million metric tons of food were given out during 1994. Figure 8-10 shows the amounts of food aid distributed globally, in comparison with the daily per capita food supply. The United States, Canada, Australia, and the nations of the European Community have been the sources of most of the donated food.

There are some less obvious factors to consider when we are talking about global food needs. One of the most serious is the debt crisis in developing countries, discussed in Chapter 7. Interest and principal payments ($180 billion per year) on the $1.8 trillion dollars now owed by these nations have imposed an enormous financial burden on them. Before 1984, there was still a net flow of funds from rich nations to poorer nations, mostly in the form of loans. Since then, however, the flow has reversed as the loans have come due, and now the poorer nations pay more money than they receive.

For example, the Philippines has a debt burden of $32 billion; principal and interest payments on this debt represent 21% of the nation's total export value of

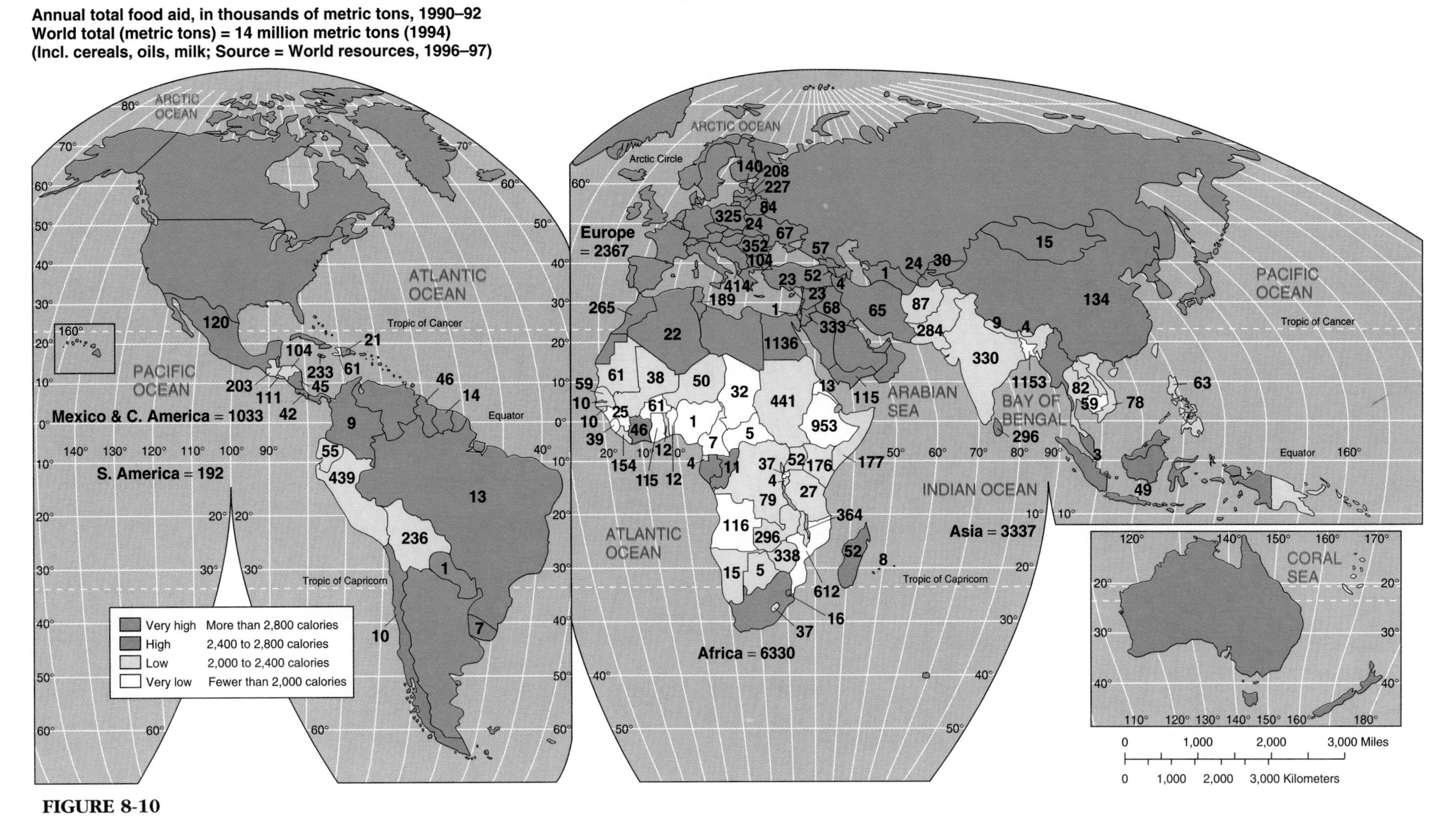

FIGURE 8-10
World daily per capita food supply (in caloric intake) and annual total food aid, 1990–92, in thousands of metric tons. World total = 14 million metric tons.

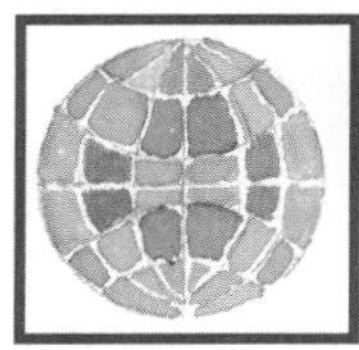

Global Perspective

World Food Summit

Under the leadership of the Food and Agricultural Organization (FAO) of the United Nations, representatives and heads of state from 100 countries met from November 13–17, 1996. Called the *World Food Summit,* the meeting was the result of two years of planning and negotiations. The objective of the meeting was to bring about a renewed commitment around the world to eradicating hunger and malnutrition, and promote conditions leading to food security for individuals, families, and countries everywhere. The need for this high-level attention was emphasized by new data from FAO indicating that 840 million people, or 18% of the population of the developing world, are malnourished or hungry. In light of continuing increases in population, rising costs of grain and declining per capita grain production, the summit is a unique opportunity for world leaders to take a new look at the meaning of sustainability.

At the start of the meeting, the delegates adopted, by acclamation, two major documents: The Rome Declaration on World Food Security, and the World Food Summit Plan of Action. These documents, in the words of the FAO, "set forth a seven-point plan stipulating concrete, political actions to ensure:

1. Conditions conducive to food security;
2. The right to access to food by all;
3. Sustainable increases in food production;
4. Trade's contribution to food security;
5. Emergency relief when and where needed;
6. The required investments to accomplish the plan;
7. Concerted efforts to achieve results by countries and organization."

One concrete objective of the plan is a 50% reduction in the number of hungry people by the year 2015. As a measure of success, this objective was criticized by some participants as being too timid, since it assumed that 420 million people would still be malnourished or hungry. The plan emphasizes the responsibilities of countries (especially developing countries) to enact reforms that would promote food security; policies that do not discriminate against agriculture or small farmers, open trade, greater efforts to bring down population growth, and investing in infrastructure.

Unlike the 1992 Earth Summit and the 1994 Population Summit, the 1996 Food Summit failed to draw many heads of state, particularly from the industrialized countries. However, if countries take the 20-page plan of action seriously, there is no doubt that significant progress will be made in holding back the serious threat of a rising tide of hunger and starvation. (http://www.fao.org/wfs/homepage.htm)

goods and services. As a consequence, half of the agricultural land in the Philippines is used to grow export crops, in spite of extensive hunger and malnutrition in that nation.

Because of the need to pay their debts to foreign creditors, debtor nations are encouraged to exploit their natural resources at an unprecedented rate. To make matters worse, prices being paid for the goods exported by developing countries are declining (Fig. 8-11), making it even more difficult for these nations to repay their debts.

Another factor in meeting global nutrition needs is trade barriers in the wealthier countries. The tariffs the European Union (EU) charges against cloth imported from developing nations are four times the tariffs it charges against cloth imported from wealthy nations. Sugar policies in the European Union and the United States guarantee high domestic prices, subsidize exports, and impose import barriers against sugar from the developing world. In general, tariffs are placed on processed goods in direct proportion to the amount of processing—there might be little duty on raw logs, more on sawn timber, and even higher duty on furniture. Discriminatory policies on clothing alone cost the developing countries an estimated $24 billion in lost earnings per year.

In light of the debt crisis and trade barriers, many observers are calling for a restructuring of the economic arrangements between rich and poor countries as the most effective way to address the food needs of the world—and at the same time to alleviate much of the environmentally destructive pressure on the poor countries to exploit their natural resources. Thus, an appropriate global goal would be for the wealthy nations to adopt policies that promote both self-sufficiency in food production and sustainable relationships between the poorer nations and their environments. Removing the trade barriers to exports from developing countries would be a large step in that direction.

Hunger, Malnutrition, and Famine

At a UN World Food Conference in 1974, delegates from all nations subscribed to the objective "that within a decade no child will go to bed hungry, that no family

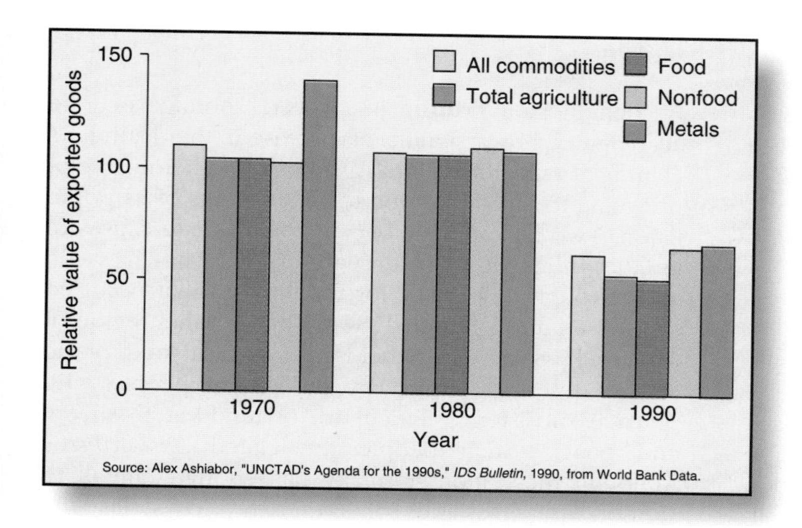

FIGURE 8-11
Declining values of exported goods sold by developing countries. The figure demonstrates the great decline in the 1980s, which has meant that these countries have less money for imports, investment, and debt repayments. (© 1990, Bread for the World Institute on Hunger and Development.)

will fear for its next day's bread, and that no human being's future and capacities will be stunted by malnutrition." More than two decades have passed since that declaration. What is the extent of hunger, malnutrition, and famine today?

Extent and Consequences of Hunger

Hunger is the general term referring to a lack of basic food required for energy and meeting nutritional needs such that the individual is unable to lead a normal, healthy life. **Malnutrition**, as we learned in Chapter 3, is the lack of essential nutrients such as amino acids, vitamins, and minerals, and **undernutrition** is the lack of adequate food energy (usually measured in calories).

Absolutely reliable figures on the worldwide extent of hunger are unavailable, mainly because few governments make any effort to document such figures. The World Health Organizaion has estimated that 840 million people are underfed and undernourished—some 18% of the population of the developing world. The regions seriously affected are southern Asia (especially Bangladesh), Latin America, and Africa. (See Fig. 8-10.) It is fair to say that hunger and malnutrition still take a terrible toll on human life and productivity. Thus, we see that, over 20 years later, the UN's 1974 objective is still unfulfilled.

The consequences of malnutrition and undernutrition vary. The effects are greatest in children and next greatest in women. Hunger can prevent normal growth in children, leaving them thin, stunted, and often mentally and physically impaired (Fig. 8-12). The UN Subcommittee on Nutrition has identified malnutrition and hunger in the children of southern Asia as the world's most serious nutritional problem, involving at least 108 million children who are underweight because of lack of food. Women in the developing world are especially vulnerable because of widespread patterns of low status and heavy labor for women in many countries.

Sickness and death are companions of hunger. According to Bread for the World Institute on Hunger and Development: "Almost 40,000 children under five die each day from malnutrition and infection. . . . The number of deaths is the same as if one hundred jumbo jets, each loaded with 400 infants and young children, crashed to Earth each day—one every 14 minutes." Hunger is often a seasonal phenomenon in rural areas

FIGURE 8-12
A malnourished child in the Denunay Feeding Center, Somalia, operated by World Vision International. Famines and food shortages in Somalia were largely a consequence of civil conflict, not drought. (Bruce Brander/Photo Researchers, Inc.)

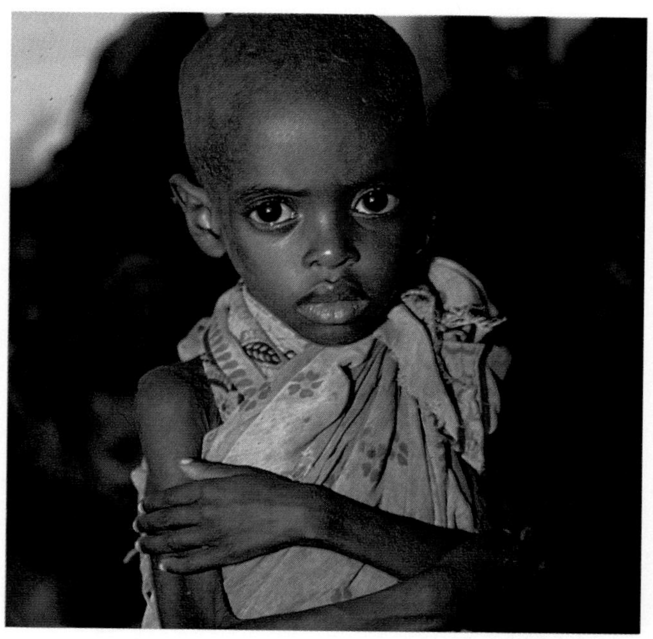

that are supported by subsistence agriculture, as people are forced to ration their stored food in order to survive until the beginning of the next harvest. Anyone who travels in the developing world cannot help but notice the fact that few people in the rural areas look well fed; most are thin and spare from a lifetime of limited access to food.

Root Cause of Hunger

In the view of most observers, *the root cause of hunger is poverty*. Our planet produces enough food for everyone alive today. Hungry and malnourished people lack either the money to buy food or adequate land to raise their own. If by some miracle world food production were to double next year, the status of most of the millions who suffer from extreme poverty and hunger would not change. Remember that any food over and above what the producers need for themselves enters the cash economy and flows in the direction of economic demand, not nutritional need.

Lack of food is only one of many consequences of poverty. Alan Durning of the Worldwatch Institute defines **absolute poverty** as "the lack of sufficient income in cash or kind to meet the most basic biological needs for food, clothing and shelter"; insufficient income is measured as income of less than $1 a day. On the basis of this definition, 1.3 billion people live in absolute poverty today. Most of these people reside in rural villages, are illiterate, spend half or more of their income on food, and represent races, tribes, or religions that suffer discrimination. They are powerless to do anything about their plight, and quite often the society in which they live is content to keep them that way. Millions are trapped in a cycle of poverty that leads to the degradation of resources and perpetuates the poverty—all made worse by the high fertility of these rural populations.

Hunger and poverty do not always go from bad to worse. A number of Asian countries, including China, Indonesia, and Thailand, significantly reduced the extent of poverty and hunger during the 1980s. Deliberate public policies and social services have greatly improved the welfare of millions in China, where food security is a matter of high national priority. Indonesia has benefited from oil exports and Green Revolution technology, and it continues to put major emphasis on rural development and social infrastructure. Clearly, it is possible for societies to address the needs of the hungry poor and to make progress in reducing the extent of absolute poverty and hunger. On the other hand, the most severe kind of hunger—famine—is found in societies that are regressing into disorder and chaos, and it is here that international responsibility comes most sharply into focus.

Famine

By definition, a **famine** is a severe shortage of food accompanied by a significant increase in the death rate. Famine is a clear signal that a society is either unable or unwilling to distribute food to all segments of its population. Two factors have been the immediate causes of famines in recent years: drought and warfare.

Drought is blamed for the famines that occurred in 1968–74 and again in 1984–85 in the Sahel region of Africa (Fig. 8-13). The Sahel is a broad belt south of the Sahara desert, occupied by 50 million people who practice subsistence agriculture or tend cattle, sheep, and goats. (Such people are called *pastoralists.)* Although the region normally has enough rainfall to support dry grasslands or savannah ecosystems, the rainfall is seasonal, undependable, and prone to failure. Making matters worse, population increases in the region have led to unsound agricultural practices and overgrazing by the expanding herds of livestock. Beginning in 1965, the region experienced 20 years of subnormal rainfall, with tragic results. Crops withered, forage for livestock declined, watering places dried up, and livestock died. Both farmers and pastoralists began abandoning their land and migrating toward urban centers, where they were often herded into refugee camps (Fig. 8-14). Unsanitary conditions in the camps and the already weakened condition of the refugees led to the spread of infectious diseases such as dysentery and cholera, and many thousands died before effective aid could be organized. The latest Sahelian famine is thought to have been responsible for 100,000 deaths; the number would have been in the millions except for the aid extended by Africans and numerous international agencies. The rains have returned to the Sahel, removing the immediate threat of famine, but the region still lacks an environmentally sustainable structure for food security.

Famines in which the common factor was not drought, but war threatened several African nations in the 1990s: Ethiopia, Somalia, Rwanda, Sudan, Mozambique, Angola, and Liberia (Fig. 8-13). Devastating and prolonged civil warfare has put millions of Africans at risk of famine. The civil wars disrupt the farmers' normal planting and harvesting and force the displacement of millions from their homes and food sources. Governments in power maintain control over food and relief supplies; relief agencies operate under dangerous conditions and frequently experience casualties. In some areas, the problem is made worse by persistent droughts. In Mozambique alone, 900,000 people died from direct military action or from indirect effects of the war there. The U.N. Food and Agricultural Organization found in a recent survey that 70% of Africans have inadequate food.

FIGURE 8-13
The geography of famine. Famines have occurred repeatedly in sub-Saharan Africa, especially in the Sahel (a band of dry grasslands that stretches across the continent). The map shows the countries where civil wars and droughts have recently brought on serious famines.

Clearly, famines from drought and war are preventable. India, Brazil, Kenya, and southern Africa have coped with droughts in recent years by mobilizing effective relief in the form of food, clothing, and medical assistance. Indeed, it was the drought of the early 1990s that accelerated the peace process in Mozambique and helped change the political landscape in southern Africa. Cooperation between South Africa and the ten nations of the Southern Africa Development Community to prevent famine was lowering barriers between South Africa and its neighbors prior to the coming of democracy to that troubled nation.

The drought and potential for famine in southern Africa were first predicted by a high-tech satellite system operated by the United States (USAID): the Famine Early Warning System (FEWS). The satellite measures trends in vegetation and rainfall in Africa. In 1994, the system alerted the world to conditions in the Horn of Africa, where more than 20 million were threatened by famine. Food aid to this region was mobilized early in order to prevent the migrations of people that usually result when crops fail. The FEWS issues regular bulletins and special reports on the World Wide Web, giving governments and relief agencies accurate and timely assessments of the status of food security in African countries. (http://gopher.info.usaid.gov/fews/fews.html)

FIGURE 8-14
A refugee camp of Rwandans in Uganda. Such camps represent the last resort from hunger and are often scenes of unthinkable human suffering and death. (Wesley Bocxe/Photo Researchers, Inc.)

Ethics

The Lifeboat Ethic of Garret Hardin

Biologist Garret Hardin has published several provocative essays addressing the worldwide food-vs.-population issue. Here we give you an opportunity to respond to Hardin's thinking.

We begin with the concept of carrying capacity—that number of a species that can be supported indefinitely without degrading the environment. For human societies, carrying capacity means the ability to meet food needs over the long term—that is, sustainably. If ecology had a decalogue, Hardin says, the first commandment would be "Thou shalt not transgress the carrying capacity." A look at the world scene reveals that numerous countries are pressing against the limits of or have exceeded their carrying capacity. This, says Hardin, is their problem, not ours, and he uses the lifeboat metaphor to show why.

Picture a number of lifeboats in the sea after a ship has sunk—some crowded with people and some in which people are riding in relatively uncrowded luxury. Each lifeboat has a limited capacity. The people in crowded boats are continually falling overboard, leaving the people on the uncrowded boats with the problem of whether to take them on board or not. Imagine an uncrowded boat with 50 on board and room for 10 more, with 100 people treading in the water and begging to be taken on board. There are several options: (1) Assume that all people have an equal right to survival, and take everyone on board. This, says Hardin, would lead to catastrophe for all. (2) Admit only 10, filling all the space on the boat. Two problems with this option are that you lose your margin of safety and you must find a basis on which to discriminate among all the people in the water. (3) Admit no more to the boat. This preserves your margin of safety and guarantees the long-term survival of the people on your boat; it is the rational solution to the lifeboat problem.

The metaphor, of course, is to be applied to the problem of food aid. Some people would argue that, if less grain went to feeding animals, there would be more available to feed the hungry in poor countries. Or, since there are often agricultural surpluses in the rich nations, we could use these surpluses to feed the hungry. The real problem, then, is a problem of food distribution, not of the quantity produced.

These arguments, says Hardin, are foolish. In giving away food, we would only be encouraging the population escalator: A population growing rapidly reaches the limits of its food capacity and is supplied with food from abroad, encouraging still further growth and necessitating still further food aid, and so on. Our hearts, says Hardin, tell us to send food, but our heads should tell us not to. We only postpone the day of reckoning, and in the end, the amount of suffering will be greater. Overpopulated, food-poor countries have transgressed the first commandment of ecology. If we want to help, we should direct our aid toward bringing population growth down, according to Hardin. What do you think? (Sample the opinion of those you speak to.)

Food Aid

As Fig. 8-10 shows, food aid is being distributed to countries all over the world, not just where famines are threatened. This raises the question, What is the proper role of food aid? Clearly, aid is vital in saving lives where famines occur. But what about the people who suffer from chronic hunger and malnutrition? The basic question is, When should food be given to those in need, instead of being distributed according to market economics?

Numerous humanitarian campaigns to end world hunger have been mounted in the last 50 years. The United States and Canada have been world leaders in giving away food (which is first purchased from farmers and therefore represents a subsidy). As noted, a number of serious famines have been moderated or averted by these efforts. As virtuous as such efforts seem on the surface, however, routinely supplying food aid in an attempt to alleviate chronic hunger in developing countries may be the worst thing to do. As Wortman and Cummings write in their book, *To Feed This World*:

> The food generosity of industrial countries, whether in their own self-interests (disposing of food surpluses) or under the mantle of alleged distributive justice, has probably done more to sap the vitality of agricultural development in the developing world than any other single factor.

The problem is, people will not pay more than they must for food. Therefore, free or very cheap foreign food undercuts the local market. In effect, local farmers must compete economically with free or low-cost imported food. When they cannot earn a profit, they stop producing and eventually enter the ranks of the poor. The cycle continues, as people who sell goods to the farmer also suffer when the farmer loses buying power. In the long run, the entire local economy deteriorates. Hence, the donation of food, while well intended, often aggravates the very conditions that it is meant to alleviate. Meanwhile, population pressures continue to build, and the magnitude of the problem increases. (See "Ethics" box on this page.)

Food aid in grains averaged 11 million metric tons per year in the 1990s. Some aid is strictly humanitarian;

Bangladesh receives 1.2 million tons per year, almost as much as it purchases on the market. The African continent receives 6.3 million tons per year, which is about 16% of the total imported. All the signs indicate that this figure will increase.

Some food aid is given for political reasons. Egypt received 1.1 million tons of grain per year during the late 1980s a result of the Camp David Peace Accord brokered by President Jimmy Carter. Egypt is the second largest recipient of food aid in the world.

Food aid will undoubtedly continue to be an international responsibility. It is at best a buffer against famine, and it will probably continue to be awarded to some nations for political reasons. As part of the solution to the chronic hunger and malnutrition among the poor, however, continued food aid is clearly counterproductive. Much more good will be accomplished by a restructuring of the economic arrangements between rich and poor nations and by the extension of loans and aid directed toward fostering self-sufficiency in food and sustainable interactions with the environment.

Building Sustainability into the Food Arena

In an ecosystem context, farmers should be viewed as herbivores who manage their producers and pastoralists as predators who manage their prey. The principles of ecosystem structure and function presented earlier in the book apply perfectly well to farming and animal husbandry. The major difference between human systems and natural ecosystems is that we do not have to allow nature to take its course—indeed, we cannot do so and expect to harvest crops instead of weeds or hope to have our livestock flourish. Perhaps for this reason, we tend to forget that our manipulations of plants and animals are nevertheless subject to ecosystem laws, and if we disregard those laws, our human systems are likely to behave counter to our best interests.

Food production and distribution, and hunger and famine, are very much a matter of how human societies interact with their environment as well as with each other. In this last section, we examine the approaches of sustainable agriculture and look once more at the socioeconomic and political dimensions of hunger.

Sustainable Agriculture

In *A Green History of the World*, Clive Ponting shows how the downfall of several past civilizations was nonsustainable farming and animal grazing. The Sumerians of Mesopotamia raised crops under intense irrigation, and in time, salinization led to the collapse of the agricultural base of their society, followed soon by the decline of the Sumerians. Overgrazing and deforestation throughout the Mediterranean basin, beginning as far back as 650 B.C. in Greece, led to soil erosion that ruined agricultural land and greatly lowered the carrying capacity for livestock, and the empires occupying the basin declined accordingly. It is Ponting's view that what happened in the past on a local scale is now occurring in global proportions.

The goal of sustainable agriculture is to maintain agricultural production while not ruining any part of the supporting system or degrading the environment. One way to measure the sustainability of an agricultural system is to apply *environmental accounting* in addition to the conventional cost accounting that balances crop yields against fertilizers, pesticides, and other costs directly absorbed by the farmer. (See Chapter 19.) This would mean measuring such natural resource costs as erosion, salinization, groundwater and surface water pollution by fertilizers and pesticides, and health costs. When this is done, resource-conserving practices will be given the credit they deserve, and real sustainability is more likely to result. Unfortunately, there is little incentive for using such an accounting method, with the result that sustainability is quite difficult to demonstrate.

There is a growing consensus that agricultural sustainability must be patterned after that of natural ecosystems in order to be successful. In Chapters 3 and 4, we presented four principles of sustainability derived from our studies of natural ecosystems. Let us apply these principles to our analysis of agricultural sustainability, differentiating where necessary between *industrialized agriculture* and *subsistence agriculture*.

1. **For sustainability, ecosystems dispose of wastes and replenish nutrients by recycling all elements**. This first principle is augmented by the tendency of wastes and nutrients to be held in place by plants and soil. Sustainable practices, therefore, emphasize soil health and stability. Chapter 9 discusses soil and the efforts needed to prevent erosion, salinization and desertification.

   Organic farming (also referred to as low-input farming) involves the regular addition of crop residues and animal manures to build up the organic matter in soil. When crops are harvested, vital mineral elements are removed from the soil. They are returned to the soil through the application of animal wastes and **green manures** (grasses or legumes that are plowed into the soil at the end of a growing season), instead of through the addition of chemical fertilizers.

   In a society that depends on subsistence agriculture, it is vitally important that the people

FIGURE 8-15
A modern farm in Schuylkill County, PA, shows a healthy diversity of crops, woodlands, and hedgerows, important in maintaining the ecological stability of the farm countryside. (W. Eastep/The Stock Market.)

not use animal dung for fuel; instead, the dung must be returned to the soil. Also, appropriate technologies can be adopted that address the problems of loss of moisture and erosion, such as the use of rock dams to hold water and protecting the forest cover around croplands. Nutrients can be cheaply added to the soil by mixing legumes with grain or root crops.

2. **For sustainability, ecosystems use sunlight as their source of energy**. Industrialized agriculture will continue to be dependent on mechanization. Subsistence agriculture, however, will do well to continue to use animal energy for working the land because the animals are fed locally and because dependence on costly fossil fuels is avoided. Wind and solar energy can be employed for many farming tasks; look for a return of the windmills once used to pump water and generate small amounts of electricity on farms all over the United States.

Sustainability is encouraged in the wealthier nations when people use small plots of land to grow their own fruits, vegetables, and small animals. This is a much more desirable use of the land than simply growing grass to mow.

3. **For sustainability, the sizes of consumer populations are maintained so that overgrazing or other overuse does not occur**. The most obvious application of this principle is in livestock management. Mismanagement of herds and overgrazing lead to the deterioration of rangeland. Therefore, sustainable livestock management all over the world must recognize the carrying capacity of rangeland ecosystems and preserve the soils and plant cover in them. In some regions of the developing world, forests and woodlands provide fodder for livestock, and in others (tropical forests) the forests are removed to make way for cattle pasture. The sustainable approach is to protect forested areas, recognizing that they are the most stable ecosystems for the site and will benefit human populations more if they are maintained as forests.

Pests are consumers, and pest control is vital to most agriculture. Chapter 10 presents the issues surrounding the use of pesticides. Alternatives to absolute dependence on pesticides are now available, including integrated pest management programs using natural predators. The selection of planting times, crop rotations, and plant residue management can provide the beneficial insects with optimal habitats. The basic strategy is to maintain biological control of pests and diseases.

One obvious application of the third principle is for human populations to come to terms with the carrying capacity of the land they occupy. This chapter and the previous two give abundant evidence that some parts of the world already have violated the principle. Clearly, efforts to lower fertility will in themselves promote sustainable agriculture by reducing the pressure put on the land to produce food. Only if these efforts are successful will the world's subsistence farmers have any hope of meeting their family's food needs in the future.

4. **For sustainability, biodiversity is maintained**. Crop rotation is a vital part of sustainable agriculture. For example, the farmer might plant three seasons of alfalfa plowed under (green manure) followed by four successive crop seasons of first wheat, then soybeans, then wheat, and then oats. In this way, weeds and insects are more easily controlled, and plant diseases do not build up in the soil. Combining crops and mixing crops, trees (agroforestry), and livestock provide a diversity of marketable products that can be an effective buffer against economic and biological risks (Figure 8-15). In more arid climates, *alley cropping*, in which rows of shade-producing trees are alternated with rows of food crops, can be adopted.

Final Thoughts on Hunger

By now it should be clear that, although food is our most vital resource, we do not treat it as a commons (free to all who need it). Indeed, the production and distribution of food is one of the most important economic enterprises on Earth. Alleviating hunger, as we have seen, is primarily a matter of addressing the absolute poverty that afflicts one of every five people on Earth. To treat food as a commons would be to treat wealth as a commons. Even though this is one of the tenets of socialist systems, it has proven to be a completely unworkable one. We are part of the world economy now, and it is a market economy—for the perfectly good reason that nothing else seems to work, given the realities of human nature.

What has not been done, however, is to bring this market economy under the discipline of sustainability. Short-term profit crowds out long-term sustainable restraint. Self-interest at every level—from the individual to the global community—generates decisions that prevent the sharing of political power, economic goods, and technology, and in the process guarantees that the environment will continue to be degraded. Why isn't land reform carried out in developing countries? Why are the rich nations content to maintain the current debt situation and keep tariffs high and developing world commodity prices low?

In 1980, a Presidential Commission on World Hunger delivered its report to President Jimmy Carter, with the major recommendation "that the United States make the elimination of hunger the primary focus of its relations with the developing world." The thrust of the report was that if we respect human dignity and have a sense of social justice, we must agree that hunger is an affront to both. The right to food must be considered a basic human right. It follows, then, that we as a nation have a moral obligation to respond to world hunger, the report concluded. Today the question is, Has that moral obligation has made its way into public policy in the ensuing years?

Our understanding of the situation is quite well developed. No new science or technology is needed in order to alleviate hunger and at the same time promote sustainability as we grow our food. The solutions lie in the realm of political and social action, at all levels of responsibility. Given the current groundswell of concern about the environment and the disappearance of the Cold War between capitalism and communism, there may never be a better time to turn things around and take more seriously our responsibilities as stewards of the planet and as our brothers' keepers.

Review Questions

1. Describe new developments the agricultural revolution brought to farming.
2. Discuss the environmental cost of each development named in Question 1.
3. What was the Green Revolution? What were its limitations?
4. Describe subsistence agriculture, and discuss its relationship to sustainability.
5. How does animal farming affect the environment?
6. What are our major prospects for increasing food production in the future?
7. Consider the benefits of eating lower on the food chain.
8. Trace the patterns in grain trade between different world regions over the last 40 years. (See Table 8–2.)
9. Describe the three levels of responsibility for meeting food needs. At each level, list several ways food security can be improved.
10. Define hunger, malnutrition, and undernutrition. What are their consequences?
11. How are hunger and poverty related?
12. Discuss the causes of famine, and name the geographical areas most threatened by it.
13. Why does food aid often aggravate poverty and hunger?
14. Define sustainable agriculture. How do each of the four principles of sustainability apply to this type of agriculture?

Thinking Environmentally

1. Although few people would argue against sending food aid to victims of famine, what conditions could be attached to the aid in order to foster self-sufficiency and prevent further dependence?
2. Imagine that you have been sent as a Peace Corps volunteer to a poor African nation experiencing widespread hunger. How would you begin to address the needs of the people of that nation?
3. Record your food intake over a two- to three-day period. Analyze the nutritional value of your diet. Which nutrients are lacking? Which are in excess? What changes in your diet would reconcile these differences?
4. Of the following methods for increasing food production, which do you feel are viable options? Why?

   clearing forests
   increasing yield on farmland already in production
   irrigating arid lands
   developing transgenic crops with biotechnology
   aquaculture

CHAPTER 9

Soil and the Soil Ecosystem

Key Issues and Questions

1. The soil environment must provide plants with water, nutrients, and air for the roots. What are the key attributes of the soil that bear on its being able (or not being able) to provide these things?
2. A dynamic interaction between mineral particles, detritus, and organisms in the soil is most important in developing the soil's key attributes. Describe this dynamic interaction and how it develops these attributes.
3. Cultural practices leading to erosion are devastating to the soil and its future productivity. What are these cultural practices? Describe the process of erosion and how it degrades each of the soil's key attributes.
4. Certain irrigation practices may be nonsustainable. Why? Describe the problems that may develop from irrigation.
5. The degradation of soil and declining productivity may become a vicious cycle leading to desertification. What is meant by desertification? Explain the sequence of events that leads to it.
6. Losses from erosion, salinization, the depletion of water supplies, and the development of farmland into residential tracts do not speak well for sustainability. Give a worldwide perspective of these issues. What would be a "greener" farm policy than present practices?

Ninety percent of the world's food comes from land-based agricultural systems, and this percentage is growing as the ocean's fish and natural ecosystems are depleted. Protecting and nurturing agricultural soils, which are the cornerstone of food production, must be a central feature of sustainability (Fig. 9-1). Yet, it is a feature that has been overlooked repeatedly in the past. The story of Easter Island in Chapter 1 is just one of many. In their book, *Topsoil and Civilization*, Carter and Dale describe how the fall of the ancient Greek, Roman, and other empires was more a result of decline of agricultural sustenance due to soil erosion than to outside forces.

Yet modern civilization seems prone to the same folly. In the United States, conversion of farmland to nonfarm uses has averaged some 1.4 million acres per year over the past decade, and the loss is continuing. The same is occurring in developing countries. Also, throughout the world agricultural soils are being degraded by erosion, buildup of salts, and other problems

Reaping the harvest in Iowa. [Photo by Martin Bond/Science Photo Library/Photo Researcher, Inc.]

that can only undercut future productivity. During the past 40 years, one third of the world's cropland (3.75 billion acres, or 1.5 billion hectares) has been abandoned due to these types of degradation. These losses have been counterbalanced by increasing production per acre by the various means discussed in Chapter 8, but, as we noted, these means of increasing production are reaching both practical and theoretical limits. Likewise, compensation by cutting more forests, plowing more grasslands, and draining more wetlands is not ecologically sustainable.

The problem is that in a cash economy food tends to flow toward where the money is, irrespective of actual social need. Thus, people with money get fed and the rest starve if those with money to purchase the food do not agree to some form of redistribution. Likewise, farmland itself goes to nonfarm uses if developers can offer more money for it than a farmer can make by growing food. And, farming methods that allow soil degradation tend to take precedence over conservation if they are significantly more profitable—even if it is only in the short term.

A major part of the problem is that most of us, now living in or near cities and getting our food at the supermarket, have lost consciousness of the fact that fertile soil remains the cornerstone of our civilization and survival. We make the assumption that others are taking care of the situation and looking out for our long-term benefit. The facts of farmland loss and soil degradation are ample evidence that this assumption is both naive and false. But we are also the citizens who will ultimately determine policy toward farmland preservation and soil conservation, or even work the soil ourselves. The encouraging fact is that there is growing interest and involvement of people in all areas, ranging from formulating and enacting farm policy to backyard gardening. But rational action and positive results depend on awareness.

Our objective in this chapter is to provide an understanding of the attributes of soil required to support good plant growth, how these attributes may deteriorate under various practices, and what is necessary to maintain a productive soil. Certain aspects of farm policy will also be presented. We begin by considering the relationships between plants and soil that sustain productivity.

Plants and Soil

Soil can make the difference between harvesting a luxuriant crop or abandoning the field to a few meager weeds. A rich soil that supports a luxuriant crop is so much more than the dirt you might get out of any hole in the ground. Indeed, agriculturists cringe when anyone refers to soil as dirt. You have already learned (Chapters 2 and 3) that detritus in an ecosystem is fed upon by various detritus feeders and decomposers. Nutrients from the detritus are thus released and reabsorbed by producers, fostering the recycling of nutrients. As we focus on a productive soil we will see that the detritus feeders and decomposers constitute a biotic community of organisms that not only facilitates the transfer of nutrients but also creates a soil environment that is most favorable to the growth of roots. In short, a productive topsoil involves dynamic interactions among the soil organisms, detritus, and the mineral particles of the soil (Fig. 9-2).

If we think of an ecosystem as a community of organisms living in a particular environment, we can see the soil as a sort of ecosystem in itself—the soil organisms living in and developing a particular soil environment. Of course, the soil ecosystem does not have its own producers; it is dependent on detritus. Therefore, the soil ecosystem is what is called a detritus-based ecosystem, and it is still basically a part of the overall ecosystem.

Let us start with a consideration of just what green plants need from the soil. Then we shall investigate how the soil ecosystem functions to provide those needs.

FIGURE 9-1
Nurturing and maintaining agricultural soils, which are the cornerstone of food production, must be a central feature of sustainability. (Paul Buckowski/Gamma-Liaison, Inc.)

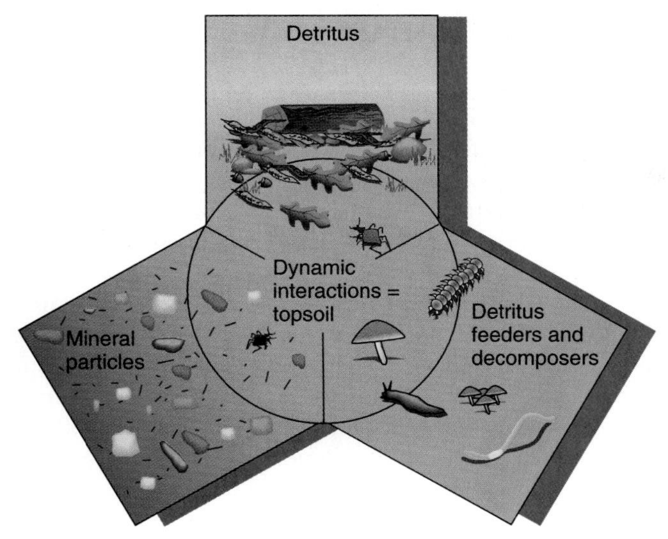

FIGURE 9-2
A productive soil is more than just "dirt." Soil maintenance involves a dynamic interaction among mineral particles, detritus, and detritus feeders and decomposers. Altering any one of these factors may have drastic effects on soil quality.

Soil for Supporting Plants

For their best growth plants need a root environment that supplies optimal amounts of the mineral nutrients, water, and air (oxygen). The pH (relative acidity) and salinity (salt concentration) of the soil are also critically important. **Soil fertility**, the soil's ability to support plant growth, often refers specifically to the presence of proper amounts of nutrients. But the soil's ability to meet all the other needs of plants is also involved in soil fertility. (See also "Earth Watch" box, p. 215.)

Mineral Nutrients and Nutrient-Holding Capacity. Mineral nutrients—phosphate (PO_4^{3-}), potassium (K^+), calcium (Ca^{2+}), and other ions—are present in various rocks along with nonnutrient elements. They initially become available to roots through the gradual chemical and physical breakdown of rock—the processes collectively referred to as **weathering**. But weathering is much too slow to support anything approaching normal plant growth. The nutrients that support plant growth in natural ecosystems are supplied mostly through the breakdown and release (recycling) of nutrients from detritus (see Fig. 3-15 for example).

But here is a problem. As free mineral nutrients are soluble in water, they may be literally washed from the soil as water moves through it, just as salt can be removed from sand by passing water through the sand. This process of removing materials from the soil by water is called **leaching**. Leaching not only lessens soil fertility; it also contributes to pollution when materials removed from the soil enter waterways. Consequently, the soil's capacity to bind and hold nutrient ions until they are absorbed by roots is just as important as the initial supply of those ions. This property is referred to as the soil's **nutrient-holding capacity** or its **ion-exchange capacity**.

In agricultural systems, there is an unavoidable removal of nutrients from the soil with each crop, because nutrients absorbed by the plants are contained in the harvested material. Therefore agricultural systems do require inputs of nutrients to replace those removed with the harvest. Nutrients are replenished with applications of **fertilizer**, material that contains one or more of the necessary nutrients. Fertilizer may be organic or inorganic. **Organic fertilizer** includes plant or animal wastes or both; manure and compost (rotted organic material) are two examples. **Inorganic fertilizers** are chemical formulations of required nutrients without any organic matter included. Inorganic fertilizers are much more prone to leaching than are organic fertilizers. Other advantages and disadvantages will be discussed later. Figure 9-3 summarizes these ideas concerning nutrients.

FIGURE 9-3
Plant-nutrient-soil relationships. The original source of mineral nutrients, such as phosphorus and potassium, required for plant growth is from weathering of parent materials. But, natural ecosystems are mainly sustained by highly efficient recycling of these nutrients. In an agricultural system, there is inevitable removal of nutrients with harvest, that can be replenished by additions of fertilizer. However, deterioration of soil may allow much of the fertilizer to be lost by leaching.

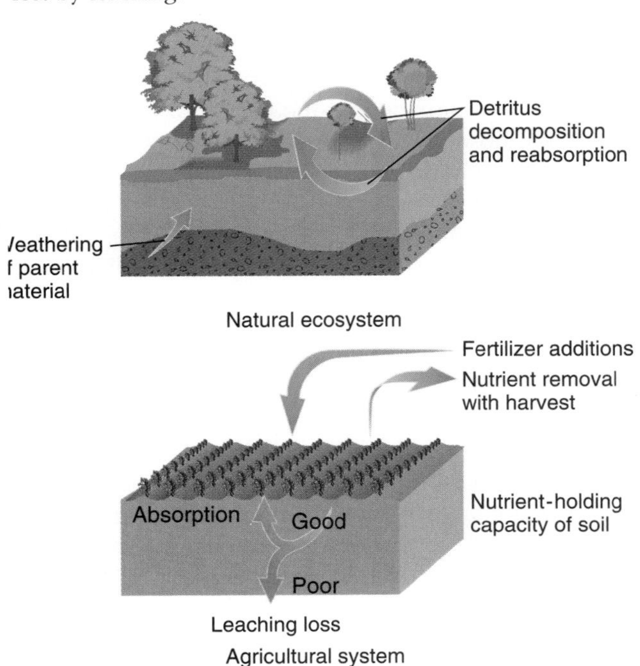

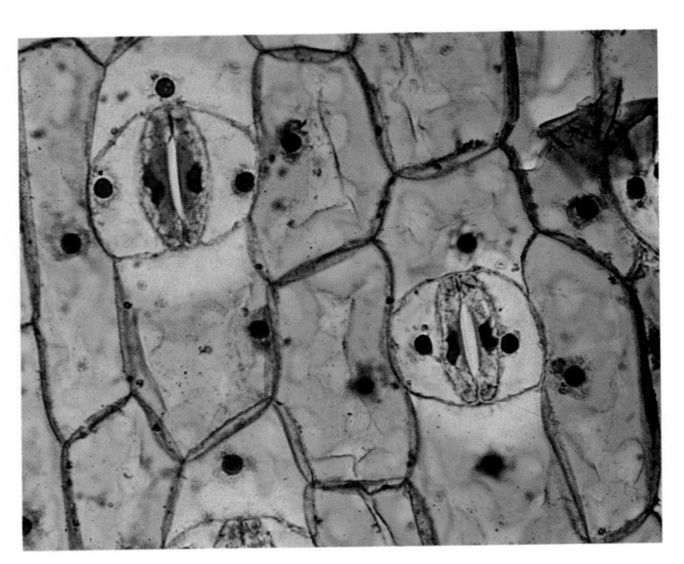

FIGURE 9-4
A microscopic view of the surface of a leaf showing individual cells and stomas. The stomas are the mouth-shaped openings between the reddish-stained cells. Stomas are essential to enable the movement of carbon dioxide into and oxygen out of the leaf during photosynthesis and the reverse of this flow during respiration. However, they also allow the escape of water vapor from the moist interior of the leaf. To balance this loss of water vapor, the plant must absorb considerable amounts of water from the soil. (Ed Reschke/Peter Arnold, Inc.)

Water and Water-Holding Capacity. A plant's need for water is self-evident. What is less conspicuous is a constant stream of water being absorbed by the roots, passing up through the plant and exiting as water vapor through microscopic pores in the leaves—a process called **transpiration** (Fig. 9-4). The pores, called *stomata* (singular, *stoma*) are essential to permit the entry of carbon dioxide and the exit of oxygen in photosynthesis; however, the plant's loss of water through the stomata is dramatic. A field of corn, for example, transpires an equivalent of a field-sized layer of water 17 inches (43 cm) deep in a single growing season. Inadequate water results first in wilting, a condition that conserves water, but also shuts off photosynthesis by closing the stomata and preventing gas exchange. Of course, if the wilted condition is too severe or too prolonged, the plants die.

Water is resupplied to the plant naturally by rainfall or artificially by **irrigation**. Three attributes of the soil are significant in this respect. First is the soil's ability to allow water to **infiltrate**, or soak in. If water runs off the ground surface it won't be useful; worse, it may cause erosion, which we shall discuss shortly.

Second is the soil's ability to hold water after it infiltrates, a feature simply called **water-holding capacity**. Poor water-holding capacity implies that most of the infiltrating water percolates on down below the reach of roots—not far in the case of seedlings and small plants—again becoming useless. What is desired is a good water-holding capacity, the ability to hold a large amount of water like a sponge, providing a reservoir on which plants can draw between rains. If the soil does not have such water-holding capacity, the plants will have to depend on frequent rains or irrigation, or suffer the consequences of drought.

The third critical attribute is **evaporative water loss** from the soil surface. It is plain that such evaporation will deplete the soil's water reservoir without serving the needs of plants. Factors that reduce evaporative water loss are significant. These aspects of the soil-water relationship are summarized in Fig. 9-5.

Aeration. Novice gardeners commonly kill plants by overwatering, or "drowning," them. Roots need to "breathe." Basically, they are living organs and need a constant supply of oxygen for energy metabolism. True, oxygen is produced in plants' stems and leaves during photosynthesis, but with the exception of some hollow-stemmed wetland plants, plants have no means of passing oxygen to their roots. Therefore, land plants depend on the soil's being loose and porous enough to allow the diffusion of oxygen into and carbon dioxide out of the soil, a property called **soil aeration**. Overwatering fills the air spaces in the soil, preventing adequate aeration. So does **compaction**, packing of the soil such as occurs with excessive foot or vehicular traffic. Compaction also reduces infiltration and increases runoff.

FIGURE 9-5
Plant-soil-water relationships. Water lost from the plant by transpiration must be replaced from a reservoir of water held in the soil. In addition to the amount and frequency of precipitation, the size of this reservoir depends on the soil's ability to allow water to infiltrate, hold water, and minimize direct evaporation.

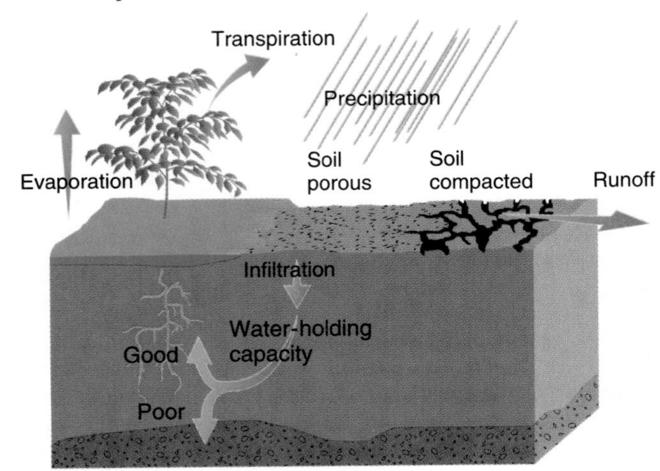

Earth Watch

Hydroponics: Growing Plants without Soil

Plants do not require any secret ingredients from the soil; all they need in order to flourish are inorganic nutrients, water, and aeration. This fact is demonstrated by the practice of **hydroponics**, the culture of plants without soil. In hydroponic systems, roots are constantly wetted with a well-aerated water solution containing optimal amounts of the required mineral nutrients. One common method is to grow plants in a container of pea-sized gravel that is constantly or periodically irrigated with a nutrient solution. The roots, supported by the gravel, grow in a film of water around the stones with plenty of air between the stones. The solution is recirculated from the bottom of the container back to drip tubes above.

Another method is to grow the plants in a trough—sections of roof eaves gutter work well—covered with black plastic except for holes for the plant stems. The trough is mounted at a slight incline so that nutrient solution constantly added at one end flows through, drains from the other end, and is recirculated. Plants are supported by strings from above, and their roots simply dangle or lie in the trough.

Because nutrients and water can be kept at optimal levels at all times, the production from hydroponic systems can be quite spectacular. Why worry about preserving topsoil if this is the case? The answer is cost. For all the equipment required, a hydroponic system will cost on the order of $100,000 per acre. When the crop is high-priced fruits and vegetables, the income may be sufficient to justify this cost. However, bread would have to sell for $10 a loaf or more to pay for hydroponically grown wheat. Thus, it seems unlikely that hydroponics will solve the food problem for the poor, but it may help them in the following way.

Urban populations consume large amounts of salad vegetables, which are often imported from great distances at great expense. At the same time, cities have large numbers of unemployed poor people. Hydroponic production on rooftops or other open areas could make the city largely self-sufficient in salad vegetables and at the same time employ a great many people, as operating hydroponic systems is labor-intensive. In this respect, hydroponic systems are a form of appropriate technology that might well be supported.

Combining a hydroponics operation with fish farming can increase efficiencies still more. The nutrient-rich water resulting from fish excrements can be used for the plants, and the plants effectively filter and clean the water for the fish. Also, waste plant material can be a component of the fish feed. John Reid, founder of Bioshelters in Amherst, Massachusetts, has developed and is marketing such a system. (Check out: http://www.magicnet.net/~itms/hydroponics.html)

Relative Acidity (pH). The term **pH** refers to the acidity or alkalinity (basicity) of any solution. A solution that is neither acidic nor alkaline is said to be neutral and has a pH of 7. The pH scale, which runs from 1 to 14, is discussed more fully in Chapter 16 as it relates to the topic of acid deposition. For now, it is important to know that most plants (as well as animals) require a pH near neutral, and most natural environments provide this.

Salt and Water Uptake. A buildup of salts in the soil makes it impossible for the roots to take in water. Indeed, if salt levels in the soil get high enough, water can be drawn out of the plant, resulting in dehydration and death. This is why highly salted soils are virtual deserts supporting no life at all. We see the importance of this problem later when we study how irrigation may lead to the accumulation of salts in soil.

Soil as an Ecosystem

Summarizing the above section, we see that to support a good crop the soil must (1) have a good supply of nutrients and nutrient-holding capacity; (2) allow infiltration and have good water-holding capacity, and resist evaporative water loss; (3) have a porous structure that permits good aeration; (4) have a pH near neutral; and (5) have a low salt content. Moreover, these attributes must be sustained. How does the natural soil ecosystem—that dynamic interaction of minerals, detritus, and soil organisms—provide and sustain these attributes? Let us begin by considering certain attributes of soil's mineral portion alone.

Soil Texture. As rock weathers, it breaks down into smaller and smaller fragments. Below the size of small stones we classify the fragments as sand, silt, and clay—called soil separates (see Table 9-1). You are familiar with sand as individual rock particles, and probably with the finer particles called silt, but it may surprise you to note that clay is actually made up of microscopic particles. Consider washing clay in water; the water immediately takes on a cloudy or muddy appearance because the clay particles are suspended in the water. The moldable "gooey" quality of clay appears when just enough water is added for the particles to slide about one another on a film of water but still cling together. On drying further, the clay particles adhere in hard

TABLE 9-1
USDA Classification of Soil Particles

Name of Particle	*Diameter (mm)*
Very Coarse Sand	2.00–1.00
Coarse Sand	1.00–0.50
Medium Sand	0.50–0.25
Fine Sand	0.25–0.10
Very Fine Sand	0.10–0.05
Silt	0.05–0.002
Clay	Below 0.002

FIGURE 9-6
Soil water- and nutrient-holding capacities increase as soil-particle size decreases. Both water and nutrient ions (represented by dots) tend to cling to surfaces, and smaller particles have relatively more surface area.

clods. These differences under varying moisture conditions is the property of *soil consistence.*

It is these sand, silt, and clay particles that constitute the mineral portion of soil. **Soil texture** refers to the relative proportions of each in a given soil. If one predominates, we speak of a sandy, silty, or clayey soil. A proportion that is commonly found consists of roughly 40% sand, 40% silt, and 20% clay. A soil with such proportions is called a **loam**. You can determine the texture of a given soil by shaking a small amount with water in a large test tube to separate the particles and then allowing them to settle. Because particles settle according to weight, sand particles settle first, silt second, and clay last. The proportion of each can then be seen.

Three basic considerations enable you to determine how infiltration, nutrient and water-holding capacities, and aeration are influenced by soil texture: (1) Larger particles have respectively larger spaces separating them than smaller particles have. (Visualize packing basketballs and golf balls in containers.) (2) Smaller particles have more surface area relative to their volume. (Visualize cutting a block in half again and again. Each time you cut it you create two new surfaces, but the total volume of the block remains the same.) (3) Nutrient ions and water molecules tend to cling to surfaces. (When you drain a nongreasy surface, it remains wet.) (See Fig. 9-6.) Soil attributes, as they vary with particle size (sand, silt, clay), are shown in Table 9-2. Do you see how they logically correspond to the considerations above?

Soil texture also affects **workability,** the ease with which a soil can be cultivated. This fact has an important bearing on agriculture. Clayey soils are very difficult to work because, with even modest changes in moisture content, they go from being too sticky and muddy to being too hard to bricklike. Sandy soils are

TABLE 9-2
Relationship between Soil Texture and Soil Properties

Soil Texture	*Water Infiltration*	*Water-Holding Capacity*	*Nutrient-Holding Capacity*	*Aeration*	*Workability*
Sand	Good	Poor	Poor	Good	Good
Silt	Medium	Medium	Medium	Medium	Medium
Clay	Poor	Good	Good	Poor	Poor
Loam	Medium	Medium	Medium	Medium	Medium

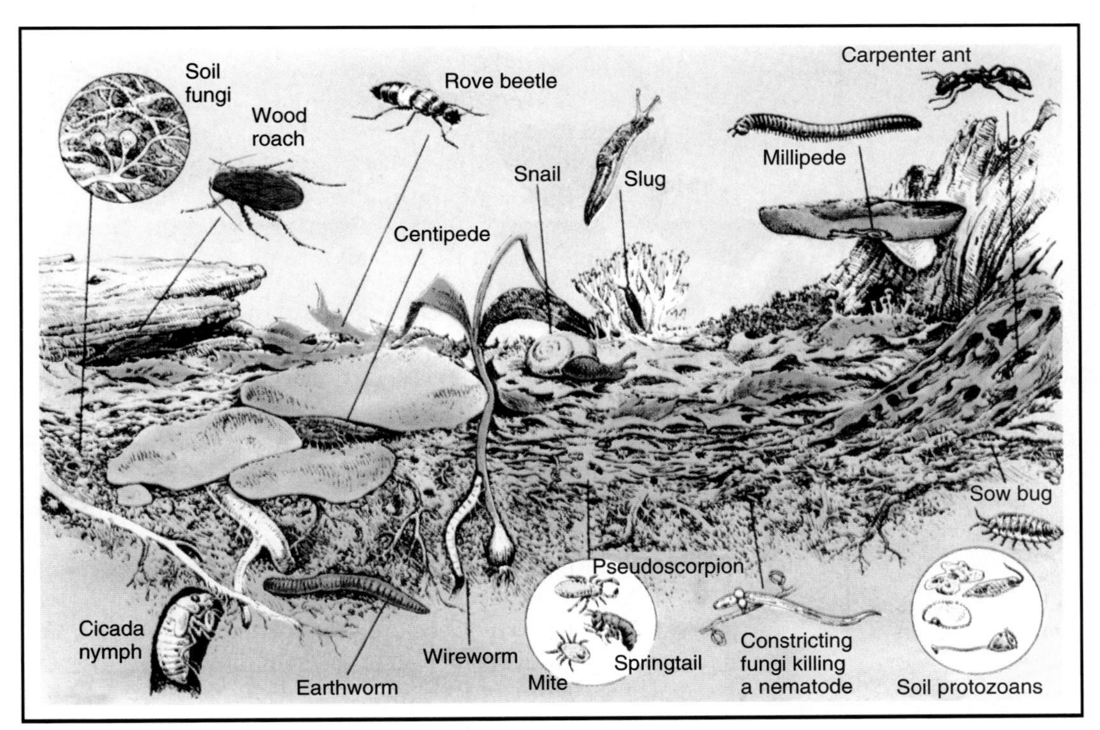

FIGURE 9-7
Soil is a detritus-based ecosystem. A host of organisms, major examples of which are shown here, feed on detritus and burrow through the soil forming a humus-rich topsoil with a loose clumpy structure. (From R. L. Smith, *Ecology and Field Biology*, 2nd ed. Copyright 1966, 1974 by Robert Leo Smith. Reprinted by permission of HarperCollins Publishers, Inc.)

very easy to work because they do not become muddy when wet, nor do they become hard and bricklike when dry.

Which is the best soil? Recall the principle of limiting factors. The poorest attribute is the limiting factor; the poor water-holding capacity of sandy soil, for example, may preclude agriculture altogether because the soil dries out so quickly. The best texture proves to be silt or loam, because such limiting factors are moderated in these two types of soil. But, the good qualities are also moderated, so this "best" is really only "medium." The organic parts of the soil ecosystem—the detritus and soil organisms—are necessary to optimize all attributes.

Detritus, Soil Organisms, Humus, and Topsoil. The accumulation of dead leaves, roots, and other detritus on and in the soil supports a complex food web, including numerous species of bacteria, fungi, protozoans, mites, insects, millipedes, spiders, centipedes, earthworms, snails, slugs, moles, and other burrowing animals (Fig. 9-7). As these organisms feed, the bulk of the detritus is consumed through their cell respiration, and carbon dioxide, water, and mineral nutrients are released as byproducts as described in Chapter 3. However, each organism leaves a certain portion undigested—that is, a certain portion resists breakdown by the organism's digestive enzymes. This residue of partly decomposed organic matter is called **humus**. A familiar example is the black or dark brown, spongy material remaining in a dead log after the center has rotted out (Fig. 9-8). **Composting** is the process of fostering the decay of organic wastes under more or less controlled conditions, and the resulting compost is the same as humus. (Composting as a means of processing and recycling various wastes is discussed further in Chapter 20.)

As organisms feed on detritus on or in the soil, they often ingest mineral soil particles as well. For example, it is estimated that as much as 15 tons per acre (37 tons per hectare) of soil pass through earthworms each year in the course of their feeding. As the mineral particles go through the worm's gut, they actually become "glued" together by the indigestible humus compounds. Thus, earthworm excrements, or **castings** as they are called, are relatively stable "clumps" of inorganic particles plus humus. The burrowing activity of organisms keeps the clumps loose. This loose, clumpy characteristic, which you may experience as softness under foot as you walk through a woods, is referred to

FIGURE 9-8
Formation of humus. Humus is the residue of organic matter, as seen in this rotted log, that remains after the bulk of organic material has been decomposed by fungi and microorganisms. Humus is resistant to the initial digestion of detritus feeders but does eventually decompose to inorganic materials. (Photograph by BJN.)

as **soil structure** (Fig. 9-9). Whereas soil texture describes the size of soil particles, soil structure refers to the arrangement of them. You can readily visualize what this loose, clumpy soil structure means for infiltration, aeration, and workability. Additionally, humus has a phenomenal capacity for holding both water and nutrients, as much as 100-fold greater than clay on the basis of weight.

This clumpy, loose, humus-rich soil, ideal for supporting plant growth, is what is commonly called **topsoil**. In natural ecosystems topsoil develops and is maintained as the uppermost layer of soil (usually between 4 to 12 inches, or 10 to 30 cm thick) because that is where most detritus and most of the activity of soil organisms occur. Below the topsoil is **subsoil**, which remains densely compacted because it lacks humus and the agitation of soil organisms. In practice, topsoil is often recognized by its dark color, since humus is black. Thus, a careful cut through a natural, undisturbed soil sample reveals a layering, referred to as the **soil profile**, of dark topsoil overlying the lighter subsoil (Fig. 9-10). Again, however, it is the loose aggregate structure of topsoil, with its water- and nutrient-holding humus, that is most significant in supporting plants.

If plants are grown on adjacent plots, one of which has had all the topsoil removed, the results are striking: The yield from plants grown on subsoil is only 10% to 15% of that from plants grown on topsoil. In other words, loss of all the topsoil would result in an 85% to 90% decline in productivity.

There are other interactions between plants and soil biota. A very significant one is the symbiotic relationship between the roots of some plants and certain fungi called **mycorrhizae**. Drawing some nourishment from the roots, mycorrhizae penetrate the detritus, absorb nutrients, and transfer them directly to the plant. Thus, there is no loss of nutrients to leaching. Another important relationship is the role of certain soil bacteria in the nitrogen cycle, as discussed in Chapter 3. Not all soil organisms are beneficial to plants. *Nematodes,* small worms that feed on living roots, are highly destructive to some agricultural crops. In a flourishing soil ecosystem, however, nematode populations may be controlled by other soil organisms, such as a fungus that forms little snares to catch and feed on nematodes (Fig. 9-11).

Soil Enrichment or Mineralization. Coming back to the context of the total ecosystem, we can now see how the above-ground portion and the soil portion of the ecosystem support each other. The bulk of the detritus, which supports the soil organisms, is from green plant producers. So green plants support the soil organisms. But then, by feeding on detritus, the soil organisms create the chemical and physical soil environment that is most beneficial to the growth of producers.

Green plants protect the soil and consequently themselves in two other important ways as well: The cover of living plants and detritus (1) protects the soil from erosion and (2) reduces evaporative water loss. Thus you can see the desirability of maintaining an organic mulch around your garden vegetables that do not maintain a complete cover themselves.

Unfortunately, the mutually supportive relationship between plants and soil can be broken all too easily. The maintenance of topsoil depends on additions of detritus sufficient to balance losses. Without continual additions of detritus, it should be obvious that soil organisms will starve, and their benefit in keeping the soil loose will be lost. However, additional consequences occur. Although resistant to digestion, humus

(a)

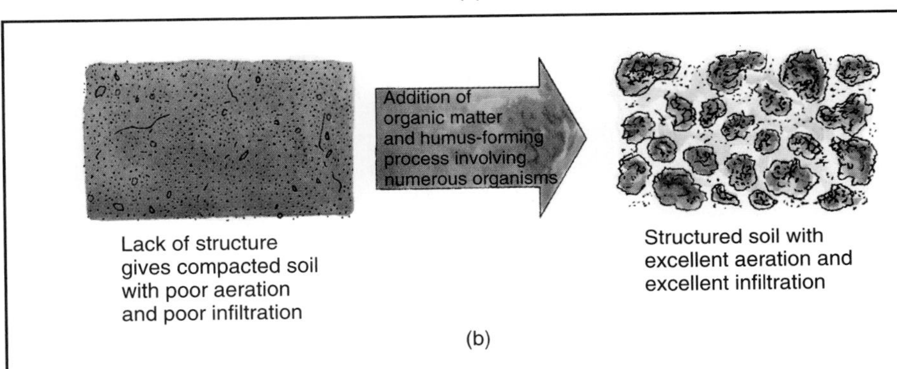

(b)

FIGURE 9-9
Humus and the development of soil structure. (a) On the left is a humus-poor sample of loam. Note that it is a relatively uniform, dense "clod." On the right is a sample of the same loam but rich in humus. Note that it has a very loose structure, composed of numerous aggregates of various sizes. (Photograph by BJN.) (b) A diagrammatic illustration of the difference.

does decompose at the rate of about 20% to 50% of its volume per year, depending on conditions. (This breakdown of humus is necessary so that the nutrients contained in humus can be released for reabsorption by plants.) As humus content declines, there is a breakdown of the clumpy aggregate structure created by soil particles "glued" together with the humus. Water and nutrient-holding capacities, infiltration, and aeration decline accordingly. This loss of humus and consequent collapse of topsoil is referred to as **mineralization** of the soil, because what is left is just the gritty mineral content—sand, silt, and clay—devoid of humus.

Thus, topsoil must be seen as a dynamic balance between detritus additions and humus-forming processes and the breakdown and loss of detritus and humus (Fig. 9-12). If additions of detritus are not sufficient, there will be a gradual deterioration of the soil. Conversely, you can see how mineralized soils can potentially be "rejuvenated" through generous additions of compost or other organic matter.

FIGURE 9-10
Soil profile. A vertical cut through soil generally reveals a layer of loose, dark topsoil overlying light-colored, compacted subsoil. The topsoil layer results from the addition of organic matter and the activity of soil organisms. (USDA photograph.)

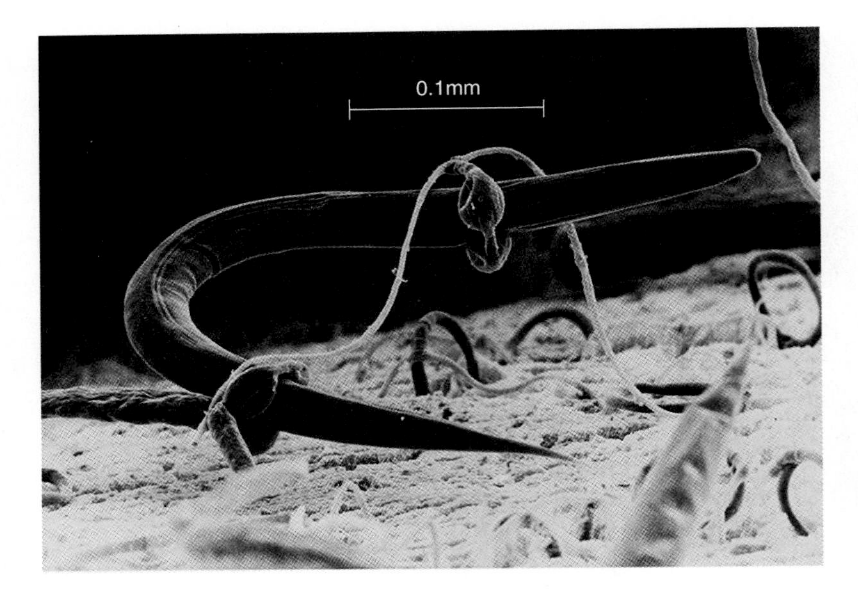

FIGURE 9-11
Soil nematode (roundworm), a root parasite, captured by the constricting rings of the predatory fungus *Arthrobotrys anchonia*. (Courtesy of Nancy Allin and O. L. Barron, University of Guelph.)

Losing Ground

In natural ecosystems the demand for additions of detritus is no problem; there is always a turnover of plant material. However, when humans come on the scene and cut forests, graze livestock, or grow crops, the soil system is at the mercy of our management or mismanagement. Contrary to knowledge and common sense, so much land is still mismanaged that sustainability of current levels of food production in many areas is in doubt. A billion and a half hectares of topsoil have already been lost. Let us examine the key factors in this loss.

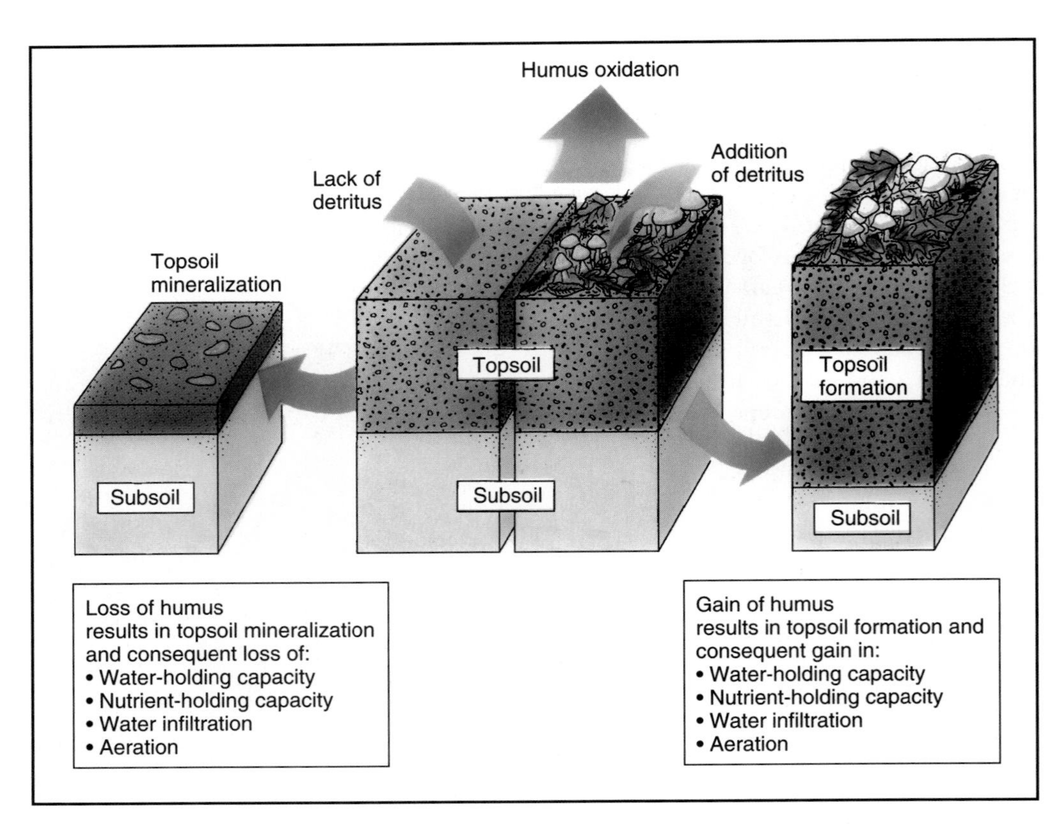

FIGURE 9-12
Topsoil must be recognized as a dynamic balance between detritus additions and humus forming processes, and the breakdown and loss of detritus and humus. If additions of detritus are not sufficient there will be a gradual deterioration of the soil. Conversely, you can see how mineralized soils can potentially be "rejuvenated" through generous additions of compost or other organic matter.

FIGURE 9-13
Bare soil is subject to severe erosion as the splash of falling rain seals the surface and resulting runoff readily carries away soil particles. The erosion of the gully in the cultivated field seen here occurred in a single rain. (USDA.)

Bare Soil, Erosion, and Desertification

How is topsoil lost? The most pervasive and damaging force is **erosion**, the process of soil and humus particles being picked up and carried away by water or wind. Erosion ensues any time soil is bared and exposed to the elements. The removal may be slow and subtle as soil is gradually blown away by wind, or it may be dramatic as gullies are washed out in a single storm (Fig. 9-13).

In natural terrestrial ecosystems other than deserts, a vegetative cover protects against erosion. The energy of falling raindrops is dissipated against the vegetation, and the water infiltrates gently into the loose topsoil without disturbing its structure. With good infiltration, runoff is minimal. Any runoff that does occur is slowed as the water moves through the vegetative or litter mat, so the water has too little energy to pick up soil particles. Grass is particularly good for erosion control because when runoff volume and velocity increase, well-anchored grass simply lies down, forming a smooth mat over which the water can flow without disturbing the soil underneath. Similarly, vegetation slows the velocity of wind and holds soil particles (Fig. 9-14a).

When soil is left bare and unprotected, however, it is highly subject to erosion. Water erosion starts with what is called **splash erosion** as the impact of falling raindrops breaks up the clumpy structure of topsoil. The dislodged particles wash into spaces between other aggregates, clogging the pores and thereby decreasing infiltration and aeration. The decreased infiltration results in more water running off carrying away the fine particles from the surface; this is called **sheet erosion**. As further runoff occurs, the water converges into rivulets and streams. These have greater volume, velocity, and energy, and hence greater capacity to pick up and remove soil. The result is the erosion of gullies, or **gully erosion**, shown in Figs. 9-13 and 9-14b. Once started, erosion can readily turn into a vicious cycle if not controlled. Eroding soil is less able to support the regrowth of vegetation and is exposed to further erosion, rendering it less able to support vegetation and so on.

Another very important and devastating feature of wind and water erosion is that both types always involve the *differential* removal of soil particles. The lighter particles of humus and clay are the first to be carried away, while rocks, stones, and coarse sand remain behind. Consequently, as erosion removes the finer materials, the remaining soil becomes progressively coarser—sandy, stony, and rocky. Such coarse soils frequently reflect past or ongoing erosion. Did you ever wonder why deserts are full of sand? The sand is what remains after the finer, lighter clay and silt particles have blown away. Often in deserts the removal of fine material by wind has left a thin surface layer of stones and coarse sand called a **desert pavement**, which protects the underlying soil against further erosion. Can you see how damaging this surface layer with vehicular traffic allows another episode of erosion to commence (Fig. 9-15)?

Recall that clay and humus are the most important components of soil for both nutrient- and water-holding capacity. As clay and humus are removed, nutrients are removed as well because they are bound to these par-

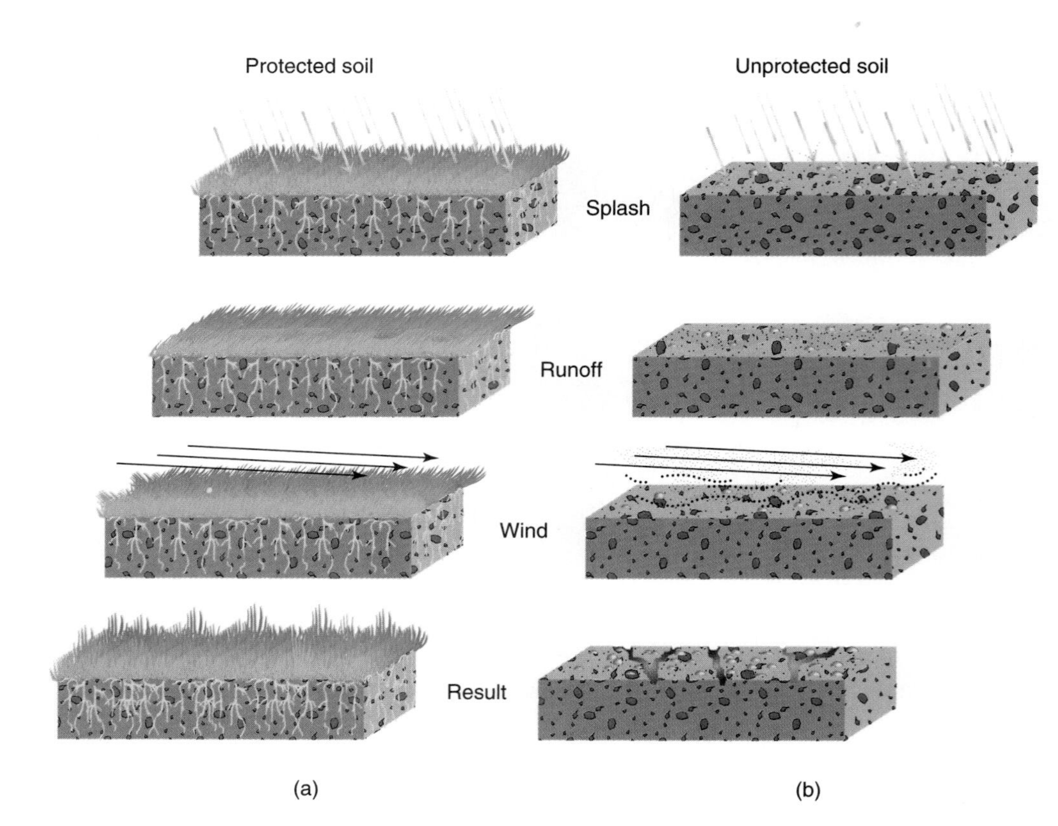

FIGURE 9-14
Erosion. (a) A vegetative cover protects soil from all forms of erosion. (b) Bare soil is extremely vulnerable to erosion. The splash of falling raindrops breaks up soil aggregates into individual particles. The finer particles of humus, clay, and silt are then readily carried away by runoff or wind, leaving only a layer of coarse sand, stones, and rocks.

ticles. The loss of water-holding capacity is even more serious, however. Regions that have sparse rainfall or long dry seasons support grass, scrub trees, or crops only insofar as soils have good water-holding capacity. As water-holding capacity is diminished by erosion of topsoil such areas become deserts both ecologically and from a standpoint of production (Fig. 9-16). Indeed, the term **desertification** is used to denote this process.

The problem is not small. Much of the world, beyond deserts, receives only 10 to 30 inches (25 to 75 cm) of rainfall a year, a minimal amount to support rangeland or nonirrigated cropland. The world over,

FIGURE 9-15
Formation of desert pavement. (a) As wind erosion removes the finer particles, the larger grains and stones are concentrated on the surface. (Adopted from Christopherson, *Geosystems* 2/e, 1994, p. 451, by permission of Prentice Hall, Upper Saddle River, New Jersey.) (b) The result is desert pavement, which protects the underlying soil from further erosion. Traffic such as off-road vehicles break up the desert pavement and initiate further erosion. (Steven Kaufman/Peter Arnold, Inc.)

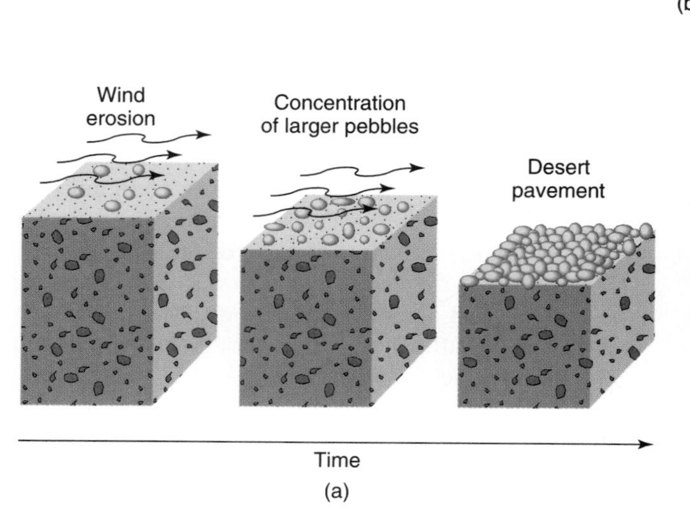

FIGURE 9-16
Desertification. Cultivation without suitable protection against erosion allows finer components of the soil to be blown away, leaving the soil increasingly coarse and stony and diminishing its water holding capacity. Persistance gradually renders the area an ecological desert as seen in this photo taken in east Africa. (Rene Grojean/Gamma-Liaison, Inc.)

some 60% of these so-called dryland types of rangelands and nonirrigated croplands have been adversely affected by erosion and desertification (Table 9-3). In recent years famines in East Africa have come into our living rooms over TV. What was less publicized was the desertification of the countryside that preceded these famines.

Unfortunately, desertification once started is not easy to reverse. With diminished productivity, the soil is left unprotected. Further erosion takes place, causing further reduction of productivity, and the vicious cycle goes on and on until it ends in a desert landscape that supports virtually no growth at all. Therefore, topsoil should be viewed as a nonrenewable resource, once destroyed it will not automatically replenish itself, at least not within a time frame that is meaningful to humans. But soil can be protected and its productivity maintained. Toward this end we will begin by taking a closer look at practices that lead to erosion and desertification.

Causing and Correcting Erosion

The major practices that expose soil to erosion are (1) overcultivation, (2) overgrazing, and (3) deforestation.

TABLE 9-3
Extent of Desertification

	Productive Dryland Types			
	Rangelands		*Rain-Fed Croplands*	
	Area (million hectares)*	*Percent Desertified*	*Area (million hectares*)*	*Percent Desertified*
Total	***2556***	***62***	***570***	***60***
Sudano-Sahelian Africa	380	90	90	80
Southern Africa	250	80	52	80
Mediterranean Africa	80	85	20	75
Western Asia	116	85	18	85
Southern Asia	150	85	150	70
Former USSR in Asia	250	60	40	30
China and Mongolia	300	70	5	60
Australia	450	22	39	30
Mediterranean Europe	30	30	40	32
South America and Mexico	250	72	31	77
North America	300	42	85	39

Source: World Resources 1989. Washington, D.C.: World Resources Institute.
* 1 hectare = 2.5 acres.
Check out: http://grid2.cr.usgs.gov/des/uncedtoc.html
http://www.mbnet.mb.ca/linkage/desert.html 

FIGURE 9-17
An important advance in reducing soil erosion is no-till planting, the apparatus for which is shown here. The "wheel" (a) opens a furrow in the soil, fertilizer is dispensed from the tank (b), seeds are dropped in from the hopper (c), and the furrow is closed by the trailing wheels (d). Weeds are controlled with herbicides. Thus all operations of field preparation, planting, and cultivating are accomplished without disturbing the protective mulch layer on the surface of the soil. (Larry Lefever/Grant Heilman Photography.)

Overcultivation. Traditionally, the first step in growing crops has been (and to a large extent still is) plowing to control weeds. The drawback is that the soil is then exposed to wind and water erosion. Further, it may remain bare for a considerable time before the newly planted crop forms a complete cover; after harvest, much of the soil again may be left exposed to erosion. Runoff and erosion are particularly severe on slopes, but on any terrain with minimal rainfall wind erosion may extract a heavy toll.

It is ironic that plowing is frequently deemed necessary to "loosen" the soil to improve aeration and infiltration, for all too often the effect is just the reverse. Splash erosion destroys the soil's aggregate structure and seals the surface so that aeration and infiltration are decreased. The weight of tractors used in plowing may add to the compaction. In addition, plowing accelerates the oxidation of humus and evaporative water loss.

Despite the harmful impacts intrinsic to cultivation, systems of crop rotations—a cash crop such as corn every third year, with hay and clover (which fixes nitrogen as well as adding organic matter) in between—has proved sustainable. However, as food or economic demands cause the abandonment of rotations, degradation and erosion exceed regenerative processes, and the result is a gradual decline in soil quality—desertification. This is the essence of overcultivation.

A technique that permits continuous cropping yet minimizes soil erosion is **no-till agriculture**, which is now routinely practiced over much of the United States. The field is first sprayed with herbicide to kill weeds; then the planting apparatus pulled behind a tractor accomplishes several operations at once. A steel disc cuts a furrow through the mulch of dead weeds, drops seed and fertilizer into the furrow, and then closes it (Fig. 9-17). At harvest, the process is repeated, and the waste from the previous crop becomes the detritus and mulch cover for the next. Thus the soil is never left exposed, erosion and evaporative water loss are reduced, and there is enough detritus including roots from the previous crop to maintain the topsoil. Of course, the drawback is the use of chemical herbicides that may have unrecognized side effects, and control of pests may become more problematic as they overwinter in the crop residues left on the surface.

Another aspect of overcultivation involves the use of inorganic versus organic fertilizer. There is no question that optimal amounts of required nutrients can be efficiently provided by suitable application of inorganic chemical fertilizer. The failing of chemical fertilizer is in the lack of organic matter to support soil organisms and build soil structure. Under intensive cultivation—a cash crop every year—nutrient content may be kept high with inorganic fertilizer, but mineralization and thus desertification proceed in any case. Then, with the soil's loss of nutrient-holding capacity, applied inorganic fertilizer is prone to simply leach into waterways, causing pollution.

This is not to say that chemical fertilizers do not have a valuable place in agriculture. Exclusive use of

(a)

(b)

FIGURE 9-18
(a) Contour farming. Cultivation up and down a slope encourages water to run down furrows and may lead to severe erosion. The problem is reduced by plowing and cultivating along the contours at a right angle to the slope. In this photograph, strip cropping is also being utilized: The light green bands are one type of crop, the dark bands are another. (USDA photograph.) (b) Shelterbelts. Belts of trees around farm fields break the wind and protect the soil from erosion. (Dan McCoy/The Stock Market.)

organic material may not provide enough of certain nutrients required to support optimal plant growth. What is required is for growers to understand the different roles played by organic material and inorganic nutrients and then to use each as necessary.

Regardless of cropping procedures and use of fertilizer, a number of other techniques are widely used to reduce erosion. Among the most conspicuous are **contour strip cropping** and establishment of **shelterbelts** (Fig. 9-18). The U.S. Natural Resource Conservation Service (NRCS, check: http://www.ncg.nrcs.usda.gov/), formerly, and still widely known as the Soil Conservation Service (SCS), which was established in response to the Dust Bowl tragedy of the early 1930s, provides information to farmers or other interested persons regarding soil or water conservation practices through a nationwide network of regional offices. (They can be found through county government telephone listings.) (Federal Extension service offices will test soil samples, provide an analysis, and make recommendations.)

However, the NRCS can only provide assistance; it has no "teeth" to enforce conservation, and most of the less-developed countries have no equivalent service. The sad fact is that erosion continues at unacceptable levels in most countries, including the United States, as seen in Table 9-4. An irony has developed relative to some of these known practices that prevent soil erosion. As equipment has increased in size, former contours and shelterbelts have given way to rows that again run up and down hills. This is why some enforcement of proper practice is necessary.

Overgrazing. Grasslands that receive too little rainfall to support cultivated crops have traditionally been used for grazing livestock. Likewise, forested slopes too

TABLE 9-4
Soil Erosion in Selected Countries

	Extent and Location	*Rate of Erosion (metric tons per hectare per year*)*
Africa		
Ethiopia	Total cropland (12 million ha)	42
Kenya	Njemps Flats	138
Madagascar	Mostly cropland (45.9 million ha)	25–250
Zimbabwe	304,000 ha	50
North & Central America		
Canada	Cultivated land New Brunswick	40
Dominican Republic	Boa watershed (9330 ha)	346
Jamaica	Total cropland (208 595 ha)	36
United States	Total cropland (170 million ha)	18
South America		
Argentina, Brazil, and Paraguay	La Plata River basin	18.8
Peru	Entire country	15
Asia		
China	Loess Plateau region (60 million ha)	11–251
India	Seriously affected cropland (80 million ha)	75
Nepal	Entire country (13.7 million ha)	35–70
Europe		
Belgium	Central Belgium	10–25
Former USSR	Total cropland (232 million ha)	11

Source: World Resources 1988–89. Washington, D.C.: World Resources Institute.
* 1 hectare = 2.5 acres.
Check: http://ingis.acn.purdue.edu.9999/cttpp/erosion.html
http://soils.ecn.purdue.edu/~wepp/

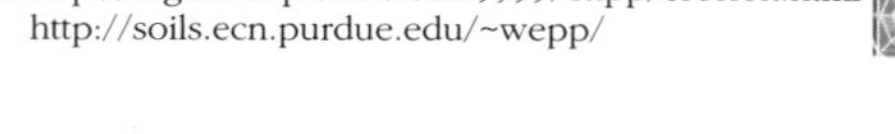

steep for cropping are commonly cleared and put into grass for grazing. Unfortunately, such lands are too often overgrazed. As grass production fails to keep up with consumption, the land becomes barren; wind and water erosion ensue, and desertification results. Data compiled by the World Resources Institute show that worldwide, 62% of rangelands suffer from desertification (Table 9-3), meaning that production has declined by 20% or more.

Overgrazing is not a new problem. In the United States in the 1800s the American buffalo (bison) were slaughtered to starve out the Native Americans and stock the rangelands with cattle. Overgrazing was rampant, leading to desertification and encroachment by hardy desert weeds such as sagebrush, mesquite, and juniper, which are not palatable to cattle. Western rangelands now produce less than 50% of the livestock forage they produced before the advent of commercial grazing. Yet, according to a U.S. General Accounting Office (GAO) study, 20% of the rangelands remain overstocked (Fig. 9-19).

The broader ecological impact of overgrazing should not go unnoticed. The World Resources Institute reports: "Overgrazing has profoundly upset the dynamics of many range ecosystems, reducing biodiversity and altering the feeding and breeding patterns of birds, small mammals, reptiles, and insects." Further, about one-third of the endangered species in the United States are in jeopardy because of overgrazing or other practices associated with raising cattle, such as predator control programs and suppression of fire. Particularly hard hit are the wooded zones along streams and rivers that are trampled to death by cattle seeking water. The resulting water pollution by sediments and cattle wastes is high on the list of factors making fish species the fastest-disappearing wildlife group in the United States.

Overgrazing occurs in many cases because the rangelands are "public lands" not owned by the people who own the animals. Where this is the case, herders who choose to withdraw their livestock from the range sacrifice income while others continue to overgraze the range. Therefore, the incentive is for all to keep grazing

FIGURE 9-19
Overgrazing. Originally grassland, the erosion, soil degradation, and loss of productivity caused by overgrazing are conspicuous in this scene of federal lands in Utah. Yet a considerable portion of such lands remain overstocked with cattle. (Rod Planck/Photo Researchers, Inc.)

despite realizing that the range is being overgrazed. This problem, known as the "tragedy of the commons," is discussed further in Chapter 19, p. 496.

In a variation of the tragedy of the commons the U.S. Bureau of Land Management (BLM, check: http://www.blm.gov/) leases huge tracts of gov ernment- (taxpayer-) owned lands for grazing rights at a nominal fee of $1.93 per animal unit (1 cow and calf pair or five sheep) per month, about a third of what they would pay a private owner for the same rights. The income to the government ($19 million in 1990) does not even cover the costs of administering the program ($50 million), and the amount of livestock permitted by BLM is considered too high by nearly all range experts. The Clinton administration made an effort to have BLM leasing fees increased in 1994, but the measure faced such strong opposition from the ranching lobby—and such apathy from nearly everyone else—that it failed. Thus, we taxpayers continue to subsidize the overgrazing and ecological destruction of our western rangelands.

The obvious solution lies in better management. Since rangelands are now producing at only a 50% level or less, their restoration could benefit both wildlife and cattle production. The Natural Resources Conservation Service now has a project that provides ranchers who own their own lands with the information and support to enable them to burn unwanted woody plants, reseed the land with perennial grass varieties that hold water, and manage cattle so that they are moved to a new location before overgrazing occurs. Results are encouraging, but there is no equivalent program for improving public (government-owned) lands in the United States or elsewhere.

Some people see a solution in decreasing the amount of beef we consume. Beef consumption in the United States has indeed declined in recent years for various reasons of health as well as the ethical issue, but it is still five to ten times higher than in most less-developed countries. Further, the people in developing countries tend to increase their beef consumption as they are able to afford to do so. Therefore world pressure is toward increasing demand for beef and the ecological pressures that go with it.

Deforestation. Forest ecosystems are extremely efficient systems both for holding and recycling nutrients and for absorbing and holding water, because they maintain and protect a very porous, humus-rich topsoil. Investigators at Hubbard Brook Forest in New Hampshire found that converting a hillside from forest to grass resulted in doubling the amount of runoff, and leaching of nutrients increased manyfold.

Much worse is what occurs if the forest is simply cut and soil is left exposed. Pounding raindrops quickly seal the soil; the topsoil becomes saturated with water and slides off the slope in a muddy mass into waterways, leaving a barren subsoil that continues to erode.

Not the least of current examples is the logging of old-growth forests of the Northwest (Fig. 9-20). Again much of the land is owned by the government (taxpayers). The U.S. Forest Service leases tracts for logging

FIGURE 9-20
Clearcutting of forests in the Pacific Northwest. The U.S. Forest Service continues to effectively subsidize the clearcutting of these lands causing tragic ecological loss, as well as financial loss, for the profit of timber interests. (Timothy Hendrix/Gamma-Liaison, Inc.)

and builds access roads for private companies to remove the timber. Like the situation of grazing rights, the government spends much more than it receives, meaning that taxpayers are subsidizing the ecological tragedy in the Northwest.

The problem is particularly acute with the cutting of tropical rain forests. The soils of equatorial regions have been subjected to heavy rains and leaching for millions of years. Parent materials are already maximally weathered, and any free nutrients have long since been leached away. Consequently, tropical soils are notoriously lacking in nutrients. All the nutrients, which support the luxuriant growth of tropical forests, are in the biomass. Nutrients are transferred directly from decaying materials back into growing vegetation by means of mycorrhizae fungi without entering the soil itself at all. When the forests are cleared, the thin layer of humus with nutrients readily washes away. Only the nutrient-poor, clayey subsoil, which is very poor for agriculture, is left.

Yet tropical forests continue to be cleared at an alarming pace—about 41.7 million acres (16.9 million hectares), or nearly 1%, per year, according to surveys conducted by the United Nations Food and Agriculture Organization (FAO). This loss is an increase from the previous rate of .6% per year. Over the past decade an area three times the size of France has been lost, and much more has been degraded by fragmentation (partial clearing and thinning). About 80% of the deforestation is for agricultural purposes despite the poor soil, and much of agriculture is directed toward the production of cash crops or conversion to grass to grow beef. Very little is managed with a clear aim of sustainable food production.

In all, less than 2% of tropical forests are under any kind of management plan for protection or harvesting of forest products on a sustainable basis, although nongovernmental organizations such as Conservation International are active in this regard. Also, the FAO has launched a Tropical Forestry Action Plan (TFAP), which at least creates a global forum to discuss and develop policies that will lead toward sustainability. Interestingly, developed countries including the United States have led the opposition to meaningful agreements restricting clearing forests. Clearly there is still a greater interest in supporting the short-term profitability of parties benefiting from the exploitation than in sustainability—an interest that might be changed if more of the public made their wishes known.

Summation. The point is that to save soil, the foundation of all terrestrial productivity, we must save the natural ecosystems that build, maintain, and nurture that soil or at least provide mechanisms that duplicate them. We observe that there is conspicuous room for improvement.

The Worldwatch Institute estimates that worldwide the loss of soil from crop, range, and deforested lands is about 23 billion tons per year. This is equivalent to all the topsoil on about 23 million acres (9.2 million hectares), an area about the size of Indiana. In all, lands suffering from or prone to desertification cover much of the globe (Fig. 9-21). Erosion is a particularly insidious phenomenon because the first 20% to 30% of the topsoil may be lost with only marginal declines in productivity, a loss that may be compensated for by additional fertilizer and favorable distribution of rains. As loss of topsoil continues, however, the decrease in productivity and increase in vulnerability to drought will become increasingly pronounced. And the declining productivity as the soil is washed away is not the total picture.

The Other End of the Erosion Problem

In addition to soil degradation and the loss of productivity resulting from erosion, we must also consider the altered pathway of water and where the **sediments**, as the eroding soil is called, go. Water, unable to infiltrate, flows over the surface immediately into streams and rivers, filling them to overcapacity and causing flooding. Eroding sediments are likewise carried into streams and rivers, where they clog channels and exacerbate flooding, fill reservoirs, kill fish, and generally upset the ecosystems of streams, rivers, bays, and estuaries. Around the world coral reefs are dying because of sediments and other pollutants carried with them. Indeed, excess sediments and nutrients resulting from erosion are recognized as the number one pollution problem of surface waters in many regions of the world. Meanwhile, groundwater resources are also depleted as

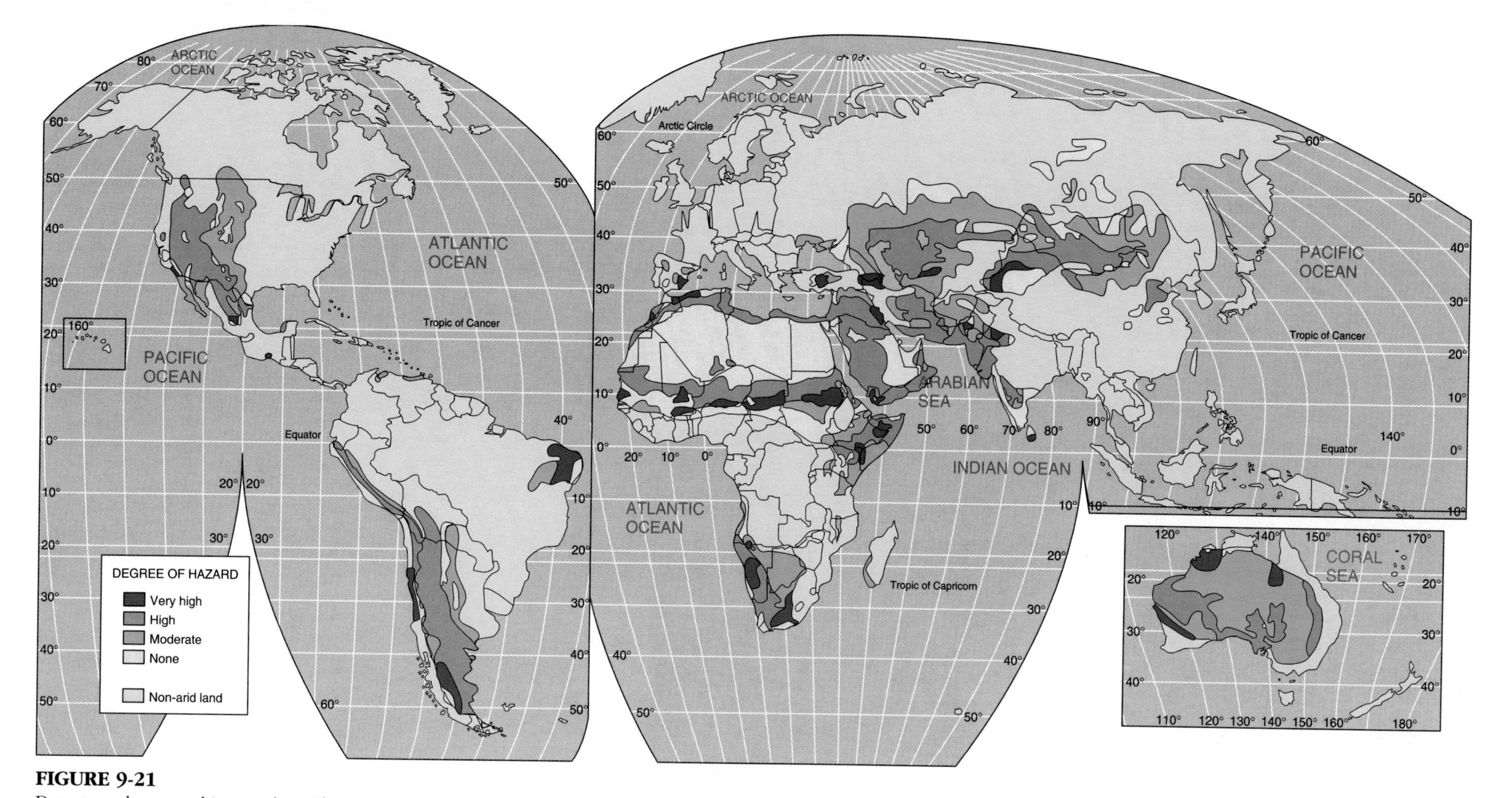

FIGURE 9-21

Deserts and areas subject to desertification. Throughout the world, overgrazing and deforestation are causing vast low-rainfall areas to degenerate into deserts. (Reprinted from "Desertification: Its Causes and Consequences," U.N. Conference on Desertification, Nairobi, 1977.)

FIGURE 9-22
Flood irrigation. The traditional method of irrigation is to flood furrows between rows with water from an irrigation canal. This method is extremely wasteful because most water either evaporates or percolates beyond the root zone, or the water table rises to waterlog the soil and prevent crop growth. (Betty Derig/Photo Researchers, Inc.)

FIGURE 9-23
Millions of acres of irrigated land are now worthless because of the accumulation of salts left behind as the water evaporates, a phenomenon known as salinization. (John M. White/Texas Agricultural Extension Service.)

the rainfall runs off rather than refilling the reservoir of water held in the soil or recharging groundwater.

Irrigation, Salinization, and Desertification

Irrigation, supplying water to croplands by artificial means, has dramatically increased crop production in regions that typically receive inadequate rainfall. Traditionally, water has been diverted from rivers through canals and flooded through furrows in fields, a technique known as **flood irrigation** (Fig. 9-22). In recent years, **center-pivot irrigation**, a procedure in which water is pumped from a central well through a gigantic sprinkler that slowly pivots itself around the well, has become much more popular (see Fig. 11-5).

Worldwide, irrigated acreage has increased dramatically in the last few decades and estimates of total irrigated land range between 555 and 620 million acres (225 to 250 million hectares). New irrigation projects continue to be built, but the expansion in acreage is now being offset in large part by another ominous trend—*salinization*. **Salinization** is the accumulation of salts in and on the soil to the point that plant growth is suppressed (Fig. 9-23). Salinization occurs because even the freshest irrigation water contains at least 200-500 parts per million (0.02 to 0.05%) of dissolved salts. As the applied water leaves by evaporation or transpiration, the salts in solution remain behind and gradually accumulate. Salinization is considered a form of desertification.

It was estimated that by the year 2000 that at least half the crop land under irrigation will have reduced productivity due to salinization (Table 9-5.) An additional 2.5 to 3.75 million acres (1.5 million hectares) are salinized each year. In the United States, the problem is especially acute in the lower Colorado River Basin area and in the San Joaquin Valley of California, areas in which a total of 400,000 acres (160,000 hectares) have

TABLE 9-5
Irrigated Land Desertified by Salinization

	Area (million hectares)*	*Percent Desertified*
Total	***131***	***30***
Sudano-Sahelian Africa	3	30
Southern Africa	2	30
Mediterranean Africa	1	40
Western Asia	8	40
Southern Asia	59	35
Former USSR in Asia	8	25
China and Mongolia	10	30
Australia	2	19
Mediterranean Europe	6	25
South America and Mexico	12	33
North America	20	20

Source: World Resources 1989. Washington, D.C.: World Resources Institute.
* 1 hectare = 2.5 acres.

been rendered nonproductive, representing an economic loss of more than $30 million per year. Adding to the problems, water supplies are fast being depleted by withdrawals for irrigation. This and related problems will be discussed further in Chapter 11.

Salinization can be avoided, even reversed, if sufficient water is applied to leach the salts down through the soil. However, unless there is suitable drainage, the soil will become a waterlogged quagmire in addition to being salinized. Drainage pipe may be installed 10–13 feet (3–4 meters) under the surface but only at great expense, and then attention must be paid to where the salt-laden water is drained. The wildlife in the Kesterson National Wildlife Reserve in California was destroyed by pollution from irrigation drainage—the area was classified as a toxic waste site after only three years of receiving salty drainage water. Problems of both salinization and depletion of water resources are exacerbated by the government's building of dam and irrigation projects and providing water to farmers at far below cost, practices that encourage the use of excessive water.

"Greening" Farm Policy

We have seen that despite the crucial importance of saving agricultural land, forests, and rangeland the world is losing ground at an alarming pace to erosion, desertification, salinization, and direct conversion of farmland to development. The losses are on a collision course with sustainability. Who is responsible for making changes?

Farm policy in the United States has traditionally focused on a single goal: increasing production. Success in attaining this goal cannot be denied; continuing surpluses speak for themselves. With the focus on increasing production, however, various practices having questionable sustainability have been allowed and even promoted through government subsidies.

The National Research Council summed it up in a 1989 report titled *Alternative Agriculture*:

> As a whole, federal policies work against environmentally benign practices and the adoption of alternative [sustainable] agricultural systems, particularly those involving crop rotations, certain soil conservation practices, reductions in pesticide use, and increased use of biological and cultural means of pest control. These policies have generally made a plentiful food supply a higher priority than protection of the resource base.

The status quo is vehemently defended by the special interest groups that benefit from subsidies and policies as they exist. Nor can farmers be expected to act in ways contrary to their economic interests. Nevertheless, changes in policy are clearly in order. A knee-jerk reaction from the public, especially in view of the building pressure to reduce government spending and move toward free trade, might be to cut all farm subsidies. But, one-fourth of net farm income is derived from subsidies of one sort or another. Cutting such subsidies cold turkey would cause economic disaster in farm communities across the nation to say nothing of disruptions in production.

The goals of sustainable agriculture are to (1) maintain a productive topsoil; (2) keep food safe and wholesome; (3) reduce the use of chemical fertilizers and pesticides; and last but far from least (4) keep farms economically viable. How can sustainable policies be implemented over the objections of special interest lobbies?

Professional agriculturists are becoming increasingly aware of the shortfalls of modern farming and are at work developing alternatives. Many options actually mimic practices of the past: contouring, crop rotation, terraces, smaller equipement, and reduced or no chemicals. In 1988 the U.S. Department of Agriculture started the Low Input Sustainable Agriculture (LISA) program, which provides funding for "alternative" methods. Agricultural colleges and universities across the country are revising curricula to emphasize sustainable agriculture. In addition, various nongovernmental organizations such as the Rodale Institute are actively engaged in researching and disseminating information regarding sustainable methodologies (see Bibliography). A small but growing number of farmers are experimenting and developing alternative systems on their own (see "Earth Watch" box, on page 234).

Ethics

Who Is Responsible For Maintaining the Soil?

The lives of all of us depend on the abundant production of agriculture. Over the years since the start of the Industrial Revolution, people have become increasingly divorced from soil and the production of food. In every developed country, countless diversified family farms have given way to huge specialized monocrop operations. In the United States, for instance, less than 2% of the population is engaged directly in crop production. From the field to storage and then to flour mills, baking companies, and supermarket shelves, grains travel an average of 1000 miles before they are consumed. Less than one penny of the dollar you pay for a loaf of bread goes to the farmer. The same is true for other foods.

With this specialization of production and its separation from consumers, who should be responsible for seeing that soil is properly protected and nurtured as a sustainable resource? Not the 98% who are off-farm consumers, surely, but not the 2% who are growers either. Far removed from the final consumers and receiving minimal compensation, most farmers consider production to be simply a business—farms are vast outdoor factories. If it fails, they will "write it off" and move on to something else.

In other words, we have unwittingly created a food production system in which virtually no one is in a position to see the whole system as an *ecosystem*, much less manage it as such.

Many people point with pride to the U.S. agricultural system, saying, "Look how efficient it is; people in this country generally spend a smaller portion of their incomes on food and yet eat better than in any other country of the world." If the system is not sustainable, however, its efficiency is a moot point.

In their book *For the Common Good*, Herman Daly and John Cobb suggest that each region of the country, a region to be no larger than a small state, should be required to be agriculturally self-sufficient (disaster-relief excluded). Such a policy would force each region to adopt a diversified agriculture, a sustainable soil policy, and would bring producers and consumers closer together. If such a policy were ever put into effect, the people of each region would take more interest in the management of their soil, because they would see a healthy soil system as crucial to their own survival and well-being.

What would be the ramifications of Daly and Cobb's plan?

- Would each region adopt a more diversified, sustainable agriculture?
- Would a more diversified agriculture facilitate recycling of animal wastes, compost, and sewage sludges back to the soil?
- Would growers develop a greater sense of responsibility toward consumers, their customers?
- Would consumers develop a greater interest in and responsibility toward how their food was grown?
- Would people face the consequences of converting more and more farmland to housing developments and office parks?
- Would prices of food skyrocket?
- Would the plant-supporting soil base be treated in a more sustainable manner?

A considerable impetus for shifting to organic sustainable farming is coming from consumers who, for various reasons, desire organically grown food. Such demand effectively provides an economic incentive for increasing numbers of growers to adopt the sustainable methods of organic farming. They find there is a lucrative market for their produce. Insofar as such food may be locally grown and distributed through local farmer's markets, it reestablishes a connection between the consumer, the grower, and the soil. An even stronger connection is made by **farm cooperatives**. In a farm cooperative, a group of consumers puts up the capital to pay a farmer to produce organically grown food. The group members then share the produce as it comes in. Effectively they are "prebuying" their produce for the season.

What we see in these initiatives is the development of what has been called an **urban-environmental coalition**—a growing number of people who, while not engaged in farming themselves, are nevertheless taking an increasing interest in how food is produced and in the sustainability of its production. The coalition is a loose network of groups such as the Sierra Club, the National Audubon Society, and the Conservation Foundation. Focusing the interests of the urban environmental coalition can bring considerable pressure to bear on policy decisions—pressure that is sufficient to override that of special interest groups. For example, during the 1970s farmers, under economic pressures to increase production, were plowing and growing crops on highly erodible land, and soil erosion was dramatic. The problem was recognized and publicized by environmentally focused organizations such as the American Farmland Trust, Rodale Press, and the Worldwatch Institute. The urban environmental coalition members lobbied Congress and won passage of the

Conservation Reserve Program in 1985. Under this program, 40 million acres (16 million hectares) of highly erodible cropland were put into a "Conservation Reserve" of forest and grass. Farmers are paid about $50 per acre ($125 per hectare) per year for land placed in the reserve. As of 1994, 36 million acres (14.5 million hectares) had been placed in the reserve, saving an estimated 700 million tons of topsoil per year from erosion—but it is also costing taxpayers over a billion dollars per year for the "deficiency payments" as they are called. Under another bill, the Food Security Act of 1985, farmers are required to develop and implement soil conservation programs in order to remain eligible for price supports and other benefits provided by the government.

Similarly, under pressure from the urban-environmental coalition the 1990 Farm Bill (which is reauthorized every five years) included measures to encourage farmers to save wetlands and reduce water-polluting runoff from their farms. In conclusion, we see that with suitable economic incentives farmers may be guided to be conservationists.

Saving agricultural land from development is one of the more intractable problems, however. It presents the classic moral dilemma of pitting the value of individual rights against the values of society as a whole. We (society) can see the importance of saving agricultural land; yet we take for granted the individual right to buy a tract of agricultural land and turn it into a housing development. Nor do we feel we can violate the right of the farmer to sell land for development when the developer can offer a price ten times as great as the land is worth for farming. Because many people prefer the aesthetics of living in a spacious rural environment, we have seen land consumption for development grow six times faster than population. In addition to squandering agricultural land, this trend leads to many other environmental problems, including air pollution, degradation of water resources, and depletion of energy resources as people drive farther and farther to meet their needs (see Chapter 24). The problem begs for solution. European countries have accepted very strict restrictions regarding development of agricultural land, as has the state of Hawaii, but suggestions of similar restrictions here are thwarted by individual rights advocates.

One notable attempt at solution is in Traverse City, Michigan. The urban-environmental coalition there won approval of a $2.5 million bond issue to protect the cherry orchards for which the region is famous. The money will be placed in a fund that will pay farmers the difference between what their land is worth for farming and what a developer may offer, in exchange for a pledge to keep the land in agriculture. The American Farmland Trust, the lead organization in this initiative, offers its expertise to other citizen groups that wish to achieve similar goals.

It is conspicuous that all these measures cost taxpayers more money. Is this a cost we must bear for sustainability, or is there a more general policy that might accomplish the goal for less cost?

Herman Daly, Professor of Economics at the University of Maryland and a founding member of the International Society of Ecological Economics, has pointed out that our government has chosen to raise revenues by taxing labor (income tax) whereas natural resources are treated as free for the taking. This system makes labor artificially expensive—we might all work for considerably less if we were not paying income tax—and resources artificially cheap. In turn, the response of all businesses to this situation has been to minimize labor and treat resources as if they were disposable. Thus we find increasing automation and mechanization displacing labor while resource exploitation and pollution—dumping wastes into the environment is another way of exploiting natural resources—run rampant. In response, we have piled on more laws and regulations in attempts to mitigate environmental impact.

Daly proposes that we might accomplish the aims of sustainability more simply if the government were to raise revenues by taxing the use and depletion of natural resources rather than labor. In relation to the present topic, suppose farmers paid no income tax but were required to pay a tax on every ton of topsoil they lost to erosion or mineralization. The farmer's final net income might be the same initially, but there would be a powerful incentive to use more labor for reducing erosion and maintaining topsoil.

Similarly there might be a transportation tax of a certain amount per mile on food products, again in exchange for a similar amount of income tax. A crazy statistic is that on average, every pound of food we eat travels 1000 miles from where it is produced to where it is consumed. Is it any wonder we lose a feeling of the connection between our food and the soil on which it is grown? In most cases we would be hard-pressed to determine where it comes from. The transportation tax on food would create an economic incentive to purchase locally grown produce; it would be cheaper because of the tax. Buying locally grown produce would, in turn, foster a closer connection between consumers, growers, and the soil they work, and it might lead to a greater incentive for nurturing the fertility of agricultural land. (See "Ethics" box, p. 232.)

Production from close-to-city diversified farms and from urban gardens (Fig. 9-24) has other ecologically practical advantages as well. First, with a diversity of both vegetables and animals it is economically practical to utilize manures and crop wastes. Likewise,

Earth Watch

An Example of a Small Diversified Farm

In the gently rolling area of southwestern Iowa is a small (160-acre) diversified farm run by Clark BreDahl and his wife, Linda, a teacher who is able to work full time on the farm during the summer. Less than half the size of the average farm in the area, the BreDahl's farm is among the top 10% in profitability and generally provides full support for the family, allowing them to save Linda's income. Crops include corn, alfalfa, and turnips to feed animals; soybeans for market; pedigreed oats sold as seed (which bring double the feed price of ordinary seed); sheep for wool; and lambs and pigs for market.

Crops are grown in 100-foot-wide contour strips and follow a five- to six-year rotation sequence of corn–soybeans–corn, oats/turnips–alfalfa. Soybeans and alfalfa, which are nitrogen fixers, and manure provide most fertilizer needs. The strips are fenced, so that animal feeding is made more efficient by letting the animals graze (and deposit manure) in the strips as feed crops are growing, allowing the use of damaged crops and crop residue as feed resources.

A unique part of the system is the use of turnips. Planted after the oats are harvested in July, turnips mature in the fall. Sheep, turned into the turnip strips in September, graze on both the turnip tops and the turnips themselves. Eating out the bulbous turnip root leaves cup-shaped depressions in the soil that fill with water, snow, and ice, helping add moisture and preventing runoff. Thus turnips provide highly nutritious fodder for the sheep into the new year, and the soil benefits in the process. Through the use of "organic methods," the soil is maintained in excellent condition, whereas soils on most other county farms are classified as moderately to severely eroded. Further crop yields per acre on the BreDahl farm are consistently above the average for the area.

The conclusion is that, with intelligent management, a family can make a comfortable living from a small diversified farm, following sustainable practices.

Source: National Research Council. *Alternative Agriculture*. 1989.

compost and treated sewage wastes can be economically utilized (see Chapter 13). Hauling expenses make such use impractical on large, specialized, distant farms.

Whether or not there are changes in the tax structure or subsidies provided, sustainability of our civilization will depend on preserving topsoil. Another aspect of sustainable agriculture is getting away from the self-defeating process of using toxic chemicals for pest control. We will turn to this topic in Chapter 10.

FIGURE 9-24
Fostering urban gardening programs could have a number of advantages in terms of providing for efficient utilization of organic wastes, enhancing incomes, and providing better nutrition for urban residents. (Ginger Chih/Peter Arnold, Inc.)

Review Questions

1. What are the main things that plant roots must obtain from the soil?
2. Describe the soil environment in terms of both content and physical structure that will best fulfill the needs of plants.
3. Define and give the relationship between soil

texture, soil structure, humus, detritus, and soil organisms.

4. Describe how each factor defined in Question 3 above is related to creating a soil environment that best supports the needs of plants.
5. What is meant by mineralization? What causes it? What are its consequences for soil?
6. How does a vegetative cover protect and nurture the soil in a way that supports the plant's growth?
7. What are the three major cultural practices that expose soil to the weather?
8. What is the impact of water and wind on bare soil? Define and describe the process of erosion in detail.
9. What is meant by desertification? Describe how the process of erosion leads to loss of water-holding capacity and, hence, to desertification.
10. What is meant by salinization, and what are its consequences? How does salinization result from irrigation?
11. What are the principles of sustainable agriculture?
12. How might government policies toward agriculture, forestry, and grazing be altered to promote sustainability?

Thinking Environmentally

1. Why is soil considered to be a detritus-based ecosystem? Describe how the aboveground portion and the belowground portion of an ecosystem act as two interrelated and interdependent ecosystems.
2. What are some long-term dangers inherent in making farms entirely commercial operations (with short-term economic interests) that must compete with other commercial enterprises such as housing development?
3. Why do human societies, past and present, seem to place so little value in maintaining topsoil? What do such attitudes bode for sustainability? What might be done to change attitudes toward soil and farming?
4. Evaluate the following argument. Erosion is always with us; mountains are formed and then erode; rivers erode canyons. You can't ever eliminate soil erosion, so it is foolish to ask farmers to do so.
5. Explain why diversified farms are necessary for the cost-effective use of manure as fertilizer. Why can't manure be used cost effectively in our present system of monoculture (single-crop) farms and feedlots? (Fifty pounds of inorganic fertilizer costs a farmer roughly $4.00 and has roughly the same nutrient content as 500 pounds of manure. Estimate and compare the cost—fuel, labor, and so on—of trucking manure 50 miles from a feedlot to a farm with the cost of using an equivalent amount of inorganic fertilizer, assuming the manure is free.)
6. Suppose you were going into farming. Describe the type of operation you would create, and give a rationale for each measure in terms of sustainability.
7. With an aim toward ending hunger in developing countries, would it be best to emphasize development of large-scale industrialized agriculture as seen in the United States or should small-scale diversified agriculture be encouraged? Give logical arguments for your recommendation.

Pests and Pest Control

Key Issues and Questions

1. For thousands of years, human enterprises have been frustrated by what we call "pests." What are pests, and why do we want to control them?
2. Different methods are used to control agricultural pests. What are the basic philosophies behind these methods?
3. As new and more effective pesticides like DDT were employed, some serious problems began to appear. What are the problems resulting from the use of chemical pesticides?
4. There are alternatives to using pesticides to control pests. How well do alternative control methods work?
5. Biotechnology has revolutionized many biological applications. How does biotechnology add to the battery of genetic methods to control pests?
6. As illustrated in the chapter opening, integrated pest management is an important way to reduce pesticide use. What is integrated pest management, and how has it worked?
7. Public policy for controlling pesticides has been criticized. What are the objectives of current policy, and what are the problems with it?
8. Large quantities of pesticides are exported to developing countries. How is pesticide export and use in developing countries regulated?

To an Indonesian rice grower in 1986, life must have seemed strange. Having been supplied in the late 1970s with the new high-yielding rice developed by the International Rice Research Institute (IRRI), the farmer had become accustomed to greatly improved yields. The government supplied pesticides at 15% of the market price, the farmer sprayed them faithfully, and rice crops flourished—for a while. Then a formerly uncommon insect called the brown planthopper began to appear in the fields, and it could not be stopped (Figure 10-1). The more the farmer sprayed, the heavier was the infestation by the hoppers, and the greater the loss of rice.

Locust in Africa. [Photo by Gianni Tortoli/Photo Researchers, Inc.]

Suddenly in 1986, rice growers were told by the government that they should no longer use the pesticides. Of 63 pesticides used on rice, 57 were banned. The IRRI scientists had found that the pesticides were killing off the predatory spiders and water striders that normally kept the planthoppers in check. They advised the government to stop trying to control the outbreaks with pesticides. In the words of one worker, spraying pesticides to control the planthopper was like pouring kerosene on a house fire! Crop advisors fanned out into the countryside and taught the farmers to protect the natural predators and spray more selectively. Once the rice farmers cut back on their use of pesticides (which now has dropped by 65% in Indonesia), the natural enemies of the planthopper returned, and now rice production is higher than ever before (Fig. 10-1). Integrated pest management had come to Indonesia.

The Need for Pest Control

Since earliest times, humans have suffered the frustration and food losses brought on by destructive pests. Genesis, the first book of the Judaic and Christian Bibles, speaks of the "thorns and thistles" thwarting Adam's attempts to eat the plants of the field, and biblical passages speak of locust invasions that destroyed crops and caused famines. To this day, both farmers and pastoralists must wage a constant battle against the insects, plant pathogens, and unwanted plants that compete with them for the biological use of crops and animals.

Defining Pests

If we adopt a dictionary definition of **pest** as "any organism that is noxious, destructive, or troublesome," we see that the term includes a broad variety of organisms that interfere with humans or with our social and economic endeavors. The principal categories of pests are:

1. Organisms that cause disease in humans or domestic plants and animals. These pests include viruses, bacteria, and such parasitic organisms as intestinal worms and flukes.
2. Organisms that annoy people and domestic animals and that may transfer disease by biting or stinging. Common examples are flies, ticks, bees, and mosquitoes.
3. Organisms that feed on ornamental plants or agricultural crops, both before and after the crops are harvested. The most notorious of these organisms are various insects, but certain worms, snails, slugs, rats, mice, and birds also fit into this category.
4. Animals that attack and kill domestic animals, such as coyotes, foxes, and raccoons.
5. Organisms that cause wood, leather, and other materials to rot and food to spoil. Bacteria and fungi—especially molds—are largely responsible for this spoilage, but in warm, moist climates, termites are the primary culprit in the destruction of wood.
6. Plants that compete with agricultural crops, forests, and forage grasses for light and nutrients. A plant in any of these roles is often referred to as a weed. Some unwanted plants poison cattle or have other serious effects; others simply detract from the appearance of lawns and gardens.

All six types of pests are highly important in human affairs. We try to bring these pests under control for three main purposes: to protect our food, to protect

FIGURE 10-1
Immature brown planthoppers, shown on the stem of a rice plant. (Nigel Cattlin/Holt Studios International/Photo Researchers, Inc.)

our health, and for convenience. Our emphasis in this chapter is on those pests that interfere with human agricultural crops, grasses, and animals.

The Importance of Pest Control

Part of the credit for human prosperity can be attributed to pest control. We would still live under extremely precarious conditions, our food supply and physical health at the mercy of all the organisms we call pests, if it were not for our ability to control them. Indeed, pests represent a major component of the environment that, until recently, kept human populations from expanding rapidly. On the credit side of the pesticide ledger, pesticides are vital elements in the prevention of the diseases that kill and incapacitate humans. Pesticides used to combat diseases such as malaria and sleeping sickness have become important public health tools, in addition to having an agricultural use.

However, nearly every garden shed, garage, or cellar in the industrialized countries holds scores of pesticides, herbicides, and repellents that simply protect us from the inconveniences brought on by certain pests. These uses are often trivial, but they are accompanied by significant health and environmental risks due to the nature of the chemicals themselves.

Crop Losses Due to Pests

Insects, plant pathogens, and weeds destroy an estimated 37% (before and after harvest) of potential agricultural production in the United States, at a yearly loss of $64 billion. Efforts to control these losses involve the use of 500,000 metric tons of **herbicides** (chemicals that kill plants) and **pesticides** (chemicals that kill animals and insects considered to be pests) annually, with a direct cost of $4 billion per year (Fig. 10-2). Many of the changes in agricultural technology, such as monoculture and the widespread use of genetically identical crops, that have boosted crop yields have also brought on an increase in the proportion of crops lost to pests—from 31% in the 1950s to 37% today. During these past 40 years, the use of herbicides and pesticides has multiplied manyfold, leading to a dependency that has disturbed those concerned with the indirect effects of the chemicals.

Different Philosophies of Pest Control

Medical practice employs two basic means of treating infectious diseases. One approach is to give the exposed patient a massive dose of antibiotics, hoping to eliminate the pathogen causing the problem or to stop the pathogen before it can get established. The other approach is to stimulate the patient's immune system

FIGURE 10-2
A crop being sprayed with pesticides to keep insects under control. Spraying is the basic technique still used in most control programs, although nowadays the pesticides in use do not persist in the environment for more than a week or two. (EPA photo.) Check: consortium for International Crop Protection at http://ipmwww.ncsu.edu/cicp/brochure.html

with a vaccine to produce long-lasting protection against any future invasion. In practice, both means are often used to keep a particular pathogen under control.

The same basic philosophies are used to control agricultural pests. The first is **chemical technology**. Like the use of antibiotics, chemical technology seeks a "magic bullet" that will eradicate or greatly lessen the numbers of the pest organism. Although it has had much success, this approach gives only short-term protection. Furthermore, the chemical often has side effects that are highly damaging to other organisms.

The second philosophy is **ecological pest management**. Like stimulating the body's immune system, this approach seeks to give long-lasting protection by developing control agents based on knowledge of the pest's life cycle and ecological relationships. Such agents, which may be either other organisms or chemicals, work in one of two ways. Either they are highly specific for the pest species being fought, or they manipulate one or more aspects of the ecosystem. Ecological pest management emphasizes the protection of people and domestic plants and animals from damage from pests, rather than eradication of the pest organism. Thus, the benefits of pest control can be obtained while the integrity of the ecosystem is maintained.

These two philosophies are combined in the approach called **integrated pest management** (IPM). IPM seeks to control pest populations using all suitable methods—chemical and ecological—in a way that brings about long-term management of pest populations and also has minimal environmental impact. This approach is increasing in usage, especially where pesticides are seen as undesirable because of health risks and in developing countries, where the cost of pesticides is prohibitive.

Promises and Problems of the Chemical Approach

Pesticides are categorized according to the group of organisms they kill. There are insecticides (for insects), rodenticides (for mice and rats), fungicides (for fungi), and so on. None of these chemicals, however, is entirely specific for the organisms it is designed to control; each poses hazards to other organisms, including humans. Therefore, pesticides are sometimes referred to as **biocides**, a name that emphasizes that they may endanger many forms of life.

Development of Chemical Pesticides and their Successes

Finding effective materials to combat pests is an ongoing endeavor. The early substances used (frequently referred to as **first-generation pesticides**) included toxic heavy metals such as lead, arsenic, and mercury. We now recognize that these substances may accumulate in soils and inhibit plant growth. Poisoning of animals and humans is also possible. In addition, these chemicals lost their effectiveness as pests became increasingly resistant to them. For example, in the early 1900s, citrus growers were able to kill 90% of injurious **scale insects** (minute insects that suck the juices from plant cells) by placing a tent over an infested tree and piping in deadly cyanide gas for a short time. By 1930, this same technique killed as few as 3% of the pests.

The next step had begun with the science of organic chemistry in the early 1800s. During the 19th century, chemists synthesized thousands of organic compounds, but, for the most part, these compounds sat on shelves because uses for them had not been found. By the 1930s, however, with agriculture expanding to meet the needs of a rapidly increasing population, and with first-generation (inorganic) pesticides failing, farmers were begging for new pesticides. In time, **second-generation pesticides**, as they came to be called, were found as a result of synthetic organic chemistry.

The DDT Story. In the 1930s a Swiss chemist, Paul Muller, began systematically testing some organic chemicals for their effect on insects. In 1938 he hit upon the chemical dichlorodiphenyltrichloroethane (DDT), a chlorinated hydrocarbon that had first been synthesized some 50 years before.

DDT appeared to be nothing less than the long-sought "magic bullet," a chemical that was extremely toxic to insects and yet seemed nontoxic to humans and other mammals. It was very inexpensive to produce. At the height of its use in the early 1960s, it cost no more than about 20 cents a pound. It was broad spectrum, meaning that it was effective against a multitude of insect pests. It was also persistent, meaning that it did not break down readily in the environment and hence provided lasting protection. This last attribute provided additional economy by eliminating both the material and labor costs of repeated treatments.

DDT quickly proved successful in controlling important insect disease carriers. During World War II, for example, the military used DDT to control body lice, which spread typhus fever among the men living in dirty battlefield conditions. As a result, World War II was one of the first wars in which fewer men died of typhus than of battle wounds. The World Health Organization of the United Nations used DDT throughout the tropical world to control mosquitoes and thereby greatly reduced the number of deaths caused by malaria. There is little question that DDT saved millions of lives. In fact, the virtues of DDT seemed so outstanding that Muller was awarded the Nobel Prize in 1948 for his discovery.

Postwar uses of DDT expanded dramatically. It was sprayed on forests to control defoliating insects such as the spruce budworm. It was routinely sprayed on salt marshes to deal with nuisance insects. It was sprayed on suburbs to control the beetles that spread Dutch elm disease. And, of course, DDT proved very effective in controlling agricultural insects. It was so effective, at least in the short run, that many crop yields increased dramatically. Growers could ignore other, more painstaking methods of pest control such as crop rotation and destruction of old crop residues. They could grow varieties that were less resistant, but more productive. They could grow certain crops in warmer or moister regions, where formerly the damage from pests had been devastating. In short, DDT gave growers more options for growing the most economically productive crop.

It is hardly surprising that DDT ushered in a great variety of synthetic organic pesticides. Examples and some characteristics of insecticides and herbicides are listed in Table 10-1. Not all of these are in use today; credit for this can be traced to Rachel Carson.

Silent Spring. Rachel Carson was a U.S. Fish and Wildlife Service biologist and accomplished science writer. In the 1950s, she began to read disturbing accounts in the scientific literature of DDT's effects on wildlife. Fish-eating birds were dying from DDT received through the food chain. Robins were dying as a result of eating worms from soil under trees sprayed

TABLE 10-1
Characteristics of Pesticides

Insecticides			
Type	***Examples***	***Toxicity to Mammals***	***Persistence***
Organophosphates	Parathion, malathion, phorate, chloropyrifos	High	Moderate (weeks)
Carbamates	Carbaryl, methomyl, aldicarb, aminocarb	Moderate	Low (days)
Chlorinated hydrocarbons	DDT, toxaphene, dieldrin, chlordane, lindane	Relatively low	High (years)
Pyrethroids	Permethrin, bifenthrin, esfenvalerate, decamethrin	Low	Low (days)

Herbicides		
Type	***Examples***	***Effects on Plants***
Triazines	Atrazine, simazine, cyanizine	Interfere with photosynthesis, especially in broadleaf plants
Phenoxy	2,4-D, 2,4,5,-T, methylchlorophenoxybutyrate (MCPB)	Plant hormonelike effects on actively growing tissue
Acidamine	Alachlor, Propachlor	Inhibit germination and early seedling growth
Dinitroaniline	Trifluralin, oryzalin	Inhibit cells in roots and shoots; prevent germination
Thiocarbamate	ethylpropylthiolcarbamate (EPTC), cycloate, butylate	Inhibit germination, especially in grasses

Source: U.S. Environmental Protection Agency, *Private Pesticide Applicator Training Manual*, 1993.

with DDT. Carson was finally galvanized into action by a letter from a friend distressed over the large number of birds killed when the friend's private bird sanctuary was sprayed for mosquito control. By 1962, she had finished a book-length documentation of the effects of the almost uncontrolled use of insecticides across the United States. The book, *Silent Spring*, became an instant best-seller. Its basic message was that if insecticide use continued as usual, there might some day come a spring with no birds—and with ominous impacts on humans as well.

Silent Spring triggered a national debate that continues today. It was immediately attacked by representatives of the agricultural and chemical industries as an unreasonable and unscientific account that, if taken seriously, they claimed it would lead to a halt of human progress. On the other side, however, the book was hailed as an unparalleled breakthrough in environmental understanding. Thirty years later, Rachel Carson is credited with the creation of the Environmental Protection Agency (EPA), which eventually banned DDT, and with stimulating the start of the environmental movement. *Silent Spring* has become a classic, and the regulation of insecticides and other toxic chemicals is in a sense a monument to Rachel Carson, who died of cancer only two years after her book was published.

Problems Stemming From Chemical Pesticide Use

Problems associated with synthetic organic pesticides can be placed in three categories:

- Development of resistance by pests
- Resurgences and secondary pest outbreaks
- Adverse environmental and human health effects

Development of Resistance by Pests. The most fundamental problem for growers is that chemical pesticides gradually lose their effectiveness. Over the years, it may become necessary to use larger and larger quantities, new and more potent pesticides, or both to obtain the same degree of control. Synthetic organic pesticides fared no better than first-generation pesticides in this respect. For example, in 1946, 1 kg (2.2 lbs) of pesticides provided enough protection to produce about 60,000 bushels of corn. By 1971, 64 kg (141 lbs) were used for the same production, and losses due to pests actually *increased* during the intervening years.

Resistance builds up because pesticides destroy the sensitive individuals of a pest population, while the more resistant individuals continue to breed, creating a new population of more resistant pests (Fig. 10-3).

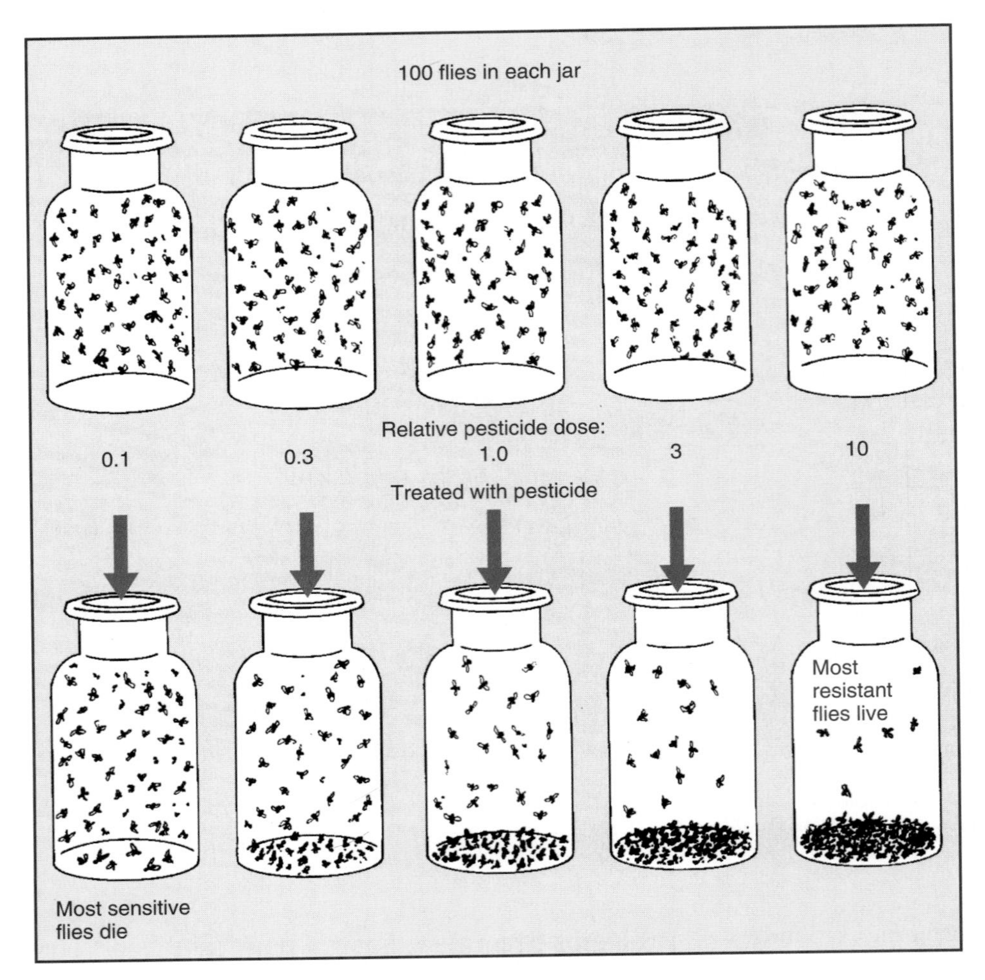

FIGURE 10-3
Genetic variation in resistance to pesticides exists in all populations. Here, the flies that were most sensitive to the pesticide died with the smallest amount of pesticide (0.1 relative dose). The flies still alive after even the heaviest dosage (10 relative dose) was applied are the most resistant members of the population. These few surviving individuals will pass on their resistance to future generations, which will be harder than ever to kill with pesticides.

Resistant insect populations develop rapidly because insects have a phenomenal reproductive capacity. A single pair of houseflies, for example, can produce several hundred offspring that may mature and reproduce themselves only two weeks later. Consequently, repeated pesticide applications result in the unwitting selection and breeding of genetic lines that are highly, if not totally, resistant to the chemicals that were designed to eliminate them. Cases have been recorded in which the resistance of a pest population has increased as much as 25,000-fold.

Over the years of pesticide use, the number of resistant species has climbed steadily. Many major pest species are resistant to all of the principal pesticides; at least 520 insects and arachnids have shown resistance to pesticides (Fig. 10-4). Interestingly, as a pest population becomes resistant to one pesticide, it also may gain resistance to other, unrelated pesticides, even though it has not been exposed to the other chemicals. (See "Earth Watch" box, p. 243.)

FIGURE 10-4
Number of species resistant to pesticides, 1908–95. Using pesticides causes selection (survival of the fittest) for those individuals that are resistant. This graph shows how an increasing number of species became resistant during the first four decades of organic pesticide use. As we continue to use pesticides, we breed insects and other pests that are increasingly resistant to the pesticides used against them. (From WorldWatch, *State of the World*, 1996.)

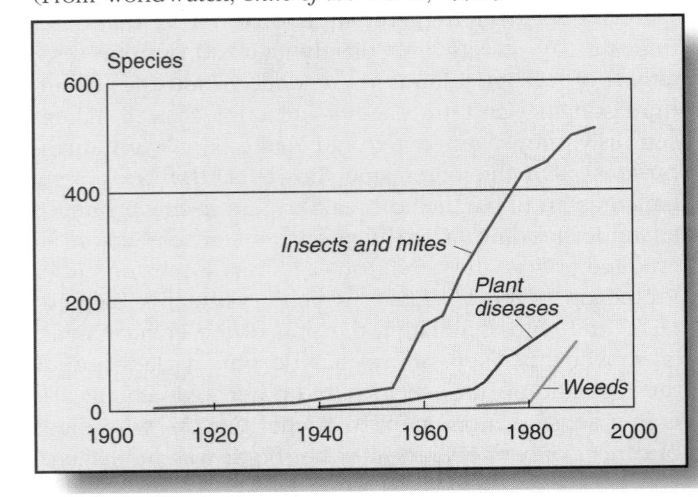

Resurgences and Secondary Pest Outbreaks. The second problem with the use of synthetic organic pesticides is that after a pest has been virtually eliminated

Earth Watch

The Ultimate Pest?

If we were to conjure up the ultimate insect pest, we might imagine the following characteristics. The ultimate insect pest would:

1. attack a broad variety of plants and fruits,
2. be resistant to all of the usual pesticides,
3. be highly prolific and have a rapid life cycle,
4. lack natural predators and parasites, and
5. become a nuisance in other ways (it would interfere with breathing, cover car windshields, and so forth).

Such a pest, if unleashed on agriculture, could cause millions of dollars in crop losses and devastate many farmers.

Unfortunately, recent news from Texas suggests that something very close to the ultimate pest might already be with us. The sweet potato whitefly (*Bemisia...*) is a tiny white insect that emerged from Florida's poinsettia greenhouses in 1986 and has become established as far away as New England as of June 1996. It has all the characteristics listed and has been dubbed the "Superbug" by farmers who have encountered it. It is known to eat at least 500 species of plants—just about everything except asparagus and onions. It thrives on roadside weeds. Total crop losses in 1991 to this insect were above $500 million and it continues to cause extensive damage. The insects swarm all over the plants, sucking them dry and leaving them withered and rotten. Pesticides have proven useless so far. Swarms become so dense that they interfere with breathing and vision. Entomologists are frantically searching for a natural predator or other pesticides to bring this pest under control. Such natural enemy research continues in 25 countries. Unfortunately, no one knows where the insect originated. Unless some breakthrough occurs, we all may be eating more onions than ever before. Check: http://ceris.purdue.edu/napis/pests/swf/index.html

---

with a pesticide, the pest population not only recovers, but explodes to higher and more severe levels. This phenomenon is known as a **resurgence**. To make matters worse, small populations of insects that were previously of no concern because of their low numbers suddenly start to explode, creating new problems. This phenomenon is called a **secondary pest outbreak**. For example, with the use of synthetic organic pesticides, mites have become a serious pest problem, and the number of serious pests on cotton has increased from 6 to 16.

At first, pesticide proponents denied that resurgences and secondary pest outbreaks had anything to do with the use of pesticides. However, careful investigations have shown otherwise, as with the brown planthopper and rice in Indonesia. Resurgences and secondary pest outbreaks occur because the insect world is part of a complex food web. Populations of plant-eating insects are frequently held in check by other insects that parasitize or prey on them (Fig. 10-5). Pesticide treatments often have a greater impact on these natural enemies than on the plant-eating insects they are meant to control. Consequently, with natural enemies suppressed, both the population of the original target pest and populations of other plant-eating insects explode.

To illustrate the seriousness of resurgences and secondary pest outbreaks, a recent study in California listed a series of 25 major pest outbreaks, each of which caused more than $1 million worth of damage. All but one involved resurgences or secondary pest outbreaks. Of course, the species appearing in secondary outbreaks quickly became resistant to pesticides, thus compounding the problem.

The chemical approach fails because it is contrary to basic ecological principles. It assumes that the ecosystem is a static entity in which one species, the pest, can simply be eliminated. In reality, the ecosystem is a dynamic system of interactions, and a chemical assault on one species will inevitably upset the system and produce other undesirable effects. The path to sustainability demands that we understand how ecosystems work and adapt our interventions accordingly.

Adverse Environmental and Human Health Effects. Perhaps of greatest concern to most people is the potential for adverse effects to human and environmental health. The story of DDT, used so widely during the 1940s and 1950s, illustrates the hazards.

In the 1950s and 1960s, ornithologists (people who study birds) observed drastic declines in populations of many species of birds that fed at the tops of food chains. Fish-eating birds such as the bald eagle and osprey (Fig. 10-6) were so affected that their extinction was feared. Investigators at the U.S. Fish and Wildlife National Research Center near Baltimore, Maryland, showed that the problem was reproductive failure; eggs were breaking in the nest before hatching. They also showed that the fragile eggs contained high concentrations of dichlorodiphenyldichloroethylene (DDE), a product of the partial breakdown of DDT by the body. DDE interferes with calcium metabolism,

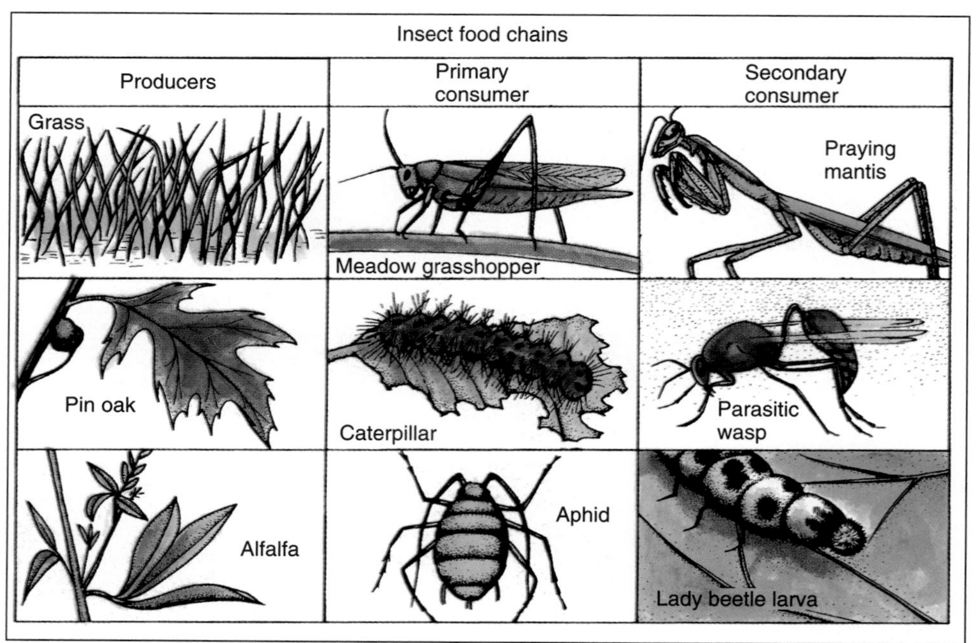

FIGURE 10-5
(a) Food chains exist among insects just as they do among higher animals. (b) Aphid lion (on right) impaling and eating a larval aphid. (Courtesy of David Pimentel, Cornell University.)

FIGURE 10-6
Bald eagle. Populations of fish-eating birds like the osprey, brown pelican, and bald eagle were decimated in the 1950s and 1960s by the effects of widespread spraying with DDT. With the banning of DDT, these populations have greatly recovered. The bald eagle was taken off the endangered species list in 1994. (Johnny Johnson/Animals Animals)

causing birds to lay thin-shelled eggs. Further study revealed that birds were acquiring high levels of DDT and DDE by **biomagnification**, the process of accumulating higher and higher doses through the food chain. (See p. 349.)

In addition, tissue assays showed that DDT was accumulating in the body fat of humans and virtually all other animals, including Arctic seals and Antarctic penguins, even though those animals were far removed from any point of DDT application. Although the evidence on DDT's health impacts is not clear, various chlorinated hydrocarbons are carcinogenic (cause cancer), mutagenic (cause mutations), and teratogenic (cause birth defects). Significantly, a recent study has shown a strong association between breast cancer in women and high DDE levels in blood serum.

Concerns about environmental and long-term health effects led to the banning of DDT in the United States and most other industrialized countries in the early 1970s. Numerous other related chlorinated hydrocarbon pesticides (e.g., chlordane, dieldrin, endrin, and heptachlor) were also banned because of their propensity for bioaccumulation in the environment and suspected potential for causing cancer.

In the years since the banning of DDT, observers have noted a marked recovery in the populations of birds that were adversely affected. However, this does not mean that the situation is under control. Because of increasing resistance, resurgences, and secondary pest outbreaks, the kinds and quantities of pesticides in use continue to grow. Approximately 70% of all cropland in the United States receives some pesticide application and for row crops use increases to 93% of that land. Globally, over 3 million tons of pesticides are used annually, but this level may decrease as more and more farmers turn to biological controls. DDT is still widely used in developing countries against malaria-bearing mosquitoes.

Many pesticides in use are toxic to people and are responsible for poisoning an estimated 45,000 persons annually in the United States. Most of the victims are farm workers or employees of pesticide companies who come in direct contact with the chemicals. Some 3000 of these victims require hospitalization, and approximately 50 of them die each year from pesticide toxicity. The World Health Organization conducted a survey to determine pesticide poisoning in humans and estimates that there are roughly 400,000 cases of acute occupational pesticide poisoning each year in developing countries of which thousands result in death. Use by untrained persons is considered to be the major cause of these poisonings.

The Pesticide Treadmill. The late entomologist Robert van den Bosch coined the term **pesticide treadmill** to describe attempts to eradicate pests with synthetic organic chemicals. It is an apt term: The chemicals do not eradicate the pests; they increase resistance and secondary pest outbreaks, which lead to the use of new and larger quantities of chemicals, which in turn lead to more resistance and more secondary outbreaks, and so on. The process is an unending cycle constantly increasing the risks to human and environmental health and is clearly not sustainable (Fig. 10-7).

FIGURE 10-7
The pesticide treadmill. The use of chemical pesticides aggravates many pest problems. The continued use of these products demands ever-increasing dosages of pesticides, which further aggravate pest problems and produce more contamination of foodstuffs and ecosystems.

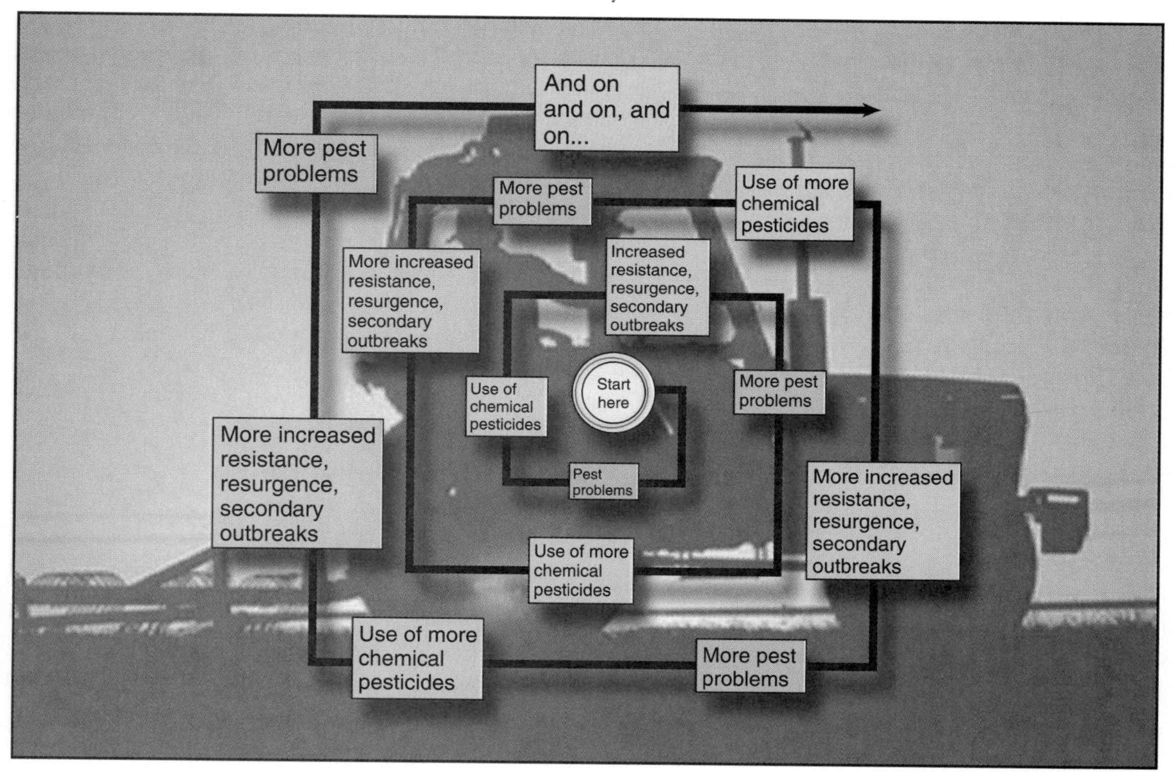

Nonpersistent Pesticides: Are They the Answer?

A key characteristic of chlorinated hydrocarbons is their persistence, that is, their slowness to break down. Lingering in the environment for years, they can contaminate many organisms and become biomagnified. This persistence results because these chemicals have such complex structures that soil microbes are unable to metabolize them. DDT, for example, has a half-life on the order of 20 years. That is, half the amount applied is still present and active 20 years later; after 20 more years, half of that amount, or one-quarter of the amount applied, remains, and so on.

Recognizing this persistence factor, the agrochemical industry has in large measure substituted *nonpersistent* pesticides for the banned compounds. For example, the synthetic organic phosphates malathion and parathion, as well as carbamates such as aldicarb and carbaryl, are now used extensively in place of chlorinated hydrocarbons (see Table 10-1). These compounds break down into simple nontoxic products within a few weeks after application. Thus, there is no danger of their migrating long distances through the environment and affecting wildlife or humans long after being applied. When used wisely, these pesticides are an important tool for farmers and gardeners. For several reasons, however, nonpersistent pesticides are not as environmentally sound as they might appear.

First of all, total environmental impact is a function not only of persistence, but also of three other important factors: toxicity, dosage applied, and location where applied. Many of the nonpersistent pesticides are far more toxic than DDT. This higher toxicity, combined with the frequent applications needed to maintain control, presents a significant hazard to agricultural workers and others exposed to these pesticides.

Second, nonpersistent pesticides may still have far-reaching environmental impacts. For example, to control outbreaks of the spruce budworm in New Brunswick, Canada, forests were sprayed with a nonpersistent organophosphate pesticide that was promoted as environmentally safe. After spraying, however, an estimated 12 million birds died. These birds may have died by direct poisoning or by the loss of their food supply, since a bird eats nearly its own weight in insects each day. In either case, after the spraying there was an eerie silence, and numerous dead warblers littered the ground.

Third, desirable insects may be just as sensitive as pest insects to these substances. Bees, for example, which play an essential role in pollination, are highly sensitive to nonpersistent pesticides. Thus, use of these compounds creates an economic problem for beekeepers, as well as jeopardizing pollination. Regular spraying of neighborhoods with malathion to control mosquitoes leads inevitably to great declines in butterflies and fireflies.

Finally, nonpersistent chemicals are just as likely to cause resurgences and secondary pest outbreaks as are persistent pesticides, and pests become resistant to nonpersistent chemicals just as quickly as they do to persistent pesticides.

Alternative Pest Control Methods

Numerous ecological and biological factors affect the relationship between a pest and its host. Ecological pest management seeks to manipulate one or more of these natural factors so that crops are protected without upsetting the rest of the ecosystem or jeopardizing environmental and human health. Since ecological pest management involves working with natural factors instead of synthetic chemicals, the techniques are referred to as **natural control** or **biological control** methods. This natural approach, unlike the chemical technology approach, depends on an understanding of the pest and its relationship with its host and with its ecosystem. The more we know about the organisms involved, the greater are our opportunities for natural control.

To illustrate, the life cycle typical of moths and butterflies is shown in Fig. 10-8. Many groups of insects have a similarly complex life cycle. The development of each stage may be influenced by numerous abiotic factors, and at each stage the insect may be vulnerable to attack by a parasite or predator. Proper completion of each stage depends on internal chemical signals provided by hormones. Locating mates, finding food, and other behaviors depend on external chemical signals. All these findings suggest ways in which pest populations may be controlled without resorting to synthetic chemical pesticides.

The four general categories of natural or biological pest control are:

- Cultural control
- Control by natural enemies
- Genetic control
- Natural chemical control

Cultural Control

A cultural control is a nonchemical alteration of one or more environmental factors in such a way that the pest

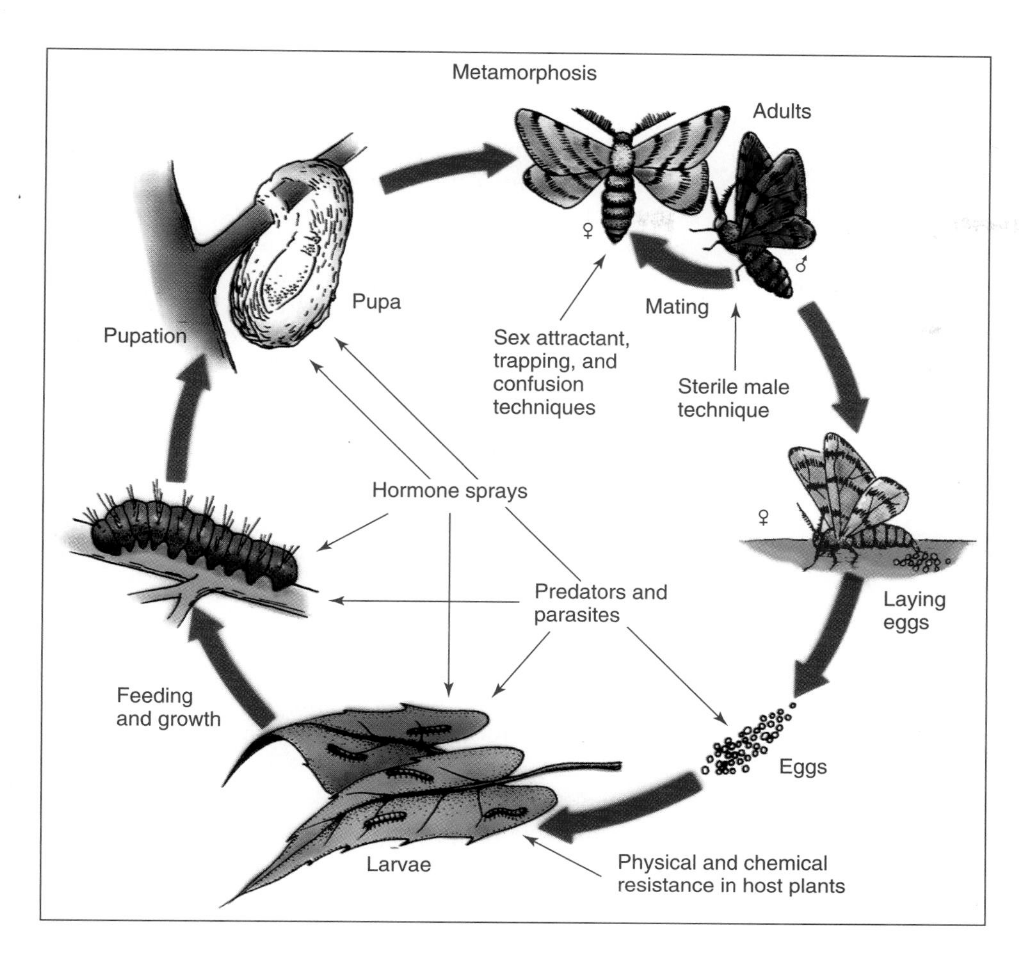

FIGURE 10-8
Life cycle of moths. Most insects have a complex life cycle that includes a larval stage and an adult stage. Biological control methods recognize the different stages and attack the insect, using knowledge of its needs and life cycle.

finds the environment unsuitable or is unable to gain access to it.

Cultural Control of Pests Affecting Humans. We routinely practice many forms of cultural control against diseases and parasitic organisms. Some of these practices are so familiar and well entrenched in our culture, that we no longer recognize them for what they are. For instance, disposing properly of sewage and avoiding drinking water from unsafe sources are cultural practices that protect against waterborne disease-causing organisms. Combing and brushing the hair, bathing, and wearing clean clothing are cultural practices that eliminate head and body lice, fleas, and other parasites. Regular changing of bed linens protects against bedbugs. Properly and systematically disposing of garbage and keeping a clean house with any cracks sealed and with good window screens are effective methods for keeping down populations of roaches, mice, flies, mosquitoes, and other pests. Sanitation requirements in handling and preparing food are cultural controls designed to prevent the spread of diseases. Refrigeration, freezing, canning, and drying of foods are cultural controls that inhibit the growth of organisms that cause rotting, spoilage, and food poisoning.

If such practices of personal hygiene and sanitation are compromised, as they usually are in any major disaster, there is very real danger of additional widespread mortality resulting from outbreaks of parasites and diseases. Also, where these practices are not broadly pursued in a society, as in some less developed countries, sickness and death are the consequences. The use of contaminated water in developing countries is responsible for infection by diseases that kill millions and make more than a billion people sick each year.

Cultural Control of Pests Affecting Lawns, Gardens, and Crops. Homeowners are prone to use excessive amounts of pesticides to maintain a weed-free lawn. Weed problems in lawns are frequently a result of cutting the grass too short. If grass is left at least 3 inches (8 cm) high, it will usually maintain a dense enough cover to keep out crabgrass and many other

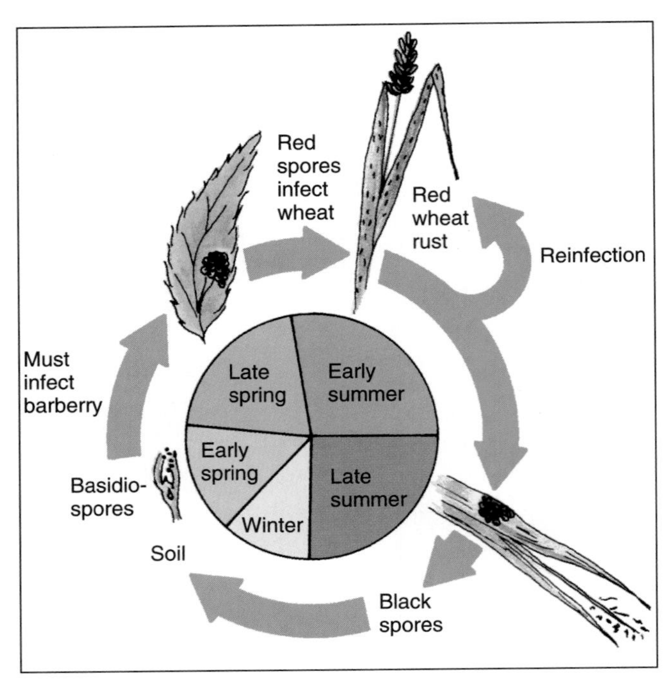

FIGURE 10-9
The life cycle of wheat rust, a parasitic fungus that is a serious pest on wheat. Since part of the life cycle requires that the rust infest barberry, an alternative host, the elimination of barberry in wheat-growing regions has been an important cultural control (along with the development of resistant varieties of wheat).

noxious weeds. Thus, monitoring the height of grass is a form of cultural weed control. Also, many homeowners allow a diversity of plants in their lawns, tolerating a variety of plants some people might consider weeds.

Some plants are particularly attractive to certain pests; others are especially repugnant. In either case, the effect may spill over to adjacent plants. A gardener may control many pests by paying careful attention to eliminating plants that act as attractants (for example, roses) and growing those that act as repellents. Marigolds and chrysanthemums are justly famous for being insect repellents.

Some parasites require an alternative host. They can be controlled by eliminating this host (Fig 10-9). Also, hedgerows, fencerows, and shelterbelts can provide refuges where natural enemies of pests (birds, amphibians, preying mantises, and so on) can be maintained.

Management of any crop residues not harvested is important. Spores of plant disease organisms and insects may overwinter or complete part of their life cycle in the dead leaves, stems, or other plant residues that remain in the fields after harvesting. Plowing under or burning the material may be very effective in keeping pest populations to a minimum. In gardens, a clean mulch of material such as grass clippings or hay will keep down the growth of weeds and protect the soil from drying and erosion.

Growing the same crop on a plot of land year after year keeps the pest's food supply continuously available. Crop rotation, the practice of changing crops from one year to the next, may provide control because pests of the first crop cannot feed on the second crop and vice versa. Crop rotation is especially effective in controlling root nematodes (roundworms that live in the soil and feed on roots) and other pests that do not have the ability to migrate appreciable distances.

For economic efficiency, agriculture has moved progressively toward monoculture—the Corn Belt, Cotton Belt, and so on. Recall the fourth principle of sustainability: Biodiversity provides stability. (Conversely, a simple system is ecologically unstable.) When a pest outbreak occurs, monoculture is most conducive to its rapid multiplication and spread. Other natural controls, even if present, may be overwhelmed by the avalanche of spreading pests. On the other hand, the spread of a pest outbreak is impeded and other natural controls may be more effective if there is a mixture of crop species, some of which are not vulnerable to attack. One approach that works well in the British Isles is to intersperse cultivated with uncultivated strips that are not treated with pesticides. Natural enemies of pests are maintained in the uncultivated strips.

Most of our pests that are hardest to control were unwittingly imported from other parts of the world, and we realize that many other species in other regions would be serious pests if introduced here. Therefore, it is important to keep would-be pests out of the country. This is a major function of the U.S. Customs Bureau and of the agriculture departments of some states. Biological materials that may carry pest insects or pathogens either are prohibited from crossing the border or are subjected to quarantines, fumigation, or other treatments to ensure that they are free of pests. The cost of such procedures is small in comparison to the costs that could be incurred if these pests gained entry and became established.

Control by Natural Enemies

The following examples illustrate the range of possibilities for controlling pests with natural enemies:

- Scale insects, potentially devastating to citrus crops, have been successfully controlled by vedalia (ladybird) beetles, which feed on them.
- Various caterpillars have been controlled by parasitic wasps (Fig. 10-10).
- Japanese beetles are controlled in part with the bacterium *Bacillus thuringiensis* (Bt), which pro-

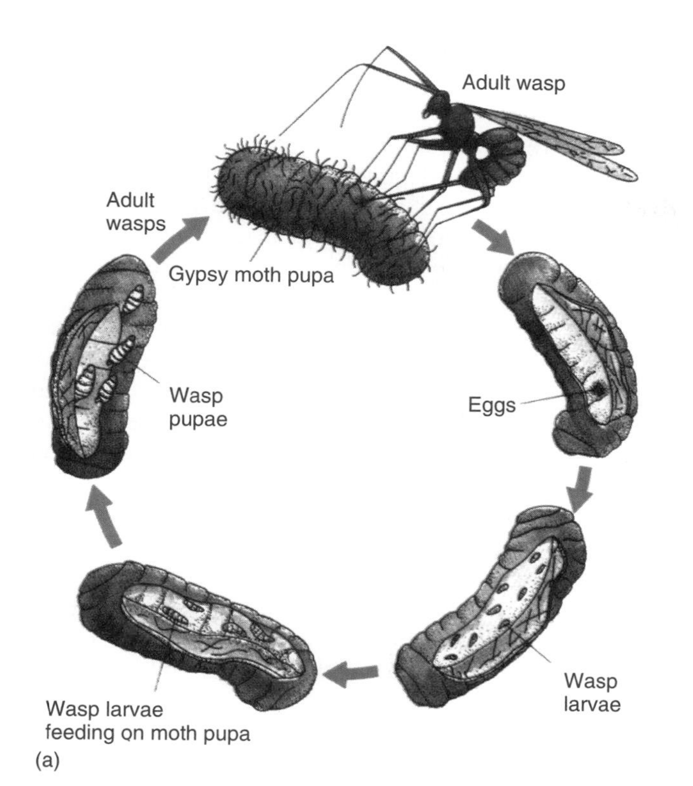

(a)

(b)

(c)

FIGURE 10-10
Parasitic wasps used to control caterpillars. (a) The life cycle of the parasitic wasp that uses the gypsy moth as its host. (b) A wasp depositing eggs in a gypsy moth larva (Dr. E. R. Degginger.) (c) Another insect parasite, a braconid wasp, lays its eggs on the pest known as the tomato hornworm, which is the larva of the sphinx moth. The wasp larvae feed on the caterpillar, and shortly before the caterpillar dies, they emerge and form the cocoons seen here. (Harry Rogers/Photo Researchers, Inc.)

duces toxic crystals when soil containing the bacterium is ingested by the larvae.

- Prickly pear cactus and numerous other weeds have been controlled by plant-eating insects (Fig. 10-11). More than 30 weed species worldwide are now limited by insects introduced into their habitats.
- Rabbits in Australia are controlled by an infectious virus.
- Aquatic weeds in the American Midwest are controlled by a vegetarian fish native to South America.
- Mealybugs in Africa are controlled by a parasitic wasp. (See "Global Perspective" box, p. 251.)

The problem with using natural enemies is finding organisms that provide control of the target species without attacking other, desirable species. Entomologists estimate that, of the 50,000 known species of plant-eating insects that have the potential of

(a)

(b)

FIGURE 10-11
Biological control of the weed, prickly pear cactus, by a cactus-eating moth in Queensland, Australia. (a) The land shown here, once settled, had to be abandoned because of prickly pear infestation. (b) Same land reoccupied following destruction of prickly pear by the moth. (Department of Lands, Queensland, Australia.)

being serious pests, only about 1% actually are. The populations of the other 99% are held in check by one or more natural enemies, so they do not do significant amounts of damage. Therefore, the first step in using natural enemies for control should be *conservation*—protecting the natural enemies that already exist. Conservation means avoiding the use of broad-spectrum chemical pesticides, which may affect natural enemies even more than they do the target pests. Eliminating or placing considerable restrictions on the use of broad-spectrum chemical pesticides will, in many cases, allow natural enemies to reestablish themselves and control secondary pests, those that became serious problems only after the use of pesticides.

However, effective natural enemies are not always readily available. In some cases, the lack of natural enemies is the result of accidentally importing the pest without its natural enemy. Quite often, effective natural enemies have been found by systematically combing the home region of an introduced pest and finding its various predators or parasites. The advantage of this approach is the great specificity of the natural enemy for its target. Of more than 100 insects introduced to new regions for weed management, none has switched its diet.

Yet the potential for utilizing natural enemies has barely been tapped. Of more than 2000 serious insect pest species, 90% remain for which effective natural enemies have not been found, mainly because no one has looked for such enemies. Finding suitable natural enemies is costly, and the profit margin for this kind of work is not as great as it is for research into pesticides. Fortunately, the U.S. Department of Agriculture (USDA) is increasing its funding of research into biological control agents. After a potential natural enemy is located, it must be propagated and carefully tested before it is released, to be sure that it will not harm other organisms. This testing often takes many years of painstaking research. When effective natural enemies are found, however, they can provide control indefinitely, saving millions of dollars per year without further expense.

Genetic Control

Most plant-eating insects and plant pathogens attack only one species or a few closely related species. This specificity implies a genetic incompatibility between the pest and species that are not attacked. Most genetic control strategies are designed to develop genetic traits in the host species that provide the same incompatibility—that is, resistance to attack by the pest. This technique has been utilized extensively in connection with plant diseases—fungal, viral, and bacterial parasites. For example, in the years 1845–47 the potato crop in Ireland was devastated by late blight, a fungal parasite. Nearly a million people starved, and another million people emigrated to escape the same fate. Nowadays, protection against such disasters is provided in large part by growing varieties of potato that are resistant to the blight. It is no overstatement to say that the world owes much of its production of potatoes, corn, wheat, and other cereal grains to the painstaking work of plant geneticists who selected and bred disease-resistant varieties.

The same potential exists for breeding plants that are resistant to insect pests. Traits that provide resistance may act as chemical barriers or physical barriers.

Control with Chemical Barriers. A chemical barrier is some chemical produced by the plant we want to protect; the substance is lethal or at least repulsive to the

Global Perspective

Wasps 1, Mealybugs 0

The cassava (manioc) plant originated in South America, but has been cultivated throughout the tropical world. Currently, it is the primary food for more than 200 million people in sub-Saharan Africa, one third of the human population in that region. It is a high-yielding crop that requires no modern technology and, accordingly, is raised by subsistence farmers everywhere.

An insect not previously seen in Africa, the mealybug, appeared in Congo in the early 1970s and spread across sub-Saharan Africa, leaving a trail of ruined harvests and hunger in its wake. Zaire and Congo, unable to afford pesticides, turned for help to the International Institute for Tropical Agriculture in Nigeria. This group formed the Biological Control Program and began to look for natural enemies of the mealybug.

Returning to the land of origin of the cassava, a researcher found the mealybug in Paraguay and observed that the insect was kept under control by natural predators and parasites. After extensive testing, researchers identified a parasitic wasp, *Epidinocarsis lopezi*, as the prime candidate for controlling the bugs. The wasp is no larger than a comma on this page. The female wasp seeks out mealybugs, paralyzes them with a sting, and deposits eggs that hatch inside the mealybug and eat their way through the insect.

Once this wasp-bug relationship was identified, the wasps were reared by the thousands on captive mealybugs and then spread across the continent over a period of eight years. The Biological Control Program has trained 400 workers to monitor the progress of the program, and all indications are that the battle has been won. It is estimated that every dollar invested in the control program has yielded $149 in crops saved from destruction. The best news is that the control is permanent and does not require the repeated application of expensive and environmentally damaging pesticides. In the wake of the project, national biological control programs have been established and are now focusing on pests of other crops of sub-Saharan Africa. (Reported by Jane Ellen Stevens, *The Boston Globe*, Jan. 3, 1994.) Check:http//www.ctpm.uq.edu.aa/programs/biocontrol.html

would-be pest. Once they have identified such a barrier, plant breeders can use selection and crossbreeding to enhance this trait in the desirable plant. The relationship between wheat and the Hessian fly provides an example. This fly lays its eggs on wheat leaves, and the larvae move down the leaves and into the main stem as they feed. The weakened stem either dies or is broken in the wind. The Hessian fly was introduced into the United States in the straw bedding of Hessian soldiers during the Revolutionary War. The fly eventually spread throughout much of the Midwest, causing widespread devastation until scientists at the University of Kansas developed a variety of wheat that produces a chemical toxic to the fly; it kills the larvae when they feed on the leaves.

Increasing resistance through breeding may not provide 100% protection, but even partial protection can make the difference between profit and loss for the grower. In addition, any degree of resistance lessens the need for chemical pesticides.

Control with Physical Barriers. Physical barriers are structural traits that impede the attack of a pest. For example, leafhoppers are significant worldwide pests of cotton, soybeans, alfalfa, clover, beans, and potatoes, but they can damage only plants with relatively smooth leaves. Hooked hairs on the leaf surfaces of some plants tend to trap and hold immature leafhoppers until they die. Similarly, alfalfa weevil larvae are fatally entrapped by glandular hairs that exude a sticky substance. Such traits can be enhanced in vulnerable plants through selective breeding.

Control with Sterile Males. Another genetic control strategy involves flooding a natural population with sterile males that have been reared in laboratories. Combating the screwworm fly provides a prime illustration. This fly, which is closely related to the housefly and looks much like it, lays its eggs in open wounds of cattle and other animals. The larvae feed on blood and lymph, keeping the wound open and festering. Secondary infections frequently occur and often lead to the death of the animal.

Early in the century, this problem became so severe that cattle ranching from Texas to Florida and northward was becoming economically impossible. In studying the situation during the 1940s, Edward Knipling, an entomologist with the USDA, observed two essential features of screwworm flies: (1) Their populations are never very large, and (2) the female fly mates just once, lays her eggs, and then dies. Knipling reasoned that if the female mated with a sterile male, no offspring would be produced. His hypothesis was correct, and today sterile males are routinely used to control this pest. After huge numbers of screwworm larvae are grown on meat in laboratories, the resulting pupae are subjected to just enough high-energy radiation to render them sterile. These sterilized pupae are then air-dropped into the infested area. Ideally, 100 sterile males are dropped for every normal female in the natural population, giving a 99% probability that wild females will mate with one of the sterile males.

This technique proved so successful that it eliminated the screwworm fly from Florida in 1958–59, and it continues to be used to control the problem in the Southwest to this day. The savings to the cattle industry are estimated at more than $300 million a year. The technique has also been used to eradicate infestations of imported pests before they could gain a strong foothold in a particular ecosystem or crop. Populations of insects used in this manner are maintained in facilities around the world so that sterile males may be called up on very short notice if the need arises. As a recent example, 1.3 billion sterilized male screwworm flies, released by FAO planes into Libya in 1991, successfully eradicated the flies from that region. If the fly had moved south of the Sahara, it could have had devastating effects on native animals and livestock, threatening food supplies there.

FIGURE 10-12
A field trial of a bioengineered potato plant. The center row shows potato plants which have the Bt gene incorporated into the plant genome. These plants are protected from the Colorado potato beetle, while surrounding rows show typical beetle devastation of non-engineered potato plants. (NatureMark Potatoes, a Unit of Monsanto Company.)

Strategies Using Biotechnology. Biotechnology has recently multiplied the potential for genetic control. More complex than basic plant breeding, genetic engineering makes it possible to introduce genes into crop plants from a variety of sources: other plant species, bacteria, and viruses. The new transgenic crops are undergoing rapid development and testing; more than 40 species of food plants have been genetically engineered and by mid-1996, 25 transgenic crops had received regulatory approval. One promising strategy is to incorporate the protein coat of a plant virus into the plant itself. When the plant "expresses" (that is, manufactures) the virus's protein coat, it becomes resistant to infection by the real virus. In this way, crop plants have been made resistant to more than a dozen plant viruses. One negative consequence is the concern that virus-resisting genes may spread to unwanted plant relatives of the crop plant, creating "superweeds."

Resistance to insects has been engineered with the use of a potent protein produced by the bacterium, *Bacillus thuringiensis* (Bt). This protein kills the larvae of a number of plant-eating insects and is harmless to mammals, birds, and most other insects. Scientists have engineered the gene into a number of plants, including cotton, potatoes, and corn (Fig. 10-12). This development alone is expected to cut in half the use of pesticides on cotton—a crop that consumes more than 10% of the pesticides used throughout the world. The protection comes from a naturally occurring protein, part of a family used for three decades by home gardeners, organic growers, and other farmers.

One unusual genetic-engineering strategy has been to give a crop species resistance to a broad-spectrum herbicide. Although this appears to encourage the use of herbicides, it is actually expected to make it possible to use less total amounts of herbicide and reduce the number of applications per season on a given acre, and to use compounds that are more environmentally benign. One such herbicide is glyphosate (Roundup®). It inhibits an amino acid pathway that only occurs in plants and not in animals, biodegrades rapidly, and can control both grasses and broadleaf plants. (Although it is nonspecific and will kill every plant, so application and clean up must be done carefully.) Genes resistant to glyphosate have been successfully introduced into cotton, canola, soybean, and corn. As a variation on this theme, genes for resistance to pesticides can be introduced into important insect predators. Farmers can then continue to use pesticides against the pest when needed, but can count on the predators being present to prevent resurgences.

Unfortunately, pests may develop the ability to overcome controls—chemical, physical, and biotechnological—in the same way they develop resistance to pesticides. This means that scientists must continually develop new resistant varieties of plants or animals to substitute for old varieties. This substitution has occurred seven times in the case of wheat and the Hessian fly. It often takes place without the public ever knowing that a potential catastrophe is being averted.

Natural Chemical Control

As in humans and other animals, each stage in the development of an insect is controlled by hormones—

FIGURE 10-13
An example of a trapping technique. In this case, adult Japanese beetles are lured into a trap by a scented bait that they find attractive. Once inside the trap, baffles prevent the beetles from escaping. These traps are available in hardware stores and are quite effective. (Jack Dermid/Photo Researchers, Inc.)

chemicals that are produced in the organism and that provide "signals" which control developmental processes and metabolic functions. In addition, insects produce many pheromones—chemicals secreted by one individual that influence the behavior of another individual of the same species.

The aim of natural chemical control is to isolate, identify, synthesize, and then use an insect's own hormones or pheromones to disrupt its life cycle. Two advantages of natural chemicals are that they are highly specific to the pest in question (they do not affect the natural enemies of the pest to any appreciable extent) and that they are nontoxic. If the affected pest is eaten by another organism, it is simply digested. Ways in which natural chemicals may be used are illustrated by the following discussion.

Scientists have discovered that caterpillar pupation is triggered by a decrease in the level of the chemical called juvenile hormone. If this chemical is sprayed on caterpillars, pupation does not occur. The larvae simply continue to feed and grow, become grossly oversized, and eventually die. One newly developed insecticide, called Mimic, is a synthetic variation of ecdysone, the molting hormone of insects. Mimic begins the molting process in insect larvae, but doesn't complete it. As a result, the larva is trapped in its old skin and eventually starves to death. Mimic is specific to moths and butterflies and is viewed as a potent agent in the future control of the spruce budworm, the gypsy moth, the beet army worm, and the codling moth—all highly devastating pests.

Adult female insects secrete pheromones that attract males for the function of mating. Once identified and synthesized, these pheromones may be used in either of two ways: the trapping technique or the confusion technique. In the trapping technique, the pheromone is used to lure males into traps or into eating poisoned bait (Fig. 10-13). In the confusion technique, the pheromone is dispersed over the field in such quantities that males become confused, cannot find the females, and thus fail to mate.

The enormous potential of natural chemicals for controlling insect pests without causing ecological damage or disruption has been recognized for at least 30 years. However, the background research and necessary testing are just now reaching fruition. Tests are extremely promising. It appears that the great expectations in this field are about to be realized, as over 800 natural chemicals have been identified, and more than 250 are being produced commercially.

Socioeconomic Issues in Pest Management

Pressures to Use Pesticides

With any method of pest control, it is important to keep in mind that a species becomes a pest only when its population multiplies to the point of causing significant damage. Natural controls are generally aimed at keeping pest populations below damaging levels, not at total eradication of the populations. By keeping pest populations down, natural controls avert significant damage while preserving the integrity and balance of the ecosystem.

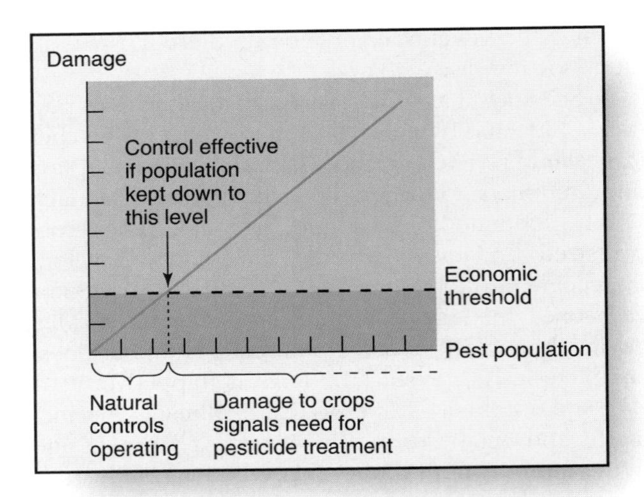

FIGURE 10-14
The economic threshold. The objective of pest control should not be to eradicate a pest totally. All that is needed is to keep population levels below the economic threshold. When the population rises into the yellow part of the graph, damage to crops is significant and pesticides are needed. Any time the population is in the blue area, however, natural controls are succeeding in keeping the pest population down; therefore, no pesticides are needed.

Therefore, the question to be asked in facing any pest species is, Is it causing significant damage? Damage should be deemed significant only when the economic losses due to the damage considerably outweigh the cost of applying a pesticide. This point is called the **economic threshold** (Fig. 10-14). If significant damage is not occurring, natural controls are already operating, and the situation is probably best left as is. Spraying with synthetic chemicals at this stage is more than likely to upset the natural balance and make the situation worse through resurgences. On the other hand, if significant damage *is* occurring, a pesticide treatment may be in order.

It makes little difference whether the threat of loss as a result of pest infestation is real or imagined, close at hand or remote; what is important is how the grower perceives the threat. Even if there is no evidence of immediate damage from a pest, a grower who believes that his or her plantings are at risk is likely to resort to **insurance spraying**, the use of pesticides "just to be safe."

Consumers also put indirect pressure on growers to use pesticides. From customers to supermarket chains to canneries, there is a tendency to select the best looking fruits and vegetables, leaving the remainder to be sold at lower prices or trashed. Like Snow White, we all tend to reach for the apparently perfect, unblemished apple. Growers know that blemished produce means less profit, so they indulge in **cosmetic spraying**—the use of pesticides to control pests that harm only the item's outward appearance. Cosmetic spraying accounts for a significant fraction of pesticide use, does nothing to increase yield or nutritional value, and results in an increase in pesticide residues remaining on the produce.

Unfortunately, the chemical companies that market the pesticides often exploit the concerns of the growers in order to increase their profits. Worldwide, the pesticide market amounts to more than $24 billion annually. Through advertising and through field representatives who are paid by commission, pesticide producers may try to convince growers that the threat of pests is much greater than it is and that spraying pesticides is good insurance. Likewise, they may emphasize the enhanced cosmetic quality that can be obtained with pesticides. In the developing world, where the use of pesticides has been growing rapidly, pesticides are often regarded as modern and progressive, symbols of the agriculture of an industrialized society.

Organically Grown Food

There is often strong public feeling against the use of chemical pesticides. Cosmetic or other unnecessary spraying persists in large part because consumers are kept ignorant about the kinds or amounts of pesticides used. Experience shows that when consumers are informed, many of them abandon the Snow White attitude. Food outlets selling **organically grown** produce (produce that is grown without synthetic chemical pesticides or fertilizers) are appearing, although the produce is not as cosmetically perfect and usually costs more than chemically treated produce (Fig. 10-15). Nonetheless, some growers now find it more profitable to rely on natural controls and sell through these specialized markets. Only 2% of United States farm production is organic, but the movement is growing rapidly. Expanding markets for certified organic foods in the U.S., Japan, and Europe have encouraged increasing numbers of growers to abandon the use of agricultural chemicals. One recent study revealed that Canadian farmers of organically grown produce were doing better than conventional Canadian farmers. Check: http//www.inform.umd.edu:8080/edres/topic/agrenv/altFarm/

Purepak is a company in California that grows, packs, and ships only organically grown vegetables. Once Purepak obtained its state-administered organic certification (an essential procedure now available in most states), the company's business took a sharp rise.

FIGURE 10-15
A roadside farm stand in Concord, Massachusetts, specializes in organic produce. Organically grown foods represent an expanding market in the United States. (Timothy Lucas/Tony Stone Worldwide.)

Growers for Purepak use machine and hand cultivation instead of herbicides to keep weeds under control. Or, if an insect infestation appears, they spray soap instead of insecticides and introduce ladybugs into the fields to prey on the insects. Purepak officials state that organic farming costs more, but as more and more growers join the movement, the costs will likely decline.

Integrated Pest Management

The approach known as **integrated pest management** (IPM) aims to minimize the use of synthetic organic pesticides without jeopardizing crops. This is made possible by addressing all the interacting sociological, economic, and ecological factors involved in protecting crops. With IPM, the crop and pests are seen as part of a dynamic ecosystem; the goal is not the eradication of pests, but maintaining crop damage below the economic threshold.

Cultural and biological control practices form the core of IPM techniques. Such practices as crop rotation, polyculture, destroying crop residues, the maintenance of predator populations, and carefully timed planting and fertilizing are basic to IPM. "Trap crops" are often used: Early strips of a crop like cotton are planted to lure existing pests, and then the pests are destroyed by hand or by the limited use of pesticides before they can reproduce. Spraying is not performed during the growing season, in order to encourage natural enemies of the pests to flourish. Pest populations are monitored, often by persons employed by local agricultural extension services or farm cooperatives, or by persons acting as independent consultants. These **field scouts** are trained in monitoring pest populations (traps baited with pheromones are used for this purpose) and other factors and in determining whether the pest population is exceeding the economic threshold. If this happens, remedial measures are taken to lower the pest population. Here, pesticides are often used, in quantities and brands that will do the least damage to predators of the pest. In addition, **pest-loss insurance**, which pays the farmer in the event of loss due to pests, has enabled insured growers to refrain from unnecessary "insurance spraying."

Making the economic benefits of natural controls known to growers is an important aspect of IPM. Because pesticides increased yields and profits when they were first used, many farmers still cling to them, believing that they offer the only way to bring in a profitable crop. In some instances, however, the rising

Ethics

The Long War against the Medfly

The program to protect crops in the United States from the Mediterranean fruit fly, or medfly, exemplifies how IPM can work. It also illustrates an ethical dilemma in which the state is caught between growers who want aerial spraying to control the medfly and citizens who strongly protest having the whole Los Angeles basin sprayed with pesticides.

Unlike the common fruit fly, which is attracted only to very ripe fruit, the Medfly lays its eggs on unripe fruits and vegetables in the field. The feeding maggots thus cause extensive damage before harvesting and during storage and transport of the fruit. Worldwide, this insect is one of the most destructive pests known. If it became established in the United States, it would present an enormous economic threat.

Medflies that were probably brought in on imported produce have invaded Florida, Texas, and California on several occasions, and whenever they have been detected, an intensive eradication campaign has been mounted. This diligence seems to have been successful in Florida and Texas, but there are signs that the medfly may now be established in California, leading to increasingly frequent but controversial sprayings. The 1990 appearance of medflies in the Los Angeles basin prompted an aerial spraying of the pesticide malathion, a spraying bitterly opposed by residents of the area.

Malathion spraying represents the last resort in a battery of IPM techniques. First, all imported produce is checked and then fumigated if officials detect any trace of the insect. Second, a network of traps baited with a sex attractant is maintained throughout the agricultural area. Monitoring these traps provides an early warning of the presence of medflies that have slipped through customs. If medflies are found in the traps, the sterile male technique is called on as the third line of defense. Stocks of sterile medflies are maintained in South America, and batches of them can be delivered on short notice and dropped over the infected area. In addition, fruit is stripped from the trees in the infected area to prevent the medfly from reproducing. Officials continue to use pheromone-baited traps to monitor the program's success. The final line of defense is aerial spraying.

Entomologist James Carey has presented convincing evidence that the medfly is now established in California and argues that strategies other than aerial spraying are needed to keep it from becoming a major pest in the state. In particular, more research is needed to study the basic biology of the medfly so that effective natural controls might be devised. Although malathion is not very toxic to humans, Los Angeles residents feel that aerial spraying is an unacceptable alternative for controlling the medfly. They argue that the benefits to the growers are not outweighed by the potential harm to the environment and to humans. As a consequence, the control program has become a very hot political issue in California. How would you feel about it if you lived in California?

costs of pesticides, along with their tendency to aggravate pest problems, have eliminated their economic advantage.

Particularly in the developing world, government agricultural policy often determines the extent to which IPM is adopted. Governments and aid agencies usually subsidize the purchase of pesticides, thus strongly encouraging growers to step onto the pesticide treadmill. By contrast, the Indonesian experience cited earlier has provided a viable IPM model for other rice-growing countries. The success of the program can be traced to close cooperation between the Indonesian government and the FAO (Fig. 10-16). FAO workers ran training sessions for farmers, weekly meetings that lasted throughout the growing season. Eventually, more than 200,000 farmers were taught about the rice agroecosystem and IPM techniques, and these farmers formed a corps who could in turn teach others. The economic and environmental benefits of the Indonesian IPM program have been remarkable: The government has saved $120 million annually by not purchasing pesticides, farmers have not had to invest in pesticides and spray equipment, the environment has been spared the application of thousands of tons of pesticide, fish are once again thriving in the rice paddies, and the health benefits of reduced pesticide use have been spread from applicators to consumers and wildlife. The FAO is sponsoring similar IPM training programs in eight other rice-growing nations and has started programs for vegetable crops in six nations.

While the aim of IPM is to reduce reliance on chemical pesticides, such pesticides still remain an integral part of the whole approach. Indeed, some entomologists feel that IPM is overly reliant on the use of synthetic organic pesticides and is thus still contributing to driving the pesticide treadmill. These entomologists advocate a renewed emphasis on the study of pest ecology and the development and implementation of

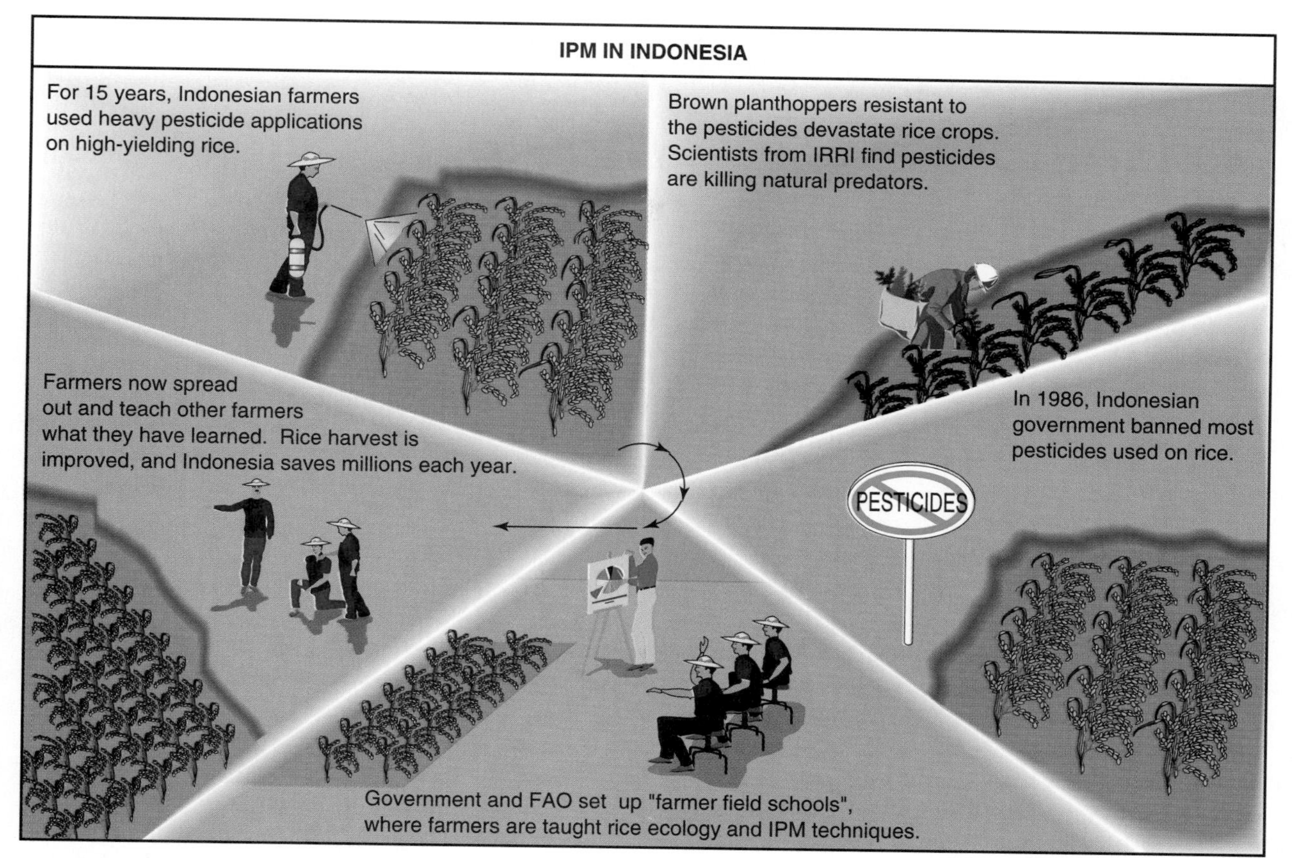

FIGURE 10-16
Integrated pest management in Indonesian rice culture. IPM has helped Indonesian farmers to bring the brown planthopper under control, after years of frustration on the pesticide treadmill.

natural controls—in short, a renewed emphasis on the concepts of ecological pest management. This is the logical pathway to a sustainable future, in which pesticides are not polluting groundwater, contaminating food, killing pollinating insects, and, in the end, creating new, more resistant pests.

Public Policy

FIFRA and Its Problems

The key legislation to control pesticides in the United States is the Federal Insecticide, Fungicide, and Rodenticide Act, commonly known as FIFRA. This law, administered by the EPA, requires manufacturers to register pesticides with the government before marketing them. The registration procedure includes testing for toxicity to animals (and, by extrapolation, to humans). From the test results, usage standards are set. For example, highly toxic compounds such as chlordane, which is used to control termites, are not authorized for use on crops. If health hazards appear after a substance has been registered and marketed, the act provides for "deregistration," whereby the pesticide may be banned from one or more uses. In 1996, President Clinton signed the Food Quality Protection Act, a strong revision relative to the exposure of children to pesticide residues. The EPA can now set tolerances to protect the young. The pesticide provisions of the Delaney Clause are replaced by a mandate to set legal limits for residues according to "reasonable certainty of no harm to public health."

Unfortunately, FIFRA has many shortcomings. The three main ones are inadequate testing, case-by-case bans, and lack of public input.

Inadequate Testing. The perils of biomagnification and the potential for long-term exposure that may cause cancer, birth defects, mutations, and other physiological disorders were not fully appreciated until the late 1960s, when bitter experience with DDT and other chemicals came to the forefront. This experience made officials

Earth Watch

The Trojan Horse Strategy

Some very clever ways are being devised to win the war against pests without resorting to chemical pesticide spraying. One is the old Trojan horse strategy. The Lyme disease tick is being fought with an environmentally benign pesticide (a product marketed as Damminix) in small tubes of cotton. Mice, which are essential hosts in the life cycle of the tick, pull cotton from the tubes and use it to build their nests—and the ticks die.

Cockroaches are now facing a variation of the Trojan horse strategy. It has long been known that a fungus found naturally in soil, *Metarhizium anisopliae*, can kill a variety of insects—but the insect must contact live spores of the fungus under moist conditions. Biologist Haim Gunner at the University of Massachusetts in Amherst reasoned that the fungus might be used to control cockroaches—especially in places where the use of pesticides would be undesirable, such as restaurants and hospitals. Gunner devised a trap 2 inches in diameter that was packed with fungus and an odor that attracts cockroaches. Roaches enter the trap, are coated with fungus, and then spread the fungus to other roaches before finally succumbing to the infection.

The trap worked well enough to encourage Gunner to found a new company, EcoScience, to develop and market the product. Now called the Bio-Path Chamber, the traps are being supplied to pest control businesses as a step in the direction of entering the retail market. Who knows how many more clever ways to use biological knowledge will be devised as fertile minds like Haim Gunner's are turned towards the age-old battle against pests?

and the public aware that many pesticides in use had never been adequately tested. Consequently, Congress amended FIFRA in 1972 to require the EPA to reevaluate and reregister all pesticide products then on the market.

When the amendment was adopted, there were already some 1400 chemicals and 40,000 formulations (mixtures of different chemicals) on store shelves. In addition, there was and still is tremendous pressure from the chemical industry to register numerous new pesticides. Needless to say, the EPA's Office of Pesticide Programs has been overwhelmed. Ongoing budget restrictions have not helped the situation.

Since the law allows existing pesticides to remain in use unless proven hazardous, many products on the market still have not been subjected to adequate tests, and existing standards may allow hazardous levels of pesticide residues in food. In 1989, the Natural Resources Defense Council (NRDC) published the results of a three-year study which concluded that many fruits and vegetables routinely contain levels of pesticide residues that, although within prescribed standards, "pose an increased risk of cancer, neurobehavioral damage, and other health problems" to children. The focus is on children both because they typically consume more fruits and vegetables per unit of body weight than do adults and because studies have shown that the young are frequently more susceptible than adults to carcinogens and neurotoxins.

Case-by-Case Bans. Each pesticide must be shown to pose a threat before it is subject to a ban. Assessing the risk of a substance is a difficult and often controversial matter. Sometimes proof of danger is so long in coming, that the damage has already been done by the time the ban is put into effect. The story of ethylene dibromide (EDB) illustrates this point.

It was known by the mid-1970s that EDB causes cancer, birth defects, and other illnesses in laboratory animals. Yet its widespread use as a soil fumigant to control root nematodes (small worms that attack roots) continued. By 1982, 4.5 million pounds of EDB were being pumped into the soil each year. In addition, EDB was increasingly used to fumigate and protect grains and other crops in storage. It could have easily been predicted that this practice would eventually result in contamination of groundwater. But EDB was not banned until 1984, after traces were found in hundreds of wells in Florida, California, Massachusetts, and other regions, and unacceptable residue levels were found in flour and other food products. The curious thing about these "finds," is that the more they looked the more they found.

Clearly, waiting until after we suffer the consequences to ban such chemicals is hardly a prudent way to protect public health. Yet this is precisely how the system operates. A few persistent halogenated hydrocarbon pesticides remain in use, and the use of some, notably lindane, is increasing. Perhaps most serious is the widespread and increasing use of herbicides. They now account for more than half of all pesticides used. Large numbers of people are being exposed to these compounds with unknown effects.

Lack of Public Input. FIFRA provides no mechanism for public input. Therefore, the legislation primarily reflects the chemical industry's interests, which are conveyed through intense lobbying efforts as opposed to public forums. The bias is obvious when one considers that the law imposes a $1000 penalty on those who misuse a pesticide and a $10,000 penalty on those who reveal a trade secret about a pesticide's formulation.

The Delaney Clause

Another law, the Federal Food, Drug and Cosmetic Act of 1958 (FFDCA), requires the EPA to set standards regarding the "safe" amounts of pesticide residues that may be left on food to be eaten by animals and humans. The Food and Drug Administration (FDA) then monitors the pesticide residues in food. Foods have been withdrawn from the market because residues of certain pesticides were above the established standard, although only a small fraction of foods tested in any given year are found in violation.

One clause of the FFDCA, the so-called Delaney clause, has attracted much attention because of its strict standard. The clause states that "no (food) additive shall be deemed to be safe if it is found to induce cancer when ingested by man or animal." Since pesticides represent one of the largest categories of toxic chemicals to which people are exposed, this clause has been applied in prohibiting many pesticides from being used on foodstuffs when those pesticides have been found to cause cancer in laboratory tests with animals. In essence, the law states that if a given pesticide presents any risk of cancer, no detectable residue may remain on the food.

The Delaney clause is in the center of a controversy over how the EPA and FDA should be protecting consumers from risk of cancer. Both of these agencies want to see the law repealed or changed to incorporate the techniques of modern risk analysis (see Chapter 17) and to reflect the great improvements made over the years in analytical instrumentation. Their basic complaint is that extremely small amounts of pesticides can now be detected on foods, amounts that can be shown to pose only "negligible risk" to human health. The federal agencies are supported in this controversy by farmers, food processors, and the agrichemical industry. On the other side of the controversy are many environmental groups and health officials, who believe that we still know far too little about how toxic chemicals such as pesticides can lead to cancer and how the risk of cancer over a lifetime of exposure to chemicals can be accurately evaluated. These groups believe that it is best to risk erring on the side of overcaution.

Pesticides in Developing Countries

The United States currently exports more than 200,000 metric tons of pesticides to developing countries each year. Some 25% of this total consists of products banned in the U.S. itself. A presidential executive order was signed in March 1981 that lifted the ban on exporting banned products to developing countries. FIFRA requires "informed permission" prior to the shipment of any pesticides banned in the United States. This permission comes from the purchaser, which can often be a foreign subsidiary of the exporting company. The EPA must then notify the government of the importing country.

Fortunately, the international community has erected a more effective system, through the cooperative work of two UN agencies: the FAO and the United Nations Environment Program (UNEP). There is now a process of prior informed consent (PIC) whereby exporting countries inform all potential importing countries of actions they have taken to ban or restrict the use of pesticides or other toxic chemicals. Governments in the importing country respond to the notifications via the UN agencies, which then disseminate all the information they receive to the exporting countries and follow up by monitoring the export practices of the exporting countries. The PIC process is presently only voluntary, but work is under way to turn the process into a legally binding instrument by 1998.

The more serious problem of unsafe pesticide use in the developing countries has also been addressed by the FAO. In spite of early opposition from the pesticide industry and exporting countries, an international "Code of Conduct" has been drawn up whereby conditions of safe pesticide use are addressed in detail. The code makes clear the responsibilities of private companies and of countries receiving pesticides in promoting their safe use. As with PIC, the code is not yet legally binding, but it has proven very useful in holding private industry and importing countries to standards of safe use. In the view of most observers, both PIC and the FAO Code of Conduct are not as strong as they should be; however, their very existence represents great progress in addressing a difficult problem in the developing countries.

New Policy Needs

The shortcomings of FIFRA are obvious. The EPA is aware of the problems and is engaged in revising its regulatory rules to address many of them. On the agenda are the following: (1) implementing a strategy to protect groundwater from infiltration by pesticides, (2) improving methods for ensuring food safety, (3) determining the extent of exposure to home and garden pesticides, (4) expanding worker protection standards to ensure the safety of all pesticide handlers, and (5) expanding the requirements for testing to include impacts on natural environments.

Some of the new policy needs will require new legislation. A "citizen suit" provision should be provided under FIFRA, so that the public can go to court to force the government to uphold the law. This provision has given the EPA added clout where it has been made part of other federal environmental statutes. The amendment giving the secretary of agriculture the opportunity for a special review before any changes are made in EPA regulations should be stricken. IPM and other elements of ecological pest management strategy need to be given more encouragement in the form of support for research and for extending their scope of application.

If further pesticide reform legislation is drafted, it will likely reflect the Clinton administration's desire to repeal or ease up on the Delaney clause requirements. The 1996 Food Quality Protection Act is a beginning. Such a move is certain to trigger a national debate over pesticides and health. Another change under consideration is to prohibit the export of any pesticides that are banned in the United States—a reversal of President Reagan's executive order.

A group of leading entomologists has called for a 50% reduction in pesticide use in the United States, supporting its recommendation with the economic and environmental benefits of such a reduction. This goal could become public policy if it is included in legislative or regulatory law. Recently, heads of the Clinton administration and the EPA, FDA, and USDA issued a joint announcement of their commitment to reduce pesticide use in the U.S. According to the EPA, a policy of "maximum feasible reduction" will apply to all pesticide users, from farmers to homeowners. Triggering this initiative was a set of reports, warning of toxic residues in the diet of American children, that came from the National Academy of Sciences (NAS) and the private, nonprofit Environmental Working Group.

Pesticide reform is in the wind. As with so many of the issues we consider in this book, significant progress in pesticide control has benefited from grassroots action and pressure from public interest groups and nongovernmental organizations. Such groups are pressing for a continued movement in the direction of ecological pest management and continued progress in keeping food free of pesticide residues. And when three governmental agencies go on record favoring a reduction in pesticide use, there is hope that we will soon make substantial movement in helping farmers to jump off the pesticide treadmill and consumers to enjoy a safer food supply.

Review Questions

1. Define pests. Why do we control them?
2. Discuss the basic philosophies of pest control.
3. What were the apparent virtues of the synthetic organic pesticide DDT?
4. What adverse environmental and human health effects can occur as a result of pesticide use?
5. Why are nonpersistent pesticides not as environmentally sound as first thought?
6. Describe the life cycle of an insect. What natural control methods can be applied at each stage?
7. Describe the four categories of natural or biological pest control. Cite examples of each and discuss their effectiveness.
8. Discuss the recent uses of biotechnology in genetic control.
9. Define the term *economic threshold* as it relates to pest control.
10. How does integrated pest management work? Give examples.
11. How do FIFRA and FFDCA attempt to control pesticides? What are the shortcomings of this type of legislation?
12. Discuss recent policy regarding the export of pesticides to developing countries.
13. What amendments to public policy would promote pest control that is environmentally safer than the methods used today?

Thinking Environmentally

1. U.S. companies export pesticides that have been banned or restricted in this country. Should this practice be allowed to continue? Support your answer.
2. Almost one third of the chemical pesticides bought in the United States are for use in houses and on gardens and lawns. What should be done by manufacturers and users to ensure their limited and prudent use?
3. Should the government give farmers economic incentives to switch from pesticide use to integrated pest management? Why or why not?
4. Investigate how bugs and weeds are controlled on your campus. Are IPM techniques being used? Organic fertilizers? Consider lobbying for their use.
5. Read or reread the ethics box entitled "The Long War against the Medfly." If you were a farmer, what methods would you advocate to control this pest? Why?

CHAPTER 11

Water, the Water Cycle, and Water Management

Key Issues and Questions

1. All water on Earth is constantly recycled, repurified, and reused. How does this recycling and repurification occur?
2. Humans have three major impacts on the water cycle. What are they, and what are their effects?
3. All the water humans use must come out of the water cycle. What are the major uses, points of withdrawal, and limitations and consequences of overdrawing water?
4. Historically, humans have addressed water problems by obtaining more water. To what degree is this not a viable option for the future?
5. Humans can reduce their water demands in numerous ways. How can demands be reduced in agriculture, industry, and domestic use?
6. Urbanization seals surfaces with pavement, increasing stormwater runoff and quickening concentration times. Discuss related problems caused by paving over soil. How should these concepts influence development?
7. There is potential for all parties' getting together to work out compromises for water usage between agriculture, cities, and natural eosystems. How great do you think this potential is?

Water is absolutely fundamental to life as we know it. It is difficult even to imagine a form of life that might exist without water. Happily, Earth is virtually flooded with water; a total volume of some 325 million cubic miles (1.4 billion cubic kilometers) covers 71% of Earth's surface. Yet it is still difficult in many locations to obtain desired amounts of water of suitable purity (see "Global Perspective" box, page 276).

All major terrestrial biota, ecosystems, and humans depend on **freshwater**, water that has a salt content of less than 0.01% (100 ppm). Ninety-seven percent of Earth's water is the salt water of oceans and seas. Then, of the 3% that is fresh water, 87% is bound up in the polar ice caps and glaciers, is inaccessible groundwater, or is in the atmosphere, leaving only 0.4% as accessible freshwater (Fig. 11-1). To be sure, evaporation from the seas and precipitation continually resupply that small percentage, as we shall describe in detail shortly. Thus, freshwater is a continually renewable resource. However, one can get no more water from a pipe than what flows through it. By the same logic, natural supplies of fresh water are limited by amounts that move through the natural system.

Thunderstorm delivers precipitation to desert lands at the Grand Canyon, Arizona. [Photo by Tom Bean/The Stock Market.]

As we have seen, precipitation patterns around the globe are far from even. Regions with abundant precipitation support lush forest ecosystems; other regions have minimal rainfall and are deserts as a result. Thus, we can visualize different volumes of flow through different natural regions (over 1 million gallons of water per acre per year in a temperate forest region; 2500 gallons or less per acre per year in desert regions). Any humans on the scene must draw on the same water for drinking, irrigating crops, and supplying industries. Hence, there is an inevitable dividing of water between the natural biota and human demands.

In high-rainfall regions there is plenty of water for both human demands and natural biota. However, in dryer regions and with growing human populations there are growing conflicts between human needs and those of the natural ecosystem. Around the world there are more examples than we can recount of ecosystems under stress or already dead because of diversions of water for human uses. Moreover, within the human arena there is growing contention between agricultural, urban, and industrial demands, and between countries which share a common water source.

Hydrologists (water experts) estimate that water shortages place a severe constraint on food production, economic development, and protection of natural ecosystems as available water drops below about 1000 cubic meters per year (750 gallons per person per day). Yet, water supply shortages loom in at least 80 countries as of 1995. Water shortages increase conflicts, public health problems, reduce food production, and endanger the environment. As populations grow, more countries will be joining the list. And, as shortages become more severe, already-contentious relations between nations sharing common supplies will become more so (Fig. 11-2). Similar contentions are growing between water-rich and water-poor areas within national boundaries as well. Finally, a considerable number of countries and regions are now living with apparent water abundance only by withdrawing groundwater faster than it is replenished, thereby depleting their supply for future generations. Conspicuously, these trends are not sustainable.

A sustainable future will depend on learning stewardship of water resources. There are abundant opportunities for sustainable development in this arena. Our objective in this chapter, then, is threefold: (1) to understand the natural water cycle, its capacities and its limitations; (2) to understand how we are overdrawing certain water resources and the consequences of this action; and (3) to understand how water must be managed if we are to achieve sustainable supplies.

The Water Cycle

Earth's **water cycle**, also called the **hydrologic cycle**, is represented in Fig.11-3. The basic cycle consists of water's rising to the atmosphere through either evaporation or transpiration and leaving it through condensation and precipitation. However, these and additional aspects bear more consideration.

Evaporation, Condensation, and Purification

As we discussed in Chapter 3, a weak attraction known as *hydrogen bonding* tends to hold water molecules (H_2O) together. Below 32°F (0°C), the kinetic energy of the molecules is so low that the hydrogen bonding is enough to hold the molecules in place with respect to one another, and the result is ice. At temperatures above freezing but below boiling, the kinetic energy of the molecules is such that hydrogen bonds keep breaking and re-forming with different molecules; the result is liquid water. As water molecules absorb energy from sunlight

FIGURE 11-1
The Earth has an abundance of water, but terrestrial ecosystems, humans and agriculture depend on accessible freshwater which constitutes just 0.4% of the total. Therefore, problems of water scarcity and purity abound.

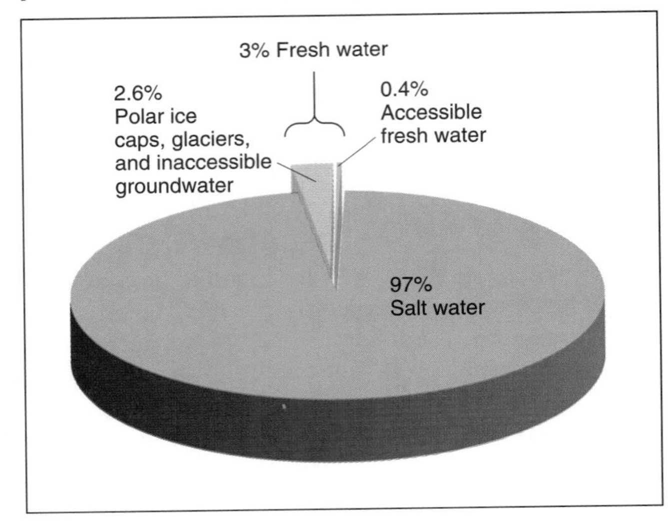

(b)

FIGURE 11-2
Human conditions range from (a) being able to luxuriate in water to (b) having to take long trecks to obtain enough to simply survive. In many cases, however, the abundance of water is illusory, since water supplies are being overdrawn to supply apparent abundance. (a) Pete Saloutos/The Stock Market. b) Peterson/ Gamma-Liaison, Inc.)

FIGURE 11-3
The water cycle. The Earth's fresh waters are replenished as water vapor enters the atmosphere by transpiration from vegetation or evaporation, leaving salts and other impurities behind. As water hits the ground, note that three additional pathways are possible.

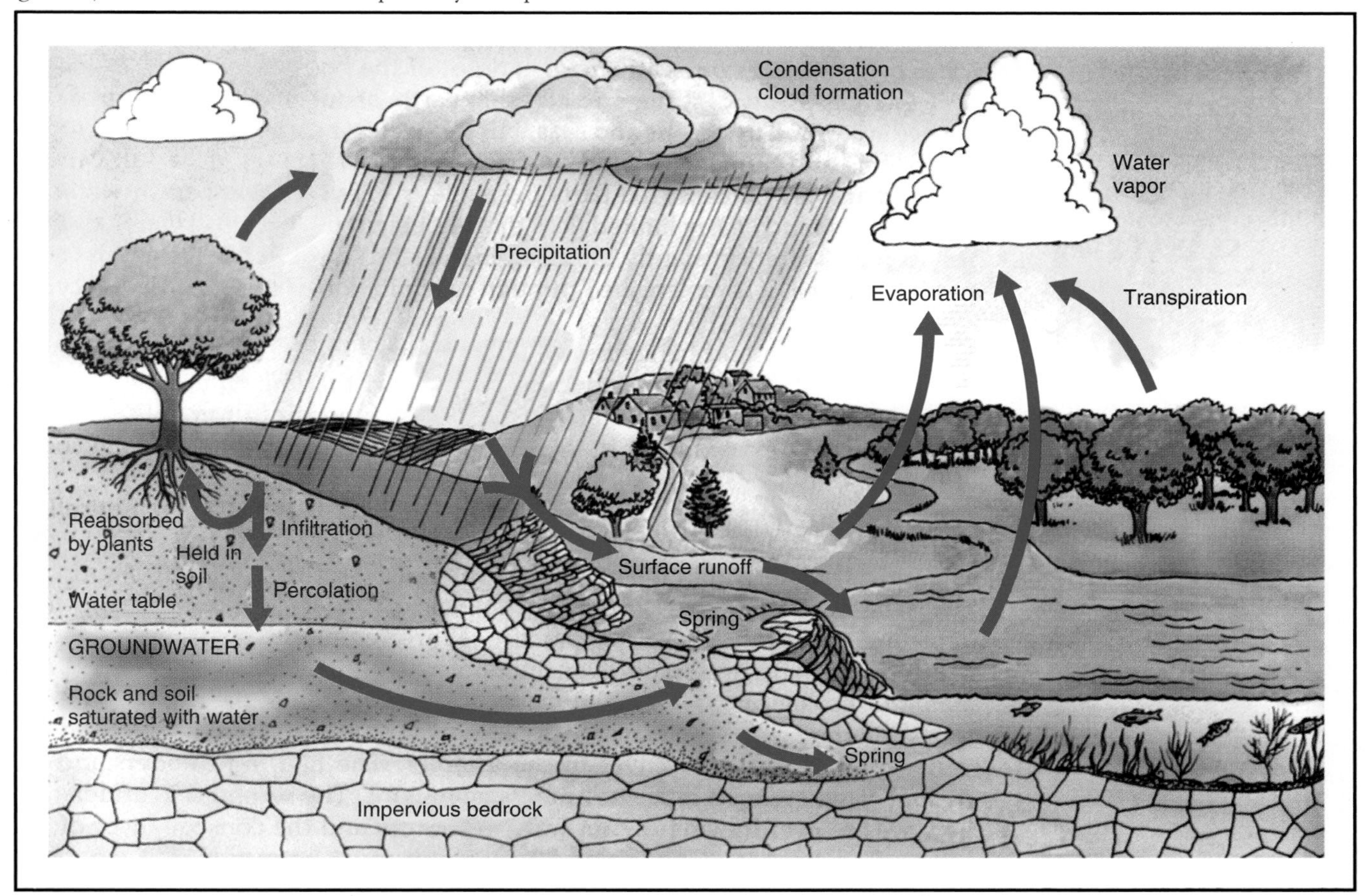

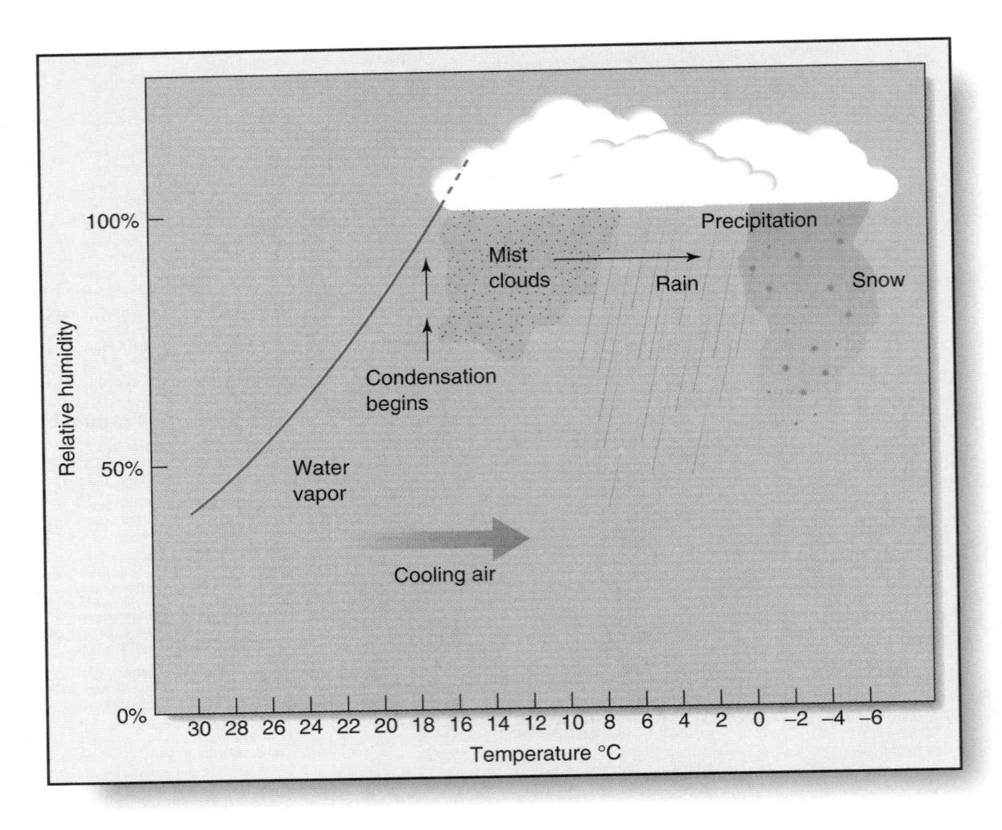

FIGURE 11-4
The amount of water vapor that air can hold increases and decreases with corresponding changes in temperature. Therefore, as warm, moist air is cooled, the amount of water it can hold decreases. Cooling air beyond the point where relative humidity (RH)—the amount of water vapor that air can hold at a given temperature—reaches 100% forces excess moisture to condense, forming clouds. Further cooling and condensation results in precipitation.

or an artificial source, the kinetic energy they gain may be enough to allow them to break away from other water molecules entirely and enter the atmosphere. This is the process we know as **evaporation**, and the water molecules are said to be in the gaseous state.

We speak of water molecules in the air as **water vapor**; the amount of water vapor in the air is **humidity**. Humidity is generally measured as **relative humidity**, the amount of water vapor as a percentage of what the air can hold *at that temperature*. For example, a relative humidity of 60% means that the air contains 60% of the maximum amount of water vapor it could hold at that temperature. The amount of water vapor air can hold increases with rising temperature and decreases with lowering temperature. Consequently, relative humidity will decrease as air warms and increase as air cools quite apart from any change in the actual amount of water vapor present. The key point is that when warm moist air is cooled, its relative humidity rises until it reaches 100%; further cooling causes the excess vapor to *condense*, because the air can no longer hold as much vapor (Fig. 11-4).

Condensation is simply water molecules rejoining by hydrogen bonding to form liquid water or ice. If the droplets form in the atmosphere, the result is fog or mist. (Masses of fog or mist in the distance are seen as clouds.) If the droplets of condensing vapor form on the cool surfaces of vegetation, the result is dew. If the temperature is below freezing as condensation occurs, the water vapor forms directly into ice crystals making up snow or frost.

Summing up, warm air readily picks up water vapor as evaporation occurs from any wet or moist surface and transpiration occurs from vegetation. Warm, moist air rises, then cools as heat is radiated and lost to outer space. When relative humidity reaches 100% and cooling continues, condensation occurs and clouds form. With continuing condensation, water droplets or ice crystals become large enough to fall as precipitation.

One very important aspect of evaporation and condensation is that these processes result in natural *water purification*. When water evaporates, only the water molecules leave the surface; salts and other solids in solution remain behind. (We noted this process when we discussed the problem of salinization in Chapter 9.) The condensed water is thus purified water—except as it picks up pollutants in the air. (The most chemically pure water for use in laboratories is obtained by distillation, a process of boiling water and recondensing the vapor.) Thus, evaporation and condensation of water vapor are the source of all natural fresh water on Earth. Fresh water from precipitation falling on the land gradually makes its way through aquifers, streams, rivers, and lakes to oceans or seas. In the process, salts from the land are constantly flushed toward locations where the only exit is by evaporation, and the salts accumulate at those points. Oceans are the prime example, but there are notable inland salt seas or lakes, such as the Great Salt Lake in Utah. Salinization of irrigated croplands is a notable human-made example.

Precipitation

The distribution of precipitation over Earth, which ranges from near zero in some areas to more than 100 inches (2.5 m) per year in others, basically depends on patterns of rising or falling air currents. As air rises, cooling and condensation occur, and precipitation results. As air descends, it tends to become warmer, causing an increase in evaporation and dryness.

Two things may cause more or less continuously rising or falling air currents over particular regions. First are global convection currents: Solar heating of Earth is most intense over and near the equator, where rays of sunlight are almost perpendicular to Earth's surface. As the air at the equator is heated by the warm soil and water, it expands and rises. Rising, it cools, and condensation and precipitation occur. Thus, equatorial regions have high amounts of rainfall. This rainfall, along with continuous warmth, supports tropical rainforests.

Rising air over the equator is just half of the convection current, however. The air must come down again. Pushed from beneath by more rising air, it literally "spills over" to the north and south of the equator and descends over subtropical regions (25° to 35° north and south of the equator), resulting in subtropical deserts (Fig. 11-5). The Sahara of Africa is the prime example.

The second situation that causes continually rising and falling air occurs where trade winds (winds that blow almost continuously from the same direction) hit mountain ranges. As the moisture-laden air in the trade winds encounters a mountain range, the air is deflected upward, causing cooling and high precipitation on the windward side of the range. As the air crosses the range and descends on the other side, it becomes warmer and increases its capacity to pick up moisture. Hence, deserts occur on the leeward side of mountain ranges. The dry region downwind of a mountain range is referred to as a **rain shadow** (Fig. 11-6). The severest deserts in the world are caused by the rainshadow effect. For example, the westerly trade winds, full of moisture from the Pacific Ocean, strike the Sierra Nevadas in California. As the winds rise over the mountains, large amounts of water precipitate out, supporting the lush forests on the western slopes. Immediately east of the southern Sierra Nevada lies Death Valley, a result of the rain shadow.

Another rain-causing event that you see in almost every television weather report is the movement of a cold front. As a cold front moves into an area, the existing warm moist air is forced upward because the cold air of the advancing front is denser. Again the rising moist air cools, causing condensation and precipitation along the leading edge of the cold front.

Water over and through the Ground

As precipitation hits the ground, it may follow either of two pathways. It may soak into the ground, **infiltration**, or it may run off the surface, **runoff**. We speak of the amount that soaks in compared with the amount that runs off as the **infiltration-runoff ratio**.

Runoff flows over the ground surface into streams and rivers, which make their way to the ocean or to inland seas. All the land area that contributes water to a particular stream or river is referred to as the **watershed** for that stream or river. All ponds, lakes, streams, rivers, and other waters on the surface of Earth are referred to as **surface waters**.

For water that infiltrates, there are two more alternatives. The water may be held in the soil; the amount held depends on the water-holding capacity of the soil, as was discussed in Chapter 9. This water, called **capillary water**, returns to the atmosphere either by way of evaporation from the soil or by transpiration as it is drawn up by plants. The combination of evaporation and transpiration is referred to as **evapotranspiration**.

The second alternative is **percolation**. Infiltrating water that is not held in the soil is called **gravitational water** because it trickles, or percolates, down through pores or cracks under the pull of gravity. Sooner or later, however, gravitational water comes to an impervious layer of rock or dense clay. There it accumulates, completely filling all the spaces above the impervious layer. This accumulated water is called **groundwater**, and its upper surface is the **water table** (Fig. 11-7). Gravitational water becomes groundwater as it hits the water table in the same way that rainwater is defined as lake water as it hits the surface of the lake. Wells must be dug to below the water table; then groundwater, which is free to move, seeps into the well and fills it to the level of the water table.

Groundwater will seep laterally as it seeks its lowest level. Where a highway has been cut through rock layers, you can frequently observe groundwater seeping out. Layers of porous material through which groundwater moves are called **aquifers**. It is often difficult to determine the location of an aquifer. Layers of porous rock are often found between layers of impervious material, and the entire formation may be folded or fractured in various ways. Thus groundwater in aquifers may be found at various depths between layers

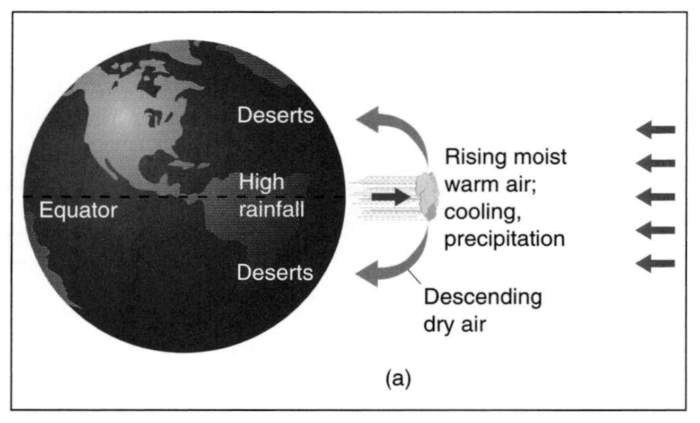

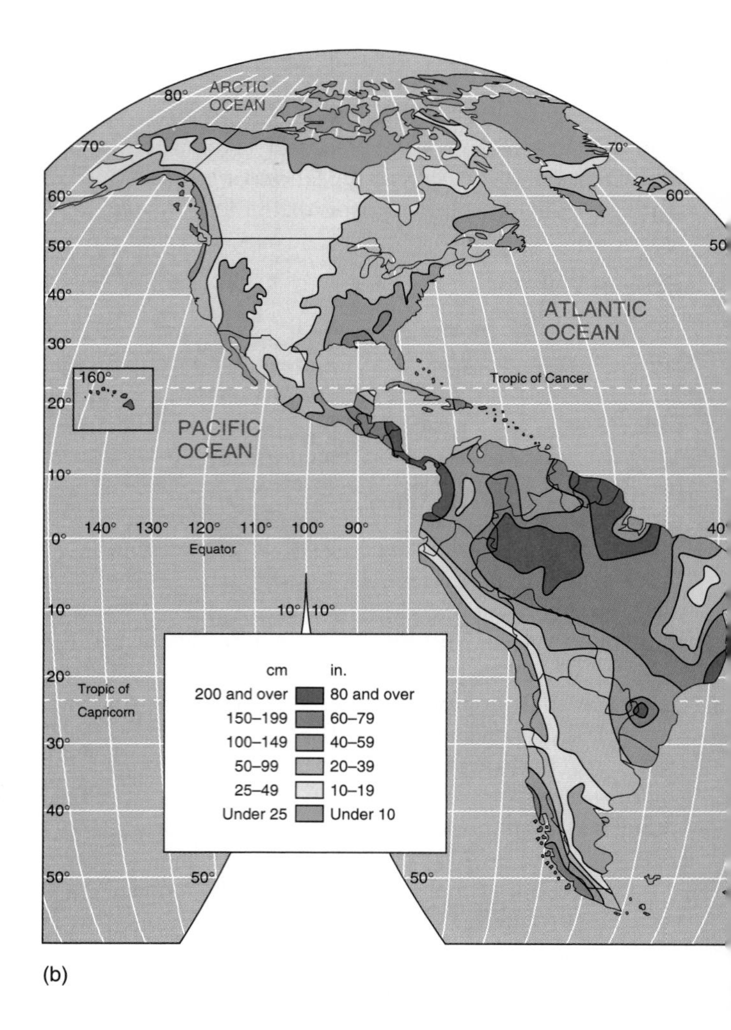

FIGURE 11-5
Equatorial rain forests and subtropical deserts. (a) Solar radiation causes maximum heating in equatorial regions and produces rising currents of moist air. As the moist air cools, there is heavy precipitation over the equatorial regions, supporting tropical rain forests. As the air descends, it becomes warmer and drier, resulting in subtropical deserts. (b) On the world map showing precipitation, note the high rainfall in equatorial regions and the regions of low rainfall to the north and south. (Robert Christopherson, *Geosystems,* 2/e. ©1994, p. 276. Reprinted by permission of Prentice Hall, Upper Saddle River, New Jersey.)

FIGURE 11-6
Rain shadow. Moisture-laden air in a trade wind cools as it rises over a mountain range, resulting in high precipitation on the windward slopes. Desert conditions arise on the leeward side as the descending air warms and tends to evaporate water from the soil. (Robert Christopherson, *Geosystems,* 2/e. ©1994, p. 218. Adapted by permission of Prentice Hall, Upper Saddle River, New Jersey.)

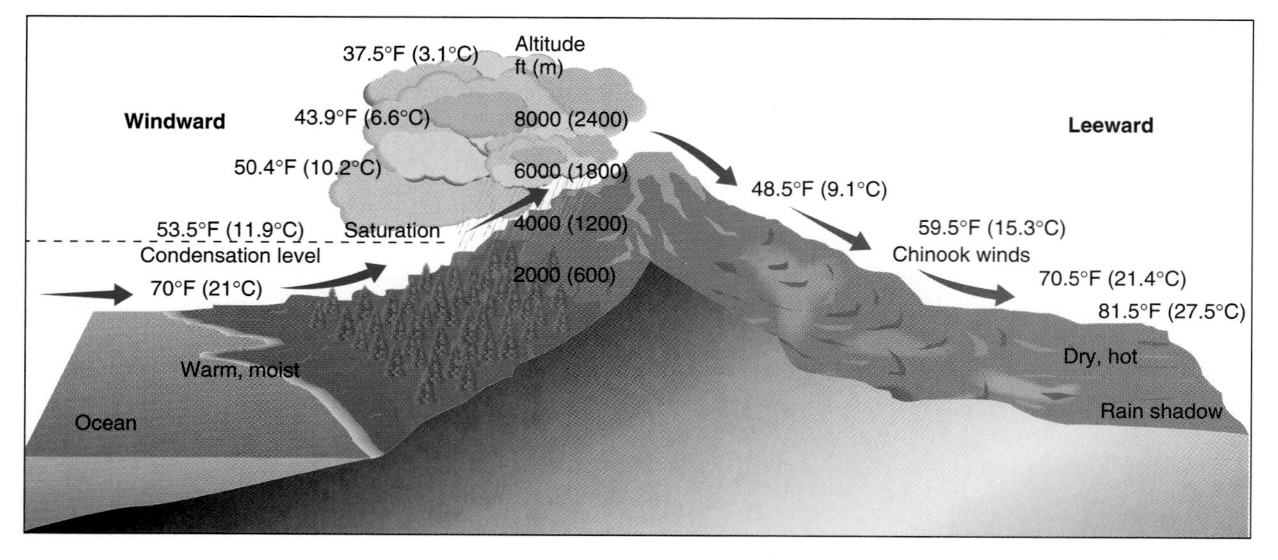

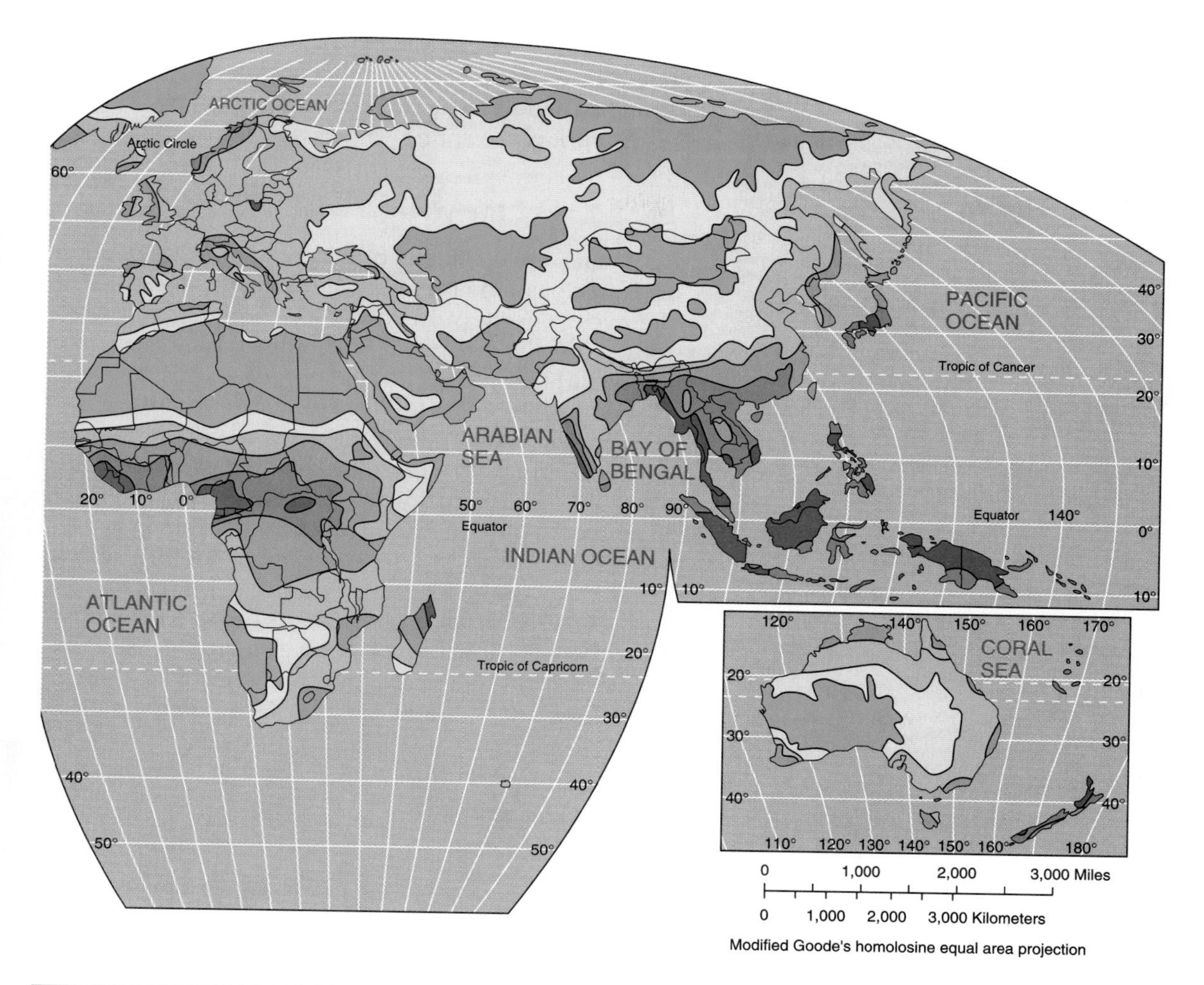

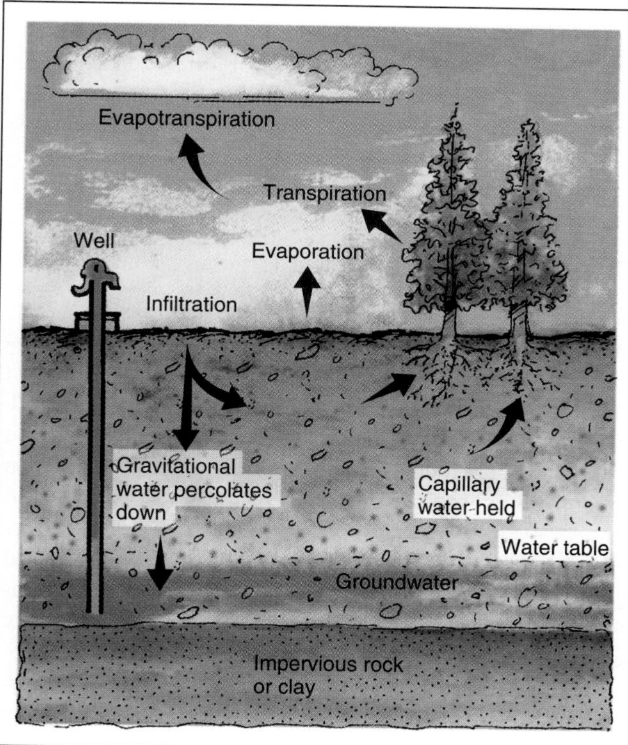

FIGURE 11-7
Pathways of infiltrating water. Water that infiltrates into the soil may be held like water in a sponge. This water may be taken up by plants and returned to the atmosphere by transpiration. When soil reaches its water-holding capacity, additional infiltrating water percolates downward under the pull of gravity and is called *gravitational* water. Water that accumulates above an impervious layer, saturating all the pore spaces in the soil, is called *groundwater*. The upper surface of the groundwater is called the *water table*. Groundwater may be withdrawn by drilling wells to below the water table, or it may percolate horizontally until it comes to a spring or seep.

of impervious rock. Also, the **recharge area**, the area where water enters an aquifer, may be many miles away from where the water leaves the aquifer.

As water percolates through the soil, debris, and bacteria from the surface are generally filtered out. However, water may dissolve and leach certain minerals. Underground caverns, for example, are the result of the gradual leaching away of limestone (calcium carbonate). In most natural situations, the minerals that leach into groundwater are not harmful. Indeed,

TABLE 11-1
Terms Commonly Used to Describe Water

Term	Definition
Water quantity	The amount of water available to meet desired demands
Water quality	The degree to which water is pure enough to fulfill the requirements of various uses
Freshwater	Water having a salt concentration below 0.01%. As a result of purification by evaporation, all forms of precipitation are fresh water, as are lakes, rivers, groundwater, and other bodies of water that have a throughflow of water from precipitation.
Saltwater	Water, typical of oceans and seas, that contains at least 3% salt (30 parts salt per 1000 parts water)
Brackish water	A mixture of fresh and salt water, typically found where rivers enter the ocean
Hard water	Water that contains minerals, especially calcium or magnesium, that cause soap to precipitate, producing a scum, curd, or scale in boilers
Soft water	Water that is relatively free of those minerals that cause soap to precipitate causing scale buildup.
Polluted water	Water that contains one or more impurities making the water unsuitable for a desired use
Purified water	Water that has had pollutants removed or rendered harmless

calcium from limestone is considered beneficial to health. Thus, groundwater is generally high-quality fresh water that is safe for drinking. A few exceptions occur where there is leaching of minerals containing arsenic or other poisonous elements that make groundwater unsafe to drink.

Drawn by gravity, groundwater may move through aquifers until it finds some opening to the surface. We observe such natural exits as springs or seeps. In a **seep**, water flows out over a relatively wide area; in a **spring**, water exits the ground as a significant flow from a relatively small opening. Since springs and seeps feed streams, lakes, and rivers, groundwater joins and becomes part of surface water. A spring will flow, however, only if the water table is higher than the spring. Whenever the water table drops below the level of the spring, the spring will dry up.

Summary of the Water Cycle

The water cycle consists of evaporation, condensation, and precipitation. There are three principal "loops" in the cycle: (1) the *surface runoff loop*, in which water runs across the ground surface and becomes part of the surface water system; (2) the *evapotranspiration loop*, in which water infiltrates, is held as capillary water, and then returns to the atmosphere by way of evapotranspiration; and (3) the *groundwater loop*, in which water infiltrates, percolates down to join the groundwater, and then moves through aquifers, finally exiting through springs, seeps, or wells, where it rejoins the surface water.

We can visualize all the land being continuously bathed with a flow of freshwater coming down as precipitation and gradually moving over and through Earth. Bodies of water—ponds, lakes, and so on—having outlets are constantly flushed by fresh water and remain fresh as long as gradual flushing continues.

Terms commonly used to describe water are given in Table 11-1.

Human Impacts on the Water Cycle

A large share of environmental problems we face stem from direct or indirect impacts on the water cycle. These impacts can be categorized into three areas: (1) changing Earth's surface, (2) pollution, and (3) withdrawals for use.

Changing the Surface of Earth

We are most concerned, perhaps, by the loss of forests and other ecosystems to various human enterprises as this loss diminishes biodiversity. However, the indirect effects of the loss on the water cycle are also profound. In most natural ecosystems there is relatively little runoff; rather, precipitation is intercepted by vegetation and infiltrates into a porous topsoil, as we described in Chapter 9, and goes on to recharge the groundwater reservoir. Then, its gradual release through springs and seeps maintains the flow of streams and rivers at relatively uniform rates. The reservoir of groundwater may be sufficient to maintain a flow during even a prolonged drought. In addition, dirt, detritus, and microorganisms are filtered out as the water percolates through

Earth Watch

Water Purification

Polluted water is defined as water that contains one or more materials that make the water unsuitable for a given use. Water purification is any method that will remove one or more such materials. Several methods may be used in combination to obtain water that is sufficiently pure for a given use. Methods that are commonly used in water purification are:

1. Settling. Soil particles and other solid material carried by flowing water may be removed by holding the water still and allowing the solids to settle. Clarified water is removed from the top. Settling may be aided by the addition of alum (aluminum sulfate). The +3 charge on aluminum ions pulls clay and other particles, which are negatively charged, into clumps that settle more readily than the individual particles do.
2. Filtration. Filtration is the passage of water through a porous material. Any materials larger than the pores will be filtered out. A bed of sand is often used for this purpose.
3. Adsorption. Certain materials bind and hold other materials on their surface. Passing water through an adsorbing material will remove certain pollutants. Activated carbon is a material commonly used in this way to remove organic contaminants from water or air.
4. Biological oxidation. Organic material (detritus and organisms) is fed upon by detritus feeders and decomposers, broken down in cell respiration, and thus removed. Passage of water through systems supporting the growth of such organisms accomplishes removal of organic material (see Chapter 13).
5. Distillation. Distillation is the evaporation and condensation of water. All materials present in the water before the evaporation step remain behind in the holding tank and are therefore not present when the water vapor is condensed.
6. Disinfection. Water is treated with chlorine or other agents that kill disease-causing organisms.

The natural water cycle includes all of these purification methods except disinfection. Sitting in lakes, ponds, or the oceans, water is subject to settling. As it percolates through soil or porous rock, it is filtered. Soil and humus are also good chemical adsorbents. As water flows down streams and rivers, detritus is removed by biological oxidation. As water evaporates and condenses, it is distilled.

Thus numerous sources of fresh water might be safe to drink were it not for human pollution. The most serious threat to human health is contamination with disease-causing organisms and parasites, which come from the excrements of humans and their domestic animals. In human settlements, you can see how these organisms can get into water and be passed on to people before any of the natural purification processes can work.

The World Health Organization estimates that 80% of the sickness, disease, and deaths of infants and children in developing countries can be attributed to contaminated water. Conversely, the greatest safeguard to human health, which we tend to take for granted in developed countries, is proper collection and treatment of sewage wastes and suitable protection and treatment of water supplies. Thus, the greatest step we could take toward improving world health would be to implement these services wherever they are not present.

soil and porous rock, resulting in groundwater that is drinkable in most cases. Similarly, streams and rivers fed by springs contain high-quality water.

As forests are cleared or land is overgrazed the pathway of the water cycle is shifted from infiltration and groundwater recharge to runoff. With runoff, the water runs into the stream or river almost immediately. This sudden influx of water into the waterway not only is likely to cause a flood but also brings along all manner of sediments and other pollutants from surface erosion.

Of course, floods are not unknown in nature. However, in many parts of the world the frequency and severity of flooding are increasing—not because precipitation is greater, but because deforested and overgrazed slopes shed water. For example, extreme flooding in Bangladesh is now common because Himalayan foothills in India and Nepal have been deforested. (This is just one example of how the actions of one country can have an environmental impact on another country.) The impacts of the eroding soil carried into the waterway are even more far-reaching as fisheries are destroyed and the usability of the water for other purposes is diminished.

Also profound and far-reaching is the fact that increased runoff necessarily means less infiltration and groundwater recharge. Thus, there may be insufficient groundwater to keep springs flowing during dry periods. Typical of deforested regions are dry, barren, and lifeless stream beds—a tragedy for both the ecosystems and the humans dependent on the flow. Wetlands function to store and release water in a manner similar to the groundwater reservoir. Therefore, destruction of wetlands has the same impacts as deforestation:

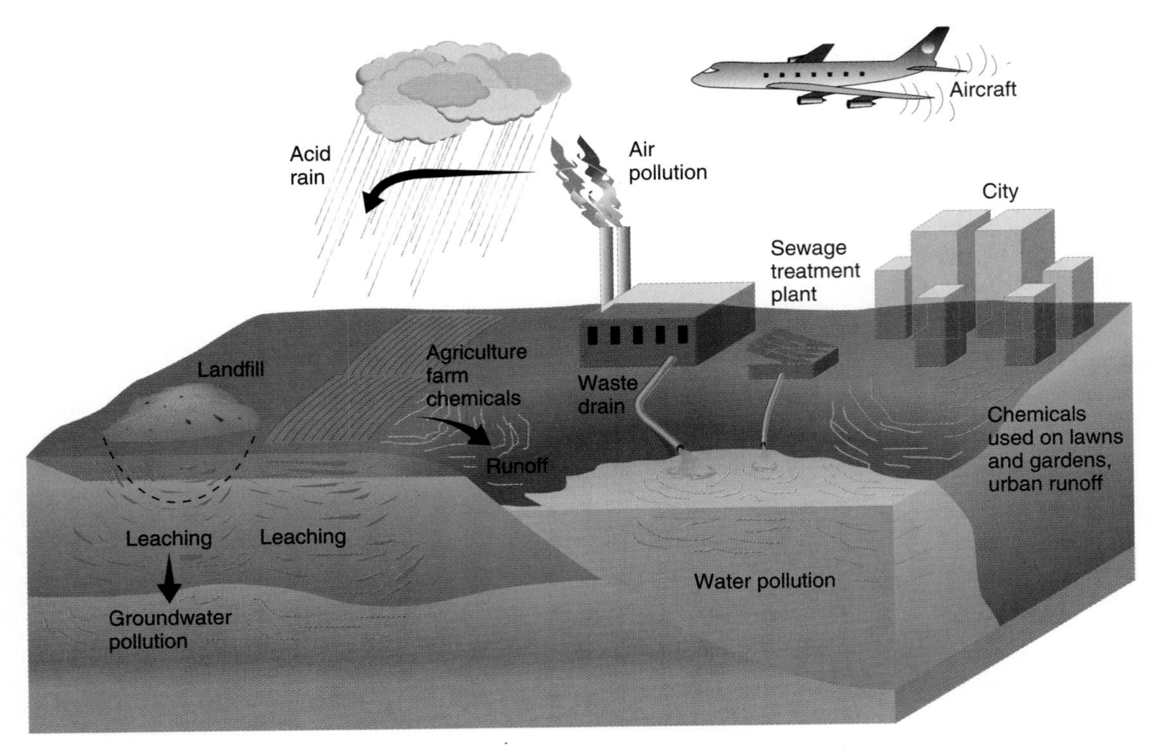

FIGURE 11-8
Human activities introduce pollution into the water cycle at numerous points as shown.

Flooding is exacerbated, and waterways are polluted during wet periods and dry up during drought periods.

Urban and suburban development provides an extreme case of altering Earth's surface as it replaces porous soil with asphalt. This problem and remedial measures will be discussed later in this chapter.

Polluting the Water Cycle

You can see that the water cycle permiates the entire biosphere. Therefore, wherever wastes are put they are inevitably introduced into the water cycle. Any smoke or fumes exhausted or evaporated into the air will come back down as contaminated precipitation. Acid rain, which is discussed in Chapter 16, is a case in point. Chemicals that we use on the soil surface, such as fertilizers, pesticides, and road salt, may either leach into groundwater or be carried into streams by runoff; the same is true of any oil, grease, or other materials we drop or spill on the ground. Any waste we bury in the ground (landfills) may eventually leach into groundwater (Fig. 11-8). (It should be added that modern landfills are constructed so as to minimize this problem; see Chapter 20.) Of course, all the water we use in the course of washing or flushing away wastes directly adds the pollutants into surface waters unless there is intervening water treatment.

Withdrawing Water Supplies

Finally, are the many problems centering around withdrawals of water for human use, not the least of which is having insufficient water to sustain human needs. We shall expand on this issue in the following sections.

Sources and Uses of Fresh Water

Human concerns regarding water can be divided into two categories: *quantitative* and *qualitative*. **Quantitative** refers to such issues as, Is there enough water to meet our needs? What are the impacts of diverting water from one point of the cycle to another? **Qualitative** refers to such issues as: Is the water of sufficient purity so as not to harm human or environmental health? In the remainder of this chapter, we shall focus primarily on the quantitative aspects, although you should bear in mind that quality is an ever-present concern. In following chapters, we shall focus on the qualitative aspects; that is, pollution.

The major uses of freshwater are given in Table 11-2. Most of the water used in homes and industries is for washing and flushing away undesired materials, and

TABLE 11-2
U.S. Demands on Freshwater

Use	*Gallons (liters) used per person per day*
Consumptive	
Irrigation and other agricultural use	700 (2800)
Nonconsumptive	
Electrical power production	600 (2400)
Industrial use	370 (1500)
Residential use	100 (400)

the water used in electrical power production is used for taking away waste heat. Such uses are termed **nonconsumptive** because the water, though now contaminated with the wastes, remains available to humans for the same or other uses if its quality is adequate or if it can be treated to remove undesired materials. In contrast, irrigation is called a **consumptive** use because the applied water is lost for further human use. It can only percolate into the ground or return to the atmosphere through evapotranspiration. Of course, in either case the water does reenter the overall water cycle, but it is gone from human control.

Worldwide, far and away the largest use of water is for irrigation (69%); second is for industry (23%); and third is for direct human consumption (8%). These percentages vary greatly from one region to another, depending on natural precipitation and degree of development (Fig. 11-9).

Humans take freshwater from whatever source they can. In some cases this means capturing precipitation directly in a rain barrel under a downspout. The major sources of freshwater, however, are surface water, namely rivers and lakes, and groundwater. Before municipal water supplies were made available, each family dipped water for its own use from a local stream or shallow well. This method is still used to a considerable extent, and women in many developing countries walk long distances each day to fetch water. Because surface water and shallow wells often receive runoff, they frequently are polluted with various wastes, including animal excrements and human sewage likely to contain pathogens (disease-causing organisms). Yet, unsafe as it is, this polluted water is the only available water for an estimated 1.2 billion poor people in less-developed countries (Fig. 11-10). It is commonly consumed without treatment—but not without consequences. According to United Nations estimates, contaminated water is responsible for 80% of the diseases in the developing world and for the deaths of 4–5 million children per year, over 1 million from simple diarrhea (see "Global Perspective" box, page 276).

In the developed countries, also, the major freshwater sources are rivers and lakes, but methods for collection, treatment, and distribution are more sophisticated. Dams are built across rivers to create reservoirs,

FIGURE 11-9
Human usage of water is divided among three major categories, as shown. The percentage used in each category varies with climate and relative development of the country. A dry-climate, less-developed region uses most of its water for irrigation (e.g., Africa), whereas moist-climate, industrialized countries (e.g., Europe) require the largest percentage for industry. (From World Resources Institute, "World Resources," WRI, 1992/1993, p. 161.)

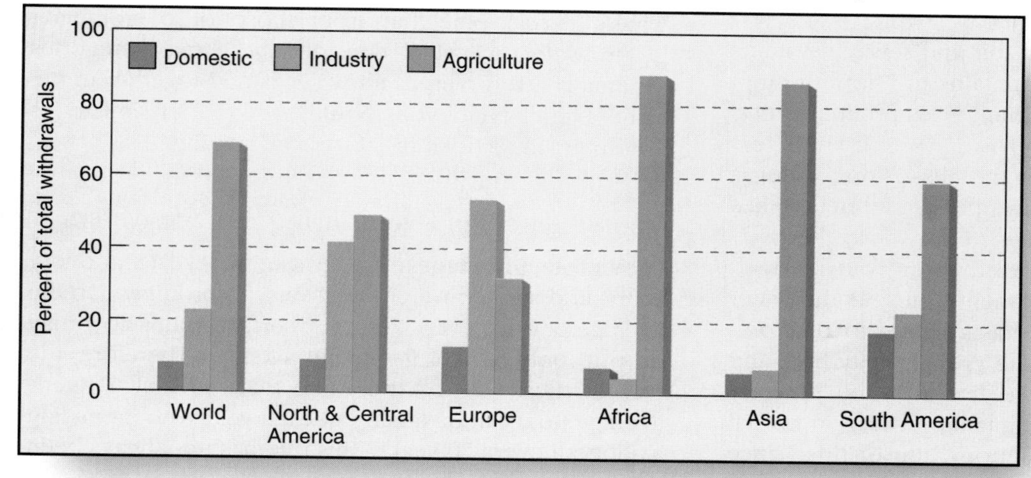

FIGURE 11-10
In many villages and cities of the developing world open, untreated wells such as this one in a Mexican village are the only water supply for people. Such sources may often be contaminated with pathogens and other pollutants. (Y. Momtiak/Photo Researchers, Inc.)

which hold water in times of excess flow and can be drawn down at times of lower flow. In addition, dams and reservoirs may provide for power generation, recreation, and flood control. Water for municipal use is piped from the reservoir to a treatment plant, where it is treated to kill pathogens and remove undesirable materials, as shown in Fig. 11-11. After treatment, water is distributed through the water system to our homes, schools, and industries. The wastewater, collected by the sewage system, is carried to a sewage-treatment plant, where it is treated before being discharged into a natural waterway (see Chapter 13). Often it is discharged into the same river from which it was withdrawn, but farther downstream.

So far as possible both water and sewage systems are laid out so that gravity maintains the flow through the system. This arrangement minimizes pumping costs and increases reliability.

On major rivers, such as the Mississippi, water is reused many times. Each city along the river takes water, treats it, uses it, and then returns it to the river. In developing nations, the wastewater is discharged often with minimal or no treatment. Thus, as the water moves downstream each city has a higher load of pollutants to contend with than the previous city had, and ecosystems at the end of the line may be severely affected by the pollution. Pollutants include industrial wastes as well as pollutants from households, since industries and residences generally utilize the same water and sewer systems.

Reservoirs created by dams on rivers are also major sources of water for irrigation. In this case no treatment is required. As noted above, the croplands are the end point for this water except insofar as it reenters the water cycle through evaporation.

Both to augment surface water supplies and to obtain water of higher quality, in the past few decades there has been an increasing trend of drilling wells and tapping groundwater. Since 1950, hundreds of cities have drilled huge wells for municipal supplies, and millions of wells have been drilled for individual households in suburbs beyond municipal supply systems. In addition, farmers have turned to the center-pivot irrigation system (Fig. 11-15). The use of such wells has increased tremendously in the last 25 years, increasing agricultural production as noted in Chapter 9, but also consuming huge amounts of groundwater. One system may use as much as 10,000 gallons (40,000 liters) per minute.

From your understanding of the water cycle, you can see that both surface water and groundwater are replenished through the cycle. Therefore, in theory at least, such water represents a sustainable or renewable (self-replenishing) resource, but it is not inexhaustible. If humans attempt to extract amounts that exceed those of natural flow, shortages will occur. Furthermore, as water is diverted from its natural pathway to human uses, there will be ecological consequences. Using and diverting water in volumes that lead to shortages or other undesirable consequences are spoken of as *overdrawing* or *overdraft of*, water resources. Let's turn our attention now to its consequences.

Overdrawing Water Resources

Because the consequences of overdrawing surface waters differ somewhat from those of overdrawing groundwater, we shall consider these two categories separately. However, as all water is tied together in the same overall cycle, you should be aware of similarities and interconnections in the two discussions.

Consequences of Overdrawing Surface Waters

Inevitable Shortages. There are wet years and dry years, and surface-water flows vary accordingly. On the average of once every 20 years, surface-water flow may drop to only 30% of its annual average. Therefore, the rule of thumb is that no more than 30% of a river's average flow can be taken out each year without risking a shortfall every 20 years. This rule has not always been

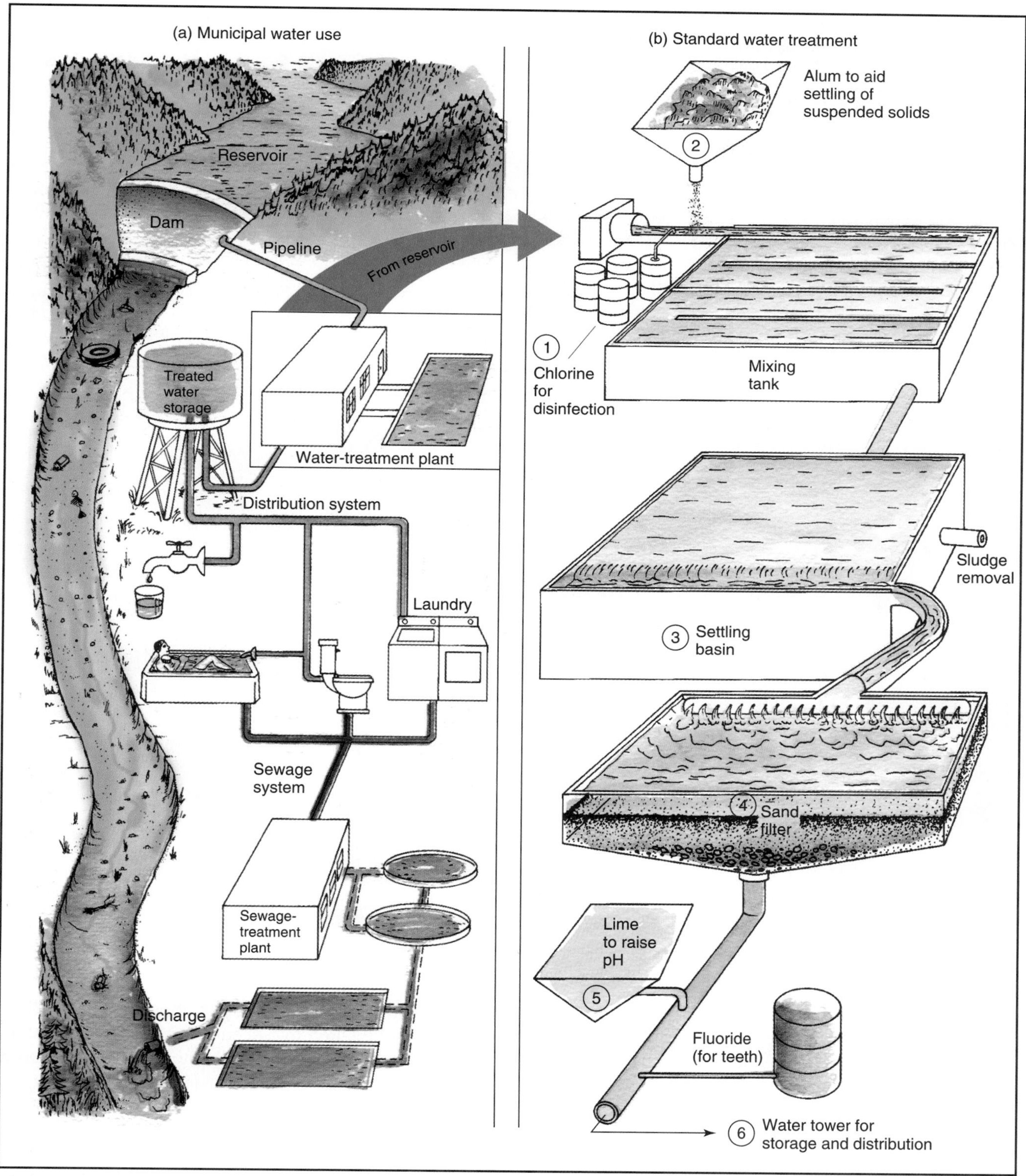

FIGURE 11-11
(a) Municipal water use. Water is often taken from a river, treated, used, then returned. (b) Water treatment. Water is piped from a reservoir to the treatment plant. At the plant, (1) chlorine is added to kill bacteria, (2) alum (aluminum sulfate) is added to coagulate organic particles, and (3) the water is put into a settling basin for several hours to allow the coagulated particles to settle. It is then (4) filtered through sand, (5) treated with lime to adjust pH, and (6) put into a storage water tower or reservoir until distribution to your home.

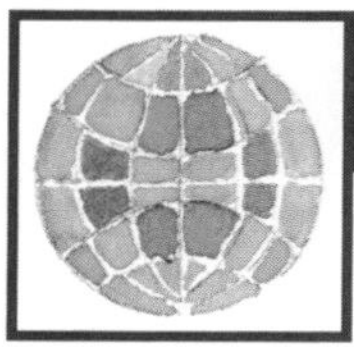

Global Perspective

People and Water

Safe drinking water is fundamental to human health and well-being. Yet much of the world's population does not have access to this essential resource, as indicated by the color coding on the map below. Furthermore, water scarcity will become a severe constraint on food production, economic development, and protection of natural ecosystems, if total annual water supplies diminish below 1000 cubic meters per person. Twenty-six countries, indicated by cross hatching on the map, are already below this threshold and, with rapid population growth, many more will cross into this category in the near future. Finally, a number of regions, including some major cities (Béijing, New Delhi, and Mexico City), are meeting current demands only by depleting groundwater reserves (blue dots), a nonsustainable solution.

Sources: "Percent of population with access to safe water," *World Bank, World Development Report*, 1994. Other data: Postal, Sandra. *Last Oasis*, Washington, DC: Worldwatch Institute, 1992.

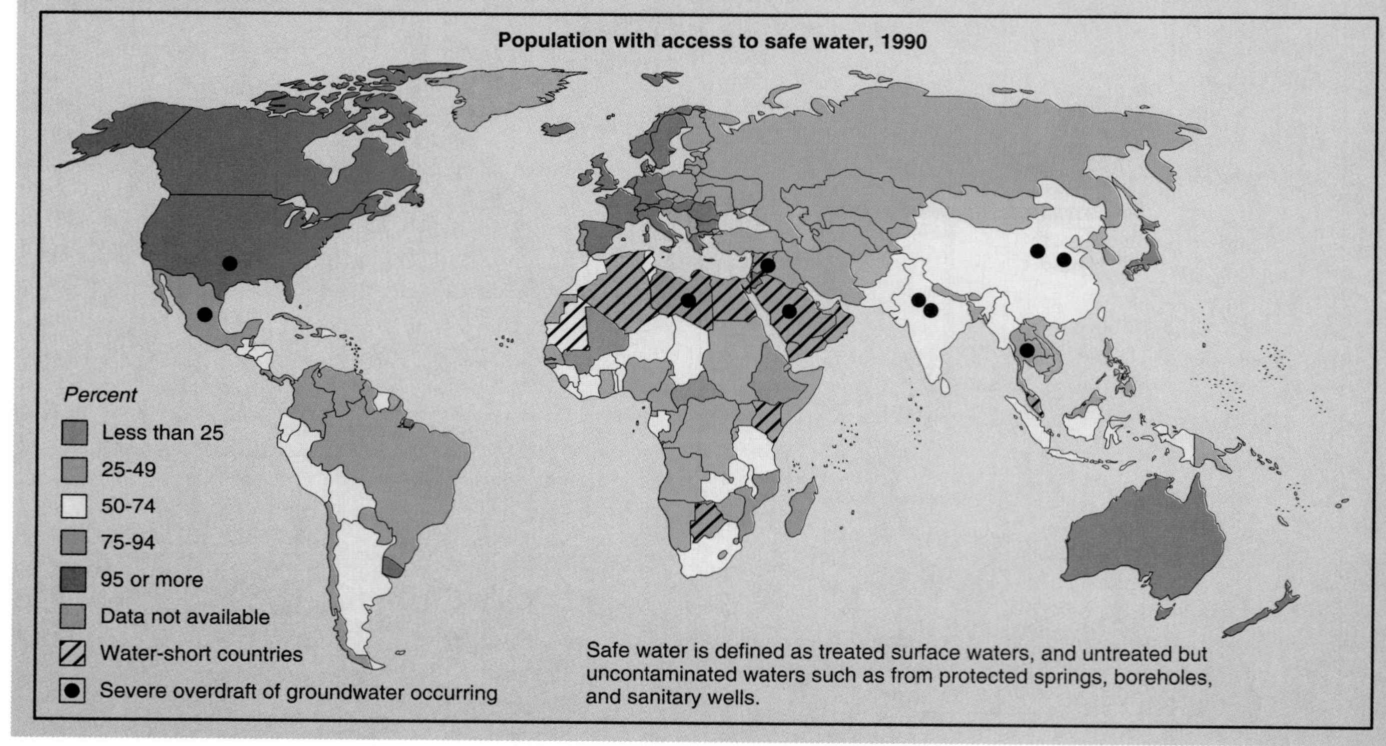

heeded, however. In some river systems in the United States, for instance, water demand has grown to 100% (and even *more than* 100%!) of average flow, making water shortages of increasing length and severity inevitable (Fig. 11-12).

Southern California is a case in point. Because of a seven-year drought, much of that part of the state entered the spring of 1992 with reservoirs down to 20% to 40% of capacity. If the drought had continued another year there would have been a severe crisis, but fortunately rains came and mitigated the situation—a narrow escape. Of course, people blamed the water shortage on the "terrible drought," but was the drought the real cause? Southern California has a desert climate, and prolonged droughts are not abnormal. Their likelihood must be taken into account in long-term planning for sustainability. California's water future must be at the forefront of growth and development planning.

Demands on the Colorado River by dry southwestern states are another example, setting the stage for inevitable shortages. Around the world there are many cases where tensions are mounting as different nations share common rivers and hence compete for the same water (see the "Ethics" box on page 278).

Ecological Effects. When a river is dammed and its flow is diverted to cities or croplands, the waterway below the diversion is deprived of that much water. The

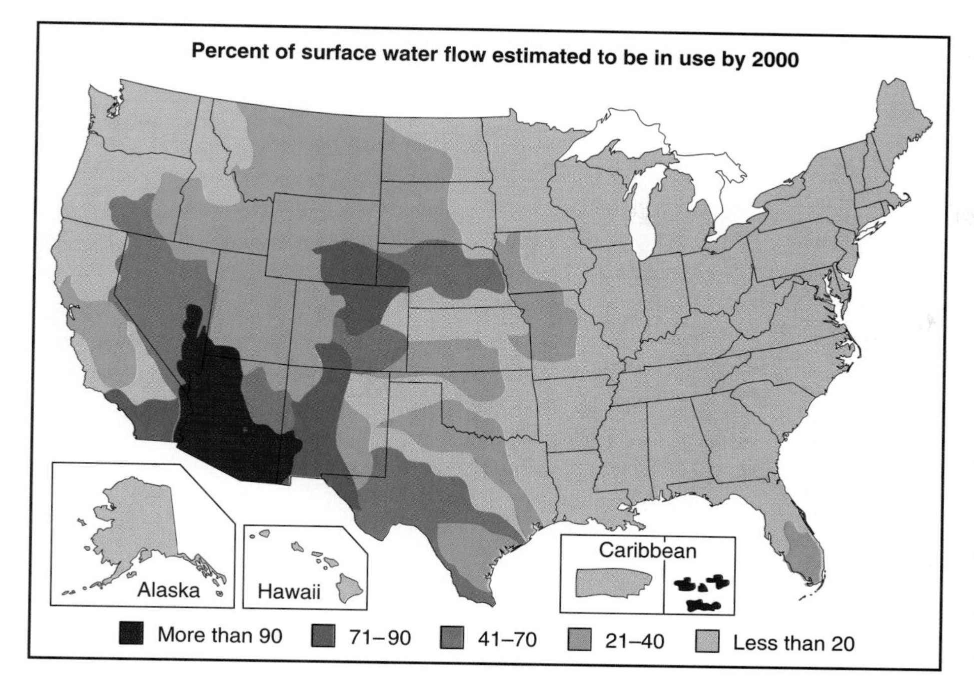

FIGURE 11-12
Droughts occur on an average of every 20 years and may reduce normal water flows by 70%. Therefore no more than 30% of the average surface-water flow can be counted on to be continuously available. By the year 2000, large areas of the United States will be above the 30% level, making severe, recurring water shortages inevitable. (U.S. Water Resources Council.)

impact on fish and other aquatic organisms is obvious, but the ecological ramifications go far beyond the river. Wildlife that depends on the water or on food chains involving aquatic organisms—and this includes virtually all wildlife—are also adversely affected. Wetlands along many rivers, no longer nourished by occasional overflows, dry up, resulting in tremendous dieoffs of waterfowl and other wildlife that depended on these habitats (Fig. 11-13). Fish such as salmon, which swim from the ocean to spawn far up rivers, are seriously affected by the reduced water level and have a problem getting around the dam, even one equipped with fish ladders. If the fish do get up the river, the hatchlings have similar problems getting back to the sea.

FIGURE 11-13
(a) Ducks and geese at Klamath, Oregon. Vast flocks of such waterfowl are supported by wetlands. (b) What happens to these birds when wetlands dry up because of water diversion or falling water table? (a) Margot Conte/Animals Animals; b) Dick Rowan/Photo Researchers, Inc.)

(a)

(b)

Ethics

Water: Who Should Get It?

Many rivers form borders between or flow through a number of states or countries, and of course having different governments involved complicates the situation. As water demands shoot up with expanding populations, the issue of who has the right to the water is leading to ever-increasing acrimony between states and between nations.

For example, in a 1922 agreement, the states of the U.S. Southwest divided the flow of the Colorado River among themselves: 7.5 million acre-feet to be shared by Colorado, Wyoming, Utah, and New Mexico—all states in which tributaries of the river originate—and 7.5 million acre-feet to be shared by Arizona, Nevada, and California—states through which the lower river flows. (An acre-foot is the amount of water that will cover 1 acre to a depth of 1 foot, about 325,000 gallons.) Another agreement gave Mexico, where the mouth of the Colorado River empties into the Gulf of California, the right to 1.5 million acre-feet, and gave Native Americans whatever amounts they need. (Their use was almost none at the time.) All went well as long as demands were below the allocated rights. As population has grown, however, each party is demanding its full share. The only problem is that the *total* average flow of the Colorado is only 13.0 million-acre feet (1930 to 1985 average). Even worse, for the past few years flow has dropped to just 9 million acre-feet.

Another water-allocation problem takes us to the other side of the world. About 90% of the tributaries of the Nile River, which flows through Egypt into the Mediterranean Sea, arise in Ethiopia, and Ethiopia is now considering ways to divert more of the water to ease its drought and famine. Egypt, which depends on the Nile water, has blocked a loan for the Ethiopian project.

The conflict between Israel and Jordan over the West Bank, a piece of territory on the western bank of the Jordan River, is as much over access to water as over territory. Many people speculate that the next war(s) in the Mideast may well be over water.

What moral and ethical principles are involved in such disputes? What are the moral dilemmas? How do you think they should be resolved? Who should be the final authorities to enforce agreements?

The problems extend to **estuaries**, which are bays in which fresh water from a river mixes with seawater. Estuaries are among the most productive ecosystems on Earth; they are rich breeding grounds for many species of fish, shellfish, and water fowl. As a river's flow is diverted to other locations, there is less fresh water entering and flushing the estuary. Consequently, the salt concentration increases, profoundly upsetting the ecology. The San Francisco Bay is a prime example. Over 60% of the freshwater which once flowed from rivers into the bay has been diverted, for irrigation in the Central Valley and for municipal use in southern California. Without the freshwater flows, salt water from the Pacific has intruded into the bay, with devastating consequences. Chinook salmon have become all but nonexistent; delta smelt and striped bass populations are down by more than 90%, and water quality for drinking and irrigation in the region is at risk. New standards will require freshwater flows to be restored to the Delta (the convergence of rivers leading into the bay), but central and southern California will have to give up some water.

Mono Lake, a 63-square-mile (163-km^2) lake in east-central California provides another example—one that also shows the potential for rectification by the public. Mono Lake has no outlet, but in the past a substantial inflow of fresh water from snowmelt off the Sierra Nevada kept its water diluted. Like other salt lakes around the world, Mono Lake supported numerous species of wildlife, including huge flocks of aquatic birds.

Beginning in the 1940s, however, much of the freshwater inflow was diverted to support water-hungry Los Angeles. Soon Mono Lake was losing more water to evaporation than it was receiving, and gradually it began to shrink. As the lake shrank the salt in its water became more concentrated, threatening ecological collapse (Fig. 11-14). In the latter 1970s, a group of students formed the Mono Lake Committee, which brought legal action under the Public Trust Doctrine, an old law protecting waterways from alteration. However, originally intended to protect navigation, the idea that destruction of the integrity of an ecosystem constituted a violation of the public trust was a new concept that had to be tried in the courts, and it was bitterly opposed by the city of Los Angeles. Nevertheless, the citizens of the Mono Lake Committee persevered through the many appeals and won the final decision in September 1994. The State Water Resources Control Board finally ruled that Los Angeles must reduce diversions of water sufficiently to allow Mono Lake, now some 40 feet below its original level, to recover by at least 15 feet (5 m) above the 1995 level. Notably, the final decision involves a complex compromise that allows Los

FIGURE 11-14
With its freshwater input diverted to Los Angeles, Mono Lake was drying up, leaving unique sculptured forms of tufa (precipitated calcium carbonate) and threatening ecological collapse. Recent court rulings requiring that a balance be struck between ecological needs and the municipal needs of Los Angeles have reduced these diversions. Whether there will be an ecological recovery remains in question. (Peter Worster, Oakland, CA)

Angeles to take variable and increasing amounts of water as the lake recovers. Thus recovery of the 15 feet biologists estimate, will require about 25 years—but at least a major ecological collapse of wildlife populations supported by the lake seems to have been averted.

And Los Angeles? The decision forces L.A. to initially reduce its takings from the Mono Lake supply by 80%, which amounts to a 12% cut in the city's total water supply. However, water-saving habits developed by citizens over previous drought years and additional conservation, and recycling schemes being implemented, are more than adequate to make up for the loss.

The problem is not limited to the United States. The southeastern end of the Mediterranean Sea was formerly flushed by water from the Nile River. Because this water is now held back and diverted for irrigation by the Aswan High Dam in Egypt, this part of the Mediterranean is suffering the ecological consequences.

The world's most dramatic example of water mismanagement is the Aral Sea, an inland sea in south-central Russia ("Earth Watch" box, page 283).

Consequences of Overdrawing Groundwater

To augment supplies of high-quality fresh water humans have increasingly turned to groundwater, and advances in drilling and pumping technology have made it convenient and economical to do so. In tapping groundwater you are tapping a large but not unlimited natural reservoir. Its sustainability ultimately depends on balancing withdrawals with rates of recharge.

In some dry regions, the groundwater found is actually water that accumulated millennia ago, when the climate in the region was wetter. Current rates of recharge are nil. The practice of tapping such pockets of ancient water is frequently spoken of as "mining fossil water" to emphasize that the resource will ultimately be depleted regardless of rates of withdrawal.

Falling Water Tables and Depletion. Rates of groundwater recharge aside, however, the simple indication that groundwater withdrawals are exceeding recharge is a falling water table, a situation that is common throughout the world (see "Global Perspective" box, page 276).

Since irrigation consumes far and away the largest amount of freshwater, depleting water resources will ultimately have its most significant impact on crop production. A prime example is the Great Plains region of Texas, Oklahoma, New Mexico, Colorado, Kansas, and Nebraska. Within this arid region some 150,000 wells tap the Ogallala aquifer to supply irrigation water to 140 million acres (56 million hectares) (Fig. 11-15). The irrigated farming in this region, which has developed since World War II, now supplies 15% of the nation's total value of wheat, corn, sorghum, and cotton and 38% of the nation's livestock about 40% overall of the grain fed to cattle. However, the withdrawal rate is about 24 million acre-feet per year whereas natural recharge is only about 3 million acre-feet and water tables are dropping rapidly. In the past 40 years, the water table has dropped 30 m (100 ft) and is lowering at 2 m per year. Irrigated farming has already come to a halt in some sections, and it is predicted that over the next 10 years another 3.5 million acres (1.4 million hectares) in this region will be abandoned or converted to dryland farming (ranching and production of forage crops) because of water depletion.

FIGURE 11-15
Exploitation of the Ogallala aquifer (a) has made this arid region of the United States into some of the most productive farmland in the country. (b) Water is applied by means of center-pivot irrigation in which water is pumped from a central well to a self-powered boom that rotates around the well, spraying water as it goes. (c) Aerial photograph shows the extent of center-pivot irrigation throughout the region, resulting in groundwater being depleted, which will bring an end to this kind of farming. (b) Earl Roberge c) F. Grohier/Photo Researchers, Inc.)

Although running out of water is the obvious eventual conclusion of overdrawing groundwater, falling water tables have other consequences before the water is entirely depleted. Let us now examine some of them.

Diminishing Surface Water. Surface waters are also affected by falling water tables. In various wetlands, for instance, the water table is essentially at or slightly above the ground surface. Dropping water tables result in such wetlands drying up, with the ecological results described earlier. Further, as water tables drop springs and seeps dry up, diminishing streams and rivers even to the point of dryness. Thus, excessive groundwater removal leads to the same effects as diversion of surface water.

Land Subsidence. Over the ages, groundwater has leached cavities in Earth. Where these spaces are filled with water, the water helps support the overlying rock and soil, but as the water table drops, this support is lost. Then there may be a gradual settling of the land, a phenomenon known as **land subsidence**. The rate of sinking may be 6–12 inches (10–15 cm) per year. In some areas of the San Joaquin Valley in California, land has settled as much as 27 feet (9 m) because of groundwater removal. Land subsidence causes building foundations, roadways, and water and sewer lines to crack. In coastal areas, subsidence causes flooding unless levees are built for protection. For example, where a 4000-square-mile (10,000-km^2) area in the Houston-Galveston Bay region of Texas is gradually sinking because of groundwater removal, coastal properties are being abandoned as they gradually are inundated by the sea. Land subsidence is also a serious problem in New Orleans, sections of Arizona, Mexico City, and many other places throughout the world.

FIGURE 11-16
Sinkhole. Removal of groundwater may drain an underground cavern until the roof, no longer supported by water pressure, collapses. The result is the sudden development of a sinkhole, such as this one, which consumed a home in Frostproof, Florida, July 12, 1991. (R. Veronica Decker/The Orlando Sentinel/Gamma-Liaison.)

Another kind of land subsidence, the occurrence of a **sinkhole**, may be sudden and dramatic (Fig. 11-16). A sinkhole results when an underground cavern, drained of its supporting groundwater, suddenly collapses. Sinkholes may be 300 feet (91 m) or more across and as much as 150 feet deep. Formation of sinkholes is particularly severe in the southeastern United States, where groundwater has leached numerous passageways and caverns through ancient beds of underlying limestone. An estimated 4000 sinkholes have occurred in Alabama alone, some of which have "consumed" buildings, livestock, and sections of highways.

Saltwater Intrusion. Another problem resulting from dropping water tables is **saltwater intrusion**. In coastal regions, springs of outflowing groundwater may lie under the ocean. As long as a high water table maintains a sufficient head of pressure in the aquifer, there is a flow of fresh water into the ocean. Thus, wells near the ocean yield fresh water (Fig. 11-17a). However, a lowering of the water table or a rapid rate of groundwater removal may reduce the pressure in the aquifer

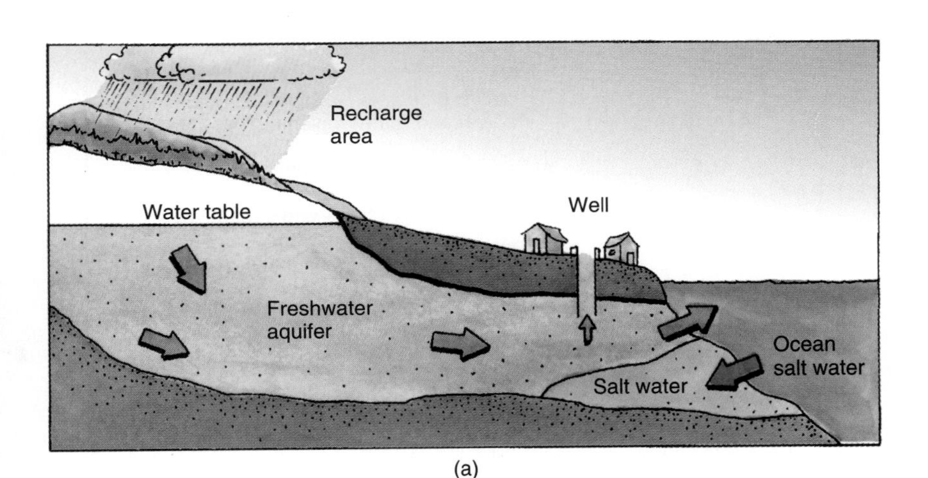

(a)

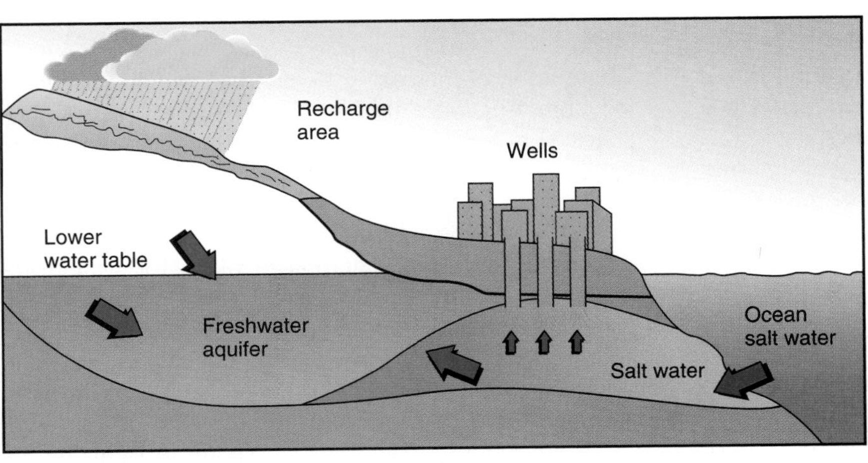

(b)

FIGURE 11-17
Saltwater intrusion. (a) Where aquifers open into the ocean, fresh water is maintained in the aquifer by the head of fresh water inland. (b) Excessive removal of water may reduce the pressure, so that salt water moves into the aquifer.

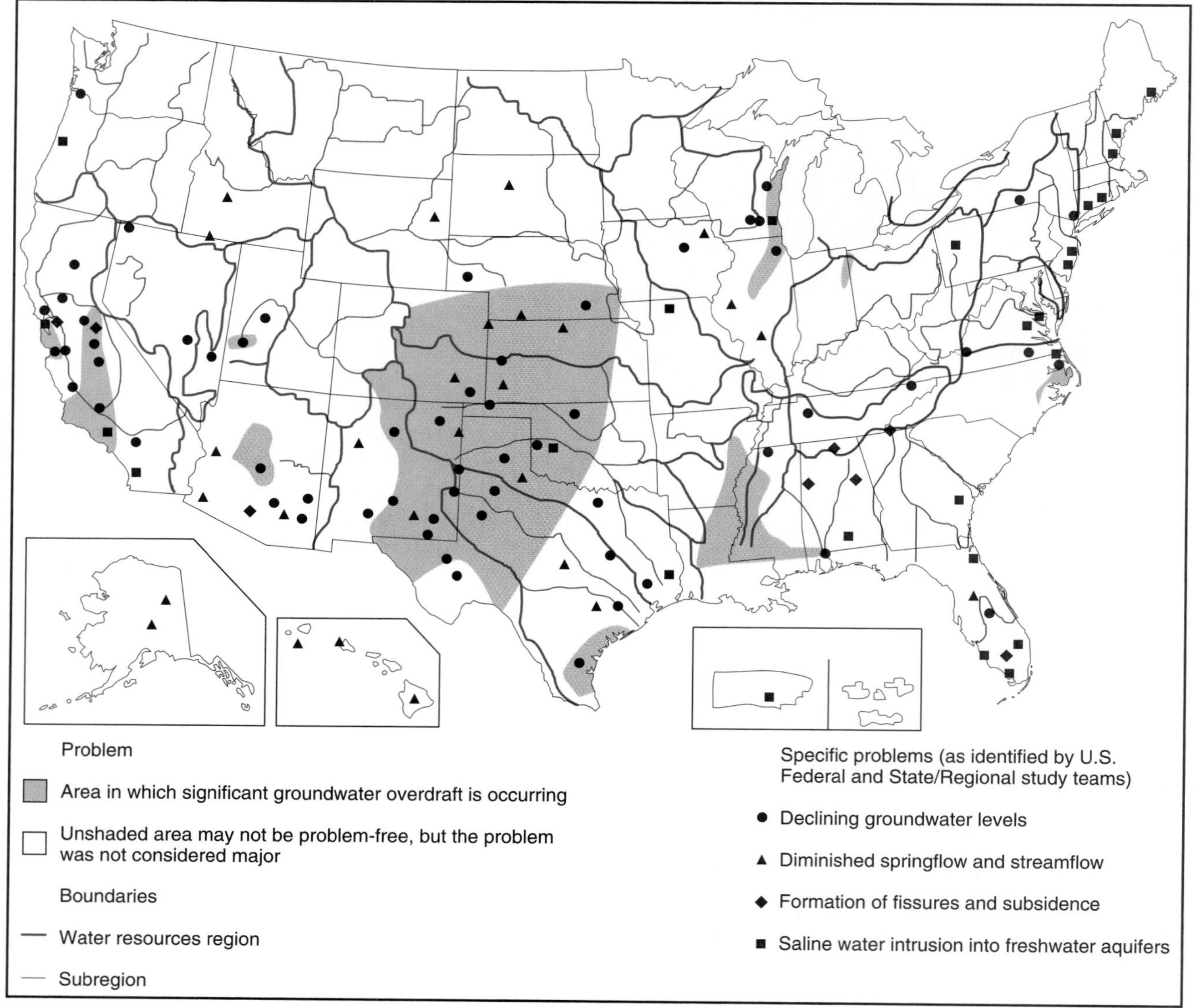

FIGURE 11-18
Declining groundwater levels and related problems in the United States. (From V. J. Pye et al., *Groundwater Contamination in the United States*, Philadelphia: University of Pennsylvania Press, 1983. Used with permission from the Academy of Natural Sciences, Philadelphia, PA.)

permitting salt water to flow back into the aquifer and hence into wells (Fig. 11-17b). Saltwater intrusion is a problem at many locations along U.S. coasts.

Figure 11-18 summarizes these effects and shows where they are occurring most severely in the United States. Of course the problems are by no means unique to this country.

Obtaining More Water

Despite the obvious and growing negative impacts of overdrawing water resources, growing populations create an ever-increasing demand for additional water for both irrigation and municipal use. In the United States, very few rivers remain undammed, and many are dammed at several points. Yet governments eager to please both people clamoring for water and businesses having economic interests in dam construction persist in offering ever more grandiose plans for water projects. One example, which is still being promoted, is a plan to dam the Yukon River in Alaska to create a reservoir that would flood nearly a third of the state and bring the water via canal all the way to the American Southwest and northern Mexico.

Increasingly, however, people are recognizing the inevitable trade-offs that occur with such projects and

Earth Watch

The Death of the Aral Sea

In the 1930s, economic planners sitting in thick-walled stone buildings in Moscow set in motion a chain of events that has almost killed an entire sea.

In order to create vast cottonfields in the drylands of Soviet Central Asia, the planners had long irrigation canals dug, fed by the waters of two rivers, the Amu Daria and the Syr Daria, that flow into the inland Aral Sea. In the statistics of the central planners, the project was a huge success. The cotton harvests grew until the Soviet Union became the world's second largest cotton exporter, after China.

But the statistics did not show the effect of diverting most of the river flow away from the Aral Sea. By 1989, the sea was receiving only one-eighth the level of water as in 1960. Its water level had dropped by 47 feet (16 meters), and its volume had shrunk by two-thirds. Once the size of North America's Lake Huron, its total area diminished by 44%.

The shores of the sea receded, leaving fishing villages tens of miles from the shore. A new desert was created around the sea, with salt strewn in massive dust storms across a vast area.

At the same time, Moscow's demands for cotton cultivation were met with saturation use of pesticides, fertilizers, and herbicides that flowed into the rivers and canals. The population around the sea, deprived of clean drinking water and living on poisoned soil, experienced rising rates of disease and infant mortality.

"The problem of the Aral Sea is very simple," said Igor Zonn, a specialist on the subject, "the Aral Sea will be dead, not soon and not completely, but it will be dead."

Dr. Zonn and fellow Russian scientist Nikita Glazorsky, head of the Institute of Geology and until recently deputy environment minister of Russia, have been working for a long time to try to save the Aral Sea. The Russian scientists say a technical solution for the sea is already well formulated. Much of the water is now wasted because of evaporation and drainage out of unlined irrigation canals and primitive irrigation technology.

At least half the (120-km^3) flow of the two rivers could be saved by rebuilding the canals and introducing new irrigation systems. It would be enough to begin to stabilize the Aral Sea at its present level, though not enough to restore it. At the same time there must be a program of health care and a concerted effort to shift the economy away from cotton cultivation, they say.

But such measures require resources and a political will that is not present. The breakup of the Soviet Union and its replacement by a loose commonwealth have given the Russian government an opportunity to dump the problem into the laps of the five former Soviet Central Asian states that form the Aral Sea's water basin. This is the source of anger for Central Asians who see the cotton monoculture imposed by Moscow as a classic example of colonial-style exploitation.

"We are left face to face with the Aral Sea problem," Kazakhstan president Nursultan Nazarbayev said, "meanwhile 97% of the cotton of the Central Asia and Kazakhstan is taken out to the European part of the Commonwealth of Independent States where the employment of 10 million workers depends on cotton use."

Russia has an ethical responsibility to help, agrees Dr. Glazorsky. "All of us live in this system, and we must solve this problem together."

By Daniel Sneider. Reprinted by permission from *The Christian Science Monitor*,

are considering them to be unacceptable. Intense counterlobbying by various environmental organizations has, in recent years, led to the rejection of a number of dam proposals, including the Yukon project (at least for the present). In addition, environmentalists won passage of the Wild and Scenic Rivers Act of 1968, which protects rivers designated as "wild and scenic" from damming and other harmful operations.

Designated wild and scenic rivers are in many ways equivalent to national parks. But like national parks, they need pubic supporters and defenders. Without such support, not only will expansion of areas under protection cease, but already-protected areas will be whittled away by various commercial interests. The primary public supporter and defender of Wild and Scenic Rivers is the association American Rivers Inc. (see Appendix A).

Bitter controversies between environmentalists and dam-building interests continue and are not limited to the industrialized world. An example currently unfolding is the construction of the Three Gorges Dam in China. For the Chinese government, this project represents a centerpiece of their effort to industrialize and join the modern age. But to others, some of whom have been jailed by the Chinese government for protesting, it is shaping up as a major ecological and social disaster. Over 1.1 million Chinese—including entire cities, farms, homes, and factories—will be displaced and relocated to make way for the 375-mile- (600 km-) long reservoir.

A lesson might be taken from the Aswan High Dam, which was constructed across the Nile River in the 1970s to bring Egypt into the modern industrial age. The loss of fisheries and productive wetlands below the dam, the loss of land flooded by the reservoir above, and unbridled population growth have largely canceled out any gains. The people in general are as bad off as they were before the dam was built.

In conclusion, does it appear that a sustainable future can be found in large-scale water exploitation and diversion projects? Fortunately, there are alternatives. They are based on the conservation and recycling of water.

Using Less Water

In the past, water has been treated as an inexhaustible resource that can be taken for granted. This viewpoint has led to extravagant and wasteful use of water. A developing-nation family living where water must be carried several miles from a well finds that one gallon per day per person is sufficient to provide for all essential needs, including cooking and washing. Yet, a typical household in a developed country consumes an average of 180 gallons (680 liters) per day per person. If all indirect uses are added this increases to a per capita use of 1600 gallons (6056 liters). Similarly, a peasant farmer may irrigate by carefully ladling water onto each plant with a dipper, while typical modern irrigation floods the whole field. This is not to say that we in the developed world should take up developing-world habits, but it does point up that water consumption can be cut back 75% or even more without people suffering any great hardship. It is through such cutbacks that we have an opportunity to meet our needs without undercutting needs of future generations or natural ecosystems.

Let's consider some specific measures that are being implemented to reduce water demands.

Irrigation

Where irrigation water is applied by traditional flood or center-pivot systems, about 60% is wasted in evaporation, percolation, or runoff. This loss can be all but eliminated by installing **drip irrigation** systems, networks of plastic pipes with pinholes that literally drip water at the base of each plant (Fig. 11-19). With such systems less than 5% of the water is wasted. They have the added benefit of retarding salinization (see Chapter 9). Although drip irrigation is being used more and more, especially in small gardens, 97% of the irrigation in the United States, and 99% throughout the world, is still done by traditional flood or center-pivot methods.

The reason for the low changeover is one of cost to farmers; it costs about $1000 per acre to install a drip system. Water for irrigation, on the other hand, is heavily subsidized by the government; the farmer pays next to nothing for it. Therefore, it makes financial sense to use the cheapest system for distributing water, even if it is wasteful. Carrying the absurdity even further, government-guaranteed crop price supports entice the farmer to grow crops such as corn and cotton under irrigation despite chronic surpluses of these crops. In effect we (taxpayers) are paying the farmer to squander water resources to grow crops that aren't needed.

FIGURE 11-19
Drip irrigation. Irrigation is the most consumptive water use. Drip irrigation offers a conservative method of applying water, dripping it on each plant through a system of plastic pipes. (Lowell Georgia/Photo Researchers, Inc.)

Municipal Systems

The water consumption of 180 gallons (680 liters) per day per person in modern homes is mostly used for washing and flushing away wastes: flushing toilets (3–5 gallons, per flush), taking showers (2–3 gallons, per minute), doing laundry (20–30 gallons, per wash), and so on. Watering lawns and filling swimming pools add to this use along with other indirect consumption.

Water conservation has long been promoted as a "good citizen" public relations measure without, it may be added, much effect. Now, numerous cities are facing the stark reality that it will be extremely expensive and in many cases impossible to increase supplies by traditional means of building more reservoirs or drilling more wells. The only practical alternative, they are discovering, is to take real steps toward reducing water consumption and wastage. A considerable number of cities have programs whereby leaky faucets will be repaired, and low-flow shower heads and water-displacement devices in toilets will be installed free of charge. New York City and Los Angeles are providing rebates of $100 to $250 dollars to people who replace an old toilet with a new model that uses only 1.6 gallons (6.5 liters) per flush. Phoenix is paying homeowners to replace their lawns with **xeroscaping**—landscaping with desert species that require no additional watering—and xeroscaping is becoming a thriving business.

Also, *gray-water* recycling systems are being adopted in some water-short areas. **Gray water**, the slightly dirtied water from sinks, showers, bathtubs, and laundry tubs, is collected in a holding tank and used for such things as flushing toilets, watering lawns, and washing cars. During the last water crisis in California many residents adopted the habit of standing in a plastic tub while taking a shower. They then used the collected water for watering gardens, which was otherwise prohibited at the time.

Going further, a number of cities are using treated wastewater (sewage water) for irrigation, both to conserve water and to reduce pollution of receiving waters (see Chapters 12 and 13). Finally, some cities are considering and developing systems to treat wastewater to a degree that it can be recycled back into the municipal system.

If the idea of reusing sewage water turns you off, recall that *all* water is recycled by nature. There is hardly a molecule of water you drink that has not moved through organisms—including humans—numerous times. A number of communities are already treating their wastewater to a such a degree that its quality surpasses what many cities take in from lakes and rivers. A major impediment to recycling wastewater is simply the cost of pumping it back to the head of the system, which is invariably uphill.

Desalting Sea Water

The world's oceans are an inexhaustible source of water, not only because they are vast but because any water removed will ultimately flow back in. Traditionally, sea water has not been used because of the costs required to remove the salt and pump it uphill to where it is used, and adequate sources of natural fresh water were available. However, with increasing water shortages and most of the world's population living near coasts, there is a growing trend toward **desalinization** (desalting) of sea water. Several thousand desalination plants already exist, primarily in Saudi Arabia, Israel, and other countries of the Mideast. The world produces 3.4 billion gallons (13 billion liters) of high-quality drinking water per day through desalination.

Two technologies are in common use for desalinization: microfiltration, or reverse osmosis, and distillation. Small plants generally use the microfiltration process in which the sea water is forced under great pressure through a membrane filter fine enough to remove the salt. Large plants, particularly where a source of waste heat is available (as from electrical power plants, for instance), generally use distillation (evaporation and recondensation of vapor). Additional efficiency is gained by using the heat given off by condensing water to heat the incoming water (Fig. 11-20). Even where waste heat is used, however, the costs of building and maintaining the plant, which is subject to corrosion from sea water, are considerable. Under the best of circumstances, the production of desalinized

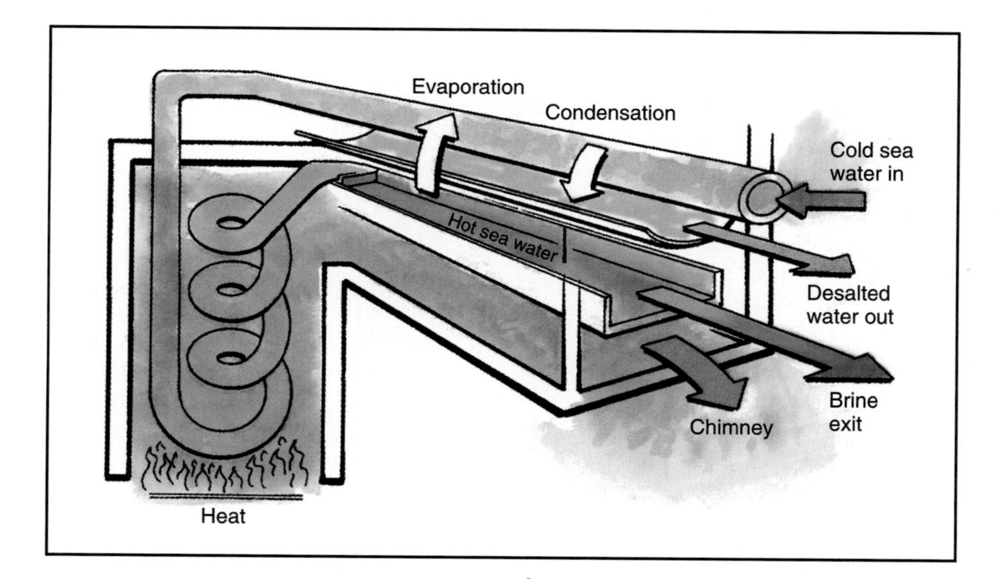

FIGURE 11-20
Desalting sea water by means of distillation. The principle of the design shown is to use the incoming cold water to condense the water vapor. Some of the heat required for vaporization is thus reused, improving the efficiency of the process.

water costs about $3 per 1000 gallons (4000 liters). This is three to six times what most city dwellers in the United States currently pay, but it is still not a high price to pay for drinking water. The high cost might cause some people to cut back on watering lawns and implement other conservation measures, but most people in the United States could afford it without undue strain.

Irrigation, however, commonly consumes 500 thousand gallons (20,000 liters) or more per acre for producing one crop. Farmers currently pay as little as 1¢ per 1000 gallons ($5 per crop for water). Thus, it is cost-effective to irrigate crops that bring in only a few hundred dollars per acre. Paying $3 per 1000 gallons of water would up the cost to $1500 per acre of crop. But, there is no way that the amounts of the cheapest desalted water made by utilizing waste heat would suffice. New plants utilizing additional energy would have to come into use. Then there would be the additional pumping costs to get the water from the coast to farmlands. As these costs are added up, to say nothing of future constraints on energy use (see Chapter 21), a future of irrigating croplands with desalinized sea water seems out of the question.

Stormwater Mismanagement and Management

You have already seen the general aspects of how changing the soil surface changes the infiltration-runoff ratio and in turn alters the rest of the water cycle. This problem is particularly acute where development results in exchanging porous, water-receiving topsoil for an asphalt pavement and rooftops, which shed nearly 100% of the water falling on them. Even the soil in lawns is much more compacted than in a natural ecosystem and sheds a high percentage of rainwater.

The traditional practice was, and in many cases still is, to channel the storm runoff down storm drains. These led to the nearest convenient off-site location to discharge the water, usually the side of valley or a natural stream bed. The ramifications of this practice are many.

Mismanagement and Its Consequences

The increase in stream flow during rains because of the sudden influx of runoff is well documented, as is the decrease in flow between rains as springs have gone dry because of reduced groundwater recharge (Fig. 11-21).

Most streams in urban and suburban areas that used to flow quietly year-round and supported a diverse biota are now little more than open storm drains alternating between surges of water when it rains and dry, dead beds when it doesn't (Fig. 11-22). Indeed, many such streams have been covered and simply incorporated into the storm-drain system. In urban areas, vast arrays of streams and their tributaries no longer exist except as underground storm drains.

The surges from increasing runoff have a number of other effects as well.

FIGURE 11-21
Curves are for similar storms on Brays Bayou in Houston, Texas, before, during, and after development in the early 1950s. Note the increasing height of the surge occurring with the storm and the decreasing volume of flow that occurs later in the cycle. Because of increasing runoff from new development in the watershed, established areas experience flooding where none occurred before. (Melissa Hayes English/Photo Researchers, Inc.; V. Van Sickle, in *Effects of Watershed Changes on Stream Flow*, W. Moore & C. Morgan, eds. Austin: U. of Texas Press, 1969.)

(a)

(b)

FIGURE 11-22
Effect of development on streamflow. Before development, this stream maintained a continuous, generally modest, flow of water throughout the year. Now, after development of the upstream watershed, the stream is dry most of time. (a) During rains there are high surges of water because of increased runoff. (b) A few hours later the stream is dry again because of depletion of groundwater. (Photographs by BJN.)

Flooding. The increased potential for flooding is self-evident. Countless communities, many of them expensive new suburban developments, have experienced flooding with increasing frequency and severity as development has led to paving more and more of the upstream watershed.

Streambank Erosion. The erosion resulting from poorly placed storm drain outlets can be horrendous (Fig. 11-23). Then effects continue downstream. Even short of flooding, the surges of water vastly accelerate erosion of the stream banks, undercutting trees and causing them to topple into the stream, which diverts water against and over the banks and causes even more erosion. While finer soil particles (clay and silt) are carried on to finally settle out in lakes and bays, coarser materials (sand, stones, and rocks) eroded from gullies below poorly placed storm drains and stream banks are deposited in the bottom of the stream channel itself. This filling of the stream channel causes the water to eat away even more at the sides, exacerbating the process. The result is that the stream channel gets wider and shallower. Indeed, a stream channel may become completely filled such that the water is diverted and simply floods the valley floor (Fig. 11-24). The results are topsoil erosion and the death of many trees because what soil is left is waterlogged. Gradually, a narrow, tree-lined stream may be converted into a broad washway of drifts of sand and gravel.

Increased Pollution. Consider all the materials used, carelessly dropped or spilled on the ground or, even

FIGURE 11-23
The gully seen in this photograph is the result of erosion caused by water exiting from a storm drain, just out of the picture in the background, funneling the runoff from a parking lot onto the hillside. The soil and stones eroded from such gullies add sediments to and clog the channels of waterways below (see Fig. 11-24). (Photograph by BJN.)

FIGURE 11-24
Sediment from upstream erosion fills and clogs stream channels, causing flooding and further erosion of the stream valley as the water is forced to find new pathways. In this photo, you can see how the stream channel has been nearly filled with fine gravel sediment, which eroded from upstream areas. (Photograph by BJN.)

worse, disposed of down storm drains. All may wash directly with runoff into adjacent natural waterways. Major categories of contaminants include:

- Nutrients from lawn and garden fertilizer
- Insecticides and herbicides used on lawns and gardens
- Bacteria from fecal wastes of pets
- Road salt and other chemicals from surface treatments or spills
- Grime and toxic chemicals from settled vehicle exhaust and other air pollution
- Oil and grease picked up from road surfaces or disposed of in storm drains
- Trash and litter carelessly discarded on the ground

How many other things can you think of? Indeed, urban runoff is now recognized as a major source of pollution for many rivers and estuaries.

On the other side of the infiltration-runoff ratio, decreased infiltration also exacerbates saltwater intrusion, land subsidence, and other problems related to falling water tables.

Channelization. What is to be done about the problems arising from increasing stormwater runoff? The old approach to streambank erosion and flooding was **channelization**. The stream bed was dredged, straightened, or gently curved, and lined with rock or concrete to prevent bank erosion (Fig. 11-25). To be sure, a channelized stream carries both water and sediment efficiently and reduces flooding and erosion in the area of the channel. However, you may also observe that it obliterates any semblance of a natural stream ecosystem. The channelized stream is simply an open storm drain. Indeed, it may be covered and made into a storm drain proper, as indicated previously.

The real defeat of channelization, however, came from recognizing that channelized sections simply served to transfer water farther downstream and create flooding there. The flooding on the Mississippi River in 1993 provides an extreme example. Millions of acres of upstream tributaries have been channelized to drain wetlands that normally would have held the water.

FIGURE 11-25
A channelized stream. To reduce flooding and streambank erosion, many streams and rivers have been channelized; that is, the water course has been straightened and lined with concrete or stone. What semblance of natural ecology remains? (Photograph by BJN.)

FIGURE 11-26
Stormwater management. Rather than letting excessive runoff from developed areas cause flooding and other environmental damage, runoff can be funneled into a retention pond as shown here. It can then drain away slowly, maintaining natural streamflow, or also can recharge ground water. The retention pond may be designed to retain a certain amount of water and thus create a pocket wildlife habitat in an otherwise urban or suburban setting. (Photograph by BJN.)

Having these areas drain efficiently exacerbated the flooding downriver.

A major impetus behind preservation of wetlands, besides protecting their natural ecology, is to maintain their water-storage capacity. In contrast, development of such wetlands invariably translates into increasing runoff, lowering water tables, and degrading water quality. It is this understanding that is behind the pressure of environmentalists to preserve wetlands, an effort that has received a serious setback by "takings legislation," legislation that basically says a landowner is free to develop wetlands—or any other ecotype—unless he or she is compensated for not developing it, a cost the government cannot afford.

Improving Stormwater Management

A number of states now require stormwater management to be a part of any new development. The management technique most commonly used is to construct a *stormwater retention reservoir* at the low extremity of the site being developed. The **stormwater retention reservoir** is simply a "pond" that receives and holds runoff from the site during storms. From the pond, the water may gradually infiltrate into the soil, or it may trickle out slowly through a standpipe mounted in the pond. Thus, the pond plays a role imitating groundwater storage and it may also create a pocket of natural wetland habitat supporting wildlife (Fig. 11-26). Large stormwater retention–flood control reservoirs may serve additionally as recreational areas with boating, fishing, and so on.

Techniques of stormwater management on small sites include such things as "wells" and trenches filled with rock that receive water and allow it to percolate; terraces that receive and hold water on the "steps" and allow it to infiltrate; and rooftops and parking lots designed to "pond" water and let it trickle away slowly. Of course, storm water collected in any sort of reservoir can act as a source of water for watering lawns, washing cars, and other nondrinking purposes.

Groping for a Better Way

Our study of the water cycle shows us that nature provides a finite flow of water through each region. When humans come on the scene, this flow of water is inevitably divided between the needs of the existing natural ecosystems and the agricultural, industrial, and domestic needs of humans. We have seen that there are many cases where the water flow is inadequate to provide the desired amounts of water to each of these categories simultaneously. Therefore, contentions build between environmentalists wishing to preserve the integrity of the natural ecosystem, farmers wanting irrigation water, and cities wanting supplies for industrial and domestic use.

We have seen vast opportunities for conserving and recycling water. The question is, How can we get ourselves and others to implement these measures short of social or ecological disaster? Laws, litigation, and the courts have been and are being used to settle disputes, but we can see that this is probably the most costly and time-consuming method; parties are left embittered and antagonistic, and wins are questionable. For example, we saw that the recovery of Mono Lake will be long and questionable despite the win by environmentalists.

We might at least imagine a better way. With suitable leadership, the contending parties might be brought together and provided with information regarding such questions as: How much water may be removed from a given source without disrupting natural ecosystems? If that amount is less than desired, what are the conservation or recycling measures available to reduce needs for irrigation and cities? Is the crop being irrigated really essential? (In a number of cases, cities are buying farm lands simply for the water rights.) How can revenues be raised and costs be shared to implement such measures? In short, how can we best balance our direct needs for water with the broader objectives of Earth stewardship and creating a sustainable future?

There are some positive precedents. In the early 1970s the Adirondack State Park in upper New York State—the source of New York City's water supply—was under severe and conflicting pressures from recreational, environmental, farming, logging, and development interests. Ecologically Sustainable Development, a firm headed by George D. Davis, was called in to negotiate. The company conducted a scientific survey to determine the region's carrying capacities for various activities. Finally, a zoning map was drawn up with rules restricting where and how the various activities could be conducted even on land that was privately owned. Everyone saw that their interests were being considered, and each "gave some to get some." The result was the first U.S. regional land-use plan. Davis is now engaged in negotiating similar agreements in and between various countries around the world.

In conclusion, the choice is between bitter disputes in which all ultimately stand to lose and finding rational compromises between the economic needs of people and the ecological needs of the land that supports us.

Review Questions

1. How does water change its state with changes in energy (heat)?
2. Define evaporation, condensation, precipitation, runoff, infiltration, transpiration, evapotranspiration, capillary water, gravitational water, percolation, groundwater, water table, aquifer, recharge area, spring, seep.
3. Use the terms defined in Question 2 to give a full description of the water cycle, including each of its three "loops." What is the water quality (purity) at different points of the cycle? Explain the reasons for the differences.
4. Why do different regions receive different amounts of precipitation?
5. Describe how groundwater and wetlands act as natural reservoirs to moderate the flow of streams and rivers in wet periods and dry periods. What impact will development of wetlands have on water quality and quantity?
6. How does changing Earth's surface, by deforestation for example, change the pathway of water? How does it affect streams and rivers? Humans? Natural ecology?
7. What are the three major categories of water use? From what point(s) in the cycle is water generally withdrawn? How is it withdrawn and distributed?
8. What are the indicators of overdrawing water supplies? What are the human, ecological, and environmental consequences of overdrawing surface waters? Of overdrawing groundwater?
9. What are the relative potentials of increasing water supplies versus decreasing demands?
10. Describe how water demands might be reduced in agriculture, industry, and households.
11. How does the change from an absorptive soil to nonporous paving change the pathway of water? What is meant by stormwater runoff?
12. What are the consequences of stormwater runoff for groundwater? For water quality in streams and rivers? For stream flow during rains? For stream flow between rains? For the natural environment and ecology?
13. Describe techniques now being implemented to manage stormwater runoff.
14. Are new policies called for to achieve sustainable water supplies? What policies would you support or promote?

Thinking Environmentally

1. Pretend you are a water molecule, and describe your travels through the many places you have been and might go in the future as you make your way around the cycle time after time. Include travels through organisms.
2. Commercial interests wish to create a large new development on what is presently wetlands. Have a debate between those representing commercial interests and environmentalists that will bring out the environmental and economic costs of development as well as the commercial values of development. Work toward negotiating a compromise.
3. Describe how many of your everyday activities, including your demands for food and other materials, add pollution to the water cycle or alters it in other ways.
4. Water is theoretically a renewable resource. Describe the various things humans are doing that are making water resources nonsustainable. Describe how such activities may be moderated to be sustainable.
5. Describe the natural system that maintains uniform streamflow despite fluctuations in weather. How do humans upset this regulation? What are the consequences?
6. Increasing numbers of people are moving to the arid Southwest, despite the fact that water supplies are already being overdrawn. If you were the governor of one of these states, what policies would you advocate regarding this situation?
7. Commercial interests wish to develop a golf course on presently forested land next to a reservoir used for city water. Describe the impacts this development might have on water quality in the reservoir.
8. Divide the class into five groups representing these various interest groups: environmentalists, farmers, city officials responsible for domestic water, industrialists, developers. The actual supply of water will support only 50% of prospective demands. Negotiate a compromise that will honor the interests of all parties.

Making a Difference—PART TWO: Chapters 6, 7, 8, 9, 10, 11

1. Think carefully about your own reproduction. What concerns will you and your mate weigh as you plan your family?
2. Become involved in the abortion debate, pro or con. Whether or not the United States supports international family planning, whether or not legal abortions remain available in your state, and other issues will be determined by votes cast by you or your representatives.
3. Become involved in and support programs promoting effective sex education and responsible sexual behavior. Consider the advantages of abstinence and monogamy as ways to avoid needing an abortion or contracting an STD.
4. Share your knowledge and energy by joining the Peace Corps or another organization engaged in appropriate technology in a Third World nation (see Food First, Appendix A).
5. Encourage sustainability by buying products that originate from appropriate technology in the Third World.
6. Demonstrate your concern for the hungry and homeless in your area by getting involved in local soup kitchens and collections of goods for the needy.
7. Stay informed on food- and hunger-related issues; join Bread for the World, a nationwide organization that seeks justice for the world's hungry by lobbying our nation's decision makers (see Appendix A for address), and write your congresspersons asking for their support of beneficial legislation.
8. Produce some of your own food by planting a vegetable garden and raising animals; practice sustainable agriculture: avoid the use of herbicides and pesticides, compost kitchen and yard wastes, and build up a humus-rich soil. Any space—even a rooftop—with 6 hours of sunlight a day is enough.
9. Contact your county soil conservation or farm

extension agent and ask for help in communicating to others, the work being done to prevent erosion and runoff of farm chemicals.

10. Inventory the pesticides and herbicides in your home, and switch to either cultural control methods or "soft" pesticides based on natural products.

11. If you are concerned about the source of your food, consider buying organically grown foods. If they are not available where you shop, ask your grocer to stock them.

12. For your community/city, investigate: Where does the water come from and how is it treated? Are water supplies adequate or being overdrawn? How is stormwater managed in your area? What policies/actions are being considered to meet future needs?

13. Observe where stormwater from your own home and yard goes (does it mostly run off or infiltrate?). Install a system, perhaps just a barrel at a downspout, to capture and use stormwater for lawn/garden watering and car washing.

14. Examine your own eating habits and opt for a nutritionally balanced diet that includes less red meat and more fruit and vegetables.

PART THREE

Pollution

An industrial plant sends a steady stream of pollutants into the air and the wind carries the pollutants away. If it didn't, there could be no field of flowers in the same scene. But where is "away"? The next county, or state, or country? The upper atmosphere? Sadly, we are learning that there is no "away," as we face the complex problems of hazardous wastes, sewage, acid deposition, global warming and loss of the ozone shield. Must we simply accept the fact that there could be no economy without polluting, or that wherever there are people, they will have to send their natural wastes to the environment for disposal?

In this Part we will take a hard look at the major forms of pollution. We will see that whether we are talking about air pollution, water pollution, pollution by toxic chemicals, or by human and animal excretions, there are responsible ways of dealing with the problems. It is possible to have people and economic enterprises without overburdening the earth, air and water with wastes. However, dealing responsibly with wastes is a deliberate and often costly matter, and it happens in the context of a human social and political system that can be slow to see the harm caused by pollution and reluctant to take costly action. As the chapters in this Part unfold, look for the changes in public policy and individual lifestyles that will signal a successful move in the right direction—toward sustainability and the environmental revolution.

Industrial Center, Detroit, Michigan. [Photo by Pete Turner/The Image Bank.]

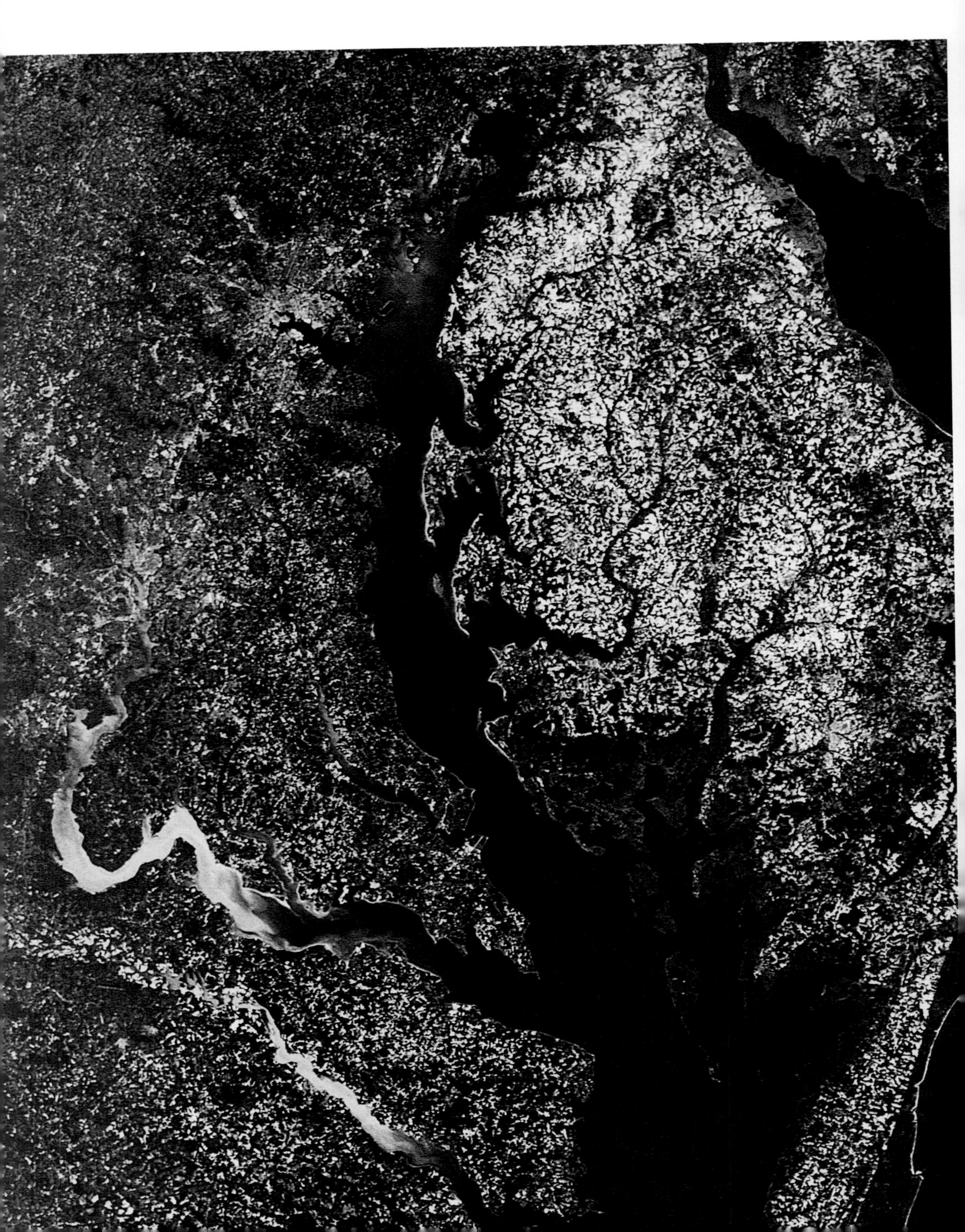

CHAPTER 12

Sediments, Nutrients, and Eutrophication

Key Issues and Questions

1. Water low in nutrients supports an ecosystem based on submerged aquatic vegetation. Water rich in nutrients supports a system based on phytoplankton. Describe these two ecosystems in more detail and how they are affected by nutrient levels.
2. Eutrophication refers to the ecosystem changes that occur with nutrient enrichment. Describe these changes, including the depletion of dissolved oxygen, and why they occur.
3. Sediments and loss of wetlands exacerbate the problem of eutrophication. Explain the effects of sediments and loss of wetlands on eutrophication.
4. Various things may be done to alleviate the symptoms of eutrophication. Describe what they are, and give their benefits and limitations.
5. Long-term control of eutrophication will depend on reducing inputs of nutrients and sediments. What are the major sources of sediment? Of nutrients? Give control strategies being implemented for each source.

Part 3 (Chapters 12–17) focuses on specific areas of concern regarding pollution. Everyone has some perception of what pollution means. Yet, pollution is many different things from many sources, contaminating air, water, or soil. A national commission developed the following definition, which we use. **Pollution** is the human-caused addition of any material or heat energy in amounts that cause *undesired alterations* to water, air, or soil. Any material that causes pollution is called a **pollutant**.

Pollution can be caused by a range of acts, from inadvertant spills and accidents to polluted discharges with criminal intent. Whatever the cause, pollution is the byproduct of economic and social activities—producing crops, creating comfortable homes, providing energy and transportation, manufacturing products, harnessing the atom and our basic biological functions (excreting wastes). Pollution problems have become more pressing over the years because both growing population and expanding per capita use of materials and energy increase the amounts of byproducts that go into the environment. Also, many materials now

Chesapeake Bay, urbanization, remaining wetlands, and sediment input from tributaries. [Image provided by EOSAT Corporation/Chesapeake Bay Foundation.]

widely used, such as aluminum cans, plastic packaging, and innumerable synthetic organic chemicals, are **nonbiodegradable.** That is, they resist attack and breakdown by detritus feeders and decomposers and consequently accumulate in the environment. An overview of the categories of pollutants that result from various activities is shown in Fig. 12-1.

It is important to note the breadth and diversity of pollution. Any part of the environment may be affected, and virtually anything may be a pollutant. The only criterion is that the addition of a pollutant results in undesirable alterations. The impact of the undesirable alteration may be largely aesthetic—hazy air obscuring a distant view or the unsightliness of roadside litter, for instance. The impact may be on ecosystems as a whole, the dieoff of fish or forests, for example. Or the impact may be on human health—toxic wastes contaminating water supplies or health problems caused by that hazy air. The impact also may range from very local—the contamination of an individual well, for instance—to global. We tend to think of pollution as the introduction of human-made materials into the environment. But undesirable alterations may be caused by introductions of too much of otherwise-natural compounds; fertilizer nutrients introduced into waterways, which we consider in this chapter, and carbon dioxide introduced into the atmosphere (Chapter 16) are examples.

Therefore, the slogan "don't pollute" is a gross oversimplification. The very nature of our existence entails the production of byproducts. Our job in remediating present and future pollution problems is parallel to the concept of sustainable development itself. It is to adapt the means of meeting our present needs so that byproducts are managed in ways that will not cause alterations that will jeopardize future generations. The general strategy in each case must be to:

1. identify the material or materials that are causing the pollution—the undesirable alteration;
2. identify the sources of those pollutants and parties responsible for the emissions;
3. develop and implement strategies to prevent those pollutants from entering the environment—pollution control—or;
4. develop and implement alternative means of meeting the need that do not produce the polluting byproduct—pollution avoidance.

We discuss this fourfold strategy with reference to each of the pollution topics we cover in this and the following chapters.

Basically, the strategy for addressing pollution is seen as finally recognizing the first basic principle of sustainability: Ecosystems dispose of wastes and replenish nutrients by recycling all elements, thereby avoiding both pollution and resource depletion. In contrast to this principle we have noted that humans have created a system that is to a large extent based on a one-way flow from resources to disposal. This one-way flow is most conspicuous as we see resources being made into various products that end up in dumps. But the concept is just as applicable in terms of the basic mineral nutrient elements such as nitrogen, phosphorus, and potassium that sustain life.

Our modern human system has been very poor at modeling the nutrient cycles seen in nature. Instead, we remove nutrients from the soil, move them through the human food chain, and discharge them into waterways from which there is no significant return. In addition, substantial amounts of nutrients applied to croplands, lawns, and gardens in fertilizers do not get into the food chain at all but wash or leach directly into waterways.

In this chapter our objective is to understand the consequences of enrichment of waterways with nutrients. The problem is called *eutrophication* (pronounced *yew-TRO-feh-KAY-shun*). Secondly, we consider ways in which we can correct the problem.

The Process of Eutrophication

The process of eutrophication is exemplified by what occurred in Chesapeake Bay during the 1970s and 80s. The Chesapeake Bay is North America's largest estuary. [See Chapter-opening photograph of the Chesapeake Bay area taken by the *Earth Resources Satellite* at an altitude of 550 miles (870 km). The gray areas at the upper left are Washington, D.C., and Baltimore, MD. Light tan and reddish rectangles are agricultural fields. The green speckles and patches are remaining forest. The rust color on the lower east side of the bay is remaining tidal wetlands. Note the extent of urbaniza-

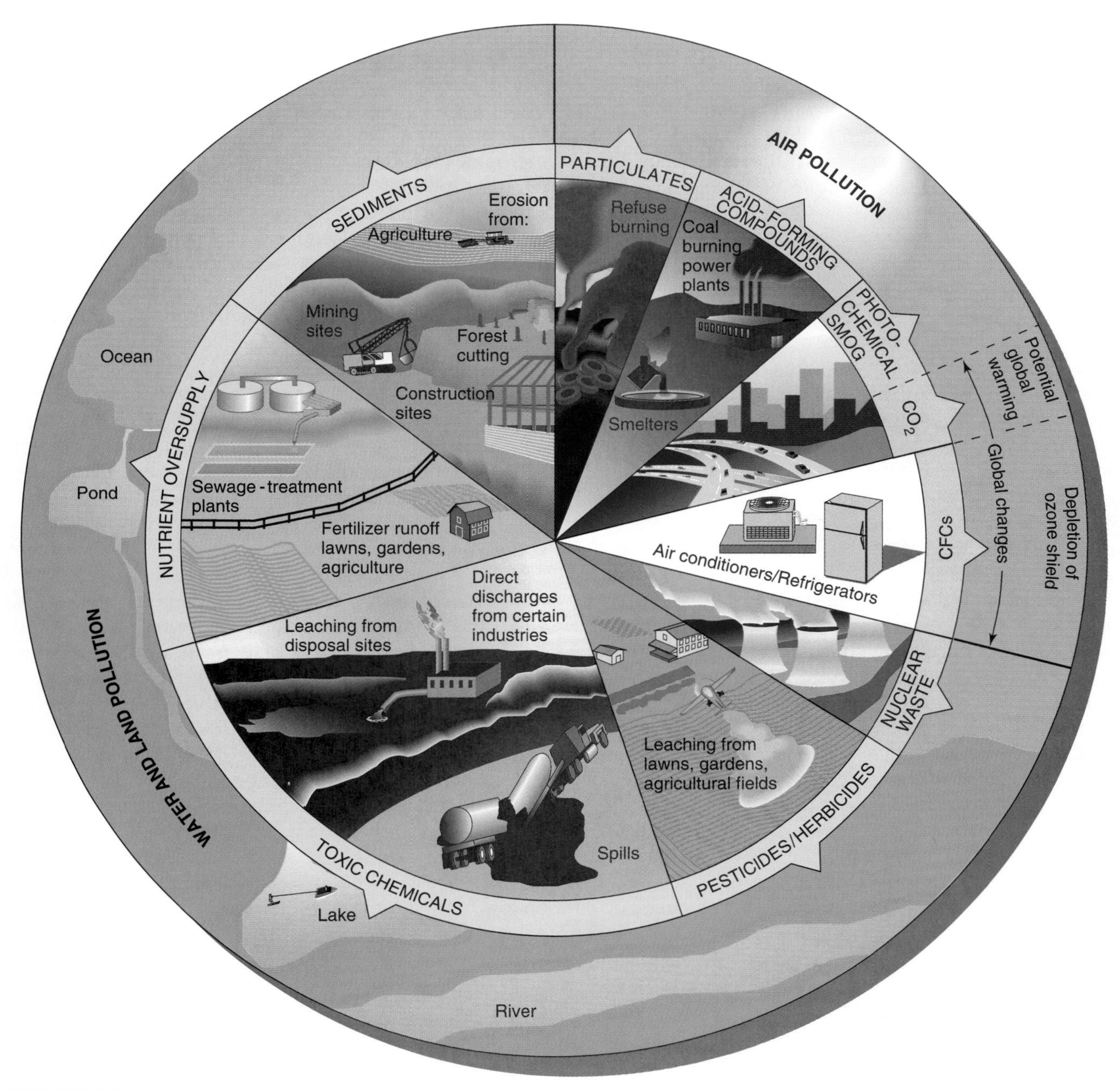

FIGURE 12-1
Pollution is a byproduct of otherwise worthy human endeavors. Major categories of pollution and activities that cause them are shown here.

tion and agriculture in the watershed. Both processes are sources of nutrients and sediments. The tan color of the two rivers in the lower left is a result of sediments eroding after a rain.]

Before the 1970s, Chesapeake Bay was fantastically productive yielding many millions of pounds of fish and shellfish and supporting vast flocks of waterfowl. Most of the food chains supporting this rich bounty had their origin in the seagrasses, 0.5 million acres (200 thousand hectares) of underwater "grass" growing on the bottom 3–6 feet (1–2 m) beneath the surface. The beds of seagrass provided food, spawning habitats, shelter for young fish and shellfish, and dissolved oxygen for them to breathe.

In the early 1970s, the seagrasses in all the major rivers and subestuaries leading into the bay started dying. By 1975, the dieback was dramatic. By 1980, the grasses were gone except in the main stem of the lower bay. Populations of the fish, shellfish, and waterfowl that had depended on the grasses soon declined also.

Even more devastating, the bottom waters in deep areas of the bay became depleted of dissolved oxygen, causing huge numbers of fish and shellfish to be suffocated. What caused the dieoff of seagrasses and the depletion of dissolved oxygen in the Chesapeake?

A team of scientists from the University of Maryland and the Virginia Institute of Marine Science, supported by grants from the Environmental Protection Agency, investigated the problem. Toxic chemicals from industry were ruled out because, although they were a problem in certain locations, they could not have been responsible for a dieback throughout the bay. Herbicides used on farmlands were suspected, but tests showed that they did not reach damaging levels except in small ditches and streams receiving drainage directly from farm fields. Then, investigations turned to the role of water clarity, and this proved to be the key. The waters of the Chesapeake had become increasingly **turbid** (murky or cloudy), and the cloudiness was persisting over extended periods of time. The increased turbidity was cutting off the light required for photosynthesis, and the seagrasses were dying as a result. What was causing the increased turbidity? It was **phytoplankton** (*phyto*, plant; *plankton*, free-floating), various forms of microscopic plants that grow and multiply freely suspended in the water. The growth of phytoplankton was stimulated by enrichment of the water with nutrients. The problem was compounded by *sediments* (mainly clay particles) in suspension. With the loss of the seagrasses, dissolved oxygen was no longer being supplied by their photosynthesis. Even more harmful were bacterial decomposers feeding on dead material; they were consuming the dissolved oxygen, making it unavailable to fish and shellfish.

The Chesapeake Bay had fallen prey to eutrophication. The process of eutrophication is not unique to the Chesapeake. In the last 40 years, many thousands of ponds, small lakes, and even some large lakes and certain rivers have suffered this fate, and the problem is continuing to spread. There are means of checking the problem, however, and remarkable recovery has occurred in some cases as a result.

Let us investigate the causes of this problem as well as the means of preventing and eventually reversing it.

Different Kinds of Aquatic Plants

To understand eutrophication, we need to consider distinct kinds of aquatic plants, namely *benthic* plants and *phytoplankton*.

Benthic plants (from *benthos*, deep) are aquatic plants that grow attached to or are rooted in the bottom. All common aquarium plants and seagrasses are examples (Fig. 12-2a). As shown in the figure, benthic plants may be divided into two categories: **submerged aquatic vegetation (SAV),** which generally grows totally under water, and **emergent vegetation**, which grows with the lower parts in water but the upper parts emerging from the water.

The most important point for understanding eutrophication is this: SAVs require water that is clear enough to allow penetration of sufficient light to support their photosynthesis. The depth to which adequate light for photosynthesis can penetrate is known as the **euphotic zone**. In very clear water, this depth may be nearly 600 feet (200 m). However, as water becomes more turbid, the euphotic zone is reduced; in extreme situations, it may be reduced to a matter of a few centimeters (2.5 cm = 1 inch). Thus increasing turbidity decreases the depth at which SAVs can survive.

A second important feature of SAVs is that they absorb their required mineral nutrients from the bottom sediments through their roots just as land plants do. They are not handicapped by water that is low in nutrients. Indeed, enrichment of the water with nutrients is counterproductive for the SAVs because it stimulates the growth of phytoplankton.

Phytoplankton consists of numerous species of algae and chlorophyll-containing bacteria (cyanobacteria, formerly referred to as blue-green algae) that grow as microscopic single cells or small groups, or "threads," of cells. Living phytoplankton are able to survive in suspension at or near the water surface (Fig. 12-2b). In extreme situations, water may become literally pea-soup green (or tea-colored, depending on the species involved), and a scum of phytoplankton may float on the surface and absorb essentially all the light. However, phytoplankton reach such densities only in nutrient-rich water because, not being connected to the bottom, they must absorb their nutrients from the water. A low level of nutrients in the water limits the growth of phytoplankton accordingly.

Upsetting the Balance by Nutrient Enrichment

Considering the requirements of phytoplankton and SAVs, you can see how the balance between them is altered when nutrient levels in the water are changed. As long as water remains low in nutrients, populations of phytoplankton are suppressed; without their presence, the water is clear, and light may penetrate to support the growth of SAVs. As nutrient levels increase, phytoplankton can grow more prolifically, making the water turbid and shading out the SAVs. Let us look at this process in somewhat more detail.

The Oligotrophic Condition. The original condition (before human impacts) of most lakes, rivers, bays, and estuaries was **oligotrophic**, the term applied to water that is *low in nutrients*, particularly phosphate or nitrogen compounds. Natural waterways' being low in nutri-

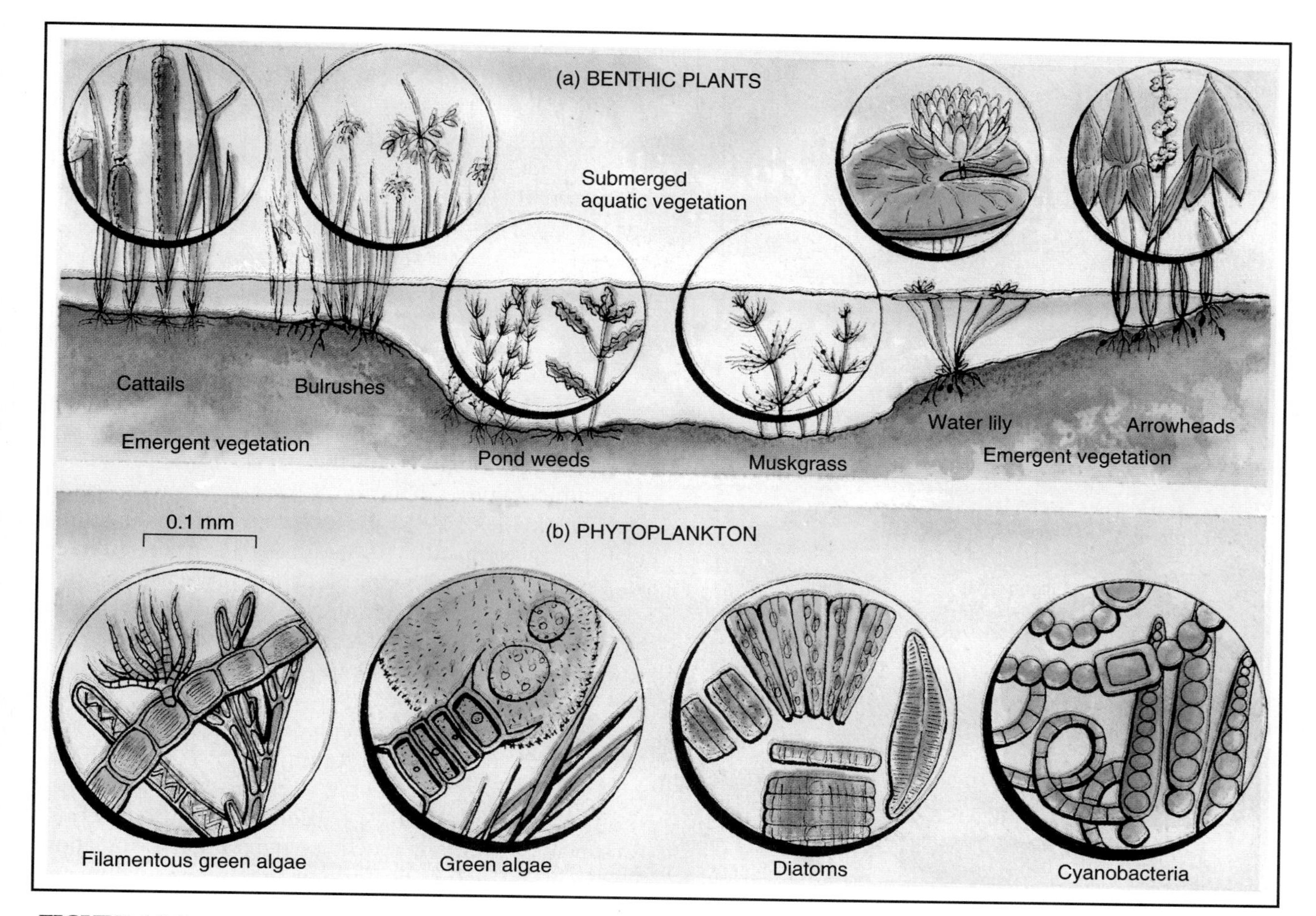

FIGURE 12-2
Two basic categories of aquatic plant life. (a) Benthic, or bottom-rooted, plants. These are subdivided into submerged aquatic vegetation (SAV) and emergent vegetation. (b) Phytoplankton, various species of plants that are either single cells, or small groups or filaments of cells, that float freely in the water. Benthic plants withdraw nutrients from sediment and hence do well in water low in nutrients if they have enough light for photosynthesis; phytoplankton depend on nutrients dissolved in the water.

ents may be contrary to your first thought. But consider what we have learned regarding the nutrient-holding capacity of topsoil and the recycling of nutrients in a natural ecosystem. Very simply, on a forested watershed, for example, nutrients are efficiently held in the cycle from soil, to trees, to detritus, and finally they are reabsorbed by trees. Very little leaching of nutrients or erosion occurs; water draining through the system and out through springs and seeps is relatively pure; concentrations of nitrogen and phosphate are near zero. Subsequently, the streams, rivers, and lakes fed by such waters are likewise low in nutrients. Another factor contributing to the low nutrient content of natural waterways is wetlands, those marshy areas of emergent vegetation adjacent to waterways or swampy areas within the watershed. Wetlands are critical in filtering and removing nutrients from water that seeps through.

Again, the low nutrient levels, by limiting the growth of phytoplankton, allow light penetration that supports the growth of SAVs, which draw their nutrients from the bottom sediments. In turn, the benthic plants support the rest of a diverse aquatic ecosystem by providing food, habitat, and *dissolved oxygen*. The importance of oxygen deserves special emphasis. Oxygen from the atmosphere is very slow to dissolve in and mix through the water. Therefore, in the absence of oxygen produced by the photosynthesis of SAVs, consumers may readily deplete the supply of dissolved oxygen and suffocate all but those bacteria and a few other organisms that can survive in the absence of oxygen. (We shall see that this is what occurs in the eutrophic condition.)

In sum, most natural bodies of water before human impacts were oligotrophic. They were characterized by having low levels of nutrients, clear water,

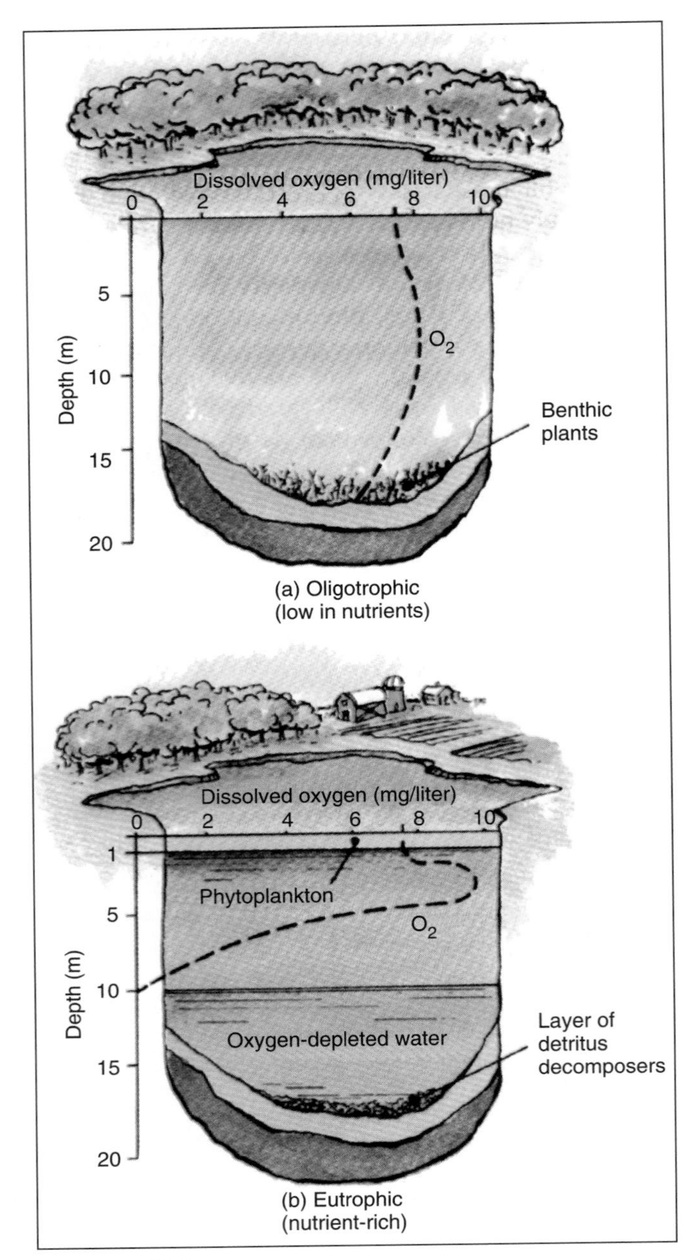

FIGURE 12-3
Dissolved oxygen levels typical of (a) oligotrophic and (b) eutrophic conditions.

and an abundance of benthic vegetation. The benthic vegetation, in turn, supported a diverse ecosystem of fish and shellfish and provided a high level of dissolved oxygen from top to bottom (Fig. 12-3a). The oligotrophic body of water is prized for its aesthetic and recreational qualities as well as for production of fish and shellfish.

Eutrophication. As the water of an oligotrophic body is enriched with nutrients, numerous changes are set in motion. First, the nutrient enrichment allows the rapid growth and multiplication of phytoplankton, causing increasing turbidity of the water. The increasing turbidity shades out the submerged aquatic vegetation. Even where light continues to penetrate to the bottom, the photosynthesis of benthic vegetation may be blocked because its leaves and stems become coated with epiphytic algae growing in the nutrient rich-water.

With the die off of SAV there is an obvious loss of food, habitat, and dissolved oxygen (DO) from their photosynthesis. However, the loss of DO becomes even more profound for the following reason. Phytoplankton are photosynthetic organisms, and they do produce oxygen as do all green plants. Because they grow near the surface, however, the surface water becomes supersaturated with oxygen, and the excess oxygen escapes to the atmosphere. On a calm, sunny day, you can often observe bubbles of oxygen entrapped by filamentous algae being released at the surface. Therefore, photosynthesis by phytoplankton does not replenish the dissolved oxygen of deeper water—except during times in the spring and fall when there is a turnover of top and bottom waters (see "Earth Watch" box, page 301). But more is involved.

Phytoplankton have remarkably high growth and reproduction rates. Under optimal conditions, phytoplankton biomass can double every 24 hours, a capacity far beyond that of benthic plants. Thus, phytoplankton soon reach a maximum population density, and continuing growth and reproduction are balanced by dieoff. Dead phytoplankton settle out, resulting in heavy deposits of detritus on the bottom.

In turn, the abundance of detritus supports an abundance of decomposers, mainly bacteria. The explosive growth of bacteria creates an additional demand for dissolved oxygen as they consume oxygen in their respiration. The result is the depletion of dissolved oxygen with the consequent suffocation of the fish and shellfish. The depletion of dissolved oxygen does not kill the bacteria, however. They have the capacity to shift to *anaerobic respiration*, alternative pathways of cell metabolism that do not require oxygen (see Chapter 3, page 66). Continuing to thrive, but using oxygen whenever it is available, such bacteria keep the water depleted in DO as long as there is detritus to support their growth. In addition, there is some direct chemical oxidation of the dead organic matter and certain other compounds, which creates an additional demand on dissolved oxygen.

Thus, a common water quality test is **biochemical oxygen demand (BOD)**, a measure of the amount of pollutants, mostly organic material, in terms of how much oxygen will be required to break it down biologically or chemically or both. The higher the BOD measure, the greater the likelihood that dissolved oxygen will be depleted in the course of breaking it down.

Earth Watch

Algal Bloom and Bust

A microbial (algae and cyanobacteria) "bloom" is not a flower! The term refers to a sudden burst of phytoplankton growth in surface waters in otherwise oligotrophic lakes. Microbial blooms typically occur in the spring and fall in lakes in temperate regions. Such blooms may be so intense that the water appears greenish, but they typically last only a week or two and then subside. Of course, as a body of water becomes enriched with nutrients, blooms become more intense and last longer until they may last the whole growing season. What causes these short-duration spring and fall blooms in otherwise oligotrophic lakes or ponds?

First, consider the winter, when temperatures and light are at their seasonal low point. The algae and other organisms from the previous year have died and dropped to the bottom as detritus. Bacteria and other detritus feeders release nutrients into water solution as they feed on and decompose the organic matter. Thus, the bottom water is enriched in nutrients. Because fresh water reaches its maximum density at 39°F (4°C), water bodies in colder climates will be stratified in the winter, with the coldest water (less than 39°F; 4°C) and ice on top. The nutrients accumulate in the deeper water layers all during the late fall, winter, and early spring.

As spring weather melts the ice and heats the cold surface water, its density increases as it warms from 32° to 39°F (0° to 4°C). As this occurs, the surface water sinks, forcing the nutrient-enriched bottom water to the surface. This phenomenon is called the spring turnover. The nutrient-rich water brought to the surface warms further under the sun until growth of phytoplankton is supported: thus the spring microbial bloom. The rapidly growing phytoplankton soon exhaust the amounts of nutrients, particularly phosphorus, in the water, and the bloom is brought to a halt. The nutrients, now in the bodies of the dead and dying phytoplankton, are carried to the bottom. As they sink, the surface water is left with nutrient levels too low to support the growth of phytoplankton, and so it becomes clear.

Now there is a summer stratification; water well above 39°F (4°C) is less dense and remains above the cooler, denser water below. Without further turnover, the surface water remains low in nutrients and, without the growth of phytoplankton, clear for the rest of the summer. Meanwhile, bottom waters are enriched again by the decomposition of the dead phytoplankton. Then come the cooling temperatures in the fall.

As the surface water cools in the fall, it becomes denser, sinks, and forces nutrient enriched water to the surface. This process is known as the fall turnover, and it stimulates another burst in growth of phytoplankton—the fall microbial bloom. Again, the bloom is brought to an end as the water is depleted of nutrients and as temperatures become too low to support growth. The winter stratification is achieved, and the same cycle will occur the following year. It should be self-evident that nutrient enrichment from cultural sources will extend the duration and intensity of these natural booms until they dominate the entire growing season.

The spring and fall turnovers are significant not only for bringing nutrient-enriched water to the surface. They also are very important in bringing oxygen-enriched water to the bottom. Because oxygen is slow to dissolve and move through the water column, without these turnovers the bottom water would be depleted of dissolved oxygen. Of course, with excessive amounts of detritus resulting from nutrient enrichment, it is.

In sum, **eutrophication** refers to this whole sequence of events—starting with nutrient enrichment, through growth and die off of phytoplankton, accumulation of detritus, growth of bacteria, and finally depletion of dissolved oxygen and suffocation of higher organisms. Thus, a **eutrophic** body of water is one characterized by *nutrient-rich* water that supports abundant growth of phytoplankton and perhaps other aquatic plants at the surface. Under the surface layer, plant growth is greatly diminished or absent because of shading; on the bottom is an accumulation of detritus. Dissolved oxygen is high at the surface because of photosynthesis of the phytoplankton, but it declines to near zero or zero at the bottom because of oxygen consumption by decomposers (Fig. 12-3b). A sample of the bottom of an oligotrophic body of water often will be a piece of "sod" of the SAVs. In contrast, a sample of the bottom of a eutrophic body is a black "ooze" (organic detritus) that smells like something dead. (The foul odors are the waste products of bacterial anaerobic respiration.)

Eutrophic bodies of water are commonly referred to as "dead." But biologically the eutrophic body is far from dead; the total production of biomass by phytoplankton, in fact, may be greater than that of the previous benthic community. Also, phytoplankton may support large populations of certain fish that are adapted to feed on it and avoid the oxygen-depleted deep water. For example, in Chesapeake Bay, populations of bay anchovy and menhaden, which are filter feeders that consume phytoplankton and other microscopic organisms, are at all-time high levels. Because they are small and oily, however, these species are not considered to be good for human food or sport fishing. It is more correct, then, to view eutrophication as an alteration of a basic abiotic factor—namely nutrient enrichment—which in turn alters the food web.

From the human perspective the eutrophic system is less than appealing for swimming, boating, and sport fishing. Therefore, it may be considered "dead" in terms of those activities. Also, if the lake is a source of drinking water, its value may be greatly impaired because phytoplankton rapidly clog water filters and may cause a foul taste. Also, some species of phytoplankton secrete into the water various toxins that may kill other aquatic life and be injurious to human health as well (see Fig. 12-11).

Eutrophication of Shallow Lakes and Ponds

In lakes and ponds where the water depth is about 3 feet (1 m) or less, eutrophication takes a somewhat different course, but the results are the same. Submerged aquatic vegetation may grow to a height of a meter or so, thus reaching the surface. Therefore, with nutrient enrichment, the SAV is not shaded out, but grows abundantly, sprawling over and often totally covering the water surface with dense mats of vegetation that make boating, fishing, or swimming impossible (Fig. 12–4). Any vegetation beneath these mats is shaded out. As the mats of vegetation die and sink to the bottom, they create a BOD that often depletes the water of dissolved oxygen, causing the death of aquatic organisms other than bacteria.

FIGURE 12-4
Eutrophication in shallow lakes and ponds. In shallow water, sufficient light for photosynthesis continues to reach the submerged aquatic vegetation. Hence, oversupply of nutrients stimulates growth so that vegetation reaches the surface (foreground of this photo) and forms mats over the surface (background). (Photograph by BJN.)

Natural versus Cultural Eutrophication

In nature, apart from human impacts, eutrophication is part of the process of aquatic succession discussed in Chapter 4 (page 100). Rates of erosion of soil and leaching of nutrients from natural ecosystems are very low, but they are not zero. Over periods of hundreds or thousands of years, bodies of water are subject to gradual enrichment with nutrients. We can observe this gradual enrichment in that otherwise oligotrophic lakes may be subject to periodic bursts of phytoplankton growth called algal blooms (described in the "Earth Watch" box on p. 301). Thus, we can speak of *natural eutrophication* as a normal process.

What humans have done is to vastly accelerate the process of nutrient enrichment. Human-produced nutrients come largely from sewage-treatment plants, poor farming practices, urban runoff, and certain other activities, as we describe shortly. We refer to the accelerated eutrophication caused by humans as **cultural eutrophication**.

The public is prone to blame any pollution problem, including eutrophication, on industrial discharges of toxic pollutants. It is very important to observe that cultural eutrophication is due to what are generally considered benign substances, namely fertilizer nutrients particularly compounds of nitrogen and phosphorus. Then, the problem may be compounded by sediments. Figure 12-5 provides a summary of cultural eutrophication.

Potential Recovery

It should be evident that the eutrophic condition will persist as long as inputs of nutrients are such that a high growth rate of phytoplankton is maintained. Even when

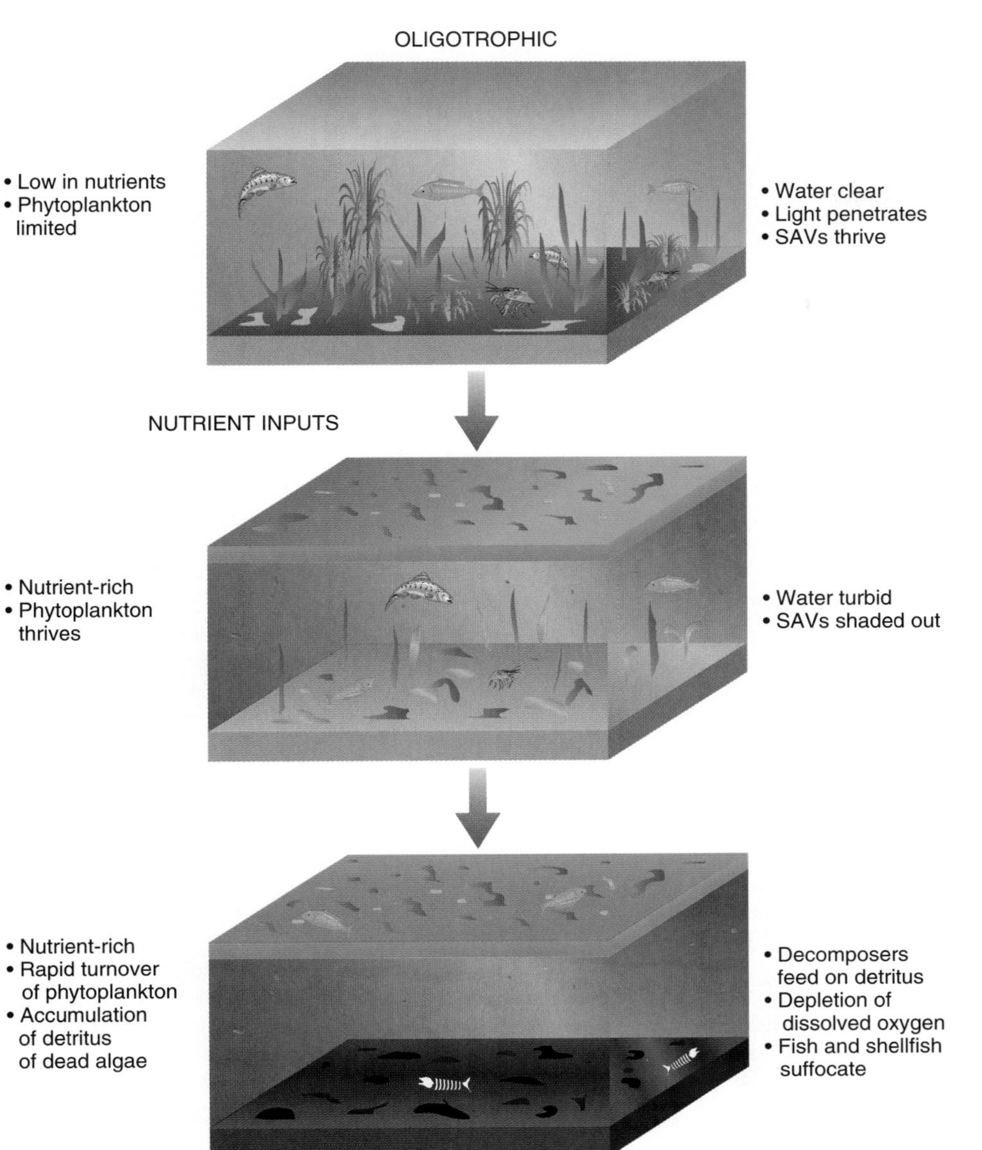

FIGURE 12-5
Summary of cultural eutrophication.

nutrient inputs are diminished, release of nutrients from the decomposing detritus may be such that eutrophication endures. However, nutrients are also gradually flushed from the system or become stabilized in the bottom sediments. As nutrient inputs are curtailed and existing nutrients are removed from the cycle, phytoplankton growth will diminish, detritus will be consumed, bacteria will die back, and dissolved oxygen levels will recover, and so may the original ecosystem if populations have not been totally exterminated. You may surmise at this point that the key to controlling eutrophication is to reduce nutrient inputs. We shall return to this topic a little later in this chapter (see "Earth Watch" box, page 304.) For now we wish to consider some additional factors contributing to eutrophication.

Additional Factors Contributing to Eutrophication

We have thus far focused on the input of nutrients as the major factor in eutrophication. However, sediments and wetlands also are extremely important. Let us consider these factors further here.

Impact of Sediments. Erosion from farmlands, deforested slopes, overgrazed rangelands, construction sites, mining sites, stream banks, and every other point is the source of sediments entering waterways. Sediments (sand, silt, and clay) have direct and extreme physical impacts on streams and rivers and eventually contribute to eutrophication.

Earth Watch

Lake Washington Recovery

One of the best ways to establish the connection between nutrients and cultural eutrophication would be to add massive amounts of nutrients to a body of water and then withdraw the nutrient input to see if the system recovered. Although no one intended it as a test of cultural eutrophication, just such an experiment was conducted on a large lake in the Seattle area—Lake Washington. The "experiment" has conclusively shown the crucial role of phosphorus in eutrophication of lakes, and it has made it clear that cultural eutrophication can be reversed.

Lake Washington is a 33.8-square-mile (76-km^2) lake east of Seattle and within commuting distance of greater metropolitan Seattle. The population began to spread outward from Seattle around the lake in the 1940s and 1950s. By 1960, 11 sewage-treatment plants close to the lake, were sending effluents into the lake at a rate of 20 million gallons (80 million liters) per day. The sewage was given state-of-the-art secondary treatment before being released to the lake or tributaries. At the same time, phosphate-based detergents were becoming widely marketed and used by homeowners. Residents began to notice changes in the clarity of the water and in the fish populations during the 1950s, and it was obvious that the lake was rapidly deteriorating in quality. In particular, blooms of phytoplankton dominated by cyanobacteria began to occur every summer, creating unpleasant smells and masses of floating cells that washed up on the shores of the lake. These cyanobacteria are symptomatic of high phosphate loading. Some species of cyanobacteria have the ability to fix atmospheric nitrogen and so can outcompete other phytoplankton species that take their nitrogen from solution.

Citizen concern led to the creation of a system that diverted the treatment-plant effluents from the lake into tidally flushed Puget Sound. The diversion was phased in gradually between 1963 and 1968. The lake began to respond positively even before the diversion was complete. The water increased in clarity as algal blooms began to diminish, and, by 1975, recovery was almost complete. Summer blooms of cyanobacteria had completely ceased, and the total algal population of the lake had diminished by 90% or more.

This "experiment" in reversing cultural eutrophication was closely followed by limnologist Tom Edmondson and co-workers from the University of Washington. Edmondson's work demonstrated beyond question the key role of phosphate in eutrophication. Before the diversion began, the lake was receiving 220 tons of phosphate per year from all sources. Under normal conditions, phosphate input is balanced by deposition to the sediments. Edmondson demonstrated that more phosphate was coming into the lake than was being deposited, with the resulting blooms of phytoplankton. By the end of the diversion, phosphate input was reduced to less than 40 tons per year, well within the ability of the lake to absorb it by deposition to the sediments.

There are several morals to this story, the most important being that cultural eutrophication can be reversed by paying attention to nutrient inputs. Also, we now know that sewage-treatment-plant effluent must never be allowed to enter a lake unless it receives tertiary treatment to remove phosphate. As a result of the work of Edmondson and other limnologists, lake management is now seen as a matter of controlling the phosphate loading into the lake from all sources: human and livestock sewage, agricultural and gardening fertilizers, street runoff, and failing septic systems. Finally, we see that with the right balance of nutrients, phytoplankton blooms can be prevented and the process of eutrophication can at least be put back on its natural schedule.

When erosion is slight, streams and rivers of the watershed run clear. They support algae and other aquatic plants that attach to rocks or take root in the bottom. These producers, plus miscellaneous detritus from fallen leaves and so on, support a complex food web of bacteria, protozoans, worms, insect larvae, snails, fish, crayfish, and other organisms. These organisms keep themselves from being carried downstream by attaching to rocks or seeking shelter behind or under rocks. Even fish that maintain their position by active swimming occasionally need such shelter to rest (Fig. 12-6a).

Sediment entering waterways in large amounts has an array of impacts. Sand, silt, clay, and humus are quickly separated by the agitation of flowing water and are carried at different rates. Clay and organic particles are carried in suspension, making the water muddy and reducing light penetration and photosynthesis. As this material settles, it coats everything and continues to block photosynthesis. It also kills the animal organisms by clogging their gills and feeding structures. Eggs of fish and other aquatic organisms are particularly vulnerable to being smothered by sediment.

Equally destructive is the **bedload**, the sand and silt, which is not readily carried in suspension but is gradually washed along the bottom. As particles roll and tumble along, they scour organisms from the rocks. They also bury and smother the bottom life and fill in the hiding and resting places of fish and crayfish. Aquatic plants and other organisms are prevented from

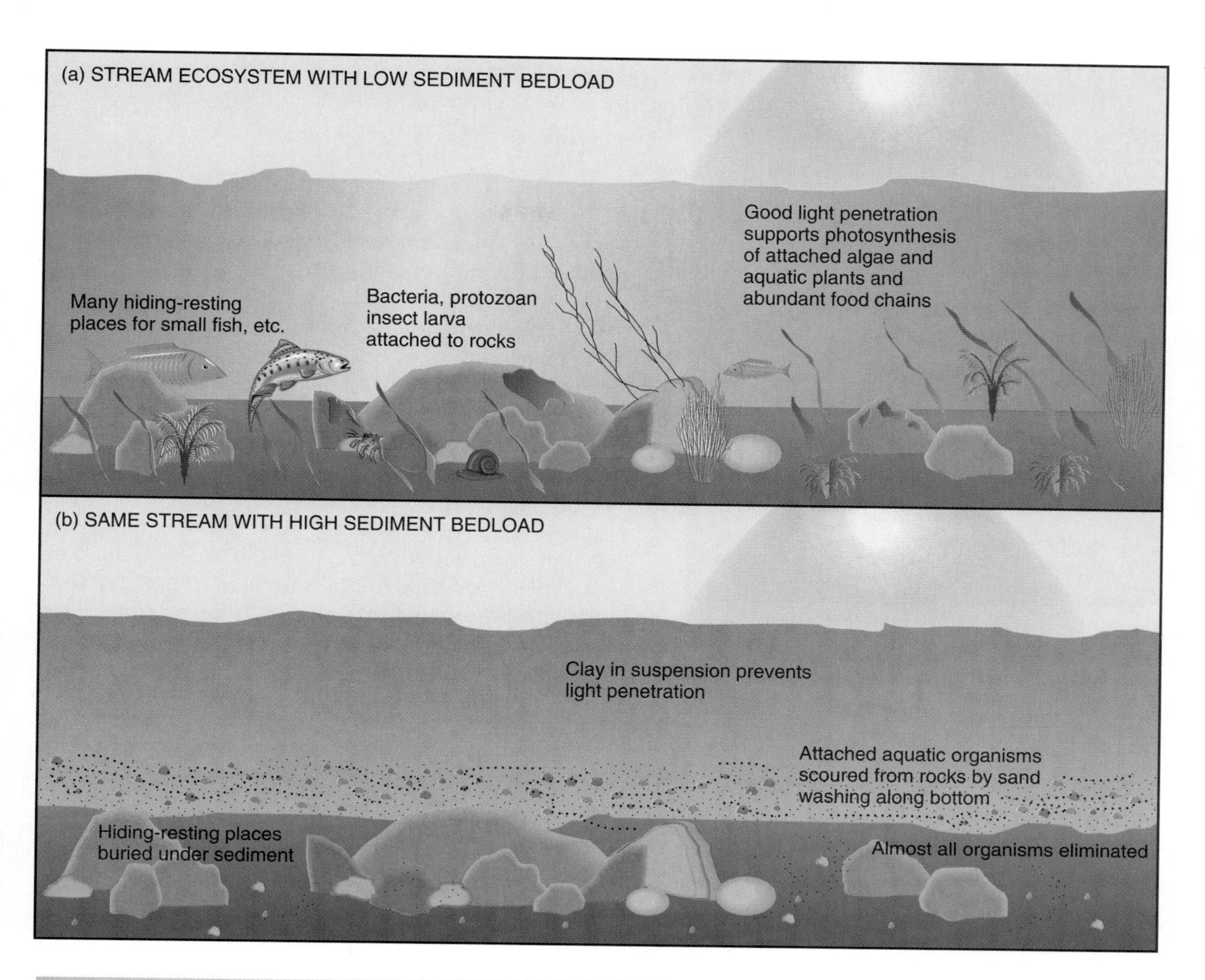

FIGURE 12-6
Negative impact of sediments on the aquatic ecosystems of streams and rivers. (a) The ecosystem of a stream that is not subjected to a large sediment bedload. (b) The changes that occur when there are large sediment inputs. (c) Platte River at Lexington, Nebraska. The river channel is choked with sediment from upstream erosion. The sandbars seen here constitute the bedload; they shift and move with high water, preventing reestablishment of aquatic vegetation. (Charles R. Belinky/Photo Researchers, Inc.)

FIGURE 12-7
Soil eroding from the land may settle out in reservoirs, gradually diminishing water-storage capacity. The level "meadow" on either side of the stream is the portion of the reservoir now filled with sediment. This filling has occurred in just the past 40 years since the reservoir was created; the muddy water at the mouth of the stream shows that the process is continuing. (Photograph is a section of Liberty Reservoir, Baltimore, MD, by Richard Adelberg, Jr., Owings Mill, MD.)

reestablishing themselves because the bottom is a constantly shifting bed of sand (Fig. 12-6b & c). In addition, the coarser sediment settles in and clogs the channel, exacerbating flooding and streambank erosion, as discussed in Chapter 11.

Sediments do not receive the attention the news media given to hazardous wastes and certain other pollution problems, but erosion is so widespread throughout the world that few streams and rivers escape the harsh impact of excessive sediment loads. The Environmental Protection Agency ranks sediments as the number one problem for U.S. streams and rivers, and they are no doubt a major problem in many other countries as well.

Sediments are responsible for the destruction of both recreational and commercial fishing in countless streams and rivers. Sediments eventually settle out in the still water of reservoirs, lakes, bays, or estuaries, but not without consequence. Many millions of cubic meters of water-storage capacity in reservoirs are lost each year because of sedimentation (Fig. 12-7). Irrigation and shipping channels are in constant need of dredging. The U.S. Natural Resources Conservation Agency estimates that the damage from sediments costs the United States over $6 billion each year. Along with the costs of dredging, there are the problems and costs of disposing of the dredged material.

Finally, sediments increase eutrophication. Particles of clay and humus in suspension contribute to the turbidity that blocks the photosynthesis of benthic plants. The same particles invariably carry nutrients that contribute to the growth of phytoplankton.

Loss of Wetlands. We have noted that wetlands are those marshy areas of emergent vegetation adjacent to waterways or swampy areas within the watershed. More technically, **wetlands** are defined as land areas that are naturally covered by shallow water at certain times and are more or less drained at other times. Depending on the depth and permanence of water, wetlands are divided into marshes, swamps, and bogs, and they may be either fresh or salt water. Also, wetlands are divided into *tidal* and *nontidal*. **Tidal wetlands** are the often broad expanses of marsh grasses and reeds along coasts and estuaries where the ground is covered by high tides but drained at low tide (Fig. 12-8). **Nontidal wetlands** are inland wetlands not affected by tides. The plant community of wetlands is typically dense stands of species of grasses, reeds, and other kinds of emergent plants and trees that are adapted to the periodic flooding (Fig. 12-9a).

Whatever their nature, wetlands are nature's most effective flood-control and water-filter device. Excess runoff is taken into wetlands and allowed to drain out

FIGURE 12-8
Tidal wetlands as seen here are immersed by high tides but drain with low tides. Such wetlands play a major role in purifying the water of adjacent estuaries or ocean, because sediments and nutrients entering with the water are largely filtered or absorbed by the wetlands as the water recedes. (Steven Krasemann/Photo Researchers, Inc.)

(a)

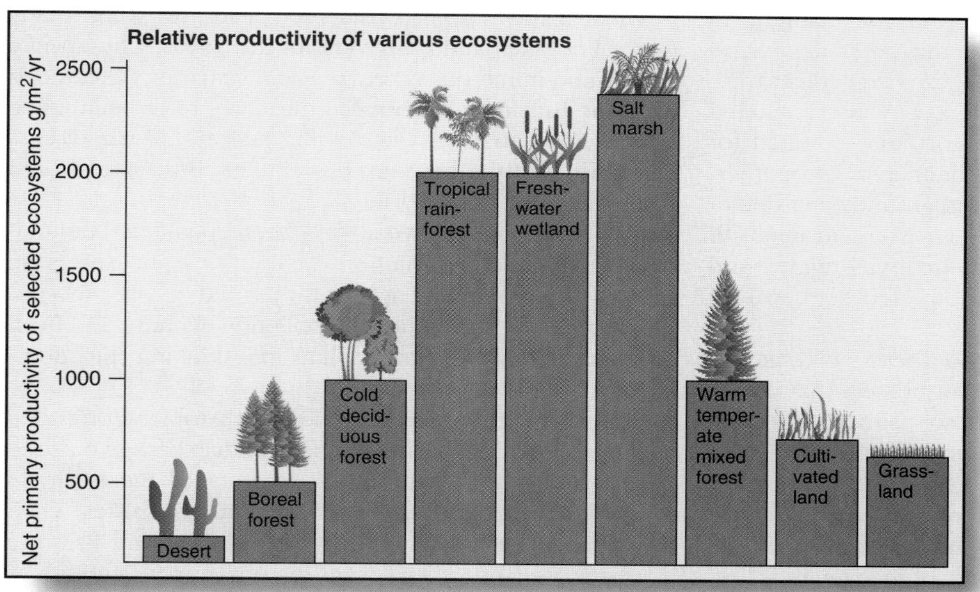

(b)

FIGURE 12-9
(a) The abundance of water and nutrients makes wetlands exceedingly rich in supporting a diversity of species. (From *Audubon*, supplement, July 1990.) (b) Wetlands, both salt- and freshwater, are the most productive of Earth's ecosystems.

slowly. In addition, wetlands hold water virtually motionless such that sediments settle. As water slowly drains through and out, nutrients are captured by adsorption to humus and eventually are reabsorbed by plants. The water draining from the wetland is almost pure. Tidal wetlands play this filtering role with each flux of the tides. With their dual role of capturing both nutrients and water, wetlands are the most biologically productive ecosystems on Earth, with food chains supporting many species of mammals, flocks of water fowl, as well as a plethora of aquatic organisms (Fig. 12-9b). Many organisms depend on wetlands for their essential breeding habitat.

These virtues of wetlands have been recognized

FIGURE 12-10 Bulkheaded shoreline. The natural shoreline of estuaries is a gentle slope covered by grasses and emergent vegetation gradually receding into the water. Much of this shoreline has now been dredged, filled, and bulkheaded to make it more usable by humans. Lost in the process are habitat, food production, and the filtering function that formerly removed nutrients and sediments. In addition, waves smashing against the bulkhead keep sediments in suspension, aggravating the problem of eutrophication. (Photograph by BJN.)

only in the past few decades. Traditionally, wetlands were viewed as wastelands or actual nuisances—breeding grounds for mosquitoes and other insects, too wet to plow, but too dry to sail ships upon. Consequently, they have been viewed as good only for changing to "better" uses. More than half of the original wetland acreage in the United States has been destroyed. Nontidal wetlands have been extensively channeled and drained for conversion to agriculture or filled for conversion to residential and commercial properties. The effects on runoff, flooding, and groundwater depletion were considered in Chapter 11. Now, an intensification of nutrient enrichment in waterways and eutrophication can be seen as another consequence of losing wetlands.

Tidal wetlands have also been extensively dredged, filled, and bulkheaded—a process that entails dredging to a depth of a meter or so, using the dredged material to build up the other portion, and stabilizing the edge with a retaining wall, or *bulkhead*. The advantage in creating more usable bayfront real estate is obvious, but what are the trade-offs? Not only are the productivity and cleansing capacity of the wetlands lost, but now waves smash against the bulkheads and create turbulence that stirs up sediments and keeps them in suspension for days and weeks at a time, cutting off light, photosynthesis, and growth of the submerged aquatic vegetation (Fig. 12-10).

Global Extent

Cultural eutrophication was first observed as a problem in farm ponds that received runoff of fertilizer or animal wastes. Then the problem spread to increasingly larger bodies of water. In larger bodies of water, it simply takes longer to build up the nutrient levels that trigger eutrophication. In the early 1970s, Lake Erie, one of the Great Lakes, was considered dead, as were many smaller lakes and reservoirs. In the 1980s, Chesapeake Bay and numerous other bays and estuaries were added to the list. Indeed, remedial measures have been taken in many situations, with some success as we discuss shortly, but the problem has continued to spread on a global scale.

The oceans have always been considered the ultimate "sink." The dilution factor of the oceans is so vast, the thinking went, that they could absorb any conceivable output of human pollution without ill effect. Now, observations are calling this thinking into question.

Among the many species of phytoplankton found in the oceans are some that are reddish and give rise to extremely toxic byproducts. Blooms of these phytoplankton are responsible for infamous **red tides**. Their numbers literally turn the water reddish, and fish, sea mammals, and other organisms unfortunate enough to be within the area of the bloom are killed by their toxic byproducts—a *biotoxin*. Like natural phytoplankton blooms in otherwise oligotrophic lakes, red tides have been observed far back in history. However, in the past decade both the locations and frequency of such events have increased dramatically (Fig. 12-11).

Moreover, coral reefs—the world's most exotic displays of benthic biodiversity—are suffering diebacks around the world. The locations of red tides and the diebacks of coral reefs are correlated with inputs of

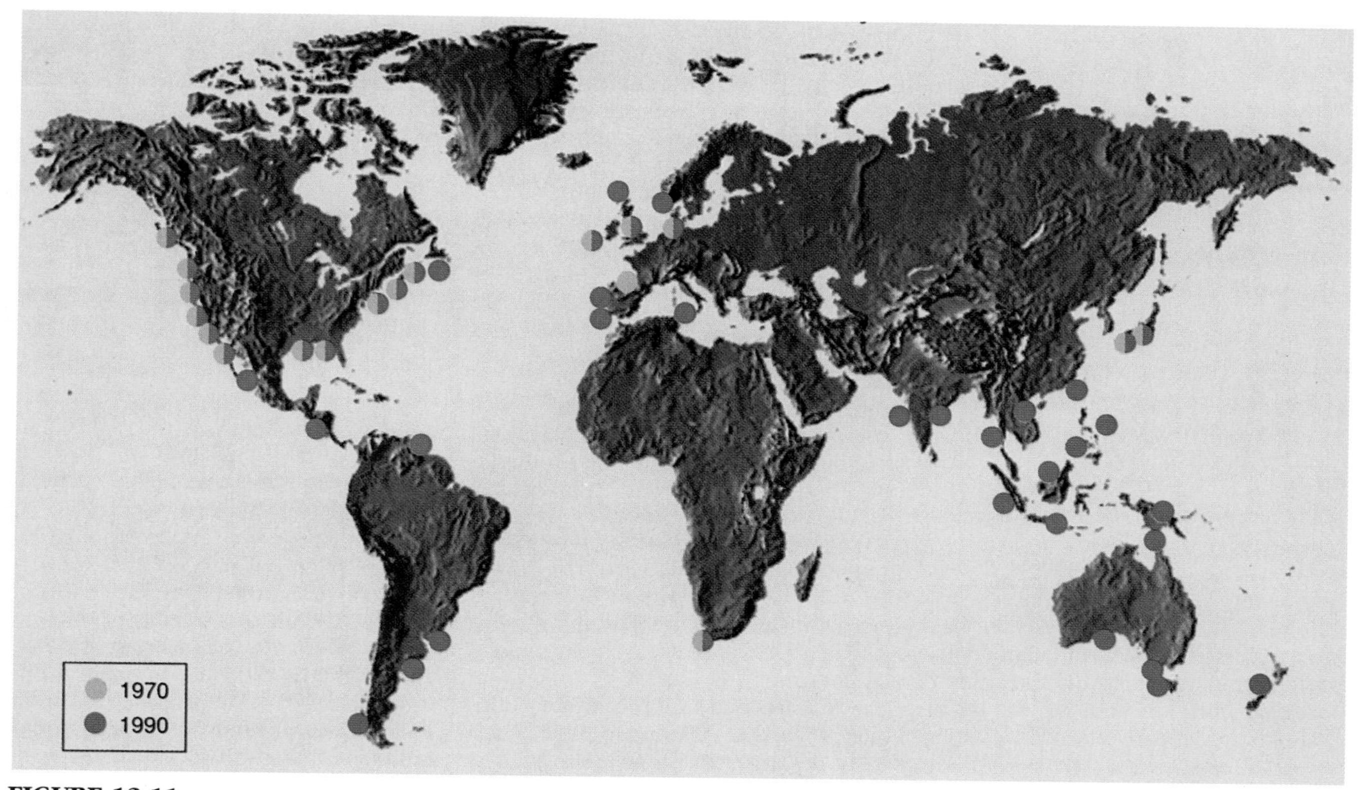

FIGURE 12-11
Outbreaks of red tide (blooms of phytoplankton), which poison fish and shellfish, affected more than twice as many areas in 1990 as in 1970. Is cultural eutrophication of oceans the cause? Some experts believe it is. (From Donald Anderson, "Red Tides," *Scientific American*, August 1994, p. 68.)

nutrients and sediments from human sources, such as the mouths of rivers carrying the runoff from poorly managed agriculture and sewage effluents.

Combating the Symptoms of Eutrophication

There are two approaches to combating the problem of eutrophication. One is to *attack the symptoms*—the growth of vegetation or the lack of dissolved oxygen or both. The other is to *get at the root cause*—excessive inputs of nutrients and sediments. Attacking the symptoms has application in certain situations where immediate remediation is the goal and costs are not prohibitive. But consider the longer-term efficacy of such methods if the causative inputs are not curtailed. Attacks on the symptoms include (1) chemical treatments, (2) aeration, (3) harvesting aquatic weeds, and (4) dredging.

Chemical Treatments

Herbicides are used in agriculture to suppress the growth of weeds. Because unwanted aquatic weeds (algae and benthic vegetation) can also be eliminated by chemical treatments, thousands of tons of chemicals were spread on U.S. ponds and lakes in the 1960s and 1970s. The results were less than inspiring. Phytoplankton—especially the cyanobacteria, which are the most obnoxious—are among the most resistant of all organisms. Therefore, amounts of chemicals sufficient to kill them also have severe effects on virtually all other aquatic organisms. When concentrations of the herbicides dissipate, cyanobacteria and other phytoplankton are among the first species to reappear.

To date, no chemical has been found that will selectively kill phytoplankton and not harm other aquatic plants and animals. Nevertheless, copper sulfate is currently being used to control the growth of phytoplankton in some water-supply reservoirs, where they would otherwise impart a bad taste to the water and cause excessive clogging of filters. Above trace

Ethics

Battle in the Everglades

Once covering 7 million acres (2.8 million hectares) of southeast Florida, the Everglades is a shallow subtropical wetland wilderness of sawgrass and tree islands blending into mangrove forests as it nears the sea. At the north is Lake Okeechobee, the third largest freshwater lake in the United States. The lake is fed by the Kissimmee River and drains into the Everglades to the south. A hundred years ago, the system absorbed the abundant rainfall of the summer wet season and allowed the slow flow of water through the sawgrass prairies to the sea. In the winter dry season, alligator holes sustained the underwater wildlife, while great concentrations of wading birds harvested fish in the shrinking wet areas. Wildlife was abundant, adapted to the cycle of summer rains and winter droughts.

Early settlers and state officials regarded the Everglades as a wasteland—without value—and began over 100 years ago to drain the wetlands and convert them into "useful" land. By 1970, the Everglades was laced by a system of 2000 miles (3000 km) of canals, Lake Okeechobee was diked to a height of 20 feet (6.6 m), and the Kissimmee River was channeled to a straight white-walled canal 48 miles (74 km) long. The changes were designed to create agricultural and urban land, control flooding, and provide water for the growing population (Florida is now the fourth largest state in the United States). There was even an attempt to build a major international airport in the heart of the Everglades. Fully half of the original Everglades has now been converted to other land uses, and only one-fifth of the original Everglades is under protection as the Everglades National Park (ENP). The ENP is considered an international treasure; the United Nations has designated it as an international biosphere reserve.

Two major cultural eutrophication problems face the Florida Everglades system. First, 200,000 acres (80,000 hectares) of wetlands in the north were converted to dairy farms, which now drain into Lake Okeechobee. Nitrogenous wastes from the dairy farms have stimulated the growth of masses of phytoplankton and water weeds, and the lake is now considered to be hypereutrophic—the most extreme phase of cultural eutrophication. Second, south of the lake, 700,000 acres (280,000 hectares) of marsh have been drained to raise sugarcane. The sugar industry has been the recipient of massive flood-control and drainage projects paid for by the federal government. Nutrients from fertilizers used on the sugarcane are draining directly into the ENP. Water is drained from other portions of the Everglades to irrigate these sugarcane fields. The influx of nutrients is converting the sawgrass prairies to a monoculture of cattails, threatening to bring on a collapse of the fragile ecosystem.

Original meandering Kissimmee River in foreground and straightened channel in background. (Garth Francis/Silver Image.)

Overriding these eutrophication concerns, the freshwater that flows southward is now almost totally controlled by the system of canals and control dams and is subject to a battle. On one side is the continued demand for water for agriculture (2 billion gallons, 8 billion liters, per day) and urban areas (more than 3 billion gallons, 12 billion liters, per day) in southern Florida. On the other side is the need for water in the ENP in a pattern that approximates the original wet and dry seasons before the canals were built. The tendency has been to treat the Everglades as a reservoir, taking water for human needs when the supply is low and channeling excess water into the marsh when rainfall is excessive. The result has been disastrous for the wildlife of the Everglades, now that the marsh is wetter than normal in the rainy season and drier than normal in the dry season. The combination of impacts on the most conspicuous wildlife—the wading birds—has reduced the population 95% since 1870. Most of the other wildlife has suffered serious declines also.

The two bases of Florida's economy are agriculture and tourism, both of which are clearly dependent on the freshwater that would normally flow south through the Everglades system. The unique wildlife of southern Florida is equally dependent on the water's being supplied according to the natural seasonal cycle.

Congress has mandated that a minimum amount of water be delivered to the ENP during times of low flow, and the state adopted in 1983 the "Save Our Everglades" initiative. This initiative has the goal of assuring "that the Everglades of the year 2000 looks and functions more like it did in 1900 than it does today." Key elements of this strategic planning initiative include cleaning up Lake Okeechobee, restoring a more natural flow of water into the Everglades National Park, and restoring major parts of the system, such as the Kissimmee River and drained lands adjacent to remaining wetlands. The price tag for the initiative runs into the hundreds of millions of dollars, and to date only a few elements of the plan have been put into place.

A governor-appointed assessment committee submitted their 1996 report to the Commission for a sustainable South Florida. Now they have a giant task in front of them if they are to maintain the megalopolis of southeastern Florida and the agricultural industry as well as the remnants of the natural system that sustains life and livelihood in South Florida. Check: http://south-florida.info-access.com/ever.html

FIGURE 12-12
To avoid the use of chemical herbicides for controlling aquatic vegetation, these residents in Columbia, Maryland, resorted to harvesting by hand. (Photograph by Bruce Fink.)

amounts, however, copper is known to be highly toxic to all organisms. Therefore, you should remain skeptical of the long-term effects of this practice. Is it sustainable?

Aeration

Depletion of dissolved oxygen by decomposers and consequent suffocation of other aquatic life is the final and most destructive stage of eutrophication. Therefore, it follows that artificial aeration of the water can avert this terminal stage. Further, in an environment with high dissolved oxygen, phosphate, a key nutrient, more readily converts to compounds that become stabilized in the sediments; thus this important nutrient is removed from the water solution.

An aeration system currently gaining popularity is to lay a network of plastic tubes with microscopic pores on the bottom of the waterway to be treated. High-pressure air pumps force from the pores microbubbles that dissolve directly into the water. (Larger bubbles would simply rise to the surface, wasting most of the pumped air.) The system is proving effective in speeding up the breakdown of accumulated detritus, improving water quality, and enabling the return of more desirable aquatic life. However, the high cost of installing and operating the system should be obvious. Nevertheless, the system has applicability in harbors, marinas, and in some water-supply reservoirs where the demand for better water quality justifies the cost.

Harvesting Excessive Plant Growth

In shallow lakes or ponds where the problem is bottom-rooted vegetation reaching and sprawling over the surface (see Fig. 12-4), harvesting the aquatic weeds may be an expedient way to improve recreational potential and aesthetics. Commercial mechanical harvesters are used, and nearby residents also have gotten together to remove the vegetation by hand. A raft made by lashing a few planks across two canoes is remarkably effective for such an effort (Fig. 12-12). The harvested vegetation makes good organic fertilizer and mulch (see Chapter 9). But even harvesting has a limited effect. The vegetation soon grows back because roots are left in the nutrient-rich sediments.

Mechanically removing phytoplankton is not at all practical. The microscopic cells would have to be filtered from the water. The volume of water to be filtered in even a modestly sized pond is overwhelming, and the effort is further frustrated by the fact that plankton cells rapidly clog the filter, precluding the passage of water.

Dredging

Dredging may be required to remove sediments blocking access for boating or shipping. Dredging, however,

TABLE 12-1
Sources of Nutrients and Sediments Causing Eutrophication

Source	*Nutrients*	*Sediments*
Agriculture/Forestry:		
Erosion from croplands	X	X
Erosion from overgrazed range	X	X
Runoff/leaching of applied fertilizer	X	
Runoff/leaching of animal wastes from feedlots, dairy barns, horse stables, etc.	X	
Erosion from logging roads	X	X
Erosion of clear-cut forests	X	X
Urban/Suburban:		
Erosion from lawns and gardens	X	X
Runoff/leaching of applied fertilizer	X	
Runoff of pet droppings	X	
Sewage Effluents:		
Discharges from sewage-treatment plants	X	
Phosphate used in detergents	X	
Seepage from individual septic systems	X	
Other Sources and Factors:		
Erosion from		
Highway construction		X
Residential and commercial construction		X
Mining sites		X
Streambank erosion (excessive runoff)	X	X
Air pollution (fallout of nitrogen compounds)	X	
Loss of wetlands to development		
Loss of natural control of runoff	X	X
Loss of filtering action	X	X

tends to increase eutrophication because it invariably stirs much settled material back into solution where it increases turbidity and stimulates the growth of phytoplankton. There is also a significant problem in finding a suitable place to dispose of the dredged material.

Long-Term Strategies for Correction

Ultimately, controlling eutrophication must involve reducing the inputs of nutrients and sediments. The first step is to identify the major sources of nutrients and sediments. Then it is a matter of developing and implementing strategies for correction. The major sources of nutrients and sediments coming into waterways and other factors involved are listed in Table 12-1. Which source or factor is most significant will depend on the human population and the land uses within the particular watershed. Therefore, each watershed must be analyzed as a separate entity, and appropriate measures must be taken to reduce the nutrients and sediments exiting from that watershed. In the following sections we discuss major control strategies that are being implemented.

First, recall the concept of limiting factors (Chapter 2); only one nutrient need be lacking to suppress growth. In natural freshwater systems, phosphorus is most commonly limiting. In marine systems, the limiting nutrient is most commonly nitrogen. Both in the environment and in biological systems, phosphorus (P) is present as phosphate (PO_4^{-3}) and nitrogen may be present as in a variety of compounds, most commonly nitrate (NO_3^-) or ammonium (NH^{+4}). Therefore, throughout the rest of this chapter we focus primarily on phosphate and nitrogen compounds.

TABLE 12-2
Progress toward Banning Phosphate from Detergents

States that ban phosphate-containing laundry detergents		
Delaware	Michigan	Pennsylvania
Georgia	Minnesota	Vermont
Indiana	New York	Virginia
Maine	North Carolina	Washington, DC
Maryland	Oregon	Wisconsin
States that restrict the level of phosphate in laundry detergents or ban phosphate detergents in some areas		
Connecticut	Illinois	Ohio
Florida	Montana	

Banning Phosphate Detergents and Advanced Sewage Treatment

We have pointed out previously that in contrast to the natural nutrient cycles, humans have created a one-directional flow of nutrients. Nitrogen, phosphate, and other nutrients withdrawn from the soil by crop plants exit with human excrements and are finally discharged into waterways. To be sure, human sewage is collected and treated (at least in developed countries) before the effluents are discharged. Nevertheless, traditional sewage-treatment practices focused only on eliminating the disease hazard and removing or breaking down the organic matter to reduce the BOD. They made no attempt to remove the nutrients. Therefore, in heavily populated areas, discharges from sewage-treatment plants are recognized as major sources of nutrients entering waterways. The levels of phosphate in effluents from sewage-treatment plants are elevated even more in areas where laundry detergents containing phosphate are used. Phosphate contained in the detergent for cleaning purposes goes through the system and out with the discharge.

In regions where eutrophication has been recognized as a problem, a key step toward prevention has been to ban the sale of phosphate detergents, or at least to regulate the maximum allowable level of phosphates. Total or partial phosphate bans are now in effect in 20 states and the District of Columbia (Table 12-2). Despite the availability of nonphosphate detergent alternatives, the detergent industry continued to favor the phosphate detergents for various economic reasons and to vigorously oppose restrictive legislation, eutrophication notwithstanding. Bans were enacted only due to the strong support of environmentalists overcoming the counter lobbying of the detergent industry. However, the critical point has now been reached, and major detergent manufacturers are shifting to general production of non-phosphate formulations, rather than making different formulations for different regions. Since this transition is still in progress, the consumer may well be advised to read labels under active ingredients. Also, the bans do not cover dishwashing detergents, and some brands are high in phosphate. You can check the labels on these also to find a brand with little or no phosphate.

In addition, there is an ongoing program aimed at upgrading sewage-treatment plants to remove nutrients from the effluents or to handle the effluents in alternative ways to avoid having the nutrients go into waterways. These processes and alternatives will be considered further in Chapter 13.

Phosphate-detergent bans and upgrading sewage treatment have brought about marked improvements in waterways that were heavily damaged by effluents from sewage-treatment plants. In a sense, however, these are the easy measures in that the target for correction, the sewage-treatment-plant effluent, is a clear concise *point-source*, and methods for correction are clearcut.

Correction becomes more difficult, but not less important, when the source is diffuse, as it is in the case of farm and urban runoff. Remediation of such *non-point sources* will involve thousands, even millions, of individual property owners' adopting new practices regarding management and use of fertilizer and other chemicals on their properties. Nevertheless, that is the challenge.

Controlling Agricultural Runoff

When the Clean Water Act (1972) was reauthorized in 1987, a new section (section 319) was added requiring states to develop management programs to address non-point, or diffuse, sources of pollution. Thus, there is a legal mandate to address the issues of agricultural and urban runoff.

Where a significant portion of the watershed is devoted to agricultural activities, the major source of nutrients and sediments is likely to be erosion and leaching of fertilizer from croplands and runoff of animal wastes from barns and feedlots. All the practices that may be used to minimize such erosion, runoff, and leaching are lumped under a single term, **best management practices (BMP)**. BMP includes all the methods of soil conservation discussed in Chapter 9. Topping the list are keeping the ground covered with

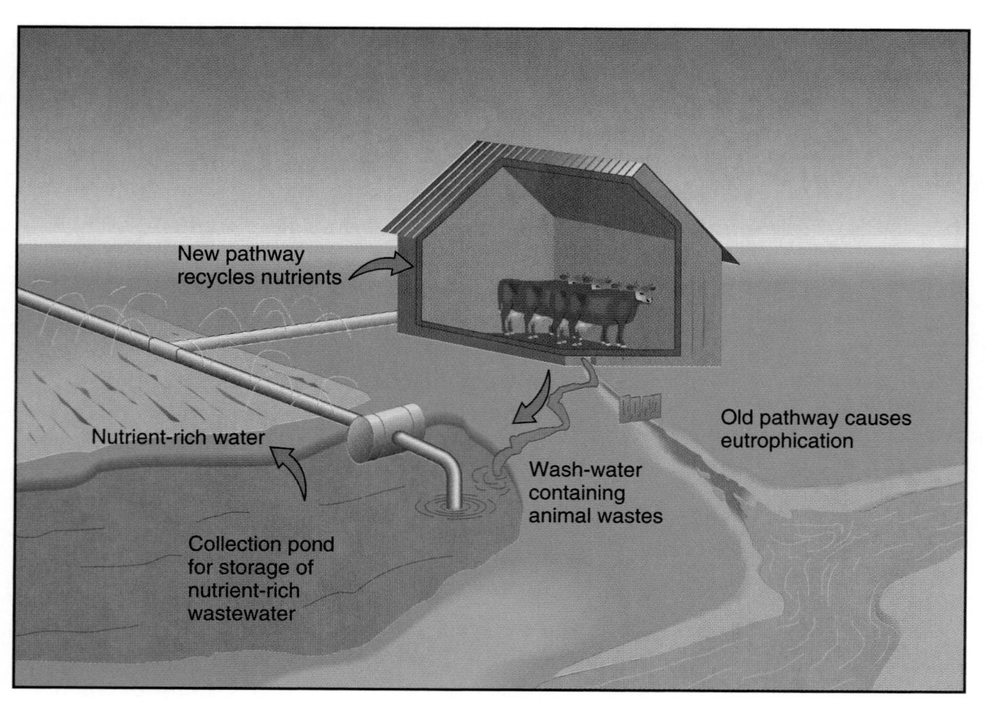

FIGURE 12-13
A collection pond for dairy barn washings. When washings from animal facilities are flushed directly into natural waterways, they contribute significantly to eutrophication. This may be avoided by collecting the flushings in ponds from which both the water and the nutrients may be recycled. Many farmers use such flushings for irrigation.

vegetation or mulch to prevent erosion, strip-cropping, using clover or other legumes for natural nitrogen addition, and using organic fertilizers (such as compost and manure) in place of inorganic fertilizers; but other things are important too.

Reestablishing riparian woodlands—the strip of trees, shrubs, and other vegetation along streams and drainage areas—has been found to be particularly helpful. The riparian woodland is very effective in filtering and absorbing nutrients from the water draining into the stream.

Where animal wastes from feedlots, dairy barns, or horse stables drain directly into waterways, ponds can be constructed to intercept the nutrient-rich runoff. Such water can then be recycled as irrigation water, returning the nutrients to the soil (Fig. 12-13). Animals allowed to wade into natural streams to drink break down the banks, accelerating streambank erosion; they also create pathways that facilitate runoff directly into the stream and incidentally drop excrements directly into the water. The alternative BMP is to provide water troughs away from the stream and actually fence the stream off for protection of the riparian woodland.

The implementation of BMP on farms has three factors in its favor. First, one has to negotiate with only a limited and fairly stable number of farmers. Second, the implementation of BMP usually has long-term economic benefits for the farmer. And third, there is an existing array of various farm subsidies that can be manipulated to provide additional economic incentives. Thus, implementation of BMP on farms is moving forward.

Controlling Urban Runoff

Homeowners tend to use three to five times more fertilizers and pesticides on lawns and gardens per unit area than farmers use on crops. Therefore, the runoff and leaching of nutrients from urban and suburban areas is considerably greater than from similar areas of agricultural land. Pet excrements also contribute significantly to the nutrient content of such runoff. Golf courses are a particular issue: Managers tend to use fertilizer intensively, and soils are compacted, thus promoting runoff. In watersheds with sizable cities, the major source of nutrients causing eutrophication may

FIGURE 12-14
The gullies in the embankment of this highway under construction attest to severe erosion. Such construction activities may be the most significant source of sediments entering waterways. (Photographs by BJN.)

be the runoff from developed areas. Of course, this source is growing with increasing population and development, and it promises to be the most difficult to control.

The same concepts of BMP may be applied to lawns and gardens as were noted in Chapter 9. However, economics is seldom a constraining factor in a homeowner's use of lawn and garden chemicals, nor is any system of subsidies to manipulate behavior practicable. Therefore, how to gain the understanding and cooperation of millions of independent homeowners in reducing their use of fertilizers and other chemicals is still a question that awaits an answer. Educational programs are being launched in some areas, but success is awaited.

Controlling Sediments from Construction and Mining Sites

We have observed that sediments cause ecological damage to streams and rivers and they exacerbate eutrophication by both carrying nutrients and adding to turbidity. Beyond erosion from agricultural lands, erosion from construction or mining sites is another major source of sediments. During construction, banks of raw, bare earth are common, and the disturbed earth is very vulnerable to erosion. It is not uncommon to see banks in construction projects deeply gullied by erosion (Fig. 12-14). Literally thousands of tons per hectare (1 hectare = 2.47 acres) may erode from a construction site.

To lessen the downstream impact of such erosion, some states have enacted legislation requiring developers to control sediments that would otherwise wash from the site. A number of techniques are applicable. The following are most widely used.

Rather than leaving the entire site bare and exposed to erosion for the duration of construction, contractors bring much of the site to final grade and restabilize it with grass immediately. Even temporary piles of dirt may be stabilized by covering them with a straw mulch and grass until needed. This approach is particularly appropriate in highway construction.

Second, **sediment traps** are constructed. These are ponds into which runoff and erosion are channeled. As sediment-laden water enters the pond, its velocity slows, sediment settles, and sediment-free water flows out over a rock dam or through a standpipe (Fig. 12-15). After construction is completed, the sediment trap may be converted to a stormwater-retention reservoir (see Chapter 11). On small sites, the lower perimeter of the site may be diked with bales of straw, which filter the runoff and remove sediment. Or a plastic fence may be used to catch water and direct it to a sediment trap. The same techniques may be applied to mining sites.

The weak link in sediment control, as in other areas, is gaining compliance. Sediment control is an added expense to developers, one they would like to minimize. Therefore, without constant inspection and enforcement, it is common to find sediment traps that are nonfunctional for lack of proper installation or

(a)

(b)

(c)

FIGURE 12-15
Sediment control at a construction site. (a) Runoff carrying sediment flows into a "pond" constructed at the lower edge of the site. Sediment settles in the "pond", while water flows over the rock dam (foreground). (b, c) Construction of a larger pond with a standpipe overflow will function both as a sediment trap during construction and as a permanent stormwater reservoir. (Photographs by BJN.)

maintenance. This is an issue where local citizen action can be effective. Save Our Streams and Trout Unlimited are two citizen organizations whose members take it upon themselves to act as "inspectors" overseeing construction projects in their areas. When they find uncontrolled erosion threatening waterways, they report it to the authorities. If the authorities fail to take action, the organizations report the situation to the local media. Media publicity frequently brings action when all else fails, because neither developers nor the government want a bad press.

With regard to numerous small patches of erosion, the best, and in many cases only, solution is for a few people to get together and do some raking, mulching, and reseeding. Government bureaucracies tend to be too large and cumbersome to deal with such small areas effectively. The U.S. Natural Resources Conservation Service (Soil Conservation Service), which maintains an office in nearly every county of the United States, does not have the workforce to do the actual work but is generally enthusiastic about providing advice and technical assistance and sometimes seed or seedlings. (See Trout Unlimited at http://flyfishing.com/tu/tu4.html

Controlling Streambank Erosion

Another important source of sediment is from the erosion of stream banks caused by excessive runoff as described in Chapter 11. Thus, stormwater control also has importance in the control of eutrophication, even more so when stormwater retention ponds (page 289) are enhanced with wetland vegetation.

Protecting Wetlands

It is only within the last few decades that we have recognized wetlands as natural treasures. We have discovered their unparalleled importance in multiple roles of filtering nutrients and sediments from water, flood control in the catchment and slow release of water, recharge of groundwater, and as centers of biological productivity and essential stopover areas for many migrating bird species.

To be sure, major portions of a number of our great cities—Washington, New York, San Francisco, and Boston, to name a few—exist only by virtue of filled wetlands, but how much is enough? When do the losses exceed the benefits? Over half of the original estimated 221 million acres of wetlands in the lower 48 states have been destroyed, and an additional 300,000 acres per year are being lost to development and agriculture. Increasingly, U.S. authorities and citizens are coming to recognize the importance of wetlands, and laws for their protection have grown accordingly.

The most encompassing law is section 404 of the Clean Water Act of 1972. Under this statute, anyone who wants to fill a wetland must get a permit from the U.S. Army Corps of Engineers. The permit is issued only on the condition that the applicant provides a plan to *mitigate* (see below) any adverse impacts. However, draining wetlands for agriculture is not restricted under this law; over 90% of the wetlands destroyed have been converted to agriculture. Still, laws protecting wetlands are gradually tightening. Over half the states now have laws paralleling or more stringent than the federal law.

Mitigation, with reference to wetlands, is a general term that refers to any activity taken to enhance, restore, or create new wetlands in exchange for those destroyed by development. Enhancing or creating new wetlands sounds good in theory, but it is fraught with problems and controversy. Basically, scientists do not yet understand the dynamics of wetland ecology well enough to build or alter a functional sustainable wetland to achieve desired results. A number of attempts have met with dismal failure.

Restoring wetlands has somewhat better prospects. When drained wetlands are rewatered by blocking the drainage, it is found that they do gradually return to their original state, and there is a small but growing movement toward wetland restoration.

The most notable example is one in Florida. In the 1960s the 98-mile, meandering Kissimmee River, which runs from Lake Kissimmee to Lake Okeechobee, was straightened and channeled to drain 20,000 acres (8000 hectares) of adjacent wetlands for agriculture. Wildlife, including wood storks, white ibises, herons, and other birds and animals, was decimated. Now, at a cost of over 10 times that of the original project, the river's meanders are in the process of being put back and the wetlands restored (see "Ethics" box, page 310).

In addition, there is a growing trend to build or reconstruct wetlands for the express purpose of treating wastewater. Wastewater, especially sewage effluents, is being put through wetlands as a means of removing nutrients, as will be described further in Chapter 13.

But despite laws and programs to protect wetlands, pressures for continued wetland destruction are also growing. Both population growth and migration toward water where most wetlands exist are creating growing pressures for development. Most of the acreage of wetlands is in private ownership. Can the state enforce regulations requiring an individual to maintain wetlands that are on his or her property? A recent Supreme Court decision said No, not without just compensation. For states to pay all property owners the equivalent of development value to not alter wetlands is financially prohibitive. Thus, the future of wetlands and the wealth of services they provide may go either way. Which way will depend largely on public attitudes and pressures.

Another factor adversely affecting tidal wetlands is the slow but steady rise of sea level, which may accelerate as a result of global warming. Rising sea level will both increase the erosion of shorelines, wetlands included, and drown out wetlands. This situation will be discussed further in Chapters 16 and 18.

Controlling Air Pollution

Table 12–1 indicates that air pollution—in particular, the fallout of nitrogen compounds from auto emissions—also contributes to eutrophication. Controlling air pollution is addressed in Chapter 15.

Summary

In summary, controlling eutrophication will require a total watershed-management approach. For a small lake, fed from a modest-sized watershed, the task may be quite straightforward. Major sources of nutrients can be identified and addressed. For large bodies of water, such as the Chesapeake Bay, the watershed of which includes some 12,000 square miles (27,000 km^2) in five states and the District of Columbia, the task is significantly greater. Nevertheless the same approach is being used. The governmental jurisdictions involved are cooperating under a Chesapeake Bay Program with the common goal of reducing nutrient inputs into the bay 40% by the year 2000. The watershed is divided into the smaller watersheds of each of the tributaries, and each jurisdiction is taking the responsibility of reducing the nutrients emanating from its tributaries.

The same need for leadership in gaining the cooperation of industry, agriculture, government, and citizens, that was described in Chapter 11, is in order here. What is the outlook? We are facing a classic issue of sustainable development. Meanwhile, several bills were introduced during the 104th Congress designed specifically to weaken key provisions of the Clean Water Act. One bill (H.R. 961) cleared a committee and passed in the House but stopped there with no comparable bill in the Senate and a certain Presidential veto ahead. H.R. 961 would have eliminated all wetlands from federal protection, placed concern for industry's interests first, and required taxpayers to pay industry and landowners to not destroy wetlands. The defeat of such an assault on the public trust can be traced to citizen action: 165 citizen and envrionmental organizations, 50 newspaper editorial boards, and some 34 Republicans joined Democrats to rally against H.R. 961. The lesson here is that no matter how successful past legislative measures may be, the public must be ever vigilant.

Review Questions

1. Describe and compare phytoplankton and submerged aquatic vegetation (SAV). Where and how does each get nutrients and light?
2. Describe and contrast oligotrophic and eutrophic ecosystems. What are the major producers, turbidity of water, and dissolved oxygen content of each?
3. Define euphotic zone and turbidity. How are the two related? What is the significance in terms of the ecosystem that may be supported?
4. Explain how oligotrophic and eutrophic ecosystems are determined by nutrient levels.
5. Explain why the natural condition (before human impacts) of waterways can be expected to be low in dissolved nutrients.
6. Describe the ecosystem changes that occur as water is enriched with nutrients.
7. Explain how dissolved oxygen becomes depleted at lower levels of eutrophic ecosystems.
8. What is meant by BOD? How is it a measure of water quality?
9. Describe the entire process of eutrophication.
10. Distinguish between natural and cultural eutrophication.
11. What are sediments? Where do they come from? What is their impact on waterways? How do they contribute to eutrophication?
12. Give four ways in which the symptoms of eutrophication can be addressed. What are the benefits and limitations of each?
13. What are the three categories of sources of nutrients entering waterways? Give specific sources under each category.
14. What methods are being implemented to reduce the nutrients coming from each source?
15. What are wetlands? Describe some different types of wetlands.
16. What role do wetlands play in preventing eutrophication? Why is it important to preserve them?

Thinking Environmentally

1. A large number of fish are suddenly floating dead on a lake. You are called in to investigate the problem. You find an abundance of phytoplankton and no evidence of toxic dumping. Suggest a reason for the fish kill.
2. A local developer plans to turn a major section of tidal wetlands on a productive estuary into a summer community and marina. Discuss the probable environmental impacts and trade-offs involved in this plan.
3. A number of regions in the United States have banned the use of phosphate-containing detergents in recent years. What harm is caused by such detergents, and what is hoped to be achieved by such bans?
4. Describe how planting trees on an eroding hillside may protect aquatic life in an estuary many miles away.
5. Relate the problem of and corrective strategies for cultural eutrophication to the first principle of sustainability.

Sewage Pollution and Rediscovering the Nutrient Cycle

Key Issues and Questions

1. Sewage is a major disease hazard and environmental pollutant. It is also a nutrient-rich organic material that may be used as fertilizer. Describe how these divergent characteristics are all aspects of sewage.
2. In natural ecosystems, nutrients are recycled. How can the human system be adapted to recycle nutrients?
3. The pollutants in sewage can be divided into four categories. What are they? Describe each.
4. Sewage treatment uses screening, settling, and biological organisms. Describe how each of these is employed in the removal of specific categories of pollutants.
5. Biosolids are a byproduct of cleaning the water. How can sludges be treated and converted into useful products?
6. Discharges of nutrient-rich water cause cultural eutrophication. What are alternative uses for such water from sewage-treatment plants?
7. A major impediment to recycling the byproducts of sewage treatment is contamination with toxic materials. What are the toxic chemicals? Where do they come from? What is being done to alleviate the problem?

Natural ecosystems avoid pollution and resource depletion by recycling all elements—the first principle of ecosystem sustainability. Yet, human societies have generally constructed systems based on a one-way flow. In particular, the pathway of fertilizer elements such as nitrogen, phosphorus, and potassium is generally from crop soils through the food chain to humans and then into waterways with the discharge of sewage effluents. The consequence of overenriching waterways with nutrients, eutrophication, is covered in Chapter 12.

But there is good news on this front. In recent years, the United States and other developed nations have recognized the problem of eutrophication and have made substantial progress in developing or modifying sewage-treatment systems that will keep nutrients in a cycle on the land and not discharge them into waterways. But, at the other extreme, particu-

Modern sewage treatment plant. Shown are primary settling tanks (circular structures at right) and activate sludge, secondary treatment tanks (rectangular structures at left.) [Photo by Gary Retherford/Photo Researchers, Inc.]

Sewage Hazards and Potential

Untreated sewage does present the significant risk of spreading disease. But it also offers the potential of being used as organic fertilizer for the benefit of agriculture.

Health Hazards of Untreated Sewage

Untreated sewage is a major public health hazard because it is a major pathway in the spread of many infectious diseases. The excrement from humans and other animals infected with certain **pathogens** (disease-causing bacteria, viruses, and other parasitic organisms) contains large numbers of these organisms or their eggs (see Table 13-1). Even after symptoms of disease disappear, an infected person or animal may still harbor low populations of the pathogen, thus continuing to act as a carrier of disease. If wastes from carriers contaminate drinking water, food, or water used for swimming or bathing, the pathogens can gain access to and infect other individuals (Fig. 13-1).

The degree to which a pathogen may spread through a population is largely determined by two factors. First, most pathogens survive at most only a few days outside a host; second, whether infection will actually occur usually depends on the number of organisms entering the body. Therefore, when hosts are sparse, relatively little transfer of pathogenic organisms occurs because considerable time elapses between elimination by one host and contact by the next and because contamination levels remain low. As host populations become denser, however, the reverse is true. This relationship between pathogens and human populations poses a particular problem for humans who live and work in dense urban areas. As long as the population is healthy, pathogens may remain below the threshold level for causing infection. If even one individual gets

TABLE 13-1
Pathogens Carried by Sewage

Disease	*Infectious Agent*
Typhoid fever	*Salmonella typhi* (bacterium)
Cholera	*Vibrio cholerae* (bacterium)
Salmonellosis	*Salmonella* species (bacteria)
Diarrhea	*Escherichia coli, Campylobacter* species (bacteria)
Infectious hepatitis	Hepatitis A virus
Poliomyelitis	Poliovirus
Dysentery	*Shigella* species (bacteria) *Entamoeba histolytica* (protozoan)
Giardiasis	*Giardia intestinalis* (protozoan)
Numerous parasitic diseases	(Roundworms, flatworms)

FIGURE 13-1
A scene along the Ganges River in India. In many places in the developing world the same waterways are used simultaneously for drinking, washing, and disposal of sewage. And in the case of the Ganges, religious practice and spiritual cleansing. A high incidence of disease, infant and childhood mortality, and parasites is the result. (T. Stoddard/Katz/Woodfin Camp & Associates.)

Earth Watch

Monitoring for Sewage Contamination

An important aspect of public health is the monitoring of water supplies and other bodies of water having contact with humans for sewage contamination. It is worth understanding how this monitoring is done.

It is exceedingly difficult, time-consuming, and costly to test for each specific pathogen that might be present. Therefore, an indirect method called the **fecal coliform test** has been developed. This test is based on the fact that huge populations of a bacterium called *E. coli* (*Escherichia coli*) normally inhabit the lower intestinal tract of humans and other animals, and large numbers of the bacterium are excreted with fecal material. In temperate regions at least, *E. coli* does not last long in the outside environment. The presumption is that when *E. coli* is found in natural waters it indicates recent and probably persisting contamination with sewage wastes. In most situations, *E. coli* is not a pathogen itself, but it is referred to as an **indicator organism**. Its presence indicates that water is contaminated with fecal wastes and that sewage-borne pathogens may be present. Conversely, the absence of *E. coli* is taken to mean that water is free from such pathogens.

The fecal coliform test shown in the figures here, detects and counts the number of coliform bacteria in a sample of water. The results indicate the relative degree of contamination and the relative risk of pathogens. To be safe for drinking, water should have an average *E. coli* count of no more than one *E. coli* per 100 milliliters (about 0.4 cups) of water. Water with as many as 200 *E. coli* per 100 mL is still considered safe for swimming. Beyond that level, a river may be posted as polluted, and swimming and other direct contact should be avoided. By comparison, raw sewage (99.9% water: 0.1% wastes) has *E. coli* counts in the millions.

The utility of the fecal coliform test seems to be largely restricted to temperate climates, however. In tropical climates, *E. coli* may persist indefinitely in natural waters, and thus it may be present where there is no sewage contamination. Conversely, certain water-borne pathogens may be present in the absence of *E. coli*. Therefore, in tropical regions the fecal coliform test is prone to give both false positive and false negative results regarding the risk of pathogens. Here, there seems to be little alternative but to perform more-definitive tests for specific pathogens. Of course, the fact remains that sewage is still a serious source of pathogens, and proper sewage handling and treatment is always in order.

(a)

(b)

(c)

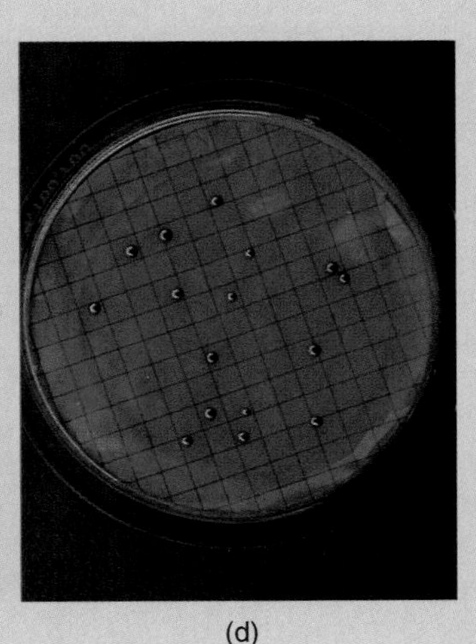

(d)

Testing water for sewage contamination—the Millipore technique. (a) A Millipore filter disc is placed in the filter apparatus. (b) A sample of the water being tested is drawn through the filter, and any bacteria present are entrapped on the filter disc. (c) The filter disc is then placed in a petri dish on a special medium that supports the growth of bacteria and will impart a particular color to fecal *E. coli* bacteria. The dish is then incubated for 24 hours at 38°C (100°F), during which time each bacterium on the disc will multiply to form a colony visible to the naked eye. (d) *Escherichia coli* bacteria, indicating sewage contamination, are identifiable as the colonies with a metallic green sheen. (Photographs by BJN, Bob Hudson, and George Waclawiw.)

sick, however, the pathogen population increases markedly and sets the stage for an escalating epidemic, especially when conditions are such that there are not suitable barriers between sewage and food and water sources.

Before the connection between disease and sewage-carried pathogens was recognized in the mid-1800s, disastrous epidemics were common in cities. For example, epidemics of typhoid fever and cholera, which killed thousands of people, were common in cities before the twentieth century. Today, public-health measures that prevent this disease cycle have been adopted throughout the developed world and to a considerable extent in the developing world. These measures involve

- purification and disinfection of public water supplies with chlorine or other agents (Chapter 11)
- sanitary collection and treatment of sewage wastes
- maintenance of sanitary standards in all facilities where food is processed or prepared for public consumption

Standards regarding the items above are set and enforced by government public health departments. A variety of other measures are enforced, as well. For example, if oyster beds are contaminated with raw sewage, health departments close them to harvesting. Implicit in all measures is monitoring for sewage contamination (see "Earth Watch" box, p. 323). Of course, our own personal hygiene, sanitation, and health precautions, such as not drinking water from untested sources and making sure foods like pork and chicken are always well cooked, remain the last and most important line of defense against disease.

Many people attribute good health in a population to modern medicine, but good health is more a result of disease prevention through public-health measures. An estimated 1.2 billion people do not have access to treated drinking water. An even greater number live in areas having poor (or no) sewage collection or treatment. This is true of large sections of developing-world cities that have grown rapidly in recent years (discussed in Chapter 6) as well as of rural areas. For example, 30% of Mexico City, 50% of Bangkok, and similar percentages of numerous other developing-world cities lack sewage-collection systems. In many cities where collection systems exist, raw sewage is still discharged into rivers. In countries of the former USSR, it is estimated that 20% of the sewage is still discharged in the raw state and another 50% receives only rudimentary treatment.

Largely because of poor sanitation regarding water and sewage, a significant portion of the world's population is chronically infected with various pathogens. More than 250 million new cases of waterborne disease are reported each year, about 10 million of them resulting in death, and about half of those deaths are among children under 5. Moreover, populations in areas where there is little or no sewage treatment are extremely vulnerable to deadly epidemics of any and all diseases spread via the sewage vector. Because of unsanitary conditions, in 1990 an outbreak of cholera in Peru killed several thousand people as it spread through Latin America. Another outbreak of cholera occurred in India in 1994.

Development of Collection and Treatment Systems

Before the late 1800s, the general means of disposing of human excrement was the outdoor privy. Seepage from the privy frequently contaminated drinking water and caused disease, especially in places where privies and wells were located near one another. In the late 1800s, Louis Pasteur and other scientists showed that sewage-borne bacteria were responsible for many infectious diseases. This important discovery led to intensive efforts to rid cities of human excrements as expediently as possible. Cities already had drain systems for storm water, but using these systems for human wastes had been prohibited. With the urgency of the situation, however, minds quickly changed. The flush toilet was introduced, and sewers were tapped into storm drains. Thus, Western civilization initiated the one-way flow of flushing sewage wastes into natural waterways.

The results of washing untreated sewage into waterways are obvious. Receiving waters that had limited capacity for dilution became essentially open cesspools of foul odors, vermin, and filth as the overload of organic matter depleted dissolved oxygen and all aquatic life suffocated. For increasing distances around or downstream from the sewage outfall, the water became unswimmable because of the sewage contamination.

In order to alleviate the problem of sewage-polluted waterways, sewage-treatment facilities were designed and constructed to treat the outflow before it entered the receiving waterway. The first treatment plants in the United States were built around 1900. However, the combined volumes of sewage and storm water soon proved impossible to handle. During heavy rains, wastewater would overflow the treatment plant and carry raw sewage into the receiving waterway. Gradually, regulations were passed requiring developers to install separate systems: **storm drains** for collecting and draining runoff from precipitation and

FIGURE 13-2
These villagers do not yet have sanitary water or sewage collection systems. Waterways through such poor areas receive direct discharges of raw sewage, yet still serve as a source of water for washing and even drinking. (Takeshi Takahara/Photo Researchers, Inc.)

sanitary sewers to receive all the wastewater from sinks, tubs, and toilets in homes and other buildings. (Note the distinction in terms; it is incorrect to speak of storm drains as sewers.)

The ideal modern system is one in which all sewage water is collected separately from storm water and fully treated to remove all pollutants before the water finally is released. But progress toward this goal has been extremely uneven. Up through the 1970s, even in the United States and other developed countries, countless communities still discharged untreated sewage directly into waterways, and for countless more the degree of treatment was minimal. Indeed, the increasing sewage pollution of waterways and beaches was the major impetus behind the passage of the Clean Water Act of 1972, and its original charge to "restore and maintain the chemical, physical, and biological integrity of the Nation's waters."

A major feature of the Clean Water Act of 1972 was to allocate many billions of dollars of federal money for installing and upgrading both collection systems and sewage-treatment plants. At first, the major thrust was to bring sewage treatment up to a standard of *secondary treatment*—a level at which essentially all the organic matter is broken down. Over 1 billion pounds of toxic pollutants and 900 million tons of untreated sewage are no longer discharged into waterways each year. It was felt that release of water free of organic matter would not degrade the environment. It was not recognized that nutrients (phosphate and nitrogen compounds) remaining in the water would cause cultural eutrophication as described in Chapter 12.

The recognition in the 1980s that effluents from sewage-treatment plants were major contributors to cultural eutrophication set off another round of upgrading sewage treatment, a round that is still very much in progress. This time the goal is to remove all pollutants, including nutrients, from the water before discharge. In the next section, we shall examine the major processes and alternatives used in sewage treatment.

In the meantime, much of the developing world still exists in the most primitive stage with regard to sewage. Innumerable poor villages and poor areas of mushrooming cities of the developing world don't even have sewer systems for collection. In such areas, it is not uncommon to find raw sewage littering the ground and overflowing gutters and streams, with children playing in the filth. Even where flush toilets and collection systems exist, the discharge of the raw sewage into waterways is still common. Many of the people living in these regions must use these badly polluted waters for bathing, laundering, and even drinking (Fig. 13-2).

Sewage Management and Treatment

In addressing the topic of sewage treatment let us first clarify which contaminants and pollutants we are talking about.

The Pollutants in Raw Sewage

Raw sewage is not only the flushings from toilets; it is also the collection from all other drains in homes and other buildings. A sewer system brings all tub, sink, and toilet drains from all homes and buildings together into larger and larger sewer pipes, just as twigs of a tree come together eventually into the trunk. This total mixture collected from all drains, which comes out at the end of the "trunk" of the collection system, is termed **raw sewage** or **raw wastewater**. Because we use such large amounts of water to flush away small amounts of dirt, especially as we stand under the shower or often just run the water with no waste at all, most of what goes down sewer drains is water. Raw sewage is about 1000 parts water for every 1 part of waste—99.9% water to 0.1% wastes.

Given our voluminous use of water, the quantity of raw sewage output is in the order of 150–200 gallons (600–800 liters) per person per day. That is, a community of 100,000 persons will produce on the order of 1.5 to 2.0 million gallons (6–8 million liters) of wastewater each day. With the addition of storm water, raw sewage is diluted still more. Nevertheless, the pollutants are sufficient to make the water black and smell foul.

FIGURE 13-3
Bar screen, the first stage of preliminary treatment. Passing the wastewater through a coarse screen consisting of parallel bars traps debris, which is then removed from the screen by a mechanical rake. This photograph shows the upper portion of the bar screens and the mechanical "rakes." The water is flowing through below floor level. (Photograph by BJN.)

The pollutants generally are divided into the four following categories, which, as we shall find, correspond to techniques used for their removal: (1) debris and grit, (2) particulate organic material, (3) colloidal and dissolved organic material, and (4) dissolved inorganic material.

Debris and Grit. Debris includes rags, plastic bags, and other objects flushed down toilets or washing through storm drains in places where they are still connected to sewers. Grit is coarse sand and gravel, and it too enters mainly through storm drains.

Particulate Organic Material. Particulate organic material includes visible particles of organic matter, originating from food wastes from home garbage-disposal units as well as fecal matter and bits of paper from toilets. Particulate organic material also includes living bacteria and other microorganisms that have begun to digest the waste, and possibly pathogenic organisms. Importantly, the particulate portion of organic material, by definition, consists of particles that will settle in still water. We shall see that its removal is based on this attribute.

Colloidal and Dissolved Organic Material. Colloidal organic material originates from the same sources as particulate organic material; the distinction is particle size. Whereas the visible particles described above will settle in still water, colloidal particles are so fine that they will not settle, at least not within any reasonable time period. Bacteria and other microorganisms, including pathogens, are also present in this category. In addition, there is dissolved organic material from soaps, detergents, shampoos, and other cleaning and washing agents.

Dissolved Inorganic Material. Dissolved inorganic material includes mainly the nitrogen, phosphorus, and other nutrients from excretory wastes plus phosphate from detergents and water softeners.

Other Contaminants. In addition to the four categories of pollutants in "standard" raw sewage, variable amounts of pesticides, heavy metals, and other toxic compounds may be found in sewage because people pour unused portions of products containing such materials down sink, tub, or toilet drains. Also, industries may discharge various toxic wastes into sewers. We shall consider the problems caused by such materials and how to cope with them at the end of the chapter. For the present, we shall assume that we are dealing with "standard" raw sewage without undue amounts of such contaminants.

Removing the Pollutants from Sewage

The challenge of sewage treatment is more than installing a technology that will do the job. It is finding one that will do the job at a reasonable cost. The following steps, which are the ones in standard use throughout the world, should be considered in this light.

Preliminary Treatment (Removal of Debris and Grit). Because debris and grit will damage or clog pumps and later treatment processes, removing them is a necessary first step and is termed **preliminary treatment**. Preliminary treatment usually involves two steps, a screening out of debris and a settling of grit. Debris is removed by letting raw sewage flow through a **bar screen**, a row of bars mounted about 1 inch (2.5 cm) apart (Fig. 13-3). Debris is mechanically raked from the screen and taken to an incinerator.

Rotating screens are also coming into use for debris removal. A screen is mounted as a conveyer belt in the flow of water (Fig. 13-4). Material caught on the screen is continually brought to the surface and removed. Such screens, which have a hole size of about one-quarter inch (0.5 cm), achieve a more nearly complete removal of debris than bar screens do. Also, in some systems, a grinder called a comminuter is used to reduce the debris to a relatively fine particle size. The particles of debris thus continue through the system to be removed at a later stage.

After passing through the screen, the water flows through a **grit-settling tank**, a swimming pool-like

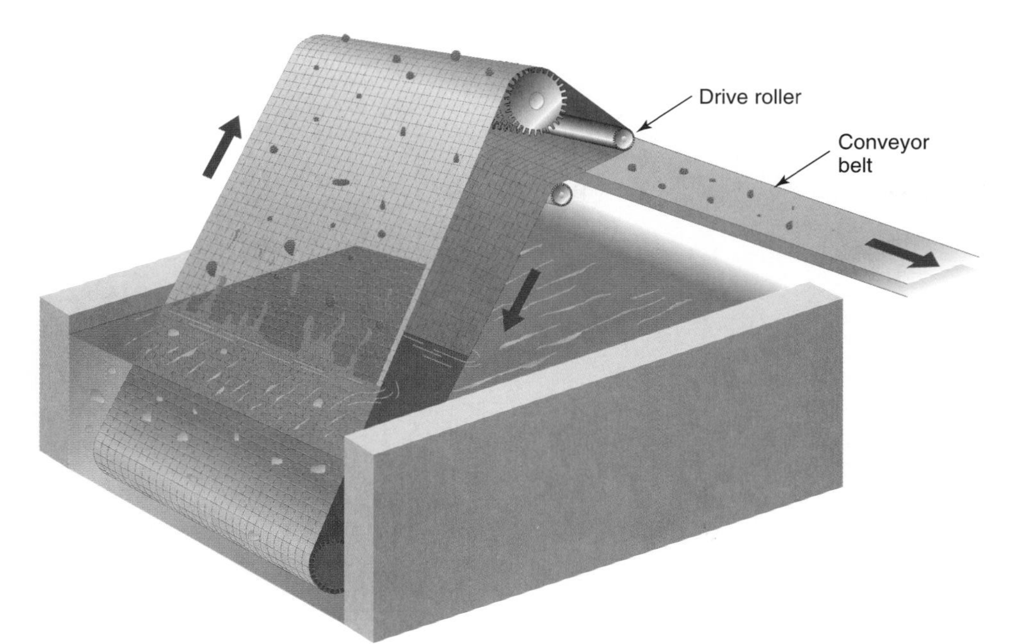

FIGURE 13-4
Rotating screen for debris removal. The screen is constructed in a manner similar to parallel bicycle chains. With the slow, continual motion of the screen, debris is collected on the screen and brought out of the water, dislodged from the screen by the cogs on the drive roller, and taken to a landfill or incinerator.

tank, where its velocity is slowed just enough to permit the grit to settle (Fig. 13-5). The settled grit is mechanically removed from these tanks and taken to landfills.

Primary Treatment (Removal of Particulate Organic Material). After preliminary treatment, the water moves on to **primary treatment**, where it flows very slowly (about 6 feet, 2 m, per hour) through large tanks called **primary clarifiers** (Fig. 13-6). Because it flows slowly through these tanks, the water is nearly motionless for several hours. The particulate organic material, about 30% to 50% of the total organic material, settles to the bottom, from where it can be removed. At the same time, fatty or oily material floats to the top,

FIGURE 13-5
Grit-settling tank, the second stage of preliminary treatment. (a) The velocity of the water flow is slowed to about 1–2 feet (0.5 m) per second, a speed that allows sand and other coarse grit to settle to the bottom while the water with other pollutants flows over the top edge (foreground). (b) Empty tank. The gates in the back wall of the tank control the velocity of the incoming water. "Plows" on the rotating arm continually push the settled grit to the side where it is mechanically removed. (Photographs by BJN.)

(a)

(b)

(a)

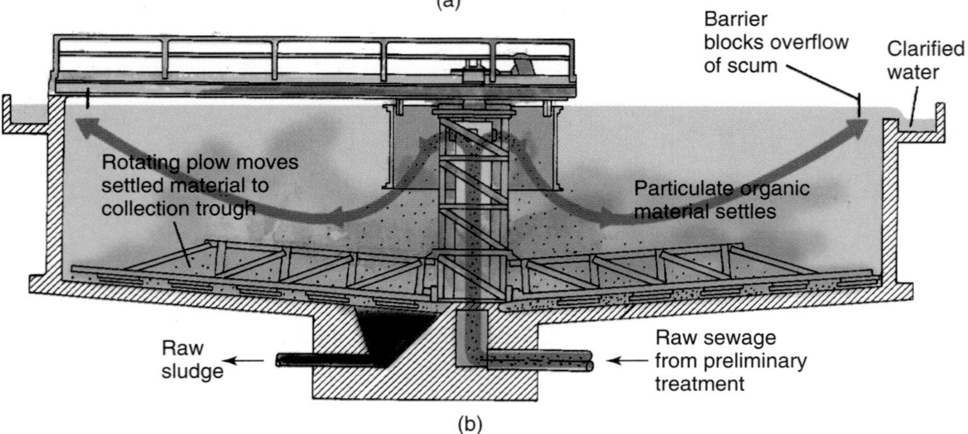

(b)

FIGURE 13-6
Primary clarifiers used for primary treatment. (a) The water enters these tanks at the center and exits over the weir at the edge. The slow velocity (about 5–10 feet, or 2–3 m, per hour) of flow through the tanks permits the particulate organic material to settle while oil and grease rise. The settled organic material is pumped from the bottom, and the oil and grease are skimmed from the surface. These combined materials constitute raw sludge. (Photograph by BJN.) (b) Cross section of clarifier. (Courtesy of Walker Process Division of C.B.I.)

where it is skimmed from the surface. All the material removed, both particulate organic material and fatty material, is combined into what is referred to as **raw sludge**; we shall consider its treatment shortly.

Note that primary treatment involves nothing more complicated than putting polluted water in a "bucket," letting material settle, and pouring off the water. Nevertheless, it removes the particulate organic matter at minimal cost.

Secondary Treatment (Removal of Colloidal and Dissolved Organic Material). Secondary treatment is also called **biological treatment** because it makes use of organisms—natural decomposers and detritus feeders (Fig. 13-7). Basically, an environment is created to enable these organisms to feed on the colloidal and dissolved organic material and break it down to carbon dioxide and water via their cell respiration. The sewage water from primary treatment is the food- and water-rich medium. The only thing that needs to be added in addition to the organisms is oxygen to enhance their respiration and growth. Recall that the oxygen consumed in the decomposition process is called the biochemical oxygen demand (BOD). Either of two systems may be used: *trickling filters* or *activated-sludge systems*. The earliest and still widely used systems are tricking filters, but the current trend is toward use of activated-sludge systems.

In a **trickling-filter system**, the water exiting from primary treatment is sprinkled onto and allowed to percolate through a bed of fist-sized rocks 6–8 feet (2–3 m) deep (Fig. 13-8). The spaces between the rocks provide for good aeration. As in a natural stream, this environment supports a complex food web of bacteria, protozoans, rotifers (organisms that consume protozoans), various small worms, and other detritus feeders attached to the rocks (Fig. 13-8b). The organic material in the water, including pathogenic organisms, is absorbed and digested by these organisms as it trickles by. Clumps of organisms that occasionally break free and wash from the trickling filters are removed by passing the water through secondary clarifiers, tanks that work in the same way as primary clarifiers. Through primary treatment and a trickling-filter system, 85% to 90% of the total organic material is removed from the wastewater.

The **activated-sludge system** is shown in Fig. 13-9. In this system, water from primary treatment enters a long tank (it could hold several tractor-trailer trucks parked end to end) that is equipped with an air-bubbling system (Fig. 13-9a). A mixture of detritus-feeding organisms, referred to as **activated sludge**, is

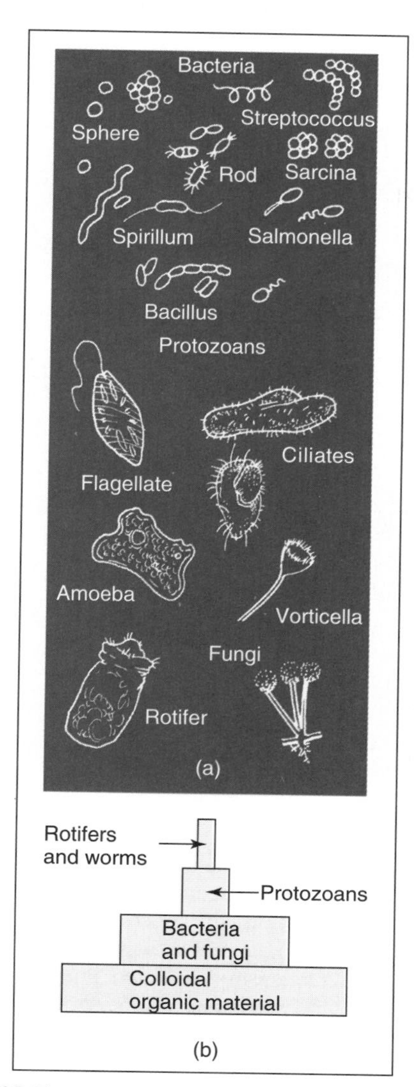

FIGURE 13-7
(a) Some of the organisms in the activated-sludge secondary treatment system. (b) These organisms represent trophic levels through which the biomass of organic material entering the system is reduced by up to 90%.

(a)

(b)

FIGURE 13-8
Trickling filters for secondary treatment. (a) The water from primary clarifiers is sprinkled onto and trickles through a bed of rocks 6–8 feet (2–3 m) deep. (b) Various bacteria and other detritus feeders adhering to the rocks consume and digest the organic material as it trickles by in the water. The water is collected at the bottom of the filters and goes on for final disinfection. (Photographs by BJN.)

added to the water as it enters the tank, and the water is vigorously aerated as it moves through the tank. Organisms in this well-aerated environment reduce the biomass of organic material, including pathogens, as they feed. As organisms feed on each other they tend to form into clumps, termed *floc*, that settle readily when the water is stilled. Thus, from the aeration tank the water is passed into a secondary clarifier tank where the organisms settle out and the water—now with better than 90% of all the organic material removed—flows on. The settled organisms are pumped back into the entrance of the aeration tank. They are the activated sludge that is added at the beginning of the process (Fig. 13-9b, c). Surplus amounts of activated sludge, which occur as populations of organisms grow, are removed and added to the raw sludge.

Recall from Chapter 3 that as organisms feed they

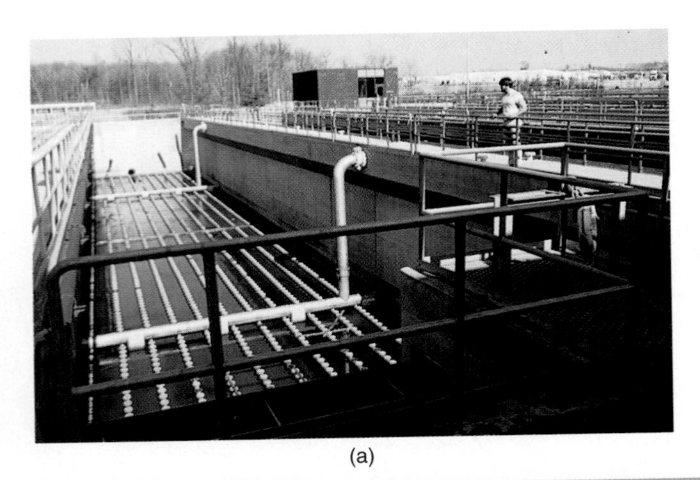

(a)

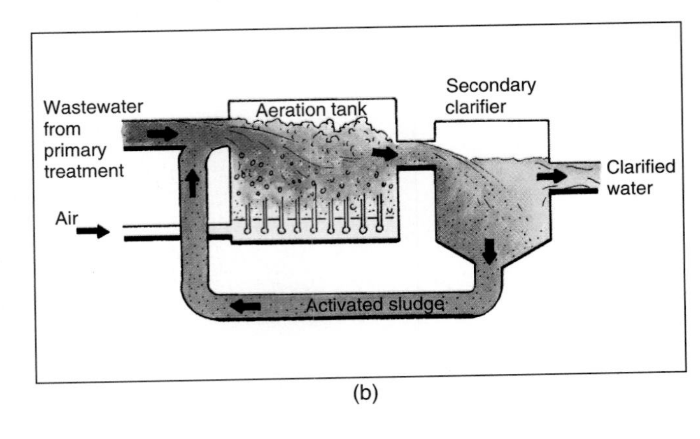

(b)

(c)

FIGURE 13-9
(a) Aeration tank used in activated-sludge treatment. Wastewater from the primary clarifiers moves through the tank and is vigorously aerated by air forced up from the tubes along the bottom. (b) In the oxygen-rich environment of the aeration tank, microorganisms consume colloidal and dissolved organic material. Organisms (activated sludge) settle out in the secondary clarifier and are returned to the aeration tank while the clarified water flows on. (c) Aeration tank in operation. The rust-colored pipe in the foreground is the inlet pipe for activated sludge being recycled from the secondary clarifier. (Photographs by BJN.)

convert some organic matter and nutrients into the biomass of their own bodies while another portion is excreted as wastes from respiration (see Fig. 3-13). Then, as food chains go toward completion, biomass is totally reduced to carbon dioxide, water, and mineral nutrients, which in this case remain in the water solution (see the biomass pyramid, Fig. 3-18).

The original secondary-treatment systems were designed and operated in a manner to let food chains go toward their limit, because elimination of organic material and its resulting BOD were considered the prime objectives. Ending up with a nutrient-rich discharge was not recognized as a problem.

Biological Nutrient Removal (Removal of Dissolved Inorganic Material). Today, with increased knowledge of the problem of cultural eutrophication, secondary activated-sludge systems are being modified and operated in a manner that achieves nutrient removal as well as detritus oxidation, a process known as **biological nutrient removal (BNR).**

Recall that in the natural nitrogen cycle (Fig. 3-16), nutrient forms of nitrogen (ammonia and nitrate) are converted by various bacteria back to nonnutritive nitrogen gas in the atmosphere. This process is called **denitrification**. For biological removal of nitrogen, then, the activated-sludge system is partitioned into zones, and the environment in each zone is controlled in a manner to promote the denitrifying process (Fig. 13-10).

With respect to phosphate, in an environment that is rich in oxygen but relatively lacking in food, the envi-

Ethics

Persisting in the Flow

Coastal cities have a reputation for dragging their heels on sewage treatment. With the enormous potential of an ocean for tidal mixing and flushing, they have become accustomed to using the dilution solution for disposing of their sewage. The city of Boston, Massachusetts, presents an intriguing case in point. (On the opposite coast, San Francisco and Los Angeles provide similar examples.) For decades, Boston sewage received only primary treatment before being discharged into Boston Harbor. Making matters worse, scum and sludge from the treatment process were also released to the harbor. The harbor had such a well-deserved reputation for gross pollution that the Republican presidential candidate George Bush used it to demolish the environmental record of Michael Dukakis (his opponent, and governor of Massachusetts) in the 1988 presidential campaign even though the Reagan/Bush administration had done little in the prior eight years on this issue.

Under federal government pressure for years to clean up the harbor, the state established the Massachusetts Water Resources Authority (MWRA) in 1984 to oversee the cleanup. The MWRA mapped out a plan for the cleanup that conforms with the federally mandated full implementation by 1999. One element of that plan was put in place in December 1991: The flow of raw sludge was stopped, and all sludge now goes to a pelletizing plant to be turned into fertilizer (see page 336). A key component of the plan is the construction of a 9.5-mile-long pipe that will carry treated effluent well out of Boston Harbor and into Massachusetts Bay. The Ocean Outfall Tunnel began carrying effluent from a new primary treatment plant in 1996. By December 1999, the secondary treatment phase of the Deer Island sewage treatment facility will be in full operation and the program will be fully implemented.

The proposed 9.5-mile outfall pipe, which is already under construction, has stirred up a firestorm of criticism from citizens and advocacy groups on Cape Cod. Their concern centers on the impact of the effluent on water quality in Cape Cod Bay, which adjoins Massachusetts Bay to the southeast. The opponents would prefer that the outfall remain in Boston Harbor, which has already been degraded by pollution and will probably always be so. They claim that because of the location of the outfall pipe, coastal water circulation may tend to carry the high-nutrient effluent into Cape Cod Bay. The Cape Codders cite numerous studies that show that nitrogen acts as a limiting factor in many coastal marine systems, and they fear that the added nitrogen from the Boston treatment plant will create conditions leading to blooms of algae, especially those that cause red tide. Red tides have been a recurring problem in recent years up and down the northeastern coast, often closing down shellfish harvesting because of the dangers of paralytic shellfish poisoning.

Resolution of this conflict may be difficult. The MWRA claims that not only will the new treatment system obviously benefit Boston Harbor, it will also improve water quality in Massachusetts and Cape Cod bays. They point out that the system as currently operating is already polluting those bays, since the effluent is eventually carried out of the harbor by tidal action. Substantial efforts by leading marine scientists are under way to assess the potential impact of the pipe, but Cape Cod residents fear that the studies (funded by the MWRA) will be biased. Their request to the MWRA for $575,000 for an independent review of the ongoing studies was recently rejected. MWRA board members felt that the request was really a form of "blackmail" on the part of the Cape Cod advocacy groups. Adding to the complexity of the conflict, Susan Tierney, the state's environmental chief and also chairwoman of the MWRA, spoke in favor of the request.

On the surface, this conflict appears to be another case of NIMBY (Not In My Back Yard). One municipality is attempting to resolve its environmental problem in an apparently effective fashion, but there are risks that the solution will create another problem many miles away. Check: http://www.history.rochester.edu/class/BOSHARB/harbor.html

ronment of zone 3, bacteria take up phosphate from solution and store it in their bodies. Thus, phosphate is removed as the excess organisms are removed from the system (see again Fig. 13-10). These organisms, with the phosphate they contain, are added to and treated with the raw sludge, ultimately producing a more nutrient-rich treated sludge (biosolids) product.

As an alternative to BNR, there are various chemical processes that may be used. One is simply to pass the effluent from standard secondary treatment through a filter of lime, which causes the phosphate to precipitate out as insoluble calcium phosphate.

Final Cleansing and Disinfection. After biological nutrient removal, the wastewater is subjected to a final cleansing by filtration through a bed of sand and by disinfection. Although few pathogens survive the combined stages of treatment and sand filtration, public-health rigors still demand that the water be disinfected before discharge into natural waterways. The most widely used disinfecting agent is chlorine gas because it is both effective and relatively inexpensive. But this treatment also introduces chlorine into natural waterways, and even minute levels of chlorine can harm aquatic animals. The hatching of trout eggs, for

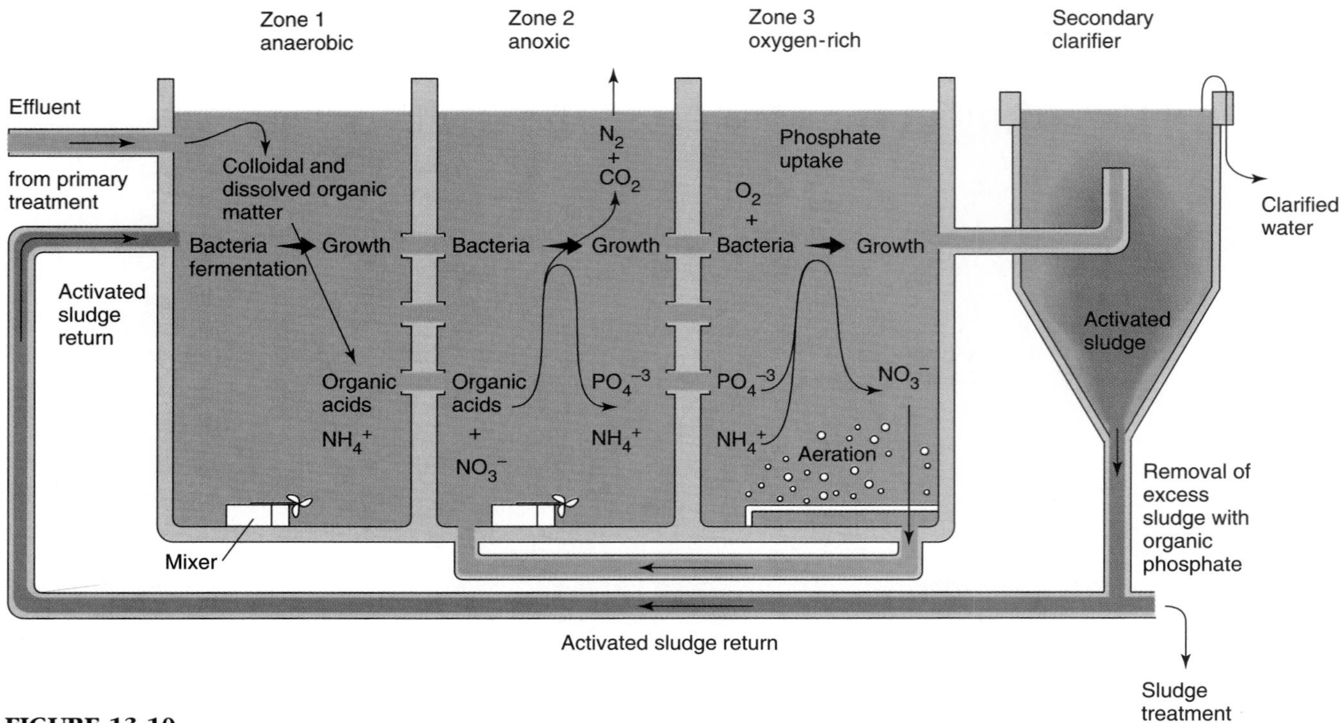

FIGURE 13-10

Biological nutrient removal (BNR). The secondary treatment, activated sludge process may be modified to remove nitrogen and phosphate while at the same time breaking down organic matter. For this BNR process the aeration tank is partitioned into three zones, only the third of which is aerated. As seen in the diagram, ammonium (NH_4^+) is converted to nitrate (NO_3^-) in zone 3. Recycled to zone 2, which is without oxygen gas (anoxic), the nitrate supplies the oxygen for cell respiration. It is converted to nitrogen gas in the process and is released into the atmosphere. Phosphate is taken up by bacteria in zone 3 and removed with the excess sludge.

instance, and development of the embryos are affected by the presence of chlorine. Also, chlorine reacts spontaneously with organic compounds to some extent to form **chlorinated hydrocarbons** (also called **organochlorides**), organic molecules with chlorine atoms attached. Many of these compounds are toxic and nonbiodegradable, and some have been identified as compounds that cause cancer, abnormal development, and reproductive problems.

Because of the negative side effects of chlorine, an additional agent may be added to convert the chlorine to a chemically inactive form.

Other disinfecting techniques are coming into use. One alternative disinfecting agent is ozone gas, which is extremely effective in killing microorganisms and in the process breaks down to oxygen gas, which actually improves water quality. But, because ozone is unstable and hence explosive, it must be generated at the point of use, a step that demands considerable capital investment and energy. Another disinfection technique is to pass the effluent though an array of ultraviolet lights mounted in the water. (Standard fluorescent lights without the white coating emit ultraviolet.) The ultraviolet radiation kills microorganisms but does not otherwise affect the water.

After these steps of treatment, the wastewater has a lower organic and nutrient content than most bodies of water into which it is discharged. In other words, discharge of such water actually contributes toward improving water quality in the receiving body. In water-short areas there is every reason to believe that the treated water itself, with little additional treatment, could be recycled into the municipal water supply system.

A summary of water treatment procedures is shown in Figure 13-11.

Bear in mind that the techniques we have described here are state-of-the-art. Many cities, even in the developed world, are still operating with antiquated systems that provide lower-quality treatment. Sizable coastal cities still persist in discharging sewage with little or no treatment directly into the ocean despite growing evidence that this practice causes ecological upset (see "Ethics" box, page 331).

On the positive side, however, alternative methods of treatment are being explored. Some of

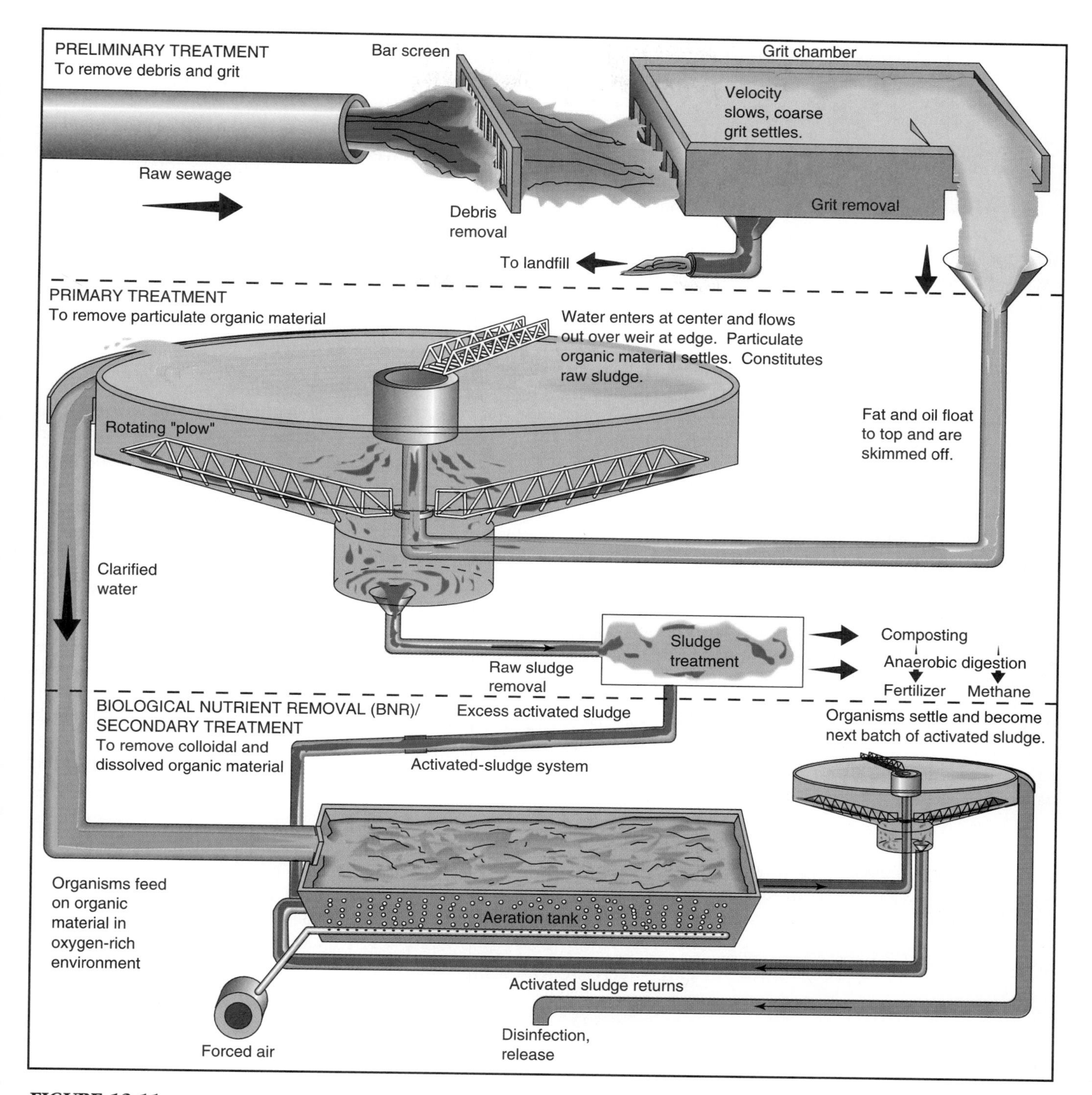

FIGURE 13-11
A summary of wastewater treatment.

these are considered shortly, after we consider treatment of the raw sludge, the organic matter removed from the water.

Sludge Treatment Options

Recall the particulate organic matter that settled out or floated to the surface of sewage water in primary treatment. This material forms the bulk of *raw sludge,* although there may be additions, as we noted, of excesses from activated sludge and BNR systems. As such, raw sludge is a black, foul-smelling, syrupy liquid with a water content of 97% to 98%. Also, the potential for the presence of pathogens is significant because raw sludge includes material directly from toilets. Indeed, it is considered a biologically hazardous material. However, as nutrient-rich, organic material it has the potential to be used as organic fertilizer (see Chapter 9) if it is suitably treated to kill pathogens and other toxic contaminants are not present.

Importantly, raw sludge is the byproduct of wastewater treatment. When raw sewage was discharged into waterways there was no sludge to be considered—"only" the consequences of polluted waterways. As sewage treatment has improved, the amount of raw sludge to be dealt with has grown in proportion. At first sludge was considered only as a material to be disposed of. It was incinerated, put into landfills, and even barged to sea and dumped, all of which create pollution. It is really only in the last decade that we have moved in a large way toward converting sewage sludge to organic fertilizer. Even so, the value of sewage sludge as fertilizer does not balance the costs of treatment and transportation. The real value must be considered in terms of recreating a sustainable nutrient cycle on the land and protecting waterways from pollution and cultural eutrophication.

At present four alternative methods for treating sludge and converting it into organic fertilizer are commonly used: (1) *anaerobic digestion*, (2) *composting*, (3) *pasteurization*, and (4) *lime stabilization*. Since this is a developing industry, it is unclear which method will prove most cost-effective and environmentally acceptable over the long run. Also, none of these methods is capable of removing toxic substances such as heavy metals and nonbiodegradable synthetic organic compounds. The presence of such toxins can preclude the use of sludge as fertilizer.

Anaerobic Digestion. Anaerobic digestion is a process of allowing bacteria to feed on the detritus in the *absence of oxygen*. The raw sludge is put into large, air-tight tanks called **sludge digesters** (Fig. 13-12). In the absence of oxygen, total breakdown of organic matter to carbon dioxide and water is impossible. But the bacteria present gain enough energy to grow through the metabolic process of fermentation or anaerobic respiration in which the organic material is broken down partially.

A major byproduct of the anaerobic processes is **biogas**, a gaseous mixture that is about two-thirds *methane*. The other one-third is made up of carbon dioxide and various foul-smelling organic compounds that give sewage its characteristic odor. Natural gas, widely used for heating and cooking, is nearly pure methane. Because of its methane content, biogas is

FIGURE 13-12
Anaerobic sludge digesters. Bacterial digestion of raw sludge in the absence of oxygen in these tanks leads to the production of methane gas, which is tapped from the tops of the tanks, and humus-like, organic matter that can be used as a soil conditioner. The "egg-shape" of the tanks facilitates mixing and digestion. (Back River STP, Baltimore. Photograph by BJN.)

FIGURE 13-13
Agricultural application of treated sludge. The treated sludge remaining after anaerobic digestion is a humus-like, nutrient-rich liquid that is an excellent soil conditioner. The vehicle shown here is specially designed for this application. Apparatus on the back injects the sludge under the soil surface, so there is no unpleasant sight or odor. (Courtesy of Ag-Chem Equipment Co. Inc., Minneapolis, Minnesota.)

flammable and can be used for fuel. In fact, it is commonly burned to heat the sludge digesters because the bacteria working on the sludge do best when maintained at about 100°F (38°C). Also, the methane can be concentrated by passing the biogas through a water column. Carbon dioxide is highly soluble in water, whereas methane is not. The nearly pure methane resulting can then be used to supplement natural gas supplies.

After four to six weeks, anaerobic digestion is more or less complete, and what remains is called **treated sludge**. It consists of the remaining organic material, which is now a relatively stable, nutrient-rich, humus-like material in water suspension. Pathogens have been largely if not entirely eliminated, so they no longer present any significant health hazard.

Treated sludge is now referred to as **biosolids** to improve its public image and it does make an excellent organic fertilizer. It may be applied directly to lawns and agricultural fields in the liquid state in which it comes from the digesters, providing the benefit of both the humus and the nutrient-rich water (Fig. 13-13). Alternatively, the sludge may be "dewatered" by means of belt presses, where the sludge is passed between rollers that squeeze out most of the water and leave the organic material as a semisolid **sludge cake** (Fig. 13-14). The sludge cake (biosolids) is easy to stockpile, distribute, and spread on fields with traditional manure spreaders.

Composting. Traditional composting is a matter of putting yard and food wastes in a well-aerated pile and letting decomposers reduce it to a stable nutrient-rich, humus-like material, as discussed in Chapter 9. The same concept is applicable to the treatment of sewage sludge. Raw sludge is mixed with wood chips or other water-absorbing material to reduce the water content. It is then placed in **windrows**, long narrow piles that allow air circulation and convenient turning with machinery (Fig. 13-15). Bacteria and other decomposers break down the organic material to rich, humus-like

FIGURE 13-14
Dewatering treated sludge. (a) Sludge (98% water) may be "dewatered" by means of belt presses as shown here. The liquid sludge is run between canvas belts going over and between rollers such that much of the water is pressed out. (b) The resulting "sludge cake" or biosolid is a semisolid humus-like material that may be used as an organic fertilizer. (Photographs by BJN.)

(a)

(b)

FIGURE 13-15
A sewage sludge composting facility near Washington, D.C. The black rectangles are the sludge being composted as described in the text. (Insert) The finished compost product is bagged, marketed, and well-received by gardeners. (Courtesy of Washington Suburban Sanitary Commission, Biosolids Operations Division.)

material. Pathogens lose out in the competition. As long as the piles are kept well aerated, the obnoxious odors typical of anaerobic respiration are negligible. After six to eight weeks of composting, the resulting humus is screened out of the wood chips. The chips may be reused, and the humus is ready for application to soil.

A technique that is gaining increasing favor is **co-composting**. In this process, raw sludge is mixed with shredded wastepaper to reduce the water content, and the entire mass is composted to a condition suitable for soil application.

Pasteurization and Drying. Raw sludge may be dewatered and the resulting sludge cake put through drying ovens that operate like oversized laundry dryers. In the dryers, the sludge is **pasteurized**, that is, heated sufficiently to kill any pathogens (exactly the same process that makes milk safe to drink). The product is dry, odorless organic pellets. Milwaukee, Wisconsin, which has a particularly rich sludge resulting from the brewing industry, has been using this process for over 60 years. The city bags and sells the pellets throughout the country as an organic fertilizer under the trade name Milorganite. In 1991, Boston, Massachusetts, started up a pasteurization and drying facility that is now the nation's largest producer of organic fertilizer. Before the plant opened, Boston and surrounding communities had been dumping some 50,000 gallons (200,000 liters) of raw sludge every day into Boston Harbor, giving the harbor the reputation of being the nation's most polluted waterway.

Lime Stabilization. In **lime stabilization**, the raw sludge is filter-pressed and the moist sludge cake is mixed with slaked lime—calcium hydroxide, $Ca(OH)_2$. The reaction of the lime with moisture generates a temperature and pH high enough to kill pathogens. The whole mixture can be applied as an organic fertilizer, the lime being another fertilizer ingredient and good for neutralizing acidic soils.

Alternative Treatment Systems

Using Effluents for Irrigation. Consider again the nutrient-rich water coming from the standard secondary treatment process. Such water is beneficial for growing plants. The problem is only that we don't want to put it

Earth Watch

The Overland Flow Wastewater Treatment System

An alternative to traditional wastewater treatment is the **overland flow system**. A facility utilizing this system was put into operation in Emmitsburg, Maryland, in 1989.

Raw wastewater, about 1 million gallons (4 million liters) per day, from all the homes and commercial establishments of the community is first put through a pond where grit and particulate organic material settle. The water is then irrigated onto the long, narrow fields you see in the photograph. These fields have about a foot (0.3 m) of rich topsoil supporting a crop of reed canary grass, a forage grass that has a voracious appetite for nitrogen and other nutrients. Below the topsoil, the subsoil is compacted clay, which is impermeable to water, and slopes gently downward away from the irrigation pipe. The wastewater applied to one side of the field thus percolates through the topsoil, across the field, and into a collecting gutter at the opposite side. During this passage, natural organisms in the topsoil break down and utilize the organic wastes and maintain the richness and aggregate structure of the topsoil, while the grass absorbs the nutrients. The water exiting into the collecting gutter is clear and nearly nutrient-free. It is collected into another reservoir and spray-irrigated onto forage crops so that none of the nutrients go to waste. The canary grass is periodically mowed and becomes feed for cattle. Thus, the nutrients make a complete cycle from wastewater to grass to beef to humans to wastewater and again back to the soil.

Overland flow wastewater treatment system. Wastewater is being irrigated onto fields. Note the lush growth of grass benefiting from the nutrients. (Photograph by BJN.)

You may ask how a small community in western Maryland happened to install a state-of-the-art ecological method of wastewater treatment. First, the town did need a new wastewater treatment facility to accommodate population growth. But it was largely the efforts of one farmer in the area, Richard Waybright, that persuaded the town officials to "go ecological." He volunteered some 200 acres (about 80 hectares) of his own land to be the final recipient of the water and the nutrients. His forward thinking has created a situation in which everyone wins. He gets free water and nutrients for irrigation, Emmitsburg meets the standards for sewage treatment that will accommodate growth with a low-cost, low-maintenance system, and everyone benefits by not having the nutrients go into Chesapeake Bay, where they would cause more eutrophication.

into waterways, where it will stimulate the growth of undesirable algae. But why not use it for irrigating plants we do want to grow? It is a way of completing the nutrient cycle. Indeed this concept has been put into practice in a considerable number of locations as a practical alternative to upgrading the treatment to remove nutrients. Again, there is the prerequisite that effluents not be contaminated with toxic materials.

For example, the nutrient-rich effluent from standard secondary treatment from St. Petersburg, Florida, was causing cultural eutrophication in Tampa Bay. Now St. Petersburg uses the effluent to irrigate 4000 acres (1600 hectares) of urban open space, from parks and residential lawns to a golf course. Revenues from the water sales help offset operating costs. Bakersfield, California, receives a $30,000 annual income from a 5000-acre (2000-hectare) farm irrigated with its treated effluent. Clayton County, Georgia, is irrigating 2500 acres (1000 hectares) of woodland with partially treated sewage. Hundreds of other similar projects are under way around the country.

A number of developing countries irrigate croplands with raw, (untreated) sewage effluents. This practice is getting the bad with the good. The crops respond

FIGURE 13-16
Orlando Easterly Wetlands Reclamation Project, Florida. Formerly cattle pasture, this is part of a mixed marsh that was created to remove nutrients from the wastewater of Orlando and to reestablish wildlife habitat. (Photograph courtesy of Post, Buckley, Schuh & Jernigan, Inc.)

well, but parasites and disease organisms can easily be transferred to farm workers and consumers. Therefore, it is important to emphasize that only treated effluents be used for irrigation.

Reconstructed Wetland Systems. It is also possible to utilize the nutrient-absorbing capacity of wetlands where suitable areas and climatic conditions exist. The project may be part of a wetlands mitigation program (see Chapter 12), or artificial wetlands may be constructed.

As one example of the former, in the 1960s and 1970s much of the land around Orlando, which was originally wetlands, was drained and converted to cattle pasture. At the time, Orlando was discharging into the James River 13 million gallons (50 million liters) per day of nutrient-rich effluent following secondary treatment. Through the Orlando Easterly Wetlands Reclamation Project, 1200 acres (480 hectares) of pastureland has been converted back to wetlands. The project involved scooping soil from and building berms (mounds of earth) around pastures to create a chain of shallow lakes and ponds. In addition, 1.2 million wetland plants ranging from bulrushes and cattails to various trees were planted. The effluent entering the upper end now percolates through the wetland for about 30 days before entering the James River virtually pure. Thus, the project has recreated a wildlife habitat (Fig. 13-16). Wetland systems can be designed for small as well as large areas and are becoming an increasingly popular alternative for small communities.

Wetlands can handle the colloidal organic matter as well as nutrients. Natural microbes exist to break the colloidal material down. However, this is *not* to say that raw sewage effluents may simply be discharged into natural wetlands. Such a practice has been found to cause significant degradation of the natural wetland. The key will always be careful management that ensures that the systems are kept in balance and not loaded beyond their ability to handle inputs.

Artificial Wetland Systems. Fully artificial wetland systems may also be constructed. An increasingly popular technique is to pass nutrient-rich effluents through shallow tanks or ponds filled with water hyacinths (Fig. 13-17). Water hyacinths are adapted to float on the surface, and they have an extensive root system that dangles down into the water and is extremely efficient at absorbing nutrients. Another variation is called the *oxidation pond system*. Here the raw sewage is directed into a series of shallow ponds, where algae and other vegetation consume the nutrients and then settle to the bottom. After passage through the series of ponds, most of the organic matter and dissolved nutrients have been removed.

Similar systems utilizing cattails and other "reeds" are also very efficient in removing nutrients. The plant material may be periodically harvested and may be used for weaving mats and baskets or for cattle feed, or it may be fermented to produce alcohol or other products. Again, climatic factors and available space are the two factors that most restrict their development.

Greenhouse Wetland Systems. Where climate is restrictive, a greenhouse wetland system may be the answer. Researchers at Ocean Arks International, Falmouth, Massachusetts, have developed a greenhouse aquatic system in which raw sewage first flows through a series of tanks, where bacteria consume the organic material (both particulate and dissolved), algae absorb the nutrients, snails eat the algae, and so on up the food chain (Fig. 13-18). The system includes clams, several species of fish, and 120 species of plants, each performing a particular role. The end of the process yields organisms harvested for food, plants for aesthetic pleasure, compost, and virtually pure water.

John Todd, president of Ocean Arks, and coworkers visualize that far from trying to put sewage treatment as far out of sight and mind as possible, their greenhouse system might become an attractive garden-plant nursery center for the community it serves. An Ocean Arks facility in Providence, Rhode Island, is beautiful, fresh-smelling, and clean. What is the potential for installing such systems in the developing world?

Overland Flow Systems. Another effective biological system that shows promise is the overland flow system. It is a sort of cross between straight irrigation and a wetland system (see "Earth Watch" box, page 337)

(a)

(b)

FIGURE 13-17
(a) Water hyacinths in artificial ponds being used for removing nutrients from sewage effluents. Hyacinths are harvested for other uses. (b) Water hyacinth plant. "Bulbs" at bases of leaves are air bladders that enable the plant to float on the surface while the roots dangle down and withdraw nutrients from the water solution.(a) François Gohier/Photo Researchers, Inc., b) Photograph by BJN.)

FIGURE 13-18
Aquatic greenhouse for sewage treatment. The Ocean Arks International greenhouse system, this purifies raw sewage using plants and animals living in a sequence of tanks through which the water passes. Another demonstration project in Frederick, Maryland, is treating 30,000 gallons a day of urban sewage. (Dann Blackwood.) Check: http://www.mbl.edu/html/OA/page1.html

Individual Septic Systems

Despite expansion of sewage-collection systems, in rural areas there always will be countless homes not connected to a municipal system. For such homes, individual septic systems are required. The traditional and still most common system is the septic tank and a drain field (Fig. 13-19). Wastewater flows into the tank, where particulate organic material settles to the bottom. The tank acts like a primary clarifier in a municipal system. Water containing colloidal and dissolved organic material as well as the dissolved nutrients flows into the drain field and gradually percolates into the soil. Organic material that settles in the tank is digested by bacteria, but accumulations still must be pumped out every two to three years. Soil bacteria decompose the colloidal and dissolved organic material that comes through the drain field. Some people establish successful vegetable gardens over septic drain fields, thus exercising the sound principle of recycling the nutrients.

Prerequisites for this traditional system are suitable land area for the drain field and a subsoil that allows sufficient percolation of water. But, if these do not exist, as is the case with many recreational retreats, certain other alternatives are available. One is the **composting toilet**.

An example of a composting toilet is the Clivus Multrum (Clivus Multrum Inc., Cambridge, Massachusetts), shown in Fig. 13-20. The Multrum receives only personal excrement and food wastes—no water other than urine. These wastes pass through a series of chambers as they decompose. After three to four years they arrive at the final chamber as a stable, nutrient-rich humus that is suitable for application on lawns and gardens. The home must have other means for disposing of bath and other gray water. Since such water is not contaminated by human excrement, however, disposing of it presents few problems. Usually it is used for watering lawns and gardens.

Impediments to Recycling Sewage Products

In spite of the ecological need to recycle nutrients on the land and the availability of technology for doing so, some impediments must be considered. The first and most serious is possible contamination of sewage with toxic chemicals. The second is simply a matter of public attitude.

Contamination with Toxic Materials

A review of the processes of sewage treatment clearly shows that they are in no way designed to or capable of removing dissolved metal ions such as lead, mercury, cadmium, and chromium, nor are they capable of breaking down nonbiodegradable synthetic organic compounds. Such elements and compounds, if present, will inevitably end up in the biosolids or composts as they adhere to organic matter, or they will remain in water solution. Because these chemicals are highly toxic, such contamination may preclude both the use of water for irrigation and the use of biosolids for fertilizer.

The source of these toxic materials is twofold, industries and individuals. Under the Clean Water Act,

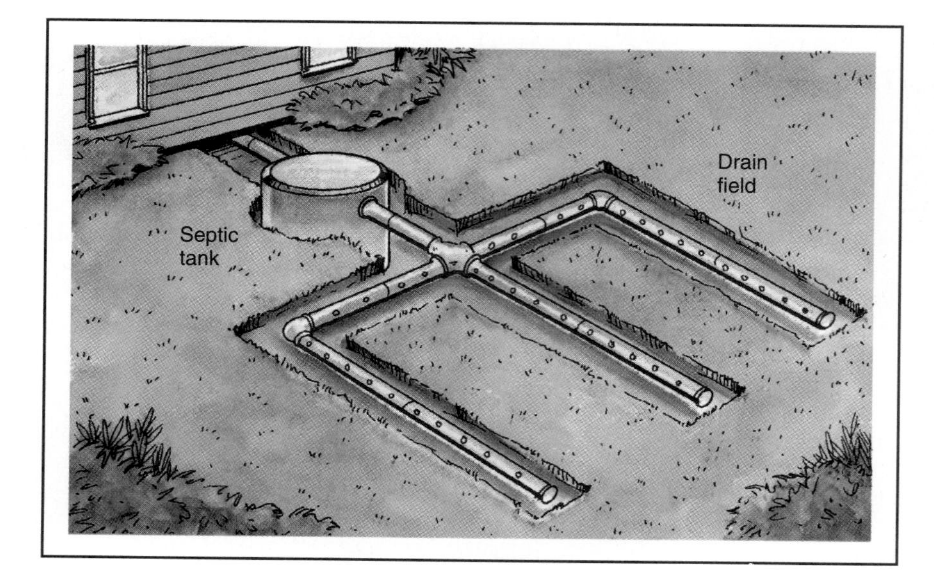

FIGURE 13-19
Sewage treatment for a private home, using a septic tank and drain field. All the pipes and the tank are normally buried underground. They are shown uncovered here only for illustration. (USDA, Soil Conservation Service.)

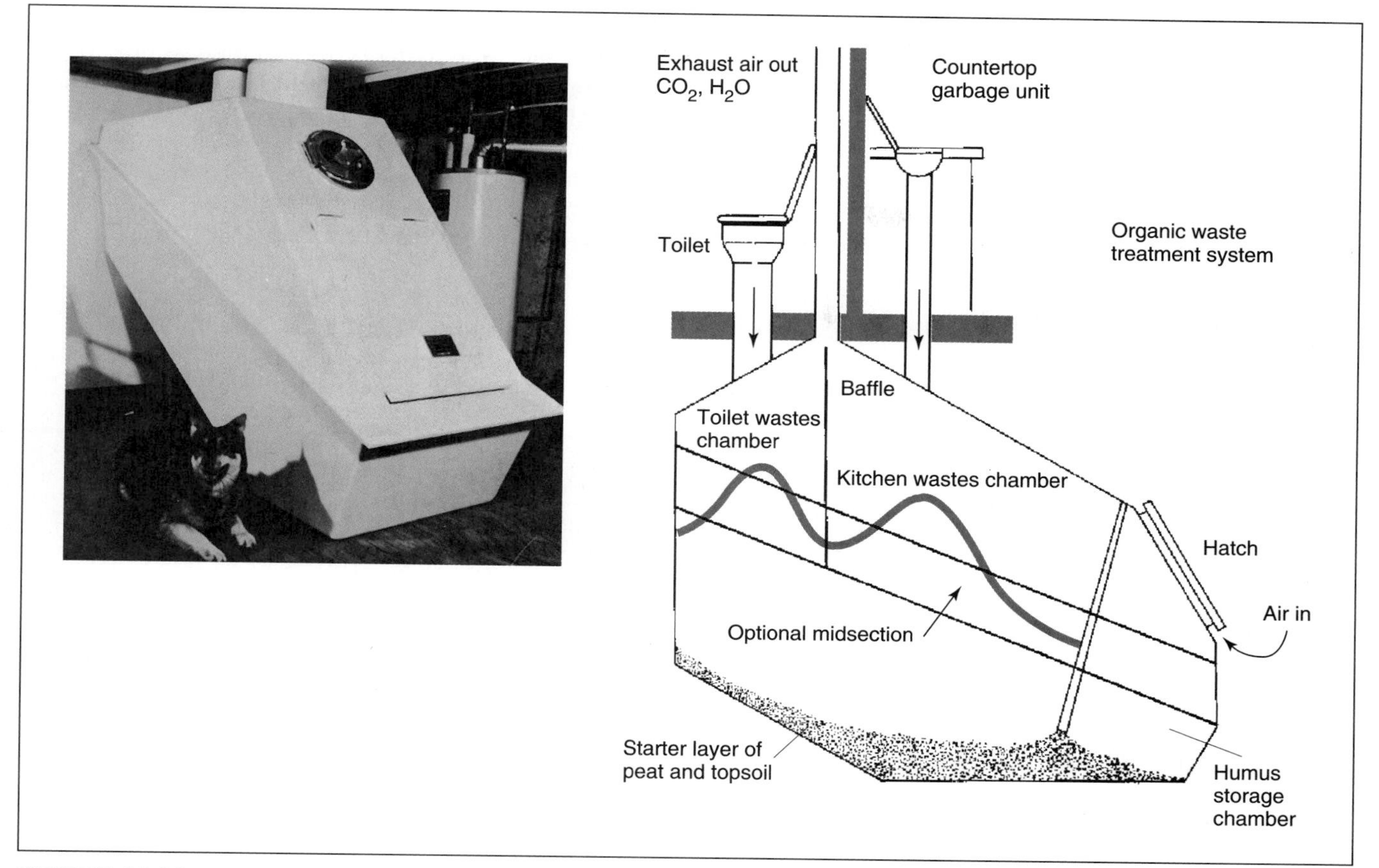

FIGURE 13-20
Composting toilet. Rather than using water to flush, excrements and food wastes are deposited into a ventilated chamber where they decompose aerobically, producing a nutrient-rich compost. The system shown here is the Clivus Multrum. (Courtesy of Clivus Multrum, Inc. Cambridge, MA)

larger industries are now required to either pretreat their wastewater to remove toxic chemicals before discharging it into municipal sewer systems, or to find alterntive means of disposal. (Further discussion of handling and disposal of hazardous wastes is the subject of Chapter 14.) This requirement has gone far toward making the products of sewage treatment usable, although there was some backsliding in the period from 1980 to 1992 as enforcement of regulations was thwarted by the White House.

Then, smaller industries, shops, individual homes, and offices that do not fall under the regualtions governing larger industries remain a problem. Individuals at home or at work are prone to discard spent or unused portions of such things as cleaning fluids, pesiticides, paints and other coatings, and photographic chemicals down drains to the municipal sewer system. Also, lead and copper may leach from water pipes. While chemicals from any single location may not be large, in total they are significant.

The situation emphasizes that we all need to be more watchful and respectful whether it is in supporting enforcement of existing regulations or being more careful with respect to what we personally put down the drain.

Public Opinion

Another impediment to upgrading sewage treatment and developing the use of byproducts is adverse public opinion. Many people feel that what goes down the drain should be "out of sight, out of mind." People fail to make any connection between spotless, sanitary bathrooms and pollution problems in waterways. Combined water and sewer bills, which include the costs for sewage treatment, often are only \$5 to \$10 a month per household in most regions. For another \$2 to \$3 per month, we could have state-of-the-art sewage treatment. Yet, many people feel adamantly that sewer charges should not be increased, and municipal referenda to raise money for sewage-treatment facilities fail again and again.

On the positive side, however, we see by the number of options that have become available, and the numerous cities and communities that have adopted them, that the public attitude is changing. For example, a few years ago so few farmers were willing to accept applications of biosolids that it was difficult to get rid of this product. Now farmers are standing in line to purchase treated sludge for application to their farms.

All in all it appears that improving sewage treatment and recycling or reuse of the water and sludge byproducts is a major growth area, offering both jobs and opportunities to participate in constructing a sustainable future.

Review Questions

1. Contrast natural ecosystems with the human system in regard to the fate of excrements from organisms.
2. What are the public-health and pollution hazards and potential benefits of sewage?
3. Give a brief history of humans' handling of sewage wastes as understanding of risks and potential benefits have been gained.
4. What is the water content of, and what are the four categories of, pollutants present in raw sewage?
5. Name and describe the facility and the process used to remove debris, grit, particulate organic matter, colloidal and dissolved organic matter, and dissolved nutrients.
6. Why is secondary treatment also called biological treatment? What is the principle involved? What are the two alternative techniques used?
7. What are the principles involved in and what is accomplished by biological nutrient removal? Where do nitrogen and phosphate go in the process?
8. Give alternatives for final disinfection of the water and the advantages and disadvantages of each.
9. Raw sludge is a byproduct of what treatment processes? What is its chemical and physical nature?
10. Name and describe four alternative methods of treating raw sludge, and give the end product(s) that may be produced from each method.
11. Rather than removing nutrients, what are alternative ways of handling effluents from sewage treatment that will still prevent cultural eutrophication?
12. Describe alternative methods for treating raw sewage.
13. How may the sewage from individual homes be handled in the absence of municipal collection systems?
14. What factors may prevent the use of water and sludges as a resource despite available treatment?
15. How do developing countries compare with developed nations in terms of sewage collection and treatment?

Thinking Environmentally

1. Arrange a tour to the sewage-treatment plant that serves your community. Contrast it with what is described in this chapter. Is the water being purified or handled in a way that will prevent cultural eutrophication? Are sludges being converted to and used as fertilizer? What improvements, if any, are in order? How can you help promote such improvements?
2. Suppose a new community of several thousand persons is going to be built in Arizona (warm, dry climate). You are called in as a consultant to design a complete sewage system including collection, treatment, and use or disposal of byproducts. Write an essay describing the system you recommend and giving a rationale for the choices involved.
3. Suppose your city or community has been disposing of raw sludge in landfills but is now proposing to compost it and use the resulting material on city parklands. The proposal is meeting considerable resistance from the public. Write a "Letter to the Editor" describing the environmental advantages of this proposal.
4. Some people favor the use of oceans as a dumping ground for sewage. Do you support or oppose this alternative? Defend your position.
5. Many Americans will continue to live in rural locations where hookups to centralized sewage collection and treatment systems are impractical. What kinds of sewage systems would you recommend for such people, particularly in regions with poor soil drainage or with frequent and heavy rains? Give a rationale for your recommendations.
6. Suppose you are the manager of a modern sewage treatment plant. Describe how your job helps protect the environment.

11

Pollution from Hazardous Chemicals

Key Issues and Questions

1. Heavy metals and synthetic organic compounds are two categories of chemicals that present particular toxic hazards. What are the attributes of these chemicals that make them hazardous?
2. Bioaccumulation and magnification are phenomena that may result in low-level exposures building up over time to cause damaging effects. Explain these two phenomena.
3. Before 1970 chemical wastes were disposed of indiscriminately. What were the consequences? What was the public response? The government response?
4. The Clean Air and Clean Water acts moved pollution from one part of the environment to another. Where did conditions improve? Worsen?
5. A number of laws have been passed to clean up and protect the public from the toxic messes that grew out of indiscriminate dumping. What are these laws and their requirements?
6. Disposal of chemical wastes has grown from almost no regulation before 1970 to highly regulated. Under what laws and by which methods are hazardous chemical wastes currently regulated?
7. Workers and the public are protected from accidental exposures to toxic chemicals by a number of laws and regulations. Name these laws and specify the regulations.
8. The direction for the future is one of pollution avoidance rather than pollution control. What is the distinction between avoidance and control? How does it change the relationship between environmentalists and businesses?

In the 1960s and 1970s, "chemical disasters" made front-page headlines. A river through Cleveland actually caught fire. In a fishing village of Minimata, Japan, an epidemic of crippling illnesses including insanity, birth defects, and deaths was due to mercury poisoning; the mercury came from wastes from a local factory through the food chain from fish to people. In Bhopal, India, an accidental release of a toxic chemical, caused by negligent operations killed an estimated 7000 people and injured more than 100,000 others. An epidemic of

Environmental Protection Agency workers sampling the contents of illegally discarded drums of chemical wastes. (See midnight dumping, page 354.) [Photo by Mugshots/The Stock Market.]

insidious illnesses—including kidney and liver dysfunction, cancer, and birth defects—in the community of Niagara Falls, New York, was caused by various toxic chemicals leaking from an abandoned dump site of chemical wastes, Love Canal.

Such tragedies have convinced the public that significant dangers are associated with the manufacture, use, and disposal of chemicals. But very few would advocate giving up the advantages of the innumerable products that derive from our modern chemical industry. Better to learn to handle and dispose of chemicals in ways that minimize the risks. Thus, over the past 25 years regulations surrounding the production, transport, use, and disposal of chemicals have strengthened. From having almost no controls at all, the chemical industry is now stringently regulated, and everyone using or handling chemicals that are deemed hazardous is affected by regulations.

Some critics are now claiming that we have gone too far. According to them, regulations are overly strict. They say costs of abiding by the regulations outweigh the benefits and are simply sapping our economic vitality. Yet, it is clear to most that not all problems have been solved. Certainly, no one wants to go back to having rivers that catch on fire and epidemics of chemical-caused illnesses. The current debate emerges: Where should we go from here? More regulation? Less regulation? Or can we find alternative ways of minimizing the chemical hazards at less cost?

This chapter focuses on chemical wastes, with the objectives of providing an understanding of the nature of the risks involved; providing an overview of laws and regulations that have come into play to protect human and environmental health and their strengths and weaknesses; and pointing out future directions toward reducing the chemical hazards at less cost.

The Nature of Chemical Risks: HAZMATs

A chemical that presents a certain hazard or risk is known as a **hazardous material (HAZMAT)**. The EPA categorizes substances on the basis of the following hazardous properties:

- **Ignitability:** Substances that catch fire readily (e.g., gasoline and alcohol).
- **Corrosivity:** Substances that corrode storage tanks and equipment (e.g., acids).
- **Reactivity:** Substances that are chemically unstable and may explode or create toxic fumes when mixed with water (e.g., explosives, elemental phosphorus [not phosphate], and concentrated sulfuric acid).
- **Toxicity:** Substances that are injurious to health when ingested or inhaled (e.g., chlorine, ammonia, pesticides, and formaldehyde).

Radioactive materials, probably the most hazardous of all, are treated as an entirely separate category and are discussed in Chapter 22.

Sources of Chemicals Entering the Environment

To understand how HAZMATs enter our environment, we need to look at various aspects of how people in our society live and work. First, consider that the materials making up almost everything we use from the shampoo and toothpaste in the morning to the TV set we watch in the evening are products of chemical technology. Our use constitutes only one step in the **total product life cycle**, a term that considers all steps from obtaining raw materials to final disposal of the product. Implicit in our use of shampoo, for example, is that raw materials were obtained and various chemicals were produced to make both the shampoo and its container. Inevitably, there are chemical wastes and byproducts in the production processes. In addition, consider the risks of accidents or spills occurring in the manufacturing process and in the transportation of the raw materials, the finished product, or the wastes. Finally, consider: What happens to the spent shampoo you rinse down the drain? What happens to the container you throw into the trash, still holding a last few drops, or perhaps most of the shampoo you didn't like after trying it?

Multiply these steps by all the hundreds of thousands of products used by billions of people, in homes, factories, and business, and you can appreciate the magnitude of the situation. At every stage—from mining raw materials through manufacturing, use, and final disposal—various chemical products and byproducts enter the environment, with potential consequences for both human and environmental health (Fig. 14-1).

The use of many products—pesticides, fertilizers, and road salts, for example—implies their direct introduction into the environment. Or the intended use may entail a fraction of the material going into the environment—the evaporation of solvents from paints and adhesives, for example. Then there are the product life

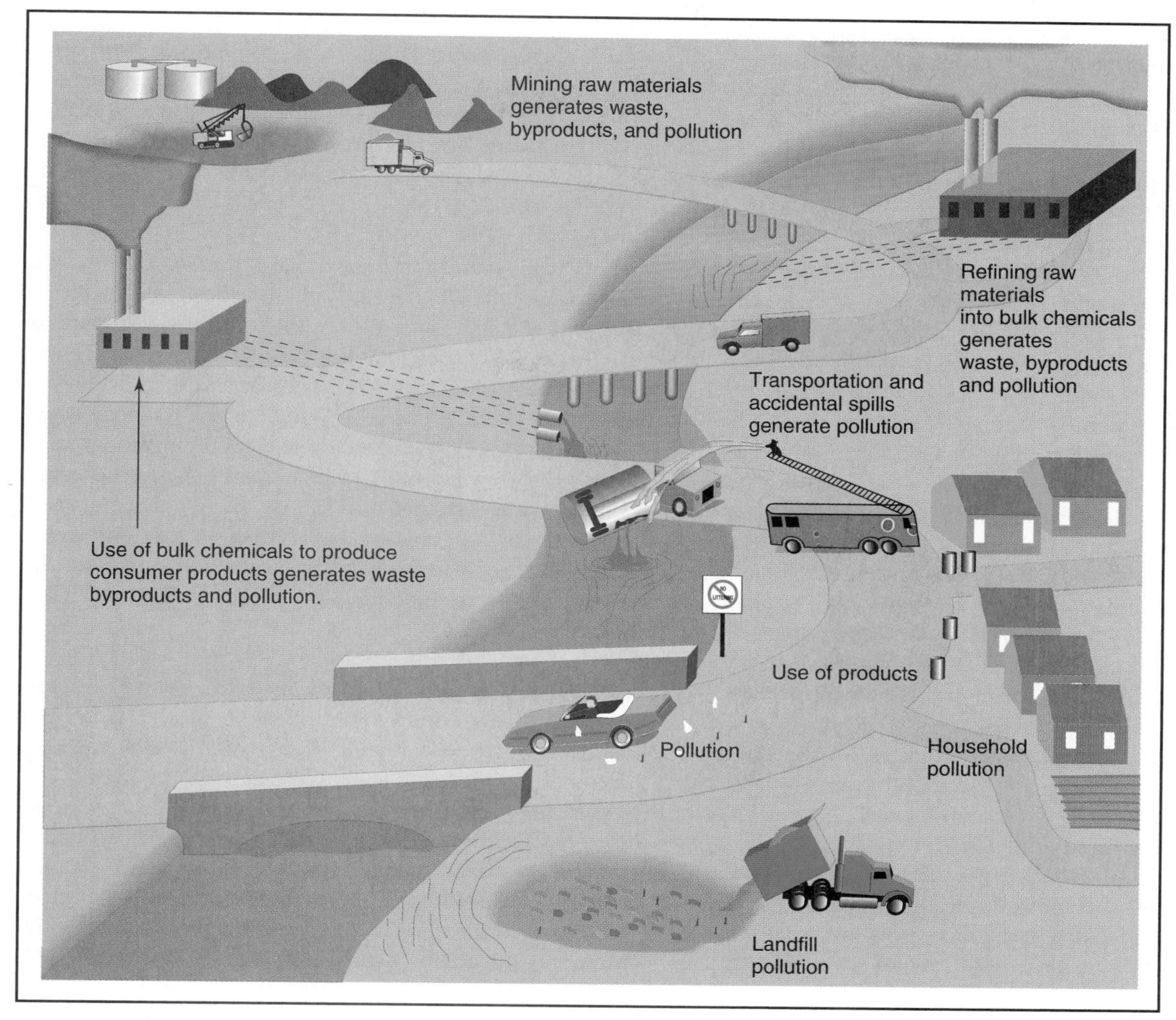

FIGURE 14-1
Total product life cycle. The life cycle of a product begins with the obtaining of raw materials and ends with final discard of the used product. At each step or in transportation between steps, wastes, byproducts, or the product itself may enter the environment, causing pollution and creating various risks to human and environmental health.

cycles of materials that are used tangentially to the desired item. Consider lubricants, solvents, cleaning fluids, cooling fluids, and so on, with whatever contaminants they may contain. Likewise, there are the product life cycles of the gasoline, coal, or other fuels that are consumed for energy. Again, in addition to the unavoidable wastes produced there is in every case the potential for accidental releases, ranging from minor leaks in storage tanks to super tanker wrecks such as the *Exxon Valdez*, which in 1989 spilled 11 million gallons of crude oil into Prince William Sound in Alaska.

Such introductions of chemicals into the environment occur in every sector from major industrial plants to small shops and individual homes. Whereas single events involving large amounts of one chemical may constitute a disaster and make headlines, the total amounts entering the environment as routine emissions from millions of homes and businesses is far greater and presents, many believe, a much greater overall health risk to society. Although the relation is difficult to prove, the steadily rising incidence of cancer (about 1% per year) is often blamed on environmental pollution.

Toxic Chemicals Presenting a Long-Term Threat

Fortunately, a large portion of the chemicals introduced into the environment are gradually broken down and assimilated by natural processes. Therefore, once these chemicals are diluted sufficiently, they pose no long-term human or environmental risk, even though they may be highly toxic in acute doses (high-level, short-term exposures). Indeed, until relatively recently it was generally assumed that "dilution was a solution to pollution."

There are two major classes of chemicals for which the dilution solution fails, however: (1) *heavy metals and their compounds* and (2) *nonbiodegradable synthetic organics*. Far from "disappearing" into the environment, these chemicals tend to be absorbed from the environment and concentrated by organisms, including humans, till they reach sometimes lethal doses. This concentration process obviously poses another dimension of risk to human and environmental health—a long-term risk from low, even minute levels.

Heavy Metals. The most dangerous heavy metals are lead, mercury, arsenic, cadmium, tin, chromium, zinc, and copper. These metals are widely used in industry, particularly in metal-working or metal-plating shops and in such products as batteries and electronics. They are also used in certain pesticides and medicines. In addition, because heavy-metal compounds can have brilliant colors, they are used in paint pigments, glazes, inks, and dyes. Thus, heavy metals may enter the environment wherever any of these products are produced, used, and ultimately discarded.

Heavy metals are extremely toxic because, as ions or in certain compounds, they are soluble in water and may be readily absorbed into the body, where they tend to combine with and inhibit the functioning of particular vital enzymes. Even very small amounts can have severe physiological or neurological consequences. The mental retardation caused by lead poisoning and the insanity and crippling birth defects caused by mercury are particularly well-known examples.

Nonbiodegradable Synthetic Organics. Synthetic organic compounds are the chemical basis for all plastics, synthetic fibers, synthetic rubber, modern paint-like coatings, solvents, pesticides, wood preservatives, and hundreds of other products. Being nonbiodegradable is an important part of what makes many such compounds useful; we wouldn't want fungi and bacteria attacking and rotting our tires, and paints and wood preservatives function only insofar as they are both nonbiodegradable and toxic to decomposer organisms.

These compounds are toxic because they are similar enough to natural organic compounds to be absorbed into the body. There they interact with particular enzymes, but their nonbiodegradability prevents them from being broken down or processed further. The result is that they upset the system. When a person ingests a sufficiently high dose, the effect may be acute poisoning and death. With low doses over extended periods, however, the effects are insidious and can be mutagenic (mutation-causing), carcinogenic (cancer-causing), or teratogenic (birth defect–causing). They may cause serious liver and kidney dysfunction, sterility, and numerous other physiological and neurological problems.

A particularly troublesome class of synthetic organics is the **halogenated hydrocarbons**, organic compounds in which one or more of the hydrogen atoms have been replaced by atoms of chlorine, bromine, fluorine, or iodine. These four elements are classed as *halogens*, hence the name *halogenated hydrocarbons* (Fig. 14-2). Of the halogenated hydrocarbons, the **chlorinated hydrocarbons** (also called **organic chlorides**) are by far the most common. Organic chlorides are widely used in plastics (polyvinyl chloride), pesticides (DDT, Kepone, and Mirex), solvents (carbon tetrachlorophenol), electrical insulation (polychlorinated biphenyls, the infamous PCBs that are often in the news), flame retardants (TRIS), and many other products. Additional *chlorinated hydrocarbons* and their health effects are listed in Table 14-1.

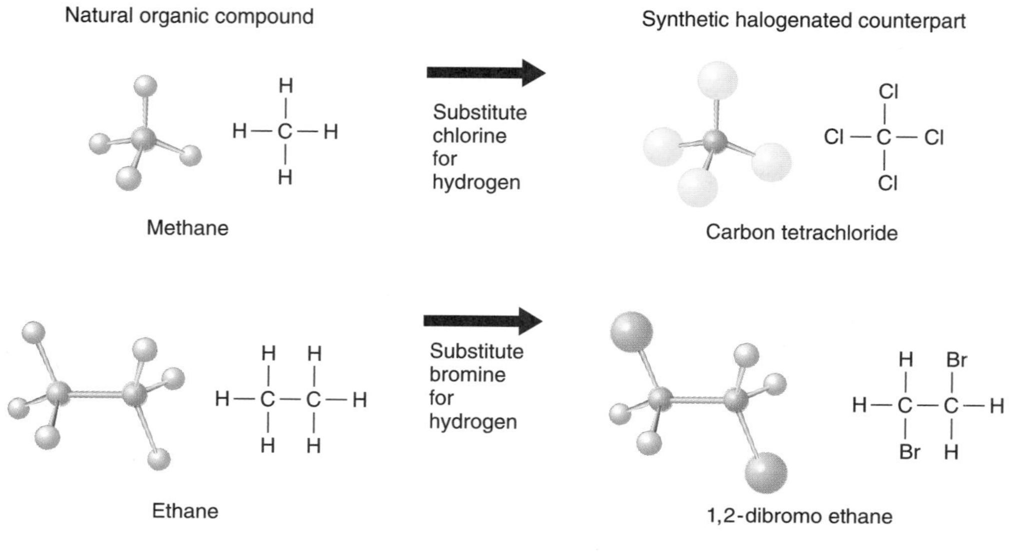

Natural organic compound

Synthetic halogenated counterpart

Substitute chlorine for hydrogen

Methane

Carbon tetrachloride

Substitute bromine for hydrogen

Ethane

1,2-dibromo ethane

FIGURE 14-2
Halogenated hydrocarbons are organic (carbon based) compounds in which one or more hydrogen atoms has been replaced by halogen atoms (chlorine, fluorine, bromine, or iodine). The most common halogenated hydrocarbons are those using chlorine, that is, chlorinated hydrocarbons. Such compounds are particularly hazardous to health because they are nonbiodegradable and they tend to bioaccumulate.

TABLE 14-1
Examples of Toxic Synthetic Organic Compounds Frequently Found in Chemical Wastes

			Known Health Effects*					
Chemical	*Mutations*	*Carcinogenic*	*Birth Defects*	*Still Births*	*Nervous Disorders*	*Liver Disease*	*Kidney Disease*	*Lung Disease*
Benzene	X	X	X	X				
Dichlorobenzene	X			X	X	X		
Hexachlorobenzene	X	X	X	X	X			
Chloroform	X	X	X		X			
Carbon tetrachloride	X		X	X	X	X		
Chloroethylene (vinyl chloride)	X	X			X	X		X
Dichloroethylene	X	X		X	X	X	X	
Tetrachloroethylene		X			X	X	X	
Trichloroethylene	X	X			X	X		
Heptachlor	X	X		X	X	X		
Polychlorinated biphenyls (PCBs)	X	X	X	X	X	X		
Tetrachlorodibenzo dioxin	X	X	X	X	X	X		
Toluene	X			X	X			
Chlorotoluene	X	X						
Xylene			X	X	X			

Source: Adapted from S. Epstein, L. Brown, and C. Pope. *Hazardous Waste in America*. Copyright © 1982 by Samuel S. Epstein, M.D., Lester O. Brown, and Carl Pope. Reprinted with permission of Sierra Club Books.
*Determined from tests on experimental animals.

Bioaccumulation and Biomagnification

The trait that makes heavy metals and nonbiodegradable synthetic organics particularly hazardous is their tendency to accumulate in organisms. Because of accumulation, small, seemingly harmless amounts received over a long period of time may reach toxic levels. This phenomenon—referred to as **bioaccumulation**—can be understood as follows.

You are familiar with the concept of a filter. As water passes through a filter, impurities too large to pass through the pores accumulate on the filter. Basically, organisms act as filters for heavy metals and synthetic organics. Heavy metals enter the body dissolved in water, but once bound to enzymes, the metals are removed from solution. Synthetic organics are highly soluble in lipids (fats or fatty compounds) but sparingly soluble in water. As they pass through cell membranes, which are lipid, they come out of water solution and enter into the lipids of the body. Thus, traces of heavy metals and synthetic organics that are absorbed with food or water are trapped and held by the body's enzymes and lipids, while the water and water-soluble wastes are passed in the urine. Since the body has no mechanism to excrete the heavy metals or synthetic organics or to metabolize them further, trace levels consumed over time gradually accumulate in the body and may sooner or later produce toxic effects.

Bioaccumulation, which occurs in the individual organism, may be compounded through a food chain. Each organism accumulates the contamination in its food, so it accumulates a concentration of contaminant in its body that is many times higher than that in its food. The next organism in the food chain effectively now has a more-contaminated food and accumulates the contaminant to yet a higher level. Essentially all the contaminant accumulated by the large biomass at the bottom of the food pyramid is concentrated, through food chains, into the smaller and smaller biomass of organisms at the top of the food pyramid. This multiplying effect of bioaccumulation that occurs through a food chain is called **biomagnification**. Figure 14-3 shows a documented example of biomagnification for the chlorinated hydrocarbon pesticide DDT, now banned in the United States but still widely used in developing countries.

One of the most distressing aspects of bioaccumulation and biomagnification is that there are no warning symptoms until contaminant concentrations in the body are high enough to cause problems. Then, it is too late to do much about it. As is often the case, bioaccumulation and biomagnification went unrecognized until serious problems brought the phenomena to light. In the 1960s, serious diebacks in the populations of many species of predatory birds, including bald eagles and ospreys, were observed. Investigations into the cause of the diebacks revealed the biomagnification

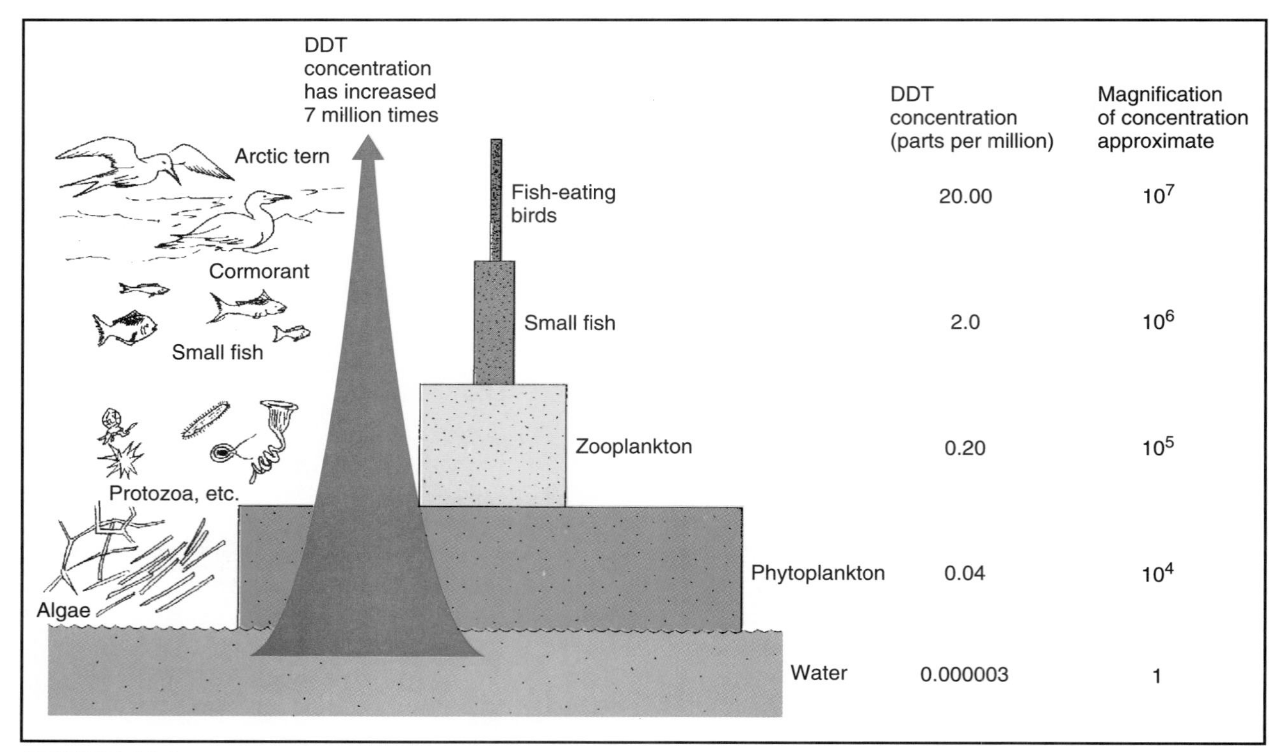

FIGURE 14-3
Biomagnification. Organisms on the first trophic level absorb the contaminant from the environment and accumulate it in their bodies (bioaccumulation). Then, each successive consumer in the food chain accumulates the contaminant to a yet higher level, as it feeds on what is a more contaminated food supply. Thus, the concentration of the contaminant in biomass is magnified manyfold throughout a food chain. Organisms at the top of the food chain are likely to accumulate toxic levels. Heavy metals and halogenated hydrocarbons are two classes of chemicals that are particularly subject to such biomagnification. Shown here is a documented case of biomagnification of the pesticide DDT, a chlorinated hydrocarbon compound.

of DDT shown in Figure 14-3. This finding led to the banning of DDT and many related pesticides in most developed countries, as described in Chapter 10.

A tragic episode in the early 1970s known as the Minamata disease revealed the potential for biomagnification of mercury and other heavy metals. The disease is named for a small fishing village in Japan where the episode occurred. In the mid-1950s, cats in Minamata began to show spastic movements, followed by partial paralysis, coma, and death. At first this was thought to be a syndrome peculiar to cats, and little attention was paid to it. However, when the same symptoms began to occur in people, concern escalated quickly. Additional symptoms such as mental retardation, insanity, and birth defects were also observed. Scientists and health experts eventually diagnosed the cause as acute mercury poisoning.

A chemical company near the village was discharging wastes containing mercury into a river that flowed into the bay where the Minamata villagers fished. The mercury, which settled with detritus, was first absorbed and bioaccumulated by bacteria and then biomagnified as it passed up the food chain through fish to cats or to humans. Cats had suffered first and most severely because they fed almost exclusively on the remains of fish. By the time the situation was brought under control, some 50 persons had died and 150 had suffered serious bone and nerve damage. Even now, the tragedy lives on in the crippled bodies and retarded minds of Minamata descendants.

Interaction between Pollutants and Pathogens

Complicating the situation of HAZMATs in the environment are interactions between pollutants and pathogens. For example, unusually high numbers of dead dolphins and other sea mammals were found washed up on beaches along the Atlantic Coast in 1991. Analysis of the dead animals showed that their tissues had bioaccumulated significant but sublethal levels of various chlorinated hydrocarbons; the immediate cause of death, however, seemed to be pathogens. Is there any relationship between these two factors, or is the presence of the pollutants only incidental? Many investigators contend that pollutants weaken the immune system, making the organism more vulnerable to infection by pathogens. It is extremely difficult to prove this hypothesis, but it creates another focus of concern.

All this knowledge concerning long-term hazards of toxic chemicals is based on relatively recent discoveries, which are based on the tragic experiences noted. Before this understanding was gained, the name of the game in HAZMAT management was essentially indiscriminate disposal.

A History of Mismanagement

Historically, chemical wastes of all kinds have been disposed of as expediently as possible. We think of today's disposal methods as superior, although many problems of land, water, and air pollution by HAZMATs remain.

Indiscriminate Disposal

From the dawn of the Industrial Age to the relatively recent past, it was common practice to exhaust all combustion fumes up smokestacks, vent all evaporating materials and solvents into the air, and flush all waste liquids and contaminated wash water into sewer systems or directly into natural waterways. Many human health problems occurred, but they either were not recognized as being caused by the pollution or were accepted as the "price of progress." Indeed, much of our understanding regarding human health effects of hazardous materials is derived from those uncontrolled exposures. For example, the expression "mad as a hatter" comes from the fact that people who made hats in the 1800s frequently became insane. The insanity, it was later found, was caused by poisoning from the mercury used in the production process.

In the 1950s, as production expanded and synthetic organics came into widespread use in the developed countries, many streams and rivers became essentially open chemical sewers, as well as sewers for human waste. These waters were not only devoid of life; they were themselves hazardous. For example, in the 1960s the Cuyahoga River, which flows through Cleveland, Ohio, carried so much flammable material that it actually caught fire and destroyed seven bridges before it burned itself out (Fig. 14-4). Worsening pollution (both chemicals and sewage) and increasing recognition of adverse health effects finally created a degree of public outrage that pushed Congress to pass the Clean Air Act of 1970 and the Clean Water Act of 1972. These acts set standards for allowable emissions into air and water and timetables for reaching those standards.

The Clean Air and Clean Water acts and their subsequent amendments remain cornerstones of environmental legislation. However, the passage of these acts in the early 1970s left an enormous loophole. If you can't vent wastes into the atmosphere or flush them into waterways, what do you do with them? Industry turned to *land disposal*, which was essentially unregulated at the time, as an expedient alternative. Indiscriminate air and water disposal became indiscriminate land disposal. Thus, in retrospect we see that the Clean Air and Clean Water acts, for all their benefits in improving air and water quality, also succeeded in transferring pollutants from one part of the environment to another. We should keep that lesson in mind as further measures are taken to control pollution. With your understanding of

FIGURE 14-4
Prior to laws and regulations to control pollution, all manner of wastes were indiscriminately discharged. In the 1960s the Cuyahoga River actually caught fire. Incidents such as this contributed to public outrage that led to the passage of the Clean Water Act of 1972. (© 1992 The Plain Dealer Publishing Company.)

the water cycle (Chapter 11), you can see how land disposal enormously increases the potential for groundwater contamination.

Methods of Land Disposal

In the early 1970s there were three primary land-disposal methods: (1) deep-well injection, (2) surface impoundments, and (3) landfills. With conscientious implementation of safeguards, each of these methods has some merit. Without adequate regulations or enforcement, however, contamination of groundwater is virtually inevitable.

Deep-Well Injection. Deep-well injection involves drilling a "well" into dry, porous material below groundwater (Fig. 14-5). In theory, hazardous waste liquids pumped into the well soak into the porous

FIGURE 14-5
Deep-well injection, a technique used for disposal of large amounts of liquid wastes. The supposition is that toxic wastes may be drained into dry, porous strata below ground, where they may reside harmlessly "forever." The theory and precautionary safety measures are listed on the left. Before the 1980s, these measures frequently were not taken. Even when they are, the potential for failure remains in actual practice (right-hand box). (Adapted with permission from Environmental Action, 1525 New Hampshire Ave. N.W., Washington, D.C. 20036.)

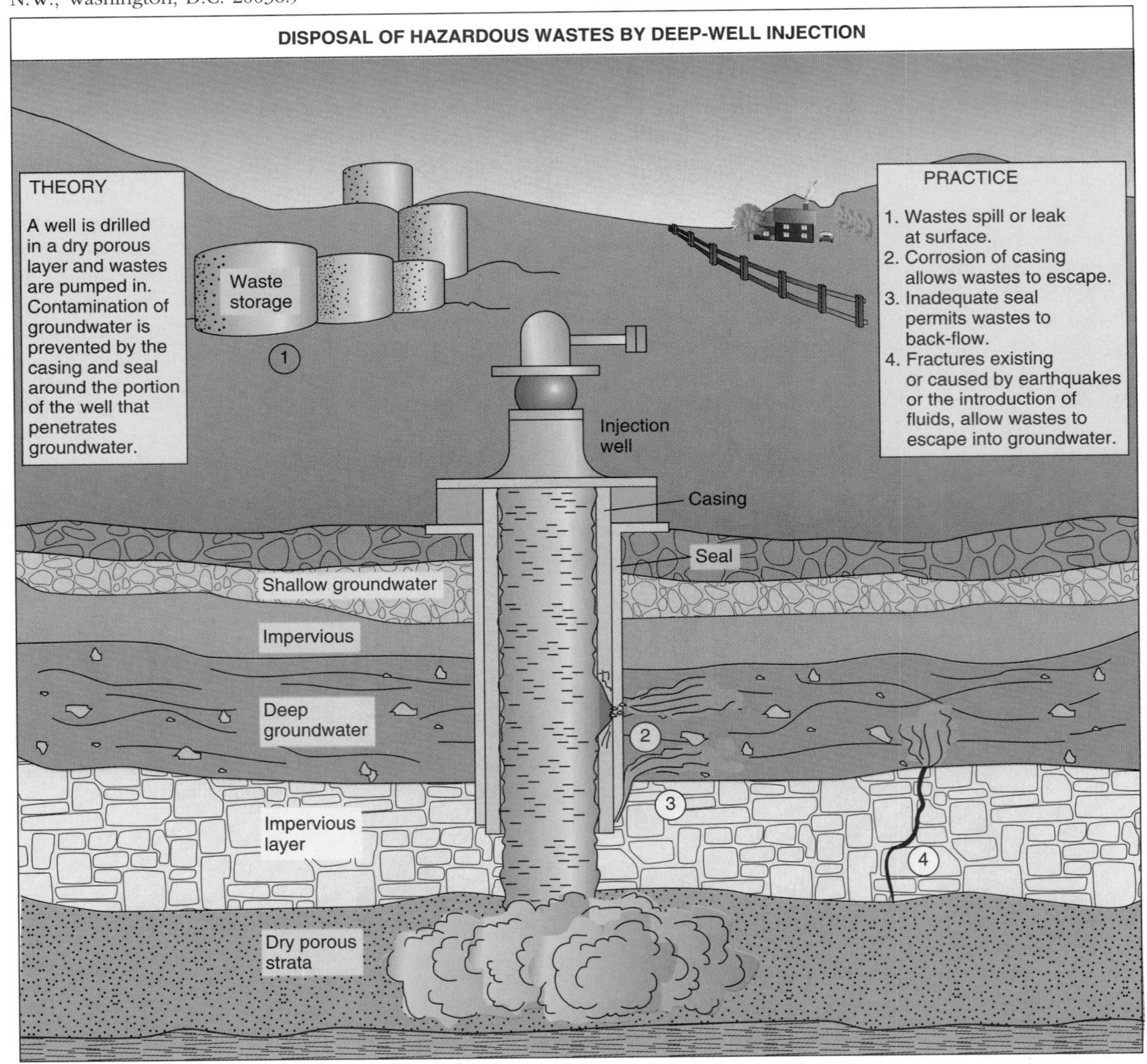

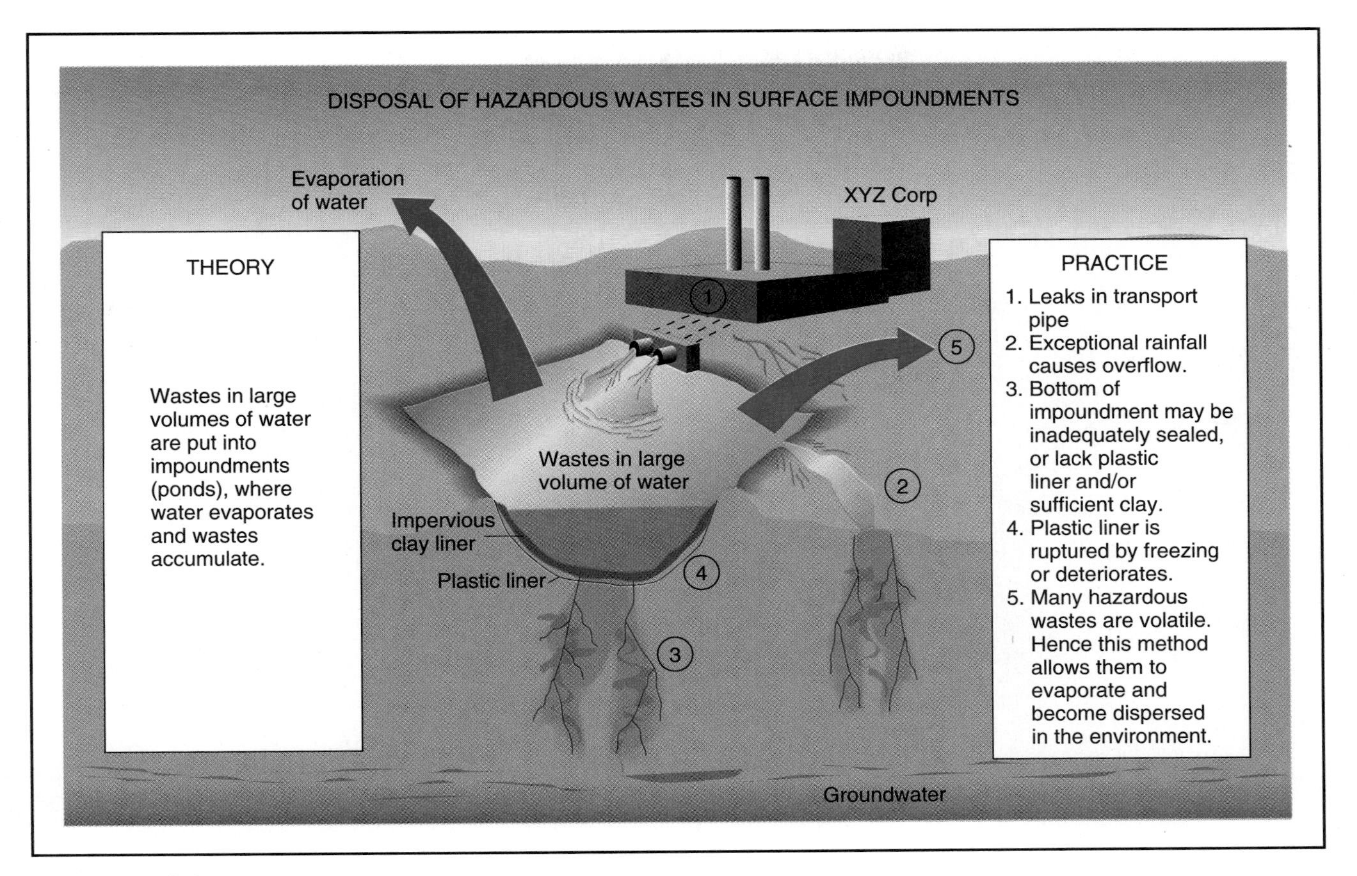

FIGURE 14-6
Surface impoundment is an inexpensive technique for disposal of large amounts of lightly contaminated liquid wastes. The supposition is that only water leaves the impoundment, by evaporation, whereas wastes remain and accumulate in the impoundment indefinitely. Before the 1980s, plastic liners were not used, and ponds were not always dug into clay. Even when these steps are taken, potential for failure remains in actual practice (right-hand box).

material and remain isolated indefinitely. In practice, however, it is almost impossible to guarantee that fractures in the impermeable layer will not eventually permit the injected wastes to escape and contaminate groundwater. Indeed, the introduction of wastes may produce enough stress to cause such fractures. Also, there are additional ways in which wastes can escape into groundwater, as Fig. 14-5 shows.

Surface Impoundments. Surface impoundments are simple excavated depressions ("ponds") into which liquid wastes are drained and held. They were the least expensive and hence most widely used way to dispose of large amounts of water carrying relatively small amounts of chemical wastes. As waste is discharged into the pond, solid wastes settle and accumulate while water evaporates (Fig. 14-6). If the pond bottom is well sealed and if evaporation equals input, impoundments may receive wastes indefinitely. However, inadequate seals may allow wastes to percolate into groundwater, exceptional storms may cause overflows, and volatile materials can evaporate into the atmosphere, adding to air pollution problems and eventually falling down with rain to contaminate water in other locations.

Landfills. When hazardous wastes are in a concentrated liquid or solid form, they are commonly put into drums and buried in landfills. If a landfill is properly lined, supplied with a system to remove leachate (material that may percolate out the bottom), and properly capped, it may be reasonably safe and is referred to as a **secure landfill** (Fig. 14-7). The various barriers are subject to damage or deterioration, however, and many experts feel it is only a question of time before contents will leach from even the most "secure" landfills.

Because early land disposal was not regulated, however, in numerous instances not even the most rudimentary precautions were taken. In many cases, deep wells were injecting wastes directly into groundwater; abandoned quarries were sometimes used as landfills with no additional precautions being taken; and surface impoundments frequently had no seals or liners whatsoever.

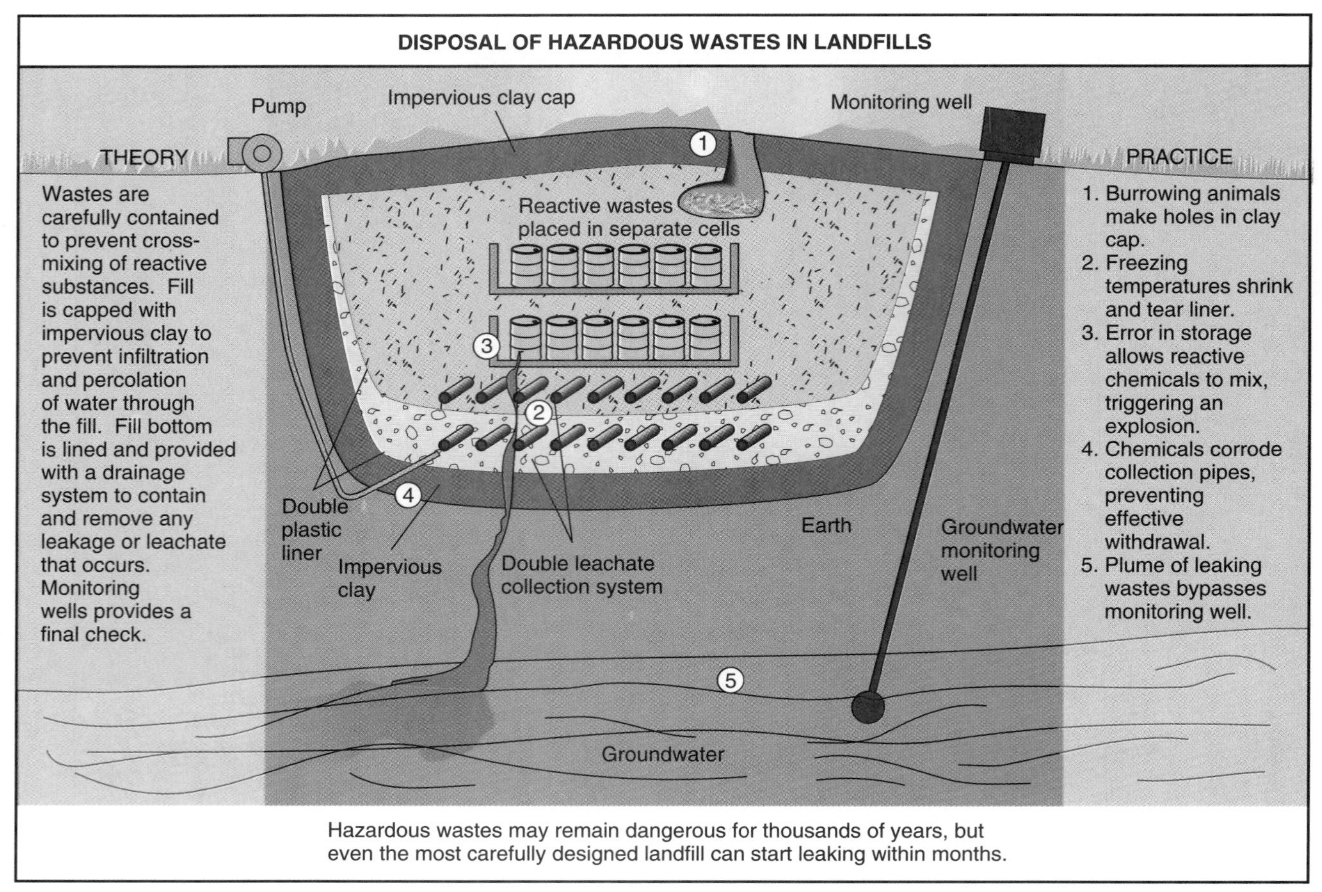

(a)

(b)

FIGURE 14-7
Landfilling, a technique widely used for disposal of concentrated chemical wastes. (a) Precautionary measures to make this method safe are listed in the Theory column. Before the 1980s, these measures frequently were not taken and even with them potential for failure remains. (Adapted from an illustration by Rick Farrell©. All rights reserved.) (b) An operating hazardous-waste landfill in Alabama. (Alon Reininger/Contact Press/Woodfin Camp Associates.)

Even worse, considerable amounts of waste failed to get to any disposal facility at all.

Midnight Dumping and On-Site Accumulation. The need for alternative methods for disposing of waste chemicals created an opportunity for a new enterprise—waste disposal. Many reputable businesses entered the field, but—again in the absence of regulations—there were also disreputable operators. As stacks of drums filled with hazardous wastes "mysteriously" appeared in abandoned warehouses, vacant lots, or municipal landfills, it became clear that some operators were simply pocketing the disposal fee and then unloading the wastes in any available location, frequently under cover of darkness, an activity termed **midnight dumping** (Fig.14-8).

FIGURE 14-8
Midnight dumping at "Valley of the Drums" near Louisville, Kentucky, about 1975. Thousands of drums of waste chemicals, many of them toxic, were simply unloaded at this site by unknown parties and left to "rot," seriously threatening the surrounding environment, waterways, and aquifers. Incidents such as this led to the passage of the Resources Conservation and Recovery Act of 1976, which outlaws such indiscriminate disposal. (Van Bucher/Photo Researchers, Inc.)

Authorities trying to locate the individuals responsible found that they had gone out of business and were nowhere to be found. Some companies simply stored wastes on their own properties and then went out of business, abandoning the property and the wastes (Fig. 14-9). As drums containing hazardous chemicals corroded and leaked, there was great danger of reactive chemicals combining and causing explosions and fires. By far the greatest long-term hazards, however, are that toxic chemicals from any form of insecure land disposal may leach into groundwater and volatile compounds may pollute the air.

Scope of the Problem

The mounting problem of unregulated land disposal of hazardous wastes was brought vividly to public attention by the episode at Love Canal, which has become a battle cry of citizens concerned about hazardous wastes.

Love Canal was an abandoned canal bed near Niagara Falls, New York. In the 1930s and 1940s, it served as a convenient burial site for thousands of drums of waste chemicals—over 20,000 tons in all. After the canal was filled and covered with earth, the land was eventually transferred to the city of Niagara Falls, and homes and a school were built on the edge of what had been the old canal. The area of covered chemicals became a playground. Over the years, children attending the school and coming in contact with "black gooey stuff" oozing out of the ground began having unusual health problems, ranging from chemical burns and skin rashes to severe physiological and nervous disorders. Even more alarming, residents began to note that an unusually high number of miscarriages and birth defects were occurring. The situation climaxed in 1978 when health authorities identified the black ooze as a potent mixture of numerous chlorinated hydrocarbons known to cause birth defects and other disorders in experimental animals.

Some $3 billion in health claims were filed against the city of Niagara Falls, several hundred times its annual operating budget, and nearly 600 families demanded relocation at state expense. Eventually the

FIGURE 14-9
In addition to midnight dumping, some companies simply stockpiled hazardous wastes on their own property with few if any of the features required for security. Hazardous waste stored in barrels near Chatfield Reservoir, Denver, Colorado, area. (Kent & Donna/Photo Researchers, Inc.)

FIGURE 14-10
Evacuation and destruction of Love Canal neighborhood. These are a few of the nearly 100 homes that the state was forced to buy and later destroy as a result of contamination from Love Canal. The area has now been cleaned up and redeveloped with new homes. (Joe Traver/Gamma-Liaison, Inc.)

state did purchase and demolish about 100 homes in the area (Fig. 14-10), but most of the residents gained only further frustration because their claims could not be proved. The difficulty is that it is essentially impossible to prove the cause of any particular disorder when multiple factors are involved. Furthermore, we simply do not have sufficient understanding regarding the long-term health effects of most bioaccumulating synthetic organic chemicals.

Investigation into who was responsible for the Love Canal disaster eventually revealed that the fault lay more with the bureaucratic processes that allowed development next to and on top of the site than with negligent disposal practices on the part of the chemical company. Nevertheless, the media attention given Love Canal did focus the public's attention on what was being done with toxic wastes and the hazards involved. (Now the old dump has been cleaned up and the area has been redeveloped with new homes.)

As both government and independent researchers began surveying the extent of the problem, bad disposal practices were found to be rampant. The World Resources Institute estimated that in the United States in the early 1980s, there existed 75,000 active industrial landfill sites, along with 180,000 surface impoundments and 200 other special facilities that were or could be possible sources of groundwater contamination. In most cases, the contaminated area was relatively small, 200 acres (80 hectares) or less, but in total the problem was immense and affected every state in the country. As studies and tests proceeded, thousands of individual wells and some major municipal wells were closed because they were contaminated with toxic chemicals.

Many cases of private-well contamination caused by careless disposal were discovered only after people experienced "unexplainable" illnesses over prolonged periods. While these incidents did not receive the media attention of Love Canal, they were nevertheless devastating to the people involved.

Another problem uncovered in the aftermath of Love Canal was that tests showed that drinking water supplies in many cities were tainted with toxic synthetic organic chemicals. In most cases, it was deemed that the trace levels detected were not dangerous, so no action was taken. However, just how much constitutes a safe versus a harmful level is a subject of considerable controversy.

Even more important than the damage already done was the recognition that this is an ongoing problem. The EPA estimates that close to 150 million tons of hazardous wastes are generated each year in the United States (close to two-thirds of a ton per person).

Thus we see that the problems concerning toxic chemical wastes can be divided into three areas:

- assuring safe drinking water, and cleaning up the "messes" already created
- regulating the handling and disposing of wastes currently being produced so as to protect public and environmental health
- looking toward future solutions

We shall address each of these areas in the following sections.

Cleaning Up the Mess

A major public health threat from land-disposed toxic wastes is contamination of groundwater that is subsequently used for drinking. The first priority is to assure that people have safe water. The second priority is to clean up or isolate the source of pollution so that further contamination does not occur.

Assuring Safe Drinking Water

To protect the public from the risk of toxic chemicals contaminating drinking water supplies, Congress passed the **Safe Drinking Water Act of 1974**. Under this act, the EPA sets standards regarding allowable levels of various toxic chemicals and requires that municipal water supplies be monitored for these chemicals on a regular basis. If specified toxic chemicals are found to exceed specified levels, the water supply is closed until adequate purification procedures or other alternatives are adopted.

There is some debate concerning the efficacy of the Safe Drinking Water Act. Most communities have never found toxic chemicals to the degree that requires remedial action. Therefore, some would like to see the act repealed because they feel that monitoring is too costly relative to the risks involved, especially for small towns supplying only a few hundred homes, which are required to do the same monitoring as large cities. Are we spending money to protect ourselves against a hazard that doesn't exist? How do you feel about relaxing the monitoring requirements for your city's water supply?

Others feel that the Safe Drinking Water Act should be expanded to cover remaining loopholes. For example, there is no provision for systematic monitoring of private wells. For these wells, contamination may not be recognized until people experience "unexplained" illness or report "funny-tasting" or "funny-smelling" water and have the water tested on their own initiative. (Your county public health department will provide this service, albeit for a fee.) If dangerous levels of contamination are found in a private well, the well may be closed but there is no compensation to the property owner. The owner must accept the burden of getting water from another source or installing purification measures that may be prohibitively expensive. In extreme cases, owners have been known to simply abandon their homes (Fig. 14-11). If only low levels of contamination deemed "safe" are found in a private well, no action is taken, and the owners are left on their own to decide what to do. Likewise, there is no provision for monitoring bottled water, which many people have turned to drinking to avoid presumed contamination of municipal water supplies. Ironically, in 1991 it came to light that numerous samples of bottled water contained higher levels of contamination than did municipal supplies.

Groundwater Remediation

If dumps, leaking storage tanks, or spills of toxic materials have contaminated groundwater and such groundwater is threatening water supplies, all is not lost. **Groundwater remediation** is a developing and growing technology. Techniques involve drilling wells, pumping out the contaminated groundwater, purifying

FIGURE 14-11 Leaching of toxic wastes from landfills and contamination of water supplies have forced the abandonment of properties in some areas. (Photograph by Thomas Busier.)

Earth Watch

The Case of the Obee Road NPL Site

The following is but one example of the almost 1300 sites on the EPA's National Priorities List (NPL) listing.

Conditions at listing (January 1987): The Obee Road Site consisted of a plume of contaminated groundwater in the vicinity of Obee Road in the eastern section of Hutchinson, Reno County, Kansas. The Kansas Department of Health and Environment (KDHE) has been investigating the area since July 1983. At that time, the state detected volatile organic chemicals, including benzene, trans-1,2-dichloroethylene, chlorobenzene, 1,1-dichloroethylene, tetrachloroethylene, trichloroethylene, vinyl chloride, and toluene, in wells drawing on a shallow alluvial aquifer. An estimated 1900 residents of suburban Obeeville obtained drinking water from private wells in the aquifer.

To protect public health, Hutchinson connected the homes of the Obeeville residents and a school that was drawing water from a contaminated well, to the Reno County Rural Water District.

Preliminary work by the state identified the source of the contamination as the former Hutchinson City landfill, which is located at the eastern edge of what is now the Hutchinson Municipal Airport. Before it was closed in about 1973, the landfill had accepted unknown quantities of liquid wastes and sludges from local industries, as well as solvents from small metal-finishing operations at local aircraft plants. The Department of Defense (DOD), which owned or maintained the airport until 1963, may also have disposed of solvents in the landfill.

Status as of 1995: Further testing and analysis showed that the original landfill had "stabilized." That is, leaching had already run its course, and there was little likelihood of further contamination from the landfill. Therefore, excavation or other cleanup of the landfill has been deemed unnecessary, but monitoring of the site to detect any further leakage will continue for the foreseeable future. Meanwhile, remediation of contaminated groundwater is ongoing.

Source: Hazardous waste site listed under the EPA Comprehensive Environmental Response, Compensation, and Liability Act of 1980 (CERCLA)("Superfund").

it, and reinjecting the purified water back into the ground or discharging it into surface waters (Fig. 14-12). Of course, cleaning up the source of the contamination is mandatory.

Superfund for Toxic Sites

Probably the most monumental task we are facing is the cleanup of the tens of thousands of toxic sites resulting from the history of careless and even criminal (midnight dumping) handling and disposal of toxic materials. Where facilities were still operating, pressures were brought to bear on the operators to clean up the sites. Many operators who could not afford to do so, however, simply declared bankruptcy, and the sites joined those already abandoned.

A major federal program aimed at cleaning up such abandoned chemical waste sites was initiated by the **Comprehensive Environmental Response, Compensation, and Liability Act of 1980**, popularly known as **Superfund**. Through a tax on chemical raw materials, this legislation provides a fund for the identification of abandoned chemical waste sites, protection of groundwater near the site, remediation of groundwater if it has been contaminated, and cleanup of the site. Funding for the program has steadily grown from about $300 million per year in the early 1980s to about $2 billion per year in the 1990s. But the cleanup job is nowhere near complete. It is estimated that it will go on for at least another 30 years at a total cost of some $300 billion. Such is the magnitude of the problem; Superfund, one of the EPA's largest ongoing programs, works as follows.

Setting Priorities. It should be obvious that resources are not sufficient to clean up all sites at once. Therefore, a system for setting priorities has been developed.

- As sites are identified—note that many abandoned sites had long since been forgotten—they are first analyzed in terms of current and potential threat to groundwater supplies by taking samples of the waste, determining waste characteristics, and testing groundwater around the site for contamination. If it is determined that no immediate threat exists, nothing more may be done.
- If a threat to human health does exist, the most expedient measures are taken immediately to protect the public. These measures may include digging a deep trench and installing a concrete dike around the site and recapping it with impervious layers of plastic and clay to prevent infiltration. Thus the wastes are isolated, at least for the short term. If the situation has gone past the threat

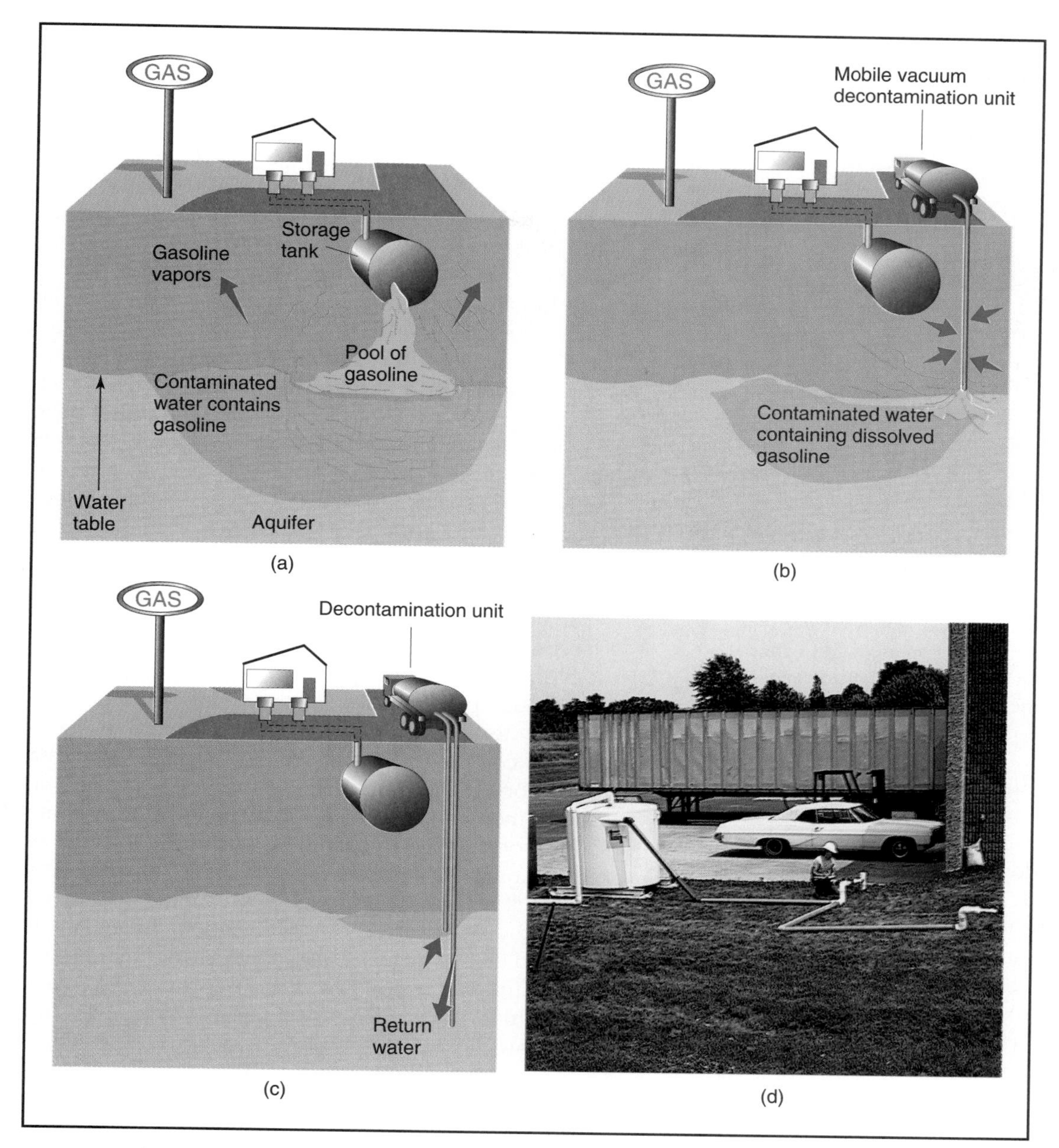

FIGURE 14-12
Groundwater remediation. (a) Typical subsurface contamination from a leaking fuel tank at a gas station. (b) After the leak has been repaired, the vacuum extraction process causes gasoline and residual hydrocarbons in the soil and on the water table to evaporate and then removes the vapors, preventing further contamination of the groundwater. (c) Contaminated groundwater is pumped out, treated, and returned to the ground. (d) Photo of remediation system. (Courtesy of Terra-Vac.)

stage and contaminated groundwater is or will be reaching wells, the remediation procedures discussed above are begun immediately.

- The worst sites (those presenting the most immediate and severe threats) are put on a **National Priorities List (NPL)** and scheduled for total cleanup. A site on this list is reevaluated in terms of the most cost-effective method of cleanup, and finally cleanup begins.

Cleanup Technology. Of course, where the chemical wastes are still contained in drums they can be picked up, treated, and managed in proper fashion. The bigger problem is the soil (often millions of tons) contaminated

FIGURE 14-13
Toxic waste mobile incinerator. The EPA is cleaning up some Superfund sites by running contaminated materials through this incinerator, which is erected at the site. (Tim Lynch/Gamma-Liaison, Inc.)

by leakage. One procedure is to excavate contaminated soil and run it through a kiln assembled on the site to burn off chemicals (Fig. 14-13). (Synthetic organic chemicals can be broken down by incineration, and heavy metals are converted to insoluble, stable oxides.) Another method is to drill a ring of injection wells around the site and a suction well in the center. Water containing a harmless detergent is injected into the injection wells and drawn to the suction well, cleansing the soil along the way. The withdrawn water is treated to remove the pollutants and is then reused for injection.

Another rapidly developing cleanup technology is **bioremediation**. In many cases, the soil is contaminated with toxic organic compounds that are biodegradable. The problem is that they do not degrade because the soil lacks organisms or oxygen or both. In bioremediation, oxygen and organisms are injected into contaminated zones. The organisms feed on and eliminate the pollutants (as in secondary sewage treatment) and then die when the pollutants are gone. Considerable research is being done to find and develop microbes that will break down certain kinds of wastes more readily. Bioremediation may be used in place of detergents to decontaminate soil. It is a rapidly expanding and developing technology of waste cleanup applicable to leaking storage tanks and spills as well as to waste-disposal sites.

Evaluating Superfund. With these mechanisms in place, it would seem that cleanup should be moving rapidly. Unfortunately it is not. Of the literally hundreds of thousands of various waste sites, slightly over 35,000 have been deemed serious enough to be given Superfund status. That is, 35,000 sites pose a risk of serious pollution in the future. It is assumed that this figure now includes most if not all of the sites in the United States that pose a significant risk. Interestingly, some of the worst that have come to light are on military bases, the result of what many characterize as totally unconscionable discard of toxic materials from military operations.

To date (1995) 1289 have been deemed to present threats serious enough to be put on the NPL list, which is growing by 25–30 sites a year. Total cleanup has been completed at only about 80 at an average cost of $25 million per site, and partial cleanup has occurred at another 200. The other NPL sites are in various stages of analysis and feasibility study. An example of a Superfund site and its progress toward cleanup is given in the "Earth Watch" box, page 358.

One reason for the high cost and slow progress is a section in the Superfund legislation that requires establishment of liability. The original intent was to identify the parties who did the dumping and make them pay for the cleanup. Thus, it was hoped that government costs for cleanup could be saved. However, the "history" of a waste disposal site may go back 40 years or more, and users may have included everything from schools and hospitals to large corporations, all of which mount legal defenses to disclaim responsibility. The result is that cleanup of a site may be delayed indefinitely by the intervening litigation. Indeed, it is estimated that as much as 50% of Superfund moneys go toward litigation rather than cleanup. Other legal delays of Superfund progress occurred because key leadership individuals in the 1980s were convicted for illegal actions associated with polluter special interest.

Another problem is the lack of any clear standards as to what constitutes clean. "How clean is clean?" has become a virtual cliché. Do you demand that chemical contamination be reduced by 90%, 99%, or 99.9%? The question has real economic significance, because each increment may double the cost of cleanup. Many feel that overly stringent standards of cleanup are costing large sums of money without providing any additional benefit to public health. This problem of finding a suitable balance between costs and benefits will be explored in more depth in Chapter 17.

Management of New Wastes

The production of chemical wastes is and will continue to be an ongoing phenomenon as long as human civilization persists. Therefore, human and environmental

Ethics

Racism and Hazardous Waste

Item: The largest commercial hazardous waste landfill in the United States is located in Emelle, Alabama. This landfill receives wastes from Superfund sites and every state in the continental United States.

Item: Kettleman, California, has been selected as the site for the state's first commercial hazardous waste incinerator.

Item: A Choctaw reservation in Philadelphia, Mississippi, was targeted to become the home of a 466-acre hazardous waste landfill.

These three places have two things in common: They are considered to be appropriate locations for hazardous waste disposal, and they are predominantly populated by people of color. African Americans make up 90% of Emelle's population; Latinos represent 95% of Kettleman's population; and the Choctaw reservation is entirely Native American. At issue is the question of environmental racism.

Several recent studies have documented the fact that all across the United States, waste sites and other hazardous facilities are more likely than not to be located in towns and neighborhoods where the majority of residents is nonwhite. If it is also true that these same towns and neighborhoods are less affluent, this only makes a stronger case for addressing the obvious injustices. The wastes involved are primarily generated by affluent industries and the affluent majority, but somehow the wastes tend to end up well away from where they were generated and in the backyards of people of color. It seems fair to assume that the siting of hazardous facilities is a matter of political power, and those with the power would like to have the sites well away from their own backyards.

Environmental racism has gotten the attention of the EPA. Former EPA Administrator William K. Reilly created the Environmental Equity Workgroup, made up of 40 professionals from within the agency. The workgroup submitted its report in February 1992 and concluded that, although there is a lack of data on actual environmental health effects by race, differences in exposure to some environmental pollutants are correlated with race and socioeconomic factors. The workgroup recommended a number of steps to be taken by the EPA to address concerns of environmental racism.

Within the nonwhite communities there are signs of a groundswell of grassroots concern about environmental racism and equity. The success of minorities in pushing for civil rights in recent decades has set a precedent that is beginning to empower people of color in seizing the initiative in this newly emerging arena. As a start, activists are turning their attention to cleaning up their own communities. Calling themselves the Toxic Avengers, a group of African American and Latino students in Brooklyn, New York, has taken on projects like battling the Radiac Research Corporation's toxic waste facility in the neighborhood, starting a recycling program, and conducting neighborhood workshops on fighting pollution. Watchdog is a grassroots, multiracial, working-class group in Los Angeles that has successfully worked toward amending the regulations of the South Coast Air Quality Management District so that the regulations reflect special concerns of minorities and low-income workers. The Good Road Coalition is an alliance of grassroots groups that successfully blocked the proposal by a Connecticut company to build a 6,000-acre landfill on the Rosebud Sioux reservation in South Dakota.

What is still uncertain about the environmental justice movement is the response of the white majority in the United States and, in particular, the very organizations that promote environmentalism. Will the nongovernmental organizations join with people of color in fighting this apparent manifestation of racism? And will businesses and local political bodies be willing to address race-related environmental inequalities? What do you think?

Source: EPA Journal, 18, no. 1 (1992).

health can be protected only if we have management procedures to handle and dispose of wastes safely so that we will not be creating more and more Superfund sites. We have already noted that the Clean Water Act (CWA) and the Clean Air Act (CAA) limit discharges into water and air. When problems regarding disposal of wastes on land became evident in the mid-1970s Congress passed the **Resource Conservation and Recovery Act of 1976 (RCRA)**, commonly spoken of as "rick-ra," in order to control land disposal. Thus, a company producing hazardous wastes in the United States today is under the regulations of these three major environmental laws, CWA, CAA, and RCRA.

It is important to understand these laws in somewhat more detail. First, let us briefly review their enactment and administration. The general process is that the U.S. Congress writes the legislation, and after it is passed it becomes the EPA's responsibility to administer and enforce. Typically, the legislation requires the EPA to set particular standards and write rules regarding how certain objectives should be met. This requirement often puts the EPA at the center of controversies, because almost invariably environmentalists feel standards and regulations are too weak and industry feels they are too stringent. The whole package of law, standards, and regulations is passed on to respective agencies in each of the states to administer to local businesses and industries, but the EPA still oversees state activities. States may also develop and apply their own standards and regulations but the law requires that state standards be at least as strong or stronger than federal standards.

Checking for compliance with all the standards and regulations necessitates a tremendous amount of collecting and analyzing of samples, as well as inspection and enforcement. Consequently, the EPA and related agencies provide a tremendous number of jobs.

The Clean Air and Water Acts

The Clean Air Act of 1970, the Clean Water Act of 1972, and their various amendments make up the basic legislation limiting discharges into the air or water. More will be said about the Clean Air Act in Chapter 15. Under the Clean Water Act any firm, including facilities such as sewage-treatment plants, discharging more than a certain volume into natural waterways must have a **discharge permit** (NPDES permit). You can't require a firm to stop polluting instantaneously, short of closing it down. The discharge permits are a means of seeing who is discharging what. The renewal of the permits then is made contingent on reducing pollutants to meet certain standards within certain time periods. Standards are being made continually stricter as technologies for pollution control improve. Some manufacturing firms discharging wastewater into municipal sewer systems are required to pretreat such water to remove any pollutant that cannot be removed by the sewage-treatment plants, namely nonbiodegradable organics and heavy metals.

Even this restriction does not end all water pollution, however. Amounts of wastes legally discharged under permits, while they may be a low percentage of total waste stream, still add up to huge amounts. Then, small firms, homes, and farms are exempt from regulation. Further, a great deal of pollution is from urban and farm runoff. While covered by the Clean Water Act, runoff is hard to control, as discussed in Chapter 12.

The Resource Conservation and Recovery Act (RCRA)

RCRA (1976) and its subsequent amendments is the cornerstone legislation designed to prevent unsafe or illegal disposal of wastes on land. RCRA has three main features. First RCRA requires that all disposal facilities such as landfills be *permitted.* The permitting process requires that they have all the safety features described in Fig. 14-7a, including monitoring wells. This demand caused most old facilities to shut down—many subsequently became Superfund sites—and new high-quality fills with safety measures to be constructed.

Second, RCRA requires that wastes destined for landfills be pretreated to convert them to forms that will not leach. Such treatment now commonly includes incineration in various kinds of facilities including cement kilns (Fig. 14-14) or biodegradation. Biodegradation refers to the use of systems similar to secondary sewage treatment as discussed in Chapter 13 and perhaps new species of bacteria capable of breaking down the synthetic organics. If treatment is thorough, there may be little or nothing to landfill. This is the ultimate objective.

For whatever is still going to disposal facilities, the third major feature of RCRA is to require "cradle-to-grave" tracking of all hazardous wastes. The generator (the company where the wastes originated) must fill out a form detailing the exact kind and amounts of waste generated. Persons transporting the waste, who are also required to be permitted, and operating the disposal facility must each sign the form, vouching that amounts of waste transferred are accurate, and copies go to the EPA. All phases are subject to unannounced EPA inspections. The generator remains responsible for any waste "lost" along the way or any inaccuracies in reporting. You can see how this provision of RCRA assures that generators will deal only with responsible parties and curtails midnight dumping.

Reduction of Accidents and Accidental Exposures

A significant risk to personal and public health lies in exposures that occur as a result of leaks, accidents, and misguided use of hazardous chemicals in the home or workplace. A considerable number of laws bear on reducing the probability of accidents and on minimizing exposure of both workers and the public should accidents occur. Examples include the following.

Leaking Underground Storage Tanks—UST Legislation. One consequence of our automobile-based transportation system is millions of underground fuel-storage tanks at service stations. Putting such tanks underground greatly diminishes the risk of explosions and fires, but it also hides leaks. Underground storage tanks have a life expectancy of about 20 years before they may spring leaks. Many thousands of tanks are crossing this threshold each year and springing leaks. Without monitoring, leaks went undetected until nearby residents began to smell fuel-tainted water flowing from their faucets. **Underground Storage Tank (UST)** legislation, passed in 1984, now requires strict monitoring of

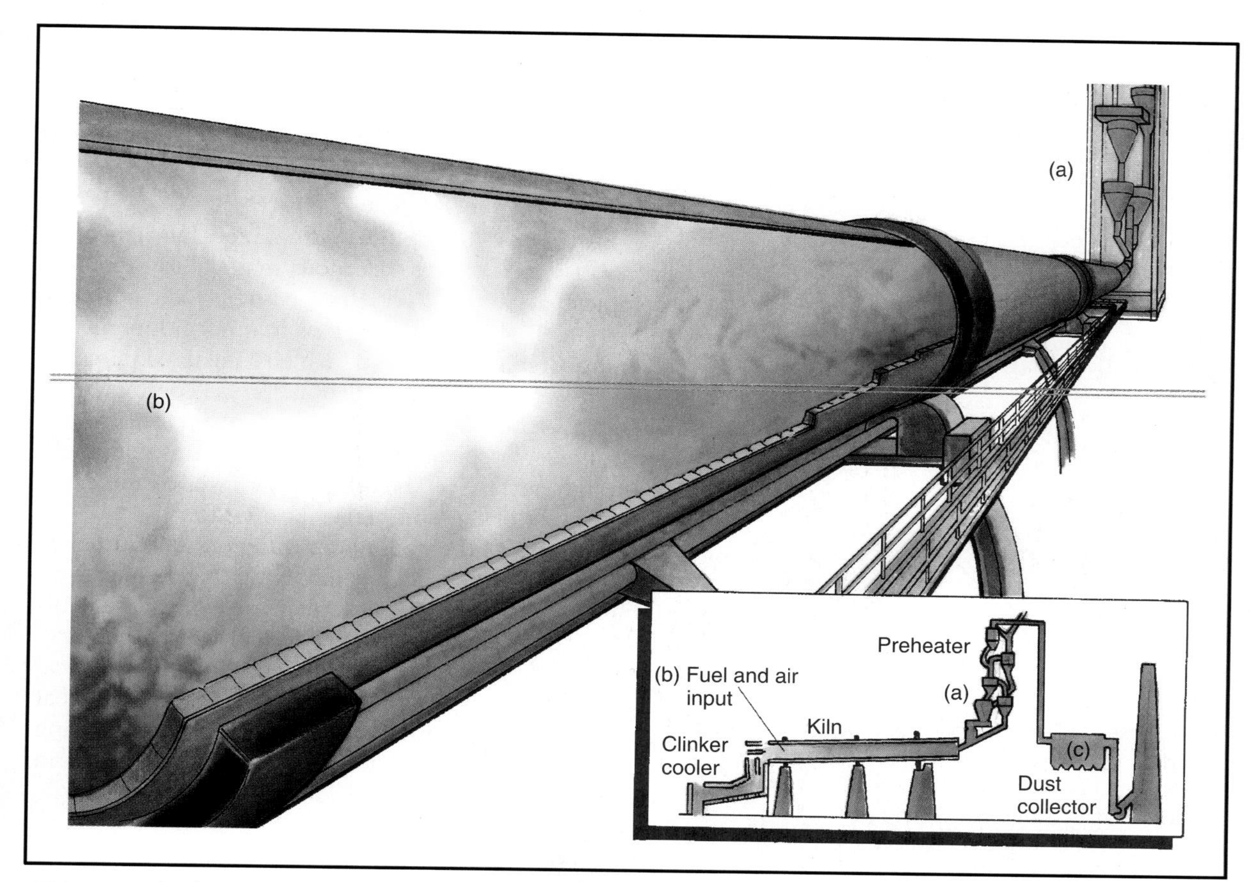

FIGURE 14-14
Cement kilns may be used to destroy hazardous wastes. A cement kiln is a huge, rotating "pipe," typically 15 feet in diameter and 230 feet long, mounted on an incline. Solid wastes fed in with the raw materials (1) are fully incinerated and made to react with cement compounds as they gradually tumble toward the combustion chamber. Flammable, liquid wastes added with the fuel and air (2) impart fuel value as they are burned. Waste dust is trapped (3) and recycled into the kiln. (Redrawn with permission. Southdown, Inc. Houston, TX 77002.)

fuel supplies, tanks, and piping so that leaks may be detected early. When leaks are detected, remediation must begin within 72 hours. New tanks must be double-walled with a monitoring device between the walls.

Department of Transportation Regulations. Transport is an area particularly prone to accidents. As modern society uses increasing amounts and kinds of hazardous materials, the stage is set for accidents to become widescale disasters. To reduce this risk **Department of Transportation Regulations (DOT Regs)** specify kinds of containers and methods of packing to be used in the transport of various hazardous materials. Such regulations are intended to reduce the risk of spills, fires, or poisonous fumes that could be generated from mixing certain chemicals in the case of an accident.

In addition, DOT Regs require that every individual container and the outside of a truck or railcar must carry a standard placard identifying the hazards (flammability, corrosiveness, potential for poisonous fumes, and so on) of the material inside (Fig. 14-15). Such placards enable police and firefighters to identify the potential hazard and respond appropriately in case of an accident. You may also see highway HAZMAT signs restricting truckers with hazardous materials to particular routes or lanes or to using a highway only during certain hours.

Worker Protection—OSHA, Worker's Right to Know. In the past it was not uncommon for industries to require workers to perform jobs that entailed exposure to hazardous materials without informing the workers of the hazards involved. This situation is now addressed by certain amendments to the **Occupational Safety and Health Act (OSHA)** known as the **hazard communication standard**, or **"worker's right to know."** Basically, the law requires businesses and industries to make information regarding hazardous

FIGURE 14-15
HAZMAT placards. These are examples of some of the placards which are mandatory on trucks and railcars carrying hazardous materials. Numbers in place of the word on the placard, or on an additional orange panel, will identify the specific material. Placards alert workers, police, and firefighters to kinds of hazards they face in the case of accidents.

materials and suitable protective equipment available. Notably, however, the responsibility to read the information and exercise proper precautions remains with the worker.

Community Protection and Emergency Preparedness—SARA, Title III. In 1984, an accident at a Union Carbide pesticide plant in Bhopal, India, caused the release of some 30–40 tons of an extremely toxic gas, methyl isocyanate (MIC). An estimated 600,000 people in communities surrounding the plant were exposed to the deadly fumes. The official death toll stands at 2500, but unofficial estimates start at 7000 and range much higher. At least 50,000 people are still suffering various degrees of visual impairment, respiratory problems, and other injuries from the exposure. Ironically, most of those deaths and injuries could have been avoided if people had known the simple protection. MIC is very soluble in water; thus a wet towel over the head would have greatly reduced exposure, and showers would have alleviated aftereffects. Unfortunately, neither people nor medical authorities had any idea of the chemical they were confronted with much less the protection or treatment. What is to prevent another such disaster?

In view of the Bhopal disaster Congress passed legislation to address the problem. For expediency in passage, it was added to a bill reauthorizing Superfund, the Superfund Amendments and Reauthorization Act of 1990 (SARA). Thus it is known as **SARA, Title III**, or "Community Right to Know."

SARA, Title III requires companies that handle in excess of 5 tons of any hazardous material to provide a "complete accounting" of storage sites, feed hoppers, and so on. This information goes to a Local Emergency Planning Committee, one of which is also required in every governmental jurisdiction. This committee is made up of officials representing local fire and police departments, hospitals, and any other groups that might be brought in case of an emergency as well as the executive officers of the companies in question.

The task of the committee is then to draw up scenarios for possible accidents involving the chemicals on site and have a contingency plan for every case. This means everything from having firefighters trained and properly equipped to fight particular kinds of chemical fires to having hospitals stocked with medicines to treat exposures to the particular chemical(s) and to having evacuation plans. Thus, there can be an immediate and appropriate response to any kind of accident.

The Toxic Substances Control Act (TSCA). In the past, new synthetic organic compounds were introduced for a specific purpose without any testing of potential side effects. For example, in the 1960s a new compound known as TRIS was found to be a very effective flame retardant and was widely used in children's sleepwear. It was only later discovered that TRIS is a potent carcinogen. Treated sleepwear was immediately withdrawn from the market. It is not known how many children (if any) developed cancer from TRIS, but the warning from this and other such cases is obvious.

Congress responded by passing the **Toxic Substances Control Act of 1976 (TSCA)**. TSCA requires that before manufacturing a new chemical in bulk, manufacturers must submit a "premanufacturing report" to the EPA in which the environmental impacts of the substance are assessed, including those that may derive from its ultimate disposal, and whether it is a carcinogen. Depending on the results of assessment, uses may be restricted or a product may be kept off the market altogether.

Laws applying to hazardous wastes are summarized in Fig. 14-16. It is significant to note that nongovernmental consumer advocate groups are a major force behind the passage of these laws as well as regulations requiring ingredient labels on all products. In short, the consumer movement in the United States is keyed to full disclosure of ingredients, complete labels, and information as to corporate compliance with the law. Citizen action does make a difference.

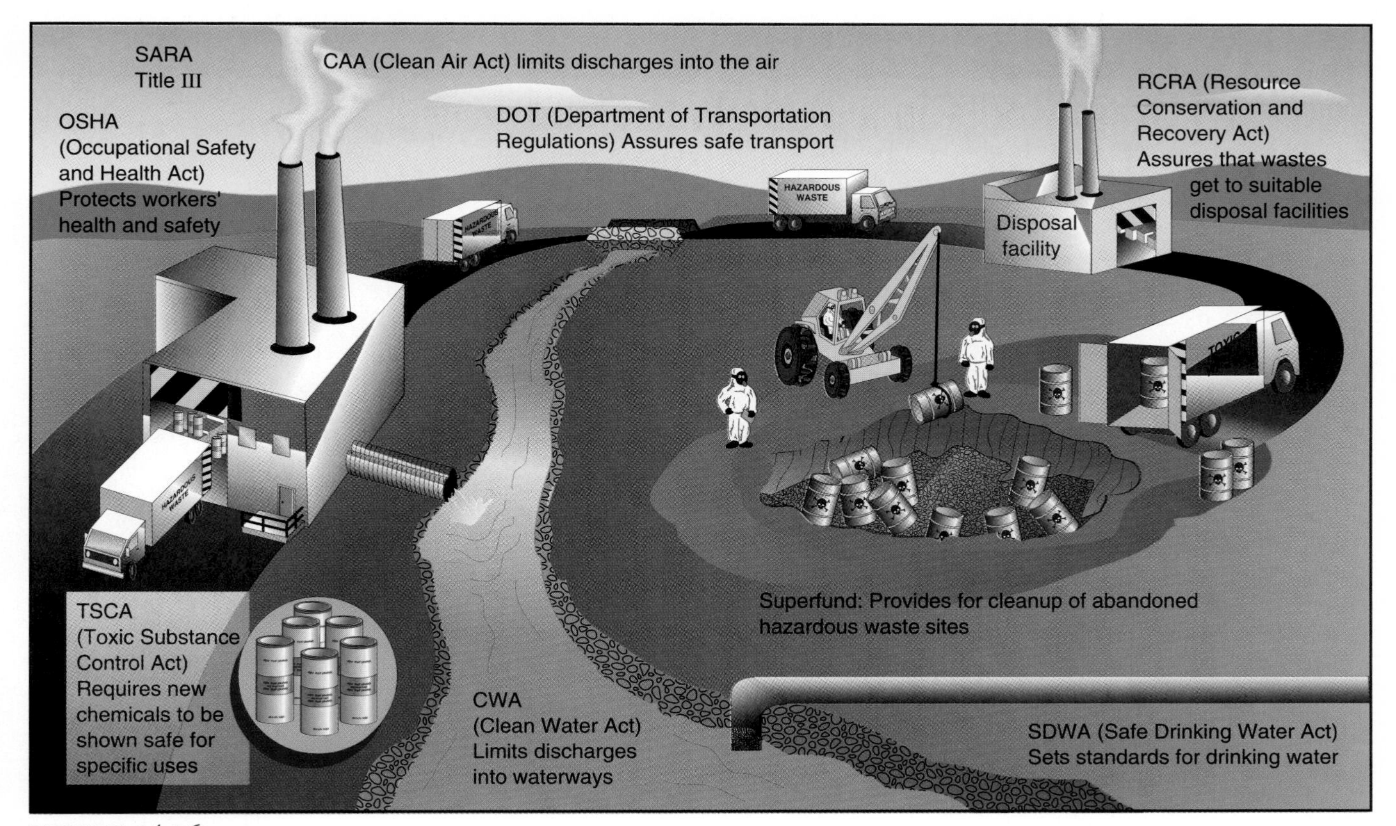

FIGURE 14-16
Summary of the major laws pertaining to the protection of workers, the public, and the environment from hazardous materials.

Looking toward the Future: Pollution Avoidance

Essentially everyone, environmentalists as well as their critics, agree that pollution control has become a horrendously complex and costly business. Consider the costs to business of installing and operating all the pollution control or treatment equipment. Then reflect on the costs to government for all the inspection, monitoring, and enforcement of all the regulations. The EPA estimates that the current yearly cost of complying with federally mandated pollution-control and cleanup programs is $115 billion.

Too Many or Too Few Regulations?

Considering all the complexities and costs of pollution control, it is not hard to see how some critics feel that business is overregulated and that we should get rid of regulations. Nor is it hard to see why some U.S. companies have moved operations to developing countries where regulations are less strict—another example of moving pollution from one part of the environment to another. Yet, which laws or regulations would you abandon? The environmental nightmare of toxic wastes that has come to light in the former USSR shows what can happen in the absence of regulations (see "Global Perspective" box, page 366).

One thing is clear, the benefit of avoided damage to ecosystems and human health far exceeds the monitary costs of regulations to date. The ratio of benefits to costs is as high as 20 to 1 by some estimates. A more complete discussion of cost-benefit ratios is in Chapter 17.

Further, our many regulations still have significant loopholes. One area of concern is the legally permitted discharges. The low concentrations that standards still allow in discharges add up. Although the figure is gradually going down, there still are more than 10 billion pounds (4.5 billion kg) of toxic chemicals legally discharged into the environment annually. Do the regulations need to be made even stricter?

A second area of concern is that companies that produce less than 100 kilograms (220 pounds) of hazardous waste per month are exempt, as are homeowners and farmers. Items such as batteries and unemptied pesticide containers go into the trash and end up in

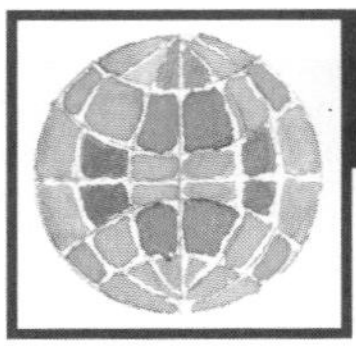

Global Perspective

Grass Roots in a Toxic Wasteland

Given the magnitude of the toxic waste problem in the United States, it is tempting to think that things might be better in a society characterized by central planning on the part of the regime in power, that environmental concerns might more easily be addressed at the highest levels of power. Recent events have put such thinking to rest. Now that the shrouds of secrecy have been lifted from the former Communist countries of Eastern Europe and the republics of the former USSR, it is apparent that central planning has been responsible for the worst kinds of environmental pollution imaginable. Pollutants are emitted from the stacks of industry and power plants with no controls, ruining thousands of square miles of forests and creating untold health problems; untreated sewage from cities fouls the rivers, destroying fish and rendering the water unfit even for industrial uses; heavy metals and toxic chemicals pour untreated into the Baltic Sea, turning the bottom into a marine desert (see map).

The truth is, the leaders of Communist regimes almost completely ignored environmental concerns in the interest of promoting economic production. In the words of Murray Feshback and Alfred Friendly, Jr., in their book *Ecocide in the U.S.S.R.*, "No other great industrial civilization so systematically and so long poisoned its air, land, water and people. . . . And no advanced society faced such a bleak political and economic reckoning with so few resources to invest toward recovery." The people of Eastern Europe and the former USSR are now faced with both a bankrupt environment and a bankrupt economy, and they are looking to the Western democracies for help in both areas.

In environmental affairs in the United States and other democratic societies, grassroots action has always preceded change in public policy. The progress in solving the hazardous waste problem—the laws passed by Congress and the implementation and enforcement of those laws—has been a direct result of grassroots pressure both on recalcitrant leaders and through electing new leaders with environmental values. All the evidence indicates that without this grassroots pressure—citizens demanding a clean, healthful environment—political leaders will promote economic and business concerns that benefit their own in-group and will ignore environmental responsibilities.

How are the environmental grass roots in the countries of Eastern Europe and the former USSR? Grassroots movements are never encouraged in totalitarian societies; they are seen as subversive elements. After the nuclear accident at Chornobyl in 1986, environmentalism achieved some legitimacy in the USSR, grassroots movements were tolerated, and some protests were successful in bringing about limited environmental changes. But now that political processes are open, environmental grassroots movements have declined, as people have become more concerned with basic economic survival. There is currently a fatalism in the response of people living in the polluted former Communist lands. They know their environment is unhealthy, but they still feel powerless to do anything about it. Such is the continuing legacy of Communism.

Thus, experience demonstrates that it is in liberal democracies, where citizens have the freedom to organize and to demand a better environment, that pollution control and prevention regulations are passed and implemented. Of course, the freedom to organize and act is only half the answer; people must take advantage of that freedom. We can only hope that the newly freed people of the former Communist nations and other peoples of the world including ourselves will take advantage of freedom and act to create a better environment.

Sources: R. Liroff, "Eastern Europe: Restoring a Damaged Environment," *EPA Journal* 16; no. 3 (1990); R. Brandt, "Soviet Environment Slips Down the Agenda," *Science*, 255 (January 3, 1992).

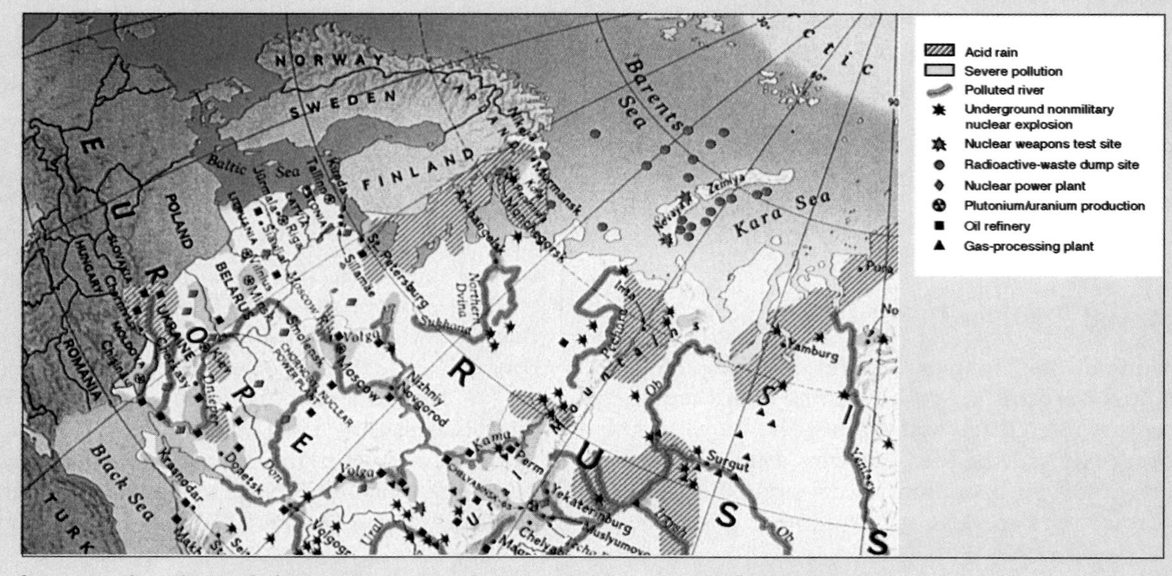

(Reprint from M. Edwards, "Lethal Legacy," *National Geographic* (August 1994), p. 80–81.)

municipal landfills, most of which don't have protective measures for toxic wastes. Can we regulate such areas without creating a police state?

In short, pollution control seems to have come to a regulatory impasse. Happily, however, a new approach is developing that may provide the best of both worlds, reducing environmental pollution while at the same time reducing costs and regulatory complexity. The concept is called *pollution avoidance* or *pollution prevention*.

Pollution Avoidance for a Sustainable Society

First, let us be clear regarding the distinction between *pollution control* and *pollution avoidance*. Pollution control refers to adding on a filter or other device at the "end of the pipe" to prevent pollutants from entering the environment. Disposal of the captured pollutants still has to be dealt with, and this entails more regulation and control. **Pollution avoidance**, on the other hand, refers to changing the production process or the materials used, or both, so that the harmful pollutants won't be produced in the first place. Take a simple example: Adding a catalytic converter to the exhaust pipe of your car is a case of pollution control. Redesigning the engine so that less pollution is produced, or even better, switching to an electric car, is a case of pollution avoidance.

Pollution avoidance often translates as better product or materials management—that is, less wastage. Thus pollution avoidance often creates a cost savings. For example, Exxon Chemical Company added simple "floating roofs" to storage tanks of its most volatile chemicals, reducing evaporative emissions by 90% and gaining a savings of $200,000 per year. That paid for the cost of the roofs in six months. As another example, Carrier, Inc., a maker of air conditioners, installed equipment for more-precise metal cutting. According to a spokesperson for the company, the new equipment greatly reduces cutting wastes, eliminates the need for a toxic cutting lubricant, and leads to the production of a better product—all results that make the company more competitive.

A second angle on pollution avoidance is to find nonhazardous substitutes for hazardous materials. For example, the electronics industry has found that a simple soap-and-water solution works for cleaning circuit boards, whereas ozone-depleting CFCs were formerly used. Clairol switched from water to foam balls for flushing pipes in hair-product production, reducing wastewater by 70% and saving $250 thousand per year. Water-based inks and paints are being substituted for those that contained synthetic organic solvents, and inks and dyes based on biodegradable organic compounds are being substituted for those based on heavy metals.

A third approach is to clean and recycle solvents and lubricants after use. Some military bases, for example, have been able to distill solvents and reuse them instead of discarding them into the environment.

Literally hundreds of additional examples might be given. Yet, INFORM, a New York–based environmental research organization, states that industry has barely scratched the surface of the potential in pollution avoidance. Many accept the estimate that vigorously pursuing pollution prevention may reduce total pollution outputs as much as 75%—perhaps more. Consider that substitution of biological controls in pest management discussed in Chapter 10 and the development of solar energy to substitute for fossil fuels (Chapter 23) are basically pollution-avoiding technologies.

Perhaps the most profound feature of pollution avoidance is that it is putting environmentalists and industrialists on the same "team." Observe that pollution control, by its very nature, tends to create an adversarial relationship, "us against them," the regulators (environmentalists) against the regulated (industry). Over the years, this adversarial relationship has cost billions in lawyer fees and court costs of parties suing one another. Now we are beginning to find environmentalists working side by side with company executives to find ways to reduce both pollution and costs through pollution avoidance, a "win-win" situation. Company executives are realizing that they too participate in the environment and depend on its life-sustaining function.

Also, many present corporation executives are themselves environmentalists from the 1960s and 1970s, and they fully recognize the problems and threats of pollution. For example, the Chemical Manufacturers Association, a trade group of chemical industry executives, has initiated what they call a **Responsible Care Program**. The aim of this program is to go beyond what is required by the law and reduce pollution as much as possible mainly through pollution avoidance.

The potential of pollution avoidance should not be taken to mean that we should relax, much less abandon, pollution-control regulations. It is undoubtedly the high cost of pollution control and the "headache" of understanding and complying with all the regulations (which are constantly changing) that is a major driving force to find a better way—i.e., pollution avoidance.

Finally, we should note that pollution avoidance is not just for business and industry. It is applicable to the individual consumer. So far as you are able to reduce or avoid using products containing harmful chemicals, you are preventing those amounts of chemicals from going into the environment. You are also reducing the byproducts resulting from producing those chemicals. Increasingly, companies are gradually beginning to produce and market **green products**, a term used for products that are more environmentally benign than their traditional counterparts. How fast and to what degree these green products displace or replace traditional products will in large part depend on how we behave as consumers. In other words, consumerism can be an extremely potent force.

In conclusion, we see that there are four ways to address the problems of chemical pollution: (1) pollution avoidance, (2) recycling, (3) treatment (breaking down or converting the material to harmless products), and (4) safe disposal. If the first three are given priority in the order listed there should be little or nothing left to dispose of, and in this respect we will find ourselves in harmony with the first principle of sustainability. There is every reason to believe that we can have the benefits of modern technology without destroying the sustainability of our environment with pollution.

Review Questions

1. The EPA divides hazardous chemicals into what four categories?
2. Define what is meant by "total product life cycle," and describe the many stages at which pollutants may enter the environment.
3. What are the two classes of chemicals that pose the most serious long-term toxic risk?
4. Describe the processes of bioaccumulation and biomagnification.
5. Why do long-term exposures to low levels of heavy metals and synthetic organics pose a risk to human and environmental health?
6. How were chemical wastes generally disposed of before 1970?
7. What two laws pertaining to the disposal of hazardous wastes were passed in the early 1970s?
8. Describe how passage of those two laws (Question 7) effectively shifted pollution from one part of the environment to another.
9. Describe three methods of land disposal that were used in the 1970s. What was a common denominator of all of them in terms of pollution?
10. What law was passed to cope with the problem of abandoned hazardous waste sites? What are the main features of the legislation?
11. What law was passed to ensure safe land disposal of hazardous wastes? What are the main features of the legislation?
12. What laws exist to protect the public against exposures resulting from accidents? What are the main features of the legislation?
13. What laws exist to protect the public against contaminated drinking water? Against new chemicals having dangerous side effects?
14. Can contaminated groundwater be cleaned up? How?
15. What methods of hazardous waste disposal or treatment may be more used in the future? Less used?
16. What is the distinction between pollution control and pollution avoidance?
17. Discuss how environmentalists and businesspeople can work together to reduce pollution. How likely is it that they will cooperate?

Thinking Environmentally

1. Do a product life cycle study for all the materials used in both manufacturing and running your car. List all the pollutants produced at all stages.
2. Before the 1970s, it was not illegal to dispose of hazardous chemicals in unlined pits, and many companies did so. Should they be held responsible today for the contamination those wastes are causing, or should the government (taxpayers) pay for the cleanup? Give a rationale for your position.
3. Our knowledge regarding bioaccumulation of toxic chemicals and their long-term health effects in humans is very limited. Why?
4. Suppose there is a proposal to build an incineration facility near your community. It is proposed that hazardous wastes currently being landfilled at the same location will be disposed of in the new facility. Would you support or oppose the proposal? Give a rationale for your position.
5. Suppose you are assigned the task of assuring that the company you work for is in compliance with environmental regulations. You would start by investigating the regulations under what particular laws? Explain how these laws might pertain to your company.
6. Should the export of hazardous wastes to Third World countries or Native American reservations be prevented? Propose legislation mandating steps or procedures that would do so.
7. Do you favor cutting back on regulations pertaining to hazardous chemicals to help balance the federal budget? What laws or regulations, if any, would you cut back? Defend your answer.

ISSSTE

Air Pollution and Its Control

Key Issues and Questions

1. Understanding air pollution begins with understanding normal atmospheric structure and function. How does the atmosphere process natural and human-caused pollution?
2. Ever since the Industrial Revolution, unhealthy smogs have plagued human cities. How are industrial and photochemical smogs generated?
3. The variety and effects of the major air pollutants have now been identified. What are the eight major air pollutants and their most serious effects?
4. Much is known now about the origins and chemistry of the air pollutants. Where do primary pollutants originate, and how do they form secondary pollutants?
5. Public policy now identifies standards for air pollutants based on their impacts on human health. What are the existing United States air pollution standards?
6. The most recent legislation addressing air pollution is the Clean Air Act of 1990. What are the main provisions of this act, and how are they related to former legislation?
7. Air inside the home and workplace may also contain serious pollutants. What are the major sources of indoor air pollution, and how does public policy address them?
8. Further improvements in air quality may require rethinking how we structure our society. What further steps could be taken to improve air quality and our way of life?

On Tuesday morning, October 26, 1944, the people of Donora, Pennsylvania (population 13,000), awoke to a dense fog. Donora lies in an industrialized valley along the Monongahela River. On the outskirts of town a sooty sign read, "Donora: Next to yours, the best town in the U.S.A." The town had a large steel mill, which used high-sulfur coke, and a zinc reduction plant that roasted ores having a high sulfur content. At the time, most homes in the area were heated with coal. The fog did not seem unusual, at least at the start. Most of Donora's fogs lifted by noon, as the sun warmed the upper atmosphere and then the land. This one didn't lift for five days.

[Photo by Steven D. Elmore/The Stock Market.]

Through Wednesday and Thursday, the air began to smell of sulfur dioxide—an acrid, penetrating odor. By Friday morning, the town's physicians began to get calls from people in trouble. At first, the calls were from elderly citizens and those with asthmatic conditions. They were having difficulty breathing. The calls continued into Friday afternoon; people young and old were complaining of stomach pain, headaches, nausea, and choking. Work at the mills went on, however. The first deaths occurred on Saturday morning. By 10 A.M. one mortician had nine bodies; two other morticians each had one. On Sunday morning the mills were shut down. Even so, the owners were certain that their plants had nothing to do with the trouble. Mercifully, the rain came on Sunday and cleared the air—but not before more than 6,000 townspeople were stricken and 20 had died. The cause? Eventually it was determined to be a combination of polluting gases and particles, a thermal inversion in the lower atmosphere, and a stagnant weather system that together gave new meaning to the term *air pollution*.

With the advent of the Industrial Revolution, the mixture of gases and particles in our atmosphere began changing rapidly, and the effects on natural ecosystems and human health have proved to be dramatic and serious. Some of the changes are regional, affecting only the air in the vicinity of the polluting sources. In this chapter, we look at these lower atmospheric problems. Chapter 16 deals with more global changes that extend into the upper atmosphere.

Atmospheric and Air Pollution Essentials

Structure of the Atmosphere

As we learned in Chapter 2, the atmosphere is a collection of gases that gravity holds in a thin envelope around Earth. It will be helpful, in our attempt to understand air pollution, to consider the basic structure of the atmosphere. The lowest layer, the **troposphere**, extends up 10 miles (16 km) and, except for local temperature inversions, gets colder with altitude (Fig. 15-1). The troposphere is well mixed vertically; pollutants can reach the top within a few days. This layer contains practically all of the water vapor and clouds; it is the site and source of our weather. Substances entering the troposphere may be washed back to Earth's surface by precipitation. Capping the troposphere is the **tropopause**, which is that altitude where temperatures reach -70°F (-59°C).

Above the tropopause is the **stratosphere**, a layer within which temperature *increases* with altitude, up to about 30 miles (50 km) above the surface of Earth. The temperature increases here primarily because the stratosphere contains ozone (O_3), a form of oxygen that absorbs high-energy (ultraviolet) radiation emitted by the sun. Because there is little vertical mixing in the stratosphere and little water vapor, substances that enter it can remain there for a long time.

Beyond the stratosphere are two more layers, the *mesosphere* and the *thermosphere*, where ozone concentration declines and where only small amounts of oxygen and nitrogen are found. Because none of the reactions we are concerned with occur in the meso-sphere or thermosphere, we shall not discuss those two layers.

Atmospheric Pollution and Cleansing

Air pollutants are substances in the atmosphere that have harmful effects. For millions of years, volcanoes, fires, and dust storms have sent smoke and other pollutants into the atmosphere. Coniferous trees and other plants emit volatile organic compounds into the air around them. However, there are mechanisms in the biosphere that remove, assimilate, and recycle these natural pollutants. First, the pollutants disperse and become diluted in the atmosphere. Then, as shown in Fig. 15-2, a naturally occurring cleanser, the hydroxyl radical (OH), oxidizes many of them to products that are harmless or that can be brought down to the ground or water by precipitation. Microorganisms in the soil further convert some of these products into harmless compounds. The complex chemistry of all these cleansing reactions is still being investigated, but we do know that such processes hold natural pollutants below toxic levels (except in the immediate area of a source, such as around an erupting volcano).

Organisms are able to deal with certain levels of pollutants without suffering ill effects. The pollutant level below which no ill effects are observed is called the **threshold level**. Above this level, the effect of a pollutant depends on both its concentration and the duration of exposure to it. Higher levels may be tolerated if the exposure time is short. Thus, for any given pollutant, the threshold level is high for shorter exposures, but gets lower as exposure time increases (Fig. 15-3). It is not the absolute amount of pollutant, but

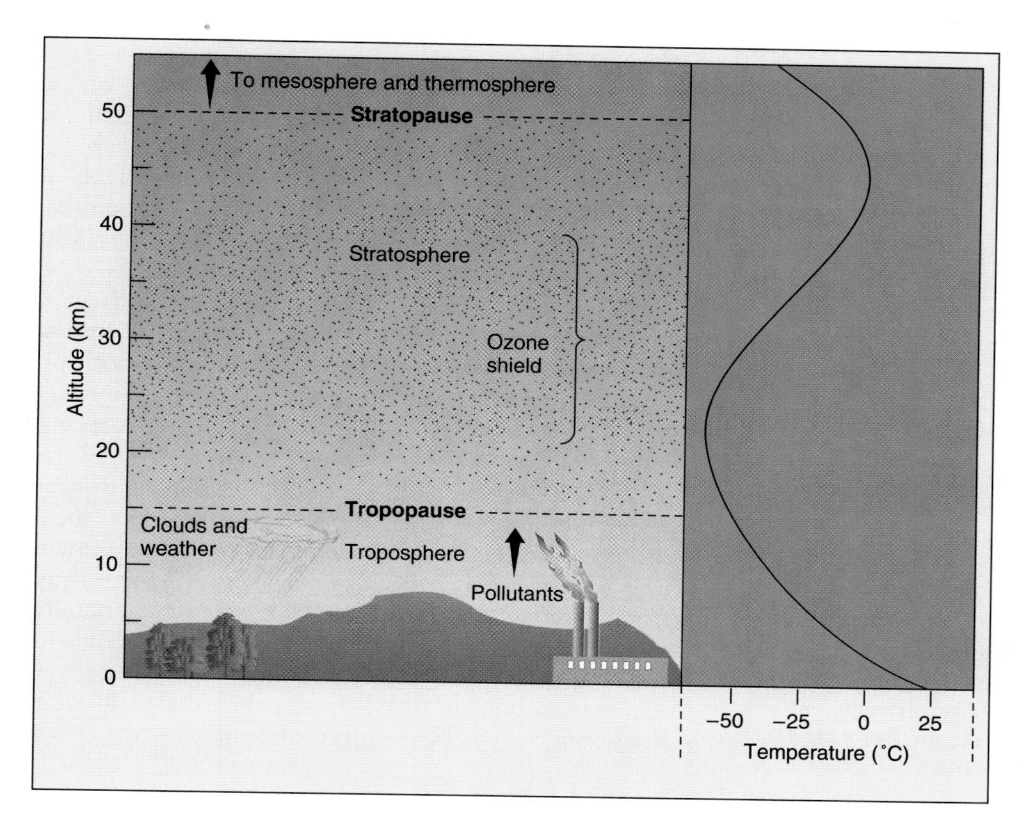

FIGURE 15-1
Structure and temperature profile of the atmosphere. Weather and most pollutants are found in the troposphere. The ozone shield is located in the stratosphere.

rather the *dose*, that is important. As you learned in Chapter 2, dose is defined as concentration multiplied by time of exposure.

Three factors determine the level of air pollution:

- Amount of the pollutants entering the air
- Amount of space into which the pollutants are dispersed
- Mechanisms that remove pollutants from the air

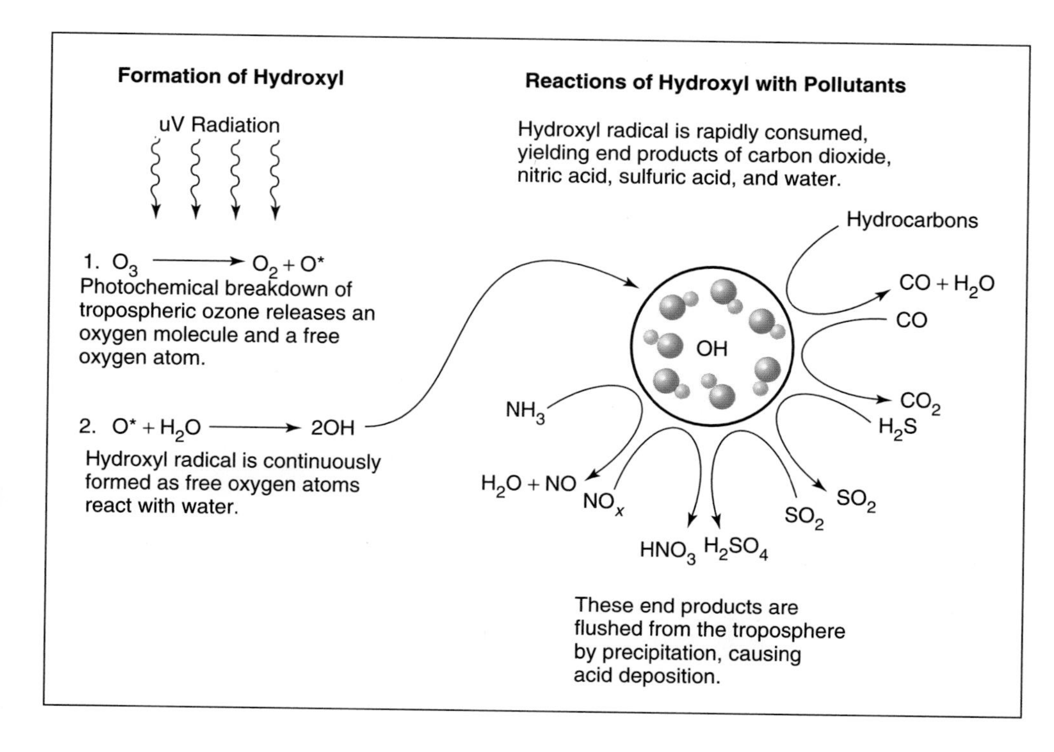

FIGURE 15-2
A simplified model of atmospheric cleansing by the hydroxyl radical. The first step, the photochemical destruction of ozone, is the major process leading to ozone breakdown in the troposphere. The second step produces hydroxyl, which reacts rapidly with many pollutants, converting them to substances that are either less harmful or brought to Earth in precipitation.

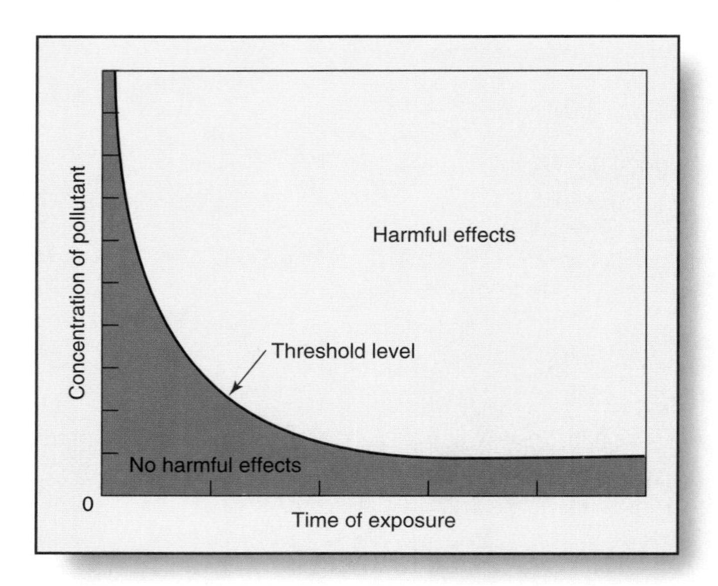

FIGURE 15-3
The threshold level diminishes with increasing exposure time. The threshold level differs for each pollutant. Also, various species and individuals may have very different thresholds for a given pollutant, and the threshold level may differ depending on the presence or absence of other pollutants or stress-causing factors.

The Appearance of Smogs

Down through the centuries, the practice of venting combustion and other fumes into the atmosphere remained the natural way to avoid their obviously noxious effects. With the Industrial Revolution of the 1800s came crowded cities and the use of coal for heating and energy, and air pollution began in earnest. In *Hard Times*, Charles Dickens commented on a typical scene: "Coketown lay shrouded in a haze of its own, which appeared impervious to the sun's rays. You only knew the town was there because there could be no such sulky blotch upon the prospect without a town." This shrouding haze became known as **industrial smog**—an irritating, grayish mixture of soot, sulfurous compounds, and water vapor. This kind of smog continues to be found wherever industries are concentrated and coal is the primary energy source.

Until recently, air pollution in cities and near certain industrial sources was considered a local problem, an inevitable outcome of economic growth and human technology. Even a short distance away from sources of pollution, air quality generally remained good. In the 1950s, however, with the mushrooming use of cars for commuting (see Chapter 24), entire metropolitan areas were enshrouded daily in a brownish haze called **photochemical smog** (Fig. 15-4). The name derives from the fact that sunlight is involved in the formation of this kind of smog.

Certain weather conditions intensify smog levels. The most significant of these conditions is a **temperature inversion**. Under normal conditions, daytime air temperature is highest near the ground, because sunlight strikes Earth and the absorbed heat radiates to the

FIGURE 15-4
A typical episode of photochemical smog. Left, early in the morning, the air is clear. Right, midmorning of the same day, the air is very hazy with smog. The haze is the result of pollutants from the exhausts of rush-hour traffic reacting in the atmosphere. The reactions are promoted by sunlight. These photographs are close to the same view over Los Angeles. (a) Tom McHugh, b) George Gerster/Photo Researchers, Inc.)

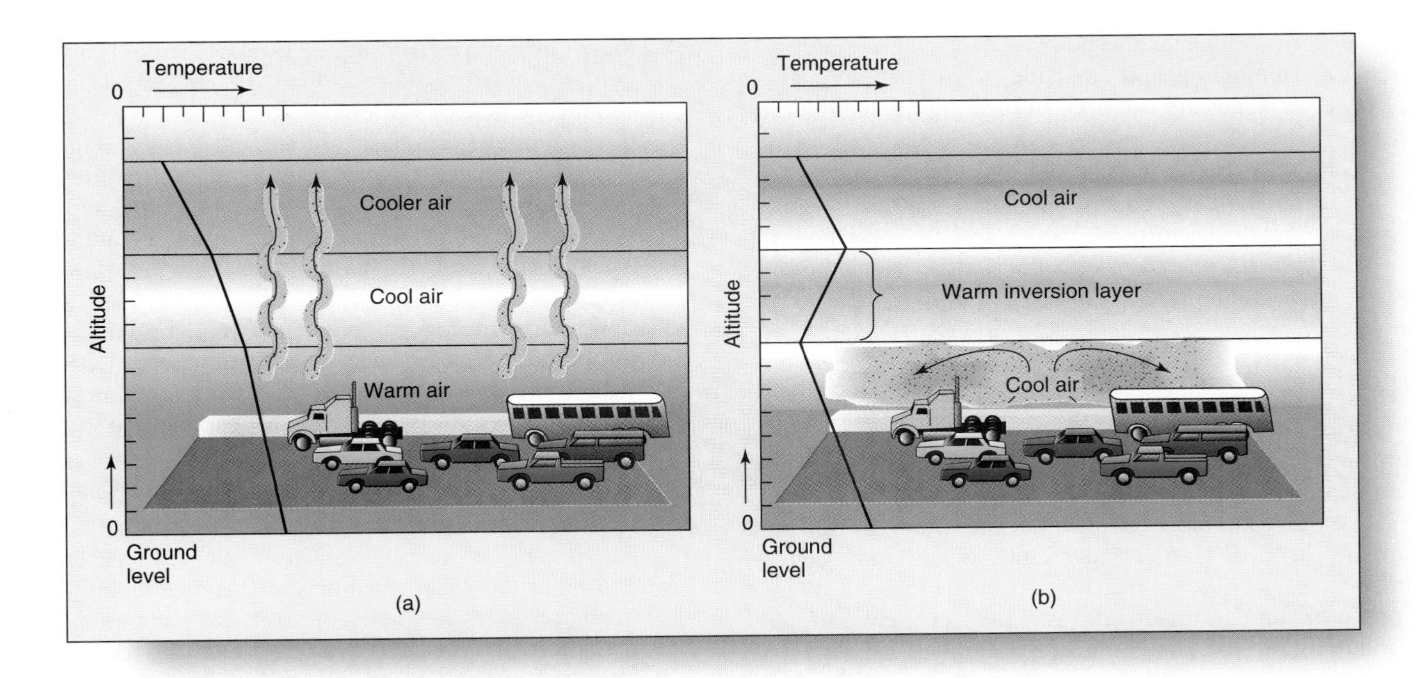

FIGURE 15-5
A temperature inversion may cause episodes of high concentrations of air pollutants. (a) Normally, air temperatures are highest at ground level and decrease at higher elevations. Because the warmer air rises, pollutants are carried upward and diluted in the air above. (b) A temperature inversion is a situation in which a layer of warmer air overlies cooler air at ground level. This blocks the normal updrafts and causes pollutants to accumulate, like cigarette smoke in a closed room.

air near the surface (Fig. 15-5a). The warm air near the ground rises, carrying pollutants upward and dispersing them at higher altitudes. At night, when the sun is no longer heating Earth, the currents cease. This condition of cooler air below and warmer air above is called a temperature inversion (Fig. 15-5b). Inversions are usually very short lived, as the next morning's sun begins the process anew, and any pollutants that accumulated overnight are carried up and away. During cloudy weather, however, the sun may not be strong enough to break up the inversion for hours or even days. Or a mass of high-pressure air may move in and sit above the cool surface air, trapping it.

When such long-term temperature inversions occur, pollutants can build up to dangerous levels, prompting local health officials to urge people with breathing problems to stay indoors. For many people, smog causes headaches, nausea, and eye and throat irritation; it may aggravate preexisting respiratory conditions such as asthma and emphysema. In some industrial cities (Donora, Pennsylvania, for example), air pollution has reached such high levels under severe temperature inversions that mortalities increased significantly. These cases became known as *air pollution disasters*. London experienced repeated episodes of inversion-related disasters in the mid-1900s; there were 4000 pollution-related deaths in one episode in 1952.

The effects of air pollution are not limited to people. In recent years, many species of trees and other vegetation in and near cities began to die back, and farmers near cities suffered damage to or even total destruction of their crops because of air pollution. Pollution-induced damage became so routine that it forced the complete abandonment of citrus growing in some parts of California and vegetable growing in certain areas of New Jersey, which had been among the most productive agricultural regions in the country. A conspicuous acceleration in the rate of metal corrosion and the deterioration of rubber, fabrics, and other materials was also noted.

The Clean Air Act

By the 1960s, it was obvious that pollutants produced by humans were overloading natural cleansing processes in the atmosphere. The unrestricted discharge of pollutants into the atmosphere could no longer be tolerated.

Under grassroots pressure from citizens, the U.S. Congress passed the **Clean Air Act of 1970**. Together with amendments passed in 1977 and 1990, this law, administered by the EPA, represents the foundation of air pollution control efforts in the United States. It calls for identifying the most widespread pollutants, setting **ambient standards**—levels that need to be achieved to protect environmental and human health—and establishing control methods and timetables to meet these goals.

Four stages are involved in meeting the mandates:

- Identifying the pollutants
- Demonstrating which pollutants are responsible for particular adverse health and/or environmental effects, so that reasonable standards may be set
- Determining the sources of the pollutants
- Developing and implementing suitable controls

These stages are adjustable. As we discover new information regarding adverse effects of pollutants, we can put more stringent standards in place and develop and implement new strategies for control.

As a result of progress under the Clean Air Act, air quality in most United States cities is markedly better now than it was in the mid-1900s. However, many problems remain unsolved. Elsewhere in the world, air pollution has become a major source of damage to forests and crops, and a rising threat to human health. Eastern Europe, the countries of the former USSR, and many cities of the developing world are experiencing air pollution reminiscent of that present in England during the early Industrial Revolution.

Major Air Pollutants and Their Impact

Major Pollutants

The following eight pollutants have been identified as most widespread and serious:

1. **Suspended particulate matter.** This is a complex mixture of solid particles and aerosols (liquid particles) suspended in the air. We see these particles as dust, smoke, and haze (Fig. 15-6). Particulates may carry any or all of the other pollutants dissolved in or adsorbed to their surfaces. The particles impair many respiratory functions, especially in individuals with chronic respiratory problems. In 1987, the EPA added a new standard for particulates, $\mathbf{PM_{10}}$, based on information that smaller particulate matter (less than 10 micrometers in diameter) has the greatest effect on health because of its capacity to be inhaled.
2. **Volatile organic compounds (VOCs).** These include materials such as gasoline, paint solvents, and organic cleaning solutions, which evaporate and enter the air in a vapor state, as well as fragments of molecules resulting from the incomplete oxidation of fuels and wastes. VOCs are prime agents of ozone formation.
3. **Carbon monoxide (CO).** This is an invisible, odorless gas that is highly poisonous to air-breathing animals because of its ability to block the delivery of oxygen to the organs and tissues.
4. **Nitrogen oxides (NO_x).** These include several nitrogen-oxygen compounds, all gases. They are converted to nitric acid in the atmosphere and are a major source of acid deposition (Chapter 16). Nitrogen dioxide is a lung irritant that can lead to acute respiratory disease in children.
5. **Sulfur oxides (SO_x), mainly sulfur dioxide (SO_2).** Sulfur dioxide is a gas that is poisonous to both plants and animals. Children and the elderly

FIGURE 15-6
A fossil-fuel power plant near Boston sends out a plume of pollutant particles and gases. Because the stack lacks electrostatic precipitators, the plume contains substantial amounts of suspended particulate matter. (Ray Pfortner/Peter Arnold, Inc.)

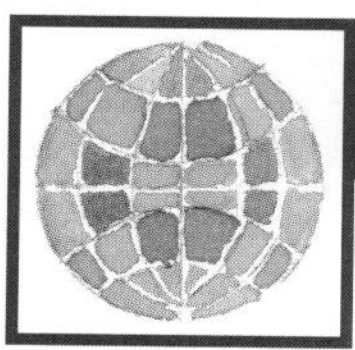

Global Perspective

Mexico City: Life in a Gas Chamber

"I'm getting out of here one of these days," said one resident of Mexico City. "The ecologists are right. We live in a gas chamber." By all accounts, Mexico City residents have the worst air in the world. The city exceeds safety limits for ozone 320 days a year, with values sometimes almost 400% over norms set by the World Health Organization. The city's 3 million vehicles (without catalytic converters and generally in disrepair) and 35,000 industries pour more than 5 million tons of pollutants into the air every day. Because it is nestled in a natural bowl, Mexico City experiences thermal inversions every day during the winter. One study showed that 40% of the city's young children suffer chronic respiratory sickness during the winter months; in January 1989, the government closed schools for a month to lessen the exposure of schoolchildren to city air. Coughs and conjunctivitis are normal to city dwellers. One controllabel source responsible for about 25% of the air pollution is leaking liquid petroleum gas (LPG) pipelines. Writer Carlos Fuentes has nicknamed the metropolis "Makesicko City"!

Mexico City's problems are indicative of a general dilemma throughout the developing world. Jobs and industrial activity, vital for a nation's economy, depend on the use of motor vehicles and on the industrial plants surrounding the city. Many automobile owners are absolutely dependent on their cars, but can ill afford to keep them repaired in order to reduce exhaust emissions. Some people see cars as status symbols and will acquire one as soon as possible—often a "clunker" discarded from the developed world. Industries in Mexico are not regulated as they are in the United States and lack the technology and capital to control emissions.

The Mexican government has initiated an air-monitoring system and, more recently, mandated emissions tests and catalytic converters for automobiles. The city has adopted a "Day without a Car" policy: Drivers are given decals that permit driving only on specific days. New air pollution laws are beginning to be enforced: One major refinery responsible for 4% of the air-quality problem was shut down by then-President Salinas in the spring of 1991, and 80 other industries were temporarily or permanently closed for violating the new laws. New subway lines are being built, and less polluting buses are being added to the fleet. Lead, sulfur dioxide, and carbon monoxide levels are declining. In the meantime, Mexico City's 22 million inhabitants are part of an unintentional experiment in demonstrating the health effects of severe air pollution.

are especially sensitive to SO_2. It is converted to sulfuric acid in the atmosphere and is also a major source of acid deposition.

6. **Lead and other heavy metals.** Lead is very dangerous at low concentrations and can lead to brain damage and death. It accumulates in the body and impairs many tissues and organs.
7. **Ozone and other photochemical oxidants.** You probably know that ozone in the upper atmosphere must be preserved to shield us from ultraviolet radiation (Chapter 16). However, ozone is also highly toxic to both plants and animals; it damages lung tissue and is implicated in many lung disorders. Therefore, ground-level ozone is a serious pollutant. The case of ozone emphasizes the fact that "a pollutant is a chemical out of place."
8. **Air toxics and radon.** Toxic chemicals in the air include carcinogenic chemicals, radioactive materials, and other chemicals (such as asbestos, vinyl chloride, and benzene) that are emitted as pollutants, but are not included in the preceding list of conventional pollutants. The Clean Air Act identifies 189 hazardous air pollutants in this category, many of which are known human carcinogens. Radon is a radioactive gas produced by natural processes within Earth. All radioactive substances have the potential to be damaging to any living matter they contact.

Adverse Effects of Air Pollution on Humans, Plants, and Materials

It is important to recognize that air pollution is not a single entity, but an alphabet soup of the foregoing materials mixed with the normal constituents of air. Further, the amount of each pollutant present varies greatly, depending on proximity to the source and various conditions of wind and weather. As a result, we are exposed to a mixture that varies in makeup and concentration from day to day—even from hour to hour—and from place to place. Consequently, the effects we feel or observe are rarely, if ever, the effects of a single pollutant; they are the combined impact of the whole mixture of pollutants acting over the total life span, and frequently these effects are *synergistic*. (That is, two or more factors combine to produce an effect greater than their simple sum.)

For example, both plants and animals may be so stressed by pollution that they become more vulnerable

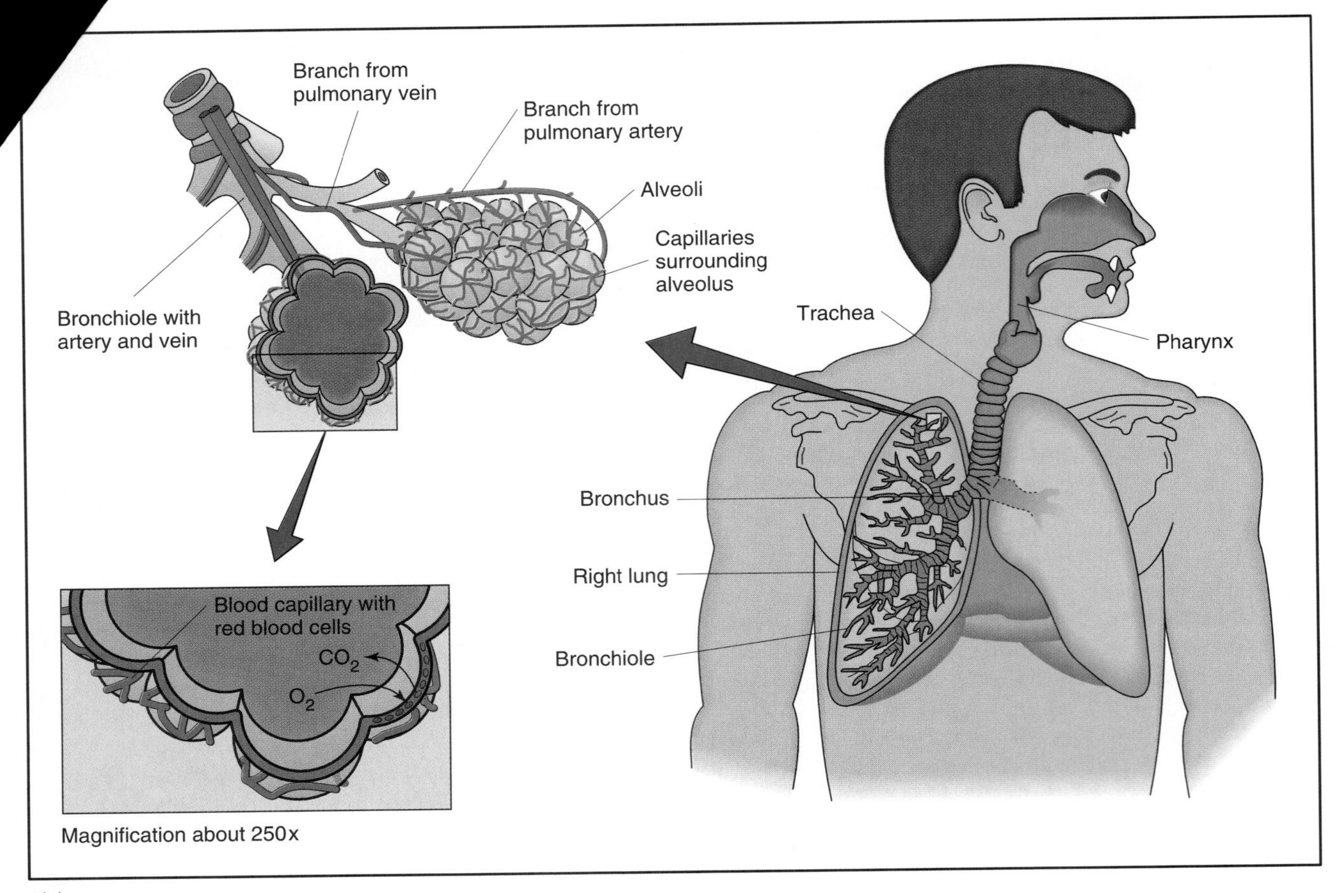

(a)

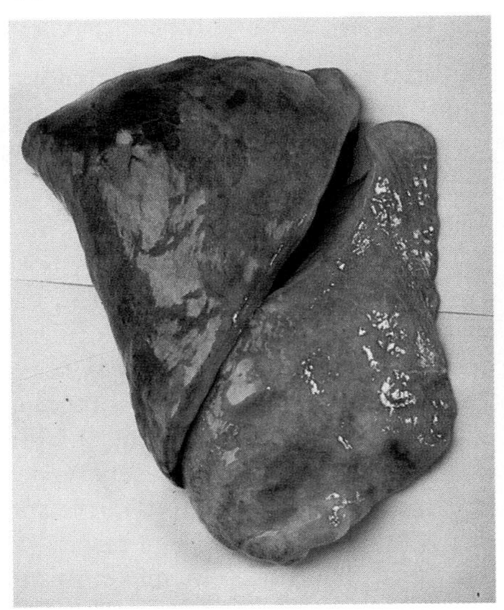

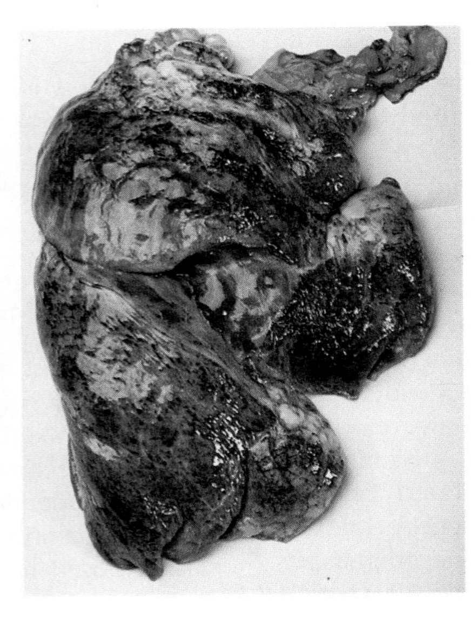

(b)

FIGURE 15-7
(a) In the lungs, air passages branch and rebranch and finally end in millions of tiny sacs called alveoli. Alveoli are surrounded by blood capillaries. As blood passes through these capillaries, oxygen from inhaled air diffuses from the alveoli into the blood. Carbon dioxide diffuses in the reverse direction and leaves the body in the exhaled air. (b) On the left, normal lung tissue; on the right, lung tissue from a person who suffered from emphysema, a chronic lung disease in which some of the structure of the lungs has broken down. Cigarette smoking and heavy air pollution are associated with the development of emphysema and other chronic lung diseases. (Matt Meadows/Peter Arnold, Inc.)

to other environmental factors, such as drought or attack by parasites and disease. Given the complexity of this situation, it is extremely difficult to determine the role of any particular pollutant in causing an observed result. Nevertheless, some significant progress has been made in linking cause and effect.

Effects on Human Health. Humans breathe 30 lb (14 kg) of air into their lungs each day. Although some of the symptoms of pollution that people suffer involve the moist surfaces of the eyes, nose, and throat, the major site of impact is the lungs (Fig. 15-7). Three categories of impact can be distinguished:

1. **Chronic:** Pollutants cause the gradual deterioration of a variety of physiological functions over a period of years.
2. **Acute:** Pollutants bring on life-threatening reactions within a period of hours or days.
3. **Carcinogenic:** Pollutants initiate changes within cells that lead to uncontrolled growth and division (cancer).

Chronic effects. Almost everyone living in areas of urban air pollution suffers from chronic effects. Long-term exposure to sulfur dioxide can lead to bronchitis (inflammation of the bronchi). Chronic inhalation of ozone and particulates can cause inflammation and, ultimately, fibrosis of the lungs, a scarring that permanently impairs lung function. Carbon monoxide reduces the capacity of the blood to carry oxygen, and extended exposure to low levels of carbon monoxide can contribute to heart disease. Chronic exposure to nitrogen oxides is known to impair the immune system, leaving the lungs open to attack by bacteria and viruses.

Those most sensitive to air pollution are small children, asthmatics, people with chronic pulmonary and heart disease, and the elderly. Asthma, an immune system disorder characterized by impaired breathing caused by constriction of air passageways, is usually brought on by contact with allergens. Many air pollutants increase the severity of asthma; the 58% increase in the incidence of asthma in the United States since 1970 is thought to be largely due to the synergistic effects of air pollutants. Studies have shown that asthmatics are extremely sensitive to levels of sulfur dioxide that are well below concentrations found in polluted cities.

Air quality is poor in countries where industrial air pollution is uncontrolled (eastern Europe and China) and where motor vehicle use is on the rise (many "megacities" in developing countries). The toll on human health and mortality in these regions can be severe. For example, the rate of infant mortality from respiratory disease in eastern Europe is nearly 20 times that in North America. The chronic effects of air pollution are reflected in morbidity and mortality data from Poland, a nation with perhaps the worst air in the world. Life expectancy for Polish men is lower than it was 20 years ago, and 30% to 45% of students are below international norms for height, weight, and other health indicators. Acute respiratory disease is responsible for some 4 million deaths a year of children under five in developing countries, second only to infant diarrhea in its impact on mortality.

One of the heavy metal pollutants, lead, deserves special attention in this discussion. For several decades, lead poisoning has been recognized as a cause of mental retardation. Researchers once thought that eating paint chips containing lead was the main source of lead contamination in humans. In the early 1980s, however, elevated lead levels in blood were shown to be much more widespread than previously expected, and they were present in adults as well as children. Learning disabilities in children, as well as high blood pressure in adults, were found to be correlated with high levels of lead in the blood. The major source of this widespread contamination was traced to leaded gasoline. (The lead in exhaust fumes may be inhaled directly or may settle on food, water, or any number of items that are put in the mouth.) This knowledge led the EPA to mandate the successful elimination of leaded gasoline, by the end of 1996. The result has been a dramatic reduction in lead concentrations in the environment—220,000 tons less lead in the environment from its elimination in gasoline alone.

Acute effects. In severe cases, air pollution reaches levels that cause death, although it should be noted that such deaths usually occur among people already suffering from severe respiratory or heart disease or both. The gases present in air pollution are known to be lethal in high concentrations, but such concentrations are not reached in outside air. Therefore, deaths attributed to air pollution are not the direct result of simple poisoning. However, intense air pollution puts an additional stress on the body, and if a person is already in a weakened condition (for example, elderly or asthmatic), this additional stress may be fatal.

Carcinogenic effects. The heavy metal and organic constituents of air pollution include many chemicals known to be carcinogenic in high doses. According to industrial reporting required by the EPA, 600,000 tons of hazardous air pollutants (air toxics) were released into the air in the United States in 1993. The presence of trace amounts of these chemicals in air may be responsible for a significant portion of the cancer observed in humans. The **Clean Air Act of 1990** says that, once an air pollutant is shown by laboratory studies to have mutagenic or carcinogenic properties, it may not be emitted if there is a lifetime risk of cancer greater than 1 in 1 million to the most exposed individual in the population.

In some cases, exposure to a pollutant can be linked directly to cancer and other health problems by way of epidemiological evidence. One pollution factor that clearly and indisputably is correlated with cancer and other lung diseases is cigarette smoking (Fig. 15-8). Cigarette smoking has been shown to be the major single cause of cancer deaths in the United States, responsible for 30% of them. Secondhand (passive) smoke is also linked to over 30,000 cancer deaths a year. Studies have shown that smokers living in polluted air experience a much higher incidence of lung disease than smokers living in clean air—demonstrating a synergistic effect. Certain diseases typically associated with

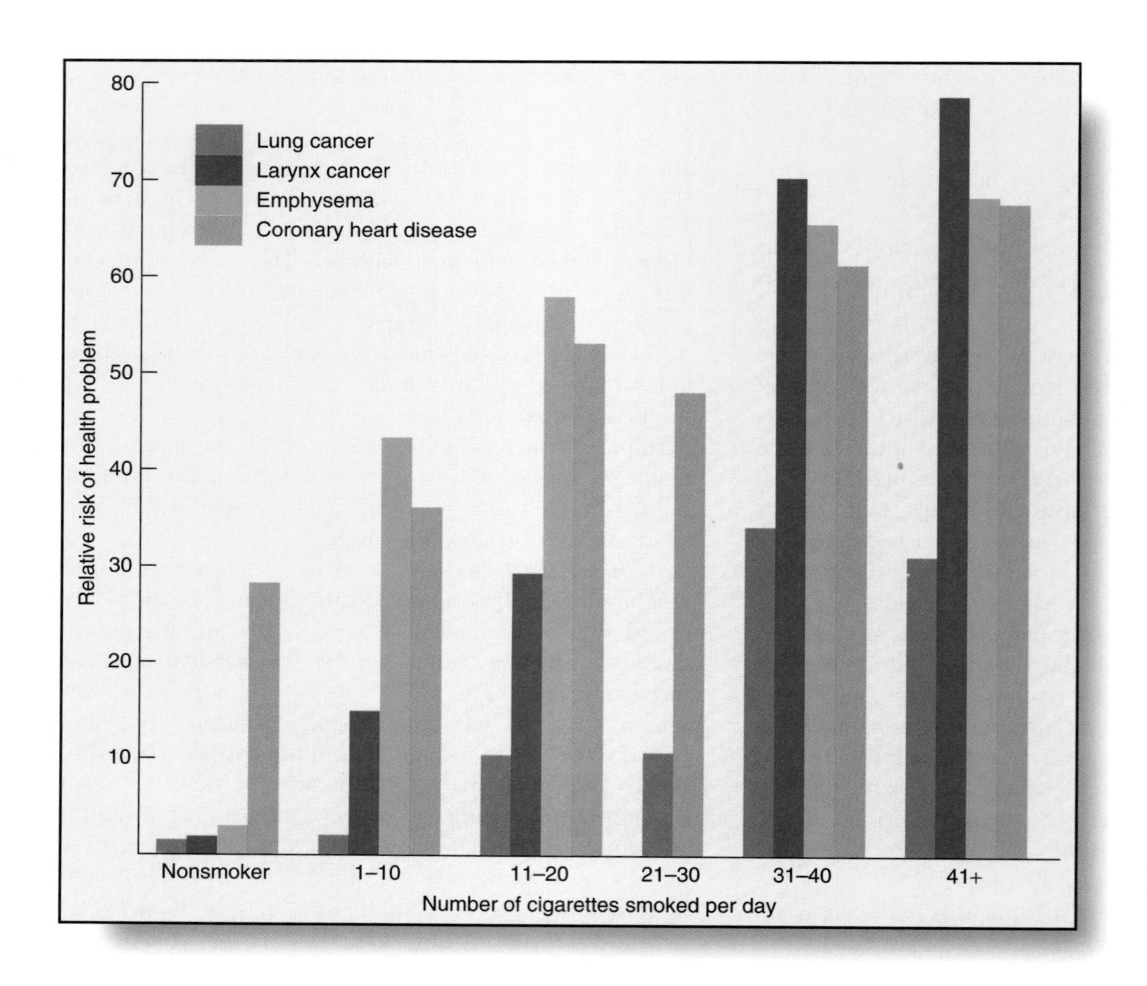

FIGURE 15-8 Cigarette smoking is strongly correlated with the development of cancer and of many chronic diseases.

occupational air pollution show the same synergistic relationship with smoking. For example, black lung disease is seen almost exclusively among coal miners who are also smokers. Smokers who are exposed to asbestos are also highly susceptable to lung disease.

Effects on Agriculture and Forests. How pollutants change vegetation is determined by growing plants in chambers where they can be subjected to any desired concentration of pollutants and the results compared with field observations. Pairs of open-top chambers set up in the field make it possible for plants in one chamber to receive filtered air while plants in the adjacent chamber receive unfiltered air, and pollutants are thus monitored. Through such experiments it is possible to determine which pollutants cause damage and to extrapolate from this information a broader picture of the effects of air pollution on agriculture, forests, and ecosystems in general.

Experiments show that plants are considerably more sensitive to gaseous air pollutants than humans are. Before emissions were controlled, it was common to see wide areas of totally barren land or severely damaged vegetation downwind from smelters or coal-burning power plants (Fig. 15-9). Here, the pollutant responsible was usually sulfur dioxide.

The dying off of vegetation in large urban areas and the damage to crops, orchards, and forests downwind of the urban centers is mainly caused by exposure to ozone and other photochemical oxidants. Crop yield reductions of 5% to 10% occur when the ozone level is well below the EPA standard of 0.12 ppm (parts per million), which was predicated on human impact alone. Estimates of crop damage by ozone range from $2 to $6 billion per year—an estimated $1 billion in California alone. It is significant that much of the world's grain production occurs in regions that receive enough ozone pollution to reduce crop yields. The same parts of the world that produce 60% of the world's food—North America, Europe, and eastern Asia—also produce 60% of the world's air pollution.

The negative impact of air pollution on wild plants and forest trees may be even greater than on agricultural crops. Open-chamber experiments in the Blue Ridge Mountains in northwestern Virginia have shown that the growth of various wild plants was substantially reduced even by an ozone level below the EPA standard. Significant damage to valuable ponderosa and Jeffrey

FIGURE 15-9
The countryside around this Butte, Montana smelter is devastated by the toxic pollutants from the industrial processes. (Joern Gerdis/Photo Researchers, Inc.)

pines occurred along the entire western foothills of the Sierra Nevada in California. U.S. Forest Service studies have concluded that ozone from the nearby Central Valley and San Francisco–Oakland metropolitan region was responsible for this damage. During the 1970s in the San Bernardino Mountains of California, an area that receives air pollution from Los Angeles, 50% of the trees died in some areas. As a result of air pollution controls, those same areas have shown significant improvement in tree growth in recent years, an encouraging sign.

Forests under stress from pollution are more susceptible to damage by insects and other pathogens than are unstressed forests. For example, the deaths of the ponderosa and Jeffrey pines in the Sierra Nevada were attributed to western pine beetles, which invade trees weakened by ozone. Even normally innocuous insects may cause mortality when combined with pollution stress. If widespread pollution worsens, reductions in tree growth and survival could occur with disastrous suddenness as threshold levels of more and more species are exceeded. As the next chapter shows, acid deposition from air pollution also has a decided impact on the growth of forest trees downwind of major urban areas.

Effects on Materials and Aesthetics. Walls, windows, and other exposed surfaces turn gray and dingy as particulates settle on them. Paint and fabrics deteriorate more rapidly, and the sidewalls of tires and other rubber products become hard and checkered with cracks because of oxidation by ozone. Metal corrosion is dramatically increased by sulfur dioxide and acids derived from sulfur and nitrogen oxides, as are weathering and deterioration of stonework (Fig. 15-10). These and other effects of air pollutants on materials increase the costs for cleaning or replacing them by hundreds of millions of dollars a year. Many of the materials damaged are irreplaceable.

A clear blue sky and good visibility—in contrast to the haze of smog—not only are a matter of health, but also have significant aesthetic value and a psychological impact on people. Can a value be put on these benefits? Many of us spend thousands of dollars and hundreds of hours commuting long distances to work so that we can live in a less polluted environment than the one in which we work. Ironically, the resulting traffic and congestion cause much of the very pollution we are trying to escape. Too many of us have come to accept pollution as the necessary cost of "progress." Is it? Can we redefine progress in terms of improving human and environmental health, rather than just increasing our gross national product? Environmentalists believe we can, but it will require political action and will exact an up-front economic cost, as we shall see.

FIGURE 15-10
The corrosive effects of acids from air pollutants are dissolving away the features of many monuments and statues, as seen in the faces of these statues in Brooklyn, New York. (Ray Pfortner/Peter Arnold, Inc.)

Pollutant Sources

In large measure, air pollutants are direct and indirect byproducts of the burning of coal, gasoline, other liquid fuels, and refuse (wastepaper, plaster, and so on). These fuels and wastes are organic compounds. With complete combustion, the byproducts of burning them are carbon dioxide and water vapor, as shown by the following equation for the combustion of methane:

$$CH_4 + 2O_2 \longrightarrow CO_2 + 2H_2O$$

Unfortunately, oxidation is seldom complete, and substances far more complex than methane are involved.

Primary Pollutants

The first six of the major pollutants listed earlier in this chapter (**particulates, VOCs, CO, NO_x, SO_x, and lead**) are called **primary pollutants** because they are the direct products of combustion and evaporation.

When fuels and wastes are burned, particles consisting mainly of carbon are emitted into the air; these are the particulates we see as soot and smoke. In addition, various unburned fragments of fuel molecules remain; these are the VOC emissions. Incompletely oxidized carbon is carbon monoxide (CO), in contrast to completely oxidized carbon, which is carbon dioxide (CO_2). Combustion takes place in the air, which is 78% nitrogen and 21% oxygen. At high combustion temperatures, some of the nitrogen gas is oxidized to form the gas nitric oxide (NO). In the air, nitric oxide immediately reacts with additional oxygen to form nitrogen dioxide (NO_2), nitrogen tetroxide (N_2O_4), or both. (These compounds are collectively referred to as the **nitrogen oxides**.) Nitrogen dioxide absorbs light and is largely responsible for the brownish color of photochemical smog.

In addition to their organic matter content, fuels and refuse contain impurities and additives, and these substances, too, are emitted into the air during burning. Coal, for example, contains from 0.2% to 5.5% sulfur. In combustion, this sulfur is oxidized, giving rise to the gas sulfur dioxide. Coal may contain heavy metal impurities, and refuse, of course, contains an endless array of "impurities."

According to EPA data, 1995 emissions in the United States of the first five primary pollutants amounted to 143 million metric tons. By comparison, in 1970, when the first Clean Air Act became law, these same five pollutants totaled 222 million metric tons. The relative amounts emitted in the U.S., as well as their major sources, are illustrated in Fig. 15-11. The EPA requires the constant monitoring of ambient pollutant concentrations in selected regions. Trends in the emissions of these primary pollutants are represented in Fig. 15-12.

The sixth type of primary pollutant, lead and other heavy metals, is discussed separately because the quantities emitted are far less than the levels for the first five. Before the EPA-directed phaseout, lead was added to gasoline as an inexpensive way to prevent engine knock. Emitted with the exhaust, the lead remained airborne and traveled great distances before settling as far away as the glaciers of Greenland. Since the phaseout, concentrations of lead in the air of cities in the United States have shown a remarkable decline (Fig. 15-12). At the same time, levels of lead in children's blood have dropped greatly. Lead concentrations in Greenland ice have also decreased significantly. All these declines indicate that lead restrictions in the United States and a few other Northern Hemisphere nations have had a global impact. Nonetheless, lead is still used in some U.S. manufacturing processes and is subject to few restrictions elsewhere in the world.

As with lead, the concentrations of toxic chemicals and radon in the air are minute compared with those of the other primary pollutants. Some of the air's toxic compounds—benzene, for example—originate with transportation fuels. Most, however, are traced to industries and small businesses. Radon, on the other hand, is produced by the spontaneous breakdown of fissionable material in rocks and soils. The radon escapes naturally to the surface and seeps into buildings through cracks in foundations and basement floors, sometimes collecting in the structures.

Secondary Pollutants

Some of the primary pollutants may undergo further reactions in the atmosphere and produce additional undesirable compounds. The latter products are called **secondary pollutants**.

Ozone and numerous reactive organic compounds are formed as a result of chemical reactions between nitrogen oxides and volatile organic carbons, with sunlight providing the energy necessary to propel the reactions. Because of the role of sunlight, these products are known collectively as **photochemical oxidants**.

The major reactions in the formation of ozone and other photochemical oxidants are shown in Fig. 15-13. Nitrogen dioxide absorbs light energy and splits to form nitric oxide and atomic oxygen. The atomic oxygen rapidly combines with oxygen gas to form ozone. If other factors are not involved, ozone and nitric oxide then react to form nitrogen dioxide and oxygen gas. A steady-state concentration of ozone results, and there is no appreciable accumulation of ozone (Fig. 15-13a). Summer concentrations of ozone in unpolluted air in North America range from 20 to 50 ppb (parts per billion).

When volatile organic compounds are present,

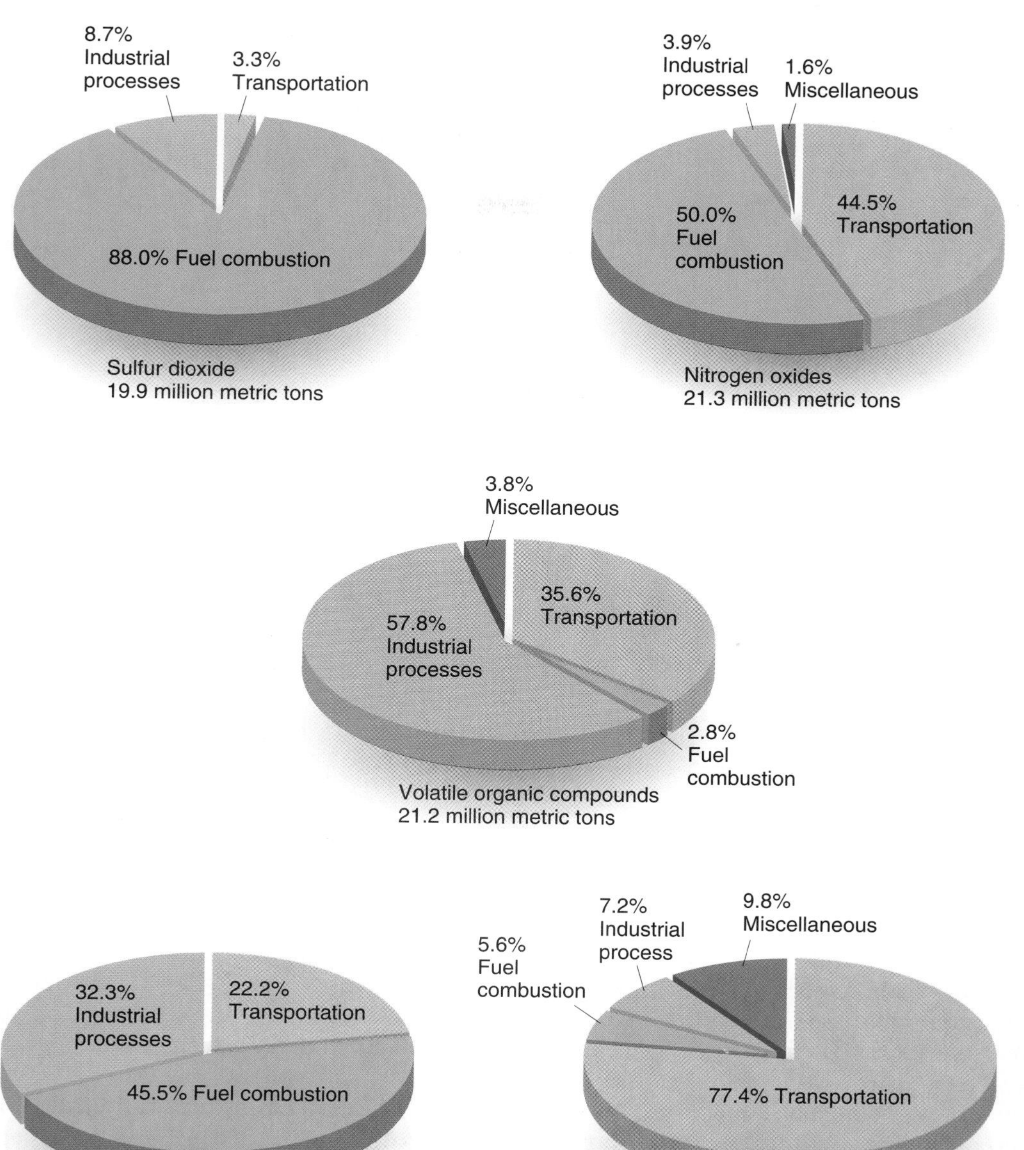

FIGURE 15-11
Emissions in the United States of five primary air pollutants, by source for 1995. Fuel combustion refers to fuels burned for electrical power generation and for space heating. Note especially the different contributions by transportation and fuel combustion, the two major sources of air pollutants. (Data from Office of Air Quality, EPA, *National Air Quality and Emissions Trends Report 1995,* October 1996.)

however, the nitric oxide reacts with them instead of with the ozone, causing several serious problems. First, the reaction between nitric oxide and VOCs leads to highly reactive and damaging compounds known as peroxyacetyl nitrates, or PANs (Fig. 15-13b). Second, numerous aldehyde and ketone compounds are produced as the VOCs are oxidized by atomic oxygen, and these compounds are also noxious. Finally, with the nitric oxide tied up in this way, the ozone tends to accumulate. Because of the complex air chemistry involved, ozone concentrations usually peak 30 to 100 miles (50 to 160 km) downwind of urban centers where the primary pollutants were generated. As a result, ozone concentrations above the health standard in the United States are often found in rural and wilderness areas.

In a sense, **sulfuric** and **nitric acids** can also be considered secondary pollutants, since they are products of sulfur dioxide and nitrogen oxides reacting with atmospheric moisture and oxidants such as hydroxyl.

Particulates take on additional potency when they adsorb other pollutants onto their surfaces. (*Ad*sorption means that the pollutants simply adhere to a surface, like flies sticking to flypaper. By contrast, *ab*sorption means "soaking in.") When inhaled, many of these contaminated particulates are fine enough to penetrate

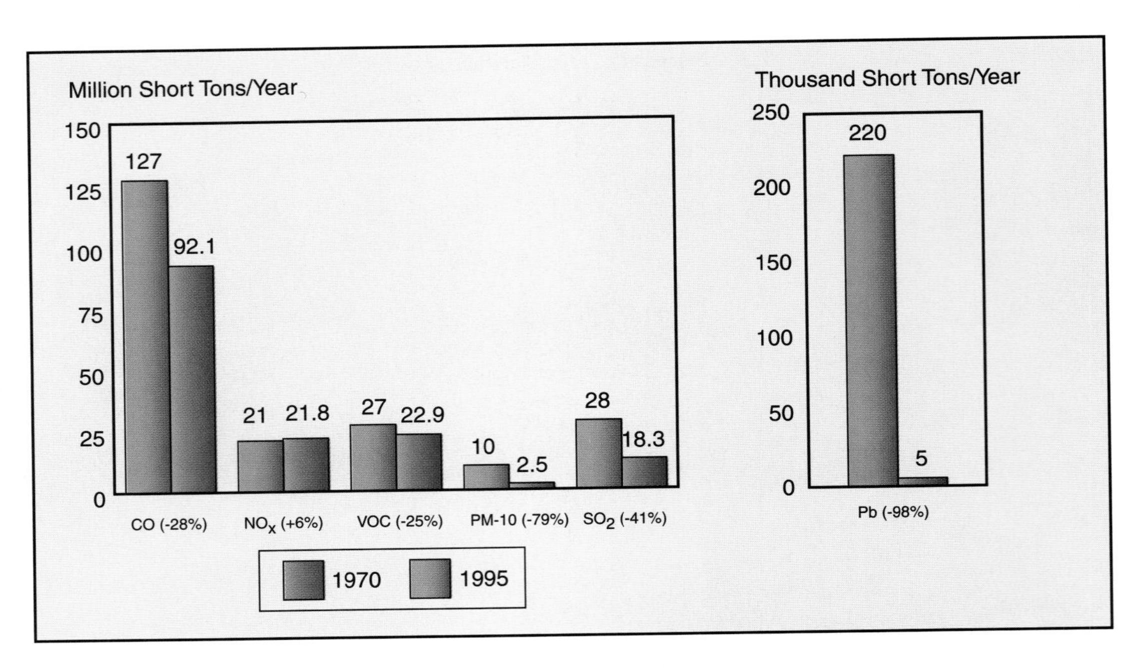

FIGURE 15-12
Trends in air pollutants in the United States, comparing 1970 and 1995 data. (From Office of Air Quality, EPA, *National Air Quality and Emissions Tends Report 1995*, October 1996, Fig. 1-1, p. 2.)

deeply into the lungs, releasing the adsorbed pollutants onto the moist lung surfaces, where they often remain trapped for life.

Sources of air pollution are summarized in Fig. 15-14. Notice that while industrial processes are the major source of particulates, transportation and fuel combustion account for the lion's share of the other pollutants. Strategies for control of air pollutants are different for these different sources.

Bringing Air Pollution Under Control

The Clean Air Act of 1970 mandated the setting of standards for four of the primary pollutants—particulates, sulfur dioxide, carbon monoxide, and nitrogen oxides—and for the secondary pollutant ozone. At the time, these five pollutants were recognized as the most widespread and objectionable; today, with the recent addition of lead, they are known as the *criteria pollutants* and are covered by the **National Ambient Air Quality Standards** (NAAQS; Table 15-1). The **primary standard** for each pollutant is based on the highest level that can be tolerated by humans without noticeable ill effects, minus a 10% to 50% margin of safety. For many of the pollutants, long-term and short-term levels are set. The short-term levels are designed to protect against acute effects, while the long-term standards are designed to protect against chronic effects. It is significant that air pollution standards are set according to human health criteria and not according to their impact on other species or on atmospheric chemistry.

In addition, **National Emission Standards for Hazardous Air Pollutants** (NESHAPS) have been issued for eight toxic substances: arsenic, asbestos, benzene, beryllium, coke oven emissions, mercury, radionuclides, and vinyl chloride. The Clean Air Act of 1990 greatly extended this section of the EPA's regulatory work by specifically naming 189 toxic air pollutants.

Control Strategies

The basic strategy of the 1970 Clean Air Act was to regulate air pollution so that the criteria pollutants remained below the primary standard levels. This approach is called a **command-and-control strategy**, because industry was given regulations to achieve a set limit on each pollutant, to be accomplished by specific control equipment. The assumption was that human and environmental health could be significantly improved by a reduction in the output of pollutants. If a particular region was in violation for a given pollu-

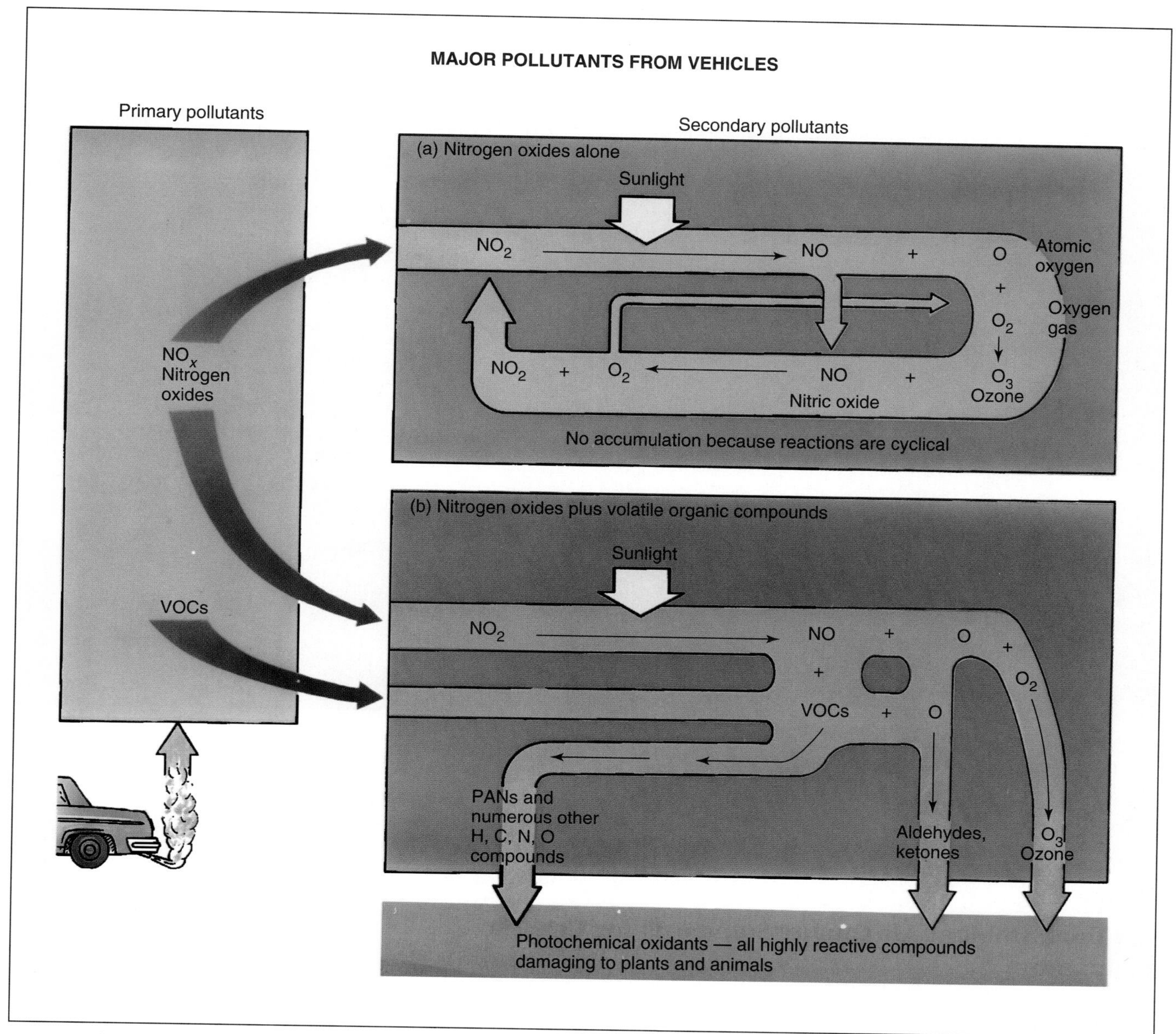

FIGURE 15-13
Formation of ozone and other photochemical oxidants. Ozone is the most injurious. (a) Nitrogen oxides, by themselves, would not cause ozone and other oxidants to reach damaging levels because reactions involving them are cyclic. (b) When VOCs are also present, however, reactions occur that lead to the accumulation of numerous damaging compounds—most significantly, ozone.

tant, a local government agency would track down the source(s) and order reductions in emissions until the area came into compliance.

Unfortunately, this strategy proved difficult to implement. Most of the regulatory responsibility fell on the states and cities, which were often unable or unwilling to enforce control. Many areas violated the standards. Even today, after over 25 years of legislated air pollution control, 93 metropolitan areas still consistently fail to meet the ozone standards, 38 do not meet the carbon monoxide standards, 83 fail to meet the particulate matter standards, and 47 fail to meet the sulfur dioxide standards. To be sure, total air pollutants have been reduced by some 24% during a time when both population and economic activity have substantially increased, but we are still sending huge quantities of pollutants into the atmosphere. (There has been one outstanding accomplishment: Emissions of lead have been reduced 98% in the past 25 years.)

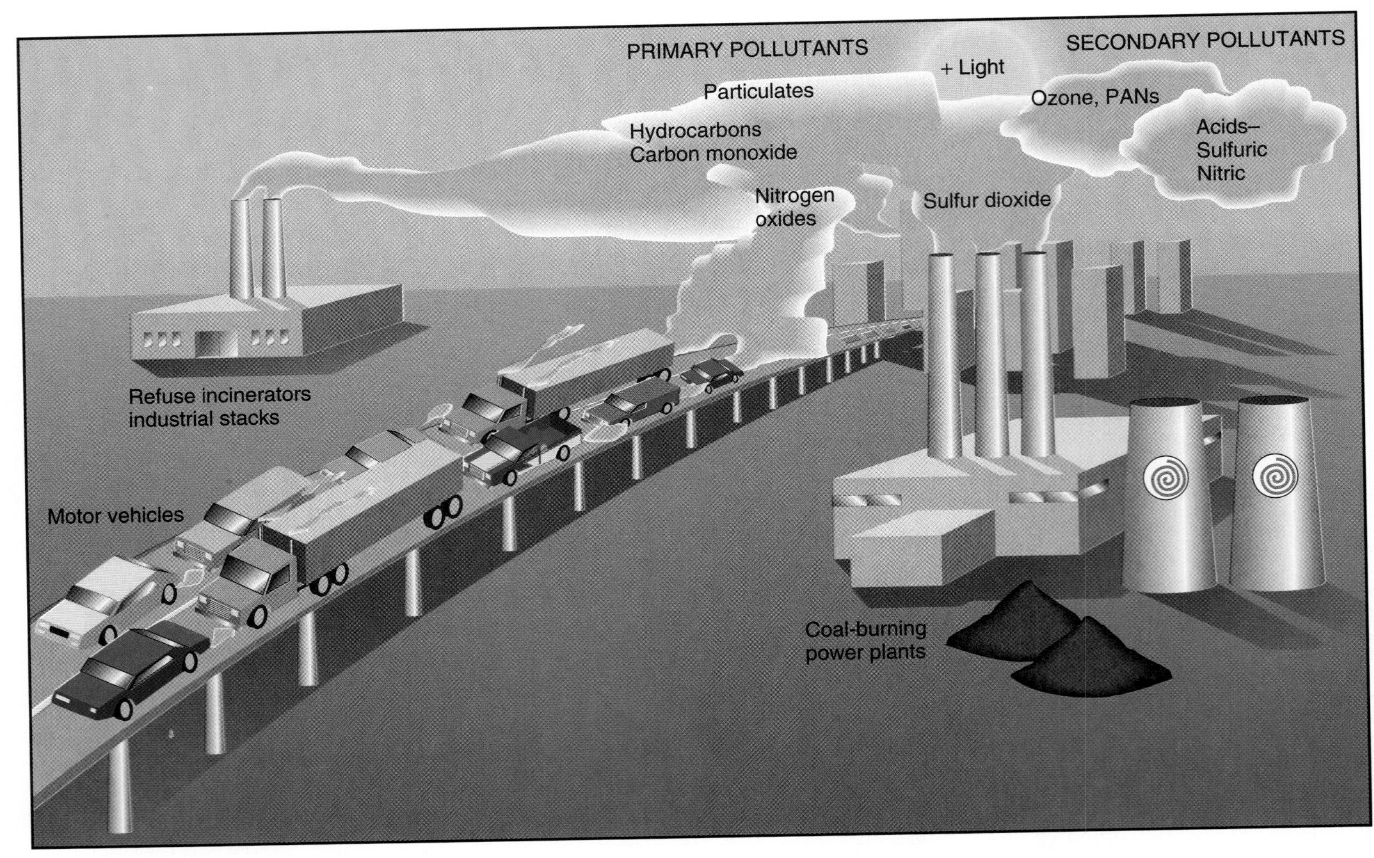

FIGURE 15-14
The prime sources of the major air pollutants.

TABLE 15-1
National Ambient Air Quality Standards for Criteria Pollutants

Pollutant	*Averaging Time**	*Primary Standard*
PM_{10} particulates†	1 year 24 hours	50 μg/m3 150 μg/m3
Sulfur dioxide	1 year 24 hours 3 hours	0.03 ppm 0.14 ppm 0.5 ppm
Carbon monoxide	8 hours 1 hour	9 ppm 35 ppm
Nitrogen oxides	1 year	0.05 ppm
Ozone	1 hour	0.12 ppm
Lead	3 months	1.5 μg/m3

* Averaging time is the period over which concentrations are measured and averaged.
† PM_{10} is the particulate fraction having a diameter smaller than or equal to 10 micrometers. It replaces Suspended Particulate Matter (SPM) as the standard. A microgram (μg) is one millionth of a gram.
Source: *EPA National Air Quality and Emissions Trends Report*, 1993.

The Clean Air Act of 1990 addresses these failures by targeting specific pollutants more directly and by more aggressively demanding compliance, through means such as the imposition of sanctions. As with the earlier law, the states do much of the work in carrying out the mandates of the 1990 act. Each state must develop a State Implementation Plan (SIP), which goes through a process of public comment before submission to the EPA for approval. The SIP is designed to reduce emissions of every NAAQS pollutant whose control standard (Table 15-1) has not been attained. One major change is a permit application process (already in place for the release of pollutants into waterways). Polluters must apply for a permit that identifies the kinds of pollutants they release, the quantities of those pollutants, and the steps they are taking to reduce pollution. Permit fees provide funds the states can use to support their air pollution control activities. The new act also provides more flexibility than the earlier command-and-control approach by allowing polluters to choose the most cost-effective way to accomplish the goals. In addition, the new law uses a market system to allocate pollution among different utilities. (See Chapter 16.)

Reducing Particulates. Prior to the 1970s, the major sources of particulates were industrial stacks and the open burning of refuse. The Clean Air Act of 1970 mandated the phaseout of open burning of refuse and required that particulates from industrial stacks be reduced to "no visible emissions."

The alternative generally taken to dispose of refuse was landfilling, a solution that has created its own set of environmental problems (discussed in Chapter 20). To reduce stack emissions, many industries were required to install filters, electrostatic precipitators, and other devices (Fig. 15-15). Unfortunately, the solid wastes removed from exhaust gases frequently contain heavy metals and other toxic substances (Chapter 14.) Although these measures have markedly reduced the levels of particulates since the 1970s (see Fig. 15-12), particulates continue to be released from steel mills, power plants, cement plants, smelters, construction sites, diesel engines, and so on. Wood-burning stoves and wood and grass fires also contribute to the particulate load, making regulation even more difficult.

Under the Clean Air Act of 1990, the 83 regions of the United States that have failed to attain the required levels must submit attainment plans. These plans must be based on **reasonably available control technology** (RACT) measures, and offending regions must convince the EPA that the standards were reached within a certain time frame.

Limiting Pollutants from Motor Vehicles. Cars, trucks, and buses release nearly half of the pollutants that foul our air. Vehicle exhaust sends out VOCs, carbon monoxide, and the nitrogen oxides that lead to ground-level ozone and PANs. Additional VOCs come from the evaporation of gasoline and oil vapors from fuel tanks and engine systems.

The Clean Air Act of 1970 mandated a 90% reduction in these emissions by 1975. This timing proved to be unrealistic, but enough improvements have been made over the years that a new car today emits 75% less pollutants than pre-1970 cars did (Fig. 15-16). This is fortunate, because driving in the United States has been increasing much more than the population has: Between 1970 and 1990, the number of vehicle miles has doubled, from 1 trillion to 2 trillion miles per year and between 1980 and 1995 the number of vehicles on the road increased over 30% (see inset data on Fig. 15-16). It is hard to imagine what the air would be like without the improvements mandated by the Clean Air Act.

The reductions in automobile emissions have been achieved with a general reduction in the size of passenger vehicles, along with a considerable array of pollution control devices. These include the computerized control of fuel mixture and ignition timing, allowing more complete combustion of fuel and decreasing VOC emissions. To this day, however, the most significant control device on cars is the **catalytic converter** (Fig. 15-17). As exhaust passes through this device, a chemical catalyst made of platinum-coated beads oxidizes most of the VOCs to carbon dioxide and water. The catalytic converter also oxidizes most of the carbon monoxide to carbon dioxide. Although newer converters reduce nitrogen oxides as well, the reduction is not impressive (Fig. 15-16).

Despite these efforts, the continuing failure to meet standards in many regions of the United States, as well as the aesthetically poor air quality in many cities, made it clear during the 1980s that further legislation was in order. In preparing for the Clean Air Act of 1990, the EPA considered three options to improve the air in our cities: (1) Tighten up emissions standards still more. (2) Encourage the development and use of cleaner burning fuels. (3) Persuade people to drive less. The new law addressed the first two options, but made only a token gesture toward promoting less driving. Following are the highlights of the sections on motor vehicles and fuels:

1. New cars sold in 1994 and thereafter must emit 30% less VOCs and 60% less nitrogen oxides than cars sold in 1990. Emission-control equipment must function properly (as represented in warranties) to 100,000 miles; on 1990 cars, this equipment need work only for 50,000 miles. Buses and trucks must meet more stringent standards. Also, the EPA is given authority to control emissions from all nonroad engines that contribute to air pollution, including lawn and garden equipment, motorboats, off-road vehicles, and farm equipment.

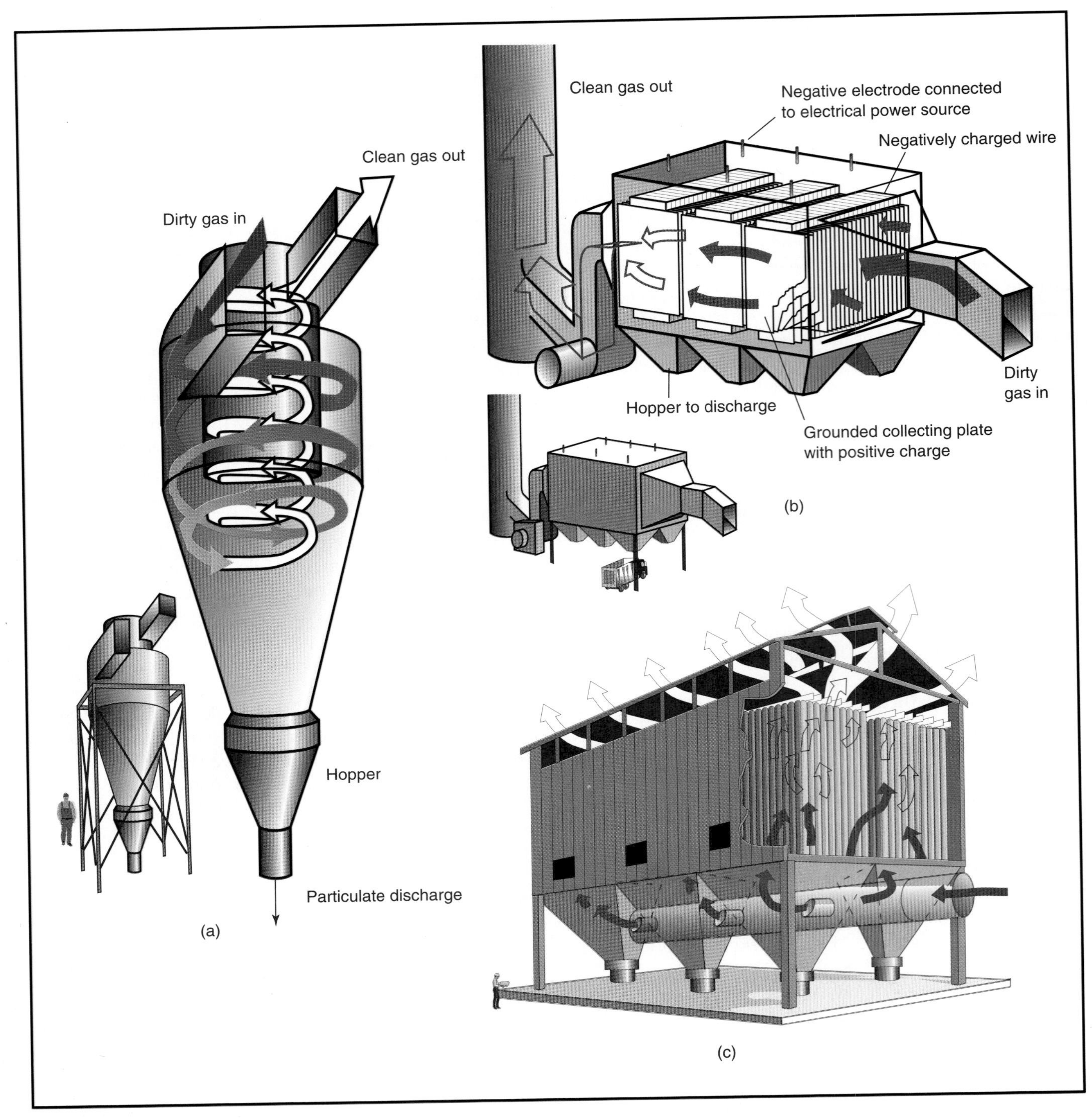

FIGURE 15-15
Devices to remove particulates from exhaust gases. (a) Cyclone precipitator. Particles are removed by centrifugal force as exhausts are swirled. (b) Electrostatic precipitator. Particles are electrically charged and then attracted to plates of the opposite charge. (c) Bag house. Exhaust gases are forced through giant vacuum-cleaner bags. None of these devices removes very fine particles or polluting gases.

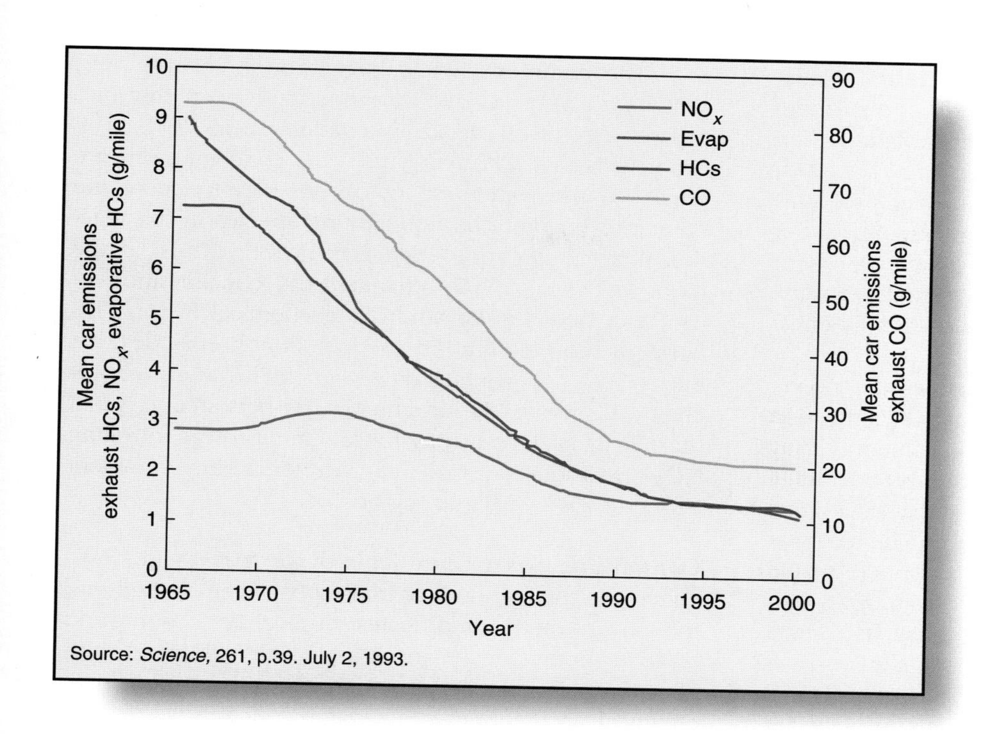

FIGURE 15-16
Average car emissions from vehicles, in grams per vehicle mile traveled, from 1965 to (estimated) 2000. Note numbers of vehicles on the road in the United States for 1980 to 1995.

2. Starting in 1992 in the 38 cities with continuing carbon monoxide problems, oxygen has been added to gasoline in the form of alcohols and ethers to stimulate more complete combustion and cut down on carbon monoxide emissions. Beginning in 1995, special "clean" fuels are required for the nine cities having the worst air pollution. These measures will result in higher gasoline prices. In 1994, the EPA ordered gasoline refiners to add grain alcohol to their products, a move calculated to improve the air in substandard cities.
3. Before 1990, only a few states and cities required that vehicle inspection stations be capable of measuring emissions accurately. The new act requires more than 40 metropolitan areas to initiate inspection and maintenance programs and other cities to improve their current programs. The result is expected to be cost effective in reducing air pollution from cars and trucks.
4. Employers in cities with high smog levels will be required to take steps to increase the number of their employees who carpool or use mass transit.

Two factors now affect fuel efficiency and consumption rates. First, the elimination of federal speed limits in early 1996 caused reduced fuel efficiencies at higher speeds. Second, the sale of sport utility vehicles (light trucks) has surged since 1990. SUVs now represent 6% of all cars and trucks on the road. In city traffic these vehicles get 12 to 16 mpg and maybe 22 mpg on the highway. Remember: as fuel efficiency decreases tailpipe emissions increase.

FIGURE 15-17
Cutaways of three kinds of catalytic converters. Top: ceramic pellets. Left: metal foil plates. Right: ceramic monolith substrates. The catalysts are platinum, palladium, and rhodium, metals that promote the oxidation of residual VOC and carbon monoxide to CO_2 and water. Catalytic converters are now required on all automobiles. (GM/AC Rochester.)

Decreasing Sulfur Dioxide and Acids. Measurements of sulfur dioxide in city air indicate that there has been great success in controlling this pollutant (Fig. 15-12). The results are highly misleading, however, because control strategies have led to other problems that are as bad, if not worse. The major source of sulfur dioxide is coal-burning electric power plants. A power plant may burn up to 10,000 tons of coal per day. Contaminated with 3% sulfur, this produces daily emissions nearing 1000 tons of sulfur dioxide.

Rather than reducing these emissions, the power plant industries lobbied for and obtained permission under the Clean Air Act to use a "dilution solution." They shut down small, inefficient plants in cities and built huge facilities in rural areas, with very tall smokestacks that ejected pollutants above the inversion layer. The idea was that pollutants ejected into the upper atmosphere would be diluted and effectively disappear. Far from disappearing, however, the sulfur dioxide from the tall stacks was converted to sulfuric acid, which has been shown to be the major source of acid precipitation (Chapter 16).

Thus, sulfur dioxide improvements in city air quality have been at the expense of declining rural air quality, and reducing sulfur dioxide at ground level has been at the expense of greatly worsening the acid deposition problem. The Clean Air Act of 1990 specifically addresses these twin problems by requiring an annual reduction of 10 million tons (roughly half) of sulfur dioxide emissions from coal-burning power plants by the year 2010.

Managing Ozone. For a long time, it has been assumed that the best way to reduce ozone levels is to reduce emissions of VOCs. The steps pertaining to motor vehicles in the Clean Air Act of 1990 address the source of about half of VOC emissions. **Point sources** (industries) account for 30% of such emissions, and **area sources** (numerous small emitters such as dry cleaners, print shops, and household products) represent the remaining 20%. RACT measures have already been mandated for many point sources, and much progress has been made through EPA, state, and local regulatory efforts to reduce emissions from these sources. Under the Clean Air Act of 1990, regions with continuing violation of ozone standards are required to further control VOC emissions from both point and area sources. The degree of control and the timing vary according to the seriousness of the violation, but the means of control will range from the prohibition of some consumer products, to annual fees paid by point sources ($5000 per ton of VOCs), to sanctions such as the withholding of federal highway funds and other grants. The last of these is an effective means of forcing recalcitrant states to address their air pollution problems.

The complex chemical reactions involving NO_x, VOCs, and oxygen have thrown some uncertainty into this strategy of emphasizing a reduction in VOCs. The problem is that *both* NO_x and VOC concentrations are crucial to the generation of ozone. Either one or the other can become the rate-limiting species in the reaction that forms ozone. (See Fig. 15-13b.) Thus, as the ratio of VOCs to NO_x changes, the concentration of NO_x can become the controlling chemical factor. This happens more commonly in air pollution episodes that occur over a period of days than in the daily photochemical smog of the urban city. The bottom line is that the EPA may have to pay more attention to controlling NO_x emissions in order to further reduce ambient ozone levels over large regions.

Controlling Toxic Chemicals in the Air. By EPA estimates, the total amount of toxic substances emitted into the air in the United States during 1993 is around 600,000 metric tons. Cancer and other adverse health effects, environmental contamination, and catastrophic chemical accidents are the major concerns associated with this category of pollutants.

Under the Clean Air Act of 1990, Congress identified 189 toxic pollutants. It directed the EPA to identify major sources of these pollutants and develop **maximum achievable control technology** (MACT) standards. Besides control technologies, the setting of these standards includes options for substituting nontoxic chemicals, giving industry some flexibility in meeting goals. State and local air pollution authorities will be responsible for seeing that industrial plants achieve the goals. The program should reduce emissions by at least 20% by the year 2005, at an estimated cost to industry of $7 billion.

Other provisions of the act include a renewed emphasis on calculating risks to public health as a result of toxic chemicals in the air from all sources (for example, radon in buildings) and more attention to the problem of accidental releases of such chemicals.

Indoor Air Pollution

After recognizing the many pollutants in outdoor air, you may be inclined to remain indoors to escape the hazards. However, air inside the home and workplace often contains much higher levels of hazardous pollutants than outdoor air does. Environmental health science has recently turned to the concept of **total exposure assessment** (TEA) in order to evaluate more accurately the impact of air pollutants on human health. To accomplish TEA, scientists measure the pollutants in the air spaces occupied by people and then calculate exposure on the basis of the time spent in those spaces. This approach of direct measurement of exposure has placed increased emphasis on indoor air pollution.

The overall indoor air pollution problem is threefold. First, increasing numbers and types of products and equipment used in homes and offices give off potentially hazardous fumes. Second, buildings have become increasingly well insulated and sealed; hence, pollutants are trapped inside, where they accumulate to potentially dangerous levels. Third, people are exposed more to indoor pollution than to outdoor pollution. The average person spends 90% of his or her time indoors, and the people who spend the most time inside are those most vulnerable to the harmful effects of pollution: small children, pregnant women, the elderly, and the chronically ill.

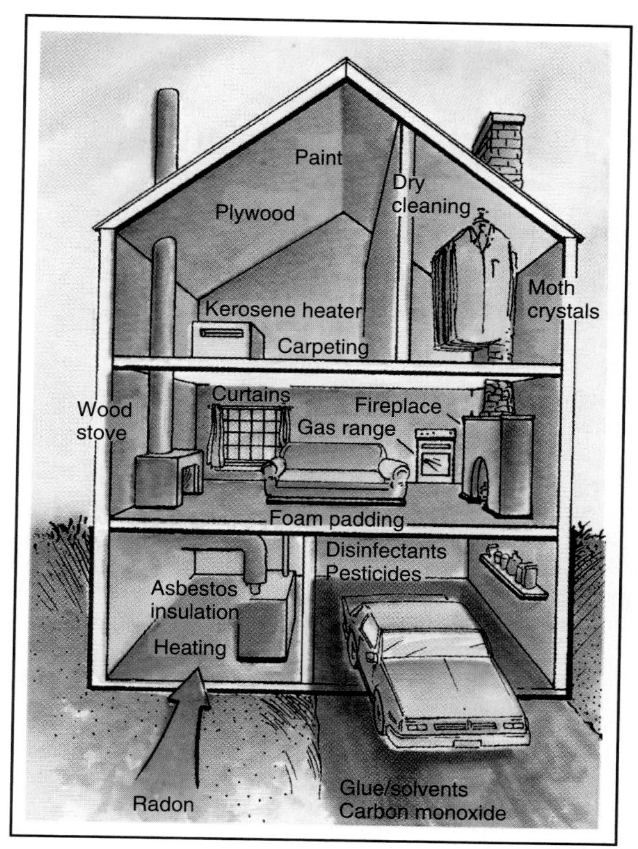

FIGURE 15-18
Indoor air pollution. Pollutants originate from many sources and can accumulate to unhealthy levels, leading to "sick building syndrome" (characterized by eye, nose, and throat irritations, nausea, irritability, and fatigue).

Sources of Indoor Pollution

The sources of indoor air pollution are numerous, as demonstrated in Fig. 15-18. They include:

- Formaldehyde and other synthetic organic compounds emanating from plywood, particle board, foam rubber, "plastic" upholstery, and no-iron sheets and pillowcases.
- A wide range of compounds from foods burned on the stove or in the oven.
- Incomplete combustion and impurities from fuel-fired heating systems such as gas or oil furnaces, kerosene heaters, and wood stoves.
- Fumes from household cleaners and other cleansing agents.
- Fumes from glues and hobby materials.
- Pesticides.
- Air fresheners and disinfectants. Most air fresheners work by either dulling the sense of smell so that you don't notice noxious odors or by introducing "high-intensity" smells that cover up odors.
- Aerosol sprays of all sorts.
- Radon. As warm air escapes from the top of a house, creating a partial vacuum, radon may be drawn through the basement floor. Trapped in the house, the gas may then reach hazardous levels. (See "Ethics" box, p. 393.) The EPA has estimated that between 7000 and 30,000 cancers per year are caused by radon.
- Asbestos. This natural mineral has fiberlike crystals. It is mined from rock and was once used for heat insulation and fire retardation material. Pipes in steam-heating systems were wrapped in asbestos, ceilings in many public buildings were covered with it, and it was used in ironing-board covers, paints, and roofing materials. In the 1960s, researchers determined that the inhalation of asbestos fibers is associated with a unique form of lung cancer that develops as long as 20 to 30 years after exposure to the material. The EPA began regulating asbestos in the mid-1970s and initiated intensive campaigns to remove it from schools. The program has moved forward slowly, however; many buildings still contain the material, and many asbestos products remain in use.
- Smoking. Smoking carries a much higher health risk than the average exposure to any of the above materials, and it may act synergistically to increase the risk of exposure to other pollutants, such as Radon gas. Smoking has also been shown to increase the health risks of nonsmokers subjected to "passive" or "secondhand" smoke (environmental tobacco smoke, or ETS). ETS contains over 4000 substances, at least 40 of which are known to be carcinogenic and many of which are potent respiratory irritants.

Public Policy on Indoor Pollution

Current EPA policies focus on two indoor sources of pollution: asbestos and tobacco smoke.

Earth Watch

Plants for Your Health

In 1989, the EPA submitted a report to Congress stating that "indoor air pollution represents a major portion of the public's exposure to air pollution and may pose serious acute and chronic health risks . . . indoor air pollution is among the nation's most important environmental health problems." Carpets, furniture coverings, wallboard, and other home furnishings are made from plastics and artificial fibers that emit VOCs. Cleaners, insecticides, glues, air fresheners, hair spray, shoe and nail polish, magic markers, oil-based paints, and any number of other commonly used products add to the pollutant load. An EPA study of one Washington, DC., nursing home detected the presence of 350 different VOCs.

In the early 1970s, Bill Wolverton, an environmental engineer working for NASA, was given the problem of keeping air clean and healthful in a spaceship. He started testing the effectiveness of house plants as "bioregenerative" agents. They proved even more effective than he had dared hope. Starting with "dangerous" levels of various VOCs, Wolverton found that some plants proved capable of reducing the pollution to nondetectable levels within 24 hours. Two of the most effective plants were spider plants and philodendrons, which are also among the easiest-to-grow house plants.

Placed in a hanging pot by a window, these plants create a cascading curtain of green. They tolerate almost any light conditions. Watering them every week or two will suffice. They are resistant to pests, and they don't produce blossoms that are allergenic. The plants are easily propagated—philodendrons from cuttings and spider plants from the numerous plantlets that are produced on runners. Among the most common and least expensive plants wherever house plants are sold, these are, indeed, plants for your health!

Asbestos. The EPA recently announced a ban on practically all uses of asbestos, to be completed by 1997. In the meantime, the agency has published guidelines for controlling old asbestos in public buildings, especially schools. The guidelines caution school administrators against assuming that removal is the only option; in most cases, asbestos can be sprayed with a sealant (encapsulation) or enclosed, for much lower expense than the cost of removal and with far less risk of exposure for the public.

Smoking. Smoking is, essentially, portable air pollution. The smoker is the primary recipient of the pollution, but the effects spread to all who breathe the smoke-clouded air and, in the case of pregnant women, to the unborn fetus. Worldwatch Institute's William Chandler maintains that "Tobacco causes more death and suffering among adults than any other toxic material in the environment." Surgeons general of the United States have issued repeated warnings against smoking since 1964, and public policy has taken them seriously by requiring warning labels on smoking materials, banning cigarette advertising on television, promoting smoke-free workplaces, requiring nonsmokers' areas in restaurants, and banning smoking on all domestic airline flights. Since the warnings began, the U.S. smoking population has gradually dropped from 40% to 26%. Unfortunately, another trend has appeared: People are likely to start smoking at a younger age—an apparent target of the tobacco industry advertising. Smoking is still the leading cause of preventable deaths in the United States; at least 420,000 die each year from smoking-related causes.

Two recent developments in public policy promise to bring major changes in the smoking environment. First, in January 1993, the EPA classified environmental tobacco smoke (ETS) as a Class A (known human) carcinogen. In the words of EPA Administrator Carol Browner, "Widespread exposure to secondhand smoke in the United States presents a serious and substantial public health risk." The EPA's action was unanimously supported by its science advisory board. Fallout from this action has been substantial: Smoke-free zones in public areas such as shopping malls, bars, restaurants, and workplaces are now common. In particular, the EPA has taken specific steps to protect children from ETS in all public places, and OSHA is promoting policies to protect workers from involuntary exposure to ETS in the workplace. The impact of these actions on public health should be substantial: One assessment of total exposure revealed that just a 2% decrease in ETS would be the equivalent, in terms of human exposure to harmful particulates, of eliminating all the coal-fired power plants in the country!

The second significant development comes from the FDA. Citing the addictive properties of nicotine, then FDA administrator David Kessler proposed in February 1994 that the agency has the authority to regulate cigarettes as a drug. At congressional hearings following the FDA's initiative, the public was entertained by tobacco company executives denying that cigarette smoking is addictive (see the ABC Case Study "Tobacco on Trial"). The hearings included allegations of a coverup of the tobacco industry's own research that revealed the addictive properties of nicotine and the further allegation that tobacco manufacturers spike their cigarettes with additional nicotine in order to "hook" smokers. Despite documented evidence of this deception, the 104th Congress (1994–1995) refused to investigate or hold further hearings. A year later, President Clinton declared nicotine an addictive drug, which automatically gives the FDA regulatory authority. The FDA's initial efforts are concentrating on curbing tobacco sales to the underaged.

For years, our approach to tobacco has been ambivalent. On the one hand, we try to protect people from the serious health effects of smoking; on the other hand, we subsidize the tobacco industry and tobacco exports. Cigarettes have even been part of the Food for Peace program!

Although the links between cigarette smoking and illness are now firmly established, the tobacco industry has managed to avoid liability for the deaths caused by smoking. As public policy moves toward banning smoking in more and more public places, the industry is fighting back with propaganda emphasizing smoking as an important freedom. (See "Ethics" box on p. 394.) Sadly as long as there are smokers, there will be a population whose addiction, illness, and mortality will remind us of the dangers of air pollution.

Taking Stock

The Cost of Controlling Air Pollution

Without question, measures taken to reduce air pollution carry an economic cost. The United States invests heavily in pollution control—sending a clear signal that human health and environmental quality rank highly in the development and implementation of public policy. There is strong citizen support for this investment as well; according to a recent *Times Mirror* poll, 89% of Americans believe that a strong economy and environmental protection can both be accomplished, while only 22% believe that the economy is more important

Ethics

Radon: The Killer in Your Home?

A deadly gas—invisible, tasteless, odorless radon—seeps into living spaces in homes, schools, and other buildings. As it is breathed into the lungs, some of the radon undergoes radioactive decay into solid chemicals that can become lodged in the lungs and, in time, initiate cancer. If this sounds like science fiction to you, you are not alone. Not everyone believes that radon is the hazard the EPA claims it is.

Radon comes from the natural breakdown of uranium in rock, soil, and water. It drew attention as a health risk when underground miners developed lung cancer at alarmingly high rates. Many studies have shown a clear link between the exposure of miners to radon and the incidence of lung cancer. Although most of the miners were exposed to much higher levels of radon than those found in homes, many scientists believe that there is no threshold level for radioactive substances. Thus, the EPA has judged that a radon concentration of 4 picocuries per liter (4pC:/L) (EPA's "action level") inhaled over a lifetime is capable of causing lung cancer in more than one person per thousand. Using risk-analysis techniques, the EPA estimates that radon causes between 7000 and 30,000 lung cancer deaths each year in the United States; smokers are at a higher risk of radon-induced lung cancer.

As a consequence of the estimates of radon's risk to health, the EPA considers it a significant indoor air health problem. The agency's strategy has been directed toward informing the public of the risk and encouraging people to test their homes and other buildings for the presence of radon (a simple procedure involving a $20 test kit). Where levels are shown to be above 4 picocuries per liter, the EPA recommends a remediation process. This usually involves sealing cracks in foundations or, in some cases, installing a ventilation system that removes radon from beneath a foundation. So far, the EPA has adopted a nonregulatory approach to radon.

Critics of the EPA's attention to radon abound. While admitting that radon does pose a health hazard, they cite the great uncertainties involved in the risk-analysis procedures used by the EPA. The critics also question the "no threshold" assumption, pointing out that humans have been exposed to low levels of radon since the beginning of the species and have likely developed resistance to such low levels of radioactivity. With an estimated $45 billion price tag for testing and remediation, the response to radon suggested by the EPA is costly and, in the judgment of the critics, largely unnecessary. So far, Congress has not taken any specific action to address radon. What do you think should be done? (Contact: Radon Hotline of the National Safety Council at 1–800–767–7236.)

Ethics

The Rights of Smokers

Smoking in public is becoming socially unacceptable in many parts of the country, forcing many smokers to retreat to rest rooms, specially designated smoking areas, or sidewalks in order to satisfy their cravings. Almost all states have laws restricting smoking in public places, and federal law now prohibits smoking on all domestic airline flights.

Although it was once considered simply a nuisance to have to breathe secondhand smoke (sidestream smoke, as it is called), evidence has shown that consistent breathing of sidestream smoke can lead to some of the same consequences experienced by smokers. One study showed that nonsmoking wives of men who smoke are twice as likely to die of lung cancer as are nonsmoking wives of men who do not smoke. Some workers (bartenders, food servers, and musicians) involuntarily inhale the equivalent of 10 or more cigarettes per day leading to increased death rates. All of this new knowledge has put smokers on the defensive, as nonsmokers adopt slogans such as "Your right to smoke stops where my nose begins" and "If you smoke, don't exhale!"

Smoking, of course, has its defenders. Smokers feel that they have a right to indulge in their habit and would like nonsmokers to understand that smokers have a physical addiction that is quite difficult to break. Some smokers have become belligerent about the restrictions and are willing to take their rights into court. The newest strategy from the smoking lobby emphasizes smoking as an issue of personal freedom, like the right to use alcohol or caffeine. However, court challenges to the restrictions on smoking have not gone well for smokers. For example, a federal court upheld a ruling that prohibited a firefighter from smoking, whether on or off the job. The judges ruled that smoking was not protected by the constitutional right to privacy. Civil libertarians object to such restrictions on the basis that a person has a right to smoke; others counter with arguments that individual liberty can be limited if there is a rational purpose to doing so. In the court case, it was determined that firefighters must be in top condition and are subjected to hazards that smoking could aggravate. The right to clean air is in the public trust and interest, whereas the right to pollute and damage health is ill founded.

In the United States, nonsmokers are in the majority, and they seem to be gaining the upper hand in the controversy over public smoking. When will the tremendous damage to society from smoking outweigh consideration of the tobacco industry? What do you think?

than the environment. Other polls indicate that most Americans are ready to pay more for cleaner gasoline. Currently, pollution control costs are estimated at $125 billion annually, about one fourth of which represents efforts to reduce air pollution. The EPA estimates that, after being fully implemented in 2005, the new Clean Air Act will cost an additional $25 billion. Although this is an enormous sum, it represents less than 1% of the estimated $7 trillion economy for that year.

Some critics—especially economists—have charged that air pollution controls are not cost effective; that is, the benefits are not nearly as great as the costs. They see lost opportunities for economic growth and tend to disregard the *avoided costs*. This despite the fact that a recent analysis of the Clean Air Act conservatively found direct benefits totaling $6.8 trillion—that is a net benefit of $6.4 trillion to the society. See the "Earth Watch" box on the next page for further details about this cost-benefit analysis. In addition, pollution control is now a major industry. Providing over 2 million jobs, it is a significant part of the national economy.

Future Directions

In the early 1970s, many cities were experiencing more than 100 days per year of pollution in the "unhealthful" range or above; Los Angeles had about 300 such days per year. The original goal of air pollution control was to achieve "good" air quality on all days by 1975. This goal was not met; however, marked progress has occurred for five of the primary air pollutants (Fig. 15-12). Certainly, the quality of the air is better than if controls had not been initiated and the total of avoided cost is significant.

However, the backsliding of the last few years is troubling. From 1970 to 1990, pollution control devices and greater fuel efficiency reduced a typical car's emissions by 75%. As old, polluting cars were replaced by new, "cleaner" cars over this period, emission levels decreased despite increasing numbers of cars and miles driven. Today, however, with most cars being "cleaner," this reduction by replacement has leveled out, and the increasing number of cars and steady erosion of fuel economy are the dominant factors. The new Clean Air Act partially addresses this problem primarily by mandating increased emission controls and cleaner burning fuels. Whether these measures will be successful in further reducing air pollution remains to be seen.

A more general approach that would benefit the whole country would be to increase the fuel efficiency standard for cars (known as corporate average fuel economy, CAFE, standards), which is now locked at 27.5 miles per gallon (mpg). Intense lobbying by the automobile and petroleum industries prevented more stringent standards of fuel efficiencies from being included in the Clean Air Act of 1990; the fuel efficiency standards were actually rolled back in the 1980s. In 1995, the actual average fuel economy of all vehicles on U.S. highways was 24.5 mpg. The sports utility vehicles (SUV), classified as light trucks, feature average efficiencies of only 18 mpg! Available yet neglected technologies could bring the average mpg to 60 by the year 2000.

California, which by all measures has the greatest problem with photochemical smog and vehicular pollutants, has taken a completely different tack to reduce pollution from vehicles. Starting in 1998, California law will require that 2% of all vehicles sold in the state must be "emission free"—in other words, powered by electricity. The required percentage will rise to 10% by 2003. Similar laws have been enacted in nine northeastern states and the District of Columbia. In response to these developments, automakers are reluctantly devoting increasing resources to the development of affordable electric cars. Entrepreneurial firms such as U.S. Electricar of California and Solectria of Massachusetts are forging ahead of the big three automakers, and already have electric cars on the market (Fig. 15-19). These cars are considerably lighter than conventional cars and quite limited in range (traveling 100 to 150 miles before needing recharging). They also lack such amenities as air conditioning and other power accessories. Nevertheless, electrical cars may well be the wave of the future, especially for short trips at low speeds, which waste energy and cause much pollution. Of course, switching from gasoline to electrical power will transfer the site of pollution emission from the moving vehicle to the power plant (Chapter 21)—a trade-off with uncertain consequences unless strategies such as wind-generated electricity are employed.

It can be argued that all this emphasis on how vehicles burn fuel is just tinkering with the mechanics of the problem. To achieve lasting progress, we need to address the fundamental way we have organized society and industry. Consider that we pour billions of dollars into expanding and improving our highways, we build malls beside every major highway corridor, and we build homes in bedroom communities that are scores of miles from major employment centers; but we put rela-

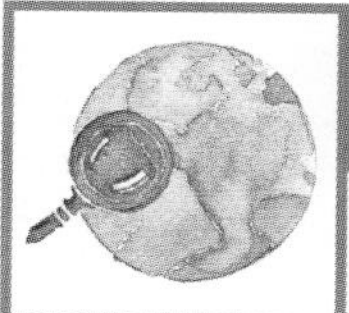

Earth Watch

Portland Takes a Right Turn

For many of us, the American life-style has come to include spending hours in the automobile each day going to and from work. Vehicle miles per year in the United States have increased much more rapidly than population, and too often the only response of state and city governments is to build more lanes on the expressways.

Portland, Oregon, had its share of expressways 20 years ago, but took a different tack when faced with the prospect of more and more commuters on the road. In response to an Oregon land-use law, Portland threw away its plans for more expressways and instead built a light rail system. This system now carries the equivalent of two lanes of traffic on all the roads feeding into downtown Portland. The result? Smoggy days have declined from 100 to 0 per year, the downtown area has added 30,000 jobs with no increase in automobile traffic, and Portland's economy has prospered.

This has clearly been a situation where everyone wins. The Portland solution seems so sensible that you have to ask why it is the exception and not the rule.

Metropolitan Area Express (MAX) in Portland, Oregon. This public transportation system has made a major contribution to the clean air and continued economic success of downtown Portland. (Courtesy Tri-Met, Portland, Oregon.)

FIGURE 15-19
The *Force*, manufactured by the Solectria Corporation, is one of many electric cars being developed for the U.S. market as demand increases for emission-free vehicles in pollution-impacted cities. (Courtesy Solectria Corporation, Wilmington, MA 01887.)

Earth Watch

The Clean Act Brings a Windfall

The Clean Air Act (1970, 1977, and 1990) has been the subject of open political warfare between those who think its cost has been too high for industry, taxpayers, labor, and consumers and those who think the health and environmental benefits were justified. Compliance affected patterns of industrial production, employment, and capital investment. Although these expenditures must be viewed as investments that generated benefits and opportunities, the dislocation in some regions was severe: such as reductions in high-sulfur coal mining and cutbacks in polluting industries such as steel. A need developed for a real cost-benefit analysis.

In 1990, Congress requested the Environmental Protection Agency to answer the question: How do the overall health, welfare, ecological and economic benefits of Clean Air Act programs compare with the costs of these programs: In response, the EPA performed the most exhaustive cost-benefit analysis of public policy ever attempted. Here is what the EPA reported in a 1996 study:

- The total direct cost to implement the Clean Air Act for all federal, state, and local rules from 1970 to 1990 was $436 billion (in 1990-value dollar). This cost was borne by businesses, consumers, and government entities in the form of higher prices for many goods, services, and some utilities.
- The mean estimate of direct benefits from the Clean Air Act from 1970 to 1990 was $6.8 trillion.
- Therefore, the net benefit of the Clean Air Act has been $6.4 trillion!

"The finding is overwhelming. The benefits far exceed the costs of the Clean Air Act in the first 20 years," said Richard Morgenstern, Associate Administrator for Policy Planning and Evaluation at the EPA. Further, the report states that "all benefits may be significantly underestimated due to the exclusion of large numbers of benefits from the monetized benefit estimate."

The benefits to society, directly and indirectly, have been widespread across the entire population. Here is a summary:

- Reduced air pollution (described in this chapter).
- Improved human health: each year, 79,000 lives saved, 18,000 fewer heart attacks, 10,000 fewer strokes, 13,000 fewer hypertension cases, and 15 million fewer respiratory cases.
- "Avoided cost": improved health has meant less debilitating disease, hospitalization, special care, and medicines.
- Less lead to harm children: in 1990, 220,000 tons of lead were not burned in gasoline because of Clean Air Act measures. Exposure to lead impairs the cognitive development of children; therefore the huge reductions in lead produced a benefit of retained IQ and the possibility of a more productive, less dependent life.
- Lower cancer rates.
- Less acid deposition.

The EPA study result should feel refreshing in our study of the environment. Society knew what to do, took action despite disruptive efforts by special interest and political partisans, and reaped about $16 in benefit for every $1 invested to control air pollution.

(Source: Adapted from R. Christopherson, *Geosystems, an Introduction to Physical Geography,* 3rd ed., by permission of Prentice Hall, Inc., ©1997, pp. 646–48.)

tively little money into improving mass transportation systems so dominant in the first half of the century.

It makes great sense, in the coming years, to fashion legislation and policies that would guide new development and redevelopment of depreciated urban areas so as to reduce average commuting distances and times and facilitate greater use of mass transit systems. At the same time, we can make development of efficient mass transit systems and fuel-efficient vehicles matters of national priority especially given the reserves of fossil fuels that will last only a few decades.

The President's Council on Sustainable Development in its 1996 report said, "the federal government should redirect policies that encourage low-density sprawl to foster investment in existing communities. It should encourage shifts in transportation toward transit and expansion of transit options rather than new highway construction."

Review Questions

1. Describe the basic structure of the atmosphere. What natural cleansing processes take place there?
2. Describe the origin of industrial and photochemical smog.
3. Name the air pollutants considered most widespread and serious.
4. What impact does air pollution have on human health?
5. Describe the negative effects of pollutants on crops, forests, and other materials. Which pollutants are mainly responsible for these effects?
6. How are primary and secondary pollutants formed?
7. Describe how the Clean Air Act of 1970 attempted to control air pollution.
8. Discuss ways in which the Clean Air Act of 1990 addresses the failures of former legislation.
9. Name the major sources of indoor air pollution. Discuss the legislative regulations that address them.
10. What further steps could be taken to improve the quality of our air?

Thinking Environmentally

1. Motor vehicles release close to half the pollutants that dirty our air. What alternatives might be introduced to encourage a decrease in our use of automobiles and other vehicles with internal combustion engines?
2. Lead paint, asbestos, tobacco, and many other products have been linked to adverse effects on human health. Research one such case that has been brought into the courts. Describe the alleged injustice and the proceedings and outcome of the trial. Do you agree with the decision? Why or why not?
3. If you were the surgeon general of the United States, would you ban smoking in all public buildings? Would such an action limit freedoms guaranteed by the constitution? Defend your answer.
4. How might traditional patterns of life change if cleaner air is not achieved? Write a short essay describing the possibilities.
5. Has radon testing been conducted in buildings on your campus? If so, find out the results. If not, consider asking the administration to authorize this testing.

Major Atmospheric Changes

Key Issues and Questions

1. The atmosphere in many regions of the world contains acidic particles and solutes, which get deposited on land and water. What is acid deposition, and where does it come from?
2. Acid deposition impacts natural ecosystems as well as industrial centers. What are its most significant effects on the environment?
3. Coping with acid deposition requires both technological and political change. What is being done in the United States to reduce acid deposition?
4. The interaction of solar radiation with atmospheric gases controls the balance of warming and cooling of Earth. How have atmospheric gases from human activities affected this balance?
5. Greenhouse gases are increasing in the troposphere. What is the probability that these gases will bring on global warming? If it occurs, what will be the major effects of global warming?
6. Global warming poses a serious threat to the world's climate. How have the countries of the world responded to the threat?
7. The stratospheric ozone shield is vital in protecting life from damaging ultraviolet radiation. How is ozone formed and destroyed in the stratosphere?
8. Chlorofluorocarbons (CFCs) and other gases destroy stratospheric ozone. What evidence confirmed these destructive processes?
9. Continued ozone destruction represents a threat to life on Earth. What is the global community doing about slowing or reversing ozone loss?

Thousands of lakes on our planet are lifeless. Tens of thousands more are threatened. Forests are dying back. The overall climate shows signs of warming, threatening the world with unprecedented climatic shifts and a rising sea level. We are warned of skin cancer because of increased exposure to ultraviolet radiation. Scientists and the media tell us that these problems are the consequences of air pollution and that the problems will get worse unless we take drastic action. This is not good news. In fact, it is such bad news that many refuse to believe it, listening instead to those who insist that civilization is robust enough to adapt to global change, or that problems of climate change are simply inventions of scientists and liberals or part of some international, new age conspiracy.

[Photo by NASA/Photo Library.]

In this chapter, we will examine three phenomena associated with major atmospheric changes: acid deposition, global warming, and statospheric ozone depletion. We will find problems in both the lower atmosphere (the troposphere) and the upper atmosphere (the stratosphere). We described the basic structure of the atmosphere in Chapter 15. (See especially Fig. 15-1.) For convenience, the characteristics of the troposphere and stratosphere are summarized in Table 16-1. The current chapter presents likely scenarios involving the three major atmospheric changes, based on the latest scientific information.

Acid deposition, global warming, and stratospheric ozone depletion have been identified by scientists as **anthropogenic**, meaning that they come from human activities. They are results of the energy-generation systems that deliver important services to modern societies. Economic well-being in the developed nations and economic progress in the developing nations are dependent on the extraction of energy from fossil fuels (coal, liquid fuels derived from crude oil, and natural gas). But as these fuels are burned and the energy from them is put to work moving vehicles, turning turbines, powering motors, and running air conditioners, pollutants are released to the air as by-products. Climatologists and atmospheric scientists tell us that air pollution from the human system has already damaged soils, forests, and lakes downwind of heavy pollution sources, and may now be altering the entire biosphere. With a future that is certain to see continued growth of the human population and greater economic development in the poorer countries, the potential for air pollution to damage the biosphere is great. By all measures, acid deposition, global warming, and stratospheric ozone depletion represent unsustainable impacts on the environment. The greatest challenge facing the world may be the challenge of protecting our atmosphere while also achieving the economic benefits that people value.

Acid Deposition

Acid precipitation refers to any precipitation—rain, fog, mist, or snow—that is more acidic than usual. Because dry acidic particles are also found in the atmosphere, the combination of precipitation and dry particle fallout is called **acid deposition**. Careful monitoring has shown that broad areas of North America, as well as most of Europe and other industrialized regions of the world, are regularly experiencing precipitation that is between 10 and 1000 times more acidic than usual. This is affecting ecosystems in diverse ways, as illustrated in Fig. 16-1.

To understand the full extent of the problem, first we must understand some principles about acids and how we measure their concentration.

Acids and Bases

Acidic properties (sour taste, corrosiveness) are due to the presence of hydrogen ions (H^+, a hydrogen atom without its electron), which are highly reactive. Therefore, an **acid** is any chemical that releases hydrogen ions when dissolved in water. The chemical formulas of a few common acids are shown in Table 16-2. Note that all of them ionize—that is, their components separate—to give hydrogen ions plus a negative ion. The higher the concentration of hydrogen ions in a solution, the more acidic the solution is.

A **base** is any chemical that releases hydroxide ions when dissolved in water. (See Table 16-2.) The bitter taste and caustic properties of all alkaline, or basic, solutions are due to the presence of hydroxide ions (OH^-, oxygen-hydrogen groups with an extra electron).

The concentration of hydrogen ions is expressed as **pH**. The pH scale goes from 0 (highly acidic) through 7 (neutral) to 14 (highly basic) (Fig. 16-2). Each of the numbers on the scale represents the negative logarithm (power of 10) of the hydrogen ion concentration, expressed in grams per liter. For example, to say that a solution has a pH of 1 means that the concentration of hydrogen ions in the solution is 10^{-1} g/L (0.1 g/L); pH = 2 means that the hydrogen ion concentration is 10^{-2} g/L, and so on.

At pH = 7, the hydrogen ion concentration is 10^{-7} (0.0000001) g/L, and the hydroxide ion concentration is also 10^{-7} g/L. This is the neutral point, where small but equal amounts of hydrogen ions and hydroxide ions are present in pure water. The pH numbers above 7 continue to express the negative exponent of hydrogen ion concentration, but they also represent an increase in hydroxide ion concentration. Because solutions of pH greater than 7 contain higher concentrations of hydroxide ions than of hydrogen ions, they are called basic solutions.

We use the same negative logarithm arrangement to express the concentration of hydroxide ions: **pOH**. There is a reciprocal relationship between pH and pOH: As the pH of a given solution goes up, its pOH goes down, and vice versa. In fact, we can state this as

TABLE 16-1
Characteristics of Troposphere and Stratosphere

Troposphere	*Stratosphere*
Extent: Ground level to 11 miles (18 km)	Extent: 11 mi to 30 mi (18 km to 50 km)
Temperature normally decreases with altitude to -70°F (-59°C)	Temperature increases with altitude to +32°F (0°C)
Much vertical mixing	Little vertical mixing, slow diffusion exchange of gases with troposphere
Substances entering may be washed back to Earth	Substances entering remain unless attacked by sunlight or other chemicals
All weather and climate take place	Isolated from the troposphere by the tropopause

a rule: For any aqueous solution, pH + pOH = 14. For example, pH = 13 means that the hydrogen ion concentration is 10^{-13} (a decimal followed by 12 zeros and a 1) g/L. However, the hydroxide ion concentration of this same solution is 10^{-1} g/L, so the pOH = 1. In a neutral solution, equal amounts of hydrogen ions and hydroxide ions are present; thus, pH = 7 and pOH = 7; and pH + pOH = 14.

Since numbers on the pH scale represent powers of ten, there is a *tenfold difference* between each unit

FIGURE 16-1
Acid deposition. Emissions of sulfur dioxide and nitrogen oxides react with the hydroxyl radicals and water vapor in the atmosphere to form their respective acids, which come back down either as dry acid deposition or, mixed with water, acid precipitation. Various effects of acid deposition are noted.

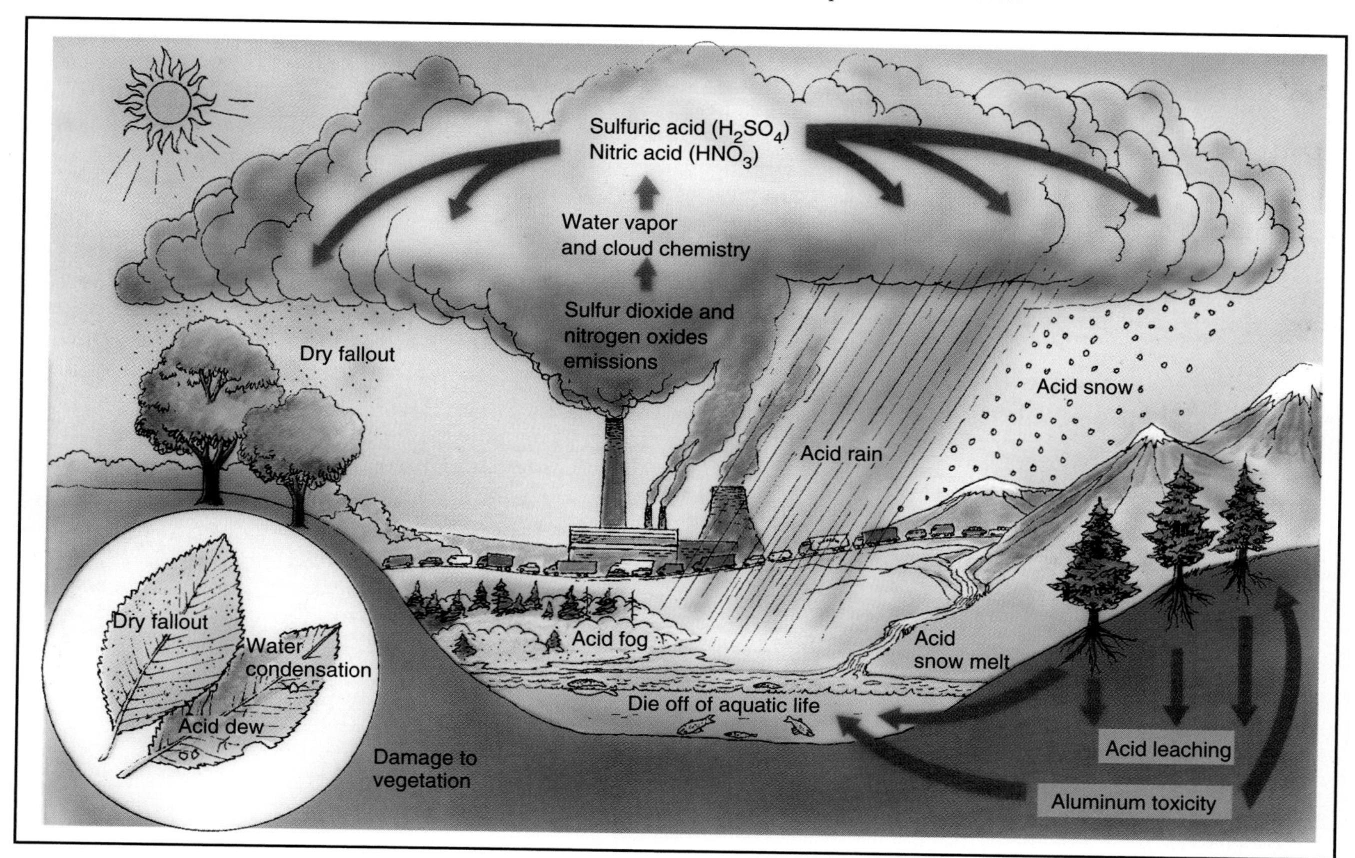

TABLE 16-2
Common Acids and Bases

Acid	*Formula*	*Yields*	*H^+ Ion(s)*	*Plus*	*Negative Ion*	
Hydrochloric acid	HCl	⟶	H^+	+	Cl^-	Chloride
Sulfuric acid	H_2SO_4	⟶	$2H^+$	+	SO_4^{2-}	Sulfate
Nitric acid	HNO_3	⟶	H^+	+	NO_3^-	Nitrate
Phosphoric acid	H_3PO_4	⟶	$3H^+$	+	PO_4^{3-}	Phosphate
Acetic acid	CH_3COOH	⟶	H^+	+	CH_3COO^-	Acetate
Carbonic acid	H_2CO_3	⟶	H^+	+	HCO_3^-	Bicarbonate
Base	***Formula***	***Yields***	***OH^- Ion(s)***	***Plus***	***Positive Ion***	
Sodium hydroxide	$NaOH$	⟶	OH^-	+	Na^+	Sodium ion
Potassium hydroxide	KOH	⟶	OH^-	+	K^+	Potassium ion
Calcium hydroxide	$Ca(OH)_2$	⟶	$2OH^-$	+	Ca^{2+}	Calcium ion
Ammonium hydroxide	NH_4OH	⟶	OH^-	+	NH_4^+	Ammonium

and the next. For example, pH 5 is ten times as acidic (has ten times as many H^+ ions) as pH 6, pH 4 is ten times as acidic as pH 5, and so on.

One easy way to measure pH is with indicator paper, which is available from any laboratory supply house. This paper contains pigments that change color when they are wetted with an acidic or a basic solution. The pH is determined by dipping a strip of indicator paper in the solution and matching the color of the wet paper with a color chart provided with the paper. Where accuracy and precision are important, however, it is necessary to use electronic instruments for measuring pH.

Extent and Potency of Acid Precipitation

In the absence of any pollution, rainfall is normally slightly acidic, with a pH of 5.6, because carbon dioxide in the air readily dissolves in and combines with water to produce an acid called carbonic acid. **Acid precipitation**, then, is any precipitation with a pH of 5.5 or less.

Unfortunately, acid precipitation is now the norm over most of the industrialized world. As Fig. 16-3 shows, for instance, the pH of rain and snowfall over a large portion of eastern North America is typically about 4.5. Many areas in this region regularly receive precipitation having a pH of 4.0 and, occasionally, as low as 3.0. Fogs and dews can be even more acidic; in mountain forests east of Los Angeles, scientists found fog water of pH 2.8—almost 1000 times more acidic than usual—dripping from pine needles. Acid precipitation has been heavy in Europe, from the British Isles to central Russia. Also it is now well documented in Japan.

Sources of Acid Deposition

Chemical analysis of acid precipitation in eastern North America and Europe reveals the presence of two acids, sulfuric acid (H_2SO_4) and nitric acid (HNO_3), in a ratio of about two to one. (In the western states and provinces, nitric acid predominates, principally formed from tailpipe emission.) As you learned in Chapter 15, burning fuels produce sulfur dioxide and nitrogen oxides, and so the source of the acid deposition problem begins to emerge. These oxides enter the troposphere in large quantities from both anthropogenic and natural sources. Once in the troposphere, they are oxidized by hydroxyl radicals (see Fig. 15-2) to sulfuric and nitric acids, which dissolve readily in water or adsorb on particles and are brought down to Earth in acid deposition. This usually occurs within a week of

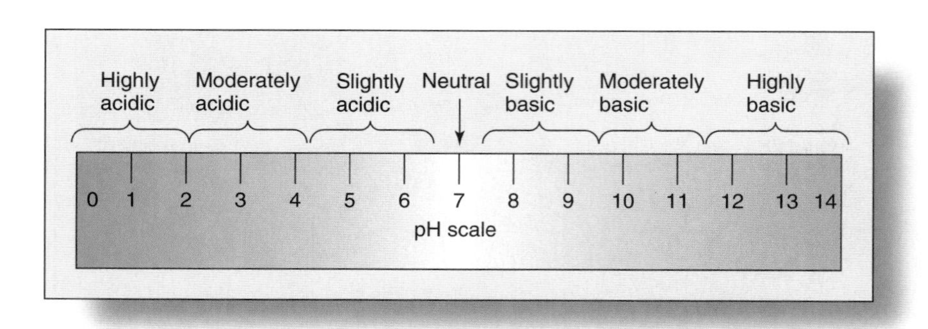

FIGURE 16-2
The pH scale.

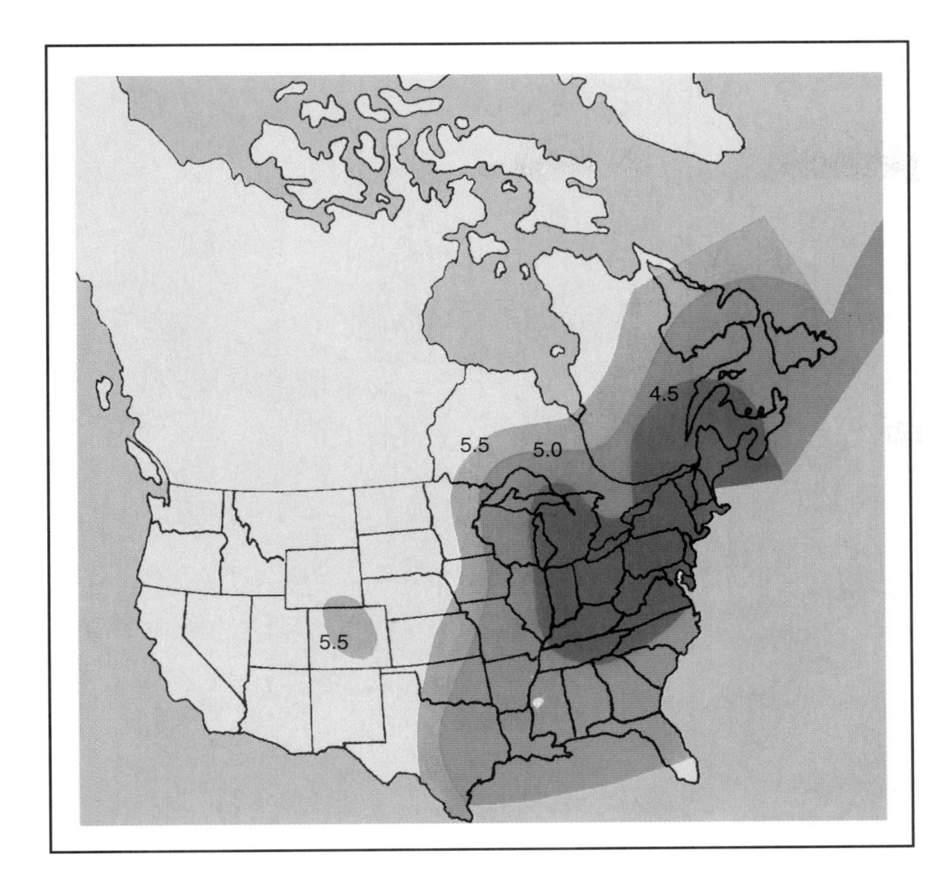

FIGURE 16-3
Regions receiving acid deposition. Monitoring the pH of deposition now reveals that acid deposition is occurring over most of the eastern United States and Canada. It is especially severe in the northeastern United States and eastern Canada. (Ontario Ministry of the Environment)

the oxides' entering the atmosphere.

We must recognize that *natural sources* contribute substantial quantities of pollutants to the air: 50 to 70 million tons per year of sulfur (from volcanoes, sea spray, and microbial processes) and 30 to 40 million tons per year of nitrogen oxides (from lightning, burning of biomass, and microbial processes). *Anthropogenic sources* are estimated at 100 to 130 million tons per year of sulfur dioxide and 60 to 70 million tons per year of nitrogen oxides. The vital difference between these two sources is that anthropogenic oxides are strongly concentrated in industrialized regions, whereas the emissions from natural sources are spread out over the globe and are a part of the global environment. Levels of the anthropogenic oxides have increased at least fourfold since 1900, while levels of the natural emissions have remained fairly constant.

As Fig. 15-11 indicates, 18.3 million tons of sulfur dioxide are released into the air annually in the United States, 88% of which are from fuel combustion (mostly from coal-burning power plants). Some 21.8 million tons of nitrogen oxides are released annually, 45% traced to transportation emissions and 50% to fuel combustion at fixed sites. For the eastern U.S., the source of over 50% of the acid deposition has been identified as the tall stacks of 50 huge coal-burning power plants (Fig. 16-4). Recall from Chapter 15 that these tall stacks were built to alleviate local sulfur dioxide pollution at ground level. The unfortunate result of the "dilution solution" is that emitting sulfur dioxide and nitrogen oxides high in the air simply provides more opportunity for them to spread hundreds of miles from their source.

Effects of Acid Deposition

Acid deposition has been recognized as a problem in and around industrial centers for over 100 years. Its impact on ecosystems, however, was noted only about 35 years ago, when anglers started noticing sharp declines in fish populations in many lakes in Sweden, Ontario, and the Adirondack Mountains of upper New York State. Scientists in Sweden were the first to identify the cause as increased acidity of the lake water and to link this increased acidity with deposition having an abnormally low pH. Since that time, as ecological damage has continued to spread, studies have revealed many ways in which acid deposition alters and may destroy ecosystems.

Impact on Aquatic Ecosystems. The pH of an environment is extremely critical because it affects the functioning of virtually all enzymes, hormones, and other proteins in the bodies of all organisms living in that

(a) (b)

FIGURE 16-4
(a) Standard smokestacks of this coal-burning power plant were replaced by new 1000-foot (330-m) stacks to aid in the dispersion of pollutants into the atmosphere. The taller stacks alleviate local air pollution problems, but create more widespread distribution of acid-generating pollutants. (b) Locations of the 50 largest sulfur dioxide emitters, all of which are utility coal-burning power plants. These facilities account for over 50% of the acid deposition falling on the eastern United States. (Grapes-Michaud/Photo Researchers, Inc.)

environment. Ordinarily, organisms are able to regulate their internal pH; however, consistently low environmental pH often overcomes this regulatory ability in many life-forms. Most freshwater lakes, ponds, and streams have a natural pH in the range of 6 to 8, and organisms are adapted accordingly. The eggs, sperm, and developing young of these organisms are especially sensitive to changes in pH. Most are severely stressed, and many die, if the environmental pH shifts as little as one unit from the optimum.

As such aquatic ecosystems are acidified, there is a rapid die off of virtually all higher organisms, either because the acidified water kills them or because it keeps them from reproducing. Figure 16-1 illustrates the fact that acid precipitation may leach aluminum and various heavy metals from the soil as the water percolates through it. Normally, the presence of these elements in the soil does not pose a problem, because they are bound in insoluble mineral compounds and therefore are not absorbed by organisms. As these compounds are dissolved by low-pH water, however, the metals are freed; they may then be absorbed and are highly toxic to both plants and animals. For example, mercury tends to accumulate in fish as lake waters become more acidic. Mercury levels are so high in the Great Lakes that many bordering states advise against eating fish caught in these waters.

In Norway and Sweden, the fish have died in at least 6500 lakes and 7 Atlantic salmon rivers. In Ontario, Canada, approximately 1200 lakes now harbor no life. In the Adirondacks, a favorite recreational region for New Yorkers, more than 200 lakes are without fish, and many are devoid of all life save resistant algae and bacteria. More than 190 New England lakes and rivers are suffering from recent acidification. The physical appearance of such lakes is deceiving. From the surface they are clear and blue, the outward signs of a healthy condition. However, a view under the surface is eerie: In spite of ample light shimmering through the clear water, there is not a sign of life in them.

During the 1980s, the EPA conducted the National Acid Precipitation Assessment Program (NAPAP), at a

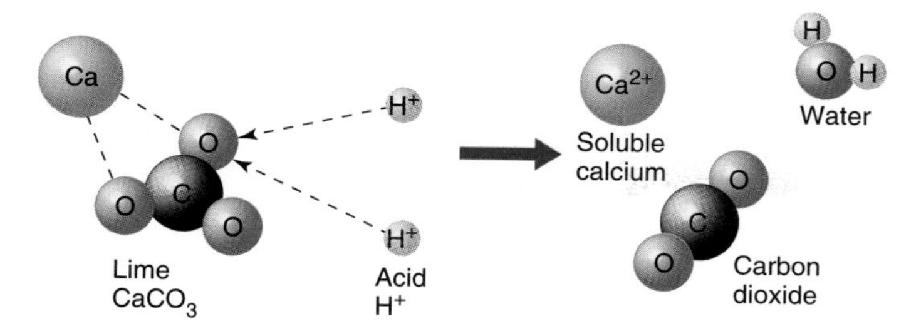

FIGURE 16-5
Buffering. Acids may be neutralized by certain nonbasic compounds called *buffers*. A buffer such as limestone (calcium carbonate) reacts with hydrogen ions as shown. Hence, the pH remains close to neutral despite the additional acid. Note, however, that the buffer is consumed by the acid. Limestone is the most widespread natural buffer.

cost of $600 million. Part of this program involved the National Surface Water Survey (NSWS), in which samples were collected from 2311 lakes (representative of a total of 28,300 lakes) and 500 streams (representative of 64,300 streams). Extrapolation from the survey results showed that acid precipitation had acidified an estimated 1180 lakes and 4670 streams, mainly in the Adirondacks, the mid-Atlantic states, New England, northern Florida, and the upper Midwest.

As Fig. 16-3 indicates, wide regions of the United States receive roughly equal amounts of acid precipitation, and yet not all areas have acidified lakes. Apparently, many areas remain healthy, whereas others have become acidified to the point of becoming lifeless. How is this possible? The key lies in the system's *buffering capacity.* Despite the addition of acid to it a system may be protected from pH change by a **buffer**—a substance that, when present in a solution, has a large capacity to absorb hydrogen ions and thus hold the pH relatively constant.

Limestone ($CaCO_3$) is a natural buffer (Fig. 16-5); its presence protects lakes from the effects of acid precipitation in many areas of the North American continent. Lakes and streams receive their water from rain and melted snow percolating through soils in their drainage basins, and if the soils are in regions of limestone rock, the lakes will contain dissolved limestone. The regions found in the NSWS survey to be sensitive to acid precipitation contain much granitic rock, which does not yield any chemical compounds that are good buffers. For these areas, the most critical time of year is the spring thaw, when accumulated winter snow melts. If the thaw is sudden, streams, rivers, and lakes are hit with what has been called "acid shock," as accumulated acids send pH levels plummeting in a sudden burst of meltwater. Making matters worse, the acid shock often coincides with the times of spawning and egg laying in aquatic animals, when they are at their most vulnerable.

Any buffer has limited capacity. Limestone, for instance, is used up by the buffering reaction and so is no longer available to react with more added hydrogen ions (Fig. 16-5). Ecosystems that have already acidified and collapsed are those which had very little buffering capacity. Those which remain healthy have greater buffering capacity. However, many of these still healthy lakes are gradually losing their buffering capacity as more and more acid continues to be deposited. In time, many healthy lakes will join the growing number of dead lakes if acid precipitation is not curbed.

Impact on Forests. Along with dying lakes, the decline of forests has been conspicuous. From the Green Mountains of Vermont to the San Bernardino Mountains of California, the die off of forest trees in the 1980s caused great concern. Red spruce forests are especially vulnerable; in New England, 1.3 million acres of high-elevation forests have been devastated. Commonly, the damaged trees lost needles and often fell prey to insects and diseases before they succumbed.

Much of the damage of acid precipitation to forests is due to chemical interactions within the forest soils. Sustained acid precipitation at first adds nitrogen and sulfur to the soils, which stimulates tree growth. In time, these chemicals impact the soils by leaching out large quantities of the buffering chemicals—usually calcium and magnesium salts. When the acid rain is no longer neutralized by these buffering salts, aluminum ions are then dissolved from minerals in the soil. The combination of aluminum—which is toxic—and the increasing scarcity of calcium—essential to plant growth—then leads to reduced tree growth. Research at the Hubbard Brook Experimental Forest in the White Mountains of New Hampshire has shown a marked reduction in calcium and magnesium in the forest soils from the 1960s on, which is reflected in the calcium of stemwood tree-rings over the same time period. The net result of these changes has been a severe decline in forest growth.

Declining forests are a serious problem in some parts of Europe, and the evidence indicates that the same kinds of soil chemical exchanges are occurring there. Because of the variations in the buffering capacity of soils, and the differing amounts of sulfur and nitrogen brought in by acid precipitation, forests are affected to varying degrees. Some forests continue to grow, whereas others are in decline. One result of the sustained acidification is sometimes a gradual shift toward more acid-tolerant species. In New England, the balsam fir is moving in to replace the dead spruce.

Impact on Humans and Their Artifacts. One of the more noticeable effects of acid deposition is the deterioration of artifacts. Limestone and marble (which is a form of limestone) are favored materials for the outsides of buildings and for monuments (collectively called **artifacts**). The reaction between acid and limestone is causing these structures to erode at a tremendously accelerated pace. Monuments and buildings that have stood for hundreds or even thousands of years with little change are now dissolving and crumbling away, as was shown in Fig. 15-10. The corrosion of buildings, monuments, and outdoor equipment by acid precipitation costs billions of dollars for replacement and repair each year in the United States. A 37-foot bronze Buddha in Kamakura, Japan, is slowly dissolving away as acid rain from Korea and China bathes the statue in rain that is more acidic than that recorded in the U.S.

Whereas the decay of such artifacts is a tragic loss in itself, it should also stand as a grim reminder of how we are dissolving away the buffering capacity of ecosystems. In addition, some officials are concerned that acid precipitation's mobilization of aluminum and other toxic elements may result in the contamination of both surface water and groundwater. Increased acidity in water also mobilizes lead from the pipes used in some old plumbing systems and from the solder used in modern copper systems.

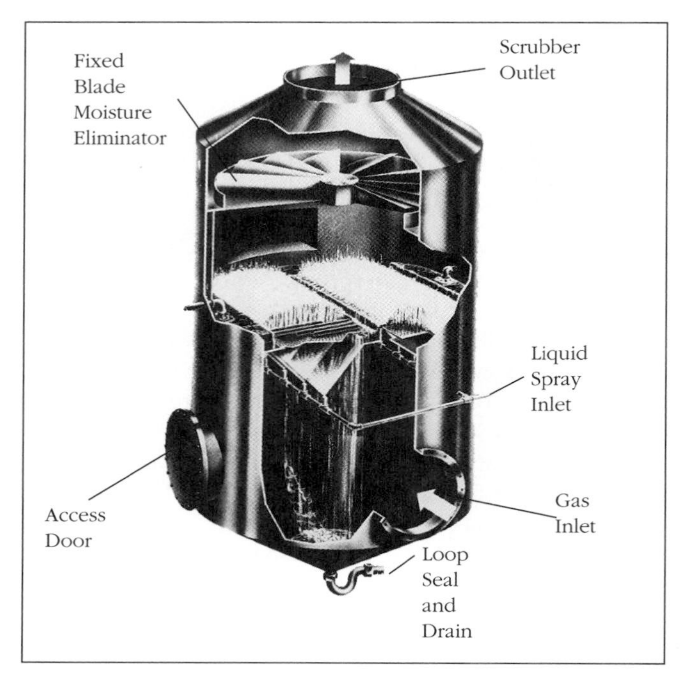

FIGURE 16-6
Scrubbers. Sulfur dioxide may be removed from flue gases by passing the flue gas through either a spray of lime slurry in a venturi scrubber or a sodium hydroxide or sodium carbonate solution in an impingment scrubber. The sulfur dioxide reacts with the alkali in the scrubbing water to form sulfite and sulfate salts. The device shown here is a cutaway view of the SLY IMPINJET gas scrubber. (Courtesy of Sly Incorporated, Cleveland, OH.)

Coping with Acid Deposition

As the song says, "What goes up must come down." The sulfur and nitrogen oxides pumped into the troposphere at the rate of 40.1 million tons per year in the United States come down as acid deposition, generally to the east of their origin because of the way weather systems flow. The deposits cross national boundaries: Canada receives half of its acid deposition from the United States and Scandinavia gets it mostly from Great Britain and other western European nations. Japan receives the windborne pollution from widespread coal burning in Korea and China. There is now a broad scientific consensus that the problem of acid deposition must be addressed at national and international levels.

Ways to Reduce Acid-Forming Emissions. Scientists have calculated that a 50% reduction in current acid-causing emissions in the United States would effectively prevent further acidification of the environment. This reduction may not correct the already bad situations, but, together with natural buffering processes, it is estimated to be capable of preventing further environmental deterioration. Because we know that about 50% of acid-producing emissions come from the tall stacks of coal-burning plants that generate electricity, control strategies are centered on these sources. Six main strategies have been proposed: (1) coal washing, (2) fluidized bed combustion, (3) fuel switching, (4) scrubbers, (5) alternative power plants, and (6) reductions in the consumption of electricity.

Economic factors do not favor the first two strategies. Coal washing to remove sulfur is costly, both economically and environmentally; large amounts of polluted water would be the outcome. In fluidized bed combustion, coal is burned in a mixture of sand and lime, churned by air forced in from underneath. The sulfur in the coal combines with the lime during combustion and is removed with the ash that is formed. When new plants are built, this may be the preferred method of emissions control, but installing the technology in existing plants would entail tearing down and rebuilding a major portion of each plant.

Oil is more expensive than coal, and switching to oil as a fuel for generating electricity would increase the dependence of the United States on foreign oil. There is, however, low-sulfur coal available in Central Appalachia and the western United States, and switching to this source represents an attractive strategy for power plants seeking to reduce their sulfur dioxide emissions without changing their technology, in spite of the higher cost of the coal and its transportation.

Scrubbers are "liquid filters" that put exhaust fumes through a spray of water containing lime (Fig. 16-6). The sulfur dioxide reacts with the lime and is precipitated as calcium sulfate ($CaSO_4$). Scrubbers have been required for all major power plants and smelters built in the United States since 1977, but the law has not required retrofitting the large, older plants in the Midwest that are the source of much of the acid rain plaguing the East and Canada. Plants in Europe and Canada and a few in the United States have demonstrated that scrubbers can be added to existing power plants relatively quickly; the scrubbers are highly effective in controlling emissions and are not prohibitively expensive.

The last two strategies in the foregoing list require major shifts in the energy policy of the United States. Currently, nuclear power represents the only alternative technology for generating electricity. It accounts for 21% of electrical power generation in the United States. However, as Chapter 22 explains, the future of nuclear energy is in serious question because of concerns about safety and the storage of nuclear waste. Reducing the consumption of electricity (conservation) makes very good sense and has already been implemented in some applications, such as appliances and insulation. Conservation and energy-efficiency strategies represent a relatively untapped potential for future reductions in air pollution, although putting it to work will involve major changes in many sectors of U.S. economic and social life, as explained in Chapter 21.

Political Developments. Although evidence of the link between power-plant emissions and acid deposition was well established by the early 1980s, no legislative action was taken until 1990. In fact, political moves blocked the publication of NAPAP annual reports for five years beginning in 1985. The problem was one of different regional interests. Western Pennsylvania and the states of the Ohio River valley, where older coal-burning power plants produce most of the electrical power, argued that controlling their sulfur dioxide emissions would make electricity in the region unaffordable. Throughout the 1980s, a coalition of politicians from these states, fossil-fuel corporation representatives of high-sulfur coal producers, and representatives from the electric power industry effectively blocked all attempts at passing legislation that would take action on acid deposition.

On the other side of the question were New York and New England, as well as most of the environmental and scientific community, which argued that it was both possible and necessary to address acid deposition and that the best way to do it was to control sulfur dioxide emissions. Since 70% of Canada's problem is from the United States they applied diplomatic pressure toward a resolution.

Title IV of the Clean Air Act of 1990. The outcome of the controversy is now history. However, we have lost two decades of action because of the political delays. **Title IV** of this act is the first law in our history to address the acid deposition problem, by mandating reductions in both sulfur dioxide and nitrogen oxide levels. The major provisions of Title IV are as follows:

1. By the year 2000, total sulfur dioxide emissions must be reduced 10 million tons below 1980 levels. This is the 50% reduction called for by scientists, and it involves the setting of a permanent cap of 8.9 million tons on such emissions.
2. The reduction will be implemented gradually. By 2000, the combined sulfur dioxide output of all the country's power plants must not exceed the permanent cap of 8.9 million tons. The utilities are required to install equipment that closely monitors their emissions of acid-generating gases.
3. In a major departure from the command-and-control approach described in Chapter 15, Title IV authorizes the EPA to use a free-market approach to regulation. Each plant is granted **emission allowances** based on formulas in the legislation. The penalty for exceeding these allowances is severe: a fee of $2000 per ton and the requirement that the utility must compensate for the excess emissions the next year. A utility may not emit more sulfur dioxide than allowances permit, but if it emits less, the difference in allowance may be sold, so that another utility may purchase the first plant's difference in place of reducing its own emissions.
4. In the future, new utilities will not receive allowances; they must buy into the system by purchasing existing allowances. Thus, there will be a finite number of allowances in existence.
5. Nitrogen oxide emissions must be reduced by 2 million tons by the year 2000. This is to be accomplished by regulating the boilers used by the utilities and by mandating the continuous monitoring of emissions.

Industry's Response to Title IV. The utilities industry has responded to the new law with three actions: (1) Many of the utilities are switching to low-sulfur coal. This will enable the targeted high-capacity utilities to achieve compliance rapidly. (2) At least 15 power plants are adding scrubbers during the first few years of compliance efforts. Many more will be forced to scrub in order to reach final compliance by 2000. Since the technology for high-efficiency scrubbing is well established, many plants will upgrade their existing scrubbers. (3) Many of the utilities are already trading in emissions allowances. At current prices, the purchase of allowances often represents a less costly way to achieve compliance than purchasing low-sulfur coal or adding a scrubber. It is too soon to know how effective the market approach will be in reducing acid deposition, but it is already helping some of the utilities keep down their costs while continuing to emit sulfur in accordance with their historic pattern. For example, the Illinois Power Company decided not to construct a $350 million scrubber because the company found that purchasing allowances to enable it to match its current emissions was less expensive.

In the wake of the passage of the Clean Air Act of 1990, Canada and the United States have signed a treaty providing that Canada will cut its sulfur dioxide emissions by half and cap them at 3.2 million tons by the year 2000. The ultimate beneficiaries will be the remaining healthy aquatic and forested ecosystems, which will now be protected from future acid deposition. In addition, it is hoped that ecosystems already harmed will be able to recover from current damage and that we will be back on the road to a sustainable interaction with the atmosphere.

Global Warming

Geochemist James W. C. White, writing in the July 15, 1993 issue of the journal *Nature*, said it well: "If the earth came with an operating manual, the chapter on climate might begin with a caveat that the system has been adjusted at the factory for optimum comfort, so don't touch the dials." Many scientists are convinced that we are madly twirling the dials of Earth's "operating system" as we release greenhouse gases into the atmosphere and put the planet in danger of unprecedented warming. Others ask for clear evidence that warming is happening before taking what will surely be costly steps to prevent it. Still others claim that we are just as likely to be on the verge of a new ice age and point to our lack of understanding of the basic forces that control climate.

Are we now on the threshold of a major human-caused change in the climate? Let us look at the evidence.

The Earth as a Greenhouse

Warming Processes. You are familiar with the way the interior of a car heats up when the car is sitting in the sun with the windows closed. This heating occurs because sunlight comes in through the windows and is absorbed by the seats and other interior objects. In being absorbed by the objects, the light energy is converted to heat energy, which is given off in the form of infrared radiation. Unlike sunlight, infrared radiation is blocked by glass and so cannot leave the car. The heat energy thus trapped causes the interior air temperature to rise. This is the same phenomenon that keeps a greenhouse warmer than the surrounding environment.

On a global scale, carbon dioxide, water vapor, and other trace gases in the atmosphere play a role analogous to the glass in a greenhouse. Therefore, they are called **greenhouse gases**. Light energy comes through the atmosphere and is absorbed by Earth and converted to heat energy at the planet's surface. The infrared heat energy radiates back upward through the atmosphere and into space. The greenhouse gases naturally present in the troposphere absorb some of the infrared radiation and reradiate it back toward the surface; other gases (N_2 and O_2) in the troposphere do not. The greenhouse gases are like a heat blanket, insulating Earth and delaying the loss of infrared to space (Fig. 16-7). Without this insulation, average surface temperatures on Earth would be 33 C° colder, and life as we know it would be impossible.

Our global climate is dependent on Earth's concentrations of greenhouse gases. If these concentrations increase or decrease significantly, our climate will change accordingly. Indeed, geological evidence indicates that Earth has undergone major climatic changes, fluctuating between ice ages (glacials) and warmings (interglacials) over tens of thousands to hundreds of thousands of years. During the height of the most recent ice age, 18,000 years ago, global temperatures were 3–5 C° colder than they are today. An analysis of gas bubbles trapped in ice from glaciers laid down during this ice age indicates that carbon dioxide concentrations then were 60% lower than those of today—an indication of cause and effect.

Global Cooling. Earth's atmosphere is also subject to cooling factors. For example, on average clouds cover 50% of Earth's surface and reflect some 21% of solar radiation away to space. This reflection of sunlight is called the **planetary albedo**, and it contributes to overall cooling. (More accurately, it prevents warming from occurring.) The effect can be readily appreciated if you think of how you are comforted on a hot day when a large cloud passes over the sun. The radiation that was heating you and your surroundings is suddenly intercepted higher up in the atmosphere, and you feel cooler.

Volcanic activity can also lead to planetary cooling. When Mount Pinatubo in the Philippines erupted in 1991, an enormous burden of particles and aerosols entered the atmosphere and contributed to a significant drop in global temperature as radiation was reflected and scattered away. This cooling effect can be

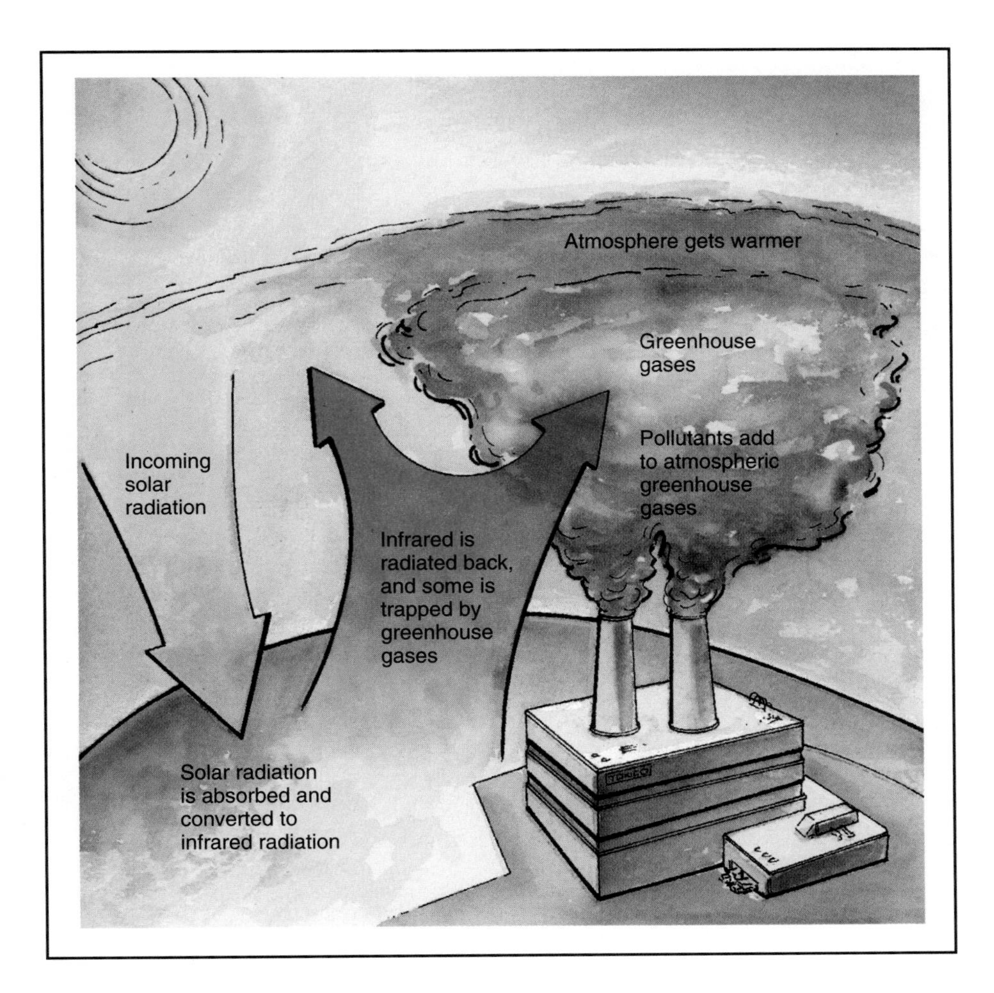

FIGURE 16-7
The greenhouse effect. Solar radiation is absorbed and converted to infrared radiation. As this radiates back through the atmosphere to outer space, some is absorbed by the greenhouse gases and insulates Earth, raising the temperature in the troposphere. Various pollutants add to the natural greenhouse gas content of the atmosphere. An increase in greenhouse gases delays more infrared losses to space, leading to global warming.

global, and it lasts until the volcanic debris is finally cleansed from the atmosphere by chemical change and deposition—a process that can take several years.

Finally, climatologists have recently found that anthropogenic sulfate aerosols (from ground-level pollution) play a significant role in canceling out some of the warming from greenhouse gases. Sulfur dioxide from industrial sources enters the atmosphere and reacts with compounds there to form a high-level aerosol—a sulfate haze. This haze reflects and scatters some sunlight and also contributes to the formation of clouds, with a concomitant increase in planetary albedo. The mean residence time of sulfates forming the aerosol is about a week; thus, the aerosol does not increase over time, as the greenhouse gases do.

Global atmospheric temperatures are a balance of the effects of factors leading to cooling and factors leading to warming. The net result varies, depending on global location. As we will see, this balance contributes much uncertainty to our predictions of what will happen in the future as greenhouse gases continue to increase. Fig. 16-8 shows a summary of the anthropogenic factors that interact to influence the temperature at any given location.

The Carbon Dioxide Story

In an article published in 1938, "The Artificial Production of Carbon Dioxide and Its Influence on Temperature," scientist G. Callendar reasoned that human beings' use of fossil fuels had the potential to increase atmospheric carbon dioxide concentrations. If that were to happen, the climate could change. This suggestion was largely forgotten until 1958, when Charles Keeling began measuring carbon dioxide levels on Mauna Loa, in Hawaii. Measurements there have continued to be recorded, and they reveal a striking increase in atmospheric levels of carbon dioxide (Fig. 16-9). The concentrations increased exponentially until the energy crisis in the mid-1970s, rose linearly for two decades, and resumed their exponential increase in the mid-1990s. The data also reveal an annual oscillation of 5 ppm, which reflects seasonal changes of photosynthesis and respiration in terrestrial ecosystems in the

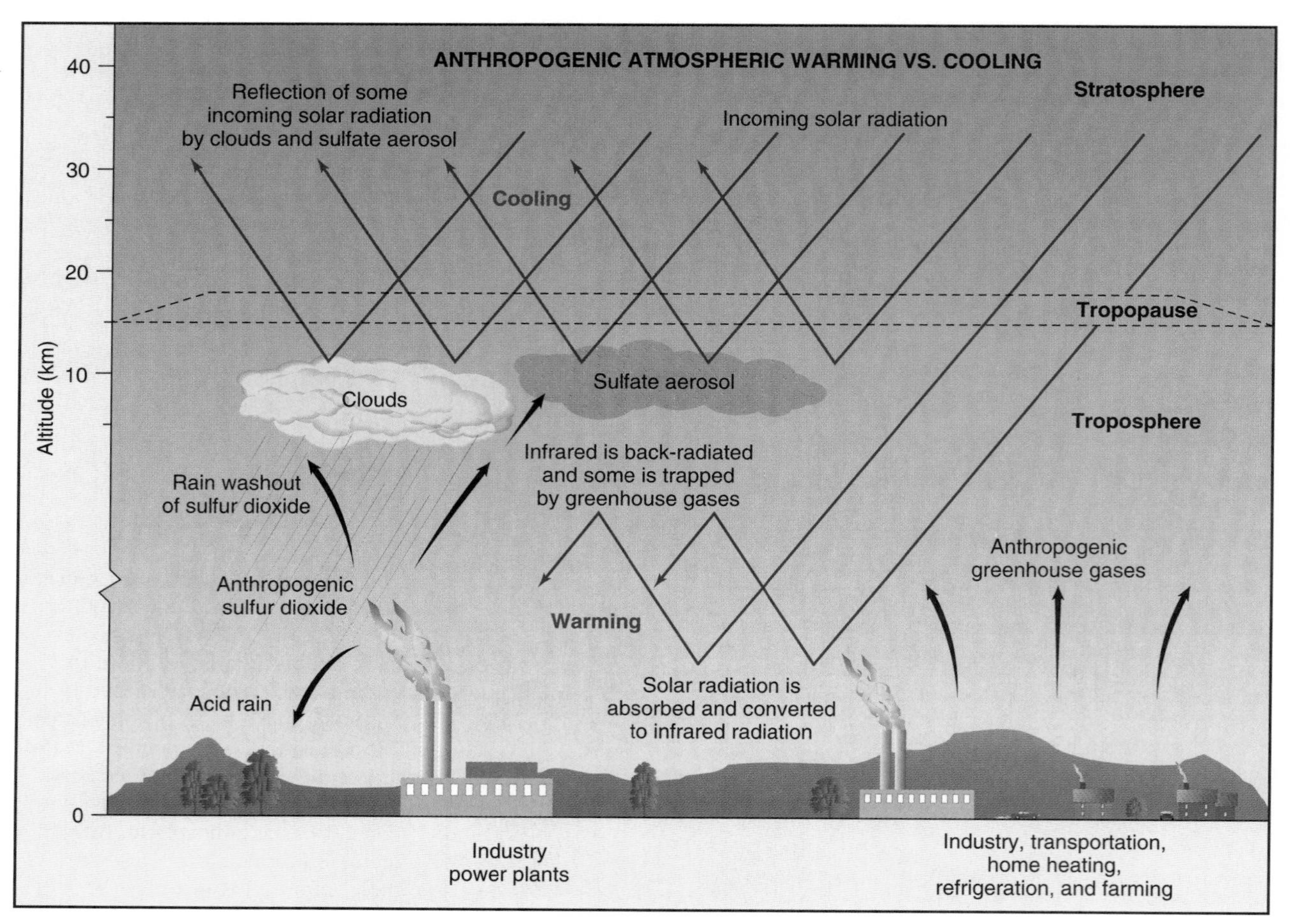

FIGURE 16-8
Anthropogenic factors involved in atmospheric warming and cooling. Some gases promote global warming; others, such as sulfur dioxide, lead to cooling. Clouds and sulfate aerosols are affecting the net temperature balance in industrial regions of the world, causing less warming of those regions than expected.

Northern Hemisphere. When respiration predominates (late fall through spring), levels rise; when photosynthesis predominates (late spring through early fall, levels fall). Most salient, however, is the relentless rise in carbon dioxide levels, ranging in recent years from 0.8 to 1.7 ppm per year. Carbon dioxide levels are now (in 1997) near 370 ppm, 35% higher than they were before the Industrial Revolution. Our insulating blanket is thicker, and there is reason to expect that this will have a warming effect.

As Callendar suggested long ago, the obvious place to look for the source of increasing carbon dioxide levels is our use of fossil fuels. Every kilogram of fossil fuel burned results in the production of about 3 kg of carbon dioxide. (The mass triples because each carbon atom in the fuel picks up two oxygen atoms in the course of burning and becoming CO_2.) Currently, 6 billion tons of fossil fuel carbon are burned each year, adding about 18 billion tons of carbon dioxide to the atmosphere. The vast majority of this and past fossil-fuel carbon dioxide has come from the industrialized countries (Fig. 16-10).

Careful calculations show that if all the carbon dioxide emitted from burning fossil fuels accumulated in the atmosphere, the concentration would rise by at least 3 ppm per year, not the 1.7 ppm or less charted in Fig. 16-9. Fortunately, the oceans are undersaturated with carbon dioxide, and there is now broad agreement that the oceans serve as a *sink* for much of the carbon dioxide emitted. There are limitations to the ocean's ability to absorb carbon dioxide, however, because only the top 300 meters of the ocean exchanges gases with the atmosphere. (The deep ocean layers do mix with the upper layers, but only very slowly—a mixing time of about 1500 years.)

The seasonal swings that are obvious in Fig. 16-9 show that the biota can influence atmospheric carbon dioxide levels. Can some of the "lost" carbon dioxide be traced to net biological uptake? Can we hope that the forests will take up increasing amounts of carbon

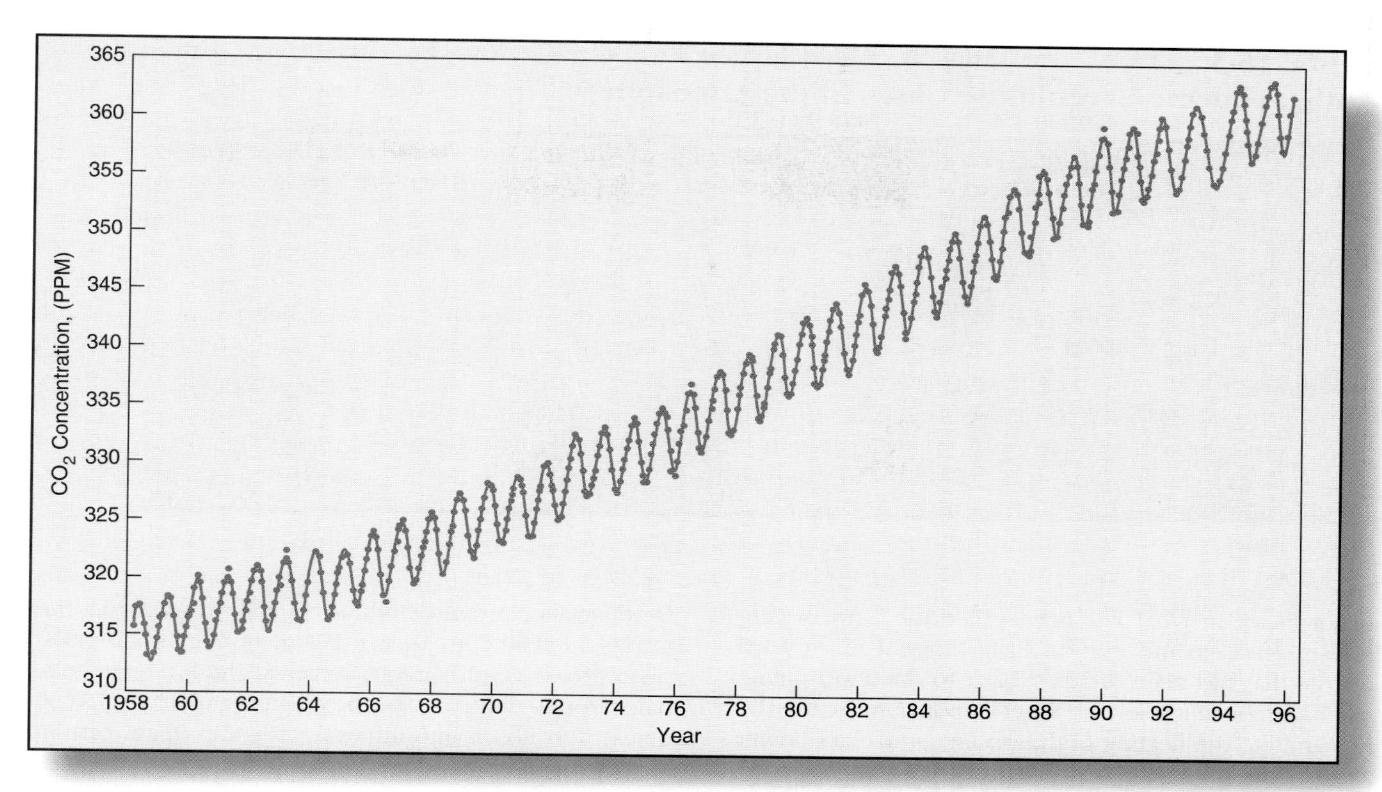

FIGURE 16-9
Atmospheric carbon dioxide concentrations from 1958 to 1996. The concentration of carbon dioxide in the atmosphere fluctuates between winter and summer because of seasonal variation in photosynthesis. The average concentration is increasing owing to human activities, namely, burning fossil fuels and deforestation. (Data for 1958 to 1974 compiled by C.D. Keeling, Scripps Institution of Oceanography, U.C. San Diego. Data for 1974 to 1996 compiled by P.P Tans and K.W. Thoning, Climate Monitoring and Diagnostics Laboratory, NOAA, Boulder, Colorado. All measurements made at Mauna Loa Observatory, Hawaii.)

dioxide as our emissions continue to rise? The evidence suggests that forests are now serving as a net *source* of carbon dioxide, not a sink! The reason is that forests are being cut and burned at a rate of 2% per year. It is estimated that the burning of forest trees is adding 1 to 2 billion tons annually to the 6 billion tons of carbon already coming from industrial processes. The net loss of forests, a serious biological concern in itself, therefore is also a cause for alarm in the context of global warming.

Other Greenhouse Gases

Several other gases also absorb infrared radiation and add to the insulating effect of carbon dioxide (Table 16-3). Some of these gases are generated from anthropogenic sources and are increasing in concentration, raising the concern that future warming will extend well beyond the calculated effects of carbon dioxide alone.

Water Vapor. Although water vapor absorbs infrared energy, the concentration of water vapor in the troposphere is quite variable. Through evaporation and precipitation, water undergoes rapid turnover in the lower atmosphere, and water vapor does not tend to accumulate over time. It does, however, appear to be a

FIGURE 16-10
Worldwide carbon dioxide emissions from fossil-fuel burning. Total emissions in 1992 were approximately 18 billion metric tons of carbon dioxide.

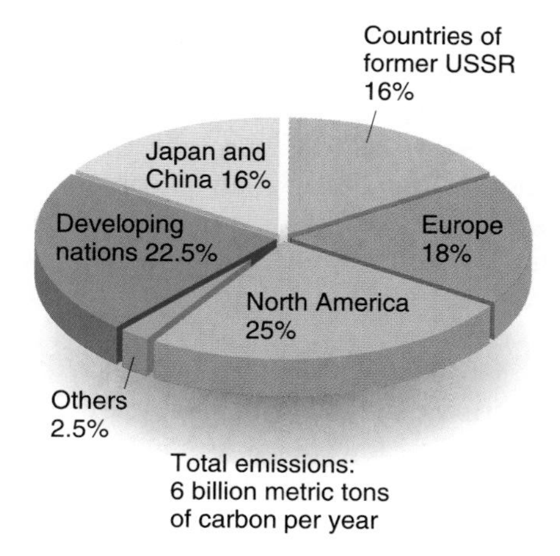

TABLE 16-3
Anthropogenic Greenhouse Gases in the Atmosphere

Gas	*Average Concentration 100 years ago (ppb)*[1]	*Approximate Current Concentration (ppb)*	*Average Residence Time in Atmosphere (years)*
Carbon dioxide (CO_2)	290,000	370,000	100
Methane (CH_4)	900	1700	10
Nitrous oxide (N_2O	285	310	170
Chlorofluorocarbons and halocarbons	0	3	60–100

[1] Parts per billion.

major factor in what has been called the "supergreenhouse effect" in the tropical Pacific ocean. As it traps energy that has been radiated back to the atmosphere, the high concentration of water vapor contributes significantly to the heating of the ocean surface and lower atmosphere in the tropical Pacific.

Methane. Methane (CH_4) is a product of microbial fermentative reactions and is also emitted from coal mines, gas pipelines, and oil wells. Because methane is generated in the stomachs of ruminants, animal husbandry is thought to be responsible for much of the increase in the gas appearing in the troposphere. Although methane is gradually destroyed in reactions with the hydroxyl radical, it is being added to the atmosphere faster than it is being broken down. The concentration of atmospheric methane has doubled since the Industrial Revolution, as revealed in core samples taken from glacial ice. In recent years, however, the rate of increase has taken a sharp downward trend: It is now only one-fourth of the 1981 rise in concentration. Scientists are puzzled by this trend and speculate that recent major repairs to natural gas pipelines in the former Soviet Union might be responsible for it.

Nitrous Oxide. Nitrous oxide (N_2O) levels are also on the increase. Sources of the gas include biomass burning and the use of chemical fertilizers; lesser quantities come from fossil-fuel burning. The emission of nitrous oxide is particularly unwelcome because its long residence time (170 years) will be a problem in not only the troposphere, where it contributes to warming, but also the stratosphere, where it contributes to the destruction of ozone.

CFCs and Other Halocarbons. Emissions of halocarbons are entirely anthropogenic in origin and are increasing more rapidly than those of any other greenhouse gas. Like nitrous oxide, halocarbons are long lived and contribute both to global warming in the troposphere and to ozone destruction in the stratosphere. Used as refrigerants, solvents, and fire retardants, halocarbons have a much greater capacity (10,000 times) for absorbing infrared radiation than carbon dioxide has. Although the rate of production of chlorofluorocarbons (CFCs) has declined since the Montreal Accord of 1987, these gases are very stable and will continue to accumulate for a few more years.

Together, these other greenhouse gases are estimated to trap about 60% as much infrared radiation as carbon dioxide does. Although the tropospheric concentrations of some of these gases are rising, it appears likely that they will gradually decline in importance, leaving CO_2 as the primary greenhouse gas to cope with in the future.

Amount of Warming and Its Probable Effects

It is certain that levels of carbon dioxide and the other greenhouse gases are increasing in the troposphere as a result of human activities. The phenomenon of the greenhouse effect is also well established. How can we be *certain*, however, that the increase in greenhouse gases will bring on a permanent increase in Earth's average temperature? The short answer is that we can't be sure. Many variables affect global temperatures, including the reflection of solar radiation by cloud cover, changes in the sun's intensity, airborne particulate matter from volcanic activity, and sulfate aerosol. But a more informed answer to the question is that all the evidence examined so far points to the *strong probability* that, as levels of greenhouse gases increase in the troposphere, global temperatures will indeed rise, and the climate will undergo major changes as a consequence—scientists have reached a consensus on this point as published by the Intergovenmental Panel on Climate Change.

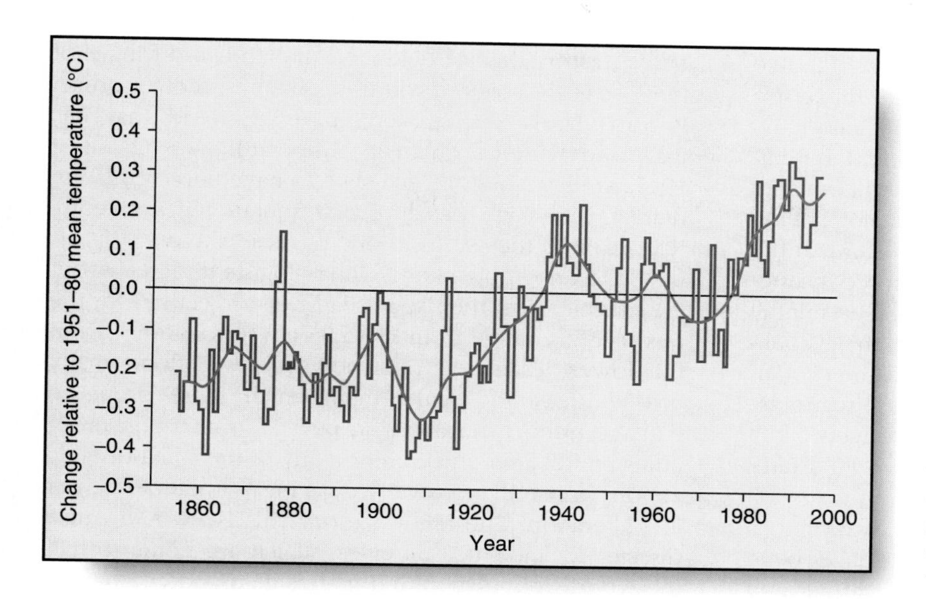

FIGURE 16-11
Annual surface temperatures for the world from 1880 to 1996, shown by the red line. The baseline, or zero point, is the 1950–80 average temperature; the blue line represents the running average for every five years. The warming trend since 1970 is conspicuous. (Courtesy of R.Ruedy, M. Sato, R. Reynolds, and J. Hansen, NASA, Goddard Institute for Space Studies, New York.)

The Contributions of Modeling. Powerful computers have made it possible to include a great number of variables in the construction of global climate models. (Computer models are simulations of real, complex situations; variables can be changed and the outcomes measured.) One early model, put together by a group of NASA scientists headed by James Hansen, got the attention of the scientific community in 1981 with its ability to track known temperature changes and link them to past and future carbon dioxide levels and global temperature changes. Hansen's model suggested that the combination of carbon dioxide and volcanic emissions was responsible for most of the observed temperature changes during the 1900s. (Fig. 16-11). A warming trend of more than 0.7° C over the last 100 years or so has coincided with an increase in carbon dioxide concentration over the same period.

Since 1981, global climate models, called *general circulation models* (GCMs), have become more numerous and more complex. Climate modeling now takes into account the global atmospheric circulation patterns and couples them to an ocean circulation model (CGCMs), in order to incorporate the significant contributions of heat exchange between the ocean and the atmosphere. Typically, a model will be run for a few simulated decades under standard conditions and then repeated with a doubled greenhouse gas concentration, to determine the climate change "caused" by the doubling. The latest information on the cooling effects of sulfates has been incorporated in current simulations, giving modelers more confidence in their results: Modeling now comes very close to simulating current-day climate fluctuations. The greatest uncertainties facing modelers today are modeling the complex behavior of clouds and incorporating the behavior of the biosphere into the model as it responds to variations in climate. What role, for example, will the planetary tree cover play as it extends farther north into the Arctic?

A number of models have been used to project the future global climate. All the projections agree that if the concentration of greenhouse gases were to double, Earth would warm up between 1.5° and 4.5 C°, with most estimates around 2.5 C°. This doesn't sound like much, but consider that temperatures were only 5 C° cooler 18,000 years ago—which was during the last ice age. At that time, ice sheets up to a mile thick stretched across North America from New York through the Great Lakes states and all of Canada. As the models simulate time into the future, the uncertainties in the models' outcomes that can be traced to variations in cloud behavior and sulfate aerosol are all dwarfed by the relentless impact of rising greenhouse-gas concentrations.

More recently, some modelers have been extending their models beyond the traditional doubling scenario. After all, these scientists reason, carbon dioxide doubling is only the beginning; there is no reason to believe that we will stop generating greenhouse gases when that level is reached. Consequently, carbon dioxide may build up to four or more times its preindustrial level, bringing with it a rise in global temperature of 7°C and a rise in sea level of about 2 meters. Although this development is a century or more away, a change of this magnitude would clearly have catastrophic effects on life on Earth.

Impacts of Future Warming. Rising global temperatures are linked to two major impacts: *regional climatic changes* and *a rise in sea level*. Both of these effects show up in all of the models and are expected to become very evident in a matter of a few decades.

Responding to these changes will likely involve unprecedented and costly adjustments. The impact on natural ecosystems could be highly destabilizing.

Warming will seriously affect rainfall and agriculture. The present-day difference between temperatures at the poles and those at the equator is a major driving force for atmospheric circulation. Greater heating at the poles than at the equator will reduce this force, changing atmospheric circulation patterns as well as rainfall distribution patterns. Further changes in climate could be initiated by alterations in the global oceanic circulation that could be brought on by greenhouse warming. Oceanographers fear that as the ocean's conveyer belt of currents goes into a stall, sudden shifts in land temperature may occur.

Another expected impact of global warming is weather change. The following changes have been observed: 1. Winters in Europe have been warmer and wetter during the last decade, as a result of differences in north Atlantic Ocean circulation. 2. The episodic warming of the southern Pacific Ocean known as "El Niño" is becoming far more common, most recently lasting an unprecedented five years (from 1990 to 1995). The El Niño phenomena influence weather by guiding jet stream and moisture patterns. 3. Thunderstorm intensity and hurricanes have been more frequent and more severe in recent years, a fact well noted by the insurance industry, which has become interested in global warming issues as huge sums for weather-related losses accrued in the 1990s.

The greatest difficulty for the agricultural community in coping with climatic change, however, is not knowing what to expect. Already farmers lose an average of one in five crops because of unfavorable weather. As the climate shifts, the vagaries of weather will become more pronounced, and crop losses are likely to increase. However, farmers are capable of rapidly switching crops and land uses, so the impact may not be as bad as some observers think.

With global warming, sea level will rise because of two factors: thermal expansion as ocean waters warm and melting of glaciers and ice fields. Sea level is already on the rise, at the annual rate of 1–1.5 mm. Most of this rise is attributed to the global warming of the last century.

The warming associated with a doubling of greenhouse gas levels is likely to be more pronounced in polar regions—as much as 10 C°—and less pronounced in equatorial regions—1 to 2 C°. The warming at the poles will have a tremendous effect when the ice melts, because so much water is stored in the world's remaining ice—enough to raise sea level by 75 m. The area of greatest concern is the Antarctic, which holds most of the world's ice. West Antarctica's ice sheet is not greatly elevated above sea level; in fact, most of it rests on land below sea level. The Ross and Weddell Seas extend outward from this ice field, and the landward fringes of these seas are covered with thick floating shelves of ice. Records show a consistent warming trend that has increased western Antarctic temperature 2.3 C° since 1945. At the same time, ice shelves have been retreating dramatically—one piece of 500 square miles recently "calved" into the sea from one of these shelves.

Oceanographers and climatologists disagree about the magnitude of the rise in sea level; projections for the next century range from 15 to 95 cm (6 to 37 in.). Even the lowest estimated rise, however, will flood many coastal areas and make them much more prone to damage from storms, forcing people to abandon properties and migrate inland. For many of the small oceanic nations, a rise in sea level would mean obliteration, not just alteration. The highest estimated rise would cause disaster for most coastal cities, which are home to half the world's population and its business and commerce. Moreover, the estimate of 15 to 95 cm extends only through the next century; the impact will certainly be greater beyond the year 2100. Are inland cities and communities ready to accommodate the billions of people that will be displaced? Are we ready to build dikes or modify all ports to accommodate the higher sea level?

Coping with Global Warming

Some would deny that preparing for the disastrous effects of global warming is necessary. They point out that Earth has a great capacity for recovering from harm and that false assumptions can skew the results of computer models.

Is Global Warming Here? What kind of evidence would convince most people that global warming was indeed occurring? For one thing, average annual temperatures should rise. There is so much natural variation in weather from year to year, that trends are not always apparent, but we would expect to see a gradual warming trend in global temperatures (while remaining aware that local temperatures do not necessarily follow globally averaged ones).

Is it significant that twelve of the hottest years on record occurred since 1980? Many scientists think so, especially since there is a likely explanation for cooler temperatures during some of the other 5 years: When Mount Pinatubo erupted in the Philippines in 1991, climatologists using their GCMs predicted that it would produce a temporary cooling trend—and they were correct. We returned to the warming trend that was so evident in the 1980s after debris from Mount Pinatuba cleared from the atmosphere. Indeed, 1994 and 1995 temperature data indicates a return to the high temperatures of the 1980s (Fig. 16-11).

Another expected impact of global warming is the retreat of glaciers. The melting and slow retreat of glac-

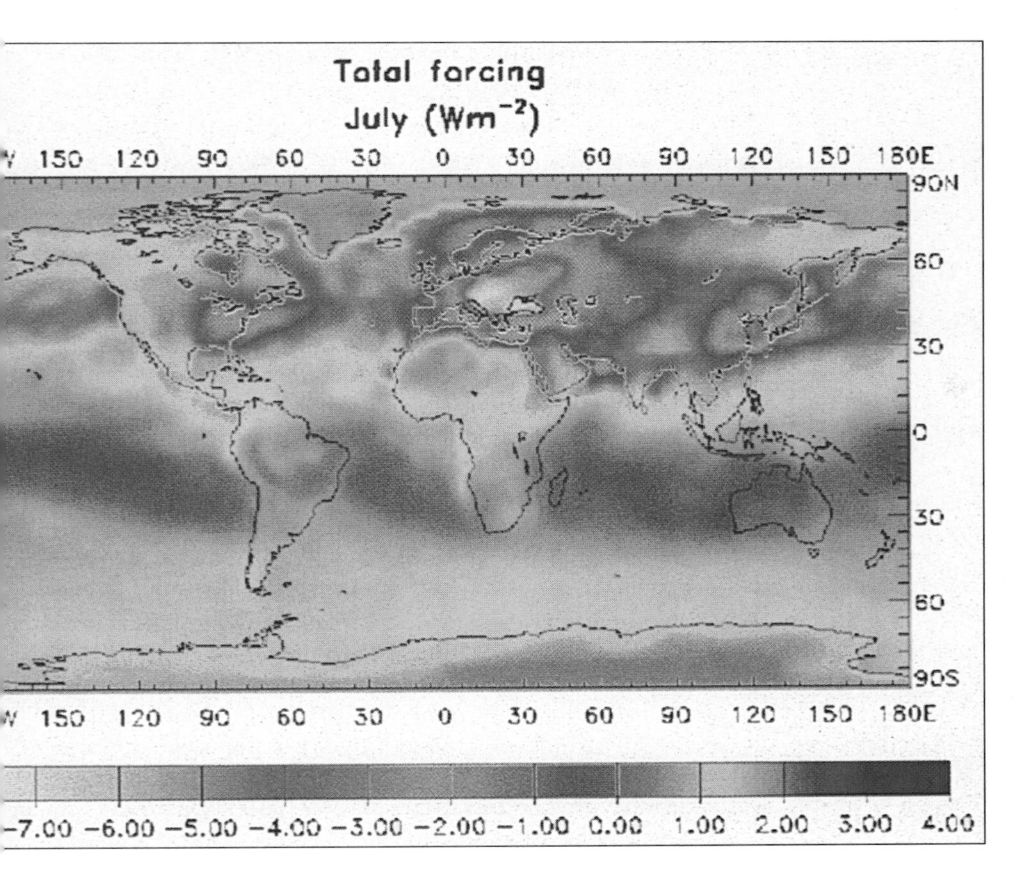

FIGURE 16-12
Global map of energy balance in midsummer in the Northern Hemisphere, showing the regional cooling that occurs in the industrialized North as a result of sulfate aerosols intercepting radiation and stimulating the formation of clouds. The map is color coded to show which regions receive more or less energy from sunlight, compared with the global average. White areas represent a large drop in incoming solar energy, blue a slight drop, green no change, yellow a slight increase, and red a major increase. (Jeffrey Kiehl/National Center for Atmospheric Research.)

iers all over the world has been followed by the World Glacier Monitoring Service. The response time of glaciers to climatic trends is slow—10 to 50 years—but the data collected clearly show a wide-scale recession of glaciers over the past 100 years, which is consistent with an increase in greenhouse gases. As already mentioned, there also is evidence that sea level is rising.

Ironically, the sulfate aerosol in the industrialized regions of the Northern Hemisphere appears to be canceling out greenhouse warming over those regions. Figure 16-12 shows the global energy balance for summer conditions and indicates clearly just where cooling effects of the aerosol occur—over the very regions most responsible for greenhouse gas emissions. However, the aerosol cooling effect is temporary. As the industrialized nations continue to reduce sulfate emissions, and greenhouse gases continue to build up, the Northern Hemisphere will likely experience its own share of warming.

Energy Use Scenarios. The potential for global warming is tied mainly to carbon dioxide emissions, which are responsible for about 60% of greenhouse warming. These emissions in turn are linked to fossil-fuel consumption. Many scenarios have been drawn up to evaluate the warming impacts of a number of energy choices (Fig. 16-13). With some variation, all the scenarios agree that, unless world cuts back on emissions, we can expect the concentrations of greenhouse gases to double from preindustrial levels during the 21st century, most probably before 2050.

Political Developments. The world's industries and transportation networks are so locked into the use of fossil fuels that massive emissions of carbon dioxide and other greenhouse gases seem sure to continue for the foreseeable future. It is possible, however, to lessen the rate at which emissions are added to the atmosphere and eventually to bring about a sustainable balance—although no one thinks that this will be easy. The means for accomplishing the goal lie in the international scientific and political arenas.

A variety of steps have been suggested to combat global warming, with the goal of *stabilizing the greenhouse gas content of the atmosphere*. Unless this goal is achieved, we will continue on an unsustainable course with unwanted consequences. Here are some of the suggestions:

- Place a worldwide cap on carbon dioxide emissions by limiting the use of fossil fuels in industry and transportation.
- Encourage the development of nuclear power, but only if cost-effectiveness, reliability, spent fuel, and high-level waste issues are resolved.

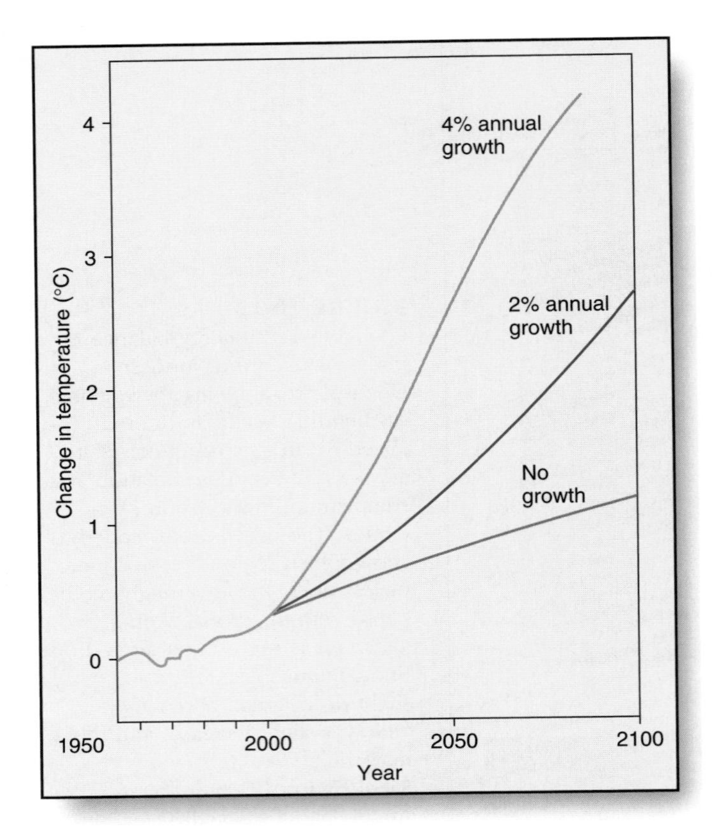

FIGURE 16-13
Warming effects of various energy-use scenarios. The no-growth curve assumes continued output of carbon dioxide remains at present levels. Current energy use is increasing at 2% per year, as indicated by the middle line. If global production of carbon dioxide increases at a rate of 4% per year, as occurred during the 1960s and 1970s, the temperature change would be dramatic (top). Source: *Science,* 213, p. 962, 28 Aug. 1981.

- Accelerate international agreements to completely phase out chlorofluorocarbons before the year 2000.
- Stop the loss of tropical forests and encourage planting of trees over vast areas now suffering from deforestation.
- Make energy conservation rules much more stringent. (Tighten building codes to require more insulation, use energy-efficient lighting, and so forth.)
- Reduce the amount of fuels used in transportation by raising mileage standards, encouraging car pooling, stimulating mass transit in urban areas, and imposing increasingly stiff carbon taxes on fuels.
- Invest in and deploy known renewable energy technologies: wind power, solar collectors, solar thermal, photovoltaics, tidal power, geothermal, among others.

Many observers advocate the so-called **tie-in**, or **no-regrets, strategy**: *Society should pursue those actions that provide widely agreed societal benefits, even if the predicted change does not materialize.* Thus, investing in more efficient use and production of energy reduces acid deposition, makes good economic sense, lowers the harmful health effects of air pollution, and lowers our dependency on foreign oil—and, of course, reduces carbon dioxide emissions.

Skeptics about global warming can be found even within the scientific community. They stress the uncertainties of the global climate models and in particular emphasize that there is much we do not know about the role of the oceans, the clouds, the biota, and the chemistry of the atmosphere. In the absence of "convincing evidence" many think that the threat of global warming may well be overplayed. Why take enormously costly steps, they ask, to prevent something that may never happen?

The Intergovernmental Panel on Climate Change (IPCC) includes scientists from all the member countries of the United Nations. The IPCC affirmed that human-produced emissions of greenhouse gases are producing a warming trend in their 1992, and subsequent 1994 and 1995 updates. The latest IPCC estimate includes a "low forecast" of 2 C° (3.6 F°) for 1990 to 2100, up to a "high forecast" of 3.5 C° (6.3 F°) world temperature rise. The IPCC consensus is summarized in Table 16-4.

Is the world taking global warming seriously? One of the five documents signed by heads of state at the United Nations Conference on Environment and Development (UNCED) "Earth Summit" in Rio de Janeiro in 1992 was the Framework Convention on Climate Change. This convention fell short of taking concrete action on global warming, largely because of opposition from the United States and a group of developing nations concerned that a limit on carbon emissions might stifle their economic growth. At issue was a proposal—endorsed by the European Community and Japan—to reduce emissions of carbon dioxide to 1990 levels by the year 2000.

The treaty embraced this goal, but called on countries to achieve the goal by voluntary actions. At the same time, the treaty commits its signatories to "stabilization of greenhouse-gas concentrations in the atmosphere at a level that would prevent dangerous anthropogenic interference with the climate system."

More than 150 countries signed the convention at the summit meeting, and sufficient legislative bodies ratified the treaty to make it legally binding on March 21, 1994. There are two major problems with its goals. First, the voluntary approach is simply not working. Although 35 industrialized countries committed to the

TABLE 16-4
IPCC Consensus on Global Climate Change

Issue	*Statement*	*Consensus*
Basic characteristics	Fundamental physics of the greenhouse effect	Virtually certain*
-	Added greenhouse gases add heat energy	Virtually certain
	Greenhouse gases increasing because of human activity	Virtually certain
	Significant reduction of uncertainty will require a decade or more	Virtually certain
	Full recovery will require many centuries	Virtually certain
Projected effects by mid-21st century	Reduction of sea ice	Very probable**
	Arctic winter surface warming	Very probable
	Rise in global sea level	Very probable
	Local details of climatic change	Uncertain***
	Increases in number of tropical storms	Uncertain

* Nearly unanimous agreement among scientists; no credible alternative.
** Roughly 9 out of 10 chance.
*** Roughly 2 out of 3 chance.
See: Houghton, John T., ed., et. al., *Climate Change 1995: The Science of Climate Change*, Working Group I; Watson, Robert T., ed., et. al., *Impacts, Adaptations, and Mitigation of Climate Change*, Working Group II; Bruce, James P., ed., et. al., *Economic and Social Dimensions of Climate Change*, Working Group III. All three IPCC volumes published by Cambridge University Press, New York, 1996.

Ethics

Staking a Claim on the Moral High Ground

The United States has rightly laid claim to the moral high ground in the matter of protecting the ozone layer, after allowing several politically forced delays. This country was the first to ban CFCs in aerosol cans and took the lead in bringing international accord on the Montreal Protocol. Our example may have motivated other nations to follow our lead in the ozone matter.

Putting these actions in the context of morality means that we are concerned about doing what is right. However, global warming is a problem that is enormously more difficult than the CFC problem. To be addressed effectively, this larger problem will require major policy changes with far-reaching consequences. Taking action on ozone-depleting chemicals is child's play compared with what lies ahead if we act to curb global warming. Scientists estimate that a 50% to 80% cut in the emissions of greenhouse gases would be needed to stabilize the global climate. As the world's leading producer of greenhouse gas emissions, the United States is in a rather conspicuous and pivotal position. What we do will carry a great deal of weight in determining the outcome of negotiations.

The dilemma is a familiar one: short-term interests versus long-term gains. In the short run, our economy is geared to growth, and achieving a steady state or even cutting back is extremely unpopular politically and economically. There is a long view, though. Twenty-first-century Americans as well as the rest of the world will bless us if we make the right choices now. But time is running out, as the pressures of population growth and continuing development stress global ecosystems.

To occupy the moral high ground, we have to deal with an existing confrontation between the industrial world and the developing countries. The industrial nations currently account for more than 70% of the greenhouse gas emissions. These have been called *luxury emissions* because they largely come from a way of life that is unavailable to the majority of the world. Most of the developing world's contributions to greenhouse gases are in the form of methane from cows and rice paddies, and carbon dioxide from burning forests—all referred to as *survival emissions*. The pathway to the moral high ground will involve our willingness to undergo a reduction in luxury emissions, as well as to transfer technology so that the developing world can both reduce its survival emissions and achieve its economic development goals. Many of the industrialized nations (Japan, Australia, and many European countries) have started down that pathway. Will the United States make good on President Clinton's promise to bring greenhouse gas emissions down to 1990 levels? Or will we see the same kind of congressional gridlock that has blocked environmental legislation and other reforms in recent years? The world is watching—will it be disappointed?

carbon dioxide reduction plan, only half are expected to meet the goal. Emissions have been increasing at a rate of 2.2 percent a year since the 1992 Earth Summit. Second, if emissions continue even at the 1990 level (and this objective is already compromised), the greenhouse gas concentration will not stabilize. Scientists estimate that we must cut emissions at least in half if we want to achieve an atmospheric concentration that will not involve major climate changes.

Since the Earth Summit, the change in leadership in the United States has led to an about-face on this issue. At first, the Clinton administration endorsed the voluntary approach to bringing emissions of carbon dioxide down to 1990 levels. More recently, stimulated by the IPCC Second Assessment, the United States is pressing for a new treaty that would require countries to reduce emissions of greenhouse gases, perhaps by a system of tradeable carbon permits similar to the system now in use for sulfur dioxide emissions in the United States. This kind of leadership from the United States is crucial if the goals of the Convention on Climate Change are to be realized. (See "Ethics" box, p. 417.)

Are we in fact on the threshold of a major human-caused change in the climate? Whatever the outcome, it is certain that we are conducting an enormous experiment—at the global level—and our children and their descendants will be living with the consequences.

Depletion of the Ozone Shield

In this third of the major atmospheric challenges traced to human technology, the work of environmental scientists again has uncovered a serious problem for which there was no prior warning. Our knowledge of the workings of the atmosphere has been appallingly poor, and one consequence of that lack of understanding is the strong possibility that ultraviolet radiation will increase in intensity all over Earth. As with global warming, however, there are skeptics who have studied the ozone-depletion phenomenon and are not convinced that it is a serious problem. Let us examine the evidence.

Nature and Importance of the Shield

Radiation from the sun includes *ultraviolet (UV) radiation*, along with visible light. UV radiation is like visible light radiation, except that UV wavelengths are slightly shorter than the wavelengths of violet light, which are the shortest wavelengths visible to the human eye. (See Fig. 16-14.) It is important to distinguish between UVB and UVA radiation. UVB wavelengths range from 280 to 320 nanometers (0.28 to 0.32 μm), UVA from 320 to 400 nanometers. Since energy is inversely related to wave-

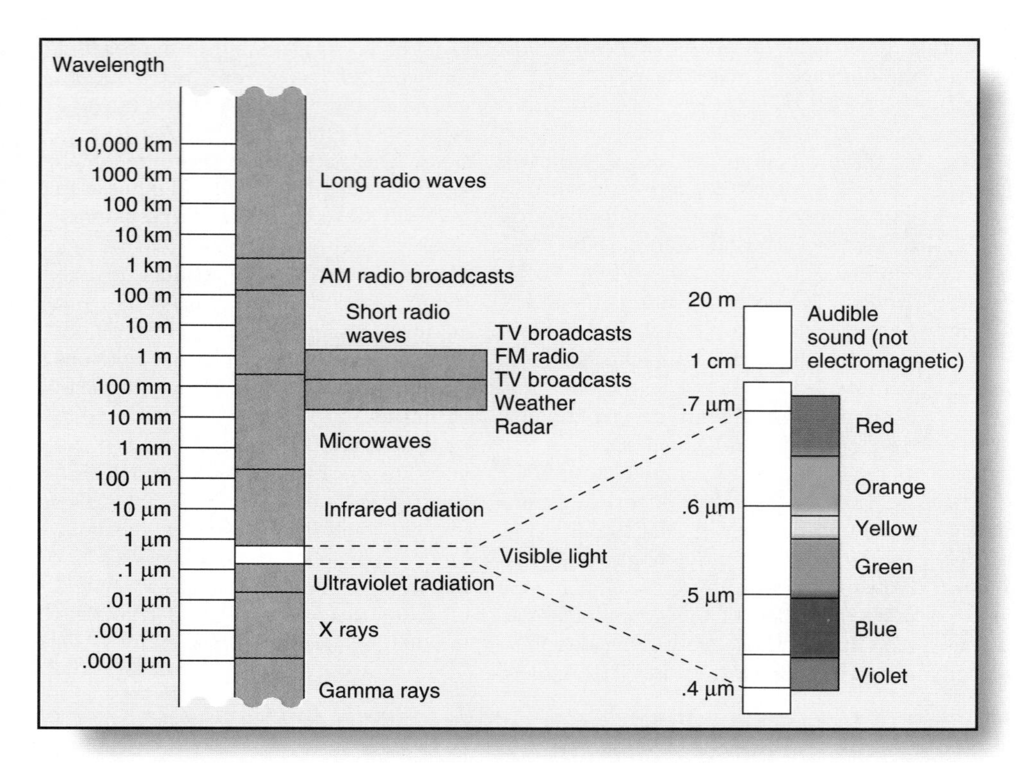

FIGURE 16-14
Ultraviolet, visible light, infrared, and many other forms of radiation are just different wavelengths of the electromagnetic spectrum.

Global Perspective

Coping with UV Radiation

People living in Chile and Australia are no strangers to the effects of the thinning ozone shield. UV alerts are given in those parts of the Southern Hemisphere during their spring months, as lobes of ozone-depleted stratospheric air move outward from the Antarctic. At times, the ozone above these regions can be less than half its normal concentration, and the consequence is, of course, greater intensities of UV radiation reaching Earth's surface. It is now predicted that one out of every three Australians will develop serious and perhaps fatal skin cancer in his or her lifetime.

A thinning ozone layer has now appeared above the Northern Hemisphere, and is getting serious attention. Concerned over the rising incidence in the United States of skin cancer and cataract surgery, the EPA, together with the National Oceanic and Atmospheric Administration and the Centers for Disease Control, has initiated a new "UV Index." The index is in the form of daily forecasts of UV exposure, issued by the Weather Service for 58 cities (see inset table). Satellite measurements of stratospheric ozone are combined with other weather patterns. The goal of the index is to remind people of the dangers of UV radiation and prompt them to take appropriate action to avoid cancer, premature skin aging, eye damage, cataracts and blindness. For at least the next decade, the dangers will be very noticeable, ranging from a greater incidence of sunburn to a likely epidemic of malignant melanoma.

Less obvious, but no less important, are the chronic effects of exposure to the sun. The "normal aging" of skin is now known to be largely the result of damage from the sun: coarse wrinkling, yellowing, and the development of irregular patches of heavily pigmented and unpigmented skin. This can happen even in the absence of episodes of sunburn. Further, a large proportion of the million cataract operations performed annually in the United States can be traced to UV exposure, another of the chronic effects of this kind of radiation.

The most serious impact of chronic exposure to the sun is skin cancer, which occurs in some 700,000 new cases each year in the United States. Three types of skin cancer are traced to UV exposure: basal cell carcinoma (BCC), squamous cell carcinoma (SCC), and melanoma. Most of the skin cancers are BCC (75% to 90%); slow growing and therefore treatable. SCC accounts for about 20% of skin cancers; this form can also be cured if treated early. However, it metastasizes (spreads away from the source) more readily than BCC and thus is potentially fatal. Melanoma is the most deadly form of skin cancer, as it easily metastasizes and is the form that causes the greatest mortality. Melanoma is often traced to occasional sunburns during childhood or adolescence, or to sunburns in people who normally stay out of the sun.

What is an appropriate response to this disturbing information? First, know when UV intensity is greatest (the hours around midday) and take precautions accordingly. Protect your eyes with sunglasses and a hat, and apply sunscreen with a Sunscreen Protection Factor (SPF) rating of 15 or higher during such times. Think twice before considering sunbathing or tanning beds, and never proceed without sunscreen that is strong enough to prevent sunburn. Always protect children with sunscreen; their years of potential exposure are many, and their skin is easily burned. If you detect a patch of skin or a mole that is changing size or color, or that is red and fails to heal, consult a dermatologist for treatment.

UV Index (EPA)

Exposure Category	*Index Value*	*Minutes to Burn for "Never Tans" (most susceptible)*	*Minutes to burn for "Rarely Burns" (least susceptible)*
Minimal	0–2	30 minutes	>120 minutes
Low	4	15 minutes	75 minutes
Moderate	6	10 minutes	50 minutes
High	8	7.5 minutes	35 minutes
Very high	10	6 minutes	30 minutes
	15	<4 minutes	20 minutes

length, UVB is more energetic and therefore more dangerous.

On penetrating the atmosphere and being absorbed by biological tissues, UV radiation damages protein and DNA molecules at the surfaces of all living things. (This is what occurs when you get a sunburn.) If the full amount of ultraviolet radiation falling on the stratosphere reached Earth's surface, it is doubtful that any life could survive; plants and animals alike would simply be "cooked." Even the small amount (less than 1%) that does reach us is responsible for all the sunburns and more than 700,000 cases of skin cancer and precancerous ailments per year in North America, as well as for untold damage to plant crops and other lifeforms.

We are spared more damaging effects from ultraviolet rays because most UVB radiation (over 99%) is absorbed by ozone in the stratosphere. (See Fig. 15-1.) For that reason, stratospheric ozone is commonly referred to as the **ozone shield**.

Stratospheric ozone is the same molecule (O_3) described in Chapter 15 as a serious air pollutant. Recall that one definition of pollution is a *chemical out of place*. Ozone is out of place in the troposphere and thus is a pollutant.

Formation and Breakdown of the Shield

The need to maintain the ozone shield requires no elaboration. However, certain anthropogenic pollutants are causing it to break down. Ozone in the stratosphere is a product of UV radiation acting on oxygen (O_2) molecules. The high-energy UV radiation first causes some molecular oxygen (O_2) to split apart into free oxygen (O) atoms, and these atoms combine with the molecular oxygen to form ozone as follows:

$$O_2 + UVB \longrightarrow O + O \quad (1)$$

$$O + O_2 \longrightarrow O_3 \quad (2)$$

Not all of the molecular oxygen is converted to ozone, however, because free oxygen atoms may also combine with ozone molecules to form two oxygen molecules:

$$O + O_3 \longrightarrow O_2 + O_2 \quad (3)$$

Finally, when ozone absorbs UVB, it is converted back to free oxygen and molecular oxygen:

$$O_3 + UVB \longrightarrow O + O_2 \quad (4)$$

Thus, the amount of ozone in the stratosphere is dynamic; there is an equilibrium due to the continual cycle of reactions of formation (Eqs. 1 and 2) and reactions of destruction (Eqs. 3 and 4). Because of seasonal changes in solar radiation, ozone concentration in the Northern Hemisphere is highest in summer and lowest in winter. Also, in general, ozone concentrations are high at the equator and diminish as latitude increases—again, a function of higher overall amounts of solar radiation. However, as we have learned in recent years, the presence of other chemicals in the stratosphere can upset the normal ozone equilibrium and promote unsustainable reactions there.

Halogens in the Atmosphere. Chlorofluorocarbons (CFCs) are a type of halogenated hydrocarbon. (See Chapter 14.) They are nonreactive, nonflammable, nontoxic organic molecules in which both chlorine and fluorine atoms replace some of the hydrogens atoms. At room temperature, CFCs are gases under normal (atmospheric) pressure; but they liquify under modest pressure, giving off heat in the process and becoming cold. When they revaporize, they reabsorb the heat and become hot. These attributes have led to the widespread use of CFCs (over a million tons per year in the 1980s) for the following purposes:

- Chlorofluorocarbons are used in virtually all refrigerators, air conditioners, and heat pumps as the heat-transfer fluid. As these machines break down or are ultimately scrapped, their CFCs generally escape into the atmosphere.
- A second major use is in the production of plastic foams. CFCs are mixed into liquid plastic under pressure. (They are soluble in organic materials.) When the pressure is released, the CFC gas causes the plastic to foam, just as the carbon dioxide in a soda causes foaming when pressure is released. After the foam is made, the CFCs escape to the air.
- CFCs are also used in the electronics industry for cleaning computer parts, which must be meticulously purified. Again, the spent CFCs may escape into the air.
- Finally, CFCs were used in widely as the pressurizing agent for aerosol cans, which, of course, release the CFCs into the air during use.

In 1972, chemist Sherwood Rowland read a report of work by James Lovelock describing the presence of CFCs in the atmosphere. In fact, Lovelock found that the amount present in the atmosphere at that time was nearly equal to the total amount ever produced by industry. Rowland and his colleague Mario Molina published a classic paper in 1974 concluding that CFCs could be damaging the stratospheric ozone layer through the release of chlorine atoms, and the result would be increased UV radiation and more skin cancer.

Rowland and Molina reasoned that, although CFCs would be stable in the troposphere (they have been found to last 60 to 100 years there), in the stratosphere they would be subjected to intense UV radiation, which would break them apart, releasing free chlorine atoms:

$$CFCl_3 + UV \longrightarrow Cl + CFCl_2 \quad (5)$$

Ultimately, all of the chlorine of a CFC molecule would be released as a result of further photochemical breakdown. The free chlorine atoms could then attack stratospheric ozone to form chlorine monoxide (ClO) and molecular oxygen:

$$Cl + O_3 \longrightarrow ClO + O_2 \quad (6)$$

Furthermore, two molecules of chlorine monoxide may react to release more chlorine and an oxygen molecule:

$$ClO + ClO \longrightarrow 2\ Cl + O_2 \quad (7)$$

Reactions 6 and 7 are called the **chlorine cycle**, because chlorine is continuously regenerated as it reacts with ozone. Thus, chlorine acts as a **catalyst**, a chemical that promotes a chemical reaction without itself being used up by the reaction. Because it lasts a long time (from 40 to 100 years), every chlorine atom in the stratosphere has the potential to cause the breakdown of 100,000 molecules of ozone. Thus, CFCs are judged to be damaging because they act as transport agents that continuously move chlorine atoms into the stratosphere. The damage can persist because the chlorine atoms are removed from the stratosphere only very slowly. Figure 16-15 shows the basic processes of ozone formation and destruction, including recent refinements of those processes that will be explained shortly.

FIGURE 16-15
Stratospheric ozone formation and destruction. UV radiation stimulates ozone production at the lower latitudes, and ozone-rich air migrates to high latitudes. At the same time, chlorofluorocarbons and other compounds carry halogens into the stratosphere, where they are broken down by UV and release chlorine and bromine. Ozone is subject to high-latitude loss during winter, as the chlorine cycle is enhanced by the polar stratospheric clouds. Midlatitude losses occur as chlorine reservoirs are stimulated to release chlorine by reacting with stratospheric sulfate aerosol.

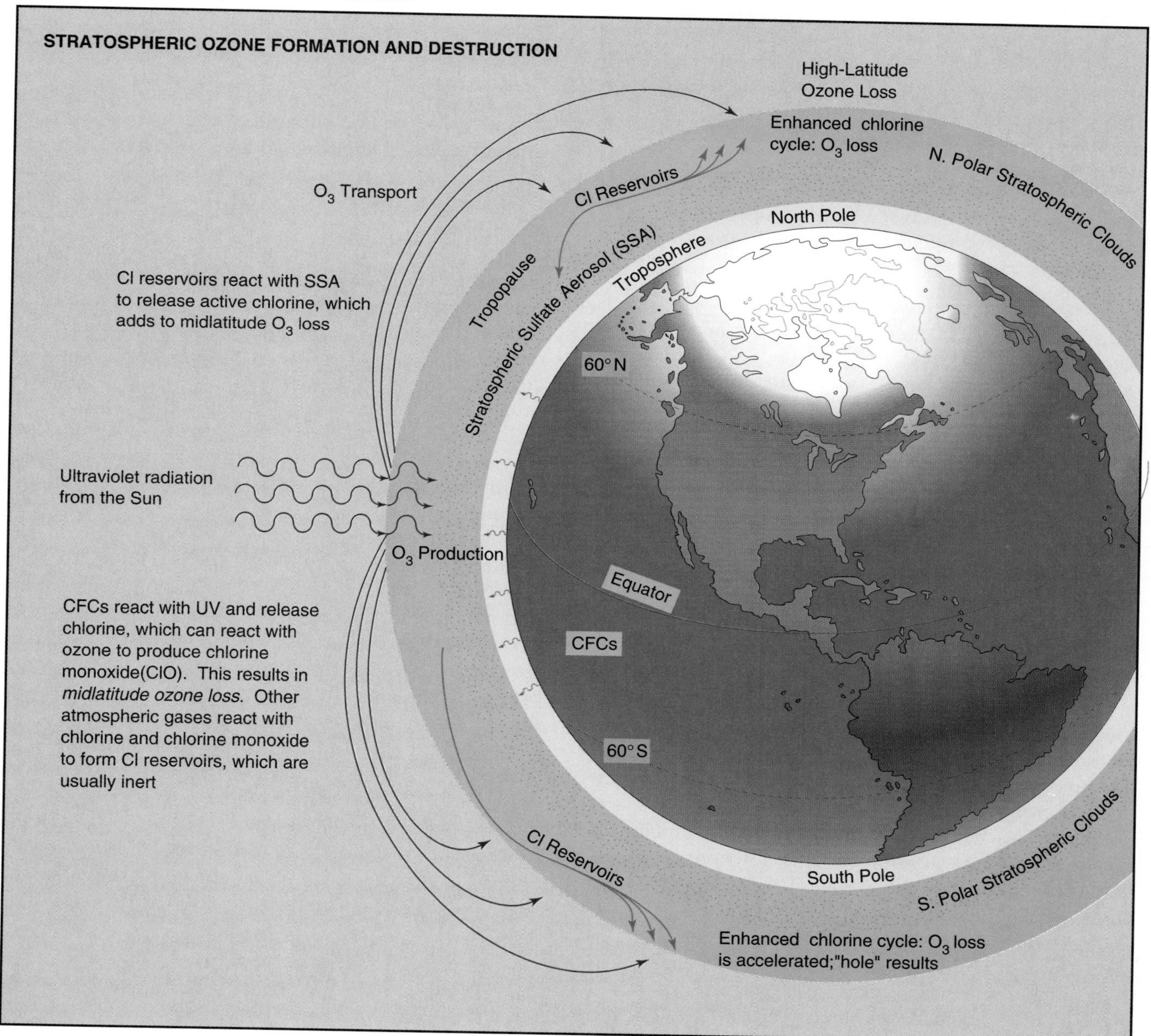

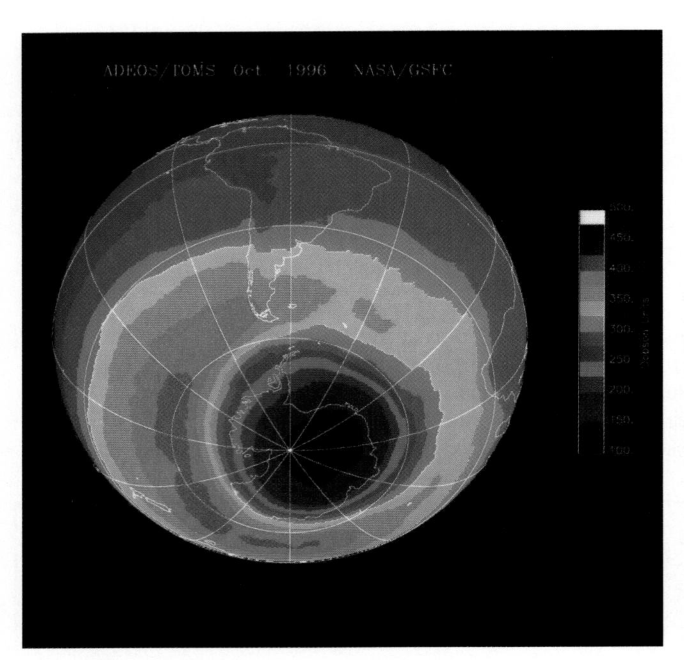

FIGURE 16-16
TOMS (Total Ozone Mapping Spectrometer) instruments aboard two new satellites recorded the largest ever ozone hole over Antarctica during September and October 1996. NASAs *Earth Probe* satellite and Japan's *Advanced Earth-Orbiting Satellite* (ADEOS) are the instrument platforms. The 1996 "hole" exceeded 25.9 million square kilometers (10 million square miles). This is an area larger than Canada, the United States, and Mexico combined. The color scale represents ozone concentrations in Dobson units, with blues and purples for amounts of less than 200. Measurements dropped as low as 111 Dobson units in 1996. (Image courtesy of Goddard Space Flight Center, NASA.)

After studying the evidence, the EPA became convinced that CFCs were a threat and, in 1978, banned their use in aerosol cans. Manufacturers in the United States quickly switched to nondamaging substitutes such as butane, and things were quiet for several years. CFCs continued to be used in applications other than aerosols, however, and critics demanded more convincing evidence of their harmfulness.

Atmospheric scientists reason that any substance carrying reactive halogens to the stratosphere has the potential to deplete ozone. These substances include halons, methyl chloroform, carbon tetrafluoride, and methyl bromide. Because of its extensive use as a soil fumigant and pesticide, methyl bromide is thought to cause as much as 10% of current stratospheric ozone loss (bromine is thought 40 times as potent as chlorine in ozone destruction).

The Ozone "Hole". In the fall of 1985, some British atmospheric scientists working in Antarctica reported a gaping "hole" (actually a thinning of one area) in the stratospheric ozone shield over the South Pole (Fig. 16-16). There, in an area the size of the United States, ozone levels were 50% lower than normal. The "hole" would have been discovered earlier by NASA satellites monitoring ozone levels, except that computers were programmed to reject data showing a drop as large as 30% as instrument "hiccups." It had been assumed by scientists that the loss of ozone, if it occurred, would be slow, gradual, and uniform over the whole planet. The "hole" came as a surprise, and if it had occurred anywhere but over the South Pole, the UV damage would have been extensive.

News of the ozone "hole" stimulated an enormous scientific research effort. The "hole" has reappeared every spring in Antarctica (our autumn) and has been intensifying—exactly what would be expected if increasing levels of CFC were responsible. In 1996, ozone levels in the "hole" reached a new low, at 40% below 1960 levels, over a 10 million square miles area, greater than all of North America.

A unique set of conditions is responsible for the ozone "hole." In the summer, gases such as nitrogen dioxide and methane react with chlorine monoxide and chlorine to trap the chlorine, forming the so-called **chlorine reservoirs** (Fig. 16-15) and so preventing much ozone depletion.

When the Antarctic winter arrives in June, it creates a vortex (like a whirlpool) in the stratosphere, which confines stratospheric gases within a ring of air circulating around the Antarctic. The extremely cold temperatures of the Antarctic winter cause the small amounts of moisture and other chemicals present in the stratosphere to form the South Polar stratospheric clouds. (See Fig. 16-15.) During winter, the cloud particles provide surfaces on which chemical reactions occur and release molecular chlorine (Cl_2) from the chlorine reservoirs.

When sunlight returns to the Antarctic in the spring, the sun's warmth breaks up the clouds; UV light then attacks molecular chlorine, releasing free chlorine and initiating the chlorine cycle, which rapidly destroys ozone. By November, the beginning of the Antarctic summer, the vortex breaks down, and ozone-rich air returns to the area. By this time, however, ozone-poor air has spread all over the Southern Hemisphere.

Shifting patches of ozone-depleted air have caused UV radiation increases of 20% above normal in Australia. Television stations there now report daily UV readings and warnings for Australians to stay out of the sun. Based on current data, estimates indicate that in Queensland, where the ozone shield is thinnest, three out of four Australians are expected to develop skin cancer.

Controversy over Ozone Depletion. Skeptics about the CFCs' role in ozone depletion abound. They claim, among other things, that probably there has always been an Antarctic ozone "hole," but we didn't use the

Earth Watch

Atmospheric Trouble from Air Traffic

Researchers have recently found unusually high concentrations of nitrogen oxides at altitudes between 5 and 9 miles above sea level. The source of these oxides is mainly air traffic: The normal cruising altitude for commercial airlines is 5–8 miles (9–13 km). This height is close enough to the tropopause to enable the gases to drift into the stratosphere. Recent work has shown that, in the stratosphere, nitrogen dioxide may break down ozone, adding to the impact of CFC emissions. Further, nitrogen oxides and water vapor contribute to the formation of the strato-spheric clouds implicated in the formation of the Antarctic ozone "hole." Studies indicate that ozone depletion over the Northern Hemisphere thus may be twice as high as what would occur with CFCs alone. Measurements above Europe have shown a 3% decrease in ozone between 6 and 10 miles up, and an 8% loss between 10 and 13 miles. These data support the conclusion that much of the measured ozone losses above the temperate zone in the Northern Hemisphere can be attributed to air traffic.

Of course, the bulk of the nitrogen oxides emitted from aircraft remain in the high troposphere, where they contribute to tropospheric ozone formation and global warming. Substances released so high in the troposphere tend to remain there at least 100 times longer than those released at ground level, due to less active cloud formation and less precipitation at those heights. When these substances finally do descend, they contribute to acid deposition. Thus, *one* activity—high-altitude air traffic—leads to depletion of the ozone layer, global warming, *and* acid deposition!

What can be done? Cruising altitudes should be kept below the tropopause, which can be as low as 6 miles during the winter. Plans to develop supersonic aircraft that cruise higher than 9 miles up should be scrapped forever, besides fuel reserves are inadequate to supply such a fleet. And air traffic should be subjected to the same scrutiny as all other forms of transportation, with attention given to the development of jet engines that emit only low levels of nitrogen oxides, as well as to limitations on the volume of traffic.

right instruments to look for it. They also claim that there is little evidence of a thinning ozone layer, and almost no evidence of ground-level UVB increases. One of the most persistent claims of the skeptics is that volcanoes have been ejecting enormous amounts of chlorine into the stratosphere for eons, with no apparent impact on the ozone shield. This chlorine, the claim goes, dwarfs any chlorine carried up into the stratosphere by CFCs.

Careful scrutiny reveals that this is a false argument. Whereas volcanoes do inject hydrochloric acid into the atmosphere, very little of it reaches the stratosphere because of the speed with which it is washed out of the atmosphere by precipitation. Furthermore, the volcanic production was highly exaggerated because of a serious calculation error by one of the skeptics (Dixy Lee Ray, former director of the Atomic Energy Commission). Despite this and other serious mistakes and flaws, the talk-radio circuit continues to repeat them.

Climatologists have, in fact, recently found evidence for ozone loss traced to volcanic eruption—but not from volcanic chlorine. Enormous volcanic eruptions, such as the 1991 Mount Pinatubo eruption, inject sulfur particles as high as the lower stratosphere. There they form a sulfate aerosol (the stratospheric sulfate aerosol pictured in Fig. 16-15) that reacts in complex ways with the chlorine reservoir chemicals normally tying up stratospheric chlorine. Active chlorine is released, which is likely one cause of ozone loss in midlatitudes. (See next section.) Climatologists point out that the volcanic impact is significant only because so much chlorine has been carried up into the stratosphere via anthropogenic sources.

More Ozone Depletion. Early in 1992, researchers monitoring the stratosphere above the Arctic found record high levels of chlorine monoxide. They concluded that the level of Arctic ozone will drop dramatically if a large polar vortex forms. If an ozone "hole" does develop above the Arctic, it will be far more serious than that above the Antarctic because ozone-depleted air will extend outward over highly populated regions of North America, Europe, and Asia. To date, however, although a winter ozone loss occurs over the Arctic, no "hole" has developed. The higher temperatures and weaker vortex formation there have so far prevented the severe losses that have become routine in the Antarctic.

Ozone losses have not been confined to Earth's polar regions, although they are most spectacular there. A worldwide network of ozone-measuring stations sends data to the World Ozone Data Center in Toronto, Canada. Reports from the center reveal winter ozone losses during the 1980s of 5% to 15% all across the temperate and tropical zones of both hemispheres. That is triple the losses that occurred in the 1970s. (Ozone trends for three regions of the world are graphed in Fig. 16-17.) Some depletion is now known to be present

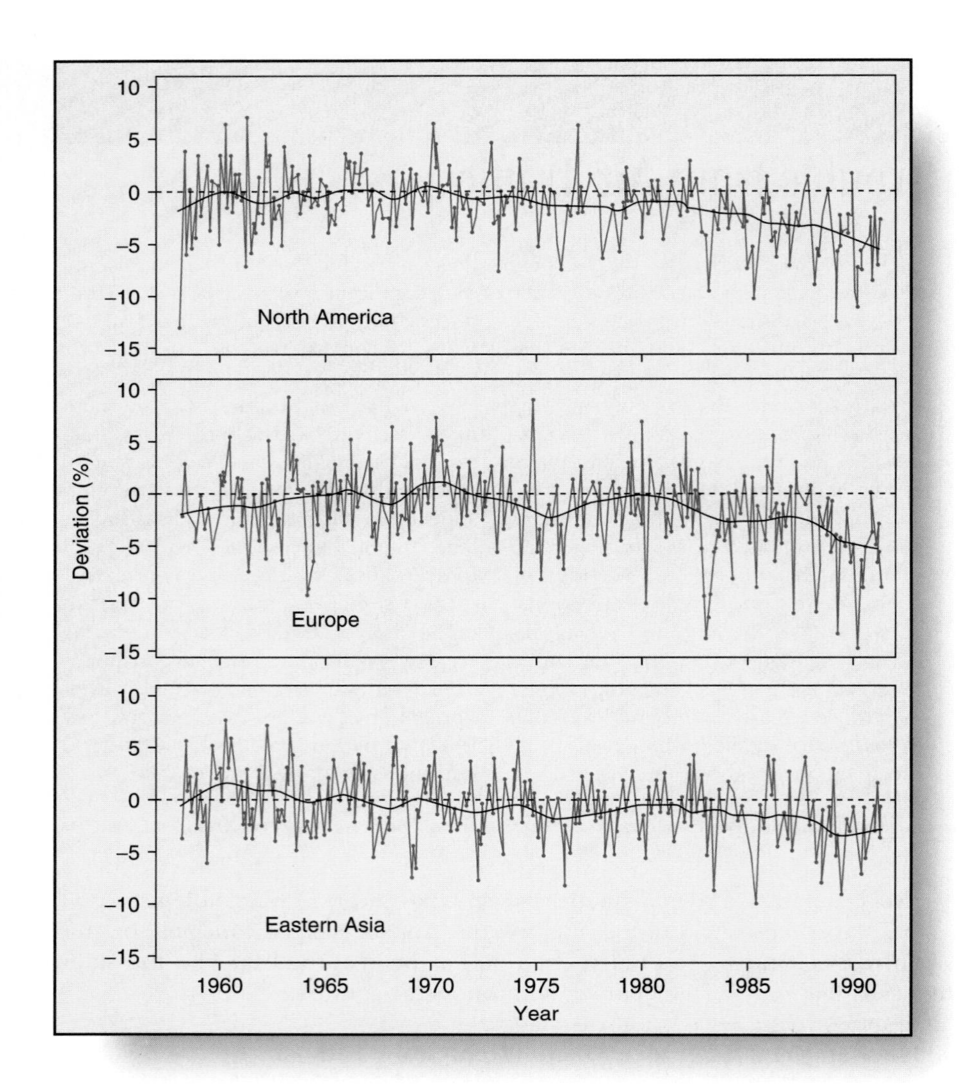

FIGURE 16-17
Average stratospheric ozone levels for three world regions since the 1950s. Seasonal and solar variations are removed; the data show clear trends in stratospheric ozone, especially the downward trend since 1980. (R. Stolarski, et. al., "Measured Trends in Stratospheric Ozone," *Science* 256 (1992): 344.)

during the summer, when UV radiation is strongest and can do the most damage to people and crops. A further loss is expected during the 1990s—some 7% to 13%, according to a recent report from a 226-member panel convened in 1994 by the World Meteorological Organization and the United Nations Environment Program. Ozone loss is expected to peak at about the year 1998, when, hopefully, the chlorine and bromine concentrations in the stratosphere will start to decline as a consequence of the international agreements that have been forged.

Is the ozone loss significant to our future? The EPA has calculated that the ozone losses of the 1980s will eventually have caused 12 million people in the United States to develop skin cancers over their lifetime and that 93,000 of these cancers will be fatal. More than 700,000 new cases of skin cancer and precancerous conditions are reported each year. The ozone losses of the 1990s are expected to allow 8% to 15% more UVB radiation than before to reach Earth.

Has there actually been a measurable increase in UVB? The skeptics argue that if the ozone loss were truly significant, we should be able to detect UV increases at Earth's surface. Canadian atmospheric scientists J. B. Kerr and C. T. McElroy made careful measurements of UVB radiation at Toronto from 1989 to 1993, with readings once or twice an hour. They found that UVB intensity increased an average of 35% per year during the winter (when UVB radiation is at its lowest) and more than 5% per year in the summer. They were able to correlate the UVB increases with a downward trend in ozone over Toronto measured at the same time (4% yearly loss in winter and 2% yearly loss in summer). The authors pointed out that their results were consistent with reports of ozone losses being measured elsewhere in the world. At the poles, where ozone losses are greater, large increases in UVB radiation have been measured by other scientists.

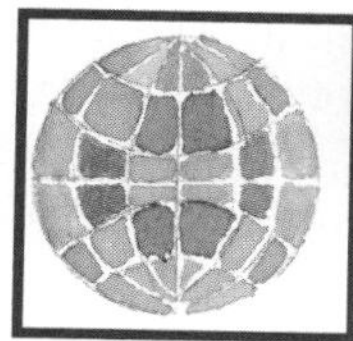

Global Perspective

Lessons from the Ozone Treaty

The 1987 Montreal Protocol and its 1990 and 1992 amendments are a remarkable and encouraging development in human affairs. Most of the countries of the world have agreed to take steps that are economically costly in order to protect a global resource—the ozone shield. This agreement was reached on the basis of research from the scientific sector, and it was a response not to an immediate problem, but to one that is expected to occur in the future. The protocol is a statement that today's generation actually has some concern about future generations. Some important lessons from the Protocol should not be missed:

- It is significant that such broad-based cooperation can be forthcoming from the collection of nations that are most accustomed to putting national self-interest at the top of their agendas.
- It is quite apparent that scientists must continue to play a role in the crafting of policy. It was their work that first drew attention to the ozone losses and the global warming complex. The development of consensus among scientists has been likewise remarkable.
- Governments must be bold enough to act even while scientific certainty is lacking. By its very nature, the scientific method cannot provide certainty short of an actual observation. At some point, delaying a decision becomes an immoral act. The costs of such delay may be enormous and far overshadow the relatively low cost of mitigation.
- Political leadership from non-governmental organizations and a few countries were crucial in bringing about the Montreal Protocol; such leadership may be just as essential in moving to a legally binding treaty on global warming.
- The United Nations, through its Environment Program, played a crucial role in the ozone accord. It has already been important in the negotiations on global warming and will likely continue to play the role of a catalyst for action.
- And, Sherwood Rowland, Mario Molina, and Paul Crutzen, were awarded the 1995 Nobel Prize for chemistry for their discoveries and subsequent fight in the political arena. The Nobel committee said, "...the 3 researchers contributed to our salvation from a global environmental problem that could have catastrophic consequences."

Coming to Grips with Ozone Depletion

The dramatic pictures of the growing "hole" in the ozone layer have galvanized a response around the world. In spite of the skepticism in the United States, scientists and politicians here and in other countries have achieved legislation designed to avert a UV disaster.

International Agreements. Under its Environmental Program, the United Nations convened a meeting in Montreal, Canada, in 1987 to address ozone depletion. Member nations reached an agreement, known as the **Montreal Protocol**, to scale CFC production back 50% by 2000; to date, 140 countries (including the United States) have signed the agreement.

The Montreal protocol was written even before CFCs were so clearly implicated in driving the destruction of ozone, and before the threat to Arctic and temperate-zone ozone was recognized. Because ozone losses during the late 1980s were greater than expected, an amendment to the protocol was adopted in June 1990. The amendment requires participating nations to completely phase out the major chemicals destroying the ozone layer, by 2000 in developed countries and by 2010 in developing countries. In the face of evidence that ozone depletion was accelerating even more, another amendment to the protocol was adopted in November 1992, moving the target date for the complete phaseout of CFCs in to 1996. Timetables for phasing out all of the suspected ozone-depleting halogens were shortened in the 1992 meeting.

Unfortunately, even with the ban, such quantities of CFCs are already present in existing cars, refrigerators and air conditioners that normal breakdown of the units will continue to contribute to atmopheric CFC levels for some years to come. On the positive side, atmospheric chlorine and bromine concentrations reached a peak in 1995 and are expected to decline to "safe" levels by about 2050.

Action in the United States. The United States is by far the leader in the production and use of CFCs and other ozone-depleting chemicals, with du Pont Chemical Company being the major producer. Following 15 years of resistance, du Pont pledged in 1988 to phase out CFC production by 2000. In late 1991, a spokesperson announced that, in response to new data on ozone loss, the company would accelerate its phaseout by three to five years. Company scientists are well along in their development of suitable substitutes.

Many of the large corporate users of CFCs (AT&T, IBM, and Northern Telcom, for example) completely phased out their CFC use by 1994. Also, du Pont spoke in oppostion to 3 bills introduced in September 1995 in the 104th Congress that were designed to terminate U.S. participation and compliance with the CFC-banning protocols.

The Clean Air Act of 1990 also addresses this problem, in Title VI, "Protecting Stratospheric Ozone." Title VI is a comprehensive program that restricts the production, use, emissions, and disposal of an entire family of chemicals identified as ozone depleting. For example, the program calls for a phaseout schedule for the hydrochlorofluorocarbons (HCFCs), a family of chemicals being used as less damaging substitutes for CFCs until nonchlorine substitutes are available. Halons—used in chemical fire extinguishers—were banned in 1994. The act also regulates the servicing of refrigeration and air-conditioning units. In addition, the EPA's hand has been substantially strengthened by Congress, and we can hope that the agency will continue taking aggressive action to protect the ozone layer.

January 1, 1996 has come and gone, and in the industrialized countries, CFCs are no longer being produced. In the wake of the ban, it might be imagined that we will be unable to find suitable substitutes and will have to shut down our compressors. The truth is, substitutes are already available, and in some cases, are even less expensive than the CFCs. The most commonly used ones are HCFCs, which still contain some chlorine, and are scheduled for a gradual phaseout. The most promising substitutes are HFCs—hydrofluorocarbons, which contain no chlorine and are judged to have no ozone-depleting potential. One unfortunate consequence of the CFC ban is a lively black market that, in Miami, is judged second only to drugs in value. Government agents have seized millions of pounds of CFCs and have brought many would-be smugglers to justice.

Final Thoughts. The ozone shield story is a remarkable episode in human history. From the first warnings in 1974 that something might be amiss in the stratosphere because of a practically inert and highly useful industrial chemical, through the development of the Montreal Protocol, and the final steps of CFC phaseout that are still occurring, the world has shown it can respond collectively and effectively to a clearly perceived threat. The scientific community has played a crucial role in this episode, first alerting the world and then plunging into intense research programs to ascertain the validity of the threat. Scientists continue to influence the political process that has forged a response to the threat. Although skeptics still stress the uncertainties in our understanding of ozone loss and its consequences, the strong consensus in the scientific community convinced the world's political leaders that action was clearly needed. This is a most encouraging development as we look forward to the forging of a sustainable society.

Review Questions

1. What are the important characteristics of the stratosphere? Of the troposphere?
2. What is the difference between an acid and a base?
3. What is the pH scale? The pOH scale? How are they related?
4. What two major acids are involved in acid deposition? Where does each come from?
5. How can a shift in environmental pH affect aquatic ecosystems?
6. In what other ecosystems can acid deposition be observed? What are its effects?
7. Discuss several strategies for controlling acid deposition. Which are considered by ecologists to be the most effective?
8. What provisions were included in the Clean Air Act of 1990 to address the problem of acid deposition?
9. Describe the heat-trapping effects of carbon dioxide.
10. Which of the greenhouse gases are the most significant contributors to global warming?
11. What are four possible impacts of rising global temperature?
12. What steps could be taken to stabilize the greenhouse gas content of the atmosphere?
13. How is the ozone shield formed? What causes its breakdown?
14. Describe four sources of CFCs entering the stratosphere.
15. How do CFCs affect the concentration of ozone in the stratosphere?
16. What causes the formation of an ozone "hole"?
17. Discuss several efforts that are currently underway to protect our ozone shield.

Thinking Environmentally

1. What arguments have the utility industries presented to delay action on acid deposition? Evaluate the validity of their concerns.
2. Compile a list of ways you are producing carbon dioxide. What steps could you take to decrease this production?
3. In your opinion, does the greenhouse effect exist? Defend your stance.
4. How would a ban on chlorofluorocarbon (CFC) production change the way you live? Give your reasoning for or against such a ban.
5. The Swedish government has made efforts to attack the symptoms of acid deposition directly, by trying to neutralize acidic lakes with lime. Discuss the pros and cons of this method.

Making a Difference

If you have an environmental action group on campus, you might consider doing what a group of students at the University of Maryland School of Law did. Through the Chicago Board of Trade, they bought an emission allowance for one ton of sulfur dioxide and retired it from use, thus guaranteeing that future sulfur dioxide emissions will be one ton less.

Pollution and Public Policy

Key Issues and Questions

1. Environmental public policy includes laws and agencies that deal with a society's interactions with the environment. How necessary are environmental public policies?
2. Environmental public policy appears in response to specific problems and needs. How is environmental public policy developed in modern societies?
3. It is argued that many environmental regulations are too costly. Do the economic effects of environmental public policy outweigh the costs?
4. It is important to craft environmental policies that make economic as well as environmental sense. Which procedures lead to sound environmental policies?
5. Cost-benefit analysis is an economic tool that is applied to environmental policies. How is the cost-effectiveness of policies measured?
6. Risk assessment is a scientific tool that the EPA is applying to its regulatory work. How is risk assessment employed in policy development?
7. Certain environmental hazards carry with them definite risks to the public. How does public policy accomplish the management of these risks?
8. The EPA is concerned that the most important risks receive their due attention, and have proposed a more rigorous comparative risk analysis. How can comparative risk analysis influence public policy?

"There was once a town in the heart of America where all life seemed to live in harmony with its surroundings. The town lay in the midst of a checkerboard of prosperous farms, with fields of grain and hillsides of orchards where, in spring, white clouds of bloom drifted above the green fields. In autumn, oak and maple and birch set up a blaze of color that flamed and flickered across a backdrop of pines. Then foxes barked in the hills and deer silently crossed the fields, half hidden in the mists of the fall mornings. . . . The countryside was, in fact, famous for the abundance and variety of its bird life, and when the flood of migrants was pouring through in spring and fall people traveled from great distances to observe them."

[Photo by Mark Boulton/Photo Researchers, Inc.]

With these words, scientist and author Rachel Carson opened the book *Silent Spring*, published in 1962 (Fig. 17-1). In the first chapter, "A Fable for Tomorrow", Carson described a town, sprayed from the air with a white powder (a pesticide), where the birds, mammals, fish, and eventually, humans began to sicken and die. Without bird songs, it was a "silent spring." Carson's book went on to describe in detail the widespread and indiscriminate use of pesticides in the United States. Her immediate target was the pesticide industry and the USDA, since the latter both promoted and regulated the use of pesticides. Even before *Silent Spring* was published, chemical companies tried in vain to discredit the book. After publication of the book, both it and Carson were vilified by the chemical and agricultural industries. Scientists were found who wrote unfavorable reviews, Carson herself was subjected to vindictive verbal attacks, and industrial sources produced books and "fact" kits to counteract the impact of *Silent Spring*. However, this book was to shock the public and the scientific community into action on many environmental fronts. Clearly, Rachel Carson had touched a nerve. Her work is credited with greatly encouraging the growth of grassroots environmental organizations. Most observers also give her credit for the birth of the Environmental Protection Agency, the premier regulatory agency responsible for environmental public policy.

Chapters 12–16 show the many serious problems we face regarding pollution. They also show the rising public concern that has led to a network of laws and agencies to deal with our interactions with the environment, both within our country and in cooperation with other nations. In this chapter, we are concerned with the ways societies make the rules that govern how we relate to our environment—in other words, environmental public policy. We look at the origins of that policy, its economic impact, the kinds of regulations that were developed, and the reasons for having such regulations. Then we turn our attention to two major tools of regulators: cost-benefit analysis, which applies economic criteria to the regulatory processes, and risk analysis, which brings scientific knowledge to bear on those processes.

Origins of Environmental Public Policy

Environmental public policy includes all of the laws and agencies in a society which deal with that society's interactions with the environment. Included are the policies that prevent or lower the pollution of air, water, and land, as well as those concerned with the use of natural resources. Public policies are developed at all levels of government: local, state, and federal. However, in a democratic society, the ultimate responsibility for environmental public policy should reside with the public—not the "authorities." Thomas Jefferson put this clearly: "I know of no safe depository of the ultimate powers of the society but the people themselves; and if we think them not enlightened enough to exercise their control with a wholesome discretion, the remedy is not to take it from them, but to inform their discretion." This is not to say that the public directly makes the rules and

FIGURE 17-1
Marine scientist and author Rachel Carson, whose courageous book *Silent Spring* broke the conspiracy of silence on the wide-scale use of pesticides and jump-started the environmental movement. (UPI/Bettmann.)

TABLE 17-1
Principal Health and Productivity Consequences of Environmental Mismanagement

Environmental problem	*Effect on health*	*Effect on productivity*
Water pollution and water scarcity	More than 2 million deaths and billions of illnesses a year are attributable to pollution; poor household hygiene and added health risks are caused by water scarcity.	Declining fisheries; rural household time (time spent fetching water) and municipal costs of providing safe water; depletion of aquifers, leading to irrversible compaction; constraint on economic activity because of water shortages.
Air pollution	Many acute and chronic health impacts: excessive levels of urban particulate matter are responsible for 300,000–700,000 premature deaths annually and for half of childhood chronic coughing; 400 million–700 million people, mainly women and children in poor rural areas, are affected by smoky indoor air.	Restrictions on vehicles and industrial activity during critical episodes; effect of acid rain on forests, bodies of water, and human artifacts.
Solid and hazardous wastes	Diseases spread by rotting garbage and blocked drains. Risks from hazardous wastes are typically local, but often acute.	Pollution of groundwater resources.
Soil degradation	Reduced nutrition for poor farmers on depleted soils; greater susceptibility to drought.	Field productivity losses in the range of 0.5–1.5% of gross national product are common on tropical soils; offsite siltation of reservoirs, river-transport channels, and other hydrologic systems.
Deforestation	Localized flooding, leading to death and disease.	Loss of sustainable logging potential and of erosion prevention, watershed stability, and carbon storage by forests. Loss of nontimber forest products.
Loss of biodiversity	Potential loss of new drugs.	Reduction of ecosystem adaptability and loss of genetic resources.
Atmospheric changes	Possible shifts in vector-borne diseases; risks from climatic natural disasters; diseases attributable to ozone depletion (300,000 more skin cancers per year; 1.7 million cases of cataracts per year).	Sea-rise damage to coastal investments; regional changes in agricultural productivity; disruption of marine food chain.

Source: "Development and the Environment," *World Development Report 1992,* The World Bank, Oxford Univ. Press, Table 1, p. 4.

regulations, but it is to say that the public should be involved in the origin of the policies and must be informed and involved as the policies are developed and administered over time.

The Need for Environmental Public Policy

The purpose of all environmental public policies is to promote the common good. Exactly what the common good consists of may be a matter of debate, but the *improvement of human welfare* would certainly be a central concern. In addition, many would include the *protection of the natural world* that provides for the continued existence of ecosystems and species—the perspective of environmental stewardship.

What are the consequences of not having effective environmental public policies? Human populations and their economic activities have the potential for doing great damage to the environment, and that damage has direct impact on present and future human welfare (Table 17-1). The impacts of pollution and the misuse of resources are most clearly seen in those parts of the world where environmental public policy is often not well established and implemented—the developing world. As the table shows, millions of deaths and wide-spread disease are directly traced to degraded environments. The costs to human welfare are imposed on health, economic productivity, and the ongoing ability of the natural environment to support human life needs. It is quite obvious that laws to protect the environment are not luxuries to be tolerated only if they do not interfere with individual freedoms or economic development. Such laws are part of the essential foundations of justice in any organized human society, and they are ignored or downgraded only at great human and environmental cost.

Relationships Between Economic Development and the Environment

Economic activity can damage the environment and human health. In addition, an unregulated economy can make intolerable inroads on natural resources. During the latter part of the 19th century in the United States, corrupt politicians allowed almost unrestrained exploitation of forests, grazing lands, and mineral deposits. These abuses began to be addressed at the turn of the century, as government rules and regulations imposed necessary limits on private enterprise. Those who complain about the present system of government regulations fail to acknowledge the past, when free enterprise meant the freedom to make fortunes at the expense of the environment.

On the other hand, economic activity in a nation also can provide the resources needed to solve the problems it creates. It is safe to say that a strong relationship exists between the level of development of a nation and the effectiveness of its environmental public policies. Figure 17-2 shows a number of environmental indicators in relation to per capita income levels. (The data were combined from studies of different countries.) Several patterns emerge:

- Many problems decline (for example, sickness due to inadequate sanitation or water treatment) as income levels rise, because the society now has the resources available to address the problems with effective technologies.
- Some problems rise and then decline (urban air pollution) as a consequence of the recognition of a problem and then the development of public policy to address it.
- Increased economic activity causes some problems to rise without any clear end in sight (municipal solid waste, CO_2 emissions).

The key to solving all of these problems brought on by economic activity is the deliberate development of effective public policies and institutions. Where this does not happen, environmental degradation is the inevitable outcome. To understand this better, let us look at a concept that is applied to public policy development as it occurs in most modern societies: the policy life cycle.

Public Policy Development: The Policy Life Cycle

Public policy regulations are developed in a sociopolitical context, usually in response to a problem. Some policies are developed at local levels to solve local problems. Cities and towns, for example, establish zoning regulations in order to protect citizens from haphazard and incompatible land uses. Thus, residential and commercial zoning districts are kept separate, conservancy zoning protects sensitive watersheds, and rural zoning protects agricultural land.

Many problems, however, are broader in their scope and must be addressed at higher levels of government. As we saw in Chapters 15 and 16, the problems of air pollution transcend local, state, and even national boundaries. The processes that contribute to air pollution are complex, involving a myriad of activities fundamental to an industrial society—from people heating their homes and driving cars, to local dry-cleaning establishments and car repair shops providing their services, to major power plants generating electricity for entire regions. When specific problems such as acid deposition and the production of ground-level ozone are addressed in a democratic society, the development of public policy often takes a predictable course, called the **policy life cycle**.

The typical policy life cycle has four stages: *recognition, formulation, implementation*, and *control* (Fig. 17-3). Each of the stages can be said to carry a certain amount of political "weight," which varies over time and is represented in the figure by the rise in the life cycle line. The thickness of the line represents controversy and political uncertainty. We can use *Silent Spring* and the formation of the Environmental Protection Agency to illustrate the policy life cycle.

Recognition Stage. Alerted in 1957 by a close friend who had experienced aerial spraying of her small bird sanctuary and the subsequent death of many birds, Rachel Carson began to turn to the technical and scientific literature to investigate what was known about the impacts of pesticides on the natural world. The evidence of fish kills, resistance of pests to DDT, massive deaths of birds, and failure of governmental agencies to respond was there in the literature. It would take several years of painstaking weaving together of the pattern of misuse, lack of regulation, and one-sided presentation of pesticides before Carson was ready to bring her book to publication. By this time, she was thoroughly aroused and angry at what she saw as a conspiracy to keep the public uninformed about pesticides. When the book was finally published in 1962

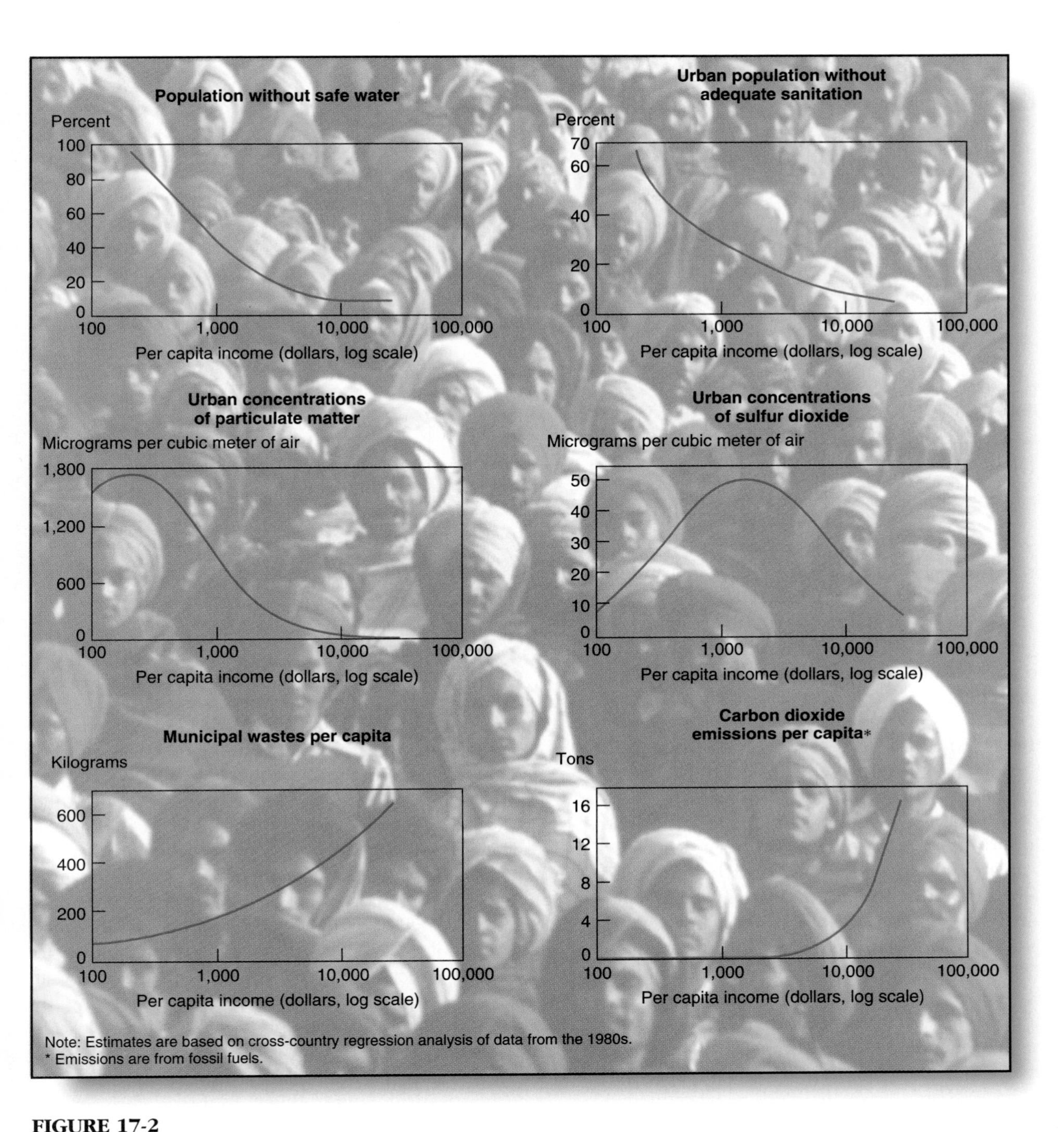

FIGURE 17-2
Environmental indicators in relation to per capita income. Some of the most serious environmental problems can be improved with income growth, others get worse and then improve, and some problems just worsen. (Graphs from "Development and the Environment," *World Development Report 1992*, The World Bank, Oxford University Press, Fig. 4, p. 11. Photo from Photo Researchers, Inc.)

(and serialized in *The New Yorker*), it ignited the firestorm of criticism already mentioned. But it also caught the public's eye, and it quickly made its way to the President's Science Advisory Committee when John F. Kennedy read the *New Yorker* version. Kennedy charged the committee with studying the pesticide problem and making appropriate recommendations for changes in public policy.

Illustrated by these events, the *recognition stage* is low in political weight. The stage begins with the early perceptions of an environmental problem, often coming as a result of scientific research. Scientists have published their findings, and the media have picked up the information and popularized it. The public is now involved, and the political process is underway. During this stage, dissension is high; opposing views on the

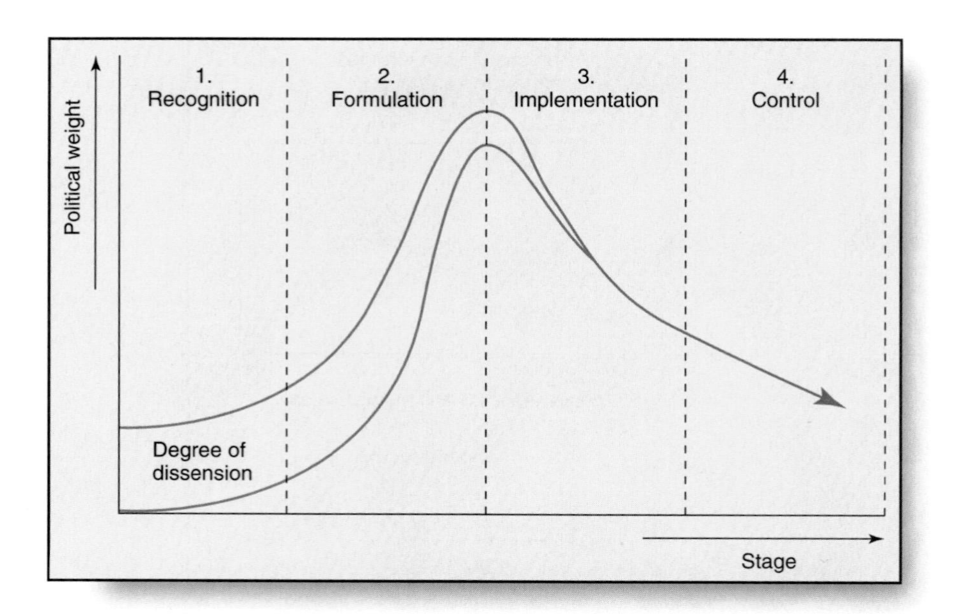

FIGURE 17-3
The policy life cycle. Most environmental issues pass through a policy life cycle in which an issue is accorded different degrees of political "weight" as it moves through the cycle. The final result is a policy that has been incorporated into the society and a problem that is under control.

problem surface as businesses or technological industries respond to the bad news that they are at the root of some new environmental problem. Eventually, the problem gets attention from some level of the government, and the possibility of addressing it with public policy is considered.

Formulation Stage. Kennedy's committee made recommendations a year later (1963) that fully supported Carson's thesis in *Silent Spring*. It became clear to policymakers during the 1960s that environmental policy was weak and often compromised by ties to the very industries creating the problems. Public debate ensued, environmentalism in the form of new organizations and legal action developed, and the executive and legislative branches of government wrestled with what to do to establish effective environmental public policy. Finally, Congress passed a bill known as the Environmental Policy Act in 1969, the first bill to recognize the interconnectedness of ecological systems and human enterprises. Shortly after that, a commission appointed by Richard Nixon to study environmental policy recommended the creation of a new agency. This agency would be responsible for dealing with air, water, solid waste, pesticide use, and radiation standards, in recognition of the interrelated web of water, soil, air, and biological life so eloquently pictured by Rachel Carson. By this time, Carson had died of cancer. Her legacy, however, was the EPA, and the new agency was given a mandate to protect the environment against pressures from other governmental agencies or industry, on behalf of the public. The year was 1970, the same year that 20 million Americans celebrated the first Earth Day.

Thus, we have the *formulation stage*—a stage of rapidly increasing political weight. The public is now aroused, and debate about policy options occurs in the corridors of power. The political battles may become fierce, as questions of regulation and who will pay for the proposed changes are addressed. The media coverage is high, and politicians begin to hear from their constituencies. Lobbyists for special interests or environmental groups put pressure on legislators to soften or harden the policy under consideration. During this stage, policymakers should be considering what may be called the *"Three E's"* of environmental public policy: **effectiveness** (the policy really accomplishes what it intends to do in improving the environment), **efficiency** (the policy accomplishes its objectives at the least possible cost), and **equity** (the policy parcels out the financial burdens fairly among the different parties involved). Often, policymakers are prone to emphasize effectiveness over efficiency and equity at this stage of development, because they are looking for a workable solution and trying to make it into a law as soon as possible.

Implementation Stage. Given its broad jurisdiction, the newly created EPA embarked idealistically on a course that would take the agency into both environmental research, which was needed in order to determine how to develop environmental protection standards, and regulatory enforcement. The 1970s were declared by President Nixon to be the "environmental decade." The EPA's first administrator, William Ruckelshaus, proposed that the agency's mission was, first and foremost, the "development of an environmental ethic." Within a few years, Congress enacted more than 20 major pieces of environmental legislation, giving the EPA increasing power to develop regulations. The agency and has become the most powerful regulatory agency in the country, able to affect everything from the design of automobiles to standards for nuclear power plants and sanitary landfills.

At this point, the policy has reached the *implementation stage*, where its real political and economic costs are exacted. The policy has been determined, and the focal point moves to a regulatory agency. During this stage, public concern and political weight are declining. By now the issue is not very interesting to the media, and the emphasis shifts to the development of specific regulations and their enforcement. Industry learns how to comply with the new regulations. Over time, greater attention may be given to efficiency and equity as all the players in the process gain experience with the policy.

Control Stage. Over time, the EPA developed into a federal bureaucracy quite vulnerable to changes in the political tides. The idealism of the early days shifted into a more pragmatic and, sometimes, insensitive attitude toward environmental concerns—at least at the highest levels of the agency. The 1980s saw the ascendency of Ronald Reagan, who seemed bent on curbing the EPA's power and effectiveness by appointing people opposed to the EPA to run it. However, after some scandals racked the agency, the Congress and the public were aroused. With changes at the top, new EPA administrators began to address pollution at the global level, while consolidating the domestic regulatory apparatus. In the 1990s the agency regained much of its former reputation for providing leadership regarding environmental issues in the nation.

The final stage in the policy life cycle is the *control stage*. By this point, years have passed since the early days of the recognition stage. Problems are rarely completely resolved, but the environment is improving as things are moving in the right direction. Policies (and their derived regulations) are broadly supported and often become imbedded in the society, although their vulnerability to political shifts continues. Regulations may become more simplified. The policymakers must now see that the problem is kept under control, and in due time, the public often forgets that there ever was a serious problem.

The policy life cycle is, of course, a simplified view of what is frequently a very complex and contentious political process. At any one time, different problems will be in different stages of the life cycle, as shown in Fig. 17-4 for a number of environmental problems in the industrialized countries. Also, countries in different stages of economic development will be in different stages of the policy life cycle for a given problem. For example, sickness and death due to polluted water are still serious problems in many developing countries; public policies have not yet caught up with the need for water treatment and sewerage. On the other hand, this problem is clearly in the control stage in most industrialized societies. The reason for this discrepancy is often traceable directly to the costs that lie behind the development and implementation of public policy.

Economic Effects of Environmental Public Policy

What are the relationships between a country's economy and its environmental public policies? Are there some policies that are too costly? How should a society parcel out its limited resources to address envi-

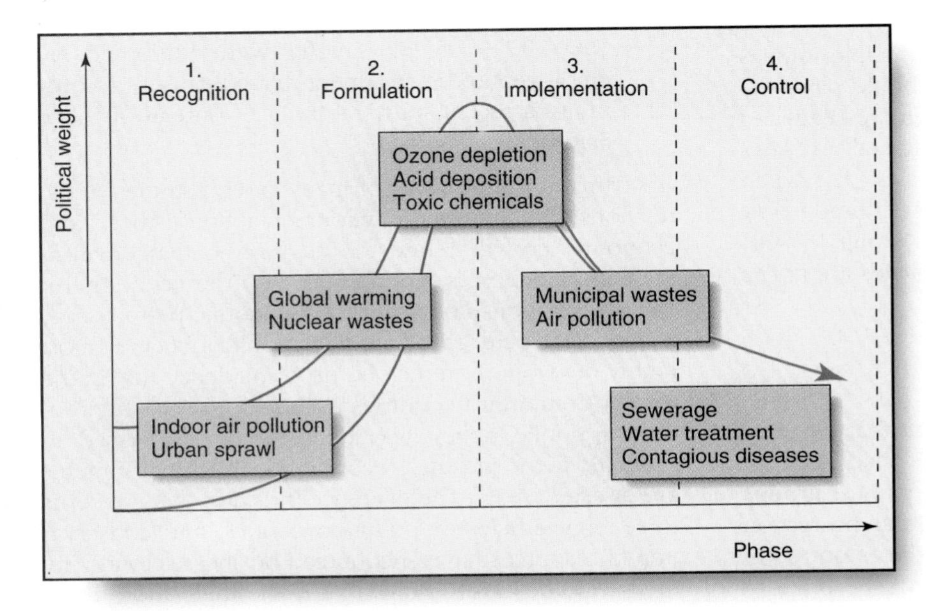

FIGURE 17-4
Some environmental problems in different stages of the policy life cycle in industrial societies. Water treatment is at the end of the cycle, while indoor air pollution and urban sprawl are still at the beginning.

ronmental problems? Are environmental regulations a drag on the economy? Let us begin to sort out these questions by looking at the relationship between environmental policies and costs.

As we have seen, policies do not just appear out of thin air; they are the result of some version of the policy life cycle. In countries undergoing economic development, the choice of environmental policies can often make the difference between rapid growth and stagnation. And in the already industrialized countries, the best policies are those that conform to the "Three E's." When they fail to do so, it is time to take a hard look at the policies and agitate for them to be changed.

Costs of Policies

Some policies have relatively little or no direct *monetary cost*—they do not require major investments of administration or resources. Thus, removing subsidies to special interests and denying special access to national resources can result in a more efficient and equitable operation of the economy and protect the environment. For example, the use of national lands for cattle grazing and timber harvesting is subsidized in the United States. As a result, the real costs—and environmental consequences—of these activities are borne not just by the special-interest groups that enjoy access to these resources, but by all taxpayers. Such policies, however, often have very real *political costs*, as powerful interests do everything they can to hold onto their privileges.

Most environmental policies involve some very real costs that must be paid by some segment of society. Here we are talking about policies that control pollution. Resources are invested in these policies because their benefits are judged to outweigh their costs—the public welfare is improved, and the environment is protected.

Who pays the costs? The equity principle implies that those benefiting from the policies should pay—the customers of businesses whose activities are regulated and the people whose health is protected by laws against polluting. (It should be recognized that when business are taxed or industries are forced to add new technologies, they will pass their costs along to consumers.) In short, the costs of public policies are borne by the public, in one way or another.

Impact on the Economy

How often have you heard the argument that environmental regulation is excessive and is bad for the economy? Many politicians and special interest groups argue that our concern for environmental protection costs hundreds of thousands of jobs, reduces our competitiveness in the marketplace, drives up the price of products and services, and in general imposes costs on the economy that are non-productive.

How sound are these arguments? This major concern was addressed by Roger Bezdek, president of Management Information Services, Inc., in *Environment* (Sept. 1993), and we refer to his article in the following analyses.

First, we must recognize that much of the evidence used to support the charges that have been made is anecdotal. Thus, "environmental restrictions on energy extraction and production have caused the loss of 400,000 jobs," according to the American Petroleum Institute. "Measures to protect the Everglades will cause the loss of 15,000 jobs," according to Florida sugar growers. Indeed, numerous cases can be cited where jobs were lost because of environmental regulations.

Anecdotal evidence is also used in favor of environmental policy. "Recycling has created 14,000 jobs in California in 1991," according to the California Planning and Conservation League Foundation. "The Clean Air Act of 1990 will generate 60,000 new jobs," according to the EPA. Many sectors of the U.S. economy most subject to environmental regulations—plastics, fabrics, etc.—have improved their efficiency and their competitiveness in the international market.

What can we conclude? Is there something more substantial to help us with this important controversy? The best evidence comes from careful studies of what is actually happening. These studies, all documented in Bezdek's article, reveal the following interesting findings:

- In general, states with the strictest environmental regulations also had the highest rates of job growth and economic performance.
- Nations with the highest environmental standards also had the most robust economies and rates of job creation.
- Only 0.1% of job layoffs were attributed by employers to environment-related causes, according to a recent study by the U.S. Bureau of Labor Statistics.

In short, the results of many careful studies of the relationship between environmental protection and economic growth refute the suggestion that environmental regulation is bad for jobs and the economy. In fact, the relationship is positive: Economic performance is highest where environmental public policy is most highly developed. (It may be no coincidence that in the former Communist countries in Europe, where environmental public policy was deliberately suppressed in order to favor industry, economies and environments are disaster areas.) The evidence indicates that concerns for energy efficiency, pollution control, and conservation of resources have encouraged businesses to modify their technologies in ways that make them more competitive, not less, in the national and international marketplace. And the evidence for jobs suggests that at least as many jobs have been created by environmental protection as have been lost.

Taken as a whole, environmental protection is a huge industry, and it has been growing more rapidly than the GNP. The graph in Figure 17-5 demonstrates environmental protection as a growth industry. A study of 1992 total expenditures for the industry revealed that it had:

- $355 billion in total industry sales
- $14 billion in corporate profits
- $63 billion in federal, state, and local government revenues
- 4 million jobs

In summary, we can draw several conclusions from our examination of the impact of environmental policy on the economy:

- Environmental public policy does not *diminish* the wealth of a nation; rather, it *transfers* wealth from polluters to pollution controllers and to less polluting companies.
- The "environmental protection industry" is a major job-creating, profit-making, sales-generating industry.
- Thus, the argument that environmental protection is bad for the economy is simply unsound: Not only is it good for the economy, but environmental public policy is responsible for a less hazardous, healthier, and more enjoyable envronment.

Implementing Environmental Public Policy

Earlier, we identified three important criteria that should be applied in formulating and evaluating environmental public policy: effectiveness, efficiency, and equity. We cannot assume that all of our environmental public policy measures up to these criteria. In reality, many of the policies now in effect could be improved. Some observers believe that a number of environmental regulations should be abolished. In light of these concerns, how might we evaluate environmental public policy? Are there ways of ensuring that the "Three E's" will indeed be incorporated into a given policy?

Adopting Good Policies

Information and knowledge are essential for formulating good environmental public policy. It is important to know the facts about hazards in the environment before attempting to protect people from those hazards. It is also important to understand the limits of exploitation of renewable resources before establishing rules for using the resources. It is no accident that environmental science now represents a major research and educational enterprise that crosses many disciplines; the task before us is formidable, and the search for knowledge and information about human interactions with the environment proceeds with increasing urgency. In the midst of this ongoing task, however, there are many

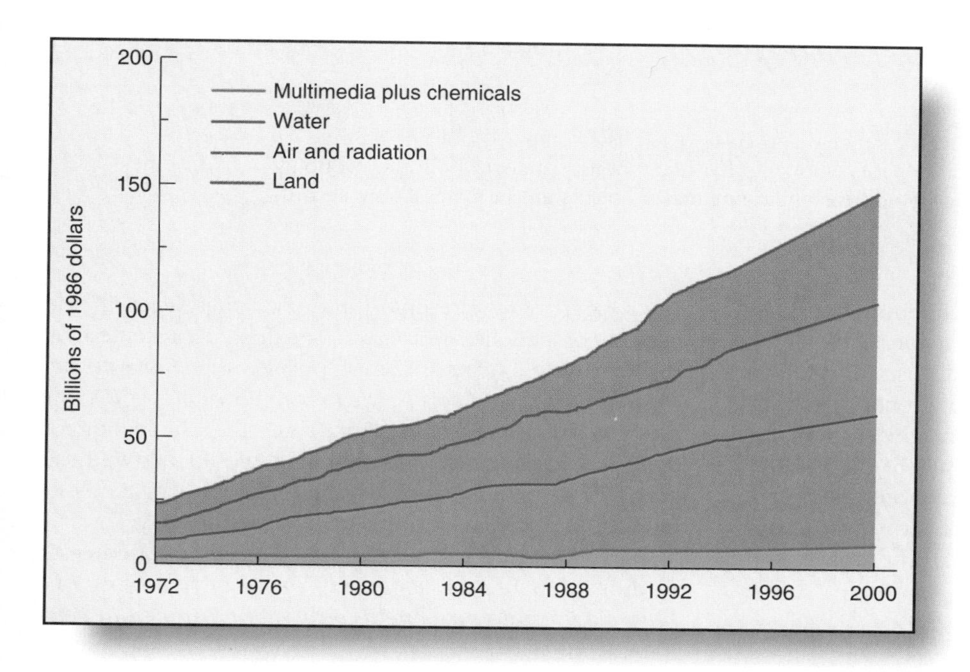

FIGURE 17-5
Total yearly costs of pollution control in the United States, assuming full implementation of existing regulatory laws. (From EPA, *Environmental Investments:* "The Cost of a Clean Environment," 1990.)

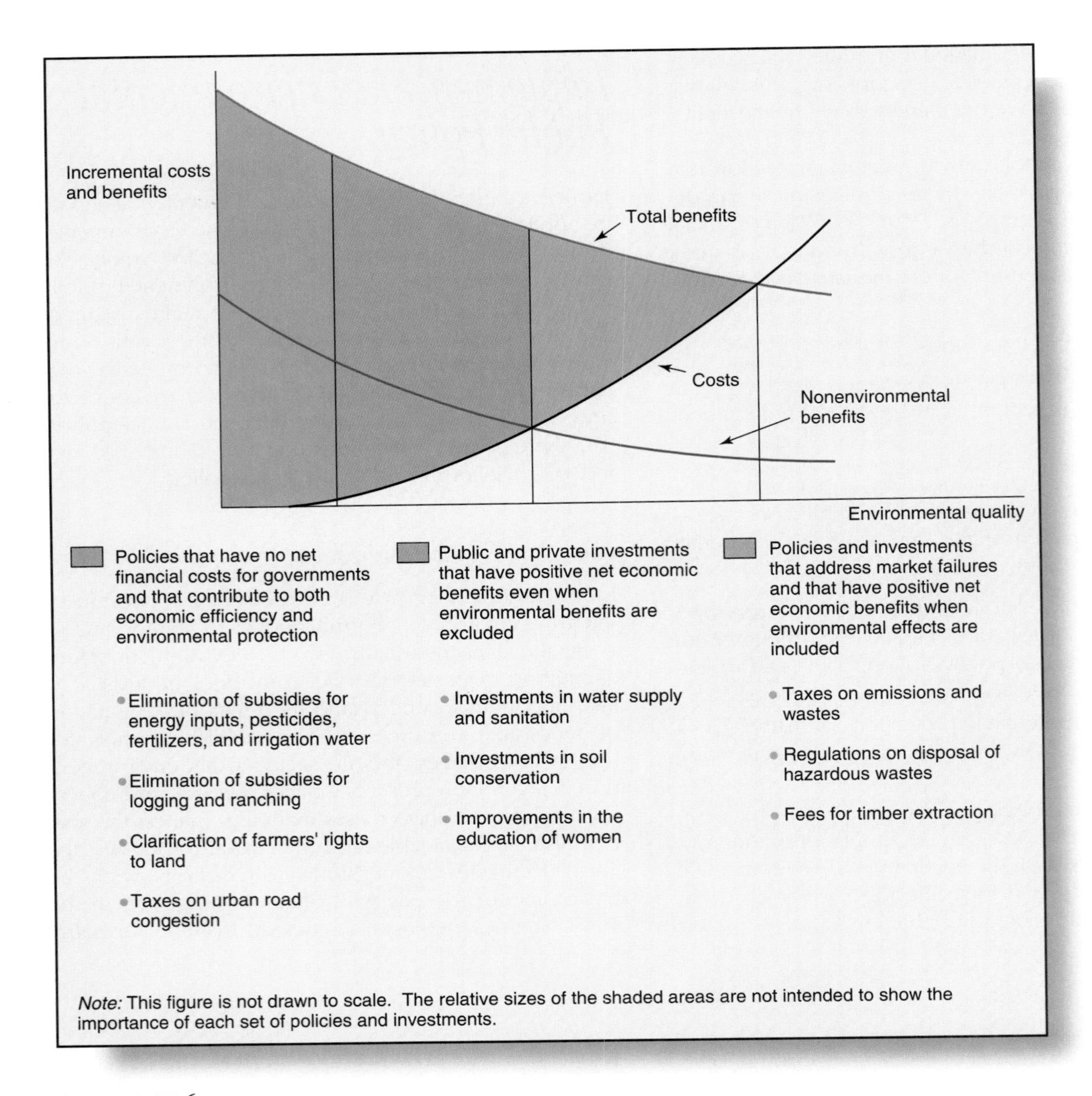

FIGURE 17-6
Environmental policies from the perspective of benefits and costs. The chart compares policies on the basis of both economic and environmental benefits, showing that many policies benefit both the economy and the environment. Policies that specifically address environmental problems are justified strictly according to cost-benefit standards.

environmental concerns that are well enough understood to enable sound policymaking to proceed.

The perspective of benefits and costs provides a useful means of looking at environmental policies. Another perspective that gives us insight into policy development is that of risk analysis, where the risks of a particular technology or hazard are compared with the benefits, and then policy is directed toward managing the risks. We look at the details of these two perspectives shortly. Let us first examine some philosophical issues.

Figure 17-6 presents three sets of policies in the context of benefits and costs. The first, represented by pink shading, shows kinds of policies that *have no net financial costs and contribute to both economic efficiency and environmental protection*. Thus, the elimination of subsidies for logging and ranching would allow these activities to be subjected to market forces; this tends to facilitate the criteria of equitability and efficiency.

The second set of policies (shaded blue) consists of those that *require some investment of resources, but have clear economic benefits even if environmental ben-*

Earth Watch

Free-Market Environmentalism

Cost-benefit analysis works well only if all of the costs and benefits of a polluting activity are included in the calculations. Let us assume this is possible. What would happen if everyone had to pay the true costs of their use of the environment as they used resources or degraded the air or water? Such charges are often called *green fees*. Oddly enough, two groups that frequently are at each other's throats—economists and environmentalists—both agree that such a policy would go a long way toward solving many of our environmental problems.

Economists point out that in a free-market economy, the market will guarantee that resources will be used in the most efficient way. This means that businesses and people who use, say, 100 gallons of gasoline and have to pay a levy that reflects the true costs of that gasoline and the pollution it produces will be highly motivated to keep their use of gasoline to an absolute minimum. They will adopt alternative modes of transportation and more efficient vehicles. Including all of the external costs of the gasoline in its price would involve adding together the cost of defending Persian Gulf shipping lanes, the costs of air pollution, and even the expected costs of coping with global warming. The green fee could be imposed as a tax on gasoline, and to ease its impact on the economy, the fee could be implemented gradually until it brings the price of gasoline to its true level.

Great, say the environmentalists. This would mean that the people who do the damage pay for their environmental impact, something environmental groups have been calling for all along. They insist that some of the revenues collected should be used to mitigate the impacts of pollution and resource use, but if this were to be accomplished, the outcome would move our society in the direction of sustainability.

Pollution fees, gasoline taxes, and user fees all have the potential for "internalizing" the external costs of environmental degradation and resource use. If such green fees could become public policy, it is possible that other taxes (e.g., personal income taxes) could be lowered, since these are often the sources of revenue that are paying much of the external costs.

The question is, Would such a policy be politically feasible? Proponents think that it may be. If people are asked if they *like* higher taxes, the answer is a highly predictable *no*. But if asked whether they would rather be taxed on their energy use, their trash production, and their purchases of high-impact goods than be taxed on their incomes and investments, they might well choose the former, especially if they can appreciate the environmental benefits that accrue from such a policy.

---

efits are not considered. Ensuring a healthful water supply and establishing sewage treatment systems would be good examples of this kind of policy.

The third set of policies (appropriately shaded green!) consists of those that are *directed toward solving specific environmental problems.* These policies target activities that are not well addressed by market forces, such as industries that pollute the air or fisheries that exploit the coastal waters. Here, the perspective of benefits and costs comes most clearly into focus. As we see later, it is not always easy to measure the benefits and costs of a given policy; however, this perspective can often be used in the process of formulating and evaluating public policy directed toward solving environmental problems. There should be no question about adopting public policies that fit into the first two categories, however; the benefits far outweigh the costs.

Policy Options: Market or Regulatory?

Let us assume that the decision has been made to develop a policy to deal with an environmental problem. Now what? The objective of environmental public policy is to *change the behavior of polluters and resource users so as to benefit public welfare and the environment.* There are two main ways to accomplish these behavior changes: by using basic **market approaches** and set prices on pollution and the use of resources, and by using a direct **regulatory approach** wherein standards are set and technologies are prescribed—the command-and-control strategy discussed in Chapter 15. Both approaches have their advantages and disadvantages. Many newer policies include elements of both.

Market-based policies have the virtues of simplicity, efficiency, and (theoretically) equity. All polluters are treated equally and will choose their responses based on economic principles of profitability. There are strong incentives to reduce the costs of using resources or of paying for the right to a certain amount of pollution. The trading of emission allowances under the Clean Air Act of 1990 represents a good example of the market approach to pollution control. User fees for the disposal of municipal solid waste is another example.

Some polluters and resource users cannot make simple market-based choices, however, because of their poverty or powerlessness. In these cases, the market approach must be modified to include concerns for the disadvantaged—distributional justice. This is especially crucial when the resources or processes in question are a matter of economic survival.

Many environmental problems are not readily amenable to market-based policies. For example, based on knowledge of the health effects of pollutants, a society may choose to set standards that reflect the health of the most vulnerable members of the population. This is the case with the basic criteria covering air pollutants (Chapter 15). To meet these standards, regulations are established, and polluters are required by law to comply with the regulations. The regulatory approach also works well with land use issues, where certain values are upheld that will not necessarily be protected by a straightforward market approach. The regulatory approach has been the most commonly used approach in environmental public policy. In the interest of addressing outstanding environmental problems, policymakers are prone to turn first to this approach because it is deemed effective.

One of the shortcomings of the regulatory approach is that it practically guarantees a certain sustained level of pollution. If a polluter is told to use a particular technology or is given a cap on emissions, the polluter has no incentive to invest in technologies or a reduction in the source that would keep pollution at lower levels than allowed.

The foregoing has been a quick look at what could be called the mine field of environmental public policymaking. It is obviously not a simple process. However, neither is it entirely opaque. Using some of the insights developed, we are able to evaluate specific public policies.

Accordingly, we now turn to the details of two specific tools that are employed in developing environmental public policy: cost-benefit analysis and risk analysis. In doing so, we will address two important questions:

- *How can we measure the cost-effectiveness of regulating pollution?*

  We can use cost-benefit analysis, a major tool of the new discipline of environmental economics.

- *How can we be sure that our public policies are providing adequate protection for humans and the environment?*

  We can apply risk analysis, which brings scientific judgment to bear on the regulatory process.

Cost-Benefit Analysis

In the first decade of strong environmental regulation—the 1970s—most environmental policies were developed with little consideration of economics. Early policies were typical command-and-control responses to air and water pollution, focusing on controlling emissions from cars and factories, releases from sewage outfalls and other point sources, and pesticides being sprayed over large areas. The policies undoubtedly had a significant economic impact on businesses, consumers, and the work force.

In reaction to the situation, and concerned that the U.S. economy in general and business in particular might be overregulated and thus unduly restricted, Ronald Reagan issued Executive Order 12291 in February 1981. This order required all executive departments and agencies to support every new *major* regulation with cost-benefit analysis. To qualify as major, a regulation had to (1) impose annual costs of at least $100 million, *or* (2) cause a significant increase in costs or prices for some sector of the economy or geographic region, *or* (3) have a significant adverse effect on competition, investment, productivity, employment, innovation, or the ability of U.S. firms to compete with foreign firms. In 1993, Bill Clinton issued Executive Order 12866, which continued most of the policies of Executive Order 12291 (the third qualification was eliminated) and added the processes of public and inter-agency review of proposed regulatory rules. The 1993 executive order uses the term "significant regulatory action," rather than "major rule," in evaluating the regulations and requires the regulatory agency to address how the proposed action will reduce risks to public health, safety, or the environment.

Both of the preceding executive orders have the effect of applying an economic tool to many decisions on public policies dealing with the environment. Let us take a closer look at this process.

A **cost-benefit analysis** compares the estimated costs of a project with the benefits that will be achieved. Such an analysis is often used as a means of rationally deciding whether to proceed with a given project. All costs and benefits are given monetary values and compared by means of what is commonly referred to as a **benefit-cost** (or cost-benefit) **ratio**. A favorable ratio for a project means that the benefits outweigh the costs. Such a project is said to be **cost effective**, and there is thus an economic justification for proceeding with it. The analysis usually involves considering several options for accomplishing the project and selecting the option with the best benefit-cost ratio. If costs are projected to outweigh benefits, the project may be revised, dropped, or shelved for later consideration.

Cost-benefit analysis of environmental issues is intended to build efficiency into policy so that society does not have to pay more than is necessary for a given level of environmental quality. If the analysis is done properly, it will take into consideration *all* of the costs and benefits associated with a regulatory option. In so doing, it must address the problem of *externalities.*

In the language of economics, an **externality** is an effect of a business process that is not included in the usual calculations of profit and loss. For example, when a business pollutes the air or water, the pollution imposes a cost on society in the form of poor health or the need to treat water before using it. This is an *external bad.* Another example, when workers improve their job performance as a result of experience and learning, the improvement is not credited on the company's ledgers. It is considered an *external good.* Amenities such as clean air, uncontaminated groundwater, and a healthy ozone layer are not privately owned. In the absence of regulatory controls, there are no direct costs to a business for degrading these amenities. (In other words, they are externalities.) Therefore, there are no incentives to refrain from polluting the air or water. By including *all* of the costs and benefits of a project or a regulation, cost-benefit analysis effectively brings the externalities into the economic accounting. Recent (1995) legislation that passed the House but did not make it past the Senate, would have forced taxpayers to pay polluters to stop bad externalities. This lost on its own poor merits.

The Costs of Environmental Regulations

The costs of pollution control include the price of purchasing, installing, operating, and maintaining pollution control equipment and implementing a control strategy. Even the banning of an offensive product costs money, because jobs are lost, new products must be developed, and machinery may have to be scrapped. In some instances, a pollution-control measure may result in the discovery of a less expensive way of doing something. In most cases, however, pollution control costs money. Thus, the effect of most regulations is to prevent an external bad by imposing economic costs that are ultimately shared by government, business, and consumers.

Pollution-control costs generally increase exponentially with the level of control to be achieved (Fig. 17-7a). That is, a partial reduction in pollution may be achieved by a few relatively inexpensive measures, but further reductions generally require increasingly expensive measures, and 100% control is likely to be impos-

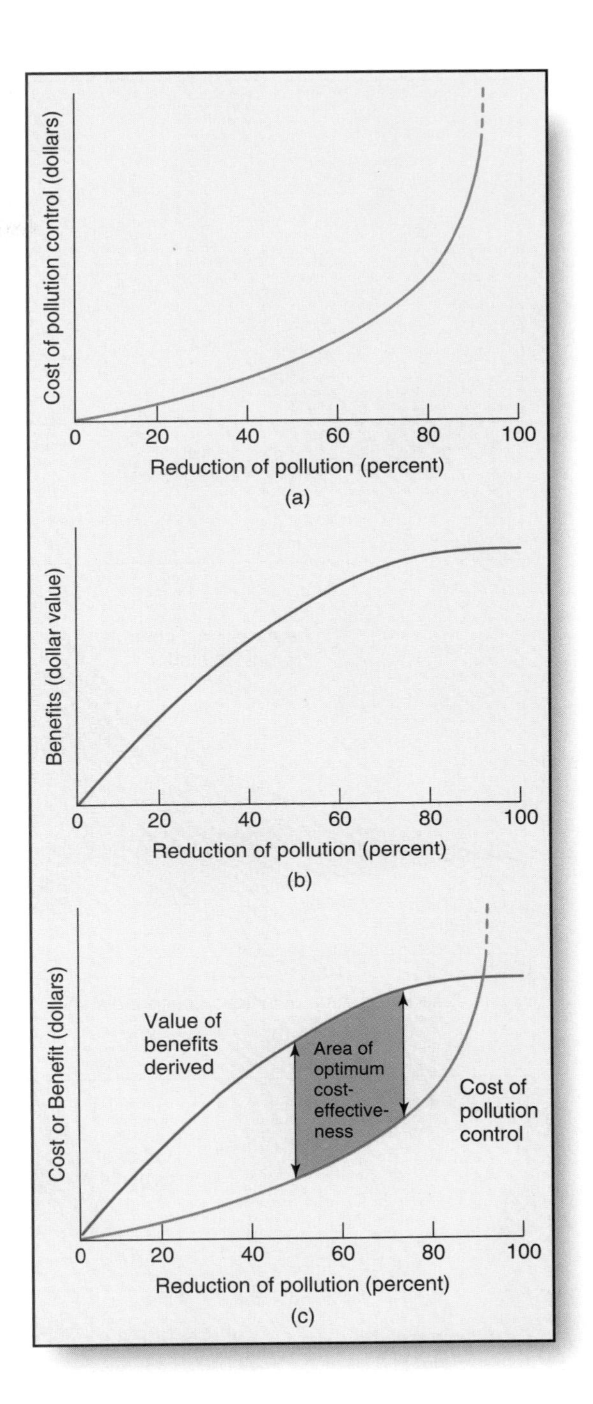

FIGURE 17-7
The cost-benefit ratio for reducing pollution. (a) The cost of pollution control increases exponentially with the degree of control to be achieved. (b) However, additional benefits to be derived from pollution control tend to level off and become negligible as pollutants are reduced to near or below threshold levels. (c) When the curves for costs and benefits are compared, we see that the optimum cost-effectiveness is achieved at less than 100% control. Expenditures to achieve maximum reduction may yield little, if any, additional benefit and hence may be cost ineffective.

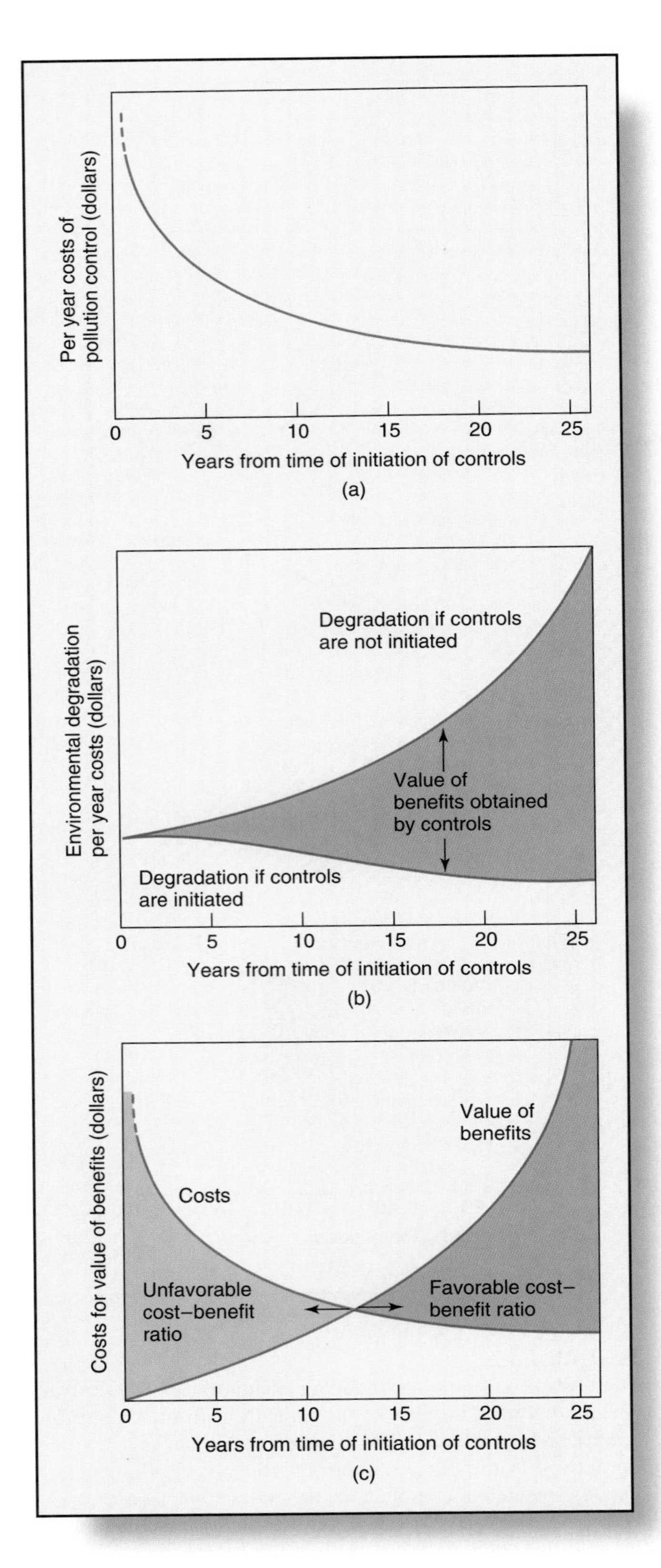

FIGURE 17-8
Evaluation of the optimum cost-effectiveness of pollution control changes with time. (a) Pollution-control strategies generally demand high initial costs. The costs then generally decline as those strategies are absorbed into the overall economy. (b) Benefits may be negligible in the short term, but they increase as environmental and human health recover from the impacts of pollution or are spared increasing degradation. (c) When the two curves are compared, we see that what may appear as cost-ineffective expenditures in the short term (5–10 years) may, in fact, be cost effective in the long term.

sible at any cost. Because of this exponential relationship, regulatory control often has to focus on stimulating new ways of reducing pollution. Indeed, the costs of pollution control constitute a powerful incentive to make substitutions, to recycle materials, or to redesign industrial processes.

In most cases, pollution-control technologies and strategies are understood and available. Thus, equipment, labor, and maintenance costs can be estimated fairly accurately. Unanticipated problems that increase costs may occur, but as technology advances and becomes more reliable, experience is gained, lower cost alternatives frequently emerge, and such unforeseen increases are negligible. The costs of pollution control are likely to be highest at the time they are initiated; then they decrease as time passes (Fig. 17-8a). The importance of this trend will become evident when we consider the time span over which costs and benefits are compared.

As we have seen in Fig. 17-5, the costs of improving the quality of the environment represent a major economic outlay. What benefits have we received in return?

The Benefits of Environmental Regulation

In Chapter 15 we analyzed the Clean Air Act's impact on society and $6.4 trillion dollar net benefit. Reduced air pollution, improved human health, avoided costs, and less harm to children from lead, all benefitted society. How does someone now argue against these regulations? The only way for special interests to weaken the Clean Air Act is if the public is not aware of these benefits in their lives.

The benefits of regulatory policies are seldom as easy to calculate as the costs. Estimating benefits is often a matter of estimating the costs of damages that *would* occur if the regulations were not imposed (Fig. 17-8b). For example, the projected environmental damage that would be brought on by a given level of sulfur dioxide emissions from a coal-fired power plant (an external bad) becomes a benefit (an external good) when a regulatory action prevents half of those emissions. Benefits include such things as improved public health, reduced corrosion and deterioration of materials, reduced damage to natural resources, preservation of aesthetic and spiritual qualities of the environment,

TABLE 17-2
Benefits That May Be Gained by Reduction and Prevention of Pollution

1. Improved human health
   Reduction and prevention of pollution-related illnesses
   Reduction of worker stress caused by pollution
   Increased worker productivity
2. Improved agriculture and forest production
   Reduction of pollution-related damage
   More vigorous growth by removal of stress due to pollution
   Higher farm profits, benefiting all agriculture-related industries
3. Enhanced commercial and/or sport fishing
   Increased value of fish and shellfish harvests
   Increased sales of boats, motors, tackle, and bait
   Enhancement of businesses serving fishermen
4. Enhancement of recreational opportunities
   Direct uses such as swimming and boating
   Indirect uses such as observing wildlife
   Enhancement of businesses serving vacationers
5. Extended lifetime of materials and less cleaning necessary
   Reduction of corrosive effects of pollution, extending the lifetime of metals, textiles, rubber, paint, and other coatings
   Cleaning costs reduced
6. Enhancement of real estate values

increased opportunities for outdoor recreation, and continued opportunities to use the environment in the future. The dollar value of these benefits is derived by estimating, for example, the reduction in health-care costs, the reduction in maintenance and replacement costs, and the economic value generated by the enhanced recreational activity. Examples of benefits are listed in Table 17-2.

The values of some benefits can be estimated fairly accurately. For example, it is well recognized that air pollution episodes cause increases in the number of people seeking medical attention. Since the medical attention provided has a known dollar value, eliminating the number of air pollution episodes provides a health benefit of that value. As another example, consider a polluted lake that is upgraded to the point where it will again support water recreation. The benefits of this are estimated by assigning a value of $3 to $5 to each anticipated swimmer-day. Such a figure is based on the fact that most people will pay that price for admission to a pool.

Many benefits, however, are difficult to estimate. Accurate cost-benefit analysis depends on assigning monetary values to every benefit, but how can a dollar value be put on maintaining the integrity of a coastal wetland, for example, or on the enjoyment of breathing cleaner air? The answer: Find out how much people are willing to pay to maintain these benefits. But how can this be done if there is no free market for the benefits? Again, economists have an answer: **shadow pricing**. This involves asking people what they *might* pay for a particular benefit if it were up to them to decide. For example, to evaluate the benefits of cleaner air in the Los Angeles basin, homeowners were asked to place a value on improving their air quality from *poor* to *fair*. The average response (in 1977) was $30 a month. The total benefit was then calculated from the number of households in the basin times the $30 average value.

The Value of Human Life. Shadow pricing becomes difficult when the analysis has to place a value on human life. Many of the pollutants to which people are exposed are hazardous; they exact a toll on health and life expectancy. To estimate the benefits of regulating such pollutants, it is necessary to calculate how many lives will be saved or how many people will enjoy better health. Finding a value for these benefits is fraught with ethical difficulties.

One approach compares how much we pay people to perform hazardous jobs with how much we pay for the same kind of work without the hazards. Another approach asks hypothetical questions, such as, what is the minimum pay you will demand to accept a risk of 1 in 50 that you will get cancer from working with chemical A? Such approaches have resulted in a range of values for human life that covers two orders of magnitude, from tens of thousands of dollars to $10 million. Any cost-benefit analysis that must factor in risk to human life requires that human lives be valued somewhere in this very broad range. Obviously, the outcome of the analysis is determined by the value selected.

Nonhuman Environmental Components. How does shadow pricing work for the nonhuman components of natural environments—for a population of rare wild flowers, for example, or a wilderness site? These things depend entirely on how willing people are to pay for their preservation. Again, monetary values must be assigned to their existence, and again, the outcome will be strongly influenced by a very subjective element in the analysis.

It is encouraging that more and more people recognize the importance of nonhuman environmental components and clearly value them highly, even if it is difficult to express their value in dollars.

Cost Effectiveness

As Fig. 17-7 (a and b) shows, significant benefit may be achieved by modest degrees of cleanup. Note how differently the cost and benefit curves behave with increasing reduction of pollution. Little additional ben-

Earth Watch

Regulation and Industry

Many people feel that industry should control its pollutants out of a good conscience. However, the competitive pressures among companies are often so severe that no company is willing to spend one penny more than is absolutely necessary to produce its products.

For example, consider two competing companies, A and B. Suppose A decides to take some pollution-control measure. Then it must either pass the costs on to customers in the form of higher prices or accept lower profits. If it raises prices, A will probably lose customers to B, who can maintain lower prices because it has no new costs to pass on. If A holds prices at their old level and so accepts lower profits, it begins to lose the financial support of investors. Thus, in its effort to be virtuous, A loses competitive advantage to B regardless of how A deals with the added costs.

Because this is the way our free-market economy works, it is necessary to institute laws and regulations that will affect all offending companies equally. Nevertheless, we see many industries, individual companies, and special-interest groups attempting to exempt themselves from regulations in order to gain an economic advantage.

This same scenario can be extended to the international level. Overseas companies without pollution control regulations (and often with lower labor costs) can sell their products in the United States for less than U.S. companies can. When this happens, the only way to establish an even playing field in U.S. markets is to erect some form of trade barrier, such as a tariff on imports.

efits are realized when cleanup begins to approach 100%, yet costs increase exponentially. This follows from the fact that living organisms—including humans—can often tolerate a threshold level of pollution without ill effect (see Fig. 15-2). Therefore, reducing the level of a pollutant below threshold levels will not yield an observable improvement.

Figure 17-7c makes it clear that, with a modest degree of cleanup, benefits can outweigh costs. At some point in the cleanup effort, the lines cross, and the costs exceed the benefits. Consequently, while it is tempting to argue that we should strive for 100% control, demanding more than 90% control may involve enormous costs with little or no added benefit. At the point when control of a particular pollutant reaches 90%, it makes more sense to allocate dollars and effort to other projects where greater benefits may be achieved for the money spent. Optimum cost-effectiveness that meets the efficiency criterion for public policy is achieved at the point where the benefit curve is the greatest distance above the cost curve.

What has been the result of cost-benefit analyses to date? Pollution of air and surface water reached critical levels in many areas of the United States in the late 1960s, and since that time huge sums of money have been spent on pollution abatement. Cost-benefit analysis shows that, overall, these expenditures have more than paid for themselves in decreased health-care costs and enhanced environmental quality. But does this mean that further expenditures of pollution control will prove equally cost effective? Or are we at the point where further expenditures will yield little, if any, benefit and the money will, in effect, be wasted? At the very least, industry, many economists, and government officials now demand more documentation of presumed benefits before consenting to further expenditures. The EPA cost-benefit analysis of the Clean Air Act is a step in this direction. Given the substantial benefits over costs ratio the EPA found, it seems clear that further delays for investing in envrionmental protection are nothing more than another method of protecting economic self-interests at the expense of the quality of the environment and society at large. To understand these views, we must take a more detailed look at cost-benefit analysis.

Problems in Comparing Costs and Benefits

The concept behind cost-benefit analysis is relatively straightforward: Simply estimate the costs that may be incurred in an environmental undertaking and the value of benefits that may be derived from it, and compare the two numbers. The difficulty lies in obtaining realistic estimates and in making objective comparisons.

Even after valid cost and benefit estimates are obtained, the comparison is often a complicated matter. We have seen that, during the initial stages of control, costs are high and observable benefits are usually few or none. As time passes, however, costs generally level off, while benefits increase and accumulate. Consequently, whether benefits outweigh costs or vice versa depends on whether one takes a short-term or a long-term view. A situation that appears to be cost ineffective in the short term may prove extremely cost effective in the long term (Fig. 17-8c). This is particularly true for problems such as acid rain or groundwater contamination from toxic wastes. In these instances, the consequences of delaying control may seriously affect large geographic areas and many millions of people, and they may be irreversible.

Those who bear the costs of pollution control and those who receive most of the benefits are frequently different groups of people. For example, industry and its shareholders may bear the costs of curtailing effluents into a river, while people who enjoy sport fishing gain the benefits. Obviously, the two parties are more than likely to reach different conclusions regarding whether benefits outweigh costs. Thus, a spokesman from the American Petroleum Institute called some recent environmental regulations unwise and inefficient and claimed that industries in the United States would be hurt by them. Specifically, he criticized the Clean Air Act of 1990, groundwater protection laws, and the ongoing resistance to opening up the Arctic National Wildlife Refuge to oil exploration.

The problem of who pays and who benefits is very complex when pollutants produced in one state or country have their greatest negative impact in another state or country. This is particularly true of acid deposition.

Progress and Public Support

In the United States, the public continues to show great concern about environmental issues. A 1996 national survey of the American voting public found that 75% of respondents viewed preserving and protecting the environment as a high priority. Half of those polled felt that there was room for improvement on environmental legislation, while most voters were unhappy with the way current environmental protection policies were handled both by the Congress and by the White House.

As we have seen, the cost of controlling pollution is high and will continue to rise. What benefits do we receive in exchange for this cost? EPA Deputy Administrator F. Henry Habicht summarized the substantial changes that have been wrought in the period 1970–90:

> Because of the EPA's efforts to implement national laws, air emissions from cars, power plants, and large industrial facilities have been curtailed sharply; hundreds of primary and secondary waste-water treatment facilities have been constructed; ocean dumping of wastes has been virtually eliminated; land disposal of untreated hazardous wastes has largely stopped; hundreds of hazardous waste sites have been identified and 52 have been cleaned up; and the production and use of substances like asbestos, DDT, PCBs, and leaded gasoline have been banned. In the aggregate, actions like these have had a measurable, positive effect on environmental quality in this country, and they have set an example for other countries around the world.

Do these benefits outweigh their costs? To answer the question, consider the phaseout of leaded gasoline as just one example. This process, occurring over an eight-year period, has cost about $3.6 billion, according to a cost-benefit report by the EPA. Benefits were valued at over $50 billion, $42 billion of which were for medical costs avoided.

Executive Order 12866 is now in effect. A cost-benefit analysis accompanies every new regulatory rule from the EPA and other federal agencies. Even if a rule is not classified as significant, the accompanying documentation must demonstrate by cost-benefit analysis that the rule does not qualify as significant. According to an EPA spokesman, the final decision on any regulatory rule is made on the basis of the legislative statute that generated the rule (that is, the basic public policy) and not strictly on cost-benefit considerations. Some statutes are more amenable than others to cost-benefit analysis, and some actually prohibit the process as a basis of decision.

In sum, cost-benefit analysis is now part of public policy in the United States, through Executive Order 12866. Apparently unhappy with, or unaware of, this policy, the Congress has recently proposed a costly and complex cost-benefit process that is heavily weighted toward industry. This "regulatory reform" legislation is warmly supported by industry and universally condemned by environmental organizations and members of the public who know about it.

Risk Analysis

All of us would like to live long and healthy lives. When confronted with the news that our days may be cut short by the effects of radiation, pesticides, secondhand smoke, or chemicals in our drinking water, we become indignant. We expect someone in authority to keep us from dying young or from spending our later years in a hospital. The high priority we place on environmental health is largely a reflection of this concern for hazards coming at us from the air we breathe, the food we eat, and the water we drink. At the grassroots level, it is this concern that is largely responsible for the billions of dollars spent on environmental protection.

In fact, our lives in this late 20th-century technological society are honeycombed with hazards (Fig. 17-9). In the context of environmental science, a **hazard** is defined as anything that can cause (1) injury, disease, or death to humans, (2) damage to personal or public property, or (3) deterioration or destruction of environ-

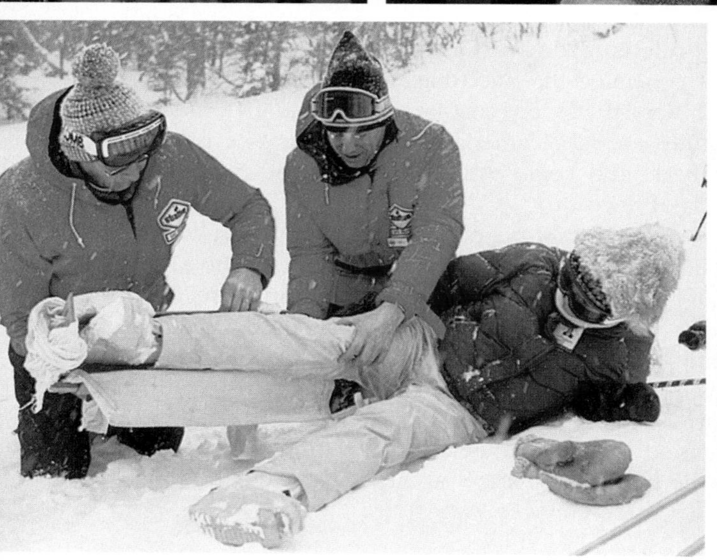

FIGURE 17-9
Hazards are an unavoidable part of life. Some are a matter of choice of lifestyle, and some are a consequence of where we live or work. All carry a risk that something unpleasant will happen. (Clockwise from top left: a) Lawrence Migdale/Photo Researchers, Inc.; b) Melissa Hayes English/Science Source/Photo Researchers, Inc.; c) David Stoecklein/The Stock Market; d/e) John Dominis/The Stock Market.)

mental components. Many hazards are a matter of personal choices we make every day. We may eat too much, get too little exercise, drive cars, sunbathe, use addictive and harmful drugs, or choose hazardous occupations.

Why do we subject ourselves to these hazards? Basically, because we derive some real or perceived *benefit* from them. Wanting the benefits, we are willing to take the *risk* that the hazard will not harm us. Other hazards, however, are a result not of personal choices, but of choices made by other people: pesticides used by farmers, chemicals emitted by power plants, and radiation from nuclear testing. Still other hazards are the result of natural causes, and no one does any choosing: earthquakes, lightning, floods, radiation from radon, and so on.

Each hazard in our lives carries with it a finite risk that something unpleasant will happen to us. Here, a **risk** is defined as the probability of suffering injury, disease, death, or other loss as a result of exposure to a hazard. Our actions each day involve choices that may subject us to higher risks.

For our own self-interest, we would like to have some knowledge and valuing of the risks we are subject to. This knowledge would enable us to make informed choices as we consider the benefits and risks of the hazards around us. We might learn, for example, that 82 million people go swimming each year, and 2600 of them drown while swimming. The risk of drowning, then, is 32 in a million. **Risk analysis**—which is what

Ethics

Living Is Risky Business

What would happen if our society decided to spend money to reduce risks in proportion to the way the experts rank the risks? How would risks from environmental pollution fare in comparison with other risks? Writing in the June, 1992, issue of *Consumer's Research*, Bernard Cohen of the University of Pittsburgh asked this question and made some interesting comparisons.

Cohen uses the measure, loss of life expectancy (LLE), to characterize the impact of a given risk. If that hazard were removed, average life expectancy would increase by the estimated LLE. Here are some interesting LLE figures:

- Heart disease (4.4 years) and cancer (3.4 years) top the disease list.
- Motor vehicle accidents (205 days), falls (28 days), and poison and fires (20 days each) are at the top of the accident list.
- Smoking (6.6 years for males, 3.9 for females) a pack of cigarettes a day
- Being overweight (52 days for each percentage point overweight!)
- All use of alcohol (1 year)
- Being poor (9 years) and unemployment (120 days for the whole population)
- Occupational exposure to toxics and pathogens (500 days)

These figures can be compared with some risks from environmental pollution:

- Pesticide residue on foods (12 days)
- Tobacco smoke (8 days)
- Regulated air pollutants (20 days)
- Releases from sewage treatment systems, incinerators, and landfills (0.4 day)
- Nuclear power (.04 day)*

Evidently, many environmental risks do not measure up to many other risks in our society. Cohen makes the point that we obviously spend money to reduce risks not on the basis of risk analysis, but on the basis of the public's risk perception. He faults the media for exaggerating environmental risks and our political system for perpetuating policies that allow us to spend millions or more per life saved from environmental risks, while spending much less on medical screening or highway safety measures that would save many more lives. What do you think?

*Assuming the electric power station only and not the entire fuel cycle.

we just did—*is the process of evaluating the risks associated with a particular hazard before taking some action in which the hazard is present.* To reduce the risk of drowning, for example, we might decide to go swimming at a beach where a lifeguard is on duty, or we might choose to swim in a calm lagoon rather than in the ocean.

Table 17-3 lists some everyday risks, expressed as the probability of dying from a given hazard. For example, the table indicates that the annual risk of dying from smoking one pack of cigarettes a day is 3.6 per 1000, meaning that in the course of a year, 3.6 out of every 1000 people who smoke a pack a day will die from smoking-related diseases.

Not very many people actually make informed choices about the hazards in their lives on the basis of such risk analysis. However, risk analysis has become an important process in the development of public policy and is being regarded as a major way of applying science to the hard problems of environmental regulation.

TABLE 17-3
Some Commonplace Hazards, Ranked according to the Degree of Risk

Hazardous Action	*Annual Risk*[a]	
Cigarette smoking, 1 pk/day	3.6 per	1 000
All cancers	2.8 per	1 000
Mountaineering (mountaineers)	6 per	10 000
Motor vehicle accident (total)	2.4 per	10 000
Police killed in line of duty	2.2 per	10 000
Air pollution, eastern U.S.	2 per	10 000
Home accidents	1.1 per	10 000
Frequent-flying professor	5 per	100 000
Alcohol, light drinker	2 per	100 000
Sea-level background radiation	2 per	100 000
Four tablespoons peanut butter/day	8 per	1 000 000
Electrocution	5.3 per	1 000 000
Drinking water containing EPA limit of chloroform	6 per	10 000 000

Source: R. Wilson and E. A. C. Crouch. "Risk Assessment and Comparisons: An introduction." *Science*, 236 [1987], 267. Copyright 1987 by the AAAS.
[a] Probability of dying.

Risk Analysis by the EPA

Risk analysis began at the EPA in the mid-1970s as a way of addressing the cancer risks associated with pesticides and toxic chemicals. Since then, the process has continued to focus largely on risks to human health. As currently performed at the EPA, there are four steps in risk

FIGURE 17-10
Laboratory mice are routinely used to test the potential of a chemical to cause cancer. They are an important source of information used to assess the presence of a hazard in food, cosmetics, or the workplace. (Courtesy the Jackson Laboratory.)

analysis: *hazard assessment, dose-response assessment, exposure assessment,* and *risk characterization*. One important limitation to the process is the fact that the analyst must work with the available information, which never seems to be sufficient. The final estimate of risk therefore includes a measure of the imprecision that always accompanies the use of imperfect information.

Hazard Assessment: Which Chemicals Cause Cancer? Hazard assessment is the process of examining evidence linking a potential hazard to its harmful effects. In the case of accidents, the link is obvious. The use of cars, for example, involves a certain number of crashes and deaths. In these cases, *historical data*, such as the annual highway death toll, are very useful for calculating risks.

In other cases the link is not so clear, because there is a time delay between first exposure and final outcome. For example, establishing a link between exposure to certain chemicals and the development of cancer some years later is often difficult. In cases where linkage is not obvious, the data may come from two sources: epidemiological studies and animal tests. An **epidemiological study** is a study that tracks how a sickness spreads through a community. Thus, in our example of finding a link between cancer and exposure to some chemical, an epidemiological study would look at all the people exposed to the chemical being examined and determine whether this population has more cancer than the general population. Data from such studies are considered to be the best data for risk analysis and have resulted in scores of chemicals being labeled as *known* human carcinogens.

The second data source, animal testing, is used when we want to find out *now* what might happen many years in the future. For example, we do not want to wait 20 years to find out that a new food additive causes cancer, so we accept evidence from **animal testing** (Fig. 17-10). A test involving several hundred animals (usually mice) takes about 3 years and costs more than $250,000. If a significant number of the animals develop tumors after being fed the substance being tested, the results indicate that the substance is either a *possible* or *probable* human carcinogen, depending on the strength of the results.

Three objections have been raised about animal testing: (1) Rodents and humans may have very different responses to a given chemical; (2) the doses used on the animals are often unrealistically high; and (3) some people are opposed on ethical grounds to the use of animals for such purposes. A point that supports the value of animal testing is that all chemicals shown by epidemiological studies to be human carcinogens are also carcinogenic to test animals.

Hazard assessment is the important first step in risk analysis. Whether it involves an analysis of accident data, epidemiological studies, or animal testing, hazard assessment tells us that we *may* have a problem that requires regulation.

Dose-Response and Exposure Assessment: How Much for How Long? When animal tests show a link between exposure to a chemical and an ill effect, the next step is to analyze the relationship between the concentrations of chemicals in the test (the **dose**) and both the incidence and the severity of the response in the test animals. From this information, projections are made about the number of cancers that may develop in humans who are exposed to different doses of the chemical. This process is **dose-response assessment**.

Following dose-response assessment is **exposure assessment**. This procedure involves identifying human groups already exposed to the chemical, learning how that exposure came about, and calculating the doses and length of time of the exposure.

Risk Characterization: How Many Will Die? The final step, **risk characterization**, is to pull together all the information gathered in the first three steps in order to determine the risk and its accompanying uncertainties. More commonly, risk is expressed as the probability of a fatal outcome due to the hazard, as in Table 17-3 for common, everyday risks.

The EPA expresses cancer risk as an "upper-bound, lifetime risk," meaning the top of the range of probabilities of the risk, calculated over a lifetime. The new Clean Air Act directs the EPA to regulate chemicals that have a cancer risk of greater than one in a million ($>1 \times 10^6$) for the people who are subject to the highest doses. That is the same standard employed by the FDA for regulating chemicals in food, drugs, and cosmetics.

Because people often have difficulty conceptualizing such data, risks may be expressed in different ways, such as reduction in life expectancy caused by engaging in a risky activity. Thus, calculations show that smoking one cigarette reduces life expectancy by 5 minutes, which is just about the amount of time it takes to smoke the cigarette!

Risk Perception

In September 1990, William K. Reilly, the EPA director at the time, received a report from a science advisory board (SAB) he had asked to evaluate different environmental risks in light of the most recent scientific data. The board consisted of 39 scientists and other experts from academia, state governments, industry, and public interest groups. Their report, entitled *Reducing Risk: Setting Priorities and Strategies for Environmental Protection*, has intensified interest in the use of science and risk analysis in setting environmental public policy.

The SAB separated risks into two categories: **health risks**, the EPA's major concern since its origin, and **ecological risks**. It is significant that the SAB recommended strongly that the EPA give at least equal attention to both kinds of risks. The board divided ecological risks into high-, medium-, and low-risk categories, as shown in Table 17-4. In performing its analyses, the board took care to weigh the *temporal* dimensions of the hazards—the length of time over which a problem is generated, understood, and corrected—and the *spatial* dimensions—the extent of geographical area affected by the problem. Thus, such problems as global warming and ozone depletion, which have long-term global implications, were ranked much higher than, say, oil spills, which have a more localized, short-term impact.

One of the most important problems raised by the scientists was the significant difference between their evaluation of risks and that of the public. Table 17-5 indicates the top 11 environmental problems from the SAB study, compared with the top 28 concerns of the public, as indicated in a Roper poll conducted in 1990. Clearly, the public perception of risks is quite different from that of the scientists. This discrepancy deserves a further look.

The U.S. public is increasingly concerned about environmental problems; most of that concern can be traced to the fear of hazards that pose a risk to human life and health. People *perceive* that their lives are more hazardous than ever before, but this is not true. In fact, our society is freer from hazards than it has ever been, as evidenced by increased longevity. Why, then, do people protest against nuclear power plants, waste sites, and pesticide residues in food when, according to experts, these hazards pose extremely small risks? The answer lies in people's **risk perceptions**—their intuitive judgments about risks. In short, people's perceptions are not consistent with the reality of the situation.

Hazard versus Outrage. The reason for the inconsistency between public perception and actual risk calculations, according to Peter Sandman of Rutgers University, is that the public perception of risks is more

TABLE 17-4
Risks to Ecology and Human Welfare, as Developed by the EPA's Science Advisory Board (not ranked within categories)

High Risk	*Medium Risk*	*Low Risk*
Alteration and destruction of habitat	Herbicides and pesticides	Oil spills
Extinction of species and loss of biodiversity	Toxics, nutrients, Biochemical Oxygen Demand (BOD), turbidity in surface waters	Groundwater pollution
Stratospheric ozone depletion	Acid deposition	Radionuclides
Global climate change	Airborne toxics	Acid runoff to surface waters
		Thermal pollution

Source: Reducing Risk: Setting Priorities and Strategies for Environmental Protection. USEPA Science Advisory Board, September 1990.

TABLE 17-5
Public Concerns vs. the EPA's Top 11 Risks

The EPA's Top 11 (not in rank order)	*Public Concerns (in rank order)*[a]
Ecological Risks	1. Active hazardous waste sites
Global climate change	2. Abandoned hazardous waste sites
Stratospheric ozone depletion	3. Water pollution from industrial wastes
Alteration of habitat	4. *Occupational exposure to toxic chemicals*
Extinction of species and loss of biodiversity	5. Oil spills
	6. *Destruction of the ozone layer*
Health Risks	7. Nuclear power plant accidents
Criteria air pollutants (e.g., smog)	8. Industrial accidents releasing pollutants
Toxic air pollutants (e.g., benzene)	9. Radiation from radioactive wastes
Radon	10. *Air pollution from factories*
Indoor air pollution	11. Leaking underground storage tanks
Contamination of drinking water	12. Contamination of coastal waters
Occupational exposure to chemicals	13. Solid waste and litter
Application of pesticides	14. *Pesticide risks to farm workers*
	15. Water pollution from agricultural runoff
	16. Water pollution from sewage plants
	17. *Air pollution from vehicles*
	18. Pesticide residues in foods
	19. *Greenhouse effect*
	20. *Contamination of drinking water*
	21. Destruction of wetlands
	22. Acid rain
	23. Water pollution from city runoff
	24. Nonhazardous waste sites
	25. Biotechnology
	26. *Indoor air pollution*
	27. Radiation from X rays
	28. *Radon in homes*

Source: "Counting on Science at EPA." *Science* 249 [1990], 616. L. Roberts. Copyright 1990 by the AAAS.
[a] Items in italics also appear on the EPA's list.

a matter of outrage than hazard. Sandman holds that, while the term *hazard* primarily expresses concern for fatalities only, the term *outrage* expresses a number of additional concerns:

1. Lack of familiarity with a technology: how nuclear power is produced and how toxic chemicals are handled, for example.
2. Involuntariness of risks: Research has shown that people who have a choice in the matter will accept risks roughly 1000 times as great as when they have no choice.
3. Memorableness of hazards: Accidents involving many deaths (Bhopal) or a failure of technology (Three Mile Island) are thoroughly imprinted into public awareness by media coverage and are not quickly forgotten.
4. Overselling of safety: The public becomes suspicious when scientists or public relations people play up the benefits of a technology and play down the hazards.
5. Morality: Some risks have the appearance of being morally wrong. If it is wrong to foul a river, the notion that the benefits of cleaning it are not worth the costs is unacceptable. You should obey a moral imperative regardless of the costs.
6. Control: People are much more accepting of a risk if they are in control of the elements of that risk, as in automobile driving, indoor air pollution, and radon.
7. Fairness: The benefits and the risks should be connected. If the benefits go to someone else, why should you accept any risk?

The public perception of risk is strongly influenced by the media, which are far better at communicating the outrage elements of a risk than the hazard. Public concern over oil spills rose precipitously following the *Exxon Valdez* oil spill in Alaska, an accident that received extensive media coverage and had a high "outrage quotient." Cigarette smoking, however, which causes *420,000 deaths a year in the United States alone* receives minimal media attention because it is not "news" and there is no outrage factor. Indeed, it has been suggested that if all the year's smoking fatalities occurred on one day, the media would have a field day, and smoking would be banned the next day!

Risk Perception and Public Policy. The most serious problem with the discrepancy between experts and the public is that, generally speaking, it is *public concern*, rather than cost-benefit analysis or risk analysis conducted by scientists, that drives public policy. The EPA's funding priorities are set largely by Congress, which reflects public concern. How important is this problem?

If public outrage is the primary impetus for public policy, some serious risks will get less attention than they deserve. In particular, risks to the environment are perceived as much less important than they really are, because of the public's preoccupation with risks to human health. As Table 17-5 indicates, two major ecological risks on the EPA's list—alteration of habitat, and extinction of species and loss of biodiversity—do not even show up on the list of public concerns. And only three of the public's top ten concerns are on the EPA's list.

This difference of opinion points to the importance of *risk communication*, a task that should not be left to the media. The value of ecosystems and their connections to human health and welfare need far greater emphasis in the public consciousness, and this responsibility falls to the scientific and educational communities, as well as to governmental agencies. Studies have shown that the most effective risk communication occurs by starting with what people already know and what they need to know and tailoring the message so that it effectively provides people with the knowledge that helps them to make an informed decision about risks.

Nonetheless, the public's concern for more than the probabilities of fatalities may have merit. Public outrage must be heard, understood, and given a reasonable response. It may not be the best source of public policy, but it reflects certain values and concerns that could easily be omitted by an "objective" risk analysis. The fact is, subjective judgments are going to play a role at every step in the risk-analysis process, from hazard assessment to risk perception to risk management.

Risk Management

Regulatory Decisions. The analysis of hazards and risks is a task that falls primarily to the scientific community. The task is incomplete, however, if no further consideration is given to the information gathered. Risk management naturally follows, and this is the responsibility of lawmakers and administrators. It is EPA policy to separate risk analysis and risk management within the agency.

Risk management involves (1) *a thorough review of the information available pertaining to the hazard in question and the risk characterization of that hazard* and (2) *a decision on whether the weight of evidence justifies a regulatory action.* Without doubt, public opinion can play a powerful role in determining these decisions. In general, however, a regulatory decision will hinge on one or more of the following considerations:

1. **Cost-benefit analysis**. As we learned earlier, a comparison of costs and benefits, if done fairly, can in many cases make a regulatory decision very clear cut.
2. **Risk-benefit analysis**. A decision can be made on the benefits versus the risks of being subjected to a particular hazard, especially where those benefits cannot be easily expressed in monetary values. The use of medical X rays is a good example. X rays carry a calculable risk of cancer, but the benefit derived when you X-ray a broken bone is much greater than the small cancer risk involved.
3. **Public preferences**. As we have seen, people have a much greater tolerance for risks that they feel are under their own control or are voluntarily accepted.

Risk management has been thoroughly incorporated into the EPA's policymaking process for at least 10 years. Its most common use has been in the design of regulations. The process has enabled the agency to target appropriate hazards for regulation and to determine where to aim the regulations—that is, at the source, at the point of use, or at the disposal of the hazardous agent. As was done for toxic chemicals and cancer, a given risk level may be adopted as a standard against which policymakers can measure new risks.

Comparative Risk Analysis. One of the advantages of risk analysis is that it provides a common language and a common procedure for comparing hazards. The basic process is understandable, even if it is fraught with uncertainties (as in the use of subjective judgments at many points). Conceivably, risk analysis may help society to choose wisely among a broad range of policy options available for reducing risks. The 1990 report of the SAB—*Reducing Risk*—addressed another potential use of risk analysis, however: **setting environmental priorities**. Indeed, much of the board's work was directed toward this goal.

Members of the group separated into different fields (human health, ecology) and defined different types of risks by which problems could be compared: cancer and noncancer health risks, impacts on materials, ecological effects, and economic impacts. They admitted that these risk types were not directly comparable, but agreed that information could be collected within each category and rankings constructed according to how serious they rated the separate hazards. One result of this ranking is Table 17-4.

As a result of the deliberations, one strong recommendation of the board was that the EPA refocus its policy orientation more toward the natural environment. Given the direct links between public concern and public policy, however, this refocusing will require convincing the public of the great value of natural ecosystems. To do so, the weight of scientific evidence must be brought to bear on the major environmental problems—which means strong governmental support of research. And the EPA will have to improve its communication with the public and the media, a large task in itself.

Currently, there is legislative pressure to apply comparative risk analysis to set national environmental policy goals. The concern behind this pressure is the rising cost of fulfilling environmental mandates, and the implication of the concern is that some risks simply do not merit the regulatory attention (and costs of compliance) they have received. Thus, scarce public and industrial resources are being wasted.

The environmental community views this new pressure with alarm, arguing that risk analysis is too imperfect a device on which to hang major policy decisions. The risk analysis would be subject to judicial review, a process that could lead to endless litigation that would paralyze regulatory agencies.

The Public and Public Policy

How is the general public involved in public policy? Another way to put this is to ask, what can you do to see that environmental concerns get the attention they deserve?

- Grassroots concern about environmental problems—often traced to the findings of environmental science—initiates the recognition stage of the policy life cycle.
- Environmental groups often fan the flames of concern by informing their constituencies—their memberships spread around the country—about the problem.
- During the policy formulation stage, members of the public should contact their legislators to inform them of their support for the environmental public policy under consideration.
- Public concern about environmental policy can play an important role in the election or defeat of political candidates.
- The public can comment on regulations when they are first proposed; the public can also bring a class action suit against polluters in some situations.
- The public is involved very directly in public policy, as people pay for the benefits they receive, through higher taxes and higher costs of the products they consume. It is easy to forget the benefits when tax time rolls around.
- Most importantly, the public should keep informed on environmental affairs; only when this happens is it likely that grassroots support will be maintained and public environmental policy will really reflect public opinion.

Review Questions

1. What are the two most important objectives of environmental public policy?
2. Review Table 17-1 and describe the effects on health and productivity associated with the following environmental problems: water pollution, air pollution, hazardous wastes, soil degradation, deforestation, loss of biodiversity, and atmospheric changes.
3. Three patterns of environmental indicators are associated with differences in the level of development of a nation: What problems decline, what problems rise and then decline, and what problems increase with level of development?
4. List the four stages of the policy life cycle, and show how *Silent Spring* and the formation of the EPA illustrate these stages.
5. What anecdotal evidence can be cited to disparage or support environmental public policies?
6. What is the conclusion of careful studies regarding the relationship between environmental policies and jobs and the economy?

7. What are the "three E's" for evaluating environmental public policy?
8. What are the advantages and disadvantages of the regulatory approach vs. a market approach to policy development?
9. How have presidential executive orders employed cost-benefit analysis in environmental regulation?
10. Define the term *cost-benefit analysis* as it relates to environmental regulation.
11. How does cost-benefit analysis address externalities?
12. List specific costs and benefits of pollution control.
13. What difficulties may be encountered when trying to apply shadow pricing to human life and the environment?
14. Discuss the cost effectiveness of pollution control.
15. Describe several problems that may arise when comparing cost and benefit estimates.
16. Define hazard, risk, and risk analysis.
17. Discuss the four steps of risk analysis used by the EPA.
18. Why do scientists and the public often come to very different conclusions regarding cost-benefit analysis?
19. What factors have been known to generate public outrage?
20. Discuss the relationship between public risk perception and public policy.
21. Describe the EPA's new strategy of comparative risk analysis.
22. Give six ways the public can become involved in public policy.

Thinking Environmentally

1. Investigate the environmental public policies of a developing country, and compare the results with the information presented in Table 17-1.
2. Suppose it was discovered that bleach commonly used for laundry was carcinogenic. Using the policy life cycle, describe a predictable course of events until the problem is brought under control.
3. Stage a debate on the following resolution: Environmental regulation is bad for the economy.
4. Consider Table 17-5. Which would you rank as the top three ecological and health risks? Which of the seven public-outrage factors might be playing a role in your rankings?
5. Imagine that you have been appointed to a risk-benefit analysis board. Explain why you approve or disapprove of widespread use of:

   genetic engineering
   drugs to slow the aging process
   nuclear power plants
   the abortion pill
6. Suppose your town wanted to spray trees to rid them of a deadly pest. How would you use cost-benefit analysis to determine whether this is a good idea?
7. What costs of our industrial processes are now treated as externalities? How could we internalize these costs?
8. Consider the complexities and limitations of cost-benefit analysis. Should we continue to support its use to determine public policy? How great a role should it play? Support your answers.

Making a Difference—PART THREE: Chapters 12, 13, 14, 15, 16, 17

1. Reduce your contribution to air pollution by amending your lifestyle to drive fewer miles: arrange to live near your workplace, car pool, use public transport, bicycle to work or school, avoid unnecessary trips.
2. Keep informed on how the Clean Air Act of 1990 is being implemented and enforced. If there are attempts to weaken it, protest to your congresspersons. Ask them to promote new standards to increase the average vehicle fuel efficiencies.

3. Be an advocate of energy conservation on all fronts, in supporting the "tie-in strategy" to combat global warming; insist on full U.S. participation in global efforts to curb greenhouse gas emissions.
4. Adopt ways to use less fuel and electricity both at home and where you work (turn thermostats back, turn off lights when leaving a room, wear sweaters, use energy-efficient light bulbs, etc.).
5. Survey the lakes and ponds in your area and determine whether cultural eutrophication is occurring. If you find cases, follow up by consulting a local environment group and get involved in combating the problem.
6. Use phosphate-free detergents for all washing. Information will be printed on the container (often in fine print!).
7. Explore sewage treatment in your community. To what extent is the sewage treated? Where does the effluent go? The sludge? Are there problems, alternatives that are ecologically more sound? Consult local environmental groups for help.
8. Read labels and become informed about the potentially hazardous materials you use in the home and workplace. Wherever possible, use nontoxic substitutes. If toxic substances are used, insure that they are disposed of responsibly.
9. Encourage your community to set up a hazardous waste collection center where you can take leftover hazardous materials.
10. Properly maintain all fuel-burning equipment to burn efficiently (oil burner, gas heater, lawnmower, outboard motor, automobile, etc.).
11. Refuse to become a smoking victim: If you are a smoker, stop; insist that others do not smoke in your presence or in your home.
12. Make sure that anyone servicing a refrigeration or air-conditioning unit for you will recapture and recycle the CFCs if the system needs to be opened.
13. Monitor hazards in your own life and take steps to reduce behavior ranked more risky by the experts. Be discriminating as you hear about hazards and risks; especially be aware of the media's tendency to exploit the *outrage* element of risks.
14. Be aware of the pressures put on regulators by business concerns to weaken regulatory rules, and insist on fairness and consistency in the application of laws passed by Congress.

PART FOUR

Resources: Biota, Refuse, Energy, and Land

The Alaska oil pipeline snakes its way down through tundra and forest to the port of Valdez, an intrusion into a beautiful and fragile wilderness that is symbolic of the dilemma facing our society. We may love the beauty of wild ecosystems and species, and turn to them for enjoyment and rest, but we also must look on much of the natural world as resources for exploitation. Can we have it both ways? Can we hope to preserve nature while we also make use of it?

In this Part we turn our attention to the resources that make it possible for some of us in the developed world to enjoy our late-20th-century lifestyle, and for those in the developing world to move slowly along the path of economic development. These resources—fossil fuels, fisheries, forests, and the like—are not evenly distributed on Earth, and their extraction and use represents a great part of the economic exchange between peoples. Wars are fought over such resources, international treaties are forged to regulate access and exchange—yet the most important actors in the picture are the individuals who through their lifestyle choices are the consumers of resources. The good news is that consumers can become stewards and bring the attitude of care and concern to the living environment that sustains us and the resources buried in Earth as minerals. Trash can be recycled, energy can be conserved, solar options can be adopted, land can be set aside for preservation. This is what the environmental revolution is all about—the transition to a sustainable society.

The Alaskan Pipeline. [Photo by Paul McCormick/The Image Bank.]

Wild Species: Biodiversity and Protection

Key Issues and Questions

1. Wild species and their habitats are threatened and endangered by human activities, a matter of concern for many people. How does public policy in the United States protect endangered species?
2. Preserving wild species may require that we find a way to show that they have value. How can we establish the value of natural species?
3. Many naturalists claim that we are losing much of the biodiversity that has enriched Earth for millions of years. What is biodiversity, and what is the current extent of it?
4. It seems certain that humans are the cause of the decline in biodiversity. What human enterprises in particular are responsible for this decline?
5. Much of the loss in wild species and biodiversity is occurring outside the United States. How has the international community acted to protect biodiversity?

Aldo Leopold was a pioneer in the field of conservation and environmental ethics. In his 1949 classic, *A Sand County Almanac,* Leopold tells the story of a wilderness trip in the southwestern United States with some friends. They were eating lunch on a hillside when they noticed an animal crossing a small river below them. They watched as the animal, evidently a female wolf, was met on the other side of the river by a half-dozen grown pups, which tumbled all over her with their greetings.

> In those days we had never heard of passing up a chance to kill a wolf. In a second we were pumping lead into the pack, but with more excitement than accuracy: how to aim a steep downhill shot is always confusing. When our rifles were empty, the old wolf was down, and a pup was dragging a leg into impassible slide-rocks.
>
> We reached the old wolf in time to watch a fierce green fire dying in her eyes. I realized then, and have known ever since, that there was something new to me in those eyes—something known only to her and to the mountain. I was young then, and full of trigger itch; I thought that because fewer wolves meant more deer, that no wolves would mean hunter's paradise. But after seeing the green fire, I sensed that neither the wolf nor the mountain agreed with such a view.

Giraffes and zebras graze on the savanna landscape of the Serengeti Plains in East Africa. [Photo by Gregory Dimijian/Photo Researchers, Inc.]

This incident not only changed Leopold's *attitude* toward shooting wolves, it also changed his *valuation* of wolves and other wild species. He began to understand the importance of wolves in keeping deer herds from overeating their food supply, and in time, Leopold articulated an ethic that stressed the *value* of natural species and the land. By value, he meant more than economic worth; he meant that wild species and their habitats have a right to exist, and protecting that right is a matter of morality. The fierce green fire in the old wolf's eyes was symbolic of wild nature, and when it went out, Leopold knew that something very precious was gone.

In this chapter, we give attention to wild species. We consider their importance to humans, examine the perspective of biodiversity, and look at the public policies that seek to protect wild species. In Chapter 19, our focus will broaden to an examination of the impact of humans on whole ecosystems, with special emphasis on human patterns of use (and misuse) of those systems and their living components. We will see that protecting wild species also means preserving their ecosystems, and vice versa.

Saving Wild Species

You saw in Chapter 17 that public indignation about a threat to our health or to the environment is often the first stage leading to environmental policy changes. As the growing human population and other influences combine to eliminate some wild animals and plants, concern is rising for preserving wild species, in the United States and in many other countries. Let us turn our attention to some of those species and to what is being done to protect them.

Game Animals in the United States

Game animals are those traditionally hunted for sport or meat. In the early days of the United States, there were no restrictions on hunting, and a number of species were hunted to near extinction (bison, wild turkey) or extinction (great auk, passenger pigeon). As game animals became scarce in the face of increasing pressure from hunting, regulations came into effect. State governments, backed up by the federal government, began to set regulations and hire wardens to enforce them.

One success story is the wild turkey. A favorite game species, this bird was hunted to the brink of extinction, but was making a slow comeback by the 1930s as a result of hunting restrictions. At that time, there was a total population of about 30,000 individuals in a few scattered states. After World War II, state and federal programs addressed the need for protecting turkey habitats. The birds were reintroduced into areas they had once inhabited, and hunting quotas were strictly limited. The turkey is now found in 49 states, and its population has risen to a total of 4.5 million as a result of these measures.

Using hunting fees as a source of revenue, state wildlife managers enhance the habitats supporting important game species and provide special areas for hunting. They monitor game populations and adjust seasons and bag limits accordingly. Game preserves, parks, and other areas where hunting is prohibited are maintained to protect habitats as well as certain breeding populations.

Common game animals, such as deer, rabbits, doves, and squirrels, are well adapted to the rural field and woods environment. Thus adapted and protected from overhunting, viable populations of these animals are being maintained. However, some problems are emerging:

1. The number of animals killed on roadways now far exceeds the number killed by hunters. Increasing numbers of animals found on roadways as rural areas are developed are a serious hazard to motorists, with scores of people killed annually from collisions with deer and moose. For example, in Ohio, 24,000 deer are hit each year by motorists.
2. Opossums, skunks, raccoons, and deer are thriving in highly urbanized areas, creating various hazards. For instance, in 1992, a rabies epidemic among "urban" raccoons was a public health risk in several states in the eastern United States.
3. Some game animals have no predators except hunters and tend to reach population densities that push them into suburban habitats, where they cannot be effectively hunted. The white-tailed deer, for example, has become a pest to gardeners and fruit nurseries; it also poses a public health risk because it is often infested with Lyme disease ticks (Fig. 18-1). Even in wilder areas, rising deer populations have reduced populations of many endangered or threatened plant species by their grazing.

FIGURE 18-1
A white-tailed deer buck grazing in a populated area of Fire Island, New York. Deer have adapted well to suburban habitats, and can often become pests there. (Heath Robbins/Gamma-Liaison, Inc.)

4. In recent years, there has been an increasing number of attacks on suburbanites by mountain lions as urbanization encroaches on the wild.

The Endangered Species Act

In colonial days, huge flocks of snowy egrets inhabited coastal wetlands and marshes of the southeastern United States. In the 1800s, when fashion dictated fancy hats adorned with feathers, egrets and other birds were hunted for their plumage. By the late 1800s, egrets were almost extinct. In 1886, the newly formed Audubon Society began a press campaign to shame "feather wearers" and end the "terrible folly." The campaign caught on, and gradually attitudes changed and new laws followed.

Florida and Texas were first to pass laws protecting plumed birds. Then, in 1900, Congress passed the **Lacey Act**, which forbids interstate commerce in illegally killed wildlife, making it more difficult for hunters to sell their kill. Since then, numerous wildlife refuges have been established to protect the birds' breeding habitats. With millions of people visiting these refuges and seeing the birds in their natural locales, attitudes have changed significantly. Today the thought of hunting these birds would be abhorrent to most of us, even if official protection were removed. Thus protected, egret populations were able to recover substantially. However, their numbers are headed down once again, as wetland habitats outside the very limited refuges continue to be destroyed and degraded by pollution.

Congress took a major step when it passed a series of acts to protect endangered species. The most comprehensive and recent of these acts is the **Endangered Species Act** (ESA) of 1973 (reauthorized in 1988). As we learned at the beginning of Chapter 4, an **endangered species** is a species that has been reduced to the point where it is in imminent danger of becoming extinct if protection is not provided (Fig. 18-2). The act also provides for the protection of **threatened species**, which are judged to be in jeopardy, but not on the brink of extinction. When a species is officially recognized as being either endangered or threatened, the law specifies substantial fines for killing, trapping, uprooting (plants), or engaging in commerce in the species or its parts. The legislation forbidding commerce includes wildlife threatened with extinction anywhere in the world.

The ESA requires the U.S. Fish and Wildlife Service, under the Department of the Interior, to draft recovery plans for protected species. Habitats must be mapped and a program for the preservation and management of critical habitats must be designed such that the species can rebuild its population. A 1995 Supreme Court decision made it clear that federal authority to conserve critical habitats extended to privately held hands. As Table 18-1 indicates, 1051 U.S. species are currently listed for protection under the act, and recovery plans are in place for 444 of those.

Many critics of the ESA believe that the act does not go far enough. A major shortcoming is that protection is not provided until a species is officially listed as endangered or threatened by the Fish and Wildlife Service and a recovery plan is established. And a species will not be listed until its population becomes dangerously low. Most ecologists are extremely concerned by the slow pace at which species are being added to the official list. Hundreds of species in the United States—over 400 in Hawaii alone—may either become extinct or else have their habitats so reduced that extinction is inevitable before they can be listed. Another source of criticism is the apparent impotence of the act: In the 20 years since its enactment, only seven species have recovered sufficiently to be either removed from the list or upgraded from endangered to threatened status.

Unfortunately, bringing about the recovery of a species may take decades or longer, as in the case of our national emblem, the bald eagle, first listed as endangered in 1978. In 1995, the Fish and Wildlife Service announced that the bald eagle has recovered well enough to be upgraded to "threatened" status, capping years of management efforts aimed at bringing this bird back in the lower 48 states. (Alaskan popula-

(a) Male Pine Barrens Tree Frog (b) Silversword (c) Red-cockaded Woodpecker (d) Karner Blue Butterfly (e) Swamp Pink (f) Whooping Cranes (g) Devil's Hole Pupfish (h) White Oryx (i) Manatee

FIGURE 18-2
Some examples of endangered species—species whose populations in nature have dropped so low that they are in imminent danger of becoming extinct unless protection is provided. (a) C. Allan Morgan/Peter Arnold, Inc.; b) Steve Kaufman/Peter Arnold, Inc.; c) Julia Sims/Peter Arnold, Inc.; d) C. C. Lockwood/ Animals Animals/Earth Science; e) Jeff Lepore/Photo Researchers, Inc.; f) A. H. Rider/Photo Researchers, Inc.; g) Tom McHugh/Steinhart Aquarium/Photo Researchers, Inc.; h) Mickey Gibson/Earth Scenes; i) Douglas Faulkner/Photo Researchers, Inc.)

TABLE 18-1
1996 Federal Listings of Threatened and Endangered U.S. Plant and Animal Species

Category	*Endangered*	*Threatened*	*Total Listings*	*Species with Recovery Plans*
Mammals	55	9	335	39
Birds	74	16	274	71
Reptiles	14	19	113	31
Amphibians	7	6	22	11
Fishes	67	40	118	73
Snails	15	7	23	18
Clams	51	6	59	45
Crustaceans	14	3	17	5
Insects	20	9	33	20
Arachnids	5	0	5	2
Plants (flowering, conifers ferns, others)	513	101	614	326
Total	835	216	1051	639 (444 plans approved)
Total endangered U.S. species		835	(322 animals, 513 plants)	
Total threatened U.S. species		216	(115 animals, 101 plants)	
Total listed U.S. species		1051	(433 animals, 614 plants)	
Total listed non-U.S. species*		565	(562 animals, 3 plants)	

*Four animals have dual status

Source: Department of the Interior, U.S. Fish and Wildlife Service, Division of Endangered Species, January 1, 1997.

tions have always been healthy.) In the early 1960s, the number of nesting pairs had declined to only 417. By 1994, the number was up to 4400, which, with an estimated 8000 juveniles, brought the bald eagle population in the lower 48 states to about 17,000 that year. Banning the insecticide DDT in 1972 undoubtedly played the key role, but the protection of its habitat under the Endangered Species Act also accelerated the eagle's comeback.

A few species have gained exceptional public attention, and heroic efforts have been mounted to save them. Efforts to save the whooping crane, for example, included virtually full-time monitoring and protection of the single remaining flock, which for years numbered only in the teens. To obtain a higher reproduction and recruitment rate, eggs were collected and artificially incubated. (When eggs are continually removed from a nest, the female can produce up to 14!) The chicks hatched in incubation and were then placed in nests of related sandhill cranes, to be raised by them as foster parents. The effort seems to have paid off; the whooping crane is now at least holding its own. The 1995 wild population of 168 birds is up from a low of 14 cranes in 1939 (146 migrate between Canada and Texas, 7 at Grays Lake in Idaho, and 15 reintroduced into Florida).

Other critics of the ESA claim that it goes too far. The most famous case in point is the northern spotted owl. This species has become a focal point in the battle to save some of the remaining old-growth forests of the Pacific Northwest (see Chapter 1, p. 8). The owl is found only in these forests and has dwindled to a population between 6000 and 8000. In June 1990, the Fish and Wildlife Service listed the owl as a threatened species. The listing prompted the service to set aside 6.9 million acres (2.8 million ha) of the old-growth forests, enough to guarantee the bird protection into the future. Neither environmentalists nor members of the timber industry were pleased with the action. Calling the plan "a legal lynching of an entire region by an out-of-control federal agency," the timber industry claimed that 33,000 jobs would be lost if the forests were protected from cutting. The Fish and Wildlife Service admitted that jobs would be lost, but stated that many of these jobs would be lost anyway because of other market forces. In fact, during the 1980s as timber yields increased, unemployment doubled because of mill modernization and other corporate moves. In contrast, the Wilderness Society claimed that the acreage was insufficient to ensure the owl's survival and that political pressure from the timber industry prevented the agency from setting aside a much larger area.

Following an attempt by the Bureau of Land Management to exempt a block of timber sales from the act and lawsuits from a number of environmental organizations charging that the Fish and Wildlife Service's plan was inadequate, a federal judge issued an injunction in 1991 banning sales of federal timber in the Northwest. The stalemate was a matter of high priority for the Clinton administration. President Clinton promised action on the issue after a Forest Conference

Earth Watch

A Beach for the Birds

The piping plover is a small shorebird well adapted to life on ocean beaches just beyond the high-tide line. The bird nests there and forages in the line of drift deposited by waves and tides. Unfortunately, this is exactly the area that receives the heaviest traffic from beach-loving humans and four-wheel-drive vehicles. Also, the nests and young of the plover are vulnerable to predators like skunks, opossums, and gulls, which are attracted to the beaches by the trash and garbage left by humans. As a result, the plover is now listed as an endangered species. The bright spot for the species is Massachusetts, where 454 pairs were recently tallied, an increase from 137 in 1989.

Ten pairs of plovers nested in 1990 on Plum Island, a part of the Parker River National Wildlife Refuge north of Boston. Because parts of the beach have been closed to protect the birds, these 10 pairs successfully raised 14 young in 1990, and refuge biologists believed that the population had the potential to increase. Accordingly, refuge manager Jack Filio closed the entire 6 miles (10 km) of beach between April and early July 1991 in order to allow the piping plovers to raise their young undisturbed. The closure met with angry protests from local business people and politicians concerned about the impact of the decision on local tourism.

The piping plover is an endangered species that breeds on outer beaches just above the high-water line. It is showing remarkable recovery in response to measures that protect it from intrusion by humans and predators. (Perry D. Slocum/Animals, Animals/Earth Science.)

"Nobody argues the point that endangered species should be protected, but at what cost to the community?" asked a spokesperson. Protestors accused the refuge manager of putting animals before people and of being unwilling to compromise. Filio pointed out that the primary purpose of the refuge is the protection of wildlife, especially endangered or threatened species. Recreational use is secondary to these concerns.

In addition to closing off the beach, refuge personnel erect fencing (called "exclosures") around and over plover nests, which effectively keep out mammalian and avian predators. As a result of these combined efforts, the nesting pairs on the island have increased to 17, with 20 young successfully fledged in 1996. Typically, the beach is reopened for recreation in phases as chicks in various nests reach the age of 35 days, when they can fly.

Events at Plum Island are symptomatic of a broader problem: The U.S. Fish and Wildlife Service has been severely criticized by some members of Congress and the General Accounting Office for allowing incompatible and harmful uses on some of the refuges it manages, as the service tries to accommodate the wishes of the public. The Parker River refuge has clearly taken a stand in favor of the wildlife it is charged with protecting.

in Oregon in April 1993, at which Vice-President Gore and five cabinet members listened both to environmentalists and to timber spokespersons. A year later, the Clinton administration acted on a plan to protect the old-growth habitat of the owl *and* to lift the ban on sales of timber—The Northwest Forest Plan. Not surprisingly, the plan was criticized by both environmentalists and the timber industry.

Critics of the ESA object to the powers the act gives to federal agencies to limit development and prohibit some activities on private lands where protected species are found. They also question the taxonomic status of some of the animals protected by the act. For example, the northern spotted owl is a subspecies of the single species designated by taxonomists as the spotted owl; the California spotted owl and the Mexican spotted owl are also subspecies of this species, and neither of them is listed as endangered or threatened. However, the Fish and Wildlife Service is on solid ground when it acts to protect subspecies and even distinct populations, as the ESA defines a species to include "any subspecies of fish or wildlife or plants, and any distinct population segment of any species of vertebrate fish or wildlife which interbreeds when mature."

The ESA was scheduled for reauthorization in 1992, but Congress decided to defer consideration of it until after the coming presidential election, fearing a veto from President Bush. With the election of Bill Clinton, environmentalists were optimistic that the ESA would get swift consideration—and were disappointed when it did not. With the shift of Congress to Republican control in 1994, the ESA seems to be in trouble. Concerns about private property rights and loss of jobs precipitated a battle between members of the 104th Congress and those who want to save the act.

In the final analysis, the ESA is a formal recognition of the importance of preserving wild species, regardless of any economic importance. Species listed as endangered or threatened have legal rights to protection under the law. The act is something of a last resort for wild species, but it embodies an encouraging attitude toward nature that has now become public policy. Unexpected oppostion to ESA repeal came from the religious community, calling the ESA analogous to a "modern Noah's Ark." Actions taken under auspices of the act have demonstrated a unique commitment to preserve wild species. With its reauthorization still pending, some changes in the ESA ar likely. One change supported by critics and supporters alike is to build incentives into the act for protection of endangered species on privately held land. Through such policies as land exchanges, tax advantages, and aid in habitat management, concerns about property rights might be reconciled. Reauthorization of a strong ESA will be a major test of our commitment to consider the value of wild species on grounds other than economic or political.

FIGURE 18-3
Clouds of passenger pigeons darkened American skies during the 18th and 19th centuries, but relentless hunting extinguished the species in the early 20th century. (Asa C. Thoresen/Photo Researchers.)

Value of Wild Species

Biological Wealth

About 1.75 million species of plants, animals, and microbes have been examined, named, and classified, but scientists estimate that between 4 and 112 million additional species have not yet been systematically explored (a working number of 13.6 million is used). The latest estimates are from a 1995 UNEP survey (*Global Biodiversity Assessment,* Cambridge University Press, 1995).

These natural species of living things, collectively referred to as **biota**, are responsible for the structure and maintenance of all ecosystems. They and the ecosystems they form represent a form of wealth—**biological wealth**—that sustains human life and economic activity. It is as if the natural world were an enormous bank account, with biological wealth capable of paying vital, life-sustaining dividends indefinitely, but only as long as the capital is maintained—through a sustainable interaction. This richness of living species is Earth's **biodiversity.**

Humankind began spending this biological wealth many centuries past. Our early ancestors were thoroughly integrated into natural ecosystems as hunter-gatherers. Their survival depended on learning the ways of the other animals and identifying plants that could be eaten. Some 10,000 years ago, humans began learning to select certain plant and animal species from the natural biota and to propagate them, and the natural world has never been the same. Over time, vast areas of forests, savannahs, and plains were converted to fields and pastures as the human population grew and human culture flourished. In the process, many living species were exploited to extinction, and others disappeared as their habitats underwent development. At least 500 plant and animal species have become extinct in the United States alone, and thousands more are at risk. We have been drawing down our biological capital, with unknown consequences.

Now, living in cities and suburbs in the industrialized world and getting all our food from supermarkets, our connections to nature seem remote. That is an illusion: Our interactions with the natural world have changed, but we are still dependent on biological wealth. Millions of our neighbors in the developing world are not so insulated from the natural world; their dependence is much more immediate, as they draw sustenance and income directly from forests, grasslands, and fisheries. However, because of overwhelming economic pressures, these people, too, are engaged in unsustainable practices by drawing down their biological capital, with consequences that are obvious and grave.

Earth Watch

Who's Afraid of the Big Bad Wolf?

The state of Alaska has between 5000 and 7000 gray wolves and not a great deal of sympathy for conservationists who want to protect Alaskan wolves. In 1992, the state's Board of Game announced a plan for an aerial wolf hunt in order to boost hunters' chances of bagging more moose and caribou. The hunt never came off, however, due to an intensive protest and a short tourism boycott in Alaska. The state then proceeded with a less publicized program of wolf control using snares. This program, too, came to a halt: In late 1994, a biologist produced a videotape of four snared wolves, one of which had chewed its leg off to a stump trying to escape the snare. Also on the tape was a scene of a state biologist firing four shots into a snared wolf with a small handgun in an attempt to kill it. After seeing the videotape played on KTUU-TV in Anchorage, Alaska's new governor, Tony Knowles, stopped the control program. The net result was the removal of 700 snare traps previously set out in a 1000-square-mile area of the state.

Alaska is the only state in which the wolf is not endangered and therefore not protected by the ESA. Prior to the act, wolves had been exterminated from all but two of the lower 48 states. Currently, gray wolf populations are on the increase; more than 2000 are now found in five states. Symptomatic of a profound shift in public attitude toward wolves, instead of states poisoning and trapping wolves and paying bounties for them, Defenders of Wildlife pays ranchers for any losses they suffer from wolf predation. The U.S. Fish and Wildlife Service began releasing captured Canadian wolves in Wyoming's Yellowstone National Park and in central Idaho. The 14 wolves released in Yellowstone in 1995 demonstrated their approval of their new home by producing 9 pups, successfully preying on the abundant elk and bison in the park, aggressively harassing the numerous coyotes in the park, and rewarding some 6000 visitors with views of their activities. Only 5 of the first-year population were killed (4 were shot, 1 was hit by a truck), leaving 18 in the wild. Encouraged by their success, the wildlife biologists added 17 more wolves to Yellowstone in 1996, and believe that this may end the need for reintroductions there.

Research biologist David Mech, probably the leading authority on wolves, believes that the key to continuing the wolf's comeback is to stop short of complete control. The conflicting interests of protectionists on the one hand and farmers and ranchers on the other will have to be worked out with a careful program of regulation of wolf populations—just as many other large animals such as bears, cougars, and coyotes are kept under control by regulations and hunting quotas. At some point, the gray wolf in the lower 48 states may be taken off the endangered list. When that happens, the wolf—in the words of David Mech—"can be accepted as a regular member of our environment, rather than as a special saint or sinner, [and] this will go a long way toward ensuring that the howl of the wolf will always be heard throughout the wild areas of the northern world."

Instrumental versus Intrinsic Value

It was not so long ago that hunters on horseback would ride out to the vast herds of bison roaming the North American prairies and shoot them by the thousands, often taking only the tongues for markets back in the east. The passenger pigeons that darkened the skies in huge flocks were ruthlessly killed at their roosts to fill a lively demand for their meat, until the species was gone (Fig. 18-3). Plume hunters decimated egrets and other shorebirds to satisfy the demands of fashion in the late 1800s. Appalled at this wanton destruction, 19th-century naturalists called for an end to the slaughter, and the U.S. public began to be sensitized to the losses that were occurring. People began to see natural species as worthy of preservation, and naturalists began to look for ways to justify their calls to conserve nature.

Just before the turn of the century, there was an emerging sense that species should not be hunted to extinction. But why? Were these early conservationists just concerned that there might not be any animals left to hunt or trees left to chop down? Their problem then, and ours now, is to establish that wild species have some *value* that makes it essential that they be preserved. If we can identify that value, then we will find it much easier to justify the action we must take to preserve them.

Philosophers who have addressed this problem inform us that two kinds of value should be considered. The first is **instrumental value**. A species or individual organism has instrumental value if its existence or use benefits some other entity. This kind of value is usually *anthropocentric*; that is, the beneficiaries are human beings. Clearly, many species of plants and animals have instrumental value to humans and will tend to be preserved (or conserved, as we would say) so that we can continue to enjoy the value we derive from them.

The second kind of value we must consider is **intrinsic value**. We assign intrinsic value to something when we agree that it has value for its own sake; that is, it does not have to be useful to us to possess value.

Ethics

Animal Rights and Wrongs

Activists who support the rights of animals have been very much in the public eye. They have publicized their cause by drawing attention to a number of particularly objectionable forms of cruelty to animals (such as leg-hold steel traps and pharmaceutical eye tests on rabbits). In Chapter 4, we considered animal rights in the light of hunting and the place of individual animals in natural ecosystems. Here, let us look at animal rights in the context of *value*.

Animal rights proponents differ from people who are concerned about preserving animal species in one main way: The former value the individual animal, whereas the latter value the whole population as a species. An animal rightist would argue that the individual animal has intrinsic value and that it is not valid for one animal (human beings) to uphold intrinsic value for individuals of its own species but not for individuals of other species. Animal rights and human rights are given parallel status on the basis that animals (at least, higher animals) have "perception, memory, a sense of the future, an emotional life together with feelings of pleasure and pain, and the ability to initiate action in pursuit of their desires and goals," as articulated by ethicist Tom Regan.

Now, if something has intrinsic value, it deserves respect, meaning that how we treat it becomes a matter of justice. Animal rightists point to the inconsistencies in our dealings with animals: We may not feed poisoned food to squirrels and pigeons in the park, but we may shoot them out of the trees; we may not bludgeon our pet dog if it fails to please us, but we may subject the eyes of rabbits to chemicals until the rabbits are blinded. In short, we recognize animal rights in some situations, but deny them in others.

Philosophers point out that, to be entirely consistent as an animal rightist, one should be against zoos, hunting, trapping, eating animals, and performing experiments on animals—as indeed many are. They believe that it is permissible to kill an animal only in self-defense, in the same way we would defend ourselves against a human intruder. The use of animals for laboratory dissection is also considered wrong, primarily because it is assumed that the animals were wrongfully caught and killed.

Most people reject this viewpoint. Few are willing to accept that animals are equal to people in terms of their intrinsic value, and although we do accord to individual animals some rights (the right not to be treated with cruelty, for example), there is no philosophical imperative to assume they have equal rights with humans. As journalist Richard Coniff pointed out in a recent *Audubon Magazine* article, there is a contradiction in arguing on the one hand that humans are merely a part of nature and should not have any more rights than other animals and on the other hand that humans have a moral obligation to give up practices that are entirely consistent with our animal nature—namely, to use other species to satisfy our needs for food and clothing.

Can we have such a high moral obligation and at the same time have no more rights or value than any other animals? What do you think?

How do we know that something has intrinsic value? That is a philosophical question, and it comes down to a matter of moral reasoning. People often disagree about intrinsic value, as illustrated by the animal-rights controversy. (See "Ethics" box, p. 465.)

As we study the problem of loss of species, we will see that many people argue that no species on Earth except *Homo sapiens* has any intrinsic value. However, we will also see that if there is no recognition of the instrinsic value of species, it is difficult to justify preserving many that are apparently insignificant or very local in distribution. In spite of the problems inherent in establishing intrinsic value for species, there is growing support in favor of preserving species that not only may be useless to humans, but also may never be seen by anyone except a few naturalists or systematists (biologists who are experts on classifying organisms).

The value of natural species can be categorized into five areas, which we examine in this chapter:

- Sources for agriculture, forestry, aquaculture, and animal husbandry
- Sources for medicines, pharmaceuticals
- Commercial value
- Recreational, aesthetic, and scientific value
- Intrinsic value

The first four categories mostly reflect instrumental value. In the case of aesthetic and scientific value, it could be argued that these sometimes represent intrinsic value.

Sources for Agriculture, Forestry, Aquaculture, and Animal Husbandry

Since most of our food comes from agriculture, we tend to believe that it is independent of natural biota. This is not true. Recall that in nature both plants and animals are

FIGURE 18-4
The winged bean, a climbing legume with edible pods, seeds, leaves, and roots. This tropical species is an example of the great potential of wild species for human use. (Courtesy ECHO, Educational Concerns for Hunger Organization.)

continuously subjected to the rigors of natural selection. Only the fittest survive. Consequently, wild populations have numerous traits for resistance to parasites, competitiveness, tolerance to adverse conditions, and other aspects of **vigor**. In addition, as we learned in Chapter 5, the gene pools of wild populations generally harbor the variations that foster adaptation to changing conditions.

Conversely, populations grown for many generations under the "pampered" conditions of agriculture tend to lose these traits because they are selected for production, not vigor. For example, a high-producing plant that lacks resistance to drought is irrigated, and the drought resistance is ignored. Also, in the process of breeding plants for maximum production, virtually all genetic variation is eliminated. Indeed, the cultivated population is commonly called a **cultivar** (for *culti*vated *var* iety), indicating that it is a highly selected strain of the original species with a *minimum* of genetic variation. When provided with optimal water and fertilizer, cultivars do give outstanding production under the specific climatic conditions to which they are adapted. With their minimum genetic variation, however, they have virtually no capacity to adapt to any other conditions. If climatic conditions change from what a cultivar is adapted to, its production may decline, and it will be unable to adapt to the new conditions because its gene pool lacks the necessary variation.

To maintain vigor in cultivars and to adapt them to various climatic conditions, plant breeders comb wild populations of related species for the desired traits. When found, these traits are introduced into the cultivar through crossbreeding or genetic engineering. Keep in mind, however, that such a trait comes from a related wild population, that is, from natural biota. If natural biota with wild populations are lost, the options for continued improvements in food plants will be greatly reduced.

Also, tremendous potential for developing *new* agricultural cultivars will be lost. From the hundreds of thousands of plant species existing in nature, humans have used perhaps 7000 in all, and modern agriculture has tended to focus on only about 30—wheat, maize (corn), and rice fulfill about 50% of global food demands. This limited diversity in agriculture makes it ill suited to production under many environmental conditions. For example, we tend to think of arid regions as being unproductive without irrigation. However, many wild species belonging to the bean family produce abundantly under dry conditions. Scientists estimate that 30,000 plant species with edible parts might be brought into cultivation. Many of these could increase production in environments that are less than ideal. For example, consider the winged bean, native to New Guinea (Fig. 18-4). This plant is a veritable supermarket, with every part edible: pods, flowers, stems, roots, and leaves. Recently introduced to many developing countries, it has already made a significant contribution to improving nutrition. Loss of biological diversity undercuts similar future opportunities.

Another area in which biodiversity has instrumental value to humans is pest control. In Chapter 10, we discussed the tremendous and invaluable opportunities to control pests through introducing natural enemies and increasing genetic resistance. Natural enemies and genes for increasing resistance can come only from natural biota. Destroying natural biota will destroy such opportunities.

Since we select species from nature for animal husbandry, forestry, and aquaculture, essentially all the same arguments can be made in connection with those important enterprises.

To use our concept of biological wealth, we can look at natural biota as a bank in which are deposited the gene pools of all the species involved. As long as natural biota are preserved, we have a rich endowment of genes in the bank, which we can draw upon as needed. Thus, natural biota are frequently referred to as a **genetic bank**. Depleting this bank cannot help but deplete our future.

Sources for Medicine

Earth's genetic bank also serves medicine, as the following example illustrates. For thousands of years, the indigenous people of the island of Madagascar used an

FIGURE 18-5
The rosy periwinkle, a plant native to Madagascar, is a source of two anticancer agents that are highly successful in treating childhood leukemia and Hodgkin's disease. (Kevin Schafer/Peter Arnold, Inc.)

obscure plant, the rosy periwinkle, in their folk medicine (Fig. 18-5). This plant grows only on Madagascar, and if it had become extinct before 1960, hardly anyone outside Madagascar would have cared. In the 1960s, however, scientists extracted two chemicals called vincristine and vinblastine, with medicinal properties, from the plant. These chemicals have revolutionized the treatment of childhood leukemia and Hodgkin's disease. Before their discovery, leukemia was almost always fatal in children; today, with vincristine treatment, there is a 95% chance of remission. These two drugs now represent a $100-million-a-year industry.

The story of the rosy periwinkle is just one of hundreds. The venom from a Brazilian pit viper (a poisonous snake) led to the development of the drug Capoten, used to control high blood pressure. Taxol, an extract from the bark of the Pacific yew, has proved to be valuable for treating ovarian, breast, and small-cell cancers. Taxol is now synthesized from the leaves of the yew tree. Bark extract from a certain rainforest tree in Samoa has been used by traditional healers there to treat patients with yellow fever; its biological effects are now being studied at the National Cancer Institute. Stories like these have created a new appreciation for the field of *ethnobotany* (the study of the relationships between plants and people). To date, some 3000 plants have been identified as having anti-cancer properties. Drug companies are now financing field studies of the medicinal use of plants by indigenous peoples and are even funding the creation of parks and reserves to promote the preservation of natural ecosystems that are home to both the people and the plants.

It is a fact that 25% of pharmaceuticals in the United States contain ingredients originally derived from native plants, representing $8 billion of annual revenue for drug companies and better health and longevity for countless people. Table 18-2 shows a number of well-established drugs that were discovered as a result of analyzing the chemical properties of plants used by traditional healers. It is likely that the search for such chemicals has barely scratched the surface.

TABLE 18-2
Modern Drugs from Traditional Medicines.

Drug	*Medical Use*	*Source*
Aspirin	Reduces pain and inflammation	*Filipendula ulmaria*
Codeine	Eases pain; suppresses coughing	*Papaver somniferum*
Ipecac	Induces vomiting	*Psychotria ipecacuanha*
Pilocarpine	Reduces pressure in the eye	*Pilocarpus jaborandi*
Pseudoephedrine	Reduces nasal congestion	*Ephedra sinica*
Quinine	Combats malaria	*Cinchona pubescens*
Reserpine	Lowers blood pressure	*Rauwolfia serpentina*
Scopolamine	Eases motion sickness	*Datura stramonium*
Theophylline	Opens bronchial passages	*Camellia sinensis*
Vinblastine	Combats Hodgkin's disease	*Catharanthus roseus*

FIGURE 18-6
Natural biota provide numerous recreational, aesthetic, and scientific values, a few of which are depicted here. (a) Lowell Georgia/Photo Researchers, Inc.; b) Explorer/Yves Gladu/Photo Researchers, Inc.; c) Mark Burnett/Photo Researchers, Inc.; d) The Stock Market; e) Jack Wilburn/Animals Animals/Earth Science; f) Earth Science.)

Commercial Value

Recreational or aesthetic activities support commercial interests: sporting goods stores, tourist and travel accommodations, and so on. **Ecotourism**—where tourists visit a place in order to observe unique ecological sites—represents the largest income-generating enterprise in many developing countries. As the amount of leisure time available to people increases, more and more money will be spent on recreation. Since some percentage of these recreation dollars will be spent on activities related to the natural environment, any degradation of that environment affects commercial interests. Examples abound of businesses folding because a lake or beach, for example, became polluted and was no longer suitable for fishing or swimming.

In addition to the indirect support resulting from recreation, natural species support a number of commercial interests directly. Commercial fishing, logging, and the trade in exotic "pets" (uncommon species of fish, reptiles, mammals, birds, and plants) are the most conspicuous examples. In one recent study, the annual contribution of wild plant and animal species to the economy of the United States was calculated to be more than $200 billion, which is about 4.5% of the gross domestic product. Their contribution to the economy of developing countries is undoubtedly much higher, because of the intense economic pressures in those countries to exploit natural resources. In many developing countries, forest plants, wild game and smaller animals, like termites and locust, are important sources of food.

Recreational, Aesthetic, and Scientific Value

The species in natural ecosystems also provide the foundation for numerous recreational and aesthetic interests, ranging from sport fishing and hunting to hiking, camping, bird watching, photography, and so on (Fig. 18-6). Interests may range from casual aesthetic enjoyment to serious scientific study. Virtually all our knowledge and understanding of evolution and ecology have come from studying wild species and the ecosystems in which they live. Pleasure and satisfaction may even be indirect. For instance, one may never see a whale, but knowing that whales and similar exciting animals exist provides a certain aesthetic pleasure. The great popularity of nature films attests to this. Further, knowing that the earth and its biosphere continue to support and maintain such wildlife provides a sense of well-being.

Recreational and aesthetic values constitute a very important source of support for maintaining wild species. These activities involve a great number of people and represent a huge economic enterprise. To cite one example, 84% of the Canadian people take part in some form of wildlife-related recreation, spending an estimated $9.4 billion. In the United States, at least 50 million people do some recreational hunting or fishing each year, in the process of which they spend over $38 billion. Very likely, the broadest public support for preserving wild species and habitats is traceable to the aesthetic and recreational enjoyment people derive from them.

Intrinsic Value

The usefulness (instrumental value) of many wild species is apparent. But what about those other species that have no obvious value to anyone—probably the majority of plant and animal species, many of which are rare or inconspicuous in the environment? Some observers believe that the most successful strategy for preserving all wild species is to emphasize the *intrinsic* value of species, rather than the unknown or uncertain ecological and economic instrumental values. Thus, we should recognize that the extinction of a species per se is an irretrievable loss of something of value.

Many observers who view wild species as having a basic right to exist claim that humans have no right to terminate a species that has existed for thousands or millions of years and that represents a unique set of biological characteristics. They argue that long-established existence carries with it a right to continued existence. Some support this view by arguing that there is value in every living thing and that one kind of living thing (e.g., human) has no greater value than any other. This argument, however, can lead to some difficulties, such as having to defend the rights of pathogens and parasites. A more common viewpoint held by ethicists is that because humans have the ability to make moral judgments, they also have a special responsibility toward the natural world, and that responsibility includes concern for other species. It should be pointed out that until recently, Western philosophers argued that only humans were worthy of ethical consideration; the Western philosophical tradition has been strongly anthropocentric.

Some ethicists find their basis for intrinsic value in religion. For example, Old Testament writings express God's concern for wild species when He created them. Jewish and Christian scholars alike maintain that by declaring His creation good (see *Genesis* 1) and giving it His blessing, God was saying that all wild things have intrinsic value and therefore deserve moral consideration. The Islamic Quran (Koran) proclaims that the environment is the creation of Allah and should be protected because it praises the Creator. This ethical concern for wild species underlies many religious traditions and represents a potentially powerful force for preserving biodiversity.

Thus, we see that even if species have no demonstrable use to humans, it can still be argued that they have a right to continue to exist. Only rarely (as in the case of parasites and pathogens) can we claim that there is any moral justification for driving other species to extinction.

Biodiversity

We saw in Chapter 5 that natural selection operating over time leads to **speciation**—the creation of new species—as well as **extinction**—the disappearance of species. Over geological time, the net balance of these processes has favored the gradual accumulation of more and more species—in other words, biodiversity. The concept of biodiversity is often extended to the genetic diversity within species, as well as the diversity of ecosystems and habitats. Its main focus, however, is the species.

How much biodiversity is there? No one knows. Today, the only two certainties are that 1.75 million species have been described and many more than that exist. Most people are completely unaware of the great diversity of species within any given taxonomic category. Groups especially rich in species are the flowering plants (270,000 species) and the insects (950,000 species), but even less diverse groups, such as birds or ferns, are rich with species that are unknown to most people (Table 18-3). Taxonomists are aware that their work in finding and describing new species is incomplete. Groups that are conspicuous or commercially important, such as birds, mammals, fish, and trees, are much more fully explored and described than, say, insects, very small invertebrates like mites and soil nematodes, fungi, and bacteria. Trained researchers who can identify major groups of organisms are few, and the task of exploring the diversity of life will require a major effort in systematic biology.

Estimates of the number of species on Earth today are based on recent work in the tropical rain forests, which hold more living species than all other habitats

TABLE 18-3
Known and Estimated Species on Earth

Species	Number of Known Species	Estimated Numbers of Species: High (000)	Low (000)	Working Number (000)	Accuracy
Viruses	4000	1000	50	400	Very poor
Bacteria	4000	3000	50	1000	Very poor
Fungi	72,000	2700	200	1500	Moderate
Protozoa	40,000	200	60	200	Very poor
Algae	40,000	1000	150	400	Very poor
Plants	270,000	500	300	320	Good
Nematodes	25,000	1000	100	400	Poor
Arthropods					
Crustaceans	40,000	200	75	150	Moderate
Arachnids	75,000	1000	300	750	Moderate
Insects	950,000	100,000	2000	8000	Moderate
Mollusks	70,000	200	100	200	Moderate
Chordates	45,000	55	50	50	Good
Others	115,000	800	200	250	Moderate
Total	**1,750,000**	**111,655**	**3635**	**13,620**	**Very poor**

Source: United Nations Environment Programme, *Global Biodiversity Assessment* (Cambridge: Cambridge University Press, 1995), Table 3.1-1, p. 118.

combined. Costa Rica, less than half the size of New York State, is estimated to contain at least 5% of all living species. The upper boundary of the estimate keeps rising as taxonomists explore the rainforests more and more. Recent estimates place the high end at 112 million species. Whatever the number, the planet's biodiversity represents an amazing and diverse storehouse of biological wealth.

The Decline of Biodiversity

Losses Known and Unknown. A recent inventory of the wild plant and animal species in the 50 states suggests that at least 9000 species are at risk of being lost. At least 500 species native to the United States are known to have become extinct since the early days of colonization. Over 100 of these are vertebrates (Fig. 18-7). Currently, the U.S. Fish and Wildlife Service and the National Marine Fisheries Service list 1051 U.S. species as endangered or threatened, categories that elicit protection and recovery plans under the ESA (Table 18-1). Thousands more are awaiting evaluation so that their status can be verified and plans made for their protection.

The species named on this long list are only part of the problem. Across North America, declines in the numbers of unlisted wild species are also occurring. Commercial landings of fish are down 42% since 1982, waterfowl populations have shown an overall decline of 30% since 1969, and many North American songbird species, such as the Kentucky warbler, wood thrush, and scarlet tanager, have been declining and have disappeared entirely from some local regions. The North American Breeding Bird Survey conducts a census of breeding birds in 2700 areas of the United States, and their latest results show that populations of 70 species of songbirds declined significantly between 1980 and 1993. Most of these birds are migrants that winter in Mexico and southward. Similarly puzzling declines in amphibian populations are occurring in North America and other parts of the world.

Worldwide, the loss of biodiversity is even greater. Most of the extinctions of the past several hundred years have occurred on oceanic islands, where small land masses limit the size of populations and human intrusions are most severe. Recently, attention has focused on the tropics, where biodiversity is richest. This richness is almost unimaginable. E. O. Wilson identified 43 species of ants on a single tree in a Peruvian rainforest, a level of diversity equal to the entire ant fauna of the British Isles. Other scientists found 300 species of trees in a single 2.5-acre (1-ha) plot and as many as 10,000 species of insects on a single tree in Peru.

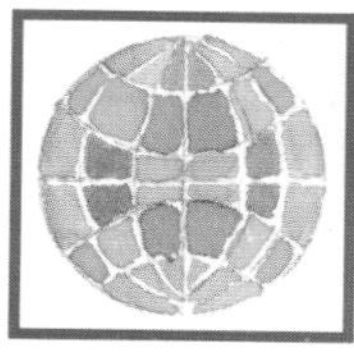

Global Perspective

The Mega-extinction Scenario

Biodiversity is under assault worldwide, and to counter this assault, a number of prominent biologists have gone on record documenting a catastrophic loss of species and the dire consequences thereof. Current losses are estimated to be as high as 17000 species a year in the tropical forests alone. At the present rates of clearing land, some scientists estimate that we will lose one fourth of all species within 50 years. If there are 14 million species (a working number), as some taxonomists suspect, this means the loss of 3.5 million of them! In light of these possibilities, calls are going out for such responses as stopping the development of all remaining natural lands in the developed world. Many advocate giving huge sums of money to the poor nations, so that they can quickly bring population growth down and convert their agriculture to sustainable, high-yield use of currently existing croplands.

This sounding of the alarm, according to some critics, has become a dogma among many biologists and is based on a number of highly questionable assumptions. Three key assumptions lie behind the mega-extinction scenario. One is that losses of habitat always mean losses of species, and often in direct proportion. However, forests cleared are not necessarily forests gone forever; many are converted to second-growth forests, which support many of the original species.

A second assumption is that the larger a geographical area, the larger is the number of species it holds. In many areas, however, the number of species levels off, even though the geographical area is large. So there may be no net loss of species, even though some of an area is lost.

The third questionable assumption is that the number of species in existence is far larger than the number of species we know about; hence, estimates of the number of extinctions are similarly high. Thus, the catastrophic extinction rates predicted by biologists are based largely on species that no one has ever seen. Also, there is a disturbing lack of agreement among observers who calculate the number of existing species; one calculation may be ten or more times another. Literally, there is an unknown genetic richness out there!

Support for the mega-extinction scenario is so strong that those who challenge it from within the biological research community risk their reputations and the normally cordial relationships expected among colleagues. Challengers do exist despite the large stakes, however, and these critics are calling for their colleagues to examine their data and base their predictions only on supportable evidence. They also ask that only realistic public-policy alternatives be advocated.

The problem is, if the dire predictions are shown not to be true, some of the real problems—tropical deforestation and loss of wetlands, for instance—will also get less than the attention they deserve. And the public will be less prone to believe other, more well-founded scientific predictions of environmental disasters.

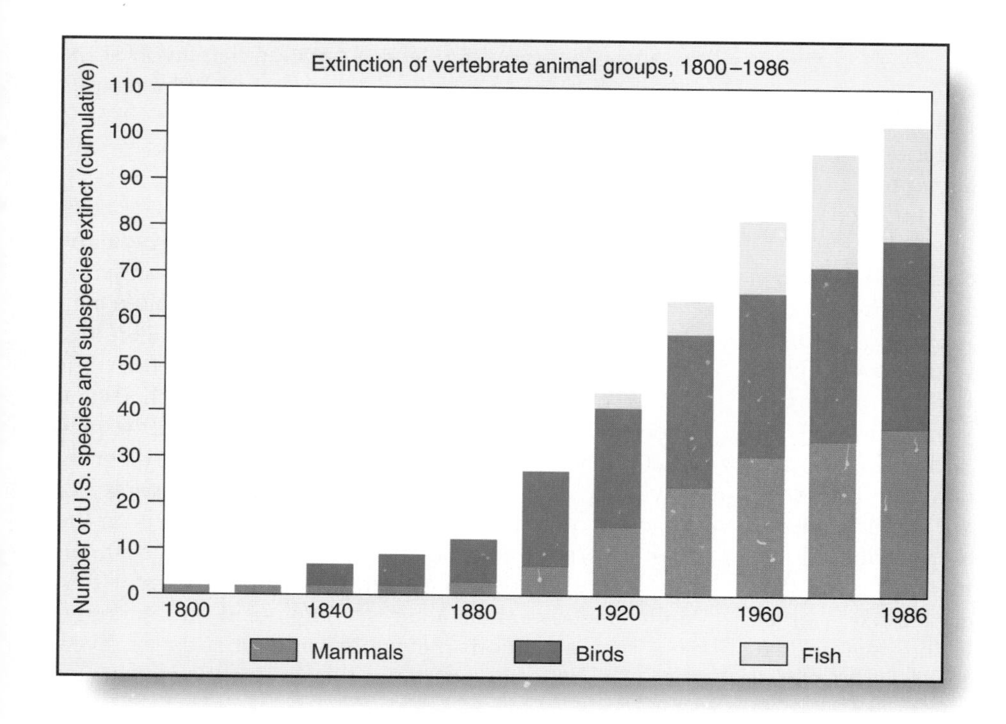

FIGURE 18-7
Extinction of vertebrate animal groups in the United States, 1800–1986. The cumulative extinction of mammals, birds, and fish is shown. (From *Environmental Trends*, 1989. Washington, DC: Council on Environmental Quality, p. 108.)

FIGURE 18-8
Heavy construction equipment removing the natural community as land is developed to make room for more housing, malls, roads, etc. (M. Kirk/Peter Arnold, Inc.)

Assuming the existence of 2 million species in the tropical forests (a conservative estimate) and a clearance rate of 1.8% per year for those forests, Wilson calculated that tropical deforestation is responsible for the loss of 4000 species a year. Other scientists have projected that as many as 17,000 species are lost per year; or, between 500,000 and 2 million species over the last half of the 20th century! Somewhere on Earth a species is going extinct about every 30 minutes—a tragic loss to the gene pool. Not all scientists agree with these calculations, however. (See "Global Perspective" box, p. 471.) The most important conclusion we can draw is that many species are in decline and some are becoming extinct. No one knows how many, but the loss is real, and it represents a continuing depletion of the biological wealth of our planet.

Reasons for the Decline

Physical Alteration of Habitats. Although multiple causes are usually the rule in losses of biodiversity, the greatest loss is caused by the physical alteration of habitats through the processes of conversion, fragmentation, and simplification.

Conversion. Natural areas are converted to farms, housing subdivisions, shopping malls, marinas, and industrial centers (Fig. 18-8). When, for example, a forest is cleared, it is not just the trees that are destroyed; every other plant and animal which occupies that destroyed ecosystem, either permanently or temporarily (e.g., migrating birds), also suffers. The idea that this wildlife will simply move "next door" and continue to live in an undisturbed section is erroneous. As we saw in Chapter 4, population balances lead to each area having all the wildlife it can support. Any loss of natural habitat can result in only one thing: a proportional reduction in all populations which require that habitat. Thus, the decline in songbird populations cited earlier has been traced to a combination of the loss of winter forest habitat in Central and South America and the increasing fragmentation of summer forest habitat in North America.

Fragmentation. For the continued survival of any natural population, the number of individuals must never fall below a *critical number*; and that requires a certain minimum area. This minimum area must be large enough to compensate for years of adverse weather. That is, more area will be required during a dry year than during a normal year. If development reduces the habitat to a point where it cannot support the critical number during an adverse year, the entire population will perish. Similarly, development (such as a highway) that fragments a territory and prevents migration between the two fragments will cause a population to perish if neither area can adequately support the critical number. Also, reducing the size of a habitat creates a greater proportion of edges, a situation that favors some species but may be detrimental to others. For example, the Kirtland's warbler is an endangered species that is highly dependent on large patches of second-growth jack pines. It is endangered because its habitat has been greatly fragmented, creating edges that favor the brown-headed cowbird, a nest parasite that can invade the forest and lay its eggs in the nest of the rare warbler.

Simplification. Human use of habitats often simplifies them. We might, for example, remove fallen logs and dead trees from woodlands for firewood, thus diminishing an important microhabitat on which several species depend. When a forest is managed for the production of a few or one species of tree, tree diversity obviously declines, and with it so does the diversity of a cluster of plant and animal species dependent on the less favored trees. Streams are sometimes "channelized"—their beds are cleared of fallen trees and riffles, and sometimes the stream is straightened out by dredging. Such alterations inevitably bring on a loss of diversity of fish and invertebrates that live in the stream.

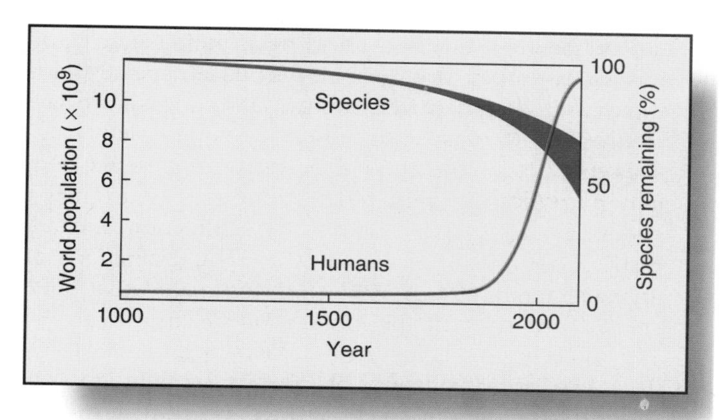

FIGURE 18-9
The inverse relationship between human population size and the survival of species worldwide. Uncertainty about the extent of species becoming extinct is reflected in the width of the species curve. (From M.E. Soulé, "Conservation: Tactics for a Constant Crisis," *Science,* 253 [1991], 744. Copyright 1991 by the AAAS.)

The Population Factor. Past losses in biodiversity can be attributed to the expansion of the human population over the globe. Continuing human population growth will bring on continued alteration of natural ecosystems and the inevitable loss of more wild species. The losses will be greatest where human population density and growth are highest—in the developing world (where biodiversity is greatest). Africa and Asia have lost almost two-thirds of their original natural habitat. People's desire for a better way of life, the desperate poverty of rural populations, and the global market for timber and other natural resources are powerful forces that will continue to draw down biological wealth.

In Kenya, where human population growth has been explosive for several decades, the conversion of savannah and woodlands to cultivation or intensive grazing by goats and cattle has driven most of the African elephant population into the existing wildlife reserves, causing a great reduction in their numbers. The other large African mammals have experienced similar reductions, as the needs of the rural population inevitably conflict with those of the large wild animals of eastern Africa. Lest we forget, as North America was gradually transformed into a great agricultural and industrial continent, similar losses in large mammal populations occurred (wolves, bison, elk, black bear, cougar, etc.).

One key to holding down the loss in biodiversity lies in bringing human population growth down. There is an inverse relationship between the size of the human population and the survival of species worldwide, illustrated in Fig. 18-9. If the human population increases to 12 billion, as some demographers believe that it will, the consequences for the natural world are enormous.

Pollution. Another major factor causing loss of biodiversity is pollution. Pollution can directly kill many kinds of plants and animals. For example, the *Exxon Valdez* oil spill killed at least 300,000 birds and 3000 sea otters (Fig. 18-10). High levels of toxic metals traced to industry and agriculture killed 150,000 grebes (waterfowl) in California's Salton Sea.

In addition, pollution destroys or alters habitats, with consequences just as severe as those caused by deliberate habitat conversions. Acid deposition and air pollution cause forests to decline, sediments and nutrients kill organisms in waterways, pesticides devastate wild bird populations, and depletion of the ozone layer increases the impact of ultraviolet radiation on wild species. Indeed, some recent experimental work with amphibians suggests that a major factor in their decline is the impact of ultraviolet radiation on their fertilized eggs.

Despite this side-effect nature, the consequences of pollution may be just as severe as those caused by deliberate conversions. Acid deposition and air pollution cause forests to die, sediments and nutrients kill species in lakes, rivers, and bays, DDT devastates wild bird populations, and depletion of the ozone layer increases the impact of ultraviolet light on wild species. The list is endless. Some scientists project that global warming from the greenhouse effect may be the greatest catastrophe to hit natural biota in 65 million years. (The fossil record reveals that a massive extinction of plants and animals occurred at that time. The probable

FIGURE 18-10
Oil-soaked seabirds killed by the *Exxon Valdez* oil spill that left 11 million gallons of crude oil spread over Prince William Sound in Alaska. (R. Levy/Gamma-Liaison, Inc.)

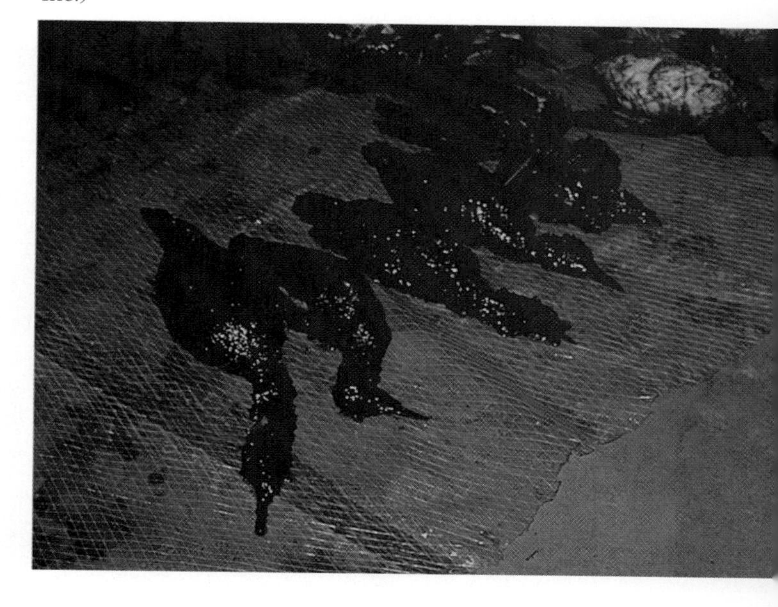

cause was the earth being hit by a major asteroid, but there is still controversy regarding this hypothesis.)

Scientists project catastrophic effects from global warming because most species adapt only very slowly and hence can adapt only to gradual changes. The greenhouse effect may cause more warming in the next 50 years than what would normally occur in the next 1000 years, a 40-fold increase in the rate of change. If this happens, numerous species will be unable to adapt so rapidly. Nor can they migrate quickly enough. Trees, for example, can spread their seeds only a few miles each generation. Seeds landing this distance from the parent must grow to mature trees and shed seeds before the population can move the next few miles, and so on. Scientists speculate that the rate of climatic warming will far outpace the ability of most species of trees to migrate northward, thus trapping them in inhospitable climates. Every species that dies out doubtlessly will take others with it. If the forest trees are unable to migrate, neither can the rest of the wildlife that depend on them for food and habitat.

Most of the more global pollution problems can be traced to the industrialized world, where energy-generating and other technologies continue to pour pollutants into the air and water at increasing rates. For this reason, it is important not to point the finger of blame primarily at the developing world, where population growth is such a problem. Global warming and ozone depletion are the legacy of the already developed nations.

Exotic Species. An **exotic species** is a species introduced into an area from somewhere else, often a different continent. Because the species is not native to the new area, it is often unsuccessful in establishing a viable population and quietly disappears. This is the fate of many pet birds, reptiles, and fish that escape or are deliberately released away from their native habitats. Occasionally, however, an introduced species finds the new environment very much to its liking. As we saw in Chapter 10, most of the insect pests and plant parasites that plague agricultural production were accidentally introduced to the region. Exotic species are major agents in driving native species to extinction and are responsible for an estimated 40% of all extinctions of animals since 1600. A 1993 study estimated that 4500 species are now in the United States; approximately 15% of these are damaging to ecosystems.

The transplantation of species by humans has occurred throughout history, to the point where most people are not aware of the distinction between native and exotic species living in their lands. Literally hundreds of weeds and forage plants were brought by the European colonists to the Americas, so that now most of the common field, lawn, and roadside plant species in eastern North America are exotics. Fully one-third of the plants in Massachusetts are alien or introduced—about 900 species! Among the animals, the most notorious have been the house mouse and the Norway rat; others include the wild boar, donkey, horse, nutria, and even the red fox, which was brought to provide better fox hunting for the early colonial equestrians. One of the most destructive exotics is the house cat. The 60 million domestic and feral cats in the United States are very efficient at catching small mammals and birds. One recent study of cat predation in Great Britain showed that cats kill 20 million birds a year there.

Deliberate introductions sanctioned by the U.S. Natural Resources Conservation Service (in the interest of "reclaiming" eroded or degraded lands) have brought us kudzu and other runaway plants such as autumn olive, multiflora rose, and amur honeysuckle. Many other exotic plants are introduced as horticultural desirables; one notorious example is purple loosestrife (Fig.

FIGURE 18-11
Purple loosestrife growing on the edge of a pond in Williams County, Ohio. This escaped horticultural species is replacing many native species in wetlands across North America. (John Lemker/Earth Scenes.)

18-11), which infests wetland habitats and replaces many plants valuable to wildlife. Another is Oriental bittersweet, an aggressive vine that chokes and shades small shrubs and trees. Often, it smothers young trees and prevents the forest from regrowing. In Oregon and California, the introduced star thistle invades and ruins pasture land, grows wildly, and is a fire hazard.

We might think that introduced species would add to the biodiversity of a region, but many introductions have exactly the opposite effect. The new species are often very successful predators that eliminate native species not adapted to their presence. For example, the brown tree snake was introduced to the island of Guam as a stowaway on cargo ships during World War II (Fig. 18-12). This snake is mildly poisonous, grows to a length of 13 feet, and eats almost anything smaller than itself. In a little over 40 years, the snakes have eliminated most of the island's birds; six of Guam's 18 native species of birds have become extinct, and populations of the remaining birds are so decimated that survivors are rarely seen or heard. Efforts at controlling the snakes have proved futile. There are no natural predators on the island, and the snake population has reached such a high density that snakes have invaded houses in search of prey (such as puppies and cats). At present, wildlife officials are concentrating their efforts on preventing the snakes from spreading to other islands in the Marianas chain.

FIGURE 18-12
The brown tree snake, accidentally introduced to Guam, has decimated bird life on the island. Often lacking in natural predators, islands are especially vulnerable to the harmful effects of exotic species. (Klaus Uhlenhut/Animals, Animals/Earth Science.)

Introduced exotic species may also drive out native species by competing with them for resources, as documented in Chapter 4. An Asian weed, hydrilla, was brought to Florida from India by a man selling plants for home aquariums. Since 1950, it has spread throughout lakes and reservoirs in the South, and federal agencies are spending $5 million a year in attempts to control it. The weed chokes intakes of power plants and waterworks, and, because of its tendency to spread over the water surface, causes fish and invertebrates to die by preventing oxygen from reaching the deep water.

One of the most recent exotic invaders is the zebra mussel, (see "Earth Watch" box, p. 476), which came to the United States in 1986 in the ballast water of ocean-going vessels. The zebra mussel is expected to cause $4 billion in damage in the Great Lakes alone over the next decade; its impact on native species is likely to be devastating.

Overuse. It should be obvious that killing whales, fish, or trees faster than they can reproduce will lead to the ultimate extinction of the species. In spite of that, overuse is another major assault against wild species. Overuse is driven by a combination of economic greed, ignorance, and desperation. The plight of birds in Europe is a good example. It is said, with some justification, that a line can be drawn across the continent north of which people watch birds and south of which they eat them. Some 700,000 "protected" birds are shot in Greece each year, Malta accounts for about 3 million a year, and Italy holds a shameful record of an astronomical 50 million birds killed and eaten each year! Most of these are small songbirds.

Another prominent form of overuse is the trafficking in wildlife and products derived from wild species—a $5 to $8 billion business annually. Much of this trade is illegal. It flourishes because some consumers are willing to pay exorbitant prices for such things as furniture made from tropical hardwoods like mahogany and for exotic pets, furs from wild animals, traditional medicines from animal parts, and innumerable other "luxuries," including polar bear rugs, baskets made from elephant and rhinoceros feet, ivory-handled knives, and reptile-skin shoes and handbags. For example, some Indonesian and South American parrots sell for up to $5000 in the United States, a panda-skin rug can bring $25,000, and the gall bladder of the North American black bear is valued as a folk medicine in Korea, where a single gall bladder can fetch as much as $2000. Such prices create a powerful economic incentive to exploit the species involved.

The long-term prospect of extinction does not curtail the activities of exploiters because the prospect of a huge immediate profit outweighs it. Even when the species is protected, the economic incentive is such that poaching and black-market trade continue (Fig. 18-13). Tiger and rhinoceros populations in the wild have declined drastically in the last two decades (a 90%

Earth Watch

Invasion of the Zebra Mussels

Since the early 1800s, the Great Lakes have become host to at least 115 exotic species of plants, fish, algae, and mollusks. A few of the invaders were deliberately introduced (silver and Chinook salmon, for example), but most came in by accident. Ships have brought in the largest percentage of exotics, especially through their ballast water.

In 1986, a ship took in ballast water from the Elbe or Rhine River in Europe and released the water in the St. Clair River near Detroit when it took on cargo. In this ballast water, there must have been a sizable quantity of the planktonic larvae of *Dreissena polymorpha*, the zebra mussel. The larvae settled to the bottom of the river and soon grew to adult size (about 1.5 inches, or 4 cm, long), whereupon they began to reproduce—with a vengeance! The mussel has now been sighted in all of the Great Lakes and has begun to spread; the Great Flood of 1993 pushed the mussels well down into the Mississippi River system.

The mussels have found biological conditions in the Great Lakes ideal. Mussel densities greater than 94,000 individuals per square meter have been found. They attach to any hard surface with strong threads (called byssal threads) and feed on plankton carried to them by normal currents. Unfortunately, they thrive particularly well on hard surfaces where water flow is continuous and so have settled in great densities on the intake pipes of municipal water supplies and thermal power stations. There they pile up and eventually block the pipes. Repairs in Detroit and Monroe, Michigan, and in Cleveland, Ohio, have already cost $300 million; the damage for the entire Great Lakes may be as much as $4 billion over the next decade.

As severe as this intake-pipe problem is, a greater fear is that the mussels will bring about catastrophic changes in the ecology of the Great Lakes. Where they reach high densities, they are able to filter virtually all of the larger plankton from the water. Indeed, in the eutrophic European waters to which they are native, the mussels have a positive impact on the water as they filter out the dense plankton and detritus. Food chains that support the Great Lakes fisheries are based on plankton, however, and it is feared that the mussels will intercept crucial quantities of the plankton and thus cut back greatly on energy flow to the higher trophic levels (the desirable fish species). The mussels also may cause the demise of native bivalve populations through competition for food and because of their habit of attaching to the shells of the other bivalves. In rivers such as the Illinois, which joins the Mississippi, high densities of the mussels have already caused oxygen levels to decline, bringing stress to other aquatic life. In Europe, several fish species prey on the zebra mussel and keep their numbers from reaching pest levels, but similar species are lacking in North America, and no one believes that importing yet another exotic species is a good idea.

In response to this invasion, in October 1990 Congress passed the Non-indigenous Aquatic Nuisance Prevention and Control Act—too late to stop the zebra mussels, but intended to prevent further such invasions by prohibiting the discharge of ballast water. The act also funds research and control programs directed toward minimizing the impact of the zebra mussels. In the meantime, biologists are waiting for the zebra mussels to reach their peak and, it is hoped, settle down to an equilibrium level where their impact is diminished. Because they are so dependent on plankton, high densities of mussels will probably continue to occur only in the most eutrophic regions of the Great Lakes area: Lake St. Clair, Lake Erie, and possibly Green Bay and Saginaw Bay. (Check: Great Lakes Sea Grant Network on the Internet)

decline for rhinos) and are on the brink of extinction. Driving this decline is the widespread but unfounded belief in Far Eastern countries that parts from these animals have medicinal or aphrodisiac properties. In just three months in 1993, customs records from South Korea revealed imports from China of 1.5 tons of tiger bones—the remains of at least 200 tigers.

It is easy to place the blame on the economic greed of the persons doing the killing. However, equal blame must be shared by the consumers who offer the "reward." It is their ignorance or insensitivity to the fact that their money is fostering the extinction of invaluable wildlife that is fueling the situation. Unfortunately, the killing shows no sign of abatement.

Particularly severe is the growing fad for exotic "pets": fish, reptiles, birds, and house plants. In many cases, these plants and animals are taken from the wild. When they are removed from the natural breeding populations, they are no better than dead, in terms of maintaining the species. At least 40 of the world's 330 species of parrots face extinction because of this fad. Numerous species of tropical fish, birds, reptiles, and plants are thus headed toward extinction because of exploitation by the pet trade.

Poor management represents another form of overuse leading to a loss of biodiversity. Forests and woodlands are overcut for firewood, grasslands are overgrazed, game species are overhunted, fisheries are

FIGURE 18-13
Products from endangered species confiscated in the breakup of a black market ring. Even though trade is illegal and the animals are protected, poaching and black market trade are still causing merciless slaughter for often trivial products. The World Wildlife Fund estimated this illegal traffic at $5 billion a year. (U.S. Fish and Wildlife Service photo. Steve Hillebrand.)

overexploited, and croplands are overcultivated. These practices not only deplete the resource in question, but often set into motion a cycle of erosion and desertification, with effects far beyond the exploited area. The problem is explored in depth in Chapter 19.

Consequences of Losing Biodiversity

What will happen as more and more rare or unknown species pass from the earth because of our activities? A few people will mourn their passing, but as one observer put it, the sun will continue to come up every morning, and the newspaper will appear on our doorstep. Currently, we seem to be getting away with it, as evidenced by the fact that many species have already become extinct as a consequence of human activities, and life goes on. However, we really do not know what we are losing when we lose species. Some ecologists have likened the loss of biodiversity to an airplane flight, where we continually pull out rivets as the plane cruises along. How many rivets can we pull out? Is this a wise activity? So far, we have gotten away with driving species to extinction, but the natural world is certainly less beautiful and more monotonous as a result.

We might someday lose what ecologists call a **keystone species,** a species whose role is absolutely vital for the survival of many other species in an ecosystem. For example, a group of insects called orchid bees play a vital role in tropical forests by pollinating trees. These bees travel great distances and are able to pollinate trees and other plants that are often widely separated from one another. If they disappear, many other plants, including major forest trees, will eventually follow. Or perhaps the keystone species in an ecosystem is a predator that keeps herbivore populations under control. Often, keystone species are the largest animals in the ecosystem, and because of their size, they have the greatest demands for unspoiled habitat. Thus, wolves, elephants, tigers, and other large animals are umbrella species for whole ecosystems. If they can survive, there will likely still be enough undisturbed habitat for the other living species.

Biodiversity will continue to decline as long as we continue to remove and constrict the natural habitats in which wild species live. This loss is definitely costly, because natural ecosystems provide vital services to human societies. Recreational, aesthetic, and commercial losses will be inevitable. The loss of wild species, therefore, will bring certain and unwelcome consequences, because it is linked directly to the degradation or disappearance of ecosystems (our focus in Chapter 19).

International Steps to Protect Biodiversity

Convention on Trade in Endangered Species. Endangered species outside the United States are only tangentially addressed by the ESA. However, under the leadership of the U.S., the Convention on Trade in Endangered Species of Wild Fauna and Flora, or CITES, was established in the early 1970s. CITES is not specifically a device to protect rare species; instead, it is an international agreement (by 118 nations) that focuses on trade in wildlife and wildlife parts. The treaty recognizes three levels of vulnerability of species, the highest being species threatened with extinction. Restrictive trade permits and agreements between exporting and importing countries are applied to those species, sometimes with the result of a complete ban on trade if the CITES nations agree.

Perhaps the best known act of CITES was to ban the international trade in ivory in 1990 in order to stop the rapid decline of the African elephant (from 2.5 million animals in 1950 to about 350,000 today). The CITES nations reconvene every two years to entertain new listings and deal with infractions of the agreement. Unfortunately, illegal poaching and inadequate or fraudulent policing, particularly in the Far East, have continued to allow a substantial amount of illegal traffic in animals and their parts. As an example of action authorized by CITES, in August 1994 the Clinton administration banned the import of some animal products from Taiwan and threatened further trade sanctions if the Taiwanese were not more diligent in preventing trade in rhino horns and tiger parts. The World Wildlife Fund, a private international organization, maintains a unit that keeps watch on wildlife trade. Sting operations in the United States and elsewhere keep the outlaw animal traders on guard; recently, a federal agent dressed as a gorilla caught five traders in a small but significant step toward curbing a business whose victims are truly innocent.

Convention on Biological Diversity. Although CITES provides for some protection of species that might be involved in international trade, it is inadequate to address broader issues pertaining to the loss of biodiversity. People of the industrialized nations have expressed a growing concern about the decline of well-loved species such as pandas, elephants, and rhinos, the rapid rate of deforestation in tropical rainforests and the feared loss of many species as a consequence, and inroads into lands occupied by indigenous peoples. It is clear that the main target of their concern is the plight of biological diversity in the developing countries, where most biodiversity is found. On the other hand, people in the developing countries regard their forests and animals as part of their natural resources; millions of rural individuals depend on access to these resources, and many governments look to their development as the key to more rapid economic growth. Thus, a problem arises wherein some of the world wants to see the protection of resources that are not in its territory, but which it values nevertheless.

With the support of the UNEP, an ad hoc working group proposed an international treaty that would resolve these differences and form the basis of action to conserve biological diversity worldwide. After several years of negotiation, the **Convention on Biological Diversity** was drafted and became one of the pillars of the 1992 Earth Summit in Rio de Janeiro (Fig. 18-14). The Biodiversity Treaty, as the convention is called, was ratified in December, 1993, and is now in force. The treaty establishes a Conference of the Parties as the agency that will provide oversight and report on its task during periodic meetings. The preamble sets forth basic guidelines for the treaty: a concern for the intrinsic value of biodiversity, its significance for human welfare, the sovereignty of a nation over its biodiversity, and the nation's obligations to protect and conserve biodiversity.

The treaty addresses the following difficult issues:

1. Funding from the industrialized world will be provided to the developing countries, enabling them to meet their obligations to protect biodiversity within their borders. The developing countries had made it clear that they needed financial help with the direct costs of protection. Further, if they were to forego, say, using land that could be farmed, they should receive some compensation for the revenue they would lose.
2. Authority to provide access to the genetic resources represented by wild species rests with the national governments within whose borders the species are found. Prior to this treaty, such resources had been considered a global commons, subject to free access.
3. All parties are obligated to make available to the country of origin the technologies (in particular, biotechnologies) that may be developed from genetic resources. This availability should be on terms better than the open market; in other words, the developing countries should receive some substantial benefit from any technological development based on their genetic resources.

FIGURE 18-14
UNCED Earth Summit in Rio de Janeiro passed the Convention on Biological Diversity over the "no" vote of President Bush. (a) Delegates in a session of the summit. (b) Maurice Strong, Secretary-General of the UNCED (far left), a representative of indigenous peoples, and other delegates. [(a) Photo by Reuters/Bettmann; (b) photo by Ricardo Funavi/Imagens Da Terra/Impact Visuals Photo and Graphics, Inc.]

Concern about the impact of the access principle on the U.S. biotechnology industry led then-President Bush to refuse to sign the treaty in Rio in 1992. President Clinton signed the treaty in June 1993, but in order to become binding, the treaty must be ratified by a two-thirds majority vote of the U.S. Senate. Continued objections raised by the agricultural and biotechnology lobbies have led the Senate to shelve ratification, leaving both the treaty and U.S. involvement in it in a state of limbo. Proponents of the treaty have claimed that these concerns are not well founded; indeed, many leaders of the supposedly endangered industries support the treaty.

Without the cooperation of the United States, the Biodiversity Treaty is unlikely to succeed. Some observers are urging President Clinton to issue executive orders that will put the U.S. in compliance with the treaty, with or without ratification by the Senate.

Further Steps

The continuing loss of habitat worldwide is largely responsible for the current trend in loss of biodiversity. Further strategies to address this loss must focus on preserving the natural ecosystems that sustain wild species. In the next chapter, we turn our attention to those ecosystems—how we depend on them, what we are doing to them, and what we must do to resist the forces that seem to be leading to a future in which all that is left of wild nature is what we have managed to protect in parks, preserves, and zoos.

Review Questions

1. What means are used to preserve game species, and what are some problems emerging from the adaptations of many game species to the humanized environment?
2. How does the Endangered Species Act (ESA) preserve rare species in the United States? What are the shortcomings of this act?
3. How does the case of the spotted owl illustrate the controversies over the ESA?
4. Define *biological wealth,* and apply the concept to human use of that wealth.
5. Define *instrumental value* and *intrinsic value* as they relate to determining the worth of natural species.
6. Describe five ways in which humans today are still dependent on natural species.
7. What is biodiversity? What are the best current estimates of its extent?
8. Describe both the known and estimated losses in biodiversity.
9. How do habitat conversion, fragmentation, and simplification affect biodiversity?
10. How do human population growth, pollution and overuse affect biodiversity?
11. What impacts can exotic species have on biodiversity?
12. What are the possible consequences of species extinctions?
13. What is CITES, and what are its limitations?
14. Give three provisions of the Convention on Biological Diversity.

Thinking Environmentally

1. Choose an endangered or threatened animal or plant species and research what is currently being done to preserve it. What dangers is it subject to? Write a protection plan for it.
2. You have been given the task of maintaining or increasing biodiversity on a small island. What steps will you take?
3. Some people argue that each individual animal has an intrinsic right to survival. Should this right extend to plants and microorganisms? Justify your answer.
4. Monies for enforcing the Endangered Species Act are limited. How should we decide which species are the most important to save?
5. Is it right to use animals for teaching and research? Defend your view.
6. Do all species have a right to exist? What about the anopheles mosquito, which transmits malaria? Tigers that kill people in India? Bacteria that cause typhoid fever? Defend your position.

Ecosystems as Resources

Key Issues and Questions

1. Forests are highly productive ecosystems that perform a host of natural services and provide lumber, fuel, and paper for the world's economy. However, in many parts of the world forests are being cleared. Why is this happening, and what can be done to make forest use sustainable?
2. The harvest of fish and shellfish has reached a plateau worldwide, and many once-abundant fisheries are no longer viable. What is causing these changes, and how can the declines be reversed?
3. Natural ecosystems perform a number of "natural services" that are vital to human interests. What are these services?
4. It is possible for a country to deplete its fisheries, degrade its soils, remove its forests, and overgraze its rangelands and have all of these activities appear as productive activities in the GNP. Should not economic systems account for natural resources and natural services?
5. Exploitation and overuse of natural systems is pervasive and nonsustainable. How do the two concepts of the tragedy of the commons and maximum sustained yield help us to understand these problems?
6. Public lands occupy 40% of the land area of the United States, and represent a vital national treasure. How effectively are public lands being protected in the United States?
7. Attempts to combat pollution, restrict overuse of natural systems, and preserve biodiversity have generated an environmental backlash known as the Wise Use Movement. What is the movement, and how does it influence public policy?

The hot African sun beat down on soil once shaded by trees. The soil dried up, and when the rains came, the soil flowed in red streams into muddy rivers and, finally, into the Indian Ocean. With it flowed the fertility desperately needed by rural families to raise their traditional crops. The forests, woodlands, and shade trees that once held the soil together with roots were removed to provide firewood for villagers, whose numbers were increasing at some of the most rapid rates of growth in the world.

This picture is all too typical of nations in East Africa, where the people and the wildlife together are greatly damaged by the loss of trees. In Kenya, however, a different story is being told, a story of belts of trees growing in abundance in urban and rural environments (Fig.

Fishermen in the North Atlantic. [Photo by Tom Stewart/The Stock Market.]

19-1). These trees are indigenous species such as baobabs, acacias, cedars, and thorns as well as fruit trees: citrus, figs, and bananas. The trees are being planted by Kenyans as part of the Green Belt movement, from seedlings raised in a thousand rural nurseries. Since the early 1980s, over ten million trees have been planted by schoolchildren and women, and the result is evident in the countryside of many communities. There is shade, woodland birds have returned, there is less dust in the air, and erosion has been reduced. These changes can be traced to the work of one remarkable woman, Wangari Maathai, who had a vision of what needed to be done and the energy and organizational skills to get the funding and launch the Green Belt movement (Fig. 19-2). Thirty other African nations now have their versions of the movement. It is significant that this movement is entirely independent of the government and foresters. In other parts of the world, similar grassroots groups are organizing to restore natural ecosystems that governments seem at best to be unable to protect and at worst to be bent on exploiting as rapidly as possible for short-term economic gain.

Unfortunately, the Green Belt story is exceptional; more often, natural ecosystems and the species in them are in trouble. This text has documented case after case where natural systems have been either lost to other uses or else degraded by pollution and unwise use. As we traced the loss of biodiversity in Chapter 18, we observed that the most effective way to save wild species is to preserve the ecosystems in which they are found—in itself an important reason to maintain natural ecosystems. Populations of wild species continue to decline, however, because ecosystems are also in decline throughout the world. Beyond the loss of species is the degradation or loss of the natural services ecosystems provide.

In this chapter we look at major ecosystems that sustain human life and economy, examine the gains and losses as we convert these ecosystems to other uses, and consider in particular those ecosystems that are in the deepest trouble. As we do this, we should be ready to apply the principle of sustainable development—development that "meets the needs of the present generation without compromising the needs of future generations," as it was first expressed by the World Commission on Environment and Development in 1987.

Ecosystems under Pressure

The decline of biodiversity is linked directly to the welfare of all Earth's ecosystems. Although human activities affect virtually all ecosystems, some are under more pressure than others.

Forests and Woodlands

Forests are the normal ecosystems in regions with year-round rainfall adequate to sustain tree growth. They are also the most productive systems the land can support, and they are self-sustaining. Forests and woodlands (ecosystems with mixed trees and grasses) perform a number of vital natural services, and they provide us with lumber, the raw material for making paper, and fuel for cooking and heating. In spite of all these valuable uses, the major threat to the world's forests is not simply *exploitation*, but rather, *total removal.*

The FAO of the United Nations released a global assessment of forest resources in 1993 and compared the data with previous information in order to assess deforestation trends. Their findings indicate that worldwide, three-fourths of the area originally covered by forests and woodlands still has some tree cover, but less

FIGURE 19-1
This Kenyan woman is cultivating seedlings in a nursery that has been planted as part of the Green Belt movement. The resulting trees provide shade and reduce erosion. (Wendy Stone/Gamma-Liaison, Inc.)

FIGURE 19-2
Wangari Maathai, founder of the Green Belt movement in Kenya. (William Campbell/Peter Arnold, Inc.)

than half of this area represents intact forest ecosystems. Most deforestation in the developed countries has been halted, but it is proceeding at a rate of 0.8% loss per year of natural forest cover in tropical forests of the developing world.

Why are forests being cleared when they could be managed for wood production on a sustainable basis? The answer is simple. Even though forests are very productive systems, it has always been difficult for humans to exploit them for food. Most of the energy in forests goes to detritus and decomposer food webs, not to the grazing food web. In contrast, grasslands have short food chains, in which herbaceous growth supports large herbivores that can yield meat and other animal products. Alternatively, natural grasses can be replaced with cultivated ones (grains) and used directly. Thus, the forests have always been an obstacle to conventional animal husbandry and agriculture. The first task the early European colonists faced when they came to the Western Hemisphere was to clear the forests so they could raise crops. Once the forests were cleared, continual grazing or plowing effectively prevented regrowth of the trees.

Clearing a forest has immediate and severe consequences for the land:

1. The overall productivity of the area is reduced.
2. The standing stock of nutrients and biomass, once stored in the trees and leaf litter, is enormously reduced.
3. Biodiversity is greatly diminished.
4. The soil is more prone to erosion and drying.
5. The hydrologic cycle is changed, as water drains off the land instead of being released by transpiration through the leaves of trees or instead of percolating into groundwater.
6. A major carbon dioxide sink, (removal of CO_2 from the air) is lost.

Unless forests are cleared and converted to other uses, they can yield a harvest of wood for fuel, paper, and building. Approximately 3.5 billion cubic feet of wood are harvested annually from the world's forests, half for fuel and half for wood and paper (Fig. 19-3). It is unreasonable to expect the countries of the world—especially the developing countries—to forego making use of their forests. However, there are ways to exploit forests, short of clearing them, that will preserve most or all of the vital ecological services they perform:

1. Sustainable logging can be carried out. Several strategies are employed: Cutting the vines before removing a desired tree often saves many surrounding smaller trees; planning a route for removing desired trees can minimize the damage done by dragging and by building roads; strip clear-cutting a 30- to 40-meter swath of trees creates gaps that can quickly be filled by seeding in from surrounding trees.
2. Plantations of trees for wood or other products (cacao, rubber, etc.) are much better than typical agricultural uses are in preserving the natural functions of forests. Although much biodiversity is lost, plantations can continue to recycle nutrients, hold the soil, and recycle water. They can also be managed in a sustainable way.
3. Extractive reserves can be created that will yield nontimber goods such as latex, nuts, fibers, and fruits. Recent calculations show that some forests are worth much more as extractive reserves than as a source of timber.
4. Forests can be preserved as part of a national heritage and put to use as tourist attractions, which can often generate much more income than logging can.
5. Forests can be put under the control of indigenous villagers, who can then collectively use the forest products in traditional ways. Given tenure over the land, the villagers tend to exercise stewardship over their forests in a way that is sustainable. Where this has been implemented, the forests have fared better than where they have been placed under state control.

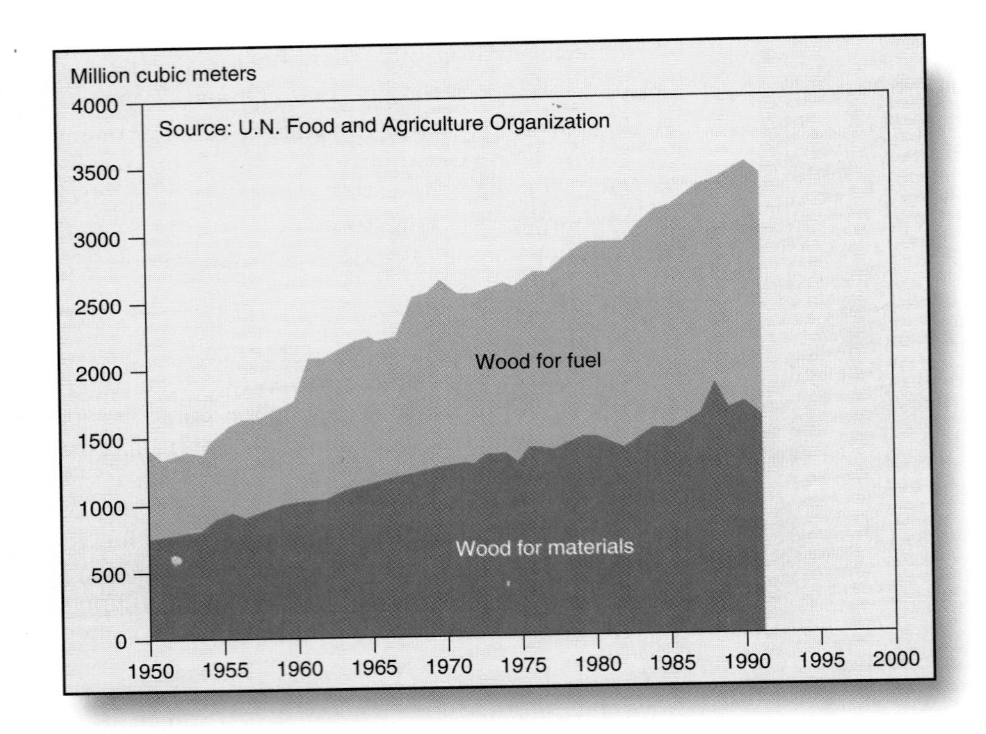

FIGURE 19-3
World wood consumption, as a consequence of its use for fuel, building material, and paper. In the United States, 43% is for lumber, 29% for paper and pulp, 17% for fuel wood, 9% for plywood and veneer, and 2% miscellaneous.
Source: U.N. Food and Agriculture Organization, *FAO Yearbook: Forest Products, 1995.*

Tropical Forests. Tropical forests are of greatest concern. They are the habitat for millions of plant and animal species, a vast number of which are still unidentified. Climatologists reason that forests are also crucial in maintaining Earth's climate, serving as a major sink for carbon and restraining the buildup of global carbon dioxide. Yet tropical forests are being destroyed at a rapid rate—over 23 million acres (9.6 million ha) a year, according to a 1993 report by the FAO. Between 1960 and 1990, 20% of the tropical forests (over 1.1 billion acres 445 million ha)—equivalent to two-fifths of the land area of the United States—have been converted to other uses.

Deforestation is being caused by a number of factors, all of which come down to the fact that the countries involved are in need of greater economic development and have rapid population growth. The huge numbers of young people in the newest generation cannot find jobs or live on land that barely supports their parents. Therefore, many governments in the developing world are encouraging the colonization of forested lands, which begins with deforestation. For example, Indonesia has embarked on a program of resettlement and intends to convert 20% of its remaining forests to agricultural production. And Brazil grants 250 acres (100 ha) of land and provides subsidies of food, money, and education to peasants who are willing to migrate to the Amazonian rain forests. Some 7 million Brazilians have moved into the rain forests since 1970, clearing the land for small-scale agriculture. Unfortunately, many tropical soils are unsuitable for agriculture and become exhausted of nutrients within a few years. Settlers then move either back to the cities or onto new land, often selling their lots to cattle ranchers.

The problem is exacerbated by the governments involved because they have huge external debts (over $108 billion in the case of Brazil). To raise money to pay the interest on their loans, they often sell the logging rights to multinational companies, which come in, wastefully destroy the forests to obtain their hardwood logs for furniture, and make no effort to replant or restore the forest. These companies see the forest as an open commons from which to get as much as they can while they can. They have no vested interest in maintaining sustainable yield.

Agricultural rights are sold to wealthy ranchers, who clear the forest and replant it in grass for grazing cattle to supply meat to industrial nations. Over the past 10 years, more than 3% of the Latin American rain forest has been converted to pastureland, another instance of the global economy driving exploitation in developing nations.

Brazil is a main focal point for concern over the rain forests. The Amazonian rain forest of Brazil covers 1.3 million square miles and is the world's largest tropical rain forest. During the 1970s and 1980s, Brazil embarked on several ambitious schemes designed to move the nation rapidly into the ranks of the industrialized nations. Loans from the World Bank and other international agencies provided capital to launch two major projects: *Grande Carajás* and *Polonoreste*. The government is also developing hydroelectric power in

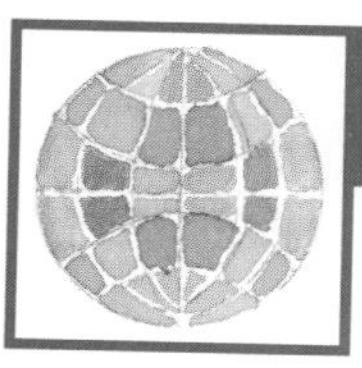

Global Perspective

Rain Forest Heroes

Acre is Brazil's westernmost state and in many ways brings to mind the old Wild West of the United States. Along with the adjoining state of Amazonas, it is a vast, sparsely settled frontier that supports large cattle ranches whose owners have turned rain forests into pastures. Like the old West, the Brazilian west has had its share of lawlessness, most apparent these days in the struggle over land ownership, which has led to thousands of deaths. It is a classic conflict: poor people who make their living from products of the forests versus wealthy ranchers who cut and burn the forests so that the land will grow grass for their cattle. The conflict has spawned its heroes, none better known than Chico Mendes. His story is proof that one person with a vision and the will to turn that vision into reality can make a difference in world affairs.

Francisco (Chico) Mendes was a *serenguero*—a rubber tapper—one of hundreds of thousands of Brazilians who harvested the latex rubber and Brazil nuts from rain forest trees. Realizing that the serengueros were fast losing their livelihood to the onslaughts of the cattle ranchers as the rain forest went up in smoke, Mendes worked tirelessly to organize the serengueros against continued cutting of the forests. Soon he was not only a serious threat to the ranchers, but also an international figure who in 1987 successfully testified in opposition to the United States' support of the Inter-American Development Bank (which was granting loans to Brazil to build roads into Acre's forests). How serious a threat Mendes was to the ranchers became clear on December 22, 1988, when, in spite of police protection, he was fatally shot as he stepped outside his house. Two years later, a rancher and his son were convicted of the killing and sentenced to 19 years in prison.

The legacy of Chico Mendes lives on, however. A key element in his crusade was **extractive reserves**—forestlands that would remain uncleared so that the serengueros could continue their way of life. Since Mendes's death, the Brazilian government has set aside 14 extractive reserves in Amazônia, covering 7.5 million acres (3 million ha) of forests in four Brazilian states; one reserve of 1 million hectares (2.5 million acres) was named the Chico Mendes Extractive Reserve. The rubber tappers' union founded by Mendes has become a vital national force, with its eye on protecting many more millions of acres. The economy of the rubber tappers is a fragile one, however, and the two major products of the rain forest—rubber and Brazil nuts—face serious competition from foreign plantation-grown products. Never-theless, the concept of Mendes's extractive reserves is spreading throughout the tropics. And the success of the rubber tappers has ignited hope for other impoverished grassroots movements in the developing world.

Judson Valentim is another Brazilian who has brought relief to the rainforests, albeit in a way very different from that of Chico Mendes: Valentim is providing aid to the group normally identified as the enemy—the ranchers. He is an agronomist with the Brazilian Institute of Ranching and Agriculture who was given the task of doing something about the failure of grasslands that had been converted from rainforest. Typically, pastures would support 20 cows on 25 acres (10 ha) of land in the first year or two after clearing. Within five years, however, the quality of the land would decline as a result of erosion, drying and hardening of the soil, and loss of fertility, so that the same area might support no more than 5 cows. This decline led to more clearing of the rainforest in order to graze a constant number of cattle. Valentim reasoned that if the land already cleared could be restored to a productive state, the pressures on the rainforest would diminish.

In the 1970s, some landowners attempted to establish rubber plantations in northern Brazil. This enterprise eventually failed, but one aspect lived on: Kudzu, the scourge of the southern United States, had been brought in to cover the ground between rubber trees. The kudzu quickly spread (no surprise!) to other areas of cleared land, and Valentim began experimenting with the plant as a source of cattle feed. The results were highly successful. Kudzu produces three times the forage that grasses do, does well during the dry season, overgrows other weeds, is resistant to pests, and, because it is a legume, restores fertility to the soil by nitrogen fixation. Cattle raised on kudzu are thriving, gaining as much as 450 pounds (200 kg) a year more than grass-fed cattle and producing more milk. Land planted to kudzu is now supporting 25 cows on 25 acres, better than freshly cleared pasture.

The ranchers of Acre adopted kudzu with enthusiasm. Valentim started a consulting service and now travels about introducing kudzu techniques and seeds to new areas. The use of kudzu is now required by law in Acre. Ranchers may graze their cattle only in previously disturbed areas and only if they plant kudzu, restrictions aimed at preventing a new round of deforestation. Valentim believes that kudzu pastures will support twice the number of cattle currently in Acre without any more clearing of the forest. The government is watching the experiment closely, but all indicators point to kudzu and Valentim's work as a substantial step toward saving the remaining Brazilian rain forests.

FIGURE 19-4
A 1988 photo from the space shuttle *Discovery* shows smoke from burning rain forests so extensive that it obscures all land features in the Amazon River basin of South America. (NASA.)

the water-rich Amazon basin; dozens of dams are planned, and those already built have flooded thousands of square miles of rain forest.

The Grande Carajás project is a huge industrial enterprise expected to involve an area in eastern Brazil the size of France and Great Britain. Its cost is projected at $62 billion. Besides the clearing needed to develop roads, towns, and industrial sites, the project has led to great losses of forests that were burned to make charcoal for fueling iron-ore smelters.

Polonoreste was a development scheme built around a highway into the Amazonian rain forest, BR-364. As settlers and ranchers moved out from the highway, the forest in the region declined from 97% cover to 80%. The fires of burning rain forests from these and other projects were vividly evident in photos from NASA's *Discovery* in 1988, alerting the world in a dramatic way to the extent of the destruction of the rain forest (Fig. 19-4). Because of the obvious destructive impacts of Polonoreste on the rain forests, the World Bank withdrew funding of BR-364. However, the Inter-American Development Bank stepped in to fund the successor to Polonoreste, called *Planafloro*, which contained $4 million for environmental protection and support for indigenous people. The road was finished in 1992 and, by recent reports, has not been accompanied by resettlement or major deforestation. International exposure of the damage to the rain forest and injustice (see "Global Perspective" box, p. 485) combined with a severe recession in Brazil to prevent the kind of damage done earlier in the road's history.

The Brazilian government has recognized and corrected serious problems with its colonization and industrial schemes. The subsidies to cattle ranchers have been halted, and the rate of deforestation (4 million acres per year in the 1980s) has consequently declined significantly, perhaps as much as 80% in the 1990s. The government has placed 7.5 million acres (3 million hectares) of rain forest into **extractive reserves**—land that is protected for native peoples who live in the forest and for other Brazilians who gather latex rubber and Brazil nuts from mature forest trees. Also, international loan agencies have become more environmentally conscious and now require that environmental impact assessments accompany requests for loans.

Although Brazil seems to have turned the corner in bringing deforestation under control, other parts of the developing world reveal a different story. In Southeast Asia, logging for the international market has intensified in recent years, promoted by Asian multinational corporations. Too often, the interests of indigenous forest dwellers have been submerged by the "development mentality" of the national governments, which view the forests as their property. In the Philippines, Indonesia, and Malaysia, logging has led to rapid deforestation or degradation of the forest, promoted by collusion between the logging industry and local politicians who receive a share of the profits. Unfortunately, these same Asian corporations are moving into small South American countries such as Guyana and Surinam and are beginning a new kind of colonialism that threatens to start a new wave of destruction of the rain forest. The Asian corporations now hold logging leases to half of Guyana's forested land, and logging has become a major part of Guyana's political economy. With only five government foresters and no commisioner, Guyana is ill prepared to monitor the logging activities. Fortunately, the situation in Guyana has been recognized by international conservation and aid agencies, and help is now being offered to bolster Guyana's forestry commission.

One of the documents signed by heads of state at the Rio de Janeiro Earth Summit of 1992 was the **Statement on Forest Principles**. Treaty negotiations leading up to the summit proved too contentious to lead to any binding agreements. The more developed countries showed great concern about the deforestation of tropical forests, whereas the less developed countries made it clear that they regarded the tropical forests as resources for exploitation. In the end, the signatories agreed to a statement of 17 nonbinding principles that stressed sustainable management. The less developed countries insisted on their sovereign right to develop their forest resources, and all of the signatories agreed to consider revising the principles in the future, to assure continuing cooperation on issues pertaining to the forest.

A recent initiative directed toward certification of wood products is the creation of the Forest Stewardship Council. This federation of NGOs (non-governmental organizations like the National Wildlife Federation), industry representatives and forest scientists has developed into a major international organization with the mission of promoting sustainable forestry by certifying forest products for the consumer market.

All of these developments suggest that global awareness of the importance of the tropical forests has reached the level of serious concern. In some cases, this concern is being translated into action that is slowing the high rate of deforestation experienced in the 1980s. In others, the outcome is uncertain. The demand for tropical wood has not diminished, and the economic rewards of exploitation will undoubtedly continue to promote unsustainable uses of the forests in many tropical countries. We address the use of forests in the United States later in the chapter, when we address the use of public lands in general.

Ocean Ecosystems

Marine Fisheries. For years, the oceans beyond a 12-mile limit were considered an international commons. By the end of the 1960s, however, numerous regions of the sea were being seriously depleted of many species by overfishing on the part of international fleets equipped with factory ships and modern fish-finding technology. As a result of agreements forged at a series of United Nations Conferences on the Law of the Sea, nations in the mid-1970s extended their limits of jurisdiction to 200 miles offshore. The United States accomplished this with the Magnuson Act of 1976. Since many prime fishing grounds are located between 12 and 200 miles from shores, this action effectively removed the fish from the international commons and placed them under the authority of a particular nation. As a result, some fishing areas recovered, while nationally based fishing fleets expanded to exploit the fisheries; however, today the global fisheries are again being overfished by fleets that are too efficient and too large. (**Fishery** refers either to a limited marine area or to a group of fish species being exploited.)

The total recorded harvest from marine fisheries and fish farming has increased remarkably since 1950, when it was just 20 million metric tons (Fig. 19-5). By 1994 it had reached 109 million metric tons (120 million short tons). Ninety-one million metric tons of this total was the catch in 1994 and is a measure of the sea's productive potential, but in a sense this obscures what is really happening. Many species and areas continue to be overfished, and when this happens, fishers turn to new areas and formerly less desirable species. For example, 70% of the rise during the 1980s is attributed to an increased catch of just four species not previously exploited; many other species yielded decreasing harvests. If 18 of the important declining species were restored to their former productivity, the catch could increase by 20 to 30 million metric tons. Therefore, scientists reason, the oceans could yield well over 100 million metric tons (well above present levels) on a sustained basis. But this could happen only if fishing for many species were temporarily eliminated or greatly reduced and the major fisheries were managed on a sustainable-yield basis. Currently, overfishing is occurring in 13 of the world's 15 major fishing areas, offering little hope that there will be any increase in the harvest. The world fishing fleet consists of more than 3 million vessels, many equipped with fish-finding sonar and underwater television cameras and aided by spotter planes and helicopters. Because of heavy government subsidies, the fleet is able to land $70 billion worth of fish at a cost of $92 billion.

Events on Georges Bank, Massachusetts, which is New England's richest fishing ground, are indicative of the trends. Cod, haddock, and flounder were the mainstay of the fishing industry for centuries. In the early 1960s, these species amounted to two-thirds of the fish population on the bank. Since 1976, fishing on Georges

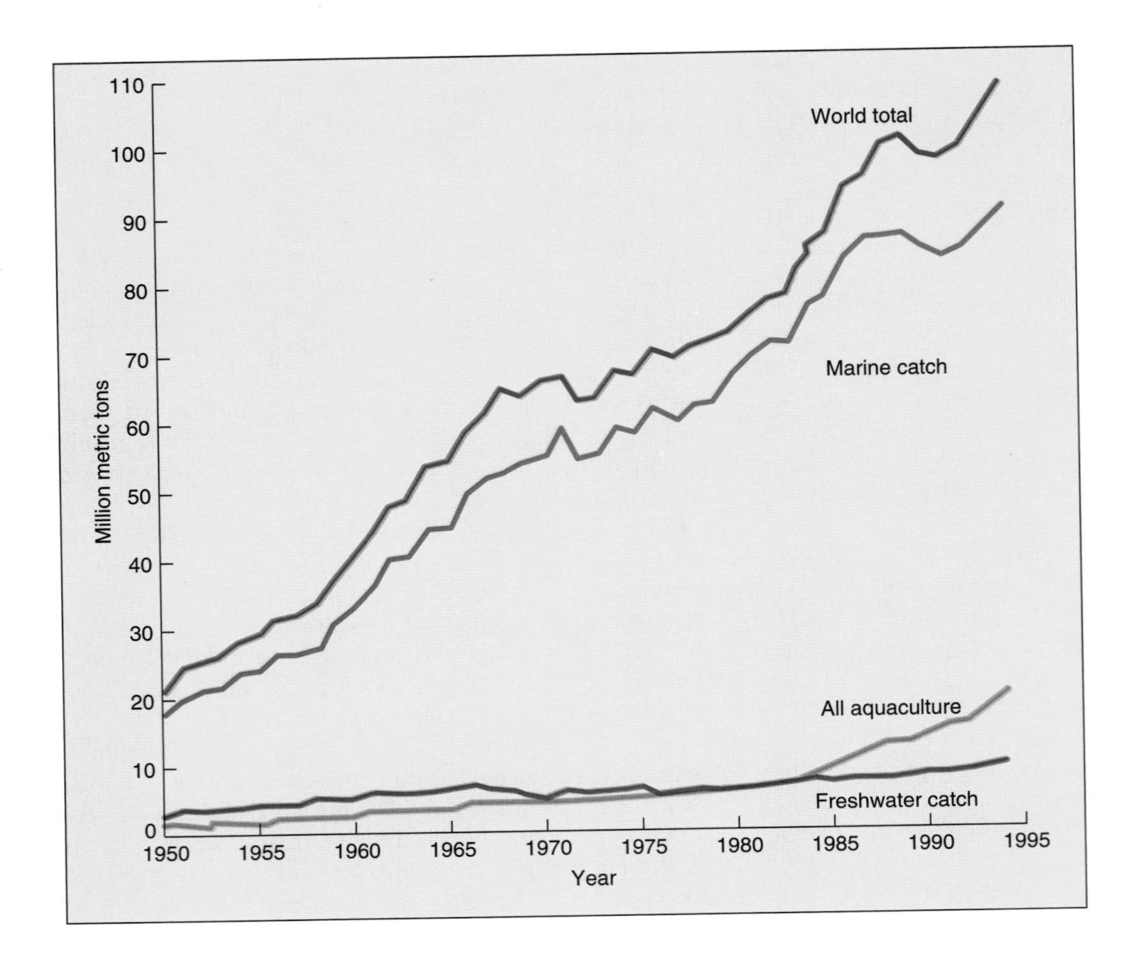

FIGURE 19-5
The global fish catch and fish farming equals world total for 1950–94. (From Worldwatch Institute, *Vital Signs,* 1996. Photo: Vanessa Vick/Photo Researchers, Inc.)

Bank has doubled in intensity, resulting in a decline in the desirable species and a rise in the so-called rough species—dogfish (sharks) and skates—which now constitute 75% of the fish population on the bank. There is concern that populations of the desirable species have fallen too low to reproduce effectively or are being disproportionately removed by predatory species, a phenomenon known as *depensation*.

The decline of the prized species on Georges Bank reflects poor management of the fishery. The Magnuson Act established eight regional management councils, made up of government officials and industry representatives. The councils are responsible for setting the management plans for their regions, while the National Marine Fisheries Service (NMFS) provides advice and scientifically based information on stock assessments, and also has the authority to reject elements of the plans submitted to them by the councils. The New England Fishery Management Council began its task by setting quotas (Total Allowable Catch, or TACs) but fishers claimed that the TACs were set too low, and successfully argued for an indirect approach that employed mesh with openings of a size that allowed smaller fish to escape. And, in less than ten years, the number of boats fishing Georges Bank doubled. The result has been disastrous, as the cod landings from Georges Bank indicate (Fig. 19-6).

With collapse of the fishery imminent, the Council has taken a more drastic approach. For 1997, all vessels are restricted to 50% of their normal days at sea; almost 5000 square miles (about 1/3 of the fishing area) is totally closed to fishing, and target TACs are set at levels which should reduce fishing mortality to 10% of the stock per year. In addition, no new vessels are allowed to enter the fishery, and the Department of Commerce has begun a buyout program that pays the fishers to scrap their boats and surrender their fishing licenses.

What will it take to restore a fishery such as Georges Bank? The Canadian answer to a very similar problem on the Grand Banks off Newfoundland was to close the fishery indefinitely in 1992. At that point, the cod population had dropped to 1/100 of its former size. The closing cost the jobs of 35,000 fishers. So far the cod population has not rebounded, and scientists believe that it will be the late 1990s before recovery occurs. The good news is that recovery *can* occur. A recent study of 128 heavily exploited fish stocks failed to show depensation occurring. Instead, the evidence indicated that stocks were ready to rebound once fishing pressure was reduced. Norwegian cod landings were dropping rapidly in the mid 1980s, and the government immediately cut quotas by 50%. Since 1990 the catch has tripled, and Norwegian cod is now supplying the North American market demand created by the decline in the Canadian and New England fisheries.

The basic problem of the fisheries is obvious to all. There are too many boats, equipped with high technology that gives the fish little chance to escape, chasing too few fish. The answer is a better system of management, where the fishing fleet is diminished in size and remaining fishers are better able to make a living without overfishing. What is needed is a limited system of access, something like the individual transferable quota (ITQ) system that has been used successfully in New Zealand. There, quotas for maximum yield are set, and individual fishers divide the yield. The system effectively gives fishers the equivalent of property rights in the fishery. These rights are transferable and allow the fishers to determine how and when they will harvest their catch.

In 1996, the Magnuson Act was reauthorized by a bill called "The Sustainable Fisheries Act." The bill requires that depleted fisheries be rebuilt and that fishing be maintained at biologically sustainable levels. The structure of the regional councils was retained, but the councils were given clear mandates to employ in their fishery management plans a number of means to see that the fish stocks were rebuilt and maintained, including ITQs, buyouts of fishing vessels, and requiring that scientific information be employed in setting yields. The bill also requires steps to be taken to assess and minimize the "bycatch" (non-target fish caught and thrown away).

International Whaling. Whales, found in the open ocean as well as coastal waters, were heavily depleted by overexploitation that was finally halted in the late 1980s. Whales once were harvested for their oil but are now prized for their meat, considered a delicacy in Japan and a few other countries. In 1974, an organization of nations with whaling interests, the International Whaling Commission (IWC), decided to regulate whaling according to the principle of maximum sustainable yield. Whenever a species of whale dropped below the optimal population for such a yield, the IWC instituted a ban on hunting that species, in order to allow the population to recover. At that time, three species (the right whale, bowhead whale, and blue whale) were at very low levels and were immediately protected. Because of difficulties in obtaining reliable data on and enforcing catch limits, the IWC took more drastic action by placing a moratorium on the harvesting of all whales in 1986. In spite of the moratorium, some limited whaling continues under the guise of "scientific research" by Japan, Norway, and Iceland, and harvest by indigenous people in Canada and Alaska.

FIGURE 19-6
Cod landings from Georges Bank, 1982–94. A new accounting system was introduced in 1995 and is still being perfected.

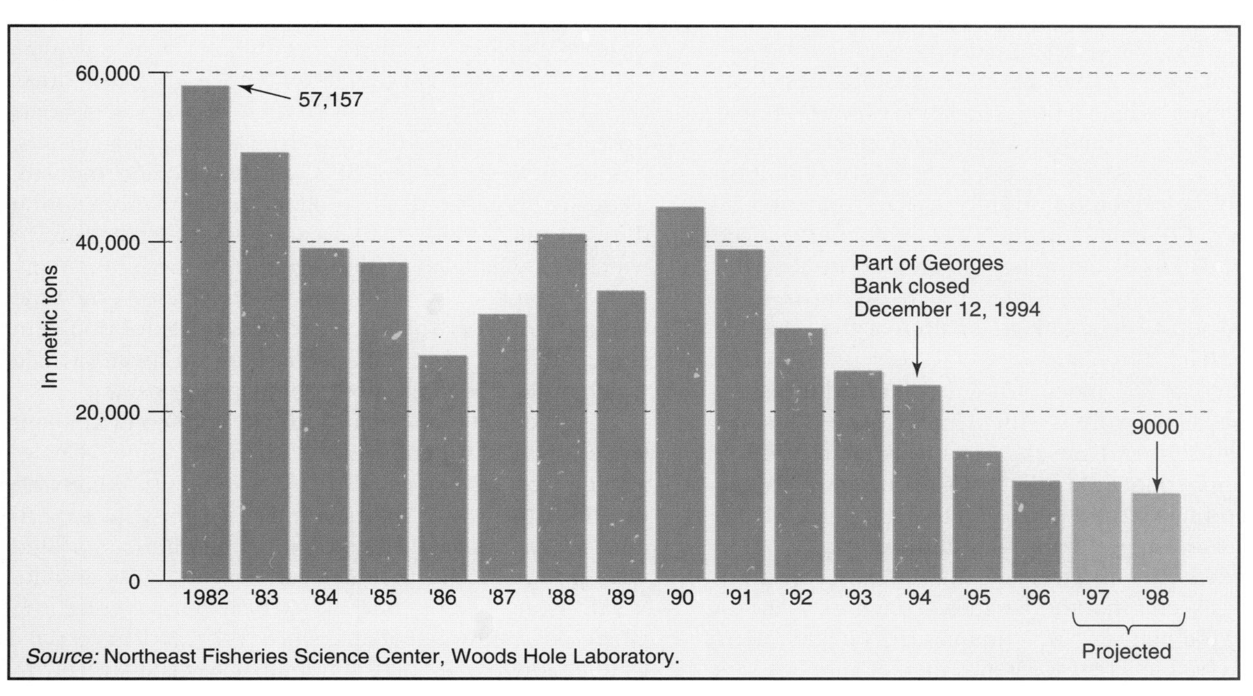

TABLE 19-1
Remaining Numbers of Large Whales

Endangered Species	*Vulnerable Species*	*Insufficient Knowledge*
Blue whale (11,700)	Bowhead whale (7,800)	Minke whale (880,000)
Northern right whale (300)	Fin whale (110,000)	Sperm whale (400,000)
	Humpback whale (10,000)	
	Sei whale (54,000)	
	Southern right whale (2,900)	
	Gray whale (18,000)	

Sources: International Union for Conservation of Nature. *Red Data Book.* 1991. Rough estimates from Congressional Research Service of Library of Congress; Schmidt, 1994 (*Science* 263:26–27).

The International Union for Conservation of Nature publishes the up-to-date status of whale populations in its *Red Data Book.* Table 19-1 lists the status of ten whale species of commercial interest. Estimates of the numbers remaining range from 300 northern right whales to approximately 880,000 minke whales. Although reliable data are still hard to acquire, many of the whale species appear to be recovering. The bowhead whale, for example, has recently been upgraded from endangered to vulnerable. In the time between the IWC moratorium and the present, however, the basis of the ethical controversy over whaling has shifted from conservation to animal rights. Many people worldwide simply believe that it is wrong to kill and eat such large and unique mammals. People from whaling nations counter with the argument that their culture includes eating whales, just as other cultures include eating cows.

There has been heavy pressure from three members of the IWC—Japan, Norway, and Iceland—to reopen whaling. The interests of these countries focus on the minke whale. Both Iceland (in 1991) and Norway (in 1993) resumed whaling for minkes, citing their rights to refuse to accept specific IWC rulings. They based their decision on information submitted by the Scientific Committee of the IWC, which judged the minke population to be numerous enough to absorb a sustainable exploitation. However, activists in the conservation group Greenpeace maintain that the scientific data on whale populations is far too uncertain to support such a judgment. Under continued pressure from antiwhaling nations (the United States, France, Australia, and others) the IWC has thus far refused to set a whaling quota on the minke. Currently, Japan, Norway, and Iceland each take up to 300 minke whales per year—the Japanese for "scientific purposes," and the Icelanders and Norwegians for human consumption. It is no secret that the Japanese "scientific" harvest ends up in the markets and commands a healthy price. Unfortunately, the Japanese retail market may also be receiving whale meat hunted illegally; a recent DNA-based test of whale meat from various Japanese markets revealed some fin and humpback whale meat along with minke.

The Norwegian decision to resume whaling has placed that country at risk of sanctions by IWC nations, including the United States, but to date, no action has been taken against Norway. At its 1994 meeting, the IWC took a new tack by passing a resolution to establish a sanctuary for whales in Antarctic waters south of 40 degrees latitude, a move that Japan strongly opposed. However, the IWC also accepted a management procedure for setting quotas that could lead to a resumption of whaling. If this happens, a clear effort towards monitoring and verifying catches will be needed to prevent further violations of international law.

One reason for the rising interest in protecting whales is the opportunity many people have had to observe whales firsthand, as whale watching has become an important tourist enterprise in coastal areas (Fig. 19-7). Stellwagen Bank, within easy reach of boats from Boston, Cape Ann, and Cape Cod, Massachusetts, has become the center of a whale-watching industry estimated to generate $17 million annually. From spring through fall, scores of boats venture offshore daily to watch the whales that congregate over the bank. Many of the humpback whales seem to enjoy entertaining the visitors and often frolic alongside the boats for hours. In some areas of the Pacific coast and Hawaii, whale watching has become more popular than fishing.

Besides its obvious aesthetic and entertainment values, whale watching is of scientific value. Whale-watching tour boats usually carry a biologist along who identifies the whale species and interprets the experience for the visitors. The biologists are often associated with groups such as the Cetacean Research Unit and the Whale Conservation Institute, which have studied the whales of Stellwagen Bank since 1979 and have published many papers on the humpback whale. Largely in

FIGURE 19-7
A humpback whale entertains a boatload of whale watchers in the Gulf of Maine off the Massachusetts coast. Whale watching has replaced whaling in New England waters that were once famous as a whaling center. (Jan Robert Factor/Photo Researchers, Inc.)

response to its importance for whales, Stellwagen Bank was designated a National Marine Sanctuary in 1993.

Coral Reefs and Mangroves. In a band from 30° north to 30° south around the equator, coral reefs occupy shallow coastal areas and form atolls, or islands, in the ocean. Coral reefs are among the most diverse and biologically productive ecosystems in the world. The coral animals live in symbiotic relationship with photosynthetic algae and therefore are found only in water more shallow than 75 meters. The corals are important in building and protecting the land shoreward of the reefs and are spectacular attractions for tourists enjoying the warm, shallow waters. In addition, because the reefs attract a great variety of fish and shellfish, they are important food sources for local people. In recent years, coral reefs have been destroyed and degraded at an alarming rate.

One puzzling phenomenon that has been spreading in the Caribbean and Eastern Pacific is *coral bleaching*, which occurs when the coral animals lose their symbiotic algae; frequently, bleaching leads to death of the coral itself. Recent studies suggest that bleaching is associated with abnormally warm temperatures in the water bathing the reefs, although other causes are being sought. Coral reefs are also being damaged by eutrophication of coastal waters. It has been shown that when nutrients—especially phosphorus—are carried to the reefs by currents in the vicinity of major developed coastlines, they encourage the growth of macroscopic algae and other submerged vegetation. The coral is shaded, the coral animals become starved for oxygen, and the reefs become brittle and stunted. Wastes from resort areas are frequently the sources of the nutrients that are polluting the coral.

Another source of damage can be traced to poverty. Islanders and coastal people in the tropics often scour the reefs for fish, shellfish, and other edible sea life. Although much of their use of the reefs is for food, some exploitation is driven by the lucrative trade in tropical fish. Local residents sometimes use dynamite and cyanide to flush the fish out of hiding in the coral, a practice that is not only very destructive to the coral, but also dangerous to the perpetrators.

Often, just inland of the coral reefs is a fringe of mangrove trees. These trees have the unique ability to root and grow in shallow marine sediments; there, they protect the coasts from damage due to storms and erosion, and form a rich refuge and nursery for many marine fish. Mangroves are under assault from two exploitative activities: logging and shrimp aquaculture. Between 1983 and the present, half of the world's 45 million acres of mangroves were cut down from 40° (e.g., Cameroon and Indonesia) to nearly 80% (e.g., Bangladesh and Philippines). Logging companies removed mangroves for paper pulp and chip boards. The most destructive practice, however, is the development of ponds for raising shrimp. In this practice, the mangrove forests are cleared and the land is bulldozed into shallow ponds, where black tiger prawns are fed with processed food. The aquaculture of prawns is now a multibillion-dollar industry, satisfying the great demand for shrimp in the developed countries. Unfortunately, in many cases the ponds become polluted and no longer support shrimp; new ponds are then created by clearing more mangroves. The massive removal of mangroves has brought on the destabilization of entire coastal areas, causing erosion, siltation of sea grasses and coral reefs, and the ruin of local fisheries. Developing nations that once considered the mangroves as useless swampland are now beginning to recognize the natural services they performed; local and international pressure to stop the destruction of mangroves is growing.

Biological Systems in a Global Perspective

Major Systems and Their Services

As we learned in the early chapters of this book, Earth is occupied by ecosystems that vary greatly in the composition of their species, but that exhibit common functions such as energy flow and the cycling of matter. The major types of terrestrial and aquatic ecosystems—the biomes—reflect the response of biotas to different climatic conditions. We can divide Earth's land area into several broad categories for the purposes of this chapter (Table 19-2): **forests and woodlands, grasslands and savannahs, croplands, wetlands**, and **desert lands and tundra** (see Fig. 2-4, pp. 26-7). We can separate the oceanic ecosystems into **coastal ocean and bays, coral reefs**, and **open ocean**. These eight systems provide all of our food; a lot of our fuel; wood for lumber and paper; leather, furs, and raw materials for fabrics; oils and alcohols; and much else. The world economy is directly dependent on their exploitation.

Ecosystems also perform a number of valuable **natural services** as they process energy and circulate matter in the normal course of their functioning (Fig. 19-8). Some of the services are general and pertain to essentially all ecosystems; others are more specific. Some of the most important natural services performed by ecosystems are as follows:

1. **Maintenance of the hydrologic cycle.** Plants absorb water from soils and release it through transpiration, returning the water to the atmosphere. Forests, grasslands, and wetlands maintain a favorable distribution and even flow of water, absorbing it when it is abundant and releasing it gradually. In doing so, they prevent much flooding.
2. **Modification of climate.** Because plants absorb considerable amounts of solar radiation and release water vapor through transpiration, ecosystems moderate temperature and help to maintain an even climate.
3. **Absorption of pollutants.** Wetlands are particularly significant in stabilizing sediments, absorbing excess nutrients and thus preventing eutrophication.
4. **Transformation of toxic chemicals.** In all ecosystems, microbes transform many toxic organic and inorganic chemicals into harmless products.
5. **Erosion control and soil building.** Plant cover and ground litter absorb the potentially destructive impact of rainfall and prevent the breakup of soil; plant roots bind the soil. Forests are especially crucial in preventing erosion on hilly terrain. All of the plants, small animals, and microorganisms in terrestrial systems contribute to the formation of the soil.

TABLE 19-2
Natural Ecosystems on the Earth's Surface

Ecosystem	*Area (million mi^2)*	*Percent of Total Land Area*
Forests and woodlands	18.2	31.7
Grasslands and savannahs	11.6	20.2
Croplands	6.2	10.7
Wetlands	2.0	3.5
Desert lands and tundra	19.6	34.0
Total land area	57.6	
Coastal ocean and bays	8.4	
Coral reefs	0.2	
Open ocean	164	
Total ocean area	172.6	

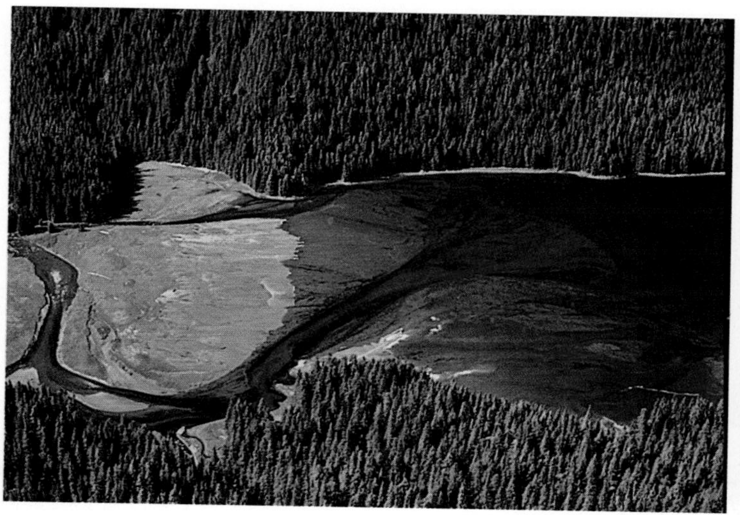

FIGURE 19-8
Natural ecosystems perform invaluable natural services, if they are treated well. On the left is an aerial view of Baranof Island, Alaska, showing a salmon-spawning stream, meadows and coniferous forest. At the right is a view of Kin-Buc landfill in New Jersey, a Superfund site where hazardous waste disposal has damaged vegetation and interrupted the normal functioning of the area biota. (a) Bud Lehnhausen/Photo Researchers, Inc. b) Ray Pfortner/Peter Arnold, Inc.)

6. **Pest management.** Natural ecosystems contain a diverse array of insect predators. Even small patches of natural habitat, such as hedgerows and woodlots, contribute to the control of pests in adjacent agricultural lands.
7. **Maintenance of the oxygen and nitrogen cycles.** Photosynthesis in green plants continually regenerates oxygen, and microbes maintain the soil's fertility through nitrogen fixation. They also prevent the buildup of potentially harmful nitrogenous compounds (for example, ammonium and nitrite).
8. **Carbon storage and maintenance of the carbon cycle.** The global carbon cycle is maintained by energy flow within natural ecosystems. Over 500 billion metric tons of carbon are stored in the standing biomass of forests, more than is found in the entire atmosphere. Much more is present in the organic matter of soils.

In their normal functioning, all natural and altered ecosystems perform some or all of the preceding natural services, free of charge. We tend to take these services for granted until the ecosystems, and thus the services, are lost. For example, deforestation in India is largely responsible for the massive siltation and flooding that are causing human tragedy and suffering in Bangladesh. Loss of wetlands, which provide sediment and nutrient control, is a major factor in the eutrophication of Chesapeake Bay (Chapter 12). And, as mentioned earlier, the removal of trees for firewood in Kenya resulted in extensive erosion of the soil and lowered the productivity of arable lands. Finally, centuries of removal of the forest and overgrazing in the Mediterranean basin have produced a climate that is much hotter and drier than it once was.

To put all this in monetary terms, scientists have calculated that it would cost more than $100,000 a year to artificially duplicate the water purification and fish propagation capacity provided by a single acre of natural tidal wetland (Fig. 19-9). And even if such sums were available for artificially carrying out these processes, the energy expenditures involved (producing and burning the requisite fuels) would probably lead to a net increase in pollution, rather than a decrease. Thus, there is no real way we can compensate for the losses of natural services that are incurred as natural ecosystems are destroyed. We simply suffer the consequent deterioration in environmental quality.

Ecosystems as Natural Resources

If the natural services performed by ecosystems are so valuable, why are we still draining wetlands and cutting down old-growth forests? The answer is revealing: *A natural area will receive protection only if the value a society assigns to its natural functions is higher than the value the society assigns to exploiting its natural resources.* When we refer to natural ecosystems and the biota in them as **natural resources**, we are consciously placing them in an economic setting and losing sight of their ecological importance. A resource is expected to produce something of economic value for its owner. Thus, even though an acre of wetland pro-

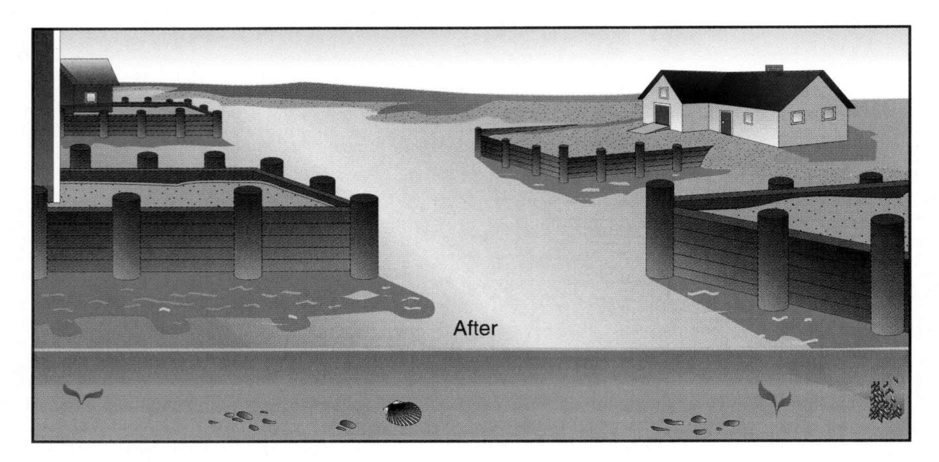

FIGURE 19-9
Tidal wetlands provide a number of valuable services, worth more than $100,000 a year for just one acre. Especially important are water purification and fish propagation. These services are lost when wetlands are bulkheaded and converted for vacation homes.

vides valuable services for Chesapeake Bay, the owner of that acre will benefit economically by draining or filling it in and selling it to a builder. Its conversion removes the acre as a functional wetland and thus represents a *loss* in natural services, but an economic *gain* to the owner.

Some natural ecosystems are maintained in a natural or seminatural state because that is how they provide the greatest economic value for their owners. For example, most of the state of Maine is owned by private corporations that periodically harvest the timber for lumber and for paper manufacturing. However, if Maine experienced a population explosion and corporations could sell their land to developers for more money than could be gained from harvesting timber, forested lands would quickly become house lots.

Finally, some natural ecosystems either are publicly owned (state and federal lands) or cannot be owned (ocean ecosystems). These ecosystems are still considered natural resources and may be subject to economically motivated exploitation. Obviously, wise exploitation of such systems will maintain the natural services they perform. However, exploitation is not always wise. One reason is that our system of economics does not take the loss of environmental assets and services into account when calculating economic wealth and progress.

Environmental Accounting. The *gross national product*, or GNP, refers to the sum of all goods and services produced in a country in a given time frame. It is the most commonly used indicator of the economic health and wealth of that country. (In the United States the comparable GDP, gross domestic product, is now used.) As a per capita index, GNP is often used to compare rich and poor countries and to assess economic progress in the developing countries. An important element in calculating GNP is accounting for the assets that are used in the production processes. Buildings and equipment, for example, are essential to production, but they gradually wear out, or depreciate. Their depreciation is charged against the value of production (called capital depreciation); that is, an accounting of capital depreciation is routinely subtracted from the production of goods and services in order to generate an accurate GNP.

The economists who invented GNP as a measuring device 50 years ago simply left out of their calculations any regard for the depreciation of natural resources—an omission that recently has received a great deal of criticism from environmental economists. Such concerns were truly regarded as externalities to the country's balance sheet! The natural resources and their associated natural services were considered a "gift from nature." Given this state of affairs, it is possible for a nation to cut down a million acres of forest and count the sale of the timber on the income side of the GNP ledger; whereas on the expense side, only the depreciation of chain saws and trucks will be seen. Completely hidden from accounting is the loss all of the natural services once performed by the forests and, incredibly, the disappearance of the million acres of forests as an economic asset! As long as this discrepancy between environmental and economic accounting remains, nations will underestimate the value of natural resources. They will be able to deplete fisheries, lose soil by intensive farming, remove forests, and degrade rangelands by overgrazing and *account for these activities as economic productivity!*

Corrections can be made in this system of accounting, and a good start toward reform would be to calculate the current market value of natural assets and consider it to be part of the stock of a nation's wealth. To this value should be added the value due to the natural services performed by the ecosystems in which the resources are found. The natural assets and the services they perform are referred to as **natural capital**. When the natural resources are drawn down, the depreciation represented by the loss of natural capital should be entered into the ledger as GNP is calculated. Interestingly, Agenda 21—one of the major documents from the 1992 Earth Summit meeting in Rio—recognized this problem and proposed "a program to develop national systems of integrated environmental and economic accounting in all countries." Let us look at an example of a pilot study in such a program.

The UN Statistical Office coordinates the accounting procedures of different countries. In 1993, the office published a new set of guidelines in a handbook, *Integrated Environmental and Economic Accounting*, which at least partly accomplishes Agenda 21's objective. In the new guidelines, countries are encouraged to perform environmental accounting by putting environmental assets and services in monetary units and keeping a parallel account of their *net domestic product* (similar to GNP, except accounting for sales and services within the borders of a nation), as it is affected by environmental accounting. Table 19-3 shows the results of an account performed for Mexico by the World Bank for 1985, the most recent year for which data were adequate. In the table, net domestic product is revised to produce two sets of environmentally adjusted net domestic products (EDP1 and EDP2). The first of these corrects for depletion of resources, the second for the loss of natural services due to environmental degradation. The environmental accounting required a downward adjustment of 13.3% for the year, a significant change that gives policymakers a better picture of their country's true economic performance.

TABLE 19-3
Environmental Accounting for Mexico, 1985

	Trillions of Pesos	***Index***
Net domestic product	42,060	100
Minus resource depletion		
Oil	1,470	3.5
Timber	164	0.4
Change in land use	764	1.8
EDP1 result	39,662	94.3
Minus environmental degradation		
Soil erosion	449	1.1
Solid wastes	197	0.5
Groundwater use	191	0.5
Water pollution	662	1.6
Air pollution	1,656	3.9
EDP2 result	36,507	86.7

From *World Bank: Measuring Environmentally Sustainable Development*, ESD Paper No. 2, 1994.

Conservation and Preservation

The beauty of a natural biota is that it is a **renewable resource**. It has the capacity to replenish itself through reproduction despite certain quantities being taken from it, and this renewal can go on indefinitely. Recall from Chapter 4 that every species has the biotic potential to increase its numbers and that, in a balanced ecosystem, the excess numbers fall prey to parasites, predators, and other factors of environmental resistance. It is difficult to find fault with activities that effectively put some of this excess population to human use. The tragedy occurs when users (hunters, fishers, loggers, and so on) take more than the excess and deplete the breeding populations, threatening or causing extinction of the species.

Conservation of natural biotas, then, does not, or at least should not, imply no use by humans whatsoever, although this may sometimes be temporarily expedient in a management program to allow a certain species to recover its numbers. Rather, the aim of conservation is to *manage or regulate use* so that it does not exceed the capacity of the species or system to renew itself. Clearly, conservation is capable of being carried out sustainably, and when sustainability is adopted in principle, conservation has a well-defined goal.

Preservation is often confused with conservation. The objective of the preservation of species and ecosystems is to *ensure their continuity, regardless of their potential utility*. Effective preservation often precludes making use of the species or ecosystems in question. For example, it is not possible to maintain old-growth (virgin) forests and at the same time harvest the trees. Thus, a second-growth forest can be *conserved* (trees can be cut, but at a rate that allows recovery of the forest) but an old-growth forest must be *preserved* (it must not be cut down at all).

There are times when conservation and preservation come into conflict. The Muriqui monkey of Brazil was once thought to require virgin forests, leading to a concern for protecting such forests for the sake of the species. Recent research, however, has shown that the monkeys actually do better in second-growth forests, which support a greater range of vegetation on which the monkeys feed. Indefinite preservation of the virgin forests would lead to a decline in the population of the Muriqui monkey, a seriously endangered species. Thus, *conservation* of the forests is essential for *preservation* of the Muriqui monkey.

Patterns of Use of Natural Ecosystems

Tragedy of the Commons. Where a resource is owned by many people in common or by no one, it is known as a **common pool resource**, or a **commons**. Examples of natural resource commons are many: federal grasslands where private ranchers graze their livestock; coastal and open-ocean fisheries used by commercial fishers; groundwater drawn for private estates and farms; nationally owned woodlands and forests burned for fuel in the developing world; and the atmosphere, which is polluted by private industry and traffic.

The exploitation of such common pool resources presents some serious problems and can lead to the eventual ruin of the resource—a phenomenon that is called the *tragedy of the commons*, after biologist Garrett Hardin's classic essay (1968) by that title. Sustainability requires that common pool resources be maintained, for the benefit of future users, not just present ones.

As described by Hardin, the original "commons" was pastureland in England provided free by the king to anyone who wished to graze cattle. In the parable, herders were quick to realize that whoever grazed the most cattle stood to benefit the most. Even if they realized that the commons was being overgrazed, those who withdrew their cattle simply sacrificed personal profits, while others went on using the commons. One herder's loss became another's gain, and the commons was overgrazed in any case. Consequently, herders would add to their herds until the commons was totally destroyed. They were locked into a system that led to their ruin.

Hardin's parable applies to a limited but significant set of problems where there is open access to the commons and where there is no regulating authority (or it is ineffective) and no functioning community. Exploitation of the commons becomes a free-for-all in which profit is the only motive. Coastal and offshore fisheries have consistently demonstrated the reality of the tragedy of the commons, as stocks of desirable fish have declined all over the world. Obviously, the tragedy can be avoided, by limiting freedom of access.

One arrangement that can mitigate the tragedy is *private ownership*. When a renewable natural resource is privately owned, access to it is restricted, and, in theory, it will be exploited in a manner that guarantees a continuing harvest for its owner(s). This theory does not hold, however, when an owner maximizes immediate income and then moves on. Most owners, however, will be in for the long run and manage the resources more responsibly.

Where private ownership is unworkable, the alternative is to *regulate access to the commons*. Regulation should allow for (1) protection, so that the benefits derived from the commons can be sustained, (2) fairness in access rights, and (3) common consent of the regulated. Such regulation can reside in the state, but it does not have to. In fact, the most sustainable approach

FIGURE 19-10
Clam diggers harvesting the soft-shelled clam, *Mya arenaria*, from a coastal Massachusetts clam flat. (Ted Levin/Animals Animals/Earth Scenes.)

to maintaining a commons may be local community control. Where the power to manage the commons resides in those who directly benefit most from its use, there are strong social ties and customs that usually function well in protecting the commons over time. And, conversely, there are many situations where state control of a commons has led to the accelerated ruin of the commons and an associated social breakdown and impoverishment of people. Protecting a commons is often a contentious enterprise.

Harvesting soft-shelled clams in New England provides a modern example (Fig. 19-10). The clam flats in any coastal town are effectively a commons belonging to the townspeople; commercial access to them is limited to town residents and policed by the local clam warden. Clamming is a way of life for some residents, but unfortunately, it is not always a dependable enterprise. Although clammers may recognize that the clams are being overharvested, any clammer who curtails his or her own take diminishes personal income, while competitors do not. His or her loss becomes their gain, and the clams are depleted in any case. Thus, while the size and number of the clams diminish, indicating overharvesting, digging continues, because the clammers have to make a living. Higher prices brought on by the shortage of clams continue to make the digging profitable despite declining harvests. In most cases, the clam warden steps in before local clamming is totally ruined and closes a number of clam flats to allow them to recover. Coastal Massachusetts towns experience regular cycles of scarcity and recovery in the soft-shelled clam industry, an indication that the commons is often difficult to maintain. The clam warden has the power to make the difference between a full-blown tragedy of the commons and a system where regulation leads to a sustainable harvest—not an easy or popular task!

Maximum Sustainable Yield. The central question in managing a renewable natural resource is, *How much continual use can be sustained without undercutting the capacity of the species or system to renew itself?* The term to describe this amount of use is **maximum sustainable yield** (MSY). This is *the highest possible rate of use that the system can match with its own rate of replacement or maintenance.*

MSY applies to more than just the preservation of natural biotas. It is also the central question in maintaining parks, air quality, water quality and quantity, soils, and, indeed, the entire biosphere. *Use* can refer to the cutting of timber, hunting, fishing, the number of park visitations, the discharge of pollutants into air or water, and so on. Natural systems can withstand a certain amount of use (or abuse, in terms of pollution) and still remain viable. However, a point exists at which increasing use begins to destroy regenerative capacity. Just short of that point is the MSY.

An important consideration in determining MSY is the **carrying capacity** of the ecosystem—the maximum population the ecosystem can support on a sustainable basis. If a population is well below the carrying capacity of the ecosystem (Fig. 19-11a), then allowing that population to grow will increase the number of reproductive individuals and, thus, the yield that can be harvested. However, as the population approaches the carrying capacity of the ecosystem, new individuals must compete with older individuals for food and living space. As a result, recruitment may fall drastically (Fig. 19-11b). When a population is at or near carrying capacity, production—and hence, sustainable yield—can be increased by thinning the population so that competition is reduced and optimal growth and reproductive rates are achieved. Thus, the MSY cannot be obtained with a population that is at the carrying capacity. Theoretically, the optimal population is just half the population at the carrying capacity (Fig. 19-11).

The matter is further complicated by the fact that carrying capacity and, hence, optimal population are not constant. They may vary from year to year as the weather fluctuates. Replacement may also vary from year to year, because some years are particularly favorable to reproduction and recruitment, while others are not. Of course, human impacts such as pollution and other forms of altering habitats adversely affect repro-

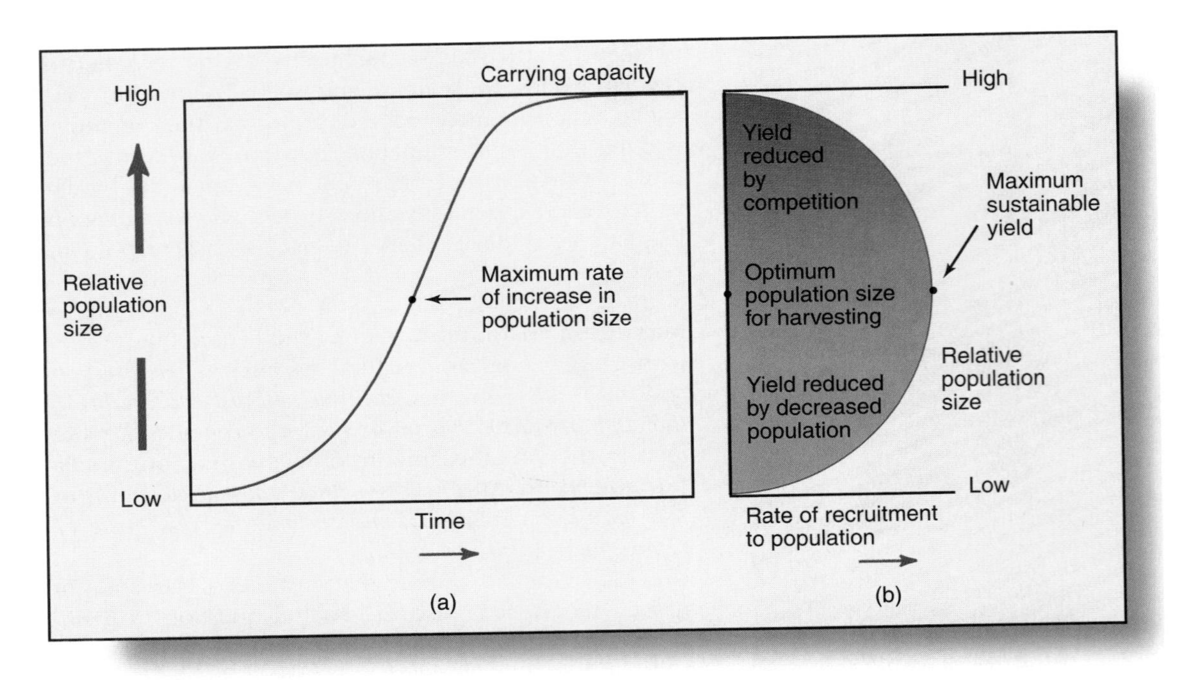

FIGURE 19-11
Maximum sustainable yield occurs not at the maximum population level, but rather at a lower, *optimal* population level. (a) The logistic curve of population size in relation to carrying capacity. The optimal population level is where the rate of population increase is at a maximum, which is well below the carrying capacity. (b) Here recruitment is plotted against population size, showing the effects of competition and decreased population levels. The maximum sustainable yield occurs where the population is at the optimal level, where the rate of increase in population is at a maximum.

ductive rates, recruitment, carrying capacity, and, consequently, sustainable yields.

For these reasons alone, managing natural populations to achieve the MSY is fraught with difficulties. Accurate estimates must be made continually of the size of the population and the recruitment rate. The usual approach is to use the estimated MSY to set a fixed quota—in fishery management, the *total allowable catch* (TAC). If the data on population and recruitment are not accurate, it is easy to overestimate the TAC, especially when there are economic pressures to maintain a high harvest quota. Earlier in the chapter, we saw how the MSY approach has failed to prevent overharvesting in many marine fisheries.

In sum, to achieve the objectives of conservation, we must be aware of both the concept of MSY and the social and economic factors that cause overuse and other forms of environmental degradation that diminish the MSY. We must then establish and enforce public policies that are effective in protecting natural resources—for example:

- Natural resources cannot be treated as an open commons because, wherever they have been, the tragedy of the commons inevitably resulted. Biotas must be put under an authority that is responsible for their sustainability and that can regulate their use.
- Regulations must be enforced.
- Economic incentives that promote the violation of regulations must be eliminated.
- Suitable habitats must be preserved.
- Habitats must be protected from pollution.

There are situations where exploitation and degradation have gone too far, and whole communities and species have been driven into a decline. In these cases, natural services and potential uses can be restored only if the habitats are restored. In recent years, the need for the restoration of endangered species and damaged ecosystems has become clear, and a new subdiscipline with that objective has appeared, namely, **restoration ecology**.

Restoration. As we have documented throughout this text, there is a global trend of destruction and degradation of natural ecosystems, accompanied by the decline and disappearance of thousands of wild species. Biodiversity in all of its manifestations is on the decline. At the same time, the global recognition of this trend (and its potentially disastrous consequences) is on the

Ethics

Stewards of Nature

With the exception of the open oceans, all natural ecosystems come under the sovereignty of the different nation-states. Some natural lands are remote from human habitation, but the vast majority are found where humans can get access to them and to the resources they hold. Indeed, it was usually the products and services of the natural lands that drew humans to them in the first place and nurtured the increases in prosperity and population that followed.

Now, in the late 20th century, essentially all natural lands are affected by human exploitation. Such exploitation is thoroughly managed in industrialized nations. In the United States, public lands are managed by federal and state agencies, such as the National Forest Service and state fish and game departments. Private lands are managed by the owners, with certain constraints placed on them by governmental authorities.

In many of the less developed countries, however, natural resources are exploited with little restraint, often by poor and needy people trying to subsist on what they can extract from the land or the sea. Concerns for conservation and preservation are usually swept aside when basic survival needs must be met. In all cases, though, whether rich nation or poor, public land or private, natural lands are exploited and managed by *people*—not by agencies. For this reason, people's *attitudes* about nature are crucial. Some attitudes lead inexorably toward the destruction of nature and the tragedy of the commons: the same sad result, regardless of whether the motive is short-term profit taking or basic survival in the face of dire human need. There is no wisdom in pursuing this course and no recognition of the rights and needs of future human generations, let alone the rights of the natural world.

There is one attitude, however, that is capable of bringing dignity to the human exploitation of nature: the attitude of the **steward**. A steward is one to whom a trust has been given; something has been placed under his or her care. Stewardship of natural lands and seas is desperately needed in this crowded and hungry world. It is an attitude which recognizes that humans do not own Earth, an attitude that sees beyond the present to the needs and rights of those yet to be born. For some people, stewardship is a biblical attitude which stems from a recognition that the land and the wild things belong to their creator, God, who has given dominion over His creation to humankind, but who always holds His stewards accountable for its care. For other people, stewardship is a matter of wisdom that stems from a deep understanding and love of the natural world and the necessary limitations on human use of that world.

It is encouraging that very often those who are given responsibility for some part of the natural world end up really *caring* for it. Frequently, those who work closest to nature not only love their work, but also love the natural world they work in. It is common to find caretakers or foresters who began as exploiters and rangers who have risked their jobs by bringing misuse to the attention of their bosses or the public.

Stewardship of the natural world is also a possibility for *you*, as you examine your own attitudes and interactions with the environment. We all have opportunities to exercise stewardship. Indeed, it may be that, unless the attitude of stewardship is widely appreciated and practiced in a society and, eventually, on a global scale, human tenure on Earth in the 20th century will be seen as the age of folly, a time when untold biological wealth was squandered beyond the point of no return.

---

rise, so that today there is a growing commitment to restore natural systems that have been lost or damaged. A great increase in restoration activity has occurred during the last 25 years, spurred on by federal and state programs and the growing science of restoration ecology.

The intent of ecosystem restoration is simple: Repair the damage so that normal functioning returns and native flora and fauna are once again present. Because of the complexity of natural ecosystems, however, the task of restoration is often difficult in practice. The restoration of a degraded or altered ecosystem, for example, is not simply a matter of "letting nature take its course." Often soils have been disturbed, pollutants have accumulated, important species have disappeared, and other—often exotic—species have achieved dominance. For these reasons, a thorough knowledge of ecosystem and species ecology is essential to success.

The ecological problems that can be ameliorated by restoration include those resulting from soil erosion, surface strip-mining, draining wetlands, coastal damage, deforestation, overgrazing, desertification, and eutrophication of lakes. For example, the global problem of desertification (described in Chapter 9) was addressed by the 1994 United Nations Convention on Desertification, an outgrowth of the 1992 Earth Summit and an earlier convention that began in 1977. About

(a)

(b)

FIGURE 19-12
The Kissimmee River in Florida (a), where a major restoration project is underway to reclaim the river once bypassed by canals (b). (Ted Levin/Animals Animals/Earth Scenes.)

one-fourth of the land area of Earth has been affected by desertification, according to the United Nations—degraded lands cover some 800 million hectares (2 billion acres). The Convention on Desertification promoted the restoration of degraded lands, with a program of replanting, applying better farming technologies, and seeking more effective aid from the developed nations.

A more limited restoration program is underway in south Florida, where the Kissimmee River is now being restored. The 103-mile-long river runs south into Lake Okeechobee and was turned into a 56-mile-long canal by the Army Corps of Engineers as part of the massive water control system in southern and central Florida. Wildlife declined, the vegetation along the river and canal was drastically changed, and the remaining river segments stagnated. Congress has authorized a 15-year project to restore a 52-mile central stretch of the river. Wetlands will be restored, the canal will be filled, and the former course of the river will once again run with flowing water (Fig. 19-12). The restoration of the Kissimmee represents the first time a major public works program is reversed—for sound ecological reasons.

As the values of natural ecosystems are recognized, efforts to restore damaged or lost ones will become increasingly important. John Berger, executive director of Restoring the Earth (a private organization dedicated to the task of restoration), states that "the business of restoration is likely to become a multibillion dollar global enterprise" because of its capacity to ameliorate many of our worst environmental problems.

Public Lands in the United States

As we saw in Chapter 18, in order to save wild species, we must protect their habitats. The time has passed when we could justify losses of ecosystems on the grounds that there were suitable substitute habitats just over the hill. With the rising human population, industrial expansion, and pressure to convert natural resources to economic gain, however, there will always be reasons to exploit natural ecosystems. The last resort for many species and ecosystems is protection by law in the form of national parks, wildlife refuges, and reserves. Worldwide, some 6930 areas representing 4.8% of the national land area on the planet, have received this kind of protection—300 areas are designated more restrictive *biosphere* reserves.. Yet in the developing world, many of these are "paper parks," where exploitation continues and protection is given only lip service.

The United States is unique among the countries in having set aside a major proportion of its landmass for public ownership. Nearly 40% of the country's land is publicly owned and is managed by state and federal agencies for a variety of purposes. The distribution of public lands, shown in Fig. 19-13, is greatly skewed toward Alaska and other western states, a consequence of historical settlement and land distribution policies. Nonetheless, although most of the East and Midwest is in private hands, there are still functioning natural ecosystems on much of those lands. (See the "Earth Watch" box, p. 503.)

Land given the greatest protection (preservation) is designated **wilderness**. Authorized by the Wilderness Act of 1964, it includes over 100 million acres at 466 locations, almost 5% of the land area of the United States. The act provides for the permanent protection of these undeveloped and unexploited areas so that natural ecological processes can operate freely. Permanent structures, roads, motor vehicles, and other mechanized transport are prohibited. Timber harvesting is excluded. Some livestock grazing and mineral development are allowed where such use existed previously; hiking and other similar activities are allowed.

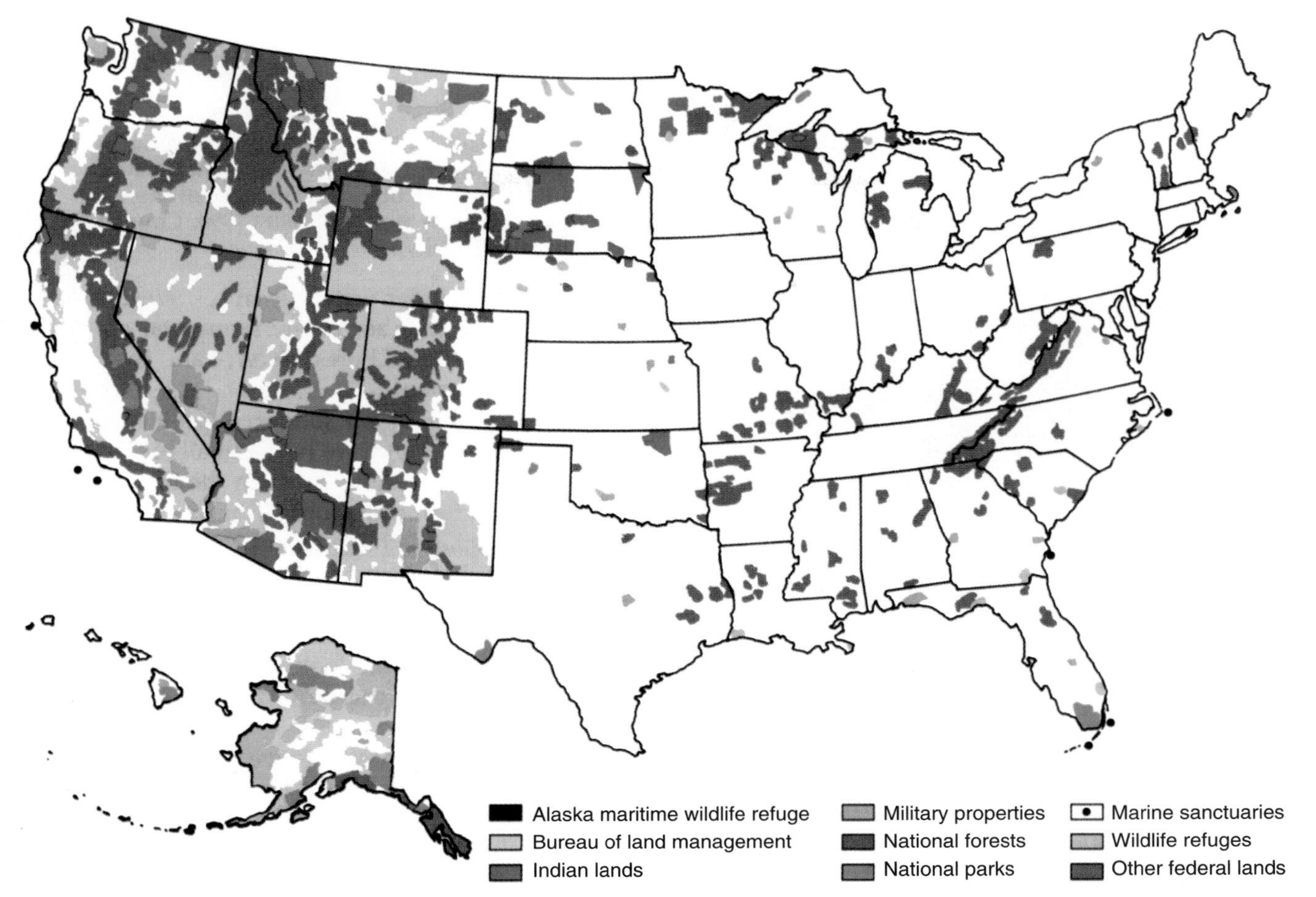

FIGURE 19-13
Distribution of public lands in the United States. (*Environmental Trends, 1989*. Council on Environmental Quality.)

The **national parks** (administered by the National Park Service) and **national wildlife refuges** (administered by the Fish and Wildlife Service) provide the next level of protection to 170 million acres. Here, the intent is to protect areas of great scenic or unique ecological significance, protect important wildlife species, and provide public access to view these scenic wonders. The dual goals of protection and providing public access often conflict with each other, because the parks and refuges are extremely popular, sometimes drawing so many visitors (273 million in 1993) that protection can be threatened by those who want to see and experience the natural sites. Because of the impacts of overuse, park rangers are under new directives to pay more attention to maintaining the parks in their natural state.

Increasingly, agencies are beginning to understand the need to manage natural sites as part of larger ecosystems. For example, the Great Smoky Mountains National Park has been made part of the Southern Appalachian Man and the Biosphere Cooperative (Fig. 19-14). Private and public land managers now have a decision-making body that can aid in the larger task of protecting natural resources. When black bears leave the park, for instance, they are likely to be shot by hunters. State wildlife managers and park officials now cooperate in setting reasonable limits on the number of bears that can be taken.

This cooperative approach is important for the continued maintenance of biodiversity, since so much of the nation's natural lands remain outside of protected areas. It may also help restrict development up to the borders of the parks and refuges, avoiding the situation where the refuge is a fairly small natural island in a sea of developed landscape.

National Forests. There is no doubt that forests in the United States represent an enormously important natural resource, providing habitat for countless wild species, as well as supplying natural services and products. U.S. forests range over 740 million acres, of which about two thirds are managed for commercial timber harvest. Almost three fourths of the managed commercial forestland is in the east and privately owned; the remainder is mainly in the west and is administered by a number of

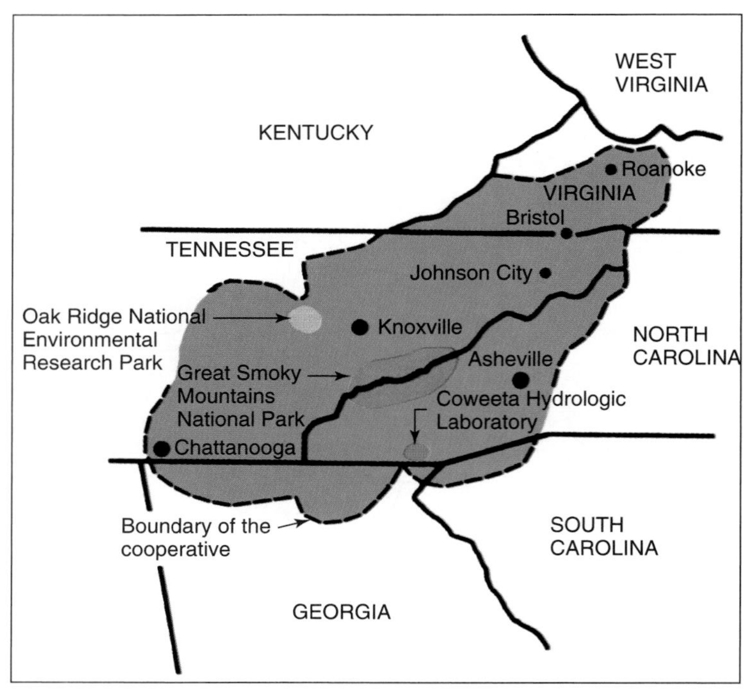

FIGURE 19-14
Southern Appalachian Man and the Biosphere Reserve, showing the relationship between the Great Smoky Mountains National Park (red) and the broader land area of the cooperative (green) that is the reserve. (*Environmental Quality, 1990.* Council on Environmental Quality.)

governmental agencies, primarily the National Forest Service and the Bureau of Land Management.

Deforestation is no longer a problem in the United States. Although we have cut all but 5% of the forests that were here when the first colonists arrived, second-growth forests have regenerated wherever forestlands have been protected from conversion to croplands and house lots. There are more trees in the U.S. today than there were in 1920. In the east, some second-growth forests have aged to the point where they look almost as they did before their first cutting, except for the lack of the American chestnut.

The Forest Service is responsible for managing the 191 million acres of national forests. The Congress has mandated management for *multiple use*, which means a combination of extraction of resources (grazing, logging, mining), recreation, and protection of watersheds and wildlife. Although the intent is to achieve a balance among these uses, in fact, multiple use creates conflicts, as groups compete to promote uses they favor. Because they see little change in management policies over the years, critics of multiple use claim that the policy does little more than justify the ongoing exploitation of public lands by private, often favored interest groups (ranchers, miners, and the timber industry). The Forest Service is specifically criticized for the way it manages logging: Government foresters select tracts of forest they judge ready for harvesting and then lease the tracts to private companies, which log and sell the timber. In reality, the national forests are losing money, and U.S. taxpayers are subsidizing the logging industry. By its own measure, the Forest Service admitted to a net loss of $50 million in its logging operations in 1992. Others place the estimates much higher, up to $300 million a year. The Clinton administration has gone on record with a commitment to ending below-cost timber sales.

President Clinton and the Congress worked out a budget compromise in the 1995 Recessions Bill that contained a highly controversial amendment—the "timber salvage rider." This rider seemed to simply allow timber companies to log forest areas already damaged by fire, insects, or disease. However, the legislation included an exemption of the timber sales from all environmental laws, even the Endangered Species Act, and included forest areas that *potentially might be damaged* in the future. This rider has allowed the timber industry to force the opening up of old timber sale proposals previously blocked because they were in violation of environmental laws. Old-growth forests that had been put off limits by the Northwest Forest Plan are now being cut, and the impact of the rider is now extending to national forests in other parts of the country. One observer has called this rider "the most anti-environmental piece of legislation ever passed by any Congress and signed by a President." A bipartisan effort has been mounted in Congress to repeal the timber salvage rider, but the timber industry will certainly oppose it.

The Forest Service is often in the middle of these controversies. The battle raging over the spotted owl and old-growth forests, discussed in Chapter 18, was resolved by the Northwest Forest Plan. Chief architect of this plan, Jack Ward Thomas was subsequently appointed by

Earth Watch

Saving the Northern Forest

Although most of the federal lands are in the west, the eastern United States boasts some impressive forested areas. One of these stretches from the eastern seaport of Machias, Maine, across to Syracuse, New York—26 million acres of forest that many refer to simply as the Northern Forest. Only 16% of the forests are publicly owned. Most of the rest is the domain of large paper companies, which view the terrain as one huge resource to be managed for paper manufacture and lumber. The land is at the same time beautiful and degraded. Pristine lakes, woods, and streams in some areas must be juxtaposed with clear-cut, bulldozed, slashed, and eroded hillsides in others. In Maine, an area the size of the state of Delaware has been swept clear of trees in the last 10 years. It was this kind of land abuse that led to the creation of the White Mountain and Green Mountain National Forests in 1911 and prompted New York to turn fields of stumps into the (now) magnificent Adirondack State Park.

There is pressure to convert some of the most scenic of these privately owned lands into condominiums and second-home development, as many easterners seek to escape to the "woods," where they can indulge themselves in "nature." Before this occurs, however, there will be a window of opportunity for conservation of a large portion of the Northern Forest. Two groups have recently made proposals that seek to do more than wring hands over the continued degradation of the forest. One of these is the Northern Forest Lands Council, which was set up by the federal government to find ways to conserve the land and develop recommendations for Congress and the states. This group, comprised of state and local officials, landowners, and environmental organizations, reported its recommendations in October 1994. Those recommendations went a long way to encourage the timber companies and property owners to keep their lands, by suggesting a system of tax incentives to conserve the forests. They also recommended more funding for state and federal land acquisition, as well as more studies of the top three environmental concerns: harmful logging practices, preservation of biodiversity, and protection of rivers and streams. One outcome of the work of the council is the National Forest Service Stewardship Act, which is under Congressional consideration. The Bill provides federal assistance to enable states to acquire land, change tax structures, and work on programs for better forest management.

Another, more radical recommendation has emerged from the grassroots conservation group, RESTORE. The group proposes that some 3.2 million privately owned acres surrounding Baxter State Park in Maine be the subject of outright purchase. This is largely untamed land, and its proximity to the scenic state park makes it part of a large natural unit that would become the nation's newest national park. The estimated price tag is between $500 million and $1 billion. Obviously aware of what the spotted owl has done for the northwestern forests, RESTORE has petitioned the U.S. Fish and Wildlife Service to consider the Atlantic salmon for endangered or threatened status; the pine marten, a member of the weasel family, is another candidate for listing. A combination of public purchase and strict land-use regulations can save the Northern Forest, but opposition to both strategies is strong. The wise use movement is active in uniting property owners by presenting the Northern Forest initiatives as a great federal land grab that will "cleanse" the area of the people who own it.

It would seem only prudent to move forward with proposals to save the Northern Forest. It is not going to remain as it is; the price will never be lower for a federal purchase. The eastern United States would enjoy a major national park, one to which some large mammals, such as elk, timber wolves, and perhaps cougars could be reintroduced.

President Clinton to head the Forest Service. Thomas inherited an agency with a mandate for change; he immediately issued a threefold directive to all employees of the service: "Tell the truth, obey the law, and practice ecosystem management!" Along with a responsibility to eliminate below-cost timber sales, Thomas inherited an agency whose professional organization, the Society of American Foresters, is promoting a **New Forestry**. This is the practice of forestry that is directed more toward protecting the ecological health and diversity of forests than toward producing a maximum harvest of logs. Thus, new forestry involves cutting trees less frequently—at 350-year intervals instead of every 60 to 80 years; leaving wider buffer zones along streams, to reduce erosion and protect fish habitats; leaving dead logs and debris in the forests, to replenish the soil; and protecting broad landscapes—up to 1 million acres—across private and public boundaries, while involving private landowners in management decisions.

Wise Use and Environmental Backlash. The use of federal lands for cattle grazing and lumbering has created a clientele accustomed to paying only low, subsidized fees for grazing and access to timber. In recent years, environmental organizations have called attention to federal lands that have been abused by overuse and poor management, and have called for greater accountability of the agencies in charge, as well as for

Earth Watch

Nature's Corporations

The spotted owl controversy pitted jobs against the preservation of the old-growth forests of the west. Timber interests maintained that preservation would result in 20,000 lost jobs. In their view, the controversy came down to this: We should continue to cut the old-growth forests for the sake of keeping loggers employed. In fact, during the 1980s when timber harvests were up, rising unemployment among loggers occurred anyway owing to mill modernization and other corporate decisions unrelated to logging!

General Motors and other major businesses facing hard times lay off thousands of workers, and no one questions their need or right to do so. Clearly, the survival of General Motors is more important than keeping all their workers employed. It is assumed that those laid off will find other employment. According to conventional wisdom, corporate America must survive if the economy is to have a chance of recovering from economic bad times.

Forested ecosystems can be viewed as nature's corporations, providing lumber and firewood for the economy. They can only do so if they are maintained sustainably, however, and not exploited just to provide jobs for loggers. (Bill Hubbell/Woodfin Camp & Associates.)

In a real sense, ecosystems are the corporations that sustain the economy of the biosphere. (See figure.) If we want those ecosystems to survive and recover, we may have to tighten our belts and withdraw some of the work force engaged in exploiting them. The maintenance of these systems is obviously more important than some jobs. Why can't laid-off loggers seek other employment the way laid-off auto workers, steel workers, or computer engineers do? Why should a natural ecosystem be "bankrupted" just to maintain the *temporary* employment of a few (temporary because the loggers will be out of a job in a few years, when the old-growth forests are finally all cut)?

With the extent of the logging cutbacks now established, the economy in the Pacific northwest is picking up. Oregon's unemployment rate of 5% is at its lowest in 25 years. Although some 15,000 wood-related jobs have been lost, there has been a net gain in employment due to rapid growth in high-technology industries and federal retraining aid to many towns and workers. Industries locate in the region because of the impact of the "second paycheck," which refers to the high quality of life that attracts workers, even at lower pay scales. The same high quality of life—the forests, streams, scenic vistas, hunting, and fishing—requires preservation of the ecosystems that were being logged over.

Clearly, it is shortsighted to assume that ecosystems are there just to provide jobs and that the jobs are more important than the ecosystems. When the old-growth forests are gone, we shall have lost more than just loggers' jobs; we shall have lost a priceless heritage and a major part of the natural world that provides us with vital services.

elimination of the subsidized uses. To these calls for action have been added the mandates of the ESA and other land use policies, directed toward more preservation of natural areas, as well as a rising tide of environmental regulation to combat pollution.

These changes have given birth in turn to an environmental backlash. As people and corporations have begun to identify concern for the environment as a serious threat to their interests, they have joined with like-minded groups to oppose environmental action. The backlash has emerged in a number of manifestations, perhaps the most prominent being the "wise use" movement. Ron Arnold, spokesman of the movement, has said, "We intend to destroy the environmental movement once and for all by offering a better alternative, the 'wise use' movement." Their rallying cry of "wise use" means that natural ecosystems should be used (by them!) and not simply preserved. Proponents of "wise use" want to see environmental legislation dismantled and promote the concept that private property owners should always be able to do what they want with their land. This movement includes many local and national organizations with a desire to use particular natural resources, usually federally owned. Thus, we have the California Desert Coalition (to protest protection of the desert), the National Wetlands Coalition (to fight changes in wetland legislation), the Public Lands Council (to maintain low grazing fees), and the Evergreen Foundation (to promote logging of federal forests). The movement labels environmentalists as "tree-worshipping pagans" who are bent on stifling the economy and putting people out of work. Their funding comes largely from the industries that have strong interests in maintaining access to public lands and resources—the paper companies, lumber industries, oil companies, manufacturers of recreational vehicles, and the like.

Another manifestation of the environmental backlash can be seen in a group of conservative, free-market "think tanks" such as the Cato Institute, the Competitive Enterprise Institute, and the Acton Institute. As one observer explains, now that world communism is dead, these organizations have discovered a new home for "socialism": the environmental movement. They maintain that environmental organizations such as the Audubon Society, the Conservation Foundation, and the Sierra Club are bent on achieving power over the national agenda in order to bring industry to its knees and return much of the American landscape to wilderness. According to them, the environmental movement is little more than fear mongering masquerading as science. They maintain that regulations promoted by environmental groups are an enormous drag on the economy, representing hundreds of billions of dollars in wasted money that could be directed toward greater productivity and creating more jobs.

A common strategy of these two groups of organizations is to minimize every environmental concern that has emerged in recent years, charging that the scientists and the environmentalists have blown things out of proportion. Global warming? Poppycock! It's not getting warmer here! Ozone holes? The holes are in the data, not the ozone layer. Extinctions of species? The few that are being lost are "nonadaptive" and thus deserve to perish. Another strategy of these groups has been to mount an assault against environmental legislation. The Republican "Contract With America," introduced with the 104th Congress in 1995, contained many bills that were seen as anti-environmental, reflecting the environmental backlash. These bills attempted to repeal or weaken existing environmental laws, cut the budgets of environmental agencies, and introduce measures such as the "takings" legislation, cost-benefit analysis and risk assessment that would seriously cripple the regulatory process. The combination of Senate filibuster, a group of environmentally sensitive Republican legislators, and the Presidential veto has so far prevented these bills from being enacted into law. Indeed, the Congress twice forced the shutdown of the Government in an attempt to push their anti-environmental agenda past President Clinton. The strategy clearly backfired when Clinton was reelected in 1996 on the strength of his stand against the Republican-dominated Congress.

Private Land Trusts

With so much land in private hands, the future of natural ecosystems in the United States is always uncertain. This is especially true of land having special appeal for recreation and aesthetic enjoyment, because here, the potential for development is great. Often, landowners and townspeople want to protect natural areas from development, but are wary of turning the land over to a governmental authority. One very creative option is the **private land trust**, a nonprofit organization that will accept either outright gifts of land or **easements**—arrangements where the landowner gives up development rights into the future, but retains ownership. The land trust may also purchase land to protect it from development. The land trust movement is growing; there were 429 land trusts in the United States in 1980, and now there are over 1100, as reported by the Land Trust Alliance, an umbrella organization in Washington serving the local and regional land trusts. Local and regional land trusts protect about 2 million acres of land, and large national land trusts such as the Nature Conservancy protect an additional 3.5 million acres.

Land trusts are proving to be a vital link in the preservation of ecosystems. The oldest trust is the Trustees of Reservations in Massachusetts, founded in 1891 and now guardian of 17,000 acres throughout the state. The Trustees, as the organization is often called, has received some ecologically prime and scenic land through the years and now maintains many of its properties for public use that is compatible with preservation. The land trusts are serving the common desires of landowners and rural dwellers to preserve the sense of place that links the present to the past. At the same time, the undeveloped land remains in its natural state, sustaining natural populations and promising to do so into the future.

Ecosystems everywhere are being exploited for human needs and profit. In addition to all the examples we have looked at in this chapter, other areas that are in trouble are wetlands drained for agriculture and recreation, rangelands overgrazed, rivers overdrawn for irrigation water, and more. Our purpose has been to highlight some of the most critical problems and to indicate steps being taken to correct them. It is certain that greater pressures will be put on natural ecosystems as the human population continues to rise. These pressures must be met with increasingly effective protective measures if we want to continue to enjoy the fruits of our various ecosystems. The problems are most difficult in the developing world, where poverty forces people to take from nature in order to survive. If natural areas are to be preserved, the needs of people must be met in ways that do not involve destroying ecosystems. People must be provided with alternatives to exploitation, a situation requiring both wise leadership and effective international aid.

Sustainability—that crucial concept once again—should be the goal of all our interactions with natural systems. The sooner we recognize the wisdom of that basic approach, the better our chances will be of having more than just a few remnants of nature left by the time the human population levels off sometime in the 21st century.

Review Questions

1. What is the extent of deforestation globally; where and how rapidly is it occurring?
2. What is the basic reason for clearing forests, and what are five consequences of that clearing?
3. Name five ways that forests can be exploited that preserve most of their vital ecological functions.
4. What factors are responsible for tropical deforestation?
5. What is the global pattern of exploitation of fisheries; how does the New England fishery exhibit the worst problems of exploitation?
6. What three countries are pressuring the IWC to reopen commercial whaling, and what is their rationale for resuming the killing?
7. What are the five major categories of terrestrial ecosystems? What are the three major categories of oceanic ecosystems?
8. Describe eight natural services that all ecosystems perform.
9. How has calculation of GNP for a country failed to accomplish environmental accounting?
10. Compare and contrast the terms *conservation* and *preservation*.
11. What is the tragedy of the commons? Give an example of a commons and how it is mistreated.
12. What are two ways to avoid a tragedy of the commons?
13. What does *maximum sustainable yield* mean? What factors complicate its application?
14. Contrast the different levels of ecosystem protection given to public lands in the United States.
15. What is the current economic problem with much of the logging of national forests, and how is the present administration proposing to change it?
16. What are two current forms of environmental backlash, and how are they attempting to influence public policy?
17. How do land trusts work, and what roles do they play in preserving natural lands?

Thinking Environmentally

1. Identify and study an area in your community where destruction or degradation of natural land or wetland is an issue. What if anything is being done to protect this land?
2. What incentives and assistance could the United States offer Brazil or Guyana to keep their tropical rain forests from further harmful development?
3. Imagine that you are the clam warden for a New

England town. What policies would you put in place to prevent a tragedy of the commons from occurring?

4. What natural resource commons do you share with those around you? Are you exploiting or conserving these commons?
5. Assume that you own 1000 acres of forest. How would you set up an accounting system for use of the forest that recognizes the natural resources and services associated with the forest?
6. See number 1; propose a program for restoration of the degraded land.
7. Does our National Park system encompass enough area, or should it be expanded? Either way, defend your answer. If you believe the system should be expanded, propose some new areas to be added.
8. Research the preservation efforts and effectiveness of one of the following conservation groups: The Sierra Club, The Appalachian Mountain Club, The Nature Conservancy, The Audubon Society.

Converting Trash to Resources

Key Issues and Questions

1. Two hundred million tons of municipal solid waste (MSW) are disposed of annually in the United States. What are the components of MSW, and how is this waste handled?
2. Sixty-one percent of MSW is disposed of in landfills. What are the problems with landfills, and how are these addressed in newer landfill sites?
3. Fifteen percent of MSW is combusted, mostly in waste-to-energy (WTE) combustion facilities. What are the advantages and disadvantages of WTE combustion?
4. The best solution to solid waste problems is to reduce waste at its source. How can the total volume of refuse be reduced?
5. More than 75% of MSW is recyclable. What role is recycling playing in waste management, and how is recycling best promoted?
6. Although most management of MSW occurs at the local level, federal regulations concerning MSW are increasing. What regulations have affected the management of municipal solid waste?
7. Much more can be done to move MSW management in a more sustainable direction. What are some recommendations to improve MSW management?

Danehy Park features three soccer fields, four baseball diamonds, two playgrounds, and a two-mile jogging trail. The 50-acre park, opened in 1990, is located close to a heavily populated area of North Cambridge, Massachusetts, and is in almost constant use in good weather (Fig. 20-1). An unusual feature of the park is a big red light in the public rest room that warns users to vacate the park if the light goes on. The light is tied in to an elaborate venting system that prevents methane from building up below ground. The system is necessary because Danehy Park is built on the former city dump. Aside from the presence of the light, and a few minor settling problems on soccer fields that have interfered with water drainage, a newcomer would never know that this was once a blight on the neighborhood—an open, burning dump in the 1950s and then a "sanitary landfill" that was closed in 1972. The park increased Cambridge's open space by 20% when it was created.

Danehy Park is a lesson in land use that is being learned only slowly in the United States—and, for that matter, in the rest of the world. The lesson is that if we could more effectively picture what 50 years can bring in the way of change, we could solve one of the most

A bulldozer-compactor works over a mountain of trash in a municipal landfill. [Photo by David R. Frazier/Photo Researchers, Inc.]

contentious problems facing local communities—what to do with municipal waste. As old dumps and landfills were closed because of environmental concerns, they created a temporary problem identified as the "solid waste crisis" in the 1970s and 1980s. Now many of those dumps and landfills are being converted into parks, golf courses, and nature preserves.

In some ways, the solid waste crisis is still with us, as we will see when we examine the problems of interstate trash movement. It is commonly said that we are running out of space to put our trash and garbage. That is a misconception; actually, we are running out of disposal space only because policymakers—local and national—refuse to make unpopular decisions about land use and trash management. To a great extent, their refusal to make hard decisions is a reflection of a fundamentally irresponsible feature of our modern society: We are happy to purchase the goods displayed so prominently in our malls and advertised in the media, but we are reluctant to accept the consequences of getting rid of them responsibly.

This chapter is about solid waste issues. We examine current patterns of disposal—landfills, combustion, and recycling—and look for solutions to our solid waste problems that can work well. The ideal would be to imitate natural ecosystems and reuse everything. Recall the first principle of sustainability: Ecosystems dispose of wastes and replenish nutrients by recycling all elements. Some solutions do well in conforming to this principle, whereas others do not—and possibly cannot.

The Solid Waste Problem

So far in this book we have talked about animal feedlot wastes (Chapter 12), sewage wastes (Chapter 13), and industrial wastes (Chapters 14 and 15). A fourth category of wastes, the focus of this chapter, is **municipal solid waste** (MSW), defined as the total of all the materials thrown away from homes and commercial establishments (commonly called trash, refuse, or garbage).

Disposing of Municipal Solid Waste

Over the years, the amount of MSW generated in the United States has grown steadily, in part because of increasing population, but more so because of changing lifestyles and the increasing use of disposable materials and excessive packaging. MSW now amounts to somewhat over 4 pounds (2 kg) per person per day. At the current (1998) U.S. population of 270 million, that is enough waste to fill 80,000 garbage trucks each day, a total of 209 million tons (190 million metric tons) per year. The *problem* can be simply stated: *We have to dispose of this stuff in the most effective and efficient way, while protecting human and environmental health.*

The refuse generated by municipalities is a mixture of materials from households and small businesses, with proportions as shown in Fig. 20-2. However, the proportions vary greatly, depending on the generator (commercial versus residential), the neighborhood (affluent versus poor), and the time of year (during certain seasons, yard wastes, such as grass clippings and raked leaves, add to the solid waste burden, often equaling all the other categories combined). It is fair to say that little attention is given to what people throw away in the trash; even if there are restrictions and prohibitions, these can be bypassed with careful packing of the trash containers. Thus, many nasty substances—paint, used motor oil, batteries, and so on—are discarded with the feeling that they are gone forever.

Customarily, local governments have assumed the responsibility for collecting and disposing of MSW. The local jurisdiction may own the trucks and employ workers, or it may contract with a private firm to provide the collection service. Alternatively, some municipalities have opted for putting all trash collection and disposal in the private sector. The collectors bill each home by volume and weight of trash. This system allows competition among collectors and gives homeowners a strong incentive to reduce the volume of trash they produce. The MSW that is collected is then disposed of in a variety of ways, and it is at the point of disposal that state and federal regulations begin to apply.

Until the 1960s, most MSW was disposed of in open, burning dumps. The waste was burned to reduce its volume and lengthen the life span of the dump site, but refuse does not burn well. Smoldering dumps produced clouds of smoke that could be seen from miles away, smelled bad, and created a breeding ground for flies and rats. Some cities turned to incinerators, or combustion facilities, as they are called today—huge furnaces in which high temperatures allow the waste to burn more completely than in open dumps. Without controls, however, incinerators were also prime sources of air pollution. Public objection and air pollution laws forced the phaseout of open dumps and many incinerators during the 1960s and early 1970s. Open dumps were then converted to landfills.

For 1994 in the United States, 61% of MSW is disposed of in approximately 6000 operating landfills, 24%

FIGURE 20-1
Thomas W. Danehy Park in Cambridge, Massachusetts, a former landfill that has been recycled into a recreational park. In addition to the playing fields, the park features a half-mile "glassphalt" pathway (built with recycled glass and asphalt), shown in the foreground. (Courtesy Cymie Payne.)

is recovered for recycling and composting, and the remainder (15%) is combusted (Fig. 20-3). The pattern is different in countries where population densities are higher and there is less open space for landfills. High-density Japan, for instance, combusts about half of its trash and recycles over half of the rest. Many Western European countries also deposit less than half of their municipal waste in landfills and combust most of the rest.

Landfills

In a **landfill**, the waste is put on or in the ground and covered with earth. Because there is no burning, and because each day's fill is covered with a few inches of earth, air pollution and vermin populations are kept down. Unfortunately, aside from those concerns and the minimizing of cost, no other factors were given real

FIGURE 20-2
The composition of municipal solid waste in 1994 in the United States. (Data courtesy of Franklin Associates, Ltd., Prairie Village, Kansas in Report No. EPA 530-R-96-011, "Characterization of Municiapal Solid Waste in the United States 1995 Update," March 1996.)

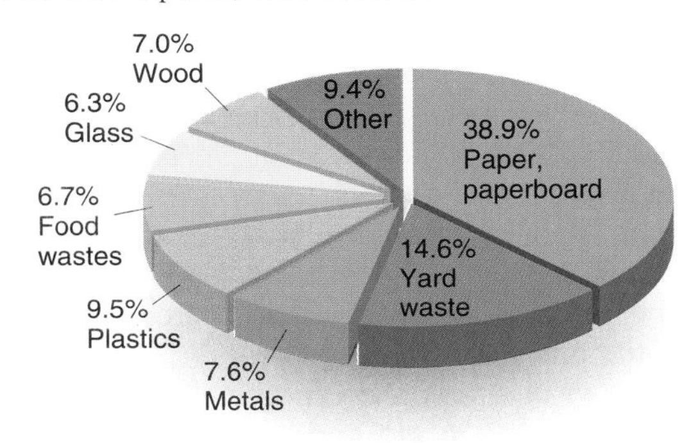

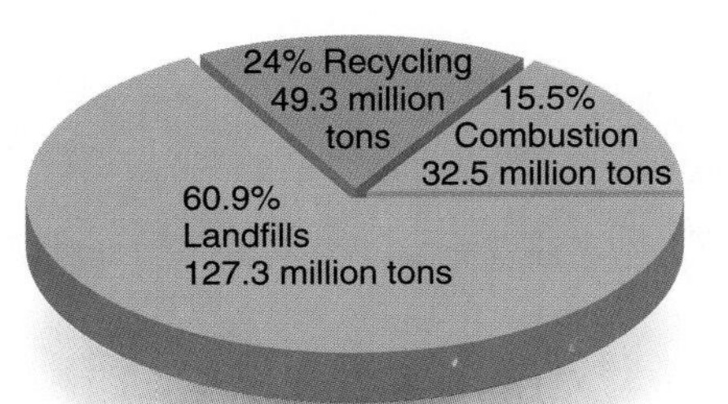

FIGURE 20-3
Categories of MSW disposal for 1994 in the United States. (Data courtesy of Franklin Associates, Ltd., Prairie Village, Kansas in Report No. EPA 530-R-96-011, "Characterization of Municiapal Solid Waste in the United States 1995 Update," March 1996.)

consideration when the first landfills were opened. Municipal waste managers generally had no understanding of or interest in ecology, the water cycle, or what products would be generated by decomposing wastes, and they had no regulations to guide them. Therefore, in general, any cheap, conveniently located piece of land on the outskirts of town became the site for a landfill. This site was frequently a natural gully or ravine, an abandoned stone quarry, a section of wetlands, or a previous dump (Fig. 20-4). Once the municipality acquired the land, dumping commenced, with no precautions taken. After the site was full, it would be covered with earth and ignored; only recently have landfills been seen as a valuable open-space resource.

Problems of Landfills. Landfills are subjected to biological and physical factors in the environment and will undergo change over time as a consequence of the operation of those factors on the waste that is deposited. Several of the changes are undesirable,

FIGURE 20-4
A landfill in Staten Island, NY, encroaching on a wetland. (Ray Pfortner/Peter Arnold, Inc.)

because they present the following problems if not dealt with effectively:

- Leachate generation and groundwater contamination
- Methane production
- Incomplete decomposition
- Settling

Leachate Generation and Groundwater Contamination. The most serious problem by far is groundwater contamination. Recall that as water percolates through any material, various chemicals in the material may dissolve in the water and get carried along, a process called *leaching*. The water with various pollutants in it is called *leachate*. As water percolates through MSW, a noxious leachate is generated that consists of residues of decomposing organic matter combined with iron, mercury, lead, zinc, and other metals from rusting cans, discarded batteries, and appliances—generously "spiced" with paints, pesticides, cleaning fluids, newspaper inks, and other chemicals. The nature of the landfill site and the absence of precautionary measures noted earlier funnel this "witches' brew" directly into groundwater aquifers.

All states have some municipal landfills that are or soon will be contaminating groundwater, but Florida is in a real crisis. Flat and with vast areas of wetlands, most of the state is only a few feet above sea level and rests on water-saturated limestone. No matter where Florida's landfills were located, they were either in wetlands or just a few feet above the water table. Since local residents rely on groundwater for 92% of their freshwater, you can guess the result: more than 200 municipal landfill sites on the Superfund list. Recall from Chapter 14 that Superfund is the federal program to clean up sites that are in imminent danger of jeopardizing human health through groundwater contamination. It will cost between $10 million and $100 million to clean up each site. So much for cheap waste disposal!

Methane Production. Because it is about two-thirds organic material, MSW is potentially subject to natural decomposition. However, buried wastes do not have access to oxygen. Therefore, their decomposition is anaerobic, and a major byproduct of this process is *biogas*, which is about two thirds methane and the rest hydrogen and carbon dioxide, a highly flammable mixture (p. 334). Produced deep in a landfill, biogas may seep horizontally through the soil and rock, enter basements, and even cause explosions if it accumulates and is ignited. Over 20 homes at distances up to 1000 feet from landfills have been destroyed, and some deaths have occurred as a result of such explosions. Also, gases seeping to the surface kill vegetation by poisoning the roots. Without vegetation, erosion occurs, exposing the unsightly waste.

A number of cities have exploited the problem by installing "gas wells" in old and existing landfills. The

Earth Watch

Trash on the Move

Long Island, New York, achieved a certain notoriety in 1987 when a garbage barge piled high with area trash cruised the Atlantic Ocean for four months in a search of a place to dump the load. Now some Long Island villages are adding to this dubious reputation as their trash is trucked 900 miles to a landfill in Taylorville, Illinois. In Tonawanda, New York, trash from Canada has become the newest item of economic exchange between Canada and the United States. Some 500,000 tons of garbage were exported into the U.S. from Ontario province in 1991. This may be Canada's most effective rebuttal to acid rain from the United States! And the small town of Welch, West Virginia, is poised to become the nation's largest recipient of out-of-state trash if a developer's plan to set up a huge landfill 3 miles from town is approved.

The underlying reason for this traffic in trash is market economics. Cities and towns are looking for the least expensive way to get rid of their MSW, and it sometimes happens that trucking the trash out of state or even out of the country is less expensive than taking it to a local incinerator or landfill. It would cost $87 a ton to dump trash from Westbury, Long Island, at a combustion facility 10 miles away in Hempstead, but Westbury can hire Star Recycling, Inc., of Brooklyn to take the trash all the way to Illinois for $69 a ton. Since the 13,000 citizens of Westbury already pay $2.1 million to get rid of their trash, they're looking for ways to keep down their costs.

In Ontario, the provincial government raised tipping fees to $136 a ton (from $18 several years ago), in order to push a new recycling program. Much lower fees in neighboring New York landfills have "drawn haulers to American dumps like sea gulls," in the words of a *New York Times* reporter. New York legislators, unhappy about the flood of Canadian garbage, are exploring the imposition of a hefty "inspection fee" of $150 a ton at the border in order to stem the tide. However, some observers fear the possibility of retaliatory action by Canada that would hurt another waste exchange: the movement of hazardous waste from the United States to Canada!

The lessons here are obvious. We are indeed in an unsustainable mode, and waste management, not simply waste removal, must become a high priority on the national agenda. Trucking trash long distances to save a few dollars makes no economic sense when the total costs of lost energy and environmental impact at the final dump sites are calculated. And it certainly makes no ecological sense. Trash on the move is the final, ridiculous consequence—and, let us hope, the death throes—of the lack of coherent public policy for managing MSW.

See: "The World of MSW," http://www.swana.org/mswweek.htm

wells tap the biogas, and the methane is purified and used as fuel. There are now 70 commercial landfill gas facilities in the United States; the largest, in Sunnyvale, California, generates enough electricity to power 100,000 homes. In a strange turn, some cities have drilled into landfills to draw out methane gas to prevent fires in the landfill. This vented gas is then simply burned off even though it is useful.

Incomplete Decomposition. The plastic components of MSW are resistant to natural decomposition. For this reason, much emphasis has been placed on developing biodegradable plastics. Serious questions remain about the degradability of these plastics, however. The term *biodegradation* refers to the complete breakdown of carbon compounds to carbon dioxide and water. All that the purported biodegradable plastics do is disintegrate into a fine polymer powder that still resists microbial breakdown.

A team of "archeologists" from the University of Arizona, led by William Rathje, has been carrying out research on old landfills. Their research has shown that even materials formerly assumed to be biodegradable—newspapers, wood, and so on—are degraded only slowly, if at all, in landfills. In one landfill, 30-year-old newspapers were recovered in a readable state; layers of telephone directories, practically intact, were found marking each year. Since paper materials are 38.9% of MSW, this is a serious matter. The reason paper and other organic materials decompose so slowly is the lack of suitable amounts of moisture; the more water percolating through a landfill, the better the biodegradation of paper materials. However, the more percolation there is, the more toxic leachate is produced!

Settling. Finally, waste settles as it compacts and decomposes. Luckily, this eventuality was recognized from the beginning, so buildings have never been put on landfills. Settling presents a problem where landfills have been converted to playgrounds and golf courses, though, because it creates shallow depressions (and sometimes deep holes) that collect and hold water. This process can be addressed by continual monitoring of the facility and the use of fill to restore a level surface.

Improving Landfills. Recognizing the foregoing problems, the EPA has upgraded siting and construction requirements for new landfills. Under current regulations:

- New landfills are sited on high ground, well above the water table. Often, the top of an existing hill is bulldozed off to supply a source of cover dirt and at the same time create a floor that is above the water table.
- The floor is first contoured so that water will drain into a tile leachate-collection system. The floor is then covered with at least 12 inches of impervious clay or a plastic liner or both. On top of this is a layer of coarse gravel and a layer of porous earth. With this design, any leachate percolating through the fill will encounter the gravel layer and then move through that layer into the leachate collection system. The clay layer or plastic liner prevents leachate from ever entering the groundwater. Collected leachate can be treated as necessary.
- Layer upon layer of refuse is positioned such that the fill is built up in the shape of a pyramid. Finally, it is capped with a layer of clay and a layer of topsoil and then seeded. The clay-topsoil cap and the pyramidal shape help the landfill to shed water. In this way, water infiltration into the fill is minimized, and less leachate is formed.
- Finally, the entire site is surrounded by a series of groundwater-monitoring wells that are checked periodically, and such checking must go on indefinitely.

These design features are summarized in Fig. 20-5. Most landfills currently in operation have the improved technologies, which protect both human health and the environment.

Although the regulations protect groundwater, the landfill pyramids may well last as long as the Egyptian pyramids. (They are not likely to become tourist attractions, however!) And if they do break down, they become a threat to the groundwater; therefore, the need for monitoring remains. Nonetheless, the creative siting and construction of landfills has the potential to address some very significant future needs, as we have seen: The abandoned landfill can become an attractive golf course, recreational facility, or wildlife preserve.

Siting New Landfills. About 1200 old landfills were scheduled to close by 1997, either because they have reached capacity or because of environmental problems. New landfills are being constructed at less than half this rate, however. This is due to the one problem associated with landfills that gets more attention than any other: **siting**. It is not that landfills take up enormous amounts of land; one landfill occupying 121 acres (Central Landfill) serves the entire state of Rhode Island, and Fresh Kill, the largest landfill in the world, serves much of New York City and its environs and is only 2400 acres in area. But as old landfills are closed, it has become increasingly difficult to locate land for the new ones needed to take their place.

People in residential communities (where MSW is generated) invariably reject proposals to site landfills anywhere near where they live. And those who already live close to existing landfills are anxious to close them down. Weary of the odor and heavy truck traffic, Staten Island, New York residents are planning legal action in an attempt to shut down the huge Fresh Kill landfill, the recipient of 13,000 tons of garbage per day. With spreading urbanization, there are few suburban areas not already dotted with residential developments. Any site selection, then, is met with protests and legal suits. This problem has been repeated in so many parts of the country (and globe!), that it has given rise to several inventive acronyms: **LULU**, **NIMBY**, and **NIMTOO**. The *LULU* is a locally unwanted land use; *NIMBY* is "Not in my backyard"; and *NIMTOO* is "Not in my term of office!" The applications of these attitudes to the landfill siting problem are obvious.

The siting problem has some undesirable consequences. First, it drives up the costs of waste disposal, as alternatives to local landfills are invariably more expensive. Second, it leads to the inefficient and equally objectionable practice of long-distance transfer of trash. (See "Earth Watch" box, p. 513), as waste generators look for private landfills anxious to receive trash. Very often, this transfer occurs across state and even national lines, leading to very real resentment and opposition on the part of citizens of the recipient state or nation. Such opposition has resulted in numerous state efforts to restrict or prohibit landfills from receiving out-of-state trash. These efforts have invariably been struck down by the courts as unconstitutional interference with interstate commerce. Five states—Illinois, Indiana, Pennsylvania, Ohio, and Virginia—import more than 1 million tons of MSW a year (Fig. 20-6). New Jersey, New York, and Missouri lead the exporting states, with New York far in the lead, at 3.8 million tons a year.

One positive impact of the siting problem is to encourage residents to reduce their waste and recycle as much as possible; this will be explored later. Another effect of the siting problem is to stimulate the use of combustion as an option for waste disposal.

Combustion: Waste to Energy

Because it has a high organic content, refuse (including the plastics portion) can be burned. Currently, more than 130 combustion facilities are operating in the United States, burning about 33 million tons of waste annually—16% of the waste stream.

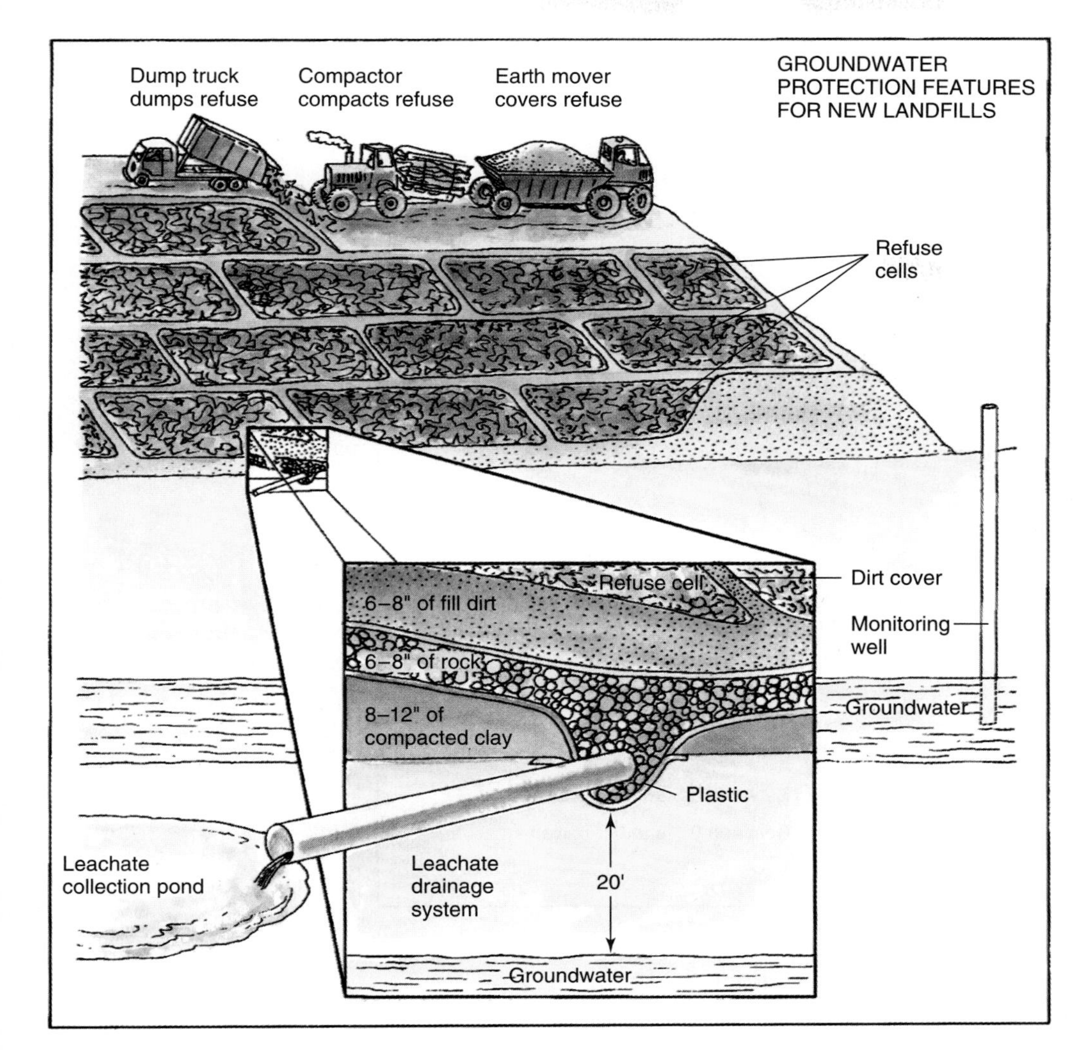

FIGURE 20-5
Features of a modern landfill with environmental safeguards. The landfill is sited on a high location well above the water table. The bottom is sealed with compacted clay or a plastic liner or both, overlaid by a rock or gravel layer with pipes to drain the leachate. Refuse is built up in layers as the amount generated each day is covered with soil, so that the completed fill has a pyramidal shape that sheds water. The fill is provided with wells for monitoring the groundwater.

Advantages of Combustion. Combustion of MSW has some advantages:

- Combustion can reduce the weight of trash by over 70% and the volume by 90%, thus greatly extending the life of a landfill (which is still required to receive the ash). Some ash goes into the landfill, and some is reused.
- Toxic or hazardous substances are concentrated into two streams of ash, which are easier to handle and control than the original MSW. The fly ash (captured from the combustion gases by air pollution control equipment) contains most of the toxic substances and can be safely put into the landfill. The bottom ash (from the bottom of the boiler) can be used as fill in some construction sites and roadbeds. Some combustion facilities process the bottom ash further, to recover metals, and then convert the remainder into concrete blocks.
- No changes are needed in trash collection procedures or people's behavior; trash is hauled to a combustion facility instead of the landfill.
- Practically all modern combustion facilities are designed to generate electricity, which is sold to offset some of the costs of disposal. The newer facilities are equipped with scrubbers and filters or electrostatic precipitators, which remove acid gases and particles to bring the emissions into compliance with Clean Air Act regulations.

Simple incineration is an outmoded method of combustion; it accomplishes a reduction in volume of the trash, but without the recovery of energy. When burned, unsorted MSW releases about half as much energy as coal, pound for pound. Almost three fourths of the currently operating combustion facilities are **waste-to-energy** (WTE) facilities equipped with modern emission control technology. Many of these facilities add **resource recovery** to their waste processing, in which many materials are separated and recov-

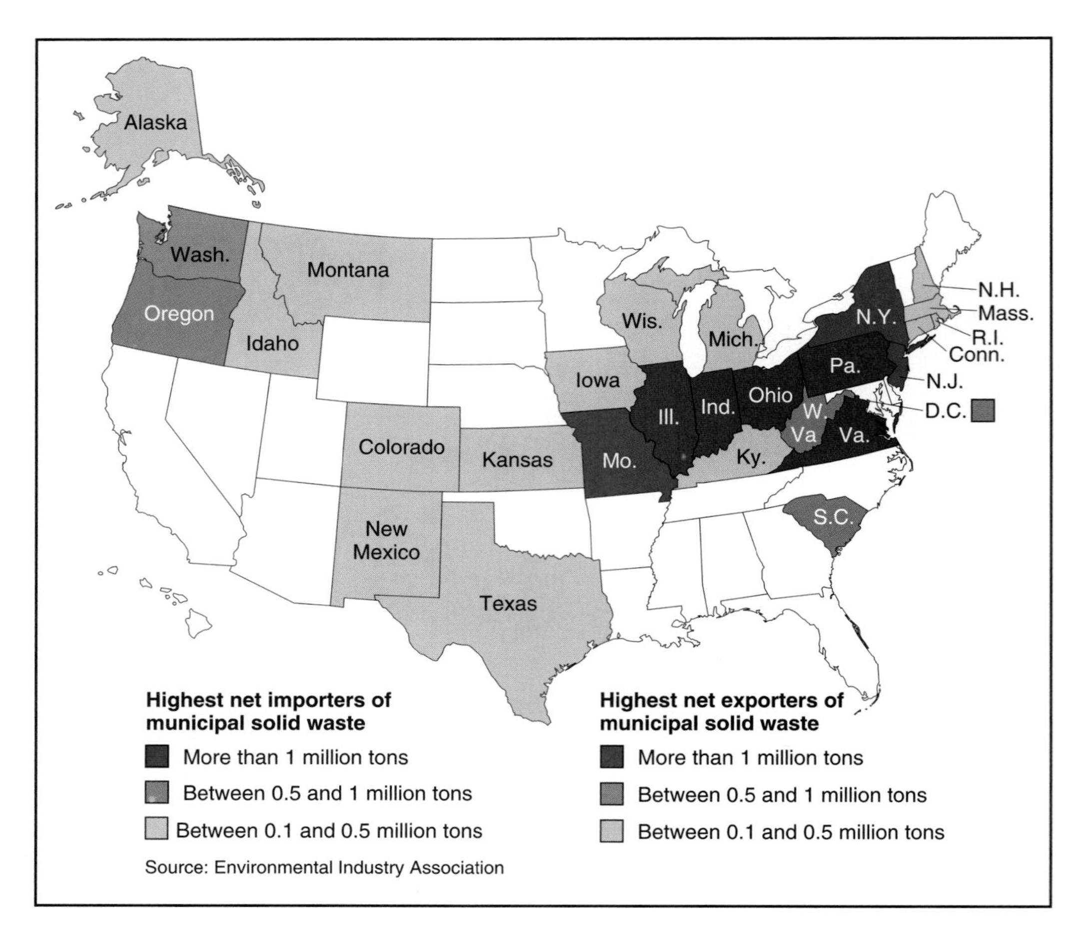

FIGURE 20-6
Net import and export of municipal solid waste, by state, in 1992. (Environmental Industry Association has a home page on the Internet.)

ered before (and sometimes after) combustion. Because the sale of the electricity offsets some of the costs of MSW disposal, the combustion facilities are usually able to compete successfully with landfills for MSW.

Drawbacks of Combustion. Combustion has some drawbacks, however:

- Trash does not burn cleanly. Despite being equipped with air pollution control devices, exhaust stacks emit toxic fumes into the air as burning oxidizes and vaporizes the assortment of metals, plastics, and hazardous materials that inevitably end up as municipal waste.
- Combustion facilities are expensive to build, and their siting has the same problem as the landfills: No one wants to live near one.
- Combustion ash is often loaded with metals and other hazardous substances and must be disposed of in secure landfills.
- To justify the cost of its operation, the combustion facility must have a continuing supply of MSW. For that reason, the facility enters into long-term agreements with municipalities, and these agreements can lessen the flexibility of the community's solid waste management options.
- Even if the combustion facility generates electricity, the process wastes both energy and materials, unless it is augmented with recycling and recovery. A number of combustion facilities compete directly with recycling for burnable materials such as newspapers and represent a major impediment to recycling in some municipalities.

An Operating Facility. Let us look at the operation of a typical modern WTE facility. The facility might serve a number of communities or a larger metropolitan area. Servicing a population of a million or more, the plant receives about 3000 tons of MSW per day. The waste comes in by rail and truck, and the communities pay tipping fees (the costs assessed at the disposal site) ranging from \$15 to \$50 per ton. Waste processing is efficient: Overall, about 80% of the MSW is burned for energy, 12% is recovered, and 8% is landfilled. The process, pictured in Fig. 20-7, is as follows:

1. Incoming waste is first inspected, and obvious recyclable and bulky materials are removed.

2. Waste is then pushed onto conveyers which feed shredders capable of reducing the width of waste particles to 6 inches or less.
3. Strong magnets remove about two-thirds of ferrous metals for recycling before combustion.
4. The waste is then blown into boilers, where light materials burn in suspension, and heavier materials burn on a moving grate.
5. The boilers produce steam, which drives turbines for generating electricity.
6. After the waste is burned, the bottom ash is conveyed to a processing facility, where further separation of metals may occur in a process that recovers brass, aluminum, gold, copper, and iron. In some facilities, this process nets $1000 a day in coins alone!
7. Combustion gases are passed through a lime-based spray dryer/absorber to neutralize sulfur dioxide and other noxious gases and then through electrostatic precipitators that remove particles. The resulting waste stream is significantly lower in pollutants than would be emitted by an energy-equivalent utility based on coal or oil combustion.
8. The fly ash and bottom ash residues are put into landfills.

An appreciation for the impact of such a facility can be gained from looking at the outcome of a year's operation. In a year, 1 million tons of MSW is processed, 40,000 tons of metal is recycled, and 650,000 megawatt-hours of electricity is generated—the equivalent of more than 80 million gallons of fuel oil, and enough electricity to power 75,000 homes. All this comes from stuff that people have thrown away!

The advantages of combustion over landfilling are obvious; the major disadvantage is that some recycling and composting opportunities are missed. This drawback can be addressed at the collection point, as we will see when we examine the recycling option. In any event, certainly substantial recovery of recyclable materials—and energy—can be accomplished with a modern WTE facility.

Costs of Municipal Solid Waste Disposal

The costs of disposing of MSW are becoming prohibitive. Increasing costs are not just a result of the new design features of landfills; more and more, they reflect the expenses of acquiring a site and providing transportation. Tipping fees now exceed $100 a ton at some landfills, and the waste collector must recover this cost, as well as transportation costs to the site. Tipping fees

FIGURE 20-7
Schematic flow for the separation of materials and combustion in a typical modern waste-to-energy combustion facility.

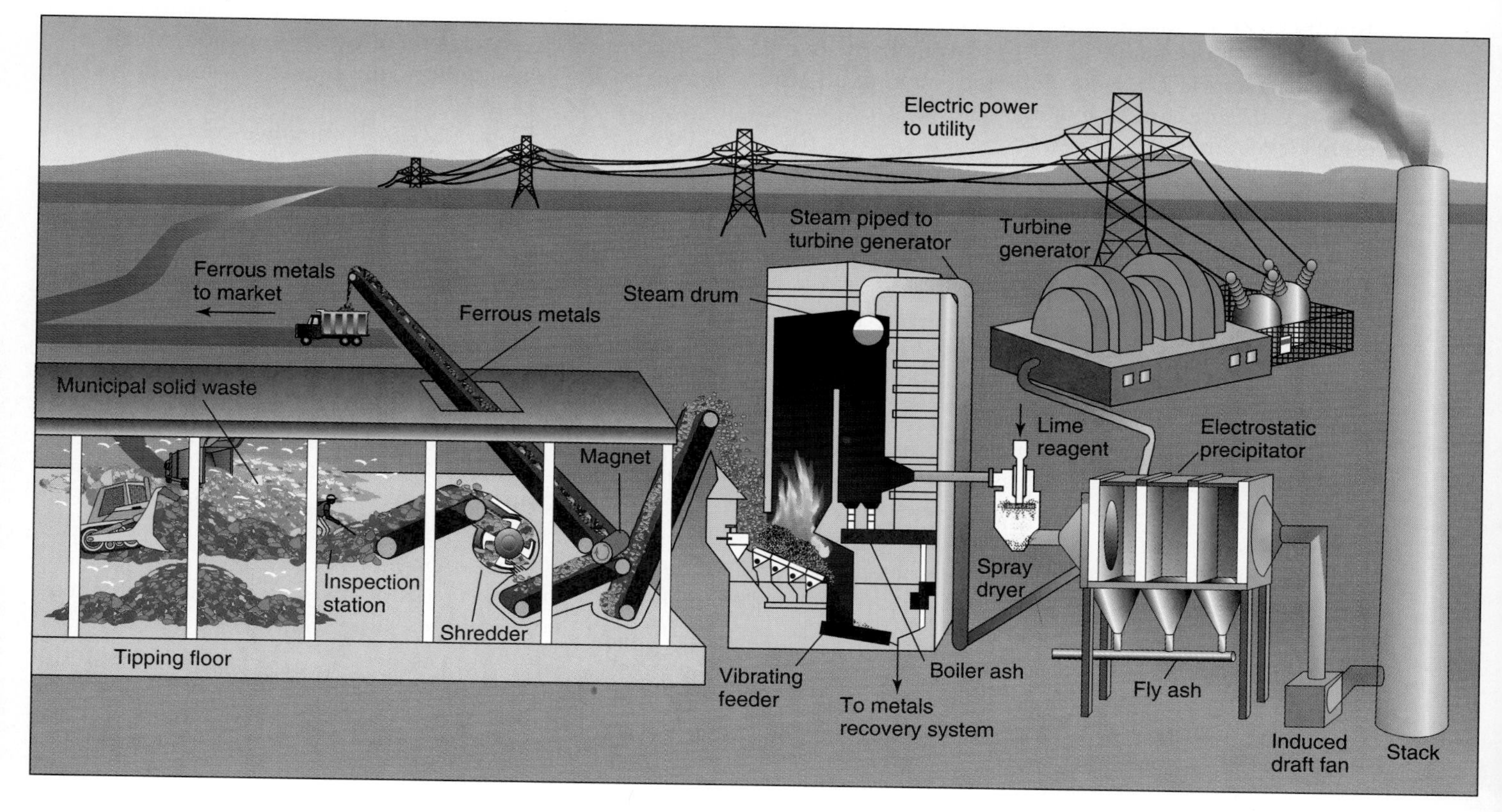

at some combustion facilities are no better; one WTE combustion facility in Saugus, Massachusetts, recently had to increase its tipping fee from $22 to $80 per ton to cover the cost of adding a mandated stack scrubber.

Getting rid of all trash is getting more expensive, and one sad consequence of this is illegal dumping. Some towns are charging up to $1.50 a bag for disposal (as we will see, this is a good thing!); it now costs $1 or more to get rid of an automobile tire and $30 or more to dispose of a refrigerator. Tires, refrigerators, yard waste, car parts, and construction waste are appearing all over the landscape. Institutions and apartments that operate with dumpsters are having to put padlocks on them in order to prevent their unauthorized use by people trying to avoid disposal costs. Many states have established a corps of environmental police, one of whose responsibilities is to track down midnight dumpers and bring them to justice.

Solutions

Reducing Waste Volume

The best strategy of all is to reduce waste at its source. This accomplishes two goals: It reduces the amount of waste that must be managed, and it conserves resources. We noted earlier that the increased amount of waste produced over the years is largely a result of changing lifestyles—which include, notably, the growing use of disposable products and excessive packaging. At the same time, however, many people are moving in the opposite direction, making an effort to reduce the amount of materials they discard. Several recent developments are as follows:

1. Concerned over the mass of disposable diapers in MSW, many families are switching to cloth diapers.
2. Environmentally concerned consumers have successfully pressured some producers to reduce their packaging. For example, the manufacturers of compact discs have agreed to reduce their unnecessarily large packages.
3. Lightening the weight of many items has reduced the amount of materials used in manufacturing. Steel cans are 60% lighter than they used to be; disposable diapers contain 50% less paper pulp, due to absorbent gel technology; and aluminum cans contain only two thirds as much aluminum as they did as 10 years ago.

An option that thus far has received too little attention is the potential for reducing the volume of waste by keeping products in use longer. Reusing items in their existing capacity is an efficient form of waste reduction. The use of returnable versus nonreturnable beverage containers is a prime case in point.

Returnable versus Nonreturnable Bottles. For many years, most beer and soft drinks were marketed by local bottlers and breweries in returnable bottles that required a deposit. Trucks delivered filled bottles to retailers and picked up empties to be cleaned and refilled. This procedure is efficient when the distance between producer and retailer is relatively short. As the distance increases, however, transportation costs become prohibitive because the consumer pays for hauling the bottles, as well as for the beverage. In the late 1950s, distributors, bent on expanding markets and growth, observed that transportation costs could be greatly reduced if they used lightweight containers that could be thrown out rather than shipped back. Thus, no-deposit, nonreturnable bottles and cans were introduced. The throwaway container is also an obvious winner for its manufacturers, who profit by each bottle or can they produce.

Through massive advertising campaigns promoting national brands and the convenience of throwaways, a handful of national distributors gained dominance during the 1950s and 1960s, and countless local breweries and bottlers were driven out of business—and with them went the returnable bottle. At the same time, nonreturnable bottle and can manufacturing grew into a multibillion-dollar industry.

The average person drinks about a quart of liquid each day. Given that there are 270 million Americans, this daily consumption amounts to some 25 billion gallons of liquid. That a significant portion of this volume should be packaged in single-serving containers that are used once and then thrown away is bizarre. It is difficult to imagine a more costly, wasteful way to distribute fluids.

Beverages in nonreturnable containers and those in returnable containers appear to be priced competitively on the market shelf, but this equality is seen to evaporate when one looks at the hidden costs of single-use containers. Nonreturnable containers constitute 6% of the solid waste stream in the United States and about 50% of the nonburnable portion; they also constitute about 90% of the nonbiodegradable portion of roadside litter. Broken bottles along roads, beaches, and park lands are responsible for innumerable cuts and other injuries, not to mention flat tires. Both the mining of the materials and the process used to manufacture these beverage containers create pollution. All of these are hidden costs that do not appear on the price tag, but we pay them with taxes for cleaning up litter, as well as with our injuries, flat tires, environmental degradation, and so on.

TABLE 20-1
States That Have Passed Bottle Laws

State	Year Passed	State	Year Passed
Oregon	1972	Iowa	1978
Vermont	1973	Massachusetts	1978
Maine	1976	Delaware	1982
Michigan	1976	New York	1983
Connecticut	1972	California	1991

In an attempt to reverse the trend, environmental and consumer groups have promoted *bottle laws*—laws that facilitate the recycling or reuse of beverage containers. Such laws generally call for a deposit on all beverage containers, both returnables and throwaways. Retailers are required to accept the used containers and pass them along for recycling or reuse.

Bottle laws have been proposed in virtually every state legislature over the last decade. In every case, however, the proposals have met with fierce opposition from the beverage and container industries and certain other special-interest groups. The reason for their opposition is obvious—economic loss to their operations—but the arguments they put forth are more subtle. The container industry contends that bottle laws will result in loss of jobs and higher beverage costs for the consumer. They also claim that consumers will not return the bottles and that the amount of litter will not diminish. In a 1996 media blitz, these same arguments were used in defeating an expansion of Oregon's successive bottle bill.

In most cases, the industry's well-financed lobbying efforts have successfully defeated bottle laws. However, some states—ten as of 1996—have adopted bottle laws of varying types despite industry opposition (Table 20-1). Their experience has proved the beverage and bottle industry's arguments false. More jobs are gained than lost, costs to the consumer have not risen, a high percentage of bottles are returned, and there is a marked reduction in the can and bottle portion of litter. In 1994, these ten states recycled 2 million tons of the total 4 million tons of glass recycled in the United States—imagine if every state had bottle-deposit laws!

A final measure of the success of bottle laws is continued public approval. Despite industry efforts to repeal bottle laws, no state that has one has repealed it. Repeated attempts have been made to get a national bottle law through Congress—to date, unsuccessfully. Opponents (the same ones that oppose state-level bottle bills) argue that such a law will threaten the newly won successes in curbside recycling, with some justification: Beverage containers represent the most important source of revenue in curbside recycling. However, since curbside recycling currently reaches only 15% of the U.S. population, a national bottle law will recover a much greater proportion of beverage containers. (States with bottle laws report 80% to 97% rates of return of containers.) A national bottle law would be labor intensive, employing tens of thousands of workers.

Other Measures.

Resale. Whenever items are reused rather than thrown away, the effect is a reduction in waste and better conservation of resources. In this respect, it is encouraging to see the growing popularity of yard sales, flea markets, consignment clothing stores and other not-new markets (Fig. 20-8).

Elimination of junk mail. We are all on bulk-mailing lists, because these lists are sold and shared widely; this guarantees that each of us will receive increasing volumes of advertising. To stay off such lists, simply inform mail-order companies and other organizations involved that you do not want your name and address shared. To get off the lists, you can write to the Mail Preference Service, Direct Marketing Association, 11 West 42nd Street, P.O. Box 3861, New York, NY 10163. This is a no-charge service that removes names from mailing lists.

An examination of our domestic wastes and their disposal reveals what a mammoth stream of material flows in one direction, from our resource base to disposal sites. Just as natural ecosystems depend on recy-

FIGURE 20-8
Volunteers preparing items for sale in a second-hand store. By reselling used clothes, such stores are reducing the volume of discarded waste. (Charles Gupton/Uniphoto Picture Agency.)

FIGURE 20-9
Bales of plastic bottles in a recycling center. (Alan L. Detrick/Photo Researchers, Inc.)

cling nutrients, the continuance of a sustainable technological society will ultimately depend on our learning to recycle or reuse not only nutrients, but virtually all other kinds of materials as well.

The Recycling Solution

In addition to reuse, recycling is another obvious solution to the solid waste problem. More than 75% of MSW is recyclable material. Of course, many people have been advocating recycling for a long time now, and various groups and individuals have been recycling paper, glass, and aluminum cans on a small scale for decades. There is an abundance of alternatives for reprocessing various components of refuse, and people are coming up with new ideas and techniques all the time. A few of the major established techniques, together with current percentages of their recovery by recycling, are as follows:

- Paper (54% recovery) can be repulped and reprocessed into recycled paper, cardboard, and other paper products; finely ground and sold as cellulose insulation; or shredded and composted.
- Glass (20% recovery) can be crushed, remelted, and made into new containers or crushed and used as a substitute for gravel or sand in construction materials such as concrete and asphalt.
- Some forms of plastic (2.2% recovery) can be remelted and fabricated into carpet fiber, outdoor wearing apparel, irrigation drainage tiles, and sheet plastic (Fig. 20-9).
- Metals can be remelted and refabricated. Making aluminum (38% recovery) from scrap aluminum saves up to 90% of the energy required to make aluminum from virgin ore. In addition, these ores are imported and are part of the mounting U.S. trade deficit. National recycling of aluminum saves energy, creates jobs, and reduces the trade deficit.
- Food wastes and yard wastes (leaves, grass, and plant trimmings—4.2% recovery) can be composted to produce a humus soil conditioner.
- Textiles (negligible recovery) can be shredded and used to strengthen recycled paper products.
- Old tires (12% recovery) can be remelted or shredded and incorporated into highway asphalt. (Asphalt can contain as much as 20% tires.)

There are two levels of recycling: **primary** and **secondary**. *Primary recycling* is a process in which the original waste material is made back into the same material—for example, newspapers recycled to make newsprint. In *secondary recycling*, waste materials are made into different products, which may or may not be recyclable—for example, cardboard from waste newspapers.

Recycling is both an environmental and an economic issue. Many people are motivated to recycle because of environmental concern, but the use of recycled materials is strongly driven by economic factors. The primary items from MSW currently being heavily recycled are cans (aluminum and steel), bottles, plastic containers, and newspapers.

Municipal Recycling. Recycling is probably the most direct and obvious way most people can become involved in environmental issues. Its popularity is on the increase; to date, at least 39 states have passed recycling mandates. EPA sources report that only 6.7% of MSW was recycled in 1960, vs. 24% in 1994. The figure is still rising in 1998. The appeal of recycling is obvious: If you recycle, you save some natural resources from being used (trees, in the case of paper), and you prevent landfills from becoming "landfulls." There is a great diversity of approaches to recycling in municipalities, from recycling centers requiring residents to drive miles to recycle to curbside recycling with sophisticated separation (Fig. 20-10). The most successful programs have the following characteristics:

1. There is a strong incentive to recycle, in the form of direct charges for general trash and no charge for recycled goods.
2. Recycling is not optional; mandatory regulations

FIGURE 20-10
Curbside recycling pickup by waste hauler using a special truck. However, this approach is reaching only 15% of the U.S. population.(Susan Greenwood/ Gamma-Liaison, Inc.)

are in place, with warnings and sanctions for violations.

3. Residential recycling is curbside, with free recycling bins distributed to households.
4. Recycling goals are clear, challenging, and feasible. Some percent of the waste stream is targeted, and progress is followed and communicated.
5. A concerted effort is made to involve local industries in the recycling process.
6. The municipality employs an experienced and committed recycling coordinator.

Municipalities experience very different recycling rates; New York City, for example, has achieved only 5% recycling, while Los Angeles diverts more than 21% of its MSW to the recycling process. The city with the highest rate, 48%, is Seattle.

Recycling has its critics, who base their arguments primarily on economics. If the costs of recycling (from pickup to disposal of recycled components) are compared with the costs of combustion or placing waste in landfills, recycling frequently comes out second best. Markets for recyclable materials fluctuate wildly, and residents often end up subsidizing the recycling effort. The shortfall between recycling costs and market value ranging from $20 to $135 a ton for some recyclables (items such as green glass and colored plastic bottles). In some cases (newspaper, glass), prices are controlled by the industry of origin. Competition between landfills and combustion facilities often lowers tipping fees, creating an even greater disincentive to recycle. Thus, critics of current recycling practices argue that, unless recycling pays for itself through the sale of recovered materials, it should not be done. However, the critics ignore the subsidies that make the use of virgin materials less expensive than recycled materials. Also, garbage collection is big business and those involved see recycling as cutting their trash business.

In spite of these economic considerations, the demonstrated support for recycling programs is strong; experience has shown that at least two thirds of households will recycle if presented with a curbside pickup program. Even when people understand that it costs more to recycle, there is strong support for the effort. (See “Ethics” box, p. 522.)

Paper Recycling. By far the most important recycled item is newspapers, because of their predominance in the waste stream—at least 13% of MSW. It is a simple matter to tie up or bag household newspapers, and the amount recovered by recycling is increasing dramatically. For the first time in our history, more paper is being recycled than put in landfills—almost 40 million tons a year to recycling, vs. 34 million tons to landfills. Since more than 25% of the trees harvested in the United States are used to make paper, recycling paper obviously saves trees. Depending on the size and type of trees, one meter stack of newspapers equals the amount of pulp from one tree.

As with the recycling of other materials, paper recycling is highly dependent on the existence of a market for the recycled products. This involves cost considerations, from the point of collection of the wastepaper to the point of use by consumers. For example, there are all types of paper, and recycling mills want specific grades; gross mixtures have little market appeal. So the processor of the recycled paper must perform some separation and packaging for specific uses.

After the wastepaper is incorporated into a final product, the market becomes a critical factor. Is there a demand for recycled paper? Several years ago, much of the recycled paper was rough and offcolor. In order to

Ethics

Recycling—the Right Thing to Do?

The town meeting is an institution in Massachusetts. Held once a year, it is a forum for decision making on the town's budget and all sorts of ancillary business (such as deciding on a leash law, changing zoning requirements, and accepting new streets). Originally, the whole town was expected to attend; nowadays, 10% of the voting population is considered a good turnout. Microphones are set up in the auditorium, and anyone in the town can express her or his opinion on matters on the docket.

In recent years, Massachusetts towns have taken a strong turn toward fiscal stringency as local property taxes have continued to escalate. Many school programs have been radically cut, town personnel have been let go, and many items on town budgets have been voted down. In the face of such stringency, it is encouraging that in almost every town and city in Boston's north shore area, recycling programs have been continued despite cutbacks in other programs. By 1994, almost all north shore towns had some kind of recycling program. Very often, the program was proposed by a nonpaid town official acting on behalf of grassroots citizens' groups. Arguments pro and con were aired at the town meeting, and, in town after town, recycling was adopted by overwhelming votes, even though at first the programs did not pay for themselves. When asked why townspeople were in favor of recycling, one town official stated that it was because "it's the right thing to do."

Thankfully, recycling is starting to pay off. For example, the town of Hamilton contracts for curbside pickup of unsorted MSW and recycled materials. Of the 3500 tons of waste picked up in 1995, 24% was recycled or composted. Even though recycling pickup is more expensive ($107/ton verses $35/ton for unsorted waste), the tipping fee for the waste—$89/ton—makes recycling a winner, since there is no tipping fee for the recycled materials. The town saved $9800 through its recycling program.

One town, Topsfield—clearly a leader in the trend—recently celebrated its 20th year of a recycling program. For towns that have not yet initiated recycling, the handwriting is on the wall. A state master plan for MSW was issued in 1990, and one goal of the plan is a 46% recycling rate by 2000. In the plan, leaf waste, large appliances, and tires were phased out of landfill dumping statewide in 1991; the ban extended to other yard waste, metals, and glass in 1992 and, in 1994, recyclable paper and plastics. Obviously, the recycling trend is not simply a matter of virtuous decision making on the part of the townspeople; they see the state mandates coming and are acting wisely. However, it should be noted that recycling in Massachusetts began as a grassroots movement. It has reached the state level and is now working back down to catch those towns that have been dragging their feet.

Recycling is one of the most obvious ways for people to demonstrate their concern for the environment. And perhaps the reason it got started is the most basic one: because it's the right thing to do. Do you agree?

---

guarantee a market for the product, many organizations—including state and federal agencies—required purchasing and using paper with recycled content. The technology for producing high-quality paper from recycled stock has improved greatly, to the point where it is virtually impossible to distinguish it from "virgin" paper. Although the mandates for "closing the loop," as the practice of creating markets for recycled products is called, are important, they may soon disappear as paper manufacture makes the use of recovered paper a standard practice.

There is often some confusion in what is meant by "recycled paper"; the key is the amount of *postconsumer* recycled paper in a given product. Much paper is "wasted" in manufacturing processes, and this is routinely recovered and rerouted back into processing—and called "recycled paper" when it is. Thus, the total recycled content of a paper can be 50%, although the actual postconsumer amount recovered by recycling might be only 10% of the total paper.

The market for used newspapers has fluctuated greatly over the past several decades. During the late 1980s, the market was saturated and municipalities often had to pay to get rid of newspapers. In 1995, discarded papers were so valuable—up to $160 a ton—thieves were stealing them off the sidewalks before the recycling trucks could pick them up. A year later, as more recycling programs came on line, the market collapsed again and many cities were once again paying as much as $25 a ton to have the newspapers hauled away (this is still less expensive than the cost of landfilling.)

There is a lively international trade as well in used paper (Fig. 20-11). Forest-poor countries in Europe and Asia purchase wastepaper from the United States and other industrial countries in the Northern Hemisphere, where there continues to be a surplus of such paper. The largest importer is Taiwan (almost 2 million tons per year). Taiwan also boasts the highest percentage of reused paper content in its paper: 98%. The United States is at a much lower 33%.

FIGURE 20-11
Bales of wastepaper being loaded for shipment overseas. (© Griffin/The Image Works.)

Plastics Recycling. There are good reasons why plastics have a bad reputation in the environmental debate. They have many uses that involve rapid throughput—for example, packaging, bottling, the manufacture of disposable diapers, and incorporation into a host of cheap consumer goods. Plastic production has increased 10% a year for the past three decades. Plastics are conspicuous in MSW and litter, the final insult being that most trash is disposed of in plastic bags manufactured just for that purpose. All of this wouldn't be so objectionable if it weren't for the fact that plastics do not decompose in the environment. Walk any beach or roadside, and you will encounter a wonderful variety of plastic items. Put plastics in landfills, and they will undoubtedly delight archeologists hundreds of years from now with their remarkable testimony to our technological prowess. They are virtually eternal, for the good reason that no microbes (or any other organisms) are able to digest plastics. Thus, when the possibility appeared of recycling at least some of the plastic in products—liquid containers—there was enthusiasm in the environmental community.

If you look on the bottom of a plastic container, you will see a number inside the little triangle of arrows that has come to represent recycling, as well as some letters (Fig. 20-12). The reason for the number is that there are so many different kinds of plastic polymers; not all of them are recyclable, and if any are recycled, they must be sorted first. The two recyclable plastics most commonly in use are high-density polyethylene (HDPE), code 2, and polyethylene terephthalate (PETE), code 1. In the recycling process, the plastics must be melted down and poured into molds, and unfortunately, some contaminants from the original containers may carry over. This makes it difficult to use the plastic again for food containers; thus, the uses for the recycled plastic are necessarily restricted. However, some new uses for recycled plastic are appearing, with more in the development stage: PETE, for example, is turned into carpet fiber and fill for outdoor wearing apparel, and HDPE becomes irrigation drainage tiles, sheet plastic, and, appropriately, recycling bins.

Critics of plastic recycling point out that if the process were purely market driven, it wouldn't happen. Recovering the plastic is more costly than starting from scratch—that is, beginning with petroleum derivatives—and manufacturers are getting involved mainly because environmentally concerned consumers demand it. Industry-supported critics also point out that plastics in landfills create no toxic leachate or dangerous biogas; also, plastics in combustion facilities burn wonderfully hot and leave almost no ash. If, however, economics were our only concern, we would have to scrap far more than plastics recycling as a means of remedying environmental problems. Perhaps we can conclude that, even though plastic recycling has its problems, continued attention to plastics in MSW will keep us moving in the right direction. As oil reserves continue to dwindle, the petroleum corporations will find it increasingly difficult to promote a once-through waste stream for plastics.

Composting

An increasingly popular way of treating yard waste and food scraps (almost 26.3% of MSW) is composting. Recall that composting involves the natural biological decomposition (rotting) of organic matter in the presence of air. Composting can be carried out in a backyard by homeowners, or it can be promoted by municipalities as a way of dealing with a large fraction of MSW. Composting has clear economic and environmental benefits over putting waste in landfills and over combustion, if the yard waste can be collected and processed independently of collecting trash. The end product is a residue of humuslike material, which can

FIGURE 20-12
Recycling symbol (representing high-density polyethylene) on the bottom of a plastic bottle. (Leonard Lessin/Peter Arnold, Inc.)

be used as an organic fertilizer and soil builder. Composting is one method of treating sewage sludge, as described in Chapter 13.

A number of companies have entered into the business of selling equipment or of building and running facilities for composting refuse. Yard waste composting facilities are on the increase: Some 2000 were added between 1990 and 1995. On a large scale, composting involves laying down the wastes in windrows up to 12 feet high and 24 feet wide. The microbes and detritus feeders (worms, grubs, etc.) will decompose the organic matter in the compost heap and greatly reduce the volume of waste. The process is speeded by aeration, which is accomplished by turning the material with large mechanical forks.

Public Policy and Waste Management

Management of MSW was entirely under the control of local governments until quite recently. It is still largely under local management, but state and federal legislation addresses some aspects of MSW disposal. Regulation at higher levels has taken some of the disarray out of the management scene, but it is fair to say that towns and cities still have the basic responsibility for managing their MSW. 1996 was one of the worst years for prices of recovered paper, aluminum, glass, and metal. Such fluctuations and market manipulations demand stronger public policies and regulation of MSW. Also, 1996 was a year of inaction in Congress and state legislatures. A new California law actually exempts all plastic food and cosmetic containers (50% of that waste stream) from the state's recycling laws!

The Regulatory Perspective

At the federal level, the following legislation has been passed:

- The first attempt of Congress to address the problem was the Solid Waste Disposal Act of 1965. The legislation gave jurisdiction over solid waste to the Bureau of Solid Waste Management, but the agency's mandate was basically financial and technical rather than regulatory.
- With the creation of the EPA in 1970, the Resource Recovery Act of 1970 gave jurisdiction to the EPA and directed attention to recycling programs and other ways of recovering resources in MSW. The act also encouraged the states to develop some kind of waste management program.
- The passage of the Resource Conservation and Recovery Act (RCRA) of 1976 saw a more regulatory ("command and control") approach to MSW, as the EPA was given power to close local dumps and set regulations for landfills. Combustion facilities were covered by air pollution and hazardous waste regulations, again under the EPA's jurisdiction. The RCRA also required the states to develop comprehensive solid waste management plans.
- The Superfund Act of 1980 (described in Chapter 14) addressed abandoned hazardous waste sites throughout the country, many of which (41%) are old landfills.
- The Hazardous and Solid Waste Amendments of 1984 gave the EPA greater responsibility to set solid waste criteria for all hazardous waste facilities. Since even household waste must be assumed to contain some hazardous substances, this meant that the EPA had to determine all landfill and combustion criteria more closely.

Largely in response to these federal mandates, the states began to put pressure on local governments to develop integrated waste management plans (see next section), with goals for recycling, source reduction, and landfill performance. State solid waste management is comprehensive and ambitious in many states, particularly in the Northeast, where landfills have been closing. States are setting goals and policies for recycling, one significant manifestation of which is the materials recycling facility (see "Earth Watch" box, p. 525). There are now more than 120 of these facilities, mostly in the Northeast. In a unique move, Rhode Island has created the Solid Waste Management Corporation, a public corporation given the responsibility for managing all aspects of the state's MSW.

Integrated Waste Management

It is not necessary to fasten onto just a single method of waste handling. Source reduction, waste-to-energy combustion, recycling, landfills, and composting all have roles to play in waste management. Different combinations of these options will work in different regions of the country. A system of having several alternatives in operation at the same time is called *integrated waste management*. Let us examine the available MSW management options, with a view toward developing recommendations that make environmental sense—that is, toward moving in a more sustainable direction.

Waste Reduction. Many observers have called the late-20th-century United States a "throwaway society," and we probably deserve that label. Setting the world record for per capita waste production, along with per capita energy consumption, is not something for which

Earth Watch

Regionalized Recycling

Currently, most recycling is on a town-by-town basis. Either the towns or their waste contractors must find markets for the recycled goods: cans, bottles, newspapers, and plastics. This problem has kept many municipalities from getting into recycling. The solution? A regionalized materials recycling facility (MRF), referred to in the trade as a "murf." Here's how the state-owned MRF in Springfield, Massachusetts, works.

Basic sorting takes place when waste is collected, either by curbside collection or by town recycling stations (sites where townspeople can bring wastes to be recycled). The waste is then trucked to the MRF and handled on three tracks—one for metal cans and glass containers, another for paper products, and a third for plastics. The materials are moved through the facility by escalators and conveyor belts, tended by workers who inspect and do further sorting. The objective of the process is to prepare materials for the recycled goods market. Glass is sorted by color, cleaned, crushed into small pebbles, and then shipped to glass companies, where it replaces the raw materials that go into glass manufacture—sand and soda ash—and saves substantially on energy costs. Cans are sorted, flattened, and sent either to detinning plants or to aluminum processing facilities. Paper is sorted, baled, and sent to reprocessing mills. Plastics are sorted into four categories, depending on their color and type of polymer, and then sold.

The facility's clear advantages are its economy of scale and its ability to produce a high-quality end product for the recycled materials market. Towns know where to bring their waste, and they quickly become familiar with the requirements for initial sorting during collection and transfer.

Currently, there is only one MRF in Massachusetts, built with state funds at a cost of $5 million and operated by the state Department of Environmental Protection. Several more are on the drawing board. After its first year in operation, the Springfield MRF was declared a success. It took in 42,886 tons of materials from 91 towns representing a population of 750,000. Depending on the community, the process has diverted 22% to 50% of the waste stream from landfills and incinerators, and saved the towns millions of dollars in disposal costs. The facility has attracted recycled wastes from Connecticut and New York while those states are setting up their own regional recycling centers.

There were some unanticipated problems at the Springfield facility during its first year of operation. Several towns reduced their trash so much, that they are not fulfilling their contractual obligations to a regional combustion facility. Landfill revenues have declined, forcing landfill operators to lower their fees in order to attract more haulers. These appear to be temporary problems that will be worked out in time, however. The state is committed to recycling almost half of MSW and combusting most of the rest by the year 2000. In theory, landfills will operate mainly as recipients of combustion ash and materials that can't be either recycled or burned.

Very likely, the future of regional recycling will be in the private sector. In order for this to work, however, the recycling market will have to be substantially strengthened. All indications point to such facilities as the wave of the future.

we should be proud (Fig. 20-13). How can we turn this around? Suppose we all had to dispose of our own trash and garbage. Very likely, that would create the strongest incentive to reduce our waste. We discussed the closed material system sealed within Biosphere 2 in Chapter 3. Imagine their challenge to recycle and reuse 100% of the material waste stream! This points to the fact that *changing personal habits* represents a powerful way to affect the MSW scene. True management of MSW begins in the home. However, in a free society like ours, changing people's behavior is always a challenge. Hopefully, if people care enough about their environment, then when they are given an opportunity to do something positive, they will do it. Thus, the role of government will be to create incentives and opportunities for people to exercise concern by doing what they can to bring about a decrease in the per capita MSW.

As a very important step in this direction, we should stop subsidizing garbage disposal. Instead of paying for trash collection and disposal through local taxes, communities should levy curbside charges for all unsorted MSW—say, $1 or more per container (Fig. 20-14). This practice is on the increase, and communities adopting it report reductions of 25% to 45% in the MSW stream. Recycling in Seattle, Washington, led to increasing the MSW that was recycled from 24% in 1988 to the current 48% after a per-bag fee was initiated.

Another policy for bringing about waste reduction is to establish a retail surcharge on goods that are troublesome to discard or recycle—tires, refrigerators, housing shingles, etc. This charge should reflect the cost of disposing of the item; Idaho, for example, adds a $1 surcharge to the cost of automobile tires. Of course, it would be important not to charge consumers again at the point of disposal! If put into practice, this policy would be an incentive to buy more durable

FIGURE 20-13
The throwaway society—the production of junk and accelerated absolescence! (Reprinted with permission of King Features Syndicate.)

goods and to refrain from buying some that are truly discretionary. Such a policy might also go a long way toward reducing midnight dumping along roadsides.

Waste Disposal. No human society can avoid generating some waste. There will always be MSW, no matter how much we reduce, reuse, and recycle it. Integrated waste management will require providing several options for waste disposal. Most experts see an increase in the percentage of waste going to WTE combustion facilities, a decrease in the percentage going to landfills, and a rise in recycling.

It will be important to break the gridlock at local and state levels on the siting of new landfills and WTE combustion facilities. Landfills will still be needed, although they should last longer than in the past. Policymakers have long known about the landfill shortage, but have opted for short-term solutions with the lowest political cost. One result of this is the long-distance hauling of MSW by truck and rail, described earlier. If regions and municipalities are required to handle their own trash locally, they will find places to site landfills, regardless of NIMBY considerations. Of course, new landfills must use the best technologies: proper lining, leachate collection and treatment, groundwater monitoring, biogas collecting, and final capping. If these are employed, a landfill will not be a health hazard. Also sites should be selected with a view to some future use that is attractive.

Another policy goal should be to encourage more WTE combustion of MSW. Again, the technologies employed should be the best available, and again, if these are used, there should be no significant threat to human health. This option seems to be the best way to

FIGURE 20-14
Pay-as-you-go trash pickup. Consumers purchase the empty bags for a set price, and only these bags are picked up by the trash hauler. (Courtesy Bob Fiore, City of Worcester, MA.)

deal with mixed waste that is nonrecyclable. Much of the energy content of the waste is converted to electricity, metals are recovered, and the waste is greatly reduced in volume.

Recycling and Reuse. Recycling is certainly the wave of the future; it should not, however, be pursued in lieu of waste reduction and reuse. The move toward more durable goods is an overlooked and underutilized option.

Many states have passed mandatory recycling laws. Such laws require that each region or municipality, under threat of loss of state funds, recycle a certain percentage of its refuse by a certain date. Massachusetts, for example, has adopted a goal of recycling 46% of all of its MSW by the year 2000.

Banning the disposal of recyclable items in landfills and combustion facilities makes very good sense. Many states have incorporated this regulation into their management program; for example, Massachusetts phased out yard waste in 1991, metals and glass in 1992, and recyclable papers and plastic in 1994. Landfill and combustion facility operators now must conduct random truck searches and are authorized to turn back trucks with significant amounts of banned items.

Enacting a national bottle-deposit law would make a giant stride toward the reuse and recycling of beverage containers and would also greatly reduce roadside litter in states lacking bottle laws.

Finally, closing the "recycling loop" remains the most significant action to be taken by governments to encourage recycling. A number of states have opted for one or more of the following approaches: (1) minimum postconsumer levels of recycled content for newsprint and glass containers; (2) requirements that state purchases of goods include recycled products, even if they are more expensive than "virgin" products; (3) requirements that all packaging be reusable or made (at least partly) of recycled materials; (4) tax credits or incentives that encourage the use of recycled materials in manufacturing; (5) assistance in the development of recycling markets.

Review Questions

1. List the components of municipal solid waste.
2. Trace the historical development of refuse disposal. What method of disposal is now most common in the United States?
3. What are the major costs and limitations of placing waste in landfills?
4. Outline the EPA's latest regulations for the construction of new landfills.
5. Why are landfills not a sustainable option for solid waste disposal?
6. What are the advantages and disadvantages of combustion?
7. How can the volume of waste in the United States be reduced?
8. Define primary and secondary recycling.
9. Discuss the characteristics of successful recycling programs.
10. Which waste materials may be composted?
11. What laws have been adopted by the federal government to control solid waste disposal?
12. What is integrated waste management?
13. What are some options for solid waste management that would encourage sustainability?

Thinking Environmentally

1. Compile a list of all the plastic items you used and threw away this week. Consider how you could reduce the length of this list.
2. How and where does your school dispose of solid waste? Is a recycling program in place? How well does it work?
3. Suppose your town planned to build a combustion facility or landfill near your home. Outline your concerns, and explain your decision for or against this site. In making your decision, consider the NIMBY problem.
4. Does your state have a bottle bill? Is it effective? If you live in a state without such a bill, explore the politics that have prevented the bill from being adopted.
5. If you live near the ocean or a large lake, walk along the beach for 20 minutes, and collect all of the items that are clearly not natural. Where did these come from? What is their volume per 100 yards of beach?

CHAPTER 21

Energy Resources: The Rise and Fall of Fossil Fuels

Key Issues and Questions

1. The development of modern civilization has depended greatly on the development of energy sources. How have the three primary fossil fuels been harnessed in recent history?
2. Much of total energy use is devoted to generating electrical power. How are fossil fuels coupled to electrical power, and what are the major environmental impacts of these processes? What are the problems of future increase in large centralized electrical-generation power plants?
3. What are the major categories of primary energy use and the energy sources that match those uses? How does an analysis of end-use energy demand help us design energy supplies?
4. The United States now imports more than 50% of the crude oil it uses. What led to this dependency on foreign oil? What are the problems of acquiring and shipping this quantity of oil?
5. Oil shortages and oil gluts have rocked the world economy in recent years. What are the reasons for these violent shifts, and what are their long-term consequences?
6. Alternative fossil fuels are available in the United States. Why are these not being put to use to meet transportation needs at present?
7. Sustainable energy options are clearly desirable for the development of a sustainable future. What are some options that meet our needs for energy with least economic environmental consequences.

In the 1970s the United States and other developed countries weathered a number of energy crises, episodes during which most gas stations put up crude signs saying, "Closed, No Gas." At the few stations having gasoline to sell, lines of cars stretched for blocks; some people slept overnight in their cars to maintain their places in line (Fig. 21-1). Highways became less congested as people conserved the gas in their tanks, uncertain about when they might find more gas for sale. Even businesses and industries that were not directly hit with fuel shortages suffered as the country's transportation system gradually slowed down. Motels, restaurants, and shopping malls suffered from a lack of customers. Thus, for a while the public received a harsh reminder that fossil fuels, particularly crude oil from which gasoline is refined, are a vulnerable (and unsustainable) energy source. At the same time, the second principle of sustainability—that sunlight is an everlasting, nonpolluting, nondepletable source—came into prominence.

Off loading of crude oil tanker in Louisiana. Over 50% of U.S. crude oil consumed must now be imported. [Photo by Craig Hammell/The Stock Market.]

Eventually, the panic abated; and here we are, more than two decades later, still using fossil fuels as if the supply were infinite. Furthermore, installation of solar energy systems and other renewable-energy systems (beyond a small vanguard of enthusiasts and some demonstration projects) seems to be languishing following the elimination of government incentives for research and development in the 1980s. Is it, as some conservatives maintain, only a myth that we are running out of fossil fuels? Is the use of fossil fuels sustainable after all? Our objective in this chapter is to gain an overall understanding of the sources of energy that support our transportation, homes, and industry and of their relative sustainability. Further, where we find energy supplies wanting, we consider options for the future.

Energy Sources and Uses

Try to imagine what life would be like without all the energy we use. There would be no transportation beyond walking, horseback, or boats powered by paddles or wind. There would be no home appliances, air conditioners, or lighting or communications equipment. There would be little manufacturing beyond what could be done by hand, on looms or on potters' wheels. Farming would depend on hand labor and on draft animals such as horses and oxen. In short, anything beyond a primitive existence is irrevocably tied to harnessing energy sources. Let's look at this matter in a little more detail.

Harnessing Energy Sources: An Overview

Throughout human history, the advance of technological civilization has been tied to the development of energy sources (Fig. 21-2). In early times (and even today in less-developed regions), the major energy source was muscle power. Some people lived in relative luxury by exploiting the labor of others—slaves, indentured servants, and minimally paid workers. Human labor was supplemented to some extent by the work of domestic animals for agriculture and transportation, by water power and wind power for milling grain, and by the sun. However, these power sources are, of course, limited by geography and climate.

FIGURE 21-1
Gas "crisis" of 1973. Because imports of foreign oil were curtailed, supplies fell below demand. The shortfall was only a few percent, but the public perceived it as a crisis, panicked, and lined up—in some cases for miles—to get gas. What will happen when even more severe shortages occur in the future? (Robert McElroy/Woodfin Camp & Associates.)

FIGURE 21-2
The development of energy sources has in large part supported the development of civilization.

By the early 1700s, inventors had already designed many kinds of machinery. The limiting factor was a continual power source to run them. The breakthrough that launched the Industrial Revolution was the development of the steam engine in the late 1700s. In a steam engine, water is boiled in a closed vessel to produce high-pressure steam, which pushes a piston back and forth in a cylinder. Through its connection to a crankshaft, the piston turns the drivewheel of the machinery. Steam engines rapidly became the power source for steamships, steam shovels, steam tractors, steam locomotives, and stationary engines to run sawmills, textile mills, and virtually all other industrial plants.

At first, the major fuel for steam engines was firewood. Then, as demands for energy increased and firewood around industrial centers became scarce, coal was substituted. By the end of the 1800s, coal had become the dominant fuel and remained so into the 1940s. In addition to being used as fuel for steam engines, coal was widely used for heating, cooking, and industrial processes. In the 1920s, coal provided 80% of all energy used in the United States, with similar percentages in other industrializing countries.

Though coal and steam engines powered the Industrial Revolution that greatly improved life for most people, they had many drawbacks. The smoke and fumes from the numerous coal fires made air pollution in cities far worse than anything seen today. Writers have recorded that often they could not see as far as a city block away because of the smoke. Coal is also notoriously hazardous to mine and dirty to handle, and burning coal results in large quantities of ashes that must be removed. As for steam engines, because of the size and bulk of the boiler, the engines were heavy and awkward to operate (Fig. 21-3). Often the fire had to be started several hours before the engine was put into operation in order to heat the boiler sufficiently.

In the late 1800s, simultaneous development of three technologies—the internal combustion engine, oil-well drilling, and the refinement of crude oil into gasoline and other liquid fuels—combined to provide an alternative to steam power. The replacement of coal-fired steam engines and furnaces with petroleum-fueled engines and oil furnaces was an immense step forward in convenience. Also, air quality was greatly improved because cities were gradually rid of the smoke and soot from burning coal. (It was only in the 1960s, with the tremendous proliferation of cars, that pollution from gasoline engines became a problem.) Further, the gasoline internal combustion engine provides a valuable power-to-weight advantage that allowed rapid advances in technology. A 100-horsepower gasoline engine weighs but a tiny fraction of what a 100-horsepower steam engine and its boiler weigh, and jet engines have an even greater power-to-weight ratio. Automobiles and other forms of transportation would be cumbersome, to say the least, without this power-to-weight advantage, and airplanes would be impossible.

Replacement of one energy technology with another is a very gradual process, however, because it is most cost-effective to use existing machinery until it wears out before replacing it. It was not until the late 1940s that crude oil surpassed coal to become the dominant energy source for the United States. Since then, however, its use has continued to grow. Crude oil currently provides about 40% of the total U.S. energy demand. The story is similar throughout the rest of the world, although the timing of events differs. We have noted that many poor regions of the world remain dependent on firewood. Coal is still the dominant fuel used in the countries of Eastern Europe and in China.

Development throughout the world, so far as it has occurred since the 1940s, has been largely predicated on technologies that consume gasoline and other fuels refined from crude oil. Thus, oil is the mainstay not only for the United States but for most other countries as well, both developed and developing. Coal has not passed from the picture however; it has become a major source of energy for generating electrical power, as we discuss shortly. In this role, coal currently provides about 22% of the total U.S. energy fuel input.

Natural gas, the third primary fossil fuel (the other two being oil and coal), is found in association with oil

FIGURE 21-3
Steam-driven tractors marked the beginning of the industrialization of agriculture. Today, tractors half the size do 10 times the work and are much easier to operate. (Tony Wells/Mason Dixon Historical Society, Inc.)

or is found during the search for oil. As it is largely methane, which produces only carbon dioxide and water as it burns, natural gas burns more cleanly than oil; thus, in terms of pollution, it is a more desirable fuel. Despite the obvious fuel potential of natural gas, at first there was no practical way to transport it from wells to consumers. Any gas released from oil fields was (and in many parts of the world still is) flared—that is, vented and burned in the atmosphere, a tremendous waste of valuable fuel. Gradually the United States constructed a network of pipelines connecting wells and consumers. With the completion of these pipelines, the use of natural gas for heating, cooking, and industrial processes escalated rapidly because of its cleanliness, convenience (no storage bins or tanks are required on the premises), and relatively low cost. Currently, natural gas provides about 25% of U.S. energy demand.

Thus, three fossil fuels—crude oil, coal, and natural gas—provide 83% of U.S. energy fuel input. The remaining 17% is provided mostly by nuclear power, water power, and renewables, which, along with coal, are used for the generation of electrical power. This changing pattern of energy sources in the United States is shown in Fig. 21-4. The picture is similar for other developed countries of the world, although the percentages differ somewhat depending on the energy resources of any given country.

Electrical Power Production

A considerable portion of total energy used is electrical power. Electrical power is called a **secondary energy source** because it depends on a **primary energy source** (coal or water power, for instance) to turn the generator.

Generators were invented in the nineteenth century. In 1831, the English scientist Michael Faraday discovered that passing a coil of wire through a magnetic field causes a flow of electrons—an electrical current—in the wire. An electric generator is basically a coil of wire that rotates in a magnetic field or that remains stationary while a magnetic field is rotated around it (Fig. 21-5). This seems like cost-free energy until we consider the First and Second Laws of Thermodynamics (Chapter 3). The phenomena described by these important energy laws prevent us from getting something for nothing, or even from breaking even. As the current flows in the wire, it creates a new magnetic field that is in opposition to the first and thus resists the movement. Also, some of the energy created is converted to thermal heat energy and lost. For every 100 calories of electrical energy produced, more than 100 calories of primary energy must be expended in turning the generator. This energy-losing proposition is justified only because electrical power is far more useful than coal, for example. Therefore, it is worth burning 300 calories of coal to obtain 100 calories of electrical power, a typical conversion ratio for electric generating power plants.

In the most widely used technique for generating electrical power, a primary energy source is employed to boil water, creating high-pressure steam that drives a **turbine**—a sophisticated "paddle wheel"—coupled to the generator (Fig. 21-6a). The combined turbine and generator are called a **turbogenerator**. Any primary energy source can be harnessed for boiling the water; coal, oil, and nuclear energy are most commonly used at present, but burning refuse, solar energy, and geothermal energy (heat from Earth's interior) may be more widely used in the future.

In addition to steam-driven turbines, gas and water (hydro-) turbines are also used. In a gas turbogenerator, the high-pressure gases produced by the combustion of a fuel (usually natural gas) drive the turbine directly (Fig. 21-6b). For hydroelectric power,

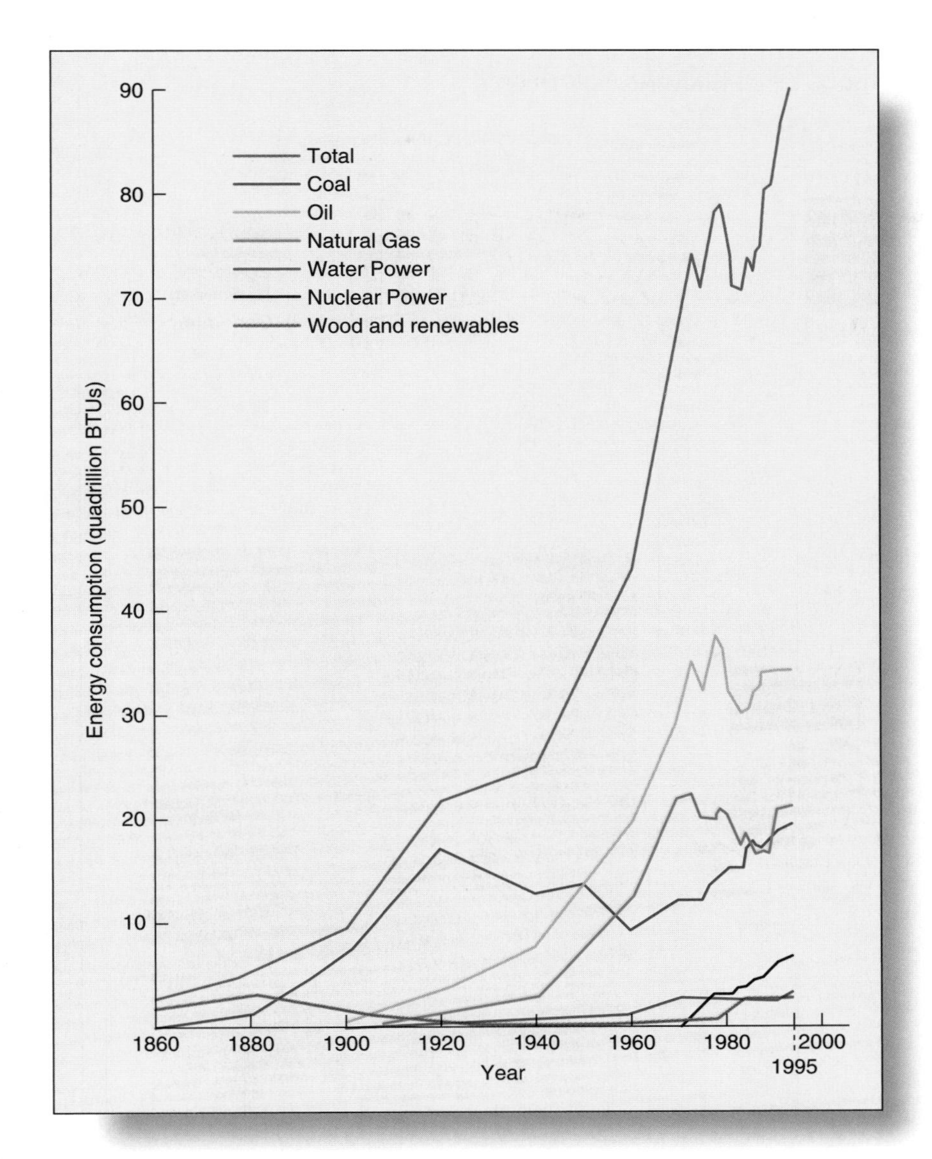

FIGURE 21-4
Energy consumption in the United States, 1860–1995, total consumption and major primary sources. Note how the mix of primary sources has changed over the years and how the total amount of energy consumed has continued to grow. Note also the skyrocketing increase in use of oil after World War II (1945) as private cars and car dependent commuting become common (see Chapter 24). Conspicuous decreases in consumption occurred in the late 1970s and early 1980s, but in recent years consumption has resumed its upward trend. (Data from *Annual Energy Review 1995,* Energy Information Administration, U.S. Department of Energy, July 1996.

water under high pressure—at the base of a dam or at the bottom of a pipe from the top of a waterfall—is used to drive a hydroturbogenerator (Fig. 21-6c). Wind turbines are also coming into use (see Fig. 23-22).

Electrical power is often promoted as being the ultimate clean, nonpolluting energy source. This is true at the point of use; using electricity creates essentially no pollution except for heat energy losses. Hidden, however, is the fact that pollution is simply transferred from one part of the environment to another: Switching to "clean" electric heat implies more demand for elec-

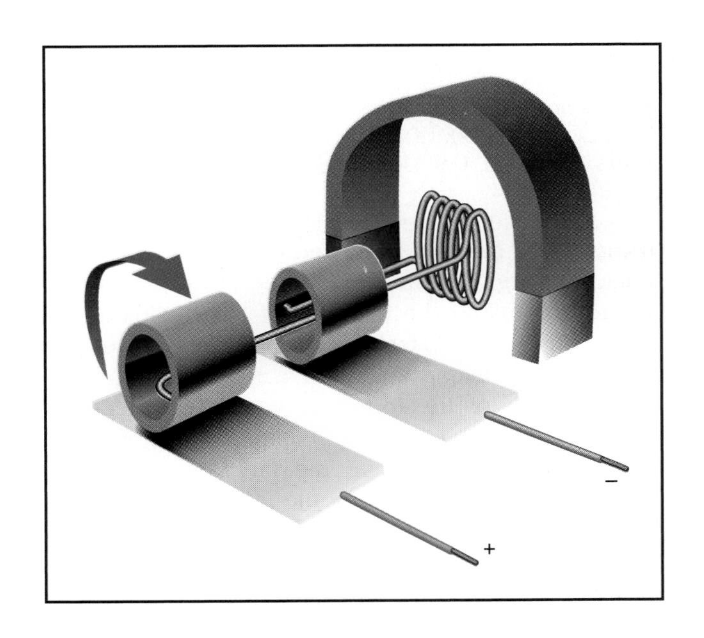

FIGURE 21-5
Principle of an electric generator. Rotating a coil of wire in a magnetic field induces a flow of electricity in the wire. In accordance with the laws of thermodynamics, more energy must go into turning the generator than is gotten out in electricity.

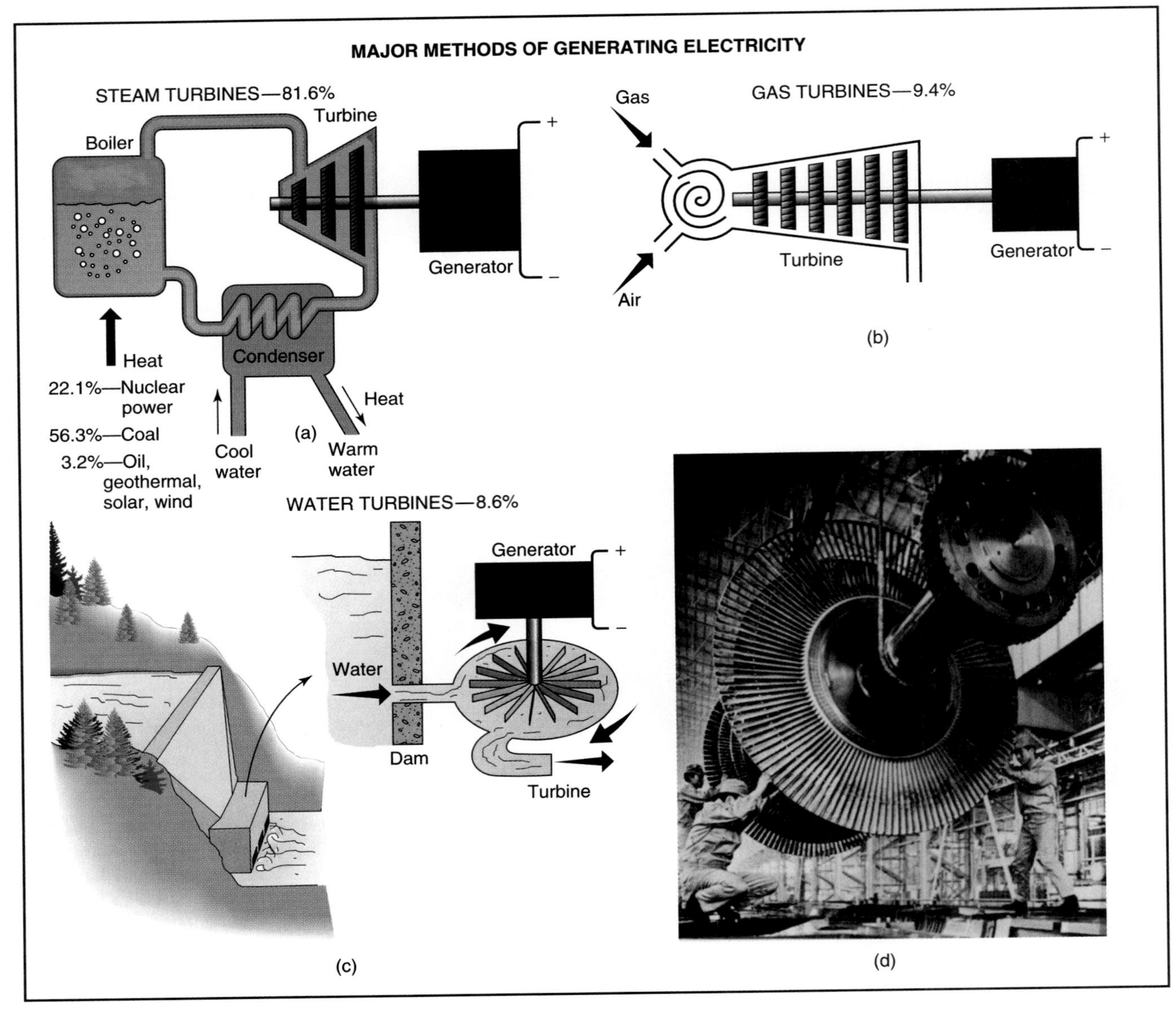

FIGURE 21-6
Electricity is produced commercially by driving generators with (a) steam turbines, (b) gas turbines, and (c) water turbines. The percentage of the U.S. electricity supply derived from each source (1995) is indicated. Small amounts of power are now also coming from solar wind energy, solar-thermal energy, and photovoltaics (solar cells). (d) Component of a steam turbine being assembled. (Data from *Statistical Abstracts of the United States*, U.S. Department of Commerce, 1995.) Photo courtesy of Counselate General of Japan, N.Y.

tricity; and it must be generated from coal, hydropower, nuclear energy, or alternative energy sources. We have already discussed the problem of sulfur dioxide emanating from the tall stacks of coal-burning power plants as being the major source of acid deposition, which affects much of the industrial world (Chapter 16). If hydroelectric power is used, creating a dam and reservoir involves displacing people, farmland, and wildlife, to say nothing of disrupting the migration of fish such as salmon. Finally, the hazards of radioactive wastes from nuclear power plants, discussed in Chapter 22, are well known. Additional adverse effects of using electrical energy apply to other parts of the fuel cycle, such as mining coal or processing uranium ore.

This problem of transferring pollution from one place to another becomes even more pronounced when we consider efficiency: Nearly 300 calories of coal, for example, must be burned to produce 100 calories of electricity. In other words, the thermal production of electricity has an *efficiency of only 30%–40%*. The 60%–70% loss is accounted for partly by some heat energy from the firebox going up the chimney and partly by a large amount of heat energy remaining in the spent steam as it comes out the end of the turbine.

FIGURE 21-7
Cooling towers. These structures that are the most conspicuous feature of both coal-fired and nuclear power plants are necessary to cool and recondense the steam from the turbines before returning it to the boiler. Air is drawn in at the bottom of the tower, over the condensing apparatus, and out the top by natural convection. In some systems, cooling may be aided by the evaporation of water that condenses in a plume of vapor as the moist air exits the top of the tower as seen in this photo. (Neal Palumbo/Gamma Liaison, Inc.)

Heat energy will go only toward a cooler place, so this heat energy cannot be recycled into the turbine. The most common practice is simply to dissipate it into the environment by means of a condenser. In fact, the most conspicuous features of coal-burning or nuclear power plants are the huge cooling towers used for this purpose (Fig. 21-7).

As an alternative to using cooling towers, water from a river, lake, or ocean can be passed over the condensing system (see Fig. 21-6a). Thus, the waste heat energy is transferred to the body of water. All the small planktonic organisms, both plant and animal, drawn through the condensing system with the water are cooked, and the warm water added back to the waterway may have deleterious effects on the aquatic ecosystem. Therefore, waste heat energy discharged into natural waterways is referred to as **thermal pollution**.

Connecting Sources to Uses

In addressing questions of whether energy resources are sufficient and where we can get additional energy, we need to consider more than just the source of energy, because some forms of energy lend themselves well to one use but not to others. For example, the U.S. transportation system of highways, cars, and trucks is virtually 100% dependent on liquid fuels. Virtually all other machines for moving, such as tractors, airplanes, locomotives and bulldozers, are likewise dependent on liquid fuels. Reserves of crude oil are needed, to support this transportation system. The nuclear industry is promoting nuclear power as the major solution to the energy shortage. However, nuclear power will do little to mitigate our demand for crude oil because it is suitable only for boiling water to drive turbogenerators, which will do essentially nothing for our transportation system. Although this situation might change with more widespread use of electric cars, their practicality for long distances still requires a technological breakthrough in the form of a low-cost, lightweight battery that will store large amounts of power.

Primary energy use is commonly divided into four categories: (1) transportation, (2) industrial processes, (3) commercial and residential uses (heating, cooling, lighting, and appliances), and (4) generation of electrical power, which is used in categories (2) and (3). The major pathways from the primary energy sources to the various end uses are shown in Fig. 21-8. Observe again that transportation is 100% dependent on petroleum, whereas nuclear power, coal, and water power are largely limited to the production of electrical power. Natural gas and oil are more versatile energy sources.

Figure 21-8 also shows the proportion of consumed energy that goes directly to waste heat energy rather than for its intended purpose. Some waste is inevitable, as dictated by the Second Law of Thermodynamics, but the current losses are much greater than necessary. The efficiency of energy use can be at least doubled—we can make cars that get twice as many miles per gallon and appliances that consume half as much power, for example. Doubling the efficiency of energy use would have the same effect as doubling the energy supply, at far less expense. Such increases in efficiency are the essence of energy conservation and are discussed in more detail later in this chapter.

In addressing our energy future we need to consider more than gross supplies and demands. We must examine how particular primary sources can be

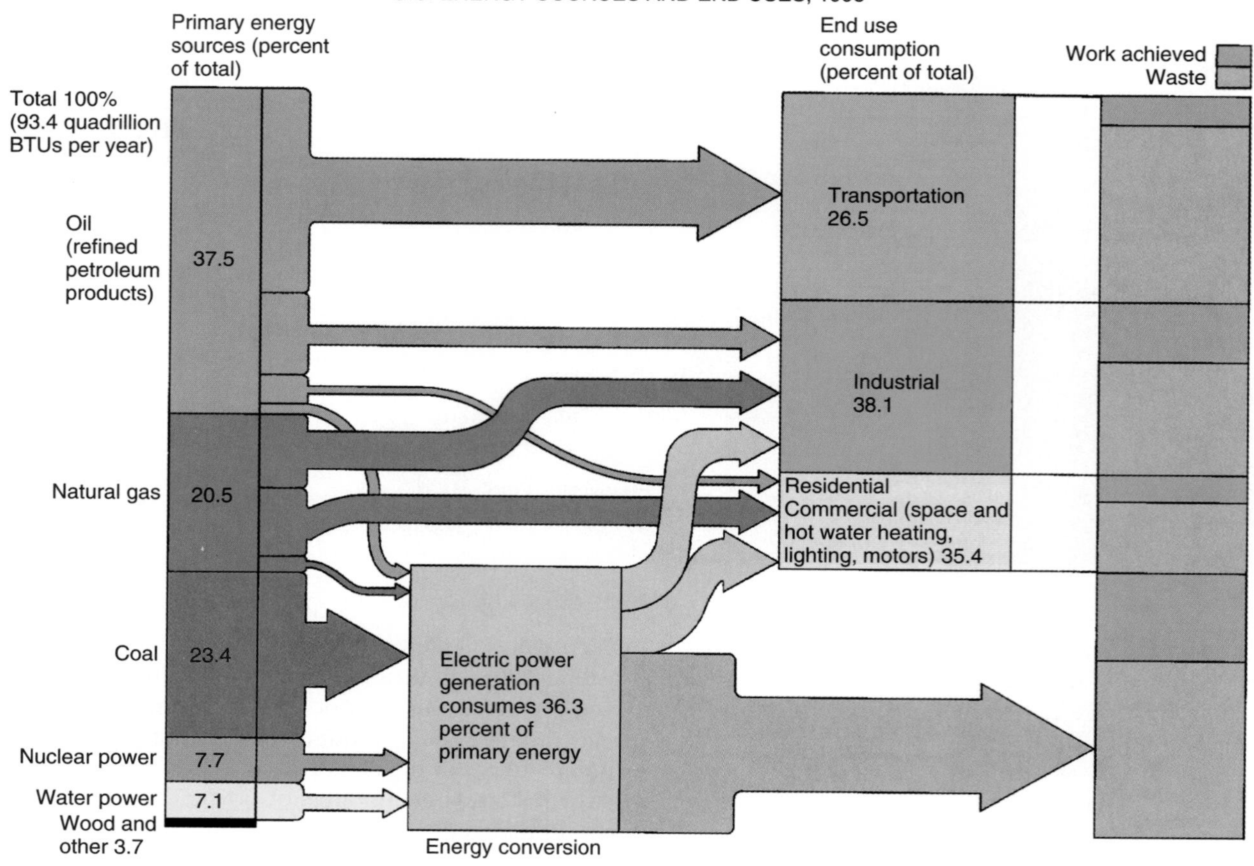

FIGURE 21-8
Pathways from primary energy sources to end uses. Only major pathways are shown. Note that end uses are connected to primary sources in specific ways. Also note the large percentage of energy that is wasted as a large portion of the energy consume is converted heat and lost. The question is: Do we base energy strategy for the future increasing supplies of primary energy or on decreasing demand through greater efficiency of use? (Based on data from Energy Information Administration, Department of Energy, December 1996.)

matched to particular end uses. Further, we need to consider the most cost-effective ways of balancing supplies and demands—by increasing supplies, or by decreasing consumption through conservation, *efficiency,* and *demand management.*

Declining Reserves of Crude Oil

The United States has abundant reserves of coal, adequate reserves for the time being of natural gas, and the potential for expansion of nuclear power (if nagging problems are resolved and nuclear economics improve). Therefore, there are no shortage crises looming in these areas now. (Note that we are setting aside the environmental concerns of coal and nuclear power for the moment.) Petroleum to fuel our transportation system is another matter. U.S. crude oil supplies fall far short of meeting demand; we are now dependent on foreign sources for over 50% of our crude oil, and this dependency is steadily growing as our reserves are diminished. These foreign sources come at increasing costs: necessary military actions, pollution of the oceans, and coastal oil spills. Let us examine this situation, starting with where crude oil comes from and how it is exploited.

How Fossil Fuels Are Formed

The reason crude oil, coal, and natural gas are called *fossil fuels* is that all three are derived from the remains of living organisms (Fig. 21-9). In the geological era 200 to 500 million years ago freshwater swamps and shallow seas, which supported abundant vegetation and phytoplankton, covered vast areas of the globe. Anaerobic conditions in the lowest layers of such bodies of water impeded the respiration of decomposers, and hence the breakdown of detritus. As a result, massive quantities of dead organic matter accu-

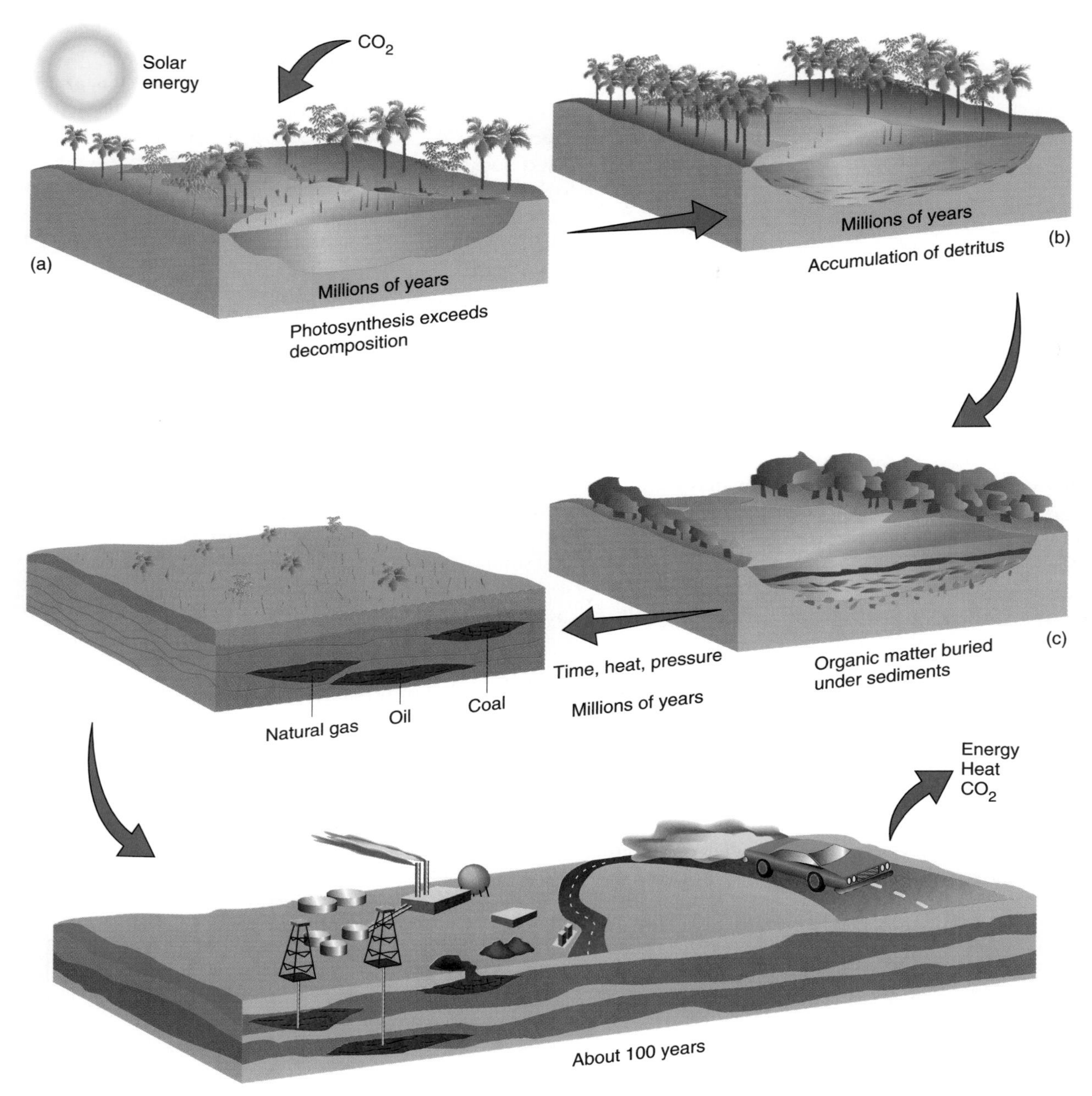

FIGURE 21-9
Energy flow through fossil fuels. Coal, oil, and natural gas are derived from biomass that was produced many millions of years ago. Deposits are finite and, since formative processes require millions of years, they are nonrenewable.

mulated. Over millions of years, this organic matter was gradually buried under layers of sediment and converted by pressure and heat to coal, crude oil, and natural gas. Which fuel was formed depended on the nature of the organic material buried, the specific conditions, and time involved. Just as anaerobic treatment (digestion) of sewage sludge yields methane gas and a residual organic sludge, a similar anaerobic process occurred with the buried vegetation. Natural gas is the trapped methane, and crude oil represents the residual "sludge." Coal is highly compressed organic matter—mostly leafy material from swamp vegetation—that decomposed relatively little.

Formation of additional fossil fuels by natural processes may be continuing to this day. However, they cannot possibly be considered renewable resources: We are using fossil fuels far faster than they ever formed. To accumulate the amount of organic matter that the world now consumes each day, 1000 years were needed. Recognizing how fossil fuels were formed, we can easily see that these resources are finite; sooner or later we will run out. The pertinent question is, When?

Crude Oil Reserves versus Production

The total amount of crude oil remaining represents Earth's oil resources. Finding and exploiting these resources is the problem. The science of geology provides information about the probable locations and extents of ancient shallow seas. On the basis of their knowledge of this science and on their field experience, geologists make educated guesses as to where oil or natural gas may be located and how much may be found. These educated guesses are the world's **estimated reserves**. Of course, such estimates may be far off the mark. There is no way to determine whether estimated reserves actually exist except by the next step, exploratory drilling.

If exploratory drilling strikes oil, further drilling is conducted to determine the extent and depth of the **oil field**. From this information, a fairly accurate estimate can be made of how much oil can be economically obtained from the field. This amount then becomes **proven reserves**. Proven reserves hinge on the economics of extraction. Thus, such reserves may actually increase or decrease with the price of oil, because higher prices justify exploiting resources that would not be worth extracting for lower prices. The final step, the withdrawal of oil or gas from the field, is called **production** in the oil business. Of course, *production* as used here is a euphemism. "It is production," says ecologist Barry Commoner, "only in the sense that a boy robbing his mother's pantry can be said to be producing the jam supply." In reality, this step is *extraction* from Earth.

Production from a given field cannot proceed at a constant rate, because crude oil is a viscous fluid held in spaces in sedimentary rock as water is held in a sponge. Initially, the field may be under so much pressure that the first penetration produces a gusher. Gushers, however, are generally short-lived, and the oil then seeps slowly from the sedimentary rock into a well from which it is withdrawn. In general, *maximum annual production is limited to about 10% of the remaining reserves*. Consider this rule in terms of maximum production from a field having 100 million barrels (one barrel equals 42 gallons) of proven reserves. As shown in Fig. 21-10, the first year's maximum production is 10 million barrels (10% of 100). Then production in subsequent years decreases as the reserve is drawn down.

Observe that there is no sudden "running out" in the sense that a car runs out of gas. Instead, the pro-

FIGURE 21-10
The 10% rule. No more than about 10% of the oil remaining in a field can be extracted in any given year. Consequently, maximum production from proven reserves invariably declines, but since there is always some oil remaining, there is no "running out" as such. The cutoff occurs when the costs of extraction equal or exceed the value of the oil extracted, a factor that varies with the price of oil. Finally, a net-yield cutoff may occur as energy expended in extraction approaches the energy yield of extracted oil.

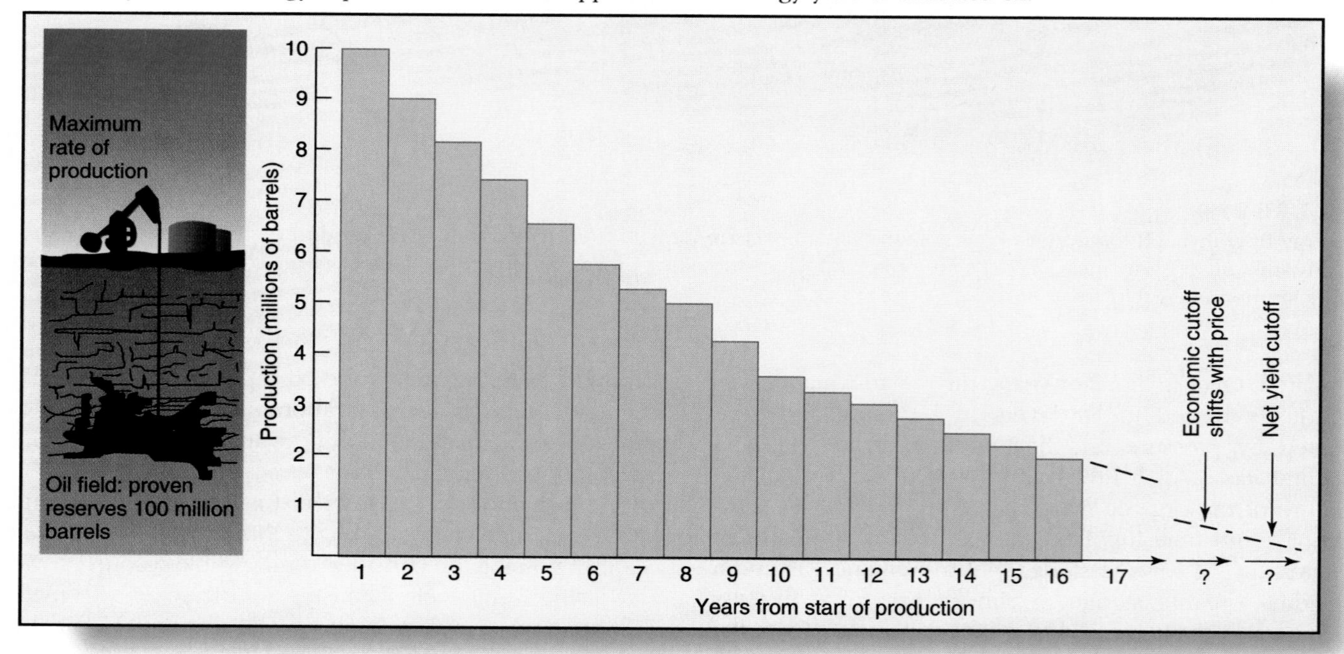

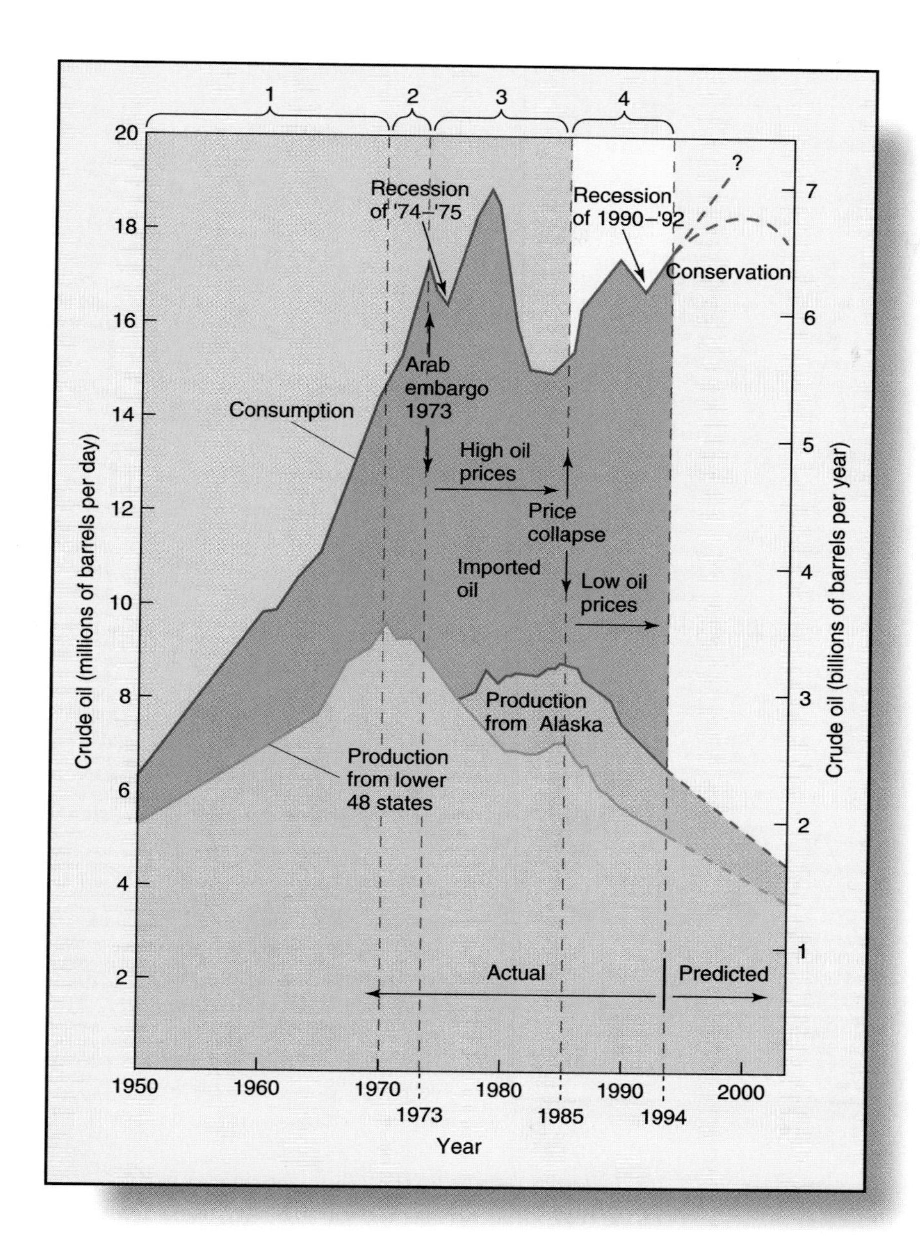

FIGURE 21-11
Oil production and consumption in the United States. Four stages can be seen: (1) Up to 1970, discovery of new reserves allowed production to parallel increasing consumption. (2) From 1970 to 1973, lack of new oil discoveries caused production to turn downward, while consumption continued to climb, causing a vast increase in oil imports and bringing on the oil crisis of 1973. (3) In the late 1970s to early 1980s, high oil prices promoted both lowered consumption and increased production—which included bringing the Alaskan oil field into production. Thus, dependence on foreign oil decreased considerably during this period. (4) From 1986, when oil prices declined sharply, until the present, consumption has again increased while production has resumed its decline, making us increasingly dependent on foreign oil. What does this bode for the future? (Data from *Statistical Abstracts of the United States*, U.S. Department of Commerce, 1995.)

duction from given proven reserves gradually diminishes according to the fact that annual production can be no more than 10% of *remaining* reserves. Production can be increased, or even kept at the first year's level, only if additional reserves are found, so proven reserves are always at least 10 times the desired rates of production.

Then, economics—the price of a barrel of oil—comes into play. No oil company will spend more money for extracting oil than they expect to make selling it. At the current price of crude oil, around $20 per barrel, it is economical to extract only about 35% of the oil resource in a given field. Further extraction involves more costly techniques that current prices cannot justify. Thus, an increase in price does indeed make more reserves available. The reverse may also occur. A practical example was experienced in the 1970s and early 1980s, when increasing oil prices justified reopening old oil fields and created an economic boom in Texas. Then an economic bust occurred in the late 1980s, as a drop in oil prices caused the fields to be shut down again.

Declining U.S. Reserves and Increasing Importation

Consideration of the relationships among production, price, and exploitation of reserves will help us understand the U.S. energy dilemma. A brief history is in order. Up to 1970, the United States was largely oil-independent. That is to say, exploration was leading to increasing proven reserves, such that production could basically keep pace with growing consumption (Fig. 21-11). But 1970 marked a significant turning point;

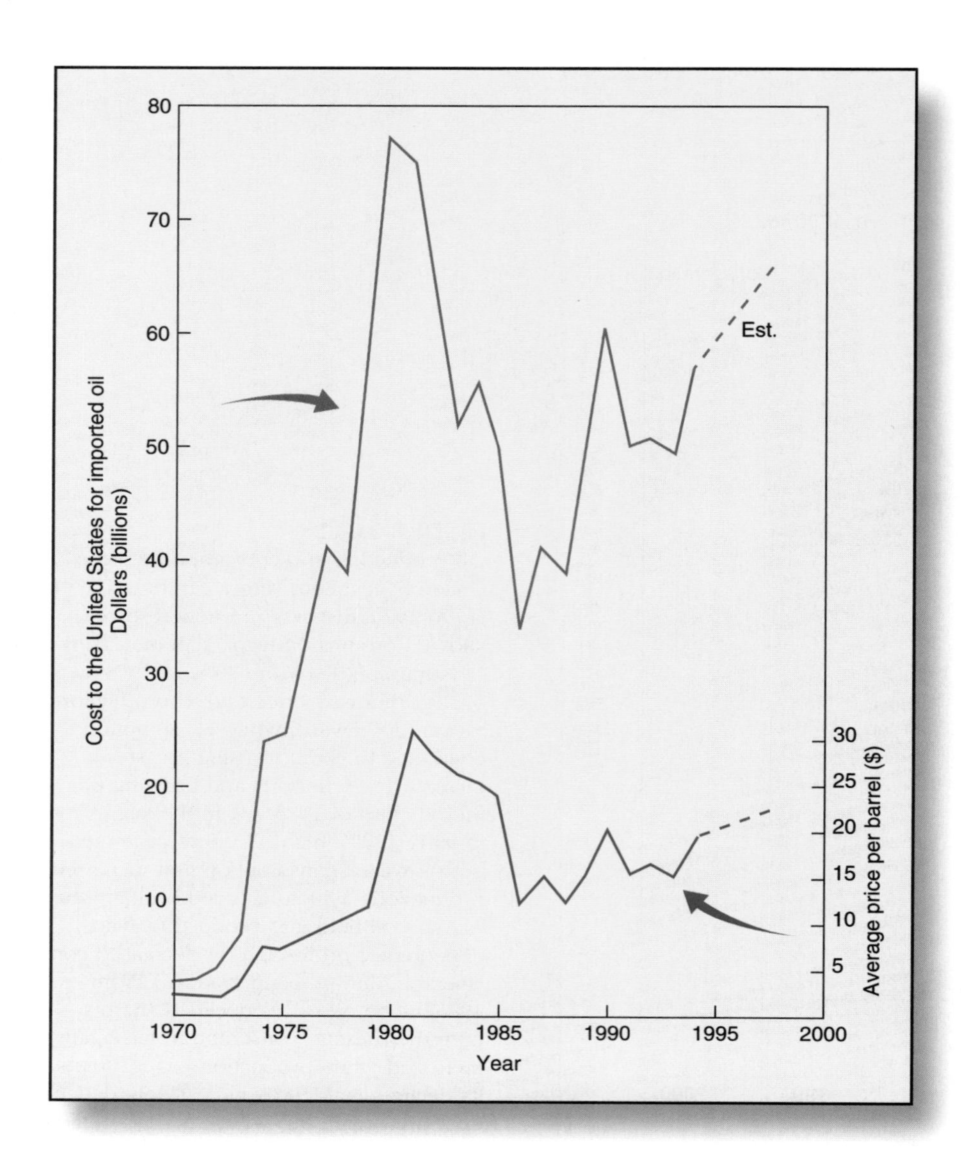

FIGURE 21-12
Annual cost to the United States of importing crude oil (upper curve). The cost fluctuates sharply with the price of foreign oil (lower curve) and the amount imported. With both the amount of oil imported and the probable rise in price, the cost to the U.S. can only increase.

new discoveries and hence reserve additions fell short, and production turned down while consumption continued its rapid growth. To fill the "energy gap" between increasing consumption and falling production, the United States became increasingly dependent on imported oil, primarily from the Arab countries of the Middle East. European countries and Japan did likewise. Since imported oil cost only about $2.30 per barrel in the early 1970s and Middle Eastern reserves were more than adequate to meet the demand, this course seemed to present few problems.

The Oil Crisis of the 1970s

Thus, in the early 1970s the United States and most other highly industrialized nations were becoming increasingly dependent on oil imports from a small group of countries known as **OPEC**, the Organization of Petroleum Exporting Countries (see "Global Perspective" box, page 541). In 1973, recognizing the dependence of industrialized nations on oil, OPEC decided to take advantage of the situation. They initiated an oil boycott, suspending oil exports. The effect was almost instantaneous, because we depend on a fairly continuous flow from wells to points of consumption. (Compared with total use, little oil was held in storage.) Spot shortages occurred, and because we are so dependent on our autos, the spot shortages quickly escalated into widespread panic. We were willing to pay almost anything to have oil shipments resumed—and pay we did. OPEC resumed shipments at a price of $10.50 a barrel, more than four times the previous price.

Then, by continuing to limit oil production all through the 1970s, OPEC was able to keep supplies tight enough to force prices higher and higher. In the early 1980s, a barrel of oil cost $30, and the United States was paying accordingly. Purchase of foreign oil became and remains the single largest factor in our balance-of-trade deficit (Fig. 21-12).

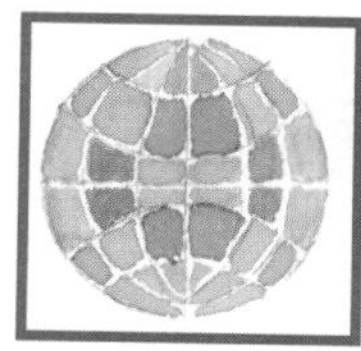

Global Perspective

Meeting the U.S. Oil Appetite

The U.S. consumption of crude oil for liquid fuels, mainly gasoline, was 17.7 million barrels per day (1 barrel = 42 gallons) in 1995 and rising, while production was less than one half that, 8.3 million barrels per day, and falling. To make up the shortfall, the United States imports oil from numerous countries around the world, both OPEC (shaded in the map) and nonOPEC. The map shows the countries of origin and the amounts in millions of barrels per year (1993) from each source. There are a number of additional minor sources not shown.

Likewise, nearly all other countries on the map, other than the exporters, both developed and developing, are net importers of crude oil. Furthermore, economic expansion and development of all countries is currently tied to increasing consumption of oil, such that there is growing demand from the exporting nations. How long can the oil-exporting nations continue to meet the world demand? Production from several countries is already in a state of decline.

Data: Millions of barrels per year (1993).
Source: Data from Statistical Abstracts of the United States, U.S. Department of Commerce, 1995.

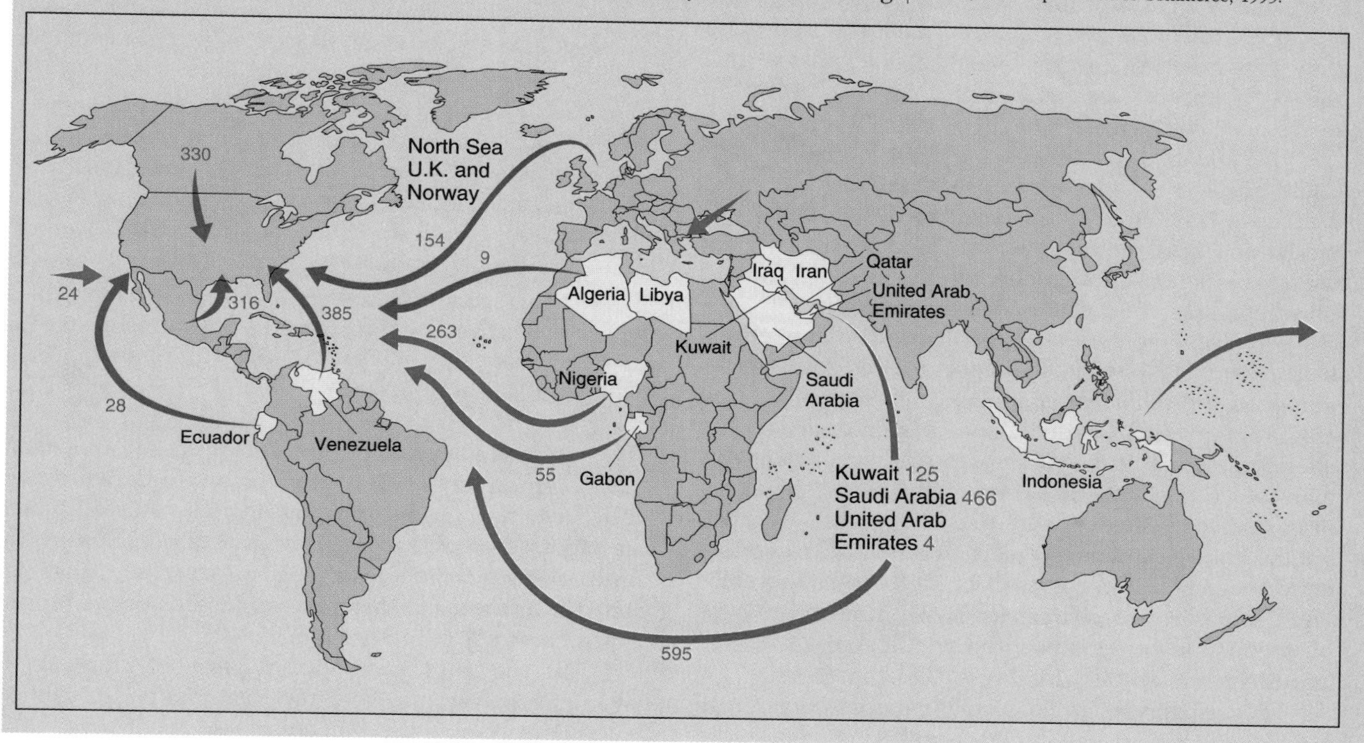

Adjusting to Higher Prices

In response to the higher prices and the recognition that oil is not forever, the United States and other industrialized countries quickly made a number of responses and adjustments during the 1970s. The most significant changes made by the United States are listed below.

To increase domestic production of crude oil:

- Exploratory drilling was stepped up.
- The Alaska pipeline was constructed. (The Alaskan oil field had actually been discovered in 1968, but it remained out of production because of its remoteness.)
- Fields that had been closed down as uneconomical when oil was only $2.30 a barrel were reopened (e.g., in Texas).

To decrease consumption of crude oil:

- Standards were set for automobile fuel efficiency. In 1973 cars averaged 13 mpg. The government mandated stepped increases so that new cars would average 27.5 mpg by 1985. Also, mpg information was required in all car advertisements.

- Other conservation goals were promoted for such things as insulation in buildings and efficiency of appliances.
- Development of alternative energy sources was begun. The government supported research and development efforts and gave tax breaks to people for installing alternative energy systems.

To protect against another OPEC boycott:

- A strategic oil reserve was created. The United States had "stockpiled," as of 1995, 592 million barrels of oil (equivalent to 75 days' of imports) in underground caverns in Louisiana.

It is noteworthy that, with the exception of the stockpile, all these steps are economically appealing only when the price of oil is high. If the price is low, the economic returns or savings may not justify the necessary investments.

Results of these actions were not immediate, nor could quick returns have been expected. For example, it takes at least three to five years to design a new car model and start production. Then another six to eight years pass before enough new models are sold and old ones retired to affect the overall highway "fleet."

As we entered the 1980s, however, the efforts were definitely having an impact. The United States remained substantially dependent on foreign oil, but oil consumption was headed down and production, up after completion of the Alaska pipeline, was holding its own (see Fig. 21-11). Elsewhere in the world significant discoveries in Mexico and the North Sea (between England and Scandinavia) made the world less dependent on OPEC oil. As a result of all these factors, plus OPEC's inability to restrain its own production, world oil production exceeded consumption in the mid-1980s, and there was an oil "glut."

What happens when supply exceeds demand? In 1986, world oil prices crashed from high $20s to close to $10 per barrel. Since that time, until now, oil prices have fluctuated in the range of $14 to $20 per barrel (see Fig. 21-12). The effect of the lower oil prices—though still higher than before the crises—was to the apparent benefit of consumers, industry, and oil-dependent developing nations, but it hardly solved the underlying problem. Indeed, it has set the stage for another crisis.

Victims of Our Success

Whereas high oil prices stimulated constructive responses, the collapse in oil prices undercut those responses:

- Exploration, which had become more costly as wells were drilled deeper and in more remote locations, was sharply curtailed. In 1982, 3105 drilling rigs were operating in the United States; in 1992 there were 660.
- Production from older oil fields, which was costing around $10–$15 per barrel to pump, was terminated. The resulting drop in production caused economic devastation in Texas and Louisiana.
- Conservation efforts and incentives were abandoned. Government standards for automobile fuel efficiency were decreased to 26 mpg in 1986 with no intention to increase them further. The requirement to include mpg information in car advertisements was dropped and the popularity of sport utility vehicles (light trucks), which are exempted from the fuel economy guidelines, surged.
- Tax incentives and other subsidies for development or installation of alternative energy sources were terminated, destroying many new businesses engaged in solar and wind energy. Grants for research and development of alternative energy sources (except for nuclear power) were sharply cut.
- The need for conservation and for development of alternative ways to provide transportation seems to have largely passed from the mind of the public.

As a result, U.S. production of crude oil is again dropping rapidly as proven reserves are drawn down. The declining production now includes Alaska, which is expected to fall to 50% of its peak production by the end of the decade. Nor does anyone see any real chance of reversing this trend as long as prices remain in the low $20 per barrel range.

At the same time, consumption of crude oil is rising again, because both the total number of cars on the highways and the average miles per year that each car is driven are increasing. Adding to that consumption, consumers are buying vehicles that are less fuel-efficient. Minivans, 4-wheel drive sport-utility vehicles, light trucks, and other less fuel efficient models now constitute 40% of the new-car market (see Fig. 21-11).

Altogether, following considerable improvement in our energy situation in the late 1970s and early 1980s, the United States has reverted to old patterns. Dependence on foreign oil has grown steadily since the mid-1980s. At the time of the first oil crisis in 1973, the United States was dependent on foreign sources for about 30% of its crude oil. Now we are even worse off: In 1994 U.S. dependency on foreign sources passed the 50% mark, and it is continuing upward at about 2 per-

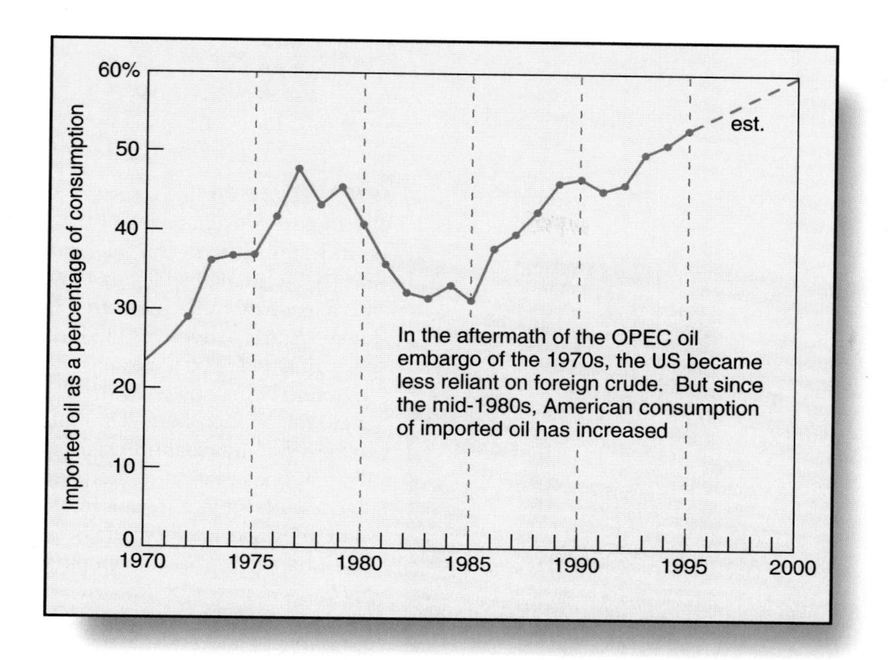

FIGURE 21-13
Imported oil as a percentage of total consumption. (Data from *Annual Energy Review,* published annually, Energy Information Administration, U.S. Department of Energy.)

centage points per year (Fig. 21-13). The United States is not alone in this situation. Much of the rest of the world, both developed and developing countries, have a growing dependency on oil from those relatively few nations with surplus production (see map in the "Global Perspective" box, page 541). What problems does this growing dependency on foreign oil present?

Problems of Growing U.S. Dependency on Foreign Oil

As the U.S. dependency on foreign oil grows again, we are faced with problems at three levels: (1) the costs of purchase, (2) the risk of supply disruptions due to political instability in the Middle East, and (3) ultimate resource limitations in any case. Let's consider each in more detail.

Costs of Purchase. The yearly cost to the United States of purchasing foreign oil is shown in Fig. 21-12. The cost has dropped considerably from its high of over $70 billion in the early 1980s, the decline being brought about both by falling oil prices and by conservation. However, the current $50 billion per year is no small number, and it is now climbing again with no downturn in sight as we become increasingly dependent on foreign oil.

This cost for crude oil represents about 65% of our current balance-of-trade deficit ($76 billion in 1993), which has manifold consequences for our economy. The price we pay at the pump is basically the same whether the oil is produced here or abroad. However, if the oil is produced at home, the money we pay for it circulates within our own economy, providing jobs and producing other goods and services. When we buy oil from a foreign country we are effectively providing "foreign aid" to that country and that much of a drain on our own economy.

Of course, the ecological costs of oil spills such as the estimated $15-billion *Exxon Valdez* incident (Fig. 21-14), and of other forms of pollution from the refining and consumption of fuels must not go unnoticed—although estimating total damage in dollars is sometimes difficult. There are also military costs, as we will describe in the next section.

Risk of Supply Disruptions. The Middle East is a politically unstable, unpredictable region of the world. As noted, it was the unexpected Arab boycott that plunged us into the first oil crisis in 1973. Recognizing this political instability, the United States maintains a military capacity to ensure our access to Middle Eastern oil. This military capacity was tested in the fall of 1990 when Saddam Hussein of Iraq invaded Kuwait, then producing 6 million barrels per day. The Persian Gulf War of early 1991 prevented further advances. Nevertheless, Kuwait's oil production was knocked out for more than a year after the war, because the Iraqis dynamited and set fire to nearly all of Kuwait's almost 700 wells before leaving (Fig. 21-15). Even with military intervention, OPEC oil may not continue flowing to U.S. consumers. The military incursion cost about $1 billion per 19 hours for the nations involved.

Military costs associated with maintaining access

FIGURE 21-14
Ecological consequences of the world's dependence on oil include accidental spills. In 1989, the release of crude oil from the tanker *Exxon Valdez* resulted in over 1500 miles of Alaskan shoreline being coated with crude oil, as seen here. This was far beyond any cleanup capacity. The total devestation to the great diversity of wildlife inhabiting shorelines and intertidal areas (see Chapter 2, chapter opening photo) will never be known. (Vanessa Vick/Photo Researchers, Inc.)

to Middle Eastern oil can be looked at as the U.S. government's subsidizing our oil consumption. The dollar amount of these "subsidies" has been calculated to be about $50 per barrel. In other words, when military "support services" are added to the $20 market price, we are actually paying about $70 per barrel.

Maintaining access to Middle Eastern oil is also a major reason for our efforts toward negotiating peace agreements between various parties in that region. Even with all this effort, however, continuing risks are present. The rise of the Islamic radicals, who threaten to take over governments of more Middle Eastern countries, may pose a great danger.

Resource Limitations. The fact that U.S. crude oil production is decreasing because of diminishing domestic reserves is not questioned by any responsible

FIGURE 21-15
Even military might may not guarantee uninterrupted access to world oil supplies. While we were the winner of the Persian Gulf War of 1991, Iraq's forces nevertheless succeeded in setting fire to more than 600 of Kuwait's oil wells before their retreat. Six million barrels per day were thus taken out of production for nearly a year. (Wesley Bocxe/Photo Researchers, Inc.)

Ethics

Trading Wilderness for Energy in the Far North

The search for new energy sources to satisfy the enormous energy appetite of the United States economy has turned northward in recent years. Discovery and development of the huge Prudhoe Bay oil field in Alaska and the building of the 800-mile (1230 km) Alaska pipeline have dramatized the stark contrasts between wilderness and modern technology. Scenes of caribou and grizzly bears are shown with a background of oil rigs and the pipeline carrying oil south to Valdez. East of Prudhoe Bay is the Arctic National Wildlife Refuge (ANWR), the largest single wilderness area in the United States. ANWR is home to 180,000 caribou and several hundred native people whose subsistence lifestyle is tied to the caribou. ANWR is also the summer breeding ground for innumerable birds and other wildlife, but it is also believed to be on top of another huge oil field that industrial interests are anxious to exploit.

Twenty-five-hundred miles to the south and east of ANWR, three major river systems flow through a vast wilderness of lakes, bogs, and forests before emptying into the James and Hudson bays. In addition to wildlife, this area of northern Canada is home to 10,000 Cree and Inuit people, 40% of whom still live off the land much as their ancestors did. But the provincial energy utility Hydro-Québec sees vast potential in this wilderness for power generation. In what is called the James Bay project, Hydro-Québec has already converted 4500 square miles (10,000 km^2) of the wilderness into a network of diversions, dams, and reservoirs, and this is just one-third the total project. Now the Crees have gone to court, demanding a full review of environmental and human costs before Hydro-Québec proceeds with the next two phases of development. Segments of the project were designed without any environmental impact assessment; other portions had been abandoned by 1996.

ANWR and James Bay have become major environmental battlegrounds of the 1990s. On one side, the energy and political interests see values in terms of energy supplies for a thriving economy and national security. On the other side, native people and environmentalists see values in terms of the beauty and biodiversity of the northern wilderness, as well as the human rights of native peoples. Those favoring development claim that it will be done in an environmentally sound way and the native people will be compensated. Those favoring preservation are not satisfied with good intentions and poor track records. They point to the *Exxon Valdez* disaster, in which 11 million gallons of crude oil were spilled in the pristine waters of Prince Edward Sound, despite all safeguards. They point out that present oil development in Alaska has introduced much of the worst of American culture to native peoples; a sense of lost identity is apparent as their traditional culture is swamped by consumer goods and pop culture. Environmentalists also question development on the basis of its sustainability. The estimated oil reserve underlying ANWR is 3.5 billion barrels; this is about six months' supply at the current rate of use in the United States. The power produced by the James Bay project will be mostly exported from Canada to the United States although several of these buyers have pulled out of the agreement. Are there ways to have adequate power without purchasing it at the expense of wilderness?

How do we balance our profligate need for energy, jobs, and economic growth, against caribou, wolves, and grizzly bears? How do we weigh the traditions and cultures of native peoples against the impacts of our own?

Sources: H. Thurston, "Power in a Land of Remembrance," *Audubon*, November–December 1991; and "Arctic Wars: The Fight for the Arctic National Wildlife Refuge," National Audubon Society film, 1990.

geologist. Nor do geologists hold out hope for major new finds. The dim hope for new U.S. finds derives from what is called the "Easter-egg hypothesis." As your searching turns up fewer and fewer eggs, you draw the logical conclusion that nearly all the eggs have been found. In terms of oil, North America is already the most intensively drilled of any continental land mass. The last major find was the Alaskan oil field in 1968. Discoveries since then have been small in comparison, with increasing numbers of dry holes in between. Indeed, most of the "new" oil in the United States is coming from innovative computer mapping of geological structures and horizontal drilling technology that enables the identification and withdrawal of oil from small isolated pockets within old fields that were previously missed.

Until recently, experts anticipated that world reserves would be drawn down to the point of limiting production on a worldwide basis by 2010–2020. That is, even foreign oil would become limited, quite apart from any action by OPEC, in the relatively near future. When this occurs, shortages will inevitably cause prices to skyrocket with manifold economic and social consequences. Some experts believe that, barring drastic changes, this scenario will still occur in the predicted time frame. However, most feel that this eventuality has been pushed into the future to 2040 or 2050.

Further drilling in the Middle East has proved those oil fields to be even more vast than previously estimated, and very significant oil fields have been identified in other parts of the world as well. These new discoveries have added to proven reserves far faster than withdrawals have diminished them. In the 1970s, proven global reserves amounted to only another 20 years' supply. Now, proven reserves are up to 50 years' of supply at current rates of use even though those rates are higher (see "Ethics" box, page 545).

This development, however, does not really change the economic and political picture for the United States. Less dependency on foreign oil will still enhance our economic security and reduce the risk of military confrontations. In the longer run it will smooth the transition to when crude oil has been used up. We can become more independent in three ways: (1) We can use the other fossil fuel resources we have, to make fuel for vehicles; (2) we can reduce our energy demand through conservation and energy efficiency; and (3) we can develop alternatives to fossil fuels, many of which are ready now. The best answer probably lies in a suitable mix of all three choices.

Alternative Fossil Fuels

Unlike crude oil, considerable reserves of natural gas and comparatively vast reserves of coal and tar sands are still available in the United States. Natural gas is gradually coming into use as a vehicle fuel, and there is some potential in this respect for coal and tar sands.

Natural Gas

With the installation of a tank for compressed gas in the trunk and some modifications of the engine fuel-intake system—costing about a thousand dollars—a car will run perfectly well on natural gas. Such cars are in widespread use in Buenos Aires, where service stations are equipped with compressed gas to refill the tank. Natural gas is a clean-burning fuel; hydrocarbon emissions are nil. Indeed, natural gas is coming into use in the United States as a vehicle fuel, and it is already being used by many buses and a number of both private and government car fleets. To be sure, natural gas reserves are not sufficient to substitute for the total oil demand, nor could they ever be a permanent solution. However, they could provide an interim measure of considerable merit.

Coal

The United States is exceptionally well endowed with coal (Fig. 21-16). Even if we tripled our current rate of coal use, these reserves could supply this country's energy needs for about 100 years. As stated earlier in

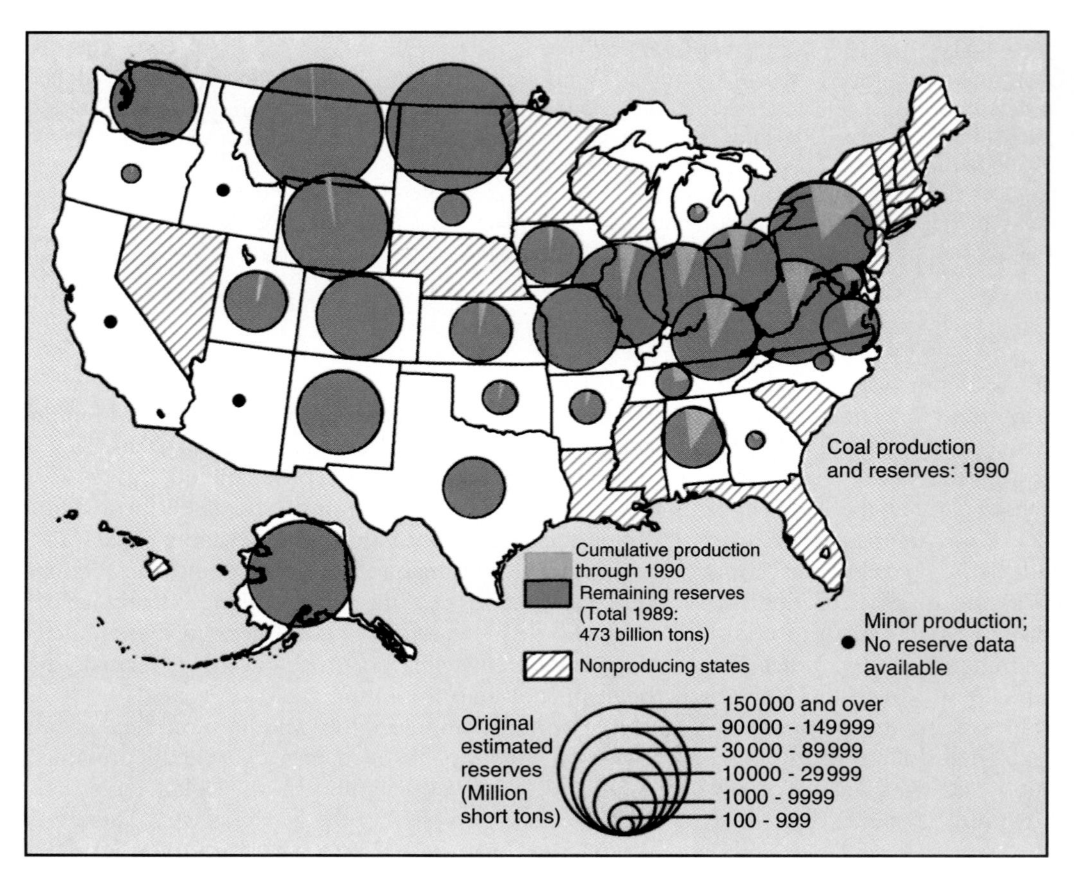

FIGURE 21-16
Major coal deposits in the United States: a solution or a potential ecological disaster? (U.S. Geological Survey.)

this chapter, however, the problem is that we cannot use coal in vehicles.

There are chemical processes that will convert coal to a liquid or gas fuel. Such coal-derived fuels are referred to as **synthetic fuels** or **synfuels**. With both government and corporate support, a great deal of research went into upscaling these chemical processes to commercial production in the late 1970s and early 1980s. The projects were all abandoned, however, because they proved too expensive, at least for the present. Profitable production of synthetic fuels could occur only if oil prices rose to $60 to $70 per barrel.

Even more troublesome, however, are the environmental impacts of using coal. Most coal deposits can be exploited practically by strip-mining. In underground mining, at least 50% of the coal must be left in place to support the mine roof. In strip-mining, gigantic power shovels turn aside the rock and soil above the coal seam and then remove the coal (Fig. 21-17). Obviously, this procedure results in total ecological destruction. Although such areas may be reclaimed—that is, graded and replanted—it takes many decades before an ecosystem resembling the original can develop. In arid areas of the West, the water limits may prevent the ecosystem's ever being reestablished. Consequently, strip-mined areas may be turned into permanent deserts. Furthermore, erosion and acid leaching from the disturbed ground have numerous adverse effects on waterways and groundwater in the surrounding area.

FIGURE 21-17
It is only economical to exploit most coal deposits by strip mining, where gigantic power shovels remove as much as 100 feet (32 meters) of rock and soil from the surface to get at the coal seam. It is hard to conceive of a more ecologically destructive activity. (Chris Jones/The Stock Market.)

The hazards of mining and burning coal are already well known from the nearly 1 billion tons that are used annually in the United States for power plants. The thought of doubling or tripling this impact to run vehicles on coal-derived fuel makes a mockery of the country's progress in environmental protection. In addition, converting coal to liquid fuel creates polluting byproducts that would be difficult and expensive to control.

Oil Shales and Tar Sands

The United States has extensive deposits of oil shales in Colorado, Utah, and Wyoming. **Oil shale** is a fine sedimentary rock containing a mixture of solid, waxlike hydrocarbons called kerogen. When shale is heated to about 1100°F (600°C) the kerogen releases hydrocarbon vapors that can be recondensed to form a black, viscous, substance similar to crude oil, which can be refined into gasoline and other petroleum products. However, it requires about a ton of oil shale to yield little more than half a barrel of oil. The mining, transportation, and disposal of wastes necessitated by an operation producing, for instance, a million barrels a day (5% of U.S. demand) would be a Herculean task, to say nothing of its environmental impact. Perhaps to our environmental advantage, oil shale, like oil from coal, has proved economically impractical for the present. Also, it requires large amounts of water, something in short supply in the west.

Tar sands are a sedimentary material containing bitumen, an extremely viscous, tarlike hydrocarbon. When tar sands are heated, the bitumen can be "melted out" and refined in the same way as crude oil. Northern Alberta, Canada, has the world's largest tar-sand deposits, and Canada is commercially exploiting them, because the cost is competitive with current oil prices. The United States has smaller, poorer tar-sand deposits, mostly in Utah, that might produce oil for $50–$60 per barrel. Again, pollution and other forms of environmental impact are drawbacks.

Global Warming—A Limiting Factor?

At the same time we are rethinking our energy future on the basis of resource availability and managing our demands, we need to consider another factor looming on the horizon—climatic change. Note again that all fossil fuels are carbon-based compounds and none can be burned without carbon dioxide (CO_2) resulting as a byproduct. The increasing atmospheric concentration of CO_2 is likely to have a warming effect (see Chapter 16, page 408). This threat of climatic change—mounting evidence indicates that it is already occurring—is likely to force a curtailment of fossil-fuel consumption long before limited resources do. World leaders meeting at the International Climate Conference in Berlin, in March of 1995, agreed to establish new limits for greenhouse gases by 1997—limits that would go into effect after the turn of the century.

Given the amounts of carbon dioxide produced—about 3 pounds for every pound of fuel consumed—CO_2 dominates the greenhouse effect, and the United States is no small contributor. Largely because of our "love affair" with basically inefficient cars and long-distance commuting, the United States, with just 4.5% of the world's population, is responsible for nearly 25% of the global CO_2 emissions. On a per capita basis this is five times the world average. We even produce more than twice as much CO_2 per capita as our major industrial competitor, Japan (Fig. 21-18). Given the agreement signed at the Berlin Climate Conference, which was a follow-up to the commitment made at the 1992 Earth Summit in Rio de Janeiro to stabilize greenhouse-gas concentrations, we are now under both moral and legal obligation to stop *increasing* fossil-fuel consumption, at the very least.

Because of their differing chemical structures, various fossil fuels produce different amounts of carbon dioxide per unit of energy delivered. Of the top three—coal, oil, and natural gas—coal produces the most, and natural gas produces the least. Synfuels would produce even more carbon dioxide than coal per unit, because of the considerable energy loss entailed in converting coal to liquid fuels. The same may be said for extracting fuel from oil shales or tar sands. Therefore, if we are at all serious about stabilizing CO_2 emissions, we have another argument for switching from crude oil to natural gas for running vehicles, and from wasteful to efficient ways, and for crossing coal, oil shales, and tar sands off the option list altogether. Further, fossil-fuel contributions to global warming are probably the most compelling reason to reduce energy demand through conservation and to examine non-fossil-fuel alternatives.

Sustainable Energy Options

In rethinking energy strategies, start with the concept that energy has no real meaning or value by itself. Its only worth is in the work it can do. Thus, what we really want is not energy but comfortable homes, transporta-

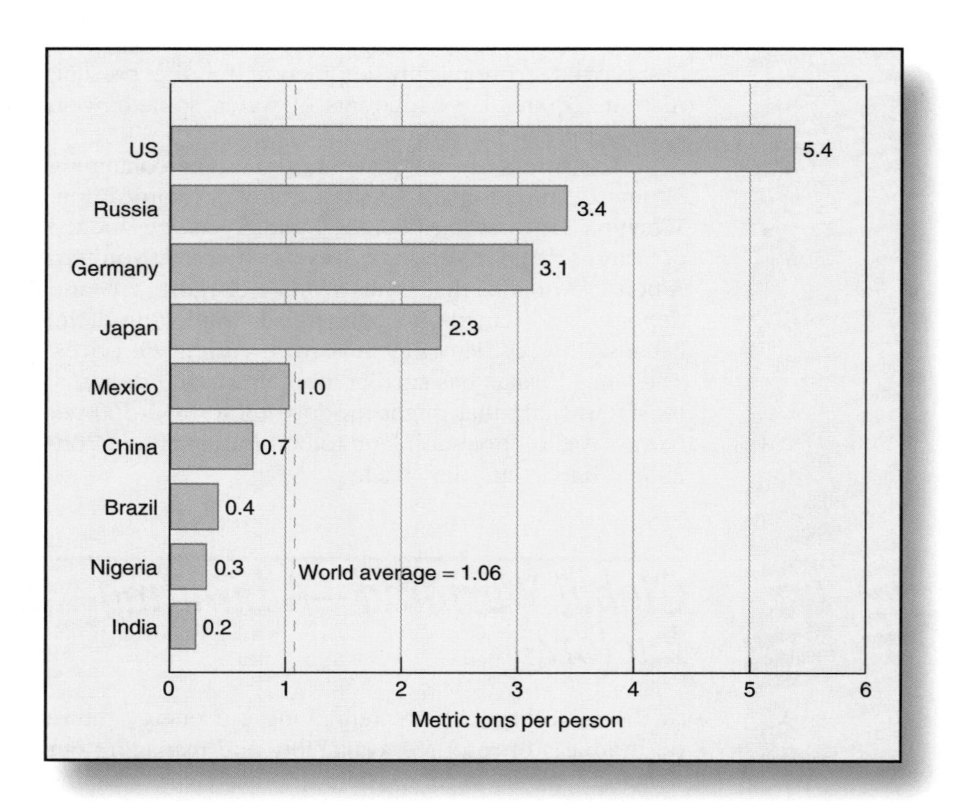

FIGURE 21-18
Carbon dioxide emissions per capita from fossil fuel burning for selected countries. (From C. Flavin and O. Tunali, "Getting Warmer," *Worldwatch*, March–April 1995, p. 17.)

Earth Watch

Cogeneration: Industrial Common Sense

In many situations a single energy source can be put to use to produce both electrical and heat energy. This process is called cogeneration. Cogeneration can be employed to conserve energy in two fundamental ways: (1) A power plant whose primary function is to generate electricity also can be used to heat surrounding homes and buildings. As one example, the Consolidated Edison electric utility has been using steam from waste heat in New York City for years. (2) A facility whose primary function is to produce heat or mechanical energy can also generate electricity, which is used by the facility and can also be sold to the local utility. The system can work for all sorts of fuels and heat engines. This is by far the most common application of cogeneration.

The arrangement makes perfect sense. Every building and factory requires both electricity and heat. Traditionally, electricity and heat come from separate sources—the electricity from power lines brought in from some central power station and the heat from an on-site heating system, usually fueled by natural gas or oil. Ordinary power plants are at best 40% efficient in their use of energy. Combining the two in cogeneration can provide an overall efficiency of fuel use as high as 80%. Besides providing savings on fuel, the cogenerating systems have at least 50% lower capital costs per kilowatt generated than do conventional utilities. This cost savings makes cogeneration quite competitive with utility-produced power.

The Santa Barbara (California) County Hospital recently installed a cogeneration system based on natural gas as the fuel source. Natural gas drives a turbine with the capacity to generate 7 MW of electricity, much of which is used by the hospital. Boilers use the waste heat from the turbine to produce high-pressure steam, which provides all of the hospital's heating and cooling needs. Any steam not required by the hospital is directed back to the turbine to increase its power. Electrical power not used by the hospital is sold to the local utility. The system has the added advantage of protecting the hospital against widescale blackouts or brownouts that may occur in the centralized utility system.

Enabling legislation known as PURPA (Public Utility Regulatory Policies Act of 1978) was essential for stimulating use of cogeneration. Under this act, a facility with cogenerative capacity is entitled to use fuels not usually available to electricity generating plants (for example, natural gas), must produce above-standard efficiency of fuel use, and is entitled to sell cogenerated electricity to the utility at a fair price. After many challenges by the utilities (which considered the procedures an interference with their business), the Supreme Court decided in favor of PURPA, and the legislation is now firmly established.

Cogeneration is becoming increasingly popular in U.S. industries as a major strategy for cutting energy costs. Given the future predictions of a downturn in availability and major price increases for petroleum, pressures favoring energy conservation will increase, and use of cogeneration will continue to grow. The Federal Energy Regulatory Commission believes that cogeneration may provide as much as 25% of our electricity by the year 2000.

Sources: E. L. Clark, "Cogeneration—Efficient Energy Source," *Annual Review of Energy* 11, 1986, p. 275; and *The Energy Sourcebook*, ed. R. Howes and A. Fainberg, American Institute of Physics, 1991.

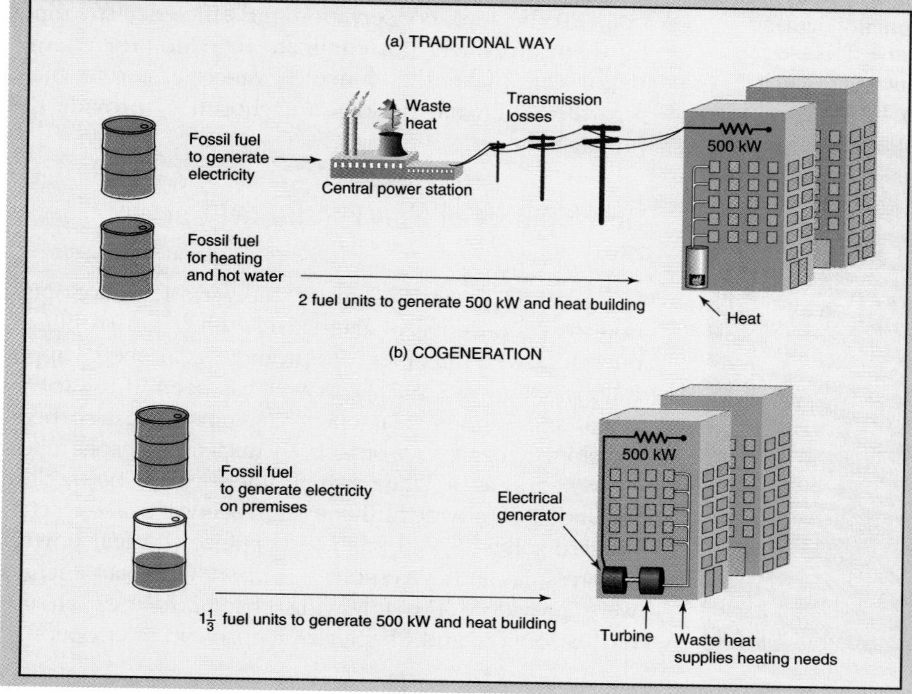

Cogeneration. (a) The traditional method of providing the energy needed for buildings. (b) In cogeneration, heating needs are provided by the heat lost from an on-site power-generating system.

tion to where we need to go, manufactured goods, and so forth. Therefore, many energy analysts are saying that we should stop thinking in terms of where we can get additional supplies of the old fuels to keep things running as they are. Instead, we should be thinking in terms of how we can satisfy these needs with minimum expenditure of energy and environmental impact.

Conservation

Imagine the discovery of a new oil field having a production potential of at least 6 million barrels a day—three times the capacity of the Alaskan field even at its height. Furthermore, assume that this new-found field is inexhaustible and that its exploitation will not adversely affect the environment. Of course, such an oil field sounds like a dream, but this is effectively what can be achieved through fossil-fuel conservation.

As noted earlier in this chapter, this "conservation reserve" has already been tapped to some extent, especially as the average fuel efficiency of cars was doubled from 13 to 26 mpg. Thus drivers of those autos are satisfying their transportation needs with only half the expenditure of energy and dollars. This and other conservation measures are already saving about $100 billion per year in oil imports. In short, advances in conservation have been extremely cost-effective as well as environmentally benign.

A number of energy experts estimate that at least 80% of the "conservation reserve" remains to be tapped. Cars are currently available that get at least double the current average of 26 mpg, and Peugeot, Toyota, Volkswagen, and Volvo all have prototypes that get from 70 to 100 mpg. A Renault prototype even attained a remarkable 124 mpg in test runs. To guarantee high sales, however, all the automakers are waiting for government mandates, consumer demand, or fuel shortages before putting these models into production. Of course, carpooling with just one person in addition to the driver effectively doubles the mpg.

FIGURE 21-19
Replacing standard incandescent light bulbs with screw-in fluorescent bulbs shown here can cut the energy demand for lighting by 75–80%. The higher cost of the fluorescent bulbs is more than offset by greater efficiency and much longer lifetime of the bulb. (Photograph by BJN.)

A technique known as **cogeneration**, described in the "Earth Watch" box, page 549, is another way of gaining large amounts of energy from the conservation reserve.

Other readily available conservation measures are underutilized: Improved insulation and double-pane windows for homes and buildings would save on heating fuel. Substitution of fluorescent lights, which are between 20% and 25% efficient (Fig. 21-19), for conventional incandescent light bulbs, which are only 5% efficient, would cut back on our electricity demands for lighting by 75%–80%. Similarly, more efficient refrigerators and other appliances could reduce energy demands for those purposes by 50%–80%. Energy-conserving light bulbs and appliances are more expensive than conventional models, but calculations show that they more than pay for themselves in energy savings. Also, some utilities are subsidizing the installation of fluorescent light bulbs and energy-efficient appliances as a cost-effective alternative to building additional power plants.

Finally, there is much to be gained by adjusting our personal habits and lifestyles to be more conserving, a topic addressed further in Chapter 24.

Many observers have pointed out that demanding more energy while failing to conserve is like demanding more water to fill a bathtub while leaving the drain open. To be sure, conservation and efficiency strategies by themselves will not eliminate demands for energy, but it can make the demands much easier to meet regardless of what options are chosen to provide the primary energy.

Development of Non-Fossil-Fuel Energy Sources

Lying before us are two major pathways for developing non-fossil-fuel energy alternatives. One is to pursue nuclear power; the other is to promote solar energy applications. Of course, nuclear power has been in use for 35 years, and various solar energy systems have also been developed. In other words, both nuclear and solar alternatives are at a stage where they could be rapidly expanded to provide further energy needs if suitable technolgical solutions and public acceptance (nuclear power) or pressure and government support (of solar energy) were provided. The following chapter focuses on the nuclear option, and Chapter 23 focuses on solar options.

Review Questions

1. How have the fuels to power homes, industry, and transportation changed from the beginning of the Industrial Revolution to the present?
2. What are the three primary fossil fuels, and what percentage does each contribute to the U.S. energy supply?
3. Electricity is a secondary energy source; how is it generated, and at what efficiencies of energy use?
4. What are the four major categories of primary energy use in an industrial society?
5. Match the energy sources that correspond to each of the four categories of primary energy use.
6. What is the distinction between estimated reserves and proven reserves of crude oil, and what factors cause the amounts of each to change?
7. Draw a graph expressing the inevitable trend of maximum production per year in the absence of new discoveries (reserve additions). Explain this trend.
8. How does the price of oil influence the amount produced?
9. What were the trends in oil consumption, in discovery of new reserves, and in U.S. production before 1970? After 1970?
10. What was done by the United States in the early 1970s to resolve the disparity between oil production and consumption?
11. What events caused the sudden oil shortages of the mid-1970s and then the return to abundant but more-expensive supplies?
12. What responses were made by the United States and other industrialized nations to the "oil crisis" of the 1970s, and what were the impacts of those responses on production and consumption?
13. Why did an oil glut occur in the mid-1980s, and what were its consequences in terms of oil prices and continuing pursuit of the responses listed in Question 12?
14. What have been the directions of U.S. production, consumption, and importation of crude oil since the mid-1980s?
15. What are the liabilities and risks of our growing dependence on foreign oil? How do these relate to the Persian Gulf War of 1991?
16. What alternative fossil fuels might the United States exploit to supplement declining reserves of crude oil? What are the economic and environmental advantages and disadvantages of each?
17. What phenomenon may restrict consumption of fossil fuels before resources are depleted?
18. To what degree has energy conservation served to mitigate energy dependency? What are some prime examples of energy conservation? To what degree may conservation alleviate future energy shortages and the cost of developing alternatives?

Thinking Environmentally

1. Statistics show that less-developed countries use far less energy per capita than developed countries. Explain why this is so.
2. The United States and many other developed and less-developed countries are going all out to expand highway systems and produce more cars and trucks—in short, to expand a transportation system that depends on liquid fuels. Discuss the sustainability of this system. Suggest alternatives.
3. Between 1980 and 1995, gasoline prices in the United States declined considerably relative to inflation. Predict what they will do in the next 10 years, and give a rationale for your prediction.
4. Suppose your region is facing power shortages (brown-outs). How would you propose solving the problem? Defend your proposed solution on both economic and environmental grounds.
5. List all the environmental impacts that occur during the "life span" of gasoline, that is, from exploratory drilling to consumption of gasoline in your car.
6. Develop what you feel should be the long-term policy for the United States to meet future transportation needs. Consider and balance the potentials and risks of continuing to increase imports, increasing exploration, developing synfuels or oil-shale deposits, conservation, and alternative modes of transportation. Suggest laws, regulations, taxes, or subsidies that might be used to implement your proposal.

Nuclear Power: Promise and Problems

Key Issues and Questions

1. Nuclear power currently generates 20% of electricity in the United States, 7% of energy overall, yet it is controversial. What is the history and current status of nuclear power?
2. The objective of nuclear power technology is to use a controlled nuclear reaction to boil water that turns a turbine that drives a generator. How does a nuclear power plant work?
3. In order to understand the hazards of nuclear power, we must understand something about radioactive materials. What are radioactive materials, and what hazards do they pose?
4. Disposal of nuclear waste is one of the major issues causing concern. What are the problems surrounding the disposal of nuclear waste?
5. The other major issue causing great concern is the possibility of nuclear accidents in nuclear plants. Of what significance are the Three Mile Island and Chornobyl´ nuclear accidents?
6. Fusion-based energy has long been considered a candidate for a pollution-free energy source of the future. What is the prognosis for fusion-based energy?
7. In light of current energy choices, nuclear power still appears to be an attractive option. What future is there for nuclear power?

"The Navajos make rugs, the Pueblos make pots, the Mescaleros make money"—quote attributed to Mescalero Apache president, Wendell Chino.

Multimillion-dollar casinos and ski resorts on Mescalero land in New Mexico testify to the truth of the preceding quote. And a new wrinkle in Apache free enterprise threatens to hold the rest of the state of New Mexico hostage to one of the most emotional issues raging throughout the United States: fear of radioactive wastes. The Apaches have offered 11 nuclear utility companies storage for 40,000 tons of high-level radioactive waste on the tribe's reservation in southern New Mexico (Fig. 22-1). The nuclear industry is faced with the prospect of closing one fourth of U.S. nuclear power plants by 1998 unless it can find someplace off-site to store spent nuclear fuel. The Apaches expect to receive $20 million in

Indian Point NPS. Buchanan, New York. [Photo by Paul Saloutos/The Stock Market.]

revenues annually in return for permitting the power plants to store their wastes in an interim storage site on their 700-square-mile reservation. A referendum in January 1995 lost. A second attempt in March 1995 resulted in negotiations. There are 3475 members in the tribe; 3200 on the reservation—less than 1000 voted. In April 1996, all negotiations and planning were suspended indefinitely. Nuclear waste remains an unsolved enigma. We have built nuclear power plants that now supply more than 20% of our electricity, but we still have not decided what to do with the rising burden of highly radioactive waste.

The Department of Energy was supposed to be accepting wastes from the nuclear utilities by 1998, but currently, the most optimistic estimate of when a facility will be ready is 2010—by which time three fourths of the nation's nuclear power plants will have exhausted their on-site storage space. New Mexico is strongly opposed to the agreement between the Apaches and the utilities, but there may be little the state can do to stop the transaction, because the Mescaleros are considered a sovereign nation. A tribal referendum has approved the basics of the deal, and the only other obstacle standing in the way of getting the site operational is licensing by the Nuclear Regulatory Commission (NRC). For the utilities, the stakes are high: billions of dollars of lost revenue if the plants have to close.

The foregoing controversy is only the most recent in the troubled history of nuclear power in the United States. In this chapter, we look into that history and add to it the perspectives of science and public policy as we sort out the many concerns about nuclear power.

Nuclear Power: Dream or Delusion?

As the industrialized nations continue their heavy use of fossil fuels, the threat of global warming gathers momentum. The Gulf War of 1991 provided graphic evidence of how developed nations dependent on overseas sources of crude oil are willing to use military force to protect those sources. In light of these two serious energy-related problems, it seems prudent to do everything possible to develop non-fossil-fuel energy sources. Nuclear power is an alternative that does not contribute to global warming, and there is sufficient uranium to fuel nuclear reactors well into the 21st century, with the possibility of extending the nuclear fuel supply indefinitely through reprocessing technologies. Thirty-two nations now have nuclear power plants either in place or under construction, and in some of these countries, nuclear power generates more electricity than any other source. Is this the key to the future of global energy—the best route to a sustainable relationship with our environment?

For hundreds of years, geologists have recognized that fossil fuels would not last forever. Sooner or later, other energy sources would be needed. That time of reckoning seemed far in the future, however. World War II, marked by both the enormous use of fossil fuels and the demonstration of the atom's awesome power, was

FIGURE 22-1
Miller Hudson, a consultant to the Mescalero Apache tribe, stands at the site in New Mexico proposed for an interim storage facility to receive spent nuclear wastes from nuclear power plants. (Robert M. Bryce.)

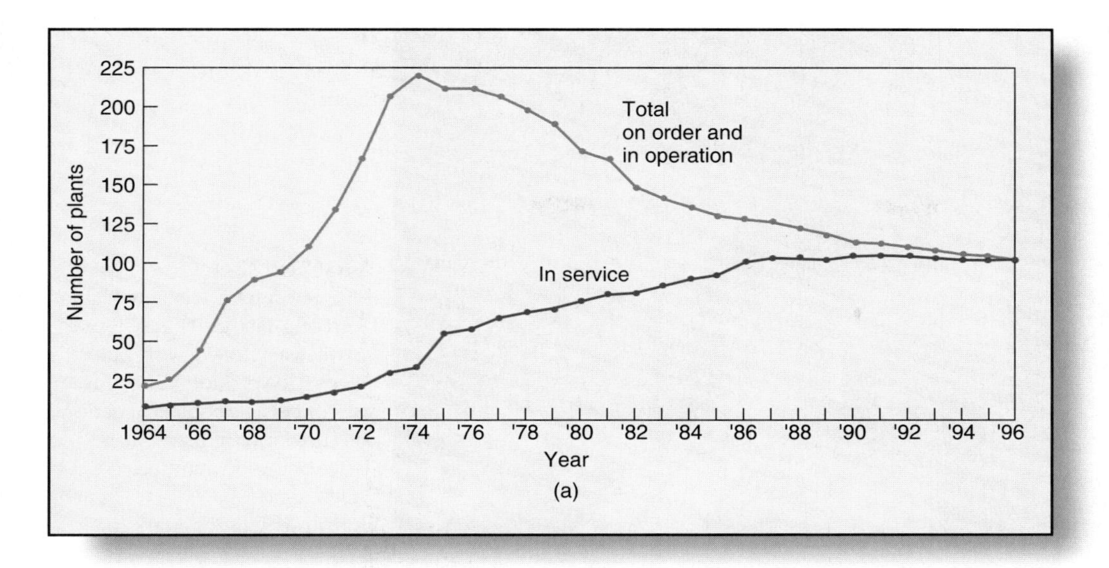

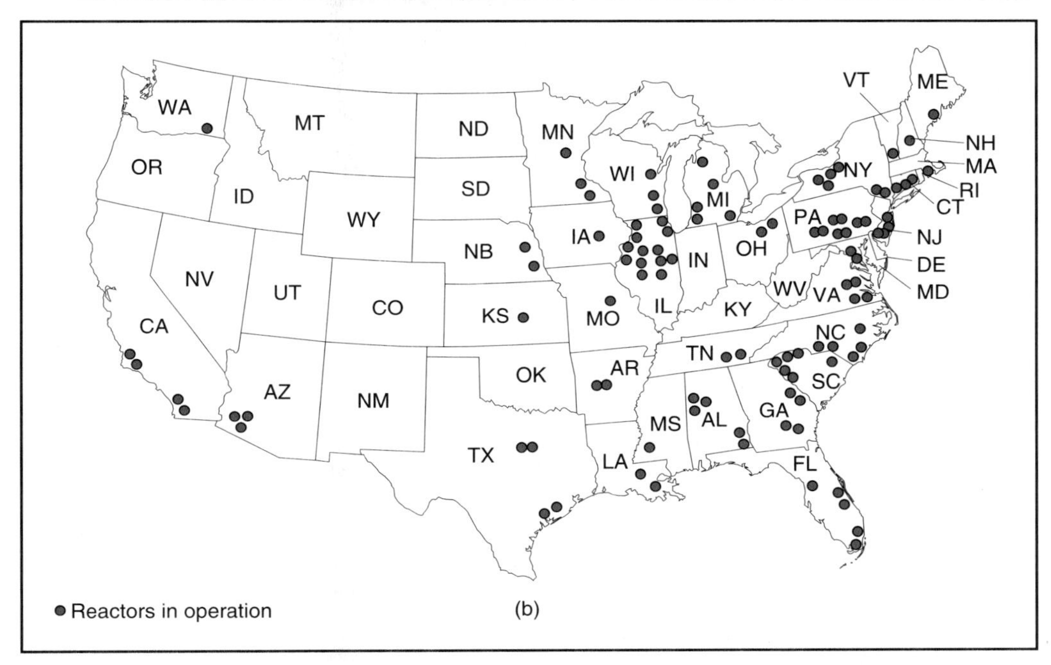

FIGURE 22-2
(a) Changing fortunes of nuclear power in the United States. Since the early 1970s, when orders for plants reached a peak, few utilities have called for new plants, and many have canceled earlier orders. (The last order not subsequently canceled was placed in 1974.) Nevertheless, the number of plants in service increased steadily as plants under construction were completed. The number of operating plants peaked at 111 and is now in decline. *Source:* U.S. Dept. of Energy.

the time of decision. The U.S. government desperately wanted to show the world that the power of the atom could benefit humankind, as well as destroy it, and embarked on a course to lead the world into the "Nuclear Age."

It was anticipated that nuclear power could produce electricity in such large amounts and so cheaply that we would phase into an economy in which electricity would take over virtually all functions, including the generation of other fuels, at nominal costs. Thus, the government moved into the research, development, and promotion of commercial nuclear power plants (along with the continuing development of nuclear weaponry). Utilizing this research, companies such as General Electric and Westinghouse constructed nuclear power plants that were ordered and paid for by utility companies, this following assurances from the federal governent that exempted these corporations and utilities from liability (Price-Anderson Act 1957). The Nuclear Regulatory Commission (NRC) (formerly the Atomic Energy Com-mission), an agency in the Department of Energy, set and enforced safety standards for the operation and maintenance of these plants, as it does today.

In the 1960s and early 1970s, utility companies moved ahead with plans for numerous nuclear power plants (Fig. 22-2). By 1975, 53 plants were operating in the United States, producing about 9% of the nation's electricity, and another 170 plants were in various stages of planning or construction. Officials estimated

FIGURE 22-3
The Shoreham Nuclear Plant on Long Island, N.Y. This plant was closed with little electricity produced because of concerns about evacuation of surrounding areas. It is now being dismantled. (DiMaggio/ Kalish/The Stock Market.)

that by 1990 several hundred plants would be on line, and by the turn of the century as many as a thousand would be operating. A number of other industrialized countries got in step with their own programs, and some less developed nations were going nuclear by buying plants from industrialized nations.

Since 1975, however, the picture has changed dramatically. Utilities stopped ordering nuclear plants, and numerous existing orders were canceled. The last order for a plant in the United States not subsequently canceled was placed in 1974. In some cases, construction was terminated even after billions of dollars had already been invested. Most striking, the Shoreham Nuclear Power plant on Long Island, New York, after being completed and licensed at a cost of $5.5 billion, was turned over to the state of New York in the summer of 1989 to be dismantled after generating electricity for only 32 hours (Fig. 22-3). The reason was that citizens and the state deemed that there was no way possible to evacuate people from the area should an accident occur. Similarly, just a few weeks earlier, California citizens voted to shut down the Rancho Seco Nuclear Power Plant located near Sacramento, which had a 15-year history of troubled operation.

At the end of 1996, there were only 110 operating nuclear power plants in the United States. When the Watts Bar nuclear plant in Tennessee came on line in February 1996, there were no more plants under construction. These plants are generating 20% of U.S. electrical power. Thus, U.S. nuclear power has peaked at 110 plants and will head downward as older plants are decommissioned. Including the United States, a worldwide total of 430 nuclear plants is operating, with an additional 55 "under construction" (many of these may never be completed). In general, the story abroad is similar to that in the United States; drastic downward revisions, if not total moratoriums, on the construction of new plants. Nuclear power generates about 17% of the world's electricity, but dependence on this source varies greatly among the 30 countries operating nuclear power plants (Fig. 22-4). Of the major industrial nations, only France and Japan remain fully committed to pushing forward with nuclear programs. Thus, France is now producing 73% of its electricity with nuclear power and has plans to go to more than 80%; Japan plans to increase its nuclear percentage from 28% to 43% by 2010.

After the catastrophic accident at Chornobyl´ in April 1986, it is not hard to see why nuclear power is being rethought. Yet, the demand for electricity is more robust than ever, and the other means of generating electricity have their own problems. Coal-powered electricity generates more of the greenhouse gases than any other form. Oil supplies are more limited, and oil is vital to transportation and home heating. Hydroelectric power is already heavily developed. Solar power technolgoy is still lagging. If the technological and economic issues can be resolved, we need not lose this source for supplying future energy needs. Are the public's safety concerns justified? In order to react intelligently to this issue, we need a clear understanding of what nuclear power is and of its pros and cons.

How Nuclear Power Works

The objective of nuclear power technology is to control nuclear reactions so that energy is released gradually as thermal infrared (heat energy). As with plants powered by fossil fuels, the heat energy is used to boil water and produce steam, which then drives conventional turbogenerators.

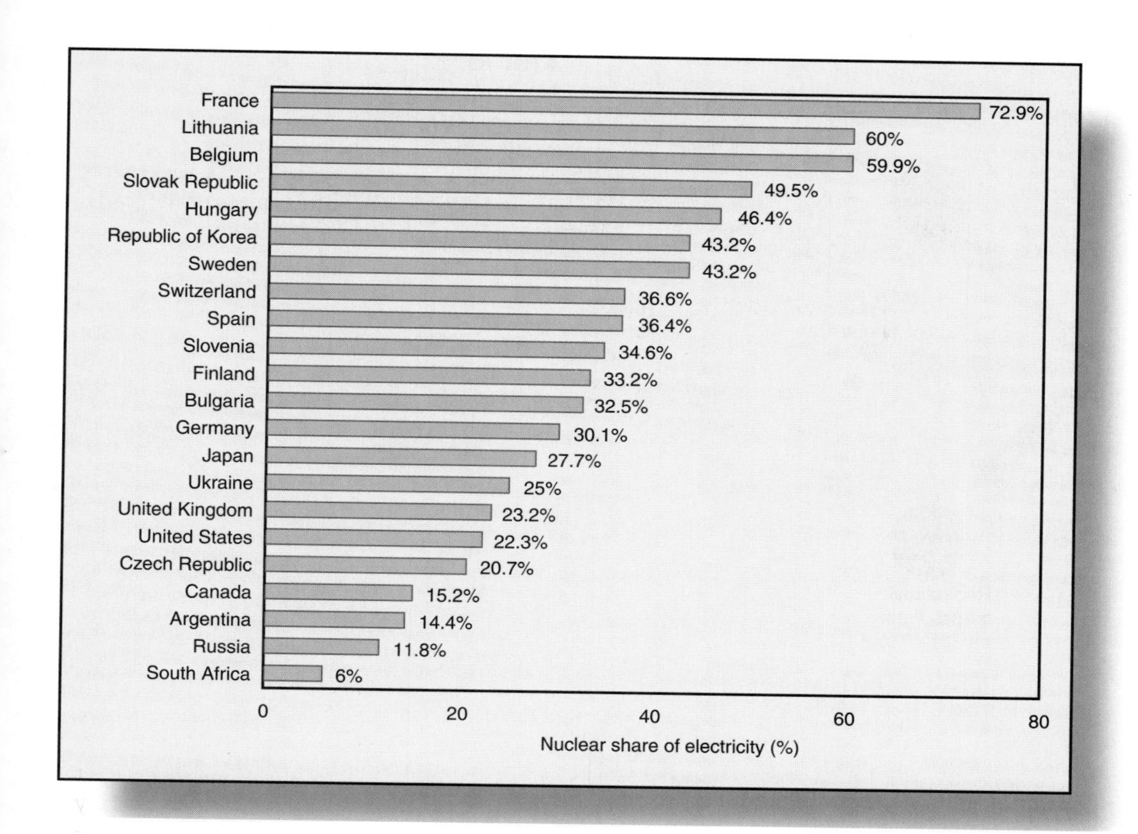

FIGURE 22-4 Percentage of electricity generated by nuclear power, in selected countries. (From International Atomic Energy Agency.)

From Mass to Energy

The release of nuclear energy is completely different from the burning of fuels or any other chemical reactions we have discussed. To begin with, materials involved in chemical reactions remain unchanged at the atomic level, although the visible forms of these materials undergo great transformation as atoms rearrange to form different compounds. Nuclear energy, however, involves changes at the atomic level through one of two basic processes: fission or fusion. In **fission**, a large atom of one element is split to produce two smaller atoms of different elements (Fig. 22-5a). In **fusion**, two small atoms combine to form a larger atom of a different element (Fig. 22-5b). In both fission and fusion, the mass of the product(s) is less than the mass of the starting material, and the lost mass is converted to energy in accordance with the law of mass-energy equivalence ($E = mc^2$), first described by Albert Einstein. The amount of energy released by this mass-to-energy conversion is tremendous. The sudden fission or fusion of a mere 1 kg of material releases the devastatingly explosive energy of a nuclear bomb.

The Fuel for Nuclear Power Plants. All current nuclear power plants utilize the fission (splitting) of uranium-235. The element uranium occurs naturally in various minerals in Earth's crust. It exists in two primary forms, or *isotopes*: uranium-238 (^{238}U) and uranium-235 (^{235}U). Like uranium, many other elements exist in more than one isotopic form. The **isotopes** of a given element contain different numbers of neutrons, but the same number of protons and electrons. Some isotopes are unstable and release particles or rays, which are called **radioactive emissions**, to be discussed shortly.

The number that accompanies the chemical name or symbol of an element is called the **mass number** of the element and is the sum of the number of neutrons and the number of protons in the nucleus of the atom. Since, by definition, all atoms of any given element must contain the same number of protons, variations in mass number represent variations in numbers of neutrons. Thus, ^{238}U contains 92 protons and 146 neutrons, while ^{235}U contains 92 protons and 143 neutrons. Whereas ^{238}U contains 3 more neutrons than ^{235}U, both isotopes contain the 92 protons that *define* the element uranium. Although all isotopes of a given element behave the same chemically, their other characteristics may differ profoundly. In the case of uranium, ^{235}U atoms will undergo fission, but ^{238}U atoms will not.

It takes a neutron hitting the nucleus at just the right speed to cause ^{235}U to undergo fission. Since ^{235}U is an unstable isotope, a small but predictable number of the ^{235}U atoms in any sample of the element undergo radioactive decay and release neutrons, among other things. If one released neutron moving at just the right speed hits another ^{235}U atom, the latter becomes ^{236}U, which is very unstable and undergoes fission immediately into lighter atoms (**fission products**). The fission reaction gives off several more neutrons and releases

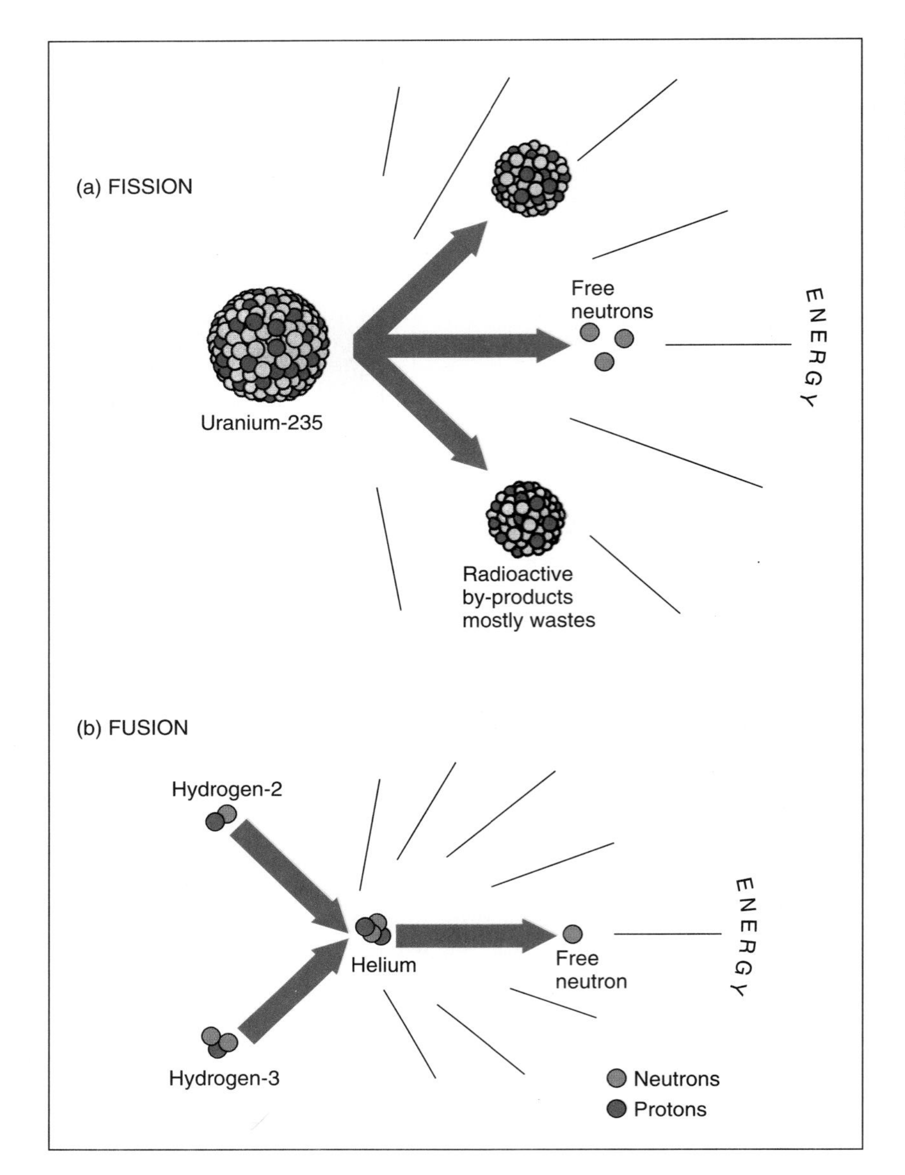

FIGURE 22-5
Nuclear energy is released from either (a) fission, the splitting of certain large atoms into smaller atoms, or (b) fusion, the fusing together of small atoms to form a larger atom. In both cases, some of the mass of the starting atom(s) is converted to energy.

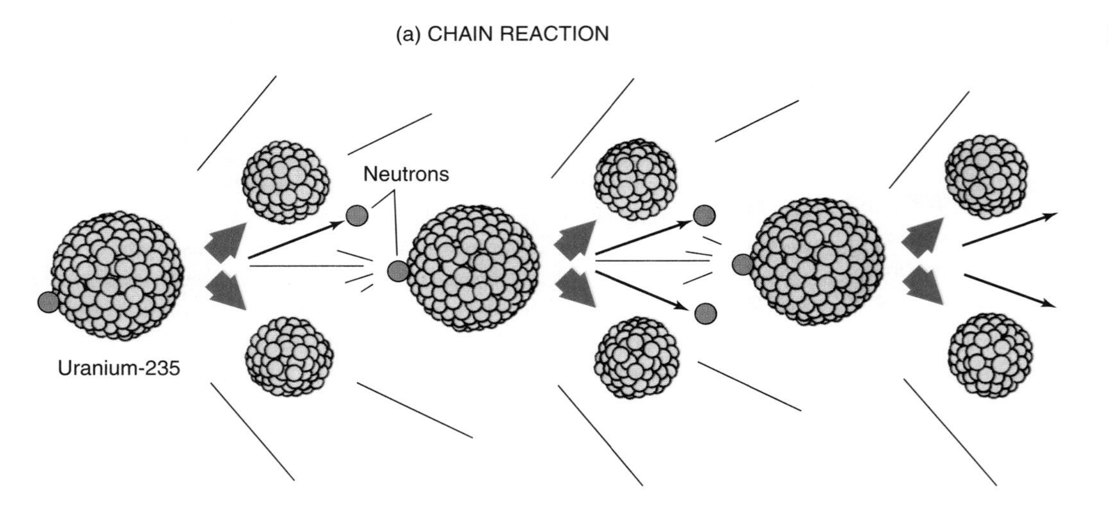

FIGURE 22-6

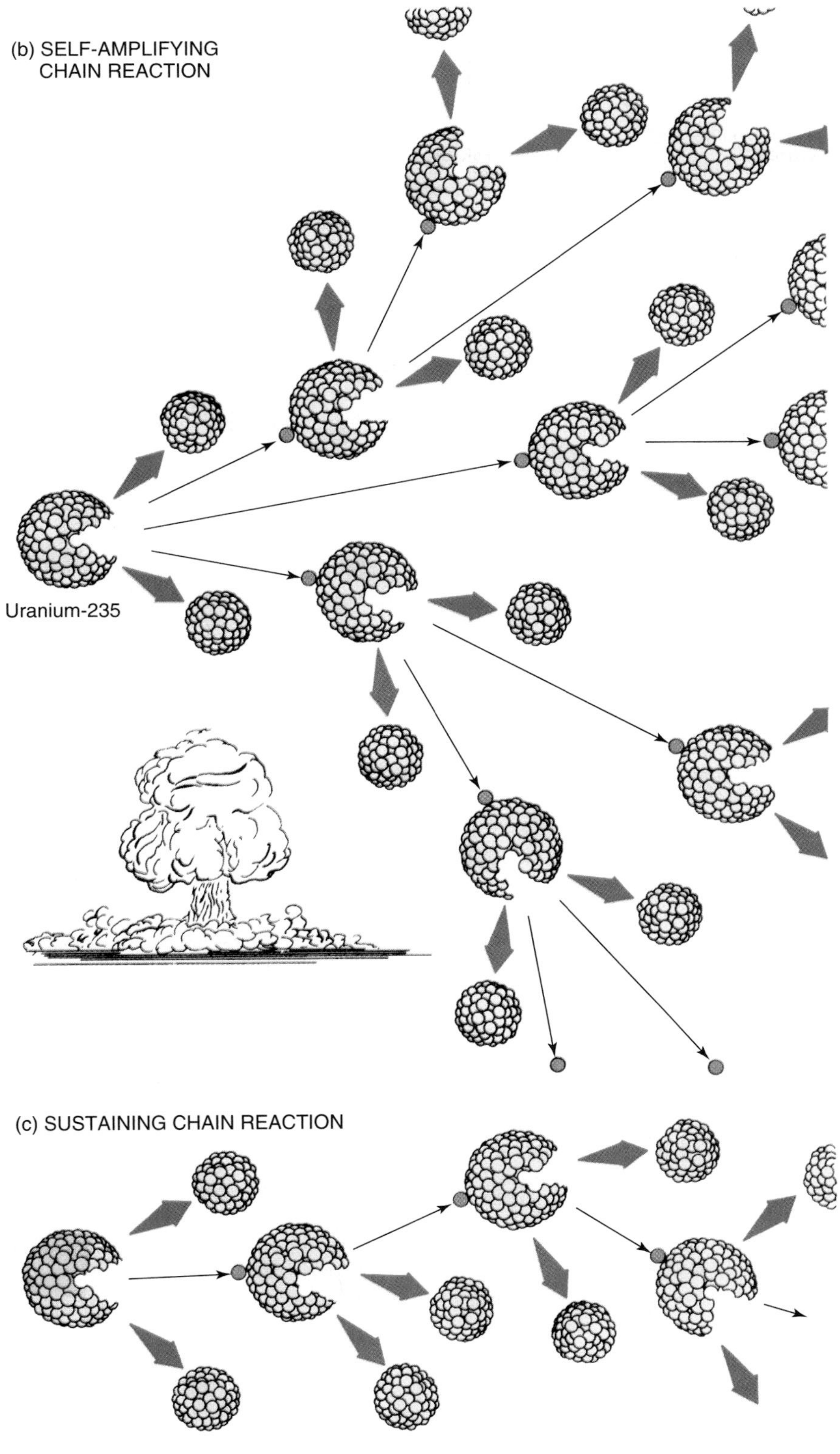

(b) SELF-AMPLIFYING CHAIN REACTION

Uranium-235

(c) SUSTAINING CHAIN REACTION

FIGURE 22-6
(a) (opposite page) A simple chain reaction. When a uranium atom fissions, it releases two or three high-energy neutrons, in addition to energy and split "halves." If another ^{235}U atom is struck by a high-energy neutron, it fissions, and the process is repeated, causing a chain reaction. (b) A self-amplifying chain reaction leading to a nuclear explosion. Since two or three high-energy neutrons are produced by each fission, each may cause the fission of two or three additional atoms. Hence, the entirety of a suitably concentrated mass of fissionable material may be caused to fission in a tiny fraction of a second, resulting in a nuclear explosion. (c) In a sustaining chain reaction, the extra neutrons are absorbed in control rods so that amplification does not occur.

a great deal of energy (Fig. 22-5a). Ordinarily, these neutrons are traveling too fast to cause fission, but if they are slowed down and then strike another ^{235}U atom, they can cause fission to occur again. In this way, more neutrons and more energy are released, with the potential to repeat the process. A domino effect, known as a chain reaction, may thus occur (Fig. 22-6a).

A chain reaction does not normally occur in nature, because ^{235}U atoms are too dispersed among other elements and more stable ^{238}U atoms, which absorb neutrons without undergoing fission. In fact, 99.3% of all uranium found in nature is ^{238}U; only 0.7% is ^{235}U. Hence, when a ^{235}U atom spontaneously decays in nature, it seldom triggers fission in another atom, and the event goes unnoticed without the aid of radiation detectors such as Geiger counters.

To make nuclear "fuel," uranium ore is mined, purified, and *enriched*. **Enrichment** involves separating ^{235}U from ^{238}U to produce a material containing a higher concentration of ^{235}U. Since ^{238}U and ^{235}U are chemically identical, enrichment is based on their slight difference in mass. The technical difficulty of enrichment is the major hurdle that prevents less developed countries from advancing their own nuclear capabilities. Enrichment has proven problematic at government plants in the United States in terms of worker exposures and materials unaccounted for (MUF).

When ^{235}U is highly enriched, the spontaneous fission of an atom can trigger a chain reaction. In nuclear weapons, small masses of virtually pure ^{235}U or other fissionable material are forced together so that the two or three neutrons from a spontaneous fission cause two or three more atoms to undergo fission; each of these in turn triggers two or three more fissions, and so on. The whole mass undergoes fission in a fraction of a second, releasing all the energy in one huge explosion (Fig. 22-6b).

The Nuclear Reactor. A nuclear reactor for a power plant is designed to sustain a continuous chain reaction (Fig. 22-6c), but not allow it to amplify into a nuclear explosion. Control is achieved by enriching the uranium to only 3% ^{235}U and 97% ^{238}U. This modest enrichment will not support the amplification of a chain reaction into a nuclear explosion.

A chain reaction can be achieved in a nuclear reactor only if a sufficient mass of enriched uranium is arranged in a suitable geometric pattern and is surrounded by a material called a *moderator*. The **moderator** slows down the neutrons which produce fission so that they are traveling at the right speed to trigger another fission. In slowing down the neutrons, the moderator gains heat. In nuclear plants in the United States, the moderator is very pure water, and the reactors are called light water reactors (LWRs). (The term *light water* denotes ordinary water, H_2O, as opposed to the substance known as heavy water, or deuterium, D_2O.) Other moderators employed in reactors of different design are graphite and deuterium.

To achieve the geometric pattern necessary for fission, the enriched uranium is made into pellets that are loaded into long metal tubes. The loaded tubes are called **fuel elements** or **fuel rods**. Many fuel rods are placed close together to form a reactor core inside a strong reactor vessel that holds the water, which serves as both moderator and heat-exchange fluid (coolant).

The chain reaction in the reactor core is controlled by rods of neutron-absorbing material, referred to as control rods, inserted between the fuel elements. The chain reaction is started and controlled by withdrawing and inserting the control rods as necessary. The nuclear reactor is simply the assembly of fuel elements, moderator-coolant, and movable control rods (Fig. 22-7). As the control rods are removed and a chain reaction is initiated, the fuel rods and the moderator become intensely hot.

The Nuclear Power Plant. In a nuclear power plant, heat from the reactor is used to boil water to provide steam for driving conventional turbogenerators. One way to boil the water is to circulate it through the reactor. In most power plants in the United States, however, a double loop is employed. The moderator-coolant water is heated to over 600°F by circulating it through the reactor, but it does not boil, because the system is under very high pressure (2100 p.s.i.). This superheated water is circulated through a heat exchanger, where it boils other unpressurized water that flows past the tubes containing the super-heated water. This produces the steam used to drive the turbogenerator.

The double-loop design of this primary system and secondary system isolates hazardous materials in the reactor from the rest of the power plant. However, it has one serious drawback: If the reaction vessel should break, the sudden loss of water from around the reactor, called a "loss-of-coolant accident" (LOCA), could result in the core overheating. The sudden loss of the moderator-coolant water would cause fission to cease, since the moderator would no longer be present. However, even though fission stops, the fuel core can still overheat because 7% of the reactor heat comes from radioactive decay in the newly formed fission products. In time, the uncontrolled decay would release enough heat energy to melt the materials in the core, a situation called a **meltdown**. Then, the molten material falling into the remaining water could cause a steam explosion. To guard against all this, there are backup cooling systems to keep the reactor immersed in water should leaks occur, and the entire assembly is housed in a thick concrete containment building (Fig. 22-8).

The fission of about 1 pound (0.5 kg) of uranium fuel releases energy equivalent to burning 50 tons of coal. Thus, one fueling of the reactor with about 60 tons of uranium oxide is sufficient to run the power plant for as long as two years. The highly radioactive spent fuel elements are then removed and replaced with new ones.

Comparison of Nuclear Power with Coal Power

In canceling plans for nuclear power plants, we did not decide to do without electricity; in effect, we opted to build coal-fired power plants instead. The United States does have abundant coal reserves, but is burning coal

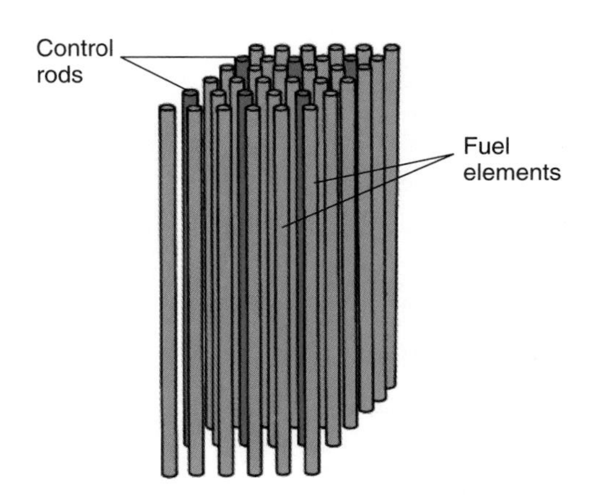

(a)

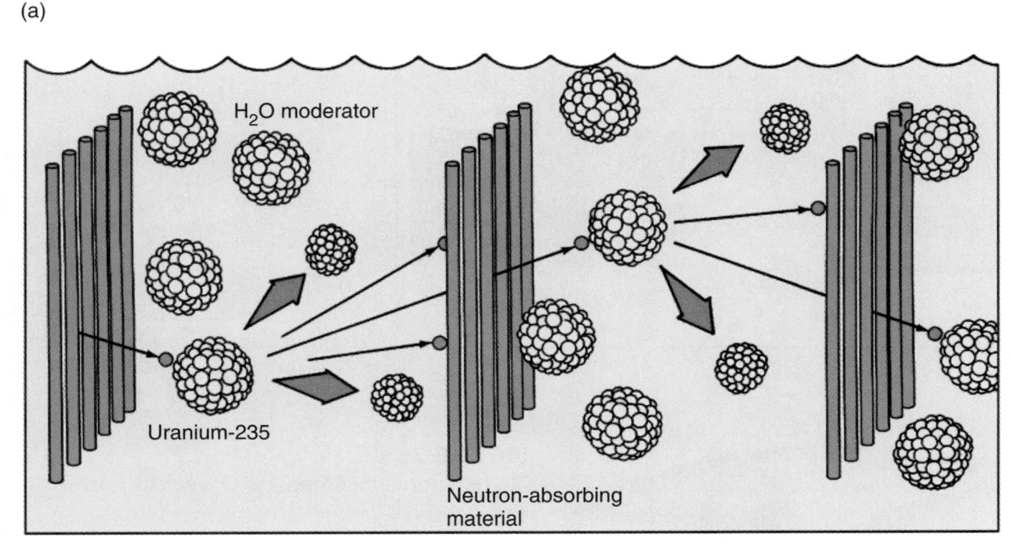

(b)

FIGURE 22-7
(a) In the core of a nuclear reactor, a large mass of uranium is created by placing uranium in adjacent tubes, called the fuel elements. The uranium is not sufficiently concentrated to permit a nuclear explosion, but it will sustain a chain reaction that will produce a tremendous amount of heat. The rate of the chain reaction is moderated by inserting or removing rods of neutron-absorbing material (control rods) between the fuel elements. The fuel and rods are surrounded by the moderator fluid, pure water. (b) Technicians ready the core housing to receive uranium fuel elements in this nuclear reactor. (Erich Hartmann/Magnum Photos, Inc.)

the course of action we wish to pursue? Nuclear power has some decided environmental advantages over coal-fired power (Fig. 22-9). Comparing a 1000-megawatt nuclear plant with a coal-fired plant of the same capacity, each operating for one year, we find the following characteristics:

- *Fuel needed.* The coal plant consumes about 3 million tons of coal. If this amount is obtained by strip mining, some environmental destruction and acid leaching will result. If the coal comes from deep

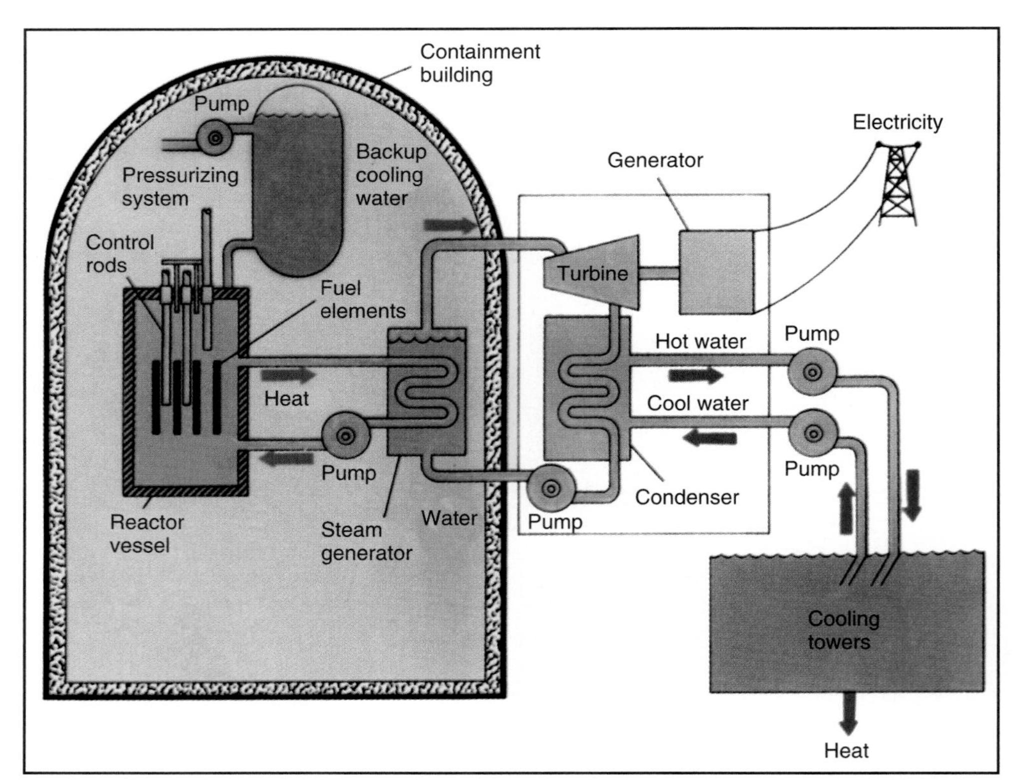

FIGURE 22-8
Schematic diagram of a pressurized nuclear power plant.

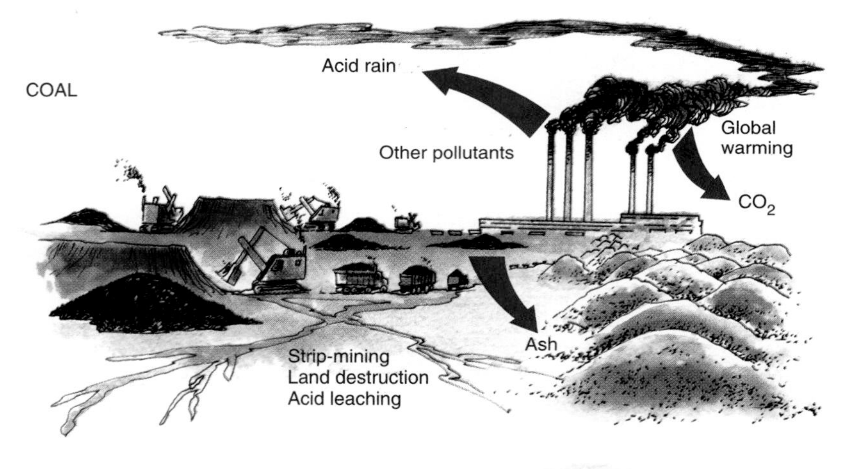

FIGURE 22-9
The environmental impacts of nuclear power and coal-fired power. The nuclear power option assumes perfect containment of radioactivity throughout the fuel cycle and that some method is finally determined for spent fuel and radioactive waste storage and disposal.

mines, there will be human costs in the form of accidental deaths and impaired health. The nuclear plant requires about 30 tons of enriched uranium obtained from mining 75,000 tons of ore, with much less harm to humans and the environment.

- *Carbon dioxide emission.* The coal plant emits over 10 million tons of carbon dioxide into the atmosphere, contributing to global warming. The nuclear plant emits none.
- *Sulfur dioxide and other emissions.* The coal plant emits over 400,000 tons of sulfur dioxide and other acid-forming pollutants, which must be captured by precipitators and scrubbers. It also releases low levels of many radioactive chemicals found naturally in the coal. The nuclear power plant produces no acid-forming pollutants, but will release low levels of radioactive waste gases.
- *Solid wastes.* The coal plant produces about 100,000 tons of ash, requiring land disposal. The nuclear plant produces about 250 tons of highly radioactive wastes requiring safe storage and ultimate safe disposal—the handling of these radioactive wastes remains an unresolved enigma.
- *Accidents.* A worst case accident in the coal plant could result in fatalities to workers and a destructive fire, a situation common to many industries. Depending on the type of nuclear plant, accidents can range from minor emissions of radioactivity to catastrophic releases that can lead to widespread radiation sickness, scores of human deaths, untold numbers of cancers, and widespread, long-lasting environmental contamination.

Of course, it is the handling of radioactive wastes and releases and the potential for accidents that have led the public to reject nuclear power. But are these problems that cannot be overcome? Or, are other energy options less problematic and less costly than nuclear?

Radioactive Materials and Their Hazards

Assessing the hazards of nuclear power necessitates our understanding of radioactive substances and their danger.

Radioactive Emissions

When uranium or any other element undergoes fission, the split "halves" are atoms of lighter elements: iodine, cesium, strontium, cobalt, or any of some 30 other elements. These newly formed atoms—called the *direct products* of the fission—are generally unstable isotopes of their respective elements. Unstable isotopes (called **radioisotopes**) become stable by spontaneously ejecting subatomic particles (alpha particles, beta particles, and neutrons), or high-energy radiation (gamma rays), or both. The particles and radiation are collectively referred to as **radioactive emissions**. Any materials in and around the reactor may also be converted to unstable isotopes and become radioactive by absorbing neutrons from the fission process (Fig. 22-10). These

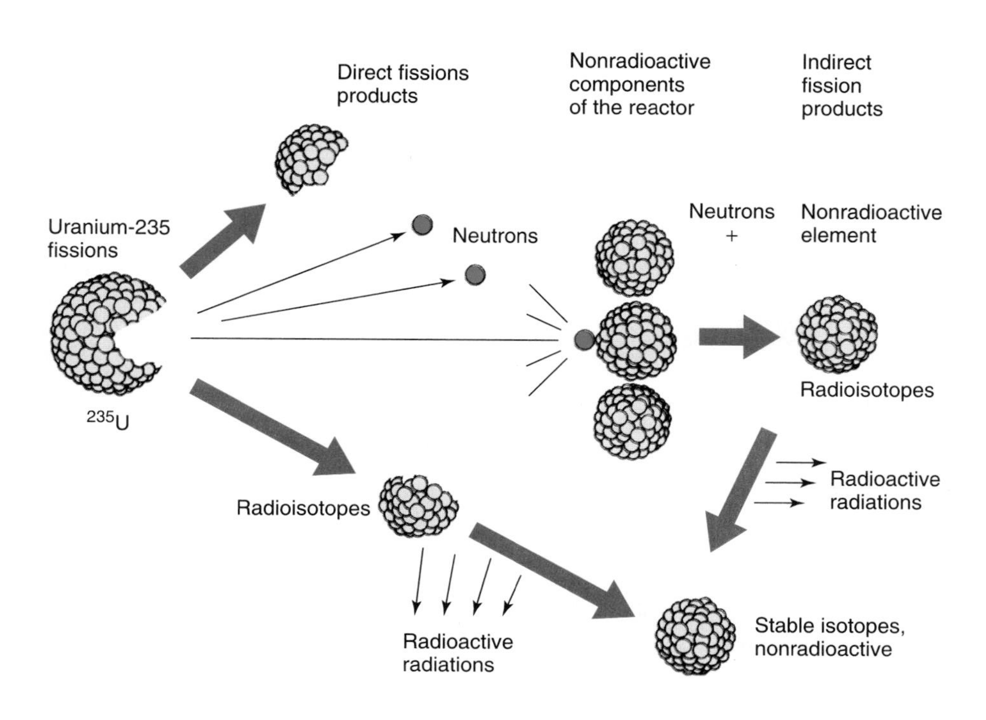

FIGURE 22-10
Radioactive wastes and radioactive emissions. Nuclear fission results in the production of numerous unstable isotopes, the radioactive wastes. They give off potentially damaging radiation until they regain a stable structure.

TABLE 22-1
Relative Doses from Radiation Sources

Source	Dose
All sources (avg.)	360 mrem/year
Cosmic radiation at sea level*	26 mrem/year
Terrestrial radiation (elements in soil)	26 mrem/year
Radon in average house	200 mrem/year
X rays and nuclear medicine (avg.)	50 mrem/year
Consumer products	11 mrem/year
Natural radioactivity in the body	39 mrem/year
Mammogram	30 mrem
Chest X ray	10 mrem
Gastrointestinal series of X rays	1.4 rem (1,400 mrem)
Continued fallout from nuclear testing	1 mrem/year
Living near a nuclear power station (assuming perfect containment)	< 1 mrem/year

*Increases 1/2 mrem for every 100 ft of elevation.

indirect products of fission, along with the direct products, are the **radioactive wastes** of nuclear power. (Radioactive fallout from nuclear explosions also consists of these direct and indirect fission products.)

Biological Effects. A major concern regarding nuclear power is that large numbers of the public may be exposed to low levels of radiation, thus elevating the risk of cancer and other disorders. Is this a valid concern? Radioactive emissions can penetrate biological tissue; their ability to do damage is measured in units called *rem*. The emissions leave no visible mark, nor are they felt, but they are capable of breaking molecules within cells. In high doses, radiation may cause enough damage to prevent cell division. Thus, in medical applications, radiation can be focused on a cancerous tumor to destroy it. However, if the whole body is exposed to such levels of radiation (over 100 rem is considered a high dose), a generalized blockage of cell division occurs that prevents the normal replacement or repair of blood, skin, and other tissues. This result is called *radiation sickness* and may lead to death a few days or months after exposure. Very high levels of radiation may totally destroy cells, causing immediate death.

In lower doses, radiation may damage DNA, the genetic material inside the cell. Cells with damaged (mutated) DNA may then begin dividing and growing out of control, forming malignant tumors or leukemia. If the damaged DNA is in an egg or a sperm cell, the result may be birth defects in offspring. The effects of exposure to radiation may go unseen until many years after the event; 10 to 40 years is typical. Other effects include a weakening of the immune system, mental retardation, and the development of cataracts.

Health effects are directly related to the level of exposure. There is broad agreement that doses between 10 and 50 rem result in an increased risk of developing cancer. Evidence for this comes from studies of patients with various illnesses who were exposed to high levels of X rays in the 1930s, when the potential harm of such radiation was not realized. People in these groups subsequently developed higher than normal rates of cancer and leukemia. Many scientists believe that no dose is without some harm; others point to the ability of living cells to repair small amounts of damage to DNA and believe that there is a threshold of radiation below which no biological effects occur. If there is such a threshold, it is virtually impossible to demonstrate, because of the many different causes of cancer, the long time it takes for cancer to develop, and the variety of sources of radiation to which people are exposed.

Sources of Radiation. Nuclear power is by no means the only potential source of radiation. There is also normal **background radiation** from radioactive materials, such as the uranium and radon gas that occur naturally in the earth's crust, and cosmic rays from outer space. For most people, background radiation is the major source of radiation exposure. In addition, we deliberately expose ourselves to radiation from medical and dental X rays, by far the largest source of human-induced exposure and, for the average person, equal to one-fifth the exposure from background sources. The average person in the United States receives a dose of about 360 millirem (*mrem*, a thousandth of a rem) per year (Table 22-1). Thus, the argument becomes one of *relative hazards*. Does or will the radiation from nuclear power significantly raise radiation levels and elevate the risk of developing cancer?

During normal operation of a nuclear power plant, the direct fission products remain within the fuel elements, and the indirect products are maintained within the containment building that houses the reactor.

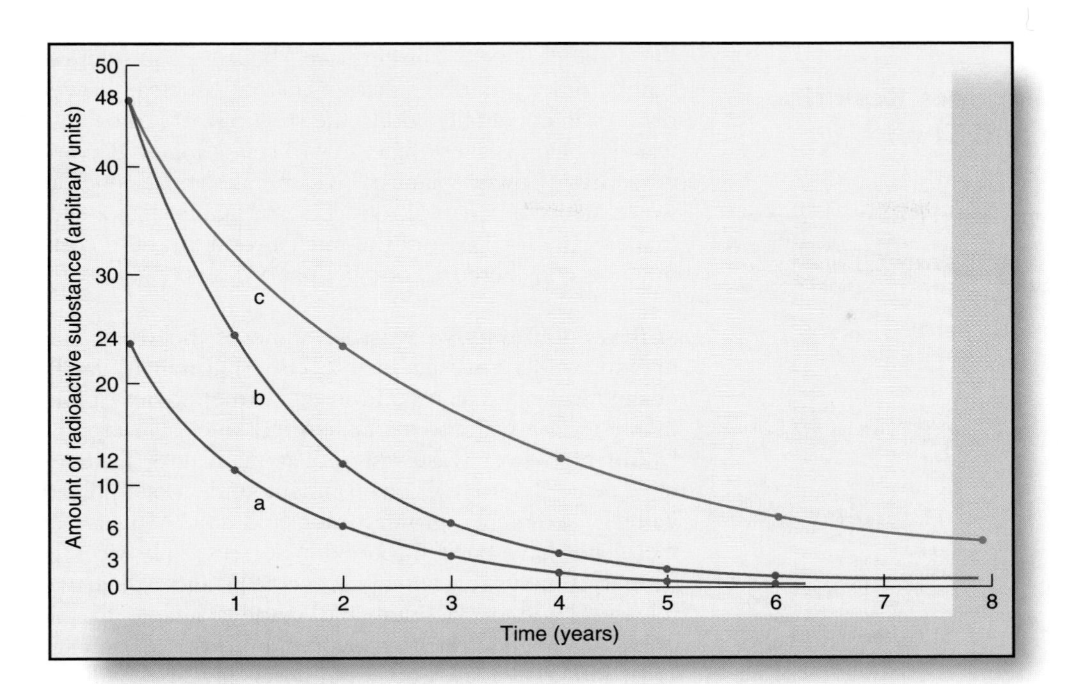

FIGURE 22-11
Radioactivity for any isotope declines as shown. Regardless of the starting amount, one half decays during each successive half-life. (a) A substance with a half-life of one year, starting with 24 units; (b) the same substance starting with 48 units; (c) a substance with a half-life of two years. The half-lives of different isotopes vary from less than one second to many thousands of years.

No routine discharges of radioactive materials (other than routine stack emissions) into the environment occur. Even very close to an operating nuclear power plant, radiation levels from the plant are lower than normal background levels. A radiation detector will pick up more radiation from Earth and concrete on a basement floor than it will when held within 150 yards of a nuclear power plant. Careful measurements have shown that public exposure to radiation from normal operations of a power plant is less than 1% of natural background.

The main concern about nuclear power, therefore, does not focus on *normal* operation. The real problems are about the storing and disposal of radioactive wastes and the potential for accidents.

Radioactive Wastes

To understand the problems surrounding nuclear waste disposal, we must understand the concept of *radioactive decay*. As unstable isotopes eject particles and radiation, they become stable and cease to be radioactive. This process is known as **radioactive decay**. As long as radioactive materials are kept isolated from humans and other organisms, the decay proceeds harmlessly.

The rate of radioactive decay is such that half of the starting amount of a given isotope will decay in a certain period of time. In the next equal period of time, half of the remainder (half of a half, which equals one fourth of the original) decays, and so on, as shown in Fig. 22-11. The time for half the amount of a radioactive isotope to decay is known as its **half-life**. The half-life of an isotope is always the same, regardless of the starting amount.

Note from Fig. 22-11 that decay of a radioisotope never equals 100%; radioactivity is reduced by only half in each half-life, and there is always an undecayed portion remaining. However, it is generally held that radiation is reduced to insignificant levels after 10 half-lives.

Each particular radioactive isotope has a characteristic half-life. The half-lives for various isotopes range from a fraction of a second to many thousands of years. Uranium fissioning results in a heterogeneous mixture of radioisotopes, the most common of which are listed in Table 22-2.

Disposal of Radioactive Wastes. Much of the radioactivity of fission wastes will dissipate in a period of months or a few years as the short-lived isotopes decay. To be safe, however, long-lived isotopes require isolation for up to 240,000 years (ten times the half-life of plutonium). Thus, the problem of nuclear waste disposal is twofold:

- *Short-term containment to allow the radioactive decay of short-lived isotopes.* Thus, in 10 years, fission wastes lose more than 97% of their radioactivity. Wastes can be handled much more easily and safely after this loss occurs.
- *Long-term containment (EPA recommends 10,000-year minimum) to provide protection from the long-lived isotopes—govenment standards require isolation for 20 half lives.*

TABLE 22-2
Common Radioactive Isotopes Resulting from Uranium Fission and Their Half-Lives

Short-Lived Fission Products	*Half-Life (days)*
Strontium-89	50.5
Yttrium-91	58.5
Zirconium-95	64.0
Niobium-95	35.0
Molybdenum-99	2.8
Ruthenium-103	39.4
Iodine-131	8.1
Xenon-133	5.3
Barium-140	12.8
Cerium-141	32.5
Praseodymium-143	13.6
Neodymium-147	11.1
Long-Lived Fission Products	***Half-Life (years)***
Krypton-85	10.7
Strontium-90	28.0
Ruthenium-106	1.0
Cesium-137	30.0
Cerium-144	0.8
Promethium-147	2.6
Addition Products of Neutron Bombardment	***Half-Life (years)***
Plutonium-239	24,000

For short-term containment, the spent fuel is first stored in deep swimming-pool-like tanks on the sites of nuclear power plants. The water in these tanks dissipates waste heat, which is still generated to some degree, and acts as a shield against radiation. After a few years of decay, the spent fuel may then be placed in dry casks in order to save space; in these casks, they are simply air cooled. (Nuclear weapons facilities also maintain various tanks for short-term containment of radioactive wastes.) There is no commercial spent-fuel reprocessing in the United States.

The development of and commitment to nuclear power went ahead without ever fully addressing the issue of ultimate long-term containment. It was generally assumed by nuclear proponents that the long-lived wastes could be solidified, placed in sealed containers, and buried in deep, stable rock formations (geologic burial) as the need for such containment became necessary (Fig. 22-12). In the meantime, the wastes from the world's commercial reactors have been accumulating at a rate of 9500 tons a year, reaching 150,000 tons at the end of 1997, all of which is stored on site at the power plants. At least one fourth of these wastes are in the United States. Furthermore, because of neutron bombardment of the reactor walls, all nuclear power plants will eventually add to the stockpile of radioactive wastes. Scientists estimate that dismantling a decommissioned power plant will generate more nuclear waste than the plant produced during its active life. Imagine the total cost of the nuclear option; these costs must be considered in our evaluation.

Military Radioactive Wastes. Some of the worst failures in handling wastes have occurred at military facilities in the United States and in the former Soviet Union, in connection with the manufacture of nuclear weapons. Liquid high-level wastes stored at many U.S. facilities have leaked into the environment and contaminated wildlife, sediments, groundwater, and soil. Activities at these sites have been shrouded in secrecy; only recently have documents revealing past accidents and radioactive releases been declassified and made available to the public. Deliberate releases of uranium dust, xenon (Xe-133), iodine (I-131), and tritium gas have been documented at Hanford, Washington; Fernald, Ohio; Oak Ridge, Tennessee; and Savannah River, South Carolina. These sites and their cleanup are now the responsibility of the Department of Energy, whose secretary, Hazel O'Leary, estimates that the cost of cleaning up and stabilizing the 20 weapons-production sites will range anywhere from $600 billion to $1.5 trillion.

Russian military weapons facilities have been even more irresponsible. The worst of these is a giant complex called Chelyabinsk-65, located in the Ural Mountains of Russia. For at least 20 years, nuclear wastes were discharged into the Techa River and then into Lake Karachay. At least 1000 cases of leukemia have been traced to radioactive contamination from the Chelyabinsk facility. Even today, standing for one hour on the shore of Lake Karachay will kill a person from radiation poisoning within a week. The lake dried up one summer, and winds blew radioactive dust across the countryside, contaminating 41,000 people. This lake is considered to be the most polluted lake on earth, a legacy of the Cold War and an enormous continuing source of radioactive contamination. Recognizing the dangers of spreading contamination, Russian authorities are now filling the lake with hollow concrete blocks and soil, with plans to have it completely filled by 1995.

The end of the Cold War has brought the welcome dismantling of nuclear weapons by the United States and the nations of the former USSR. Thousands of nuclear weapons are being dismantled, many as a result of agreements between President Clinton and Russian President Boris Yeltsin. The agreements include a joint closure of all remaining plutonium weapons production reactors by the year 2000, funds to aid Russia in destroying missle silos and dismantling nuclear submarines and bombs, and vast cuts in nuclear arsenals of

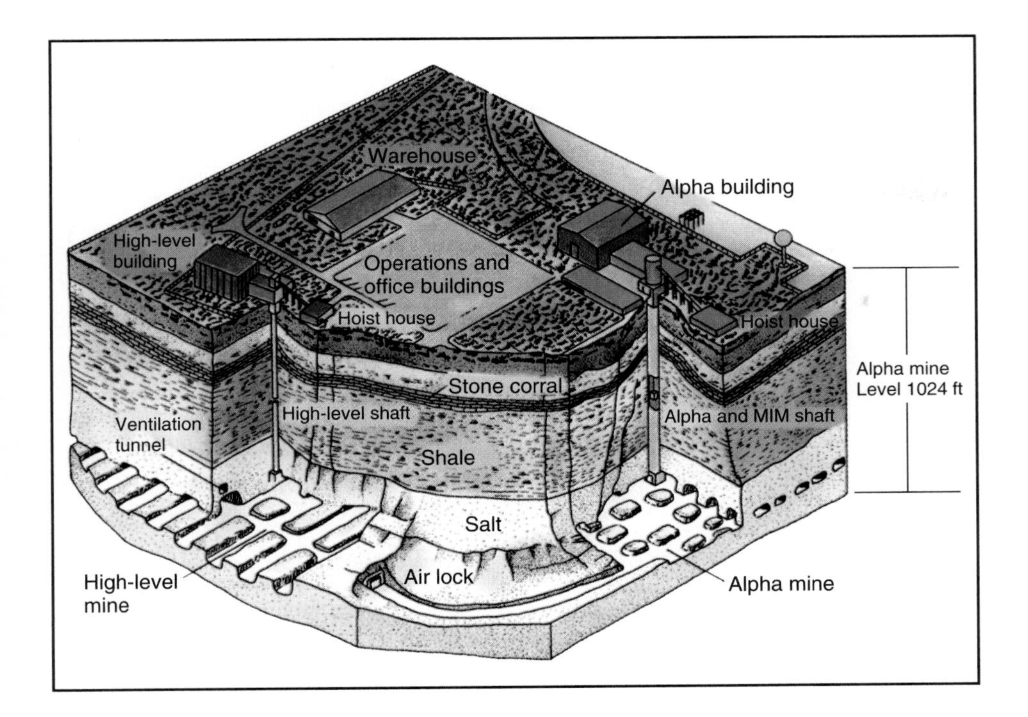

FIGURE 22-12
Disposal of radioactive wastes from nuclear power plants. These wastes must be isolated from the environment for thousands of years. Elaborate plans have been made for their disposal, but will the plans ensure that the wastes will stay where they are put?

the two former enemies. The radioactive components of the weapons—at least 100 tons of weapons-grade plutonium—must be handled with great care and disposed of where they are safe from illegal access and where they will pose no danger to human health for centuries to come—a daunting task. Several attempts to smuggle weapons-grade plutonium have occurred. A suitcase containing 363 grams of plutonium and 200 grams of lithium 6 was confiscated in Munich in August of 1994. The closing down of much of the nuclear warfare enterprise is certainly a welcome development, but the problems of security and of disposal of the wastes will remain as a long-term legacy of the Cold War.

High-level Nuclear Waste Disposal. The United States and most other countries using nuclear power have decided on geologic burial for the ultimate consignment of nuclear wastes, but no nation has developed plans to the point of actually carrying out the burial. Many nuclear nations have not even been able to find a site that may be suitable for receiving the wastes. Where sites have been selected, many questions about safety have surfaced. The basic problem is that no rock formation can be guaranteed to remain stable and dry for tens of thousands of years. Everywhere scientists look, there is evidence of volcanic or earthquake activity or groundwater leaching within the last 10,000 years or so, which is to say that it may occur again in a similar period of time. If such events did occur, the still radioactive wastes could escape into the environment and contaminate water, air, or soil, with consequent effects on humans and wildlife.

In the United States, efforts to locate a long-term containment facility, which have been going on for about 25 years, have been hampered by the *NIMBY* syndrome. (See Chapter 20.) A number of states, under pressure from citizens, have passed legislation categorically outlawing the disposal of nuclear wastes within their boundaries. In the meantime, the need to select and develop a long-term repository has become increasingly critical. With the current lack of public support for nuclear power, it is proving extremely difficult for the United States or any other nation to establish a broadly acceptable long-term nuclear waste facility. (See "Ethics" box, page 568.) The problem is more political than technical.

At the end of 1987, Congress called a halt to debate and arbitrarily selected a remote site, Yucca Mountain in southwestern Nevada, to be the nation's nuclear dump. Not surprisingly, Nevadans have fought this selection, passing a law in 1989 that prohibits anyone from storing high-level radioactive waste in the state. The federal government has the power to override state prohibitions, though, and intensive study and evaluation of the Nevada site are now underway. To date, studies have shown that the storerooms will be 1000 feet (300 m) above current groundwater levels, presumably safe from invasion by groundwater. However, earthquakes or volcanic activity could change conditions in a short time, and the site lies in a zone of fault lines and has a history of volcanic activity only 5000 years ago. The Yucca Mountain facility could begin receiving wastes from commercial and military facilities all over the country in the year 2010 although these plans are on hold at present. It should not escape notice that this plan means that thousands of tons of

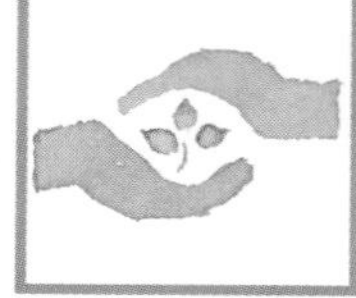

Ethics

Showdown in the New West

In a scenario reminiscent of the Civil War, two western states have become the focus of a controversy over states' rights. On one side is the federal government, in the form of the Department of Energy, which is trying to solve the problem of what to do with nuclear wastes and thinks it has a reasonable solution. On the other side are politicians in the states designated as repository sites, who are responding to the voices of citizens of those states. At stake may be the whole future of nuclear energy in the United States.

The Department of Energy has constructed a $1.5 billion Waste Isolation Pilot Plant in salt caves 2150 feet beneath the desert in southwestern New Mexico, 26 miles from the city of Carlsbad. The plant is on federal land and has passed all necessary safety reviews. It was ready to open for business in October 1991. "Business," in this case, was for the plant to become the repository for up to a million barrels of plutonium wastes from nuclear weapons plants and laboratories around the country, as well as the plutonium from dismantled nuclear warheads. (No commercial nuclear wastes will go there.) Eager to prove that nuclear wastes can be safely moved and placed in such a repository, the energy agency wants to bring a few thousand barrels to the New Mexico site in order to demonstrate the usefulness of the facility. The main objection to the New Mexico site is the prospect of accidents on the narrow roads in the area. There was talk of barricades in the streets to block the trucks if the energy department starts them rolling.

New Mexico sued the Department of Energy on the grounds that it was in violation of federal law prohibiting the opening of such a repository without congressional approval. The state of Texas and four environmental groups joined New Mexico in bringing suit. In October 1992, Congress gave approval to the department to begin testing the facility's ability to store the plutonium wastes—with one significant catch: The EPA was ordered to strengthen the radiation standards protecting the public from the stored wastes. The EPA wants the energy department to show that the facility will function acceptably for the next 10,000 years, and the department has turned its attention to that daunting task. Thus, the situation is presently a stalemate, and some observers believe that the plant will not open for business before 1999, if at all.

Opposition to long-term storage of nuclear wastes is perhaps even stronger in Nevada, which has the dubious distinction of being selected for the nation's only repository for *commercial* nuclear wastes, at Yucca Mountain. The site is a barren ridge in the desert about 100 miles northwest of Las Vegas. In 1989, the Nevada legislature passed a bill making it unlawful for any agent or agency to store high-level radioactive waste in the state. All sorts of surveys have been carried out to test the mood of Nevadans regarding long-term storage of wastes in their state. The results have been uniformly negative: Three-fourths of Nevadans agreed that the state should continue to do all it can to prevent the establishment of the Yucca Mountain site. Clearly, ordinary Nevadans perceive the risks of such a site as enormous and unacceptable, in contrast to the opinions of scientists who have performed risk calculations for the site. In this case, however, the state's attempts to pose a legal blockade to the Department of Energy have failed; the Supreme Court has ruled that Nevada must process applications for permits to continue work on the site.

The dilemma is plain. Public and political support for any form of nuclear energy is at an all-time low, and people in these two western states are especially indignant at their states' becoming the repository of nuclear wastes from around the nation. Yet we need to do something with the legacy of nuclear energy other than to pass the problem along to the next generation. What do you think should be done?

---

radioactive wastes will be shipped by rail and truck through congested areas across the whole country. It is also significant that no other sites are being considered.

Many observers are suggesting that the Department of Energy establish an intermediate storage facility, called a *monitored retrievable storage* (MRS) facility, for receiving spent fuel from the nuclear plants. Earlier, we mentioned that many of the utilities feel pressured to establish such a facility on their own; hence the pact with the Mescalero Apaches of New Mexico. An MRS facility would function as a safe repository until the permanent storage facility at Yucca Mountain is ready for use. Several circumstances support this proposal: (1) Beginning in 1998, the Department of Energy will be required by law to accept spent nuclear fuel from power plants; (2) Electricity customers have paid more than $10 billion into the nuclear waste fund for the purpose of spent-fuel storage and disposal; (3) The MRS approach has been successful in other countries. For example, Sweden's MRS facility is located in a rock cavern, receives spent fuel rods after a year's storage on site, and will hold the rods for 30 to 40 years until they can be transported to a deep geologic repository. An increasing number of utilities will be storing spent fuel in dry-cask, above-ground sites at the power plants, due to limits on their capacity to keep the spent fuel rods in cooling pools. Without some interim solution such as the MRS to the waste storage problem, 70 nuclear power plants will be out of room for storage by the time 2010 rolls around.

FIGURE 22-13
An aerial view of the Chernobyl´ reactor on April 29, 1986, three days after the explosion. The disaster was the worst nuclear power accident ever in the former USSR, directly killing at least 33 people and putting countless numbers of people in the surrounding country at risk for future cancer deaths. (Shone-Zoufarov/Gamma-Liaison, Inc.)

The Potential for Accidents

Prior to 1986, the scenario for a worst case nuclear power plant disaster was a matter of speculation. Then, at 1:24 A.M. local time on April 26, 1986, events at a nuclear power plant in Ukraine made such speculation irrelevant (Fig. 22-13). Since that day, Chornobyl´ has served as a horrible example of nuclear energy gone awry.

Chornobyl´. While conducting a test of stand-by-diesel generators, engineers disabled the safety systems, withdrew the control rods, shut off the flow of steam to the generators, and decreased the flow of coolant water in the reactor. They did not allow for the radioactive heat energy generated by the fuel core and, lacking coolant, the reactor began to heat up. The extra steam produced could not escape and had the effect of rapidly boosting the energy production of the reaction. In an attempt to quell the reactor the engineers quickly inserted the carbon-tipped control rods. The carbon tips acted as moderators, slowing down the neutrons to a speed where they triggered more fission reactions, and the result was a splitsecond power surge to 100 times the maximum level. Steam explosions then blew the 2000 ton top off the reactor; the reactor melted down, and a fire was ignited in graphite, which burned for days. At least 90 million curies of radioactivity in the form of fission products were released in a plume that rained radioactive particles over thousands of square miles.

As radioactive fallout settled, 135,000 people were evacuated and relocated. The reactor was eventually sealed in a sarcophagus of concrete and steel. A barbed-wire fence now surrounds a 1000-square-mile exclusion zone around the reactor site. The soil remains contaminated with radioactive compounds, yet 2000 Ukrainian workers are bused in daily to work at the two remaining reactors in the Chornobyl´ complex; the Ukrainian government does not want to lose the power still being generated.

Only two engineers were killed by the expolsion, itself, but 31 of the personnel brought in to contain the reactor in the aftermath of the explosion died of radiation in a few months. Over a broad area down wind of the disaster, buidings and roadways were washed down to flush away radioactive dust. Even with these precautions, however, many people in or near the evacuation zone were exposed to dangerous levels of radiation, especially the short-lived radioisotope iodine-131. Because iodine collects in the thyroid gland, the radioactive iodine has been responsible for great increases in the incidence of thyroid cancer in Ukraine and neighboring Belarus (over 700 cases in children have occurred since 1986). The long-term effects are estimated to range from 140,000 to 475,000 cancer deaths worldwide from the accident.

Are we in the United States in danger of such an explosion occurring at a nuclear power plant? Nuclear scientists argue that the answer is no because U.S. power plants have a number of design features that should make a repeat of Chornobyl´ impossible. The Chornobyl´ reactor used graphite as a moderator, rather than water as in LWRs. Further, LWRs are incapable of developing a power surge more than twice normal power, well within the designed containment capacity of the reactor vessel. And, in LWRs, there are more backup systems to prevent the core from overheating, and the reactors are housed in a thick concrete-walled containment building designed to withstand explosions such as the one that occurred in Chornobyl´. The Russian reactor had no containment building.

However, LWRs are not immune to accidents, the most serious being a complete core meltdown as a result of total loss of coolant. This has never happened, but there was a close call at Three Mile Island.

Earth Watch

A Test of Fire

A drunk driver going the wrong way plowed head-on into a flatbed truck on Interstate 91 in Springfield, Massachusetts, early in the morning of December 16, 1991. The truck's diesel fuel burst into flames, and the fire consumed 12 wooden shipping crates on the trailer. Rescuers pulled the occupants from both vehicles in time to save their lives. What made this accident something more than a "normal" head-on collision was the fact that the truck was carrying 12 steel and lead casks of nuclear fuel destined for the Vermont Yankee nuclear power plant in Vernon, Vermont.

There were some anxious moments when state police and fire fighters arrived on the scene and saw the symbol on the side of the truck showing that it carried a radioactive cargo; however, all but one of the casks withstood the impact and the fire and came out intact. The one that didn't suffered a small break, but none of the ceramic uranium dioxide pellets inside it spilled. Even if they had, the pellets were harmless, because they had not yet been irradiated. If they had been spent nuclear fuel pellets, it would have been a different matter, but as we have seen, spent nuclear fuel is not transported but kept on site. However, if the western nuclear repository sites are ever activated (See "Ethics" box, p. 568), much more highly radioactive cargo will be on the highways.

Opponents of nuclear energy were upset because the pertinent towns had not been notified that nuclear fuel was coming through. State public health officials replied that although the state issues permits for the movement of such cargos, federal regulations prohibit notification of cities and towns on the route because of a desire to avoid possible terrorist interception.

This accident is a stern reminder that human error must always be factored into any assessment of the risks of a technology. Plans to move nuclear waste should be based on the assumption that the truck will have a head-on collision at any moment. The next test of fire could have far more serious consequences.

Three Mile Island. On March 28, 1979, the Three Mile Island nuclear power plant near Harrisburg, Pennsylvania, suffered a partial meltdown as a result of a series of human and equipment failures and a flawed design. The steam generator (see Fig. 22-8) shut down automatically because of a lack of power in its feedwater pumps, and eventually a pressure valve on top of the generator opened in response to the ensuing buildup of pressure. Unfortunately, the valve remained stuck in the open position and drained coolant water from the reactor vessel. There were no sensors to indicate that this pressure-operated relief valve was open. Operators responded poorly to the emergency, shutting down the emergency cooling system at one point and shutting down the pumps in the reactor vessel. One instrument error compounded the problem: Gauges told operators that the reactor was full of water when, actually, it needed water badly. The core was uncovered for a time and suffered a partial meltdown, and a small amount of radioactive gas was released to the atmosphere.

The drama held the whole nation—particularly the 300,000 residents of metropolitan Harrisburg poised for evacuation—in suspense for several days. The situation was eventually brought under control and no injuries occurred, but it could have been much worse if meltdown had been complete. The reactor was so badly damaged, and so much radioactive contamination occurred inside the containment building, that the continuing cleanup is proving to be as costly as building a new power plant. There are no plans to restart the reactor.

Real costs of the accidents at Three Mile Island and Chornobyl´ must also be reckoned in terms of public trust. Public confidence in nuclear energy already was declining, but plummeted after the two accidents. The accidents pointed to human error as a highly significant factor in nuclear safety, and human error is something the public understands well. Nuclear proponents suffered a serious loss of credibility with Three Mile Island, but Chornobyl´ was their worst nightmare come true—a full catastrophe, just as the antinuclear movement had predicted might someday occur.

Safety and Nuclear Power

As a result of Three Mile Island and other, lesser incidents in the United States, the Nuclear Regulatory Commission has upgraded safety standards not only in the technical design of nuclear power plants, but also in maintenance procedures and in the training of operators. Thus, proponents contend, nuclear plants were designed to be safe in the beginning, and now they are safer than ever. It is estimated that as a result of new procedures instituted after the accident at Three Mile Island, nuclear plants are now six times safer than before. Some proponents of nuclear power claim that we now have the technology to build *inherently safe* nuclear reactors—reactors designed in such ways that any accident would result in an automatic quenching of the chain reaction and suppression of heat from nuclear decay. In reality, however, there is no such thing as an inherently safe reactor, since the concept implies no release of radioactivity under any circumstances—an impossible expectation.

Instead, nuclear scientists are proposing a new generation of nuclear reactors with built-in *passive safety* features, rather than the *active safety* features found in current reactors. The distinction is important, and it represents a major change in the direction of thinking within the nuclear community. **Active safety** relies on operator-controlled actions, external power, electrical signals, and so forth. As accidents have shown, operators may override such safety factors, and electricity, valves, and meters can fail or give false information. **Passive safety**, on the other hand, involves engineering devices and structures that make it virtually impossible for the reactor to go beyond acceptable levels of power, temperature, and radioactive emissions.

In anticipation of a possible renewed interest in nuclear power, engineers are now designing LWRs that include passive safety features and much simpler, smaller power plants—the so-called advanced light water reactors (ALWRs). For example, one passive safety feature would be to position a cold-water reservoir so that, in the event of a LOCA, the water would drain by gravity to the reactor core (Fig. 22-14). In addition, the design will make it impossible for operators to inactivate the passive safety systems. Another feature being planned is to build reactors as modular units small enough to conduct heat from nuclear decay outward into the soil; in this manner, the reactors could not possibly experience a core meltdown. One such modular reactor, built in Germany, suffered no damage to its core when tested against a LOCA. Of course, all this increases costs. Nuclear power already is more expensive than any other option to generate electricity, when government subsidies are considered.

The motive behind the new designs is restoration of the public's confidence in nuclear energy. Previously, nuclear proponents had emphasized the very low probabilities of accidents; as we have seen, though, improb-

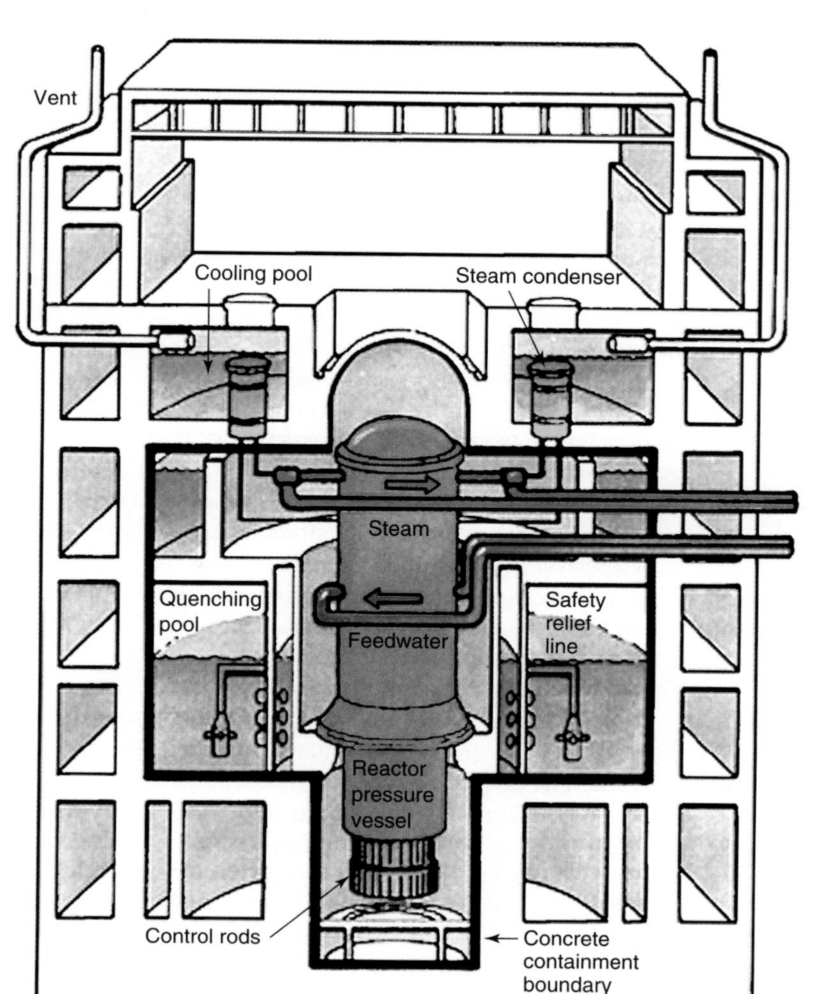

FIGURE 22-14
Advanced light water reactor, showing passive safety features. The core is surrounded by three concentric structures: a reactor pressure vessel, in which heat from the reactor directly boils water into steam; a concrete chamber (outlined with heavy black line) and water pool, which contains and quenches steam vented from the reactor in an emergency; and a concrete building, which acts as a secondary containment vessel and shield. Any excessive pressure in the reactor will automatically open valves that release steam into a quenching pool, reducing the pressure. Water from the quenching pool can, if necessary, flow downward to cool the core. Evaporation from a pool on top of the containment building limits the buildup of containment pressure. (From "Advanced Light-Water Reactors", by M. W. Golay and N. E. Todreas. Copyright © 1990 by *Scientific American*, Inc. All rights reserved.)

FIGURE 22-15
Laser fusion. In this experimental instrument at Lawrence Livermore Laboratory, 30 trillion watts of optical power are focused onto a tiny pellet of hydrogen smaller than a grain of sand located in the center of a vacuum chamber. For less than a billionth of a second, the fusion fuel is heated and compressed to temperatures and densities near those found in the sun. (Alexander Tsiaras/Science Source/Photo Researchers, Inc.)

able events can happen, and when they happen to nuclear power plants, the consequences are awesome. The new emphasis is on convincing the public of the fundamental safety of ALWR designs by demonstrations and explanations that nontechnical people can understand.

Economic Problems with Nuclear Power

As Fig. 22-2 shows, in the United States utilities already were turning away from nuclear power considerably before the disaster at Chornobyl´. The reasons are mainly economic.

First, projected future energy demands were overly ambitious; the slower growth rate in the demand for electricity has postponed orders for all types of power plants. Second, increasing safety standards for the construction and operation of nuclear power plants have caused the costs of nuclear power plants to increase at least fivefold, even after inflation is considered. Also, adding to the rise in costs is the withdrawal of government subsidies to the nuclear industry. Third, public protests have frequently delayed the construction or startup of a new power plant. Such delays increase costs still more, because the utility is paying interest on its investment of several billion dollars even when the plant is not producing power. As these costs are passed on, consumers become yet more disillusioned with nuclear power. Finally, safety systems may protect the public, but they do not prevent an accident from being financially ruinous to the utility. Since radioactivity prevents straightforward cleanup and repair, an accident can convert a multibillion-dollar asset into a multibillion-dollar liability in a matter of minutes, as Three Mile Island demonstrated. Thus, nuclear power involves a financial risk that utility executives are reluctant to take.

Another factor that promises to increase the cost of nuclear-generated electricity is a shorter-than-expected lifetime for nuclear power plants. Originally, it was thought that nuclear plants would have a lifetime of about 40 years. It now appears that their lifetime will be considerably less. Worldwide, 60 nuclear plants have been closed, after an average lifetime of 17 years. This shorter lifetime substantially increases the cost of the power produced, because the cost of the plant must be repaid in the shorter period.

The lifetimes of nuclear power plants are shorter than originally expected because of two problems: *embrittlement* and *corrosion*. **Embrittlement** occurs when some of the neutrons from fission bombard the reactor vessel and other hardware. Gradually, this neutron bombardment causes the metals to become brittle enough that they may crack under thermal stress—for example, when emergency coolant waters are introduced in the event of a LOCA. When the reactor vessel becomes too brittle to be considered safe, the plant must either be shut down or repaired at great cost. Thus, utility executives shut down Yankee Rowe in Massachusetts in 1992, eight years before its license expired, rather than spend over $100 million to repair the embrittled reactor vessel.

Corrosion is a normal consequence of steam generation. Very hot, pressurized water flows from the core into the steam generator through thousands of 3/4-inch-diameter pipes immersed in water (See Fig. 22-8). Water inside and outside these pipes contains corrosive chemicals which, over time, cause cracks to develop in some of the pipes. If the main line conveying steam from the generator to the turbine were to break, the sudden increase in pressure in the generator could cause several cracked pipes to break at once. If this happened, radioactive moderator-coolant water would be released and overload safety systems, forcing the plant to vent radioactive gas to the outside. Cracked pipes are "repaired" basically by plugging them; in 1993 alone, over 10,000 worn or cracked pipes were plugged in pressurized LWRs, reducing the efficiency of the reactors and cutting their total power. Using a new high-tech probe, officials discovered in March 1995 that up to half of the steam generator pipes in the Maine Yankee plant had developed cracks, many of them going the full circumference of the pipe, with some penetrating 80% of the pipe's thickness. As a result, the plant was shut down for a year, and the owners of 70 other pressurized water nuclear reactors were notified by the Nuclear Regulatory Commission that they, too, could be facing similar cracking problems.

Closing down, or *decommissioning*, a power plant can be extremely costly. The estimated price tag for decommissioning Yankee Rowe, a small nuclear plant by current standards, is $368 million; $56 million of this is needed to construct a temporary storage facility until locations are found to receive the high-level and low-level wastes resulting from the decommissioning. Faced with these costs, some utilities are opting to repair older plants in spite of the high costs of the repairs. Of the plants operating in the United States, it is estimated that at least ten will be decommissioned by the year 2000, and most of the rest will be in the ten years thereafter.

More Advanced Reactors

Uranium—especially ^{235}U—is not a highly abundant mineral on Earth. At the height of optimism about nuclear energy in the 1960s, when as many as 1000 plants were envisioned by the turn of the century, it was foreseen that shortages of ^{235}U would develop. Breeder reactors, which utilize chain reactions, were seen as the solution to this problem.

Breeder Reactors

Recall that when a ^{235}U atom fissions, two or three neutrons are ejected. Only one of these neutrons hitting another ^{235}U atom is required to sustain a chain reaction; the others are absorbed by something else. The breeder reactor is designed so that the extra neutrons are absorbed by (nonfissionable) ^{238}U. When this occurs, the ^{238}U is converted to plutonium (^{239}Pu), which is fissionable. The ^{239}Pu can be purified and used as a nuclear fuel, just as ^{235}U is. Thus, the breeder converts nonfissionable ^{238}U into a useful nuclear fuel. Consequently, since there are generally two neutrons in addition to the one needed to sustain the chain reaction, the breeder may produce more fuel than it consumes. Because 99.7% of all uranium is ^{238}U, converting this to ^{239}Pu through breeder reactors effectively increases the nuclear fuel reserves over one-hundred-fold.

Breeder reactors present all of the problems and hazards of standard fission reactors, plus a few more. If a meltdown occurred in a breeder, the consequences would be much more serious than in an ordinary fission reactor because of the large amounts of ^{239}Pu, which has an exceedingly long half-life of 24,000 years. In addition, because plutonium can be purified and fabricated into nuclear weapons more easily than ^{235}U can, the potential for the diversion of breeder fuel to weapons production is greater. Hence, the safety and security precautions needed for breeder reactors are greater.

With its scaled-down nuclear program, the United States currently has enough uranium stockpiled. Thus, there is no urgency for the United States to develop breeders. However, in the United States and elsewhere, small breeder reactors are operated for military purposes. The only commercial breeder reactor—the Ferm:1 near Detroit—suffered a catastrophic partial meltdown in October 1966. France is currently the only nation with a commercial breeder reactor, the Superphénix. With no oil fields, France is determined to achieve energy independence with nuclear power. French officials point out that one gram of plutonium equals a ton of oil; together with a new waste fuel reprocessing facility, the French can now produce 16 tons of plutonium a year.

Fusion Reactors

The vast energy emitted by the sun and other stars comes from fusion (Fig. 22-5b). The sun, as well as other stars, is composed mostly of hydrogen. Solar energy is the result of fusion of this hydrogen into helium. Scientists have duplicated the process in the

Earth Watch

Radiation Phobia?

It is a familiar scene to TV viewers: Angry people with placards picket a nuclear power plant, and some are dragged off by police for obstructing access to the plant. The protesters are convinced that nuclear power is dangerous and a threat to health, even when a plant is operating normally. Often, the ranks of the protesters include medical doctors and research scientists along with the more familiar members of environmental protest organizations. What is it about nuclear power that ignites such strong feelings? David Willis, professor emeritus of radiation biology at Oregon State University, believes that many protesters are motivated by fear of something they do not understand—that they in fact have a radiation phobia. This fear is made worse by the way the media portray radiation. For example, the word *radiation* is very often preceded by the adjective *deadly*. Yet only a few people suffer ill effects from radiation, and those effects are almost always due to occupational or medical accidents involving carelessness in handling sources of radiation.

Perhaps the fear occurs because we are not able to perceive radiation by our normal senses of smell, taste, touch, vision, and hearing. We hear that there is radiation all around us in the form of cosmic rays—the background radiation—but there is nothing we can do about it, and life seems to go on. Willis points out that, yes, ionizing radiation can induce cancer in humans, but the kinds of cancer induced are identical with cancers that arise spontaneously from all sorts of causes, mostly unknown. The higher the radiation dose, the greater the tendency to develop cancer. As the dose declines, however—as in small exposures to radiation that might be an occupational hazard for some lines of work—radiation effects simply disappear in the statistical data showing cancer development in a given population.

To illustrate his point, Willis asks us to consider an airplane descending into a fog bank as it comes in to land. Above the fog, the airplane's descent is quite visible to another airplane, but when it descends into the fog, it disappears from view even though it is still there. By analogy, the airplane represents the cancers induced from ionizing radiation. Its descent represents the lessening number of cancers induced by diminishing doses of radiation. The fog bank represents the cancers normally occurring from a variety of causes or spontaneously. Any effects of low levels of radiation are completely masked by the large fog of cancers naturally occurring. If some cancers are induced by low-level radiation, the number is insignificantly low compared with the natural rate.

Willis asserts that many opponents of nuclear power are afraid of something that actually carries a very low risk, especially when compared with other hazards in our environment. In spite of numerous studies, no linkage has been shown between the incidence of cancer and the presence of a nuclear facility. Willis has debated the issue with nuclear opponents and finds that many of them will not listen to a reasoned explanation of the health effects of ionizing radiation and the demonstratedly low risks involved with nuclear power. The result is apparent: Nuclear power is on the wane and may never recover.

hydrogen bomb, but hydrogen bombs hardly constitute a useful release of energy. The aim of fusion technology is to carry out fusion in a controlled manner in order to provide a practical heat source for boiling water to power steam turbogenerators.

Since hydrogen is an abundant element on Earth (there are two atoms of it in every molecule of water) and helium is an inert, nonpolluting, nonradioactive gas, hydrogen fusion is promoted as the ultimate solution to all our energy problems—that is, pollution-free energy from a virtually inexhaustible resource, water. However, the dream is still a long way from reality. Indeed, a fusion plant has not yet been proven possible, much less practical.

In the present state of the art, fusion power is still an energy consumer rather than a producer. The problem is that it takes an extremely high temperature—some 100 million degrees Celsius—and pressure to get hydrogen atoms to fuse. In the hydrogen bomb, the temperature and pressure are achieved by using a fission bomb as an igniter—an unthinkable way to initiate a sustained, controlled fusion reaction

A major technical problem is how to contain the hydrogen while it is being heated to the tremendously high temperatures required for fusion into helium. No material known can withstand these temperatures without vaporizing; however, two techniques are being tested. One is the Tokamak design, in which ionized hydrogen is contained within a magnetic field while being heated to the necessary temperature. The second is laser fusion, in which a tiny pellet of frozen hydrogen is dropped into a "bull's-eye" where it is hit simultaneously from all sides by powerful laser beams (Fig. 22-15, page 572). The laser beams simultaneously heat and pressurize the pellet to the point of fusion.

Some fusion has been achieved in both types of devices, and in late 1994, a Tokamak facility at the Princeton Plasma Physics Laboratory in Princeton, N. J., achieved 10.7 megawatts of fusion power in 0.27 second of reaction. As yet, however, the break-even point has not been reached: More energy is required to run the magnets or lasers than is obtained by the fusion. (It took 39.5 megawatts of energy to sustain the reaction at Princeton.) The most optimistic workers in the field believe that, with sufficient money for research—$10 billion—the break-even point might be reached in the late 1990s. Even if this goal is achieved, however, it is still a long way from a practical commercial fusion-reactor power plant. Developing, building, and testing such a plant would require at least another 20 to 30 years and many more billions of dollars. Additional plants would require additional years. Thus, fusion is, at best, a very long-term option. Many scientists believe that fusion power will always be the elusive pot of gold at the end of the rainbow.

Even if the break-even point is reached, fusion energy will still fail to be either clean or an unlimited resource. Current designs do not use regular hydrogen (^{1}H), but rather the isotopes deuterium (^{2}H) and tritium (^{3}H)—the *d-t* reaction. The Princeton Tokamak employed a 50–50 mixture of these isotopes. Fusion of deuterium alone (a *d-d* reaction) demands much greater temperatures and pressures than the *d-t* reaction and has a low energy yield. Deuterium is a naturally occurring nonradioactive isotope and can be extracted in almost any desired amounts from the hydrogen in seawater. Tritium, however, is an unstable radioactive isotope that must be produced. Current plans call for the production of tritium by a breeding reaction in which the element lithium is bombarded with neutrons. The neutrons will be produced by the fusion of deuterium and tritium, the essential energetic reaction. The overall reaction is as follows:

$$\underset{\text{(Deuterium isolated from water)}}{^{2}\text{H}} + \underset{\text{Tritium}}{^{3}\text{H}} \longrightarrow \underset{\text{Helium}}{^{4}\text{He}} + \underset{\text{Neutron}}{\text{n}} + \underset{\text{(fusion reaction)}}{\text{energy}}$$

$$\uparrow \qquad\qquad\qquad \downarrow$$

$$\underset{\text{Tritium}}{^{3}\text{H}} + \underset{\text{Helium}}{^{4}\text{He}} \longleftarrow \underset{\text{Neutron}}{\text{n}} + \underset{\text{Lithium}}{^{6}\text{Li}}$$

Lithium is not an abundant element, and it could easily become the limiting factor in the wide-scale use of fusion reactors. Also, tritium is radioactive and therefore hazardous. Moreover, it is gaseous and difficult to contain. As a result, fusion reactors could easily become a source of radioactive tritium leaking into the environment, unless effective (and costly) designs prevent leaks. On the plus side, tritium has a short half-life and emits only weak beta particles (electrons), and the amounts needed in the reactor are so small as to be dangerous to the public only if a massive loss were to occur.

One distinct advantage of fusion over fission is the absence of spent fuel wastes, which we have identified as one of the major problems with current nuclear power. However, the reactor hardware will be embrittled and made highly radioactive by the constant bombardment from neutrons. Thus, there will be the cost of constantly replacing reactor components and the problem of disposing of components that have been made radioactive. This problem is considered to be about as great as that posed by an equal-power fission reactor. Finally, fusion reactors promise to be the source of unprecedented thermal pollution. A steam turbogenerator is only 30% to 40% efficient, and if half the power produced must be fed back into the reactor to sustain the fusion process, the overall reactor is only 15% to 20% efficient. That is to say, 80 to 85 units of heat energy will be dissipated into the environment for every 15 to 20 units of electrical energy produced. At best, all this boils down to the fact that fusion power will be exceedingly expensive if it is achieved at all.

Cold Fusion?

In the spring of 1989, there was great excitement regarding fusion power when two scientists announced that they had achieved the fusion of hydrogen atoms at room temperature, called *cold fusion*. The claim was that when palladium electrodes were used to pass an electric current through heavy water (water in which the hydrogen atoms have been replaced by deuterium), some of the deuterium diffused into the electrodes and fused, which generated heat in addition to that from the electric current being used. However, no neutrons were detected in the reaction, and most scientists thought that the reactions were ordinary chemical reactions and not nuclear fusion. Attempts to reproduce the experiments in other laboratories around the world have failed, and the scientific community has generally rejected the claims of cold fusion.

Fusion research may continue in an effort to increase our understanding of physical processes. However, it seems likely that solar technologies will be producing power less expensively and with fewer risks before fusion reactors become a reality.

The Future of Nuclear Power

The long-term outlook for energy is discouraging, as fossil fuels are being used up rapidly, and alternatives such as solar power are in their infancy. Using nuclear power seems to be an inevitable choice. However, many are opposed to nuclear energy and it is the most costly option; worldwide, public opposition to nuclear energy is higher now than it has ever been and raises the question of whether nuclear energy has any future once the currently operating plants have lived out their life spans.

Opposition

Opposition to nuclear power is based on several premises:

- People have a general distrust of technology they do not understand, especially when that technology carries with it the potential for catastrophic accidents or the hidden, but real, capacity to induce cancer. (See "Earth Watch" box, p. 574).
- Many observers are critical of the way nuclear technology is being managed. They are aware that the same agency (the Nuclear Regulatory Commission) responsible for licensing and safety regulations is also a strong supporter of the commercial nuclear industry. This concern is warranted. The NRC has come under fire recently in connection with the Millstone nuclear plants in Connecticut. Concerned that safey rules were routinely being ignored in the facility, two employees tried to get corrective action from the plant's managers. The management balked, and it then took three years for the employees to get action from the NRC. During that time they were subjected to intimidation from their superiors. When the NRC finally stepped in, it became evident that the Commission had for years been routinely waiving safety rules in order to let nuclear plants avoid costly shutdowns. The Millstone plant was shut down indefinitely in 1996 in order to address a host of safety violations.
- The nuclear industry has repeatedly presented nuclear energy as extremely safe, using arguments based on the low probabilities of accidents occurring. However, when accidents occur, probabilities become realities, and the arguments are moot.
- There remains the crucial problem of nuclear waste. All parties agree that the waste must be placed somewhere safe, but there the agreements end, and siting a long-term nuclear waste repository has become an apparently impossible political as well as technological problem in country after country.

Completely aside from these objections, there is the basic mismatch between nuclear power and the energy problem. As we have emphasized, the United States' main energy problem involves an eventual shortage of crude oil for transportation purposes, yet nuclear power produces electricity, which is not used for transportation. If we were moving toward a totally electric economy that included even electric cars, nuclear-generated electricity could be substituted for oil-based fuels. Unfortunately, electric cars have not yet proved practical, and the outlook for them in the near future remains uncertain. Consequently, nuclear power simply competes with coal-fired power plants in meeting limited demands for electricity. The fact is that, given the escalated costs and additional financial risks of nuclear power plants, coal is cheaper, and the United States does have abundant coal reserves.

However, there are still the environmental problems of mining and burning coal, including acid precipitation and global warming. If these were factored into the price of coal, we would find that price considerably higher than it is now. Of course, costs such as long-term confinement of nuclear wastes and decommissioning of power plants have yet to be factored into the cost of nuclear power.

At the bottom line, we find that one of the most touted energy options, nuclear power, is also a feared technology. The public simply does not trust the safety of the reactors and the plans for waste disposal. Opponents believe that the technology poses long-term risks to both human and environmental health and will in no way solve our most critical energy problem: fuel for transportation. Also, although the United States has relatively abundant reserves of uranium, it is not a renewable resource. Opting to meet the energy problem by all-out exploitation of this resource would assuredly lead to another energy resource crisis and increased imports. Over the course of history, that is a very short period. The cost of the nuclear option is an important factor.

Rebirth of Nuclear Power?

What would it take to revitalize the nuclear power option? If nuclear energy is to have a brighter future, it may be because we have found continued use of fossil fuels to be so damaging to the atmosphere that we have placed limits on their use, but have not been able to successfully develop adequate alternative energy sources. If that day comes, observers agree that a number of changes will have to be made:

- Reactor safety concerns will have to be addressed, perhaps by promoting only smaller, advanced light water reactor designs with built-in passive safety features.
- Manufacturing philosophy will have to change, to favor standard designs and "mass production" of the smaller reactors instead of the custom-built reactors presently in use in the United States. Recently, the Department of Energy selected two designs for future development, one of which—the Westinghouse advanced pressurized water reactor—is a 600-megawatt unit similar to the one pictured in Fig. 22-14.

- The framework for licensing and monitoring must be streamlined, but without sacrificing safety concerns. This has largely been accomplished with the National Energy Policy Act of 1992.
- The political problems of siting new reactors must be resolved; perhaps the best sites will be on the grounds of present or closed reactors, taking advantage of the existing infrastructure and familiarity with (and acceptance of) a nuclear facility.
- The waste dilemma must be resolved, perhaps by lowering what some deem the unrealistic policy demands that the site must somehow be proven safe for the next 10,000 years.
- Political leadership will be required to accomplish these developments. Currently, nuclear power would be more of a liability than an asset for any politician who would forge a policy for our energy future because of the need for large government subsidies. Perhaps only a broad-based change in the public perception of nuclear power would bring the necessary political leadership to the surface, but it is hard to picture such a change happening, given current nuclear economics and unresolved problems.

Review Questions

1. Compare the current outlook for nuclear power with that of the 1960s and 1970s.
2. Describe how energy is produced in a nuclear reaction.
3. How do nuclear power plants and nuclear reactors work?
4. What environmental advantages does nuclear power offer over coal energy?
5. How are radioactive wastes produced, and what are the associated hazards?
6. Describe the two stages of nuclear waste disposal.
7. What problems are associated with the long-term containment of nuclear waste?
8. Describe what happened at Chornobyl´ and Three Mile Island.
9. What features might make a nuclear power plant inherently safe?
10. Discuss economic reasons that have caused many utilities to opt for coal-burning rather than nuclear-powered plants.
11. How do breeder and fusion reactors work? Does either one offer promise for alleviating our energy shortage?
12. Explain why nuclear power does little to address our shortfall in crude oil production.
13. Discuss changes in nuclear power that might brighten its future.

Thinking Environmentally

1. Discuss the advantages and disadvantages of nuclear power. Are we overly concerned or not concerned enough about nuclear accidents? Could an accident like Chernobyl happen in the United States?
2. Would you rather live next door to a coal-burning plant or a nuclear power plant? Defend your choice.
3. Many people feel that nuclear power is a fading dream. Discuss this statement from a historical perspective.
4. Alvin Weinberg, one of the scientists who developed the first fission reactor, believes that we have no choice other than to opt for nuclear power. He believes that, in order to maintain ourselves at our present numbers and affluence, we must commit ourselves to nuclear power and the proper operation of nuclear plants. Do you agree or disagree? Why?
5. After reading the "Ethics" box entitled "Showdown in the New West," what do you think should be done with our nuclear waste?

Solar and Other Renewable Energy Sources

Key Issues and Questions

1. The total amount of solar energy reaching Earth is enormous. What is the potential for harnessing this energy?
2. Solar water-heating and space-heating systems for buildings represent well-developed technologies. What is preventing the more widespread adoption of these forms of solar heating?
3. Solar energy via photovoltaic cells and solar-tough collectors is used to produce electrical power. What are the current applications of these technologies, and what is their promise for the future?
4. There is a great need to develop solar energy sources that can be coupled to fuel for transportation. What is the potential for fuel from solar hydrogen production?
5. Water, fire, and wind have provided energy for centuries. What are sustainable ways of expanding these options in the near future?
6. The options for energy into the twenty-first century seem to be "business as usual"; using the current mix, which is highly dependent on diminshing fossil fuels, or deliberately forcing a transition to primarily solar energy sources. What policies either favor or discourage movement in the direction of renewable energy sources?

Silent panels of solar cells are providing electricity in both developed and developing countries (Fig. 23-1). Throughout Israel and other countries in warm climates around the world it is now commonplace to have hot water heated by the Sun (Fig. 23-2). Even in temperate climates many people have discovered that standard furnaces can be made obsolete with suitable designs to utilize the Sun's warmth. In the desert northeast of Los Angeles are "farms" with rows of trough-shaped mirrors tipped toward the Sun (Fig. 23-3). These reflectors are focusing the Sun's rays to boil water or synthetic oil and drive turbogenerators. In the hills east of San Francisco, regiments of "windmills" (more properly called *wind turbines*) standing in rows up the slopes and over the crests of the hills are producing electrical power equivalent to that produced by a large coal-fired power plant (refer to Fig. 23-22). Wind turbines are also becoming commonplace across northern Europe and sprouting up in India, Mexico, Argentina, New Zealand, and other countries around the world.

A photovoltaic power station in Israel. Photovoltaic cells convert sunlight into electrical power. [Photo by Peter Ginter/The Stock Market.]

The inertia of "business as usual" has kept the world on a track of growing dependence on fossil fuels, with nuclear power retaining its powerful advocates as well. In the meantime, however, the above examples demonstrate how use of energy from the Sun and wind has been making quiet but steady progress, to the point where these renewable sources of energy are becoming cost-competitive with traditional energy sources. Thus, the world has at its disposal the potential to move toward a nonpolluting, inexhaustible energy economy based on using the Sun's current energy output. In short, we have the capability of abiding by the second principle of sustainability—running on solar energy.

Our objective in this chapter is to give you a greater understanding of the ways in which sunlight may be utilized to provide various energy needs. We also consider some other sustainable energy options. Finally, we examine how we can overcome the hurdles that prevent us from moving more rapidly toward a sustainable energy future.

Principles of Solar Energy

First, let's consider some general parameters of solar energy. Then we will turn to the practical ways in which solar energy is captured and used to meet our needs.

Solar energy originates with thermonuclear fusion reactions occurring in the Sun. Importantly, all the chemical and radioactive polluting byproducts of the reactions remain behind on the Sun. The solar energy reaching Earth is radiant energy, ranging from ultraviolet light, which is largely screened out by ozone in the stratosphere, through visible light, to infrared (heat energy). The most energy is in the visible-light part of the spectrum (Fig. 23-4).

The total amount of solar energy reaching Earth is vast almost beyond belief. By one calculation, just 30 days of sunshine striking Earth have the energy equivalent of the total of all the planet's fossil fuels, both known and unknown. If all the solar energy falling on the paved areas of the United States were captured and used, it would more than suffice for all our energy needs.

Moreover, utilizing some of this solar energy will not change the basic energy balance of the biosphere. Solar energy absorbed by water or land surfaces is converted to heat energy and eventually lost to outer space. Even that absorbed by vegetation and used in photosynthesis is ultimately given off again in the form of heat energy as the food is broken down by various consumers (as we learned in Chapter 3). Similarly, if

FIGURE 23-1
Panels of photovoltaic cells, which convert sunlight into electrical power, are coming into widespread use in both developed and developing countries. (a) A modern, electricity self-sufficient home in the U.S. Note solar panels on upper part of roof (H.R. Bramaz/Peter Arnold, Inc.) (b) A small solar electric power supply for a medical clinic on the Amazon River in Brazil. (Ricardo Beliel/Gamma Liaison, Inc.)

(a)

(b)

FIGURE 23-2
In warm climates, solar hot water heaters are becoming commonplace. Note the solar water-heating panels on each of the homes in this development in Southern California. (Tom McHugh/Photo Researchers, Inc.)

humans were to capture and obtain useful work from solar energy, it would still ultimately be converted to and lost as heat in accordance with the second law of thermodynamics (see page 59). The overall energy balance would not change.

The problem of using solar energy is one of taking a diffuse source—one falling evenly over a wide area—and concentrating it into an amount and form, such as fuel or electricity, that we need for heat and to run vehicles, appliances, and other machinery. Then there is the obvious problem: What does one do when the Sun is not shining? These problem areas may be categorized as: (1) collection, (2) conversion, and (3) storage. Also, in the final analysis, overcoming these hurdles must be cost-effective.

Reflect on how the natural ecosystems handle these three aspects: Leaves collect light over a wide area. By photosynthesis, light energy is both converted

FIGURE 23-3
A solar–thermal power plant in southern California. Sunlight striking the parabolic shaped mirrors is reflected onto the central pipe, where it heats a fluid that is used, in turn, to boil water and drive turbogenerators. (Gabe Palmer/The Stock Market.)

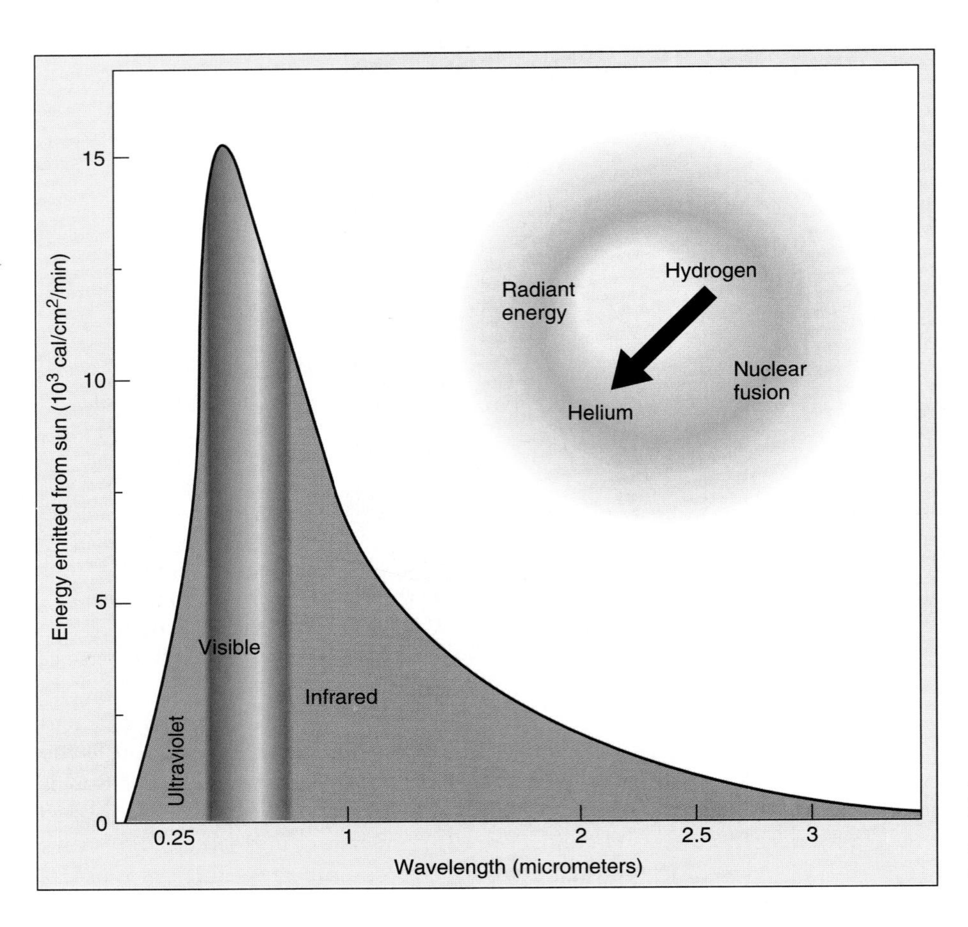

FIGURE 23-4
The solar energy spectrum. The greatest output of solar energy is in the visible-light part of the spectrum.

to and stored as chemical energy—namely, glucose and other forms of organic matter—in the plant. The plant biomass then "fuels" the rest of the ecosystem.

Putting Solar Energy to Work

In the following sections you see how we can overcome or, even better, sidestep these hurdles in various ways to cost-effectively meet our needs from solar energy.

Solar Heating of Water

As noted in the introduction to this chapter, solar hot-water heating is already popular in warm, sunny climates. A solar collector for heating water consists of a thin, broad "box" with a glass or clear plastic top and a black bottom with water tubes embedded in it (Fig. 23-5). Such collectors are called **flat-plate collectors**. Faced to the sun, the black bottom gets hot as it absorbs sunlight—similar to how black pavement heats up in the Sun—and the clear cover prevents the heat from escaping. (Recall the greenhouse effect described in Chapter 16.) Water circulating through the tubes is thus heated and goes to the tank where it is stored.

The heated water may be actively moved by means of a pump—*active systems*. Or natural convection currents may be used—*passive systems*. Passive systems, as you can guess, are more economical.

For a passive solar water-heating system, the system must be mounted so that the collector is lower than the tank. Thus, heated water from the collector moves by natural convection into the tank while cooler water from the tank descends into the collector (Fig. 23-6). There are no pumps to buy, maintain, or remember to turn on and off. The size and number of collectors, of course, is adjusted to the need. The greatest economy is gained by careful management of requirements for hot water, so that the size of the tank and collectors can be minimized.

In temperate climates, where water in the system might freeze, the system may be adapted to include a heat exchange coil within the hot water tank. Then antifreeze fluid is circulated between the collector and the tank. In the United States, there are an estimated 800,000 solar hot-water systems operating, but this number is still only about 0.5% of the total of all hot water heaters.

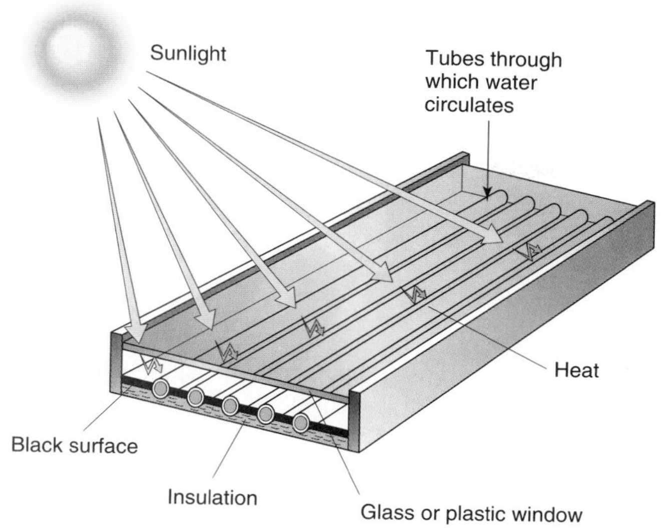

FIGURE 23-5
The principle of a flat-plate solar collector. As it is absorbed by a black surface, sunlight is converted to heat. A clear glass or plastic window over the surface allows the sunlight to enter but traps the heat. Air or water is heated as it passes over and through tubes embedded in the black surface.

Solar Space Heating

Flat-plate collectors like those used in water heating can be used for space heating. Indeed, the collectors for space heating may be even less expensive, homemade devices—because it is necessary only to have air circulate through the collector box. Again, efficiency is gained if the collectors are mounted to allow natural convection to circulate the heated air into the space to be heated (Fig. 23-7).

The greatest efficiency in solar space heating, however, is gained by designing the building to act as its own collector. Such designs can be found in numerous sources (see Bibliography); however, the basic principle is to have Sun-facing windows. In the winter, sunlight can come in and heat the interior (Fig. 23-8a). At night, insulated drapes or shades are pulled to trap the heat inside. The well-insulated building, with appropriately made doors and windows, acts as its own best heat-storage unit. (On the other hand, heating a poorly insulated building by any means is like trying to keep a leaky bucket full of water. It is an exercise in incredible waste.) Beyond good insulation, other systems of storing heat such as in tanks of water or masses of rocks have proved cost-ineffective. Excessive

FIGURE 23-6
(a) Solar water heater. In nonfreezing climates, simple water-convection systems may suffice. Where freezing occurs, an antifreeze fluid is circulated. (b) Solar panels (flat black collectors) and hot water tanks on the roof of a hotel in Barbados. (Photograph by BJN.)

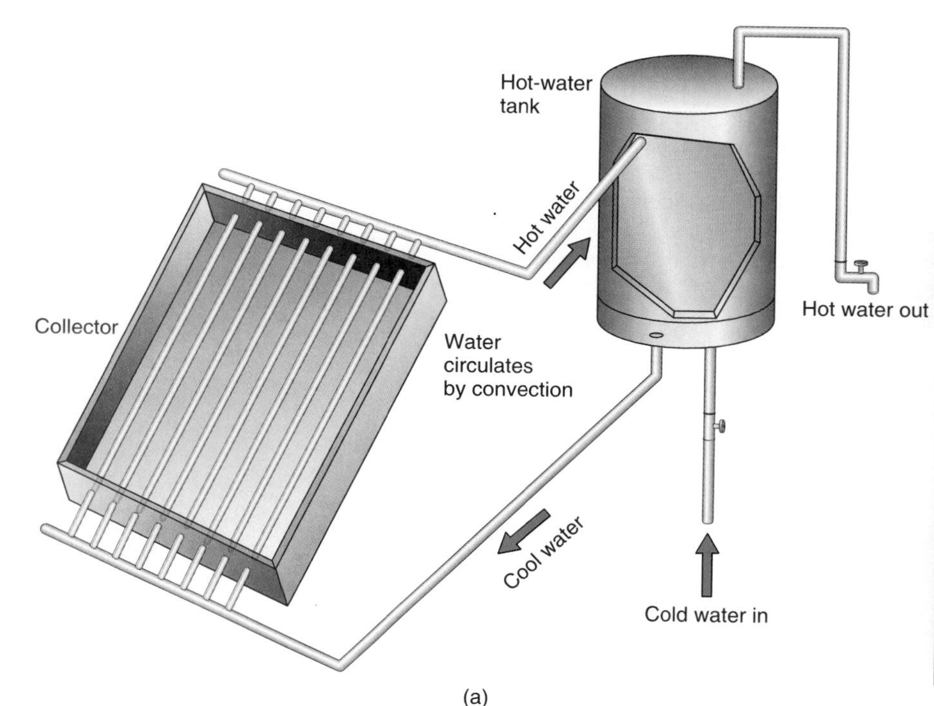

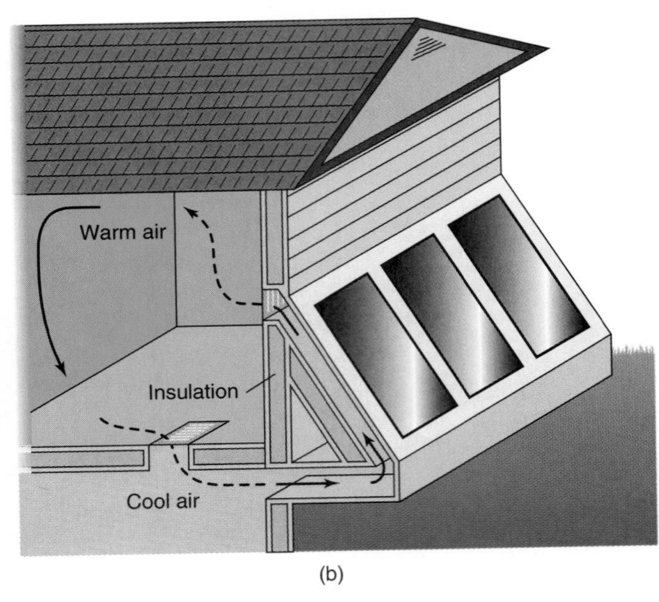

(a) (b)

FIGURE 23-7
(a) Many homeowners could save on fuel bills by adding homemade solar collectors as shown here. (Photograph by BJN.) (b) Air heated in the collector moves into the room by passive convection.

heat load in the summer can be avoided by using an awning or overhang to shield the window from the high summer Sun. Good insulation helps here, too (Fig. 23-8b).

Along with improved insulation, appropriate landscaping can contribute to the heating and cooling efficiency of both solar and nonsolar designs. In particular, deciduous trees or vines on the sunny side of a building will block much of the excessive summer heat, letting the desired winter heat pass through. An evergreen hedge on the shady side will provide protection from the cold (Fig. 23-9).

A common criticism of solar heating is that a backup heating system is still required for periods of inclement weather. Good insulation is a major part of the answer to this criticism. People with well-insulated solar homes find that they have relatively little need for backup heating. When backup is needed, a small wood stove or gas heater suffices. In any case, the criticism concerning the need for backup heating misses the point. Remember that the objective of solar heating is to reduce our dependency on traditional fuels. Even if solar heating and improved insulation reduced the demand for conventional fuels by a mere 20%, this

FIGURE 23-8
Passive solar heating. In contrast to expensive and complex active solar systems, solar heating *can* be achieved by suitable architecture and orientation of the home at little or no additional cost. (a) The fundamental feature is large, Sun-facing windows that permit sunlight to enter during winter months. Insulating drapes or shades are drawn to hold in the heat when the Sun is not shining. (b) Suitable overhangs, awnings, and deciduous plantings will prevent excessive heating in the summer. (From B. Anderson, with M. Riordan, *The Solar Home Book*, Harrisville, N.H.: Brick House Publishing Co., 1976, p. 87.)

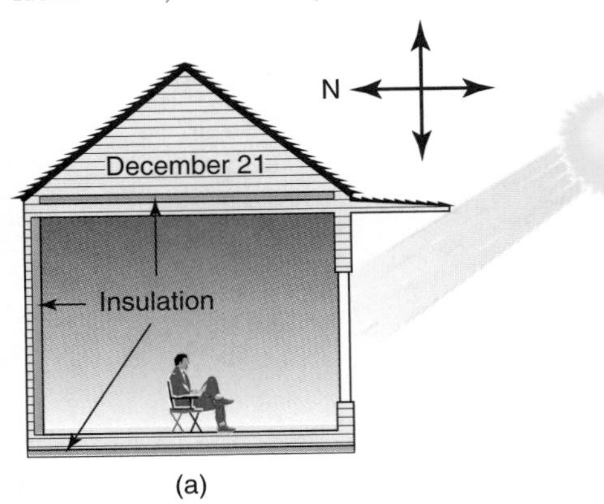

(a) (b)

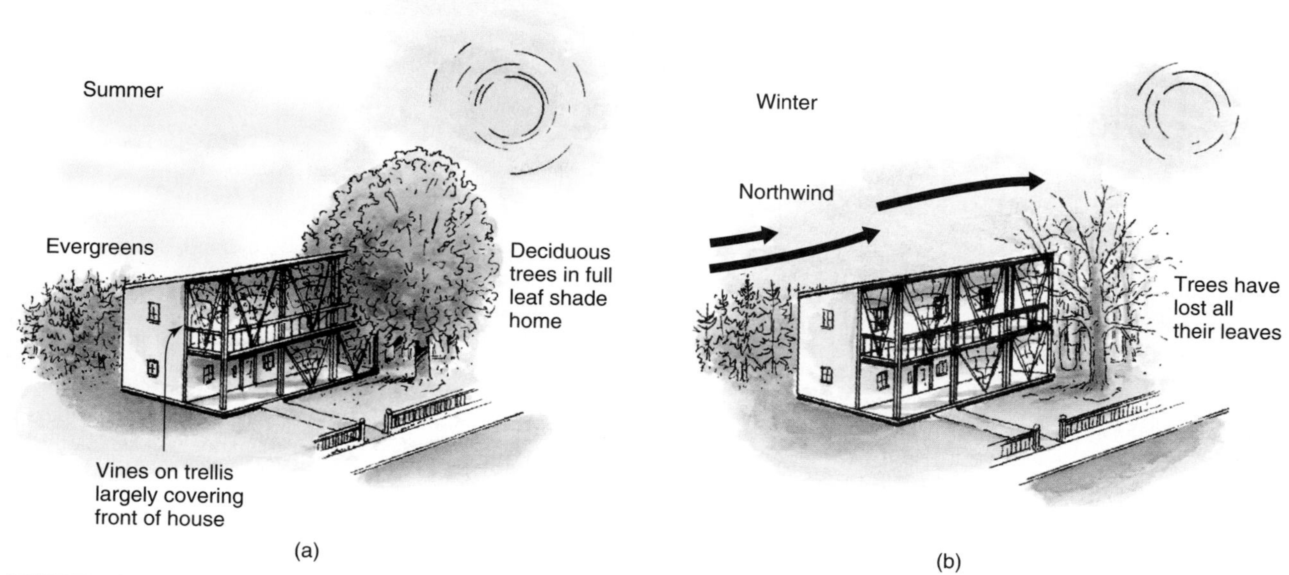

FIGURE 23-9
Landscaping may be an important adjunct to solar heating and cooling. (a) In summer, the house may be shaded with deciduous trees or vines. (b) In winter, leaves drop, and the bare trees allow the house to benefit from sunlight. Evergreen trees on the opposite side protect and provide insulation from cold winds.

would still represent sustainable savings of 20% of the traditional fuel and its economic and environmental costs. With any systematic effort, the savings could be much greater.

The Center for Renewable Resources estimates that in almost any climate, a well-designed passive solar home can reduce energy bills by 75% with an added construction cost of only 5% to 10%. About 25% of our primary energy is used for space and water heating. If solar heating were optimally used in this sector it would save considerably on the oil, natural gas, and electrical power currently used.

What's Delaying Solar Homes?

It is disappointing that, more than 25 years after recognizing the energy crisis and the virtues of solar energy, we are still building homes in traditional ways with traditional furnaces and hot-water heaters, ignoring the potential benefits of the Sun. Insulation is somewhat improved, but it is often poorly installed. Thus, we are continuing to build a larger and larger stock of energy-inefficient homes and buildings that will continue to demand far more energy than necessary. The Center for Renewable Resources concludes, "The chief barrier to more widespread use of passive solar designs is ignorance. Many builders and consumers are unaware of the potential benefits of passive solar design. . . . [Further, these benefits] are often ignored by policymakers and have received meager government support."

In the 1980s, intensive advertising campaigns by utility and oil companies purporting that solar energy is impractical and cost-ineffective did much to maintain this state of ignorance. These fossil-fuel interests successfully lobbied the government to cancel research and end incentive programs for renewable energy. Now, in the 1990s, the major factor in discouraging the use of solar heating is relatively low energy prices, which are a disincentive to making any change. But the opportunity is open.

Solar Production of Electricity

Solar energy can also be used to produce electrical power, thus providing an alternative to coal and nuclear power. Currently, two methods are proving to be economically viable: *photovoltaic cells* and *solar-trough collectors*.

Photovoltaic Cells. A solar cell, more properly called a **photovoltaic**, or **PV, cell** looks like a simple wafer of material with one wire attached to the top and one to the bottom (Fig. 23-10). As sunlight shines on this "wafer," it puts out an amount of electrical current roughly equivalent to a flashlight battery. Thus, PV cells accomplish the collection of light and its conversion to electrical power in one step. One PV cell is no more than about 2 inches (5 cm) in diameter. However, almost any amount of power can be produced by wiring cells together in panels (Fig. 23-11).

Photovoltaic cells are already in common use in pocket calculators, watches, and numerous toys. Panels

Earth Watch

Another Payoff of Solar Energy

In addition to being a nonpolluting renewable energy source, photovoltaic, solar-trough, and wind-powered energy-generating facilities have another advantage: They can be installed quickly and added to the utility system in relatively small increments. To understand this advantage, let's draw a comparison with a nuclear power plant.

A nuclear power plant can be cost-effective only if it is large—about 1000 MW capacity, enough to power a million average homes. Such a plant, even assuming public acceptance, requires in the order of 10 to 15 years from the time of the decision to build to getting it into operation, and it costs around $5 to $10 billion. Therefore, a utility taking this route must borrow billions of dollars and keep that money tied up 10 to 15 years before a return on the investment is realized by selling power from the new plant. Even then, if the 1000 MW of additional power is not needed when the new plant comes on line, the return on the investment may be meager indeed. In the late 1970s and early 1980s a number of utility companies went bankrupt (or nearly so) because they had huge amounts of money tied up in nuclear power plants but a slack demand for the power produced. Clearly, building large power plants (nuclear or coal) involves considerable financial risk on the part of both utility and consumers, as consumers are the ones who ultimately pay.

On the other hand, solar or wind facilities can be operational within a few months of the decision to build; they can be installed in small increments, and additional modules can be added as demand requires. The utility is not placed in the position of having to guess at what power demands will be in 10 or 15 years. Through solar or wind, a utility can add capacity as it is needed, making relatively small investments at any given time and having those investments start paying back almost immediately. You can see that this approach involves much less financial risk both for the utility and for consumers.

of PV cells are providing power for rural homes, irrigation pumps, traffic signals, radio transmitters, lighthouses, offshore oil-drilling platforms, and other installations that are distant from power lines. Rural electrification projects based on photovoltaic cells are beginning to spread throughout the developing world where centralized power is not available. It is cheaper to install a PV system on a rural home than it is to run in a power line 1 mile.

FIGURE 23-10
The thin wafer of material with wires attached is a photovoltaic cell. Converting light to electrical energy, this cell provides enough energy to run the small electric motor needed to turn these fan blades. (Photograph by BJN.)

It is not hard to imagine a future in which every home and building has its own source of pollution-free, sustainable electrical power from an array of PV panels on the roof (Fig. 23-12). Some even speculate that vehicles might be electrically propelled along highways from PV cells embedded in the pavement or along the shoulders of the road. Such a future is not entirely out of the question, but it must await further development in the technology of manufacturing high-efficiency, low-cost cells.

The deceptively simple appearance of PV cells belies a very sophisticated materials science and technology. Each cell consists of two very thin layers of material. The lower layer has atoms with single electrons in the outer orbital; such electrons are easily lost. The upper layer has atoms lacking one electron from their outer orbital; such materials readily gain electrons. The kinetic energy of light striking this "sandwich" dislodges electrons from the lower layer, creating an electrical potential between the two layers. This potential provides the energy for an electrical current through the rest of the circuit. Electrons from the lower side flow through a motor or other electrical device back to the upper side. Thus, with no moving parts, solar cells convert light energy directly to electrical power.

Because there are no moving parts, solar cells do not wear out, but their present life span is in the range

FIGURE 23-11
Photovoltaic cells can be wired together to obtain any amount of power desired. The array seen here is used to power an irrigation system on a farm in Nebraska. No backup or storage of power is needed in this situation because the plants' need for water corresponds to light intensity, which causes evapotranspiration. (David Copp/Woodfin Camp & Associates.)

of 20 years because of deterioration due to exposure to the weather. The major material used in PV cells is the element silicon, one of the most abundant elements on Earth, so there is little danger that production of PV cells will ever suffer because of limited resources. The cost of these cells lies mainly in their sophisticated design and construction.

The cost of PV power (cents per kilowatt-hour) will be the cost of the PV cells divided by the total amount of power they may be expected to produce over their lifetime. Then the cost of PV power must be compared with that of other power alternatives. The first PV cells had a cost factor several hundred times that of electricity from conventional power stations transmitted through the **power grid** (the network of power lines taking power from generating stations to customers), so they have been used mainly in areas far from the grid. It is easy to see why PV power had its first significant application in the 1950s, in the solar panels of space satellites. This application in satellites, in fact, started the development cycle rolling. As more-efficient cells and less-expensive production techniques evolved, costs came down. As costs came down, applications, sales, and potential markets expanded, creating the incentive for further development, and this cycle is still nowhere near its end (Fig. 23-13).

In 1995, PV power was still more costly than electricity from the power grid. Although the technology does not share in the same subsidies and incentives that are given to fossil fuels and nuclear. In addition, PV cells do not store any power; batteries or other backup must be part of the PV package. Still, in locations where there is not easy access to the power grid, a PV package may well be less expensive than extending power lines from the grid, particularly when only small amounts of power are needed.

The world's electric utilities install some 60,000 megawatts (MW) of new generation capacity annually, whereas that of PV installations is about 81 MW. Clearly, there is a vast potential market for PV cells if costs are lowered sufficiently, and there is reason to believe that they will be. Additional breakthroughs toward increasing the efficiency of PV cells have been demonstrated at the laboratory stage, and a number of major multina-

FIGURE 23-12
Georgetown University's Intercultural Center (Washington, D.C.) supports an array of 4400 photovoltaic modules providing a power output of 300 kw under full Sun. Such rooftop arrays may become commonplace in the future as production costs of photovoltaic cells are reduced. (U.S. Department of Energy.)

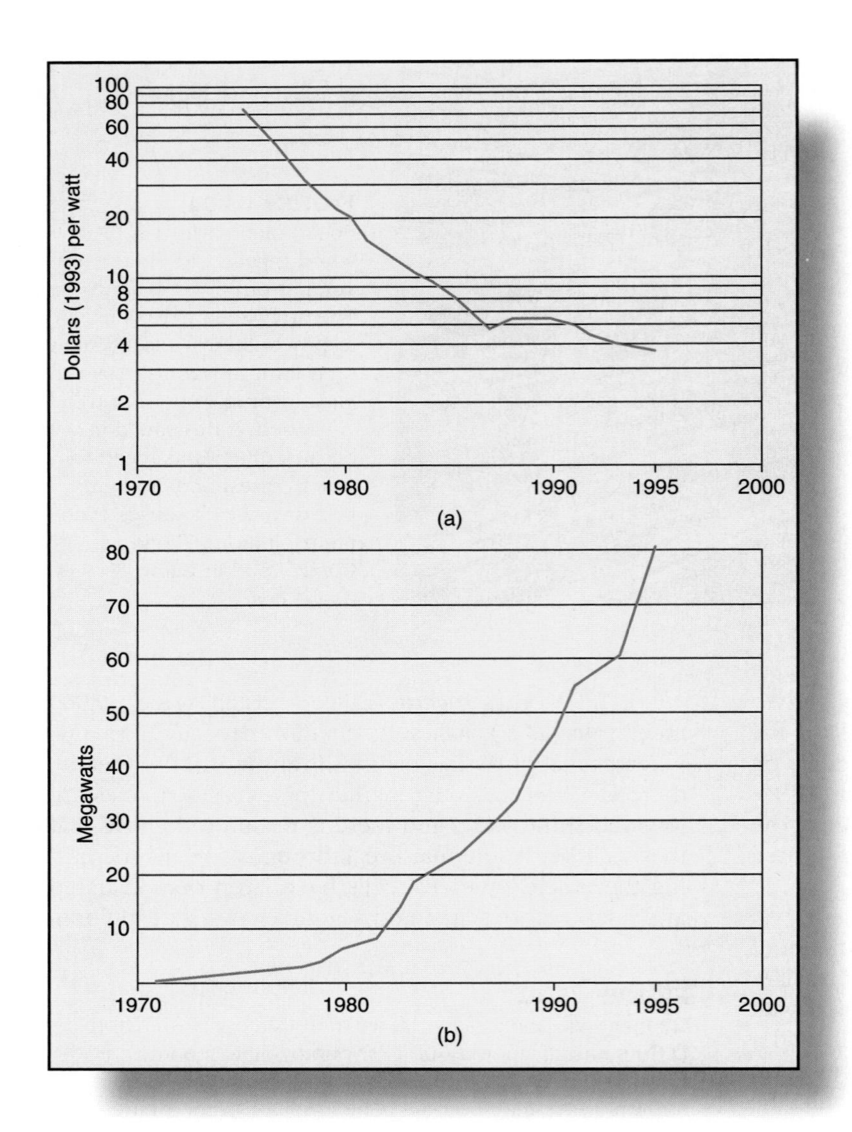

FIGURE 23-13
(a) Research and development has brought the price of photovoltaic cells down about 50-fold in the past 20 years. (b) As prices have come down, sales have increased dramatically. Note the surge in sales for 1994 and 1995—to 81.4 megawatts. Increased sales potential provides the incentive for further reduction in costs. Thus, both trends are likely to continue in the years ahead. (From C. Flavin and N. Lenssen. *Power Surge: Guide to the Coming Energy Revolution*. New York: W. W. Norton, 1994, figs. 8-1, 8-2; and *Vital Signs 1996*, by Lester Brown, Worldwatch Institute, 1996, p. 59.)

tional corporations are developing mass-production techniques, another factor that will reduce costs significantly. Some predict that within 10 years PV power may be competitive with conventional sources.

In moving into PV power, batteries may not be necessary after all because about 70% of the overall demand for electrical power is during daytime hours. Thus, potential savings may be achieved by using solar panels just for daytime needs and continuing to rely on conventional sources at night. In particular, the demand for air conditioning, which, after refrigeration, is the second largest power consumer, is well matched to energy from PV cells. In addition, solar-powered air conditioners could operate independently from the rest of the electrical system, thus avoiding the costs of interconnecting systems. Such air conditioners may be expected to come on the market within the next few years.

What will utility companies do as PV power becomes competitive? A demonstration PV power plant has been built in southern California (Fig. 23-14). However, the more promising future for PV power would appear to be installation of home-sized systems on rooftops, thus avoiding additional land and transmission costs. In this effort the Idaho Power Company is already purchasing, installing, and maintaining PV systems for off-grid homeowners. Customers pay for the system and service on an installment basis. According to the power company, this is a more cost-effective alternative than expanding the utility's own generating capacity and grid.

Experimentation with vehicles run on photovoltaic cells is also under way. The basic challenge for this usage is that all the light falling on a car still represents less than a single horsepower of energy. (Imagine running your car with a small lawnmower engine!) Of course, the PV cells can charge batteries while the car is sitting in the Sun, but a good measure of sitting-in-the-sun time is needed in comparison with travel time. Even ultra-efficient designs have very limited speed and performance (Fig. 23-15).

FIGURE 23-14
The world's first photovoltaic (solar cell) power plant, located near Bakersfield, California. The array of 220 34-foot (11-m) panels produces 6.5 MW at peak, enough for 2400 homes. (T. J. Florian/Rainbow.)

Solar-Trough Collectors. The future includes centralized solar power plants based on the principle of using the Sun's energy to boil water to produce steam for driving a conventional turbogenerator. A system that is proving cost-effective is the **solar-trough** collector system, which is so named because the collectors are long, trough-shaped reflectors tilted toward the Sun (Fig. 23-16). The curvature of the "trough" is such that sunlight hitting the collector is all reflected onto a pipe running down the center. Oil or other fluid circulating through the pipe is thus heated to very high temperatures. The heated fluid is passed through a heat exchanger to boil water and produce steam for driving a turbogenerator. This solar-trough concept was pioneered by an inventor named Charles Abbott in the 1930s (Fig. 23-17). Sixty years later it is finally coming into its own.

An Israel-based company, Luz International (which has now been taken over by another Israel-based company, Solel), built nine solar-trough facilities in southern California, with a combined capacity of 350 MW, about one-third the capacity of a large nuclear power plant. The most recent Luz facility, which converts a remarkable 22% of incoming sunlight to electrical power, is producing power at a cost of 8 cents per kilowatt-hour, barely more than the cost at coal-fired facilities.

Two additional methods of harnessing thermal energy from the Sun have been tested but, at least for the present, have been abandoned as cost-ineffective. The first is what was commonly called a "power tower" constructed in the desert east of Los Angeles. An array of sun-tracking mirrors was used to focus the sunlight falling on several acres of land onto a boiler mounted on a tower in the center (Fig. 23-18). The intense heat generated steam for driving a conventional turbogenerator.

The second is solar ponds. An artificial pond is partially filled with brine (very salty water), and fresh

FIGURE 23-15
The Sunraycer, built by General Motors, is an experimental car powered by solar energy. Limitations in speed and maximum range of solar cars prevent them at this time from offering a viable alternative to internal combustion engines. (Sygma.)

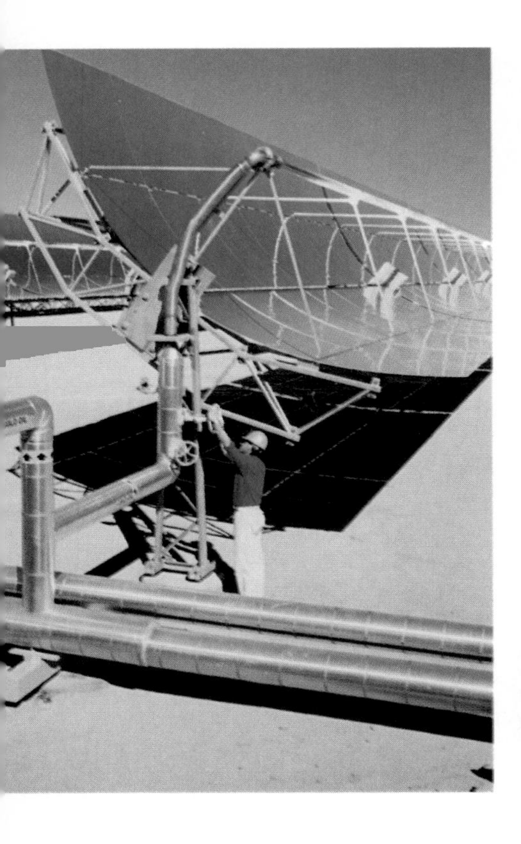

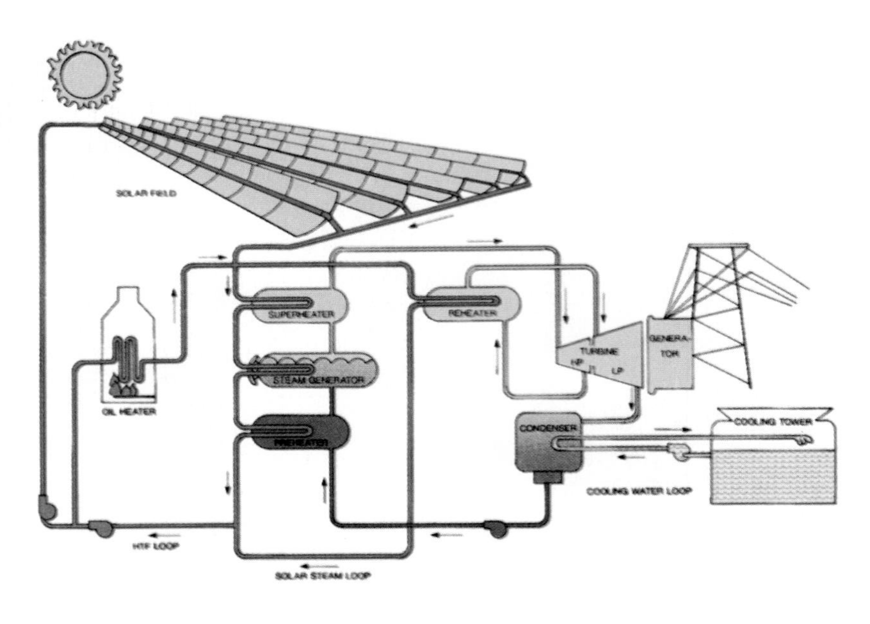

FIGURE 23-16
Several solar-trough facilities are now in operation in southern California. (a) The curved reflector focuses sunlight on and heats oil in the pipe. (b) The heated oil is used to boil water and generate steam for driving a conventional turbogenerator. (Courtesy of Luz International.)

water is placed over the brine. Because it is much denser than fresh water, the brine remains on the bottom, and little or no mixing occurs. Sunlight passes through the fresh water but is absorbed and converted to heat in the brine. The freshwater then acts as an insulating blanket and holds the heat in. The hot brine solution can be circulated through buildings for heating, or it can be converted to electrical power by vaporizing fluids with low boiling points and using the vapors to drive low-pressure turbogenerators. Because the pond also acts as a very effective heat-storage unit, it will supply power continuously. Israel pioneered the

FIGURE 23-17
The solar-trough concept of harnessing solar energy was pioneered by Charles Abbott in the 1930s. Shown here is Abbott, on right, with his 1/2-horsepower "solar boiler." This machine was used to provide power for the National Broadcasting Company to broadcast news of this event to the United States and Canada in 1936. (UPI/Bettmann.)

FIGURE 23-18
A power tower for producing electrical power from sunlight. Sun-tracking mirrors are used to focus a broad area of sunlight onto a boiler mounted on the tower in the center. The steam produced is used to drive a conventional turbogenerator. This facility in southern California is now out of service, since upkeep proved too costly. (Courtesy of Southern California Edison Co.)

concept of solar ponds, but development efforts for the present have been put on hold as cost-ineffective.

The Promise of Solar Energy

One can make a sound argument that solar-produced power is already considerably cheaper because the hidden costs of air pollution, strip-mining, and nuclear waste disposal are not included in the cost of power from traditional sources. Likewise, criticism that solar power requires an exorbitant amount of land to "harvest" the sunlight rings hollow when one considers the thousands of acres that are devastated by strip-mining of coal each year or that might be made uninhabitable by a nuclear accident.

Again, the fact that the Sun provides power only during the day is countered with the argument that 70% of the electrical demand occurs in daytime hours when industries, offices, and stores are in operation. All the solar power that can conceivably be built in the next 20 years or more can be used to supply the daytime demand, saving traditional fuels to provide nighttime demand. In the long run, we might envision the nighttime load being carried by forms of indirect solar energy such as wind power or water power (discussed later in the chapter).

You observed in Fig. 21-8 that about 75% of our electrical power is currently generated by coal-burning and nuclear power plants. Therefore, development of solar electrical power can be seen as gradually reducing the need for coal and nuclear power. Environmentalists consider solar electrical power to be extremely promising as a means of mitigating the acid rain, carbon dioxide, and other environmental impacts of burning coal and of countering the risks of nuclear power. It still doesn't address our greatest resource concern, the fact that our oil reserves needed to fuel transportation are dwindling, but solar energy may offer an answer here as well.

Solar Production of Hydrogen—The Fuel of the Future

Conventional cars can be run on hydrogen gas (H_2) as a fuel in the same manner as they are now beginning to be run on natural gas (methane, CH_4) (see Chapter 21). This fact has been amply demonstrated. Furthermore, there is no carbon dioxide or hydrocarbon pollution produced in burning hydrogen. The only byproduct is water vapor.

$$2\,H_2 + O_2 \longrightarrow 2\,H_2O + \text{energy}$$

(Small amounts of nitrogen oxides may be produced, however, because the burning still uses air, which is nearly four-fifths nitrogen.)

If hydrogen gas is such a great fuel, why are we not using it? The answer is that there is virtually no hydrogen gas on Earth. Any hydrogen gas in the atmosphere has long since been ignited by lightning and burned to form water. And, although there are many soil bacteria that produce hydrogen in fermentation reactions, other bacteria are quick to use the hydrogen because it is an excellent energy source. Thus, there are abundant amounts of the *element* hydrogen, but it is all combined with oxygen in the form of water (H_2O) or other low-energy compounds.

Although we can re-isolate hydrogen from water, this process *uses* energy. Recall that the First Law of Thermodynamics dictates that energy cannot be created; "You can't get something for nothing." At least as much energy is required to convert water into hydrogen and oxygen as is produced in the burning reaction. In fact, it requires *more* energy to convert water into hydrogen and oxygen because of inevitable losses (the Second Law of Thermodynamics). Therefore, the use of hydrogen as a fuel must await an energy source that is itself suitably cheap, abundant, and nonpolluting. Do these specifications sound like solar energy?

Nature, long ago, developed a method of splitting water into hydrogen and oxygen using light energy. This is what happens in photosynthesis. (You will recall that the oxygen from photosynthesis is released into the atmosphere, and the hydrogen is attached to carbon dioxide to form sugars. See page 60.) Scientists, however, have not yet developed a means of mimicking the photosynthetic reaction on a scale sufficient to produce commercial amounts of hydrogen.

However, hydrogen can be produced with very simple equipment by means of *electrolysis of water*. **Electrolysis** of water works as follows: A direct current passed through water causes water molecules to dissociate. Hydrogen bubbles come off at the negative electrode, while oxygen bubbles come off at the positive electrode. You can demonstrate this process for yourself with a battery, some wire, and a dish of water. In addition, a small amount of battery acid in the water facilitates the conduction of electricity and the release of hydrogen.

Now, taking this concept to the ultimate of using solar energy to produce hydrogen for vehicular fuel, the World Resources Institute proposes that we build arrays of solar-trough or photovoltaic generating facilities in the deserts of the southwestern United States, where land is cheap and sunlight plentiful. The electrical power would be used to produce hydrogen gas from water by electrolysis. Conveniently, Texas is the hub of a network of natural gas pipelines that were originally constructed to transport natural gas from the Texas oil fields. Most of the pipelines throughout the nation are now underutilized because those fields have been largely depleted. The hydrogen gas produced in the Southwest could be transported throughout the nation by way of these pipelines.

As vehicles are already being adapted to run on natural gas, hydrogen could be phased in by mixing gradually increasing proportions of hydrogen gas with natural gas. Thus there could be a smooth transition from present fuel through natural gas to hydrogen. (The details of the proposal and supporting data from exhaustive analysis may be found in the World Resources Institute report "Solar Hydrogen: Moving Beyond Fossil Fuels.")

World Resources Institute's economic analysis projects that the cost of solar-produced hydrogen gas would be equivalent to gasoline costing $1.65 to $2.35 per gallon. This may not sound cheap, but it is less than what gasoline will cost as inevitable constraints come into play (see Chapter 21). Consumers in European nations and Japan are already paying much more. Germany is constructing what will be the world's first solar-hydrogen plant near Nüremberg. It will utilize photovoltaic cells.

An alternative to burning hydrogen in conventional internal combustion engines uses the hydrogen in *fuel cells* to produce electricity and power the vehicle with motors. Fuel cells are devices in which hydrogen is recombined with oxygen chemically in a manner that produces an electrical potential rather than burning (Fig. 23-19). Because fuel cells create much less waste heat than conventional engines do, there is a more efficient transfer of energy from the hydrogen to the vehicle.

Indirect Solar Energy

Water, fire, and wind have provided energy for humans throughout history. We are all familiar with hydroelectric dams; burning firewood is common; and photos of wind turbines in California have appeared far and wide. We group these age-old sources of energy together under "indirect solar energy" because a moment's reflection shows us that energy from the Sun is the driving force behind each. The question is, What is the potential to expand these options from the past into major sources of sustainable energy for the future?

Hydropower

It was discovered early in technological history that the force of falling water could be used to turn paddle wheels, which in turn would drive machinery to grind grain, saw logs into lumber, and do other laborious tasks. The modern culmination of this concept is huge hydroelectric dams, where water under high pressure at the bottom of the reservoir behind the dam drives turbogenerators as it flows through (Fig. 23-20). The amount of power generated is proportional to both the height of water behind the dam—that is what provides the pressure—and the volume of water that flows through.

About 8% of the electrical power generated in the United States currently comes from hydroelectric dams, most of it from about 300 large dams concentrated in the Northwest and Southeast.

Whereas water power is basically a nonpolluting, renewable energy source, harnessing it by means of hydroelectric dams still involves tremendous ecological, social, and cultural trade-offs:

1. The reservoir created behind the dam inevitably drowns farmland or wildlife habitat, and perhaps towns or land of historical, archaeological, or cultural value. Glen Canyon Dam (on the border

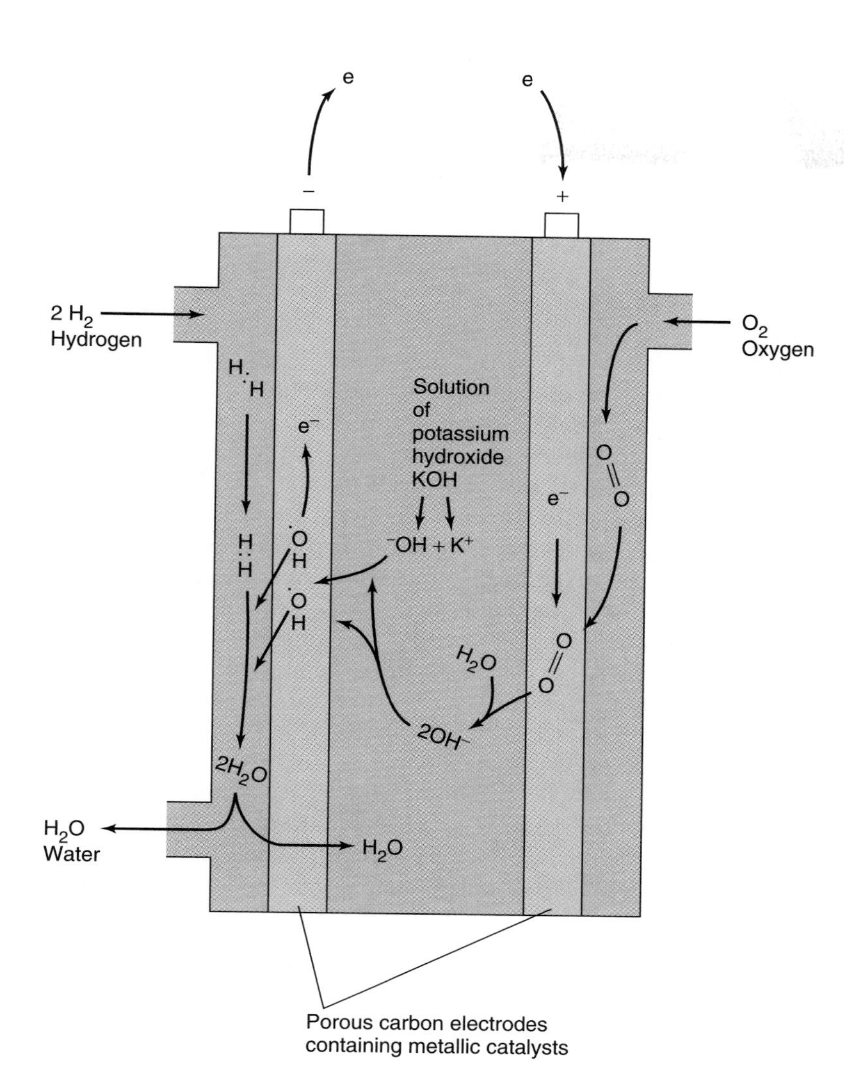

FIGURE 23-19
A hydrogen-oxygen fuel cell accomplishes the reaction of hydrogen and oxygen through stages, as shown in the diagram, such that their high potental energy creates an electric potential. In principle a fuel cell is like an ordinary battery, but the reactants are continuously supplied from an external tank instead of being self-contained within the battery. The only byproduct is water, the same as when burning hydrogen and oxygen.

between Arizona and Utah) drowned one of this world's most spectacular canyons. A number of dams have obliterated sacred lands of Native Americans.

2. Migration of fish, even when fish ladders are provided, is impeded or prevented. Federal surveys show that fish habitat is suffering in 68% of the nation's streams because of damming. Salmon fishing, one of the great industries of the Northwest, has been heavily impacted there because of this factor.
3. Changing from a cold-flowing river to a warm-water reservoir can have unforeseen ecological consequences. The reservoir behind the Aswan High Dam in Egypt has fostered the spread of a parasitic worm that causes a debilitating disease. The reservoir has also increased humidity over a widespread area, which is now accelerating the deterioration of ancient monuments and artifacts that have stood virtually unchanged for many centuries.
4. Ecological consequences occur below the dam as well. Because water flow is regulated according to the need for power, dams play havoc downstream; water levels may go from near flood levels to virtual dryness and back to flood levels in a single day. Other ecological factors are also affected because sediments with nutrients settle in the reservoir, and smaller amounts reach the river's mouth.

As if such trade-offs were not enough, thoughts of greatly expanding water power in the United States are nullified by the fact that few sites conducive to large dams remain. There are already 65,000 U.S. dams; only 2% of the nation's rivers remain free-flowing, and many of these are now protected by the Wild and Scenic Rivers Act of 1968, a law that effectively gives certain scenic rivers the status of national parks.

In short, proposals for new dams are embroiled in controversy over whether the projected benefits justify the ecological and sociological trade-offs. In 1991, the Environmental Protection Agency, after protracted dis-

FIGURE 23-20
Hoover Dam. About 8.6% of the electrical power used in the United States comes from large hydroelectric dams such as this. Water flowing through the base of the dam drives turbines. (Lowell Georgia/Photo Researchers, Inc.)

putes, finally denied permits for the proposed Two Forks Dam in Colorado on environmental grounds. The James Bay project in northern Quebec is creating even more controversy. If completed—one-third of it is done—the project will consist of 215 dams and dikes, as well as nine river diversions, and will affect fish and wildlife throughout an area the size of New England, New York, and Pennsylvania combined. It will largely displace two tribes of Native Americans. Most of the power produced will be exported to the United States, although the state of New York withdrew orders in 1992 owing to the success of conservation efforts. Such controversies can be seen in many developing countries as well; one is the Three Gorges Dam in China.

Finally, it is questionable just how sustainable these hydroelectric projects really are. Reservoirs gradually fill with sediments, and dredging them is not economically practical. Many are not expected to last more than another 50 years. Even now, in years of drought like 1992, the use of hydropower for electricity is curtailed sharply in western states.

Wind Power

Wind power, which was all but forgotten in the era of cheap oil (1930–1970) is now looking more promising. Throughout much of history wind power—in addition to propelling sailing ships—was widely used for grinding grain, hence the term *windmill*. Then, wind-driven propellers went on to perform other tasks. Until the 1940s, most farms in the United States used windmills for pumping water and generating small amounts of electricity. In the thirties and forties, windmills fell into disuse, however, as transmission lines brought abundant lower-cost power from central generating plants. Not until the energy crisis and rising energy costs of the seventies did wind begin to be seriously considered again as a potential source of sustainable energy.

Many different designs of wind machines were proposed and tested, but what has proven most practical is the age-old concept of wind-driven propeller blades. The propeller shaft is geared directly to a generator. (Wind driving a generator is more properly called a **wind turbine** than a windmill.)

The United States, always intrigued with bigness and on the basis of mathematical calculations showing that efficiency increases with size, embarked on building some truly monstrous wind turbines—blades as much as 300 feet (100 m) from tip to tip mounted on 200-foot (70 m) towers (Fig. 23-21). All of these huge machines suffered frequent and repeated mechanical failures because parts could not withstand the tremendous stresses, and all have since been abandoned.

Meanwhile, Denmark, also hard-pressed by the energy crisis, started deploying and has refined the development of more modestly sized wind turbines with a blade diameter in the range of 45 feet (15 m) (Fig. 23-22). Most of the wind turbines now installed in California and elsewhere in the world are imported from Denmark.

As reliability and efficiency of wind turbines have improved, the cost of wind-generated electricity has come down. **Wind farms**, arrays of 50 to several thousand such machines, are now producing pollution-free, sustainable power for as little as 7 cents per kilowatt-hour, cheaper than traditional sources. Moreover, the amount of wind that potentially can be tapped is immense. The American Wind Energy Association calculates that wind farms located throughout the Midwest could meet the electrical needs for the entire country, while the land beneath the turbines could still be used

FIGURE 23-21
MOD 2 wind turbine in Washington State. Blades, 300 feet (100 m) from tip to tip; tower, 200 feet (70 m) tall; capacity, 2.5 MW in winds 14 to 45 mph. Such wind turbines have now been abandoned because severe stresses on parts caused maintenance problems. (U.S. Department of Energy.)

for farming. This situation is similar throughout the world.

There are currently some 17,000 wind turbines generating 1500 MW (supplanting the need for two nuclear power plants) operating in California, and over 50,000 operating elsewhere in the world. Even so, this is only the beginning.

There are still some drawbacks with wind power, however. First, as it is an intermittent source, problems of backup or storage must be considered. Second is the aesthetic consideration. One or two windmills can be charming, but a landscape covered with them can be visually tiresome, to say nothing of their continual whirring and whistling. Third, they are a hazard to birds, although the single *Exxon Valdez* accident in 1989 killed 300,000 birds. Location of wind farms on migratory routes could be a problem for some avian populations.

Biomass Energy, or Bioconversion

There is nothing like a new name to put pizzazz into an old concept. Thus "burning firewood in a stove" has become **biomass energy**, or **bioconversion**. Actually, these terms include any means of deriving energy from present-day photosynthesis—as opposed to that of millions of years ago, as is the case with fossil fuels.

In addition to burning wood in a stove, major methods of bioconversion include: obtaining energy from the burning of municipal waste paper and other

FIGURE 23-22
Wind farm at Altamont Pass, an area east of San Francisco. Such arrays of wind turbines are proving to be an economically competitive way of generating electrical power. The land under the wind turbines can still be used for agriculture. (Tom McHugh/Photo Researchers, Inc.)

Ethics

Transfer of Energy Technology to the Developing World

The nations of the industrialized Northern Hemisphere have achieved their level of development using energy technologies based largely on fossil fuels (and, to a lesser extent, nuclear energy). Their development took place during a time when fossil fuels were inexpensive; only as the expense of those fuels has risen have the nations of the North begun to get serious about technologies to make energy use more efficient. As we saw in Chapter 21, these nations still have a long way to go in conserving energy; just putting into place efficient technologies already developed could save three-fourths of energy expenditures in the United States.

The same traditional fossil-fuel–based technologies have been adopted by the developing nations of the Southern Hemisphere, except that they are lagging behind in efficiency. On the whole, the industrialized North is three times more efficient in energy use than the developing South. As we have seen in Chapter 21, however, the fossil-fuel–based energy pattern of the twentieth century will have to be largely replaced with renewable energy systems in the next century. If the United States and other rich nations are willing to play a significant role in helping the poor nations develop, we should not promote the same wasteful development path the North has taken. Instead, there is the opportunity to engage in some "leap-frogging" technology transfer. The North can put its development dollars into such technologies as electrification of rural areas via photovoltaics and efficient public transport systems for the cities.

The solar route is especially attractive for many of the climates in the developing world, where an estimated 2 billion people lack electricity. A model program is the work of Richard Hansen, whose nonprofit development group Enersol helped establish the Asociacion de Desarrollo de Energia Solar (ADES) in the Dominican Republic. ADES has established a revolving low-interest loan fund and the necessary equipment to help rural people electrify their homes. Rooftop photovoltaic units the size of two pizza boxes produce enough power for five electric lights, a radio, and a television set. Similar programs have been started in Sri Lanka and Zimbabwe.

It is entirely possible that if this strategy of development aid is followed, the nations of the South may end up pointing the way to a sustainable energy future for the rest of the world. The Rio Declaration, signed at the Earth Summit in June 1992, commits the industrial countries to providing greater levels of development aid to the poor countries. It also stipulates that the development should be sustainable. As the details of this commitment get fleshed out, it will be interesting to see if solar energy receives the attention it deserves. If it does, there is much the donor countries can learn from the results. What do you think?

organic waste; methane production from anaerobic digestion of manure and sewage sludges; and alcohol production from fermenting grains and other starchy materials. Let's look further at the potential of each of these techniques.

Burning Firewood. Where forests are ample relative to population, firewood can be a sustainable energy resource, and indeed wood has been a main energy resource over much of human history. In the United States, wood stoves have enjoyed a tremendous resurgence in recent years—about 5 million homes rely entirely on wood for heating and another 20 million use it for partial heating, so much so that air pollution from wood stoves has become a problem, and some communities are finding it necessary to apply restrictions.

However, using firewood is hardly novel, for over a billion people in developing countries still depend on it as their only source of fuel for cooking. Recall that such regions are suffering from the severe ecological effects of deforestation as well as the human deprivation caused by running out of wood (see Chapter 9). Therefore, any blanket recommendation to burn wood because in theory it is renewable is environmentally irresponsible. Using firewood must be coupled with programs of reforestation to be sure that the process is sustainable.

Burning Municipal Waste. Facilities to generate electrical power from the burning of municipal wastes (waste-to-energy conversion) were discussed in Chapter 20. Many sawmills and woodworking companies are now burning wood wastes, and a number of sugar refineries are burning the cane wastes to supply all or most of their power. You will recall, however, that the primary virtue in this method was in finding a productive or less-costly way to dispose of wastes. It is *not* a cost-effective means of producing power, nor can the power from this source ever meet more than a few percent of total electrical needs. Also, such schemes must be weighed against the potential for recycling or for other uses of such wastes. Wood wastes, for example, are made into pressed "plywood."

Producing Methane. You will recall from Chapter 13 that anaerobic digestion of sewage sludge yields *biogas*, which is two-thirds methane, plus a nutrient-rich treated sludge that is a good organic fertilizer. Animal manure can be digested likewise. Where these three aspects—manure disposal, energy production, and fertilizer creation—can be combined in an efficient cycle, great economic benefit can be achieved.

This concept is demonstrated on the Mason-Dixon Dairy farm located near Gettysburg, Pennsylvania (Fig. 23-23). Manure from 2000 cows that are fed in a barn drops through a grated floor and slowly flows by gravity into anaerobic digesters. The biogas produced (purification of the methane has proved unnecessary) fuels engines that drive generators supplying not only all the electrical power for the dairy, but also considerable excess that is sold to the utility company. Waste heat from the engines is used to heat the digesters and the buildings. The nutrient-rich sludge remaining after digestion is recycled back to the land, in order to maintain the fertility of the fields growing feed for the cows.

With the savings and proceeds from selling energy, as well as sale of dairy products and additional savings on fertilizer, the Mason-Dixon Dairy is probably the most profitable dairy in the country. If all dairy farms and cattle-feeding operations in the United States followed this example, we could be getting nearly as much electrical power from cows as we are currently getting from nuclear power at about one-fifth the cost, and reducing pollution (eutrophication) in the bargain.

In China, millions of small farmers maintain a simple digester in the form of a sealed pit into which they put agricultural wastes. The biogas produced is used as a source of cooking fuel. This concept could be expanded throughout the developing world to provide an alternative to fuelwood.

Producing Alcohol. Alcohol is produced by the fermentation of starches or sugars. The usual starting material is grain, sugar cane, or various sugary fruits

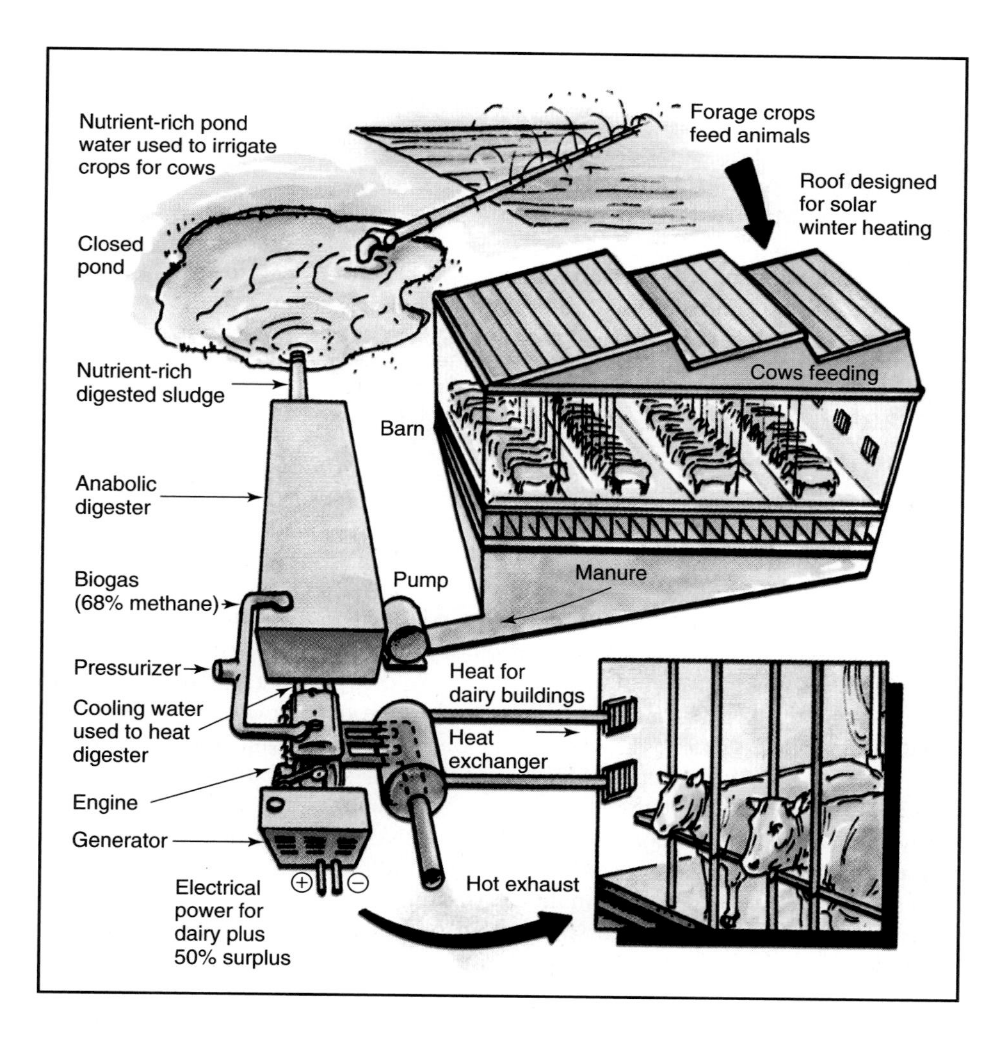

FIGURE 23-23
The total power needs for the Mason-Dixon Dairy, located near Gettysburg, Pennsylvania, are obtained as a byproduct of cow manure, and nutrients are recycled in the process. Excess power, nearly half of what is produced, is sold to the local utility.

Earth Watch

Another Look at Conservation

The promise of electrical power and alternative fuels from solar sources should not cause us to overlook the importance of energy conservation and efficiency. If total energy use continues to grow with population growth as it has in the past, even massive installation of solar alternatives may barely keep pace with growing demand. It is difficult to see how solar facilities can be deployed rapidly enough to allow an orderly phaseout of traditional fossil fuels and nuclear power while keeping abreast of growing demand. Thus, regardless of which alternatives are chosen, conservation will remain a most important element in energy policy. Conservation is also exceptionally cost-effective. Whatever the source of power or fuels, the cost is less if fewer kilowatts or gallons need to be produced.

Most people still think of conservation in terms of turning the heat down, turning off lights, and so on. Thus, conservation still has the reputation among many of necessitating "freezing in the dark." Importantly, the greatest energy savings are not to be achieved by measures that cause inconvenience or discomfort; rather, they are to be achieved by making heating, lighting, and transportation systems more efficient so that we can still have the same comfort, convenience, and transportation but use less energy in the process. By pursuing conservation through greater energy efficiency, the world can support a growing economy without demanding increasing amounts of energy.

Amory and Hunter Lovins, founders of the Rocky Mountain Institute (RMI) located near Aspen, Colorado, are recognized as world authorities on energy conservation. According to their analysis, universal application of today's best energy-saving technologies—high-mileage vehicles and more efficient appliances, motors, lights, windows, and insulation—could reduce U.S. oil and electricity consumption by 75%, saving $300 billion annually with no loss of services.

Many critics feel that this analysis is overly optimistic, but numerous companies are taking advantage of RMI's consulting services and realizing tremendous savings. For example, the *Boston Globe* is achieving a savings of $350,000 per year through installing a high-efficiency lighting, heating, and cooling system recommended by RMI. The World Bank headquarters in Washington, D.C., is now saving $500,000 per year on a one-time investment of $100,000 for more efficient lighting. The greatest success, however, is the following story.

In 1988, Northeast Utilities proposed building a new power plant to meet Connecticut's growing demand. The Conservation Law Foundation (CLF), an environmental group, using information from RMI, challenged the proposal in court claiming that power needs could be met through conservation. CLF won! The court ordered the utility to work with CLF to start a program of conservation. The program developed involves the utility's spending the money that would have been spent on the new power plant for retrofitting homes, businesses, and factories with energy-efficient lighting, appliances, and insulation and designing more energy-efficient new buildings. A key ingredient was the state's changing utility regulations so that the utility could earn a higher rate of return on its money invested in conservation than on its money spent for traditional power production. This is a win-win situation in which the utility is earning more on its investments, consumers are saving money on power, and the environment is not suffering the impacts of yet another power plant. Other utilities have now joined in the program, and the concept is spreading.

With growth in power demand stemmed by implementing conservation, we can move toward meeting demand with power from solar sources.

such as grapes. Yes, this is exactly the process used in the production of alcoholic beverages. The only new part of the concept is that instead of drinking the brew, we distill it and put it in our cars. *Gasohol*, a mixture of gasoline and alcohol, has been promoted and marketed in the Midwest since the late 1970s.

However, the enthusiastic support of gasohol in the Midwest is not motivated purely by the desire to save fossil fuel. This product provides another market for grain surpluses and thereby improves the economy of the region. Where grain surpluses do not exist—and this includes most of the world—alcohol as a fuel is a dubious goal at best. For example, Brazil promoted the large-scale production of alcohol from sugarcane in the 1980s. Sugarcane production rose to record levels, while food crops declined by 10%, despite widespread malnutrition and a rapidly growing population. Brazil has now abandoned its promotion of alcohol fuel.

Another problem with alcohol as fuel is pollution. Alcohol is promoted as clean-burning, and it is. However, producing the alcohol generates much pollution because inexpensive, dirty-burning fuels such as soft coal are used to distill the alcohol.

Finally, proposals that trees or other crops be grown specifically for biomass energy—either for direct burning or for conversion to alcohol or methane—must be weighed against the impacts of soil erosion, fertilizers, and pesticides that such production generally

FIGURE 23-24
One of the 11 geothermal units operated by the Pacific Gas & Electric Company at The Geysers in Sonoma and Lake counties, California. The field was generating 2000 MW in 1988, but power output is now falling off because the field is running out of water. (Photograph courtesy of Pacific Gas & Electric Company.)

entails. Also, energy crops will invariably compete with food crops because either may be grown on the same land.

Ocean Thermal Energy Conversion

Over much of the world's oceans, a thermal gradient of about 20° C (36° F) exists between surface water heated by the Sun and colder deep water. **Ocean thermal energy conversion (OTEC)** is the name of an experimental technology for using this temperature difference to produce power. It involves using the warm surface water to heat and vaporize a low-boiling-point liquid such as ammonia. The increased pressure of the vaporized liquid would drive turbogenerators. The ammonia vapor leaving the turbines would then be recondensed by cold water pumped up from as much as 300 feet (100 m) deep and returned to the start of the cycle.

Various studies show that OTEC power plants show little economic promise—unless, perhaps, they can be coupled with other, cost-effective operations. For example, in Hawaii a shore-based OTEC plant uses the cold, nutrient-rich water pumped from the ocean bottom to cool buildings and supply nutrients for vegetables in an aquaculture operation in addition to cooling the condensers in the power cycle. Even so, interest in duplicating such operations is minimal at present.

Additional Renewable Energy Options

Two additional, theoretically sustainable energy options have strong support from certain sectors and deserve consideration: geothermal power and tidal power.

Geothermal Energy

In various locations in the world, such as the northwestern corner of Wyoming (now Yellowstone National Park), one finds springs that yield hot, almost boiling water. Natural steam vents and other thermal features are also found in such areas. They occur where the hot molten rock of Earth's interior is close enough to the surface to heat groundwater, as may occur in volcanic regions. Using such naturally heated water or steam to heat buildings or drive turbogenerators is the basis of **geothermal energy**. Frequently, additional wells are drilled nearby to enhance the release of hot water or steam.

In 1993, geothermal energy was being used to generate electrical power equivalent to seven large nuclear or coal-fired power stations in countries as widely diverse as Nicaragua, the Philippines, Kenya, Iceland, and New Zealand. Today, the largest single facility is in the United States at a location known as The Geysers, 70 miles (110 km) north of San Francisco (Fig. 23-24). As impressive as this application for power generation is, nearly double the amount of geothermal energy used for power generation is being used to directly heat homes and buildings largely in Japan and China.

Many experts feel that the potential for geothermal energy has been barely tapped, although its development will probably remain restricted to regions such as Japan, the Philippines, Iceland, and New Zealand, which have suitable underlying volcanic geology. Another concern is that recent experience has shown that geothermal energy may not always be sustainable. The Geysers in California was steadily expanded. In 1988 it reached about 2000 MW (equivalent to two large nuclear power plants), and it was projected that by the year 2000 the facility would be producing 3000 MW. In fact, output from The Geysers has been steadily declining since 1988; in 1991 it was producing just 1500 MW, and it is now projected that by the end of the decade this number may be down to half the peak 1988 level. The problem is that, although the source of geothermal heat may be unlimited, the amount of groundwater isn't. Tapping the hot steam is depleting the groundwater.

Geothermal power presents pollution problems as well. The hot steam and water brought to the surface are frequently heavily laced with salts and other contaminants, particularly sulfur compounds leached from minerals in the bedrock. These contaminants are highly corrosive to turbines and other equipment, and they cause air pollution if the steam is released into the atmosphere. Sulfur dioxide pollution from a geothermal plant can be equivalent to that from a plant burning high-sulfur coal. Hot brines from geothermal sources released into streams or rivers may be ecologically disastrous.

Tidal Power

A phenomenal amount of energy is inherent in the twice-daily rise and fall of ocean tide, and many imaginative schemes have been proposed for capturing this eternal, pollution-free source of energy. The most straightforward idea is to build a dam across the mouth of a bay and mount turbines in the structure. The incoming tide flowing through the turbines generates power. As the tide shifts, the blades can be reversed, so the outflowing water continues to generate power.

Tantalizing as this idea sounds, it is fraught with a fundamental problem: In most regions of the world the maximum difference between high and low tide is a matter of only 16 to 20 in. (40 to 50 cm). A head of pressure 20 inches or less is not enough to drive turbines efficiently. There are about 30 locations in the world where shoreline topography generates tides high enough—18 feet (6 m) or more—for this kind of use. Large tidal power plants already exist at three of those places—in France, Canada, and in Russia. The only suitable location in North America is the Bay of Fundy, where the Annapolis Tidal Generating Station has operated since 1984, with 20 MW capacity.

Thus, tidal power has potential only in certain localities, and even there it would not be without adverse environmental impacts. The dams would trap sediments, impede migration of fish and other organisms, and alter the circulation and mixing of salt and fresh water in estuaries, perhaps having other unforeseen ecological effects.

In summary, the solar and other renewable energy alternatives we have discussed in this chapter are diagramatically shown in Fig. 23-25.

Policy for a Sustainable Energy Future

In the last three chapters, we studied our planet's energy resources, requirements, and management options. A review of the energy situation as a whole suggests that there is no reason to fear "running out" of energy. World reserves of fossil fuels, even crude oil, are adequate for at least the next 20 to 50 years. Then, there are combinations of solar and wind options that can be developed and phased in over that time frame. And there is nuclear power if we should choose to go that route.

The major question confronting us is: Should we be taking any specific steps to promote or even force the transition to an economy based on solar energy? Or should we simply eliminate all subsidies for energy to level the playing field and let free-market forces sort out the situation?

Some policies of promoting the transition away from dependence on crude oil and toward solar energy were implemented in the 1970s. They included:

- Subsidies to producers of solar energy or solar energy products
- Subsidies to people for solar installations in homes and other structures
- Additional taxes on traditional fuels, making them more expensive relative to solar energy
- Mandates to increase fuel efficiency of cars
- Additionally, rationing of fuel and denial of new hookups to traditional power supplies were strongly considered.

All of these policies, except for a modest tax on gasoline, were phased out in the early 1980s effectively shifting to the alternative policy—"business as usual"—leaving things as they are—that is with an economic bias toward traditional energy. Subsidies and incentives for fossil fuels and nuclear remained in force. There is some reason to think that economic forces may suffice, as we see photovoltaic cells and wind energy making headway in today's marketplace. But, the "business as usual" view overlooks the fact that the field of competition between solar and traditional sources of energy is

FIGURE 23-25
Renewable energy resources. At the present time, direct solar heating of space and hot water, photovoltaic cells, solar-trough collectors, wind power, production of hydrogen from solar or wind power, and production of methane from animal manure and sewage sludges seem to offer the greatest potential for supplying sustainable energy with a minimum of environmental impact. However other alternatives shown should not be ruled out.

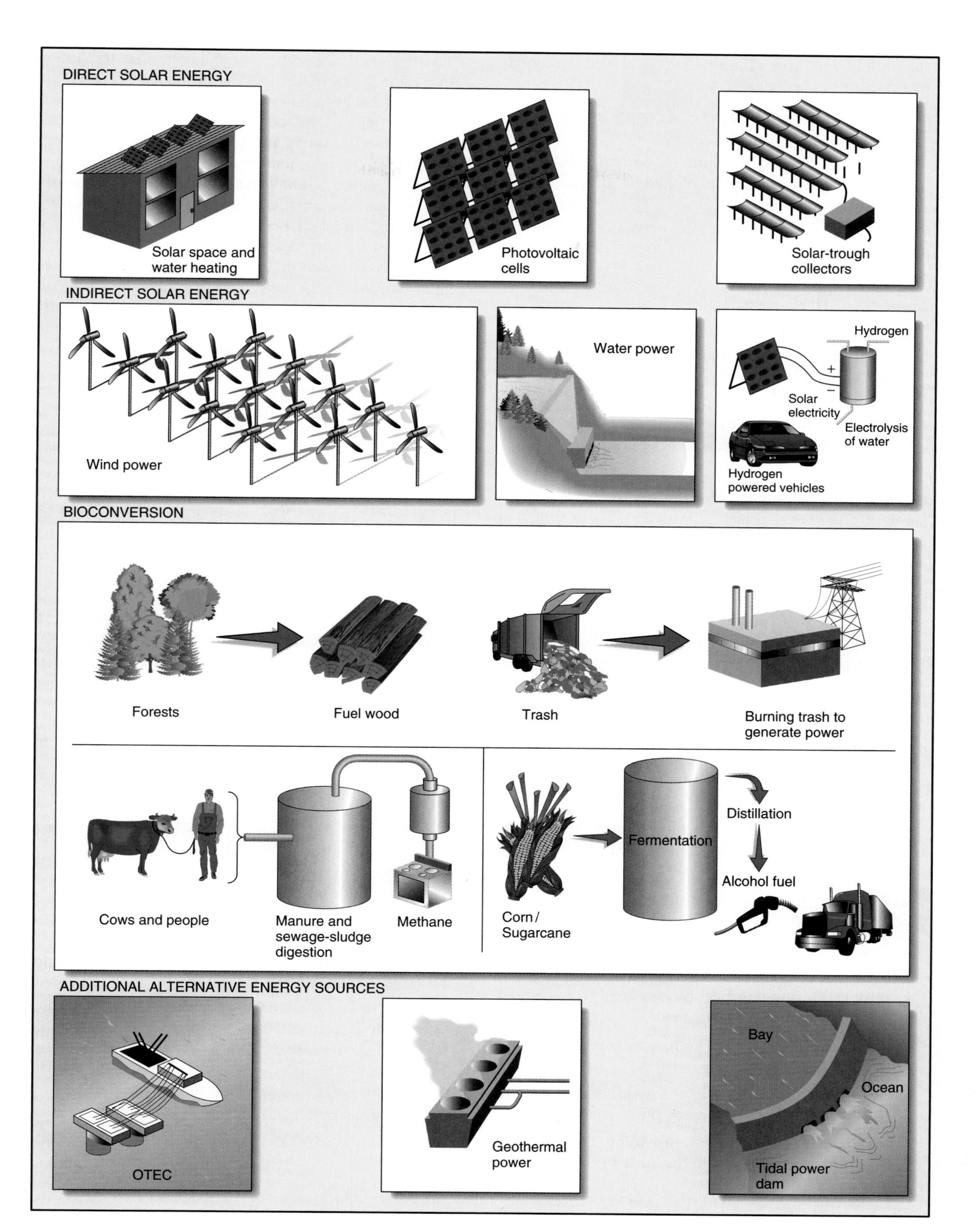
DIRECT SOLAR ENERGY
Solar space and water heating
Photovoltaic cells
Solar-trough collectors
INDIRECT SOLAR ENERGY
Wind power
Water power
Hydrogen
Solar electricity
Electrolysis of water
Hydrogen powered vehicles
BIOCONVERSION
Forests
Fuel wood
Trash
Burning trash to generate power
Cows and people
Manure and sewage-sludge digestion
Methane
Corn / Sugarcane
Fermentation
Distillation
Alcohol fuel
ADDITIONAL ALTERNATIVE ENERGY SOURCES
OTEC
Geothermal power
Bay
Ocean
Tidal power dam

steeply tilted in favor of traditional fuels. The traditional fuel industry is heavily subsidized by:

- Depletion allowances (tax writeoffs as a resource is depleted)
- Leasing of public lands at bargain basement prices, encouraging exploitation of fossil fuel reserves
- Military support to assure access to oil in the Middle East

Moreover, substantial costs are discounted or overlooked. These costs include:

- Damage to the environment and human health from the pollution caused by burning fossil fuels
- Environmental destruction from strip-mining, oil spills, and other accidents
- The continuing and increasing economic drain as we import growing amounts of crude oil

Finally, we are under the moral and legal "Earth Summit" agreement with other nations to stabilize emissions of greenhouse gases—the major one of which is carbon dioxide from fuel burning (see Chapter 16, page 408).

In view of these facts, what should our policy be? Should the subsidies for solar energy be reinstated? Should mandates for conservation be reimposed? Should all the hidden costs of fossil fuel consumption be put "up front" on the price of fuel? The subjects for debate are nearly endless.

It should not escape your notice that the United States stands out as the only industrialized country that has seen fit to keep gasoline prices relatively low. Fuel in all other highly developed countries is so heavily taxed that it costs consumers $3 to $5 per gallon—U.S. prices, adjusted for inflation, remain at the 1952 price level.

A number of environmental groups promote the idea of a carbon tax, a tax levied on all fuels according to the amount of carbon dioxide that is produced in their consumption. Such a tax, proponents believe, would provide both incentives for using solar sources, which would not be taxed, and disincentives for consumption of fossil fuels. This one tax would avoid the bureaucratic nightmare of administering many separate regulations. A number of European countries have already adopted such a tax.

Many people ask, "Why don't they develop solar?" We should see that the obstacles are no longer technological or even economic. The main obstacle is political and the reluctance to upset our traditional energy-using habits. Each one of us, to a greater or lesser degree, contributes to that inertia. The question therefore needs to be rephrased as, "Why don't *I* develop solar?" Each one of us—individuals, consumers, and voters—has numerous opportunities to help promote conservation, as well as solar and other renewable energy alternatives—to have a demand/consumer-driven energy future rather than a supply/producer-driven one. In Chapter 24, we examine in more detail the connections between lifestyle and sustainability.

Review Questions

1. How does the energy reaching Earth from the Sun compare with fossil-fuel energy and basic energy needs?
2. What are the three basic problems of harnessing solar energy?
3. How do active and passive solar hot-water heaters work?
4. How can a building best be designed to become a passive solar collector for heat?
5. What are the barriers to more widespread adoption of solar homes?
6. How does a photovoltaic cell work, and what are some present applications of photovoltaic cells?
7. What is the potential for providing more energy from photovoltaic cells in the near future?
8. Describe what the solar-trough system is, how it works, and its potential for providing power.
9. Respond to the criticisms that solar power requires too much land and doesn't work at night.
10. How can hydrogen gas be produced using solar energy?
11. How may hydrogen meet the needs for fuel for transportation in the future?
12. What is the potential for developing more hydroelectric power in North America, and what would be the environmental impacts of such development?
13. How is wind power being harvested, and what is the future potential for wind farms?
14. What are four ways of converting biomass to useful energy?

15. What is the potential for and environmental impact of each of the four methods you listed in your answer to Question 14?
16. What is meant by geothermal energy, and how may it be harnessed?
17. What has been the recent experience with geothermal power in California, and how has it changed the outlook for geothermal energy?
18. What is the potential for developing tidal power in the United States?
19. What policies were moving the United States towards solar energy in the 1970s?
20. How is the traditional fossil-fuel–based system subsidized?
21. What are the "hidden costs" of continuing to rely on fossil-fuel energy systems?
22. What policies might move us in a direction of a sustainable energy future?

Thinking Environmentally

1. A sustainable society will ultimately be one that gains all its energy needs from what source(s)? Consider all energy sources, both traditional and alternative, and discuss each in terms of long-term sustainability. What conclusions do you reach? Defend your conclusions.
2. Design what you feel should be the energy policy for the United States based on the total range of energy options, including fossil fuels and nuclear power as well as various solar alternatives. Which energy sources should be promoted, and which should be discouraged? Give a rationale for each of your recommendations.
3. Suggest and give a rationale for laws, taxes, subsidies, and so forth that might be used to bring to fruition the policy you suggested for Question 2.
4. Various solar energy alternatives seem to promise a sustainable energy future. Yet, the U.S. government seems locked into a policy of maintaining dependence on fossil fuels and nuclear power. Does this mean that solar advocates are wrong? Discuss the economic and political forces that might be at work in maintaining the status quo.
5. Consider all the energy needs in your daily life. Describe how each might be provided by one or another solar option.

CHAPTER 24

Lifestyle and Sustainability

Key Issues and Questions

1. Before World War II cities had a compact integrated structure. Why? How did the increasing reliance on automobiles after World War II begin to change this structure?
2. Urban trends since World War II are marked by exurban migration, urban sprawl, urban blight, and a growing dependence on cars. Describe these trends and how they are all related.
3. Urban sprawl is at the root of many environmental problems. List and describe these problems.
4. Exurban migration set into motion a vicious cycle of urban decline. Describe the factors involved in this cycle. What is economic exclusion?
5. A sustainable future will depend on both reining in urban sprawl and rehabilitating cities. How can these objectives be accomplished?
6. Most development since the 1950s has focused on moving cars. Future development must focus on communities and people. Describe the distinction.

All the environmental problems we have discussed, from the loss of biodiversity through various forms of pollution to depletion of energy resources, have their origins in people and their lifestyles. Recall the relationship we presented in Chapter 6 (page 146) showing that negative environmental impact is a function of population size multiplied by the consumptiveness of lifestyle divided by environmental regard. Preceding chapters can be seen from the perspective of environmental regard. Concern for our environment has provided the impetus for better pollution control, sewage treatment, waste management, wildlife protection, and more benign energy sources. Yet, is this enough? We do not need to look far to see more and more farms, open space, wildlife habitat, and biodiversity giving way to new highways, shopping malls, industrial parks, and housing developments (Fig. 24-1). Since the mid-1980s, new development has consumed an average of 1.4 million acres of agricultural land per year and an equal amount of other natural areas in the United States alone. The problem exists in most other countries as well. Is such a pathway of development sustainable?

The old, inner cities are blighted with poverty, unemployment, drug use, crime, and violence. Family life is difficult in the inner city and these destitute neighborhoods. How can

Development in the years since World War II has emphasized a lifestyle that is profligate in its use of space and consumption of energy, water, and other resources. Shown here is a section of a Los Angeles neighborhood. [Photo by Pacific Gas and Electric Co.]

mothers and fathers support their children? What kind of adults will these children become as they grow up in this environment of impoverishment and social chaos? What does this situation say about social justice? What does it bode for sustainability?

Many people think that both of these problems, expanding development into more and more open space on the one hand and urban blight on the other, are products of population growth. However, statistics show that outward development has occurred much faster than population growth. For example, between 1965 and 1990, the population of the New York metropolitan region grew by just 5%, while developed land area increased by 61%. Also, statistics show that the inner cities suffering from blight have actually *decreased* in population.

Our objective in this chapter is to show how these two problems, exurban development and urban blight, actually are related. Both stem from a profound shift in lifestyle that has occurred mainly since the end of World War II. Further, we shall consider the potential for changing lifestyle so as to bring about sustainability.

From Urban Structure to Urban Sprawl

The dominant, and environmentally most provocative, feature of life in the United States is our near-total dependence on cars. We have developed an urban-suburban layout in which the locations where we live, work, go to school, and shop are widely separated. Thus, we have come to rely on private cars, which we drive an average of 200 miles (300 km) a week simply to meet our everyday needs, and multilane highways are often congested with traffic. We shall refer to this lifestyle that revolves around going everywhere by car as a *car-dependent lifestyle*.

This far-flung urban-suburban network of residential areas, shopping malls, industrial parks, and other facilities loosely laced together by multilane highways is referred to as **urban sprawl**. The word *sprawl* is used because the perimeters of the city have simply been extended outward into the countryside, one development after the next, with little plan as to where the expansion is going and no notion as to where it will stop. That the sprawl is continuing is all too evident. Almost everywhere we go near urban areas, we are confronted by farms and natural areas giving way to new developments, new highways being constructed, and old roadways being upgraded and expanded (Fig. 24-2).

FIGURE 24-1
We do not need to look far to see more and more farms, open space, wildlife habitat, and biodiversity giving way to new highways, shopping malls, industrial parks, and housing developments. Is such "progress" sustainable? (Joan Baron/The Stock Market.)

The Origins of Urban Sprawl

Most of you reading this text have grown up in the midst of urban sprawl and America's car-dependent lifestyle. Therefore it may be difficult to imagine that things were ever different. However, urban sprawl really started only in the late 1940s when, after World War II, private ownership of cars became common.

Even though cars had been developed around the turn of this century, they were unaffordable to most people until Henry Ford developed assembly-line production in the 1920s. Just as that technological advance came about, however, ownership was sharply curtailed first by the Great Depression and then by World War II, at which time all production facilities were given over to the war effort. Therefore, until the end of World War II, a relatively small percentage of people owned cars.

Cities had developed in ways that allowed people to meet their needs by means of the transportation available—mainly walking. (For most people, owning a horse in a city was as untenable then as now because

FIGURE 24-2
The hallmark of development over the past several decades has been the construction of new highways. New highways, however, enable people to commute to work from more distant locations. Thus they support further development and sacrifice of open space. (David Pollack/The Stock Market.)

of the expense and impracticality of maintaining one in the limited space of an urban property.) Every few blocks had a small grocery, a pharmacy, and other stores, as well as professional offices integrated with residences. Often buildings had a store at the street level and residences above (Fig. 24-3). Numerous schools were scattered throughout the city as were city parks for outdoor recreation or relaxation. Thus, walking distances were generally short, and bicycling was even more convenient.

For more-specialized needs people boarded public transportation—horse-drawn or electric trolleys, cable cars, and, later, buses—from neighborhoods to the "downtown" area, where big department stores, specialty shops, and offices were located. Public transportation did not change the compact structure because people still needed to walk to the transit line. At the outer ends of the transit lines, cities gave way abruptly to farms, which provided most of the food for the city, and open country. The small towns and villages surrounding cities, the original suburbs, were compact for the same reasons and mainly served farmers in the immediate area. This pattern held until the end of World War II; then it began to change dramatically.

FIGURE 24-3
Before the widespread use of automobiles, cities had an integrated structure. A wide variety of small stores and offices on ground floors, with residences on upper floors, placed everyday needs within walking distance. This arrangement still exists in certain areas of some cities. (Geri Engberg/The Stock Market.)

FIGURE 24-4
Prime agricultural land is being sacrificed for development around most U.S. cities and towns. Location of developments is determined more by the availability of land than by any overall urban planning. The scattered nature of development that results is a major contributing factor to urban sprawl. (Photography by Grant Heilman.)

Despite the many advantages of urban life, many people found cities less-than-pleasant places to live. Especially in industrial cities, poor housing, inadequate sewage systems, inadequate refuse collection, pollution from home furnaces and industry, and generally congested, noisy conditions were much the norm. A decrease in services during World War II aggravated those problems. Hence, many people had the desire—the "American Dream"—to live in their own house on their own piece of land away from the city.

The Automobile. Because of a shortage of consumer goods during the war, there was a pent-up demand for a shopping spree when the war ended, and many civilians and returning veterans had accumulated considerable savings during the war. When mass production of cars commenced at the end of the war, people flocked to buy them. With private cars, people were no longer restricted to live within walking distance of their workplaces or of transit lines. They could move out of their cramped city apartments and into homes of their own outside the city. Their cars would allow them to drive back and forth easily to their jobs, shopping, recreation, and so on.

Developers responded quickly to the demand for private homes. They bought farms wherever they could and put up houses. The government aided this trend by providing low-interest mortgages through the Veterans Administration and the Federal Housing Administration, and interest payments on mortgages were made tax-deductible (rent continued to be non-tax-deductible). Property taxes in the suburbs were much lower than in the city. These financial factors meant that, for the first time in history, making monthly payments on one's own home in the suburbs was cheaper than paying rent for equivalent or less living space in the city.

Thus, the mushrooming development around cities did not proceed according to any plan; rather, it happened wherever developers could acquire land (Fig. 24-4). In fact, planning was more than merely neglected; it was actively prevented by the fact that cities were surrounded by a maze of more or less autonomous local jurisdictions (towns, townships, counties, municipalities). No governing body existed to devise, much less enforce, an overall plan. Local governments were simply thrown into a catch-up role of trying to provide schools, sewers, water systems, and other public facilities—and most of all, roads—to accommodate the uncontrolled growth.

Highways. The influx of commuters into previously rural areas soon resulted in traffic congestion, creating a need for new and larger roads. To raise money to build and expand highways, Congress passed the Highway Revenue Act of 1956, which created the **Highway Trust Fund**. This legislation placed a tax on gasoline and earmarked the revenues to be used exclusively for building new roads. (The fund still exists, but was modified in 1991 to allow a portion of the money to be used for other forms of transportation, as will be described later.)

Perhaps you can already see how the Highway Trust Fund perpetuates development. A new highway not only alleviates existing congestion; it also encourages development at farther locations because time, not distance, is the limiting factor for commuters. (Listen to how people describe how far they live from work; it is almost always put in terms of minutes, not distance.) The average person is willing to spend 20 to 40 minutes each way on a daily commute. Given this time limit, people who have to walk would have to live within 1 or 2 miles (1.5–3.0 km) of their work. With cars and expressways, people can live 20 to 40 miles (30 to 60 km) from work and still get there in the same amount of time.

Therefore, new highways that are intended to

(a)

(b)

FIGURE 24-5
The new highway–traffic congestion cycle. (a) Growing traffic congestion creates the need for new and upgraded highways. (b) However, upgraded highways encourages development at more distant locations, thus creating more traffic congestion and hence the need for more highways. Is this snowball effect sustainable? (a) David Lawrence/The Stock Market. b) Steve Elmore/The Stock Market.)

reduce congestion actually foster development of open land and commuting by more drivers from distant locations (Fig. 24-5). Soon traffic conditions are as congested as ever. Average commuting *distance* has doubled since 1960, but average commuting *time* has remained about the same. The increase in commuting distance, however, requires more fuel, which generates more money for the Highway Trust Fund. Thus, the whole process is repeated in a continuing cycle (Fig. 24-6). Thus, government policy, far from curtailing urban sprawl, has become a party to supporting and promoting it.

Residential developments are followed (or sometimes led) by shopping malls and, more recently, industrial parks and office complexes. This trend has not lessened dependence on cars. Indeed, such commercial centers are often situated in such a way that the only possible access is by car, as is clearly seen in the aerial photograph of the mall shown in Fig. 24-7; this has only changed the direction of commuting. Whereas in the early days of suburban sprawl the major traffic flow was into and out of cities, it is now between suburban centers. Multilane highways connecting suburban centers are perpetually congested with traffic going in *both* directions.

In broad perspective then, urban sprawl is a

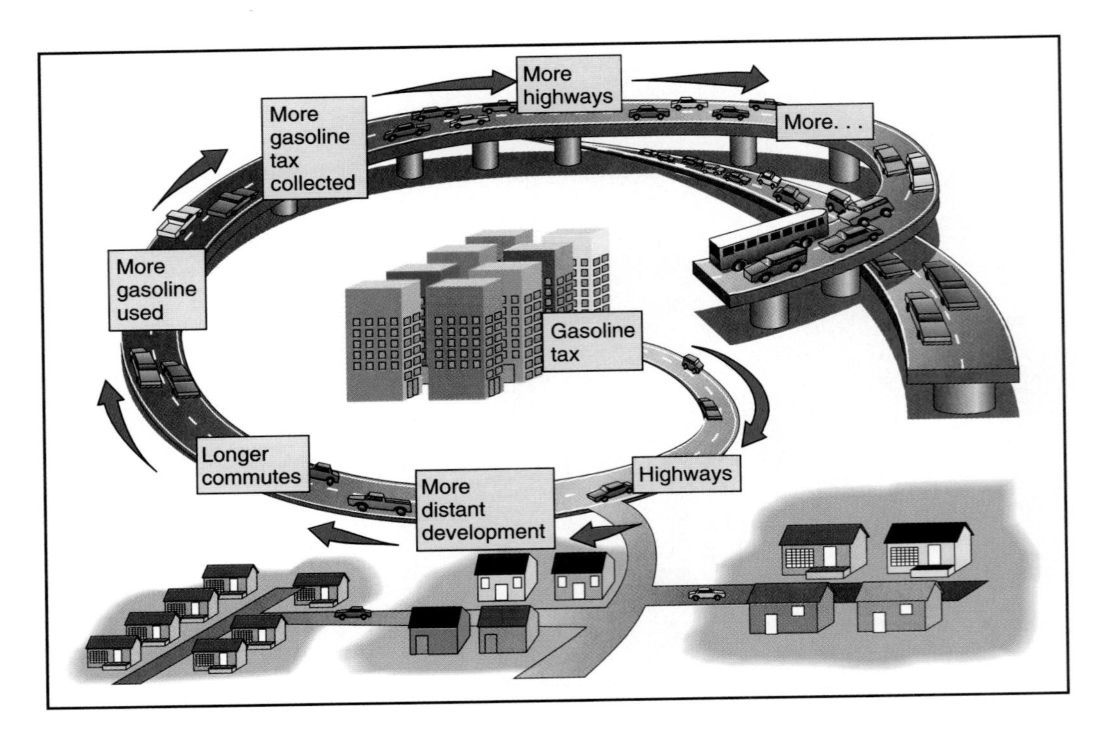

FIGURE 24-6
The development cycle spawned by the Highway Trust Fund.

process of **exurban migration**—that is, a relocation of residences, shopping areas, and workplaces from their traditional spots in the city to outlying areas. Population growth, although it has occurred, has played a relatively minor part in the development of urban sprawl. The populations of many U.S. cities, excluding the suburbs, have actually declined in the last several decades as a result of exurban migration (Table 24–1). Exurban migration is continuing in a leap-frog fashion as people from older suburbs move to **exurbs** (communities farther from cities than suburbs).

The "love affair" with cars is not just a U.S. phenomenon. Around the world, in both developed and developing countries, people aspire to own cars and adopt the car-dependent lifestyle. As a result, urban sprawl is occurring around many developing world cities as people become affluent enough to own cars. Love of the car-dependent lifestyle, however, is not suf-

FIGURE 24-7
Favored locations for new shopping malls and industrial parks have good access to major traffic arteries. Such locations, far removed from any housing, assure that the only access by shoppers and workers is by automobile, forcing car dependency, an athema to the concept of sustainability. Shown here is White Marsh shopping mall, northeast of Baltimore. (Photograph by Richard Adelberg, Jr., Owings Mills, MD.)

TABLE 24-1
Decline in Population of American Cities 1950–1992

City	Population (thousands)					Percent change
	1950	1960	1970	1980	1992	1950–1992
Baltimore, MD	950	939	905	787	726	−23
Boston, MA	801	697	641	563	552	−31
Buffalo, NY	580	533	463	358	323	−44
Cleveland, OH	915	876	751	574	503	−45
Detroit, MI	1850	1670	1514	1203	1012	−45
Louisville, KY	369	391	362	298	271	−26
Minneapolis, MN	522	483	434	371	363	−30
New York (Bronx), NY	1451	1425	1472	1169	1195	−18
Oakland, CA	385	365	362	339	373	−03
Philadelphia, PA	2072	2003	1949	1688	1553	−25
St. Louis, MO	857	705	662	453	384	−55
Washington, D.C.	802	764	757	638	585	−27

(Data from Statistical Abstracts of the United States U.S. Bureau of Census, 1994.)

ficient to make it sustainable. In fact, this lifestyle is at the crux of a number of the unsustainable trends described in earlier chapters.

Environmental Impacts of Urban Sprawl

The environmental impacts of urban sprawl can be put into the following categories:

- Depletion of energy resources
- Air pollution
- Water pollution and degradation of water resources
- Loss of landscapes and wildlife
- Loss of agricultural land

Depletion of Energy Resources. Shifting to a car-dependent lifestyle has entailed an ever-increasing demand for energy resources. From 1945 to 1980 U.S. oil consumption nearly quadrupled while population grew by just 60%, a nearly threefold increase in per capita consumption. In addition, individual suburban homes require 1.5 to 2 times more energy for heating and cooling than comparable attached city dwellings. We have amply described the costs and risks of becoming increasingly dependent on foreign sources of oil for fuel (see Chapter 21).

Air Pollution. Despite improvements in pollution control, many cities still fail to meet desired air-quality standards (see Chapter 15). Vehicles are responsible for an estimated 80% of the air pollution in metropolitan regions. Likewise, the threefold increase in per capita energy consumption exacerbates the potential for global warming as burning fossil fuels produces carbon dioxide. Automobiles likewise are implicated in depletion of the stratospheric ozone shield because a major source of CFCs entering the atmosphere is that which escapes from vehicle air conditioners (see Chapter 16).

Water Pollution and Degradation of Water Resources. All the highways, parking lots, driveways, and other paved areas associated with urban sprawl lead to a substantial increase in runoff, as well as decreased infiltration over large regions. Even suburban lawns increase runoff because they are more compacted than the original soils. Recall the consequences of increasing runoff—increased flooding, streambank erosion, and decreased water quality, to name a few—and the consequences of decreasing infiltration—depletion of groundwater, drying of springs and waterways, saltwater encroachment, and others (see Chapter 11). Also, water quality is degraded by the runoff of fertilizer, chemicals, crankcase oil, pet droppings, and so on.

Loss of Landscapes and Wildlife. New highways are frequently routed through parklands or along stream valleys because such open areas provide the least expensive rights of way. Countless city parks have been eliminated in the course of highway development. The result is the sacrifice of aesthetic, recreational, and wildlife values in the very places where they are most important: metropolitan areas. Highway planners argue that a highway through a large park will take a relatively small portion of the total area. If a park is to provide humans with a measure of peace and tranquillity, however, it is doubtful that two halves split by a pollution- and noise-generating highway will still add up to the whole.

Where does wildlife go when natural habitat is destroyed for development? Many assume that the

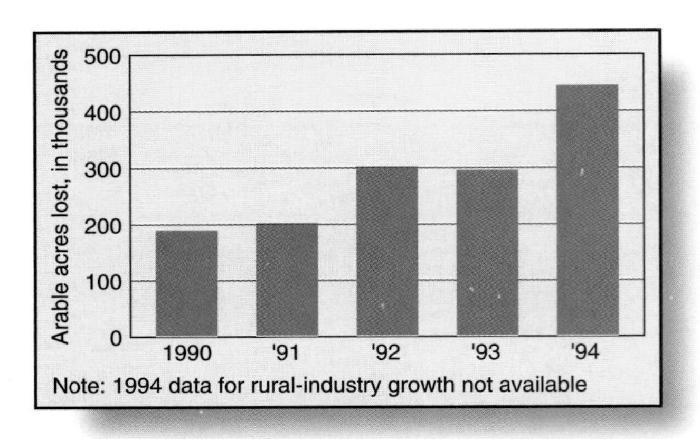

FIGURE 24-8
China has no agricultural land to spare, yet it is allowing land to be taken for development. As a result, China will become increasingly dependent on food imports. The trend in other developing countries is similar. (From "Feeding the Masses: China's Development Squeezes Peasant Farming, Increases Need for Imported Grain," *Wall Street Journal*, March 10, 1995, p. 4A.)

animals will simply find another place to live. Recall from Chapter 4, however, that in natural ecosystems the biotic (reproductive) potential of each species keeps it at the maximum population that can be supported by the ecosystem. Additional numbers, if added, will result in additional deaths due to one or another of the factors of environmental resistance. Therefore, the excess individuals of every species displaced by development are inevitably assigned to death even if similar habitat exists. Of course, if no similar habitat exists, the result is self-evident.

Furthermore, many species of wildlife need certain minimum unbroken areas to maintain viable populations, and unbroken "lanes" of habitat through which to migrate. Wildlife biologists are finding marked declines in countless species ranging from birds to amphibians as natural areas are fragmented by highways and new developments. The loss is even more conspicuous when wildlife is blocked from a source of drinking water, as frequently happens when highways run along rivers or stream valleys. Also, expanded and improved highways lead to increasing numbers of roadkills. Today much more wildlife is killed by vehicles than by hunters.

Loss of Agricultural Land. Finally, and perhaps most serious in the long run, is the loss of prime agricultural land. You may have heard this problem minimized by arguments that the world has plenty of land. However, 65% of the world's land is deserts, tundra, or mountains, none of which can be farmed (although about half of it supports forests). Another 24% is considered marginal land; it is excessively dry (savannahs and grasslands) or wet (wetlands), but given suitable inputs, it might be used for agriculture. Only the remaining 11% is prime agricultural land.

In the United States alone, sprawling development eats up 1.4 million acres (0.6 million hectares) per year of prime agricultural land. For the time being, it is considered that the United States has adequate agricultural land and can tolerate this loss. However, most of the food for cities used to be locally grown on small, diversified, family farms surrounding the city. With most of these farms turned into housing developments it is estimated that food now travels an average of 1000 miles (1500 km) from where it is produced—mostly on huge commercial farms—to where it is eaten. The loss is not just the locally grown produce but also the social interactions and ties with the farm community.

The loss of agricultural land to development is a worldwide phenomenon, and in countries having no land to spare, the consequences are becoming significant. For example, China became food-independent with the Green Revolution. However, with its push toward modernization and industrialization it is experiencing mounting losses of prime agricultural land to development (Fig. 24-8). This loss of agricultural land, coupled with growing population and a growing demand for meat, is making China dependent on imports again. Some question whether world surplus capacity can actually meet China's expected demands in the years ahead. Considering the additional degradation of agricultural soil because of erosion and salinization and the fact that irrigation is being curtailed by limited water resources (see Chapters 9 and 11), the squandering of agricultural land for development does not seem to be a responsible direction for any nation to be taking.

These environmental impacts resulting from or exacerbated by urban sprawl and development are summarized in Fig. 24-9.

Social Consequences of Exurban Migration

Now let us take a look at the other end of the exurban migration, the city from which people are moving. Here we find that exurban migration is the major factor underlying the *urban decay*, or *blight*, that has occurred over the last 50 years.

Economic and Ethnic Segregation

Historically, U.S. cities have included people with a wide diversity of economic and ethnic backgrounds.

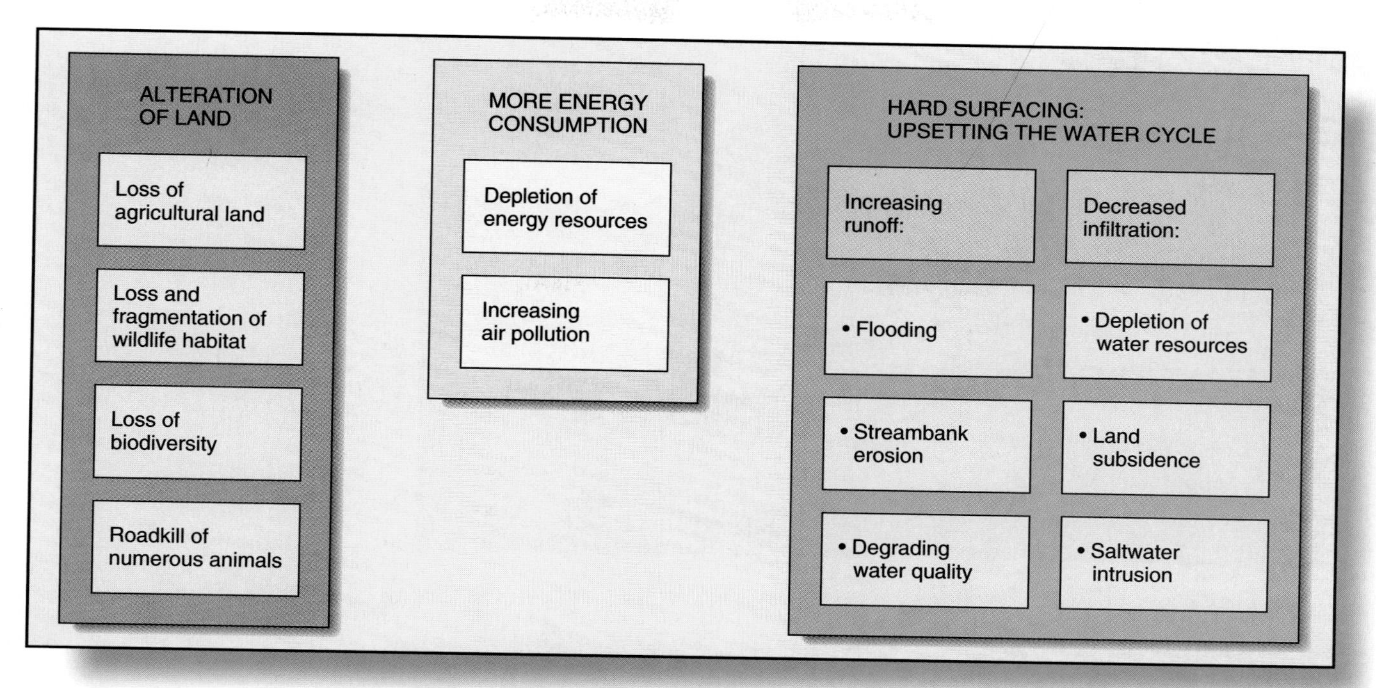

FIGURE 24-9
A summary of the environmental side effects of urban sprawl.

However, moving to the suburbs required some degree of affluence—at least the ability to manage a down payment and mortgage on a home and the ability to buy and drive a car. Therefore, exurban migration, for the most part, excluded the poor, the elderly, and the handicapped. The poor were largely African Americans and other minorities because a history of discrimination had kept them from education and well-paying jobs. Discriminatory lending practices by banks and sales practices by real estate agents kept minorities out of the suburbs even when money was not a factor. Civil rights laws passed in the 1960s made such practices illegal, but there is much evidence that they still persist in various forms. Even without the racial segregation, the economic segregation continues to exist and is even intensified.

In short, exurban migration and urban sprawl have led to segregation of the population into groups sharing common economic, social, and cultural backgrounds. Moving into the new suburban and exurban developments were the economically advantaged, whereas many areas of the cities and older suburbs were effectively abandoned to the economically depressed, who in large part were ethnic minorities, the handicapped, and elderly (Fig. 24-10).

People moving from rural areas to urban regions split according to the same pattern: based on their economic and ethnic status. Likewise, new immigrants are segregating along the same economic lines.

The Vicious Cycle of Urban Blight

Affluent people moving to the suburbs set into motion a vicious cycle of exurban migration and urban blight, which still continues. To understand how this downward cycle occurs, we must note several points concerning local governments. First, local governments (usually city or county, though the particular entities vary from state to state) are responsible for providing public schools, maintenance of local roads, police and fire protection, refuse collection and disposal, public water and sewers, welfare services, libraries, and local parks. Second, a local government's major source of revenue to pay for these services is local property taxes, a yearly tax proportional to the market value of the property and home or other buildings on it. If property values increase or decrease, the property taxes are adjusted accordingly. Third, in most cases, the central city is a governmental jurisdiction separate from the surrounding suburbs.

Because of these three characteristics of local government, exurban migration has the following consequences: By the economics of supply and demand, property values in the suburbs escalate with the influx of affluent newcomers. Thus, suburban jurisdictions enjoy increasing tax revenue with which they can improve and expand local services. At the same time, property values in the city decline because of decreasing demand. Also, property taxes create a powerful dis-

FIGURE 24-10
Exurban migration, the driving force behind urban sprawl, has also led to the segregation of the population along economic and racial lines. In (a) an area of suburbia and (b) an area of inner-city Baltimore, you see the contrast. (Photographs by BJN.)

incentive toward maintaining property. Often, landlords allow property to deteriorate to reduce their taxes while keeping rents high to maximize their income. Many properties end up being abandoned by the owner for nonpayment of taxes. The city "inherits" such abandoned properties, but they are a liability for the city rather than a source of revenue (Fig. 24-11). The declining tax revenue resulting from declining property

FIGURE 24-11
As exurban migration progresses, there is no one to buy urban properties. Many hundreds of thousands of urban properties such as these in Philadelphia now stand abandoned in U.S. cities. (Joe Sohm/The Stock Market.)

values is referred to as an **eroding**, or **declining, tax base**, and it has been a serious handicap for most U.S. cities since the exurban migration started in the late 1940s. Adding to the problem is the fact that many of those remaining in the city are disadvantaged people requiring public assistance of one sort or another. Thus cities bear a disproportionate burden of welfare obligations as well as the declining tax revenues.

The eroding tax base forces city governments to cut local services or increase the tax rate or, generally, both. Hence, the property tax on a home in a city is often two to three times greater than on a comparably priced home in the suburbs, a difference that amounts to $2000 to $3000 per year in the tax bill on an average-priced home. At the same time, schools, refuse collection, street repair, libraries, parks and other services and facilities deteriorate from neglect. This situation of increasing taxes and deteriorating services causes more people to leave the city and become part of urban sprawl even if the car-dependent lifestyle is not their primary choice. However, it is still only the relatively more affluent individuals who can make this move and thus the whole process of exurban migration, declining tax base, and deteriorating real estate, worsening schools and other services and facilities is perpetuated in a vicious cycle (Figure 24-12). This downward spiral

FIGURE 24-12
The vicious cycle of exurban migration and urban decay.

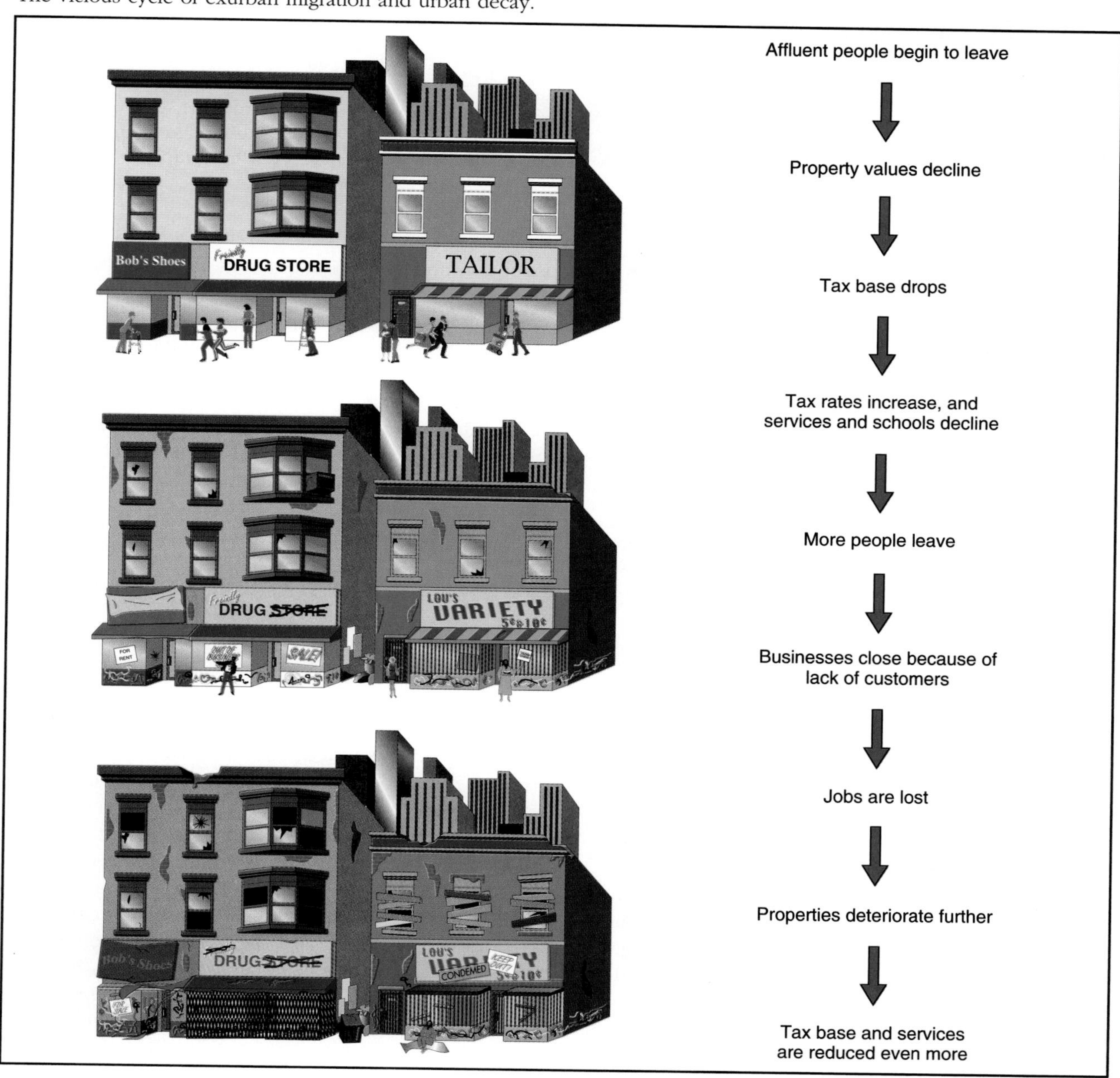

FIGURE 24-13
Many cities' ills are because of too *little* population, not too *much*. When the exurban migration causes population to drop below a certain level, businesses are no longer supported. The result is abandonment, the bane of urban blight. Shown here is a section of Baltimore only a few blocks from the redeveloped Inner Harbor shown in Fig. 24-14. (Photograph by BJN.)

of conditions is referred to as **urban blight** or **urban decay**.

Economic Exclusion of the Inner City

The exurban migration includes more than just individuals and families. The exiting of affluent people from the city removes the purchasing power necessary to support stores, professionals, and other businesses. Faced with declining business, these enterprises are forced either to go out of business or to move to the new locations in the suburbs. Either way, vacant store fronts are the result (Fig. 24-13), and people remaining in the city lose convenient access to goods, services, and, most important, *jobs*, because each business also represents employment opportunities.

Unemployment rates run at 50% or more in depressed areas of inner cities, and what jobs do exist are mostly at the minimum wage. The new jobs being created are in the shopping malls and new industrial parks in the outlying areas that are largely inaccessible for lack of public transportation. Therefore, people remaining in the inner city are not only poor from the outset, but, the cycle of exurban migration and urban decay has now led to their **economic exclusion** from the mainstream. Is it any wonder that drug dealing, crime, violence, and other forms of maladaptive social behavior are rife and getting worse in such areas?

To be sure, in the last two decades many cities have redeveloped core areas. Shops, restaurants, hotels, convention and entertainment centers, office buildings, residences, and places for walking and relaxing are all combining to bring a new spirit of life and hope, as well as the practical assets of taxes and employment, back into cities (Fig. 24-14). This is an encouraging new trend, so far as it goes. However, walk two blocks in any direction from the sparkling, revitalized core of most U.S. cities and you will find urban blight continuing unabated. The Los Angeles riot that occurred in May 1992 exemplifies the potential for urban explosion.

Moving toward Sustainable Cities

Some envision a future in which the entire human population is spread out over the landscape in low-density housing, and this is indeed the direction in which most current development seems headed. However, given the current and projected population, such a scenario is insupportable. Solar energy and electric automobiles may alleviate the constraints of limited fossil-fuel supplies, air pollution, and global warming, but they will not eliminate the constraints of water resources, pollution from urban runoff, and the need to preserve agricultural land, natural habitats, and biodiversity. Nor can we tolerate the social consequences of the economic exclusion of the significant portion of society that is being left behind. Also, people in the suburbs are

FIGURE 24-14
Inner Harbor, Baltimore. High population densities help *make* a city, not destroy it. Redeveloping key areas of cities with heterogeneous mixtures of office buildings, shopping, restaurants, residences, and recreational facilities, all of which attract people, has helped to bring new life back into cities. (Photograph by BJN.)

increasingly frustrated with having to commute through congested traffic 1 to 3 hours per day and then searching for parking spaces.

The good news is that more and more people in all walks of life are recognizing that the environmental consequences of urban sprawl and the social consequences of urban blight are two sides of the same coin. There is increasing awareness that a sustainable future will depend on both reining in urban sprawl and revitalizing cities. The only possible way to sustain the global population is by having viable, resource-efficient cities, leaving the countryside for agriculture and natural ecosystems (See "Earth Watch" box, page 619.) The key word is *viable*, which means livable. No one wants to or should be required to live in the conditions that have come to typify urban blight.

What Makes Cities Livable?

Livability is a general index based on people's response to the question, "Do you like living here, or would you rather live somewhere else?" Crime; pollution; recreational, cultural, and professional opportunities; and many other social and environmental factors are summed up in the subjective answer.

Though many people assume that the social ills of the city are an outcome of high population densities, in fact crime rates and other problems in U.S. cities have climbed while populations have dwindled as a result of the exurban migration. As reported in the Worldwatch Institute's study titled *Shaping Cities: The Environmental and Human Dimensions*:

> There is no scientific evidence of a link between population density per se and social ills. . . . Copenhagen and Vienna—two cities widely associated with urban charm and livability—each have relatively high "activity density" (the number of residents and jobs in the city), 47 and 72 people per hectare (1 hectare = 2.5 acres) respectively. By contrast, low-density cities such as Phoenix (13 people per hectare) often are dominated by unwelcoming, car-oriented commercial strips and vast expanses of concrete and asphalt.

Renowned city planner William Whyte goes even further. His studies show that wherever urban population density drops below a certain point, there are no longer enough people to support restaurants, department stores, and so on, and the area "dies" as a result.

Looking at livable cities around the world, we find that the common denominators are (1) maintaining a high population density; (2) preserving a heterogeneity of residences, businesses, stores, and shops; and (3) keeping layouts on a human dimension, where people can incidentally meet, visit, or conduct business over coffee at a sidewalk café, or stroll on a promenade through an open area. In short, the space is designed for and devoted to people (Fig. 24-15). In contrast, development of the past 50 years has focused on accommodating automobiles and traffic. Two-thirds of the land in cities that have grown up in the era of the automobile is devoted to moving, parking, or servicing cars, and such space is essentially alien to the human psyche (Fig. 24-16). William Whyte remarked, "It is difficult to design space that will not attract people. What is remarkable is how often this has been accomplished."

The world's most livable cities are not those with "perfect" auto access between all points; they are ones that have taken measures to reduce outward sprawl,

FIGURE 24-15
The key to livable cities is having a heterogeneity of residences, businesses, and stores, and in keeping layouts on a human dimension where people can incidentally meet or conduct business over coffee in a sidewalk café, or stroll on a promenade through an open area. In short, the space is designed for and devoted to people. New vitality is being brought back into cities by using these concepts. Shown here is Quincy Market, Boston. (Geri Engberg/The Stock Market.)

reduce automobile traffic, and improve access by foot and bicycle in conjunction with mass transit. For example, Geneva, Switzerland, prohibits automobile parking at workplaces in the city's center, forcing commuters to use the excellent public transportation system. Copenhagen bans all on-street parking in the downtown core. Paris has removed 200,000 parking places in the downtown area. Curitiba, Brazil, with a population of 1.5 million, is cited as the most livable city in all of Latin America. The achievement of Curitiba is due almost entirely to the efforts of Jaime Lerner who, serving as mayor since the early 1970s, has guided development in terms of mass transit rather than cars. The space saved by not building highways and parking lots has been put into parks and shady walkways, causing green area per inhabitant to increase from 4.5

FIGURE 24-16
In contrast to what is required for livability, city development of the last several decades has focused on moving and parking cars, and creating homogeneity. Note the sterility of parking lots surrounding buildings in the foreground of this photo of Los Angeles. (Saloutos/The Stock Market.)

Earth Watch

Clustered versus Detached Development

The most common type of suburban development is a subdivision of property into equal lots and the building of a single (detached) home on each lot. The entire property is thus converted into houses, lawns, and roadways, and any semblance of the natural ecology is lost. An alternative development plan is to cluster the dwellings in attached units, leaving a major portion of the property as open space or for other uses. The figure shows two alternative plans for a specific piece of property. The size, number, and cost of the homes in each is approximately equal.

Studies show that cluster development, in contrast to detached housing, achieves the following savings:

- Farmland and forest losses are reduced by 70%.
- Stormwater runoff and pollution are reduced by 50%.
- Infiltration and recharge of groundwater is increased by 50%.
- Fuel required for space heating is reduced by 25%.
- Air pollution from space heating is reduced by 25%.
- Cost of installing and maintaining utility lines, (water, gas, electric) is reduced by 50%.
- Cost of school busing is reduced by 38%.
- Cost of refuse collection is reduced by 40%.
- Vehicle fuel consumption and pollution are reduced by 17%.
- Servicing by public transportation becomes more cost effective.

When communities of clustered dwellings are integrated with stores, professional offices, and schools they provide even greater savings in vehicle use, fuel costs, traffic congestion, and pollution since people can do many of their errands by walking. Building such communities, as opposed to the typical suburban development, can go a long way toward reining in urban sprawl.

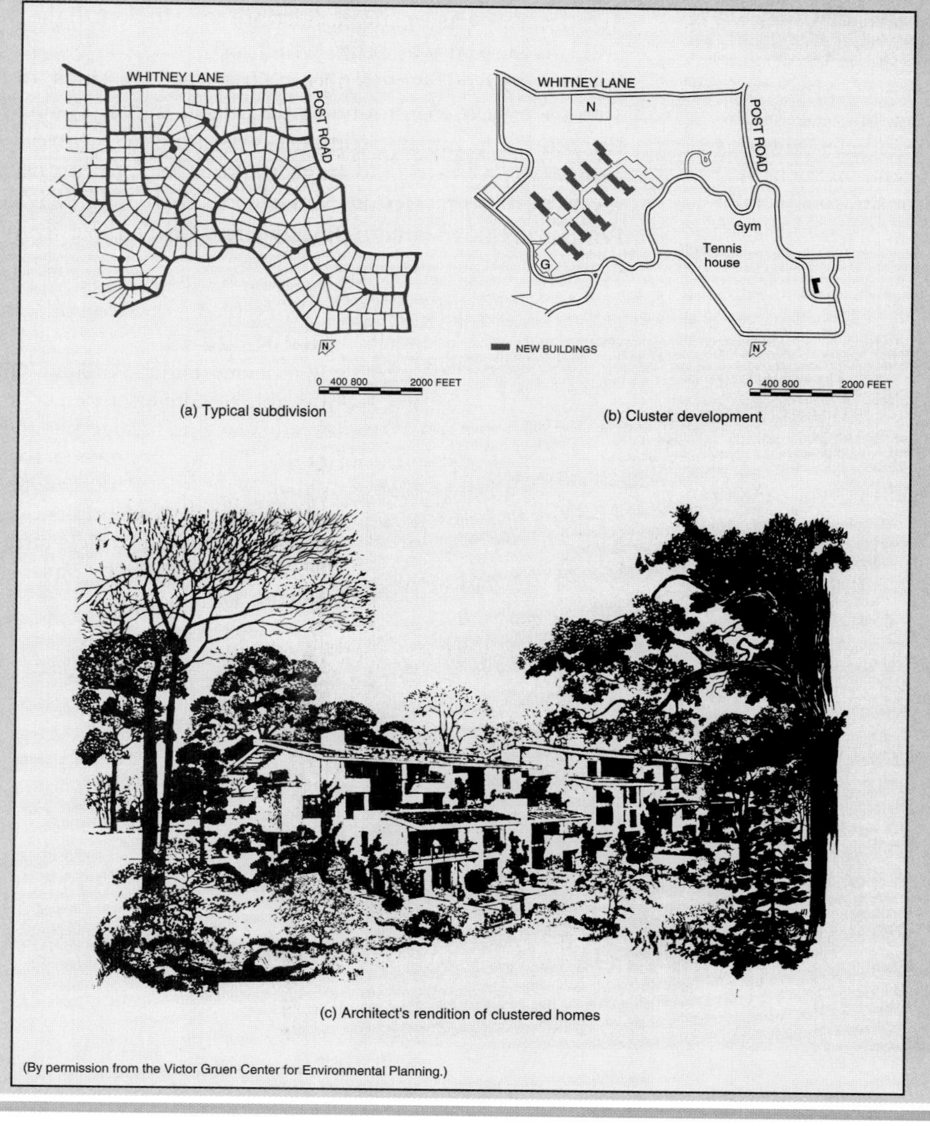

Alternative plans for development of the same property. These planning techniques combine environmental sciences, landscape architecture, and architecture.

square feet (0.5 m^2) in 1970 to 4500 square feet (50 m^2) today.

In Tokyo, millions of people ride bicycles (Fig. 24-17), either all the way to work or to stations from which they catch fast, efficient subways or the "bullet train" to their destination. By sharply restricting development outside certain city limits, Japan has maintained population densities within cities and along metropolitan corridors that ensure the viability of commuter trains. Japan's cities have maintained a heterogeneous urban structure that mixes small shops, professional offices, and residences in such a way that a large portion of the population meets their needs without cars. In maintaining an economically active city it is probably no coincidence that street crime, vagrancy, and begging are virtually unknown in the vast expanse of Tokyo despite the seeming congestion.

Portland, Oregon, is one of the few U.S. cities that have taken more than token steps to curtail automobile use. The first step was to encircle the city with an urban growth boundary, a line outside of which new development was prohibited. Thus, compact growth rather than sprawl was ensured. Second, an efficient light rail and bus system was built, which now carries 43% of all commuters to downtown jobs. (In most U.S. cities, only 10% to 25% of commuters ride public transit systems.) By reducing traffic, Portland was able to convert a former expressway and huge parking lot into the now-renowned Tom McCall Waterfront Park (Fig. 24-18). Portland is now ranked among the world's most livable cities.

FIGURE 24-17
The bicycle parking lot at Nakano Station. Most people in Tokyo do not own automobiles but ride bicycles or walk to stations, from which they take fast, efficient, inexpensive subways to reach their destinations. (Kim Newton/Woodfin Camp & Associates.)

FIGURE 24-18
By reducing traffic, Portland, OR, was able to convert a former expressway and a huge parking lot into the now-renowned Tom McCall Waterfront Park. Due to such planning, Portland is now ranked among the world's most livable cities. (Photo by Janis Miglavs, Portland, OR.)

Reining In Urban Sprawl

The obvious way to curtail urban sprawl might appear to be: Simply pass laws that preclude development of exurban properties. Indeed, a number of European countries and Japan do have such laws. However, laws cannot be passed without the support of the public. In the United States the strong sentiment toward the right of a property owner to develop property as he or she feels fit has made such restrictions, with few exceptions, politically impossible to pass or financially impossible to implement. (see Chapter 18, page 462).

Yet, many outdated laws and regulations that exacerbate urban sprawl might well be changed. Examples include archaic zoning laws that require such things as large lot sizes, detached housing, and the separation of residential and commercial or industrial areas. Most such laws were passed by pressures from the first exurban migrants in the 1940s and 1950s as a means of preventing "undesirable elements" from invading the new exurban neighborhoods. By demanding such separation and low-density residential areas, however, these regulations now assure a car-dependent lifestyle. Mass transit (bus or rail lines), a much-promoted idea, is unworkable where residential densities are low because there are not enough riders between any two locations to make it viable. It makes no sense, economically or ecologically, to run a 40-passenger bus with only a few passengers.

Furthermore, the requirement for separation of different functions led architects and developers themselves to focus on specialized aspects—homes, shopping malls, industrial parks, while government transportation departments focused on providing highways. Lost in the process was any sense of creating integrated communities for people.

Worldwatch's *Shaping Cities* points out that a key to developing sustainable communities is changing zoning laws to allow stores, light industries, professional offices, and higher-density housing to be integrated. A second point stressed is the need to provide affordable housing around industrial parks and shopping malls for the people who work and shop in those places. Such housing need not be unattractive or lacking in privacy. A third point emphasized is the need to provide safe bicycle and pedestrian access between residential areas and workplaces. It is assumed that employers will provide a convenient, safe place to keep a bicycle; at present, they seldom do, although they do spend a great deal of money providing parking spaces for automobiles. These steps will allow more people to walk or bicycle to their jobs and will reduce the space needed for highways and parking. The effectiveness of these steps in reining in urban sprawl is illustrated by the following example. Fairfax County, Virginia, next to Washington, D.C., is now almost 100% covered by sprawl-type development. It is estimated that if Fairfax had *allowed* the kind of development described above, the county would still be 70% open space and would be far less congested and less polluted.

There is considerable indication that what is prescribed above is beginning to occur. Zoning laws are being changed in many localities, and a new generation of architects and developers are beginning to focus on creating integrated communities as opposed to disassociated facilities. Early indications in the marketplace—the final determining factor—are that the new communities are a great success.

As further steps toward reining in urban sprawl (short of outright land-use restrictions) many environmentalists support placing a substantial tax on gasoline. All other industrialized countries have already done this to the amount of $1.50–$3.00 per gallon. The assumption is that a higher cost for gasoline will discourage people from moving farther and farther out and will promote energy conservation in general. Actually, early in its term the Clinton administration proposed such a tax (50 cents per gallon). Overwhelming negative public reaction diminished the increase to a negligible 4 cents per gallon.

In addition, there have been repeated attempts to break the cycle created by the Highway Trust Fund by allowing revenues to be used for purposes other than construction of more highways. The first significant inroad was made in 1991 with the passage of the Intermodal Surface Transportation Efficiency Act, a bill environmentalists had been promoting for many years. Under this act almost half of the money levied by the Highway Trust Fund is now eligible for use on other modes of transportation, including cycling and walking as well as mass-transit facilities. The legislation also requires states to hire bicycle and pedestrian coordinators to oversee programs and to establish long-range pedestrian-bicycle plans. The act also requires that states spend a total of $3 billion on "transportation enhancements," which include pedestrian and bicycle facilities. Whether this funding, which is now technically available, actually gets used to provide bicycle-pedestrian facilities in your area will still depend on citizen lobbying. Finally, the efficacy of private land trusts should be reconsidered (see page 505.)

Refocusing on Cities

Developing new integrated communities in the suburbs may slow urban sprawl, but it obviously does not stop it, nor does it address the "other side of the coin," the unacceptable problems and social injustice of urban blight. Not only do the social problems need to be addressed for their own sake, but American cities provide a vast opportunity for environmentally sound redevelopment and accommodating increasing popula-

Earth Watch

Rooftop Gardens for Livability and Food

For years, workers at ECHO (Educational Concerns for Hunger Organization, a nonprofit organization helping to meet food needs of the poor) have been developing techniques for growing vegetables on rooftops at ECHO's model facilities in Fort Myers, Florida. Even though it would be possible to use hydroponic technology to accomplish this, ECHO staff decided to avoid traditional hydroponics because of the dependence on water or air pumps to keep roots aerated—technology that could be expensive and energy-dependent. Instead, they found that they could grow luxuriant vegetables in a "stagnant water" system only 1 inch deep. The shallow depth allowed oxygen to diffuse effectively to the roots.

ECHO workers also found that they could grow almost any vegetable in a shallow (3–4 inch) layer of compost or slightly decomposed refuse. No soil was used, and the beds were kept shallow to keep weight down (rooftops are often too weak to support heavy weights). In one developing country application of the technology, rooftops on a school and a hospital in Port-au-Prince, Haiti, were utilized for raising high-nutrient crops to feed students and patients.

Water and nutrients were fed to the vegetables with a wick made from a sheet of polyester cloth. The cloth was spread out on the cement roof, vegetable plants were spaced out on the cloth, and a few inches of pine needles were used to fill in around the plants for lightweight support and protection against drying. A 5-gallon bucket was then filled with a complete, soluble fertilizer, a 3/8-inch hole was drilled in the lid, and the bucket was turned upside down on a clear section of the cloth. Water and nutrients simply wicked out to the vegetables through the cloth. The workers found that if pine needles were unavailable, they could substitute crushed soda cans to occupy the space between plants and inhibit evaporation of water. They were even able to raise field corn with this system!

The lessons to be learned are straightforward: Plants really need light, water, and nutrients, but, surprisingly, little or no soil. These needs can be met in virtually any developing world city, by the literally acres of flat "land" in the form of rooftops in the city. This kind of urban "farming" also has the potential to rehabilitate cities in the industrialized countries, giving people the opportunity to raise some of their own food and enjoy living in the company of green plants. (For information write ECHO, 17430 Durrance Rd., Fort Myers, Florida 33917.)

---

tions. Recall that over the years since World War II central cities have declined in population and now have many tens of thousands of abandoned properties, both residential and commercial. Additional space will be available as car-space is converted to people-space. However, few people are going to be enticed to come back into the city—or even stay if they can leave—unless the problems of crime, violence, and blight in general can be solved. The problem is how to reverse the cycle of urban decline.

Revitalizing Urban Economies and Rehabilitating Cities. From our earlier discussion, we can see that the urban ills of poverty and unemployment stem mainly from the economic exclusion caused by exurban migration. Conversely, professionals now recognize that if we are to reverse the downward cycle of urban blight we must find ways of bringing economic vitality back into blighted neighborhoods. Indeed, the shortfalls of earlier approaches can be seen in the context of their failure to address the underlying problem of economic exclusion. For example, huge public housing projects built in the 1950s and early 1960s to provide decent housing for the poor are now being torn down, themselves victims of urban decay. Welfare, intended to help the needy, is now coming under attack as having simply fostered greater dependency. But is welfare the cause of the underlying problem, or is the cause the economic exclusion that has occurred? Finally, huge urban redevelopment projects have revitalized the core areas of many cities. However, for all their virtues they have failed in general to reverse urban decay because the projects themselves have tended to be economically exclusive. Increasing real estate values engendered by development or condemnation of properties for redevelopment have only forced the poor aside; and jobs created fall short of the inner city's needs.

The conclusion is that revitalizing urban economies and rehabilitating cities will require coordinated efforts on the part of all sectors of society, including the poor themselves, who must become active participants. In fact, such efforts are beginning to occur. What is termed a "sustainable communities movement" or an "urban recovery movement" is taking root in cities around the country. Chattanooga, Tennessee, is now regarded as a prototype of what can occur with such a movement.

Twenty-five years ago Chattanooga, which straddles the Tennessee River, was a decaying industrial city with horrendous levels of pollution. Employment was falling, and residents were fleeing to the suburbs, leaving abandoned properties and increasing crime. Then, a nonprofit organization launched "Vision 2000" the first step of which was to bring people from all walks of life together to build a consensus about what the city *could* be like. Literally thousands of ideas were gradually distilled into somewhat over a hundred specific projects. Then with the cooperation of all sectors—including government, business, lending institutions, and average persons—work on the projects began, providing employment for more than 8000 people. Among the projects were the following:

With the support of the Clean Air and Clean Water acts, local government clamped down hard on industries to control pollution.

- Low-cost housing was renovated.
- A new industry to build pollution-control equipment was spawned as was another industry to build electric buses, which are now serving the city without noise or pollution.
- A recycling center employing mentally handicapped adults to separate materials was built.
- An urban greenway demonstration farm, which school children may visit, was created.
- A zero-emissions industrial park utilizing-pollution avoidance principles was built.
- The river was cleaned up, and parks were created along the riverbanks.
- Theaters, museums, and a freshwater aquarium were renovated or built.
- All facilities and a renovated business district were made pedestrian-friendly and accessible (Fig. 24-19).

With these and numerous other projects, many of them still ongoing, Chattanooga has moved its reputation from one of the worst to one of the best places to live.

An example of another urban rehabilitation project, this one in a large city, is the Sandtown project in Baltimore. Sandtown is a 72-block area of about 3000 decrepit housing units (some 600 abandoned). Of the 12,000 residents, over 50% are unemployed, and most of the rest have incomes below the poverty line. Crime and drugs are rife. The project involves the cooperative leadership of Habitat for Humanity, a worldwide charitable organization devoted to renovating housing for the poor; developer James Rouse, chairman of the Enterprise Foundation; Baltimoreans United in Leadership Development (BUILD); Mayor Kurt Schmoke; and others. The plan is to do much more than make Sandtown's housing stock fit and affordable. Residents will be trained in the process and will do most of the work (Fig. 24-20). At the same time, residents will be trained to staff their own neighborhood schools, health-care facilities, and to provide police protection, and other services. The essence of a vibrant economy is everyone providing goods or services which they can exchange for the benefit of all. You can see how this is the goal.

The Federal Role. What role in inner-city improvement, if any, should the federal government play? In

FIGURE 24-19
With numerous projects to reduce pollution and traffic and to provide attractions and amenities for people, Chattanooga, TN, has changed its reputation from one of the worst to one of the best places to live in the U.S. Note the amount of green space in the downtown area in contrast to Los Angeles (Fig. 24-16). (Chattanooga Convention and Visitors Bureau.)

FIGURE 24-20
Habitat for Humanity volunteers. By using charitable contributions and volunteer labor, Habitat for Humanity creates quality homes—over 8000 built throughout the world by 1994—that needy families can afford to buy with no-interest mortgages, enabling the money to be recycled to build more homes. Prospective home buyers participate in the construction and may gain valuable skills in the process. (Karl Gehring/Gamma-Liaison, Inc.)

December of 1994 the U.S. Department of Housing and Urban Development (HUD) awarded six $100 million grants to each of six cities—Atlanta, Baltimore, Chicago, Detroit, New York, and Philadelphia-Camden. These cities were the winners among some 500 competing applicants. The criterion for these awards was not who could cry most loudly for need, but demonstrating that there were already solid commitments of coordinated help from other sectors within the city to bring about improvements in a designated zone referred to as the **empowerment zone**. For example, Detroit, which has lost 50% of its population and 40% of its job base in the last 40 years, now has a commitment from the business community led by General Motors, Ford, and Chrysler to create 3275 jobs and invest $1.8 billion, and there is a commitment from banks to lend additional millions in the zone. (Another additional handicap on the economy of blighted areas has been the refusal of banks to make loans in such areas.) There were additional commitments from utilities, insurance companies, law firms, and so on to make services more affordable to people and new businesses in the zone. Universities and the media were also engaged to offer and promote education, job training, counseling, and so on. In all, the private sector in Detroit is committing some $20 for every dollar of federal money.

In other words, federal grants acted mainly as an incentive for cities to organize their own resources in addressing the problem of blight. The key ingredient for success is to bring all the people of the community together to share in the conception of a vision as to what they want their community or city to be like. Then the "pieces begin to fall into place." Of course, success cannot be guaranteed. However, to argue that nothing will work and to do nothing is to argue for a future of continuing rise in crime and social anarchy.

Making Communities Sustainable

The same factors that underlie livability also lead to sustainability. The reduction of auto traffic and greater reliance on foot and public transportation reduce energy consumption and pollution. Urban heterogeneity can facilitate recycling of materials. Housing can be retrofitted with passive solar heating of space and hot water. Use of landscaping can provide cooling, as described in Chapter 23. A number of cities are developing vacant or cleared areas into garden plots (Fig. 24-21), and rooftop hydroponic gardens are becoming popular. Such gardens will not make cities agriculturally self-sufficient, of course, but they add to livability, provide an avenue for recycling compost and nutrients removed from sewage, give a source of fresh vegetables, and have the potential of providing income for many unskilled workers (see "Earth Watch" box page 622). If urban sprawl is curbed, relatively close-in farms could

FIGURE 24-21
Urban garden plots on formerly vacant lots in Philadelphia. Urban gardening is becoming recognized as having sociological, economic, environmental, and aesthetic benefits. (Blaine.)

provide most of the remaining food needs for the city.

This discussion of urban issues may seem to have strayed far from nature and environmental science. However, there is a close connection—the decay of our cities is hastening the breakdown of our larger environment. As the growing human population spreads outward from the old cities, we are encroaching on all other life. Thus, without the creation of sustainable human communities, there is little chance for sustainability of the rest of the biosphere.

By making urban areas more appealing and economically stable, we can not only improve the lives of those who choose to remain or must remain in cities, but also spare surrounding areas. Our parks, wildernesses, and farms will not be replaced by exurbs and shopping malls, but saved for future generations (see "Ethics" box, page 627).

Epilogue

What is to be the destiny of humankind on Planet Earth? In the introduction (Chapter 1) we presented the concept of sustainability and the contrast between sustainability and various unsustainable global trends. In each chapter we have developed this theme further. In each topic covered we have shown how certain practices and technologies are not sustainable, but then and most important, we described current steps and progress toward sustainability.

Is the glass half empty or half full? One can be depressed by the bad news, but one should also be encouraged and inspired by the literally millions of people in all walks of life who are acutely aware of the problems and who are making outstanding efforts to bring about solutions. Every pathway toward solution that we have mentioned represents the work of thousands of dedicated professionals and volunteers ranging from scientists and engineers through businesspeople, lawyers, and public servants. Still, the forces aligned with the traditional, nonsustainable directions are formidable. Thus the outcome is still unsure. Will only time tell? Lester Brown, president of WorldWatch Institute states:

> Until now the Environmental Revolution has been viewed by society much like a sporting event—one where thousands of people sit in the stands watching, while only a [relative] handful are on the playing field actively attempting to influence the outcome of the contest. Success in this case depends on erasing the imaginary sidelines that separate spectators from the participants so we can all get involved. Saving the planet is not a spectator sport.

Indeed, we are all involved whether we recognize it or not. Simply by our existence on the planet, everything we do—the car we drive, the products we use, the wastes we throw away, virtually every choice we make and action we take—has a certain environmental impact and a certain consequence for the future. Therefore, it is not a matter of having an effect but of what and how great that effect will be. It is a matter of each of us asking ourselves, Will I be part of the problem or part of the solution? The outcome will depend on how each of us responds to the challenges ahead.

There are four levels on which we may participate to work toward a sustainable society (Fig. 24–22):

- Individual lifestyle changes
- Political involvement
- Membership and participation in nongovernmental environmental organizations
- Career choices

Lifestyle changes may involve such things as switching to a more fuel-efficient car or using a bicycle for short errands; recycling paper, cans, and bottles; retrofitting your home with solar energy; starting a backyard garden and composting and recycling food and garden wastes into your soil; choosing low-impact recreation such as canoeing rather than speed boating; living closer to your workplace, and any number of additional things.

Political involvement ranges from supporting and voting for particular candidates to expressing your support for particular legislation through letters or phone calls. The outcome when environmentally concerned people are not sufficiently active in this regard is highlighted by the results of the 1994 congressional election. The election brought in a large number of people whose primary objective appears to be to dis-

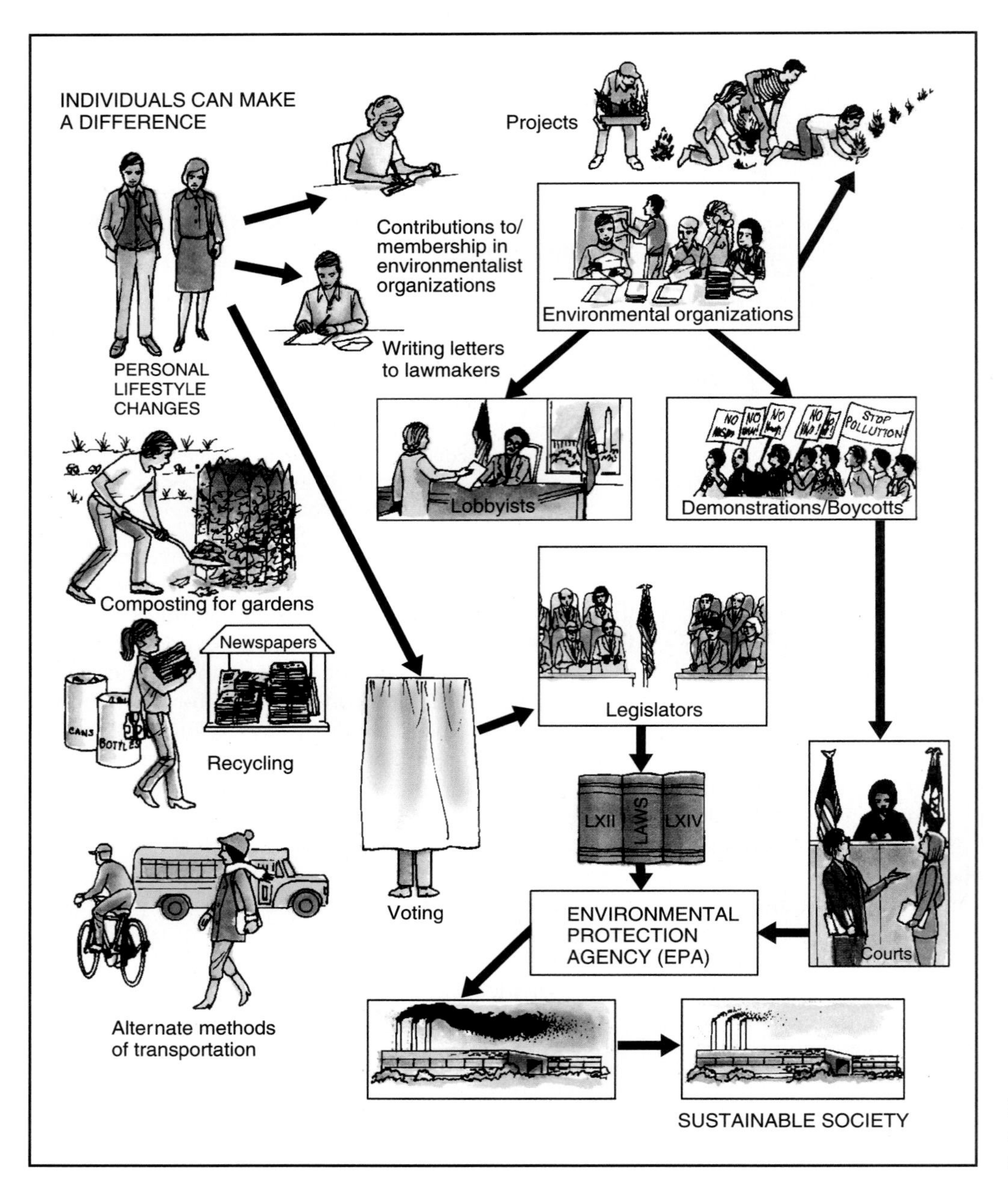

FIGURE 24-22 Avenues of individual action.

mantle most of the environmental legislation of the past 25 years. Of course, such dismantling is being done under the guise of balancing the budget and reducing burdensome government regulations—worthy objectives. Before jumping on such a bandwagon, however, consider the rebuff to the 104th Congress in the 1996 election results. Citizens let it be known that they want clean air, water, and other protective regulation.

Membership in nongovernmental environmental organizations can enhance both lifestyle change and political involvement. As a member of an environmental organization you will receive and may help disseminate information making you and others more aware of particular environmental problems and things you can do to help. Specifically, you will be informed regarding environmentally significant legislation so that you may focus your political efforts at the most effective time and place. Also, your membership and contribution serve to support lobbying efforts of the organization. A lobbyist representing only him- or herself has relatively little impact on the legislators. On the other hand, if the lobbyist represents a million-member organization that can follow up with that many phone calls and letters (and ultimately votes) the impact is considerable. Finally, where enforcement of existing law has been the weak link, some organizations such as Public Interest Research Groups, the Natural Resources Defense Council, and Environmental Defense Fund, have been very influential in bringing polluters or the government to court to see that the law is upheld. Again, this can be

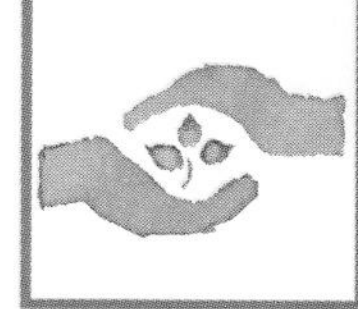

Ethics

A Sense of Justice

This chapter documented all too fully the human costs of urban blight caused by people and businesses moving to the suburbs. The car-dependent lifestyle that characterizes our country is so thoroughly embedded in our economy and social structure that it is almost bizarre to try to imagine how to live differently. Yet, as we have seen, there will be no environmental sustainability unless we make major moves away from our current urban-suburban divergence and rehabilitate our cities. There is another important reason for doing this, however, and that is *justice*.

Justice has to do with the distribution of things that are scarce or in relatively short supply: jobs, education, good housing, food, consumer goods, and the like. A sense of *distributive justice* is part of our human makeup; we all believe that we should have a share of those things that are in short supply. History has proved that distributive justice is not accomplished when there is a free-for-all scramble for scarce things. It is accomplished only when law and order exist and people cooperate to maintain that order. Thus, an orderly society is one where people perceive that at least they are being dealt with justly. They then have a stake in maintaining order. However, if people believe they are being treated unjustly, their stake in the orderly society is more tenuous. Sooner or later they become convinced that things are not going to change, and then they are prone to take action that is disruptive to that order—such as the riots we have seen in cities.

As we have seen, the exurban migration has left the inner city to minority groups, the elderly, and the handicapped, and society has become divided along economic lines. The distribution of goods and services has become similarly divided, leading to the downward spiral of urban decay. The pathway to making the cities more livable is going to require a commitment on the part of all segments of society to distributive justice. It will take major structural changes to correct the injustices that exist. New programs and new sources of funding must be put in place.

It is important, however, that enhancing distributive justice also be put in a moral context. The case can be made that those who "have it all" are benefiting from an unjust distribution of goods and services that leaves others out. We will probably not make much progress until the outrage of the disadvantaged minorities is augmented by those who are both well off and fair-minded enough to care about the plight of those who have been left out. The responsibility to contribute to distributive justice should be proportionate to the degree of well-being—in other words, those who have more should be more willing to work toward a just society. The moral imperative becomes even greater when we realize that we are also talking about environmental justice—as we make cities more livable we are also creating a more sustainable society for ourselves and the other creatures with whom we share the planet.

done only with the support of members.

Finally, you may choose to devote your career to implementing solutions to environmental problems. Environmental careers go far beyond the traditional occupation of wildlife or park management. There are any number of lawyers, journalists, teachers, scientists, engineers, medical personnel, entertainers, and others focusing their talents and training on environmental issues or hazards. There are innumerable business and job opportunities in pollution control, recycling, waste management, ecological restoration, city planning, environmental monitoring and analysis, nonchemical pest control, production and marketing of organically grown produce, and so on. Some developers concentrate on rehabilitation and reversal of urban blight as opposed to contributing more to urban sprawl. Some engineers are working on the development of pollution-free vehicles to help solve the photochemical smog dilemma of our cities. Indeed, it is difficult to think of a vocation that cannot be focused on promoting solutions to environmental problems.

Bear in mind that we are entering a new millennium, not just in terms of the calendar flipping from 1999 to 2000, but in terms of a major shift in world view from seeing nature as resources to be exploited to seeing nature as life's supporting structure that demands our stewardship. (See "Ethics" box, page 13.) Such major paradigm shifts have never occurred without controversy and strife, but once they occur they usher in a new age of social, cultural, and scientific progress. The key is to have the faith in yourself that you can help make it happen.

> Thou art even as a finely tempered sword concealed in the darkness of its sheath and its value hidden from the artificer's knowledge. Wherefore come forth from the sheath of self and desire that thy worth may be made resplendent and manifest unto all the world.
>
> Bahá'u'lláh, *Hidden Words*

Review Questions

1. What were cities like before the advent of cars? How did people run errands? How was running errands facilitated by the layout of the city?
2. How did the structure of cities begin to change after World War II? What factors were responsible for the change?
3. Why are the terms *car-dependent* and *urban sprawl* used to describe our current suburban lifestyle and urban layout?
4. What federal laws and policies support urban sprawl?
5. Describe five categories of environmental consequences stemming from the car-dependent lifestyle and urban sprawl.
6. What is meant by exurban migration? How is it related to suburban sprawl?
7. What services do local governments provide, and what is their prime source of revenue for these services?
8. What is meant by "erosion of a city's tax base"? Why does it follow from exurban migration? What are the results?
9. How do exurban migration, urban sprawl, and urban decay become a vicious cycle?
10. What is meant by "economic exclusion"? How is it related to the problems of crime and poverty in cities?
11. What has been the focus of development over the past 50 years? How have communities and people been neglected?
12. How can urban sprawl be reined in?
13. What has been the failing feature of past efforts to mitigate urban blight?
14. What opportunities or advantages do cities offer in terms of accommodating growing populations?
15. What is the focus of modern approaches to rehabilitating cities? Give examples of successes.
16. What are the common denominators of successful efforts to rehabilitate cities?
17. What features or principles of sustainability can be built into the redevelopment of cities?

Thinking Environmentally

1. Interview an older person (65 years or older) about what her or his city or community was like before 1950 in terms of meeting people's needs for shopping, recreation, getting to school, work, and so on. Was it necessary to have a car?
2. Do a study of your region. What aspects of urban sprawl and urban blight are evident? What environmental and social problems are evident? Are they still going on? Are efforts being made to correct them?
3. Identify nongovernmental organizations in your region that are working toward reining in urban sprawl or working toward preserving or bettering the city. What specific projects are under way in these areas? What roles are local governments playing in the process? How can you become involved?
4. Suppose you are a planner or developer. Design a community that is people-oriented and that integrates the principles of sustainability and self-sufficiency. Consider all the fundamental aspects: food, water, energy, sewage, solid waste, and how people get to school, shopping, work, and recreation.

ABC Video Case Studies, Volume III

Each of the following 9 ABC Video Case Studies will bring you face to face with a significant environmental issue somewhere in the world, if not in your own backyard. With the cooperation and courtesy of Prentice Hall and ABC News, we selected video segments from such award-winning ABC news programs as *Nightline*, *20/20*, and *World News Tonight*. Presented below are a written overview, or abstract, for each video segment, along with several questions designed to encourage you to focus on the relevant issues and controversies. Each written case study has a chapter and/or topic reference that serves to emphasize the relevance of the case study to specific sections of the text. We hope that each of these case studies will foment class discussion and debate, for environmental science is a dynamic field that should always accommodate diverse opinions and perspectives. People must talk about the environmental issues of our day; opinions and perspectives should be exchanged and examined.

Prentice Hall is making all of these ABC video segments available to your instructor for classroom use on two convenient long-playing video cassettes. We encourage you to explore these cases as you use this text.

ABC CASE STUDY
Reviving the Biosphere

The Sonoran desert of southeastern Arizona is known for its giant Saguaro cactus and a remarkable diversity of desertland biological life. It is also known as the location of Biosphere 2, an experiment in closed-system ecosystem living—sort of what might be expected if we were to send a module to colonize space. In 1991, eight men and women were sealed in the "Biosphere," and charged with growing all their food and recycling all their wastes within the 2.5 acre sealed structure. The results of their two year stay are controversial, and the private financing that built the structure and maintained it through the sojourn of the eight "biospherians" has been replaced by Columbia University. The news segment discusses the new objectives of Biosphere, and suggests that the structure now has a more serious mission: education and research. Students are taking courses at the facility, and experiments are being conducted to test, for example, the impact of a doubled CO2 concentration on an isolated ecosystem within Biosphere. *Nebel/Wright reference: Earth Watch, "Biosphere 2," p. 75, Chapter 2.*

QUESTIONS

1. What was the basic problem with the Biosphere 2 experiment?
2. According to the Earth Watch box, what is the important lesson learned from Biosphere 2?

ABC CASE STUDY
World Food Prize Doctors

One of the remarkable achievements of the Green Revolution took place in the International Rice Institute in the Philippines. Two agricultural scientists—Henry Beachell and Gurdev Khush—led the way toward the development of two new strains of rice that have provided more food for hundreds of millions of hungry peopleAccording to the Earth Watch box, what is the important lesson learned from Biosphere 2?IR 8 and IR 36. These strains have doubled the rice yield, now accounting for 70% of the rice grown in the world. The news segment highlights their work as they are presented with the World Food Prize in October, 1996. Beachell, a retired U.S. Department of Agriculture scientist, and Khush, who grew up amid the poverty of India, have clearly made a difference in the lives of millions of people through their dedicated work. *Nebel/Wright reference: Chapter 8, Green Revolution, pp. 190–191; Chapter 10, IPM in Indonesia, pp. 237–238.*

QUESTIONS

1. What motives were offered to account for the work of Beachell and Khush?
2. What particular problems in reice culture do IR 8 and IR 36 address?

ABC CASE STUDY
Drinking Water Safety

Access to safe drinking water is taken for granted in most of the developed world, and especially in the United States. The news in this ABC report is a wakeup call to citizens and politicians,telling us that all is not well with our drinking water. The report refers to two recent comprehensive surveys that indicate that during 1993 and 1994 one of every five Americans drank water from their tap that was contaminated with fecal material, radiation, lead or dangerous parasites. As the 104th Congress was considering an overhaul that would have weakend the Clean Water Act, the public is reminded of a 1994 episode where the drinking water in Las Vegas was contaminated by the parasite cryptosporidium. The infestation left 37 dead and at least 200 very sick. The Las Vegas episode exemplifies a larger problem—old and ailing water treatment facilities in many parts of the country allow pathogens like cryptosporidium to get through the treatment process. *Nebel/Wright reference: Water treatment is discussed in Chapter 11, pp. 273–274; Chapter 13, pp. 322–324 deals with fecal contamination of water.*

QUESTIONS

1. What subpopulation has proven to be particularly vulnerable to the cryptosporidium parasite, and why?
2. Evaluate water treatment procedures (Ch. 11) and explain what should be done to prevent parasites like cryptosporidium from getting into drinking water.

ABC CASE STUDY
GM's Electric Car

A number of states have enacted laws that require automobile sellers to begin to market *zero emission vehicles* (ZEVs), which are electric cars. California, once the trend-setter, recently waffled on a rule requiring sales to begin in 1998, putting off the day of reckoning to 2003. Nevertheless, the automobile manufacturers have gotten the message, and are now racing to outdo several smaller entrepreneurial firms in establishing a position of dominance once the market for the ZEVs heats up. The newscast unveils GM's EV1, a trim, attractive vehicle that is getting attention wherever it has appeared. Right now, the vehicles are only available for lease, although other manufacturers have ZEVs for sale (e.g., Solectria's Force model (see Fig. 15-19) which sells for $34,000). The major problem is still the limited range of the cars, which are powered by conventional lead-acid batteries. However, the ZEVs are a clear indicator of the future, and in the words of the newscast, "Detroit finally seems to have made an electric car that won't be mistaken for a golf cart!" *Nebel/Wright reference: Chapter 15, p. 396.*

QUESTIONS

1. What are the advantages and disadvantages of a ZEV?
2. Would you consider buying an electric automobile? In the vernacular of car salespersons, "What would it take to get you in this car?"

ABC CASE STUDY
Global Warming

The goal of the Framework Convention on Climate Change, a treaty forged at the 1992 Earth Summit, is stabilization of greenhouse gas concentrations in the atmosphere. The IPCC scientists have reached a consensus on the fact of a human-caused global warming. The treaty calls for the industrial nations to reduce their carbon emissions to 1990 levels by 2000, but the nations are to accomplish this by voluntary actions. In the United States, the world's largest carbon emitter, doing something about global climate change is turning out to be an exercise in rhetoric rather than action. This newsclip focuses on the frustration being generated by this issue—and blame is cast in various directions. Looking on are the British, who are making progress cutting their emissions by switching from coal to natural gas, and the Chinese, who are on a sharp rise in their emissions as they industrialize. The fact is, our economy is so strongly tied to fossil fuel use, and our political process is so directed towards maintaining the economy at all costs, that it is hard to imagine any kind of effective action by the United States to cut carbon emissions. Yet this must happen if we are to prevent accelerating global climate change, and the question is, when will we have the moral courage to do what must be done? *Nebel/Wright reference: Chapter 16, pp. 416-419; see also Ethics: Staking a Claim on the Moral High Ground, p. 417.*

QUESTIONS

1. How can the "no-regrets strategy" be put into play to lower our carbon emissions, and what benefits would come from this move?
2. Try to imagine a scenario where the United States does take action on lowering our emissions. What will we expect from the Executive Branch of government, and what will we expect from the Legislative Branch?

ABC CASE STUDY *Yellowstone's Wolves*

Yellowstone National Park—the crown jewel of our national park system—has seen its share of controversy in recent years. Perhaps the most contentious issue has been the reintroduction of the gray wolf into the park, after many decades of absence. Although the park is large, wolves range over many square miles, and ranchers near Yellowstone fear that they will lose livestock to wolf predation. In spite of these fears, wolves were reintroduced in 1995 and successfully preyed on the abundant deer, elk and bison in the park. This news segment presents the second round of wolf releases into the park. Unfortunately, the releases coincided with some recent sheep losses attributed to wolf predation, and the old fears were once again evident. The story continues, however, as recent news from the park has revealed that 14 more pups were born in 1996. Wildlife biologists believe that they may not need to release any more wolves to the park—the population is now a viable one. Now the Fish and Wildlife Service is looking toward the northeast, where the last wolf was exterminated in 1899. Adirondack State Park in New York and upstate Maine have been selected as areas that could support wolf populations. *Nebel/Wright reference: Chapter 18, Earth Watch box "Who's Afraid of the Big Bad Wolf?", p. 464.*

QUESTIONS

1. According to the newscast, what happens when a wolf does prey on livestock?
2. Which animal do you think is more dangerous, the wolf or the cougar? Why?

ABC CASE STUDY *Mollie Beattie*

The value of wild species is a matter of perspective. To some, wild and threatened species are of little consequence; there are more important things in life to worry about, and besides, who would miss spotted owls or the Karner blue butterfly? However, there are some people who care very much about wild species, people who believe in the intrinsic value of wild creatures and who will go to great lengths to protect them. Such a person is Mollie Beattie. This newsclip salutes Mollie Beattie, the first woman to become director of the US Fish and Wildlife Service. Beattie has valiantly defended the Endangered Species Act as it has come under fire from Congress. And she has spearheaded the remarkable efforts to restore wolves to Yellowstone National Park. There is a sad note to this salute to Molly Beattie, however. At the time of this news brief, Mollie Beattie was in a Vermont hospital, battling a malignant brain tumor. She had just resigned from the Fish and Wildlife Service, and was fighting for her life. Bruce Babbitt, Secretary of the Interior, states that her legacy is going to be the Endangered Species Act, rescued from the brink of oblivion and renewed. Her story is a reminder that people can make a difference, that commitment to a worthy cause is still possible, and that humans can indeed act as stewards of wild nature, caring for things that cannot care for themselves. The story does not have a happy ending. Mollie Beattie died on June 27, 1996 in Townsend, Vermont. Yet, her memory lives on at Yellowstone, where one of the new wolf packs was named after her. *Nebel/Wright reference: Chapter 18, pp. 459–465; Chapter 19, Ethics box, Stewards of Creation (p. 499).*

QUESTIONS

1. How does Mollie Beattie explain the importance of protecting endangered species?
2. What is the difference between the intrinsic value and the instrumental value assigned to wild species?

ABC CASE STUDY
Lobbyists and Environmental Law

Our government is often portrayed as a liberal democracy where the will of the majority of the people prevails, and our senators and representatives are responsive to that will. Recent events in the Congress have shown that this portrayal is at best a simplification, and at worst, a sham. The people put the Republican party in power in the 104th Congress, and then something unexpected happened. The Congress decided to attempt to rewrite the nation's environmental protection laws, something the people hadn't asked for, and turned to the special interest groups most affected by those laws for help. The result? Lobbyists for industry were submitting their proposals, and those proposals were showing up in the bills introduced to Congress. The newsclip highlights this process as it reveals the wielding of special interest politics in Washington. The good news for the environment is that something happened on the way to the White House, and the Republican Party paid the price for their attempt to water down protection of the environment as Bill Clinton was re-elected in 1996. *Nebel/Wright reference: Chapter 19, p. 505.*

QUESTIONS

1. Can you trace how the "Contract With America" became an attack on environmental public policy?
2. What are the consequences for the environment of "special interest" politics becoming the major mode of political policy-making in Washington? Who represents environmental interests if that happens?

ABC CASE STUDY
Chornobyl, Ten Years Later

The massive explosions and meltdown of the Chornobyl reactor on April 26, 1986, released 200 times as much radiation as the Hiroshima atomic bomb. The highest radiation still is found in the soil of the exclusion zone that surrounds the reactor site. This newscast takes viewers to that exclusion zone, which now is home to an array of wildlife. Amazingly, the animals and plants are thriving in an environment which continually bombards them with radiation. A study of field mice has indicated that the mice are coping with the radiation by rapidly adapting through gene mutations. In essence, they are evolving, and at a far more rapid rate than any mammalian species ever has. Genetic changes have also been found in the children of families living downwind from the accident, although the newscast does not identify the kinds of changes found. One change known to have been caused by Chornobyl is the highly elevated incidence of thyroid cancer in children in Ukraine and Belarus. The Chornobyl explosion can be seen as a huge experiment in testing the effects of radiation on natural ecosystems and on humans—and it is hopefully the last experiment of this kind we ever see. *Nebel/Wright reference: Chapter 22, pp. 569-570.*

QUESTIONS

1. How can the results shown in this study of mice be reconciled with the great concerns most people have about nuclear energy?
2. Why would radiation conditions continually improve for the mice in the ten years since the Chornobyl disaster? (see text, p. 565)

ABC Video Case Studies, Volumes I and II

Also being made available to instructors who adopt the 6th edition of Environmental Science are the three cassettes of videos from the 4th and 5th edition Video Case Studies. Most of these videos are still quite relevant to current environmental concerns. The following are the titles of the Case Studies as they appear at the end of the 4th and 5th editions, which refer to the videos available in Volume I and II:

Volume I

Famine and Relief Efforts in Somalia
Hazardous Wastes in the Wrong Place
Environmental Concerns under Communism
Global Warming and the Scientific Community
Break in the Ozone Shield
The Place of Risk Analysis in Pollution Cleanup
Council on Competitiveness: Bush's Shadow Government
Biodiversity: What Can Be Done to Protect It?
Challenge on the Right: Environmental Movement as Subversive
Mismanaging the National Forests
Who Wants the Trash?
Oil and the Arctic National Wildlife Refuge
The Case for Energy Conservation
Fuel-Efficient Cars: Detroit versus Japan
Oil on the Sound: Alaskan Oil Spill
The Nuclear Waste Dilemma
Energy Alternatives in Our Future

Volume II

The World's Population Explosion
Assignment Rwanda: Profiles in Caring
Nature's Warning: Environmental Poisons and Sterility
Tobacco on Trial
Ozone and UV Radiation
Resoring Rivers for Wildlife
Environmental Backlash and the Scientific Community
Empty Nets: Collapse of the Fisheries Turning Landfills into Nature Preserves
Nuclear Waste and Apaches

APPENDIX A

Environmental Organizations

This is a list of nongovernmental organizations active in environmental matters. Included here are national organizations as well as some small, specialized ones. These organizations offer a variety of fact sheets, brochures, newsletters, publications, educational materials, and annual reports. Requests for information should be specific and are best made by phone to determine what information is available. Some organizations have internship positions available for those wishing to do work for an environmental group. A more complete listing of environmental organizations can be found in the *Conservation Directory* put out by National Wildlife Federation (address below). This directory includes local, regional, national and international organizations. The cost is $18.00 plus $4.50 for shipping and handling. Also included at the end of this Appendix is a section on environmental information available via Internet, and E-mail addresses and World Wide Web sites for selected environmental organizations.

AMERICAN FARMLAND TRUST. Focuses on preservation of American farmland. 1920 N Street, N.W., Suite 400, Washington, D.C. 20036. (202)659-5170.

AMERICAN LUNG ASSOCIATION. Research, education, legislation, lobbying, advocacy: indoor and outdoor air pollution effects and means of control. 1740 Broadway, New York, NY 10019. (212)315-8700.

AMERICAN RIVERS, INC. Mission is "to preserve and restore America's rivers' systems and to foster a river stewardship ethic." Education, litigation, lobbying: wild and scenic rivers, hydropower relicensing. Information on specific rivers available. 801 Pennsylvania Avenue, S.E., Suite 400, Washington, D.C. 20003. (202)547-6900.

BREAD FOR THE WORLD. National advocacy group that lobbies for hunger-related legislation. 1100 Wayne Avenue, Suite 1000, Silver Springs, MD 20910. (301) 608-2400.

CENTER FOR ENVIRONMENTAL INFORMATION. Provides training for corporations and individuals on environmental risks, offers environmental law courses, generates publications, and has library open to the public. 46 Prince Street, Rochester, NY 14607. (716)271-3550.

CENTER FOR SCIENCE IN THE PUBLIC INTEREST. Research and education: alcohol policies, food safety, health, nutrition, organic agriculture. 1875 Connecticut Avenue, N.W., Suite 300, Washington, D.C. 20009. (202)332-9110.

CHESAPEAKE BAY FOUNDATION. Research, education, and litigation: environmental defense and management of Chesapeake Bay and surrounding area. 162 Prince Georges Street, Annapolis, MD 21401. (301)268-8816.

CLEAN WATER ACTION PROJECT. Lobbying, education, research: water quality. 1320 18th Street, N.W., Suite 300, Washington, D.C. 20036. (202)457-1286.

COMMON CAUSE. Lobbying: government reform, energy reorganization, clean air. 2030 M Street, N.W., Washington, D.C. 20036. (202)833-1200.

COMMUNITY TRANSPORTATION ASSOCIATION OF AMERICA. Provides technical assistance for rural and specialized transportation systems. 725 15th Street, N.W., Suite 900, Washington, D.C. 20005. (800)527-8279.

CONCERN, INC. Environmental education. 1794 Columbia Road, N.W., Washington, D.C. 20009. (202)328-8160.

CONGRESS WATCH. Lobbying: consumer health and safety, pesticides. 215 Pennsylvania Avenue, S.E., Washington, D.C. 20003. (202)546-4996.

CONSERVATION INTERNATIONAL. Education and research: rain forests. 1015 18th Street, N.W., Suite 1000, Washington, D.C. 20036. (202)429-5660.

CRITICAL MASS ENERGY PROJECT. Research and education: alternative energy and nuclear power. 215 Pennsylvania Avenue, S.E., Washington, D.C. 20003. (202)546-4996.

DEFENDERS OF WILDLIFE. Research, education and lobbying: endangered species. 1244 19th Street, N.W., Washington, D.C. 20036. (202)659-9510.

EARTHWATCH EXPEDITIONS, INC. Environmental research is encouraged by organizing teams to make expeditions to various locations all over the world. Team members contribute to the expenses and spend several weeks to months on site. 680 Mount Auburn Street, P.O. Box 403, Watertown, MA 02272. (617)926-8200.

ENVIRONMENTAL ACTION, INC. Lobbying, education, grass roots organizing: energy efficiency and conservation, toxics reduction, right-to-know laws, transportation, solid waste, solar energy, deposit legislation. Offer internships. 6930 Carroll Avenue, Suite 600, Takoma Park, MD 20912. (301)891-1100.

ENVIRONMENTAL DEFENSE FUND. Research, litigation, and lobbying: cosmetics safety, drinking water, energy, transportation, pesticides, wildlife, air pollution, cancer prevention, radiation. 257 Park Avenue South, New York, NY 10010. (212)505-2100.

ENVIRONMENTAL LAW INSTITUTE. Training, educational workshops, and seminars for environmental professionals, lawyers, and judges on institutional and legal issues affecting the environment. 1616 P Street, N.W., Suite 200, Washington, D.C. 20036. (202)328-5150.

FREEDOM FROM HUNGER. Develop programs for the elimination of hunger worldwide. 1644 DaVinci Court, Davis, CA 95617. (916)758-6200.

FRIENDS OF THE EARTH. Research, lobbying: all aspects of energy development, preservation, restoration, and rational use of the earth. 218 D Street, S.E., Washington, D.C. 20036. (202)544-2600.

GREENPEACE, USA, INC. "An international environmental organization dedicated to protecting the planet through nonviolent, direct action providing public education, scientific research, and legislative lobbying. Greenpeace campaigns to free the Earth of nuclear and toxic pollution, to protect marine ecology, and to end atmospheric destruction." 1436 U Street, N.W., Washington, D.C. 20009. (202)462-1177.

HABITAT FOR HUMANITY INTERNATIONAL. Fosters home building and ownership for the poor. Education on housing issues. 121 Habitat Street, Americus, GA 31709-3498. (912)924-6935.

INSTITUTE FOR LOCAL SELF-RELIANCE. Research and education: appropriate technology for community development. 2425 18th Street, N.W., Washington, D.C. 20009. (202)232-4108.

IZAAK WALTON LEAGUE OF AMERICA, INC. Research, education, endowment grants: conservation, air and water quality, streams. 1401 Wilson Boulevard, Level-B, Arlington, VA 22209. (703)528-1818.

LAND TRUST ALLIANCE. Works with local and regional land trusts to enhance their work. 900 17th Street, N.W., Suite 410, Washington, D.C. 20006. (202)785-1410.

LEAGUE OF CONSERVATION VOTERS. Political arm of the environmental community. Works to elect candidates to the U.S. House and Senate who will vote to protect the nation's environment, and holds them accountable by publishing the *National Environmental Scorecard* each year, which can be ordered for $6 and is free to students. 1707 L Street, N.W., Suite 550, Washington, D.C. 20036. (202)785-8683.

LEAGUE OF WOMEN VOTERS OF THE U.S. Education and lobbying, general environmental issues. Publications on groundwater, agriculture and farm policy, and other topics available. 1730 M Street, N.W., Suite 1000, Washington, D.C. 20036. (202)429-1965.

MONITOR CONSORTIUM OF CONSERVATION AND ANIMAL WELFARE ORGANIZATIONS. Lobbying and networking: conservation issues, animal issues, endangered species and their habitats. 1506 19th Street, N.W., Washington, D.C. 20036. (202)234-6576.

NATIONAL AUDUBON SOCIETY. Research, lobbying, education, litigation, and citizen action: broad-based environmental issues. 700 Broadway, New York, NY 10003. (212)546-9100.

NATIONAL PARK FOUNDATION. Education, land acquisition, management of endowments, grant making. National Parks, 1101 17th Street, N.W., Suite 1102, Washington, D.C. 20036. (202)785-4500.

NATIONAL PARKS AND CONSERVATION ASSOCIATION. Research and education: parks, wildlife, forestry, general environmental quality. 1776 Massachusetts Avenue, N.W., Suite 200, Washington, D.C. 20036. (202)223-6722.

NATIONAL RESOURCES DEFENSE COUNCIL. Research and litigation: water and air quality, land use, energy, pesticides, toxic waste, 40 West 20th Street, New York, NY 10011. (212)727-2700.

NATIONAL WILDLIFE FEDERATION. Research, education, lobbying: general environmental quality, wilderness, and wildlife. 1400 16th Street, N.W., Washington, D.C. 20036. (202)797-6800.

NATURE CONSERVANCY. "Preserve plants, animals, and natural communities that represent the diversity of life on Earth by protecting the land and the water they need to survive." 1815 North Lynn Street, Arlington, VA 22209. (703)841-5300.

OXFAM AMERICA. Funding agency for projects to benefit the "poorest of the poor" in South America, Africa, India, Central America, the Caribbean, and the Philippines. Provide whatever resources are needed. 115 Broadway, Boston, MA 02116. (617)482-1211.

PLANNED PARENTHOOD FEDERATION OF AMERICA. Education, services, and research: fertility control, family planning. 810 7th Avenue, New York, NY 10019. (212)541-7800.

THE POPULATION INSTITUTE. Education, research, and speaking engagements: population control. 107 2nd Street, N.E., Washington, D.C. 20002. (202)544-3300.

POPULATION REFERENCE BUREAU, INC. Organization engaged in collection and dissemination of objective population information. Excellent publications. 1875 Connecticut Avenue, N.W., Suite 520, Washington, D.C. 20009. (202)483-1100.

RACHEL CARSON COUNCIL, INC. Publication and distribution of information on pesticides and toxic substances, educational conferences, and seminars. 8940 Jones Mill Road, Chevy Chase, MD 20815. (301)652-1877.

RAIN FOREST ACTION NETWORK. Information and educational resources: world's rain forests. 450 Sansome Street, Suite 700, San Francisco, CA 94111. (415)398-4404.

RAINFOREST ALLIANCE. Education, medicinal plants project, timber project to certify "smart wood," and news bureau in Costa Rica. 279 Lafayette Street, Suite 512, New York, NY 10012. (212)941-1900.

RENEW AMERICA. "A nationwide clearinghouse for environmental solutions by seeking out and promoting successful programs. We offer positive, constructive models to help communities meet environmental challenges." Twenty different environmental categories addressed. 1400 16th Street, N.W., Suite 710, Washington, D.C. 20036. (202)232-2252.

RESOURCES FOR THE FUTURE Research and education: conservation of natural resources, environmental quality. 1616 P Street, N.W., Washington, D.C. 20036. (202)328-5000.

SCIENTISTS INSTITUTE FOR PUBLIC INFORMATION. Education: scientists of full range of disciplines provide information for the public. 355 Lexington Avenue, New York, NY 10017. (212)661-9110.

SIERRA CLUB. Education and lobbying: broad-based environmental issues. 730 Polk Street, San Francisco, CA 94109. (415)776-2211.

TRUST FOR PUBLIC LAND. Works with citizen groups and government agencies to acquire and preserve open space. 116 New Montgomery, 4th Floor, San Francisco, CA 94105. (415)495-4014.

U.S. ENVIRONMENTAL PROTECTION AGENCY, PUBLIC INFORMATION CENTER. Provides general information about environmental topics. 401 M Street, S.W., Washington, D.C. 20460. (202)260-7751.

U.S. Public Interest Research Group. Research, education, and lobbying: alternative energy, consumer protection, utilities regulation, public interest. 215 Pennsylvania Avenue, S.E., Washington, D.C. 20003. (202)546-9707.

Water Environment Federation. Research, education, and lobbying. 601 Wythe Street, Alexandria, VA 22314. (703)684-2400.

Wilderness Society. Research, education and lobbying: wilderness, public lands. 900 17th Street, N.W., Washington, D.C. 20006. (202)833-2300.

World Resources Institute. Research and publish reports on environmental issues that are sent to educators, policy makers, and organizations. 1709 New York Avenue, N.W., Washington, D.C. 20006. (800)822-0504.

World Wildlife Fund. Preservation of wildlife habitats and protection of endangered species. 1250 24th Street, N.W., Washington, D.C. 20036. (202)293-4800.

WorldWatch Institute. Research and education: energy, food, population, health, women's issues, technology, the environment. 1776 Massachusetts Avenue, N.W., Washington, D.C. 20036. (202)452-1999.

Zero Population Growth, Inc. Public education, lobbying, and research: population. 1400 16th Street, N.W., Suite 320, Washington, D.C. 20036. (202)332-2200.

Environmental Information on the Internet

Without a doubt, one of the most exciting and profound revolutions in information exchange is the system referred to as the Internet. For many years the Internet was used as a means of personal communication, via E-Mail. Once a user has computer access to the Internet, a few strokes of the keyboard make possible rapid communication over vast distances at little or no cost. E-Mail continues to be a major function of the Internet, as people all over the world keep in touch with each other via E-Mail messages. You can send an E-Mail message to the President of the United States, or to your cousin in Albuquerque. For many years, the Internet could also be used for transferring files, for accessing information from other computers, or for conducting ongoing "discussions" on topics of interest.

Access to the Internet can be secured through many commercial services, some local and some very widespread, like America OnLine and Compuserve. All you need is a computer, a modem to connect to a telephone line, and browsing software, which is provided by your service. Most colleges and universities, indeed, public and private schools now have their own connections to the Internet, allowing anyone with a computer to hook up to their internal network and access the Internet.

By far the most important and revolutionary function of the Internet, however, is the World Wide Web (referred to as "the Web"). This medium has literally exploded into "cyberspace" in the last few years. The Web links millions of computers via electronic communication. For students of environmental science, the Web has become a fantastic information resource. Because it also involves graphics, sound and movies, the Web has greatly enhanced communication possibilities over the former text-only uses of the Internet. It is a multimedia extravaganza that defies the imagination. Individuals, organizations, businesses, virtually anyone with a computer can establish a Web site, or "Home Page." The Home Page can contain photos, graphics, text, even movies, and can provide links to other Home Pages in a never-ending series of connections. Your college, university or high school very likely has a Home Page, with an index to different departments, informative materials, and other resources within the institution's Home Page. Each Home Page has a unique address, which must use a specific protocol—the URL (universal resource locator) notation. For example, the URL for the EPA is <http://www.epa.gov/> (everything within the brackets is the URL). It is important to get the URL right; every letter and symbol is imortant, even capitalization.

Two extremely vital kinds of software have opened up the World Wide Web to users. The first is the "browser," a program that is able to read the URL address of a Home Site and also read the coded language used in setting up Web pages (HTML). Microsoft Internet Explorer, **Mosaic** and **Netscape** are the three most heavily used browsers. They are quite user-friendly, making possible "surfing the net" with menu-driven options and all sorts of ways of manipulating the options available.

The other program that has contributed so much to the Web revolution is the "search tool," a program that accepts **keywords** and lists web sites according to the degree of fidelity of the site with your topic. Search tools available on Netscape include **Yahoo, Alta Vista, Lycos, Excite,** and **WebCrawler.** You must be fairly specific with your keywords, as general terms (for example, population) will generate many thousands of web locations, or "hits." To access a location, you simply click on the listed site, and you are quickly there - if the site's server computer is on line, of course. The search tools have put information retrieval and research on a new level of accessibility. Would you like to see the text of a bill introduced to Congress? You can have it in a few minutes! Do you want to know what the weather is like in Nairobi, Kenya? No problem! Another option is to use the **subject directories** available on search tools like Yahoo. These are menu-driven, and after you open a category (say, Entertainment) you will keep choosing from successive menus until you reach the specific subject you are seeking.

The Web is extremely well connected to environmental organizations and concerns. Below we list some web addresses that are useful; they are organized according to types of organization. The list is far from comprehensive, but it should provide a start. A few web sites are uniquely constructed to serve the environmental community. One such site is the **Envirolink Network** (<http://www.envirolink.org/>). This network provides access to environmental activist organizations, mailing lists, information on specific environmental issues, and many links to other organizations. Note that **all web addresses have the prefix http://** We will just list what comes after this in the addresses below.

A note of caution: the information you glean on the Web may be accurate and useful, or it may be false and misleading; there is no guarantee of the validity of entries on Home Pages. Organizations put information on their web page for a reason—they want you to know what they have to say. Beware of bias and misinformation. Don't let this stop you from surfing the web, however. It's an enormously interesting and useful resource.

I. General Resources

Campus Green Vote (geared towards getting college student involved in campus projects, political activism)	www.cgv.org/cgv/
Center for Environmental Philosophy (many links, Environmental Ethics journal info. home page for many organizations)	www.cep.unt.edu/
Christian environmental web site and the Evang. Envir. Network)	(contains link to the Global Stewardship Conference www.cesc.montreat.edu/
E Magazine (net version, links)	www.emagazine.com/
Econet (directory of env. issues)	www.igc.org/igc/econet/
Ecomall (consumer-oriented, eco-business)	www.ecomall.com/
eNet Digest (weekly reviews of web sites of envir. interest)	www.enetdigest.com/
Envirolink Network (see above)	www.envirolink.org
Environmental Professional's Home Page (full scope of government sites and envir. issues)	www.clay.net/govag.html
Greennet (a UK-based site with international and developing country emphasis, many links)	www.gn.apc.org/
National Institute for the Environment (includes the National Library for the Environment, and other information services)	www.cnie.org/index.shtml
The Best Environmental Resources Directory (lists other environmental directories and Internet sites)	www.ulb.ac.be/ceese/cds/html
World Conservation Monitoring Centre	www.wcmc.org.uk

II. Non-Governmental Organizations

American Rivers	www.amrivers.org/amrivers/
Answers to Rush Limbaugh	www.econet.apc.org/igc/rush.html
Bread for the World	www.bread.org/
Clean Water Action	www.essential.org/cwa/
Common Cause	www.commoncause.org/index.html
Defenders of Wildlife	www.defenders.org
Freedom From Hunger	www.freefromhunger.org/
Friends of the Earth	www.essential.org/FOE.html
Greenpeace	www.greenpeace.org
Habitat for Humanity Internat.	www.habitat.org/
League of Conservation Voters	www.econet.apc.org/lcv/scorecard.html
National Audubon Society	www.audubon.org/audubon/
National Wildlife Federation	www.nwf.org/nwf/
Natural Resources Defense Council	www.nrdc.org
Nuclear Energy Institute	www.nei.org/
Oxfam America	www.web.net/oxfamgft/index.htm
Rainforest Action Network	www.ran.org/ran/
Rainforest Alliance	www.rainforest-alliance.org/
Resources for the Future	www.rff.org/
Sierra Club	www.sierraclub.org
Trust for Public Land	www.igc.apc.org/tpl/
Wise Use Movement inform.	www.wiseuse.com
WorldWatch Institute	www.worldwatch.org
World Wildlife Fund	www.worldwildlife.org/
Zero Population Growth	www.zpg.org/zpg/

III. International Agencies and Concerns

Canadian Forestry Management	www.nofc.forestry.ca/
Developing country population inform.	www.macroint.com/dhs/
Global information center for maps, relief operations	www.info/usaid.gov/ofda/reliefweb/
Index to many articles on population	www.popindex.princeton.edu/
U.N. Development Program	www.undp.org/
U.N. Envir. Program, Biodiversity Convention	www.unep.ch/biodiv.html
U.N. Environment Program	www.unep.org/
U.N. Environment Program Journal	www.ourplanet.com
U.N. Food and Agric. Organization	www.fao.org/
U.N. FAO, World Food Summit inform.	www.fao.org/wfs/homepage.htm
The World Bank	www.worldbank.org/
World Hunger Relief Foundation	www.estatebond.com/hunger1.htm
U.S. and Internat. web sites on pop.	www.pitt.edu/HOME/GHNet/poprepro.htm

IV. U.S. Governmental Organizations

Agency for International Development	www.info.usaid.gov/
Center for Disease Control	www.cdc.gov/
Congress	thomas.loc.gov/
Dept. of Agriculture, Natural Resources Conservation Service	www.ncg.nrcs.usda.gov/
Dept. of Commerce, National Oceanic and Atmospheric Administration	www.esdim.noaa.gov/
Department of Energy	www.doe.gov/
EPA	www.epa.gov/
Fish & Wildlife Service	www.fws.gov/law/law5.html
Geological Survey, Water Resources inform.	h2o.er.usgs.gov/
National Aeronautic and Space Admin.	spsogsfc.nasa.gov/
NASA's ozone web page	spsogsfc.nasa.gov/NASA_FACTS/ozone/ozone.html
National Marine Fisheries Service	kingfish.sssp.nmfs.gov/nmfs_pubs.html
Nuclear Regulatory Commission	www.nrc.gov/
The Whitehouse web site	www.whitehouse.gov

APPENDIX B

Units of Measure

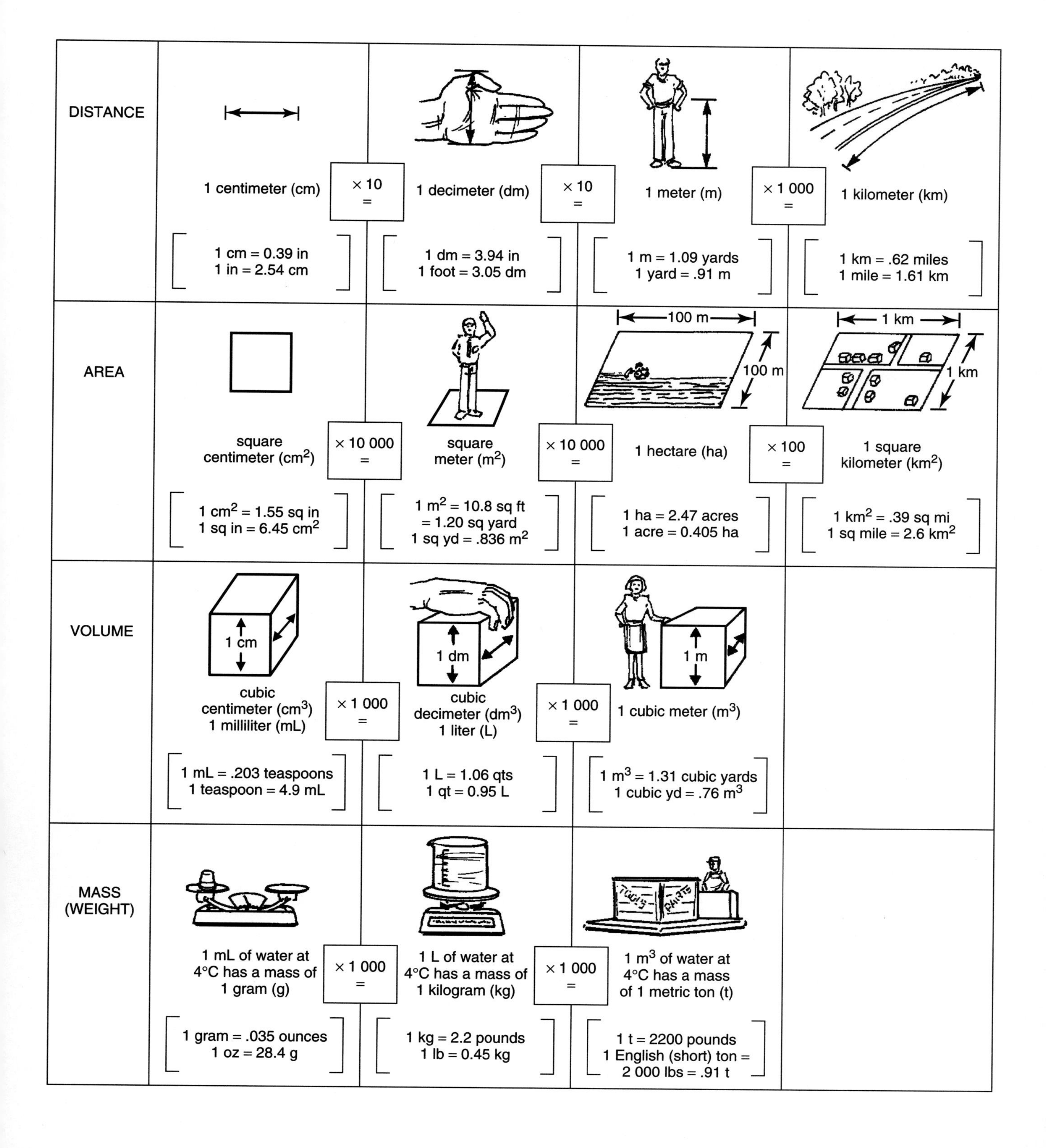

DISTANCE	1 centimeter (cm)	× 10 =	1 decimeter (dm)	× 10 =	1 meter (m)	× 1 000 =	1 kilometer (km)
	1 cm = 0.39 in 1 in = 2.54 cm		1 dm = 3.94 in 1 foot = 3.05 dm		1 m = 1.09 yards 1 yard = .91 m		1 km = .62 miles 1 mile = 1.61 km
AREA	square centimeter (cm^2)	× 10 000 =	square meter (m^2)	× 10 000 =	1 hectare (ha)	× 100 =	1 square kilometer (km^2)
	1 cm^2 = 1.55 sq in 1 sq in = 6.45 cm^2		1 m^2 = 10.8 sq ft = 1.20 sq yard 1 sq yd = .836 m^2		1 ha = 2.47 acres 1 acre = 0.405 ha		1 km^2 = .39 sq mi 1 sq mile = 2.6 km^2
VOLUME	cubic centimeter (cm^3) 1 milliliter (mL)	× 1 000 =	cubic decimeter (dm^3) 1 liter (L)	× 1 000 =	1 cubic meter (m^3)		
	1 mL = .203 teaspoons 1 teaspoon = 4.9 mL		1 L = 1.06 qts 1 qt = 0.95 L		1 m^3 = 1.31 cubic yards 1 cubic yd = .76 m^3		
MASS (WEIGHT)	1 mL of water at 4°C has a mass of 1 gram (g)	× 1 000 =	1 L of water at 4°C has a mass of 1 kilogram (kg)	× 1 000 =	1 m^3 of water at 4°C has a mass of 1 metric ton (t)		
	1 gram = .035 ounces 1 oz = 28.4 g		1 kg = 2.2 pounds 1 lb = 0.45 kg		1 t = 2200 pounds 1 English (short) ton = 2 000 lbs = .91 t		

Energy Units and Equivalents

1 Calorie, food calorie, or kilocalorie—The amount of heat required to raise the temperature of one kilogram of water one degree Celsius (1.8°F).

1 BTU (British Thermal Unit)—The amount of heat required to raise the temperature of one pound of water one degree Fahrenheit.

1 Calorie = 3.968 BTU's
1 BTU = 0.252 calories

1 therm = 100,000 BTU's
1 quad = 1 quadrillion BTU's

1 watt standard unit of electrical power

1 watt-hour (wh) = 1 watt for 1 hr. = 3.413 BTU's

1 kilowatt (kw) = 1000 watts

1 kilowatt-hour (kwh) = 1 kilowatt for 1 hr. = 3413 BTU's

1 megawatt (Mw) = 1,000,000 watts

1 megawatt-hour (Mwh) = 1 Mw for 1 hr. = 34.13 therms

1 gigawatt (Gw) = 1,000,000,000 watts or 1,000 megawatts

1 gigawatt-hour (Gwh) = 1 Gw for 1 hr. = 34,130 therms

1 horsepower = .7457 kilowatts; 1 horsepower-hour = 2545 BTU's

1 cubic foot of natural gas (methane) at atmospheric pressure = 1031 BTU's

1 gallon gasoline = 125,000 BTU's

1 gallon No. 2 fuel oil = 140,000 BTU's

1 short ton coal = 25,000,000 BTU's

1 barrel (oil) = 42 gallons

APPENDIX C

Some Basic Chemical Concepts

Atoms, Elements, and Compounds

All matter, whether gas, liquid, or solid, living or nonliving, organic or inorganic, is comprised of fundamental units called **atoms**. Atoms are extremely tiny. If all the world's people, about 5,700,000,000 of us, were reduced to the size of atoms, there would be room for all of us to dance on the head of a pin. In fact, we would only occupy a tiny fraction (about Z\z/,///) of the pin's head. Given the incredibly tiny size of atoms, even the smallest particle which can be seen with the naked eye consists of billions of atoms.

The atoms comprising a substance may be all of one kind; or they may be of two or more different kinds. If the atoms are all of one kind, the substance is called an **element**. If the atoms are of two or more different kinds bonded together, the substance is called a **compound**.

Through countless experiments, chemists have ascertained that there are only 96 distinct kinds of atoms which occur in nature. They are listed in Table C-1 with their chemical symbols. By scanning Table C-1, you can see that a number of familiar substances such as aluminum, calcium, carbon, oxygen, and iron are elements; that is, they are a single distinct kind of atom. However, most of the substances with which we interact in every-day life, such as water, stone, wood, protein, and sugar, are not on the list. Their absence from the list is indicative that they are not elements; rather they are compounds, which means they are actually comprised of two or more different kinds of atoms bonded together.

Atoms, Bonds, and Chemical Reactions

In chemical reactions, atoms are neither created, nor destroyed, nor is one kind of atom changed into another. What occurs in chemical reactions, whether mild or explosive, is simply a rearrangement of the ways in which the atoms involved are bonded together. An oxygen atom, for example, may be combined and recombined with different atoms to form any number of different compounds, but a given oxygen atom always has been, and always will be, an oxygen atom. The same can be said for all the other kinds of atoms. In order to understand how atoms may bond and rearrange to form different compounds, it is necessary to first have some concepts concerning the structure of atoms.

Structure of Atoms

In every case, an atom consists of a central core called the nucleus (not to be confused with the cell nucleus). The nucleus of the atom contains one or more **protons** and, except for hydrogen, one or more **neutrons** as well. Surrounding the nucleus are particles called **electrons**. Each proton has a positive (+) electric charge and each electron has an equal but opposite negative (−) electric charge. Thus, the charge of the protons may be balanced by an equal number of electrons making the whole atom neutral. Neutrons have no charge.

Atoms of all elements have this same basic structure consisting of protons, electrons, and neutrons. The distinction among atoms of different elements is in the number of protons. The atoms of each element have a characteristic number of protons which is known as the **atomic number** of the element (see Table C-1). The number of electrons characteristic of the atoms of each element also differs corresponding to the number of protons. The general structure of the atoms of several elements is shown in Figure C-1.

The number of protons and electrons, i.e., the atomic number of the element, determines the chemical properties of the element. However, the number of neutrons may also vary. For example, most carbon atoms have six neutrons in addition to the six protons as indicated in Figure C-1. But some carbon atoms have eight neutrons. Atoms of the same element which have different numbers of neutrons are known as **isotopes** of the element. The total number of protons plus neutrons is used to define different isotopes. For example, the usual isotope of carbon is referred to as carbon-12 while the isotope noted above is referred to as carbon-14. The chemical reactivity of different isotopes of the same element is identical. However, certain other properties may differ. Many iso-

TABLE C-1
The Elements

Element	*Symbol*	*Atomic Number*	*Element*	*Symbol*	*Atomic Number*
Actinium	Ac	89	Neodymium	Nd	60
Aluminum	Al	13	Neon	Ne	10
Americium	Am	95	Neptunium	Np	93
Antimony	Sb	51	Nickel	Ni	28
Argon	Ar	18	Niobium	Nb	41
Arsenic	As	33	Nitrogen	N	7
Astatine	At	85	Nobelium	No	102
Barium	Ba	56	Osmium	Os	76
Berkelium	Bk	97	Oxygen	O	8
Beryllium	Be	4	Palladium	Pd	46
Bismuth	Bi	83	Phosphorus	P	15
Boron	B	5	Platinum	Pt	78
Bromine	Br	35	Plutonium	Pu	94
Cadmium	Cd	48	Polonium	Po	84
Calcium	Ca	20	Potassium	K	19
Californium	Cf	98	Praseodymium	Pr	59
Carbon	C	6	Promethium	Pm	61
Cerium	Ce	58	Protoactinium	Pa	91
Cesium	Cs	55	Radium	Ra	88
Chlorine	Cl	17	Radon	Rn	86
Chromium	Cr	24	Rhenium	Re	75
Cobalt	Co	27	Rhodium	Rh	45
Copper	Cu	29	Rubidium	Rb	37
Curium	Cm	96	Ruthenium	Ru	44
Dysprosium	Dy	66	Samarium	Sm	62
Einsteinium	Es	99	Scandium	Sc	21
Erbium	Er	68	Selenium	Se	34
Europium	Eu	63	Silicon	Si	14
Fermium	Fm	100	Silver	Ag	47
Fluorine	F	9	Sodium	Na	11
Francium	Fr	87	Strontium	Sr	38
Gadolinium	Gd	64	Sulfur	S	16
Gallium	Ga	31	Tantalum	Ta	73
Germanium	Ge	32	Technetium	Tc	43
Gold	Au	79	Tellurium	Te	52
Hafnium	Hf	72	Terbium	Tb	65
Helium	He	2	Thallium	Tl	81
Holmium	Ho	67	Thorium	Th	90
Hydrogen	H	1	Thulium	Tm	69
Indium	In	19	Tin	Sn	50
Iodine	I	53	Titanium	Ti	22
Iridium	Ir	77	Tungsten	W	74
Iron	Fe	26	Unnilennium	Une	109
Krypton	Kr	36	Unnilhexium	Unh	106
Lanthanum	La	57	Unniloctium	Uno	108
Lawrencium	Lr	103	Unnilseptium	Uns	107
Lead	Pb	82	Uranium	U	92
Lithium	Li	3	Vanadium	V	23
Lutetium	Lu	71	Xenon	Xe	54
Magnesium	Mg	12	Ytterbium	Yb	70
Manganese	Mn	25	Yttrium	Y	39
Mendelevium	Md	101	Zinc	Zn	30
Mercury	Hg	80	Zirconium	Zr	40
Molybdenum	Mo	42			

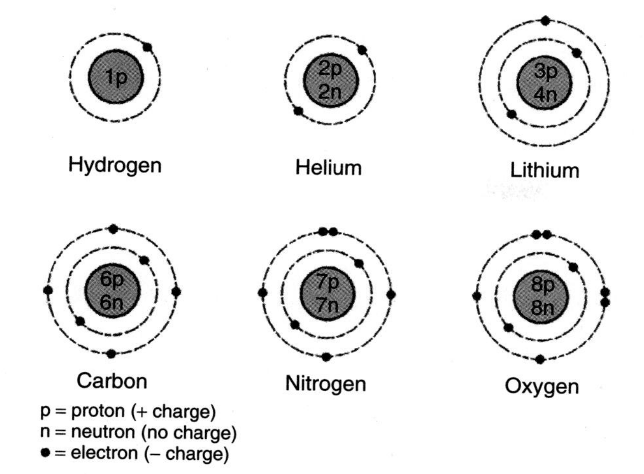

FIGURE C-1
Structure of atoms. All atoms consist of fundamental particles: protons (P), which have a positive electric charge, neutrons (n), which have no charge, and electrons, which have a negative charge. Protons and neutrons are located in a central core, the nucleus. The positive charge of the protons is balanced by an equal number of electrons, which occupy various levels or orbitals around the nucleus. The uniqueness of each element is given by its atoms having a distinct number of protons, its atomic number.

topes of various elements prove to be radioactive as is carbon-14.

Bonding of Atoms

The chemical properties of an element are defined by the ways in which its atoms will react and form bonds with other atoms. By examining how atoms form bonds, we shall see how the number of electrons and protons determines these properties. There are two basic kinds of bonding: (1) **covalent bonding**, and (2) **ionic bonding**.

In both kinds of bonding, it is first important to recognize that electrons are not randomly distributed around the atom's nucleus. Rather, there are, in effect, specific spaces in a series of layers, or **orbitals**, around the nucleus. If an orbital is occupied by one or more electrons but not filled, the atom is unstable; it will tend to react and form bonds with other atoms to achieve greater stability. A stable state is achieved by having all the spaces in the orbital filled with electrons. But, it is also important to keep the charge neutral, i.e., the total number of electrons equal to that of the protons.

Covalent Bonding

These two requirements, filling all the spaces and keeping the charge neutral, may be satisfied by adjacent atoms sharing one or more pairs of electrons as shown in Figure C-2. The sharing of a pair of electrons holds the atoms together in what it is called a **covalent bond**.

Covalent bonding, by satisfying the charge-orbital requirements, leads to discrete units of two or more atoms bonded together. Such units of two or more covalently bonded atoms are called **molecules**. A few simple but important examples are shown in Figure C-2.

A chemical formula is simply a shorthand description of the number of each kind of atom in a given molecule. The element is given by the chemical symbol and a subscript following the symbol gives the number present, no subscript being understood as one. A molecule with two or more different kinds of atoms may also be called a compound, but a molecule comprised of a single kind of atom, oxygen (O_2) for example, is still defined as an element.

Only a few elements, namely carbon, hydrogen, oxygen, nitrogen, phosphorus and sulfur, have configurations of electrons which lend readily to the formation of covalent bonds. But, carbon specifically, with its ability to form four covalent bonds, can produce long, straight or branched chains, or rings (Fig. C-3). Thus, an infinite array of molecules can be formed by using covalently bonded carbon atoms as a "backbone" and filling in the sides with atoms of hydrogen or other elements. Thus, it is covalent bonding among atoms of carbon and these few other elements that produces all natural organic molecules, those molecules that comprise all the tissues of living things, and also synthetic organic compounds such as plastics.

Ionic Bonding

Another way in which atoms may achieve a stable electron configuration is to gain additional electrons to complete the filling of an orbital, or lose electrons which are over a completed orbital. In general, the maximum number of electrons that can be gained or lost by an atom is three. Therefore, an element's atomic number determines whether one or more electrons will be lost or gained. If an atom's outer orbital is one to three electrons short of being filled, it will always tend to gain additional electrons. Conversely, if an atom has one to three electrons over its last complete orbital it will always tend to give them away.

Of course, gaining or losing electrons results in the number of electrons being greater or less than the number of protons, and the atom consequently having an electric charge. The charge will be one negative for each electron gained or one positive for each electron lost (Fig. C-4). A covalently bonded group of atoms may acquire an electric charge in the same way. An atom or group of atoms which has acquired an electric charge in this way is called an **ion**, positive or negative. Ions are designated by a superscript following the chemical symbol giving the number of positive or negative charges. Absence of superscripts indicates that the atom or molecule is neutral. Some important ions are listed in Table C-2.

Since unlike charges attract, positive and negative ions tend to join and pack together in dense clusters in such a way as to neutralize the overall electric charge. This joining together of ions through the attraction of their opposite charges is called **ionic bonding**. The result is the formation of hard, brittle, more or less crystalline substances of which all rocks and minerals are examples (Fig. C-5).

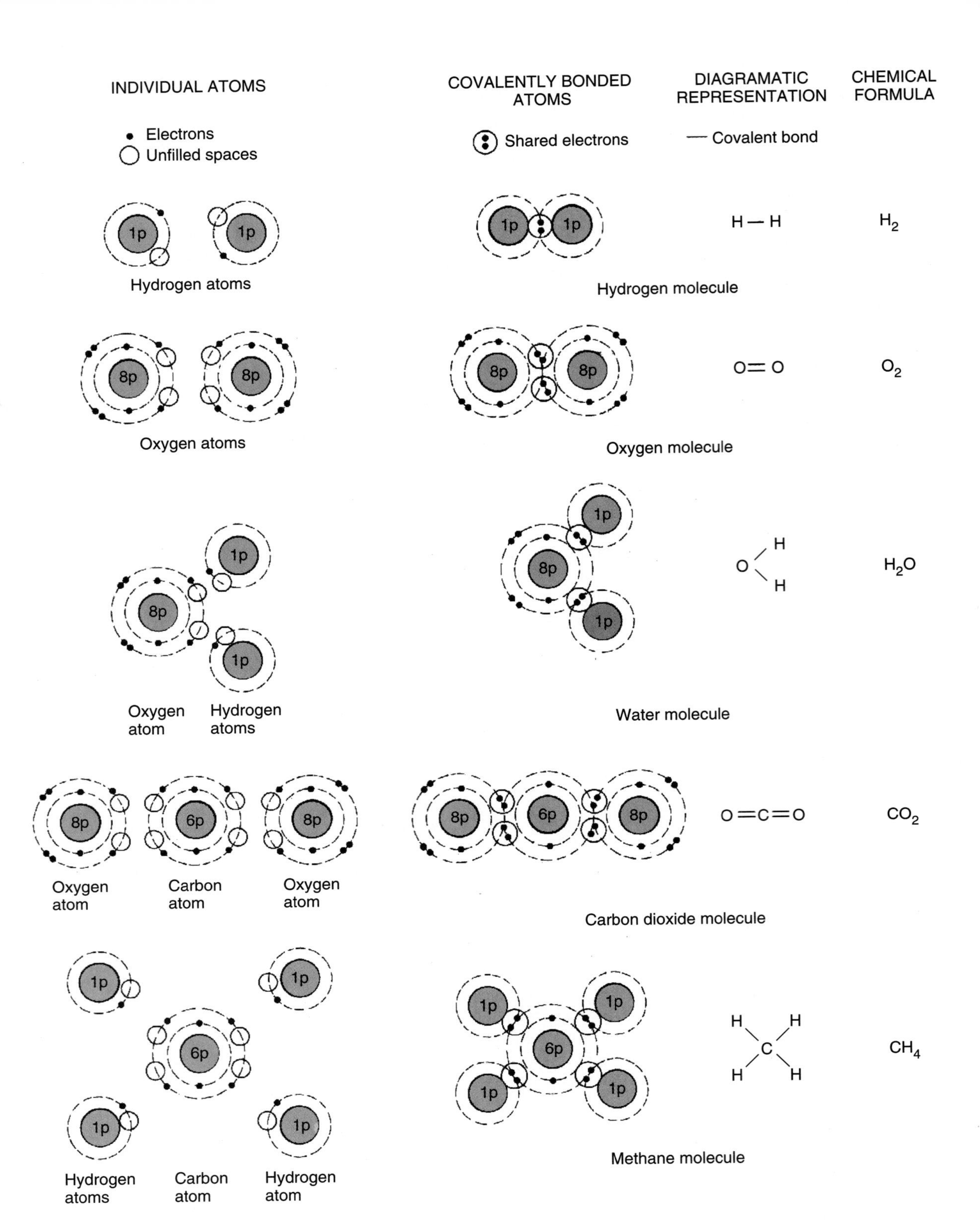
INDIVIDUAL ATOMS
COVALENTLY BONDED ATOMS
DIAGRAMATIC REPRESENTATION
CHEMICAL FORMULA
Electrons
Unfilled spaces
Shared electrons
— Covalent bond
1p
Hydrogen atoms
Hydrogen molecule
H — H
H2
8p
Oxygen atoms
Oxygen molecule
O = O
O2
Oxygen atom
Hydrogen atoms
Water molecule
H2O
6p
Oxygen atom
Carbon atom
Oxygen atom
Carbon dioxide molecule
O = C = O
CO2
Hydrogen atoms
Carbon atom
Hydrogen atom
Methane molecule
CH4

FIGURE C-2

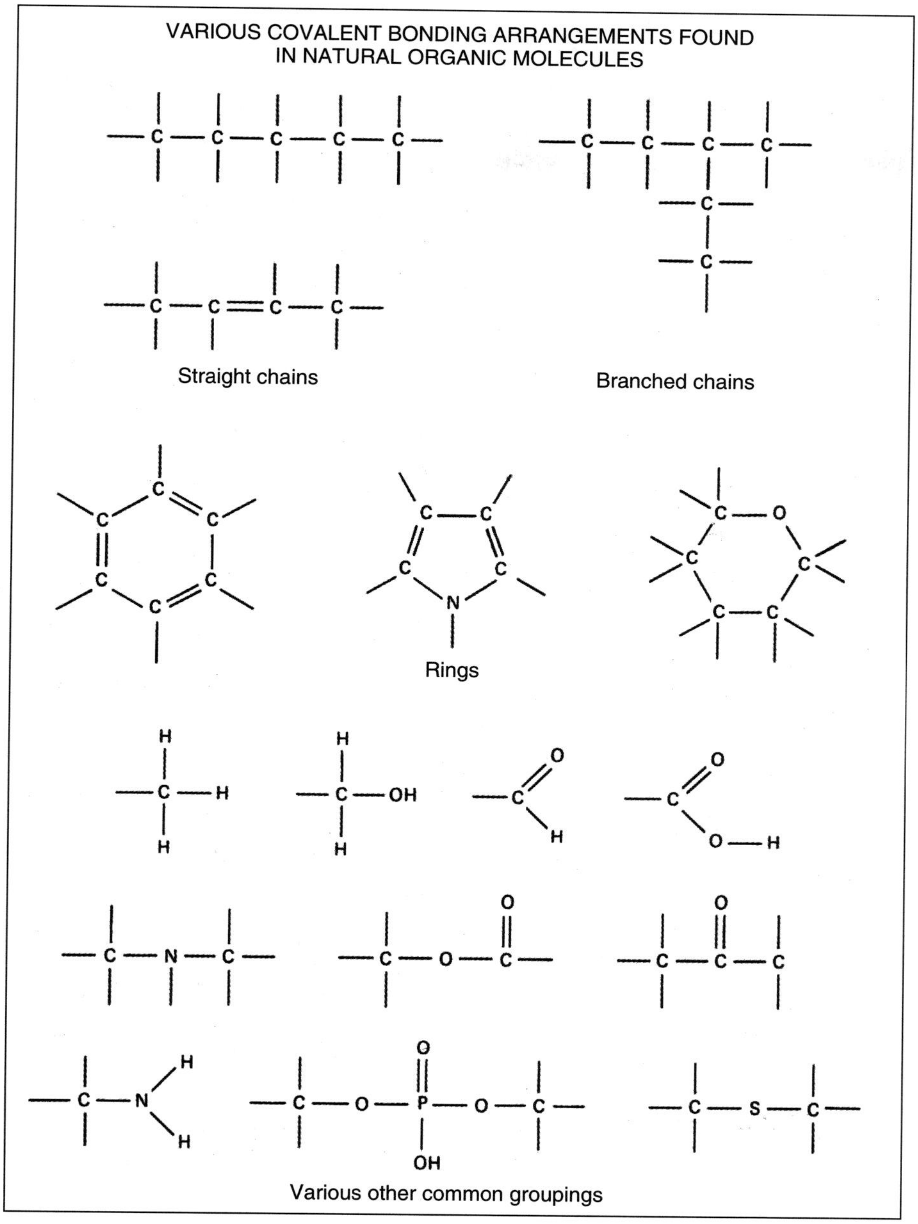

VARIOUS COVALENT BONDING ARRANGEMENTS FOUND IN NATURAL ORGANIC MOLECULES

Straight chains

Branched chains

Rings

Various other common groupings

FIGURE C-3
Covalent bonding and organic molecules. The ability of carbon and a few other elements to readily form covalent bonds leads to an infinite array of complex molecules, organic molecules, which constitute all living things. A few major kinds of groupings are shown here. Note that each element forms a characteristic number of bonds: carbon, 4; nitrogen, 3; oxygen, 2; hydrogen, 1; sulfur, 2; phosphorus, 5. Bonds (dashed lines) left "hanging" indicate attachments to other atoms or groups of atoms.

It is significant to note that whereas covalent bonding leads to discrete molecules, ionic bonding does not. Any number and combination of positive and negative ions may enter into an ionically bonded cluster to produce crystals of almost any size. The only restriction is that the overall charge of positive ions is balanced by that of negative ions. Thus, ionicly bonded substances are properly called compounds but not molecules. When chemical formulas are used to describe such compounds, they define the ratio of various elements involved, not specific molecules.

Chemical Reactions and Energy

While atoms themselves do not change, the bonds between atoms may be broken and reformed with different atoms producing different compounds and/or molecules. This is essentially what occurs in all chemical reactions. What determines whether a given chemical reaction will occur or not? We noted above that atoms form bonds because they achieve a greater stability by doing so. But some bonding arrangements may provide greater overall stability than others. Consequently

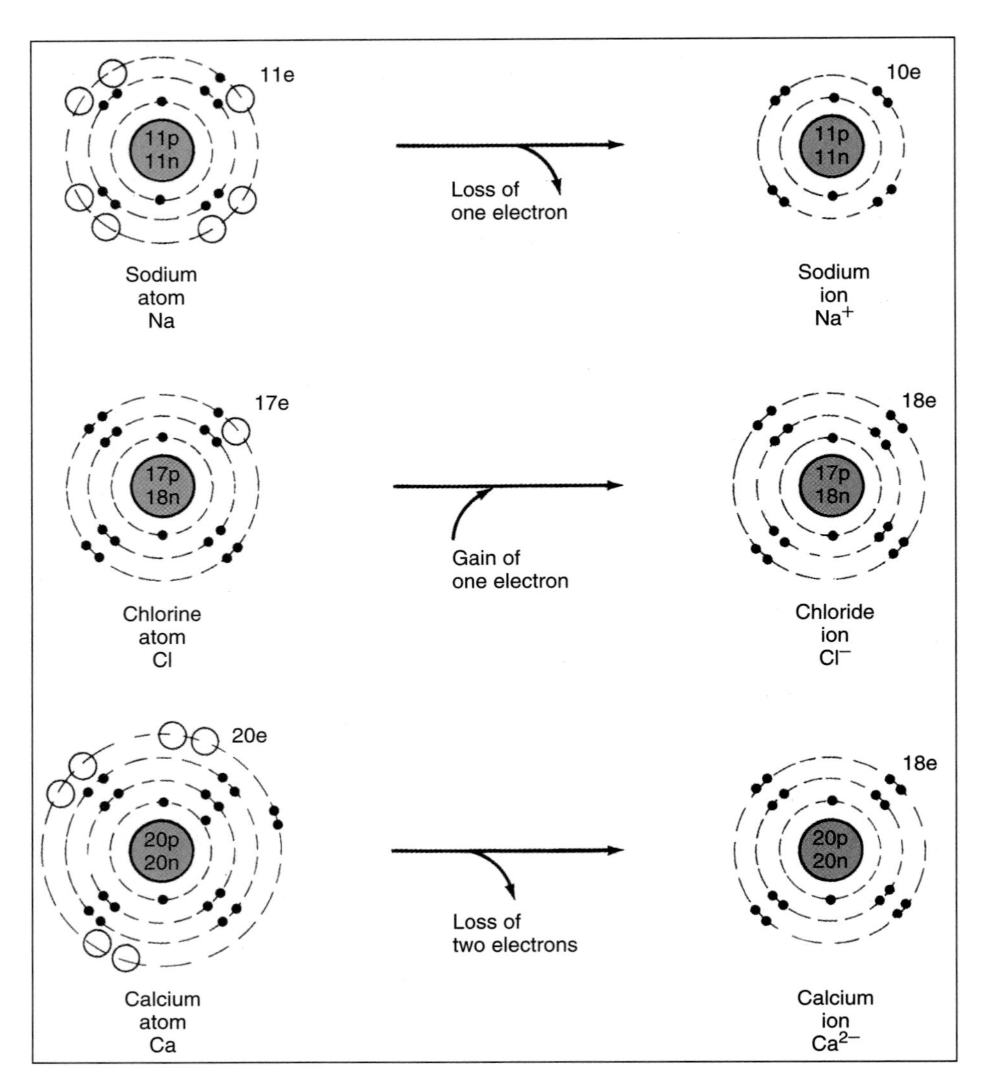

FIGURE C-4
Formation of ions. Many atoms will tend to gain or lose one or more electrons in order to achieve a state of complete (electron-filled) orbitals. In doing so they become positively or negatively charged ions as indicated.

TABLE C-2
Ions of Particular Importance to Biological Systems

Negative (−) Ions		*Positive (+) Ions*	
Phosphate	PO_4^{3-}	Potassium	K^+
Sulfate	SO_4^{2-}	Calcium	Ca^{2+}
Nitrate	NO_3^-	Magnesium	Mg^{2+}
Hydroxyl	OH^-	Iron	Fe^{2+}, Fe^{3+}
Chloride	Cl^-	Hydrogen	H^+
Bicarbonate	HCO_3^-	Ammonium	NH_4^+
Carbonate	CO_3^{2-}	Sodium	Na^+

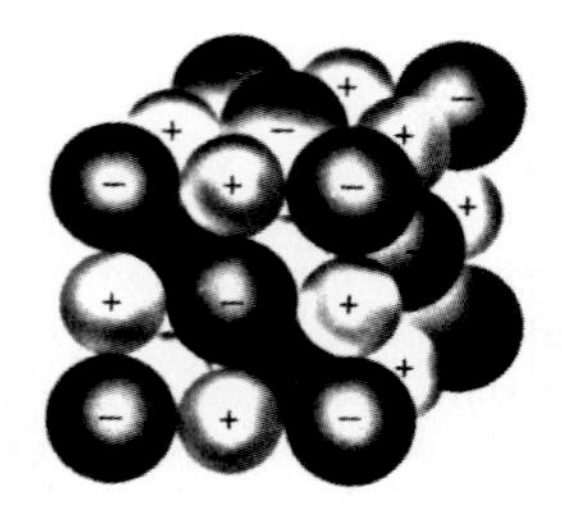

FIGURE C-5
Positive and negative ions bond together by their mutual attraction.

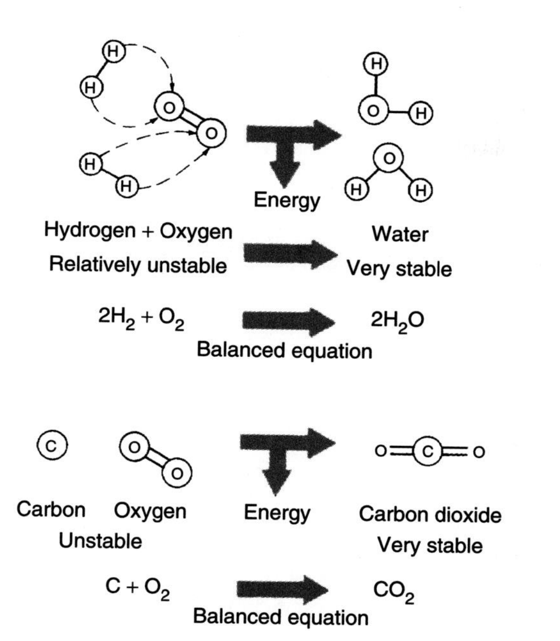

FIGURE C-6
Some bonding arrangements are more stable than others. Chemical reactions will go spontaneously toward more stable arrangements, releasing energy in the process. But, reactions may be driven in the opposite direction with suitable energy inputs.

substances with relatively unstable bonding arrangements will tend to react to form one or more different compounds which have more stable bonding arrangements. Common examples are the reaction between hydrogen and oxygen to produce water, and the reaction between carbon and oxygen to produce carbon dioxide (Fig. C-6).

Additionally, energy is always released in the process of gaining greater overall stability as indicated in Figure C-6. Thus, energy being released from a chemical reaction is synonymous with the atoms achieving more stable bonding arrangements. Thus, it may be said that chemical reactions always tend to go in a direction that releases energy as well as one which gives greater stability.

However, chemical reactions can be made to go in a reverse direction. With suitable energy inputs and under suitable conditions, stable bonding arrangements may be broken and less stable arrangements formed. As described in Chapter 2, this is the basis of photosynthesis occurring in green plants. Light energy is brought to bear on splitting the highly stable hydrogen-oxygen bonds of water and forming less stable carbon-hydrogen bonds thus creating high-energy organic compounds.

APPENDIX D

Science and the Scientific Method

Many environmental issues are embroiled in controversies that are so polarized that no middle ground seems possible. On the one hand are persons who argue from presumably sound facts and proven theories. On the other are persons who dismiss the theories and mistrust scientists and their motives, particularly when findings and conclusions of scientists conflict with corporate interests and traditional ways of doing things. Many people are understandably left confused. It is our objective in this appendix to give a brief overview of the nature of science and the scientific method so that you can evaluate for yourself the two sides of such controversies.

What is Science?

In its essence, science is simply a way of gaining knowledge; the way is called *the scientific method.* The term *science* further refers to all the knowledge gained through that method. What is the scientific method?

Basic Assumptions of the Scientific Method

First, the scientific method rests on certain basic assumptions that most of us accept without argument. The first assumption is that what we perceive with our basic five senses represents the existence of an objective reality; that is, our perceptions are not some kind of mirage or dream. The second assumption is that this "objective reality," which ranges from the atomic to the cosmic, functions according to certain basic principles and natural laws that remain consistent through time and space.

The third assumption, which follows directly from the second, is that every result has a cause, and every event, in turn, will cause other events. This is to say that if the universe functions according to certain basic principles and natural laws, each event can be seen as resulting from a prior event and in turn it will cause further reactions in a predictable manner according to the principles and laws. In other words, we assume that events do not occur without reason and that there is an explainable cause behind every happening. The fourth and final assumption is that through our powers of observation, manipulation, and reason, we can discover and understand the basic principles and natural laws by which the universe functions.

"The proof of the pudding is in the taste." Although assumptions, by definition, are premises that cannot be proved, the fact is that the assumptions underlying science have served us well and are borne out with everyday experience. For example, we suffer severe consequences if we do not accept our perception of fire as real. All our experience confirms that gravity is a predictable force acting throughout the universe and that it is not subject to unpredictable change. (Weightlessness in orbit is not a change in gravity; it is effectively a state of free fall.) The same can be said for any number of other phenomena that we observe. If our car fails to start, we know that there is a logical reason and call a mechanic to fix it. We do not want to believe, as an act of faith, that the wheels are securely bolted; this is something we want to *know*.

Thus, whether we are conscious of the fact or not, we all accept the basic assumptions of science in the conduct and understanding of our everyday lives. Scientists and scientific investigations only extend the boundaries of everyday experience, deepen our understanding of cause-effect relationships, and provide greater appreciation for the principles and natural laws that seem to determine the behavior of all things, from the outcome of a chemical reaction to the functioning of the biosphere.

Observation

In previous schooling you may have learned that the scientific method consists of the sequence: question, hypothesis, test (experiment), theory. This sequence is an oversimplification in that it both fails to describe what is involved in these steps and it omits what is really the most fundamental aspect of science.

The foundation of all science and scientific discovery is *observation* (seeing, hearing, smelling, tasting, feeling). Indeed, many branches of science such as natural history (where and how various plants and animals live, mate, reproduce, etc.), astronomy, anthropology, and evolutionary biology are based entirely on observation because experimentation is either inappropriate or impossible. For example, experimentation is obviously counterproductive if you want to discover what plants and animals do in nature, and it is simply impossible to conduct experiments on stars or past events.

Likewise, many of the data and conclusions of other sciences such as zoology, botany, geology, comparative anatomy, and taxonomy (classification of organisms) are based on nothing more (or less) than the careful observing and chronicling of things and events by persons taking the pains and time to do so. Even experimentation, as we will discuss shortly, is conducted to gain another window of observation. Therefore, in all science careful *observation* is the keystone. How can we be sure that observations are accurate?

Of course, not every reported observation is accurate for reasons ranging from honest misperceptions to calculated mischief. Therefore, an important aspect of science and trait of scientists is to be skeptical of any new report until it is confirmed or verified. Such confirmation usually involves other investigators repeating and checking out the observations of the first investigator and validating (or invalidating) the accuracy. As observations are confirmed by additional investigators they gain the status of scientific facts. In other words, *facts* are things or events that have been confirmed by more than one observer and remain open to be reconfirmed by additional people. Things or events that do not allow such confirmation, UFOs for example, remain in the realm of speculation from a scientific standpoint.

Various observations by themselves, like the pieces of a puzzle, may be put together into a larger picture or hypothesis and theory. To give a simple example, we observe that water evaporates and leads to moist air, that water from moist air condenses on a cool surface, and we observe clouds and precipitation. Putting these observations together logically, we derive a hypothesis of the water cycle. Water evaporates and then condenses as air is cooled; condensation forms clouds and precipitation follows. Water thus makes a cycle from the surface of Earth into the atmosphere and back to Earth. Note how this simple example incorporates the four assumptions described above—there is an objective reality; it operates according to principles; every result has a cause; and we can discover the principles. Also note how the example broadens our everyday experiences of the evaporation of water and falling of rain into an understanding of a cycle involving both.

Thus the essence of science and the scientific method may be seen as a process making observations and logically integrating those observations into a larger picture of how the world works. Another good example is taking the numerous observations of what plants and animals need and what they produce and integrating those observations into a picture of how they function together to create a sustainable ecosystem as described in Chapters 2 and 3. To be sure, many areas of science get more complex and difficult to comprehend. Still, the basic process—constructing a logically coherent picture of causes and effects from basic observations—is the same.

Where then does experimentation, that additional hallmark of science, fit in?

Experimentation

Experimentation is basically setting up situations to make more-systematic observations regarding causes and effects. For example, the number of chemical reactions one can readily observe in nature is limited. However, in the laboratory it is possible to purify elements or compounds, mix them together in desired combinations, and carefully observe and measure how they react (or fail to react). From the way chemicals react chemists have constructed a coherent cause-effect picture, the atomic theory, and they have determined the attributes of each element. Similarly, biologists put plants or animals into determined situations where they can carefully observe and measure their responses to particular conditions or treatments. (Again note that experimentation is necessarily limited to things that lend themselves to artificial manipulation. In many cases, such as stars, geological events, atmospheric events, and events that have occurred in the past, one has only observations to work with in the construction of the broader picture.)

Some experimentation may be more or less spontaneous and random—the childhood inclination to "do this and see what happens." And sometimes valuable information may be obtained in this way if careful accounts are kept so that one has an accurate record of causes and effects. However, to solve a particular problem—What is the cause of this event?—a more systematic line of experimentation is used. This is where the sequence question, hypothesis, test (experiment), and theory comes in. Let's consider the observation of the die off of submerged aquatic vegetation in Chesapeake Bay.

The question was, What has caused the die off? The next step is to make educated guesses as to the cause. Each educated guess as to the cause is a *hypothesis*. Each hypothesis is then tested through making further field observations or conducting experiments to determine if the hypothesis really accounts for the observed effect. In this particular case, the first hypothesis or presumed cause was industrial wastes. However, this hypothesis was disproved by the observations that the die off occurred in locations where no industrial wastes were present. A second hypothesis, herbicides being used on farmlands, was rejected as the primary cause by laboratory testing and field measurement that showed that herbicide levels were generally too low to cause damage.

However, the third hypothesis, reduction of light due to increased turbidity, stood up to tests of measuring photosynthesis at various light intensities. Additional observations concerning the causes for the increased turbidity filled in the picture and brought what started out as a hypothesis to the status of the proven cause.

Development and Establishment of Theories

We have already noted in the case of the water cycle how various specific observations (evaporation, precipitation, etc.) may fit together to give a logically coherent cause-effect picture of a broader phenomenon, i.e., the water cycle. At first, the broader picture is termed a *hypothesis* because it has not been observed precisely; it is really a tentative explanation as to how various observations are related. The word *hypothesis* is really a *hypo-theory*, or less than a theory. The distinction between a hypothesis and a theory is only in the body of information covered. A hypothesis generally is a tentative explanation for a specific event, whereas a theory is an understanding of how things happen and behave in the real world, as part of broad, general principles. A general theory represents the governing laws we see in nature.

In any case, theories may be tested and confirmed (or denied) and perfected so they are logically consistent with all observations. Further, through cause-effect reasoning, theories will generally suggest additional aspects: "If this is so, it should follow that . . ." If the outcome predicted by the theory is indeed found, it provides strong evidence for the correctness of the theory. Predictions require experiments and testing and more data gathering and observation. When theories reach a state of providing a logically consistent framework for all relevant observations (facts) and can be used to reliably predict outcomes, they represent a correct interpretation. For example, we have never seen atoms as such, but innumerable observations and experiments are coherently explainable by the concept that all gases, liquids, and solids consist of various combinations of only slightly more than a hundred kinds of atoms. Hence, we fully accept the *atomic theory of matter.*

Sometimes people will argue that because a theory is not proven fact, one theory is as good as another. That notion is *false*! One theory may have overwhelming supporting evidence, whereas much evidence may contradict another theory. In evaluating a theory, ask: What is the supporting evidence? Is there more, or less, evidence supporting an alternative theory?

Principles and Natural Laws

The second assumption underlying science—that the universe functions according to certain basic principles and natural laws that remain consistent through time and space—cannot be proved, but every observation and test bear it out. All our observations, whether direct or through experimentation, demonstrate that matter and energy do not behave randomly or even inconsistently but in precise and predictable ways without exception. We refer to these principles by which we can define and precisely predict the behavior of matter and energy as *natural laws*. Examples are the Law of Gravity, the Law of Conservation of Matter, and the various Laws of Thermodynamics. Our technological success in space exploration and many other fields is in no small part due to our recognition of these principles and precise calculations based on them. Conversely, trying to make something work in a manner contrary to a principle invariably results in failure.

The Role of Instruments in Science

Complex instrumentation is another hallmark of science and often gives it an aura of mystery. Yet, regardless of complexity, all scientific instruments perform one of three basic functions. First, they may extend our powers of observation—telescopes, microscopes, X-ray machines, and CAT scans, for example. Second, instruments are used to quantify observations; that is, they enable us to measure exact quantities. For example, we may feel cold, but a thermometer enables us to measure and quantify exactly how cold it is. Comparisons, communication, and verification of different observations and events would be impossible if it were not for such measurement and quantification. Third, instruments such as growth chambers and robots help us achieve conditions and perform manipulations required to make certain observations or to perform experiments.

All instruments used in science are themselves subjected to testing and verification to be sure that they are giving an accurate representation of what is there as opposed to creating illusions.

Scientific Controversies

With the scientific method capable of coming to objective conclusions, why does so much controversy still prevail? There are four main reasons. First, we are continually confronted by new observations—the hole in the ozone shield, for instance, or the dieback of certain forests. There is considerable time before all the hypotheses regarding the cause can be adequately tested. During this time, there may be honest disagreement as to which hypothesis is most likely. Such controversies are gradually settled by further observations and testing, but this process leads into the second reason for continuing controversy.

Phenomena such as the hole in the ozone shield or the loss of forests do not lend themselves to simple tests or experiments. Therefore, it is difficult and time-consuming to prove the causative role of one factor or to rule out the involvement of another. Gradually, different lines of evidence will support one hypothesis and exclude another and enable the issue to be resolved.

When is there enough evidence to say unequivocally that one hypothesis is right and another wrong? Deciding that there is enough evidence to be convincing involves subjective judgment. The biases or vested interests of a person may affect the amount of information that person requires to be convinced. The Tobacco Institute, a lobbying association for the tobacco industry, provides a prime example. It continually makes the point that the connection between smoking and ill health effects is not proved and that more studies are necessary. By harping on the lack of absolute proof and simply ignoring the overwhelming amount of evidence supporting the connection between smoking and ill health effects, the tobacco lobby has succeeded in keeping the issue controversial and thereby has delayed restrictions on smoking. Thus, the third reason for controversy is that there are many vested interests who wish to maintain and promote disagreement because they stand to profit by doing so. The need to keep a watch for this kind of behavior in evaluating the two sides of a controversy is self-evident. (Internal tobacco industry memoranda now show that they have perjured themselves in the public forum.)

The fourth reason for controversy, which may be seen as a generalization of the third reason, is that subjective value judgments, as well as subjective judgments of facts, may be involved. This is particularly true in environmental science because it deals with the human response to environmental issues. For example, there is virtually no controversy regarding nuclear power as long as it is considered at the purely scientific level of physics. However, when it comes to the environmental level of deciding whether to promote or shun the further use of nuclear energy to generate electrical power, controversy arises, and it stems from the fact that different people have different subjective feelings regarding the relative risks and benefits involved.

Evaluating Information

We have seen that the scientific method is a way of gaining understanding that starts with basic observations and develops them into a logically coherent picture of broader phenomena through logical cause-effect reasoning. The soundness in this approach is seen in the scientific and technological progress it has made possible. However, the human mind (and its inclination to gain advantage as opposed to the truth) is also extremely adept at starting with a desired conclusion and working in the opposite direction. This backward approach often involves:

- Simply propounding the conclusion without regard to supporting evidence.
- Putting forth "observations" that have not been and cannot be verified.
- Pointing to certain facts that support the conclusion while ignoring or even denying other facts that contradict the conclusion.
- Propounding as "causes" phenomena or events that have no logical connection to the effect.
- Committing any number of other logical fallacies.

The danger is that any action or policy based on such unfounded conclusions will almost certainly end in failure if not disaster. Too often technological optimism, fueled by profit motive, runs ahead of evidence and experience. Therefore it is important that we distinguish between such "wishful thinking" and conclusions that are soundly based. Whether we are scientists or not, we can use facets of the scientific method to judge the relative validity of alternative viewpoints and also to develop our own capacity for logical reasoning. The basic questions to ask are:

- What are the basic observations (facts) underlying the conclusion (theory)?
- Can the observations be satisfactorily verified?
- Do the conclusions follow logically from the observations?
- Does the conclusion account for all observations? (If the conclusion is logically inconsistent with any observations, it must be judged as questionable at best.)
- Does the conclusion or predicted outcome violate any natural laws? (If it does the conclusion is fundamentally flawed. Stupidity, ignorance, fraud, and deceit come before exceptions to natural laws.)

Science and Value Judgments

By its nature, science can deal only with things and phenomena that can be verifiably observed with the basic five senses and quantitatively measured in some way. This limitation obviously excludes many areas of human experience such as justice love, values, purpose, and spirituality. Some people think of science as "Godless" on account of its failure to address such aspects. It is important to emphasize that the inability of science to address such issues in no way denies their existence any more than describing one side of a coin denies the existence of the other. It only remains to us as total human beings to bring the two sides together. The majesty of the universe, mystery of subatomic space, and complexity of DNA and our genetic makeup should actually motivate the spiritual journey—we are surrounded by awesome handiworks!

By providing an understanding of how things work in terms of causes having predictable effects, science can predict the likely outcomes of different courses of action. Still the course we choose to take depends on the priority we give to different values.

For example, our scientific studies give us an understanding of how ecosystems function and how they interact to make the planetary biosphere. Our studies have also revealed certain principles that underlie sustainability (Table 4–1, p. 104). However, how we decide to act in light of such information depends on the way we weigh short-term versus long-term satisfaction and how we balance personal desires against benefits to society at large. People who adhere to the cornucopian world view basically put values of maintaining the status quo foremost. In supporting this position they maintain that exploitation of Earth's resources as in the past can go on without consequence. Looking at this view in light of understanding how the biosphere works, however, reveals that it is based heavily on wishful thinking, denial of many facts, and various logic gaps. Some religious groups have now spoken out that this view is "destroying the creation."

On the other hand, if our primary value really is a sustainable future, we should be willing to use our scientific understanding and move toward implementing the principles of sustainability in our human system.

Bibliography and Additional Reading

General References

Council on Environmental Quality. *Environmental Quality: 23rd Annual Report.* Washington, D.C.: Council on Environmental Quality, 1993. Annual report on the state of the environment in the United States, with policies, issues, and data.

Daly, Herman E., and John B. Cobb Jr. *For the Common Good: Redirecting the Economy toward Community, the Environment, and a Sustainable Future.* Boston: Beacon Press, 1989.

Garbarino, James. *Toward a Sustainable Society.* Chicago: The Noble Press, 1992.

Gore, Al. *Earth in the Balance: Ecology and the Human Spirit.* Boston: Houghton Mifflin, 1992.

Government of Canada. "The State of Canada's Environment." Ottawa, 1991.

Matthews, Jessica Tuchman, ed. *Preserving the Global Environment: The Challenge of Shared Leadership.* New York: W. W. Norton, 1991.

Myers, N., ed. *Gaia: An Atlas of Planet Management.* Garden City, NY: Anchor and Doubleday, 1993.

Parker, Jonathan, and Chris Hope. "The State of the Environment: A Survey of Reports from around the World." *Environment* 34 (Jan/Feb 1992), pp. 19–20, 39–45.

Sullivan, Thomas F. P., ed. *Environmental Law Handbook*, 14th ed. Rockville, MD: Government Institutes, 1995.

Switzer, Jacqueline Vaughn. *Environmental Politics: Domestic and Global Dimensions.* New York: St. Martin's Press, 1994.

Wild, Russell, ed. *The Earth Care Annual, 1992.* Emmaus, PA: Rodale Press, 1992.

World Bank. *World Development Report.* New York: Oxford University Press, 1994. An annual report of the World Bank.

World Bank Group. *Making Development Sustainable.* Wash-ington, D.C.: The World Bank, 1994.

World Resources Institute. *World Resources 1994–95.* New York: Oxford University Press, 1994. An annual compilation of data and reports covering a wide range of environmental issues.

WorldWatch Institute. *State of the World: A Worldwatch Institute Report on Progress toward a Sustainable Society.* Washington, D.C.: Worldwatch Institute. (A collection of articles from the institute covering an array of environmental issues, published annually since 1984.)

WorldWatch Institute. *Vital Signs, The Trends That Are Shaping Our Future.* Washington, D.C.: WorldWatch Institute (A presentation of data showing trends that will affect our future, published annually since 1992.)

Chapter 1. Introduction: Environmental Science and Sustainability

Alpert, J. et al. "The Sustainable Biosphere Project of SCOPE." *Ambio* 24(2) (Mar 1995), pp. 133–134.

Brown, Lester R. "Launching the Environmental Revolution." Pages 174–190 in *State of the World 1992.* Washington, D.C.: WorldWatch Institute, 1992.

Cairns, John, Jr. "Ecosocietal Restoration: Reestablishing Humanity's Relationship with Natural Systems." *Environment* 37 (June 1995), pp. 4–9.

Chaney, Ed, et al. "People Who Make a Difference." *National Wildlife* 30 (Oct/Nov 1992), pp. 34–41.

Conniff, Richard. "Easter Island Unveiled." *National Geographic* 183 (Mar 1993), pp. 54–79.

Diamond, Jared. "Easter's End." *Discover* 16 (Aug 1995), pp. 62–69.

Diamond, Jared. "Playing Dice with Megadeath." *Discover* 11 (Apr 1990), pp. 54–59.

Drew, Lisa. "25 Messages from Wildlife." *International Wildlife* 33 (Apr/May 1995), pp. 8–19.

Dunlap, Riley E., et al. "Of Global Concern: Results of the Health of the Planet Survey." *Environment* 35 (Nov 1993), pp. 6–15, 33–39.

Evans, Diane L., et al. "Earth from Sky." *Scientific American* 271 (Dec 1994), pp. 70–75.

Glick, Daniel. "Having Owls and Jobs Too." *National Wildlife* 33 (Aug/Sept 1995), pp. 8–13.

Glick, Daniel. "Windows on the World." *National Wildlife* 32 (Feb/Mar 1994), pp. 4–13.

Greence, George. "Caring for the Earth." *Environment* 36 (Sept 1994), pp. 25–28.

Halpern, Sue. "Losing Ground." *Audubon* 94 (July/Aug 1992), pp. 70–79.

Helvarg, David. "The War against the Green: The Wise Use Movement, the New Right and Anti-Environmental Violence." Sierra Club Books, 1994.

Holloway, Marguerite. "Nurturing Nature." *Scientific American* 270 (Apr 1994), pp. 98–108.

Homer-Dixon, Thomas F., et al. "Environmental Change and Violent Conflict." *Scientific American* 268 (Feb 1993), pp. 38–45.

Kates, Robert W. "Sustaining Life on the Earth." *Scientific American* 271 (Oct 1994), pp. 114–125.

Kenny-Gilday, Cindy, et al. "The Question of Sustainable Development." *Nature Conservancy* 45 (Jan/Feb 1995), pp. 10–15.

Loehr, Raymond. "Looking Ahead to the Planet's Future." *EPA Journal* 21 (Winter 1995), pp. 22–25.

Meadows, Donella, et al. *Beyond the Limits.* Post Mills, VT: Chelsea Green Publishing, 1992.

Mittlo, Cindy, and Alisa Gravitz. "Green Businesses Hold a New Vision." *Co-op America's National Green Pages* (Premiere Issue 1993), pp. 6, 7, 73.

Munda, G., et al. "Qualitative Multicriteria Evaluation for Environmental Management." *Ecological Economics* 10 (July 1994), pp. 97–112.

Myers, Norman. "What's Ailing the Globe?" *International Wildlife* 24 (Mar/Apr 1994), pp. 34–41. Top ecologist Norman Myers offers his list of Earth's top 10 environmental problems.

National Commission on the Environment. *Choosing a Sustainable Future.* Washington, D.C.: Island Press, 1993.

Oelschlaeger, Max. *Caring for Creation.* New Haven, CT: Yale University Press, 1994.

Ojima D. S., et al. Special Section Articles: "The Global Impact of Land-use Change" and following articles. *BioScience* 44 (May 1994), pp. 300–356.

Oppenheimer, Michael. "Context, Connection, and Opportunity in

Environmental Problem Solving." *Environment* 37 (June 1995), pp. 10–15, 34–38.
Orians, Gordon H. "Ecological Concepts of Sustainability." *Environment* 32 (Nov 1990), pp. 21–24.
Rees, William E. "The Ecology of Sustainable Development." *The Ecologist* 20 (Jan/Feb 1990), pp. 18–23.
Ruckelshaus, William D. "Toward a Sustainable World." *Scientific American* (Sept 1989), pp. 166–175.
Rudloe, Anne, and Jack Rudloe. "Sea Turtles: In a Race for Survival." *National Geographic* 185 (Feb 1994), pp. 94–121.
Sawhill, John C. "Building a Sustainable Future." *Nature Conservancy* (Jan/Feb 1993), pp. 5–8.
Schmidheiny, Stephan. *Changing Course*. Cambridge, MA: MIT Press, 1992.
Schneider, Stephen H. "Can We Repair the Air?" *Discover* 13 (Sept 1992), pp. 28–32.
Sitarz, Daniel, ed. *Agenda 21: The Earth Summit Strategy to Save Our Planet*. Boulder, CO: Earth Press, 1993.
Smith, Courtland L. "Assessing the Limits to Growth." *BioScience* 45 (July/Aug 1995), pp. 478–483.
"Special Section: Population, Consumption & Environment." *The Amicus Journal* 15 (Winter 1994).
"Sustainability." *EPA Journal* 18 (Sept/Oct 1992).
"The 26th Environmental Quality Index." *National Wildlife* 32 (Feb/Mar 1994), pp. 38–44.
Van Dyk, Jere. "The Amazon." *National Geographic* 187 (Feb 1995), pp. 2–39.
Weber, Peter. "Failing Coasts." *WorldWatch* 7 (Mar/Apr 1994), pp. 20–29.
World Commission on Environment and Development. *Our Common Future*. New York: Oxford University Press, 1987.
Wright, Richard T. *Biology through the Eyes of Faith*. San Francisco: Harper/Collins, 1989.
Young, John E. "Mining the Earth." *WorldWatch Paper 109*. Washington, D.C.: WorldWatch Institute, 1992.
Young, Louise B. "Easter Island: Scary Parable." *World Monitor* (Aug 1991), pp. 40–45.

Chapter 2. Ecosystems: Units of Sustainability

Acharya, Anjali. "The Fate of the Boreal Forests." *WorldWatch* 8 (May/June 1995), pp. 20–29.
Ahmadjian, Vernon. "Lichens Are More Important Than You Think." *BioScience* 45 (Mar 1995), p. 124.
Breining, Greg. "Rising from the Bogs." *Nature Conservancy* 42 (July/Aug 1992), pp. 24–29.
Bruemmer, Fred. "Colors of My Arctic." *International Wildlife*, 22 (Mar/Apr 1992), pp. 4–11.
Carey, John. "Cutting through the Mystery of Fog." *National Wildlife* 32 (Feb/Mar 1994), pp. 52–59. The moody mist is crucial to the survival of some species, deadly to others.
Chadwick, Douglas. "Ndkoki–the Last Place on Earth." *National Geographic* 188 (July 1995), pp. 22–45.
Chadwick, Douglas H., and Jim Brandenburg. "The American Prairie." *National Geographic* 184 (Oct 1993), pp. 90–119.
Cloud, Preston. "The Biosphere." *Scientific American* 249 (Sept 1983), pp. 176–189.
Degre, Alain. "Heart of the Kalahari." *International Wildlife* 22 (Nov/Dec 1992), pp. 44–51.
Diamond, Jared. "The Great Leap Forward." *Discover* (May 1989), pp. 50–60.
Dybas, Cheryl Lyn. "Ocean Forests." *Sea Frontiers* 41.2(1995), pp. 6–8.
Elder, Jane, et al. "21 Ecoregions." *Sierra* 79 (Mar/Apr 1994).
Glick, Daniel. "Windows on the World." *National Wildlife* 32 (Feb/Mar 1994), pp. 4–13.
Johanson, D. C., and T. D. White. "Systematic Assessment of Early African Hominids." *Science* 203 (Jan 1979), pp. 321–328.
Lemonick, Michael D. "One Less Missing Link." *Time* (Oct. 3, 1994), pp. 68–69.
Ralston, Jeannie. "Our National Shores." *Audubon* 96 (May/June 1994), pp. 38–59.
Risser, Paul G. "The Status of the Science Examining Ecotones." *BioScience* 45 (May 1995), pp. 318–325.
Roach, Mary. "Aliens in the Treetops." *International Wildlife* 24 (Nov/Dec 1994), pp. 4–11.
Solbrig, Otto T., and Michael D. Young. "Toward a Sustainable and Equitable Future for Savannas." *Environment* 34 (Apr 1992), pp. 6–15.
Stevens, Jane E. "The Antarctic Pack-ice Ecosystem." *BioScience* 45 (Mar 1995), pp. 128–133.
Terborgh, John. "Why American Songbirds Are Vanishing." *Scientific American* (May 1992), pp. 98–104.
Wiewandt, Thomas A. "Desert Flowers—Now You see 'em, Now You Don't." *Smithsonian* 25 (Mar 1995), pp. 78–93.

Chapter 3. Ecosystems: How They Work

Ballafante, Ginia. "New Wave Fish Farming." *Garbage* 3 (Jan/Feb 1991), pp. 16–17.
Bazzaz, Fakhri, and Eric D. Fajer. "Plant Life in a CO_2-Rich World." *Scientific American* 266 (Jan 1992), pp. 68–77.
Berner, Robert A., and Antonio Lasaga. "Modeling the Geochemical Carbon Cycle." *Scientific American* (Mar 1989), pp. 74–81.
Brill, W. F. "Biological Nitrogen Fixation." *Scientific American* 236 (Mar 1977), pp. 19–21.
Doubilet, David. "The Desert Sea." *National Geographic* (Nov 1993), pp. 60–87.
Durning, Alan B. "Junk Food, Food Junk." *WorldWatch* (Sept/Oct 1991), pp. 7–9.
Frieden, Earl. "The Chemical Elements of Life." *Scientific American* (July 1972), pp. 52–60.
Galston, Arthur W. "Photosynthesis as a Basis for Life Support on Earth and in Space." *BioScience* 42 (Aug 1992), pp. 490–493.
Holmes, Hannah. "Eating Low on the Food Chain." *Garbage* (Jan/Feb 1992), pp. 32–39.
Kirshner, Robert P. "The Earth's Elements." *Scientific American* 271 (Oct 1994), pp. 58–65.
Kreeyer, Karen Young, and John Gamble. "Plankton Pyramid." *Sea Frontier* 41.2 (1995), pp. 32–35.
Lemonick, Michael. "The Last Frontier." *Time* (Aug 14, 1994), pp. 52–60.
Nelson, Mark, et al. "Using a Closed Ecological System to Study Earth's Biosphere." *BioScience* 43 (Apr 1993), pp. 225–236.
Rifkin, Jeremy. *Beyond Beef, The Rise and Fall of the Cattle Culture*. New York: Dutton, Penguin Group, 1992.
Rounick, J. S., and J. J. Winterbourn. "Stable Carbon Isotopes and Carbon Flow in Ecosystems." *BioScience* 36 (Mar 1986), pp. 171–177.
Scientific American 223 (Sept 1970). Issue devoted to articles describing nutrient cycles and energy flow in ecosystems.
Scrimshaw, Nevin S. "Iron Deficiency." *Scientific American* 265 (Oct 1991), pp. 46–52.
Sieloff, D. A. *Biosphere 2: A World in Our Hands*. Oracle AZ: Biosphere Press, 1995. The classroom biosphere and other lessons about how Earth's biosphere works.
Sims, Grant. "Shedding Light on Life." *National Wildlife* 32 (Dec/Jan 1994), pp. 4–11.
Sunquist, Fiona. "Blessed Are the Fruit-Eaters." *International Wildlife* 22 (May/June 1992), pp. 4–11.
Zimmer, Carl. "The Case of the Missing Carbon." *Discover* (Dec 1993), pp. 38–39.
Zimmer, Carl. "More Productive, Less Diverse." *Discover* (Sept 1994), pp. 24–25.

Chapter 4. Ecosystems In and Out of Balance

Alcover, Josep Antoni, and Miquel McMinn. "Predators of Vertebrates on Islands." *BioScience* 44 (Jan 1994), pp. 12–18.
Allen, William H. "Reintroduction of Endangered Plants." *BioScience*

44 (Feb 1994), pp. 65–69.
Arrow, Kenneth, et al. "Economic Growth, Carrying Capacity, and the Environment." *Science* 268 (Apr 1995), pp. 520–521.
Askins, Robert A. "Hostile Landscapes and the Decline of Migratory Songbirds." *Science* 267 (Mar 1995), pp. 1956–1958.
Baskin, Yvonne. "Forests in the Gas." *Discover* (Oct 1994), pp. 116–121.
Baskin, Yvonne. "There's a New Wildlife Policy in Kenya: Use It or Lose It." *Science* 265 (Aug 1994), pp. 733–734.
Begley, Sharon. "Why Trees Need Birds." *National Wildlife* 33 (Aug/Sept 1995), pp. 42–45.
Bright, Chris. "Biological Invasions: The Spread of the World's Most Aggressive Species." *World Watch* 8 (Jul/Aug 1995), pp. 10–19.
Burke, William K. "Return of the Native: The Art and Science of Environmental Restoration." *E Magazine* (July/Aug 1992), pp. 38–44.
Carroll, Dan. "Subduing Purple Loosetrife." *The Conservationist* (Aug 1994), pp. 6–9.
Castello, John D., et al. "Pathogens, Patterns, and Processes in Forest Ecosystems." *BioScience* 45 (Jan 1995), pp. 16–24.
Cheater, Mark. "Alien Invasion." *Nature Conservancy* 42 (Sept/Oct 1992), pp. 24–29. Exotic weeds are more than just a nuisance—they pose one of the major threats to preserves, parks, and refuges.
Cherfas, Jeremy. "Disappearing Mushrooms: Another Mass Extinction?" *Science* 254 (Dec 6, 1991), p. 1458.
Conniff, Richard. "Fuzzy-Wuzzy Thinking about Animal Rights." *Audubon* (Nov 1990), pp. 120–135.
Conover, Adele. "Bamboo and Katydid: Their Surprising 'Inside' Story." *Smithsonian* 25 (Oct 1994), pp. 120–131.
Drew, Lisa. "Whose Home Is the Range Anyway?" *National Wildlife* 32 (Dec/Jan 1994), pp. 12–19. The latest research is confirming that in the West's fragile public lands, cattle are bad news for wildlife.
Falkowski, Paul G., et al. "Population Control in Symbiotic Corals." *BioScience* 43 (Oct 1993), pp. 606–611.
Gillis, Anna Marie. "Fiddling with Foaling." *BioScience* 44 (July/Aug 1994), pp. 443–450. One team's quest for a contraceptive to manage wild horse populations humanely seems near realization.
Hansen, A. J., et al. "Conserving Biodiversity in Managed Forests." *BioScience* 41 (June 1991), p. 382.
Harrison, George H. "Is There a Killer in Your House?" *National Wildlife* 30 (Oct/Nov 1992), pp. 10–13.
Hauserman, Julie. "Landscape Architects Help Design a Statewide Greenway Networld." *Landscape Architecture* (July 1995), pp. 56–61.
Hedgpeth, John. "Foreign Invaders." *Science* 261 (July 1993), pp. 34–35.
Holloway, Marguerite. "Nurturing Nature." *Scientific American* 270 (Apr 1994), pp. 98–108. Florida's Everglades are serving as testing ground—and battlefield—for an epic attempt to restore an environment damaged by human activity.
Homer-Dixon, Thomas F., et al. "Environmental Change and Violent Conflict." *Scientific American* 268 (Feb 1993), pp. 38–45.
Klein, David R. "The Introduction, Increase, and Crash of Reindeer on St. Matthew Island." *Journal of Wildlife Management* 32 (Apr 1968), pp. 350–367.
Kunzig, Robert. "These Woods Are Made for Burning." *Discover* 10 (Mar 1989), pp. 87–93.
Kuznik, Frank. "What Difference Does the Dogwood Make?" *National Wildlife* 31 (Apr/May 1993), pp. 46–51.
Line, Les. "Curse of the Cowbird." *National Wildlife* 32 (Dec/Jan 1994), pp. 40–45.
Line, Les. "Silence of the Songbirds." *National Geographic* 183 (June 1993), pp. 68–91.
Ludyanskiy, Michael L., et al. "Impact of the Zebra Mussel, a Bivalve Invader." *BioScience* 43 (Sept 1993), pp. 533–544.
Mairson, Alan. "The Everglades: Dying for Help." *National Geographic* 185 (Apr 1994), pp. 2–35.
Malecki, Richard A., et al. "Biological Control of Purple Loose-strife." *BioScience* 43 (Nov 1993), pp. 680–686.
McLaren, B. E., and Peterson, R. O. "Wolves, Moose, and Tree Rings on Isle Royale." *Science* 266 (Dec 1994), pp. 1555–1558.
McLean, Herbert. "Fighting Fire with Fire." *American Forests* (Jul/Aug 95), pp. 13–18, 56–63.
Mills, Edward L., et al. "Exotic Species and the Integrity of the Great Lakes." *BioScience* 44 (Nov 1994), pp. 666–676.
Newhouse, Joseph R. "Chestnut Blight." *Scientific American* (July 1990), pp. 106–111.
Norris, Ruth. "Can Ecotourism Save Natural Areas?" *National Parks* 66 (Jan/Feb 1992), pp. 30–34.
Poiani, Karen A., and W. Carter Johnson. "Global Warming and Prairie Wetlands." *BioScience* 41 (Oct 1991), pp. 611–618.
Sawhill, John C., et al. "Biodiveristy." *Nature Conservancy* 44 (Jan/Feb 1994).
Shafer, Craig L. "Values and Shortcomings of Small Reserves." *BioScience* 45 (Feb 1995), pp. 80–88.
Si Silvestro, Roger. "Rescue from the Twilight Zone." *Inter-national Wildlife* 24 (Jan/Feb 1994), pp. 4–13. Scientists from Morocco to the Philippines are devising last-ditch strategies to save earth's disappearing wildlife.
Smith, Val H. "Resource Competition between Host and Pathogen." *BioScience* 43 (Jan 1993), pp. 21–31.
Travis, John. "Invader Threatons Black, Azov Seas." *Science* 262 (Nov 1993), pp. 1366–1367.
Turbak, Gary. "A Pleasant Bird Is the Pheasant." *National Wildlife* 30 (Oct/Nov 1992), pp. 14–21.
Weingrod, Carmi. "On the Horns of a Dilemma." *National Parks* 68 (May/June 1994), pp. 28–32. Olympic National Park managers must make some hard choices about mountain goats, a popular but destructive non-native species.
Williams, Ted. "The Imperiled Northern Forest." *Audubon* 96 (May/June 1994), pp. 26–33.
Wood, Daniel B. "Public, Private Sectors Join to Save Hawaii's Fragile Ecosystems." *The Christian Science Monitor*, Dec 20, 1994, pp. 10–11.

Chapter 5. Ecosystems: Adapting to Change—or Not

Allegre, Claude J., and Stephen H. Schneider. "The Evolution of the Earth." *Scientific American* 271 (Oct 1994), pp. 66–75.
Amabile-Cuevas, Carlos F., et al. "Antibiotic Resistance." *American Scientist* 83 (July/Aug 1995) pp. 320–329.
Barlow, Connie, and Tyler Volk. "Thinking of Biology: Gaia and Evolutionary Biology." *BioScience* 42 (Oct 1992), pp. 686–693.
Baskin, Yvonne. "Ecologists Dare to Ask: How Much Does Diversity Matter?" *Science* 264 (Apr 1994), pp. 202–203.
Begley, Sharon. "A Question of Breeding." *National Wildlife* 29 (Feb/Mar 1991), pp. 12–17.
Benton, J. "Diversification and Extinction in the History of Life." *Science* 268 (Apr 1995), pp. 52–58.
Bloch, Nini. "Cost of Living." *Earthwatch* (May/June 1994), pp. 16–17. A shrinking gene pool has farmers resurrecting traditional livestock breeds.
Blumenschine, Robert J., and John A. Cavallo. "Scavenging and Human Evolution." *Scientific American* 267 (Oct 1992), pp. 90–97.
Broecker, Wallace S., and George Denton. "What Drives Glacial Cycles?" *Scientific American* 262 (Jan 1990), p. 48.
Chadwich, Douglas H. "The Endangered Species Act." *National Geographic* 187 (Mar 1995), pp. 2–41.
Coppens, Yves. "East Side Story: The Origin of Humankind." *Scientific American* 270 (May 1994), pp. 88–95.
Dalziel, Ian W. D. "Earth before Pangea." *Scientific American* 272 (Jan 1995), pp. 58–63.
de Duve, Christian. "The Beginning of Life on Earth." *American Scientist* 83 (Sept 95) pp. 428–437.
Gasser, Charles S., and Robert T. Fraley. "Transgenic Crops." *Scien-*

tific American (June 1992), pp. 62–69.
Gittleman, John L. "Are the Pandas Successful Specialists or Evolutionary Failures?" *BioScience* 44 (July/Aug 1994), pp. 456–464.
Gore, Rick, and O. Louis Mazzatenta. "Explosion of Life: The Cambrian Period." *National Geographic* 184 (Oct 1993), pp. 120–136.
Gould, Stephen Jay. "The Evolution of Life on the Earth." *Scientific American* 271 (Oct 1994), pp. 84–91.
Grant, Peter R. "Natural Selection and Darwin's Finches." *Scientific American* 265 (Oct 1991), pp. 82–87.
The Harvard Working Group on New and Resurgent Diseases. "New and Resurgent Diseases: The Failure of Attempted Eradication." *The Ecologist* 25 (Jan/Feb 1995), pp. 21–26.
Hill, Geoffrey E. "Ornamental Traits as Indicators of Environmental Health." *BioScience* 45 (Jan 1995), pp. 25–31.
Horgan, John. "Eugenics Revisited." *Scientific American* 268 (June 1993), pp. 122–131.
Kerr, Richard A. "Evolution Made Visible." *Science* 267 (Jan 1995), pp. 30–34.
Kerr, Richard A. "Huge Impact Tied to Mass Extinction." *Science* 257 (Aug 1992), pp. 878–880.
Lemonick, Michael D. "The Killers All Around: New Viruses and Drug-resistant Bacteria Are Reversing Human Victories over Infectious Disease." *Time* (Sept 12, 1994), pp. 60–69.
Levinton, Jeffrey S. "The Big Bang of Animal Evolution." *Scientific American* 267 (Nov 1992), pp. 84–93.
Likens, Gene. "Human-Accelerated Environmental Change." *BioScience* 41 (Mar 1991), p. 130.
Lovelock, James, and Lynn Margulis. "Rethinking Life on Earth." *Earthwatch* 11 (Sept/Oct 1992), pp. 20–29.
May, Robert M. "How Many Species Inhabit the Earth?" *Scientific American* 267 (Oct 1992), pp. 42–49.
McNeely, Jeffrey A., et al. "Strategies for Conserving Biodiversity." *Environment* 32 (Apr 1990), pp. 16–20, 36–40.
Mellon, Margaret. "Altered Traits." *Nucleus* 15 (Fall 1993), pp. 4–6, 12.
Novacek, Michael, et al. "Dinosaurs." *Natural History* (June 1995). Special section.
Powledge, Fred. "The Food Supply's Safety Net." *BioScience* 45 (Apr 1995), pp. 235–243.
Powledge, Tabitha M. "The Genetic Fabric of Human Behavior." *BioScience* 43 (July 1993), pp. 362–366.
Rhoades, Robert E., and L. Johnson. "The World's Food Supply at Risk." *National Geographic* (Apr 1991), pp. 74–105.
Ryan, John, C. "Life Support: Conserving Biological Diversity." *WorldWatch Paper 108*. Washington, D.C.: WorldWatch Institute, 1992.
Savage-Rumbaugh, Sue, and Roger Lewin. "Ape at the Brink." *Discover* 15 (Sept 1994), pp. 90–99.
Sereno, Paul C. "Dinosaurs and Drifting Continents." *Natural History* (Jan 1995), pp. 40–47.
Shell, Ellen Ruppel. "Waves of Creation." *Discover* 14 (May 1993), pp. 54–61. Abrupt climate shifts force all creatures to change their ways—or die.
Shreeve, James. "*Erectus* Rising." *Discover* 15 (Sept 1994), pp. 80–89.
Trefil, James. "Life on Earth: Was It Inevitable?" *Smithsonian* 25 (Feb 1995), pp. 32–41.
Weinberg, Steven. "Life in the Universe." *Scientific American* 271 (Oct 1994), pp. 44–51.
Wills, Christopher. "Escape from Stupidworld." *Discover* 14 (Aug 1993), pp. 54–78. Our brainpower might not just set us apart—it might also set us up for extinction.
Wills, Christopher. "Is Evolution Over for Us?" *Discover* 13 (Aug 1992), pp. 22–24.
Wilson, Allan C., and Rebecca L. Cann. "The Recent African Genesis of Humans." *Scientific American* 266 (Apr 1992), pp. 66–73.
Wilson, Edward O. "The Diversity of Life." *Discover* 13 (Sept 1992), pp. 45–68.
Wilson, Edward O., et al., eds. *Biodiversity*. Washington, D.C.: National Academy Press, 1988.
Zimmer, Carl. "Coming onto the Land." *Discover* 16 (June 1995), pp. 118–127.

Chapter 6. The Global Human Population Explosion: Causes and Consequences

Bilsborrow, Richard, E., and Pamela F. Delargy. "Land Use, Migration, and Natural Resource Deterioration: The Experience of Guatemala and the Sudan." Pages 125–147 in *Resources, Environment, and Population*. New York: Oxford University Press, 1991.
Brown, Lester R. "Feeding China." *WorldWatch* 7 (Sept/Oct 1994), pp. 10–22.
Brown, Lester, R., and Jodi L. Jacobson. "Our Demographically Divided World." *WorldWatch Paper 74*. Washington, D.C.: WorldWatch Institute, 1986.
Caputo, Robert. "Tragedy Stalks the Horn of Africa." *National Geographic* 184 (Aug 1993), pp. 88–212.
Cobb, Charles E., Jr. "Bangladesh: When the Water Comes." *National Geographic* 183 (June 1993), pp. 118–134.
Daily, Gretchen C., and Paul R. Ehrlich. "Population, Sustainability, and Earth's Carrying Capacity." *BioScience* 42 (Nov 1992), pp. 761–71.
Daily, Gretchen, et al. "Optimum Human Population Size." *Population and Environment* 15(6) (July 1994), pp. 469–475.
Dasgupta, Partha S. "Population, Poverty and the Local Environment." *Scientific American* 272 (Feb 1995), pp. 40–45.
Davis, Kingsley. "Population and Resources: Fact and Interpretation." Pages 1–21 in *Resources, Environment and Population*. New York: Oxford University Press, 1991.
Denniston, Derek. "High Priorities: Conserving Mountain Ecosystems and Cultures." *WorldWatch Paper 123*. Washington, D.C.: WorldWatch Institute, 1995.
Dilworth, D. "Two Perspectives on Sustainable Development." *Population and Environment* 15(6) (July 1994), pp. 441–467.
Durning, Alan. *How Much Is Enough?* W. W. Norton, New York, 1992.
Durning, Alan B., and Holly B. Brough. "Taking Stock: Animal Farming and the Environment." *WorldWatch Paper 103*. Washington, D.C.: WorldWatch Institute, 1991.
Ehrlich, Anne, and Paul Ehrlich. "The Population Explosion: Why Isn't Everyone As Scared As We Are?" *The Amicus Journal* 12 (Winter 1990), pp. 22–29.
Gardner, Gary. "Third World Debt Is Still Growing." *WorldWatch* 8 (Jan/Feb 1995), pp. 37–40.
Haub, Carl. *The UN Long-Range Population Projections: What They Tell Us*. Washington, D.C.: Population Reference Bureau, 1993.
Haub, Carl. "World Growth Rate Slows, But Numbers Build Up." *Population Today* 22 (Nov 1994), pp. 1–2.
Homer-Dixon, Thomas F., et al. "Environmental Change and Violent Conflict." *Scientific American* 268 (Feb 1993), pp. 37–38.
Horiuchi, Shiro. "World Population Growth Rate." *Population Today* 21 (June 1993), pp. 6–7.
Kalish, Susan. "International Migration: New Findings on Magnitude, Importance." *Population Today* 22 (Mar 1994), pp. 1–2.
Kane, Hal. "A Deluge of Refugees." *WorldWatch* 5 (Nov/Dec 1992), pp. 32–33.
Keyfitz, Nathan. "The Growing Human Population." *Scientific American* 261 (Sept 1989), pp. 118–127.
Lutz, Wolfgang. "The Future of World Population." *Population Bulletin* 49 (June 1994).
Mabogunje, Akin L. "The Environmental Challenges in Sub-Saharan Africa." *Environment* 37 (May 1995), pp. 4–9.
Mazur, Laurie Ann, ed. *Beyond the Numbers*. Washington D.C.: Island Press, 1995.
McFalls, Joseph A. *Population: A Lively Introduction*, rev. 2nd ed. Washington, D.C.: Population Reference Bureau, 1995.
McGranahan, Gordon, and Jacob Songsore. "Wealth, Health, and the Urban Household: Weighing Environmental Burdens in Accra, Jakarta, and Sao Paulo." *Environment* 36 (July/Aug 1994), pp. 4–11.
McMarry, John. "Bombay." *National Geographic* 187 (Mar 1995), pp. 42–67.
Newland, Kathleen. "Refugees: The Rising Flood." *WorldWatch* 7 (May/June 1994), pp. 10–20.

Passerini, Edward. "The Curve of the Future." Dubuque, IA: Kendal/Hunt Publishers, 1992.
Pimentel, David, et al. "Natural Resources and an Optimum Human Population." *Population and Environment* 15(5) (May 1994), pp. 347–370.
Platt, Anne E. "Dying Seas." *WorldWatch* 8 (Jan/Feb 1995), pp. 10–19.
Platt, Anne E. "The Resurgence of TB." *WorldWatch* 7 (July/Aug 1994), pp. 31–34.
Poleman, Thomas T. "Population: Past Growth and Future Control." *Population and Environment* 17(1) (Sept. 1995), pp. 19–40.
Popline. A bimonthly publication of the Population Institute, 110 Maryland Avenue, N.E., Washington, D.C. 20002.
Population Reference Bureau, Inc. "1992 World Population Data Sheet," "Population Today," and other informational and educational materials. Washington, D.C.: Population Reference Bureau.
Robey, Bryant, et al. "The Fertility Decline in Developing Countries." *Scientific American* 269 (Dec 1993), pp. 60–67.
Sachs, Aaron. "Child Prostitution: The Last Commodity." *WorldWatch* 7 (July/Aug 1994), pp. 25–30.
Saltz, I. S. "Income Distribution in the Third World: Its Estimation via Proxy Data." *American Journal of Economics and Sociology* 54.1 (Jan 1995), pp. 15–32.
Smil, Vaclav. *China's Environmental Crisis.* New York: M. E. Sharpe, 1993.
"Special Section: Population, Consumption & Environment." *The Amicus Journal,* 15 (Winter 1994).
Teisch, Jessica, and Alex de Sherbinin. "Population Doubling Time." *Population Today* 23 (Feb 1995), p. 3.
United Nations Population Fund. *Population and the Environment: The Challenges Ahead.* New York: United Nations Population Fund, 1991.
United Nations World Food Council. *Hunger and Malnutrition in the World: Situation and Outlook.* New York: U.N. World Food Council, June 5–8, 1991.
Weber, Peter. "Net Loss: Fish, Jobs, and the Marine Environment." *WorldWatch Paper 120.* Washington, D.C.: WorldWatch Institute, 1994.
Weiss, Robin, et al. "AIDS: A Worldwide Emergency." *The UNESCO Courier* (June 1995), pp. 6–39.
World Resources Institute. *World Resources 1994–95.* New York: Oxford University Press, 1994.

Chapter 7. Addressing the Population Problem

Ashford, Lori S. "New Perspectives on Population: Lessons from Cairo." *Population Bulletin* 50 (Mar 1995).
Benjamin, Medea, and Andrea Freedman. *Bridging the Global Gap: A Handbook to Linking Citizens of the First and Third Worlds."* Washington, D.C.: Seven Locks Press, 1989.
Bongaarts, John. "Can Growing Human Population Feed Itself?" *Scientific American* 270 (Mar 1994), pp. 36–42.
Chege, Nancy. "An African Success Story." *WorldWatch* 6 (July/Aug 1993), pp. 5–7.
Chen, Lincoln C., et al. "Women, Politics, and Global Management." *Environment* 37 (Jan/Feb 1995), pp. 4–9.
Clancy, Paul. "Putting People First." *Calypso Log* (Oct 1994), pp. 4–5.
Dale, H. E. "Fostering Environmentally Sustainable Development: Four Parting Suggestions for the World Bank." *Ecological Economics* 10 (Aug 1994), pp. 183–188.
Donaldson, Peter, J., and Amy Ong Tsui. "The International Family Planning Movement." *Population Bulletin* 45 (Nov 1990).
Fisher, Julie. "Third World NGOs: A Missing Piece to the Population Puzzle." *Environment* 36 (Sept 1994), pp. 6–11.
Fornos, Werner. "Population Politics." *Technology Review* (Feb/Mar 1991), pp. 43–51.
French, Hilary F. "The World Bank: Now Fifty, But How Fit?" *WorldWatch* 7 (July/Aug 1994), pp. 11–18.
Gibbons, Ann. "Small Is Beautiful: Microlivestock for the Third World." *Science* 253 (26 July 1991), p. 378.
Hanford, Heather. "Nepal Moves Mountains with Literacy." *WorldWatch* 5 (Nov/Dec 1992), pp. 9, 32.
Holloway, Marguerite. "Trends in Women's Health: A Global View." *Scientific American* 271 (Aug 1994), pp. 76–83.
Jacobson, Jodi L. "Gender Bias: Roadblock to Sustainable Development." *WorldWatch Paper 110.* Washington, D.C.: WorldWatch Institute, 1992.
Jacobson, Jodi L. "India's Misconceived Family Plan." *WorldWatch* 4 (Nov/Dec 1991), pp. 18–25.
Jordan, Andrew. "Financing the UNCED Agenda: The Controversy over Additionality." *Environment* 36 (Apr 1994), pp. 16–20, 26–33.
Kammen, Daniel M. "Cookstoves for the Developing World." *Scientific American* 273 (July 1995), pp. 72–75.
Kane, Hal. "Growing Fish in Fields." *WorldWatch* 6 (Sept/Oct 1993), pp. 20–27.
Kelber, Mim. "The Women's Environment and Development Organization." *Environment* 36 (Oct 1994), pp. 43–45.
Levine, Robert A., et al. "Women's Schooling and Child Care in the Demographic Transition: A Mexican Case Study." *Population and Development Review* 17 (Sept 1991), pp. 459–496.
Lipschutz, Ronnie D. "Redefining National Security." *E Magazine* (Mar/Apr 1991), pp. 54–55.
Livernash, Robert. "The Future of Populous Economies: China and India Shape Their Destinies." *Environment* 37 (Jul/Aug 1995), p. 60.
Livernash, Robert. "The Growing Influence of NGOs in the Developing World." *Environment* 34 (June 1992), pp. 12–20.
McIntosh, Alison C. and Jason L. Finkle. "The Cairo Conference on Population and Development." *Population and Development Review* 21(2) (June 1995), pp. 223–260.
McNicoll, Geoffrey, and Mead Cain, eds. *Rural Development and Population, Institutions and Policy.* Food and Agriculture Organization of the United Nations report. New York: Oxford University Press, 1990.
Mittlo, Cindy, and Alisa Gravitz. "Green Businesses Hold a New Vision." *Co-op America's National Green Pages* (Premiere Issue 1993), pp. 6, 7, 73.
Mortimore, Michael, and Mary Tiffen. "Population Growth and a Sustainable Environment: The Machakos Story." *Environment* 36 (Oct 1994), pp. 10–20.
Mosley, W. Henry, and Peter Cowley. "The Challenge of World Health." *Population Bulletin* 46 (Dec 1991), pp. 1–39.
Nepal, S. K. and K. E. Weber. "Prospects for Coexistence: Wildlife and Local People." *Ambio* 24(4) (June 1995), pp. 238–245.
Norse, David. "A New Strategy for Feeding a Crowded Planet." *Environment* 34 (June 1992), pp. 6–11.
Pearce, David, et al. "Debt and the Environment." *Scientific American* 272 (June 1995), pp. 52–57.
Population Institute. "Bangladesh Defying 'Transition' Theory." *Popline* 16 (Jan/Feb 1994), p. 3.
Population Institute. "1994 ICPD Program of Action Highlights." *Popline* (Sept/Oct 1994), p. 3.
Press, Robert. "Big Payoff for Small Loans." *The Christian Science Monitor,* Jan 7, 1992, p. 10.
Reid, Walter V. C. "Sustainable Development: Lessons from Success." *Environment* 31 (May 1989), pp. 7–9, 29–35.
Repetto, Robert. "Economic Aid and the Environment." *EPA Journal* 16 (July/Aug 1990), pp. 20–22.
Robey, Bryant, et al. "The Fertility Decline in Developing Countries." *Scientific American* 269 (Dec 1993), pp. 60–67.
Roudi, Nazy. "Peru's Family Planning Success." *Population Today* 19 (June 1991), p. 4.
Sachs, Aaron. "Male Responsibility." *WorldWatch* 7 (Mar/Apr 1994), pp. 12–19.
Sachs, Ignacy, et al. "Development: The Haves and the Have-Nots." *The UNESCO Courier* (Mar 1995), pp. 9–31.
Sen, Gita. "The World Programme of Action: A New Paradigm for

Population Policy." *Environment* 37 (Jan/Feb 1995), pp. 10–15.
Serageldin, Ismail, and Andrew Steer, eds. *Making Development Sustainable: From Concepts to Action*. Washington D.C.: The World Bank, 1994).
Stone, Laurie. "Solar Cooking in Developing Countries." *Solar Today* 8 (Nov/Dec 1994), pp. 27–30.
Swerdlow, Joel L. "Burma, the Richest of Poor Countries." *National Geographic* 188 (July 1995), pp. 70–97.
United Nations World Food Council. "Focusing Development Assistance on Hunger and Poverty Alleviation." New York: U.N. World Food Council, June 5–8, 1991.
United Nations World Food Council. "Food-Security Implications of the Changes in the Political and Economic Environment." New York: U.N. World Food Council, June 5–8, 1991.
United Nations World Food Council. "Meeting the Developing Countries' Food Production Challenges of the 1990s and Beyond." New York: U.N. World Food Council, June 5–8, 1991.
Wright, R. Michael. "Sharing the Wealth." *National Parks* 68 (Jan/Feb 1994), pp. 24–25. The needs of the rural poor, who live surrounded by biological riches, must be included in plans for preservation.
Youth, Howard. "Farming in a Fish Tank." *WorldWatch* 5 (May/June 1992), pp. 5–7.

Chapter 8. The Production and Distribution of Food

Aliteri, Miguel A., and Susanna B. Hecht. *Agroecology and Small Farm Development*. New York: CRC Press, 1990.
Bread for the World. *Hunger 1994. Transforming the Politics of Hunger. Fourth Annual Report on the State of World Hunger.* Washington, D.C.: Bread for the World Institute, 1994.
Brown, Lester R. "Facing Food Insecurity." *State of the World 1994*. Washington, D.C.: WorldWatch Institute, 1990.
Capute, Robert. "Tragedy Stalks the Horn of Africa." *National Geographic* 184 (Aug 1993), pp. 88–212.
Durning, Alan B., and Holly W. Brough. "Taking Stock: Animal Farming and the Environment." *WorldWatch Paper 103*. Washington, D.C.: WorldWatch Institute, 1991.
Edwards, Clive A., et al., eds. *Sustainable Agricultural Systems*. Ankeny, IA: Soil and Water Conservation Society, 1990.
Faeth, Paul. "Building the Case for Sustainable Agriculture: Policy Lessons from India, Chile, and the Philippines." *Environment* 36 (Jan/Feb 1994), pp. 16–20, 34–39.
French, Hilary F. "Costly Tradeoffs: Reconciling Trade and the Environment." *WorldWatch Paper 113*. Washington, D.C.: WorldWatch Institute, 1993.
Gasser, Charles S., and Robert T. Fraley. "Transgenic Crops." *Scientific American* 266 (June 1992), pp. 62–69.
Jackson, Wes. "Nature as the Measure for a Sustainable Agriculture." Chapter 4 in *Ecology, Economics, Ethics: The Broken Circle*. New Haven, CT: Yale University Press, 1991.
Kane, Hal. "The Hour of Departure: Forces That Create Refugees and Migrants." *WorldWatch Paper 125*. Washington, D.C.: WorldWatch Institute, 1995.
Kurlansky, Mark. "On Haitian Soil." *Audubon* (Jan/Feb 1995), pp. 50–57.
Moffat, Anne Simon. "Developing Nations Adapt Biotech for Own Needs." *Science* 265 (July 8, 1994), pp. 186–187.
Moffett, George. "'Super Rice' May Ease World Food Crisis." *The Christian Science Monitor*, Oct 26, 1994, p. 3.
Mortimore, Michael, and Mary Tiffen. "Population Growth and a Sustainable Environment: The Machakos Story." *Environment* 36 (Oct 1994), pp. 10–20, 28–32.
Paarlberg, Robert L. "The Politics of Agricultural Resource Abuse." *Environment* 36 (Oct 1994), pp. 6–9, 33–42.
Plucknett, Donald L., et al. "International Agricultural Research for the Next Century." Special issue. *BioScience* 43 (July/Aug 1993).
Ponting, Clive. "Historical Perspectives on Sustainable Development." *Environment* 32 (Nov 1990), pp. 4–9, 31–33.
Reganold, John P., et al. "Sustainable Agriculture." *Scientific American* 262 (June 1990), pp. 112–120.
Rhoades, Robert E., and L. Johnson. "The World's Food Supply at Risk." *National Geographic* (Apr 1991), pp. 74–105.
Schmidt, Karen. "Genetic Engineering Yields First Pest-Resistant Seeds." *Science* 265 (Aug 5, 1994), p. 739.
Tefft, Sheila. "A Shrinking Rice Bowl in China: Rising Food Prices Spur Unease." *The Christian Science Monitor*, Jan 19, 1995, pp. 1, 6.
White, Peter T., and Robb Kendrick. "Rice, the Essential Harvest." *National Geographic* 185 (May 1994), pp. 48–79.
World Resources Institute. "Food and Agriculture." Chapter 6 in *World Resources 1994–95*. New York: Oxford University Press, 1994.

Chapter 9. Soil and the Soil Ecosystem

Adger, W. N. "Aggregate Estimate of Environmental Degradation for Zimbabwe: Does Sustainable National Income Ensure Sustainability?" *Ecological Economics* 11 (Nov 1994), pp. 93–104.
American Farmland (Spring 1991). Issue devoted to conservation for a sustainable agriculture.
Baker, Beth. "Washington Watch: Environmental aspects of the Farm Bill." *BioScience* 45 (June 1995), p. 393.
Belcher, Linford. "Curbing Erosion." *American Farmland* (Summer 1992), pp. 16–17.
Berton, Valerie. "Battling Sprawl in Berks County, PA." *American Farmland* (Spring 1993), pp. 6–7.
Berton, Valerie. "Low-Input Farming." *American Farmland* (Winter 1993), pp. 14–15.
Bidwell, Dennis P. "Private Options for Farmland Protection." *American Farmland* (Winter 1994), pp. 18–20.
Biswas, A. K. "Environmental Sustainability of Egyptian Agriculture: Problems and Perspectives." *Ambio* 24(1) (Feb 1995), pp. 16–20.
Bongaarts, John. "Can Growing Human Population Feed Itself?" *Scientific American* 270 (Mar 1994), pp. 36–42.
Bray, Francesca. "Agriculture for Developing Nations." *Scientific American* 271 (July 1994), pp. 30–37.
Brookfield, Harold and Christine Padoch. "Appreciating Agrodiversity: A Look at the Dynamism and Diversity of Indigenous Farming Practices." *Environment* 36 (June 1994), pp. 6–11, 37–43.
Buringh, P. Pages 69–83 in *Food and Natural Resources*, ed. David Pimentel and C. W. Halls, San Diego, CA: Academic Press, 1989.
Carter, Vernon G., and T. Dale. *Topsoil and Civilization*. Norman, OK: University of Oklahoma Press, 1974.
Clemings, Russell. "Mirage." *Earthwatch* (June 1991), pp. 14–21. Description of salinization.
Collier, George A., et al. "Peasant Agriculture and Global Change." *BioScience* 44 (June 1994), pp. 398–407.
"Composting." *BioCycle* (Jan 1993).
Dodd, Jerrold L. "Desertification and Degradation in Sub-Saharan Africa." *BioScience* 44 (Jan 1994), pp. 28–34.
Durning, Alan B., and Holly B. Brough. "Taking Stock: Animal Farming and the Environment." *Worldwatch Paper 103*. Washington, D.C.: WorldWatch Institute, 1991.
Edwards, J. H., et al. "Applying Organics to Agricultural Land." *BioCycle* 34 (Oct 1993), pp. 48–50.
Ellis, William S. "Africa's Stricken Sahel." *National Geographic* 172 (Aug 1987), pp. 40–179.
Ettlin, Lauren, and Bill Stewart. "Yard Debris Compost for Erosion Control." *Journal of Composting & Recycling* 34 (Dec 1993), pp. 46–57.
Faeth, Paul. "Building the Case for Sustainable Agriculture: Policy Lessons from India, Chile, and the Phillippines." *Environment* 36 (Jan/Feb 1994), pp. 16–20, 34–39.
Garbriel, Clifford J. "AIBS News: Research in Support of Sustainable Agriculture." *BioScience* 45 (May 1995), pp. 346–353.
Giampietro, Mario. "Sustainability and Technological Devel-opment

in Agriculture." *BioScience* 44 (Nov 1994), pp. 677–689.
Hulme, Mike, and Mick Kelly. "Exploring the Links between Desertification and Climate Change." *Environment* 35 (July/Aug 1993), pp. 4–11.
Izac, A.-M. N., and J. J. Swift, "On Agricultural Sustainability and Its Measurement in Small-Scale Farming in Sub-Saharan Africa." *Ecological Economics* 11 (Nov 1994), pp. 105–126.
Jackson, Wes. "Listen to the Land." *The Amicus Journal* 15 (Spring 1993), pp. 32–41.
Justin, E. L. "Taxpayer-Subsidized Resource Extradition Harms Species." *BioScience* 45.7 (Jul/Aug 1995), pp. 446–455.
Kashmanian, Richard M., et al. "Building Support for Composting in Agriculture." *Journal of Composting & Recycling* 35 (Dec 1994), pp. 67–70.
Lipske, Mike. "Cutting Down Canada." *International Wildlife* 24 (Mar/Apr 1994), pp. 10–17.
Mann, R.D. "Time Running Out: The Urgent Need for Tree Planting in Africa." *The Ecologist* 20 (Mar/Apr 1990), pp. 48–53.
Mihaly, Mary. "Farming for the Future." *Nature Conservancy* 44 (Sept/Oct 1994), pp. 25–29.
Milton, Suzanne J., et al. "A Conceptual Model of Arid Rangeland Degradation." *BioScience* 44 (Feb 1994), pp. 70–76.
Mollison, Bill. *Permaculture: A Practical Guide for a Sustainable Future*. Washington, D.C.: Island Press, 1990.
Paarlberg, Robert L. "The Politics of Agricultural Resource Abuse." *Environment* 36 (Oct 1994), pp. 6–9.
Pimentel, David, et al. "Environmental and Economic Costs of Soil Erosion and Conservation Benefits." *Science* 267 (Feb 24, 1995), pp. 1117–1123.
Pretty, J. N. et al. "Agricultural Regeneration in Kenya: The Catchment Approach to Soil and Water Conservation." *Ambio* 24(1) (Feb 1995), pp. 7–15.
Reganold, John P., et al. "Sustainable Agriculture." *Scientific American* 262 (June 1990), pp. 112–120.
Riggle, David, and Hannah Holmes, et al. "Earthworms and Composting." *Journal of Composting & Recycling* 35 (Oct 1994), pp. 58–67.
Robbins, Grace Jones, ed. *Alternative Agriculture*. Washington, D.C.: National Academy Press, 1989.
Rominger, Rich. "USDA Aims to Make Conservation a Farm Priority." *American Farmland* (Winter 1994), pp. 16–17.
Runnels, Curtis N. "Environmental Degradation in Ancient Greece." *Scientific American* 272 (Mar 1995), pp. 96–99.
Soule, Judith D., and Jon Piper. *Farming in Nature's Image*. Washington, D.C.: Island Press, 1992.
Speth, J. G. *Towards an Effective and Operational International Convention on Desertification*. New York: International Negotiating Committee, International Convention on Desertification, United Nations, 1994.
Tanji, Kenneth, et al. "Selenium in the San Joaquin Valley." *Environment* 28 (July/Aug 1986), pp. 6–39.
Troeh, F. R., et al. *Soil and Water Conservation*. Englewood Cliffs, NJ: Prentice Hall, 1991.
Troeh, F. R., and L. M. Thompson, *Soils and Soil Fertility*, 5th ed. New York: Oxford University Press, 1993).
Wilken, Elena. "Assault of the Earth." *WorldWatch* 8 (Mar/Apr 1995), pp. 20–27.
Zimmer, Carl. "How to Make a Desert." *Discover* (Feb 1995), pp. 51–56.

Chapter 10. Pests and Pest Control

Barinaga, Marcia. "Entomologists in the Medfly Maelstrom." *Science* 247 (Mar 9, 1990), pp. 1168–1169.
Beard, Jonathan D. "Bug Detectives Crack the Tough Cases." *Science* 254 (Dec 13, 1991), pp. 1580–1581.
Carson, Rachel. *Silent Spring*. Boston: Houghton Mifflin, 1962.
Culotta, Elizabeth. "Biological Immigrants under Fire." *Science* 254 (Dec 6, 1991), pp. 1444–1447.
DeBach, Paul. *Biological Control by Natural Enemies*. London: Cambridge University Press, 1974.
Friends of the Earth. *How to Get Your Lawn and Garden off Drugs*. Ottawa, Ontario: Friends of the Earth, 1990.
Gardner, Gary. "Preserving Agricultural Resources." *State of the World 1996*. Washingtont, D.C. WorldWatch Institute, 1996
Hussey, N. W., and N. Scopes. *Biological Pest Control*. Ithaca, NY: Cornell University Press, 1986.
Hynes, H. Patricia. *The Recurring Silent Spring*. New York: Pergamon Press, 1989.
Kimm, Victor J. "The Delaney Clause Dilemma." *EPA Journal* 19 (Jan/Mar 1993), pp. 39–41.
Kling, James. "Could Transgenic Supercrops One Day Breed Superweeds?" "*Science* 274 (11 October 1996), pp. 180-181.
Lewis, Jack. "Rachel Carson." *EPA Journal* 18 (May/June 1992), pp. 60–62.
Malecki, Richard T. "Biological Control of Purple Loosestrife." *BioScience* 43 (Nov 1993), pp. 680–686.
Marco, Gino J., et al., eds. *Silent Spring Revisited*. Washington, D.C.: American Chemical Society, 1987.
Meyerhoff, Al, and Clausen Ely. "The Delaney Clause: Point/Counterpoint." *EPA Journal* 19 (Jan/Mar 1993), pp. 42–45.
Natural Resources Defense Council. *A Report on Intolerable Risk: Pesticides in Our Children's Food*. Washington, D.C.: Natural Resources Defense Council, 1989.
Paarlberg, Robert L. "Managing Pesticide Use in Developing Countries." Chapter 7 in *Institutions for the Earth*, ed. Peter M. Haas, Robert O. Keohane, and Marc A. Levy. Cambridge. MA: MIT Press, 1993.
Perring, Thomas M., et al. "Identification of a Whitefly Species by Genomic and Behavioral Studies. *Science* 259 (Jan 1, 1993), pp. 74–77.
Pimental, David. "The Dimensions of the Pesticide Question." Chapter 5 in *Ecology, Economics, Ethics: The Broken Circle*. New Haven, CT: Yale University Press, 1991.
Pimental, David. "Reducing Pesticide Use through Alternative Agricultural Practices: Fungicides and Herbicides." Chapter 23 in *Pesticide Interactions in Crop Production: Beneficial and Deleterious Effects*, ed. J. Altman. Boca Raton, FL: CRC Press, 1993.
Pimental, David , et al. "Environmental and Economic Costs of Pesticide Use." *BioScience* 42 (Nov 1992), pp. 750–760.
Revkin, Andrew C. "March of the Fire Ants." *Discover* 10 (Mar 1989), pp. 71–76.
Schulten, Gerard G. M. "Integrated Pest Management in Developing Countries." Pages 205–212 in *Environmentally Sound Technology for Sustainable Development*. New York: United Nations, 1992.
Smith, Miranda, ed. *"The Real Dirt: Farmers Tell About Organic and Low-Input Practices in the Northeast."* Burlington, VT: Northeast Organic Farming Association, 1994.
Stevens, Jane Ellen. "In Africa, Imported Wasps Defeat Imported Pests." *Boston Globe,* Jan 3, 1994, pp. 67, 70.
Stewart, Doug. "Luck Be a Ladybug." *National Wildlife* 32 (June/July 1994), pp. 30–33.
Stone, Richard. "Researchers Score Victory over Pesticides—and Pests—in Asia." *Science* 256 (May 29, 1992), pp. 1272–1273.
Stone, Richard. "Environmental Estrogens Stir Debate." *Science* 265 (July 15, 1994), pp. 308–310.
Strobel, Gary. "Biological Control of Weeds." *Scientific American* 265 (July 1991), pp. 72–78.
United Nations Environment Program. "The Conamination of Food." *UNEP/GEMS Environment Library No 5*. Nairobi, Kenya, 1992.
United Nations Environment Program. *"Highlights of the Biennium: 1994-1995."* Nairobi, Kenya, 1995.
Wargo, John, *"Our Children's Toxic Legacy. How Science and Law Fail to Protect Us from Pesticides."* New Haven: Yale University Press. 1996.
Wolff, M. S. et al. "Blood Levels of Organochlorine Residues and Risk of Breast Cancer." *Journal of the National Cancer Institute* 85 (1993), pp. 648–652.
World Resources Institute. "Food and Agriculture." Chapter 6 in *World Resources 1994–95*. New York: Oxford University Press, 1994.

Chapter 11. Water, the Water Cycle, and Water Management

Barker, Mary L., and Dietrich Soyez. "Think Locally, Act Globally? The Transnationalization of Canadian Resource-Use Conflicts." *Environment* 36 (June 1994), pp. 12–20, 32–36.
Baylery, Peter. B. "Understanding Large River-Floodplain Ecosystems." *BioScience* 45 (Mar 1995), pp. 153-158.
Boucher, Norman. "Back to the Everglades." *Technology Review* (Aug/Sept 1995), pp. 24–35.
Brown, Lester R. "The Aral Sea: Going, Going . . ." *WorldWatch* 4 (Jan/Feb 1991), pp. 20–27.
Carrier, Jim. "Water and the West: The Colorado River." *National Geographic* (June 1991), pp. 2–35.
Daniel, John. "Dance of Denial." *Sierra* 78 (Mar/Apr 1993), pp. 64–73. The Pacific salmon swim toward extinction.
Dulle, Kevin J., and Ted H. Streakfull. "Greening of Ground Water." *Civil Engineering* 65 (Apr 1995), pp. 62–65.
Falkenmark, Malin, and Carl Widstrand. "Population and Water Resources: A Delicate Balance." *Population Bulletin* 47 (Nov 1992).
Gardner, Gary. "A Tale of Three Aquifers." *WorldWatch* 8 (May/June 1995), pp. 30–36.
Gillis, Anna Marie. "Israeli Researchers Planning for Global Climate Change on the Local Level." *BioScience* 42 (Sept 1992), pp. 587–589. Greening the desert is not a sensible option given limited water.
Gleick, Peter H. *Water in Crisis: A Guide to the World's Fresh Water Resources*. New York: Oxford University Press, 1993.
Gleick, Peter H. "Water, War, and Peace in the Middle East." *Environment* 36 (Apr 1994), pp. 6–15, 35–41.
Gore, James A., and F. Douglas Shields, Jr. "Can Large Rivers Be Restored?" *BioScience* 45 (Mar 1995), pp. 142–152.
Gorman, James. "Wetlands? Wetlands? Whatever Happened to the Swamps?" *Audubon* 94 (May/June 1992), pp. 82–83.
Grossi, Ralph. "Wetlands and the Farm." *American Farmland* (Summer 1992), pp. 4–7, 18.
Gruchow, Paul. "Rite of Spring." *Nature Conservancy* 44 (Mar/Apr 1994), pp. 24–29. Sandhill cranes herald a new season at Nebraska's Platte River.
Harger, Cindy. "A Spoonful of Sugar Makes the Everglades Go Down." *Conservation* 90 (June 8, 1990), pp. 7–10.
Jackson, Donald Dale. "Saving the Last Best River." *Audubon* 96 (May/June 1994), pp. 76–88, 120–121.
Johnson, Barry L., et al. "Past, Present, and Future Concepts in Large River Ecology." *BioScience* 45 (Mar 1995), pp. 134–141.
Kourik, Robert. "Drip Irrigation." *Garbage* (May/June 1991), pp. 46–51.
Kourik, Robert. "Greywater: Why Throw It Away?" *Garbage* (Jan/Feb 1990), pp. 41–45.
Kusler, Jon A. "Wetlands." *Scientific American* 270 (Summer 1994), pp. 64B–70.
Kwai-cheong, Chau. "The Three Gorges Project of China: Resettlement Prospects and Problems." *Ambio* 24(2) (Mar 1995), pp. 98–102.
Lee, Fred and Ann Jones Lee. "Issues in Managing Urban Stormwater Rushoff Quality." *Water Engineering and Management* (May 1995), pp. 51–53.
Ligon, Franklin K. "Downstream Ecological Effects of Dams." *BioScience* 45 (Mar 1995), pp. 183–192.
Mairson, Alan. "Great Flood of '93." *National Geographic* 185 (Jan 1994), pp. 42–81.
Markus, R. Michael, et al. "Aquifer Recharge Enhanced with Rubber Dam Installations." *Water Engineering and Management* (Jan 1995), pp. 37–40.
McGivney, Annette. "Troubled Waters: America's Endangered Rivers." *E Magazine* 4 (Sept/Oct 1993), pp. 30–37.
Melody, Ingrid. "Solar Water Desalinization." *Solar Today* 4 (Nov/Dec 1990), pp. 14–17.
Misch, Ann. "India's Wells Run Dry." *WorldWatch* (May/June 1991), pp. 9–10.
Mitchell, John G. "Our Disappearing Wetlands." *National Geographic* (Oct 1992), pp. 3–45.
Monks, Vicki. "Engineering the Everglades." *National Parks* 65 (Sept/Oct 1990), pp. 32–36.
"Mono Lake: Saved." *National Wildlife* 33 (June/July 1995), pp. 36–37.
Okum, Daniel A. "A Water and Sanitation Strategy for the Developing World." *Environment* 33 (Oct 1991), pp. 16–20.
Palmer, Tim. *Lifelines: The Case for River Conservation*. Washington, D.C.: Island Press, 1994.
Postel, Sandra. *Last Oasis*. New York: W. W. Norton, 1992.
Postel, Sandra. "Rivers Drying Up." *WorldWatch* 8 (May/June 1995), pp. 9–19.
Postel, Sandra. "Water Tight." *WorldWatch* 6 (Jan/Feb 1994), pp. 19–25. Innovative conservation plans prove that saving water—by acknowledging its true value—makes economic and environmental sense.
Ridgeway, James. "Watch on the Danube." *Audubon* 94 (July/Aug 1992), pp. 46–64.
Schueler, Donald. "That Sinking Feeling." *Sierra* 75 (Mar/Apr 1990), pp. 42–51.
Schwarzbach, David A. "Promised Land. (But What about the Water?)" *The Amicus Journal* 16 (Summer 1995), pp. 35–39.
Seraydarian, Harry. "San Francisco Bay, Beset by Freshwater Diversion." *EPA Journal* 16 (Nov/Dec 1990), pp. 20–22.
Sparks, Richard E. "Need for Ecosystem Management of Large Rivers and Their Floodplains." *BioScience* 45 (Mar 1995), pp. 168–182.
Starr, Joyce R. "Water Resources: A Foreign-Policy Flash Point." *EPA Journal* (July/Aug 1990), pp. 34–36.
Stutz, Bruce. "Water and Peace." *Audubon* 96 (Sept/Oct 1994), pp. 64–77.
Tenbergen, B. et al. "Harvesting Runoff: The Minicatchment Technique—An Alternative to Irrigated Tree Plantations in Semiarid Regions." *Ambio* 24(2) (Mar 1995), pp. 72–76.
Tennebaum, David. "Rethinking the River." *Nature Conservancy* 44 (July/Aug 1994), pp. 10–15.
Terrazas, Michael. "Saltwater Intrusion: Florida's Underground Movement." *American City and County* (Feb 1995), pp. 46–59.
Vesilind, Pritt J. "Middle East Water—Critical Resource." *National Geographic* 183 (May 1993), pp. 38–72.
Viessman, Warren. "Water Management." *Environment* 32 (May 1990), pp. 11–15.
"Water: The Power, Promise, and Turmoil of North America's Fresh Water." *National Geographic* (Special Edition, Vol. 184(5A) 1993).
Wood, Daniel B. "Clock Starts Ticking on California's Water Wars." *The Christian Science Monitor*, Jan 5, 1994, pp. 12–13.
Wood, Daniel B. "Hopi Water Dispute Tests Federal Resolve on Protection." *The Christian Science Monitor*, Apr 20, 1994, p. 11.

Chapter 12. Sediments, Nutrients, and Eutrophication

Anderson, Donald M. "Red Tides." *Scientific American* 271 (Aug 1994), pp. 62–69.
Baker, William, C., and Tom Horton. "Runoff and the Chesapeake Bay." *EPA Journal* 16 (Nov/Dec 1990), pp. 13–16.
Baskin, Yvonne. "Losing a Lake." *Discover* 15 (Mar 1994), pp. 72–81. Lake Victoria, Africa's largest body of water, is in trouble.
Bell, P.R.F. and I. Elmetri. "Ecological Indicators of Large-scale Eutrophication in the Great Barrier Reef Lagoon." *Ambio* 24(4) (June 1995), pp. 208–215.
Burke, Maria. "Phosphorus Fingered as Coral Killer." *Science* 263 (Feb 1994), p. 1086.
Carmichael, Wayne W. "The Toxins of Cyanobacteria." *Scientific American* 270 (Summer 1994), pp. 78–86.
Dennison, William C., et. al. "Assessing Water Quality with Submersed Aquatic Vegetation." *BioScience* 43 (Feb 1993), pp. 86–94.
Dugan, Eugene, ed. *Wetlands in Danger: A World Conservation Atlas*. New York: Oxford University Press, 1993.
EPA Journal (Nov/Dec 1991). Issue devoted to problems of nutrient, sediment, and chemical runoff from agricultural and urban areas.

Foster, Catherine. “Nipping Algal Blooms in the Bud.” *The Christian Science Monitor*, Jan 26, 1994, p. 14.
Gup, Ted. “Getting at the Roots of a National Obsession: How to Maintain a Chemical-Free Lawn.” *National Wildlife* 29 (June/July 1991), pp. 18–24.
Horton, Tom. “Chesapeake Bay—Hanging in the Balance.” *National Geographic* 183 (June 1993), pp. 2–35.
Horton, Tom, and William Eichbaum. *Turning the Tide*. Washington, D.C.: Island Press, 1991.
Hudson, Kar. “Bottom-up Approach Is Winning Pollution Battle.” *American City and County* (June 1995), pp. 30–49.
Hunnsaker, Carolyn T. “Hierarchical Approaches to the Study of Water Quality in Rivers.” *BioScience* 45 (Mar 1995), pp. 193–205.
Johnson, H. J. “Backyard Composting Education Programs.” *BioCycle* (Jan 1995), pp. 75–77.
Kusler, Jon. “Wetlands Delineation: An Issue of Science or Politics?” *Environment* 34 (Mar 1992), pp. 6–11, 29–37.
Lanza, Guy R. and J.K.G. Silvey, “Interactions of Reservoir Microbiota: Eutrophication—Related Environmental Problems.” In *Microbial Processes in Reservoirs*. Dordrecht: Kluwer Academic Publishers Group, 1985.
Makarewicz, Joseph C., and Paul Bertram. “Evidence for the Restoration of the Lake Erie Ecosystem.” *BioScience* (Apr 1991), pp. 216–224.
Mitchell, John G. “Our Disappearing Wetlands.” *National Geographic* (Oct 1992), pp. 3–45.
Platt, Anne E. “Dying Seas.” *WorldWatch* 8 (Jan/Feb 1995), pp. 10–19.
Salvesen, David. *Wetlands: Mitigating and Regulating Development Impacts*, 2nd ed. Washington, D.C.: The Urban Land Institute, 1994.
Schlosser, Isaac J. “Stream Fish Ecology: A Landscape Perspective.” *BioScience* 41 (Nov 1991), pp. 704–712.
Turner, R. Eugene, and Nancy N. Rabalais. “Changes in Mississippi River Water Quality This Century.” *BioScience* 41 (Mar 1991), pp. 140–147.
Weber, Peter. “Oceans in Crises.” *E Magazine* (May/June 1994), pp. 36–43.
Wheeler, Timothy B. “Pa. Legislature Passes Bill to Curb Farm Pollution.” *Baltimore Sun*, May 6, 1993, p. 1.
Williams, Ted. “The Wetlands—Protection Farce.” *Audubon* 97 (Mar/Apr 1995), pp. 30–36.

Chapter 13. Sewage Pollution and the Nutrient Cycle

Anderson, Geoffrey, and Karen Smith. “Mixed Paper Teams Up with Biosolids.” *Journal of Composting & Recycling* (Mar 1994), pp. 61–65.
Bencivenga, Jim. “Florida Wetlands Filter City Waste.” *The Christian Science Monitor*, Mar 5, 1992, pp. 16–17.
Briscoe, John. “When the Cup Is Half Full: Improving Water and Sanitation Services in the Developing World.” *Environment* 35 (May 1993), pp. 6–15.
Clark, Sarah L. “Building Support for Sludge Reuse.” *BioCycle* (Dec 1992), pp. 74–79.
Fabrikant, Reva, and Ray Kearney. “Cocomposting in Los Angeles Optimizes Resource Management.” *BioCycle* (Dec 1994), pp. 58–60.
Gillette, Becky. “The Green Revolution in Wastewater Treatment.” *BioCycle* (Dec 1992), pp. 44–45.
Goldstein, Nora, and Robert Steuteville. “Annual Survey: Biosolids Composting.” *BioCycle* (Dec 1993), pp. 48–57.
Graff, Gordon. “The Chlorine Controversy.” *Technology Review* (Jan 1995), pp. 54–60.
Hammer, Donald A., ed. *Constructed Wetlands for Wastewater Treatment*. Chelsea, MI: Lewis Publishers, 1989.
Hazan, Terry C. “Fecal Coliforms as Indicators in Tropical Waters: A Review.” *Toxicity Assessment: An International Journal* 3 (1988), pp. 461–477.
Hochrein, Peter, and Thomas Outerbridge. “Anaerobic Digestion for Soil Amendment and Energy.” *BioCycle* 33 (June 1992), pp. 63–64.
Lisk, Ian. “Tertiary-Treated Wastewater Irrigates Monterey Peninsula's Recreational Acreage.” *Water Engineering and Management* (July 1995), pp. 18–23.
Logsdon, Gene. “Innovative Waste Treatment in a Midwest Cornfield.” *BioCycle* (Jan 1992), pp. 46–48.
Maritato, M. C., E. R. Algeo, and R. E. Keenan. “Potential Human Health Concerns from Composting.” *BioCycle* (Dec 1992), pp. 70–73.
McDonald, George J. “Applying Sludge to Agricultural Land—Within the Rules.” *Water Engineering and Managment* (Feb 1995), pp. 38–40.
Meister, John. “Waste Not, Want Not: Putting Wastewater to Use.” *American City and County* (June 1995), pp. 32–43.
Okum, Daniel A. “A Water and Sanitation Strategy for the Developing World.” *Environment* 33 (Oct 1991), pp. 16–20.
Panetta, Daniel. “Sustainability and Wastewater Treatment.” *Solar Today* 6 (Sept/Oct 1992), pp. 18–20.
Parten, Susan. “Using Waste Woods Chips to Treat Septage.” *BioCycle* 35 (Apr 1994), pp. 74–76.
Platt, Anne E. “Duckweed: Cleaning Water at the Grassroots.” *WorldWatch* 6 (Nov/Dec 1993), pp. 8–9, 30.
Ross, Elizabeth. “Boston Recycles Sludge into Fertilizer.” *The Christian Science Monitor*, Jan 22, 1992, p. 15.
Scherer, Ron. “Sydney Pipes Sewage out to Sea.” *The Christian Science Monitor*, Jan 22, 1992, p. 14.
Smith-Vaago, Linda. “New Possibilities for Wastewater Treatment.” *Water Engineering and Management* 138 (Mar 1991), pp. 27–29.
“UV Effect on Wastewater.” *Water Engineering and Management* (Dec 1990), pp. 15–23.
Waterman, Melissa. “Pollution-Prevention Tactics in the Gulf of Maine.” *EPA Journal* 16 (Nov/Dec 1990), pp. 17–19.
Wittrup, Larry. “Biogas Project Advances in California.” *BioCycle* (Apr 1995), pp. 48–49.

Chapter 14. Pollution from Hazardous Chemicals

Amato, Ivan. “The Slow Birth of Breen Chemistry.” *Science* 259 (Mar 1993), p. 294.
Anderson, John V., and Bimleshwar P. Gupta. “Solar Detoxification of Hazardous Waste.” *Solar Today* 4 (Nov/Dec 1990), pp. 10–13.
Atlas, Ronald M., and Carl E. Cerniglia. “Bioremediation of Petroleum Pollutants.” *BioScience* 45 (May 1995), pp. 332–38.
Bell, Lauren. “Once Aflame and Filthy, a River Shows Signs of Life.” *National Wildlife* 28 (Feb/Mar 1990), p. 24.
Bellafante, Ginia. “Bottled Water: Fads and Facts.” *Garbage* (Jan/Feb 1990), pp. 46–51.
Bellafante, Ginia. “Minimizing Household Hazardous Waste.” *Garbage* (Mar/Apr 1990), pp. 44–49.
Browner, Carol M., et al. “Clean Water Agenda: Remaking the Laws That Protect Our Water Resources.” *EPA Journal* 20 (Summer 1994), pp. 6–41.
Bullard, Robert D., ed. *Unequal Protection: Environmental Justice and Communities of Color*. San Francisco, CA: Sierra Club Books, 1994.
Busse, Phil, and Julia Gorton. “The Great Lakes.” *Sierra* 78 (Mar/Apr 1993), pp. 78–79.
Chepesiuk, Ron. “From Ash to Cash: The International Trade in Toxic Waste.” *E Magazine* 11(July/Aug 1991), pp. 31–37, 63.
Clean Water Action News. “Mercury Contaminates the Land of 10,000 Lakes.” *Environment* 33 (May 1991), pp. 21–22.
Dibb, Sue. “Swimming in a Sea of Oestrogens: Chemical Hormone Disrupters.” *The Ecologist* 25 (Jan/Feb 1995), pp. 27–31.
Edwards, Mike. “Soviet Pollution.” *National Geographic* 186 (Aug 1994), pp. 70–99.
EPA Journal (July/Aug 1991). Issue devoted to hazardous waste cleanup.
Frederick, Robert J., and Margaret Egan. “Environmentally Compatible Applications of Biotechnology.” *BioScience* 44 (Sept 1994),

pp. 529–535.
Geiser, Ken. "The Greening of Industry." *Technology Review* (Aug/Sept 1991), pp. 64–72.
Gillette, Becky. "Constructed Wetlands for Industrial Wastewater." *BioCycle* 35 (Nov 1994), pp. 80–83.
Glass, David J. "Waste Management—Biological Treatment of Hazardous Wastes." *Environment* 33 (Nov 1991), pp. 5, 43–45.
Goodno, James. "Bye-Bye Subic Bay; Hello Toxic Mess." *E Magazine* (Mar/Apr 1994), pp. 42–43.
Gore, Al, et al. "Making Economic and Environmental Sense: How Green Technology Works Better, Costs Less." *EPA Journal* 20 (Fall 1994), pp. 6–43.
Grumbly, Thomas P. "Lessons from Superfund." *Environment* 37 (Mar 1995), pp. 33–35.
Holmes, Hannah, and Bill Breen. "Getting the Lead Out." *Garbage* (Nov/Dec 1993), pp. 26–31.
Karliner, Joshua. "Profiting from Pollution." *The Ecologist* 24 (Mar/Apr 1994), pp. 59–60.
Kindler, Janusz, and Stephen F. Lintner. "An Action Plan to Clean Up the Baltic." *Environment* 35 (Oct 1993), pp. 6–15.
Kotov, Vladimi, and Elena Nikitina. "Russia in Transition: Obstacles to Environmental Protection." *Environment* 35 (Dec 1993), pp. 10–20.
Krieg, E. J. "A Socio-Historical Interpretation of Toxic Waste Sites: The Case of Greater Boston." *American Journal of Economics and Sociology* 54 (Jan 1995), pp. 1–13.
Lipske, Michael. "Cracking Down on Mining Pollution." *National Wildlife* 33 (June/July 1995), pp. 20–24.
Luoma, Jon R. "Havoc in the Hormones." *Audubon* 97 (May/June 1995), pp. 60–67.
McGivney, Annette. "Troubled Waters: America's Endangered Rivers." *E Magazine* 4 (Sept/Oct 1993), pp. 30–37.
Mentasti, E. and C. Ramel. "Spread of Toxic Substances and Environmental Pollution." *Ambio* 24(4) (June 1995), pp. 250–251.
Misch, Ann. "The Amazon: River at Risk." *WorldWatch* 5 (Jan/Feb 1992), pp. 35–37.
Misch, Ann. "Chemical Reaction." *WorldWatch* 6 (Mar/Apr 1993), pp. 10–17.
Moore, Curtis A. "Down Germany's Road to a Clean Tomorrow." *International Wildlife* 22 (Sept/Oct 1992), pp. 24–28.
Mukerjee, Madhusree. "Persistently Toxic." *Scientific American* (June 1995), pp. 16–18.
Murphy, Steven. "Law—The Basic Convention on Hazardous Wastes." *Environment* 35 (Mar 1993), pp. 42–44.
Nriagu, Jerome O. "Global Metal Pollution: Poisoning the Biosphere?" *Environment* 32 (Sept 1990), pp. 6–11.
Olsen, Roger L., and Michael C. Kavanaugh. "Can Groundwater Restoration Be Achieved?" *Water Environment and Technology* (Mar 1993), pp. 42–47.
Olsson, A. and A. Bergman. "A New Persistent Contaminant Detected in Baltic Wildlife: Bis(4-chlorophenyl) Sulfone." *Ambio* 24(2) (March 1995), pp. 119–123.
Penfield, Wendy. "Message from the Belugas." *International Wildlife* 20 (May/June 1990), pp. 40–44.
"Pollution Prevention: It's a Whole New Way of Doing Business." *EPA Journal* 19 (July/Sept 1993).
Probstein, Ronald F., and Edwin R. Hicks. "Removal of Contaminants from Soils by Electric Fields." *Science* 260 (Apr 1993), pp. 498–503.
"Resettlement of Love Canal." *E Magazine* (Jan/Feb 1991), p. 16.
Ritter, Don, et al. "How Clean Is It?" *Environment* 37 (Mar 1995), pp. 10–15.
Roy, Kimberly. "Thermochemical Reduction Process Offers Absolute Detoxification." *Hazmat World* 3 (Oct 1990), pp. 50–51.
Schwartz, Joel, and Ronnie Levin. "Lead: Example of the Job Ahead." *EPA Journal* 18 (Mar/Apr 1992), pp. 42–44.
Selcraig, Bruce. "Border Patrol." *Sierra* 79 (May/June 1994), pp. 58–64.
Shulman, Seth. "Operation Restore Earth—Cleaning Up after the Cold War." *E Magazine* (Mar/Apr 1994), pp. 36–43.
Simons, Marlise. "Virus Linked to Pollution Is Killing Hundreds of Dolphins in Mediterranean." *The New York Times International,* (Oct 28, 1990), pp. 2–3.
Smith, D. R. and A. R. Flegal. "Lead in the Biosphere: Recent Trends." *Ambio* 24(1) Feb 1995), pp. 21–23.
Stewart, Doug. "Will This Lake Stay Superior?" *National Wildlife* 31 (Aug/Sept 1993), pp. 4–11.
Szekely, Julian, and G. Trapaga. "From Villain to Hero." *Technology Review* (Jan 1995), pp. 30–37.
United States Environmental Protection Agency. *Bioremediation in the Field.* Cincinnati: Center for Environmental Research Information, 1991.
van Voorst, Bruce. "Toxic Dumps: The Lawyer's Money Pit." *Time* (Sept 13, 1993), pp. 63–64.
Wilson, Elizabeth K. "Ground Water Cleanup." *Chemical and Engineering News* (July 3, 1993), pp. 19–23.
Wolkomir, Richard, and Robb Kendrick. "Where in the World Is Hector Villa?" *Smithsonian* 25 (May 1994), pp. 26–37.
Young, John E. "For the Love of Gold." *WorldWatch* 6 (May/June 1993), pp. 19–26.

Chapter 15. Air Pollution and Its Control

"Air Pollution in the World's Megacities." *Environment* 36 (Mar 1994), pp. 4–13, 25–37.
American Medical Association. *Journal of the American Medical Association* 225 (Feb 28, 1986). Issue devoted to effects of smoking tobacco.
Baldauf, Scott. "EPA Nudges the Northeast toward More Electric Cars." *The Christian Science Monitor*, Sept 19, 1994, p. 7.
Bartecchi, Carl E. et al. "The Global Tobacco Epidemic." *Scientific American* 272 (May 1995), pp. 44–51.
Brune, William H. "Stalking the Elusive Atmospheric Hydroxyl Radical." *Science* 256 (May 22, 1992), pp. 1154–1155.
Chameides, W. L., et al. "Growth of Continental-Scale Metro-Agro-Plexes, Regional Ozone Pollution, and World Food Production. *Science* 264 (April 1, 1994), pp. 74–77.
Chivian, E., et al. *Critical Condition: Human Health and the Environment.* Cambridge, MA: MIT Press, 1993.
Cole, Leonard A. "Radon: The Silent Killer?" *Garbage* (Spring 1994), pp. 22–28.
Corcoran, Elizabeth. "Cleaning Up Coal." *Scientific American* 264 (May 1991), pp. 107–116.
Eastham, Tony R. "High-Speed Rail: Another Goden Age?" *Scientific American* 273 (September 1995), pp. 100–101.
Eisenstein, Paul A. "Land Yachts May Be Extinct, But Gas Guzzlers Thrive in US." *The Christian Science Monitor,* April 16, 1996.
EPA Journal (Oct/Dec 1993). Most of issue is devoted to problem of indoor air pollution.
Evarts, Eric C. "Electric Cars Power Up (Quietly) to Hit the Road." *The Christian Science Monitro* (May 21, 1996).
Findlayson-Pitts, Barbara J., and James N. Pitts. *Atmospheric Chemistry: Fundamentals and Experimental Techniques.* New York: John Wiley & Sons, 1986.
French, Hilary F. "Clearing the Air: A Global Agenda." *WorldWatch Paper 94.* Washington, D.C.: WorldWatch Institute, 1990.
Global Environmental Monitoring System. "An Assessment of Urban Air Quality." *Environment* 31 (Oct 1989), pp. 6–13, 26–37.
Graedel, Thomas E., and Paul J. Crutzen. "The Changing Atmosphere." *Scientific American* 261 (Sept 1989), pp. 58–68.
Kerr, Richard. "Hydroxyl, the Cleanser That Thrives on Dirt." *Science* 253 (Sept 13, 1991), pp. 1210–1211.
LaFranchi, Howard. "Mexico City Struggles to Clear the Air." *The Christian Science Monitor*, Oct 19, 1994, p. 12.
Lents, James M., and William J. Kelley. "Clearing the Air in Los Angeles." *Scientific American* (Oct 1993), pp. 32–39.
Mumme, Stephen P. "Clearing the Air: Environmental Reform in Mexico." *Environment* 33 (Dec 1991), pp. 7–11, 26–30.
Nero, Anthony V., Jr. "Controlling Indoor Air Pollution." *Scientific American* 258 (May 1988), pp. 42–48.
Page, Stephen. "Radon Risks Are Real." *Garbage* (Spring 1994), pp. 29–31.

Petrick, William. "The Ozone Below." *Audubon* (Sept/Oct 1994), pp. 14, 22–23.

The President's Council on Sustainable Development. *Sustainable America: A New Consensus."* Washington, D.C.: President's Council on Sustainable Development.

Schelling, Thomas C. "Addictive Drugs: The Cigarette Experience." *Science* 255 (Jan 24, 1992), pp. 430–433.

Seinfeld, John. "Urban Air Pollution: State of the Science." *Science* 243 (Feb 10, 1989), pp. 745–752.

Sillman, Sanford. "Tropospheric Ozone: The Debate over Control Strategies." *Annual Review of Energy and the Environment.* 18 (1993), pp, 31–56.

Smith, Kirk R. "Fuel Combustion, Air Pollution Exposure, and Health: The Situation in Developing Countries." *Annual Review of Energy and the Environment.* 18 (1993), pp. 529–566.

Sperling, Daniel. "The Case for Electric Vehicles." *Scientific American* 275 (November 1996), pp. 54–59.

Thompson, Anne M. "The Oxidizing Capacity of the Earth's Atmosphere: Probable Past and Future Changes." *Science* 256 (May 22, 1992), pp. 1157–1165.

United Nations Environment Program. "Urban Air Pollution." *UNEP/GEMS Environment Library No. 5.* Nairobi, Kenya, 1991.

U.S. Department of Health and Human Services. *The Health Consequences of Smoking.* Rockville, MD: U.S. Dept. of Health and Human Services, Annual.

U.S. Environmental Protection Agency. *A Citizen's Guide to Radon*, 2nd ed. Washington, D.C.: U. S. Environmental Protection Agency, 1992.

U.S. Environmental Protection Agency. "Clean Air Act Amendments of 1990: Detailed Summary of Titles." Washington, D.C.: U.S. Environmental Protection Agency, 1990.

U.S. Environmental Protection Agency. *Home Buyer's and Seller's Guide to Radon.* Washington, D.C.: U.S. Environmental Protection Agency, 1993.

U.S. Environmental Protection Agency. "The New Clean Air Act." *EPA Journal* 17 (Jan/Feb 1991). Issue on the Clean Air Act of 1990.

Wager, Janet S. "Where the Rubber Meets the Road." *Nucleus* 18: (Fall, 1996), pp. 4–5, 8.

Whiteman, Lily. "Trades to Remember: The Lead Phasedown." *EPA Journal* (May/June 1992), pp. 38–39.

Chapter 16. Major Atmospheric Changes

Abbatt, Jonathan P. D., and Mario J. Molina. "Status of Strato-spheric Ozone Depletion." *Annual Review of Energy and the Environment.* 18 (1993), pp. 1–29.

Baker, Lawrence A., et al. "Acidic Lakes and Streams in the United States: The Role of Acidic Deposition." *Science* 252 (May 24, 1991), pp. 1151–1154.

Bazzaz, Fakhri A., and Eric D. Fajer. "Plant Life in a CO_2-Rich World." *Scientific American* 266 (Jan 1992), pp. 68–74.

Benedick, Richard E. *Ozone Diplomacy: New Directions in Safeguarding the Planet.* Cambridge, MA: Harvard University Press, 1991.

Broeker, Wallace S. "Chaotic Climate." *Scientific American* 273 (November 1995), pp. 62–68.

Charlson, Robert J., and Tom M. L. Wigley. "Sulfate Aerosol and Climatic Change." *Scientific American* (Feb 1994), pp. 48–57.

Decker, Jonathan P. "Miami Ice: Freon Smuggling Rivals Contraband in Drugs." *The Christian Science Monitor* (October 23, 1995), p.3.

Edgerton, Lynne T. *The Rising Tide: Global Warming and World Sea Levels.* Covelo, CA: Island Press, 1991.

Environmental Canada. *Downwind: The Acid Rain Story.* Ottawa, Ontario: Ministry of Supply and Services, 1981.

Fahnestock, Mark. "An Ice Shelf Breakup." *Science* 271 (9 February, 1996), pp. 775–776.

Flavin, Christopher and Odil Tunali. "Climate of Hope: New Strategies for Stabilizing the World's Atmospher." *WorldWatch Paper* 130 (June 1996). Worldwatch Institute.

Graedel, Thomas E., and Paul J. Crutzen. "The Changing Atmosphere." *Scientific American* 261 (Sept 1989), pp. 58–68.

Hordijk, Leen. "A Model Approach to Acid Rain." *Environment* 30 (Mar 1989), pp. 17–20, 40–41.

Houghton, J. T., et al., eds. *Climate Change: The IPCC Scientific Assessment.* New York: Cambridge University Press, 1990.

Kauppi, Pekka E. et al. "Biomass and Carbon Budget of European Forests, 1971–1990." *Science* 256 (Apr 3, 1992), pp. 70–74.

Kempton, Willett, and Paul P. Craig. "European Perspectives on Global Climate Change." *Environment* 35 (Apr 1993), pp. 16–20, 41–45.

Kerr, J. B., and C. T. McElroy. "Evidence for Large Upward Trends of Ultraviolet-B Radiation Linked to Ozone Depletion." *Science* 262 (Nov 12, 1993), pp. 1032–1034.

Kerr, Richard. "It's Official: First Glimmer of Greenhouse Warming Seen." *Science* 270 (8 December 1995), pp. 1565–1566.

Kerr, Richard A. "Antarctic Ozone Hole Fails to Recover. *Science* 266 (Oct 14, 1994), p. 217.

Kerr, Richard A. "U.S. Climate Tilts toward the Greenhouse." *Science* 268 (Apr 21, 1995), pp. 363–364.

Kowalok, Michael E. "Research Lessons from Acid Rain, Ozone Depletion, and Global Warming." *Environment* 35 (July/Aug 1993), pp. 12–20, 35–38.

Levine, Joel S. et al. "Biomass Burning: A Driver for Global Change." *Environmental Science and Technology* 29 (1995), pp. 120–125.

Likens, G.E., et al. "Long-term Effects of Acid Rain: Response and Recovery of a Forest Ecosystem. *Science* 272 (12 April 1996), pp. 244–246.

Minnis, P., et al. "Radioactive Climate Forcing by the Mount Pinatubo Eruption." *Science* 259 (Mar 5, 1993), pp. 1411–1415.

Mohnen, Volker A. "The Challenge of Acid Rain." *Scientific American* 259 (Aug 1988), pp. 30–38.

Montzka, Stephen, et al. "Decline in the Tropospheric Abundance of Halogen from Halocarbon: Implications for Stratospheric Ozone Depletion. *Science* 272: (31 May 1996), pp. 1318–1322.

Muller, Frank. "Mitigating Climate Change: The Case for Energy Taxes." *Environment* 38 (March 1996), pp. 12–20, 36–43.

Myers, Norman. "Environmental Refugees in a Globally Warmed World." *National Wildlife* 43 (Dec 1993), pp. 752–761.

National Academy of Sciences. *Acid Deposition: Long-Term Trends.* Washington, D.C.: National Academy Press, 1986.

National Academy of Sciences. *Policy Implications of Greenhouse Warming.* Washington, D.C.: National Academy Press, 1991.

Nerem, R.S. "Global Mean Sea Level Variations from TOPEX/POSEIDON Altimeter Data." *Science* 268 (5 May 1995), pp. 708–710.

Nordhaus, William D. "An Optimal Transition Path for Control-ling Greenhouse Gases." *Science* 258 (Nov 20, 1992), pp. 1315–1319.

Oerlemans, Johannes. "Quantifying Global Warming from the Retreat of Glaciers." *Science* 264 (Apr 8, 1994), pp. 243–245.

Rodriguez, Jose M. "Probing Stratospheric Ozone." *Science* 261 (Aug 27, 1993), pp. 1128–1129.

Rowland, F. Sherwood. "Chlorofluorocarbons and the Depletion of Stratospheric Ozone." *American Scientist* 77 (1989), pp. 36–45.

Rowland, F. Sherwood. "President's Lecture: The Need for Scientific Communication with the Public." *Science* 260 (June 11, 1993), pp. 1571–1576.

Rowlands, Ian H. "The Fourth Meeting of the Parties to the Montreal Protocol: Report and Reflection." *Environment* 35 (July/Aug 1993), pp. 25–34.

Rubin, Edward S., et al. "Realistic Mitigation Options for Global Warming." *Science* 257 (July 10, 1992), pp. 148–149, 261–266.

Schindler, D. W. "Effects of Acid Rain on Freshwater Ecosystems." *Science* 239 (Jan 8, 1988), pp. 149–153.

Schwartz, S. E. "Acid Deposition: Unraveling a Regional Phenomenon." *Science* 243 (Feb 10, 1989), pp. 753–761.

Service, Robert R. "Uncovering Threats to the Ozone Layer Brings Rewards." *Science* 270 (20 October 1995), pp. 381–382.

Stix, Gary. "Green Policies: Insureers Warm to Climate Change." *Scientific American* 274 (February 1996), pp. 27–28.

Stolarski, Richard, et al. "Measured Trends in Stratospheric Ozone."

Science 256 (Apr 17, 1992), pp. 342–349.
Tabazadeh, A., and R. P. Turco. "Stratospheric Chlorine Injection by Volcanic Eruptions: HCl Scavenging and Implications for Ozone." *Science* 260 (May 21, 1993), pp. 1082–1086.
Tolbert, Margaret A. "Sulfate Aerosols and Polar Stratospheric Cloud Formation." *Science* 264 (Apr 22, 1994), pp. 527–528.
Toon, Owen B., and Richard P. Turco. "Polar Stratospheric Clouds and Ozone Depletion." *Scientific American* 264 (June 1991), pp. 68–74.
Torrans, Ian M., et al. "The 1990 Clean Air Act Amendments: Overview, Utility Industry Responses, and Strategic Implications." *Annual Review of Energy and the Environment.* 17 (1992), pp. 211–237.
White, Robert M. "Our Climatic Future: Science, Technology, and World Climate Negotiations." *Environment* 33 (March 1991), pp. 18–20, 38–41.

Chapter 17. Pollution and Public Policy

Aggleton, Peter, et al. "Risking Everything? Risk Behavior, Behavior Change, and AIDS." *Science* 265 (July 15, 1994), pp. 341–345.
Ahearne, John F. "Integrating Risk Analysis into Public Policymaking." *Environment* 35 (Mar 1993), pp. 16–20, 37–40.
Alper, Joe. "Protecting the Environment with the Power of the Market." *Science* 260 (June 25, 1993), pp. 1884–1885.
Arrow, Kenneth J., et al. "Is There a Role for Benefit-Cost Analysis in Environmental, Health, and Safety Regulation?" *Science* 272 (12 April, 1996) pp. 221–222.
Bezdek, Roger H. "Environment and Economy: What's the Bottom Line?" *Environment* 35 (Sept 1993), pp. 6–11, 25–32.
Bhagwate, Jagdish, and Herman E. Daly. "Debate: Does Free Trade Harm the Environment?" *Scientific American* (Nov 1993), pp. 41–57.
Brown, Phil. "Popular Epidemiology Challenges the System." *Environment* 35 (Oct 1993), pp. 16–20, 32–41.
Bullard, Robert D. "Overcoming Racism in Environmental Decision-Making." *Environment* 36 (May 1994), pp. 10–19, 39–44.
Clarke, Lee. *Acceptable Risk? Making Decisions in a Toxic Environment.* Berkeley: University of California Press, 1989.
Cohen, Bernard L. "How to Assess the Risks You Face." *Consumer's Research* (June 1992), pp. 11–16.
Council on Environmental Quality. "Making the Environment Count: Costs, Benefits, Goals and Tools." Chapter 2 in *Environmental Quality: 21st Annual Report.* Washington, D.C.: Council on Environmental Quality, 1990.
Cruz, Wilfrido et al. "Greening Development: Environmental Implications of Economic Policies." *Environment* 38 (June 1996), pp. 6–11, 31–38.
Graham, J. D., et al., eds. *In Search of Safety: Chemicals and Cancer Risk.* Cambridge, MA: Harvard University Press, 1988.
Graham, John D. "Edging Toward Sanity on Regulatory Risk Reform." *Issues in Science and Technology*, Summer, 1995, pp. 61–66.
Kempton, William. "Will Public Environmental Concern Lead to Action on Global Warming?" *Annual Review of Energy and the Environment.* 18 (1992), pp. 217–245.
Kneese, Allan V. *Measuring the Benefits of Clean Air and Water.* Washington, D.C.: Resources for the Future, 1984.
Krupnick, Alan J., and Paul R. Portney. "Controlling Urban Air Pollution: A Benefit-Cost Assessment." *Science* 252 (Apr 26, 1991), pp. 522–528.
Kunreuther, Howard, and Ruth Patrick. "Managing the Risks of Hazardous Waste." *Environment* 33 (Apr 1991), pp. 12–15, 31–36.
Morgan, M. Granger. "Risk Analysis and Management." *Scientific American* (July 1993), pp. 32–41.
Nash, James A. "Moral Values in Risk Decisions." Ch. 14, pp. 195–212 in *Handbook for Environmetnal Risk Decision Making: Values, Perceptions and Ethics,* ed. C. Richard Cothern. Boca Raton, CRC Press/Lewis Publishers, 1996.
"Profiles in Risk Assessment." *EPA Journal* 19 (Jan/Mar 1993). Issue about risk assessment.
Repetto, Robert, and Roger C. Dower. "Green Fees: Charging for Environmentally Damaging Activities." *Environmental Science and Technology.* 27 (1993), pp. 214–216.
Russell, Milton, and Michael Gruber. "Risk Assessment in Environmental Policy-Making." *Science* 236 (Apr 17, 1987), pp. 286–290.
Sandman, Peter M. "Risky Business." *Rutgers Magazine* (April/ June, 1989), pp. 36, 37.
"Setting Environmental Priorities: The Debate about Risk." *EPA Journal* 17 (Mar/Apr 1991). Issue about risk analysis.
Stavins, Robert. "Harnessing Market Forces to Protect the Environment." *Environment* 31 (Jan/Feb 1989), pp. 4–7, 28–38.
Stavins, Robert, and Bradley W. Whitehead. "Pollution Charges for Environmental Protection: A Policy Link between Energy and Environment." *Annual Review of Energy and the Environment.* 17 (1992), pp. 187–210.
U.S. Environmental Protection Agency. *Reducing Risk: Setting Priorities and Strategies for Environmental Protection.* Washington, D.C.: Science Advisory Board, U.S. EPA, 1990.
Vernon, Raymond. "Behind the Scenes: How Policymaking in the European Community, Japan, and the United States Affects Global Negotiations." *Environment* 35 (June 1993), pp. 13–20, 35–43.
Wilson, Richard, and E. A. C. Crouch. "Risk Assessment and Comparisons: An Introduction." *Science* 236 (Apr 17, 1987), pp. 267–270.
Winsemius, Pieter, and Ulrich Guntram. "Responding to the Environmental Challenge." *Business Horizons* (Mar/Apr, 1992), pp. 12–20.
The World Bank. *World Development Report 1992: Development and the Environment.* New York: Oxford University Press, 1992.

Chapter 18. Wild Species: Diversity and Protection

Blaustein, Andrew R. and David B. Wake. "The Puzzle of Declining Amphibian Populations *Scientific American* 272 (April 1995), pp. 52–57.
Blum, Elissa. "Making Biodiversity Conservation Profitable." *Environment* 35 (May 1993), pp. 16–20, 38–45.
Butler, William A. "Incentives for Conservation." Chapter 11 in *Ecology, Economics, Ethics: The Broken Circle.* New Haven, CT: Yale University Press, 1991.
Council on Environmental Quality. "Linking Ecosystems and Biodiversity." Chapter 4 in *Environmental Quality: 21st Annual Report.* Washington, D.C.: Council on Environmental Quality, 1990.
Cox Paul Alan, and Michael J. Balick. "The Ethnobotanical Approach to Drug Discovery." *Scientific American* (June 1994), pp. 82–87.
Culotta, Elizabeth. "Is Marine Biodiversity at Risk?" *Science* 263 (Feb 18, 1994), pp. 918–919.
Eisner, Thomas, Jane Lubchenco, Edward O. Wilson, David S. Wilcove and Michael J. Bean. "Building a Scientifically Sound Policy for Protecting Endangered Species." *Science* 268 (1 September 1996), pp. 1231–1232.
Government of Canada. *The State of Canada's Environment.* Especially Chapter 6 "Wildlife: Maintaining Biological Diversity." Ottawa, Ontario: Government of Canada, 1991.
Graham, Frank, Jr. "Winged Victory." *Audubon* (July/Aug 1994), pp. 36–41.
Hargrove, Eugene C. *Foundations of Environmental Ethics.* Englewood Cliffs, NJ: Prentice Hall, 1989.
Karr, Paul. "The Scourge of Suburbia." *Massachusetts Audubon* 31 (Fall 1991), pp. 5–7.
Kumar, Ashok, et al. "How We Busted the Tiger Gang." *International Wildlife* 24 (May/June 1994), pp. 38–43.
Lazell, James. "The Evils of Exotics." *Massachusetts Audubon* 31 (Fall 1991), pp. 12–13.
Mann, Charles C. "Are Ecologists Crying Wolf?" *Science* 253 (Aug 16, 1991), pp. 736–738.

Mann, Charles C., and Mark L. Plummer. "California vs. Gnatcatcher." *Audubon* (Jan/Feb 1995), pp. 38–48.
Mann, Charles C., and Mark L. Plummer. "Is Endangered Species Act in Danger?" *Science* 267 (Mar 3, 1995), pp. 1256–1258.
May, Robert M. "How Many Species Inhabit the Earth?" *Scientific American* (Oct 1992), pp. 42–48.
McNeely, Jeffrey A., et al. "Strategies for Conserving Biodiversity." *Environment* 32 (Apr 1990), pp. 16–20, 36–41.
Mech, L. David. "The Howling: Dealing with the Wolf Boom." *Earthwatch* (Nov/Dec 1994), pp. 25–31.
Milstein, Michael. "Gray Wolves Reclaiming Their Land." The Boston Globe, April 15, 1996.
Myers, Norman. "Biological Diversity and Global Security." Chapter 2 in *Ecology, Economics, Ethics: The Broken Circle*. New Haven, CT: Yale University Press, 1991.
Myers, Steven Lee. "Wild Turkeys Roar Back from Near-Extinction." *The New York Times* Nov 24, 1991, p. 5.
Power, Thomas Michael. "The Wealth of Nature." *Issues in Science and Technology* (Spring, 1996), pp. 48–54.
Prance, G.T. "Biodiversity." pp. 183–193 in *Encylopedia of Environmental Biology*. New York: Academic Press.
Raustiala, Kal and David G. Victor. "Biodiversity Since Rio: The Future of the Convention on Biological Diversity." *Enviroment* 38 (May 1996), pp. 16–20, 37–43.
Reid, Walter V., and Kenton R. Miller. *Keeping Options Alive: The Scientific Basis for Conserving Biodiversity*. Washington, D.C.: World Resources Institute, 1989.
Roberts, Leslie. "Zebra Mussel Invasion Threatens U.S. Waters." *Science* 249 (Sept 21, 1990), pp. 1370–1372.
Rolston, Holmes, III. *Environmental Ethics: Duties to and Values in the Natural World*. Philadelphia: Temple University Press, 1988.
Rudloe, Anne, and Jack Rudloe. "Sea Turtles: In a Race for Survival." *National Geographic* 186 (Feb 1994), pp. 94–121.
Ryan, John C. "Life Support: Conserving Biological Diversity." *WorldWatch Paper 108*. Washington, D.C.: WorldWatch Institute, 1992.
Sharp, Bill, and Elaine Appleton. "Plight of the Plovers." *National Parks* 68 (Mar/Apr 1994), pp. 35–38.
Solbrig, Otto T. "The Origin and Function of Biodiversity." *Environment* 33 (June 1991), pp. 16–20, 34–39.
Soulé, Michael E. "Conservation: Tactics for a Constant Crisis." *Science* 253 (1991), pp. 744–750.
Steinfells, Peter. "Evangelical Group Defends Laws Protecting Endangered Species as a Modern 'Noah's Ark.'" *The New York Times* (Jan. 31, 1996).
Tear, Timothy, et al. "Status and Prospects for Success of the Endangered Species Act: A Look at Recovery Plans." *Science* 262 (Nov 12, 1993), pp. 976–977.
Terborgh, John. "Why American Songbirds Are Vanishing." *Scientific American* 266 (May 1992), pp. 98–104.
United Nations Environment Program. "*Global Biodiversity Assessment.*" Cambridge, UK: Cambridge University Press, 1995.
U.S. Environmental Protection Agency. *Ecology and Manage-ment of the Zebra Mussel and Other Introduced Aquatic Nuisance Species*. EPA/600/3-91/003. Washington, D.C.: U.S. EPA, 1991.
U.S. Fish and Wildlife Service. *Endangered and Threatened Species Recovery Program: Report to Congress*. Washington, D.C.: U.S. Dept. of the Interior, 1990.
Williams, Ted. "Invasion of the Aliens." *Audubon* (Sept/Oct 1994), pp. 24–32.
Wilson, Edward O. "Biodiversity, Prosperity, Value." Chapter 1 in *Ecology, Economics, Ethics: The Broken Circle*. New Haven, CT: Yale University Press, 1991.
Wilson, Edward O., ed. *Biodiversity*. Washington, D.C.: National Academy Press, 1988.
The World Bank. *World Development Report 1992: Development and the Environment*. New York: Oxford University Press, 1992.
World Resources Institute. "Biodiversity." Chapter 8 in *World Resources 1994–95*. Oxford University Press, New York, 1994.
Youth, Howard. "Flying into Trouble: The Global Decline of Birds, and What It Means." *WorldWatch* (Jan/Feb 1994), pp. 10–19.

Chapter 19. Ecosystems as Resources

Allen, Scott. "George's Bank: A Comeback Is No Sure Thing." *Boston Globe*, Dec 19, 1994, pp. 29, 33.
Alper, Joe. "How to Make the Forests of the World Pay Their Way." *Science* 260 (June 25, 1993), pp. 1895–1896.
Barker, Mary L., and Dietrich Soyez. "Think Locally, Act Globally? The Transnationalization of Canadian Resource-Use Conflicts." *Environment* 36 (June 1994), pp. 12–20, 32–36.
Bartlett, Ellen. "Culling the Herd." *Boston Globe* (July 24, 1995), pp. 25, 27. On killing elephants in Kruger National Park, South Africa.
Berger, John J., ed. *Environmental Restoration: Science and Strategies for Restoring the Earth*. Washington, D.C.: Island Press, 1990.
Bohnsack, James A. "Marine Reserves: They Enhance Fisheries, Reduce Conflicts, and Protect Resources." *Oceanus* (Fall 1993), pp. 63–71.
Brown, Barbara. "Coral Bleaching." *Scientific American* (Jan 1993), pp. 64–70.
Carothers, Andre. "Don't Beat the Retreat." *E Magazine* (Jan/Feb 1995), p. 64.
Colchester, Marcus. "The New Sultans: Asian Loggers Move in on Guyana's Forests. *The Ecologist* 24 (Mar/Apr 1994), pp. 45–52.
Deal, Carl. "The Greenpeace Guide to Anti-Environmental Organizations." Berkeley, CA: Odonian Press, 1992.
Denniston, Derek. "High Priorities: Conserving Mountain Ecosystems and Cultures." *WorldWatch Paper 123*. Washington, D.C. WorldWatch Institute, 1995.
Derr, Mark. "Redeeming the Everglades." *Audubon* (Sept/Oct 1993), pp. 48–56, 128–131.
Dixon, R. K., et al. "Carbon Pools and Flux of Global Forest Ecosystems." *Science* (Jan 14, 1994), pp. 185–190.
Durning, Alan Thein. "Guardians of the Land: Indigenous Peoples and the Health of the Earth." *WorldWatch Paper 112*. Washington, D.C.: WorldWatch Institute, 1992.
Durning, Alan Thein. "Saving the Forests: What Will It Take?" *WorldWatch Paper 117*. Washington, D.C.: WorldWatch Institute, 1993.
Fisher, Ron, et al. *The Emerald Realm: Earth's Precious Rain Forests*. Washington, D.C.: National Geographic Society, 1990.
Goldsmith, Edward, et al. *Imperiled Planet: Restoring Our Endangered Ecosystems*. Cambridge, MA: MIT Press, 1990.
Hardin, Garrett, and John Baden, eds. *Managing the Commons*. San Francisco: W. H. Freeman, 1977. Contains Hardin's original paper on the tragedy of the commons and many other papers related to the concept of the commons.
Howard, Malcolm. "Is US, Land Trusts Grow at Rapid Rate to Curb Development." *The Christian Science Monitor* (January 4, 1996), pp. 10, 11.
Jordan, William R., III, et al. *Restoration Ecology: A Synthetic Approach to Ecological Research*. Cambridge: Cambridge University Press, 1987.
Knickerbocker, Brad. "Activists' New Tune in the West: This Land Is Our Land." *The Christian Science Monitor*, Feb 7, 1995, pp. 1, 5.
Knickerbocker, Brad. "Beauty Fuels Growth Spurt in Northwest." *The Christian Science Monitor* (19 March 1996), pp. 10, 11.
Knickerbocker, Brad. "'New Forestry:' A Kinder, Gentler Approach to Logging." *The Christian Science Monitor*, June 20, 1995, pp. 10, 11.
Knickerbocker, Brad. "U.S. Eyes Opening 'High Risk' Forests." *The Christian Science Monitor*, Mar 16, 1995, p. 4.
Kusler, Jon A. "Wetlands." *Scientific American* (Jan 1994), pp. 64–70.
Kusler, Jon A., and Mary E. Kentula. *Wetland Creation and Restoration: The Status of the Science*. Washington, D.C.: Island Press, 1990.
McNeely, Jeffrey A., et al. *Conserving the World's Biological Resources: A Primer on Principles and Practice for Development Action*. Washington, D.C.: World Resources Institute, 1989.
Myers, Norman. "The World's Forests and Human Populations: The Environmental Interconnections." Pages 237–251 in *Resources, Environment and Population: Present Knowledge, Future*

Options. New York: Oxford University Press, 1991.
Myers, R.W. et al. "Population Dynamics of Exploited Fish Stocks at Low Population Levels." *Science* 269 (25 August 1995), pp. 1, 18.
Norse, Elliott A. *Ancient Forests of the Pacific Northwest*. Washington, D.C.: Island Press, 1990.
Oelschlager, Max. *The Idea of Wilderness: From Prehistory to the Age of Ecology*. New Haven, CT: Yale University Press, 1991.
Paarlberg, Robert L. "A Domestic Dispute: Clinton, Congress and International Enviromental Policy." *Environment* 38 (Oct. 1996), pp. 16–20, 28–33.
Polson, Shiela, "Common ground Found in the North Woods." *The Christian Science Monitor* (19 March 1996), pp. 10, 11.
Prugh, Thomas, et al. *"Natural Capital and Human Economic Survival."* Sunderland, MA: Sinauer Associates, Inc. 1995.
Repetto, Robert. "Accounting for Environmental Assets." *Scientific American* (June 1992), pp. 94–100.
Revkin, Andrew. *The Burning Season: The Murder of Chico Mendes and the Fight for the Amazon Rain Forest*. Boston: Houghton Mifflin, 1990.
Ross, Michael R. *Recreational Fisheries of Coastal New England*. Amherst, MA: University of Massachusetts Press, 1991.
Rutzler, Klaus, and Ilka C. Feller. "Caribbean Mangrove Swamps." *Scientific American* 274 (March 1996), pp. 94–99.
Sadler, A.E. ed. *"The Environment: Opposing Viewpoints."* San Diego: Greenhaven Press, 1996. The book contains arguments between "wise use" advocates and their opponents.
Safina, Carl. "The World's Imperiled Fish."*Scientific American* 273 (November 1995), pp. 46–53.
Schmidt, Karen. "Scientists Count a Rising Tide of Whales in the Sea." *Science* 263 (Jan 7, 1994), pp. 25–26.
Skole, David, and Compton Tucker. "Tropical Deforestation and Habitat Fragmentation in the Amazon: Satellite Data from 1978 to 1988." *Science* 260 (June 25, 1993), pp. 1905–1910.
Skore, Mari. "Whaling: A Sustainable Use of Natural Resources or a Violation of Animal Rights?" *Environment* 36 (Sept 1994), pp. 12–20, 30–31.
U.S. Fish and Wildlife Service. *Restoring America's Wildlife, 1937–1987*. Washington, D.C.: U.S. Dept. of the Interior, 1987.
Vincent, Jeffrey R. "The Tropical Timber Trade and Sustainable Development." *Science* 256 (June 19, 1992), pp. 1651–1655.
Wallace, Aubrey. "EcoHeroes: Twelve Tales of Environmental Victory." San Francisco: Mercury House, 1993.
Weber, Peter. "Abandoned Seas: Reversing the Decline of the Oceans." *WorldWatch Paper 116*. Washington, D.C.: WorldWatch Institute, 1993.
Weber, Peter. "Net Loss: Fish, Jobs, and the Maine Environemtns." *WorldWatch Paper* 120. Washington, D.C.: WorldWatch Institute, 1994.
Williams, Ted. "Whose Woods Are These? Endgame in the Northern Forest." *Audubon* (May/June 1994), pp. 26–33.
World Resources Institute. *World Resources 1994–1995*. New York: Oxford University Press, 1994. Excellent resource containing separate chapters on different habitats and extensive tables.

Chapter 20. Converting Trash to Resources

Boener, Christopher, and Kenneth Chilton. "False Economy: The Folly of Demand-Side Recycling." *Environment* 36 (Jan 1994), pp. 6–15, 32–33.
Buchholz, Rogene A., et al. "The Politics of Recycling in Rhode Island." *In Managing Environmental Issues: A Casebook*. Englewood Cliffs, NJ: Prentice Hall, 1992.
Bukro, Casey. "Have Trash, Will Travel." *Garbage* (Fall 1994), pp. 14–16.
The Composting Council of Canada. "Emphasis on Home Composting." *BioCycle* 34 (June 1993), p. 40.
Connett, Paul H. "The Disposable Society." Chapter 7 in *Ecology, Economics, Ethics: The Broken Circle*. New Haven, CT: Yale University Press, 1991.
EPA Journal (July/Aug 1992). Whole issue devoted to recycling.
Gaige, C. David, and Richard T. Halil Jr. "Clearing the Air about Municipal Waste Combustors." *Solid Waste & Power* (Jan/Feb 1992), pp. 12–17.
Grassy, John. "Bottle Bills." *Garbage* 4 (Jan/Feb 1992), pp. 44–48.
Grove, Noel, and Jose Azel. "Recycling." *National Geographic* 186 (July 1994), pp. 92–113.
Gutin, JoAnn C. "Here's Another Look at Plastics." *E Magazine* (May/June 1994), pp. 28–35.
Hay, Jonathan, et al. "Understanding the Total Waste Stream." *BioCycle* 34 (July 1993), pp. 40–42.
Hilts, Michael E. "WTE: Building a Record of Dependable Waste Disposal and Environmental Safety." *Solid Waste Technologies/Industry Sourcebook 1994*. pp. 12–16.
Kharbanda, O.P., and E. A. Stallworthy. *Waste Management: Toward a Sustainable Society*. New York: Auburn House, 1990.
Levenson, Howard. "Wasting Away: Policies to Reduce Trash, Toxicity and Quantity." *Environment* 32 (March 1990), pp. 10–15, 31–36.
O'Leary, Philip R., et al. "Managing Solid Waste." *Scientific American* 259 (Dec 1988), pp. 36–42.
Platt, Brenda, et al. *Beyond 40 Percent: Record-Setting Recycling and Composting Programs*. Washington, D.C.: Island Press, 1991.
Preer, Robert. "Sealed and Buried, Closed Dumps Get a Life." *Boston Globe*, Oct 10, 1993, pp. 28, 30.
Reidy, Chris. "Economics of Recycling Paper Take a Tumble." *The Boston Globe,*, July 24, 1996, pp. 1, 16.
Russell, Gerard F. "Trash Bag Fee Buys Recyclers: Worcester's Rules Pay Off in Compliance." *Boston Globe* Jan 13, 1994, pp. 25, 31.
Scherer, Ron. "Staten Island Has Had Its Fill of Nation's Largest Landfill." *The Christian Science Monitor* (April 26, 1996), pp. 1,4.
Seidman, Ethan, et al. "An Inside Look at Paper Recycling." *Garbage* (Sept/Oct 1993), pp. 30–37.
Stavins, Robert N., and Bradley W. Whitehead. "Polluton Charges for Environmental Protection: A Policy Link between Energy and the Environment." *Annual Review of Energy and the Environment* 17 (1992), pp. 187–210.
Tyson, James L. "Your Old Newspapers Now Worth a Bundle." *The Christian Science Monitor*, Oct 14, 1994, p. 8.
United Nations Environment Program. "Promoting Waste Recycling, Part I and II. *"Industry and Environment,* Vol. 17, Nos. 2 and 3 (1994) feature articles on recycling in the international arena.
U.S. Environmental Protection Agency. "Characterization of Municipal Solid Waste in the United States, 1994 Update." Report No. EPA 530-S-94-042. Washington, D.C.: U.S. EPA, 1994.
U.S. Environmental Protection Agency. *The Solid Waste Dilemma, An Agenda for Action*. EPA/530-SW-89-019. Washington, D.C.: U.S. EPA, Office of Solid Waste, 1989.
Wolf, Nancy, and Ellen Feldman. *America's Packaging Dilemma*. Covelo, CA: Island Press, 1990.
Young, John E. "Discarding the Throwaway Society." *WorldWatch Paper 101*. Washington, D.C.: WorldWatch Institute, 1991.
Young, John E. "Mining the Earth." *WorldWatch Paper 109*. Washington, D.C.: WorldWatch Institute, 1992.
Young, John E. "The Sudden New Strength of Recycling." *WorldWatch 8* (July/August 1995), pp. 20–25.
Young, William. "A Tree Grows on Fresh Kills." *Garbage* (Summer 1994), pp. 60–61.

Chapter 21. Energy Resources: The Rise (and Fall?) of Fossil Fuels

Asmus, Peter. "Saving Energy Becomes Company Policy." *The Amicus Journal* 14 (Winter 1993), pp. 38–42.
Ayres, Ed. "Breaking Away." *WorldWatch* 6 (Jan/Feb 1994), pp. 10–18.
Baumgarten, Fred. "America's Arctic Refuge—At the Crossroads." *Audubon Activist* (Apr 1991), pp. 1, 4.
Cooper, George A. "Directional Drilling." *Scientific American* 270 (May 1994), pp. 82–87.

Corcoran, Elizabeth. "Cleaning Up Coal." *Scientific American* 264 (May 1991), pp. 106–117.
Cullen, Robert. "The True Cost of Coal." *Atlantic Monthly* (Dec 1993), pp. 38–52.
Decicco, John, and Marc Ross. "Improving Automotive Efficiency." *Scientific American* 271 (Dec 1994), pp. 52–59.
Dillin, John. "Natural Gas Seeps into Spotlight." *The Christian Science Monitor*, March 8, 1991), p. 3.
Dillin, John. "As US Oil Output Falls, Reliance on Imports Grows." *The Christian Science Monitor*, Jan 28, 1994, p.2.
Drew, Lisa. "Truth and Consequence along Oiled Shores." *National Wildlife* 28 (June/July 1990), pp. 34–43.
"Energy for Planet Earth." *Scientific American* (Sept 1990). Special issue.
Feldman, David Lewis. "Revisiting the Energy Crisis: How Far Have We Come?" *Environment* 37 (May 1995), pp. 16–20.
Fickett, Arnold. "Efficient Use of Electricity." *Scientific American* 263 (Sept 1990), pp. 65–74.
Flavin, Christopher. "The Bridge to Clean Energy." *WorldWatch* 5 (Aug 1992), pp. 10–18.
Flavin, Christopher. "Building a Bridge to Sustainable Energy." Pages 3–45 in *State of the World*. New York: W. W. Norton, 1992.
Flavin, Christopher, and Tunali Odil. "Getting Warmer." *WorldWatch* 8 (Mar/Apr 1995), pp. 10–19.
Freedman, David H. "Batteries Included." *Discover* 13 (Mar 1992), pp. 90–99.
Gross, Daniel. "The College Eco-Olympics." *E Magazine* (Mar/Apr 1994), pp. 32–33.
Harris, Mark. "Energy-Efficient Appliances: Lowering the Cost of Plugging In." *E Magazine* (Nov/Dec 1990), pp. 58–61.
Hirst, Eric. "A Bright Future: Energy Efficiency Programs at Electric Utilities." *Environment* 36 (Nov 1994), pp. 10–15.
Hubbard, Harold M. "The Real Cost of Energy." *Scientific American* 264 (Apr 1991), pp. 36–43.
Hwang, Roland J. "All Charged Up." *Nucleus* 15 (Fall 1993), pp. 1–3.
Jackson, Julia. "Perspectives on Petroleum." *Geotimes* (July 1995), pp. 17–19.
Landay, Jonathan S. "US Sees Threat in Buildup by Iran Near Shipping Lanes." *The Christian Science Monitor,* Mar 23, 1995, pp. 1–8.
Lenssen, Nicholas. "All the Coal in China." *WorldWatch* 6 (Mar/Apr 1993), pp. 22–29.
Lenssen, Nicholas. "Empowering Development: The New Energy Equation." *WorldWatch Paper 111*. Washington, D.C.: WorldWatch Institute, 1992.
Lewis, Thomas A. "The States Seize Power." *National Wildlife* 31 (Dec/Jan 1993), pp. 16–19.
Lofstedt, Ragnar E. "Hard Habits to Break: Energy Conservation Patterns in Sweden." *Environment* 35 (Mar 1993), pp. 10–15.
Lovins, Amory B. "Energy, People, and Industrialization." Pages 95–124 in *Resources, Environment, and Population*. New York: Oxford University Press, 1991.
Masters, C. D., et al. "Resource Constraints in Petroleum Production Potential." *Science* 253 (July 12, 1991), pp. 146–152.
Miller, William H. "Balance Sought: Energy, Environment, Economy." *Industry Week* (Apr 1, 1991), pp. 62–65, 68–70.
Muckleston, Keith W. "Salmon vs. Hydropower: Striking a Balance in the Pacific Northwest." *Environment* (Jan/Feb 1990), pp. 10–15.
Oppenheimer, Ernest J. "Raise the Gasoline Tax." *The Christian Science Monitor*, June 10, 1993, p. 19.
Pendleton, Scott. "Looking for Oil." *The Christian Science Monitor*, June 20, 1994, pp. 9–11.
Roodman, David Malin. "The Obsolescent Incandescent." *WorldWatch* 6 (May/June 1993), pp. 5–7.
Ross, Mark, et al. *Converting a Gasoline-Powered Vehicle to Run on Natural Gas*. Cambridge, MA: UCS Publications, 1994.
Starr, C., et al. "Energy Sources: A Realistic Outlook." *Science* (May 15, 1992), pp. 981–986.
Steiner, Rick. "Probing an Oil-Stained Legacy." *National Wildlife* 31 (Apr/May 1993), pp. 4–11.
Wager, Janet S. "How Much Are Fossil Fuels Really Costing You?" *Nucleus* 15 (Summer 1993), pp. 4–7.
Wilbanks, Thomas J. "Improving Energy Efficiency: Making a 'No-Regrets' Option Work." *Environment* 36 (Nov 1994), pp. 16–20.
Zuchermann, Wolfgang. *End of the Road*. Post Mills, VT: Chelsea Green Publishing, 1992.

Chapter 22. Nuclear Power: Promise and Problems

Anspaugh, Lynn R., et al. "The Global Impact of the Chernobyl Reactor Accident." *Science* 242 (Dec 16, 1988), pp. 1513–1519.
Balter, Michael. "Children Become the First Victims of Fallout." *Science* 272 (19 April 1996), pp. 357–360.
Barkenbus, Jack N. and Charles Forsberg. "Internationalizing Nuclear Safety: The Pursuit of Collective Responsiblility." *Annual Review of Energy and the Environment* 20 (1995), pp. 83–118.
Bryce, Robert. "Nuclear Waste's Last Stand: Apache Land." *The Christian Science Monitor,* Sept 2, 1994, pp. 6, 7.
Chandler, David L. "Dismantling Yankee Rowe." *Boston Globe,* June 5, 1995, pp. 25, 27.
Cochran, Thomas B., et al. "Radioactive Contamination at Chelyabinsk-65, Russia." *Annual Review of Energy and the Environment* 18 (1993), pp. 507–528.
Conn, Robert W., et al. "The International Thermonuclear Experimental Reactor." *Scientific American* 266 (Apr 1992), pp. 103–110.
Cowen, Robert C. "Scientists Announce Fusion Breakthrough at Princeton U. Lab." *The Christian Science Monitor*, Nov 7 1994, p. 5.
Energy Information Administration. *International Energy Outlook 1991: A Post-War Review of Energy Markets.* DOE/EIA-0484(91). Washington, D.C.: U.S. Dept. of Energy, 1991.
Flavin, Christopher. "Reassessing Nuclear Power: The Fallout from Chernobyl." *WorldWatch Paper 75.* Washington, D.C.: WorldWatch Institute, 1987.
Flynn, James and Paul Slovic. "Yucca Mountain: A Crisis for Policy: Propects for America's High-Level Nuclear Waste Program. *Annual Review of Energy and the Environment* 20 (1995), pp. 83–118.
Forsberg, C. W., and A. M. Weinberg. "Advanced Reactors, Passive Safety, and Acceptance of Nuclear Energy." *Annual Review of Energy* 15 (1990), pp. 133–152.
Friedlander, Gerhart, et al. *Nuclear and Radiochemistry*, 3d ed. New York: John Wiley and Sons, 1981.
Furth, Harold P. "Fusion." *Scientific American* 273 (September 1995), pp. 174–177.
Golay, Michael W., Neil E. Todreas. "Advanced Light-Water Reactors." *Scientific American* 262 (Apr 1990), pp. 82–89.
Haefele, Wolf. "Energy from Nuclear Power." *Scientific American* 263 (Sept 1990), pp. 137–144.
Holdren, J. P. "Safety and Environmental Aspects of Fusion Energy." *Annual Review of Energy and the Environment* 16 (1991), pp. 235–258.
Lenssen, Nicholas. "Chernobyl, Firsthand." *WorldWatch* (Jan/Feb 1994), p. 8.
Lenssen, Nicholas. "Nuclear Waste: The Problem That Won't Go Away." *WorldWatch Paper 106.* Washington, D.C.: WorldWatch Institute, 1991.
Lichtenstein, Kenneth, and Ira Helford. "Radiation and Health: Nuclear Weapons and Nuclear Power." In *Critical Condition: Human Health and the Environment.* Cambridge, MA: MIT Press, 1993, pp. 93–122.
Makhijani, Arjun, Howard Hu and Katherine Yik, eds. "*Nuclear Wasteland: A Global Guide to Nuclear Weapons Production and Its Health and Environmental Effects.*" Cambridge, MA: The MIT Press, 1995.
McCafferty, nell. "Life After Chernobyl." *Audubon* 98 (May–June, 1996), pp. 66–75.
Medvedev, Zhores A. *The Legacy of Chernobyl.* New York: W. W.

Norton, 1990.
Morgan, M. Granger. "What Would It Take to Revitalize Nuclear Power in the United States?" *Environment* 35 (Mar 1993), pp. 6–9, 30–33.
Pooley, Eric. "Nuclear Warriors." Time (March 4, 1996), pp. 47–54.
Razavi, Hossein, and Fereidun Fesharaki. "Electricity Generation in Asia and the Pacific: Historical and Projected Patterns of Demand and Supply." *Annual Review of Energy and the Environment* 16 (1991), pp. 275–294.
Rochlin, Gene I., and Alexandra von Meier. "Nuclear Power Operations: A Cross-Cultural Perspective." *Annual Review of Energy and the Environment* 19 (1994), pp. 153–187.
Rosen, Yereth. "USSR Leaves Radioactive Legacy." *The Christian Science Monitor*, Aug 26, 1993, p. 8.
Rossin, A. D. "Experience of the U.S. Nuclear Industry and Requirements for a Viable Nuclear Industry in the Future." *Annual Review of Energy* 15 (1990), pp. 153–172.
Shcherbak, Yuri M. "Ten Years of the Chornobyl Era." *Scientific American* 274 (April 1996)
Slovic, Paul, et al. "Lessons from Yucca Mountain." *Environment* 33 (Apr 1991), pp. 7–11, 28–31.
Slovic, Paul, et al. "Perceived Risk, Trust, and the Politics of Nuclear Waste." *Science* 254 (Dec 13, 1991), pp. 1603–1608.
Spotts, Peter N. "Nuclear Power Plants Losing Steam." *The Christian Science Monitor*, Apr 4, 1995, p. 4.
Spotts, Peter N. "The Road Ahead for Nuclear Safety." *The Christian Science Monitor*, Mar 23, 1995, p. 4.
Stone, Richard. "The Explosions that Shook the World." *Science* (19 April 1996), pp. 352–354.
Taubes, Gary. "No Easy Way to Shackle the Nuclear Demon." *Science* 263 (Feb 4, 1994), pp. 629–631.
Vogelsang, W. F.,and H. H. Barschall. "Nuclear Power." Pages 127–152 in *The Energy Sourcebook: A Guide to Technology, Resources, and Policy*. New York: American Institute of Physics, 1991.
Wheelwright, Jeff. "For Our Nuclear Wastes, There's Gridlock on the Way to the Dump." *Smithsonian* (May 1995), pp. 40–51
Whipple, Chris G. "Can Nuclear Waste Be Stored Safely at Yucca Mountain?" *Scientific American* 274 (June 1996), pp. 40–44.
Williams, Phil and Paul N. Woessner. "The Real Threat of Nuclear Smuggling." *Scientific American* 274 (January 1996), pp. 40–44.
Zorpette, Glenn. "Hanford's Nuclear Wasteland." *Scientific American* 274 (May 1996), pp. 88–97.

Chapter 23. Solar and Other Renewable Energy Resources

Aitken, Donald, and Paul Bony. "Passive Solar Production Housing and the Utilities." *Solar Today* 7 (Mar/Apr 1993), pp. 23–26.
Aitken, Donald, and Paul Neuffer. "Low Cost, High Value Passive Solar." *Solar Today* 9 (Mar/Apr 1995), pp. 18–20.
Anderson, Bruce, with Michael Riordan. *The New Solar Home Book*. Andover, MA: Brick House Publishing, 1987.
Awerbuch, Shimon. "The Case for Public Investment in Photovoltaic Technology." *Solar Today* 8 (July/Aug 1994), pp. 24–26.
Bain, Richard L., and Jim Jones. "Renewable Electricity from Biomass." *Solar Today* 7 (May/June 1993), pp. 21–23.
Charters, W.W.S. "Solar Energy: A Viable Pathway toward ecologically sustainable development." *Solar Energy* 53 (Oct 1994), pp. 311–314.
Crowley, Larry A. "Utility-Owned Remote PV Systems." *Solar Today* 7 (Mar/Apr 1993), pp. 29–30.
Denniston, Derek. "Second Wind." *WorldWatch* 6 (Mar/Apr 1993), pp. 33–35.
Dobb, Edwin. "Solar Cooker." *Audubon* 94 (Nov/Dec 1992), pp. 100–109.
Dostrovsky, Israel. "Chemical Fuels from the Sun." *Scientific American* 265 (Dec 1991), pp. 102–107.
Evans, Lori C. "Wind Energy in Europe." *Solar Today* 6 (May/June 1992), pp. 32–34.
Flavin, Christopher, and Nicholas Lenssen. "Beyond the Petroleum Age: Designing a Solar Economy." *WorldWatch Paper 100*. Washington, D.C.: WorldWatch Institute, 1990.
Flavin, Christopher, and Nicholas Lenssen. "Powering the Future toward a Sustainable Electricity Industry." *Solar Today* 8 (Sept/Oct 1994), pp. 26–28.
Flavin, Christopher, and Nicholas Lenssen. *Power Surge*. New York: W. W. Norton, 1994.
Greenberg, David A. "Modeling Tidal Power." *Scientific American* 257 (Nov 1987), pp. 128–133.
Haas, R. "The Value of Photovoltaic Electricity for Society." *Solar Energy* 54 (Jan 1995), pp. 25–32.
Haggard, Kenneth. "Straw Bale Passive Solar Construction." *Solar Today* 7 (May/June 1993), pp. 17–20.
Harvey, Hal. "Innovative Policies to Promote Renewable Energy." *Solar Today* 7 (Nov/Dec 1993), pp. 23–25.
Herber, B. P., and J. T. Raga. "An International Carbon Tax to Combat Global Warming." *American Journal of Economics and Sociology* 54.3 (July 1995), pp. 257–267.
Kerr, Richard A. "Geothermal Tragedy in the Commons." *Science* 253 (July 12, 1991), pp. 134–135.
Kozloff, Keith Lee. "Renewable Energy Technology: An Urgent Need, A Hard Sell." *Environment* 36 (Nov 1994), pp. 4–9.
Lehman, Peter, and Christine Parra. "Hydrogen Fuel from the Sun." *Solar Today* 8 (Sept/Oct 1994), pp. 20–25.
Lenssen, Nicholas. "Third World PVs Hit the Roof." *WorldWatch* 5 (May/June 1992), pp. 7–8.
Lindley, David, et al. "Wind Energy." *SunWorld* 19 (June 1995), pp. 2–13.
Lipske, Michael. "Playing for Power in Quebec's North." *International Wildlife* 21 (May/June 1991), pp. 10–17.
Lotker, Michael, and David Kearney. "Solar Thermal Electric Performance and Prospects: The View from Luz." *Solar Today* (May/June 1991), pp. 10–13.
Lynd, Lee R., et al. "Fuel Ethanol from Cellulosic Biomass." *Science* 251 (Mar 15, 1991), pp. 1318–1323.
Melody, Ingrid. "Florida's Hydrogen Research." *Solar Today* 7 (Sept/Oct 1993), pp. 14–16.
Muckleston, Keith W. "Salmon vs. Hydropower: Striking a Balance in the Pacific Northwest." *Environment* 32 (Jan/Feb 1990), pp. 10–15.
Nelson, Erik, and Cecile Leboeuf. "Sunrayce 93." *Solar Today* 8 (Jan/Feb 1994), pp. 21–22.
Ogden, Joan M., and Robert H. Williams. *Solar Hydrogen, Moving beyond Fossil Fuels*. Washington, D.C.: World Resources Institute, 1989.
Pimentel, David. *Grains for Food or Fuel*. Energy Policy Research and Information Program, Publication Series No. 81-3. West Lafayette, IN: Purdue University, 1981.
Pimental, David. "Renewable Energy: Economic and Environmental Issues." *BioScience* 44 (Sept 1994), pp. 536–547.
Probhu, Edan. "SCE's Innovative Solar Neighborhood Program." *Solar Today* 9 (July/Aug 1995), pp. 22–29.
Rawlings, Lyle and Mark Kapner. "The New PV Solar Homes." *Solar Today* 8 (Jan/Feb 1994), pp. 26–28.
Roodman, David Malin, and Nicholas Lenssen. "Blueprint for Better Buildings." *Solar Today* 9 (July/Aug 1995), pp. 34–37.
Selengut, Stanley. "Sustainable Development in Paradise." *Solar Today* 8 (Sept/Oct 1994), pp. 16–19.
Shapiro, Andrew M. *The Homeowner's Complete Handbook for Add-On Solar Greenhouses and Sunspaces*. Emmaus, PA: Rodale Press, 1985.
Sheffer, Marcue B. "Solar Water Heating in Pennsylvania." *Solar Today* 8 (Jan/Feb 1994), pp. 12–15.
Shugar, Daniel S., et al. "Photovoltaic Grid Support: A New Screening Methodology." *Solar Today* 7 (Sept/Oct 1993), pp. 21–24.
Smith, Charles. "Revisiting Solar Power's Past." *Technology Review* (July 1995), pp. 38–47.
Thayer, Burke Miller. "Passive Solar Rowhouses." *Solar Today* 9 (July/Aug 1995), pp. 38–41.

Thomas, C. E. "Solar Hydrogen: A Sustainable Energy Option." *Solar Today* 7 (Sept/Oct 1993), pp. 11–13.
Wager, Janet S. "Renewables in the Midwest." *Solar Today* 8 (Mar/Apr 1994), pp. 16–18.

Chapter 24. Lifestyle and Sustainability

Bacow, Adele Fleet. *Designing the City.* Washington, D.C.: Island Press, 1995.
Berton, Valerie. "Farming on the Edge." *American Farmland* (Summer 1993), pp. 10–14, 16–17.
Berton, Valerie. "Reurbanizing America." *American Farmland* (Winter 1995), pp. 12–13.
Bidwell, Dennis P. "Private Options for Farmland Protection." *American Farmland* (Winter 1994), pp. 18–20.
"A Blueprint for Farmland Protection." *American Farmland* (Fall 1993), pp. 18–20.
The Conservation Foundation. "Will We Live in Accidental Cities or Successful Communities?" In *Conservation Foundation Letter.* Washington, D.C.: The Conservation Foundation, 1987.
Earth Works Group. *The Next Step: 50 More Things You Can Do to Save the Earth.* Kansas City, MO: Andrews and McMeel, 1991.
Environmental Careers Organization. *The New Complete Guide to Environmental Careers.* Washington, D.C.: Island Press, 1993.
French, Hilary F. "After the Earth Summit: The Future of Environmental Governance." *WorldWatch Paper 107.* Washington, D.C.: WorldWatch Institute, 1992.
Fuller, Millard, and Diane Scott. *No More Shacks.* Waco, TX: Word Books, 1986.
Gangloff, Deborah, et al. "Urban Forests." Special section. *American Forests* (May/June 1995), pp. 30–37.
Grauch, Sarah. "Egypt's Riddle: A Beltway—Or Sphinx and Pyramids." *The Christian Science Monitor,* Jan 11, 1995, pp.1–4.
Gravitz, Alisa, et al. "Creating a Sustainable Society" and following articles. *Co-op America Quarterly* (Summer 1992), pp. 11–24.
Hair, Jay D., et al. "NWF at Work." *International Wildlife* 24 (May/June 1994), pp. 25–28.
Hamcock, David. "Creativity Helps Cites Find Development Dollars." *American City and County* (May 1995), pp. 40–52.
Heuer, Robert. "Growth Management in the Midwest." *American Farmland* (Fall 1993), pp. 4–7.
Hiss, Tony, et al. "Special Section: Cities." *The Amicus Journal* 14 (Summer 1992), pp. 11–36.
Jackson, Wes. "Listen to the Land." *The Amicus Journal* 15 (Spring 1993), pp. 32–41.
Leonetti, Carol, and Jim Motavalli. "Seriously Green." *E Magazine* 6 (July/Aug 1995), pp. 36–41.
Lerner, Steve. "Suburban Wilderness." *The Amicus Journal* 16 (Spring 1994), pp. 14–17.
Lowe, Marcia D. "Alternatives to the Automobile: Transport for Livable Cities." *WorldWatch Paper 98.* Washington, D.C.: WorldWatch Institute, 1990.
Lowe, Marcia D. "China's Shrinking Cropland." *WorldWatch* 2 (July/Aug 1989), pp. 9–36.
Lowe, Marcia D. "Reclaiming Cities for People." *WorldWatch* 5 (Aug 1992), pp. 19–25.
Lowe Marcia D. "Shaping Cities: The Environmental and Human Dimensions." *WorldWatch Paper 105.* Washington, D.C.: WorldWatch Institute, 1991.
Lyman, Francesca. "Reinventing Suburbia." *The Amicus Journal* 14 (Summer 1992), pp. 18–23.
Margolis, Mac. "A Third-World City That Works." *World Monitor* 5 (Mar 1992), pp. 42–51.
Mills, Judy. "Rails-to-Trails: An Exercise in Linear Logic." *Smithsonian* 21 (Apr 1990), pp. 132–145.
Moore, Curtis. "Greenest City in the World." *International Wildlife* 24 (Jan/Feb 1994), pp. 38–43.
Morris, David. *Getting from Here to There: Building a Rational Transportation System.* Washington, D.C. Institute for Local Self-Reliance, 1992.
Newcomb, Duane. *Small Space, Big Harvest.* Rockland, CA.: Prima.
Newman, Morris. "New Urbanism Meets Nature." *Landscape Architecture* (Aug 1995), pp. 60–63.
Ojima D. S., et al. Special Section Articles: "The Global Impact of Land-use Change" and following articles. *BioScience* 44 (May 1994), pp. 300–356.
Oppenheimer, Michael. "Context, Connection, and Opportunity in Environmental Problem Solving." *Environment* 37 (June 1995), pp. 10–15.
Pirro, Joe F. "Two-Wheel Highway." *The Amicus Journal* 16 (Spring 1994), pp. 11–13.
Poole, William. "In Land We Trust." *Sierra* 77 (Mar/Apr 1992), pp. 52–58.
Rawcliffe, Peter. "Making Inroads: Transport Policy and the British Environmental Movement." *Environment* 37 (Apr 1995), pp. 16–20.
Renew America. "Environmental Success Index 1992." Washington, D.C.: Renew America, 1992. A compilation of noteworthy environmental projects throughout the country, with contact names and telephone numbers.
Renew America. "National Awards for Environmental Sustainability." *Renew America* (Jan 1995).
Renner, Michael. "Creating Sustainable Jobs in Industrial Countries." Pages 138–154 in *State of the World 1992.* New York: WorldWatch Institute, W. W. Norton, 1992.
Rockwell, Fulton. "Inner City Composting Yields Diversion and Vegetables." *Journal of Composting & Recycling* 35 (Aug 1994), pp. 74–77.
Rudig, Wolfgang. "Green Party Politics around the World." *Environment* 33 (Oct 1991), pp. 6–9.
Russell, Dick, et al. "Special Section: Green Architecture." *The Amicus Journal* 15 (Summer 1993), pp. 14–23.
Sachs, Aaron. "The Nature of a City." *WorldWatch* 6 (Sept/Oct 1993), pp. 35–37.
Sessions, Kathy. "Building the Capacity for Change." *EPA Journal* 19 (Apr–June 1993), pp. 15–19.
Shane, Scott. "City seeks Third World Remedies to Cure Its Ills." *Baltimore Sun* June 6, 1994, pp. 1, 8.
Sudol, Frand, and Allen Dresdner. "The Sustainable City." *American City and County* (Mar 1995), pp. 6–11.
Thayer, Burke Miller. "Esperanza de Sol: Sustainable, Affordable Housing." *Solar Today* 8 (May/June 1994), pp. 20–23.
Whyte, William H. *City. Rediscovering the Center.* New York: Doubleday, 1988.
Zorc, Anne. "Sustainable Communities from the Grassroots." *Co-op America Quarterly* 5 (Spring 1993), pp. 15–18.

Glossary

abiotic. Pertaining to factors or things that are separate and independent from living things; nonliving.

absolute poverty. The lack of sufficient income in cash or exchange items for meeting the most basic human needs for food, clothing, and shelter.

acid. Any compound that releases hydrogen ions when dissolved in water. Also, a water solution that contains a surplus of hydrogen ions.

acid deposition. Any form of acid precipitation and also fallout of dry acid particles. (See **acid precipitation**)

acid precipitation. Includes acid rain, acid fog, acid snow, and any other form of precipitation that is more acidic than normal, i.e., less than pH 5.6. Excess acidity is derived from certain air pollutants, namely sulfur dioxide and oxides of nitrogen.

activated sludge. Sludge made up of clumps of living organisms feeding on detritus that settles out and is recycled in the process of secondary wastewater treatment.

activated sludge system. A system for removing organic wastes from water. The system uses microorganisms and active aeration to decompose such wastes. The system is used most as a means of secondary sewage treatment following the primary settling of materials.

active safety features. Those safety features of nuclear reactors that rely on operator-controlled reactions, external power sources, and other features that are capable of failing. (See **passive safety features**)

adaptation. An ecological or evolutionary change in structure or function that produces better adjustment of an organism to its environment and hence enhances its ability to survive and reproduce.

adsorption. The process whereby chemicals (ions or molecules) stick to the surface of other materials.

aeration. *Soil:* The exchange within the soil of oxygen and carbon dioxide, necessary for the respiration of roots. *Water:* The bubbling of air or oxygen through water to increase the dissolved oxygen.

age structure. Within a population, proportions of people who are old, middle-aged, young adults, and children.

air pollution disaster. Short-term situation in industrial cities in which intense industrial smog brings about significant increase in human mortality.

air toxics. A category of air pollutants including radioactive materials and other toxic chemicals that are present at low concentrations but are of concern because they often are carcinogenic.

alga, pl. **algae.** Any of numerous kinds of photosynthetic plants that live and reproduce entirely immersed in water. Many species, the planktonic forms, exist as single or small groups of cells that float freely in the water. Other species, the "seaweeds," may be large and attached.

algal bloom. A relatively sudden development of a heavy growth of algae, especially planktonic forms. Algal blooms generally result from additions of nutrients, whose scarcity is normally limiting.

alleles. The two or more variations of a gene for any particular characteristic, e.g., blue and brown are alleles of the gene for eye color.

alternative farming. Farming methods designed to minimize the use of agricultural chemicals.

ambient standards. Air quality standards (set by the EPA) stating that outside average air should always maintain a level of purity. That is, certain levels of pollution should not be exceeded in order to maintain environmental and human health.

anaerobic. Oxygen-free.

anaerobic digestion. The breakdown of organic material by microorganisms in the absence of oxygen. The process results in the release of methane gas as a waste product.

anaerobic respiration. Respiration carried on by certain bacteria in the absence of oxygen. Methane, which can be used as fuel gas (it is the same as natural gas), may be a byproduct of the process.

anthropogenic. Referring to pollutants and other forms of impacts on natural environments that can be traced to human activities.

appropriate technology. Technology that seeks to increase the efficiency and productivity of hand labor without displacing workers. That is, it seeks to enable people to improve their well-being without disrupting the existing social and economic system.

aquaculture. A propagation and/or rearing of any aquatic (water) organism in a more or less artificial system.

aquifer. An underground layer of porous rock, sand, or other material that allows the movement of water between layers of nonporous rock or clay. Aquifers are frequently tapped for wells.

artificial selection. Plant and animal breeders' practice of selecting individuals with the greatest expression of desired traits to be the parents of the next generation.

asbestos fibers. Crystals of asbestos, a natural mineral, that have the form of minute strands.

atom. The fundamental unit of all elements.

autotroph. Any organism that can synthesize all its organic substances from inorganic nutrients, using light or certain inorganic chemicals as a source of energy. Green plants are the principal autotrophs.

background radiation. Radioactive radiation that comes from natural sources apart from any human activity. We are all exposed to such radiation.

bacterium, pl. **bacteria.** Any of numerous kinds of microscopic organisms that exist as simple, single cells that multiply by simple division. Along with fungi, they constitute the decomposer component of ecosystems. A few species cause disease.

balanced herbivory. A diversified plant community held in balance by various herbivores specific to each plant species.

bar screen. A set of iron bars about an inch apart used to screen debris out of wastewater.

base. Any compound that releases hydroxyl ions (OH^-) when dissolved in water. A solution that contains a surplus of hydroxyl ions.

bedload. The load of coarse sediment, mostly coarse silt and sand, that is gradually moved along the bottom of a riverbed by flowing water rather than being carried in suspension.

benefit-cost analysis. An analysis and/or comparison of the value benefits in contrast to the costs of any particular action or project. (See **cost-benefit ratio**)

benthic plants. Plants that grow under water attached to or rooted in the bottom. For photosynthesis, they depend on light penetrating the water.

best management practice. Farm management practices that serve best to reduce soil and nutrient runoff and subsequent pollution.

bioaccumulation. The accumulation of higher and higher concentrations of potentially toxic chemicals in organisms. It occurs in the case of substances that are ingested but cannot be excreted or broken down (nonbiodegradable substances).

biochemical oxygen demand (BOD). The amount of oxygen that will be absorbed or "demanded" as wastes are being digested or oxidized in both biological and chemical processes. Potential impacts of wastes are commonly measured in terms of the BOD.

biocide. Applies to any pesticide or other chemical that is toxic to many, if not all, kinds of living organisms.

bioconversion. The use of biomass as fuel. Burning materials such as wood, paper, and plant wastes directly to produce energy, or converting such materials into fuels such as alcohol and methane.

biodegradable. Able to be consumed and broken down to natural substances such as carbon dioxide and water by biological organisms, particularly decomposers. Opposite: **nonbiodegradable**.

biodiversity. The diversity of living things found in the natural world. The concept usually refers to the different species but also includes ecosystems and the genetic diversity within a given species.

biogas. The mixture of gases—about two-thirds methane, one-third carbon dioxide, and small portions of foul-smelling compounds—resulting from the anaerobic (without air) digestion of organic matter. The methane content enables biogas to be used as a fuel gas.

biological control. Control of a pest population by introduction of predatory, parasitic, or disease-causing organisms.

biological treatment. See **secondary treatment**.

biological wealth. The combination of commercial, scientific, and aesthetic values imparted to a region by its biota.

biomagnification. Bioaccumulation occurring through several levels of a food chain.

biomass. Mass of biological material. Usually the total mass of a particular group or category; for example, biomass of producers.

biomass energy, biomass fuels. Energy or fuels such as alcohol and methane produced from current photosynthetic production of biological materials. (See **bioconversion**)

biomass pyramid. Refers to the structure that is obtained when the respective biomasses of producers, herbivores, and carnivores in an ecosystem are compared. Producers have the largest biomass, followed by herbivores and then carnivores.

biome. A group of ecosystems that are related by having a similar type of vegetation governed by similar climatic conditions. Examples include prairies, deciduous forests, arctic tundra, deserts, and tropical rain forests.

bioremediation. Refers to the use of microorganisms for the decontamination of soil or groundwater. Usually involves injecting organisms and/or oxygen into contaminated zones.

biosolids. Organic material removed from sewage effluents in the course of treatment. Formerly referred to as sludge.

biosphere. The overall ecosystem of Earth. It is the sum total of all the biomes and smaller ecosystems, which ultimately are all interconnected and interdependent through global processes such as water and atmospheric cycles.

biota. Refers to any and all living organisms and the ecosystems in which they exist.

biotic. Living or derived from living things.

biotic community. All the living organisms (plants, animals, and micro-organisms) that live in a particular area.

biotic potential. Reproductive capacity. The potential of a species for increasing its population and/or distribution. The biotic potential of every species is such that, given optimum conditions, its population will increase. (Contrast **environmental resistance**)

biotic structure. The organization of living organisms in an ecosystem into groups such as producers, consumers, detritus feeders, and decomposers.

birth control. Any means, natural or artificial, that may be used to reduce the number of live births.

BOD. See **biochemical oxygen demand**.

borrowed time. Time preceding a predictable and inevitable collapse or failure of a system during which nothing is done to avert the end result despite awareness of it.

bottle law (bottle bill). A law that provides for the recycling or reuse of beverage containers, usually by requiring a returnable deposit at the purchase of the item.

breeder reactor. A nuclear reactor that in the course of producing energy also converts nonfissionable uranium-238 into fissionable plutonium-239, which can be used as fuel. Hence, a reactor that produces as much nuclear fuel as it consumes or more.

broad-spectrum pesticides. Chemical pesticides that kill a wide range of pests. They also kill a wide range of nonpest and beneficial species; therefore, they may lead to environmental upsets and resurgences. The opposite of narrow-spectrum pesticides and biorational pesticides.

BTU (British thermal unit). A fundamental unit of energy in the English system. The amount of heat required to raise the temperature of 1 pound of water 1 degree Fahrenheit.

buffer. A substance that will maintain the pH of a solution by reacting with the excess acid. Limestone is a natural buffer that helps to maintain water and soil at a pH near neutral.

buffering capacity. Refers to the amount of acid that may be neutralized by a given amount of buffer.

calorie. A fundamental unit of energy. The amount of heat required to raise the temperature of 1 gram of water 1 degree Celsius. All forms of energy can be converted to heat and measured in calories. Calories used in connection with food are kilocalories, or "big" calories, the amount of heat required to raise the temperature of 1 liter of water 1 degree Celsius.

capillary water. Water that clings in small pores, cracks, and spaces against the pull of gravity, like water held in a sponge.

carbon monoxide. A highly poisonous gas, the molecules of which consist of a carbon atom with one oxygen attached. Not to be confused with nonpoisonous carbon dioxide, a natural gas in the atmosphere.

carbon tax. A tax levied on all fossil fuels in proportion to the amount of carbon dioxide that is released as they burn.

carcinogenic. Having the property of causing cancer, at least in animals and by implication in humans.

carnivore. An animal that feeds more or less exclusively on other animals.

carrying capacity. The maximum population of a given species that an ecosystem can support without being degraded or destroyed in the long run. The carrying capacity may be exceeded, but not without lessening the system's ability to support life in the long run.

castings. The humus-rich pellets resulting from earthworm activity.

catalyst. A substance that promotes a given chemical reaction without itself being consumed or changed by the reaction. Enzymes are catalysts for biological reactions. Also catalysts are used in some pollution control devices, e.g., the **catalytic converter**.

catalytic converter. The device used by U.S. automobile manufacturers to reduce the amount of carbon monoxide and hydrocarbons in the exhaust. The converter contains a catalyst that oxidizes these compounds to carbon dioxide and water as the exhaust passes through.

cell. The basic unit of life; the smallest unit that still maintains all the attributes of life, Many microscopic organisms consist of a single cell. Large organisms consist of trillions of specialized cells functioning together.

cell respiration. The chemical process that occurs in all living cells wherein organic compounds are broken down to release energy required for life processes. Higher plants and animals require oxygen for the process as well and release carbon dioxide and water as waste products, but certain microorganisms do not require oxygen. (See **anaerobic respiration**)

cellulose. The organic macromolecule that is the prime constituent of plant cell walls and hence the major molecule in wood, wood products, and cotton. It is composed of glucose molecules, but because it cannot be digested by humans its dietary value is only as fiber, bulk, or roughage.

center pivot irrigation. An irrigation system consisting of a spray arm several hundred meters long supported by wheels pivoting around a central well from which water is pumped.

CFCs. See **chlorofluorocarbons**.

chain reaction. Nuclear reaction wherein each atom that fissions (splits) causes one or more additional atoms to fission.

channelization/channelized. The straightening and deepening of stream or river channels to speed water flow and reduce flooding. A waterway so treated is said to be channelized.

chemical barrier. In reference to genetic pest control, a chemical aspect of the plant that makes it resist pest attack.

chemical energy. The potential energy that is contained in certain chemicals; most importantly, the energy contained in organic compounds such as food and fuels, which may be released through respiration or burning.

chemical technology. Applied to the control of agricultural pests, refers to the use of pesticides and herbicides to control or eradicate the pests.

chemosynthesis. The ability of some microorganisms to utilize the chemical energy contained in certain inorganic chemicals such as hydrogen sulfide for the production of organic material. Such organisms are producers.

chlorinated hydrocarbons. Synthetic organic molecules in which one or more hydrogen atoms have been replaced by chlorine atoms. They are extremely hazardous compounds because they tend to be nonbiodegradable and therefore to bioaccumulate; many have been shown to be carcinogenic. Also called organochlorides.

chlorination. The process of adding chlorine to drinking water or sewage water in order to kill microorganisms that may cause disease.

chlorofluorocarbons (CFCs). Synthetic organic molecules that contain one or more of both chlorine and fluorine atoms, and implicated in ozone destruction.

chlorophyll. The green pigment in plants responsible for absorbing the light energy required for photosynthesis.

Clean Air Act of 1970. Amended in 1977 and 1990, this is the foundation of U.S. air pollution control efforts.

Clean Water Act of 1972. The cornerstone federal legislation addressing water pollution.

clearcutting. Cutting every tree, leaving the area completely clear.

climate. A general description of the average temperature and rainfall conditions of a region over the course of a year.

climax ecosystem. The last stage in ecological succession. An ecosystem in which populations of all organisms are in balance with each other and with existing abiotic factors.

clone. A group of genetically identical individuals derived from the asexual propagation a single individual.

clustered development. The development pattern in which homes and other facilities are arranged in dense clusters on a relatively small portion of the land considered for development, allowing the rest of the land to remain open.

co-composting. A technique of composting sewage sludge and shredded paper together.

cogeneration. The joint production of useful heat and electricity. For example, furnaces may be replaced with gas turbogenerators that produce electricity while the hot exhaust still serves as a heat source. An important avenue of conservation, it effectively avoids the waste of heat that normally occurs at centralized power plants.

command-and-control strategy. The basic strategy behind most air and water pollution public policy. It involves setting limits on pollutant levels and specifying control technologies that must be used to accomplish those limits.

commons, common pool resources. Resources (usually natural ones) owned by many people in common, or, as in the case of the air or the open oceans, owned by no one but open to exploitation.

compaction. Packing down. *Soil:* Packing and pressing out air spaces present in the soil. Reduces soil aeration and infiltration and thus reduces the capacity of the soil to support plants. *Trash:* Packing down trash to reduce the space that it requires.

compliance schedule. A timetable for reducing pollutants by certain amounts by certain dates. Such schedules are arrived at through negotiations between companies and regulatory agencies.

composting/compost. The process of letting organic wastes decompose in the presence of air. A nutrient-rich humus, or compost, is the resulting product.

composting toilet. A toilet that does not flush wastes away with water but deposits them in a chamber where they will compost. (See **composting**)

compound. Any substance (gas, liquid, or solid) that is made up of two or more different kinds of atoms bonded together. (Contrast **element**)

Comprehensive Environmental Response, Compensation, and Liability Act of 1980. See **Superfund**.

condensation. The collecting of molecules from the vapor state to form the liquid state, as for example, water vapor condenses on a cold surface and forms droplets. Opposite: **evaporation.**

confusion technique. Pest control method in which a quantity of sex attractant is applied to an area so that males become confused and are unable to locate females. The actual quantities of pheromones applied are very small because of their extreme potency.

conservation. The management of a resource in such a way as to assure that it will continue to provide maximum benefit to humans over the long run. Conservation may include various degrees of use or protection, depending on what is necessary to maintain the resource over the long run. *Energy:* Saving energy. It entails not only cutting back on use of heating, air conditioning, lighting, transportation, and so on but also increasing the efficiency of energy use. That is, developing and instigating means of doing the same jobs, e.g., transporting people, with less energy.

consumers. In an ecosystem, those organisms that derive their energy from feeding on other organisms or their products.

consumptive water use. Use of water for such things as irrigation, where the water does not remain available for potential purification and reuse.

containment building. Reinforced concrete building housing the nuclear reactor. Designed to contain an explosion should one occur.

contour farming. The practice of cultivating land along the contours across rather than up and down slopes. In combination with strip cropping it reduces water erosion.

control group. The group in an experiment that is the same as and is treated like the experimental group in every way except for the particular factor being tested. Only by comparison with a control group can one gain specific information concerning the effect of any test factor.

controlled experiment. An experiment with adequate control groups. (See **control group**)

control rods. Part of the core of a nuclear reaactor; the rods of neutron-absorbing material that are inserted or removed as necessary to control the rate of nuclear fissioning.

Convention on Biological Diversity. The Biodiversity Treaty signed by 158 nations at the Earth Summit in Rio de Janeiro in 1992 calling for various actions and cooperative steps between nations to protect the world's biodiversity.

cooling tower. A massive tower designed to dissipate waste heat from a power plant (or other industrial process) into the atmosphere.

cornucopianism. Dominant world view that embodies the assumption that all parts of the environment are natural resources to be exploited for the advantage of humans.

cosmetic damage. Damage to the surface of fruits and vegetables that affects appearance but does not otherwise affect taste, nutritional quality, or storability.

cosmetic spraying. Spraying of pesticides to control pests that damage only the surface appearance.

cost-benefit analysis. See **benefit-cost analysis**.

cost-benefit ratio/benefit-cost ratio. The value of the benefits to be gained from a project divided by the costs of the project. If the ratio is greater than 1, the project is economically justified; if the ratio is less than 1, the project is not economically justified.

cost-effective. Pertaining to a project or procedure that produces economic returns or benefits that are significantly greater than the costs.

covalent bond. A chemical bond between two atoms, formed by sharing a pair of electrons between the two atoms. Atoms of all organic compounds are joined by covalent bonds.

credit associations. Associated with microlending. Groups of poor people with no collateral to assure loans forming an association to assure each other's loans.

criteria pollutants. Certain pollutants the level of which is used as a gauge for the determination of air (or water) quality.

critical level. The level of one or more pollutants above which severe damage begins to occur and below which few if any ill effects are noted.

critical number. The minimum number of individuals of a given species that is required to maintain a healthy, viable population of the species. If a population falls below its critical number its extinction will almost certainly occur.

crop rotation. The practice of alternating the crops grown on a piece of land. For example, corn one year, hay for two years, then back to corn. (Contrast **monocropping**)
crude birth rate. Number of births per 1000 individuals per year.
crude death rate. Number of deaths per 1000 individuals per year.
crystallization. The joining together of molecules or ions from a liquid (or sometimes gaseous) state to form a solid state.
cultivar. A cultivated variety of a plant species. All individuals of the cultivar are genetically highly uniform.
cultural control. A change in the practice of growing, harvesting, storing, handling, or disposing of wastes that reduces the susceptibility or exposure to pests. For example, spraying the house with insecticides to kill flies is a chemical control; putting screens on the windows to keep flies out is a cultural control.
cultural eutrophication. The process of natural eutrophication accelerated by human activities. (See **eutrophication**)

DDT (dichlorodiphenyltrichloroethane). The first and most widely used of the synthetic organic pesticides belonging to the chlorinated hydrocarbon class.
debt crisis. Refers to the fact that many less-developed nations are so heavily in debt that they may not be able to meet their financial obligations, e.g., interest payments. Their failure to meet such obligations could have severe economic impacts on the entire world.
declining tax base. The loss of tax revenues that occurs when affluent taxpayers and businesses leave an area and property values subsequently decline. Also referred to as eroding tax base.
decommissioning. Refers to the inevitable need to take nuclear power plants out of service after 25–35 years because the effects of radiation will gradually make them inoperable.
decomposers. Organisms whose feeding action results in decay or rotting of organic material. The primary decomposers are fungi and bacteria.
deep-well injection. A technique used for the disposal of liquid chemical wastes that involves putting them into deep dry wells where they permeate dry strata.
demographic transition. The transition from a condition of high birth rate and high death rate through a period of declining death rate but continuing high birth rate finally to low birth rate and low death rate. This transition may result from economic development.
demography/demographer. The studies of population trends (growth, movement, development, and so on). People who perform such studies and make projections from them.
denitrification. The process of converting nitrogen compounds present in soil or water back to nitrogen gas in the atmosphere. It is a natural process conducted by certain bacteria (see text discussion of the nitrogen cycle), and it is now utilized in the treatment of sewage effluents.
density-dependent. Refers to population balancing, factors such as parasitism that increase and decrease in intensity corresponding to population density.
deoxyribonucleic acid. See **DNA**.
Department of Transportation regulations (DOT Regs). Regulations intended to reduce the risk of spills, fires, and poisonous fumes by specifying the kinds of containers and methods of packing to be used in transporting hazardous materials.
desalinization. Processes that purify seawater into high-quality drinking water via distillation or microfiltration.
desertification. Declining land productivity caused by mismanagement. Overgrazing and overcultivation allowing erosion and salinization are the major causes.
desertified. Land for which productivity has been significantly reduced (25% or more) because of human mismanagement. Erosion is the most common cause.
desert pavement. A covering of stones and coarse sand protecting desert soils from further wind erosion. The covering results from the differential erosion of finer material.
detritus. The dead organic matter, such as fallen leaves, twigs, and other plant and animal wastes, that exists in any ecosystem.
detritus feeders. Organisms such as termites, fungi, and bacteria that obtain their nutrients and energy mainly by feeding on dead organic matter.
deuterium (^2H). A stable, naturally occurring isotope of hydrogen. It contains one neutron in addition to the single proton normally in the nucleus.
developed countries. Industrialized countries—United States, Canada, Western European nations, Japan, Australia, and New Zealand—in which the gross domestic product exceeds $7000 per capita.
developing countries. All free-market countries in which the gross domestic product is less than $7000 per capita. Includes nations of Latin America, Africa, and Asia excepting Japan.
development rights. Legal documents that grant permission to develop a given piece of property. They must be owned by the developer before development can occur. They can be bought and sold apart from the property itself.
differential reproduction. Refers to the fact that within a population certain individuals reproduce much more than others.
diffuse sources. Widespread sources of pollution such as agricultural and urban runoff. Also called **nonpoint sources**. (Contrast **point sources**)
dioxin. A synthetic organic chemical of the chlorinated hydrocarbon class. It is one of the most toxic compounds known to humans, having many harmful effects, including induction of cancer and birth defects, even in extremely minute concentrations. It has become a widespread environmental pollutant because of the use of certain herbicides that contain dioxin as a contaminant.
direct solar energy. See **solar energy**.
discharge permit. (Technically called NPDES permit.) A permit that allows a company to legally discharge certain amounts or levels of pollutants into air or water.
disinfection. The killing (as opposed to removal) of microorganisms in water or other media where they might otherwise pose a health threat. For example, chlorine is commonly used to disinfect water supplies.
dissolved oxygen (DO). Oxygen gas molecules (O_2) dissolved in water. Fish and other aquatic organisms depend on dissolved oxygen for respiration. Therefore concentration of dissolved oxygen is a measure of water quality.
distillation. A process of purifying water or other liquids by

boiling the liquid and recondensing the vapor. Contaminants remain behind in the boiler.

district heating. The heating of an entire community or city area through circulating heat (e.g., steam) from a central source; particularly, utilizing waste heat from a power plant or from incineration of refuse.

diversion. Taking some or all of the flow of a natural waterway and carrying it to other places for uses such as municipal water supplies or irrigation.

DNA (deoxyribonucleic acid). The natural organic macromolecule that carries the genetic or hereditary information for virtually all organisms.

DO. See **dissolved oxygen**.

domestic solid wastes. Wastes that come from homes, offices, schools, and stores, as opposed to wastes that are generated from agricultural or industrial processes.

dose. A consideration of the concentration of a hazardous material times the length of exposure to it. For any given material or radiation, effects correspond to the product of these two factors.

doubling time. The time it will take a population to double in size, assuming the continuation of current rate of growth.

drift-netting. The practice of harvesting marine fish and squid by laying down miles of gill nets across the open seas. The nets collapse around larger organisms and kill many whales, dolphins, seals, marine birds, and turtles.

drip irrigation. Supplying irrigation water through tubes that literally drip water onto the soil at the base of each plant.

easement. In reference to land protection, an arrangement whereby a landowner gives up development rights into the future but retains ownership.

ecological pest management. Control of pest populations through understanding the various ecological factors that provide natural control and so far as possible utilizing these factors as opposed to using synthetic chemicals.

ecological regard. Taking into consideration the environmental impact, direct and indirect, of one's actions and lifestyle. Adjusting actions and lifestyle to minimize their impact as much as possible.

ecological restoration. See **restoration ecology**.

ecological risk. Any factor that may cause undetermined damage or upset to the existing natural ecological system.

ecological succession. Process of gradual and orderly progression from one ecological community to another.

ecologists. Scientists who study ecology, i.e., the ways in which organisms interact with each other and with their environment.

ecology. The study of any and all aspects of how organisms interact with each other and with their environment.

economic exclusion. The cutting of access of certain ethnic or economic groups to jobs, quality education, and other opportunities and thus preventing them from entering the economic mainstream of society—a condition that prevails in poor areas of cities.

economic threshold. The level of pest damage that, to be reduced further, would require an application of pesticides that is more costly than the economic damage caused by the pests.

ecosystem. A grouping of plants, animals, and other organisms interacting with each other and with their environment in such a way as to perpetuate the grouping more or less indefinitely. Ecosystems have characteristic forms such as deserts, grasslands, tundra, deciduous forests, and tropical rain forests.

ecotone. A transitional region between two ecosystems that contains some of the species and characteristics of the two adjacent ecosystems and also certain species characteristic of the transitional region.

ecotourism. The enterprises involved in promoting tourism of unusual or interesting ecological sites.

electrolysis. The use of electrical energy to split water molecules into their constituent hydrogen and oxygen atoms. Hydrogen gas and oxygen gas result.

electrons. Fundamental atomic particles that have a negative electrical charge but virtually no mass. They surround the nuclei of atoms and thus balance the positive charge of protons in the nucleus. A flow of electrons in a wire is synonymous with an electrical current.

element. A substance that is made up of one and only one distinct kind of atom. (Contrast **compound**)

embrittlement. Becoming brittle. Pertains especially to the reactor vessel of nuclear power plants gradually becoming prone to breakage or snapping as a result of continuous bombardment by radiation. It is the prime factor forcing the decommissioning of nuclear power plants.

emergency response teams. Teams of people, generally associated with police or fire departments, specially trained to handle accidents involving hazardous materials.

emergent vegetation. Aquatic plants whose lower parts are under water but whose upper parts emerge from the water.

emission allowance/standards. See **discharge permit**.

endangered species. A species the total population of which is declining to relatively low levels, a trend that if continued will result in extinction.

Endangered Species Act. The federal legislation that mandates protection of species and their habitats that are determined to be in danger of extinction.

energy. The ability to do work. Common forms of energy are light, heat, electricity, motion, and chemical bond energy inherent in compounds such as sugar, gasoline, and other fuels.

enrichment. With reference to nuclear power, signifies the separation and concentration of uranium-235 so that, in suitable quantities, it will sustain a chain reaction.

entomologist. A scientist who studies insects, their life cycles, physiology, behavior, and so on.

entropy. Refers to the degree of disorder: increasing entropy means increasing disorder.

environment. The combination of all things and factors external to the individual or population of organisms in question.

environmental impact. Effects on the natural environment caused by human actions. Includes indirect effects through pollution, for example, as well as direct effects such as cutting down trees.

environmental impact statement. A study of the probable environmental impacts of a development project. The National Environmental Policy Act of 1968 (NEPA) requires such studies prior to proceeding with any project receiving federal funding.

environmental movement. Refers to the upwelling of public awareness and citizen action regarding environmental issues that began during the 1960s.

environmental regard. A factor that may moderate negative environmental impacts, such as suitable attention to conservation or recycling.

environmental resistance. The totality of factors such as adverse weather conditions, shortage of food or water, predators, and diseases that tend to cut back populations and keep them from growing or spreading. (Contrast **biotic potential**)

environmental science. The branch of science concerned with environmental issues.

environmentalism. Embodies the assumption that what we generally view as natural resources are products of the natural environment and can be maintained only insofar as the natural environment is maintained.

environmentalist. Has come to include any person who believes that sustainability of civilization hinges on preserving natural aspects of the biosphere, namely, freedom from pollution and maintenance of biodiversity.

EPA. U.S. Environmental Protection Agency. The federal agency responsible for control of all forms of pollution and other kinds of environmental degradation.

epidemiological study. Determination of causes of disease (e.g., lung cancer) through the study and comparison of large populations of people living in different locations or following different lifestyles and/or habits (e.g., smoking versus nonsmoking).

epiphytes. Air plants that are not parasitic but "perch" on tree branches, where they can get adequate light.

erosion. The process of soil particles' being carried away by wind or water. Erosion moves the smaller soil particles first and hence degrades the soil to a coarser, sandier, stonier texture.

estimated reserves. See **reserves**.

estuary. A bay open to the ocean at one end and receiving fresh water from a river at the other. Hence, mixing of fresh and salt water occurs (brackish).

ETS (environmental tobacco smoke). "Second-hand" tobacco smoke to which nonsmokers are exposed when in the presence of smokers.

euphotic zone. The layer or depth of water through which there is adequate light penetration to support photosynthesis.

eutrophic. Refers to a body of water characterized by nutrient-rich water supporting abundant growth of algae and/or other aquatic plants at the surface. Deep water has little or no dissolved oxygen.

eutrophication. The process of becoming eutrophic.

evaporation. Molecules leaving the liquid state and entering the vapor or gaseous state as, for example, water evaporates to form water vapor. Opposite: **condensation**.

evapotranspiration. The combination of evaporation and transpiration.

evolution. The theory that all species now on Earth descended from ancestral species through a process of gradual change brought about by natural selection.

evolutionary succession. The succession of different species that have inhabited Earth at different geological periods, as revealed through the fossil record. The process whereby new species come in through the process of speciation while other species pass into extinction.

experimental group. The group in an experiment that receives the experimental treatment in contrast to the **control group**, used for comparison, which does not receive the treatment. Synonym: test group.

exotic species. A species introduced to a geographical area where it is not native.

exponential increase. The growth produced when the base population increases by a given percentage (as opposed to a given amount) each year. It is characterized by doubling again and again, each doubling occurring in the same period of time. It produces a J-shaped curve.

externality. Any effect of a business process not included in the usual calculations of profit and loss. Pollution of air or water is an example of a *negative* externality—one that imposes a cost on society that is not paid for by the business itself.

extinction. The death of all individuals of a particular species. When this occurs, all the genes of that particular line are lost forever.

extractive reserves. As now established in Brazil, forest lands that are protected for native peoples and others who harvest natural products of the forests, such as latex and Brazil nuts.

exurban migration. Refers to the pronounced trend since World War II of relocating homes and businesses from the central city and older suburbs to more-outlying suburbs.

exurbs. New developments beyond the traditional suburbs but from which most residents still commute to the associated city for work.

famine. A severe shortage of food accompanied by a significant increase in the local or regional death rate.

FAO. Food and Agriculture Organization of the United Nations.

farm cooperatives. An association of consumers who jointly own and manage a farm for the production of produce specifically for their own consumption.

fecal coliform test. A test for the presence of *Escherichia coli,* the bacterium that normally inhabits the gut of humans and other mammals. A positive test indicates sewage contamination and the potential presence of disease-causing microorganisms carried by sewage.

fermentation. A form of respiration carried on by yeast cells in the absence of oxygen. It involves a partial breakdown of glucose (sugar) that yields energy for the yeast and the release of alcohol as a byproduct.

fertility rate. See **total fertility rate**.

fertilizer. Material applied to plants or soil to supply plant nutrients, most commonly nitrogen, phosphorus, and potassium but may include others. Organic fertilizer is natural organic material such as manure, which releases nutrients as it breaks down. Inorganic fertilizer, also called chemical fertilizer, is a mixture of one or more necessary nutrients in inorganic chemical form.

field capacity. A measure of the maximum volume of water that a soil can hold by capillary action, i.e., against the pull of gravity.

field scouts. Persons trained to survey crop fields and determine whether applications of pesticides or other pest management procedures are actually necessary to avert significant economic loss.

FIFRA. Federal Insecticide, Fungicide, and Rodenticide Act; the key U.S. legislation to control pesticides.

filtration. The passing of water (or other fluid) through a filter to remove certain impurities.

fire climax ecosystems. Ecosystems that depend on the recurrence of fire to maintain the existing balance.

first basic principle of ecosystem sustainability. Resources are supplied and wastes are disposed of by recycling all elements.

first-generation pesticides. Toxic inorganic chemicals that were first used to control insects, plant diseases, and other pests. Included mostly compounds of arsenic and cyanide and various heavy metals such as mercury and copper.

First Law of Thermodynamics. The fact based on irrefutable observations that energy is never created or destroyed but may be converted from one form to another, e.g., electricity to light. Also called the Law of Conservation of Energy. (See also **Second Law of Thermodynamics**)

fishery. Fish species being exploited, or a limited marine area containing commercially valuable fish.

fission. The splitting of a large atom into two atoms of lighter elements. When large atoms such as uranium or plutonium fission, tremendous amounts of energy are released.

fission products. Any and all atoms and subatomic particles resulting from splitting atoms in nuclear reactors. All or most such products are highly radioactive.

flat-plate collector. A solar collector that consists of a stationary, flat, black surface oriented perpendicular to the average sun angle. Heat absorbed by the surface is removed and transported by air or water (or other liquid) flowing over or through the surface.

flood irrigation. Technique of irrigation in which water is diverted from rivers through canals and flooded through furrows in fields.

food aid. Food of various forms that is donated or sold below cost to needy people for humanitarian reasons.

food chain. The transfer of energy and material through a series of organisms as each one is fed upon by the next.

food security. For families, the ability to meet the food needs of everyone in the family, providing freedom from hunger and malnutrition.

food web. The combination of all the feeding relationships that exist in an ecosystem.

fossil fuels. Energy sources, mainly crude oil, coal, and natural gas, that are derived from prehistoric photosynthetic production of organic matter on Earth.

fourth principle of ecosystem sustainability. Biodiversity must be maintained.

freshwater. Water that has a salt content of less than 0.05% (500 parts per million).

fuel assembly. The assembly of many rods containing the nuclear fuel, usually uranium, positioned close together. The chain reaction generated in the fuel assembly is controlled by rods of neutron-absorbing material between the fuel rods.

fuel elements. The pellets of uranium or other fissionable material that are placed in tubes, which, with the control rods, form the core of the nuclear reactor.

fuel rods. See **Fuel elements**.

fungus, pl. **fungi.** Any of numerous species of molds, mushrooms, brackets, and other forms of nonphotosynthetic plants. They derive energy and nutrients by consuming other organic material. Along with bacteria they form the decomposer component of ecosystems.

fusion. The joining together of two atoms to form a single atom of a heavier element. When light atoms such as hydrogen are fused, tremendous amounts of energy are released.

gasohol. A blend of 90% gasoline and 10% alcohol, which can be substituted for straight gasoline. It serves to stretch gasoline supplies.

gene pool. The sum total of all the genes that exist among all the individuals of a species.

genes. Segment of DNA that codes for one protein, which in turn determines a particular physical, physiological, or behavioral trait.

genetic bank. The concept that natural ecosystems with all their species serve as a tremendous repository of genes that is frequently drawn upon to improve domestic plants and animals and to develop new medicines, among other uses.

genetic control. Selective breeding of the desired plant or animal to make it resistant to attack by pests. Also, attempting to introduce harmful genes—for example, those that cause sterility—into the pest populations.

genetic engineering. The artificial transfer of specific genes from one organism to another.

genetic makeup. Refers to all the genes that an individual possesses and that determine all of the individual's inherited characteristics.

genetics. The study of heredity and the processes by which inherited characteristics are passed from one generation to the next.

genetic variation. An expression of the range of genetic (DNA) differences that occur among individuals of the same species.

gentrification. The trend seen in modern society of people moving into more or less isolated communities with others of similar economic, ethnic, and social backgrounds.

geothermal. Refers to the naturally hot interior of Earth. The heat is maintained by naturally occurring nuclear reactions in Earth's interior.

geothermal energy. Useful energy derived from the naturally hot interior of Earth.

global warming. The term given to the possibility that Earth's atmosphere is gradually warming because of the greenhouse effect of carbon dioxide and other gases. Global warming is thought by many to be the most serious global environmental issue facing our society. (See also **greenhouse effect** and **greenhouse gases**)

glucose. A simple sugar, the major product of photosynthesis. Serves as the basic building block for cellulose and starches and as the major "fuel" for the release of energy through cell respiration in both plants and animals.

gravitational water. Water that is not held by capillary action in soil but percolates downward by the force of gravity.

graying. The increasing average age in populations in developed and many developing countries that is occurring because of decreasing birth rates and increasing longevity.

gray water. Wastewater, as from sinks and tubs, that does not contain human excrements. Such water can be reused without purification for some purposes.

greenhouse effect. An increase in the atmospheric temperature caused by increasing amounts of carbon dioxide and certain other gases that absorb and trap heat radiation, which normally escapes from Earth.

greenhouse gases. Gases in the atmosphere that absorb infrared energy and contribute to the air temperature. These gases are like a heat blanket and are important in insulating Earth's surface. They include carbon dioxide, water vapor, methane, nitrous oxide, and chlorofluorocarbons and other halocarbons.

green manure. A legume crop such as clover that is specifically grown to enrich the nitrogen and organic content of soil.

green revolution. Refers to the development and introduction of new varieties of wheat and rice (mainly) that increased yields per acre dramatically in some countries.

grit chamber. Part of preliminary treatment in wastewater-treatment plants; a swimming pool like tank in which the velocity of the water is slowed enough to let sand and other gritty material settle.

grit-settling tank. See **grit chamber**.

gross domestic (national) product per capita. The total value of all goods and services exchanged in a year in a country, divided by its population. A common indicator for the average level of development and standard of living for a country.

groundwater. Water that has accumulated in the ground, completely filling and saturating all pores and spaces in rock and/or soil. Groundwater is free to move more or less readily. It is the reservoir for springs and wells and is replenished by infiltration of surface water.

groundwater remediation. The repurification of contaminated groundwater by any of a number of techniques.

growth momentum. Refers to the fact that the human population will continue to grow for some time even after the fertility rate is reduced to 2.0 because there is currently such an excessive number of children moving into the reproductive age brackets.

gully erosion. Gullies, large or small, resulting from water erosion.

habitat. The specific environment (woods, desert, swamp) in which an organism lives.

habitat alteration. Any change in a natural habitat that may occur because of changing drainage, pollution, or direct impacts.

half-life. The length of time it takes for half of an unstable isotope to decay. The length of time is the same regardless of the starting amount. Also refers to the amount of time it takes compounds to break down in the environment.

halogenated hydrocarbon. Synthetic organic compound containing one or more atoms of the halogen group, which includes chlorine, fluorine, and bromine.

hard water. Water that contains relatively large amounts of calcium and/or certain other minerals that cause soap to precipitate. (Contrast **soft water**)

hazard. Anything that can cause (1) injury, disease, or death to humans; (2) damage to property; or (3) degradation of the environment.

hazard assessment. The process of examining evidence linking a particular hazard to its harmful effects.

hazardous materials (HAZMAT). Any material having one or more of the following attributes: ignitability, corrosivity, reactivity, toxicity.

heavy metals. Any of the high atomic weight metals such as lead, mercury, cadmium, and zinc. All may be serious pollutants in water or soil because they are toxic in relatively low concentrations and they tend to bioaccumulate.

herbicide. A chemical used to kill or inhibit the growth of undesired plants.

herbivore/Herbivorous. An organism such as rabbit or deer that feeds primarily on green plants or plant products such as seeds or nuts. Such an organism is said to be herbivorus. Synonym: **primary consumer**.

herbivory. The feeding on plants that occurs in an ecosystem. The total feeding of all plant-eating organisms.

heterotroph/Heterotrophic. Any organism that consumes organic matter as a source of energy. Such an organism is said to be heterotrophic.

Highway Trust Fund. The monies collected from the gasoline tax designated for construction of new highways.

hormones. Natural chemical substances that control development, physiology, and/or behavior of an organism. Hormones are produced internally and affect only that individual. Hormones are coming into use in pest control. (See also **pheromones**)

host. In feeding relationships, particularly parasitism, refers to the organism that is being fed upon, i.e., supporting the feeder.

host-parasite relationship. The combination of a parasite and the organism upon which it feeds.

host-specific. Referring to insects, fungal diseases, and other parasites that are unable to attack species other than their particular host.

human system. The entire system that humans have created for their own support, consisting of agriculture, industry, transportation and communications networks, etc.

humidity. The amount of water vapor in the air. (See also **relative humidity**)

humus. A dark brown or black, soft, spongy residue of organic matter that remains after the bulk of dead leaves, wood, or other organic matter has decomposed. Humus does oxidize, but relatively slowly. It is extremely valuable in enhancing physical and chemical properties of soil.

hunger. A general term referring to the lack of basic food required for meeting nutritional and energy needs, such that the individual is unable to lead a normal, healthy life.

hunter-gatherers. Humans surviving by hunting wild game and gathering seeds, nuts, berries, and other edible things from the natural environment.

hybrid. A plant or animal resulting from a cross between two closely related species that do not normally cross.

hybridization. Cross-mating between two more or less closely related species.

hydrocarbon emissions. Exhaust of various hydrogencarbon compounds due to incomplete combustion of fuel. They are a major contribution to photochemical smog.

hydrocarbons. *Chemistry:* Natural or synthetic organic substances that are composed mainly of carbon and hydrogen. Crude oil, fuels from crude oil, coal, animal fats, and vegetable oils are examples. *Pollution:* A wide variety of relatively small carbon-hydrogen molecules resulting from incomplete burning of fuel and emitted into the atmosphere. (See **volatile organic compounds**)

hydroelectric dam. A dam and associated reservoir used to produce electrical power by letting the high-pressure water behind the dam flow through and drive a turbogenerator.

hydroelectric power. Electrical power that is produced from hydroelectric dams or, in some cases, natural waterfalls.

hydrogen bonding. A weak attractive force that occurs between a hydrogen atom of one molecule and, usually, an oxygen atom of another molecule. It is responsible for holding water molecules together to produce the liquid and solid states.
hydrogen ions. Hydrogen atoms that have lost their electrons. Chemical symbol, H^+.
hydrological cycle. (See **water cycle**)
hydroponics. The culture of plants without soil. The method uses water with the required nutrients in solution.
hydroxyl radical. The hydroxyl group (OH^-) missing the electron. It is a natural cleansing agent of the atmosphere. It is highly reactive and readily oxidizes many pollutants on contact and thus contributes to their removal.
hypothesis. An educated guess concerning the cause of an observed phenomenon that is then subjected to experimental tests to prove its accuracy or inaccuracy.

indicator organism. An organism, the presence or absence of which indicates certain conditions. For example, the presence of *Escherichia coli* indicates that water is contaminated with fecal wastes and pathogens may be present; the absence indicates that the water is free of pathogens.
indirect products. Air pollutants that are not contained in emissions but are formed when compounds in emissions undergo various reactions in the atmosphere.
indirect solar energy. (See **solar energy**)
industrialized agriculture. Using fertilizer, irrigation, pesticides, and energy from fossil fuels to produce large quantities of crops and livestock with minimal labor for domestic and foreign sale.
industrial smog. The grayish mixture of moisture, soot, and sulfurous compounds that occurs in local areas where industries are concentrated and coal is the primary energy source.
infant mortality. The number of babies that die before age 1, per 1000 babies born.
infiltration. The process in which water soaks into soil as opposed to running off the surface.
infiltration-runoff ratio. The ratio of the amount of water soaking into the soil to that running off the surface. The ratio is obtained by dividing the first amount by the second.
infrared radiation. Radiation of somewhat longer wavelengths than red light, the longest wavelengths of the visible spectrum. Such radiation manifests itself as heat.
infrastructure. The sewer and water systems, roadways, bridges, and other facilities that underlie the functioning of a city and that are owned, operated, and maintained by the city.
inherently safe reactor. In theory, a nuclear reactor that is designed in such a way that any accident would be automatically corrected and no radioactivity released.
inorganic compounds/molecules. *Classical definition:* All things such as air, water, minerals, and metals that are neither living organisms nor products uniquely produced by living things. *Chemical definition:* All chemical compounds or molecules that do not contain carbon atoms as an integral part of their molecular structure. (Contrast **organic compounds**)
inorganic fertilizer. See **fertilizer**.
inorganic molecules. See **inorganic compounds/molecules**.
insecticide. Any chemical used to kill insects.
instrumental value. Based on the belief that living organisms or species are worthwhile if their existence or use benefits people; the degree to which they benefit humans. (Contrast **intrinsic value**)
insurance spraying. Spraying of pesticides that is done when it is not really needed in the belief that it will insure against loss due to pests.
integral urban house. A house in an urban setting that utilizes ecological principles, including water and materials conservation and recycling, solar energy, and intensive cultivation of food plants insofar as possible.
integrated pest management (IPM). Two or more methods of pest control carefully integrated into an overall program designed to avoid economic loss from pests. The objective is to minimize the use of environmentally hazardous, synthetic chemicals. Such chemicals may be used in IPM, but only as a last resort to prevent significant economic losses.
integrated waste management. The approach to municipal solid waste that provides for several options for dealing with wastes, including recycling, composting, waste reduction, and landfilling and incineration where unavoidable.
intrinsic value. Based on the belief that living organisms or species are worthwhile in their own right; they do not have to be useful to have value. (Contrast **instrumental value**)
inversion. See **temperature inversion**.
ion. An atom or group of atoms that has lost or gained one or more electrons and consequently has acquired a positive or negative charge. Ions are designated by + or − superscripts following the chemical symbol.
ion-exchange capacity. See **nutrient-holding capacity**.
ionic bond. The bond formed by the attraction between a positive and a negative ion.
IPM. See **integrated pest management**.
irrigation. Any method of artificially adding water to crops.
isotope. A form of an element in which the atoms have more (or less) than the usual number of neutrons. Isotopes of a given element have identical chemical properties, but they differ in mass (weight) as a result of the additional (or lesser) neutrons. Many isotopes are unstable and radioactive. (See **radioactive decay, radioactive emissions,** and **radioactive materials**)

juvenile hormone. The insect hormone that, at sufficient levels, preserves the larval state. Pupation requires diminished levels; hence artificial applications of the hormone may block development.

keystone species. A species whose role is essential for the survival of many other species in an ecosystem.
kinetic energy. The energy inherent in motion or movement, including molecular movement (heat) and movement of waves, hence radiation including light.

landfill. A site where wastes (municipal, industrial, or chemical) are disposed of by burying them in the ground or placing them on the ground and covering them with earth. Also used as a verb meaning to dispose of a material in such a way.

land subsidence. The phenomenon whereby land gradually sinks. It may result from removing groundwater or oil, which is frequently instrumental in supporting the overlying rock and soil.

land trust. Land that is purchased and held by various organizations specifically for the purpose of protecting its natural environment and biota that inhabit it.

larva, pl. **larvae.** A free-living immature form that occurs in the life cycle of many organisms and that is structurally distinct from the adult. For example, caterpillars are the larval stage of moths and butterflies.

Law of Conservation of Energy. See **First Law of Thermodynamics**.

Law of Conservation of Matter. Law stating that in chemical reactions, atoms are neither created, changed, nor destroyed; they are only rearranged.

Law of Limiting Factors. Also known as Liebig's Law of Minimums. A system may be limited by the absence or minimum amount (in terms of that needed) of any required factor. (See **limiting factor**)

leachate. The mixture of water and materials that are leaching.

leaching. The process in which materials in or on the soil gradually dissolve and are carried by water seeping through the soil. It may result in the removal of valuable nutrients from the soil, or it may carry buried wastes into groundwater, thereby contaminating it.

legumes. The group of land plants that is virtually alone in its ability to fix nitrogen; includes such common plants as peas, beans, clovers, alfalfa, and locust trees but no major cereal grains. (See **nitrogen fixation**)

lethal mutation. A genetic alteration that results in such severe abnormalities that the organism cannot survive.

Liebig's Law of Minimums. See **Law of Limiting Factors**.

limiting factor. A factor primarily responsible for determining the growth and/or reproduction of an organism or a population. The limiting factor may be a physical factor such as temperature or light, a chemical factor such as a particular nutrient, or a biological factor such as a competing species. The limiting factor may differ at different times and places.

limits of tolerance. The extremes of any factor, e.g., temperature, that an organism or a population can tolerate and still survive and reproduce.

lipids. A class of natural organic molecules that includes animal fats, vegetable oils, and phospholipids, the last being an integral part of cellular membranes.

litter. In an ecosystem, the natural cover of dead leaves, twigs, and other dead plant material. This natural litter is subject to rapid decomposition and recycling in the ecosystem, whereas human litter, such as bottles, cans, and plastics, is not.

loam. A solid consisting of a mixture of about 40% sand, 40% silt, and 20% clay.

longevity. The average life span of individuals of a given population.

LULU. An acronym standing for "locally unwanted land use." Expresses the difficulty in siting a facility that is necessary but which no one wants in the immediate locality.

macromolecules. Very large, organic molecules such as proteins and nucleic acids that constitute the structural and functional parts of cells.

MACT (maximum achievable control technology). The best technologies available for reducing the output of especially toxic industrial pollutants.

malnutrition. The lack of essential nutrients such as vitamins, minerals, and amino acids. Malnutrition ranges from mild to severe and life-threatening.

mariculture. The propagation and/or rearing of any marine (saltwater) organism in more or less artificial systems.

marine environment. An ocean environment that supports a distinctive array of seaweeds, plankton, fish, shellfish, and other marine organisms depending on temperature, water depth, nature of the bottom, and concentrations of nutrients and sediments.

mass number. The number that accompanies the chemical name or symbol of an element or isotope. It represents the number of neutrons and protons in the nucleus of the atom.

materials recycling facility (MRF). A processing plant where regionalized recycling is facilitated. Recyclable municipal solid waste, usually presorted, is prepared in bulk for the recycling market.

matter. Anything that occupies space and has mass. Refers to any gas, liquid, or solid. (Contrast **energy**)

maximum sustainable yield. The maximum amount of a renewable resource that can be taken year after year without depleting the resource. It is the maximum rate of use or harvest that will be balanced by the regenerative capacity of the system—for example, the maximum rate of tree cutting that can be balanced by tree regrowth.

meltdown. The event of a nuclear reactor's getting out of control or losing its cooling water so that it melts from its own production of heat. The melted reactor would continue to produce heat and could melt its way out of the reactor vessel and eventually down into groundwater, where it would cause a violent eruption of steam that could spread radioactive materials over a wide area.

metabolism. The sum of all the chemical reactions that occur in an organism.

methane. A gas, CH_4. It is the primary constituent of natural gas. It is also produced as a product of fermentation by microbes. Methane from ruminant animals is thought to be responsible for the rise in atmospheric methane, of concern because methane is one of the greenhouse gases.

microbe. A term used to refer to any microscopic organism, primarily bacteria, viruses, and protozoans.

microclimate. The actual conditions experienced by an organism in its particular location. Owing to numerous factors such as shading, drainage, and sheltering, the microclimate may be quite distinct from the overall climate.

microfiltration. A process for purifying water in which water is forced under very high pressure through a membrane that is fine enough to filter out ions and molecules in solution; used by small desalination plants to filter salt from seawater. Also called reverse osmosis.

microlending. The process of providing very small loans (usually $50–$100) to poor people to facilitate their starting a small enterprise to become economically self-sufficient.

microorganism. Any microscopic organism, particularly bacteria, viruses, and protozoans.

midnight dumping. The wanton illicit dumping of materials, particularly hazardous wastes, frequently under the cover of darkness.

Minamata disease. A "disease" named for a fishing village in Japan where an "epidemic" was first observed. Symptoms, which included spastic movements, mental retardation, coma, death, and crippling birth defects in the next generation, were found to be the result of mercury poisoning.

mineral. Any hard, brittle, stonelike material that occurs naturally in Earth's crust. All consist of various combinations of positive and negative ions held together by ionic bonds. Pure minerals, or crystals, are one specific combination of elements. Common rocks are composed of mixtures of two or more minerals.

mineralization. The process of gradual oxidation of the organic matter (humus) present in soil that leaves just the gritty mineral component of the soil.

mixture. Means there is no chemical bonding between the molecules of the element involved. For example, air contains (is a mixture of) oxygen, nitrogen, and carbon dioxide.

mobilization. In soil science, the bringing into solution of normally insoluble minerals. Presents a particular problem when the elements of such minerals have toxic effects.

moderator. In a nuclear reactor, the moderator is any material that slows down neutrons from fission reactions so that they are traveling at the right speed to trigger another fission. Water and graphite represent two types of moderators.

molecule. A specific union of two or more atoms. The smallest unit of a compound that still has the characteristics of that compound.

monocropping. The practice of growing the same crop year after year on the same land. (Contrast **crop rotation** and **polyculture**)

monoculture. The practice of growing a single crop over very wide areas, for example, thousands of square kilometers of wheat, and only wheat, grown in the Midwest.

Montreal Protocol. An agreement made in 1987 by a large group of nations to cut back the production of chlorofluorocarbons by 50% by the year 2000 in order to protect the ozone shield. A 1990 amendment calls for the complete phaseout of these chemicals by 2000 in developed nations and by 2010 in less-developed nations.

MRF. See **materials recycling facility**.

municipal solid waste. The entirety of refuse or trash generated by a residential and business community. The refuse that a municipality is responsible for collecting and disposing of, distinct from agricultural and industrial wastes.

mutagenic. Causing mutations.

mutation. A random change in one or more genes of an organism. Mutations may occur spontaneously in nature, but their number and degree are vastly increased by exposure to radiation and/or certain chemicals. Mutations generally result in a physical deformity and/or metabolic malfunction.

mutualism. Refers to a close relationship between two organisms in which both organisms benefit from the relationship.

mycelia. The threadlike feeding filaments of fungi.

mycorrhizae, sing. **mycorrhiza.** The mycelia of certain fungi that grow symbiotically with the roots of some plants and provide for additional nutrient uptake.

NASA. National Aeronautics and Space Administration.

national forests. Administered by the National Forest Service, these are public forest and woodlands that are managed for multiple uses, such as logging, mineral exploitation, livestock grazing, and recreation.

national parks. Administered by the National Park Service, national parks are lands and coastal areas of great scenic, ecological, or historical importance. They are managed with the dual goals of protection and providing public access.

national priorities list (NPL). A list of the chemical waste sites presenting the most immediate and severe threats. Such sites are scheduled for cleanup ahead of other sites.

natural. Describes a substance or factor that occurrs or is produced as a normal part of nature apart from any activity or intervention of humans. Opposite of artificial, synthetic, human-made, or caused by humans.

natural capital. The natural assets and the services they perform are referred to as natural capital.

natural chemical control. The use of one or more natural chemicals such as hormones or pheromones to control a pest.

natural control methods. Any of many techniques of controlling a pest population without resorting to the use of synthetic organic or inorganic chemicals. (See **biological control, cultural control, genetic control, hormones,** and **pheromones**)

natural enemies. All the predators and/or parasites that may feed on a given organism. Organisms used to control a specific pest through predation or parasitism.

natural increase. The number of births minus the number of deaths in a given population. It does not consider immigration and emigration.

natural laws. Derivations from our observations that matter, energy, and certain other phenomena apparently always act (or react) according to certain "rules."

natural organic compounds. See **organic compounds/ molecules**.

natural rate of change. The percent of growth (or decline) of a given population during a year. It is found by subtracting the crude death rate from the crude birth rate and changing the result to a percent. It does not include immigration or emigration.

natural resources. As applied to natural ecosystems and species, this term indicates that they are expected to be of economic value and may be exploited. Likewise the term applies to particular segments of ecosystems such as air, water, soil, and minerals.

natural selection. The process whereby the natural factors of environmental resistance tend to eliminate those members of a population that are least well adapted to cope and thus, in effect, select those best adapted for survival and reproduction.

natural services. Functions performed free of charge by natural ecosystems such as control of runoff and erosion, absorption of nutrients, and assimilation of air pollutants.

natural succession. See **ecological succession**.

net energy yield. The amount of energy produced minus the amount that is expended in production and transmission to consumers.

neutron. A fundamental atomic particle found in the nuclei of atoms (except hydrogen) and having one unit of atomic mass but no electrical charge.

niche (ecological)**.** The total of all the relationships that bear on how an organism copes with both biotic and abiotic factors it faces.

NIMBY. Acronym for "not in my back yard." NIMBY refers to a common attitude regarding undesirable facilities such as incinerators, nuclear facilities, and hazardous waste treatment plants, whereby people do everything possible to prevent the location of such facilities nearby.

NIMTOO. An acronym for "not in my term of office."

nitric acid (HNO_3). One of the acids in acid rain. Formed by reactions between nitrogen oxides and the water vapor in the atmosphere.

nitric oxide. See **nitrogen oxides**.

nitrogen dioxide. See **nitrogen oxides**.

nitrogen fixation. The process of chemically converting nitrogen gas (N_2) from the air into compounds such as nitrates (NO_3^-) or ammonia (NH_3) that can be used by plants in building amino acids and other nitrogen-containing organic molecules.

nitrogen oxides (NO_x). A group of nitrogen-oxygen compounds formed when some of the nitrogen gas in air combines with oxygen during high-temperature combustion; they are a major category of air pollutants. Along with hydrocarbons, they are a primary factor in the production of ozone and other photochemical oxidants that are the most harmful components of photochemical smog. They also contribute to acid precipitation (see **nitric acid**). Major nitrogen oxides are nitric oxide, NO; nitrogen dioxide, NO_2; nitrogen tetroxide, N_2O_2.

nitrous oxide. A gas, N_2O. Nitrous oxide comes from biomass burning, fossil fuel burning, and the use of chemical fertilizers. It is of concern because in the troposphere it is a greenhouse gas and in the stratosphere it contributes to ozone destruction.

NOAA. National Oceanic and Atmospheric Administration.

nonbiodegradable. Not able to be consumed and/or broken down by biological organisms. Nonbiodegradable substances include plastics, aluminum, and many chemicals used in industry and agriculture. Particularly dangerous are human-made nonbiodegradable chemicals that are also toxic and tend to accumulate in organisms, i.e., nonbiodegradable synthetic organic compounds. (See **biodegradable and bioaccumulation**)

nonconsumptive water use. Use of water for such purposes as washing and rinsing where the water, albeit polluted, remains available for further uses. With suitable purification, such water may be recycled indefinitely.

nonpersistent. Refers to chemicals that break down readily to harmless compounds, as, for example, natural organic compounds break down to carbon dioxide and water.

nonpoint sources. Sources of pollution such as general runoff of sediments, fertilizer, pesticides, and other materials from farms and urban areas as opposed to specific points of discharge such as factories. Also called **diffuse sources**. (Contrast **point sources**)

nonrenewable resources. Resources such as ores of various metals, oil, and coal that exist as finite deposits in Earth's crust and that are not replenished by natural processes as they are mined. (Contrast **renewable resources**)

nontidal wetlands. Inland wetlands not affected by tides.

no-till agriculture. The farming practice in which weeds are killed with chemicals (or other means) and seeds are planted and grown without resorting to plowing or cultivation. The practice is very effective in reducing soil erosion.

NPL. See **National Priorities List**.

NRCS. Natural Resources Conservation Service, formerly the SCS (U.S. Soil Conservation Service).

nuclear power. Electrical power that is produced by using a nuclear reactor to boil water and produce steam, which, in turn, drives a turbogenerator.

Nuclear Regulatory Commission (NRC). The agency within the Department of Energy that sets and enforces safety standards for the operation and maintenance of nuclear power plants.

nucleic acids. The class of natural organic macromolecules that function in the storage and transfer of genetic information.

nucleus. *Biology:* The large body contained in most living cells that contains the genes or hereditary material, DNA. *Physics:* The central core of atoms, which is made up of neutrons and protons. Electrons surround the nucleus.

nutrient. *Animal:* Material such as protein, vitamins, and minerals required for growth, maintenance, and repair of the body and material such as carbohydrates required for energy. *Plant:* An essential element in a particular ion or molecule that can be absorbed and used by the plant. For example, carbon, hydrogen, nitrogen, and phosphorus are essential elements; carbon dioxide, water, nitrate (NO_3^-), and phosphate (PO_4^{3-}) are the respective nutrients.

nutrient cycle. The repeated pathway of particular nutrients or elements from the environment through one or more organisms back to the environment. Nutrient cycles include the carbon cycle, the nitrogen cycle, the phosphorus cycle, and so on.

nutrient-holding capacity. The capacity of a soil to bind and hold nutrients (fertilizer) against their tendency to be leached from the soil.

observations. Things or phenomena that are perceived through one or more of the basic five senses in their normal state. In addition, to be accepted as factual, the observations must be verifiable by others.

ocean thermal energy conversion (OTEC). The concept of harnessing the temperature difference between surface water heated by the sun and colder deep water to produce power.

oil field. The area in which exploitable oil is found.

oil shale. A natural sedimentary rock that contains a material, kerogen, that can be extracted and refined into oil and oil products.

oligotrophic. Refers to a lake the water of which is nutrient-poor. Therefore, it will not support phytoplankton, but it will support submerged aquatic vegetation, which get nutrients from the bottom.

omnivore. An animal that feeds more or less equally on both plant material and other animals.

OPEC. Organization of Petroleum Exporting Countries.

optimal range. With respect to any particular factor or combination of factors, the maximum variation that still supports optimal or near-optimal growth of the species in question.

optimum. The condition or amount of any factor or combination of factors that will produce the best result. For example, the amount of heat, light, moisture, nutrients, and so on that will produce the best growth. Either more or less than the optimum is not as good.

optimum population. The population of a resource that will provide the maximum sustainable yield. The yield is reduced at higher or lower populations.

organically grown. Generally refers to produce grown

without the use of hard chemical pesticides or inorganic fertilizer. However, as of yet there are no official standards defining the use of the term.

organic compounds/molecules. *Classical definition:* All living things and products that are uniquely produced by living things, such as wood, leather, and sugar. *Chemical definition:* All chemical compounds or molecules, natural or synthetic, that contain carbon atoms as an integral part of their molecular structure. Their structure is based on bonded carbon atoms with hydrogen atoms attached. They can be either biodegradable to nonbiodegradable. (Contrast **inorganic compounds**)

organic fertilizer. See **fertilizer**.

organic gardening/farming. Gardening or farming without the use of inorganic fertilizers, synthetic pesticides, or other human-made materials.

organic molecules. See **organic compounds/molecules**.

organic phosphate. Phosphate (PO_4^{-3}) bonded to an organic molecule.

organism. Any living thing—plant, animal, or microbe.

organochlorides. See **chlorinated hydrocarbons**.

OSHA. Occupational Safety and Health Administration. Promulgates regulations concerning measures that must be taken to protect workers.

osmosis. The phenomenon whereby water diffuses through a semipermeable membrane toward an area where there is more material in solution (where there is a relatively lower concentration of water). Has particular application regarding salinization of soils where plants are unable to grow because of osmotic water loss.

outbreak. A population explosion of a particular pest. Often caused by an application of pesticides that destroys the pest's natural enemies.

overgrazing. The phenomenon of animals' grazing in greater numbers than the land can support in the long run. There may be a temporary economic gain in the short run, but the grassland (or other ecosystem) is destroyed, and its ability to support life in the long run is vastly diminished.

overland flow system. An alternative method of wastewater treatment that involves allowing water to percolate through a field of grass or other vegetation.

oxidation. Chemical reaction process that generally involves breakdown through combining with oxygen. Both burning and cellular respiration are examples of oxidation. In both cases, organic matter is combined with oxygen and broken down to carbon dioxide and water.

ozone. A gas, O_3, that is a pollutant in the lower atmosphere but necessary to screen out ultraviolet radiation in the upper atmosphere. May also be used for disinfecting water.

ozone hole. First discovered over the Antarctic, this is a region of stratospheric air that is severely depleted of its normal levels of ozone during the Antarctic spring because of CFCs from anthropogenic (human-made) sources.

ozone shield. The layer of ozone gas (O_3) in the upper atmosphere that screens out harmful ultraviolet radiation from the sun.

PANs (peroxyacetylnitrates). A group of compounds present in photochemical smog that are extremely toxic to plants and irritating to eyes, nose, and throat membranes of humans.

parasites. Organisms (plant, animal, or microbial) that attach themselves to another organism, the host, and feed on it over a period of time without killing it immediately but usually doing harm to it. Commonly divided into *ectoparasites,* those that attach to the outside, and *endoparasites,* those that live inside their hosts.

parent material. The rock material, the weathering and gradual breakdown of which is the source of the mineral portion of soil.

particulates. (See **PM-10** and **suspended particulate matter**.)

parts per million (ppm). A frequently used expression of concentration. It is the number of units of one substance present in a million units of another. For example, 1 g of phosphate dissolved in 1 million grams (= 1 ton) of water would be a concentration of 1 ppm.

passive safety features. Those safety features of nuclear facilities that involve processes that are not vulnerable to operator intrusion or electrical power failures. Passive safety features enhance the degree of safety of nuclear reactors. (See **active safety features**)

passive solar heating system. A solar heating system that does not use pumps or blowers to transfer heated air or water. Instead, natural convection currents are used or the interior of the building itself acts as the solar collector.

pasteurization. The process of applying heat to kill pathogens.

pastoralist. One involved in animal husbandry, usually in subsistence agriculture.

pathogen. An organism, usually a microbe, that is capable of causing disease. Such an organism is said to be pathogenic.

PCBs (polychlorinated biphenyls). A group of widely used industrial chemicals of the chlorinated hydrocarbon class. They have become serious and widespread pollutants, contaminating most food chains on Earth, because they are extremely resistant to breakdown and are subject to bioaccumulation. They are known to be carcinogenic.

percolation. The process of water seeping through cracks and pores in soil or rock.

permafrost. The ground of arctic regions that remains permanently frozen. Defines tundra, since only small herbaceous plants can be sustained on the thin layer of soil that thaws each summer.

persistent. Refers to pesticides or other chemicals that are nonbiodegradable and very resistant to breakdown by other means. Such chemicals therefore remain present in the environment more or less indefinitely.

pesticide. A chemical used to kill pests. Pesticides are further categorized according to the pests they are designed to kill—for example, herbicides kill plants, insecticides kill insects, fungicides kill fungi, and so on.

pesticide treadmill. Refers to the fact that use of chemical pesticides simply creates a vicious cycle of "needing more pesticides" to overcome developing resistance and secondary outbreaks caused by the pesticide applications.

pest-loss insurance. Insurance that a grower can buy that will pay in the event of loss of crop due to pests.

petrochemical. A chemical made from petroleum (crude oil) as a basic raw material. Petrochemicals include plastics, synthetic fibers, synthetic rubber, and most other synthetic organic chemicals.

pH. Scale used to designate the acidity or basicity (alkalinity) of solutions or soil, expressed as the logarithm of the concentration of hydrogen ions (H^+). pH 7 is neutral; values decreasing from 7 indicate increasing acidity; values increasing from 7 indicate increasing basicity. Each unit from 7 indicates a tenfold increase over the preceding unit.

pheromones. A chemical substance secreted externally by certain members of a species that affects the behavior of other members of the same species. The most common examples are sex attractants, which female insects secrete to attract males. Pheromones are coming into use in pest control. (See also **hormones**)

phosphate. An ion composed of a phosphorus atom with four oxygen atoms attached. PO_4^{3-}. It is an important plant nutrient. In natural waters it is frequently the limiting factor. Therefore, additions of phosphate to natural water are frequently responsible for algal blooms.

photochemical oxidants. A major category of air pollutants, including ozone, that are highly toxic and damaging especially to plants and forests. Formed as a result of interactions between nitrogen oxides and hydrocarbons driven by sunlight.

photochemical smog. The brownish haze that frequently forms on otherwise clear sunny days over large cities with significant amounts of automobile traffic. It results largely from sunlight-driven chemical reactions among nitrogen oxides and hydrocarbons, both of which come primarily from auto exhausts.

photosynthesis. The chemical process carried on by green plants through which light energy is used to produce glucose from carbon dioxide and water. Oxygen is released as a byproduct.

photovoltaic cells. Devices that convert light energy into an electrical current.

physical barrier. A genetic feature on a plant, such as sticky hairs, that physically blocks attack by pests.

phytoplankton. Any of the many species of algae that consist of single cells or small groups of cells that live and grow freely suspended in the water near the surface. Given abundant nutrients, they may become so numerous as to give the water a green "pea soup" appearance and/or form a thick green scum over the surface.

plankton. Any and all living things that are found freely suspended in the water and that are carried by currents as opposed to being able to swim against currents. It includes both plant (phytoplankton) and animal (zooplankton) forms.

plant community. The array of plant species, including numbers, ages, distribution, that occupies a given area.

PM-10. The new standard criterion pollutant for suspended particulate matter. PM-10 refers to particles smaller than 10 micrometers in diameter. Such particles are readily inhaled directly into the lungs.

point sources. Specific points of origin of pollutants, such as factory drains or outlets from sewage-treatment plants. (Contrast **nonpoint sources**)

pollutant. A substance the presence of which contaminates air, water, or soil.

pollution. Contamination of air, water, or soil with undesirable amounts of material or heat. The material may be a natural substance, such as phosphate, in excessive quantities, or it may be very small quantities of a synthetic compound such as dioxin that is exceedingly toxic.

pollution avoidance, pollution prevention. A strategy of encouraging development of techniques that would not generate pollutants.

polyculture. The growing of two or more species together. (Contrast **monoculture**)

poor. Economically unable to afford adequate food and/or housing.

population. A group within a single species, the individuals of which can and do freely interbreed. Breeding between populations of the same species is less common because of differences in location, culture, nationality, and so on.

population density. The numbers of individuals per unit of area.

population explosion. The exponential increase observed to occur in a population when or if conditions are such that a large percentage of the offspring are able to survive and reproduce in turn. Frequently leads, in turn, to overexploitation, upset, and eventual collapse of the ecosystem.

population momentum. Refers to the fact that a rapidly growing human population may be expected to grow for 50–60 years after replacement fertility (2.1) is reached because of increasing numbers entering reproductive age.

population profile. A bar graph that shows the number of individuals at each age or in each 5-year age group.

population structure. Refers to the proportion of individuals in each age group. For example, a population may be made up predominantly of young people, old people, or a more or less even distribution of young and old.

potential energy. The ability to do work that is stored in some chemical or physical state. For example, gasoline is a form of potential energy; the ability to do work is stored in the chemical state and is released as the fuel is burned in an engine.

ppm. See **parts per million**.

pOH. The negative logarithm of the concentration of hydroxyl ions (OH^-). Like pH, the scale ranges from 0 to 14, each unit representing a tenfold increase over the preceding unit. The lower the pOH, the higher the concentration of hydroxyl ions.

precipitation. Any form of moisture condensing in the air and depositing on the ground.

predator. An animal that feeds on another.

predator-prey relationship. A feeding relationship existing between two kinds of animals. The predator is the animal feeding on the prey. Such relationships are frequently instrumental in controlling populations of herbivores.

preliminary treatment. The removal of debris and grit from wastewater by passing the water through a coarse screen and grit-settling chamber.

prey. In a feeding relationship, the animal that is killed and eaten by another.

primary pollutants. The air pollutants that are emitted into the air as direct byproducts of combustion or other process as opposed to those (secondary air pollutants) that form as a result of various chemical reactions occurring in the atmosphere.

primary consumer. An organism such as a rabbit or deer that feeds more or less exclusively on green plants or their products, such as seeds and nuts. Synonym: **herbivore**.

primary energy sources. Fossil fuels, radioactive material, and solar, wind, and water and other energy sources that exist as natural resources.

primary standard. The maximum tolerable level of a pollutant. The standard is intended to protect human health.

primary succession. See **succession**.

primary treatment. The process that follows preliminary sewage treatment. It consists of passing the water very slowly through a large tank, so that the particulate organic material in the water can settle out. The settled material is **raw sludge**.

private land trust. A tract of land that is acquired and put into a protected status by a group of private individuals without government funding or support.

producers. In an ecosystem, those organisms, mostly green plants, that use light energy to construct their organic constituents from inorganic compounds.

production. In the oil industry, refers to the withdrawing of oil reserves.

profligate growth. Growth characterized by extravagant and wasteful use of resources.

property taxes. Taxes that the local government levies on privately owned properties, generally a few dollars per hundred dollars of property value. This is the major source of revenue for local governments.

protein. The class of organic macromolecules that is the major structural component of all animal tissues and that functions as enzymes in both plants and animals.

proton. Fundamental atomic particle with a positive charge, found in the nuclei of atoms. The number of protons present equals the atomic number and is distinct for each element.

protozoan, pl. **protozoa.** Any of a large group of microscopic organisms that consist of a single, relatively large complex cell or in some cases small groups of cells. All have some means of movement. Amoebae and paramecia are examples.

proven reserves. See **reserves**.

punctuated evolution. "Step" model of evolution in which there is little change while an ecosystem is in a balanced state but a shift alters selective pressures and sets into motion fairly rapid charges in almost all, if not all, species in the ecosystem until a new balance is reached.

qualitative. Refers to issues involving purity.

quantitative. Refers to issues involving numbers.

RACT (reasonably available control technology). Applied to the goals of the Clean Air Act, EPA-approved forms of technology that will reduce the output of industrial air pollutants. (See also **MACT**)

radioactive decay. The reduction of radioactivity that occurs as an unstable isotope (radioactive substance) gives off radiation and becomes stable.

radioactive emissions. Any of various forms of radiation and/or particles that may be given off by unstable isotopes. Many such emissions have very high energy and can destroy biological tissues or cause mutations leading to cancer or birth defects.

radioactive materials. Substances that are or that contain unstable isotopes and that consequently give off radioactive emissions. (See **isotope** and **radioactive emissions**)

radioactive wastes. Waste materials that are or that contain or are contaminated with radioactive substances. Many materials used in the nuclear industry become wastes because of their contamination with radioactive substances.

radioisotope. An isotope of an element that is unstable and may tend to gain stability by giving off radioactive emissions. (See **isotope** and **radioactive decay**)

radon. A radioactive gas produced by natural processes in Earth that is known to seep into buildings. It can be a major hazard within homes and is a known carcinogen.

rain shadow. The low-rainfall region that exists on the leeward (downwind) side of mountain ranges. It is the result of the mountain range's causing the precipitation of moisture on the windward side.

range of tolerance. The range of conditions within which an organism or population can survive and reproduce, for example, the range from the highest to lowest temperature that can be tolerated. Within the range of tolerance is the optimum, or best, condition.

raw sludge. The untreated organic matter that is removed from sewage water by letting it settle. It consists of organic particles from feces, garbage, paper, and bacteria.

raw wastewater. (See **raw sludge**)

reactor vessel. Steel-walled vessel that contains the nuclear reactor.

recharge area. With reference to groundwater, the area over which infiltration and resupply of a given aquifer occurs.

recruitment. With reference to populations, the maturation and entry of young into the adult breeding population.

relative humidity. The percentage of moisture in the air compared with how much the air can hold at the given temperature.

remediation. The return to the original uncontaminated state. (See also **bioremediation** and **groundwater remediation**)

renewable energy. Energy sources, namely solar, wind, and geothermal, that will not be depleted by use.

renewable resources. Biological resources such as trees that may be renewed by reproduction and regrowth. Conservation to prevent overcutting and protection of the environment are still required, however. (Contrast **nonrenewable resources**)

replacement capacity. The capacity of a system to recover to its original state after a harvest or other form of use of biological resources.

replacement fertility/level. The fertility rate that will just sustain a stable population.

reproductive strategy. The particular methodologies seen in nature to enhance the chance of subsequent generations: for example, producing massive numbers of young but offering no care or protection vs. producing few young and caring for them.

reserves. The amount of a mineral resource (including oil, coal, and natural gas) remaining in Earth that can be exploited using current technologies and at current prices. Usually given as proven reserves, those that have been positively identified, and estimated reserves, those that have not yet been discovered but that are presumed to exist.

Resources Conservation and Recovery Act of 1976 (RCRA). The cornerstone legislation to control indiscriminate land disposal of hazardous wastes.

respiration. See **cell respiration**.

restoration ecology. The branch of ecology devoted to restoring degraded and altered ecosystems to their natural state.

resurgence. The rapid comeback of a population especially of pests after a severe dieoff, usually caused by pesticides, and the return to even higher levels than before the treatment.

reuse. The practice of reusing items as opposed to throwing them away and producing new items, as, for example, bottles can be collected and refilled (**recycling**).

reverse osmosis. See **microfiltration**.

riparian woodlands. The strip of woods that grow along natural watercourses.

risk. The probability of suffering injury, disease, death, or other loss as a result of exposure to a hazard.

risk analysis. The process of evaluating the risks associated with a particular hazard before taking some action. Often called risk assessment.

risk characterization. The process of determining a risk and its accompanying uncertainties after **hazard assessment**, dose-response assessment, and exposure assessment have been accomplished.

risk management. The task of regulators, involving reviewing the risk data and making regulatory decisions based on the evidence. The process often is influenced by considerations of costs and benefits as well as by public perception.

risk perception. Nonexperts' intuitive judgments about risks, which often are not in agreement with the level of risk as judged by experts.

runoff. That portion of precipitation that runs off the surface as opposed to soaking in.

Safe Drinking Water Act of 1974. Legislation to protect the public from the risk that toxic chemicals will contaminate drinking water supplies. Mandates regular testing of municipal water supplies.

salinization. The process whereby soil becomes saltier and saltier until finally the salt prevents the growth of plants. It is caused by irrigation because salts brought in with the water remain in the soil as the water evaporates.

saltwater intrusion, saltwater encroachment. The phenomenon of seawater's moving back into aquifers or estuaries. It occurs when the normal outflow of freshwater is diverted or removed for use.

sand. Mineral particles 0.2–2.0 mm in diameter.

sanitary sewer. Separate drainage system used to receive all the wastewater from sinks, tubs, and toilets.

SARA (Title III). Superfund Amendments and Reauthorization Act section that promulgates community Right-to-Know requirements.

savanna. A type of grassland usually dotted with trees supported by a wet season and dry season and frequent natural fires, typical of subtropical regions, particularly in Africa.

secondary air pollutants. Air pollutants resulting from reactions of primary air pollutants while resident in the atmosphere. These include ozone, other reactive organic compounds, and sulfuric and nitric acids. (See **ozone, PANs,** and **photochemical oxidants**)

secondary consumer. An organism such as a fox or coyote that feeds more or less exclusively on other animals that feed on plants.

secondary energy source. A form of energy such as electricity that must be produced from a primary energy source such as coal or radioactive material.

secondary pest outbreak. The phenomenon of a small, and therefore harmless, population of a plant-eating insect suddenly exploding to become a serious pest problem. Often caused by the elimination of competitors through pesticide use.

secondary succession. See **succession**.

secondary treatment. Also called biological treatment. A sewage-treatment process that follows primary treatment. Any of a variety of systems that remove most of the remaining organic matter by enabling organisms to feed on it and oxidize it through their respiration. Trickling filters and activated-sludge systems are the most commonly used methods.

second basic principle of ecosystem sustainability. Ecosystems run on solar energy, which is exceedingly abundant, nonpolluting, constant, and everlasting.

second-generation pesticides. Synthetic organic compounds used to kill insects and other pests. Started with the use of DDT in the 1940s.

Second Law of Thermodynamics. The fact based on irrefutable observations that in every energy conversion (e.g., electricity to light) some of the energy is converted to heat and some heat always escapes from the system because it always moves toward a cooler place. Therefore, in every energy conversion, a portion of energy is lost. Therefore, since energy cannot be created (First Law) the functioning of any system requires an energy input.

secure landfill. A landfill with suitable barriers, leachate drainage, and monitoring systems such that it is deemed secure against contaminating groundwater with hazardous wastes.

sediment. Soil particles, namely sand, silt, and clay, carried by flowing water. The same material after it has been deposited. Because of different rates of settling, deposits generally are pure sand, silt, or clay.

sedimentation. The filling in of lakes, reservoirs, stream channels, and so on with soil particles, mainly sand and silt. The soil particles come from erosion, which generally results from poor or inadequate soil conservation practices in connection with agriculture, mining, and/or development. Also called siltation.

sediment trap. A device for trapping sediment and holding it on a development or mining site.

seep. Where groundwater seeps from the ground over some area as opposed to a spring, which is the exit as a single point.

selective breeding. The breeding of certain individuals because they bear certain traits and the exclusion from breeding of others.

selective pressure. A fundamental mechanism of evolution. An environmental factor that causes individuals with certain traits, which are not the norm for the population, to survive and reproduce more than the rest of the population. The result is a shift in the genetic makeup of the population. For example, the presence of insecticides provides a selective pressure to increase pesticide resistance in the pest population.

sex attractant. A natural chemical substance (pheromone) secreted by the female of many insect species that serves to attract males for the function of mating. Sex attractants may be used in traps or for the **confusion technique** to aid in the control of insect pests.

shadow pricing. In cost-benefit analysis, a technique used to estimate benefits where normal economic analysis is ineffective. For example, people could be asked how much they might be willing to pay monthly to achieve some improvement in their environment.

sheet erosion. The loss of a more or less even layer of soil from the surface due to the impact and runoff from a rainstorm.

shelterbelts. Rows of trees around cultivated fields for the purpose of reducing wind erosion.

silt. Soil particles between the size of sand particles and clay particles; namely, particles 0.002–0.2 mm in diameter.

siltation. See **sedimentation**.

sinkhole. A large hole resulting from the collapse of an underground cavern.

slash-and-burn agriculture. The practice, commonly exercised throughout tropical regions, of cutting and burning vegetation to make room for agriculture. The process is highly destructive of soil humus and may lead to rapid degradation of soil.

sludge cake. Treated sewage sludge that has been dewatered to make it a moist solid.

sludge digesters. Large tanks in which raw sludge (removed from sewage) is treated through anaerobic digestion by bacteria.

smog. See **industrial smog** and **photochemical smog**.

soft water. Water with little or no calcium, magnesium, or other ions in solution that will cause soap to precipitate (form a curd that makes a "ring" around the bathtub). (Contrast **hard water**)

soil. A dynamic system involving three components: mineral particles, detritus, and soil organisms feeding on the detritus.

soil aeration. See **aeration**.

soil erosion. The loss of soil caused by particles' being carried away by wind and/or water.

soil fertility. Soil's ability to support plant growth; often refers specifically to the presence of proper amounts of nutrients. The soil's ability to fulfill all the other needs of plants is also involved.

soil profile. A description of the different, naturally formed layers within a soil.

soil structure. The composition of soil in terms of particles (sand, silt, and clay) stuck together to form clumps and aggregates, generally with considerable air spaces in between. Structure affects infiltration and aeration. It develops as organisms feed on organic matter in and on the soil.

soil texture. The relative size of the mineral particles that make up the soil. Generally defined in terms of the sand, silt, and clay content.

solar cells. See **photovoltaic cells**.

solar energy. Energy derived from the sun. Includes direct solar energy (the use of sunlight directly for heating and/or production of electricity) and indirect solar energy (the use of wind, which results from the solar heating of the atmosphere, and biological materials such as wood, which result from photosynthesis).

solar-trough collectors. Reflectors in the shape of a parabolic trough, which reflect the sunlight onto a tube of oil at the focal point. The oil thus heated is used to boil water to drive a steam turbine.

solid waste. The total of materials discarded as "trash" and handled as solids, as opposed to those that are flushed down sewers and handled as liquids.

solubility. The degree to which a substance will dissolve and enter into solution.

solution. A mixture of molecules (or ions) of one material in another. Most commonly, molecules of air and/or ions of various minerals in water. For example, seawater contains salt in solution.

specialization. With reference to evolution, the phenomenon whereby species become increasingly adapted to exploit one particular niche but, thereby, are less able to exploit other niches.

speciation. The evolutionary process whereby populations of a single species separate and, through being exposed to different forces of natural selection, gradually develop into distinct species.

species. All the organisms (plant, animal, or microbe) of a single kind. The "single kind" is determined by similarity of appearance and/or by the fact that members do or potentially can mate and produce fertile offspring. Physical, chemical, or behavioral differences block breeding between species.

splash erosion. The compaction of soil that results when rainfall hits bare soil.

springs. Natural exits of groundwater.

standards. Air or water quality levels set by the federal or state government; the maximum levels of various pollutants that are to be legally tolerated. If levels go above the standards, various actions may be taken.

standing biomass. That portion of a population that is not available for consumption but must be conserved to maintain the productive potential of the population.

starvation. The failure to get enough calories to meet energy needs over a prolonged period of time. It results in a wasting away of body tissues until death occurs.

sterile male technique. Saturating an infested area with males of the pest species that have been artificially reared and sterilized by radiation. Matings between normal females and sterile males render the eggs infertile.

steward/stewardship. A steward is one to whom a trust has been given. In reference to natural lands, stewardship is an attitude of active care and concern for nature.

stomata sing. **stoma.** Microscopic pores in leaves, mostly on the undersurface, that allow the passage of carbon dioxide and oxygen into and out of the leaf and that also permit the loss of water vapor from the leaf.

storm drains. Separate drainage systems used for collecting and draining runoff from precipitation.

stormwater. In cities, the water that results directly from rainfall, as opposed to municipal water and sewage water piped to and from homes, offices, and so on. The extensive hard surfacing in cities creates a vast amount of stormwater runoff, which presents a significant management problem.

stormwater management. Policies and procedures for handling stormwater in acceptable ways to reduce the problems of flooding and erosion of stream banks.

stormwater retention reservoirs. Reservoirs designed to hold stormwater temporarily and let it drain away slowly in order to reduce problems of flooding and stream bank erosion.

stratosphere. The layer of Earth's atmosphere between 10 and 30 miles above the surface that contains the ozone shield. This layer mixes only slowly; pollutants that enter may remain for long periods of time. (See also **troposphere**)

strip cropping. The practice of growing crops in strips alternating with grass (hay) at right angles to prevailing winds or slopes in order to reduce erosion.

strip mining. The mining procedure in which all the earth covering a desired material such as coal is stripped away with huge power shovels in order to facilitate removal of the desired material.

submerged aquatic vegetation (SAV). Aquatic plants rooted in bottom sediments growing under water depend on light's penetrating through the water for photosynthesis.

subsistence farming. Farming that meets the food needs of the farmers and their families but little more. It involves hand labor and is practiced extensively in the developing world.

subsoil. In a natural situation, the soil beneath topsoil. In contrast to topsoil, subsoil is compacted and has little or no humus or other organic material, living or dead. In many cases, topsoil has been lost or destroyed as a result of erosion or development, and subsoil is at the surface.

succession. The gradual, or sometimes rapid, change in the species that occupy a given area, with some species invading and becoming more numerous while others decline in population and disappear. Succession is caused by a change in one or more abiotic or biotic factors that benefits some species at the expense of others. *Primary succession:* The gradual establishment, through a series of stages, of a **climax ecosystem** in an area that has not been occupied before, e.g., a rock face. *Secondary succession:* The reestablishment, through a series of stages, of a climax ecosystem in an area from which it was previously cleared.

sulfur dioxide (SO_2). A major air pollutant, this toxic gas is formed as a result of burning sulfur. The major sources are burning coal (coal-burning power plants) that contains some sulfur and refining metal ores (smelters) that contain sulfur.

sulfuric acid (H_2SO_4). The major constituent of acid precipitation. Formed when sulfur dioxide emissions react with water vapor in the atmosphere. (See also **sulfur dioxide**)

Superfund. The popular name for the Comprehensive Environmental Response, Compensation, and Liability Act of 1980. This act is the cornerstone legislation that provides the mechanism and funding for the cleanup of potentially dangerous hazardous waste sites to protect groundwater.

surface impoundments. Closed ponds that used to be used to collect and hold liquid chemical wastes.

surface water. Includes all bodies of water, lakes, rivers, ponds, and so on that are on Earth's surface in contrast to groundwater, which lies below the surface.

suspended particulate matter (SPM). A category of major air pollutants consisting of solid and liquid particles suspended in the air. (See also **PM-10**)

suspension. With reference to materials contained in or being carried by water, materials kept "afloat" only by the water's agitation that settle as the water becomes quiet.

sustainability. Refers to whether a process can be continued indefinitely without depleting the energy or material resources on which it depends.

sustainable agriculture. Agriculture that maintains the integrity of soil and water resources such that it can be continued indefinitely. Much of modern agriculture is depleting these resources and hence, is, not sustainable.

sustainable development. Development that provides people with a better life without sacrificing or depleting resources or causing environmental impacts that will undercut future generations.

sustainable society. A society that functions in a way so as not to deplete energy or material resources on which it depends.

sustainable yield. The taking of a biological resource (e.g., fish or forests) that does not exceed the capacity of the resource to reproduce and replace itself.

symbiosis. The intimate living together or association of two kinds of organisms.

synergism. The phenomenon in which two factors acting together have a very much greater effect than would be indicated by the sum of their effects separately—as, for example, modest doses of certain drugs in combination with modest doses of alcohol may be fatal.

synfuels, synthetic fuels. Fuels similar or identical to those that come from crude oil and/or natural gas, produced from coal, oil shale, or tar sands.

synthetic. Human-made as opposed to being derived from a natural source. For example, synthetic organic compounds are those produced in chemical laboratories, whereas natural organic compounds are those produced by organisms.

synthetic organic compounds. See **organic compounds**.

tar sands. Sedimentary material containing bitumin that can be "melted out" using heat and then refined in the same way as crude oil.

taxonomy. The science of identification and classification of organisms according to evolutionary relationships.

tectonic plates. Huge slabs of rock that make up Earth's crust.

temperature inversion. The weather phenomenon in which a layer of warm air overlies cooler air near the ground and prevents the rising and dispersion of air pollutants.

teratogenic. Causing birth defects.

terracing. The practice of grading sloping farmland into a series of steps and cultivating only the level portions in order to reduce erosion.

territoriality. The behavioral characteristic exhibited by many animal species, especially birds and mammalian carnivores, to mark and defend a given territory against other members of the same species.

texture. With reference to solids, the sizes of the particles, sand, silt, and/or clay, that make up the mineral portion.

theory. A conceptual formulation that provides a rational explanation or framework for numerous related observations.

thermal pollution. The addition of abnormal and undesirable amounts of heat to air or water. It is most significant with respect to discharging waste heat from electric generating plants, especially nuclear power plants, into bodies of water.

third basic principle of ecosystem sustainability. Large biomasses cannot be supported at the end of long food chains. The size of consumer populations is maintained such that overgrazing does not occur.

third world. See **developing countries**.

threatened species. A species the population of which is declining precipitously because of direct or indirect human impacts.

threshold level. The maximum degree of exposure to a pollutant, drug, or other factors that can be tolerated with no ill effect. The threshold level will vary depending on the species, the sensitivity of the individual, the length of exposure, and the presence of other factors that may produce synergistic effects.

tidal wetlands. Areas of marsh grasses and reeds along coasts and estuaries where the ground is covered by high tides but drained at low tide.

tie-in strategy. In connection with global warming, the idea that society should take actions that not only deal with global warming but also have other beneficial effects. For example, energy conservation not only reduces carbon dioxide emissions but also saves money, reduces acid deposition, and lowers our dependency on foreign oil.

topsoil. The surface layer of soil, which is rich in humus and other organic material, both living and dead. As a result of the activity of organisms living in the topsoil, it generally has a loose, crumbly structure as opposed to being a compact mass. In many cases, because of erosion, development, or mining activity, the topsoil layer may be absent.

total fertility rate. The average number of children that would be born alive to each woman during her total reproductive years if she followed the average fertility at each age.

total product life cycle. Consideration of all steps from the obtaining of raw materials through the manufacture, use, and finally disposal of a product. Consideration of byproducts and pollution resulting from each step.

total watershed planning. A consideration of the entire watershed, and planning development and other activities so as to maintain the overall water flow characteristics of the area.

trace elements. Those essential elements that are needed in only very small amounts.

traditional farming. Current farming methods involving intensive use of fertilizers, pesticides, and other chemicals.

tragedy of the commons. The overuse or overharvesting and consequent depletion and/or destruction of a renewable resource that tends to occur when the resource is treated as a commons, that is, when it is open to be used or harvested by any and all with the means to do so.

trait. Any physical or behavioral characteristic or talent that an individual is born with.

transpiration. The loss of water vapor from plants. Water evaporates from cells within the leaves and exits through stomata.

trapping technique. The use of sex attractants to lure male insects into traps.

treated sludge. Solid organic material that has been removed from sewage and treated so that it is nonhazardous.

trickling filter system. System in which wastewater trickles over rocks or a framework coated with actively feeding microorganisms. The feeding action of the organisms in a well-aerated environment results in the decomposition of organic matter. Used in secondary or biological treatment of sewage.

tritium (^3H). An unstable isotope of hydrogen that contains two neutrons in addition to the usual single proton in the nucleus. It does not occur in significant amounts naturally but is human-made.

trophic level. Feeding level with respect to the primary source of energy. Green plants are at the first trophic level, primary consumers at the second, secondary consumers at the third, and so on.

troposphere. The layer of Earth's atmosphere from the surface to about 10 miles in altitude. The *tropopause* is the boundary between the troposphere and the stratosphere above. This layer is well mixed and is the site and source of our weather, as well as the primary recipient of air pollutants. (See also **stratosphere**)

turbid. Refers to water purity; means cloudy.

turbine. A sophisticated "paddle wheel" driven at a very high speed by steam, water, or exhaust gases from combustion.

turbogenerator. A turbine coupled to and driving an electric generator. Virtually all commercial electricity is produced by such devices. The turbine is driven by gas, steam, or water.

turnover rate. The rate at which a population is replaced by the next generation.

ultraviolet radiation. Radiation similar to light but with wavelengths slightly shorter than violet light and with more energy. The greater energy causes it to severely burn and otherwise damage biological tissues.

undernutrition. A form of hunger in which there is a lack of adequate food energy as measured in calories. Starvation is the most severe form of undernutrition.

urban decay. General deterioration of structures and facilities such as buildings and roadways, and also the decline in quality of services such as education, that has occurred in inner city areas as growth has been focused on suburbs and exurbs.

urban environmental coalition. The coming together of urban residents who may be members of various environmental organizations to solve particular environmental problems facing the city.

urban sprawl. The rapid expansion of metropolitan areas through building housing developments and shopping centers farther and farther from urban centers and lacing them together with more and more major highways. Widespread development that has occurred without any overall land-use plan.

UST legislation. Amendments to the Resources Conservation and Recovery Act of 1976, passed in 1984 to address the mounting problem of leaking underground storage tanks (USTs).

vigor. Applied to crop plants or animals, refers to traits for hardiness to disease, drought, cold, and other adverse factors or conditions.

vitamin. A specific organic molecule that is required by the body in small amounts but that cannot be made by the body and therefore must be present in the diet.

volatile organic compounds (VOCs). A category of major air pollutants present in the air in vapor state, including fragments of hydrocarbon fuels from incomplete combustion and evaporated organic compounds such as paint solvents, gasoline, and cleaning solutions. They are major factors in the formation of photochemical smog.

water cycle. The movement of water from points of evaporation through the atmosphere, through precipitation, and through or over the ground, returning to points of evaporation.

water-holding capacity. The ability of a soil to hold water so that it will be available to plants.

waterlogging. The total saturation of soil with water. Results in plant roots' not being able to get air and dying as a result.

watershed. The total land area that drains directly or indirectly into a particular stream or river. The watershed is generally named from the stream or river into which it drains.

watershed management. See **total watershed planning**.

water table. The upper surface of groundwater. It rises and falls with the amount of groundwater.

water vapor. Water molecules in the gaseous state.

weathering. The gradual breakdown of rock into smaller and smaller particles, caused by natural chemical, physical, and biological factors.

wetlands. Areas that are constantly wet and are flooded at more or less regular intervals. Especially, marshy areas along coasts that are regularly flooded by tide.

wetland systems. A biological aquatic system (usually a restored wetlands) to remove nutrients from treated sewage wastewater and return it, virtually pure, to a river or stream. Wetland systems are sometimes used when using treated wastewater for irrigation is not feasible.

Wilderness Act of 1964. Federal legislation that provides for the permanent protection of undeveloped and unexploited areas so that natural ecological processes can operate freely. Most uses are excluded from such areas, which now total 90 million acres in the United States.

wind farms. Arrays of numerous, modestly sized wind turbines for the purpose of producing electrical power.

windrows. Piles of organic material extended into long rows to facilitate turning and aeration to enhance composting.

wind turbines. "Windmills" designed for the purpose of producing electrical power.

work. Any change in motion or state of matter. Any such change requires the expenditure of energy.

workability. With reference to soils, the relative ease with which a soil can be cultivated.

World Bank. A branch of the United Nations that acts as a conduit to handle loans to developing countries.

world view. A set of assumptions that a person holds regarding the world and how it works.

xeroscaping. Landscaping with drought-resistant plants that need no watering.

yard wastes. Grass clippings and other organic wastes from lawn and garden maintenance.

zones of stress. Regions where a species finds conditions tolerable but suboptimal. Where a species survives but under stress.

Index

Boldface type indicates definition of indexed terms, none of which comes from Figures, Tables, Bibliography, Appendices, or Glossary.